...er: Peter Marshall
...Development Editor: Elizabeth Howe Marsh
...velopment Editor: Tommy Moorman
...tions: Tommy Moorman
...ing: John Britch
...ing Assistant: Bailey James
...Editor: Mike Jones
...ger of Digital Development: Amanda Dunning
...Project Editor: Elizabeth Geller
...ector: Diana Blume
...and Text Design: Tom Carling, Carling Design Inc.
...Editor: Jennifer Atkins
...Researcher: Julia Phelan
...ction Manager: Paul Rohloff
...osition and Layout: Sheridan Sellers
...g and Binding: RR Donnelley

...photo: Cloud Gate, 2004, © Anish Kapoor. © 2014 Anish Kapoor/ARS,
...York/DACS, London. Image by Thomas Jackson.

...y of Congress Control Number: 2014952562

...nt Edition: ISBN-13: 978-1-4641-3595-8
                   ISBN-10: 1-4641-3595-9

...5, 2012, 2010 by W. H. Freeman and Company
...hts reserved

...d in the United States of America

...h Printing

...Freeman and Company
...adison Avenue, New York, NY 10010
...dmills, Basingstoke RG21 6XS, England

...whfreeman.com

WHAT IS *Life*
A GUIDE TO BIOLOGY

THIRD
EDITIO

Jay Phelan

University of California, Los Angeles

W. H. FREEMAN
& COMPANY

A Macmillan Education Imprint

BRIEF CONTENTS

CONTENTS

PART 1 The Facts of Life

1 • Scientific Thinking

Your best pathway to understanding the world

2 • Chemistry
Raw materials and fuel for our bodies

StreetBIO: KNOWLEDGE YOU CAN USE

Melt-in-your-mouth chocolate may not be such a sweet idea.   78

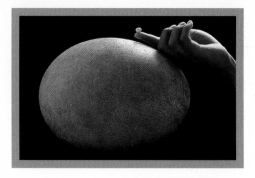

3 · Cells

The smallest part of you

StreetBIO: KNOWLEDGE YOU CAN USE

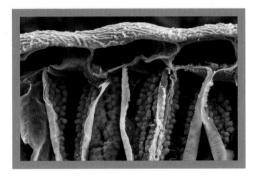

4 • Energy

135

From the sun to you in just two steps

StreetBIO: KNOWLEDGE YOU CAN USE

PART 2 Genetics, Evolution, and Behavior

5 · DNA, Gene Expression, and Biotechnology (177)
What is the genetic code, and how is it harnessed?

StreetBIO: KNOWLEDGE YOU CAN USE

Mixing aspirin and alcohol can lead to metabolic interference and unexpected inebriation.   224

6 · Chromosomes and Cell Division

Continuity and Variety

StreetBIO: KNOWLEDGE YOU CAN USE

Can you select the sex of your baby? (Would you want to?)   270

7 · Genes and Inheritance 277
Family resemblance: how traits are inherited

StreetBIO: KNOWLEDGE YOU CAN USE

Can a gene nudge us toward novelty-seeking (and spicy foods)? 308

8 • Evolution and Natural Selection

315

Darwin's dangerous idea

StreetBIO: KNOWLEDGE YOU CAN USE

Evolution: what it is and what it is not . . . 358

9 • Evolution and Behavior

Communication, cooperation, and conflict in the animal world

StreetBIO: KNOWLEDGE YOU CAN USE

How to win friends and influence people   398

PART 3 Evolution and the Diversity of Life

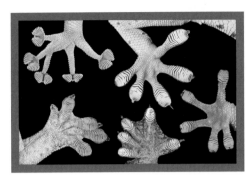

10 · The Origin and Diversification of Life on Earth

(405)

Understanding biodiversity

StreetBIO: KNOWLEDGE YOU CAN USE

11 · Animal Diversification
Visibility in motion

StreetBIO: KNOWLEDGE YOU CAN USE

Where are you from? "Recreational genomics" and the search for clues to your ancestry in your DNA   486

12 • Plant and Fungi Diversification

493

Where did all the plants and fungi come from?

StreetBIO: KNOWLEDGE YOU CAN USE

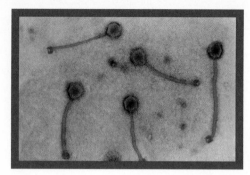

13 · Evolution and Diversity Among the Microbes (533)

Bacteria, archaea, protists, and viruses: the unseen world

StreetBIO: KNOWLEDGE YOU CAN USE

The five-second rule: how clean is that food you just dropped?   564

PART 4 Ecology and the Environment

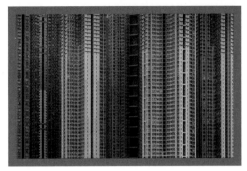

14 · Population Ecology
Planet at capacity: patterns of population growth

StreetBIO: KNOWLEDGE YOU CAN USE

Life history trade-offs and a mini-fountain of youth: what is the relationship between reproduction and longevity? 602

15 · Ecosystems and Communities

609

Organisms and their environments

StreetBIO: KNOWLEDGE YOU CAN USE

Life in the dead zone: in boosting plant productivity on farms, we've created a "dead zone" in the Gulf of Mexico bigger than Connecticut.   642

16 • Conservation and Biodiversity

Human influences on the environment

StreetBIO: KNOWLEDGE YOU CAN USE

The perils of (exotic) pets! 680

Dear Reader,

How many days do you wake up to breaking news about a scary-sounding virus, or a potential cause of cancer, or newly identified genes that make you better at math? In a world of fast progress and easy access to information, it can be difficult to know how much confidence we should have about such reports.

My mission is to help you become more aware of the beauty and the utility of biology, and to help you evaluate the sometimes conflicting messages about science topics and science-related issues. If you could learn anything from reading this book, I hope it would be this: *Biology is about you, and it touches your life every day in dozens of ways. It's creative. And it's fun.*

There are two versions of this third edition of *What Is Life? A Guide to Biology*. One of them, *What Is Life? A Guide to Biology with Physiology,* includes all sixteen chapters of the other version, with an additional ten chapters on plant and animal physiology. It's not always possible to include these additional chapters in a one-term course, but they provide a rich introduction to the importance of biology, with particular significance for human health.

In these pages, you'll find an overview of the key themes in biology as well as detailed information and stories about meaningful topics. I hope you will find answers to questions you're curious about, and will be spurred to ask many more. You'll also find many Red **Q** questions, such as:

- Do megadoses of vitamin C reduce cancer risk?
- An onion has five times as much DNA as a human. Why doesn't that make onions more complex than humans?
- Why doesn't natural selection lead to the production of perfect organisms?
- Why are big, fierce animal species so rare in the world?

The Red **Q**s point toward passages that help uncover the answers. Often, the answer may not be apparent—but look again and think some more. Sometimes you know more than you realize. And sometimes it's possible to transfer the things you learn in one context to another, helping you to recognize new connections. Understanding and developing these abilities will help you tackle novel problems, serving you well long after you may have forgotten this or that specific fact.

At the heart of scientific thinking is a determination to ask and answer questions about the world. This process of inquiry is carried out in diverse and creative ways. Within each chapter of *What Is Life?* you'll find a section called **This Is How We Do It**. In these sections we explore the diverse ways that scientists approach problems, and how they go about finding answers. Example topics include *Why do we yawn?* and *Does sunscreen use reduce skin cancer risk?*

At the end of each chapter, you'll find a section called **StreetBIO: Knowledge You Can Use**. These sections unpack some questions and issues that are particularly practical, such as *How clean is that food you just dropped?*

There's much more to biology than just words. Flip through *What Is Life?* and look at the **photographs**. Images can do much more than simply illustrate ideas; they can inspire and provide an alternative hook for remembering and understanding concepts. They can also challenge you to see ideas in new ways. I have hand-picked every photo, with a goal of provoking and engaging, while helping you make connections between complex ideas.

You'll also notice brief quotes from a variety of literary sources. There is a rich tradition of scientific imagery, references, and metaphors throughout literature. It is my hope that you will recognize that as your scientific literacy increases, your experience and appreciation of literature also will be richer.

In a world of information overload, it is more important than ever to learn how to distill ideas, examples, and implications, forming hierarchies of importance. I don't want you to lose sight of the big picture. In organizing each chapter, I have broken down the topics into **discrete, manageable sections**. And at the end of each, I provide a

Take-Home Message that concisely and precisely highlights and reinforces the section's most important ideas.

Also included in the book are four-page illustrated summaries of each chapter, integrated with multiple-choice and short-answer questions. This **Review & Rehearse (R&R)** "mash-up" of content and quizzing reflects important insights from educational research: integrating testing to hone your retrieval abilities, while reviewing the concepts themselves, enhances your learning—taking it beyond the simple recognition that comes from just revisiting the material.

The multiple-choice questions are just a tiny sampling of the thousands of questions available to you in the online adaptive quizzing system **PrepU**, and each is marked with a difficulty thermometer, which reflects the difficulty of the questions based on more than 30 million responses from students nationwide.

Increasingly, the information you consume includes graphs. It's essential to understand how to read and interpret such figures. To help you, I've included an exercise within each chapter called **Graphic Content**. This critical thinking challenge will help you become adept at reading and analyzing visual displays of information, while identifying subtle assumptions, biases, and even manipulations.

This is just a sample of some of the features in *What Is Life?* I hope that you find this book stimulates new ways of thinking about and understanding the world.

Sincerely,

Jay Phelan

P.S. About the cover: I want to convey that biology isn't something that exists far away, separate from our personal lives. Rather, it intersects with our lives and is a central part of our world.

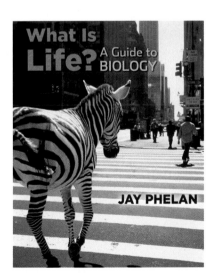 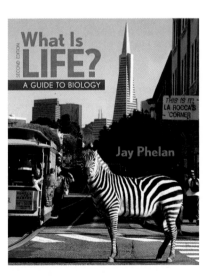 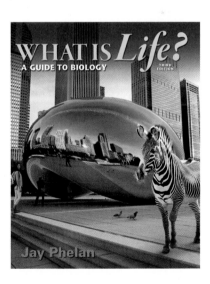

About the Author

Jay Phelan teaches biology at UCLA, where he has taught introductory biology to more than 11,000 majors and non-majors students over the past seventeen years. He is the recipient of more than a dozen teaching awards, including UCLA's highest teaching honor, the Distinguished Teaching Award, in 2011. He received his Ph.D. in evolutionary biology from Harvard in 1995, a master's degree in environmental studies from Yale, and a bachelor's degree from UCLA. His primary area of research is evolutionary genetics, and his original research has been published in *Evolution*, *Experimental Gerontology*, and the *Journal of Integrative and Comparative Biology*, among other journals. His research has been featured on *Nightline*, CNN, the BBC, and National Public Radio; in *Science Times* and *Elle*; and in more than a hundred newspapers.

Jay lectures frequently on a variety of topics in education, including the nurturing of critical thinking skills in undergraduate students and the use and efficacy of online adaptive assessment systems. His research in these areas has been published in the *International Encyclopedia of Education* and numerous journals.

With economist Terry Burnham, Jay is co-author of the international bestseller *Mean Genes: From Sex to Money to Food—Taming Our Primal Instincts*. Written for the general reader, *Mean Genes* explains in simple terms how knowledge of the genetic basis of human nature can empower individuals to lead more satisfying lives.

(The Daily Bruin.)

To Julia

Q: HOW DOES *WHAT IS LIFE?* SO THOROUGHLY CAPTIVATE NON-MAJORS? IT WAS CREATED WITH THEM IN MIND.

Engaging Examples Showcase Biology in Everyday Life

What Is Life? A Guide to Biology threads fascinating, relevant, contemporary examples of the science throughout each chapter.

Brief Sections Make the Material Manageable

Each chapter is broken down into a series of short, accessible sections.

Clear and Consistent Illustrations

Fresh and easy-to-understand figures bring the concepts to life. Collaboratively developed by the author and the scientific illustrator, the text and illustrations are seamlessly integrated, effective learning tools.

From an evolutionary perspective, every living species is successful.

FIGURE 11-2 Two equally successful organisms: the earthworm and the tiger.

Vivid Photos Capture the Story of Biology

Striking images appear as unit openers and are combined with illustrations of biological processes, concepts, and experimental techniques to engage the imagination of the student.

Intriguing, Often Surprising Q Questions Motivate Readers

Q Questions spark students' interest and encourage critical thinking. **Q** Animations (interactive versions of these questions) are available in LaunchPad.

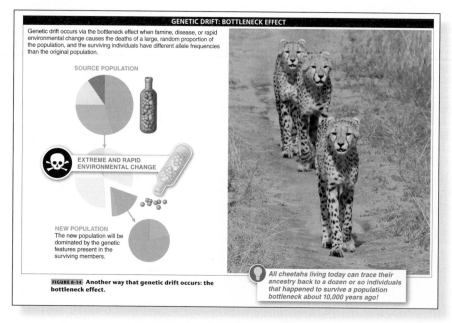

GENETIC DRIFT: BOTTLENECK EFFECT

Genetic drift occurs via the bottleneck effect when famine, disease, or rapid environmental change causes the deaths of a large, random proportion of the population, and the surviving individuals have different allele frequencies than the original population.

SOURCE POPULATION

EXTREME AND RAPID ENVIRONMENTAL CHANGE

NEW POPULATION
The new population will be dominated by the genetic features present in the surviving members.

FIGURE 8-14 Another way that genetic drift occurs: the bottleneck effect.

All cheetahs living today can trace their ancestry back to a dozen or so individuals that happened to survive a population bottleneck about 10,000 years ago!

Q Mammals get bigger and bigger the more they eat. Why don't insects?

Take-Home Messages Provide a Quick Summary

Each section of the chapter includes a concise, memorable summary of key ideas.

TAKE-HOME MESSAGE 14·8

Because constraints limit evolution, life histories are characterized by trade-offs between investments in growth, reproduction, and survival.

StreetBIOs Make Biology Memorable

StreetBIO: KNOWLEDGE YOU CAN USE features are found in every chapter, and demonstrate the practicality and fun of biology.

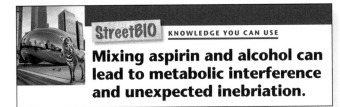

StreetBIO KNOWLEDGE YOU CAN USE

Mixing aspirin and alcohol can lead to metabolic interference and unexpected inebriation.

End-of-Chapter Study Tools

Each chapter includes **Key Terms**, an **R&R** visual summary with embedded questions, and a **Graphic Content** feature that gives students practice in thinking critically about visual displays of data.

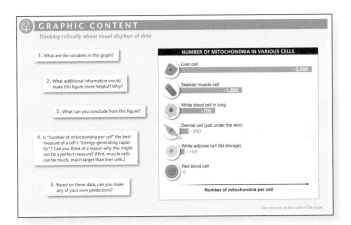

Q: WHAT'S NEW IN THE NEW EDITION? INNOVATIVE NEW STUDY TOOLS

This Is How We Do It

In each chapter, Jay Phelan highlights an intriguing question—for example, *Does sunscreen use reduce skin cancer risk?*—and shows how scientists have approached the problem and thought it through. It's an effective new way to guide students through the process of science and develop their science literacy skills.

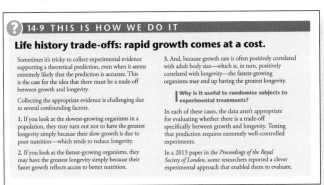

? 14·9 THIS IS HOW WE DO IT

Life history trade-offs: rapid growth comes at a cost.

R&R (Review and Rehearse)

Each chapter now concludes with a four-page visual summary that represents the key ideas, with a brief recap and central illustration from each section. **R&R** also includes short-answer and multiple-choice questions, making it ideal for chapter review, for exam preparation, or as assignments.

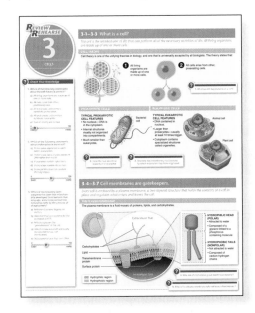

NEW! FOR THE INSTRUCTOR AND FOR THE CLASSROOM

From the front of the classroom to the top of the bestseller list, award-winning educator Jay Phelan knows how to tell the story of how scientists investigate the big questions about life. He is also a master at using biology as a springboard for developing the critical thinking skills and scientific literacy that are essential to students through college and throughout their lives.

Phelan's dynamic approach to teaching biology is the driving force behind *What Is Life?*—the most successful new non-majors biology textbook of the millennium. The rigorously updated new edition brings forward the features that made the book a classroom favorite (chapters anchored to intriguing questions about life, spectacular original illustrations, innovative learning tools). The third edition also includes enhanced art and full integration with its own dedicated version of LaunchPad—W. H. Freeman's breakthrough online course space. LaunchPad fully integrates an interactive e-Book, all student media, and a wide range of assessment and course management features, in a new interface in which power and simplicity go hand in hand.

LaunchPad

Developed with extensive feedback from instructors and students, W. H. Freeman's new online course space offers:

- **Pre-built units for each chapter**, curated by experienced educators, with media for that chapter organized and ready to assign or customize to suit your course.

- **All online resources for the text in one location**, including an interactive e-Book, LearningCurve adaptive quizzing (see below), Bio 101 Tutorials, **Q** Animations, graph-reading and data analysis activities, assessment questions (written by the author), all instructor resources, and more.

- **Intuitive and useful analytics**, with a Gradebook that lets you see how your class is doing individually and as a whole.

- **A streamlined interface** that lets you build an entire course in minutes.

LearningCurve

In a game-like format, LearningCurve adaptive and formative quizzing provides an effective way to get students involved in the coursework. It offers:

- A unique learning path for each student, with quizzes shaped by each individual's correct and incorrect answers.

- A Personalized Study Plan, to guide students' preparation for class and for exams.

- Feedback for each question with live links to relevant e-Book pages, guiding students to the reading they need to do to improve their areas of weakness.

Acknowledgments

As a new graduate student at Harvard, I heard from experienced teaching fellows that if you were interested in learning how to be an effective teacher, it was essential to seek out extraordinary mentors. Based on word of mouth, I became involved with E. O. Wilson's course in Evolutionary Biology and Irven DeVore's course in Human Behavioral Biology. Both were known to be unusually provocative, challenging, and entertaining classes for non-science majors. I aggressively pursued teaching positions in both classes—which I held onto tightly for twelve semesters. Working under these legendary educators, I was set on a course that inspired and prepared me to write this book.

The two courses were quite different from each other, but at their core both were built on two beliefs that are central to this book and to my thinking about education: (1) Biology is creative, interesting, and fun. (2) Biology is relevant to the daily life of every person. There was a palpable sense that, in teaching non-science majors especially, we had a responsibility to provide our students with the tools to thrive in a society increasingly permeated by scientific ideas and issues, and that one of our most effective strategies would be to convey the excitement we felt for biology and the enormous practical value it has to help us understand the world. I thank Professors Wilson and DeVore for all that they have shared with me.

My development as a scientist and, particularly, my appreciation for rigorous and methodical critical thinking have been shaped by the kind support and wise guidance of Richard Lewontin. I have also been fortunate to have as a long-time mentor and collaborator Michael Rose, who has instilled in me a healthy skepticism about any observation in life that is not fivefold replicated. And for almost daily insightful input on matters relating to scientific content, teaching, writing, and more, I thank Terry Burnham.

There are many other friends and colleagues I wish to thank for helping me with *What Is Life?*

In researching and writing the book and in developing the numerous courses I teach, I have benefited from more than a decade of perceptive and valuable contributions, too numerous to list, from Glenn Adelson, Alon Ziv, and Michael Cooperson. I am tremendously appreciative of all they have done for me.

For a project covering so many topics and years, it is essential to have a close group of trusted, tolerant, and knowledgeable colleagues. I am grateful to Alicia Moretti, Harold Owens, Greg Graffin, Brian Swartz, Greg Laden, Jeff Egger, Andy Tobias, Elisabeth Tobias, Joshua Malina, Melissa Merwin-Malina, Bill U'ren, Chris Bruno, and Michelle Richmond, who have offered advice, guidance, and support, far beyond the call of duty.

Numerous colleagues at UCLA provided assistance and support, including Steve Strand, Cliff Brunk, Fred Eiserling, Victoria Sork, Deb Pires, Lianna Johnson, Gaston Pfluegl, Frank Laski, Jeff Thomas, and Tracy Knox. As a result of their commitment to excellence in the UCLA Department of Life Science Core Education, I have been able to acquire a wealth of experiences that have helped me continue improving as a teacher.

I owe a tremendous debt to Sara Tenney, without whose encouragement and support this project could never have been begun or completed.

W. H. Freeman is an extremely author-centric publisher. Throughout the process of creating this book, Liz Widdicombe, John Sargent, and Susan Winslow have been tremendously supportive. I am grateful for their welcoming me into their publishing family. Publisher Peter Marshall has been a tenacious, versatile, and skillful manager of the entire team. I am very fortunate to have such a wise leader and friend overseeing all aspects of this project. The team of editors that worked with me on this book improved it immeasurably. Most importantly, development editor Beth Marsh oversaw every aspect of the writing and production of the book, attending to issues of content, production, and design while making insightful contributions throughout and expertly managing the thousand details necessary to put everything together. I could not have completed this book without Beth's commitment and guidance. I must also convey my gratitude to development editor Jane Tufts, whose meticulous attention to detail, commitment to accuracy, and almost obsessive drive to create a thorough and readable book are apparent on every page.

It is impossible to teach biology without illustrations. My deepest gratitude goes to Tommy Moorman for creating such innovative and effective figures for the book. Tommy's vision for an elegant and beautiful art program completely integrated with the text is apparent on every page. Working with him to develop each illustration in this book has been (and continues to be) one of my most enjoyable and satisfying professional collaborations. Thanks also go to Alison Kendall, Tamara Lau, and Erin Daniel for assisting with the creation of the illustrations. For the design of the book, I thank Tom Carling. And for excellent assistance with photo research, thanks to Julia Phelan, Jennifer Atkins, and Christine Buese.

For creating the innovative media and print materials that accompany the book, I am thankful for the contributions of Mike Jones and for the extensive input of Amanda Dunning and Elaine Palucki. I thank all of the contributors and advisors who helped create the student and instructor resources; your efforts have been invaluable. I thank Jennifer Warner and Meredith Norris for their work on the Student Success Guide. I also appreciate the contributions of Troy Williams and the PrepU team. Sheri Snavely provided significant input in developing pedagogical strategies throughout the book; I also appreciate her thoughtful and wise advice at nearly every step in the publishing process. Copyeditor Linda Strange helped to ensure consistency, accuracy, and readability throughout the text. The rest of the life sciences editorial team at W. H. Freeman, too, have been knowledgeable and supportive, particularly Anna Bristow, Lisa Samols, Susan Moran, Kate Parker, and Elaine Palucki. For their efficiency and commitment to producing a beautiful book, I am most grateful to the W. H. Freeman production team: Sheridan Sellers, Elizabeth Geller, Paul Rohloff, Diana Blume, and Catherine Woods.

The people on the marketing team at W. H. Freeman have contributed enormously in helping with the challenging task of introducing this book to students and instructors across the country. John Britch, Shannon Howard, and Todd Elder have been enthusiastic and dedicated in creating materials and strategies to assist instructors in evaluating the ways in which *What Is Life?* can aid them as they develop their own courses and strategies for success.

Finally, I thank my family—Kevin Phelan, Patrick Phelan, Erin Enderlin, and my parents—for their unwavering support and interest as I wrote this book. Reading draft after draft and following each revision, they made valuable contributions at every stage. I thank Charlie, Jack, and Sam, too. Most of all, for her generous and passionate support of this project from day one, her substantive contributions to both the content and presentation of ideas, and so much more, I thank Julia.

Contact the author with your feedback.

The content of this book has been greatly improved through the comments of reviewers and students. Your comments, suggestions, and criticism are also welcome; they are essential in guiding its ongoing evolution. Please contact the author at jay@jayphelan.com. I'm serious.

We thank the many reviewers who aided in the development of this text.

Third Edition Development

Mike Aaron, Shelton State Community College
Jyoti Abraham, Bacone College
Andrew Accardi, Central Carolina Technical College
Christa Adam, Prairie State College
Patricia Adams, Middlesex Community College
Adjoa Ahedor, Rose State College
Laura Almstead, University of Vermont
Dennis Ancinec, San Diego Mesa College
Eric Anderson, East Carolina University
Taylor Anderson-McGill, College of the Canyons
Corrie Andries, Central New Mexico Community College
Megan Anduri-Flynn, California State University, Fullerton
Jon Aoki, University of Houston–Downtown
Chander Arora, Los Angeles Valley College
Joe Arruda, Pittsburg State University
Momchil Atanassov, Texas Tech University
Kim Atwood, Cumberland University
Yael Avissar, Rhode Island College
Rao Ayyagari, Lindenwood University
Caryn Babaian, Bucks County Community College
Dave Bachoon, Georgia College and State University
Nina Baghai-Riding, Delta State University
Joseph Bailey, University of Tennessee
Michael Bailey, Anderson University
Robert Bailey, Central Michigan University
Ellen Baker, Santa Monica College
Andrew Baldwin, Mesa Community College
Sarah Bales, Moraine Valley Community College
Marilyn Banta, Texas State University
Don Bard, Cabrillo College
John Barone, Columbus State University
Verona Barr, Heartland Community College
Susan Barrett, Massasoit Community College; Bristol Community College
Lois Bartsch, Metropolitan Community College
Thomas David Bass, University of Central Oklahoma
Tonya Bates, University of North Carolina–Charlotte

David Baumgardner, Texas A&M University
Greg Beaulieu, University of Victoria
Bette Beck, Kent State University
Diane Beechinor, Northeast Lakeview College
Mark Belk, Brigham Young University
Rebecca Benard, Case Western Reserve University
Chipley Bennett, Spartanburg Community College
Morgan Benowitz-Fredericks, Bucknell University
Tiffany Bensen, University of Mississippi
Christine Bezotte, Elmira College
Cynthia Bida, Henry Ford Community College
Charles Biles, East Central University
Subhasis Biswas, University of Medicine & Dentistry of New Jersey
Donna Bivans, Pitt Community College
Benjie Blair, Jacksonville State University
Shean Blair, Yakima Valley Community College
Lisa Ann Blankinship, University of North Alabama
Lanh Bloodworth, Florida State College at Jacksonville
Judy Bluemer, Morton College
Lisa Boggs, Southwestern Oklahoma State University
Cheryl Boice, Florida Gateway College
Charles Booth, Eastern Connecticut State University
Aiwei Borengasser, Pulaski Technical College
Karen Borgstrom, Moraine Valley Community College
Susan Bornstein-Forst, Marian University
Bruno Borsari, Winona State University
Anthony Botyrius, York College of Pennsylvania
Lisa Boucher, University of Nebraska
Colin Bradshaw, Medgar Evers College
Linda Brandt, Henry Ford Community College
Susan Brantley, University of North Georgia; Gainesville State College
James Bray Jr, Blackburn College
Mimi Bres, Prince George's Community College
Robert Brewer, Cleveland State Community College
Randy Brewton, University of Tennessee, Knoxville

Peggy Brickman, University of Georgia
Heather Brown, Seattle University
Wendy Brown, Danville Area Community College
Carole Browne, Wake Forest University
Steven Brumbaugh, Green River Community College
Lisa Bryant, Arkansas State University–Beebe
Pam Bryer, Bowdoin College
Michael Bucher, College of San Mateo
Diep Burbridge, Long Beach City College
Jamie Burchill, Troy University
Stephanie Burdett, Brigham Young University
Rebecca Burton, Alverno College
Jack Buser, Iowa Central Community College
Greg Butcher, Centenary College of Louisiana
David Butler, Montana State University–Billings
Nancy Butler, Kutztown University
Pamela Byrd-Williams, Los Angeles Valley College
David Byres, Florida State College at Jacksonville
William Caire, University of Central Oklahoma
Karen Campbell, Albright College
Britt Canada, Western Texas College
Cassandra Cantrell, Western Kentucky University
Laurel Carney-Zelko, Joliet Junior College
Tate Carter, Brigham Young University–Idaho
Dale Casamatta, University of North Florida
Aaron Cassill, University of Texas at San Antonio
Emma Castro, Victor Valley College
Maitreyee Chandra, Diablo Valley College
Jeannie Chapman, University of South Carolina Upstate
Rebekah Chapman, Georgia State University
Kerry Cheesman, Capital University
Jianguo Chen, Claflin University
Samuel Chen, Moraine Valley Community College
Xiaomei Cheng, Mount St. Mary College
Catherine Chia, University of Nebraska
Mark Chiappone, Miami Dade College–Homestead Campus
Genevieve Chung, Broward College
Patricia Clark, Indiana University–Purdue University at Indianapolis
Kimberly Cline-Brown, University of Northern Iowa

Lisa Cobb, Cumberland University
Randy Cohen, California State University, Northridge
Ronald Coleman, California State University, Sacramento
Karen Conzelman, Glendale Community College
Sean Cooney, Montgomery College
Jennifer Cooper, University of Akron
Sarah Cooper, Arcadia University
Bryan Coppedge, Tulsa Community College
Jaimee Corbet, Paradise Valley Community College
Andrea Corbett, Cleveland State University
Clay Corbin, Bloomsburg University
Susan Cordova, Central New Mexico Community College
David Corey, Midlands Technical College
Andrew Corless, Vincennes University
Anthony Cornett, Polk State College
Richard Cowart, Coastal Bend College
Keith Crandall, George Washington University
Michael Crandell, Carl Sandburg College
Jerry Cronin, Salem Community College
Stephen Cronin, Ave Maria University
Jan Crook-Hill, University of North Georgia
Roni Crotty, Tarrant County College
James Crowder, Brookdale Community College
Karen Curto, University of Pittsburgh
Don Dailey, Austin Peay State University
Karen Dalton, Community College of Baltimore County
Hattie Dambroski, Nomandale Community College
Melody Danley, University of Kentucky
Juville Dario-Becker, Central Virginia Community College
Douglas Darnowski, Indiana University Southeast
Pradeep Dass, Appalachian State University
Rachel Davenport, Texas State University
Renee Dawson, University of Utah
Denise Deal, Nassau Community College
Carolyn Dehlinger, Keiser University
Leigh Delaney-Tucker, University of South Alabama
Craig Denesha, Spartanburg Community College
Phil Denette, Delgado Community College
Daniela Derderian, Wayland Baptist University
Mary Dettman, Seminole State College of Florida
Gregg Dieringer, Northwest Missouri State University
Jessica DiGirolamo, Broward College
Mary Dion, St. Louis Community College–Meramec
Wendy Dixon, California State Polytechnic University, Pomona
Tiffany Doan, State College of Florida, Manatee-Sarasota
Pamela Dobbins, Shelton State Community College
Cathy Dobbs, Joliet Junior College
Danielle Dodenhoff, California State University, Bakersfield
Melinda Donnelly, University of Central Florida
Robert Dotson, Tulane University
Kristin Dragos, Marist College
Sondra Dubowsky, McLennan Community College
Dani DuCharme, Waubonsee Community College
Erastus Dudley, Huntingdon College
Lisa Duich Perry, Chaminade University of Honolulu
Jacquelyn Duke, Baylor University
Cindy Duong, California State University, Fullerton
Joseph Eagan, SUNY Adirondack
George Ealy, Keiser University

Kari Eamma, Tarrant County College–Northeast
Angela Edwards, Trident Technical College
Stan Eisen, Christian Brothers University
Jennifer Ellington, Belmont Abbey College
Whitney Elmore, Middle Georgia State College
Ray Emmett, Daytona State College
Miles Engell, North Carolina State University
Amy Erickson, Louisiana State University in Shreveport
Debra Erikson, Allen Community College
Ana Escandon, Los Angeles Harbor College
Marirose Ethington, Genesee Community College
Greg Farley, Chesapeake College
Maureen Farley, North Shore Community College
Gerald Farr, Texas State University–San Marcos
Myriam Feldman, Lake Washington Institute of Technology
Michael Ferkin, University of Memphis
Victor Fet, Marshall University
Steven Fields, Winthrop University
Kerri Finlay, University of Regina
Mary Beth Finn, Herzing University
Thomas Finnegan, Dunwoody College of Technology
Linda Flora, Delaware County Community College
Patricia Flower, San Diego Miramar College
April Ann Fong, Portland Community College, Sylvania Campus
Donald Fontes, Community College of Rhode Island
James Forbes, Hampton University
Charlene Forest, Brooklyn College of City University of New York
Reza Forough, Bellevue College
Brandon Foster, Wake Technical Community College
Michael Fox, SUNY–The College at Brockport
Ellen France, Georgia College and State University
Bettina Francis, University of Illinois
Barbara Frank, Idaho State University–Idaho Falls
Jennifer Fritz, University of Texas at Austin
Jonathan Frye, McPherson College
Susannah Fulton, Shasta College
Cynthia Galloway, Texas A&M Kingsville
Chunlei Gao, Middlesex Community College
Kelley Gaske, South University
Clayton (Ed) Gasque, University of Wisconsin–Stevens Point
Chad Gatlin, Indian Hills Community College
Tom Gehring, Central Michigan University
Carrie Geisbauer, Moorpark College
Zoe Geist, St. Louis Community College
Deborah Gelman, Pace University
Bagie George, Georgia Gwinnett College
Patricia Geppert, University of Texas at San Antonio
Philip Gerard, University of Maine, Rockland College
Kadrin Getman, Blue Ridge Community College
Julie Gibbs, College of DuPage
Diane Gibson, Kentucky Community Technical College, Hazard
Philip Gibson, Gwinnett Technical College
Paul Gier, Huntingdon College
Catherine Gilbert, University of California, Merced
Richard Gill, Brigham Young University
Amy Glaser, Erie Community College
Melissa Glenn, Broome Community College
Mary Gobbett, University of Indianapolis
Jan Goerrissen, Orange Coast College
Andrew Goliszek, North Carolina A&T State University
Brad Goodbar, College of the Sequoias

Amy Goode, Illinois Central College
Dalton Gossett, Louisiana State University in Shreveport
Tamar Goulet, University of Mississippi
Becky Graham, Ohio Dominican University
Jen Grant, University of Wisconsin–Stout
Sherri Graves, Sacramento City College
Madoka Gray-Mitsumune, Concordia University
Melissa Greene, Northwest Mississippi Community College
Suzanna Gribble, Grove City College
John Griffis, Joliet Junior College
Carole Griffiths, Long Island University
Brad Griggs, Piedmont Technical College
Tim Grogan, Valencia College–Osceola Campus
Deann Grossi, The Illinois Institute of Art
Laine Gurley, Harper College
Carla Guthridge, Cameron University
Ghislaine Guyot Jackson, Keiser University
Sue Habeck, Tacoma Community College
Nick Hackett, Moraine Valley Community College
Cheryl Hackworth, West Valley College
Mark Haefele, Community College of Denver
William Hairston, Harrisburg Area Community College
Kristy Halverson, University of Southern Mississippi
Michael Hamed, Virginia Highlands Community College
Pamela Hanratty, Indiana University–Bloomington
Katy Hansen, Bismarck State College
Shari Harden, Metropolitan Community College–Blue River
Joyce Hardy, Chadron State College
Janelle Hare, Morehead State University
John Harley, Eastern Kentucky University
Olivia Harriott, Fairfield University
Katherine Harris, Hartnell College
J. Scott Harrison, Georgia Southern University
Jay Hatch, University of Minnesota
Keith Hench, Kirkwood Community College
Krista Henderson, California State University–Fullerton
Scottie Henderson, Cerritos College
J. L. Henriksen, Bellevue University
Deborah Henry, Coastline Community College
John Hnida, Peru State College
Gloria Hoffman, Morgan State University
Kirsten Hokeness, Bryant University
Amy Hollingsworth, University of Akron
Tina Hopper, Missouri State University
Jane Horlings, Saddleback College
Vanessa Hormann, Broward College
Harold Horn, Lincoln Land Community College
Kris Horn, DeVry University
Anne-Marie Hoskinson, University of Colorado–Boulder
Laurie Host, Harford Community College
Laura Houston, Northeast Lakeview College
Steve (Robert) Howard, Middle Tennessee State University
Carina Howell, Lock Haven University
Kenny Hudiburg, Fort Scott Community College
Joan Hudson, Sam Houston State University
Vicki Huffman, Potomac State College of West Virginia University
Kimberly Hurd, Bakersfield College
Joseph Husband, Florida State College at Jacksonville
Cynthia Hutton, Northland Pioneer College
Jeba Inbarasu, Metropolitan Community College, South Omaha Campus
Anthony Ippolito, DePaul University
Virginia Irintcheva, Black Hawk College

Kamal Islam, Ball State University
Robert Iwan, Inver Hills Community College
Evelyn Jackson, University of Mississippi
Kristin Jacobson, Illinois Central College
Kathleen Janech, College of Charleston
David Jenkins, University of Alabama at Birmingham
Dianne Jennings, Virginia Commonwealth University
Jamie Jensen, Brigham Young University
Bradley Jett, Oklahoma Baptist University
Carl Johansson, Fresno City College
Mitrick Johns, Northern Illinois University
Keith Johnson, Bradley University
Kelly Johnson, Somerset Community College
Kristy Johnson, The Citadel
Robert Johnson, Pierce College
Staci Johnson, Southern Wesleyan University
Tanganika Johnson, Southern University and A&M College
Gail Jones, Texas Christian University
Gregory Jones, Santa Fe College
Ken Jones, Dyersburg State Community College
Kevin Jones, Charleston Southern University
Jacqueline Jordan, Clayton State University
Glenn Kageyama, California State Polytechnic University, Pomona
Ron Kaltreider, York College of Pennsylvania
Lisa Kaplan, Quinnipiac University
Arnold Karpoff, University of Louisville
Judy Kaufman, Monroe Community College
Jerry Kavouras, Lewis University
Joanna Kazmierczak, Community College of Allegheny County
Dawn Keller, Hawkeye Community College
Kristopher Kelley, University of Louisiana at Monroe
Tonya Kerschner, Butler Community College
Moshe Khurgel, Bridgewater College
Amine Kidane, Columbus State Community College
Michael Kiel, Marywood University
Brandon King, California State Polytechnic University, Pomona
Kevin Kinney, DePauw University
Dennis Kitz, Southern Illinois University, Edwardsville
Mark Knauss, Georgia Highlands College
Jennifer Kneafsey, Tulsa Community College
Nighat Kokan, Cardinal Stritch University
Olga Kopp, Utah Valley University
Mary Korte, Concordia University Wisconsin
Anna Koshy, Houston Community College
Melissa Kosinski-Collins, Brandeis University
Karen Koster, University of South Dakota
Peter Kourtev, Carnegie Mellon University
Karen Kowalski, Tidewater Community College
Dennis Kraichely, Cabrini College
Brian Kram, Prince George's Community College
Rene Kratz, Everett Community College
David Krauss, City University of New York
Joe Kremer, Alvernia University
Tim Kreps, Bridgewater College
Nadine Kriska, University of Wisconsin–Whitewater
Dana Kurpius, Elgin Community College
Kim Lackey, University of Alabama
Katherine LaCommare, Lansing Community College
Troy Ladine, East Texas Baptist University
Archana Lal, Independence Community College
Kevin Lam, Alexander College
Ellen Lamb, University of North Carolina at Greensboro
Kirkwood Land, University of the Pacific
Gary Lange, Saginaw Valley State University

Patrick Larkin, Santa Fe College
Liz Lawrence, Miles Community College
Tyler Lawson, Baldwin-Wallace University
Brenda Leady, University of Toledo
Lorraine Leiser, Southeast Community College
Laurie Len, El Camino College
Andrea LeSchack, Trident Technical College
Suzanne Lewis, Truckee Meadows Community College
Harvey Liftin, Broward College
Tammy Liles, Bluegrass Community and Technical College
Debra Linton, Central Michigan University
Kathryn Lipson, Western New England University
Robert Lishak, Auburn University
Cynthia Littlejohn, University of Southern Mississippi
Jason Locklin, Temple College
Suzanne Long, Monroe Community College
David Loring, Johnson County Community College
David Luther, George Mason University
Margaret Lutze, DePaul University
Kimberly Lyle-Ippolito, Anderson University
Cyrus MacFoy, Montgomery College
William Mackay, Edinboro University of Pennsylvania
Tom Maida, Santa Fe College
Margaret Major, Georgia Perimeter College
Charles Mallery, University of Miami
Cindy Malone, California State University, Northridge
Mark Manteuffel, St. Louis Community College
Lisa Maranto, Prince George's Community College
Brian Maricle, Fort Hays State University
Marlee Marsh, Columbia College
Paul Marshall, Middlesex Community College
David Martin, Tacoma Community College, Centralia College
Sharon Marusiak, Cuyahoga Community College
Roy Mason, Mt. San Jacinto College
James McCandless, Methodist University
John McDonald, University of Delaware
Tara McGoey, Canadore College
Joan McKearnan, Anoka Ramsey Community College
Karen McLellan, Indiana University–Purdue University Fort Wayne
Margaret McMichael, Baton Rouge Community College
Brock McMillan, Brigham Young University
Daniel Meer, Cardinal Stritch University
Karen Meisch, Austin Peay State University
Sandi Melkonian, Eastern Florida State College
Maryanne Menvielle, California State University, Fullerton
John Mersfelder, Sinclair Community College
Jennifer Metzler, Ball State University
Heather Miceli, Johnson and Wales University
James Mickle, North Carolina State University
Jessica Miles, Palm Beach State College
Craig Milgrim, Grossmont College
Sheila Miracle, Southeast Kentucky Community & Technical College
Jeanne Mitchell, Truman State University
Par Mohammadian, Los Angeles Mission College
Robert Moldenhauer, St. Clair County Community College
Charles Molnar, Camosun College
Thomas Montagno, Mount Wachusett Community College
Scott Moody, Ohio University
Vertigo Moody, Santa Fe College
Brenda Moore, Truman State University

Frances Moore, Patrick Henry Community College
Jeanelle Morgan, University of North Georgia
Erin Morrey, Georgia Perimeter College
Esther Muehlbauer, Queens College
Tim Mulkey, Indiana State University
John Mull, Weber State University
Catherine Murphy, Ocean County College
Christine Mutiti, Georgia College and State University
Elizabeth Nash, Long Beach City College
Rocky Nation, Southern Wesleyan University
Dana Newton, College of The Albemarle
Necia Nicholas, Calhoun Community College
Zia Nisani, Antelope Valley College
Louise Mary Nolan, Middlesex Community College
Meredith Norris, University of North Carolina at Charlotte
Deanna Noyes, Dallas Baptist University
Richard Nuckels, Austin Community College
Judith Ogilvie, St. Louis University
Olumide Ogunmosin, Houston Community College
Katrina Olsen, University of Wisconsin–Oshkosh
Laura O'Riorden, Tallahassee Community College
Joshua Osborn, Henry Ford College
Cassandra Osborne, Western State Colorado University
Mary O'Sullivan, Elgin Community College
Kathleen Page, Bucknell University
David Palmer, Motlow State Community College
Laura Palmer, Penn State Altoona
Joshua Parker, Clayton State University
Ann Paterson, Williams Baptist College
Bruce Patterson, University of Arizona, Tucson
James Peliska, Ave Maria University
Kathleen Pelkki, Saginaw Valley State University
Clayton Penniman, Central Connecticut State University
Marc Perkins, Orange Coast College
Micah Perkins, Owensboro Community and Technical College
Beverly Perry, Houston Community College
Irene Perry, University of Texas of the Permian Basin
John Peters, College of Charleston
Angela Petitclerc, Bishop's University
Stacy Pfluger, Angelina College
Susan Phillips, Brevard Community College
Don Plantz, Mohave Community College
John Pleasants, Iowa State University
Michael Plotkin, Mt. San Jacinto College
Angela Porta, Kean University
Dan Porter, Amarillo College
Kumkum Prabhakar, Nassau Community College
Saroj Pramanik, Morgan State University
Heather Prestridge, Blinn College
Shaun Prince, Lake Region State College
Dorothy Puckett, Kilgore College
Charles Pumpuni, Northern Virginia Community College
Chittur Radhakrishnan, Santa Fe College
Jeffery Ray, University of North Alabama
Mario Raya, MassBay Community College
Samir Raychoudhury, Benedict College
Amy Reber, Georgia State University
Timothy Redman, SUNY Sullivan
Nick Reeves, Mt. San Jacinto College
Rob Reinsvold, University of Northern Colorado
Erin Rempala, San Diego City College
Ashley Rhodes, Kansas State University
Eugenia Ribeiro-Hurley, Fordham University
Stanley Rice, Southeastern Oklahoma State University
Brenden Rickards, Gloucester County College
Allison Rober, Ball State University

Leah Roesch, Emory University
Karen Rogowski, Baruch College
Peggy Rolfsen, Cincinnati State Technical and Community College
Chris Romero, Front Range Community College–Larimer Campus
Lori Rose, Sam Houston State University
Sadie Rosenthal, Cascadia Community College
Judy Rosovsky, Johnson State College
Hiranya Roychowdhury, New Mexico State University–Doña Ana Community College
Yelena Rudayeva, Palm Beach State College
Lynn Rumfelt, Gordon College
Heather Rushforth, University of North Carolina at Greensboro
Lynette Rushton, South Puget Sound Community College
Michael Rutledge, Middle Tennessee State University
Kim Sadler, Middle Tennessee State University
Stephen Salek, Fayetteville State University
Sydha Salihu, West Virginia University
Shamili Sandiford, College of DuPage
Georgianna Saunders, Missouri State University
Michael Sawey, Texas Christian University
Michael Scanlon, Cornell University
John Schampel, Phoenix Community College
Anna Schmidt, University of Wisconsin–Platteville
Emily Schmitt, Nova Southeastern University
Doreen Schroeder, University of St. Thomas
Malcolm Schug, University of North Carolina at Greensboro
Jennifer Scoby, Illinois Central College
Craig Scott, Clarion University
Michael Scott, Lincoln University
Erik Scully, Towson University
Robin Searles-Adenegan, University of Maryland University College
Pramila Sen, Houston Community College
Carla Serfas, Henry Ford Community College
Brian Seymour, Edward Waters College
Wallace Sharif, Morehouse College
Elizabeth Sharpe-Aparicio, Blinn College
Dave Sheldon, St. Clair County Community College
Marilyn Shopper, Johnson County Community College
Mark Shotwell, Slippery Rock University
Allan Showalter, Ohio University
Eric Shows, Jones County Junior College
Jack Shurley, Idaho State University
Amanda Simmons, Technical College of the Lowcountry
Jeffrey Simmons, Mount St. Mary's University
Derek Sims, Hopkinsville Community College
Michael Small, Texas State University
Jennifer Smith, Triton College
Rob Smith, McHenry County College
Tom Smith, Brigham Young University
William Smith, Wake Forest University
Anna Bess Sorin, University of Memphis
Kathryn Sparace, Tri-County Technical College
Salvatore Sparace, Clemson University
Ashley Spring, Brevard Community College; Eastern Florida State College
William Sproat, Walters State Community College
Brooke Stabler, University of Central Oklahoma
Sonja Stampfler, Kellogg Community College
Sharon Standridge, Middle Georgia State College
Wendy Stankovich, University of Wisconsin–Platteville
John Starnes, Somerset Community College
Nancy Staub, Gonzaga University

Ernest Steele, Morgan State University
Thomas Stege, Austin Community College
Mikel Stevens, Brigham Young University
Jon Stoltzfus, Michigan State University
Lisa Strong, Northwest Mississippi Community College
Irina Stroup, Shasta College
Julie Stupi, Lakeland Community College
Aaron Sullivan, Houghton College
Susan Sullivan, Louisiana State University at Alexandria
Qiang Sun, University of Wisconsin–Stevens Point
Kirby Swenson, Middle Georgia College
Joyce Tamashiro, University of Puget Sound
David Tapley, Salem State University
Tatiana Tatum Parker, St. Xavier University
Kimberly Taugher, Diablo Valley College
Mike Taylor, Santiago Canyon College
Shervia Taylor, Southern University
Franklyn Te, Miami Dade College
Don Terpening, Ulster County Community College
Pamela Thinesen, Century College
Jeffrey Thomas, Queens University of Charlotte
Anna Thompson, Feather River College
Jamey Thompson, Hudson Valley Community College
Rita Thrasher, Pensacola State College
Nina Thumser, Edinboro University of Pennsylvania
Candace Timpte, Georgia Gwinnett College
Jonathan Titus, SUNY Fredonia
Anne Tokazewski, Burlington County College
Robert Tompkins, Belmont Abbey College
Willetta Toole-Simms, Azusa Pacific University
Michelle Tremblay, Georgia Southern University
Dan Trubovitz, San Diego Miramar College
Anh-Hue Tu, Georgia Southwestern State University
Marsha Turell, Houston Community College
Corey Turner, Lone Star College
Muatasem Ubeidat, Southwestern Oklahoma State University
Sophia Ushinsky, Concordia University
Jagan Valluri, Marshall University
Tom Vance, Navarro College
Pete van Dyke, Walla Walla Community College
Eric Vaught, Northwest Arkansas Community College
Jose Vazquez, New York University
Leticia Vega, Barry University
Aggie Veld, Olivet Nazarene University
William Velhagen, New York University
Uma Vithala, Des Moines Area Community College
Jennifer Vlk, Elgin Community College
Ann Vogelmann, Augustana College
Amie Voorhees, Fresno City College
Walt Wagner, Rowan University
Katherine Warpeha, University of Illinois Chicago
Kathy Webb, Bucks County Community College
Colby Weeks, Brigham Young University–Hawaii
Charles Welsh, Duquesne University
Michael Wenzel, Folsom Lake College
Alexander Werth, Hampden-Sydney College
Alicia Whatley, Troy University
Clay White, Lone Star College–CyFair
Jennifer Wiatrowski, Pasco-Hernando Community College
Rachel Wiechman, West Liberty University
Torelen Winbush, Westwood College
Donnell Young, Waubonsee Community College
John Zamora, Middle Tennessee State University
Lara Zerkowski, Albright College
Ted Zerucha, Appalachian State University
Brenda Zink, Northeastern Junior College

Contributors and Advisors

Instructor's Manual—Classroom Catalysts
Verona A. Barr, Heartland Community College
Stephen M. Gómez, Central New Mexico Community College
Paul H. Marshall, Northern Essex Community College

eBook and Non-Major's Biology Study Tools
Christine Lama
Mark S. Manteuffel, St. Louis Community College
Jeffrey Thomas, Queens University

Exploring Biology: Case Studies
Michelle Cawthorn, Georgia Southern University
Jennifer L. Holzman, Emory University

Figure Conversion Engine
Ann Aguanno, Marymount Manhattan College

Hands-On Biology: Laboratories for Distance Learning
Mimi Bres, Cassandra Moore-Crawford, and Arnold Weisshaar, Prince George's Community College (all)

Image Bank Advisor
Jane J. Henry, Baton Rouge Community College

Keynote Lecture Presentation
Michael C. Bucher, College of San Mateo

PowerPoint Lecture Outlines
Kristen L. Curran, University of Wisconsin–Whitewater
Danielle DuCharme, Waubonsee Community College
Jennifer Lange, Chabot College
Mark S. Manteuffel, St. Louis Community College

Prep-U Instructor's Test Bank and Prep-U for Students
Glenn Adelson, Lake Forrest College
Jay Phelan, University of California, Los Angeles
Alon Ziv, Prep-U

Q Animations
Anne Bunnell, East Carolina University
Erastus (Topher) Dudley, Huntingdon College
Johnny El-Rady, University of South Florida
Julie V. Gibbs, College of DuPage
Tina Hopper, Missouri State University
Tracy Hurley, Sumanas, Inc.
Eric Stavney
Sheela Vemu, Northern Illinois University
Carol L. Wymer, Morehead State University
Sophia Ushinsky, Stetson University

Student Worksheets
Michael C. Bucher, College of San Mateo

Active Learning Lecture Outlines
Kristen L. Curran, University of Wisconsin-Whitewater
Mark S. Manteuffel, St. Louis Community College

Bio 101 Tutorials
Amy Horner-Reber, Georgia State University
Kavita Oommen, Georgia State University

Biology Connections Articles
Amy Coombs
Paul H. Marshall, Northern Essex Community College

1 | Scientific Thinking

YOUR BEST PATHWAY TO UNDERSTANDING THE WORLD

More than just a collection of facts, science is a process for understanding the world.

A beginner's guide: what are the steps of the scientific method?

Well-designed experiments are essential to testing hypotheses.

Scientific thinking can help us make wise decisions.

On the road to biological literacy: what are the major themes in biology.

1·1–1·3
More than just a collection of facts, science is a process for understanding the world.

A walk through Carnarvon National Park, Australia.

1·1 What is science? What is biology?

You are already a scientist. You may not have realized this yet, but it's true. Because humans are curious, you have no doubt asked yourself or others questions about how the world works and wondered how you might find the answers.

- Does the radiation released by cell phones cause brain tumors?

- Do large doses of vitamin C reduce the likelihood of getting a cold?

These are important and serious questions. But you've probably also pondered some less weighty issues, too.

- Why is morning breath so stinky? And can you do anything to prevent it?

- Why is it easier to remember gossip than physics equations?

And if you really put your mind to the task, you will start to find questions all around you whose answers you might like to know (and some whose answers you'll learn as you read this book).

- Which parent determines a baby's sex? Why?

- Why do so few women get into barroom brawls?

- What is "blood doping," and does it really improve athletic performance?

- Why is it so much easier for an infant to learn a complex language than it is for a college student to learn biology?

Still not convinced you're a scientist? Here's something important to know: science doesn't require advanced degrees or secret knowledge dispensed over years of technical training. It does, however, require an important feature of our species: a big brain, as well as curiosity and a desire to learn. But curiosity, casual observations, and desire can take you only so far.

Explaining how something works or why something happens requires methodical, objective, and rational observations and analysis that are not clouded with emotions or preconceptions. **Science** is not simply a body of knowledge or a list of facts to be remembered. It is an intellectual activity, encompassing observation, description, experimentation, and explanation of natural phenomena. Put another way, science is a pathway by which we can come to discover and better understand our world.

Later in this chapter, we explore specific ways in which we can most effectively use scientific thinking in our lives. But first let's look at a single powerful question that underlies scientific thinking:

How do you know that is true?

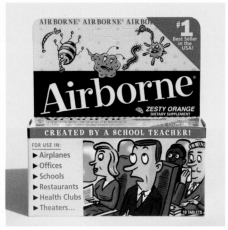

FIGURE 1-1 Some products claim to improve our health, but how do we know whether they work?

Once you begin asking this question—of others and of yourself—you are on the road to a better understanding of the world.

The following two stories about popular and successful products show the importance of questioning the truth of many "scientific" claims you see on merchandise packages or read in a newspaper or on the internet.

Dannon yogurt. According to the Federal Trade Commission (FTC), a U.S. government agency with the mission of consumer protection, the Dannon Company claimed in nationwide advertisements that its Activia yogurt relieves irregularity and helps with "slow intestinal transit time." Dannon also claimed that its DanActive dairy drink helps prevent colds and flu (**FIGURE 1-1**). The FTC charged that the ads were deceptive because there was no substantiation for the claims and, further, that the claims had been clinically proven to be false. In an agreement finalized in 2011, Dannon agreed to pay $21 million in fines and to stop making those claims unless the company gets reliable scientific evidence demonstrating the claims.

Airborne. For more than 15 years, the product Airborne has been marketed and sold to millions of customers. On the packaging and in advertisements, the makers originally asserted that Airborne tablets could ward off colds and boost your immune system (see Figure 1-1). Not surprisingly, Airborne quickly became a great success; it has generated more than $500 million in revenue. Then some consumers posed a reasonable question to the makers of Airborne: *How do you know that it wards off colds?*

Q Can we trust the packaging claims that companies make?

To support their claims, the makers of Airborne pointed to the results of a "double-blind, placebo-controlled study" conducted by a company specializing in clinical drug trials. We'll discuss exactly what those terms mean later in the chapter; for now we just need to note that as a result of a class-action lawsuit, it became clear that no such study had been conducted and that there was *no* evidence to back up Airborne's claims. The Airborne company removed the claims from the packaging and agreed to refund the purchase price to anyone who had bought Airborne. It also removed any reference to its "clinical trials," with the company's CEO saying that people "are really not scientifically minded enough to be able to understand a clinical study."

Are you insulted by the CEO's assumption about your intelligence? You should be. Did you or your parents fall for Airborne's false claims? Possibly. But here's some more good news: you can learn to be skeptical and suspicious (in a good way) of product claims. You can learn exactly what it means to have scientific proof or evidence for something. And you can learn this by learning what it means to think scientifically.

Scientific thinking is important in the study of a wide variety of topics: it can help you understand economics, psychology, history, and many other subjects. Our focus in this book is **biology,** the study of living things. Taking a scientific approach, we investigate the facts and ideas in biology that are already known and study the process by which we come to learn new things. As we move through the book, we explore the most important questions in biology.

- What is the chemical and physical basis for life and its maintenance?

- How do organisms use genetic information to build themselves and to reproduce?

- What are the diverse forms that life on earth takes, and how has that diversity arisen?

- How do organisms interact with each other and with their environment?

In this chapter, we explore how to think scientifically and how to use the knowledge we gain to make wise decisions. Although we generally restrict our focus to biology, scientific thinking can be applied to nearly every endeavor, so in this chapter we use a wide range of examples—

including some from beyond biology—as we learn how to think scientifically. Although the examples vary greatly, they all convey a message that is key to scientific thinking: it's okay to be skeptical.

Fortunately, learning to think scientifically is not difficult—and it can be fun, particularly because it is so empowering. **Scientific literacy,** a general, fact-based understanding of the basics of biology and other sciences, is increasingly important in our lives, and literacy in matters of biology is especially essential.

TAKE-HOME MESSAGE 1·1

Through its emphasis on objective observation, description, and experimentation, science is a pathway by which we can discover and better understand the world around us.

1·2 Biological literacy is essential in the modern world.

A brief glance at any magazine or newspaper will reveal just how much scientific literacy has become a necessity (FIGURE 1-2). Many important health, social, medical, political, economic, and legal issues pivot on complex scientific data and theories. For example, why are unsaturated fats healthier for you than saturated fats? And why do allergies strike children from clean homes more than children from dirty homes? And why do new agricultural pests appear faster than new pesticides?

As you read and study this book, you will be developing **biological literacy,** the ability to (1) use the process of scientific inquiry to think creatively about real-world issues that have a biological component, (2) communicate these thoughts to others, and (3) integrate these ideas into your decision making. Biological literacy doesn't involve just the big issues facing society or just abstract ideas. It also matters to you personally. Should you take aspirin when you have a fever? Are you using the wrong approach if you

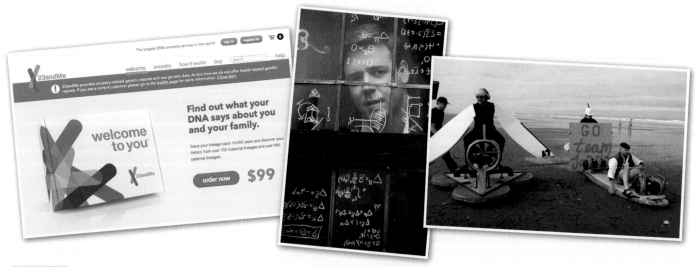

FIGURE 1-2 **In the news.** Every day, news sources report on social, political, medical, and legal issues related to science.

try to lose weight and, after some initial success, you find your rate of weight loss diminishing? Is it a good idea to consume moderate amounts of alcohol? Lack of biological literacy will put you at the mercy of "experts" who may try to confuse you or convince you of things in the interest of (their) personal gain. Scientific thinking will help you make wise decisions for yourself and for society.

TAKE-HOME MESSAGE 1·2

Biological issues permeate all aspects of our lives. To make wise decisions, it is essential for individuals and societies to attain biological literacy.

1·3 Scientific thinking is a powerful approach to understanding the world.

It's a brand new age, and science, particularly biology, is everywhere. To illustrate the value of scientific thinking in understanding the world, let's look at what happens in its absence, by considering some unusual behaviors in the common laboratory rat.

Rats can be trained, without much difficulty, to push a lever to receive a food pellet from a feeding mechanism (**FIGURE 1-3**). When the mechanism is altered so that there is a 10-second delay between the lever being pushed and the food pellet being dispensed, however, strange things start to happen. In one cage, the rat will push the lever and then, very methodically, run and push its nose into one corner of the cage. Then it moves to another corner and again pushes its nose against the cage. It repeats this behavior at the third and fourth corners of the cage, after which the rat stands in front of the feeder and the pellet is dispensed. Each time the rat pushes the lever it repeats the nose-in-the-corner sequence before moving to the food tray.

In another cage, with the same 10-second delay before the food pellet is dispensed, a rat pushes the lever and then proceeds to do three quick back-flips in succession. It then moves to the food tray for the food pellet when the 10 seconds have elapsed. Like the nose-in-the-corner rat, the back-flip rat will repeat this exact behavior each time it pushes the lever.

In cage after cage of rats with these 10-second-delay food levers, each rat eventually develops its own peculiar series of behaviors before moving to the food dish to receive the pellet. Why do they do this? Because it seems to work! They have discovered a method by which they can get a food pellet. To some extent, the rats' behaviors are reasonable. They associate two events—pushing the lever and engaging in some sequence of behaviors—with another event: receiving food. In a sense, they have taken a step toward understanding their world, even though the events are not actually related to each other.

Q Why do people develop superstitions? Can animals be superstitious?

Humans can also mistakenly associate actions with outcomes in an attempt to understand and control their world. The irrational belief that actions or circumstances that are not logically related to a course of events can influence its outcome is called **superstition** (**FIGURE 1-4**). For example, Nomar Garciaparra, a former major league baseball player, always engaged in a precise series of toe taps and adjustments to his batting gloves before he would bat.

Thousands of different narratives, legends, fairy tales, and epics from all around the globe exist to help people understand the world around them. These stories explain everything from birth and death to disease and healing.

FIGURE 1-3 **"In the absence of the scientific method . . ."** Rats develop strange, superstition-like behaviors if there is a 10-second delay between when they push a lever and when food is delivered.

FIGURE 1-4 **Superstitions abound.** As comforting as myths and superstitions may be, they are no substitute for really understanding how the world works.

As helpful and comforting as stories and superstitions may be (or seem to be), they are no substitute for understanding achieved through the process of examination and discovery called the **scientific method.**

The scientific method usually begins with someone observing a phenomenon and proposing an explanation for it. Next, the proposed explanation is tested through a series of experiments. If the experiments reveal that the explanation is accurate, and if others complete the experiments with the same result, then the explanation is considered to be valid. If the experiments do not support the proposed explanation, then the explanation must be revised or alternative explanations must be proposed and tested. This process continues as better, more accurate explanations are found.

While the scientific method reveals much about the world around us, it doesn't explain everything. There are many other methods through which we can gain an understanding of the world. For example, much of our knowledge about plants and animals does not come from the use of the scientific method, but rather comes from systematic, orderly observation, without the testing of any explicit hypotheses. Other disciplines also involve

understandings of the world based on non-scientific processes. Knowledge about history, for example, comes from the systematic examination of past events as they relate to humans, while the "truths" in other fields, such as religion, ethics, and even politics, often are based on personal faith, traditions, and mythology.

Scientific thinking can be distinguished from these alternative ways of acquiring knowledge about the world in that it is **empirical.** Empirical knowledge is based on experience and observations that are rational, testable, and repeatable. The empirical nature of the scientific approach makes it self-correcting: in the process of analyzing a topic, event, or phenomenon with the scientific method, incorrect ideas are discarded in favor of more accurate explanations. In the next sections, we look at how to put the scientific method into practice.

TAKE-HOME MESSAGE 1·3

There are numerous ways of gaining an understanding of the world. Because it is empirical, rational, testable, repeatable, and self-correcting, the scientific method is a particularly effective approach.

1·4–1·10
A beginner's guide: what are the steps of the scientific method?

Eugenie Clark (at left), a pioneering investigator of shark behavior since the 1940s.

1·4 Thinking like a scientist: how do you use the scientific method?

"Scientific method"—this term sounds like a rigid process to follow, much like following a recipe. In practice, however, the scientific method is an adaptable process that can be done effectively in numerous ways. This flexibility makes the scientific method a powerful process that can be used to explore a wide variety of thoughts, events, or phenomena, not only in science but in other areas as well.

The basic steps in the scientific method are:

Step 1. Make observations.

Step 2. Formulate a hypothesis.

Step 3. Devise a testable prediction.

Step 4. Conduct a critical experiment.

Step 5. Draw conclusions and make revisions.

Once begun, though, the process doesn't necessarily continue linearly through the five steps until it is concluded (**FIGURE 1-5**). Sometimes, observations made in the first step can lead to more than one hypothesis and several testable predictions and experiments. And the conclusions drawn from experiments often suggest new observations, refinements to hypotheses, and, ultimately, increasingly precise conclusions.

An especially important feature of the scientific method is that its steps are self-correcting. As we continue to make new observations, a hypothesis about how the world works might change (**FIGURE 1-6**). If our observations do not support our current hypothesis, that hypothesis must be given up in favor of one that is not contradicted by any observations. This may be the most important feature of the scientific method: *it tells us when we should change our minds.*

Q What should you do when something you believe in turns out to be wrong?

THE SCIENTIFIC METHOD

STEP 1	STEP 2	STEP 3	STEP 4	STEP 5
Make observations.	Formulate a hypothesis.	Devise a testable prediction.	Conduct a critical experiment.	Draw conclusions and make revisions.

 The scientific method rarely proceeds in a straight line. Conclusions, for example, often lead to new observations and refined hypotheses.

FIGURE 1-5 **The scientific method: five basic steps and one flexible process.**

❝ If science proves some belief of Buddhism wrong, then Buddhism will have to change. ❞

— THE 14TH DALAI LAMA,
New York Times, December 2005

FIGURE 1-6 **Hold the fries.** We apply an understanding of science when we choose foods from the menu that have fewer calories and less saturated fat.

FIGURE 1-7 **"With your own two eyes . . ."?** How reliable is eyewitness testimony in criminal courts?

Because the scientific method is a general strategy for learning, it needn't be used solely to learn about nature or scientific things. In fact, we can analyze an important criminal justice question using the scientific method:

- How reliable is eyewitness testimony in criminal courts?

For more than 200 years, courts in the United States have viewed eyewitness testimony as unassailable. Few things are seen as more convincing to a jury than an individual testifying that she can identify the person she saw commit a crime (FIGURE 1-7). But is eyewitness identification always right? Can the scientific method tell us whether this perception—or some other commonly held idea—is supported by evidence? As we describe how to use the scientific method to answer questions about the world, it will become clear that the answer is a resounding *yes*. In the coming sections of this chapter, we also look at how the scientific method can be used to address a variety of issues.

In addition to our criminal justice question, we'll answer two additional questions:

- Does echinacea reduce the intensity or duration of the common cold?

- Does shaving hair from your face, legs, or anywhere else cause it to grow back coarser or darker?

> ## TAKE-HOME MESSAGE 1·4
>
> The scientific method (observation, hypothesis, prediction, test, and conclusion) is a flexible, adaptable, and efficient pathway to understanding the world, because it tells us when we must change our beliefs.

1·5 Step 1: Make observations.

Scientific study always begins with observations: we simply look for interesting patterns or cause-and-effect relationships. This is where a great deal of the creativity of science comes from. In the case of eyewitness testimony, DNA technologies have made it possible to assess whether tissue such as hair or blood from a crime scene came from a particular suspect. Armed with these tools, the U.S. Justice Department recently reviewed 28 criminal convictions that had been overturned by DNA evidence.

It found that in most of the cases, the strongest evidence against the defendant had been eyewitness identification. The observation here is that many defendants who are later found to be innocent were initially convicted based on eyewitness testimony.

Opportunities for other interesting observations are unlimited. Using the scientific method, we can (and will) also answer our two other questions.

STEP 1   STEP 2   STEP 3   STEP 4   STEP 5

STEP 1: MAKE OBSERVATIONS

OBSERVATION
To many people, consuming echinacea extract seems to reduce the intensity or duration of symptoms of the common cold.

FIGURE 1-8 **The first step of science: making observations about the world.**

Many people have claimed that consuming extracts of the herb echinacea can reduce the intensity or duration of symptoms of the common cold (**FIGURE 1-8**). We can ask: how do you know this is true?

- Does taking echinacea reduce the intensity or duration of the common cold?

Some people have suggested that shaving hair from your face, legs, or anywhere else causes the hair to grow back coarser and darker. Is this true?

- Does hair that is shaved grow back coarser or darker?

Using the scientific method, we can answer these questions.

TAKE-HOME MESSAGE 1·5

The scientific method begins by making observations about the world, noting apparent patterns or cause-and-effect relationships.

1·6 Step 2: Formulate a hypothesis.

Based on observations, we can develop a **hypothesis** (*pl.* **hypotheses**), a proposed explanation for observed phenomena. What hypotheses could we make about the eyewitness-testimony observations described in the previous section? We could start with the hypothesis "Eyewitness testimony is always accurate." We may need to modify our hypothesis later, but this is a good start. At this point, we can't draw any conclusions. All we have done is summarize some interesting patterns we've seen in a possible explanation.

To be most useful, a hypothesis must accomplish two things.

1. It must establish an alternative explanation for a phenomenon. That is, it must be clear that if the proposed explanation is not supported by evidence or further observations, a different hypothesis is a more likely explanation.

2. It must generate testable predictions (**FIGURE 1-9**). This characteristic is important because we can evaluate the validity of a hypothesis only by putting it to the test. For example, we could disprove the "Eyewitness testimony is always accurate" hypothesis by demonstrating that, in certain circumstances, individuals who have witnessed a crime might misidentify someone as the criminal when asked to select the suspect from a lineup.

Researchers often pose a hypothesis as a negative statement, proposing that there is no relationship between two factors, such as "Echinacea has no effect on the duration and severity of cold symptoms." Or "There is no difference in the coarseness or darkness of hair that grows after shaving." A hypothesis that states a *lack* of relationship between two factors is called a **null hypothesis.** Both types of hypothesis are equally valid, but a null hypothesis is easier to disprove. This is because a single piece of evidence or a

STEP 1 STEP 2 STEP 3 STEP 4 STEP 5

STEP 2: FORMULATE A HYPOTHESIS

HYPOTHESIS
Echinacea reduces the duration and severity of the common cold.

FIGURE 1-9 **Hypothesis: the proposed explanation for a phenomenon.**

single new observation that contradicts a null hypothesis is sufficient for us to reject it and conclude that an alternative hypothesis must be considered. So, once you have one piece of solid evidence that your null hypothesis is not true, you gain little by collecting further data. Conversely, it is impossible to prove that a hypothesis is absolutely and permanently true: all evidence or further observations that support a hypothesis are valuable, but they do not rule out

the possibility that some future evidence or observation might show that the hypothesis is not true. They simply give us more confidence about our hypothesis.

> **❝** Scientific issues permeate the law. I believe [that] in this age of science we must build legal foundations that are sound in science as well as in law. The result, in my view, will further not only the interests of truth but also those of justice.**❞**
>
> — U.S. SUPREME COURT JUSTICE STEPHEN BREYER, at the annual meeting of the American Association for the Advancement of Science, February 1998

For our two additional observations, we could state each of our hypotheses in two different ways:

Hypothesis: Echinacea reduces the duration and severity of the symptoms of the common cold.

Null hypothesis: Echinacea has no effect on the duration or severity of the symptoms of the common cold.

Hypothesis: Hair that is shaved grows back coarser and darker.

Null hypothesis: There is no difference in the coarseness or color of hair that is shaved relative to hair that is not shaved.

TAKE-HOME MESSAGE 1·6

A hypothesis is a proposed explanation for an observed phenomenon.

1·7 Step 3: Devise a testable prediction.

Not all hypotheses are created equal. For a hypothesis to be useful, it must generate a prediction. That is, it must suggest that under certain conditions we will be able to observe certain outcomes. Put another way, a good hypothesis helps us make predictions about novel situations. This is a powerful feature of a good hypothesis: it guides us to knowledge about new situations.

All of this is rather abstract. Let's get more concrete with the three hypotheses we are considering. Keep in mind that when you do not understand some aspect of the world, any one of several possible explanations could be true. In devising a testable prediction from a hypothesis, the goal is

to propose a situation that will give a particular outcome if your hypothesis is true, but will give a different outcome if your hypothesis is not true.

Hypothesis: Eyewitness testimony is always accurate.
Prediction: Individuals who have witnessed a crime will correctly identify the criminal regardless of whether multiple suspects are presented one at a time or all at the same time in a lineup.

This is a good, testable prediction because, if our hypothesis is true, then our prediction will always be true. On the other hand, if one method of presenting suspects consistently causes

STEP 1   STEP 2   STEP 3   STEP 4   STEP 5

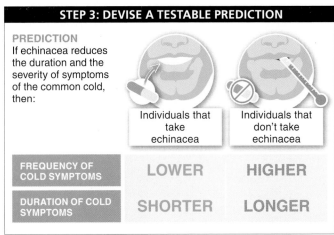

STEP 3: DEVISE A TESTABLE PREDICTION

PREDICTION
If echinacea reduces the duration and the severity of symptoms of the common cold, then:

	Individuals that take echinacea	Individuals that don't take echinacea
FREQUENCY OF COLD SYMPTOMS	LOWER	HIGHER
DURATION OF COLD SYMPTOMS	SHORTER	LONGER

FIGURE 1-10 **Coming up with a testable prediction.** For a hypothesis to be useful, it must generate a testable prediction.

incorrect identification of the criminal, our hypothesis cannot be true and must be revised or discarded.

Hypothesis: Echinacea reduces the duration and severity of the symptoms of the common cold.

Prediction: If echinacea reduces the duration and severity of the symptoms of the common cold, then individuals taking echinacea should get a cold less frequently than those not taking it, and when they do get sick, their illness should not last as long (**FIGURE 1-10**).

Hypothesis: Hair that is shaved grows back coarser and darker.

Prediction: If shaving leads to coarser, darker hair growing back, then if individuals shaved one leg (or eyebrow) only, the hair that re-grows should become darker and coarser than the hair on the other leg (or eyebrow).

As you begin to think scientifically, you will find yourself making a lot of "if . . . then" types of statements: "*If* that is true," referring to some hypothesis or assertion someone makes, or perhaps a claim made by the manufacturer of a new health product, "*then* I would expect . . . ," proposing your own prediction about a related situation. Once you've made a testable prediction, the next step is to go ahead and test it.

TAKE-HOME MESSAGE 1·7

For a hypothesis to be useful, it must generate a testable prediction.

1·8 Step 4: Conduct a critical experiment.

Once we have formulated a hypothesis that generates a testable prediction, we conduct a **critical experiment,** an experiment that makes it possible to decisively determine whether a particular hypothesis is correct. There are many crucial elements in designing a critical experiment, and Section 1-11 covers the details of this process. For now, it is important just to understand that, with a critical experiment, if the hypothesis being tested is not true, we will make observations that compel us to reject that hypothesis.

In this step of the scientific method, a bit of cleverness can come in handy. Suppose we were to devise a critical experiment to test our hypothesis "Eyewitness testimony is always accurate." First, we stage a mock crime such as a purse snatching in front of a group of observers who do not know the crime is staged. Next, we ask observers to identify the criminal. To one group of observers we might present six "suspects" all at once in a lineup. To another

group of observers we might present the six suspects one at a time. The beauty of this experiment is that we actually know who the "criminal" is. With this knowledge, we can evaluate with certainty whether an eyewitness's identification is correct or not.

This exact experiment was done, but with an additional, devious little twist: the researcher did not include the actual "criminal" in any of the lineups or one-by-one presentations of the six "suspects." If eyewitness testimony is always accurate, however, this slight variation should not matter. We would predict that the observers in both groups (the lineup group and the one-at-a-time group) would indicate that the criminal was not present. In the next section, we'll see what happened. Right now, let's devise critical experiments for our other hypotheses.

Hypothesis: Echinacea reduces the duration and severity of the symptoms of the common cold. The critical experiment

WHAT IS SCIENCE?   SCIENTIFIC METHOD   EXPERIMENTAL DESIGN   DECISION MAKING   THEMES IN BIOLOGY

11

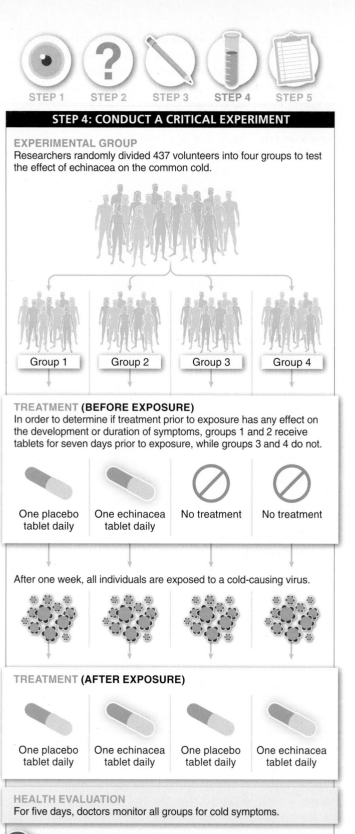

STEP 1 STEP 2 STEP 3 STEP 4 STEP 5

STEP 4: CONDUCT A CRITICAL EXPERIMENT

EXPERIMENTAL GROUP
Researchers randomly divided 437 volunteers into four groups to test the effect of echinacea on the common cold.

Group 1 | Group 2 | Group 3 | Group 4

TREATMENT (BEFORE EXPOSURE)
In order to determine if treatment prior to exposure has any effect on the development or duration of symptoms, groups 1 and 2 receive tablets for seven days prior to exposure, while groups 3 and 4 do not.

One placebo tablet daily | One echinacea tablet daily | No treatment | No treatment

After one week, all individuals are exposed to a cold-causing virus.

TREATMENT (AFTER EXPOSURE)

One placebo tablet daily | One echinacea tablet daily | One placebo tablet daily | One echinacea tablet daily

HEALTH EVALUATION
For five days, doctors monitor all groups for cold symptoms.

💡 *A critical experiment makes it possible to conclusively determine whether a hypothesis is correct.*

FIGURE 1-11 "When you need to know . . ."

FIGURE 1-12 **Does shaving hair cause it to grow back coarser or darker?**

for the echinacea hypothesis has been performed, and it is about as close to a perfect experiment as possible. Researchers began with 437 people who volunteered to be exposed to viruses that cause the common cold. Exposure to the cold-causing viruses was a bit unpleasant: all of the volunteers had cold viruses (in a watery solution) dripped into their noses. The research subjects were then secluded in hotel rooms for five days, and doctors examined them for the presence of the cold virus in their nasal cavity and for any cold symptoms (**FIGURE 1-11**). (It is actually quite rare to expose human subjects to illness, and they always must be informed of the risks and then must give, by signing a form, what is called "informed consent.")

Before the cold virus treatment, the volunteers were randomly divided into four groups. In two of the groups, each individual began taking a pill each day for a week prior to exposure to the virus. Those in one group received echinacea tablets, while those in the other took a **placebo,** a pill that looked identical to the echinacea tablet but contained no echinacea or other active ingredient. Neither the subject nor the doctor administering the tablets (and later checking for cold symptoms) knew what they contained. In the other two groups, the individuals did not begin taking the tablets until the day they were exposed to the cold virus. Again, one group got the echinacea and the other group got the placebo.

Q Does shaving or cutting hair make it grow back more thickly?

Hypothesis: Hair that is shaved grows back coarser and darker. A critical experiment does not have to be complex or high-tech. All that matters is that it can decisively determine whether or not a hypothesis is correct (**FIGURE 1-12**). This point is illustrated by an experiment published in the scholarly journal *Archives of Facial Plastic*

Surgery. Researchers decided to test the hypothesis that shaving hair causes it to grow back coarser or darker. A group of volunteers had one of their eyebrows, selected at random, completely shaved off. The subjects were then evaluated for eyebrow re-growth over the course of the next six months, with observers also analyzing photographs of the individuals.

In the next section we'll see how our hypotheses survive being confronted with the results of these critical experiments and how we can move toward drawing conclusions.

TAKE-HOME MESSAGE 1·8

A critical experiment is one that makes it possible to decisively determine whether a particular hypothesis is correct.

1·9 Step 5: Draw conclusions, make revisions.

Once the results of the critical experiment are in, they are pulled apart, examined, and analyzed. Researchers look for patterns and relationships in the evidence they've gathered from their experiments; they draw conclusions and see whether their findings and conclusions support their hypotheses. If an experimental result is not what you expected, that does not make it a "wrong answer." Science includes a great deal of trial and error, and if the conclusions do not support the hypothesis, then you must revise your hypothesis, which often spurs you to conduct more experiments. This step is a cornerstone of the scientific method because it demands that you must be open-minded and ready to change what you think.

Q Is eyewitness testimony in courts always right?

The results of the purse-snatching experiment were surprising. When the suspects (which, as you'll recall, did not include the actual "criminal") were viewed together in a lineup, the observers/witnesses erroneously identified someone as the purse snatcher about a third of the time. When the suspects were viewed one at a time, the observers made a mistaken identification less than 10% of the time.

In this or any other experiment, it does not matter whether we can imagine a reason for the discrepancy between our

hypothesis and our results. What is important is that we have demonstrated that our initial hypothesis—"Eyewitness testimony is always accurate"—is not supported by the data. Our observations suggest that, at the very least, the accuracy of an eyewitness's testimony depends on the method used to present the suspects. Based on this result, we might adjust our hypothesis to: "Eyewitness testimony is more accurate when suspects are presented to witnesses one at a time."

We can then devise new and more specific testable predictions in an attempt to further refine our hypothesis. In the case of eyewitness testimony, further investigation suggests that when suspects or pictures of suspects are placed side by side, witnesses compare them and tend to choose the suspect that *most resembles* the person they remember committing the crime. When viewed one at a time, suspects can't be compared in this way, and witnesses are less likely to make misidentifications.

Hypothesis: Echinacea reduces the duration and severity of the symptoms of the common cold. In the echinacea study, the results were definitive (**FIGURE 1-13**). Those who took the echinacea were just as likely to catch a cold as those who took placebo, and, once they caught the cold, the symptoms lasted for the

Q Does echinacea help prevent the common cold?

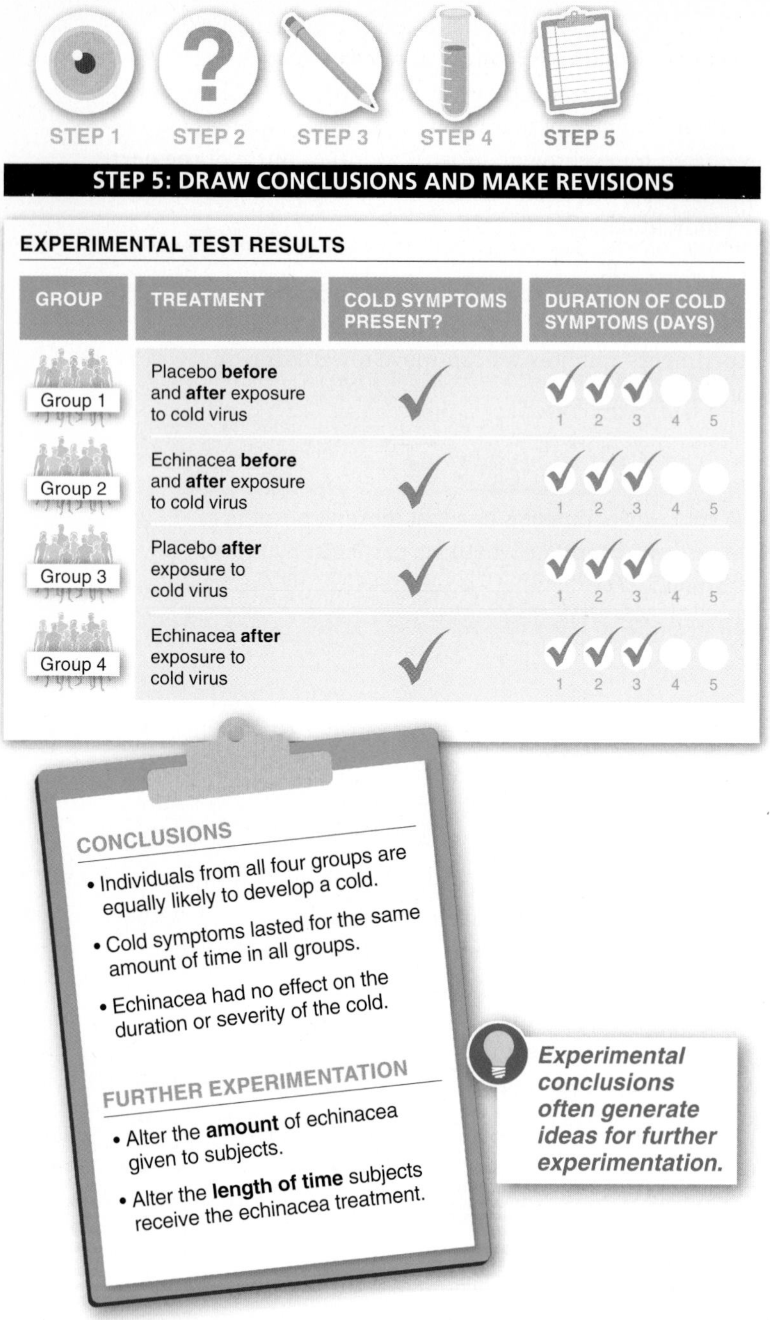

STEP 5: DRAW CONCLUSIONS AND MAKE REVISIONS

EXPERIMENTAL TEST RESULTS

GROUP	TREATMENT	COLD SYMPTOMS PRESENT?	DURATION OF COLD SYMPTOMS (DAYS)				
Group 1	Placebo **before** and **after** exposure to cold virus	✓	✓ 1	✓ 2	✓ 3	4	5
Group 2	Echinacea **before** and **after** exposure to cold virus	✓	✓ 1	✓ 2	✓ 3	4	5
Group 3	Placebo **after** exposure to cold virus	✓	✓ 1	✓ 2	✓ 3	4	5
Group 4	Echinacea **after** exposure to cold virus	✓	✓ 1	✓ 2	✓ 3	4	5

CONCLUSIONS

- Individuals from all four groups are equally likely to develop a cold.
- Cold symptoms lasted for the same amount of time in all groups.
- Echinacea had no effect on the duration or severity of the cold.

FURTHER EXPERIMENTATION

- Alter the **amount** of echinacea given to subjects.
- Alter the **length of time** subjects receive the echinacea treatment.

Experimental conclusions often generate ideas for further experimentation.

FIGURE 1-13 **Using experimental conclusions to make revisions.**

same amount of time. In short, echinacea had no effect at all. Several similar studies have been conducted, all of which show that echinacea does not have any beneficial effect. As one of the researchers commented afterward, "We've got to stop attributing any efficacy to echinacea."

Although it seems clear that our initial hypothesis that echinacea stops people from catching colds and reduces the severity and duration of cold symptoms is not correct, further experimentation might involve altering the amount of echinacea given to the research subjects or the length of time they take echinacea before exposure to the cold-causing viruses.

Hypothesis: Hair that is shaved grows back coarser and darker. In the hair-shaving experiment, the observers discovered that, after six months, it was impossible to distinguish which eyebrow had been shaved. This was not a surprise to dermatologists evaluating the study, because all of the living parts involved in hair growth are below the surface of the skin. (Plucking hairs can damage the root of the hair, potentially affecting future hair growth, but that is another story that awaits further study.)

The outcomes in all of these studies show that after the results of a critical experiment have been gathered and interpreted, it is important not just to evaluate the initial hypothesis but also to consider any necessary revisions or refinements to it. This revision is an important step; by revising a hypothesis, based on the results of experimental tests, we can explain the observable world with greater and greater accuracy.

TAKE-HOME MESSAGE 1·9

Based on the results of experimental tests, we can revise a hypothesis and explain the observable world with increasing accuracy. A great strength of scientific thinking, therefore, is that it helps us understand when we should change our minds.

1·10 When do hypotheses become theories, and what are theories?

It's an unfortunate source of confusion that, among the general public and in the popular media, the word "theory" is often used to refer to a hunch or a guess or speculation—that is, something we are not certain about. In fact, to scientists, the word means nearly the opposite: a hypothesis of which they are most certain. To reduce these common misunderstandings, we examine here the ways in which scientists describe our knowledge about natural phenomena.

Hypothesis As we have seen, hypotheses are at the very heart of scientific thinking. A hypothesis is a proposed explanation for a phenomenon. A good hypothesis leads to testable predictions. Commonly, when non-scientists use the word "theory"—as in, "I've got a theory about why there's less traffic on Friday mornings than on Thursday mornings"—they actually mean that they have a hypothesis.

Theory A **theory** is an explanatory hypothesis for natural phenomena that is exceptionally well supported by the empirical data. A theory can be thought of as a hypothesis that has withstood the test of time and is unlikely to be altered by any new evidence. Like a hypothesis, a theory is testable; but because it has already been repeatedly tested and

no observations or experimental results have contradicted it, a theory is viewed by the scientific community with nearly the same confidence as a fact. For this reason, it is inappropriate to describe something as "just a theory" as a way of asserting that it is not likely to be true.

Theories in science also tend to be broader in scope than hypotheses. In biology, two of the most important theories (which we explore in more detail in Chapters 3 and 8) are the *cell theory*, that all organisms are composed of cells and all cells come from preexisting cells, and the *theory of evolution by natural selection*, that species can change over time and all species are related to one another through common ancestry.

> ### TAKE-HOME MESSAGE 1·10
> Scientific theories do not represent speculation or guesses about the natural world. Rather, they are hypotheses—proposed explanations for natural phenomena—that have been so strongly and persuasively supported by empirical observation that the scientific community views them as very unlikely to be altered by new evidence.

1·11–1·14
Well-designed experiments are essential to testing hypotheses.

Controlled experiments increase the power of our observations. Here, laboratory-grown plants are measured.

1·11   Controlling variables makes experiments more powerful.

From our earlier discussion of critical experiments, you have a sense of how important it is to have a well-planned, well-designed experiment. In performing experiments, our goal is to figure out whether one thing influences another thing: if an experiment enables us to draw a correct conclusion about that cause-and-effect relationship, it is a good experiment.

In our initial discussion of experiments, we just described what the experiment was, without examining why the researchers chose to perform the experiment the way they did. In this section, we explore some of the ways to maximize an experiment's power, and we'll find that, with careful planning, it is possible to increase an experiment's ability to discern causes and effects.

First, let's consider some elements common to most experiments.

1. Treatment: any experimental condition applied to the research subjects. It might be the shaving of one of an individual's eyebrows, or the pattern used to show "suspects" (all at once or one at a time) to the witness of a staged crime, or a dosage of echinacea given to an individual.

2. Experimental group: a group of subjects who are exposed to a particular treatment—for example, the individuals given echinacea rather than placebo in the experiment described above. It is sometimes referred to as the "treatment group."

3. Control group: a group of subjects who are treated identically to the experimental group, with one exception—they are not exposed to the treatment. An example would be the individuals given placebo rather than echinacea.

4. Variables: the characteristics of an experimental system that are subject to change. They might be, for example, the amount of echinacea a person is given, or a measure of the coarseness of an individual's hair. When we speak of "controlling" variables—the most important feature of a good experiment—we are describing the attempt to minimize any differences (which are also called "variables") between a control group and an experimental group other than the treatment itself. That way, any differences between the groups in the outcomes we observe are most likely due to the treatment.

Let's look at a real-life example that illustrates the importance of considering all these elements when designing an experiment.

Stomach ulcers are erosions of the stomach lining that can be very painful. In the late 1950s, a doctor reported in the *Journal of the American Medical Association* that stomach ulcers could be effectively treated by having a patient swallow a balloon connected to some tubes that

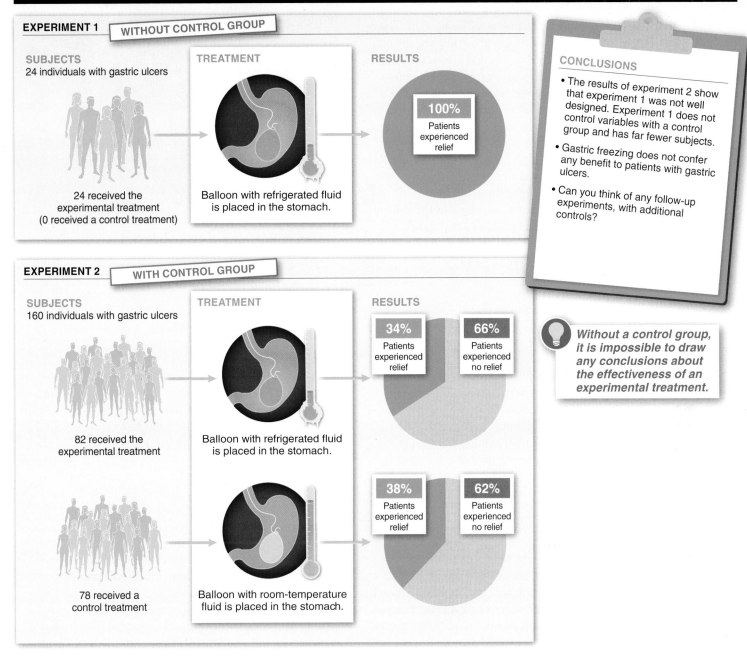

EXPERIMENT 1 WITHOUT CONTROL GROUP

SUBJECTS
24 individuals with gastric ulcers

24 received the
experimental treatment
(0 received a control treatment)

TREATMENT

Balloon with refrigerated fluid
is placed in the stomach.

RESULTS

100%
Patients
experienced
relief

CONCLUSIONS
• The results of experiment 2 show
that experiment 1 was not well
designed. Experiment 1 does not
control variables with a control
group and has far fewer subjects.
• Gastric freezing does not confer
any benefit to patients with gastric
ulcers.
• Can you think of any follow-up
experiments, with additional
controls?

EXPERIMENT 2 WITH CONTROL GROUP

SUBJECTS
160 individuals with gastric ulcers

82 received the
experimental treatment

TREATMENT

Balloon with refrigerated fluid
is placed in the stomach.

RESULTS

34%
Patients
experienced
relief

66%
Patients
experienced
no relief

Without a control group,
it is impossible to draw
any conclusions about
the effectiveness of an
experimental treatment.

78 received a
control treatment

Balloon with room-temperature
fluid is placed in the stomach.

38%
Patients
experienced
relief

62%
Patients
experienced
no relief

FIGURE 1-14 No controls. Gastric freezing and stomach ulcers: a poorly designed experiment and a well-controlled "do-over."

circulated a refrigerated fluid. He argued that by super-cooling the stomach, acid production was reduced and the ulcer symptoms were relieved. He had convincing data to back up his claim: in all 24 of his patients who received this "gastric freezing" treatment, their condition improved (**FIGURE 1-14**). As a result, the treatment became widespread for many years.

Although there was a clear hypothesis ("Gastric cooling reduces the severity of ulcers") and some compelling

observations (all 24 patients experienced relief), this experiment was not designed well. In particular, there was no clear group with whom to compare the patients who received the treatment. In other words, who is to say that just going to the doctor or having a balloon put into your stomach doesn't improve ulcers? The results of this doctor's experiment do not rule out these interpretations.

A few years later, another researcher decided to do a more carefully controlled study. He recruited 160 ulcer patients and gave 82 of them the gastric freezing treatment. The other 78 received a similar treatment in which they swallowed the balloon but had room-temperature water pumped in. The latter was an appropriate control group because the subjects were treated exactly like the experimental group, with the exception of only a single difference between the groups—whether they experienced gastric freezing or not. The new experiment could test for an effect of the gastric freezing, while controlling for the effects of other, lurking variables that might affect the outcome.

Surprisingly, although the researcher found that for 34% of those in the gastric freezing group their condition improved, he also found that 38% of those in the control group improved. These results indicated that gastric freezing didn't actually confer any benefit when compared with a treatment that did not involve gastric freezing. Not surprisingly, the practice was abandoned.

A surprising result from the gastric freezing study—and many other studies, as we'll see—was that they demonstrated the **placebo effect,** the frequently observed, poorly understood phenomenon in which people respond favorably to *any* treatment. The placebo effect highlights the need for an appropriate control group. We want to know whether the treatment is actually responsible for any effect seen; if the control group receiving the placebo or sham treatment has an outcome much like that of the experimental group, we can conclude that the treatment itself does not have an effect.

Another pitfall to be aware of in designing an experiment is to ensure that the persons conducting the experiment don't influence the experiment's outcome. An experimenter can often unwittingly influence the results of an experiment. This phenomenon is seen in the story of a horse named Clever Hans. Hans was considered clever because his owner claimed that Hans could perform remarkable intellectual feats, including multiplication and division. When given a numerical problem, the horse would tap out the answer number with his foot. Controlled experiments, however, demonstrated that Hans was only able to solve problems when he could see the person asking the question and when that person knew the answer (**FIGURE 1-15**). It turned out that the questioners, unintentionally and through very subtle body language, revealed the answers.

The Clever Hans phenomenon highlights the benefits of instituting even greater controls when designing an experiment. In particular, it highlights the value of **blind experimental design,** in which the experimental subjects do not know which treatment (if any) they are receiving, and **double-blind experimental design,** in which neither

FIGURE 1-15 **Math whiz or ordinary horse?** The horse Clever Hans was said to be capable of mathematical calculations, until a controlled experiment demonstrated otherwise.

the experimental subjects nor the experimenter know which treatment a subject is receiving.

Another hallmark of an extremely well-designed experiment is that it combines the blind/double-blind strategies we've just described in a **randomized,** controlled, double-blind study. In this context, "randomized" refers to the fact that, as in the echinacea study described above, the subjects are randomly assigned into experimental and control groups. In this way, researchers and subjects have no influence on the composition of the two groups.

The use of randomized, controlled, double-blind experimental design can be thought of as an attempt to imagine all the possible ways that someone might criticize an experiment and to design the experiment so that the results cannot be explained by anything other than the effect of the treatment. In this way, the experimenter's results either support the hypothesis or invalidate it—in which case, the hypothesis must be rejected. If multiple explanations can be offered for the observations and evidence from an experiment, then it has not succeeded as a critical experiment.

Suppose you want to know whether a new drug is effective in fighting the human immunodeficiency virus (HIV), the virus that leads to AIDS. Which experiment would be better, one in which the drug is added to HIV-infected cells in a test tube under carefully controlled laboratory conditions, or one in which the drug is given to a large number of HIV-infected individuals? There is no definitive answer. In laboratory studies, it is possible to control nearly every environmental variable. In their simplicity, however, lab studies may introduce difficulties in coming to broader conclusions about a drug's effectiveness under real-world conditions. For example, complex factors in human subjects, such as nutrition and stress, may interact with the experimental drug and influence its effectiveness. These interactions will not be present and taken into account in the controlled lab study.

Good experimental design is more complex than simply following a single recipe. The only way to determine the quality of an experiment is to assess how well the variables that were not of interest were controlled, and how well the experimental treatment tested the relationship of interest.

TAKE-HOME MESSAGE 1·11

To draw clear conclusions from experiments, it is essential to hold constant all those variables we are not interested in. Control and experimental groups should differ only with respect to the treatment of interest. Differences in outcomes between the groups can then be attributed to the treatment.

? 1·12 THIS IS HOW WE DO IT

Is arthroscopic surgery for arthritis of the knee beneficial?

Sometimes we may think that we "know" something when in fact there is no strong evidence supporting that belief. It can be particularly difficult to question ideas that are widely accepted. That's when the power of a randomized, well-controlled study can help us to gain a better understanding of our world—and to know when we should change our mind.

Consider, for example, the efficacy of a common knee surgery for arthritis. In this arthroscopic surgery, surgeons make several small incisions and insert a tiny, flexible tube into the knee in order to see inside the joint. As a treatment for arthritis, the surgeon may then remove or sand down debris and damaged cartilage. She may also flush out the debris and other materials from the knee joint.

Until very recently, this type of surgery was performed on more than 650,000 people in the United States each year—at a cost of $5,000 for each surgery, or over $3 billion a year in all. Some researchers asked a basic and reasonable question: Is this surgery actually beneficial for patients?

How could you determine whether a particular type of surgery is effective?

The general approach that the researchers took was straightforward. From a large group of volunteers who suffered from osteoarthritis of the knee, some received arthroscopic surgery, *while others received a placebo surgery.* The researchers then evaluated individuals' knee function and degree of pain relief over the course of the two years following surgery.

Wait a minute . . . what?

Yes. After recruiting 180 volunteer patients who were candidates for arthroscopic surgery for their arthritis, the researchers randomly assigned them to one of three groups. Two of the groups would undergo arthroscopic surgery, while the third group would receive the placebo surgery. The participants did not know which group they were in, but all of them did understand that they might undergo only the placebo surgery. They all signed an "informed consent" form stating: "I realize that I may receive only placebo surgery. I further realize that this means that I will not have surgery on my knee joint. This placebo surgery will not benefit my knee arthritis."

How does general scientific literacy—particularly among non-scientists such as the volunteers in this study—help in advancing our knowledge and understanding about a particular phenomenon?

The patients were randomly assigned to one of the three groups. This occurred only after they were in the operating room, at which point the surgeon (one surgeon performed all the operations) was given an envelope indicating which treatment each patient was to receive. The treatments included:

1. *Arthroscopic surgery with debridement:* The surgeon made the incisions and irrigated the knee joint with 10 liters of fluid, and then shaved, trimmed, and smoothed (that is, debrided) the cartilage.

2. *Arthroscopic surgery with lavage:* The surgeon made the incisions and irrigated the knee joint with 10 liters of fluid, as in the first group, but did not

debride. Any debris that could be flushed out was removed (the process of "lavage").

3. *Placebo surgery:* The surgeon made the same three incisions as in the first two groups, and then asked for all instruments and manipulated the knee as if performing arthroscopic surgery with debridement. The surgeon also splashed saline to simulate lavage.

In all three treatments, patients spent the night in the hospital, cared for by nurses who did not know which treatment group they were in.

| **How did the researchers decide whether the arthroscopic surgery was effective?**

The researchers evaluated the effectiveness of the surgery at seven points over the next two years. The evaluations included patients' self-reports of knee pain and body pain and researchers' measurements of knee function.

| **Why do you think these two different measures were used?**

Pain scores were on a scale of 0–100 (higher scores indicating greater pain), based on a 12-item evaluation called the Knee-Specific Pain Scale. The graph shows the mean results for the three groups.

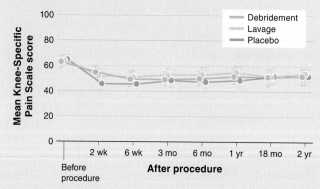

Knee function measures were also on a scale of 0–100 (higher scores indicating more limited function), based on measurements of walking 100 feet and climbing up and down a flight of stairs—using a scale called the AIMS2 Walking-Bending Subscale. The second graph shows the mean results for this measurement.

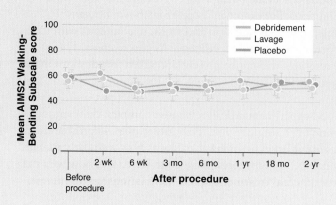

| **What is the take-home message from these two graphs?**

At no point following surgery was pain reduced or knee function improved in patients receiving either type of arthroscopic surgery when compared with those undergoing the placebo surgery. At two years, for example, the pain scores were:

Patient Group	Mean Pain Score
1. Debridement	51 ± 23
2. Lavage	54 ± 24
3. Placebo	52 ± 24

| **What conclusions can you draw from these results?**

Publishing their results in the *New England Journal of Medicine,* the researchers summarized their conclusions succinctly: "This study provides strong evidence that arthroscopic lavage with or without débridement is not better than and appears to be equivalent to a placebo procedure in improving knee pain and self-reported function."

TAKE-HOME MESSAGE 1·12

In a well-controlled experiment, researchers demonstrated that arthroscopic knee surgery for osteoarthritis was no more beneficial for patients—in terms of knee pain and knee functioning—than a placebo surgery.

1·13 Repeatable experiments increase our confidence.

Q Can science be misleading? How can we know?

In 2005, a study showed that some patients being treated for HIV infection who were also taking an epilepsy drug called valproic acid had significantly reduced numbers of HIV particles in their blood. Newspaper headlines announced this finding—"AIDS Cure Possible, Study Suggests"—and raised many people's hopes. Two years later, however, a study of people who were already taking valproic acid and anti-HIV drugs concluded that those taking valproic acid had not benefited at all from the drug (**FIGURE 1-16**). It's not certain why the later study didn't produce the same results as the first study, but the second study did address two serious shortcomings of the first study—namely, the first study was much smaller, involving only four patients, and no control group was used. This pair of studies reveals the importance of repeatability in science.

A powerful way to demonstrate that observed differences between a treatment group and a control group truly reflect the effect of the treatment is for the researchers to conduct the experiment over and over again. Even better is to have other research groups repeat the experiment and get the same results. Researchers describe this desired characteristic of experiments by saying that an experiment must be "reproducible" and "repeatable."

An experiment that can be done over and over again by a variety of researchers to give the same results is an effective defense against biases (which we discuss in the next section) and reflects a well-designed experiment.

Q Do megadoses of vitamin C reduce cancer risk?

Experiments whose results cannot be confirmed by repeated experiments or by experiments performed by other researchers are the downfall of many dramatic claims. Even though chemist Linus Pauling had not one, but two Nobel prizes, he was never able to convince the scientific community that megadoses of vitamin C are an effective treatment for cancer. Every time other individuals tried to repeat Pauling's studies, properly matching the control and treatment groups, they found no difference. Vitamin C just isn't effective against cancer—a conclusion that the vast majority of the medical community now believes. The scientific method is profoundly egalitarian: more important than a scientist's credentials are sound and reproducible results.

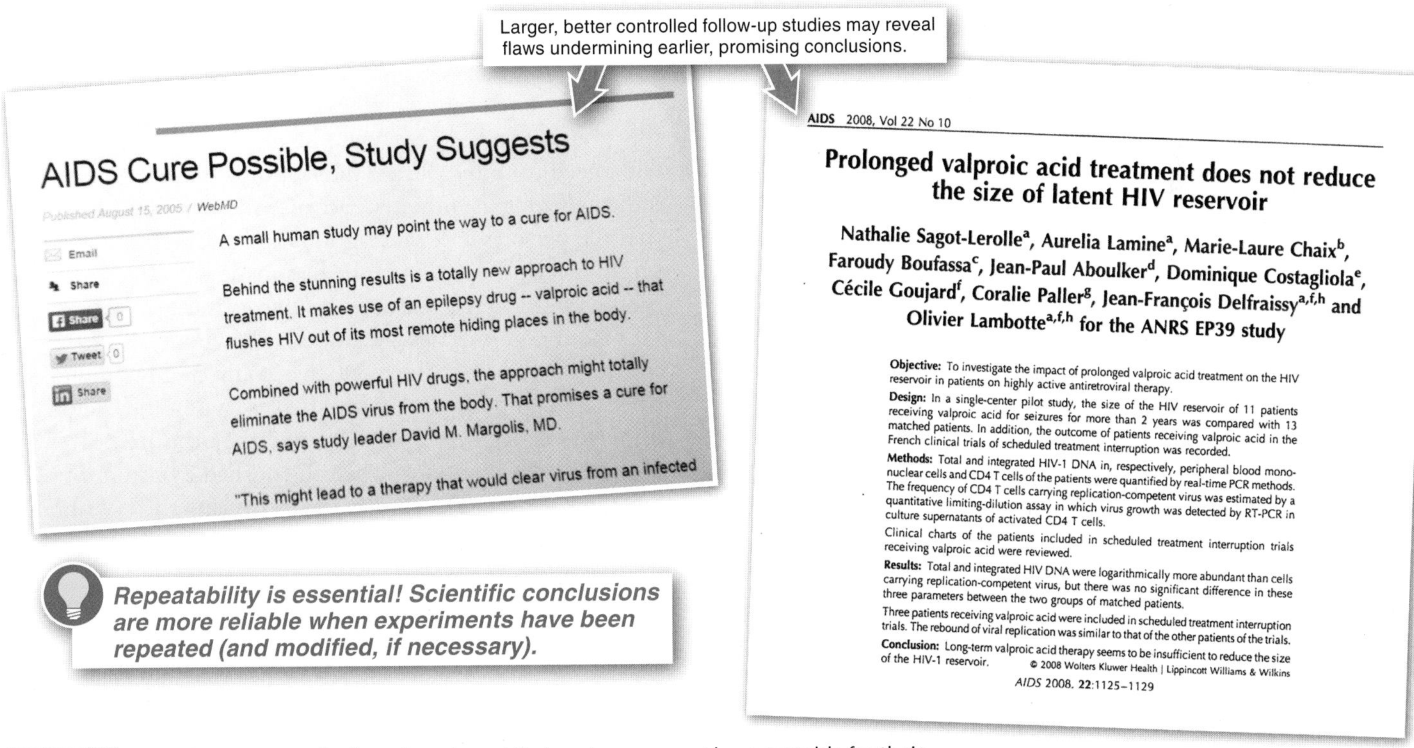

Larger, better controlled follow-up studies may reveal flaws undermining earlier, promising conclusions.

AIDS Cure Possible, Study Suggests

Published August 15, 2005 / WebMD

A small human study may point the way to a cure for AIDS.

Behind the stunning results is a totally new approach to HIV treatment. It makes use of an epilepsy drug -- valproic acid -- that flushes HIV out of its most remote hiding places in the body.

Combined with powerful HIV drugs, the approach might totally eliminate the AIDS virus from the body. That promises a cure for AIDS, says study leader David M. Margolis, MD.

"This might lead to a therapy that would clear virus from an infected

AIDS 2008, Vol 22 No 10

Prolonged valproic acid treatment does not reduce the size of latent HIV reservoir

Nathalie Sagot-Lerolle[a], Aurelia Lamine[a], Marie-Laure Chaix[b], Faroudy Boufassa[c], Jean-Paul Aboulker[d], Dominique Costagliola[e], Cécile Goujard[f], Coralie Paller[g], Jean-François Delfraissy[a,f,h] and Olivier Lambotte[a,f,h] for the ANRS EP39 study

Objective: To investigate the impact of prolonged valproic acid treatment on the HIV reservoir in patients on highly active antiretroviral therapy.

Design: In a single-center pilot study, the size of the HIV reservoir of 11 patients receiving valproic acid for seizures for more than 2 years was compared with 13 matched patients. In addition, the outcome of patients receiving valproic acid in the French clinical trials of scheduled treatment interruption was recorded.

Methods: Total and integrated HIV-1 DNA in, respectively, peripheral blood mononuclear cells and CD4 T cells of the patients were quantified by real-time PCR methods. The frequency of CD4 T cells carrying replication-competent virus was estimated by a quantitative limiting-dilution assay in which virus growth was detected by RT-PCR in culture supernatants of activated CD4 T cells. Clinical charts of the patients included in scheduled treatment interruption trials receiving valproic acid were reviewed.

Results: Total and integrated HIV DNA were logarithmically more abundant than cells carrying replication-competent virus, but there was no significant difference in these three parameters between the two groups of matched patients. Three patients receiving valproic acid were included in scheduled treatment interruption trials. The rebound of viral replication was similar to that of the other patients of the trials.

Conclusion: Long-term valproic acid therapy seems to be insufficient to reduce the size of the HIV-1 reservoir. © 2008 Wolters Kluwer Health | Lippincott Williams & Wilkins

AIDS 2008. **22**:1125–1129

💡 *Repeatability is essential! Scientific conclusions are more reliable when experiments have been repeated (and modified, if necessary).*

FIGURE 1-16 **Once is not enough.** Experiments and their outcomes must be repeatable for their conclusions to be valid and widely accepted.

When a study is repeated (also referred to as "replicating" a study), sometimes a tiny variation in the experimental design can lead to a different outcome; this can help us isolate the variable that is primarily responsible for the outcome of the experiment. Alternatively, when experiments are repeated and the same results are obtained, our confidence in them is increased.

TAKE-HOME MESSAGE 1·13

Experiments and their outcomes must be repeatable for their conclusions to be considered valid and widely accepted.

1·14 We've got to watch out for our biases.

In 2001, the journal *Behavioral Ecology* changed its policy for reviewing manuscripts that were submitted for publication. Its new policy instituted a double-blind process, whereby neither the reviewers' nor the authors' identities were revealed. Previously, the policy had been a single-blind process in which reviewers' identities were kept secret but the authors' identities were known to the reviewers. In an analysis of papers published between 1997 and 2005, it turned out that after 2001, when the double-blind policy took effect, there was a significant increase in the number of published papers in which the first author was female (**FIGURE 1-17**). Analysis of papers published in

Q Can scientists be sexist? How would we know?

a similar journal that maintained the single-blind process over that period revealed no such increase.

This study reveals that people, including scientists, may have biases—sometimes subconscious—that influence their behavior. It also serves as a reminder of the importance of proper controls in experiments. If knowing the sex of the author influences a reviewer's decision on whether a paper should be published, it is also possible that researchers' biases can creep in and influence their collection of data and analysis of results. (Even the decision about what—and what not—to study can be influenced by our biases.)

It can be hard to avoid biases. Consider a study that required precise measurement of the fingers of the left and right hands—comparing the extent to which people in different groups were physically symmetrical. The researcher noted that when she measured individuals from the group she predicted would be more symmetrical, she felt a need to re-measure if the reading on her digital ruler indicated a big asymmetry. Because she was thinking that the person's left- and right-hand fingers *should* be symmetrical, she assumed she had made an error. She felt no such need to re-measure for subjects in the other group, for whom asymmetries confirmed her hypothesis. To control for this bias, the researcher connected her digital ruler to a computer and, without ever having the number displayed on the screen, transmitted the measurement directly to the computer when she pushed a button. In this way, she was able to make each measurement without introducing any regular bias.

PERCENTAGE OF PAPERS PUBLISHED WITH FEMALE FIRST AUTHOR

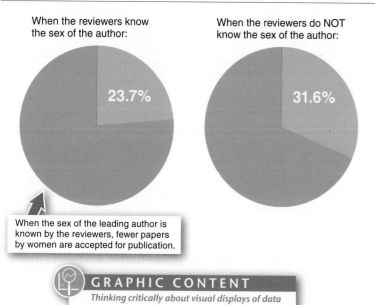

When the reviewers know the sex of the author:

23.7%

When the reviewers do NOT know the sex of the author:

31.6%

When the sex of the leading author is known by the reviewers, fewer papers by women are accepted for publication.

GRAPHIC CONTENT
Thinking critically about visual displays of data
Turn to p. 33 for a closer inspection of this figure.

FIGURE 1-17 Bias against female scientists? The journal *Behavioral Ecology* accepted more papers from female authors when the reviewers were not aware of the author's sex.

TAKE-HOME MESSAGE 1·14

Biases can influence our behavior, including our collection and interpretation of data. With careful controls, it is possible to reduce the impact of biases.

1·15–1·18
Scientific thinking can help us make wise decisions.

What can you believe? Reading labels is essential to evaluating products and the claims about them.

1·15 Visual displays of data can help us understand and explain phenomena.

"Let's look at the data." Using points and lines and symbols and a variety of other graphic elements to display measured quantities is a powerful tool commonly used throughout the sciences. Such visual displays of data can serve a range of purposes, which relate to both the presentation of and the exploration of the data.

Whether making a point, illustrating an idea, or facilitating the testing of a hypothesis, visual displays of data typically have one feature in common: they condense large amounts of information into a more easily digested form. In doing so, they can help readers think about and compare data, ultimately helping them to synthesize the information and see useful patterns.

There is an almost infinite variety of ways to display data, including maps, tables, charts, and graphs. Graphs are particularly prevalent in biology, and a few forms are used most frequently. These include bar graphs, line graphs, and pie charts (**FIGURE 1-18**).

Visual displays of data generally have a few common elements. Most have a title, for example, which usually

FIGURE 1-18 **Presenting what we have observed, precisely and concisely.**

COMMON VISUAL DISPLAYS OF DATA USED IN BIOLOGY

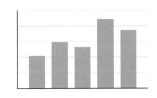

BAR GRAPH
Rectangular bars are used to represent data, each with a height that is proportional to the value being represented.

LINE GRAPH
A line or curve may be used to connect data points or to illustrate trends across many data points.

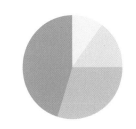

PIE CHART
"Slices" are used to represent data, in which each slice is a proportion of the whole.

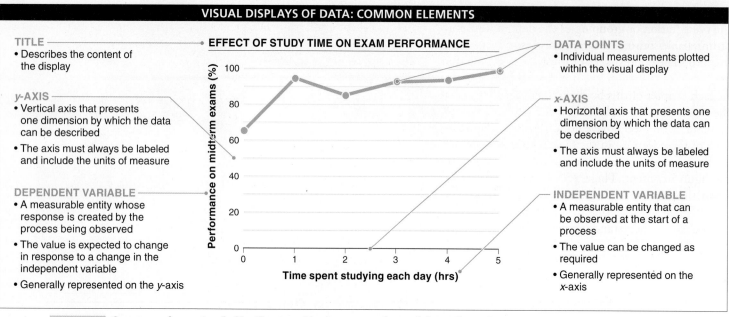

VISUAL DISPLAYS OF DATA: COMMON ELEMENTS

TITLE
- Describes the content of the display

EFFECT OF STUDY TIME ON EXAM PERFORMANCE

DATA POINTS
- Individual measurements plotted within the visual display

y-AXIS
- Vertical axis that presents one dimension by which the data can be described
- The axis must always be labeled and include the units of measure

x-AXIS
- Horizontal axis that presents one dimension by which the data can be described
- The axis must always be labeled and include the units of measure

DEPENDENT VARIABLE
- A measurable entity whose response is created by the process being observed
- The value is expected to change in response to a change in the independent variable
- Generally represented on the y-axis

INDEPENDENT VARIABLE
- A measurable entity that can be observed at the start of a process
- The value can be changed as required
- Generally represented on the x-axis

FIGURE 1-19 **Common elements of effective graphical presentations of data.** Shown here: an example of a line graph.

appears at the top and describes the content of the display. Bar graphs and line graphs include axes, usually a horizontal axis, also called the *x*-axis, and a vertical axis, called the *y*-axis. Each axis has a scale, generally labeled with some gradations, indicating one dimension by which the data can be described.

The *x*-axis of a line graph, for example, might describe the number of hours a student spends studying for a class, in which case it would be labeled "Time spent studying each day (hrs)," while the *y*-axis might be labeled "Performance on midterm exams (%)." Depending on the size of the data set (the total number of observations collected in the experiment that relate an individual's performance on a midterm and the number of hours that student spent studying each day), individual data points are included on the graph, with additional information conveyed by the shape, color, or pattern of the data points (**FIGURE 1-19**). One data point might reflect that one student spent 2 hours studying per day and scored 85% on a midterm exam, while another spent 1 hour per day and scored 94% on the midterm exam. A line or curve may be used to connect data points or to illustrate a relationship between the two variables, and the axes must always be labeled and include the units of measure.

Rather than displaying individual data points, a bar graph has rectangular bars, each with a height proportional to the value being represented—maybe hours spent studying in a particular class. In a pie chart, each "slice" is a proportion

of the whole—for instance, each slice representing the portion of a student's total number of study hours spent on each different topic every day. A legend may be included for the graph, identifying which information is represented by which bar or data point or pie slice.

One of the most common functions of visual displays of information is to present the relationship between two variables, such as in a graph. These variables may be described as independent and dependent variables. An **independent variable** is some entity that can be observed and measured at the start of a process, and whose value can be changed as required. A **dependent variable** is one that can also be observed and measured, but whose response is created by the process being observed and depends on the independent variable. The dependent variable is generally represented by the *y*-axis and is expected to change in response to a change in the independent variable, represented on the *x*-axis. The number of hours of sleep a student gets each night, for example, could be thought of as an independent variable, while some measure of academic performance—maybe grade point average—would be a dependent variable.

Visual displays of data can be simple and straightforward, but they can also have features that reduce their effectiveness, or even cause them to be downright misleading. These difficulties can arise from ambiguity in the axis labels or scales, or incomplete information on how each data point was collected (and how the points might have varied), or biases or hidden assumptions in the presentation or grouping of the data, or unknown or unreliable sources of data, or an insufficient or inappropriate context given for the data presentation.

In each chapter of this book, you will see that one of the visual displays of data is labeled with the "Graphic Content" icon (you'll have noticed this on Figure 1-17). For each of these figures, in the Check Your Knowledge section at the end of the chapter you'll find several questions about the figure's content. These questions will guide you in evaluating aspects of the figure (such as: "What is its main point?") and in extracting information from it. They will also help you understand when it might be warranted to

approach a data display with more than the usual amount of skepticism, and when greater trust is reasonable.

TAKE-HOME MESSAGE 1·15

Visual displays of data, which condense large amounts of information, can aid in the presentation and exploration of the data. The effectiveness of such displays is influenced by the precision and clarity of the presentation, and it can be reduced by ambiguity, biases, hidden assumptions, and other issues that reduce a viewer's confidence in the underlying truth of the presented phenomenon.

1·16 Statistics can help us in making decisions.

In Section 1-13 we saw that researchers repeatedly found, through experimentation, that megadoses of vitamin C do not reduce cancer risk. If you put yourself in the researchers' shoes, you might wonder how you'd figure out that the vitamin C did not reduce cancer risk. Perhaps there were 100 individuals in the group receiving megadoses of vitamin C: some of them developed cancer and some of them did not. And among the 100 subjects in the group not receiving the megadoses, some of them developed cancer and some did not. How do you decide whether the vitamin C actually had an effect? This knowledge comes from a branch of mathematics called **statistics,** a set of analytical and mathematical tools designed to help researchers gain understanding from the data they gather. To understand statistics, let's start with a simple situation.

Suppose you measured the height of two people. One is a woman who is 5 feet 10 inches tall. The other is a man who is 5 feet 6 inches tall. If these were your only two observations of human height, you might conclude that female humans are taller than males. But suppose you measured the height of 100 women and 100 men chosen randomly from a population. Then you can say, "for the 100 men, the average height is 5 feet 9.5 inches, and for the 100 women, the average height is 5 feet 4 inches." Better still, the data can illuminate for you not only the average but also some measure of how much variation there is from one individual to another. Statistical analysis can tell you not only that the average height for a man in this study is 5 feet 9.5 inches but that two-thirds of the men are between 5 feet 6.5 inches tall (3 inches less than the average) and 6 feet 0.5 inches tall (3 inches more than the average).

You will often see this type of range stated as "5 feet 9.5 inches ± 3 inches" ("plus or minus 3 inches"). Similarly, the data might show for the female subjects "5 feet 6 inches ± 3 inches," indicating that two-thirds of the women are between 5 feet 3 inches and 5 feet 9 inches tall. As we discussed in Section 1-11, on experimental design, and as this example shows, larger numbers of participants are better than smaller numbers if you want to draw general conclusions about natural phenomena such as the height of men and women (**FIGURE 1-20**).

Using data to describe the characteristics of individuals participating in a study is useful, but often we want to know whether data support (or do not support) a hypothesis. If the scientific method is to be effective in helping us understand the world, it must help us make wise decisions about concrete things. For example, suppose we want to know whether having access to a textbook helps a student perform better in a biology class. Statistics can help us answer this question. After conducting a study, let's say we find that students who had access to a textbook scored an average of 81% ± 8% on their exams, while those who did not scored an average of 76% ± 7%.

In this example, it is difficult to distinguish between the two possible conclusions.

Possibility 1: Students having access to a textbook *do perform better* in biology classes. In other words, our sampling of this class reveals a true relationship between the two variables, textbook access and class performance.

HUMAN HEIGHT AND STATISTICS

Don't get fooled! Making one or two observations isn't enough to draw general conclusions about natural phenomena, such as height.

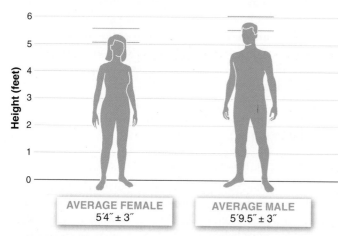

FIGURE 1-20 Drawing conclusions based on limited observations is risky. Measuring a greater number of people will generally help us draw more accurate conclusions about human height.

AVERAGE FEMALE
5′4″ ± 3″

AVERAGE MALE
5′9.5″ ± 3″

Possibility 2: Students having access to a textbook *do not perform better* in biology classes. The variation in the scores for the two groups may be too large to allow us to conclude that there's any effect of having access to a textbook. Instead, the difference in average scores for the two groups might mean that more of the high-performing students just happened, through random chance, to be in the group given access to a textbook.

But what if the students with access to a textbook scored, on average, 95% ± 5%, while those without access scored only 60% ± 5%? In this case, we would be much more confident

that there is a significant effect of having access to a textbook (FIGURE 1-21), because even with the large variation in the scores seen in each group, the two averages are still very different from each other.

Statistical methods help us to decide between these two possibilities and, equally important, to state how confident we are that one or the other is true. The greater the difference between two groups (95% vs. 60% is a greater difference than 81% vs. 76%), and the smaller the variation in each group (± 5% in the 95% and 60% groups, vs. ± 8% and ± 7% in the 81% and 76% groups), the more confident we are of the conclusion that there is a significant effect of the treatment (having access to a textbook, in this case). In other words, in the case where the groups of students scored 95%

INTERPRETING STATISTICS: A HYPOTHETICAL STUDY

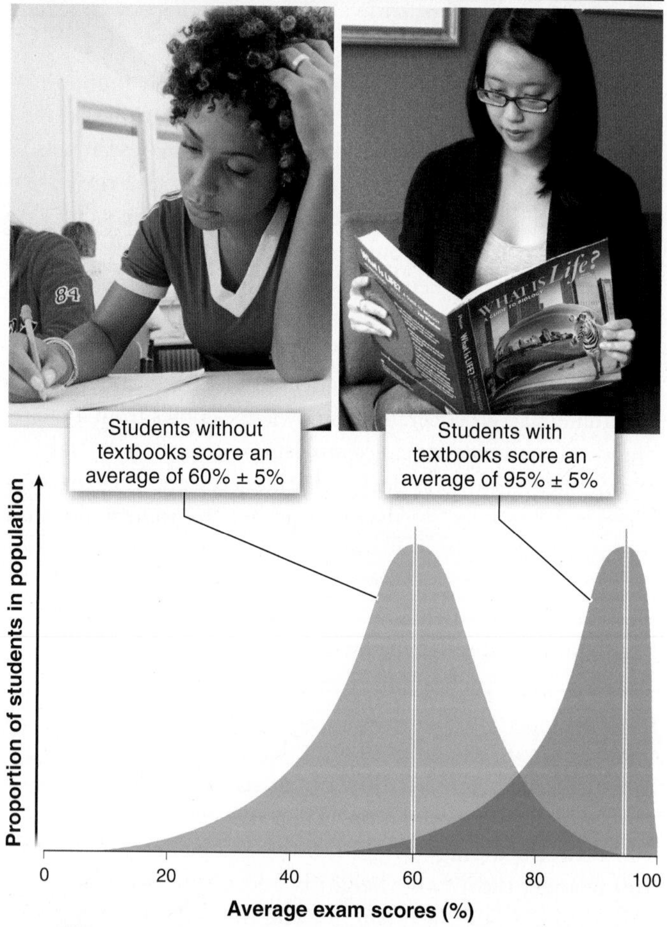

Students without textbooks score an average of 60% ± 5%

Students with textbooks score an average of 95% ± 5%

Proportion of students in population

Average exam scores (%)

Statistics can quantify and summarize large amounts of information, making it possible to draw more accurate conclusions.

FIGURE 1-21 Drawing conclusions based on statistics. Statistical analyses can help us evaluate the significance of our observations.

or 60%, depending on whether they did or did not have access to a textbook, it is possible that having a textbook did not actually improve performance and that this observed difference was just the result of chance. But this conclusion is very, very unlikely.

Statistics can also help us identify relationships (or lack of relationships) between variables. For example, we might note that when there are more firefighters at a fire, the fire is larger and causes more damage. This is a **positive correlation,** meaning that when one variable (the number of firefighters) increases, so does the other (the severity of the fire). Should we conclude that firefighters make fires worse? No. While correlations can reveal relationships between variables, they don't tell us *how* the variables are related or whether change in one variable *causes* change in another. (You may have heard or read the phrase "correlation is not causation," which refers to this sort of situation.) Before drawing any conclusions about more firefighters causing larger fires, we need to know about the type of fire and its size when the firefighters arrived, because those factors will

significantly influence the ultimate amount of damage. To estimate the effect of the number of firefighters on the amount of damage, we would need to compare the amount of damage from fires of similar size that are fought by different numbers of firefighters.

Ultimately, statistical analyses can help us organize and summarize the observations that we make and the evidence we gather in an experiment. These analyses can then help us decide whether any differences we measure between experimental and control groups are likely to be the result of the treatment, and how confident we can be in that conclusion.

> ## TAKE-HOME MESSAGE 1·16
> Because much variation exists in the world, statistics can help us evaluate whether any differences between a treatment group and a control group can be attributed to the treatment rather than random chance.

1·17 Pseudoscience and misleading anecdotal evidence can obscure the truth.

One of the major benefits of the scientific way of thinking is that it can prevent you from being taken in or fooled by false claims. There are two types of "scientific evidence" that frequently are cited in the popular media and are responsible for people erroneously believing that links between two things exist, when in fact they do not.

1. **Pseudoscience,** in which individuals make scientific-sounding claims that are not supported by trustworthy, methodical scientific studies

2. **Anecdotal observations,** in which, based on just one or a few observations, people conclude that there is or is not a link between two things

Pseudoscience is all around us, particularly in the claims made on the packaging of consumer products and food (**FIGURE 1-22**). Beginning in the 1960s, for example, consumers encountered the assertion by the makers of a sugarless gum that "4 out of 5 dentists surveyed recommend sugarless gum for their patients who chew gum." Maybe the statement is factually true, but the

Pseudoscience can manipulate you with deliberately misleading claims, or with "scientific sounding" language sufficiently ambiguous to avoid charges of deception.

FIGURE 1-22 **Pseudoscientific claims are often found on food products.**

general relationship it implies may not be. How many dentists were surveyed? If the gum makers surveyed only five dentists, then the statement may not represent the proportion of *all* dentists who would make such a recommendation. And how were the dentists sampled? Were they at a shareholders meeting for the sugarless gum company? What alternatives were given—perhaps gargling with a tooth-destroying acid? You just don't know. That's what makes it pseudoscience.

Pseudoscience capitalizes on a belief shared by most people: that scientific thinking is a powerful method for learning about the world. The problem with pseudoscience is that the scientific bases for a scientific-sounding claim are not clear. The claims generally sound reasonable, and they are persuasive in convincing people to purchase one product over another. But one of the beauties of real science is that you never have to just take someone's word about something. Rather, you are free to evaluate people's research methods and results, and then decide for yourself whether their conclusions and claims are appropriate.

We are all familiar with anecdotal evidence. Striking stories that we read or hear, or our own experiences, can shape our views of cause and effect. We may find compelling parallels between suggestions made in horoscopes and events in our lives, or we may think that we have a lucky shirt, or we may be moved by a child whose cancer went into remission after eating apricot seeds. Yet, despite lacking a human face, data are more reliable than anecdotes, primarily because they can illustrate a broader range of observations, capturing the big picture.

Anecdotal observations can seem harmless and can be emotionally powerful. But because they do not include a sufficiently large and representative set of observations of the world, they can lead people to draw erroneous conclusions, often with disastrous consequences. One important case of anecdotal evidence being used to draw general conclusions involves autism, a developmental disorder that impairs social interaction and communication, and the vaccination for measles, mumps, and rubella (commonly called the MMR vaccine) that is given to most children.

Q Does the measles, mumps, and rubella vaccine cause autism?

In 1998, the prestigious medical journal *The Lancet* published a report by a group of researchers that described a set of symptoms (diarrhea, abdominal pain, bloating) related to bowel inflammation in 12 children who exhibited the symptoms of autism. The parents or physicians of 8 of the children in the study said that the behavioral symptoms of autism appeared shortly after the children received MMR vaccination. For this reason, the authors of the report recommended further study of a possible link between the MMR vaccine, the bowel problems, and autism.

In a press conference, one of the paper's authors suggested a link between autism and the MMR vaccine, recommending single vaccines rather than the MMR triple vaccine until it could be proved that the MMR vaccine did not trigger autism. This statement caused a major health scare that only recently has begun to subside. Noting that there had been a significant increase in the incidence of autism in the 1990s and early 2000s, the press and many people took the claims made by the researcher at the press conference as evidence that the MMR vaccine causes autism. Over the course of the next few years, the number of children getting the MMR vaccine dropped significantly, as parents sought to reduce the risk of autism in their children (**FIGURE 1-23**).

Unfortunately, this is a notable case of poor implementation of the scientific method, with numerous flaws leading to an incorrect conclusion. Most important among these is that the study was small (only 12 children), the sample was carefully selected rather than randomized (that is, the researchers selected the study participants based on the symptoms they showed), and no control group was included for comparison

A misguided fear of "catching" autism has caused some parents to decline immunizations for their children.

FIGURE 1-23 Bad science became headline news.

(for example, children who had been vaccinated but did not exhibit autism symptoms). On top of this, the study, even flawed as it was, did not actually find or even report a link between autism and the MMR vaccine. That purported link came only from the unsupported, suggestive statements made by one of the study's authors at the press conference.

On their own, the design flaws in the study would be sufficient to invalidate the one author's claims about the link between autism and the MMR vaccine, but there are even more weaknesses in the research. It later was discovered that before the paper's publication, the study's lead author (the one who spoke at the press conference) had received large sums of money from lawyers seeking evidence to use in lawsuits against the MMR vaccine manufacturers. It also came to light that he had applied for a patent for a vaccine that was a rival to the most commonly used MMR vaccine. In the light of these initially undisclosed biases, 10 of the paper's 12 authors published a retraction of their original interpretation of their results. And in 2011, the *BMJ* (*British Medical Journal*) published an article declaring that the "article linking MMR vaccine and autism was fraudulent."

In the years since the original paper was published, many well-controlled, large-scale studies have conducted critical experiments on this subject, and all have been definitive in their conclusion that there is no link between the MMR vaccine and autism. Here are a few of these studies.

1. A study of all children born in Denmark between 1991 and 1998 found no difference in the incidence of autism among the 440,655 children who were vaccinated with the MMR vaccine and the 96,648 children who were not vaccinated.

2. A 2005 study in Japan showed that after use of the MMR vaccine was stopped in 1993, the incidence of autism continued to increase.

3. A study of 1.8 million children in Finland who were followed up for 14 years after getting the MMR vaccine found no link at all between the occurrence of autism and the vaccine.

At this point, the consensus of the international scientific community is that there is no scientific evidence for a link between the MMR vaccine and autism.

> "Science is a way to call the bluff of those who only pretend to knowledge. It is a bulwark against mysticism, against superstition, against religion misapplied to where it has no business being. If we're true to its values, it can tell us when we're being lied to."
>
> —**CARL SAGAN**, in *The Demon-Haunted World: Science as a Candle in the Dark*, **1997**

So what explains the observation that there are more autism cases now than in the past? It seems that this is a function of the better identification of autism by doctors and changes in the process by which autism is diagnosed. Another reason for the perceived link between autism and the MMR vaccine was simply the coincidence that most children receive the vaccine at around 18-19 months of age, which happens to be the age at which the first symptoms of autism are usually noticed. In the end, what we learn from this is that we must be wary that we do not generalize from anecdotal observations or let poorly designed studies obscure the truth.

TAKE-HOME MESSAGE 1·17

Pseudoscience and anecdotal observations often lead people to believe that links between two phenomena exist, when in fact there are no such links.

1·18 There are limits to what science can do.

The scientific method is a framework that helps us make sense of what we see, hear, and read in our lives. There are limits, however, to what science can do.

The scientific method will never prove or disprove the existence of God. Nor is it likely to help us understand the mathematical elegance of Fermat's last theorem or the beauty of Shakespeare's sonnets. As one of several approaches to the acquisition of knowledge, the scientific method is, above all, empirical. It differs from non-scientific approaches such as mathematics and logic, history, music, and the study of artistic expression in that it relies on *measuring* phenomena in some way. The generation of value judgments and other types of non-

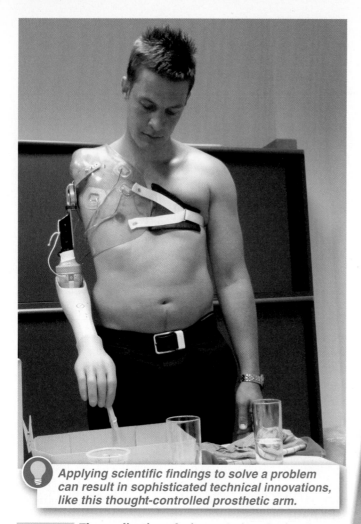

Applying scientific findings to solve a problem can result in sophisticated technical innovations, like this thought-controlled prosthetic arm.

FIGURE 1-24 **The application of science.** Andrew Garthwaite, a soldier who lost an arm, learns how to use a mind-controlled prosthetic limb.

quantifiable, subjective information—such as religious assertions of faith—falls outside the realm of science. Despite all of the intellectual analyses that the scientific method gives rise to and the objective conclusions it makes possible, it does not, for example, generate moral statements and it cannot give us insight into ethical problems. What "is" (i.e., what we observe in the natural world) is not necessarily what "ought" to be (i.e., what is morally right). It may or may not be.

Further, much of what is commonly considered to be science, such as the construction of new engineering marvels or the heroic surgical separation of conjoined twins, is not scientific at all. Rather, these are technical innovations and developments. While they frequently rely on sophisticated scientific research, they represent the application of research findings to varied fields such as manufacturing and medicine to solve problems (**FIGURE 1-24**).

As we begin approaching the world from a more scientific perspective, we can gain important insights into the facts of life, yet must remain mindful of the limits to science.

TAKE-HOME MESSAGE 1·18

Although the scientific method may be the most effective path toward understanding the observable world, it cannot give us insights into the generation of value judgments and other types of non-quantifiable, subjective information.

1·19
On the road to biological literacy: what are the major themes in biology?

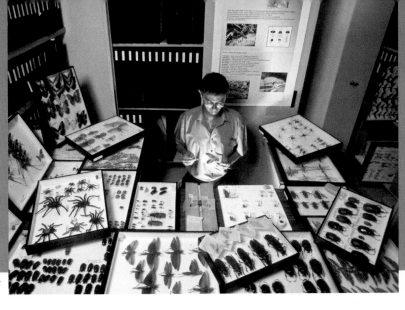

Contemplating unity and diversity. An entomologist organizes and categorizes specimens.

1·19 What is life? Important themes unify and connect diverse topics in biology.

Biology is, literally, the study of life. But what is life? And how is it distinguished from non-life? Spend a few moments trying to define exactly what life is and you'll realize that it is not easily described with a simple definition. A useful approach is to consider the characteristics shared by all living organisms and living systems:

- **A complex, ordered organization consisting of one or more cells.** Cells carry out the functions necessary for life.

- **The use and transformation of energy to perform work.** Organisms can perform many reactions and activities, by acquiring, using, and transforming energy.

- **Sensitivity and responsiveness to the external environment.** Living organisms are able to respond to stimuli—such as light, moisture, or another organism.

- **Regulation and homeostasis.** Organisms are able to maintain relatively constant internal conditions that may differ from the external environment.

- **Growth, development, and reproduction.** Organisms can grow and develop, and they carry information relating to these and other processes that they can pass on to offspring.

- **Evolutionary adaptation leading to descent with modification over time.** Populations have the capacity to change over time. As a consequence of organisms' ability to reproduce and of evolutionary change, populations may become better adapted to their environments.

In this guide to biology, as we explore the many facets of biology and its relevance to life in the modern world, you will find two central and unifying themes recurring throughout.

Hierarchical organization. Life is organized on many levels within individual organisms, including atoms, cells, tissues, and organs. And in the larger world, organisms themselves are organized into many levels: populations, communities, and ecosystems within the biosphere.

The power of evolution. Evolution, the change in genetic characteristics of a population over time, accounts for the diversity of organisms and the unity among them.

These central unifying themes connect the diverse topics in biology, which include the chemical, cellular, and energetic foundations of life; the genetics, evolution, and behavior of individuals; the staggering diversity of life and the unity underlying it; and ecology, the environment, and the links between organisms and the world they inhabit. Let's continue our exploration of life!

TAKE-HOME MESSAGE 1·19

"Life" is not easily described with a simple definition. The characteristics shared by all living organisms include complex and ordered organization; the use and transformation of energy; responsiveness to the external environment; regulation and homeostasis; growth, development, and reproduction; and evolutionary adaptation leading to descent with modification.

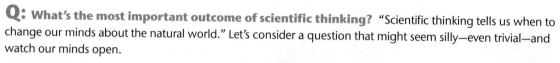

StreetBIO   KNOWLEDGE YOU CAN USE

Rainy days and Mondays

Q: **What's the most important outcome of scientific thinking?** "Scientific thinking tells us when to change our minds about the natural world." Let's consider a question that might seem silly—even trivial—and watch our minds open.

Q: **It often seems to rain more on weekends. Could it be true?** Before you read on, write down your one-word answer (and a justification for it, if you'd like).

Q: **Well, does it rain more on weekends?** At first, it seems that the answer must be no. How could the weather even "know" what day it is? Weather, after all, existed long before humans, calendars, and Monday mornings.

Rather than relying on common sense, let's think scientifically.

My hypothesis: It rains more on weekends.

Restated as a null hypothesis: The amount of rainfall does not differ across the days of the week.

A testable prediction: The amount of rainfall should not differ depending on what day of the week it is.

And now the data . . .

In a 1998 study published in the journal *Nature*, researchers used 17 years of data to analyze rainfall along the eastern coast of North America. Here's what they found: there was 22% more rain on Saturdays than on Mondays! (And this difference was statistically significant.)

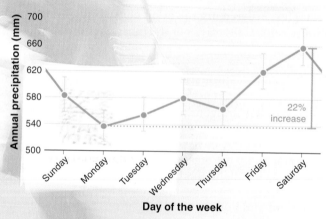

Q: **How can this be?** The researchers also tracked the pollutants carbon monoxide and ozone, and found that both gradually increased from Monday through Friday. This pattern, they suggested, is probably related to accumulations from driving, which is greater during the week than on weekends. They suggested that the weekly pollution cycle causes increased cloud formation and rain on weekends (while the reduced pollution on Mondays leads to reduced rainfall). In other words, human activities are influencing the weather!

One good hypothesis deserves another. If their suggestion is true, we should expect the "Rainy Saturday/Dry Monday" phenomenon to occur only in the vicinity of human activities. As a test of this hypothesis, the researchers measured daily rainfall patterns over the oceans in the northern hemisphere, away from large human populations. Here they found that one-seventh of the rain fell on each day, with no days rainier than others.

Q: **The take-home message here?** Scientific thinking rewards open minds with satisfying answers.

Check Your Knowledge

GRAPHIC CONTENT

Thinking critically about visual displays of data

1. What does the green portion of the pie chart represent? What does the blue portion represent? What does the "whole pie" represent?

2. Why are there two pie charts?

3. What can you conclude from this figure?

4. What might be an alternative explanation (other than bias against female scientists) for the phenomenon shown here?

5. How big was the increase in number of papers published with a female first author following the institution of a double-blind review policy?

6. Why isn't the percentage of papers published with a female first author 50%, even after the new review policy? Is that cause for concern? What data would help you to answer that question?

7. Do these graphs prove a general bias against female scientists?

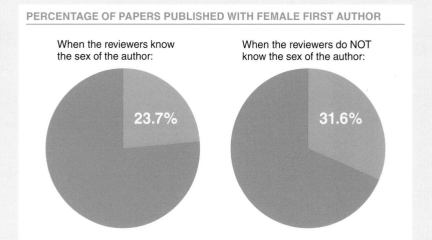

PERCENTAGE OF PAPERS PUBLISHED WITH FEMALE FIRST AUTHOR

When the reviewers know the sex of the author: 23.7%

When the reviewers do NOT know the sex of the author: 31.6%

See answers at the back of the book.

Key Terms in Scientific Thinking

anecdotal observation, p. 27
biological literacy, p. 4
biology, p. 3
blind experimental design, p. 18
control group, p. 16

critical experiment, p. 11
dependent variable, p. 24
double-blind experimental design, p. 18
empirical, p. 6

experimental group, p. 16
hypothesis (pl. hypotheses), p. 9
independent variable, p. 24
null hypothesis, p. 9

placebo, p. 13
placebo effect, p. 18
positive correlation, p. 27
pseudoscience, p. 27
randomized, p. 18
science, p. 2
scientific literacy, p. 4

scientific method, p. 6
statistics, p. 25
superstition, p. 5
theory, p. 15
treatment, p. 16
variable, p. 16

ABOUT THE CHAPTER OPENING PHOTO

Beginning in 1960 (and continuing for more than 50 years), Jane Goodall has studied the social and family interactions of chimpanzees in the wild, in Gombe Stream National Park, Tanzania.

REVIEW & REHEARSE

1

SCIENTIFIC THINKING

? Check Your Knowledge

All questions are not created equal ...
We've determined the difficulty of these questions based on more than four hundred million responses from students nationwide using the online adaptive quizzing system Prep-U.

The difficulty meter below each question indicates the proportion of students answering the question incorrectly.

How can this help you?
We've included questions with a wide range of difficulties. It's okay if you get some wrong. By noting the difficulty level of those giving you trouble, you can better assess where you stand as you master the material.

Smarter than the average quiz

1. Science is:

a) a field of study that requires certain "laws of nature" to be taken on faith.

b) both a body of knowledge and an intellectual activity encompassing observation, description, experimentation, and explanation of natural phenomena.

c) a process that can be applied only within the scientific disciplines, such as biology, chemistry, and physics.

d) the only way to understand the natural world.

e) None of the above are correct.

0 — 8 — 100
EASY · · · HARD

2. All of the following are elements of biological literacy except:

a) the ability to use the process of scientific inquiry to think creatively about real-world issues having a biological component.

b) reading the most important books in biology.

c) the ability to integrate into your decision making a consideration of issues having a biological component.

d) the ability to communicate with others about issues having a biological component.

e) All of the above are elements of biological literacy.

0 — 48 — 100
EASY · · · HARD

1·1–1·3 More than just a collection of facts, science is a process for understanding the world.

Through objective observation, description, and experimentation, science helps us to discover and better understand the world around us.

WHAT IS SCIENCE?

Science is not simply a body of knowledge or a list of facts to be remembered. It is an intellectual activity, encompassing observation, description, experimentation, and explanation of natural phenomena.

WHAT IS BIOLOGY?

Biology is the study of living things.

? 1. In nationwide advertisements, the Dannon Company claimed that its Activia yogurt relieves irregularity and helps with "slow intestinal transit time." Dannon also claimed that its DanActive dairy drink helps prevent colds and flu. These claims were based on no evidence. How would you design an experiment to try to validate these statements?

BIOLOGICAL LITERACY IS ESSENTIAL IN THE MODERN WORLD

Biological issues permeate all aspects of our lives. To make wise decisions, it is essential for individuals and societies to attain biological literacy.

? 2. Describe two examples of biological literacy.

? 3. Scientific thinking can be distinguished from alternative ways of acquiring knowledge about the world in that it is empirical. What does empirical mean, and how does it relate to the study of biology?

1·4–1·10 A beginner's guide: what are the steps of the scientific method?

The scientific method is a flexible, adaptable, and efficient pathway to understanding the world.

THE SCIENTIFIC METHOD: FIVE BASIC STEPS AND ONE FLEXIBLE PROCESS

The scientific method consists of five basic steps. Once begun, though, the process doesn't necessarily continue linearly through the five steps until it is concluded.

STEP 1: MAKE OBSERVATIONS
The scientific method begins by making observations about the world, noting patterns or cause-and-effect relationships.

STEP 2: FORMULATE A HYPOTHESIS
A hypothesis is a proposed explanation for an observed phenomenon.

STEP 3: DEVISE A TESTABLE PREDICTION
For a hypothesis to be useful, it must generate a testable prediction.

STEP 4: CONDUCT A CRITICAL EXPERIMENT
A critical experiment is one that makes it possible to decisively determine whether a particular hypothesis is correct.

STEP 5: DRAW CONCLUSIONS AND MAKE REVISIONS
Based on the results of experimental tests, we can revise a hypothesis and explain the observable world with increasing accuracy.

A "mash-up" of the chapter summary and review questions? YES.

But it's not an accident. This feature is built on a growing body of evidence from cutting-edge research about improving learning.

You'll find that increased time and effort may be required when incorporating testing in your review of the chapter content. But you should consider this a "desirable difficulty." Using testing to hone your retrieval abilities actually enhances your learning of the concepts. By improving your depth of encoding the material, testing takes your learning beyond the simple recognition of the content that comes from just revisiting the material.

In short: testing yourself shouldn't be a separate activity from your studying. It's a powerful tool to improve learning and long-term retention.

? 4. An especially important feature of the scientific method is that its steps are self-correcting. What does self-correcting mean in this context?

? 5. Describe something that you have observed in the world around you that you could study using the scientific method. Describe something that you could not study scientifically.

? 6. A researcher hypothesizes that the more a person exercises, the less acne he or she will have. What would the null hypothesis be in this situation?

? 7. What would be a reasonable prediction for the hypothesis: "Eating fresh fruit reduces the likelihood that you will get sick"?

? 8. Describe the key features of a critical experiment.

9. Many claims have been made concerning the health benefits of green tea. Suppose you read a claim that alleges drinking green tea causes weight loss. You are provided with the following information about the studies that led to this claim.

   • People were weighed at the beginning of the study.
   • People were asked to drink two cups of green tea every day for 6 weeks.
   • People were weighed at the end of the study.
   • People who drank green tea for 6 weeks lost some weight by the end of the study.
   • It was concluded that green tea is helpful for weight loss.

   This study obviously had some holes in its design. Assuming no information other than that provided above, indicate at least four things that could be done to improve the experimental design.

10. Following an experimental conclusion, what is a likely next step? Why?

HYPOTHESIS vs. THEORY

HYPOTHESIS
A hypothesis is a proposed explanation for a phenomenon. (And a good hypothesis leads to testable predictions.)

THEORY
A theory is an explanatory hypothesis for a phenomenon that is exceptionally well supported by the empirical data. (And can be thought of as a hypothesis that has withstood the test of time and is unlikely to be altered by any new evidence.)

11. Compare and contrast "theory" and "hypothesis."

3. Superstitions are:
   a) held by many humans, but not by any non-human species.
   b) just one of many possible forms of scientific thinking.
   c) true beliefs that have yet to be fully understood.
   d) irrational beliefs that actions not logically related to a course of events influence its outcome.
   e) proof that the scientific method is not perfect.

   0 ——**11**———————— 100
   EASY   HARD

4. To be useful in the scientific method, an observation must be:
   a) definite.
   b) measurable.
   c) proven to be true.
   d) hypothetical.
   e) All of the above are correct.

   0 ——————**36**—————— 100
   EASY   HARD

5. Empirical results:
   a) rely on intuition.
   b) are generated by theories.
   c) are based on observation.
   d) cannot be replicated.
   e) must support a tested hypothesis.

   0 —————**33**——————— 100
   EASY   HARD

6. To be useful, a hypothesis will:
   a) generate a testable prediction.
   b) lead you to the conclusions you feel confident are true.
   c) establish many overlapping explanations for a phenomenon.
   d) be based on evidence that cannot be falsified by scientific experimentation.
   e) be deduced from a critical experiment.

   0 —**12**————————— 100
   EASY   HARD

7. Which of the following statements is correct?
   a) A hypothesis that does not generate a testable prediction is not useful.
   b) Common sense is usually a good substitute for the scientific method when trying to understand the world.
   c) The scientific method can be used only to understand scientific phenomena.
   d) It is not necessary to make observations as part of the scientific method.
   e) All of the above are correct.

   0 ————**27**————————— 100
   EASY   HARD

1·11–1·14 Well-designed experiments are essential to testing hypotheses.

To draw clear conclusions from experiments, variables not of interest should be held constant, outcomes must be repeatable, and biases should be minimized.

CONTROLLING VARIABLES

To draw clear conclusions from experiments, it is essential to hold constant all those variables we are not interested in. Control and experimental groups should differ only with respect to the treatment of interest. Differences in outcomes between the groups can then be attributed to the treatment.

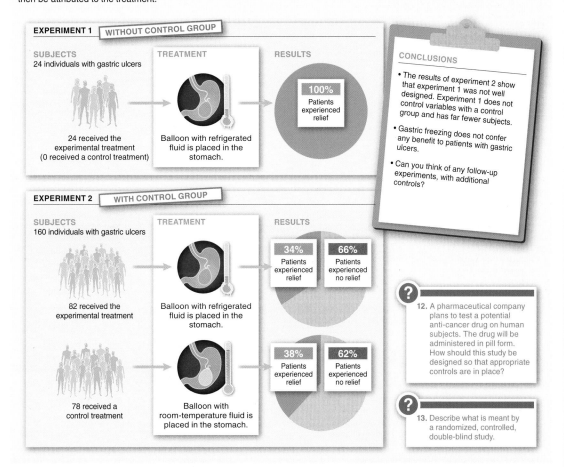

EXPERIMENT 1 WITHOUT CONTROL GROUP

SUBJECTS
24 individuals with gastric ulcers

24 received the experimental treatment (0 received a control treatment)

TREATMENT
Balloon with refrigerated fluid is placed in the stomach.

RESULTS
100% Patients experienced relief

EXPERIMENT 2 WITH CONTROL GROUP

SUBJECTS
160 individuals with gastric ulcers

82 received the experimental treatment

78 received a control treatment

TREATMENT
Balloon with refrigerated fluid is placed in the stomach.

Balloon with room-temperature fluid is placed in the stomach.

RESULTS
34% Patients experienced relief / 66% Patients experienced no relief

38% Patients experienced relief / 62% Patients experienced no relief

CONCLUSIONS

• The results of experiment 2 show that experiment 1 was not well designed. Experiment 1 does not control variables with a control group and has far fewer subjects.

• Gastric freezing does not confer any benefit to patients with gastric ulcers.

• Can you think of any follow-up experiments, with additional controls?

12. A pharmaceutical company plans to test a potential anti-cancer drug on human subjects. The drug will be administered in pill form. How should this study be designed so that appropriate controls are in place?

13. Describe what is meant by a randomized, controlled, double-blind study.

REPEATABLE EXPERIMENTS

Experiments and their outcomes must be repeatable for their conclusions to be considered valid and widely accepted.

AVOIDING BIASES

Biases can influence our behavior, including our collection and interpretation of data. With careful controls, it is possible to minimize such biases.

14. Biases can influence our behavior, including our collection and interpretation of data. Which type of experimental design can help us eliminate biases?

8. The placebo effect:

a) is the frequently observed phenomenon that people tend to respond favorably to any treatment.

b) reveals that sugar pills are more effective than actual medications.

c) reveals that experimental treatments cannot be proven effective.

d) demonstrates that most scientific studies cannot be replicated.

e) is an urban legend.

0 —— (32) —— 100
EASY · HARD

9. Which of the following correctly describes a double-blind test?

a) The researchers apply two-layered blindfolds to prevent the subjects from seeing the treatment drug.

b) Neither the researchers nor the study participants know who is receiving the drug and who is receiving the placebo.

c) The researchers know who is receiving the drug and who is receiving the placebo, but they do not know what the supposed effects of the drug should be.

d) The researchers do not know who receives the drug or the placebo, but the participants do.

e) None of the above are correct.

0 —— (15) —— 100
EASY · HARD

10. In controlled experiments:

a) one variable is manipulated while others are held constant.

b) all variables are dependent on each other.

c) all variables are held constant.

d) all variables are independent of each other.

e) all critical variables are manipulated.

0 —— (24) —— 100
EASY · HARD

11. If a researcher uses the same experimental setup as in another study, but with different research subjects, the process is considered:

a) an uncontrolled experiment.

b) intuitive reasoning.

c) extrapolation.

d) replication.

e) exploration.

0 —— (40) —— 100
EASY · HARD

12. An independent variable:

a) can cause a change in a dependent variable.

b) is generally less variable than a dependent variable.

c) is plotted on the *y*-axis in a line graph.

d) can be controlled less well than a dependent variable.

e) is typically more important than a dependent variable.

0 —— (10) —— 100
EAS · HARD

1·15–1·18 Scientific thinking can help us make wise decisions.

Visual displays of data can help readers think about and compare data, ultimately helping them to synthesize the information and see useful patterns.

COMMON VISUAL DISPLAYS OF DATA USED IN BIOLOGY

There is an almost infinite variety of ways to display data, including maps, tables, charts, and graphs. Graphs are particularly prevalent in biology, and a few forms are used most frequently.

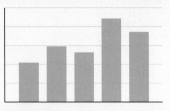

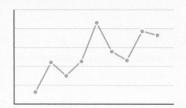

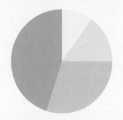

BAR GRAPH
Rectangular bars are used to represent data, each with a height that is proportional to the value being represented.

LINE GRAPH
A line or curve may be used to connect data points or to illustrate trends across many data points.

PIE CHART
"Slices" are used to represent data, in which each slice is a proportion of the whole.

COMMON ELEMENTS OF EFFECTIVE GRAPHICAL PRESENTATIONS OF DATA

Shown here: an example of a line graph.

TITLE
• Describes the content of the display

y-AXIS
• Vertical axis that presents one dimension by which the data can be described
• The axis must always be labeled and include the units of measure

DEPENDENT VARIABLE
• A measurable entity whose response is created by the process being observed
• The value is expected to change in response to a change in the independent variable
• Generally represented on the *y*-axis

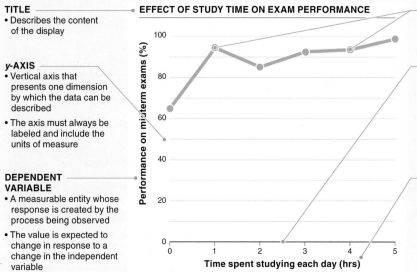

DATA POINTS
• Individual measurements plotted within the visual display

x-AXIS
• Horizontal axis that presents one dimension by which the data can be described
• The axis must always be labeled and include the units of measure

INDEPENDENT VARIABLE
• A measurable entity that can be observed at the start of a process
• The value can be changed as required
• Generally represented on the *x*-axis

STATISTICS

Statistics can quantify and summarize large amounts of data, making it possible to draw more accurate conclusions. Because much variation exists in the world, statistics can help us evaluate whether differences between a treatment group and control group can be attributed to the treatment rather than random chance.

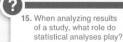

15. When analyzing results of a study, what role do statistical analyses play?

16. You notice that all of the male students who signed up for French tutoring have blue eyes. What conclusions can you draw from this observation?

PSEUDOSCIENCE AND ANECDOTAL OBSERVATIONS

Pseudoscience, in which individuals make scientific-sounding claims that are not supported by trustworthy, methodical scientific studies, and anecdotal observations often lead people to believe that links between two phenomena exist, when there are no such links.

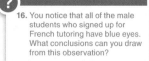

17. You have heard that people from Scandinavia usually have blonde hair and fair skin. Your roommate is from Sweden and he has darker hair and more olive skin. Does this disprove the idea that Scandinavians are fair and blonde? Why or why not?

1·19 On the road to biological literacy: what are the major themes in biology?

Although the diversity of life on earth is tremendous, the study of life is unified by the themes of hierarchical organization and the power of evolution.

HIERARCHICAL ORGANIZATION

Life is organized on many levels within individual organisms, including atoms, cells, tissues, and organs. And in the larger world, organisms themselves are organized into many levels: populations, communities, and ecosystems within the biosphere.

THE POWER OF EVOLUTION

Evolution, the change in genetic characteristics of individuals within populations over time, accounts for the diversity of organisms, but also explains the unity among them.

18. What are the two major unifying themes in this guide to biology?

WHAT'S THIS?

When I was a student, one of my professors had a policy of allowing students to bring a single 8½" x 11" sheet of paper into our exams. We were allowed to write whatever we wanted on this page beforehand and use it during the exam. I spent hours trying to distill the course material to its essential ideas and information. I mastered the art of tiny writing. And I was always very proud of the summary documents that I crafted.

It was with those little "cheat-sheets" in mind that I created the "Review & Rehearse" R&R guides. Please understand that they are not meant to be a replacement for the chapter itself. Rather, they are a prompt, to help you recall, review, and contemplate the most important material from the chapter after you have spent time reading and studying it. I hope they help you!

—Jay Phelan

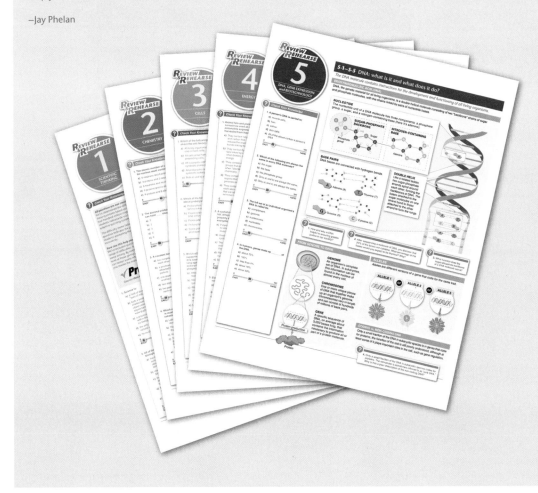

13. Statistical methods make it possible to:

a) prove that any hypothesis is true.

b) determine how likely it is that certain results have occurred by chance.

c) unambiguously learn the truth.

d) reject any hypothesis.

e) test non-falsifiable hypotheses.

0 — 23 — 100
EASY — HARD

14. Anecdotal evidence:

a) is the basis of scientific thinking.

b) tends to be more reliable than data based on observations of large numbers of diverse individuals.

c) is a necessary part of the scientific method.

d) is often the only way to prove important causal links between two phenomena.

e) can seem to reveal links between two phenomena, but the links do not actually exist.

0 — 43 — 100
EASY — HARD

15. A relationship between phenomena that has been established on the basis of large amounts of observational and experimental data is referred to as:

a) a theory.

b) a fact.

c) an assumption.

d) a conjecture.

e) an experimental control.

0 — 30 — 100
EASY — HARD

16. Which of the following issues would be least helped by application of the scientific method?

a) developing more effective high school curricula

b) evaluating the relationship between violence in video games and criminal behavior in teens

c) determining the most effective safety products for automobiles

d) formulating public policy on euthanasia

e) comparing the effectiveness of two potential antibiotics

0 — 31 — 100
EASY — HARD

17. What is the meaning of the statement "Correlation does not imply causation"?

a) Just because two variables vary in a similar pattern does not mean that changing one variable causes a change in the other.

b) It is not possible to demonstrate a correlation between two variables.

c) When a change in one variable causes a change in another variable, the two variables are not necessarily related to each other.

d) It is not possible to prove the cause of any natural phenomenon.

e) Just because two variables vary in a similar pattern does not mean that they have a relationship to each other.

0 — 49 — 100
EASY — HARD

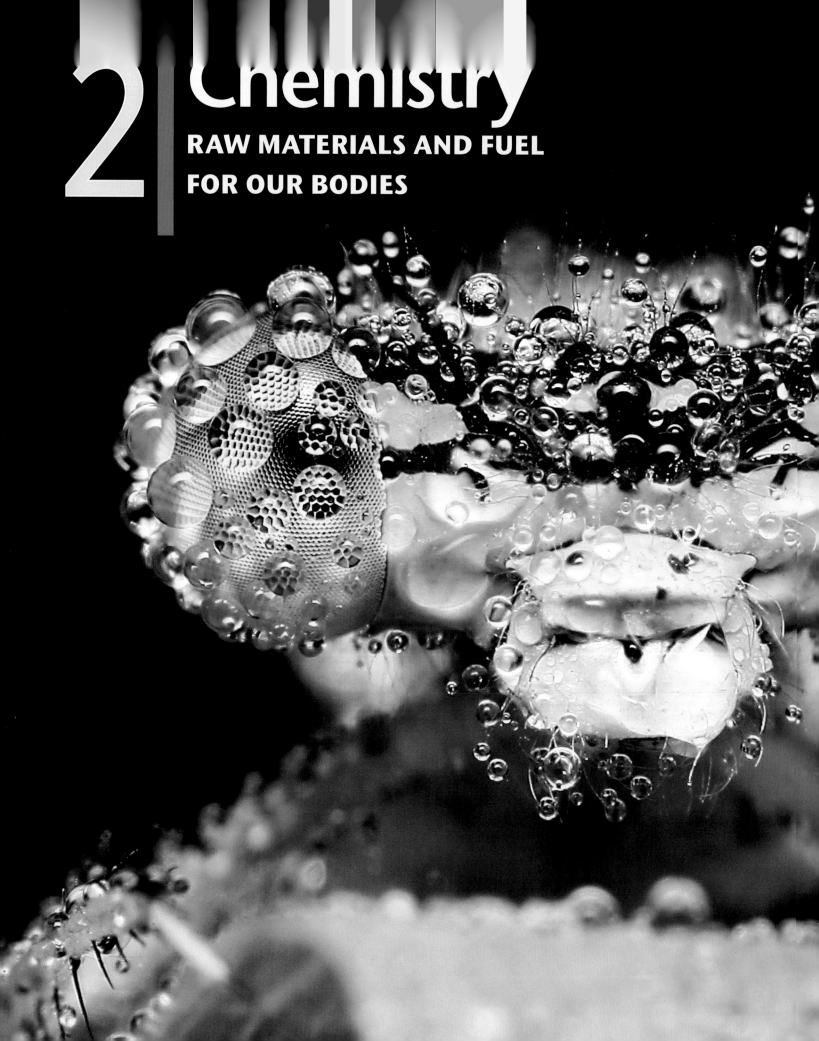

2 Chemistry

RAW MATERIALS AND FUEL
FOR OUR BODIES

Atoms form molecules through bonding.

Water has features that enable it to support all life.

Carbohydrates are fuel for living machines.

Lipids store energy for a rainy day.

Proteins are versatile macromolecules that serve as building blocks.

Nucleic acids store information on how to build and run a body.

2·1–2·3
Atoms form molecules through bonding.

Classical atomic models: a nucleus as a packed cluster of protons and neutrons, orbited by electrons.

2·1 Everything is made of atoms.

A little bit of chemistry goes a long way in the study of biology and in understanding a great deal about your everyday life. Will eating those beans in your soup keep you up all night in gastrointestinal distress? Just knowing whether the beans are lentils or lima beans—each contains slightly different types of sugar molecules—will give you an answer. Will the butter you spread on your toast sabotage your efforts to lose weight? Understanding something about the carbon-hydrogen connections in the fat molecules can help you decide.

The chemistry that is most important in biology revolves around a few important elements, which are introduced at the end of this section. An **element** is a substance

that cannot be broken down chemically into any other substances. Gold, carbon, and copper are elements you might be familiar with. Whatever the element, if you keep cutting it into ever smaller pieces, each of the pieces behaves exactly the same as any other piece. The smallest piece of pure gold will still have the softness, reflectivity, and malleability characteristic of that element (FIGURE 2-1).

If you could continue cutting, you would eventually separate the gold into tiny pieces that could no longer be divided without losing their gold-like properties. These individual component pieces of an element are called **atoms.** An atom is a bit of matter that cannot

Gold

Carbon

Copper

FIGURE 2-1 **Familiar elements.**

 Elements are substances that cannot be broken down chemically into any other substances.

be subdivided any further without losing its essential properties. The word "atom" is from the Greek for "indivisible."

Everything around us, living or not, can be reduced to atoms. All atoms—whether of the element gold or some other element such as oxygen or aluminum or calcium—have the same basic structure. At the center of an atom is a **nucleus,** which is usually made up of two types of particles, called protons and neutrons. **Protons** are particles that have a positive electrical charge and **neutrons** are particles that have no electrical charge. The amount of matter in a particle is its **mass;** protons and neutrons have approximately the same mass (**FIGURE 2-2**).

Whirling in a cloud around the nucleus of every atom are negatively charged particles called **electrons.** An electron weighs almost nothing—less than one-twentieth of one percent of the weight of a proton. (The mass of an atom—its **atomic mass**—is made up of the combined mass of all of its protons and neutrons; for our purposes here, electrons are so light that their mass can be ignored.)

Particles that have the same charge repel each other; those with opposite charges are attracted to each other. Because all electrons have the same charge, the electrons in an atom repel each other. But because they are negatively charged, they are attracted to the positively charged protons in the nucleus. This attraction holds electrons close enough to the nucleus to keep them from flying away, while the energy of their fast movement keeps them from collapsing into the nucleus. When the number of protons and electrons is equal, the charges in the atom are balanced.

Atoms are tiny. Enlarge an atom by a billion times and it would only be the size of a grapefruit. Paradoxically, most of the space taken up by an atom is empty. That is, because the nucleus is very small and compact, the electrons zip about relatively far from the nucleus. If the nucleus were the size of a golf ball, the electrons would be anywhere from half a mile to six miles away.

Elements differ in their number of protons

What distinguishes one element, such as chlorine, from another, such as neon or oxygen? The number of protons in an atom's nucleus determines what element it is. As a rule, atoms of different elements have a different number of protons in the nucleus. A chlorine atom has 17 protons, a neon atom has 10 protons, and an oxygen atom has 8 protons. Each element is given a name (and an abbreviation, such as O for oxygen and C for carbon) and an **atomic number** that corresponds to how many protons it has (**FIGURE 2-3**).

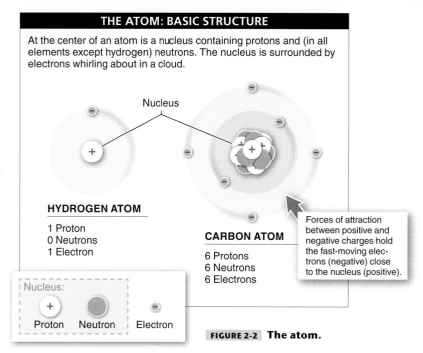

THE ATOM: BASIC STRUCTURE

At the center of an atom is a nucleus containing protons and (in all elements except hydrogen) neutrons. The nucleus is surrounded by electrons whirling about in a cloud.

Nucleus

HYDROGEN ATOM

1 Proton
0 Neutrons
1 Electron

CARBON ATOM

6 Protons
6 Neutrons
6 Electrons

Nucleus:

Proton   Neutron   Electron

Forces of attraction between positive and negative charges hold the fast-moving electrons (negative) close to the nucleus (positive).

FIGURE 2-2   **The atom.**

The mass of an atom is often about double the element's atomic number. This is the case when the number of neutrons in the nucleus is equal to the number of protons, because protons and neutrons have approximately the same

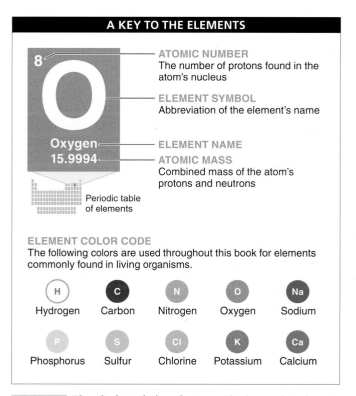

A KEY TO THE ELEMENTS

8

O

Oxygen
15.9994

Periodic table of elements

ATOMIC NUMBER
The number of protons found in the atom's nucleus

ELEMENT SYMBOL
Abbreviation of the element's name

ELEMENT NAME

ATOMIC MASS
Combined mass of the atom's protons and neutrons

ELEMENT COLOR CODE
The following colors are used throughout this book for elements commonly found in living organisms.

H — Hydrogen
C — Carbon
N — Nitrogen
O — Oxygen
Na — Sodium

P — Phosphorus
S — Sulfur
Cl — Chlorine
K — Potassium
Ca — Calcium

FIGURE 2-3   **The vital statistics of atoms.** This key explains how to read a periodic table of elements. A full periodic table of the elements is at the back of the book.

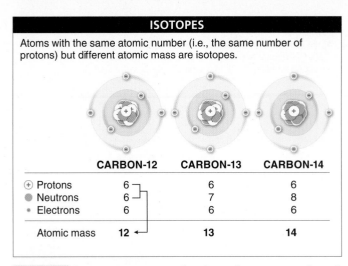

ISOTOPES

Atoms with the same atomic number (i.e., the same number of protons) but different atomic mass are isotopes.

	CARBON-12	CARBON-13	CARBON-14
⊕ Protons	6	6	6
● Neutrons	6	7	8
• Electrons	6	6	6
Atomic mass	**12**	13	14

FIGURE 2-4 **Isotopes are atoms that have the same number of protons but a different number of neutrons.**

mass. The element oxygen, for example, has the atomic number 8 reflecting that it has 8 protons, and because it has 8 neutrons it has an atomic mass of 16, simply the mass of the 8 protons and the mass of the 8 neutrons added together. (As noted above, we can ignore the mass of the electrons.)

Atoms of the same element don't always have the exact same number of neutrons in their nucleus and electrons circling around it. They sometimes acquire or lose components. For example, an atom may have extra neutrons or fewer neutrons than the number of protons. Atoms with the same number of protons but different numbers of neutrons are called **isotopes.** An atom's charge doesn't change in an isotope, because neutrons have no electrical charge, but the atom's mass changes with the loss or addition of another particle in the nucleus (**FIGURE 2-4**). Carbon, for example, has six protons and so usually has a mass of 12. Occasionally, though, a rare carbon atom has an extra neutron or two and an atomic mass of 13 or 14. These isotopes are called carbon-13 (^{13}C) and carbon-14 (^{14}C) and are referred to as "heavy" carbon. In nature, we frequently see mixtures of several isotopes for a given element. So although a sample of pure carbon is predominantly ^{12}C atoms (with 6 protons and 6 neutrons), some ^{13}C and ^{14}C atoms are present in the sample, too.

Most elements and their isotopes have perfectly stable nuclei that remain unchanged virtually forever, never losing or gaining neutrons, protons, or electrons. A few atomic nuclei are not so stable, however, and break down spontaneously sometime after they are created. These atoms are **radioactive,** and in the process of decomposition they release, at a constant rate, a tiny, high-speed particle carrying a lot of energy. (The particle may be a proton,

neutron, or electron; sometimes just energy is released and no particle.) For example, uranium-238 (which has 92 protons and 146 neutrons in its nucleus) is a radioactive element. It spontaneously loses a particle containing 2 protons and 2 neutrons, turning it into an isotope of a different element altogether—thorium-236, in this case. Thorium is radioactive as well, one in a long chain of isotopes, each decaying into another radioactive element, until finally producing the stable element lead (with an atomic mass of 206). Radioactive atoms turn out to be useful in determining the age of fossils (see Section 8-18), in medical imaging and cancer treatment, and in generating vast amounts of energy.

All the known elements can be arranged in a scheme, in the order of their atomic number, called the **periodic table** (see Figure 2-3). A copy of the periodic table is provided at the back of the book. So far, about 90 elements have been discovered that are present in nature, and about 25 others can be made in the laboratory. Everything you see around you is made up of some combination of the 90 or so naturally occurring elements.

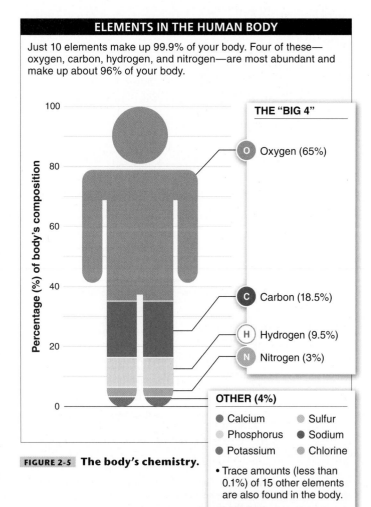

ELEMENTS IN THE HUMAN BODY

Just 10 elements make up 99.9% of your body. Four of these—oxygen, carbon, hydrogen, and nitrogen—are most abundant and make up about 96% of your body.

Percentage (%) of body's composition

THE "BIG 4"

O Oxygen (65%)

C Carbon (18.5%)

H Hydrogen (9.5%)

N Nitrogen (3%)

OTHER (4%)

● Calcium ● Sulfur
○ Phosphorus ● Sodium
● Potassium ○ Chlorine

FIGURE 2-5 **The body's chemistry.**

• Trace amounts (less than 0.1%) of 15 other elements are also found in the body.

Of all the elements found on earth, only 25 are found in your body. The "Top 10" most common of these make up 99.9% of your body mass, but they are not present in equal measures (**FIGURE 2-5**). The "Big 4" elements—oxygen, carbon, hydrogen, and nitrogen—predominate and make up more than 96% of your body mass. With knowledge about the Big 4, you can understand a huge amount about nutrition and physiology (how your body works), so we'll focus on the properties of these four elements later in this chapter.

TAKE-HOME MESSAGE 2·1

Everything around us, living or not, is made up of atoms, the smallest unit into which material can be divided without losing its essential properties. All atoms have the same general structure. They are made up of protons and neutrons in the nucleus, and electrons, which circle far and fast around the nucleus.

2·2 An atom's electrons determine whether (and how) the atom will bond with other atoms.

Most substances are not pure elements but are compounds made of atoms from different elements that are joined by bonds. It is an atom's electrons that determine whether (and how) it bonds with other atoms. Electrons move so quickly that it is impossible to determine, at any given moment, exactly where an electron is. Electrons are not just moving about haphazardly, though. Speeding around the nucleus, they tend to stay within a prescribed area called an "electron shell." An atom may have several shells, each shell occupied by its own set of electrons. Within a shell, the electrons stay far apart because their negative charges repel one another. (Electron shells aren't actual physical tracks but, rather, are simplified placeholders for the different possible energy levels of the electrons.)

The first electron shell is closest to the nucleus and can hold two electrons (**FIGURE 2-6**). If an atom has more than two electrons, as most atoms do, the other electrons are arranged in other shells. The second shell is a bit farther away from the nucleus and can hold as many as eight electrons. There can be up to seven shells in total, holding varying numbers of electrons.

Atoms become less reactive and more stable when their outermost shell is filled to capacity. Atoms with a completely filled outer shell behave like loners, neither reacting nor combining with other atoms. On the other hand, when atoms have outer shells with vacancies, they are likely to interact with other atoms, giving, taking, or sharing electrons to achieve that desirable state: a full outer shell of electrons. In fact, based on the number of electron vacancies in the outermost shell of an atom, it's possible to predict how likely that atom will be to bond, and even which other atoms its likely bonding partners will be.

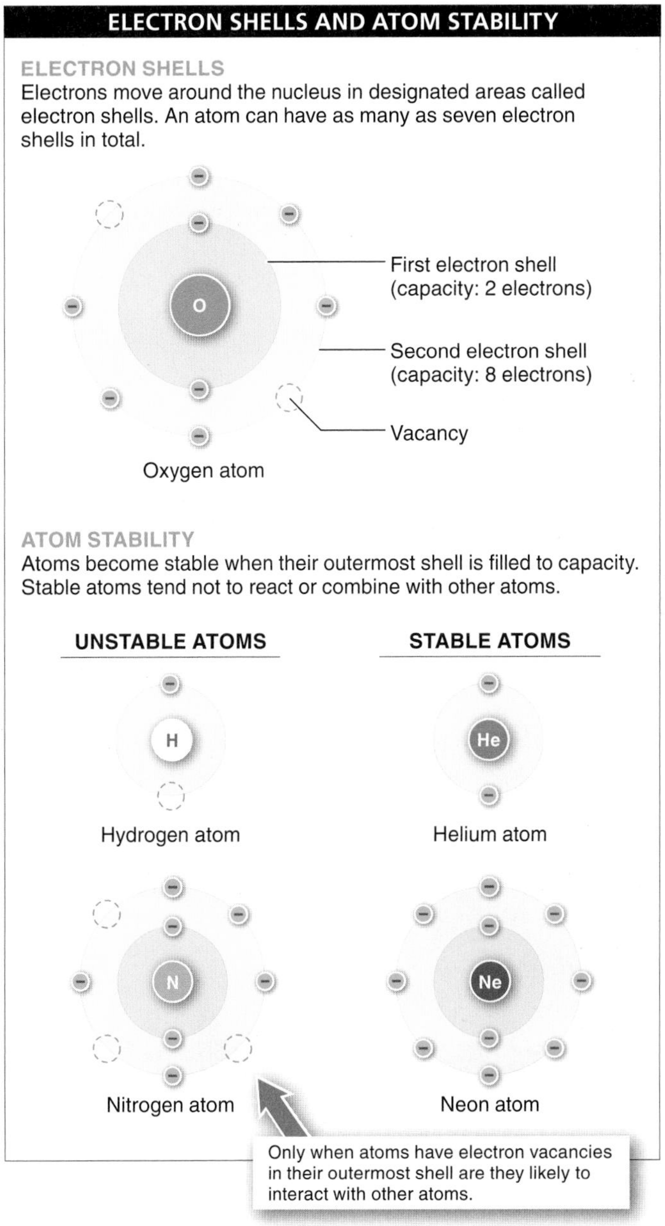

FIGURE 2-6 **Electron interactions.** The chemical characteristics of an atom depend upon the number of electrons in its outermost shell.

ELECTRON SHELLS AND ATOM STABILITY

ELECTRON SHELLS
Electrons move around the nucleus in designated areas called electron shells. An atom can have as many as seven electron shells in total.

First electron shell (capacity: 2 electrons)

Second electron shell (capacity: 8 electrons)

Vacancy

Oxygen atom

ATOM STABILITY
Atoms become stable when their outermost shell is filled to capacity. Stable atoms tend not to react or combine with other atoms.

UNSTABLE ATOMS	STABLE ATOMS

Hydrogen atom

Helium atom

Nitrogen atom

Neon atom

Only when atoms have electron vacancies in their outermost shell are they likely to interact with other atoms.

Let's take a brief look at one element—carbon. Carbon's electron configuration gives it considerable versatility when it comes to bonding with other atoms and making important compounds. Carbon has 6 electrons overall: 2 in the first electron shell and 4 in the second electron shell. Because the second electron shell has a capacity of 8 electrons, carbon can share the 4 electrons in its outermost shell. Having 4 electrons to share gives a carbon atom the ability to bond with other atoms in a large number of different ways—including in four different directions—and makes a huge variety of complex molecules possible (**FIGURE 2-7**).

THE VERSATILITY OF CARBON

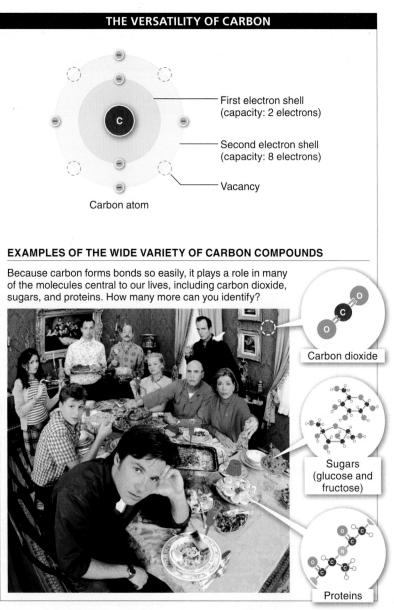

First electron shell
(capacity: 2 electrons)

Second electron shell
(capacity: 8 electrons)

Vacancy

Carbon atom

EXAMPLES OF THE WIDE VARIETY OF CARBON COMPOUNDS

Because carbon forms bonds so easily, it plays a role in many of the molecules central to our lives, including carbon dioxide, sugars, and proteins. How many more can you identify?

Carbon dioxide

Sugars
(glucose and
fructose)

Proteins

FIGURE 2-7 Carbon can bond with other atoms in many different ways.

IONS ARE CHARGED ATOMS

An atom that loses one or more electrons becomes positively charged, while an atom that acquires electrons becomes negatively charged. This transfer of electrons is driven by the fact that atoms with full outer electron shells are more stable.

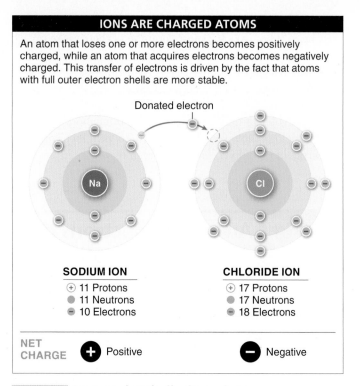

Donated electron

SODIUM ION	CHLORIDE ION
⊕ 11 Protons	⊕ 17 Protons
● 11 Neutrons	● 17 Neutrons
⊖ 10 Electrons	⊖ 18 Electrons

NET CHARGE **+** Positive **−** Negative

FIGURE 2-8 Ions are electrically charged atoms.

Carbon atoms most commonly bond with oxygen, hydrogen, nitrogen, and other carbon atoms. Chains formed of carbon atoms bonded to other carbon atoms are very common—and each carbon bound to two other carbon atoms can still bind with one or two additional atoms. These carbon chains are present in most organic molecules.

Typically, an atom has the same number of electrons as protons. Sometimes an atom may have one or more extra electrons or may lack one or more electrons relative to the number of protons. An atom with extra electrons becomes negatively charged, and an atom lacking one or more electrons is positively charged. Such a charged atom is called an **ion** (**FIGURE 2-8**). Due to their electrical charge, ions behave very differently from the atoms that give rise to them. As we see later in the chapter, ions are more likely to interact with other, oppositely charged ions.

TAKE-HOME MESSAGE 2·2

The chemical characteristics of an atom depend on the number of electrons in its outermost shell. Atoms are most stable and least likely to bond with other atoms when their outermost electron shell is filled to capacity.

2·3 Atoms can bond together to form molecules or compounds.

When you eat a meal, it is like filling your car's tank with gasoline; you have a source of energy that can be used to fuel activities like running, thinking, building muscle, and maintaining the machinery of life. That energy initially comes from the sun and is captured and stored by plants. When we eat plants or eat other animals that eat plants, we ingest the energy stored in the plant material. But how exactly is energy stored in plants?

Groups of atoms held together by bonds are called **molecules.** It takes a certain amount of energy to break a bond between two atoms. The amount of this energy, called the **bond energy,** depends on the atoms involved. When chemical reactions occur—such as when animals (including humans) consume and digest molecules in their diet—some existing bonds are usually broken and some new bonds are usually formed. If the bond energy of the new bonds formed is less than the bond energy of the starting materials, the excess energy is released—and can be used to fuel the body's activities. In a sense, molecules are created as a short-term storage of energy that can be harnessed later.

Before looking at specific types of bonds that hold molecules together, let's look at how molecules are illustrated. In this book you will most often see molecular structures represented by the ball-and-stick and space-filling models.

DIFFERENT WAYS OF REPRESENTING MOLECULAR STRUCTURE

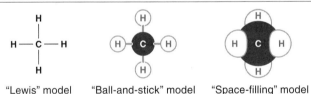

"Lewis" model   "Ball-and-stick" model   "Space-filling" model

There are three principal types of bonds that hold atoms together. The type of bonding that any atom is likely to take part in depends almost entirely on the number of electrons in its outermost shell.

Covalent bonds When two atoms share electrons, a strong **covalent bond** is formed. The simplest example of a covalent bond is the bonding of two hydrogen atoms to form a hydrogen molecule, H_2 (**FIGURE 2-9**), the simplest of all molecules. A hydrogen atom has an atomic number of 1: it has a single proton in its nucleus and a single electron circling around the nucleus in the first shell. Because the atom is most stable when the first shell has two electrons, two hydrogen atoms can each achieve a complete outermost shell by sharing

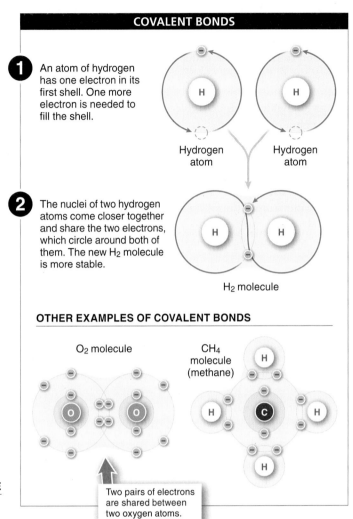

COVALENT BONDS

1 An atom of hydrogen has one electron in its first shell. One more electron is needed to fill the shell.

Hydrogen atom   Hydrogen atom

2 The nuclei of two hydrogen atoms come closer together and share the two electrons, which circle around both of them. The new H_2 molecule is more stable.

H_2 molecule

OTHER EXAMPLES OF COVALENT BONDS

O_2 molecule CH_4 molecule (methane)

Two pairs of electrons are shared between two oxygen atoms.

FIGURE 2-9 **Covalent bonds: strong chemical bonds formed through electron sharing.**

electrons. The nuclei come close together (but not too close, because they are positively charged and their mutual repulsion would destabilize the molecule), and the two electrons circle around both of the nuclei, almost in a figure 8. The new H_2 molecule is very stable because, now that both atoms have two electrons in their outermost shell, they are no longer likely to bond with other atoms. The sharing of two electrons between two atoms forms a *single* covalent bond.

Oxygen, with an atomic number of 8, has two electrons in its innermost shell and six in its outermost shell. Consequently it needs to gain two electrons to fill its outermost shell. Sometimes two oxygen atoms join together to form O_2. Each oxygen atom shares a pair of electrons with the other, filling the outermost shell of both. The sharing of two pairs of electrons between two atoms is called a **double bond** (see Figure 2-9). O₂ is the most common form in which we find oxygen in the world.

Carbon is a particularly "extroverted" atom. It has an atomic number of 6, meaning that two electrons fill its first shell and four remain to occupy the second shell. Because four electron vacancies are left, carbon can, and frequently does, form four covalent bonds, joining up with other atoms in a wide variety of molecules. Methane, the chief component of natural gas, is formed when one atom of carbon covalently bonds with four atoms of hydrogen (see Figure 2-9). It has the chemical formula CH_4.

Ionic bonds Atoms can also bond together without sharing electrons. When one atom transfers one or more of its electrons completely to another, each atom becomes an ion, since each has an unequal number of protons and electrons. The atom gaining electrons becomes negatively charged, while the atom losing electrons becomes positively charged. An **ionic bond** occurs when the two oppositely charged ions attract each other. Unlike covalent bonds, in ionic bonds each electron circles around a single nucleus. Ions of two or more elements linked by ionic bonds form an **ionic compound.** Because the ions attracted to each other are of equal and opposite charge, the compound is neutral—that is, it has no charge (**FIGURE 2-10**). A common ionic compound you are familiar with is table salt (NaCl), composed of sodium and chloride ions.

> ❝In the last third of his life, there came over Laszlo Jamf . . . a hostility, a strangely personal hatred, for the covalent bond . . . That something so mutable, so soft, as a sharing of electrons by atoms of carbon should be at the core of life, *his* life, struck Jamf as a cosmic humiliation. *Sharing?* How much stronger, how everlasting was the ionic bond—where electrons are not shared, but *captured. Seized!* and held! polarized plus and minus, these atoms, no ambiguities . . . how he came to love that clarity: how stable it was, in such mineral stubbornness!❞❞
>
> — THOMAS PYNCHON, in *Gravity's Rainbow*
>
> [Was Laszlo Jamf mistaken in his understanding of the relative strengths of ionic and covalent bonds?]

Hydrogen bonds Ionic and covalent bonds link two or more atoms together within an ionic compound or a molecule. **Hydrogen bonds,** on the other hand, are important in bonding molecules together. A hydrogen bond is formed between a hydrogen atom in one molecule and another atom, often an oxygen or nitrogen atom, in another molecule (or even in another part of the same

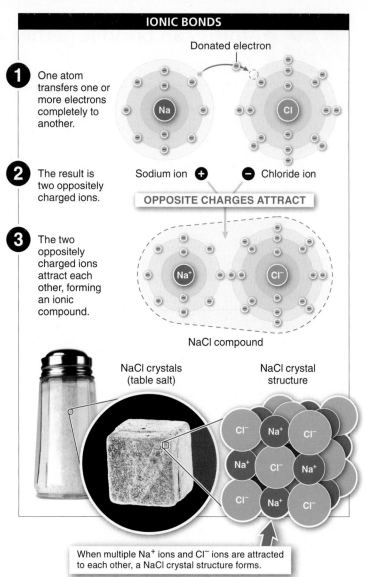

IONIC BONDS

1 One atom transfers one or more electrons completely to another.

Donated electron

2 The result is two oppositely charged ions.

Sodium ion ⊕   ⊖ Chloride ion

OPPOSITE CHARGES ATTRACT

3 The two oppositely charged ions attract each other, forming an ionic compound.

Na⁺   Cl⁻

NaCl compound

NaCl crystals (table salt)   NaCl crystal structure

When multiple Na⁺ ions and Cl⁻ ions are attracted to each other, a NaCl crystal structure forms.

FIGURE 2-10 Ionic bonds: transfer of electrons from one atom to another. The resulting charged atoms (ions) attract each other.

molecule). This bond is based on the attraction between positive and negative charges.

The atoms taking part in hydrogen bonds are not necessarily ions, so where do the electrical charges come from? The hydrogen atom is already covalently bonded to another atom in the same molecule and shares its electron. That electron circles both the hydrogen nucleus and the nucleus of the other atom, but the electron is not shared equally. It's as if the nuclei of the atoms are playing tug-of-war with the electrons. Because the other atom always has more than the one proton found in the hydrogen nucleus, it is more positively charged. As a result, the electron contributed by hydrogen spends more of its time near the other, more positively charged nucleus than near the hydrogen nucleus. Having an extra electron nearby causes the larger atom to be slightly negatively charged, while the hydrogen atom becomes slightly positively charged (**FIGURE 2-11**).

In a sense, molecules with slightly charged atoms become like a magnet, with distinct positive and negative sides (see Figure 2-11), and are said to be "polar." Polar molecules are attracted to other polar molecules, lining up in particular orientations such that the positive regions of one molecule are near the negative regions of another. A hydrogen atom with a slight positive charge can become attracted to an atom in the neighboring molecule with a slight negative charge; these attractions are called hydrogen bonds. Although hydrogen bonds are only about one-thirtieth as strong as covalent or ionic bonds, a multitude of hydrogen bonds form in water and are responsible for many of the unique characteristics that make water one of the most

HYDROGEN BONDS

1 In a water molecule, oxygen's eight positively charged protons attract electrons more readily than does the single proton in the hydrogen atoms. As a result, the electrons are pulled toward the oxygen side of the molecule, making it slightly negative in charge, while the hydrogen side is slightly positive.

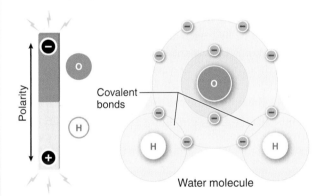

2 Hydrogen bonds are formed between the slightly positively charged hydrogen atoms of one molecule and the slightly negatively charged oxygen atom of another.

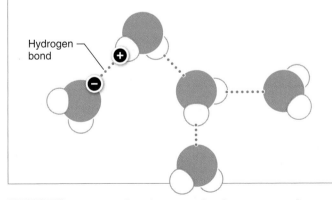

FIGURE 2-11 **Hydrogen bonds: attraction between a polar atom or molecule and a hydrogen atom.** As we'll see later in the chapter, the hydrogen bonding between H₂O molecules gives water some of its characteristic properties.

SUMMARY: THREE TYPES OF BONDS

COVALENT BOND
A strong bond formed when atoms share electrons in order to become more stable, forming a molecule.
BOND STRENGTH: STRONG

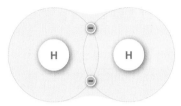

H₂ molecule

IONIC BOND
An attraction between two oppositely charged ions, forming an ionic compound.
BOND STRENGTH: STRONG

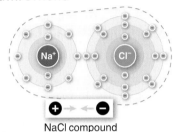

NaCl compound

HYDROGEN BOND
An attraction between the slightly positively charged hydrogen atom of one molecule and the slightly negatively charged atom of another.
BOND STRENGTH: WEAK

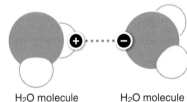

H₂O molecule   H₂O molecule

FIGURE 2-12 **The properties of covalent, ionic, and hydrogen bonds compared.**

important molecules for life on earth. See **FIGURE 2-12** for a review of the three types of bonds discussed in this section.

TAKE-HOME MESSAGE 2·3

Atoms can be bound together in three different ways. Covalent bonds are formed when atoms share electrons. In ionic bonds, one atom transfers its electrons to another and the two oppositely charged ions are attracted to each other, forming an ionic compound. Hydrogen bonds, which are weaker than covalent and ionic bonds, are formed from the attraction between a hydrogen atom and another atom with a slight negative charge.

2·4–2·7
Water has features that enable it to support all life.

Three drops of water bead on a blade of grass.

2·4 Hydrogen bonds make water cohesive.

Every so often, a tadpole or small fish gets an unexpected—and life-ending—surprise from above as a giant spider moves quickly across the surface of the water, reaches down with its two front legs, and plucks the animal from the water. The spider, known as the fishing spider (*Dolomedes triton*), injects its prey with venom, carries it back to shore—moving across the water's surface as if across solid ground—and eats it. How can these spiders walk on water?

The fishing spider, like numerous other insects such as the water strider, makes use of the fact that water molecules have tremendous cohesion. That is, they stick together with

unusual strength. This molecular cohesiveness is due to hydrogen bonds between the water molecules.

Each water molecule is V-shaped (**FIGURE 2-13** ; see also Figure 2-11). The hydrogen atoms are at the ends of the two arms and the oxygen is at the bottom end of the V, between the two hydrogen atoms. Oxygen's strongly positively charged nucleus pulls the circling electrons toward itself and holds on to them for more than its fair share of the time. Consequently, the oxygen at the bottom of the V has a slight negative charge and the other end of the water molecule, containing the hydrogen atoms, has a slight positive charge.

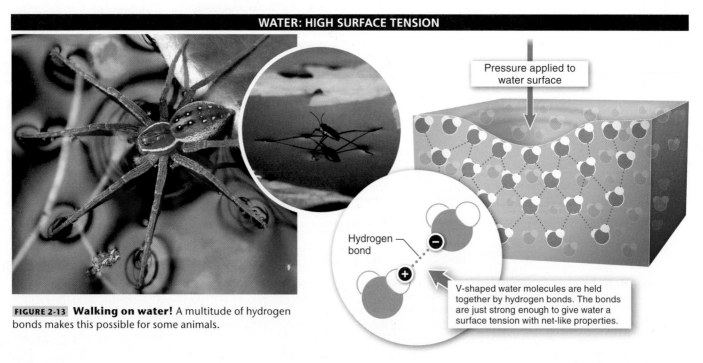

WATER: HIGH SURFACE TENSION

Pressure applied to water surface

Hydrogen bond

V-shaped water molecules are held together by hydrogen bonds. The bonds are just strong enough to give water a surface tension with net-like properties.

FIGURE 2-13 **Walking on water!** A multitude of hydrogen bonds makes this possible for some animals.

Because of their unequally shared electrons, water molecules are polar (see Figure 2-11). Consequently, large numbers of water molecules orient themselves so that the negative side of one molecule is near the positive side of another. Hydrogen bonds form between the relatively positively charged hydrogen atoms and the relatively negatively charged oxygen atoms of *adjacent* water molecules.

Hydrogen bonds, as we've noted, are much weaker than covalent and ionic bonds, and they don't last very long.

Nonetheless, to the fishing spider, the cumulative effect of all the hydrogen bonds in water is to link together all the water molecules in the stream just enough to give the water a surface tension with some net-like properties.

TAKE-HOME MESSAGE 2·4

Water molecules easily form hydrogen bonds, giving water great cohesiveness.

2·5 Water has unusual properties that make it critical to life.

All life on earth depends on water; organisms are made up mostly of water and require it more than any other molecule. Hydrogen bonding among water molecules gives water several important properties that contribute to its crucial role in the biology of all organisms.

1. Cohesion. We saw in the previous section how the connection of water molecules through hydrogen bonds makes water cohesive, resulting in, for example, high surface tension. The cohesiveness of water molecules also makes it possible for tall trees to exist (**FIGURE 2-14**). Leaves need water. Molecules of water in the leaves are continually lost to the atmosphere through evaporation or are used up in the process of photosynthesis. To get more water, plants must pull it up from the soil. The problem for many plants, such as the giant sequoia trees, is that the soil may be 300 feet below the leaf. However, water molecules can pull up adjacent water molecules to which they have hydrogen-bonded. The chain of linked molecules extends all the way down to the soil, where another water molecule is pulled in via the roots each time a water molecule evaporates from a leaf far above.

2. Large heat capacity. Walking across a sandy beach on a hot day, you can feel how easily sand heats up. By comparison, the water is cool as you step from the beach into the ocean, because water resists warming. It takes a lot of energy to change the temperature of water even a small amount. Why? Again, we must look to hydrogen bonding for our answer.

The temperature of a substance is a measure of how quickly all of the molecules are moving. The molecules move more quickly when energy is added in the form of heat. When we heat water, the added energy doesn't immediately increase the movement of the individual water molecules. Rather, it disrupts some of the hydrogen bonds between the molecules (**FIGURE 2-15**). As quickly as they can be

disrupted, though, hydrogen bonds form again somewhere else. And since the water molecules themselves don't increase their movement, the temperature doesn't increase. The net effect is that even if you release a lot of energy into

WATER: STRONG COHESIVENESS

Because of the cohesive properties of water, trees such as the giant sequoia are able to transport water molecules from the soil to their leaves 300 ft. above.

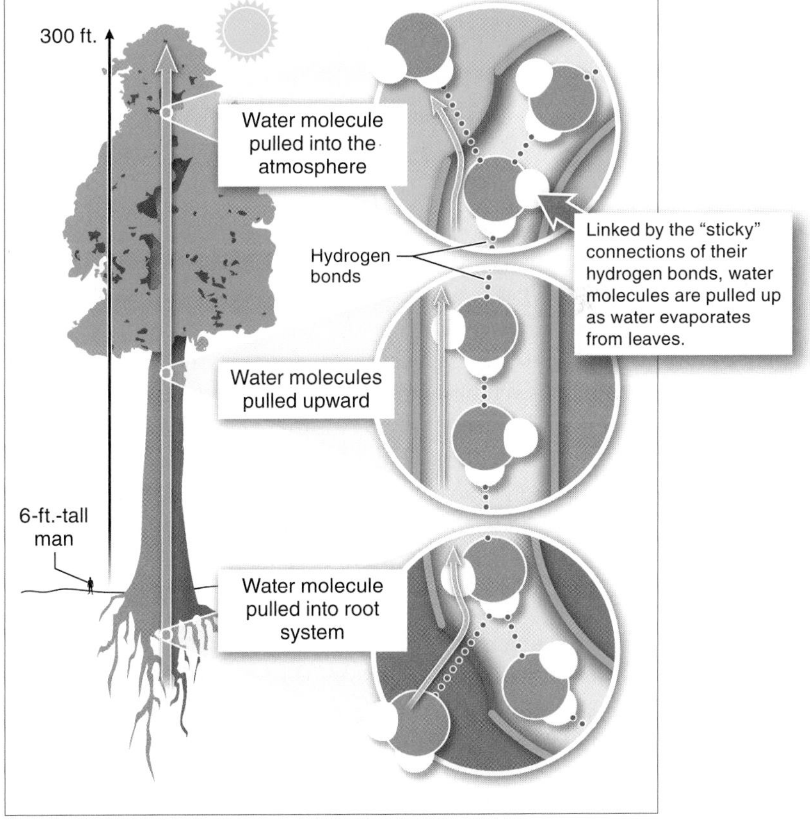

300 ft.

Water molecule pulled into the atmosphere

Hydrogen bonds

Linked by the "sticky" connections of their hydrogen bonds, water molecules are pulled up as water evaporates from leaves.

Water molecules pulled upward

6-ft.-tall man

Water molecule pulled into root system

FIGURE 2-14 Like a giant straw. Hydrogen bonds cause water molecules to "stick" together, which is why they can be pulled up through the giant sequoia.

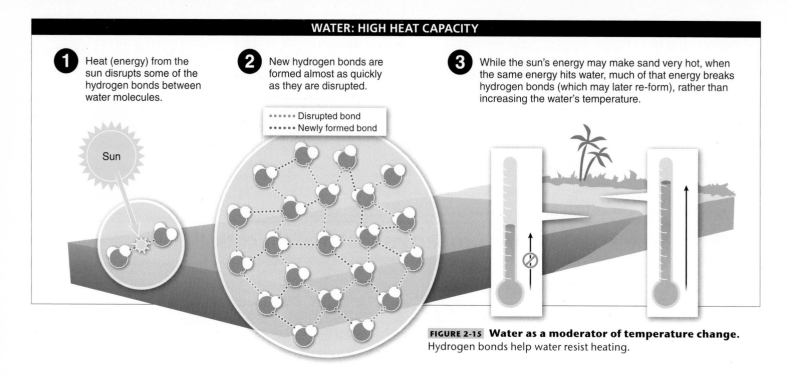

1 Heat (energy) from the sun disrupts some of the hydrogen bonds between water molecules.

2 New hydrogen bonds are formed almost as quickly as they are disrupted.

3 While the sun's energy may make sand very hot, when the same energy hits water, much of that energy breaks hydrogen bonds (which may later re-form), rather than increasing the water's temperature.

••••• Disrupted bond
••••• Newly formed bond

Sun

FIGURE 2-15 Water as a moderator of temperature change. Hydrogen bonds help water resist heating.

water, the temperature doesn't change much. For this reason, because so much of your body is water, you are able to maintain a relatively constant body temperature.

Q Why do coastal areas have milder, less variable climates than inland areas?

Large bodies of water, especially oceans, can absorb huge amounts of heat from the sun during warm times of the year, reducing temperature increases on the coastland. During cold times of the year, the ocean cools slowly, giving off heat that reduces the temperature drop on shore.

3. Low density as a solid. Ice floats. This is unusual because most substances *increase* in density when frozen. As the cooling molecules slow down, they pack together more and more efficiently—and densely. Consequently, the solid sinks. Water, however, becomes less dense and, as you might expect by now, this unusual property is due to hydrogen bonding. As the temperature drops and water molecules slow down, each V-shaped water molecule bonds with four partners, via hydrogen bonds, forming a crystalline lattice in which the molecules are held slightly farther apart than in the liquid, causing ice to be less dense than water (**FIGURE 2-16**).

4. Good solvent. If you put a pinch of table salt into a glass of water, it will quickly dissolve. This means that all the charged sodium (Na^+) and chloride (Cl^-) ions that were ionically bonded together become separated from one another. The sodium and chloride ions were initially attracted to each other because they carry a slight opposing charge. Water is able to pry them apart because, as a polar molecule, it, too, carries charges. The positively charged sodium ions are attracted to the negatively charged side of

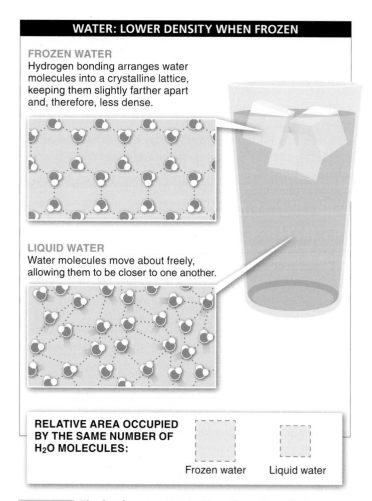

FROZEN WATER
Hydrogen bonding arranges water molecules into a crystalline lattice, keeping them slightly farther apart and, therefore, less dense.

LIQUID WATER
Water molecules move about freely, allowing them to be closer to one another.

RELATIVE AREA OCCUPIED BY THE SAME NUMBER OF H_2O MOLECULES:

Frozen water   Liquid water

FIGURE 2-16 The lattice structure of ice allows it to float.

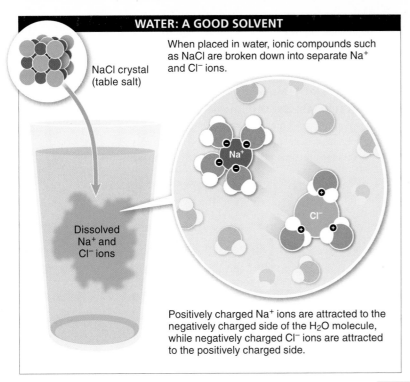

WATER: A GOOD SOLVENT

NaCl crystal (table salt)

When placed in water, ionic compounds such as NaCl are broken down into separate Na^+ and Cl^- ions.

Dissolved Na^+ and Cl^- ions

Positively charged Na^+ ions are attracted to the negatively charged side of the H_2O molecule, while negatively charged Cl^- ions are attracted to the positively charged side.

FIGURE 2-17 **Solutions.** Water pries apart ionic bonds, dissolving ionic compounds.

the water molecule, and the negatively charged chloride ions are attracted to the positively charged side (**FIGURE 2-17**). The ionic bonds holding together the ions are broken, and each ion becomes surrounded by water moleculses. Many

substances are, like water, polar and consequently dissolve easily in water.

Nonpolar molecules (such as oils) have neither positively charged regions nor negatively charged regions. Consequently, the polar water molecules are not attracted to them. Instead, when an oil is poured into a container of water, the oil molecules remain in clumps that never dissolve.

Because so much salt is dissolved in the oceans, many of the water molecules have their positively charged sides facing Cl^- ions and, simultaneously, many molecules of water have their negatively charged sides facing Na^+ ions. Consequently, the orderly lattices of hydrogen bonds found in ice cannot form in salt water, and it does not freeze well.

Q Why don't oceans freeze as easily as freshwater lakes?

TAKE-HOME MESSAGE 2·5

The hydrogen bonds between water molecules give water several of its most important characteristics, including cohesiveness, reduced density as a solid, the ability to resist temperature changes, and broad effectiveness as a solvent for ionic and polar substances.

2·6 Living systems are highly sensitive to acidic and basic conditions.

There's a lot more going on in water than meets the eye. Most of the molecules are present as H_2O, but at any instant some of them break into two parts: H^+ and OH^-. In pure water, the amount of H^+ and OH^- must be exactly the same, since every time a molecule splits, one of each type of ion is produced. But in some fluids containing other dissolved materials, this balance is lost: the fluid can have more H^+ or more OH^-.

The amount of H^+ or OH^- in a fluid gives it some important properties. In particular, the amount of H^+ in a solution is a measure of its acidity and is called **pH.** The greater the number of free hydrogen ions floating around, the more acidic the solution is.

Pure water is in the middle of the pH scale, with a pH of 7.0. Any fluid with a pH below 7.0 has more H^+ ions (and fewer OH^- ions) and is considered an **acid.** Any fluid with a pH above 7.0 has fewer H^+ ions (and more OH^- ions) and is considered a **base.** The pH scale, like the Richter scale for earthquakes, is logarithmic, although in the case of pH, the *lower* the number the *greater* the acidity: a decrease of 1 on the pH scale represents a 10-fold increase in the hydrogen ion concentration (**FIGURE 2-18**). A decrease of 2 represents a 100-fold increase. This means that a cola, with a pH of

HYDROGEN IONS and HYDROXIDE IONS

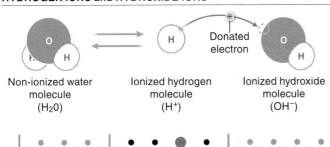

Non-ionized water molecule (H_2O)

Donated electron

Ionized hydrogen molecule (H^+)

Ionized hydroxide molecule (OH^-)

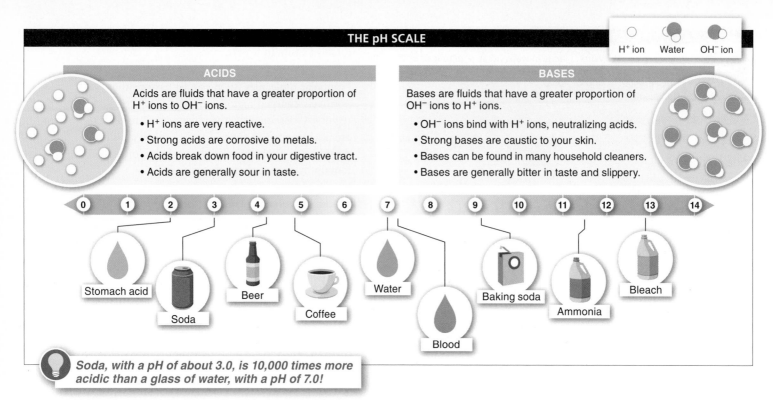

H⁺ ion   Water   OH⁻ ion

ACIDS

Acids are fluids that have a greater proportion of H⁺ ions to OH⁻ ions.

- H⁺ ions are very reactive.
- Strong acids are corrosive to metals.
- Acids break down food in your digestive tract.
- Acids are generally sour in taste.

BASES

Bases are fluids that have a greater proportion of OH⁻ ions to H⁺ ions.

- OH⁻ ions bind with H⁺ ions, neutralizing acids.
- Strong bases are caustic to your skin.
- Bases can be found in many household cleaners.
- Bases are generally bitter in taste and slippery.

0 1 2 3 4 5 6 7 8 9 10 11 12 13 14

Stomach acid   Soda   Beer   Coffee   Water   Blood   Baking soda   Ammonia   Bleach

Soda, with a pH of about 3.0, is 10,000 times more acidic than a glass of water, with a pH of 7.0!

FIGURE 2-18 The pH scale is a way of referring to the acidic, basic, or chemically neutral quality of a fluid.

about 3.0, is 10,000 times (!) more acidic than a glass of water, with a pH of 7.0.

H⁺ ions are essentially free-floating protons. Acids can donate their H⁺ ions to other chemicals. In fact, H⁺ is a very reactive little ion. Its presence gives acids some unique properties. For instance, the hydrogen ions in acids can bind with atoms in metals, causing them to corrode. That's why you can dissolve nails by dumping them in a bucket of acid (or cola).

Q Is acid rain likely to have any impact on populations of microbes living in lakes, streams, and soil? Why?

Your stomach produces large amounts of hydrochloric acid (HCl dissolved in water) and has a pH between 1 and 3. (HCl dissolved in water is acidic because most of the hydrogen ions split off from the chlorine, raising the H⁺ concentration of the fluid.) The acid in your stomach helps to kill most bacteria that you ingest. It also greatly enhances the breakdown of the chemicals in the food you eat and the efficiency of digestion and absorption. You may have learned firsthand of the high acidity of your stomach fluids if you have experienced heartburn or the sour taste of vomit.

Acids have a higher concentration of H⁺ than of OH⁻. Conversely, bases have a higher concentration of OH⁻ than of H⁺. Baking soda is a common basic substance. Some bases are called "antacids" because the OH⁻ ions in bases can bind with excess H⁺ ions in acidic solutions, neutralizing the acid. Base-containing products such as Alka-Seltzer and Milk of Magnesia reduce the unpleasant feeling of heartburn and acid indigestion that sometimes arise from the overproduction of acids by the stomach. Bases are commonly used in household cleaning products and generally have a bitter taste and a soapy, slippery feel.

The pH of blood is usually 7.4. Given that most cellular reactions produce or consume H⁺ ions, you might expect there to be great swings in the pH of our blood. Unfortunately, our bodies can't tolerate such swings. Most of the chemical reactions in our blood or cells stop proceeding properly if the pH swings up or down, even by less than half a point. Fortunately, our bodies contain some chemicals that act like bank accounts for H⁺ ions (**FIGURE 2-19**). Called **buffers,** these chemicals can quickly absorb excess H⁺ ions to keep a solution from becoming too acidic, and they can quickly release H⁺

BUFFERS IN BLOOD STABILIZE pH

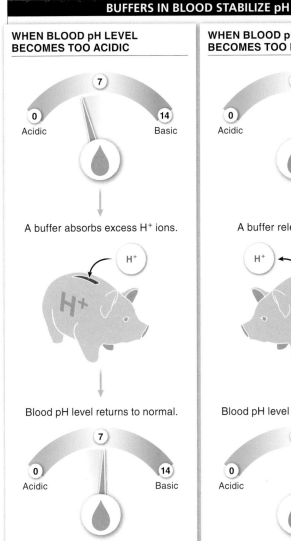

WHEN BLOOD pH LEVEL BECOMES TOO ACIDIC

7
0 — Acidic
14 — Basic

A buffer absorbs excess H⁺ ions.

H^+

H^+

Blood pH level returns to normal.

7
0 — Acidic
14 — Basic

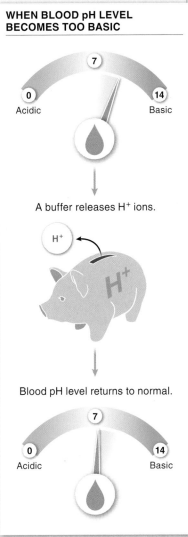

WHEN BLOOD pH LEVEL BECOMES TOO BASIC

7
0 — Acidic
14 — Basic

A buffer releases H⁺ ions.

H^+

H^+

Blood pH level returns to normal.

7
0 — Acidic
14 — Basic

FIGURE 2-19 Maintaining a constant internal environment. Buffers in blood prevent potentially damaging pH swings.

ions to counteract any increases in OH^- concentration. Buffers are chemicals that act to resist changes in pH.

TAKE-HOME MESSAGE 2·6

The pH of a fluid is a measure of how acidic or basic the solution is and depends on the concentration of dissolved H^+ ions present; the lower the pH, the more acidic the solution. Acids, such as vinegar, can donate protons to other chemicals; bases, including baking soda, bind with free protons.

② 2·7 THIS IS HOW WE DO IT

Do anti-acid drugs impair digestion and increase the risk of food allergies?

Your stomach is a pretty extreme environment compared with the rest of your body—at least when it comes to acidity. In the previous section, we saw that hydrochloric acid produced in your stomach creates an extremely acidic environment, with a pH between 1 and 3.

How does the stomach's acidic environment aid in digestion?

Your stomach's high acidity aids in breaking down chemicals in the food you eat—particularly proteins. (It also helps your body fight against bacterial infections.) This is important because proteins that pass through the stomach undigested can trigger an allergic response, and your body will treat them as a sort of foreign invader and health risk, mounting an immune response to them.

But for millions of people, the acidity in their stomach can lead to discomfort. For a wide variety of complaints—including ulcers, heartburn, and acid reflux—many people take medications (called antacids) that neutralize stomach acids or medications that reduce the stomach's production of acids or the major protein-digesting enzymes. In fact, one of these medications, Prilosec, is one of the best-selling drugs in the United States, with sales of more than $700 million in 2013.

> **Does reducing stomach acidity have health consequences? How could we evaluate whether acidity-reducing medications increase the potential for food allergy?**

Researchers have suspected that anti-acid medications might reduce protein digestion, and that undigested proteins stimulate allergies. To test this idea, they explored two experimental strategies.

1. First they evaluated the effectiveness of protein-digesting enzymes at several different levels of acidity. For this, they simply combined these enzymes and common dietary proteins in test tubes, at body temperature.

The result? At the typical stomach pH of 2, they found that the proteins in milk are completely digested within 1 minute. Conversely, at less acidic pH values of 3 and 5, the milk proteins were largely undigested, even after 1 hour.

> **From these results, what could the researchers conclude about the role of stomach acidity in digesting proteins?**

2. In their second experimental strategy, the researchers observed 152 people who were seeking treatment for an ulcer, before and after a three-month anti-acid treatment. They evaluated each participant for signs of allergic reactions to a variety of proteins.

The result? Before the three-month anti-acid treatment, 10% of the subjects exhibited signs of having had an immune response to at least 1 of the 19 dietary proteins the researchers tested. After three months of anti-acid treatment, 26% of the subjects exhibited signs of having had an immune response to at least 1 of the 19 tested dietary proteins.

> **From these results, what could the research team conclude about the role of anti-acid medications in causing normal dietary proteins to stimulate an allergic response?**

The researchers also made observations on 50 control subjects, who were not seeking anti-ulcer treatment. (What do you think was the purpose of these controls?) As with the other subjects prior to anti-acid treatment, just 10% exhibited signs of having had an immune response to at least 1 of the 19 tested dietary proteins.

From their results, the researchers concluded that anti-acid drugs "hamper the biological gate-keeping function of the stomach."

TAKE-HOME MESSAGE 2·7

Dietary proteins are not digested by the stomach when the stomach's pH is increased as a result of medications taken to treat ulcers, heartburn, and other digestive problems. This can put people who take anti-acid medications at risk for developing allergic responses to common foods.

2·8 – 2·11
Carbohydrates are fuel for living machines.

Worldwide, more corn is produced (by weight) than any other grain.

2·8 Carbohydrates include macromolecules that function as fuel.

Hardly a day goes by without an item in a magazine or newspaper or on TV talking about whether carbohydrates are good or bad for us. Sports drinks such as Gatorade are filled with carbs, and in the days before a big game or race, athletes often "carbo-load" by eating bowls of pasta. Yet other dietary supplements and "power" bars tout that they are low-carb and are effective in weight loss by causing the body to resort to using fat for energy. Meanwhile, nutritionists, doctors, and many diet programs exhort people to increase the amount of fiber—another type of carbohydrate—in their diet. What exactly are they all talking about?

Carbohydrates are one type of **macromolecule**—a large molecule made up from smaller building blocks or subunits. Four types of macromolecule are essential to the building and functioning of living organisms: carbohydrates, lipids, proteins, and nucleic acids. **Carbohydrates** are molecules that contain carbon, hydrogen, and oxygen: they are the primary fuel for running all of the cellular machinery and also form much of the structure of cells in all life forms. Sometimes they contain atoms of other elements, but they must have carbon, hydrogen, and oxygen to be considered a carbohydrate (**FIGURE 2-20**). Additionally, a carbohydrate generally has approximately the same number of carbon atoms as it does H_2O units. For instance, the best-known carbohydrate, glucose, has the composition $C_6H_{12}O_6$ (6 carbons and, as a little math will show, 6 H_2O units;

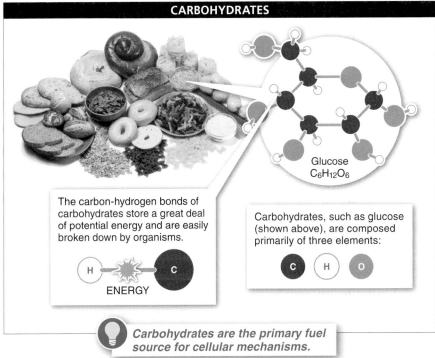

CARBOHYDRATES

The carbon-hydrogen bonds of carbohydrates store a great deal of potential energy and are easily broken down by organisms.

ENERGY

Carbohydrates, such as glucose (shown above), are composed primarily of three elements:

Glucose
$C_6H_{12}O_6$

Carbohydrates are the primary fuel source for cellular mechanisms.

FIGURE 2-20 **All carbohydrates have a similar structure and function.**

notice that $6 \times H_2 = H_{12}$ and $6 \times O = O_6$). A carbohydrate called maltose has the composition $C_{12}H_{22}O_{11}$.

Carbohydrates are classified into several categories, based on their size and their composition. The simplest carbohydrates are the **monosaccharides,** or **simple sugars.** The simple sugars contain anywhere from three to six carbon atoms and, when they are broken down, the products usually are not carbohydrates. Two common monosaccharides are glucose, found in the sap and fruit of many plants, and fructose, found primarily in fruits and vegetables, as well as in honey. Fructose is the sweetest of all naturally occurring sugars. The suffix *-ose* tells us that a substance is a carbohydrate.

SOME COMMON MONOSACCHARIDES

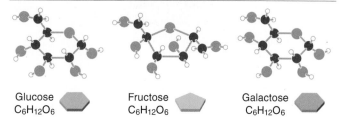

Glucose
$C_6H_{12}O_6$

Fructose
$C_6H_{12}O_6$

Galactose
$C_6H_{12}O_6$

Carbohydrates function well as fuels because of their many carbon-hydrogen bonds; as those bonds are broken down

and other, more stable bonds (primarily between carbon and oxygen) are formed, a great deal of energy is released that organisms can use.

In the next section we investigate glucose, the chief carbohydrate used by organisms to fuel their activities.

TAKE-HOME MESSAGE 2·8

Carbohydrates are the primary fuel for running all cellular machinery and also form much of the structure of cells in all life forms. Carbohydrates contain carbon, hydrogen, and oxygen, and generally have the same number of carbon atoms as they do H_2O units. The simplest carbohydrates, including glucose, are monosaccharides or simple sugars. They contain from three to six carbon atoms. As the chemical bonds of carbohydrates are broken down and other more stable bonds are formed, a great deal of energy is released that can be used by the organism.

2·9 Glucose provides energy for the body's cells.

The carbohydrate of most importance to living organisms is glucose. This simple sugar is found naturally in most fruits, but most of the carbohydrates that you eat, including table sugar (called sucrose) and the starchy carbohydrates found in bread and potatoes, are converted into glucose in your digestive system. The glucose then circulates in your blood at a concentration of about 0.1%. Circulating glucose, also called "blood sugar," has one of three fates (**FIGURE 2-21**).

1. Fuel for cellular activity. Once it arrives at and enters a cell, glucose can be used as an energy source. Through a series of chemical reactions, the relatively high-energy bonds in the glucose molecule are converted into lower-energy bonds of other molecules (a process explained in detail in Chapter 4). The change from higher-energy to lower-energy bonds releases energy, and organisms use the released energy to fuel cellular activity, including the muscle contractions that enable you to move and the nerve activities that enable you to think.

2. Stored temporarily as glycogen. If there is more glucose circulating in your bloodstream than is necessary to meet your body's current energy needs, the excess glucose

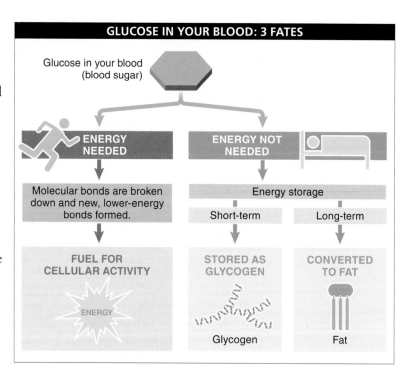

FIGURE 2-21 What happens to sugar in your blood? Depending upon whether energy is needed, glucose can be released for immediate use or stored on a short-term or long-term basis.

FIGURE 2-22 **Some competitive athletes carbo-load to maintain high energy levels.**

blood glucose than is necessary, so that much of the excess glucose is stored as glycogen.

Glycogen also plays a role in the initial rapid weight loss people experience when dieting. If you reduce your caloric intake such that your body is burning more calories than you are consuming, your body must use stored energy. The first, most accessible molecules that can be broken down for energy in the absence of sufficient sugar in your bloodstream are glycogen molecules in your muscles and liver. Large amounts of water are bound to glycogen. As that glycogen is removed from your tissue, so, too, is the water. This loss of water leads to the initial dramatic weight loss that occurs before your body resorts to using stored fat, and the rate of weight loss then slows considerably (FIGURE 2-23).

can be temporarily stored in various tissues, primarily your muscles and liver. The stored glucose molecules are linked together to form a large web of molecules called **glycogen.** When you need energy later, the glycogen can easily be broken down to release glucose molecules back into your bloodstream. Glycogen is the primary form of short-term energy storage in animals.

3. Converted to fat. Finally, glucose circulating in your bloodstream can be converted into fat, a form of long-term energy storage.

Q What is "carbo-loading"? "Carbo-loading" is a method by which athletes can, for a short time, double or triple the usual amount of glycogen stored in their muscles and liver, increasing the store of fuel available for extended exertion and delaying the onset of fatigue during an endurance event (FIGURE 2-22).

Carbo-loading is usually done in a depletion phase and a loading phase. Six or seven days before competition, the athlete depletes glycogen in the muscles through a super-low carbohydrate intake and exhaustive exercise. Then two days before the competition, the athlete eats a super-high carbohydrate diet and reduces exercise to achieve a higher

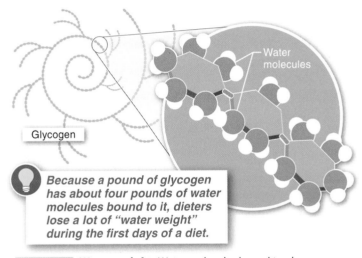

Glycogen

Water molecules

💡 *Because a pound of glycogen has about four pounds of water molecules bound to it, dieters lose a lot of "water weight" during the first days of a diet.*

FIGURE 2-23 **Water weight.** Water molecules bound to glycogen account for much of the weight lost early in a diet.

TAKE-HOME MESSAGE 2·9

Glucose is the most important carbohydrate to living organisms. Glucose in the bloodstream can be used as an energy source, can be stored as glycogen in the muscles and liver for later use, or can be converted to fat.

2·10 Many complex carbohydrates are time-release packets of energy.

In contrast to the simple sugars, some carbohydrates contain more than one sugar unit or building block. When two simple sugars are bonded together they form a **disaccharide;** the joining of glucose and galactose, for example, makes the disaccharide lactose, which is the sugar found in milk. Much larger numbers of simple sugars may be joined together—sometimes as many as 10,000. In this case, the resulting carbohydrate is called a **polysaccharide,** or a **complex carbohydrate** (**FIGURE 2-24**). Depending on how the simple sugars are bonded together, polysaccharides may function as "time-release" stores of energy or as structural materials that may be completely indigestible by most animals. An example of such a structural material is the polysaccharide cellulose—the primary component of plant cell walls.

Like simple sugars, many disaccharides and polysaccharides are important sources of fuel for cells. Unlike simple sugars, however, disaccharides and polysaccharides must undergo some preliminary processing before the energy can be released from their bonds. Let's look at what happens when we eat sucrose, common table sugar. Sucrose is the primary carbohydrate in plant sap and is composed of two simple sugars, glucose and fructose, linked together. To use sucrose, the body must first break the bond linking the glucose and fructose. Only then can the individual monosaccharides be broken down into their component atoms and the energy that was stored in their chemical bonds be harvested and used.

In plants, the primary form of energy storage is a complex carbohydrate called **starch,** found in roots and other tissues (see Figure 2-24). Starch consists of a hundred or more glucose molecules joined together in a line. Grains such as barley, wheat, and rye are high in starch content, and corn and rice are more than 70% starch. Although it is composed of glucose molecules linked together, starch does not taste sweet. Because of its molecular shape, it does not stimulate the sweetness receptors on the tongue. The glycogen that stores energy in your muscles and liver is also a complex carbohydrate, so it is sometimes referred to as "animal starch" (although it has a more branched structure than starch and carries more glucose units linked together).

Many complex carbohydrates are like "time-release" fuel pellets. The glucose molecules become available slowly, as the bonds between glucose units are broken. As a demonstration,

Q Before heading to the library for a long study session, students would be wise to consume oatmeal rather than fresh fruits. Why?

put a piece of potato or bread into your mouth and let it rest on your tongue. Initially, it does not taste sweet. After a few minutes, you will begin to taste some sweetness as the chemicals in your saliva break the starch down into glucose.

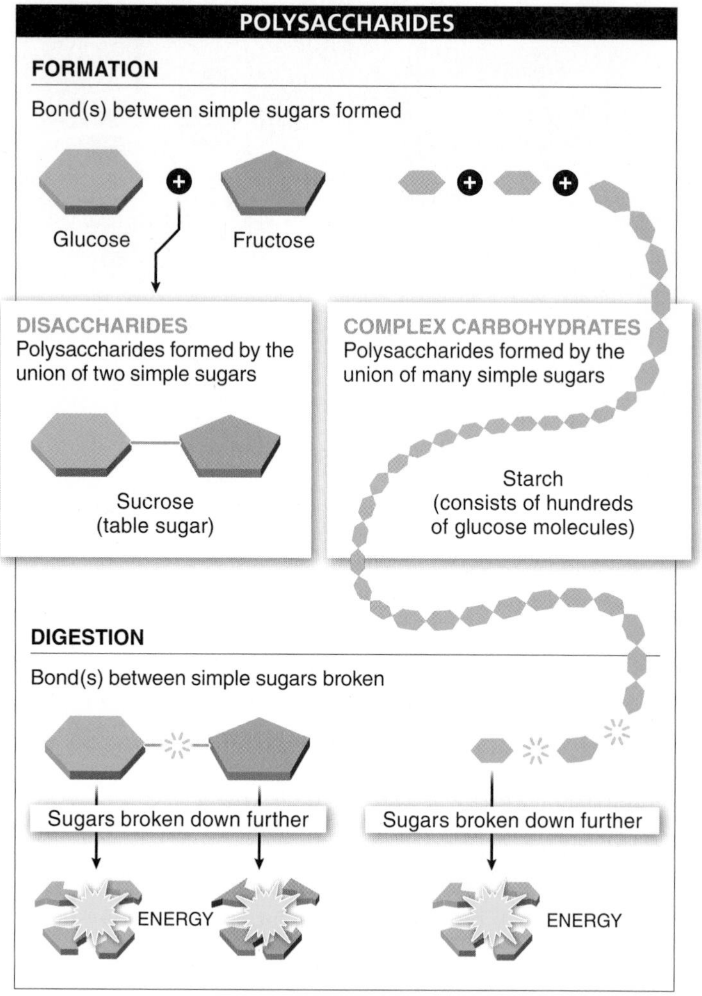

The relative amounts of complex carbohydrates and simple sugars in foods cause them to have different effects when you eat them. Oatmeal (along with rice and pasta), for example, is rich in complex carbohydrates. Fresh fruits, on the other hand, are rich in simple sugars, such as fructose. Consequently, although fruit will give a quick burst of energy as the sugars

POLYSACCHARIDES

FORMATION
Bond(s) between simple sugars formed

Glucose + Fructose

DISACCHARIDES
Polysaccharides formed by the union of two simple sugars

Sucrose (table sugar)

COMPLEX CARBOHYDRATES
Polysaccharides formed by the union of many simple sugars

Starch (consists of hundreds of glucose molecules)

DIGESTION
Bond(s) between simple sugars broken

Sugars broken down further — ENERGY

Sugars broken down further — ENERGY

FIGURE 2-24 Chains of sugars. Polysaccharides are made from simple sugars bonded together.

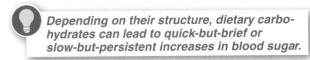

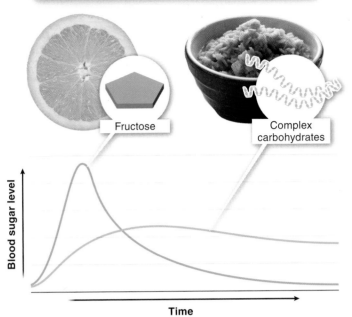

Depending on their structure, dietary carbo-hydrates can lead to quick-but-brief or slow-but-persistent increases in blood sugar.

Fructose

Complex carbohydrates

Blood sugar level

Time

FIGURE 2-25 **Short-term versus long-term energy.** Simple sugars and complex carbohydrates differ in the way they make energy available to you.

are almost immediately available, the fuel will soon be gone from the bloodstream. Simple sugars in the oatmeal will become available only gradually, as the complex carbohydrates of the oats are broken down into their simple sugar components (**FIGURE 2-25**).

TAKE-HOME MESSAGE 2·10

Multiple simple carbohydrates are sometimes linked together into more complex carbohydrates. Types of complex carbohydrates include starch, which is the primary form of energy storage in plants, and glycogen, which is a primary form of energy storage in animals.

2·11   Not all carbohydrates are digestible.

Despite their general importance as a fuel source for humans, not all carbohydrates can be broken down by our digestive system. Two different complex carbohydrates—both indigestible by humans—serve as structural materials for invertebrate animals and plants: **chitin** (pronounced KITE-in) and **cellulose** (**FIGURE 2-26**). Chitin forms the rigid outer skeleton of most insects and crustaceans (such as lobsters and crabs). Cellulose forms a huge variety of plant structures that are visible all around us. We find cellulose in trees and the wooden structures we build from them, in cotton and the clothes we make from it, in leaves and in grasses. In fact, it is the single most abundant compound on earth.

Surprisingly, cellulose is almost identical in composition to starch. Nonetheless, because of one small difference in the chemical bond between the simple sugar units, cellulose has a slightly different three-dimensional structure. Even tiny differences in the shape of a molecule can have a huge effect on its behavior. In this case, the difference in shape

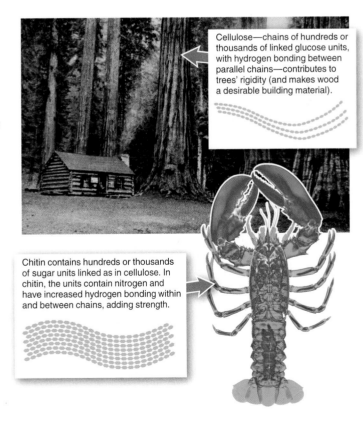

Cellulose—chains of hundreds or thousands of linked glucose units, with hydrogen bonding between parallel chains—contributes to trees' rigidity (and makes wood a desirable building material).

Chitin contains hundreds or thousands of sugar units linked as in cellulose. In chitin, the units contain nitrogen and have increased hydrogen bonding within and between chains, adding strength.

FIGURE 2-26 **Carbohydrates can serve as structural materials.**

Cellulose in our diet—fiber—is indigestible, but it still aids in digestion and has numerous health benefits.

FIGURE 2-27 **Fiber.** It's not digestible, but it's still important for our diet.

makes cellulose a sturdy structural material and, unlike starch, also makes it impossible for humans to digest. Consequently, the cellulose we eat passes right through our digestive system unused.

Although it is not digestible, cellulose is still important to human diets. The cellulose in our diet is known as "fiber" (**FIGURE 2-27**). It is also appropriately called "roughage" because, as the cellulose of celery stalks and lettuce leaves passes through our digestive system, it scrapes the wall of the digestive tract. Its bulk and the scraping stimulate the more rapid passage of food and the unwanted, possibly harmful products of digestion through our intestines. That is why fiber reduces the risk of colon cancer and other diseases (but it is also why too much fiber can lead to diarrhea).

Unlike humans, termites have microorganisms living in their gut that are able to break down cellulose. That's why they can chew on wood and, with the help of the cellulose-digesting boarders in their gut, can break down the cellulose and extract usable energy from the freed glucose molecules.

TAKE-HOME MESSAGE 2·11

Some complex carbohydrates, including chitin and cellulose, cannot be digested by most animals. Such indigestible carbohydrates in the diet, called fiber, aid in digestion and have many health benefits.

2·12–2·14
Lipids store energy for a rainy day.

Protected from the cold: lipids help insulate seals.

2·12 Lipids are macromolecules with several functions, including energy storage.

Lipids are a second group of macromolecules important to all living organisms. Lipids, just like carbohydrates, are made primarily from atoms of carbon, hydrogen, and oxygen, but the atoms are in different proportions. The lipids in your diet, for example, tend to have significantly more energy-rich carbon-hydrogen bonds than carbohydrates, and so contain significantly more stored energy (**FIGURE 2-28**).

What exactly is a lipid? That's not as easy to answer as you might expect. Lipids come in a wide variety of structures.

They don't have any unique subunits (such as the simple sugars that make up disaccharides and polysaccharides) or particular ratios of atoms that serve as defining features. Consequently, lipids are defined based on their physical characteristics. Most notably, lipids do not dissolve in water and are greasy to the touch—think of salad dressings.

Lipids are insoluble in water because they tend to have long chains consisting only of carbon and hydrogen

The Emperor penguin (*Aptenodytes forsteri*) benefits from the insulating effect of fat and is able to thrive in the extreme cold of Antarctica.

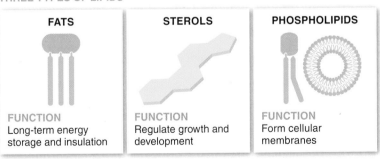

INTRODUCTION TO LIPIDS

TYPICAL FEATURES OF LIPIDS
- Nonpolar molecules that do not dissolve in water
- Greasy to the touch
- Can be a significant source of energy storage

THREE TYPES OF LIPIDS

FATS	STEROLS	PHOSPHOLIPIDS
FUNCTION Long-term energy storage and insulation	**FUNCTION** Regulate growth and development	**FUNCTION** Form cellular membranes

FIGURE 2-28 Lipids serve many roles in the body.

ATOMS   WATER   CARBOHYDRATES   LIPIDS   PROTEINS   NUCLEIC ACIDS

Q Why does a salad dressing made with vinegar and oil separate into two layers shortly after you shake it?

atoms. In contrast to water, these chains of carbon and hydrogen atoms are nonpolar—there are no regions of positive or negative charge. Thus they do not form weak bonds with water and cannot dissolve in water. Nonpolar molecules (or parts of molecules) tend to minimize their contact with water and are considered **hydrophobic** ("water-fearing"). Lipids cluster together when mixed with water, never fully dissolving. Molecules that readily form hydrogen bonds with water, on the other hand, are considered **hydrophilic** ("water-loving").

One familiar type of lipid is *fat,* the type most important in long-term energy storage and insulation. (Penguins and walruses can maintain relatively high body temperatures, despite living in very cold habitats, due to their thick layer of insulating fat.) Lipids also include *sterols,* which include *cholesterol* and many of the sex hormones that play regulatory roles in animals, and *phospholipids,* which form the membranes that enclose cells.

TAKE-HOME MESSAGE 2·12

Lipids are insoluble in water and greasy to the touch. They are valuable to organisms for long-term energy storage and insulation, in membrane formation, and as hormones.

2·13 Fats are tasty molecules too plentiful in our diets.

All fats have two distinct components: they have a "head" region and two or three long "tails" (**FIGURE 2-29**). The head region is a small molecule called **glycerol.** It is linked to "tail" molecules known as **fatty acids.** A fatty acid is simply a long hydrocarbon—that is, a chain of carbon atoms, often a dozen or more, linked together and with one or two hydrogen atoms attached to each carbon atom.

The fats in most foods we eat are **triglycerides,** which are fats having three fatty acids linked to the glycerol molecule. For this reason, the terms "fats" and "triglycerides" are often used interchangeably. Triglycerides that are solid at room temperature are generally called "fats," while those that are liquid at room temperature are called "oils."

Fat molecules contain much more stored energy than do carbohydrate molecules. That is, the chemical breakdown of fat molecules releases significantly more energy. A single gram of carbohydrate stores about 4 calories of energy, while the same amount of fat stores about 9 calories—not unlike the difference between a $5 bill and a $10 bill. Because fats store such a large amount of energy, a strong taste preference for fats over other energy sources has evolved in animals (**FIGURE 2-30**). Organisms evolving in an environment of uncertain food supply will build the largest surplus by consuming molecules that hold the largest amount of energy in the smallest mass. This feature helped the earliest humans to survive, millions of years ago, but today puts us in danger from the health risks of obesity now that fats are all too readily available.

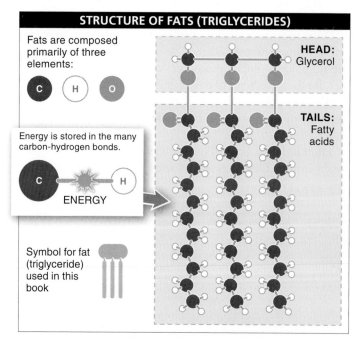

STRUCTURE OF FATS (TRIGLYCERIDES)

Fats are composed primarily of three elements:

C H O

HEAD: Glycerol

Energy is stored in the many carbon-hydrogen bonds.

C — H
ENERGY

TAILS: Fatty acids

Symbol for fat (triglyceride) used in this book

FIGURE 2-29 Triglycerides have glycerol heads and fatty acid tails.

An important distinction is made between "saturated" and "unsaturated" fats (**FIGURE 2-31**). These terms refer to the hydrocarbon chain in the fatty acids. If each carbon atom in the hydrocarbon chain of a fatty acid is bonded to two hydrogen atoms, the fat molecule carries the maximum number of hydrogen atoms and is said to be a **saturated fat.** Most animal fats, including those found in meat and eggs, are saturated. They are not essential to your health

FIGURE 2-30 Animals (including humans) prefer the taste of fats.

Because fats store such large amounts of energy, a strong taste preference for fats over other energy sources has evolved in animals.

SATURATED FATS vs. UNSATURATED FATS

SATURATED FATS

In saturated fats, each carbon in the hydrocarbon chain is bound to two hydrogen atoms.

Straight fatty acids can be packed together tightly. As a result, saturated fats are solid at room temperature.

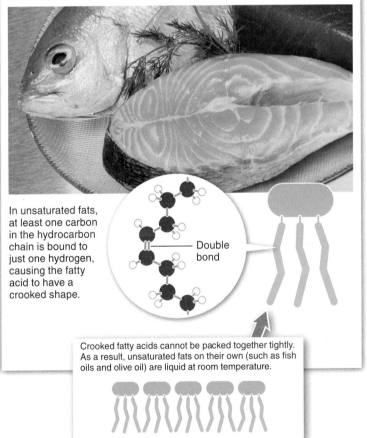

UNSATURATED FATS

In unsaturated fats, at least one carbon in the hydrocarbon chain is bound to just one hydrogen, causing the fatty acid to have a crooked shape.

Double bond

Crooked fatty acids cannot be packed together tightly. As a result, unsaturated fats on their own (such as fish oils and olive oil) are liquid at room temperature.

FIGURE 2-31 **Degrees of saturation.** Fatty acids (and thus the fats that contain them) can be unsaturated or saturated.

ATOMS   WATER   CARBOHYDRATES   LIPIDS   PROTEINS   NUCLEIC ACIDS

and, because they accumulate in your bloodstream and can narrow the vessel walls, they can contribute to heart disease and strokes.

An **unsaturated fat** is one in which some of the carbon atoms are bound to only a single hydrogen (and are connected to each other by a double bond). Most plant fats are unsaturated. Unsaturated fats may be *mono-unsaturated* or *polyunsaturated*. A mono-unsaturated fatty acid hydrocarbon chain has only one pair of neighboring carbon atoms in an unsaturated state—that is, has only one double bond. A *polyunsaturated* fatty acid hydrocarbon chain has more than one pair of carbons in an unsaturated state—there's more than one double bond. Unsaturated fats are still high in calories, but because they can lower cholesterol, they are generally preferable to saturated fats. Foods high in unsaturated fats include avocados, peanuts, and olive oil. Relative to other animals, fish tend to have less saturated fat.

Q How will the "chewy-ness" of a cookie differ depending on whether you make it with butter or vegetable oil as the lipid? Which is better for your health?

The shapes of unsaturated fat molecules and saturated fat molecules are different. When saturated, the hydrocarbon tails of the fatty acids all line up very straight and the fat molecules can be packed together tightly. The tight packing causes the fats, such as butter, to be solid at room temperature. When unsaturated, the fatty acids have kinks in the hydrocarbon tails and the fat molecules cannot be packed together as tightly (see Figure 2-31). Consequently, unsaturated fats, such as canola oil and vegetable oil, do not solidify as easily and are liquid at room temperature.

The ingredient list for many snack foods includes "partially hydrogenated" vegetable oils. The hydrogenation of an oil means that hydrogen atoms have been added to a liquid, unsaturated fat so that it becomes more saturated. Adding hydrogen atoms can create a food with a more desirable texture, since increasing a fat's degree of saturation changes its consistency and makes it more solid at room temperatures. By attaining just the right degree of saturation, it is possible to create foods, such as chocolate, that are near the border of solid and liquid and "melt in your mouth" (**FIGURE 2-32**). Unfortunately, hydrogenation also makes the food less healthful because saturated fats increase the risk of heart disease. They are less reactive—your body is less likely to break them down—and so are more likely to accumulate in your blood vessels.

Hydrogenation of unsaturated fats is doubly problematic from a health perspective because it also creates **trans fats,**

HYDROGENATION

Hydrogenation can improve a food's taste, texture, and shelf-life, but it also makes the food less healthful.

Hydrogenation is the artificial addition of hydrogen atoms to an unsaturated fat in order to make the fat more saturated.

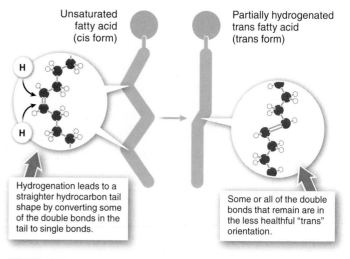

Unsaturated fatty acid (cis form)

Partially hydrogenated trans fatty acid (trans form)

Hydrogenation leads to a straighter hydrocarbon tail shape by converting some of the double bonds in the tail to single bonds.

Some or all of the double bonds that remain are in the less healthful "trans" orientation.

FIGURE 2-32 Hydrogenation improves a food's taste, texture, and shelf-life (but at a cost).

the "trans" referring to the unusual orientation of some or all of the double bonds that remain following the addition of hydrogen atoms. This orientation differs from that in other unsaturated dietary fats—which have their double bonds in an orientation called "cis." Trans fats in your diet cause your body to produce more cholesterol, further raising the risk of heart disease, and they also reduce your body's production of a type of cholesterol that protects against heart disease.

Q Olestra is a recently developed "fake fat" chemical that gives foods the taste of fat, without adding the calories of fats. What chemical structure might make this possible?

Because of the well-documented links between dietary fats and heart disease, many people are trying to reduce their fat intake. "Fake fats" make this possible. They are designed to be similar to fats in taste and texture, but have one big difference: they cannot be digested by humans. One such "fake fat" is olestra. Olestra, instead of being a triglyceride fat with three fatty acid tails linked to a glycerol molecule, has eight separate hydrocarbon fatty acids attached to a molecule of sucrose. The fatty acids in this octopus-like molecule stimulate their usual taste buds on your tongue, telling your brain that you are eating a fat.

The complex shape of the molecule, however, prevents your body's digestive chemicals from grabbing onto it and breaking it down. As a consequence, it passes through your digestive system without being digested. It's not a perfect solution, however. Olestra reduces absorption of some vitamins, and in some people causes abdominal cramping.

TAKE-HOME MESSAGE 2·13

Fats, including the triglycerides common in the food we eat, are one type of lipid. Characterized by long hydrocarbon tails, fats effectively store energy in the many carbon-hydrogen and carbon-carbon bonds. Their caloric density is responsible for humans' preferring fats to other macromolecules in the diet, and is also responsible for their association with obesity and illness in the modern world.

2·14 Cholesterol and phospholipids are used to build sex hormones and membranes.

Not all lipids are fats, nor do lipids necessarily function in energy storage. A second group of lipids, called the **sterols,** plays an important role in regulating growth and development (**FIGURE 2-33**). This group includes some very familiar lipids: cholesterol and the steroid hormones such as testosterone and estrogen. All of these molecules are variations on one basic structure formed from four interlinked rings of carbon atoms.

Cholesterol is an important component of most cell membranes. For this reason, it is an essential molecule for living organisms. Cells in our liver produce almost 90% of the circulating cholesterol by transforming the saturated fats in our diet. Cholesterol is also present in foods and has a bad reputation in most Western cultures that is mostly well deserved. When we ingest too much cholesterol (present in animal-based foods such as egg yolks, red meat, and cream) and high levels of cholesterol circulate in our bloodstream, the cholesterol can attach to blood vessel walls and cause them to thicken. In turn, this thickening can lead to high blood pressure, a major contributor to strokes and heart attacks. For these reasons, nutritionists advise limiting the consumption of foods high in cholesterol and saturated fats.

The steroid hormones estrogen and testosterone are built through slight chemical modifications to cholesterol.

These hormones are among the primary molecules that direct and regulate sexual development, maturation, and sperm and egg production. In both males and females, estrogen influences memory and mood, among other traits. Testosterone has numerous effects, one of which

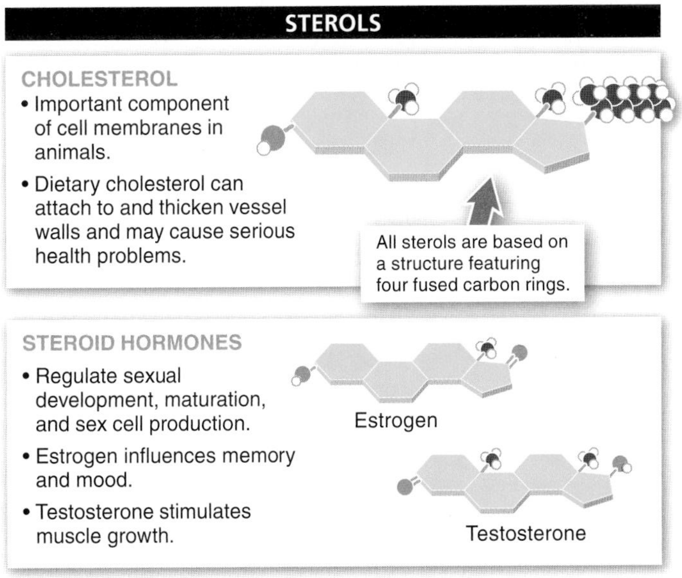

STEROLS

CHOLESTEROL
• Important component of cell membranes in animals.
• Dietary cholesterol can attach to and thicken vessel walls and may cause serious health problems.

All sterols are based on a structure featuring four fused carbon rings.

STEROID HORMONES
• Regulate sexual development, maturation, and sex cell production.
• Estrogen influences memory and mood.
• Testosterone stimulates muscle growth.

Estrogen

Testosterone

FIGURE 2-33 Not all lipids are for energy storage. Cholesterol, estrogen, and testosterone are all lipids.

is to stimulate muscle growth. As a consequence, athletes (particularly bodybuilders) have often been found to take synthetic variants of testosterone to increase their muscularity (**FIGURE 2-34**). But the use of these supplements is often accompanied by dangerous side effects, including extreme aggressiveness ("'roid rage"), high cholesterol, and, following long-term use, cancer. As a consequence, nearly all athletic organizations have banned their use.

Phospholipids and waxes are also lipids. **Phospholipids** are the major component of the membrane that surrounds the contents of a cell and controls the flow of chemicals into and out of the cell (**FIGURE 2-35**). They have a structure similar to fats, but with two differences: they contain a phosphorus atom (hence *phospho*lipids) and they have two fatty acid chains rather than three. We explore the significant role of phospholipids in cell membranes in the next chapter.

PHOSPHOLIPIDS

HYDROPHILIC HEAD (attracted to water)

Phosphate group

HYDROPHOBIC TAILS (not attracted to water)

Fatty acids

PHOSPHOLIPIDS IN WATER

Hydrophilic heads

Water

Hydrophobic tails

Phospholipids align so that their hydrophilic heads extend toward the water, while their hydrophobic tails are directed away from the water.

Symbol for phospholipid used in this book

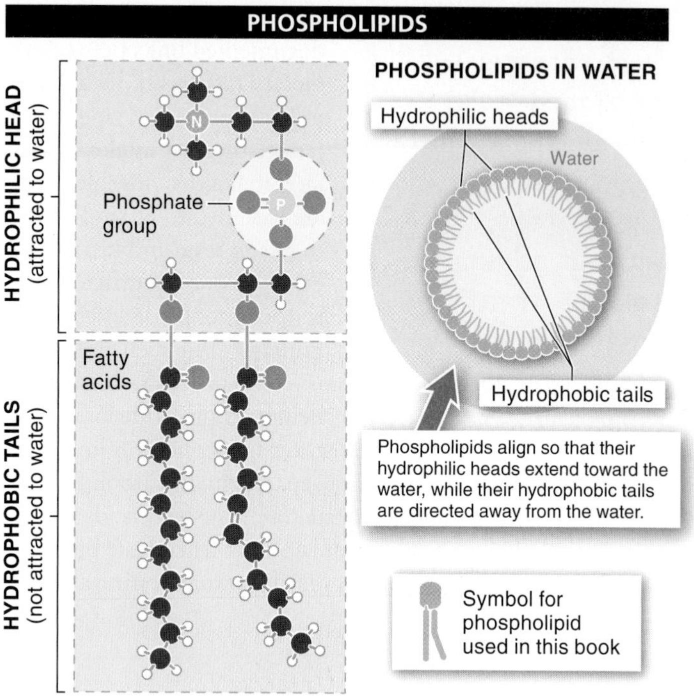

FIGURE 2-35 **Dual nature.** Phospholipids have a head region that is attracted to water and a tail that is not.

Waxes resemble fats but have only one long-chain fatty acid linked to the glycerol head of the molecule. Because the fatty acid chain is highly nonpolar, waxes are strongly hydrophobic; that is, these molecules do not mix with water but repel it. Their water resistance accounts for their presence as a natural coating on the surface of many plants and in the outer coverings of many insects. In both cases, the waxes prevent the plants and animals from losing the water essential to their life processes. Many birds, too, have a waxy coating on their wings, keeping them from becoming water-logged when they get wet.

TAKE-HOME MESSAGE 2·14

Cholesterol and phospholipids are lipids that are not fats. Both are important components in cell membranes. Cholesterol also serves as a precursor to steroid hormones, important regulators of growth and development.

FIGURE 2-34 **Dangerous bulk.** Steroids can increase muscularity, but with serious health consequences.

2·15–2·19
Proteins are versatile macromolecules that serve as building blocks.

Two newborn Cattle Egret chicks; their feathers are built from proteins.

2·15 Proteins are bodybuilding macromolecules.

You can't look at a living organism and not see proteins (**FIGURE 2-36**). Inside and out, **proteins** are the chief building blocks of all life. They make up skin and feathers and horns. They make up muscles and are a significant component of bone. In your bloodstream, proteins fight invading microorganisms and stop you from bleeding to death from a shaving cut. Proteins control the levels of sugar and other chemicals in your bloodstream and carry oxygen from one place in your body to another. And in just about every cell in every living organism, proteins called

---

PROTEIN DIVERSITY

Proteins perform a variety of different functions. They all, however, are built the same way and from the same raw materials in organisms.

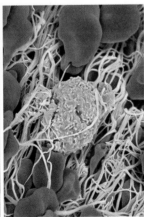
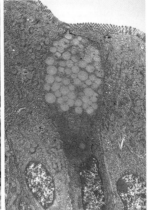

Wing feathers on a Scarlet Macaw	*Blood clot*	*Goblet cell (pink and blue) in the mucosal lining of the small intestine*	*Heart muscle cells*	*A model of hemoglobin molecules carrying oxygen*
STRUCTURE Hair, fingernails, feathers, horns, cartilage, tendons	**PROTECTION** Help fight invading micro-organisms, coagulate blood	**REGULATION** Control cell activity, constitute some hormones	**CONTRACTION** Allow muscles to contract, heart to pump, sperm to swim	**TRANSPORTATION** Carry molecules such as oxygen around your body

FIGURE 2-36 **Proteins everywhere!** Proteins are the chief building blocks of all organisms.

ATOMS   WATER   CARBOHYDRATES   LIPIDS   PROTEINS   NUCLEIC ACIDS

enzymes initiate and assist every chemical reaction that occurs.

Although proteins perform several very different types of functions, all are built in the same way and from the same raw materials in all organisms. In the English language, every sentence is made up of words and every word is formed from one or more of the 26 letters of the alphabet. With 26 letters we can write anything, from sonnets to cookbooks to biology textbooks. Proteins, too, are constructed from a sort of alphabet. Instead of 26 letters there are 20 molecules, known as **amino acids.** Unique combinations of these 20 amino acids are strung together, like beads on a string, and the resulting protein has a unique structure and chemical behavior.

Let's look more closely at the structure of the amino acids in the protein alphabet. They all have the same basic two-part structure: one part is the same in all 20 amino acids, and the other part is unique, differing in each of the 20 amino acids.

Amino acids contain the same familiar atoms as carbohydrates and lipids—carbon, hydrogen, and oxygen—but differ in an important way: they also contain nitrogen. At the center of every amino acid is a carbon atom, with its four covalent bonds (**FIGURE 2-37**). One bond attaches the carbon to a group of atoms called a **carboxyl group,** which is a carbon atom bonded to two oxygen atoms (to one by a single bond and to the other by a double bond). The second bond attaches the central carbon to a single hydrogen atom. The third bond attaches the central carbon to an **amino group,** which is a nitrogen atom bonded to hydrogen atoms (usually two or three). These components—the central carbon with its attached hydrogen atom, carboxyl group, and amino group—are the foundation that identifies a molecule as an amino acid. As multiple amino acids are joined together, the unit formed by these three components forms the "backbone" of the protein.

The fourth bond of the central carbon atom attaches to a side chain. This side chain is the unique part of each of the 20 amino acids. In the simplest amino acid, glycine, for

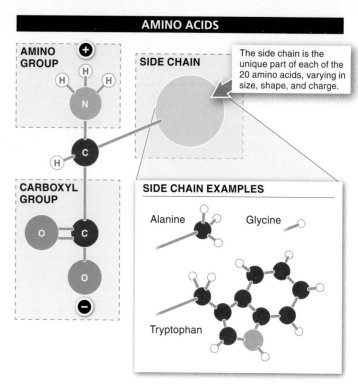

FIGURE 2-37 **Amino acid structure.** Amino acids are made up of a central carbon atom attached to a hydrogen atom, an amino group, a carboxyl group, and a side chain.

example, the side chain is simply a hydrogen atom. In other amino acids, the side chain is a single CH_3 group or three or four such groups. Most of the side chains include both hydrogen and carbon atoms, and a few include nitrogen or sulfur atoms. The side chain determines an amino acid's chemical properties, such as whether the amino acid molecule is polar or nonpolar.

TAKE-HOME MESSAGE 2·15

Unique combinations of 20 amino acids give rise to proteins, the chief building blocks of the physical structures that make up all organisms. Proteins perform myriad functions, from assisting chemical reactions to causing blood clotting to building bones to fighting microorganisms.

2·16 Proteins are an essential dietary component.

The atoms present in the proteins we eat—especially the nitrogen atoms—are essential to the growth, repair, and replacement that take place in our bodies. As we

digest protein, breaking it down into its amino acids, our bodies use these amino acids for various building projects. Proteins also store energy in their bonds and, like

ESSENTIAL AMINO ACID CONTENT OF COMMON FOODS

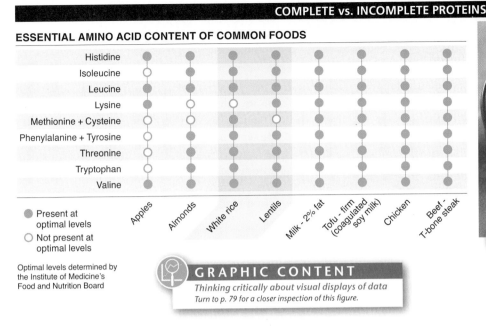

	Apples	Almonds	White rice	Lentils	Milk - 2% fat	Tofu - firm (coagulated soy milk)	Chicken	Beef - T-bone steak
Histidine	●	●	●	●	●	●	●	●
Isoleucine	○	●	●	●	●	●	●	●
Leucine	●	●	●	●	●	●	●	●
Lysine	●	○	○	●	●	●	●	●
Methionine + Cysteine	○	○	●	○	●	●	●	●
Phenylalanine + Tyrosine	○	●	●	●	●	●	●	●
Threonine	○	●	●	●	●	●	●	●
Tryptophan	○	●	●	●	●	●	●	●
Valine	●	●	●	●	●	●	●	●

● Present at optimal levels
○ Not present at optimal levels

Optimal levels determined by the Institute of Medicine's Food and Nutrition Board

GRAPHIC CONTENT
Thinking critically about visual displays of data
Turn to p. 79 for a closer inspection of this figure.

Traditional dishes in many cultures combine proteins, bringing together all essential amino acids.

FIGURE 2-38 **All proteins are not created equal.** Some foods have "complete proteins" with all the essential amino acids. Other foods have "incomplete proteins," and we must consume proteins from multiple sources to get all the essential amino acids.

carbohydrates and lipids, they can be used to fuel living processes.

The amount of protein we need depends on the extent of the building projects under way. Most individuals need 40–80 grams of protein per day. Bodybuilders, however, may need 150 grams or more a day to achieve the extensive muscle growth stimulated by their training; similarly, the protein needs of pregnant or nursing women are very high.

Contrary to the impression you might get from food labels, all proteins are not created equal. Every different protein has a different composition of amino acids. And while our bodies can manufacture certain amino acids as they are needed, many other amino acids must come from our diet. Those that we must get from our diet—about half of the 20 amino acids—are called "essential amino acids." For this reason, we shouldn't just speak of needing "*x* grams of protein per day." We need to consume all of the essential amino acids every day.

Many foods, containing "complete proteins," have all of the essential amino acids. Animal products such as milk, eggs,

Q Food labels indicate how many grams of protein are contained in a food item. Why is this information only partially helpful for effectively guiding your protein intake?

fish, chicken, and beef tend to provide complete proteins. Vegetables, fruits, and grains often contain "incomplete proteins," which do not have all the essential amino acids. If you consume only one type of incomplete protein in your diet, you may be deficient in one or more of the essential amino acids. But two incomplete proteins that are "complementary proteins," when eaten together, can provide all the essential amino acids.

Traditional dishes in many cultures often include such pairings. Examples are corn and beans in Mexico and rice and lentils in India (**FIGURE 2-38**).

TAKE-HOME MESSAGE 2·16

Twenty amino acids make up all the proteins necessary for growth, repair, and replacement of tissue in living organisms. Of these amino acids, about half are essential for humans: they cannot be synthesized by the body so must be consumed in the diet. Complete proteins contain all essential amino acids, while incomplete proteins do not.

2·17 A protein's function is influenced by its three-dimensional shape.

Proteins are formed by linking individual amino acids together with a **peptide bond,** in which the amino group of one amino acid is bonded to the carboxyl group of another. Two amino acids joined together form a *dipeptide,* and several amino acids joined together form a *polypeptide.* The sequence of amino acids in the polypeptide chain is called the **primary structure** of the protein and can be compared to the sequence of letters that spells a specific word (**FIGURE 2-39**).

Amino acids in a polypeptide chain don't remain in a simple straight line like beads on a string. The chain begins to fold as side chains come together and hydrogen bonds form between various atoms in the chain. The two most common patterns of hydrogen bonding between amino acids cause a segment of the chain to either twist in a corkscrew-like shape or form a zigzag folding pattern. The distribution of corkscrews and zigzags within a protein gives a protein its **secondary structure.**

The secondary structure itself continues to fold and bend, bringing together amino acids that then form bonds such as hydrogen bonds or covalent sulfur-sulfur bonds (see Figure 2-39). Eventually, the protein folds into a unique and complex three-dimensional shape called its **tertiary structure.**

Some protein molecules have a **quaternary structure** in which two or more polypeptide chains are held together by hydrogen bonds and other non-peptide bonds between amino acids in the different chains. An example of a protein with quaternary structure is hemoglobin, the protein molecule that carries oxygen from the lungs to the cells where it is needed. Hemoglobin is made from four polypeptide chains, two "alpha" chains and two "beta" chains.

Some proteins are attached to other types of macromolecules. *Lipoproteins,* for example, circulate in the bloodstream carrying fats. They are formed when molecules of cholesterol and a triglyceride (both lipids) combine with a protein. *Glycoproteins* are combinations of carbohydrates and proteins. These are found on the surfaces of nearly all animal cells and play a role in helping the immune system distinguish between your own cells and foreign cells. (We learn more about glycoproteins in the next chapter, which discusses cells.)

The overall shape of a protein molecule determines its function—how it behaves and the other molecules it interacts with. For proteins to function properly, they must retain their three-dimensional shape. When their shapes are deformed, they usually lose their ability to function. We can see proteins deforming when we fry an egg. The heat

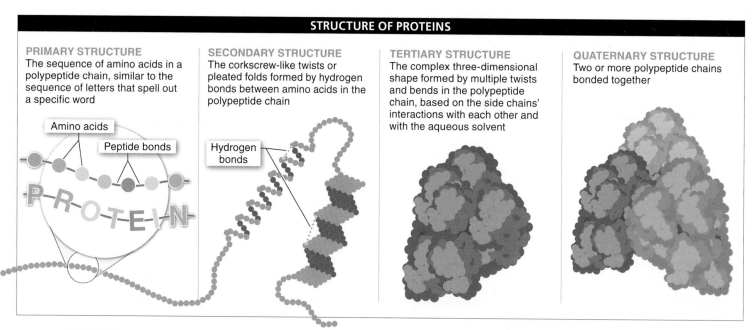

STRUCTURE OF PROTEINS

PRIMARY STRUCTURE
The sequence of amino acids in a polypeptide chain, similar to the sequence of letters that spell out a specific word

Amino acids
Peptide bonds

SECONDARY STRUCTURE
The corkscrew-like twists or pleated folds formed by hydrogen bonds between amino acids in the polypeptide chain

Hydrogen bonds

TERTIARY STRUCTURE
The complex three-dimensional shape formed by multiple twists and bends in the polypeptide chain, based on the side chains' interactions with each other and with the aqueous solvent

QUATERNARY STRUCTURE
Two or more polypeptide chains bonded together

FIGURE 2-39 **The several levels of protein structure.** The functions of proteins are influenced by their three-dimensional shape.

Q Egg whites contain a lot of protein. Why does beating them change their texture, making them stiff?

breaks the hydrogen bonds that give the proteins their shape. The proteins in the clear egg white unfold, losing their secondary and tertiary structure. This disruption of protein folding is called **denaturation** (**FIGURE 2-40**).

Almost any extreme environment will denature a protein. Take a raw egg, for instance, and crack it into a dish containing baking soda or rubbing alcohol. Both chemicals are sufficiently extreme to turn the clear protein opaque white, as in fried egg whites.

Hair is a protein whose shape most of us have modified at one time or another. Styling hair—whether curling or straightening it—involves altering some of the hydrogen bonds between the amino acids that make up the hair protein, changing its tertiary structure. When your hair gets wet, the water is able to disrupt some of the hydrogen bonds, causing some amino acids in the protein to form hydrogen bonds with the water molecules instead. Thus, if you style your hair while it's wet, you can change your hair's shape— making it straighter or, if you manipulate it around curlers, making it curlier. The hair can then hold this shape when it dries, because as the water evaporates the hydrogen bonds to water are replaced by hydrogen bonds between amino acids of the hair protein. Once your hair gets wet again, however, unless it is combed, brushed, or wrapped in a different style, it will return to its natural shape.

Whether your hair is straight or curly or somewhere in between depends on your hair protein's amino acid sequence and the three-dimensional shape it confers

FIGURE 2-41 **Curly or straight?** Proteins determine it!

(**FIGURE 2-41**). This amino acid sequence is something you're born with (that is, it's genetically determined). The chains are more or less coiled, depending on the extent of covalent and hydrogen bonding between different parts of the coil. Many hair salons make use of the ability to alter covalent bonds to change hair texture semi-permanently. They are able to do this in three simple steps. First, the bonds are broken chemically. Second, the hair is wrapped around curlers to hold the polypeptide chains in a different position. And third, chemicals are put on the hair to create new covalent bonds between parts of the polypeptide chains. The hair thus becomes locked in a new position. (New hair will continue to grow with its genetically determined texture, of course, requiring the procedure to be repeated regularly.)

Q Why do some people have curly hair and others have straight hair?

TAKE-HOME MESSAGE 2·17

A protein's particular amino acid sequence determines how it folds into a particular three-dimensional shape. This shape determines many of the protein's properties, including which molecules it will interact with. When a protein's shape is deformed, the protein usually loses its ability to function.

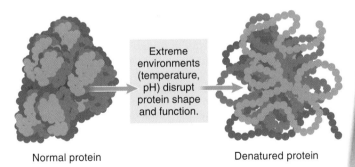

Normal protein → Extreme environments (temperature, pH) disrupt protein shape and function. → Denatured protein

FIGURE 2-40 **Denaturation.** When proteins are unfolded, they lose their function.

2·18 Enzymes are proteins that speed up chemical reactions.

Protein shape is particularly critical in **enzymes,** molecules that help initiate and accelerate the chemical reactions in our bodies. Enzymes emerge unchanged—in their original form—when the reaction is complete and thus can be used again and again. Here's how they work.

Think of an enzyme as a big piece of popcorn. Its tertiary or quaternary structure gives it a complex shape with lots of nooks and crannies. Within one of those nooks is a small area called the **active site** (**FIGURE 2-42**). Based on the chemical properties of the atoms lining this pocket, the active site provides a place for the participants in a chemical reaction, the reactants or **substrate** molecules, to nestle briefly.

Enzymes are very choosy: they bind only with their appropriate substrate molecules, much like a lock that can be opened with only one key (see Figure 2-42). Only the substrate molecules are of the correct size and shape to fit into the active-site groove. Moreover, the exposed atoms in the active site are positioned to form weak interactions with specific atoms in the substrates. Once a substrate molecule is bound to the active site, a reaction can take place—and usually does so very quickly.

The chemical reactions that occur in organisms can either release energy or consume energy. But in either case, a certain minimum energy—a little "push"—is needed to initiate the reaction, called **activation energy.** And although enzymes don't alter the amount of energy released by a reaction, they act as catalysts by lowering the activation energy, which causes the reaction to occur more quickly. For example, enzymes may lower the activation energy by holding substrate molecules in an orientation that stresses bonds that need to break or brings together atoms that need to bond.

By virtue of their catalytic capacities, enzymes are at the heart of the chemistry of living organisms. Taken together, all of the chemical reactions in a living organism are its *metabolism.*

Increasingly complex molecules are synthesized or degraded in a series of sequential reactions called a "metabolic pathway." Each step of these metabolic pathways is

ENZYMES FACILITATE CHEMICAL REACTIONS

Enzymes can help to bring about chemical reactions in a variety of ways. The enzyme lactase, for example, breaks down the milk sugar lactose into two simple sugars that can be used for energy.

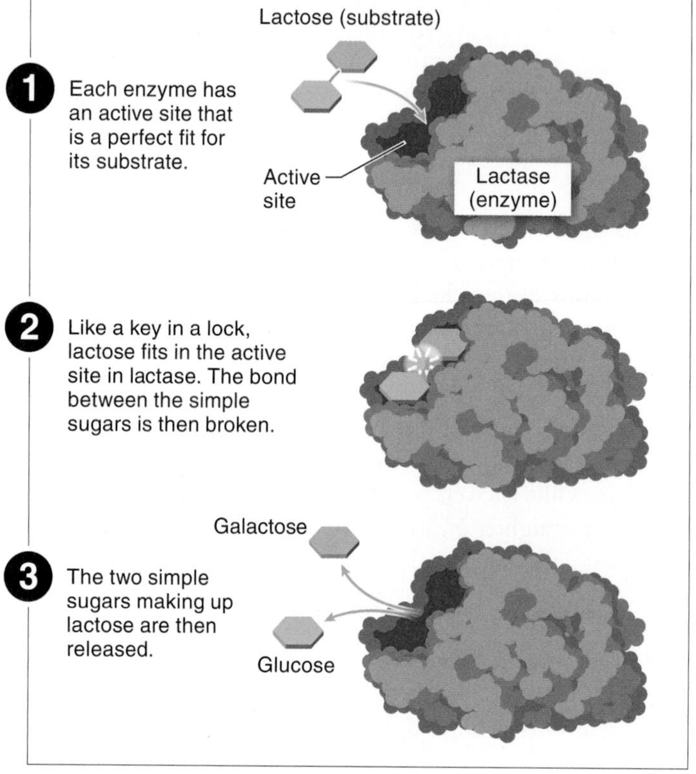

1 Each enzyme has an active site that is a perfect fit for its substrate.

Lactose (substrate)

Active site

Lactase (enzyme)

2 Like a key in a lock, lactose fits in the active site in lactase. The bond between the simple sugars is then broken.

3 The two simple sugars making up lactose are then released.

Galactose

Glucose

FIGURE 2-42 **Lock and key.** Enzymes are very specific about which molecules and reactions they will catalyze.

catalyzed by an enzyme produced in the body. Proteins are the building blocks with which living organisms are built, but since nearly all enzymes are proteins, proteins can also be thought of as the *builders* of bodies, too.

TAKE-HOME MESSAGE 2·18

Enzymes are proteins that help initiate and speed up chemical reactions. They aren't permanently altered in the process, but rather can be used again and again.

2·19 Enzymes regulate reactions in several ways (but malformed enzymes can cause problems).

If not for enzymes, it is possible that nothing would ever get done. At least not in living organisms. Enzymes don't alter the outcome of reactions, but without the chemical "nudge" they supply—often increasing reaction rates to millions of times their uncatalyzed rate—the processes necessary to sustain life could not occur.

The rate at which an enzyme catalyzes a reaction is influenced by several chemical and physical factors (FIGURE 2-43). These include:

1. Enzyme and substrate concentration. For a given amount of substrate, an increase in the amount of enzyme increases the rate at which the reaction occurs. Similarly, for a given amount of enzyme, an increase in the substrate concentration increases the reaction rate. In both cases, once all of the enzyme molecules are bound to substrate, or vice versa, additional enzyme or substrate no longer increases the reaction rate.

2. Temperature. Because increasing the temperature increases the speed of movement of molecules, reaction rates generally increase at higher temperatures. Reaction rates continue to increase only up to the optimum temperature for an enzyme. At temperatures above the optimum, reaction rates decrease as enzymes lose their shape or even denature. Enzymes from different species can have widely differing optimum temperatures.

3. pH. As with temperature, enzymes have an optimum pH. Above or below this pH, excess hydrogen or hydroxide ions interact with amino acid side chains in the active site. These interactions disrupt enzyme function (and sometimes structure) and decrease reaction rates.

4. Presence of inhibitors or activators. One of the most common ways that cells can speed up or slow down their metabolic pathways is through the binding of other chemicals to enzymes. This binding can alter enzyme shape in a way that increases or decreases the enzyme's activity. **Inhibitors** reduce enzyme activity and come in two types. **Competitive inhibitors** bind to the active site, blocking substrate molecules from the site and thus from taking part in the reaction. **Noncompetitive inhibitors** do not compete for the active site but, rather, bind to another part of the enzyme, altering its shape in a way that

FIGURE 2-43 Getting the job done. Enzyme activity is influenced by physical factors such as temperature and pH and by chemical factors such as enzyme and substrate concentrations.

ENZYME ACTIVITY

The rate at which an enzyme catalyzes a reaction is influenced by several chemical and physical factors.

ENZYME AND SUBSTRATE CONCENTRATION
Reaction rates increase with increased amounts of enzyme (or substrate), but only up to the point at which all of the enzyme molecules are bound to substrate. At that point, additional enzyme (or substrate) no longer increases the reaction rate.

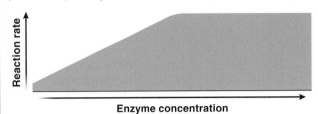

TEMPERATURE
Reaction rates generally increase at higher temperatures, but only up to the optimum temperature for an enzyme. At temperatures above the optimum, reaction rates decrease as enzymes can lose their shape or even denature.

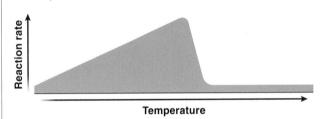

pH
Reaction rates generally increase as pH nears the optimum level for an enzyme. Above or below this pH, enzyme function can be disrupted and reaction rates decrease.

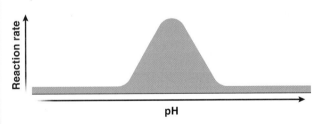

PRESENCE OF INHIBITORS OR ACTIVATORS
Reaction rates increase in the presence of activators and decrease in the presence of inhibitors.

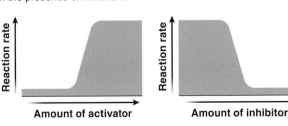

Sometimes a protein "word" is misspelled—that is, the sequence of amino acids is incorrect. If an enzyme is altered even slightly, the active site may change, which can cause the enzyme to no longer function.

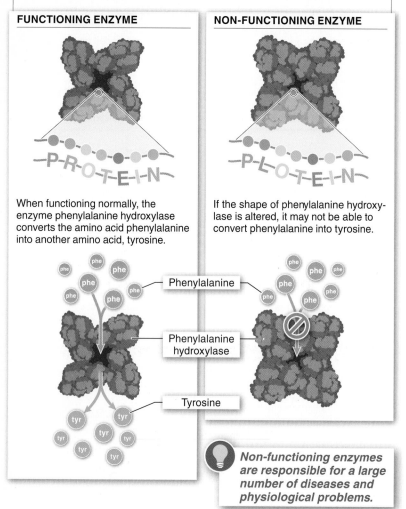

FUNCTIONING ENZYME

When functioning normally, the enzyme phenylalanine hydroxylase converts the amino acid phenylalanine into another amino acid, tyrosine.

Phenylalanine

Phenylalanine hydroxylase

Tyrosine

NON-FUNCTIONING ENZYME

If the shape of phenylalanine hydroxylase is altered, it may not be able to convert phenylalanine into tyrosine.

💡 *Non-functioning enzymes are responsible for a large number of diseases and physiological problems.*

FIGURE 2-44 **When a protein is "misspelled."**

changes the structure of the active site, thus reducing or blocking its ability to bind with substrate. Often, it is the very product of a metabolic pathway that acts as an inhibitor of enzymes early in the pathway, effectively shutting off the pathway when enough of its end product has been produced.

Just as a molecule can bind to an enzyme and inhibit the enzyme's activity, so can some cellular chemicals act as **activators.** Instead of their binding to the enzyme "turning it off," their binding to the enzyme "turns it on," altering the enzyme's shape or structure so that it can now catalyze a reaction.

Sometimes a cell produces a protein "word" that is misspelled—that is, the sequence of amino acids is

incorrect. If an enzyme is altered even slightly, the active site may change and the enzyme will no longer function (**FIGURE 2-44**). Slightly modified, non-functioning enzymes are responsible for a large number of diseases and physiological problems (see Section 5-9). An example is the body's inability to break down the amino acid phenylalanine (in a condition known as phenylketonuria).

One health issue influenced by enzyme function is the condition called lactose intolerance. Normally, during digestion, the lactose in milk is broken down into its component parts, glucose and galactose (see Figure 2-42). These simple sugars are then used for energy. But some people, when they become adults, are unable to break the bond linking the two simple sugars because they no longer produce the enzyme lactase that assists in this process. Consequently, any lactose in their diet passes through their stomach and small intestine undigested. Then, when it reaches the large intestine, bacteria living there consume the lactose. The problem is that, as the bacteria break down the lactose, they produce some carbon dioxide and other gases. These gases are trapped in the intestine and lead to severe discomfort. Interestingly, in regions of the world with long traditions of pastoralism (raising and consuming livestock), lactose intolerance is much rarer than in other parts of the world. Only about 10% of people from Denmark or Sweden have lactose intolerance, but among people from China, which has historically been largely non-pastoral, the vast majority of adults (more than 80% according to numerous published studies) are lactose intolerant.

Q *Why do some adults get sick when they drink milk?*

The unpleasant symptoms of lactose intolerance can be avoided by not consuming milk, cheese, yogurt, ice cream, or any other dairy products, but they can also be avoided by taking a pill containing the enzyme lactase. It doesn't matter how the enzyme gets into your digestive system; as long as it's there, the lactose in the milk can be broken down.

TAKE-HOME MESSAGE 2·19

Enzyme activity is influenced by physical factors such as temperature and pH, as well as by chemical factors, including enzyme and substrate concentrations. Inhibitors and activators are chemicals that bind to enzymes and, by blocking the active site or altering the shape or structure of the enzyme, can change the rate at which the enzyme catalyzes reactions.

2·20 – 2·22
Nucleic acids store information on how to build and run a body.

Spirals Time—Time Spirals: *a sculpture inspired by DNA's double helix structure.*

2·20 Nucleic acids are macromolecules that store information.

We have examined three of life's macromolecules: carbohydrates, lipids, and proteins. We turn our attention now to the fourth: **nucleic acids,** macromolecules that store information. There are two types of nucleic acids: **deoxyribonucleic acid (DNA)** and **ribonucleic acid (RNA).** Both play central roles in directing the production of proteins in living organisms, and by doing so play a central role in determining all of the inherited characteristics of an individual.

Nucleic acids and are made up of individual units called **nucleotides.** All nucleotides have three components: a molecule of sugar, a phosphate group (containing a phosphorus atom bound to four oxygen atoms), and a nitrogen-containing molecule (**FIGURE 2-45**). In both types of nucleic acids, nucleotides are linked in a series to form a ribbon-like strand that is the backbone of the nucleic acid molecule: a sugar molecule is attached to a phosphate group, which is attached to another sugar, which is attached to another phosphate, and so on. Attached to each sugar, and protruding from the backbone, is one of the nitrogen-containing molecules called DNA or RNA **bases** (so named because of their chemical structure). A 10-unit nucleic acid strand therefore would have 10 bases, one attached to each

FIGURE 2-45 **The molecules that carry genetic information.** The nucleic acid shown here is DNA.

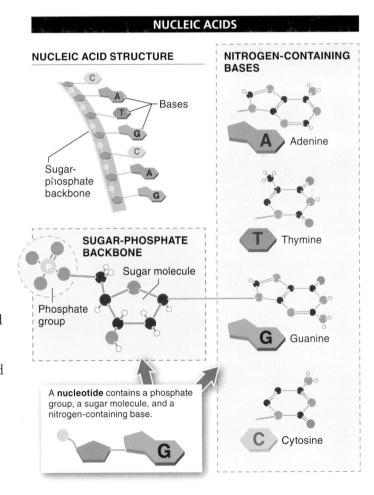

NUCLEIC ACIDS

NUCLEIC ACID STRUCTURE

Bases

Sugar-phosphate backbone

SUGAR-PHOSPHATE BACKBONE

Phosphate group

Sugar molecule

A **nucleotide** contains a phosphate group, a sugar molecule, and a nitrogen-containing base.

G

NITROGEN-CONTAINING BASES

A Adenine

T Thymine

G Guanine

C Cytosine

sugar within the sugar-phosphate-sugar-phosphate backbone. But the base attached to each sugar is not always the same. It can be one of several different bases. For this reason, a nucleic acid is often described by its sequence of bases.

Nucleic acids store information in the order of bases attached at each position in the molecule's backbone. At each position in a molecule of DNA, for example, the base can be any one of four possible bases: adenine (A), thymine (T), guanine (G), or cytosine (C). Just as the meaning of a sentence is determined by which letters are strung together, the information in a segment of DNA is determined by its sequence of bases. One DNA segment may have the sequence adenine, adenine, adenine, guanine, cytosine,

thymine, guanine—abbreviated as AAAGCTG. Another DNA segment may have the sequence CGATTACCCGAT. Because the information differs in each case, so, too, does the polypeptide for which the sequence codes, as we'll see.

TAKE-HOME MESSAGE 2·20

The nucleic acids DNA and RNA are macromolecules that store information in their unique sequences of bases contained in nucleotides, their building-block molecules. Both nucleic acids play central roles in directing protein production in organisms.

2·21 DNA holds the genetic information to build an organism.

A molecule of DNA has two strands, each a sugar-phosphate-sugar-phosphate backbone with a base sticking out from each sugar molecule. The two strands wrap around each other, each turning in a spiral. Although each strand has its own sugar-phosphate-sugar-phosphate backbone and sequence of bases, the two strands are connected by the bases protruding from them.

You can picture a molecule of DNA as a ladder. The two sugar-phosphate-sugar-phosphate backbones are like the long vertical sides of the ladder that give it height. A base sticking out represents a rung on the ladder. Or, more accurately, half a rung. The bases protruding from each strand meet in the center and bind to each other (via hydrogen bonds). DNA differs slightly from a ladder, though, in that it has a gradual twist. The two spiraling strands together are said to form a **double helix** (**FIGURE 2-46**).

The two intertwining spirals fit together because only two combinations of bases pair up together. The base A always pairs with T, and C always pairs with G. Consequently, if the base sequence of one of the spirals is CCCCTTAGGAACC, the base sequence of the other must be GGGGAATCCTTGG. That is why researchers working on the Human Genome Project describe only one sequence of nucleotides when presenting a DNA sequence—even though that DNA is double-stranded in our bodies. With that one sequence, we can infer the identity and order of the bases in the complementary sequence, and thus we know the exact structure of the nucleic acid.

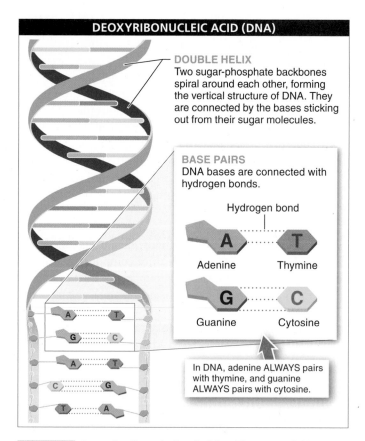

DEOXYRIBONUCLEIC ACID (DNA)

DOUBLE HELIX
Two sugar-phosphate backbones spiral around each other, forming the vertical structure of DNA. They are connected by the bases sticking out from their sugar molecules.

BASE PAIRS
DNA bases are connected with hydrogen bonds.

Hydrogen bond

A — T
Adenine   Thymine

G — C
Guanine   Cytosine

In DNA, adenine ALWAYS pairs with thymine, and guanine ALWAYS pairs with cytosine.

FIGURE 2-46 A gradually twisting ladder. The rungs of the DNA ladder are nucleotide base pairs, and the sides of the ladder are made up of two sugar-phosphate backbones.

The sequences of nucleotide bases containing the information about how to produce a particular protein have anywhere from a hundred to several thousand bases. In a human, all of the DNA in a cell, containing all of the instructions for every protein that a human must produce, contains about three billion base pairs. Almost all of this DNA is in the cell's nucleus.

TAKE-HOME MESSAGE 2·21

DNA is shaped like a ladder in which the long, vertical sides of the ladder are made from a sequence of sugar-phosphate-sugar-phosphate molecules and the rungs are pairs of nucleotide bases. The sequence of nucleotide bases contains the information about how to produce a particular protein.

2·22 RNA is a universal translator, reading DNA and directing protein production.

The process of building a protein from a DNA sequence is not a direct one. Rather, it incorporates a middleman, RNA, that is also a nucleic acid (**FIGURE 2-47**). Although both RNA and DNA are built from nucleotides, the RNA nucleotide differs from the DNA nucleotide in three important ways. First, the sugar portion of the nucleotide differs slightly, containing an extra atom of oxygen. Second, while RNA has the bases A, G, and C, instead of thymine (T) it has a similar base called uracil (U). And third, unlike DNA, RNA is single-stranded. The sugar-phosphate-sugar-phosphate backbone is still there, as are the bases that protrude from each sugar, but the bases do not bind with bases in another RNA strand to form the ladder-like structure we see in DNA.

When the cell needs to synthesize a protein, a short strip of RNA is produced using a segment of a DNA strand as a model. The RNA nucleotides are therefore complementary to the DNA nucleotides, so the RNA molecule contains all the information present in the order of nucleotides of that DNA segment. The RNA moves to another part of the cell and then directs the linking together of amino acids to form a polypeptide chain that folds into a three-dimensional protein. We explore this in greater detail in Chapter 5.

Whether we're looking at the nucleotides that make up RNA and DNA or the lipids used to build sex hormones and cell membranes, we see a recurring theme in the construction of biological macromolecules: from relatively simple sets of building blocks linked together, infinitely complex molecules can be formed. Complex webs of one simple sugar, bonded together as glycogen, for instance, provide fuel for organisms. Similarly, sequences of amino acids of 20 different types, joined together, specify the structure of all the proteins found in every species on earth.

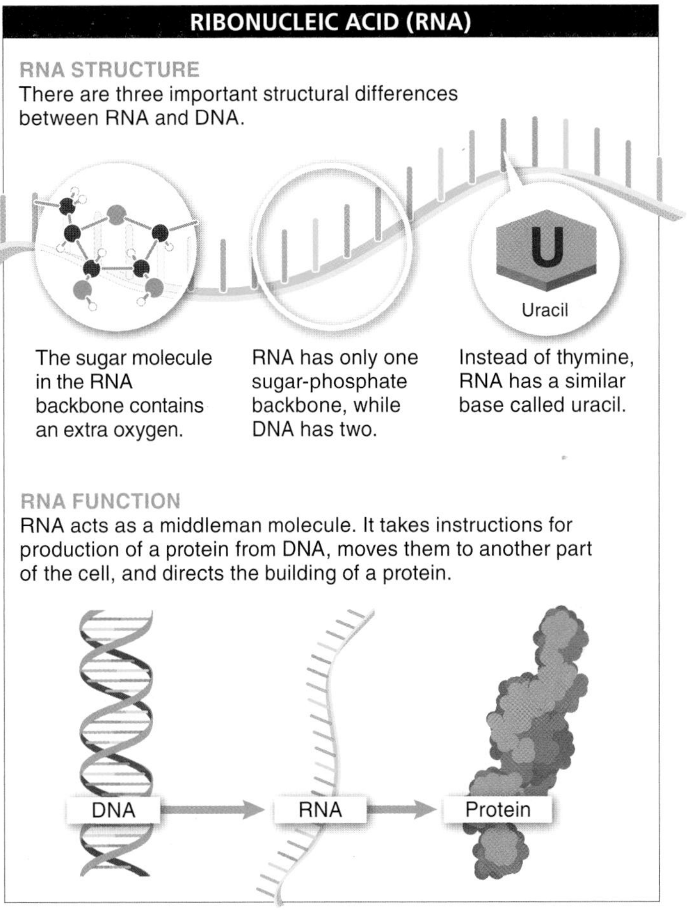

RIBONUCLEIC ACID (RNA)

RNA STRUCTURE
There are three important structural differences between RNA and DNA.

Uracil

The sugar molecule in the RNA backbone contains an extra oxygen.

RNA has only one sugar-phosphate backbone, while DNA has two.

Instead of thymine, RNA has a similar base called uracil.

RNA FUNCTION
RNA acts as a middleman molecule. It takes instructions for production of a protein from DNA, moves them to another part of the cell, and directs the building of a protein.

DNA → RNA → Protein

FIGURE 2-47 **The middleman between DNA and protein.** The structure of RNA.

TAKE-HOME MESSAGE 2·22

RNA acts as a middleman molecule—taking the instructions for protein production from DNA to another part of the cell, where, in accordance with the RNA instructions, amino acids are linked together into proteins.

Melt-in-your-mouth Chocolate May Not Be Such a Sweet Idea

Food chemists have figured out how to make chocolate that melts in your mouth. Is that a good thing?

Q: **Why are some fats "liquidy," like oil, and others solid?** The less saturated a fat is, the more "liquidy" it is at room temperature. Most animal fats are saturated and are solid at room temperature. Best example: butter. Most plant fats are polyunsaturated and are liquid at room temperature. Best example: vegetable oils.

Q: **Can oils be made more solid?** It's possible to increase the saturation of plant fats. Just heat them up and pass hydrogen bubbles through the liquid. In creating partially hydrogenated plant oils, this process reduces the number of carbon-carbon double bonds and makes the oil more solid. (It's easy, it's cheaper than just using butter, and it increases foods' shelf-life.)

Q: **Does that improve their taste?** By precisely controlling the level of saturation in plant fats, it is possible to create foods that are solid but have such a low melting point that they quickly melt on contact with the warmth of your mouth. This seems great, but . . .

Q: **Is there a downside?** The saturation of vegetable fats creates trans fats, due to the position taken by the newly added hydrogen atoms in the molecule. Trans fats increase levels of LDL ("bad") cholesterol and decrease HDL ("good") cholesterol, narrowing blood vessel walls and increasing the risk of heart disease and strokes.

Conclusion: Partially hydrogenated vegetable oils can give food a perfect texture and a pleasing feel in your mouth. But the creaminess comes with a high cost when it comes to your health.

Check Your Knowledge

Thinking critically about visual displays of data

1. In this figure, what does a blue dot indicate? What does a white dot indicate?

2. What can you conclude from this figure?

3. Why is there shading behind the data for white rice and lentils?

4. Should a food be avoided if it doesn't contain all of the essential amino acids at optimal levels? Why or why not?

5. Would a meal containing apples and white rice contain all of the essential amino acids? What about a meal of lentils and almonds?

6. Can you ascertain what "optimal levels" means in this figure?

7. What additional information would make this figure more helpful? Why?

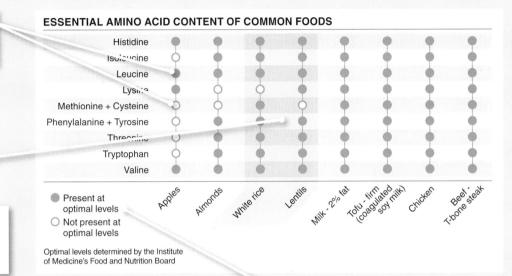

ESSENTIAL AMINO ACID CONTENT OF COMMON FOODS

Rows: Histidine, Isoleucine, Leucine, Lysine, Methionine + Cysteine, Phenylalanine + Tyrosine, Threonine, Tryptophan, Valine

Columns: Apples, Almonds, White rice, Lentils, Milk - 2% fat, Tofu - firm (coagulated soy milk), Chicken, Beef - T-bone steak

● Present at optimal levels
○ Not present at optimal levels

Optimal levels determined by the Institute of Medicine's Food and Nutrition Board

See answers at the back of the book.

Key Terms in Chemistry

acid, p. 51
activation energy, p. 72
activator, p. 74
active site, p. 72
amino acid, p. 68
amino group, p. 68
atom, p. 40
atomic mass, p. 41
atomic number, p. 41
base, p. 51
base (of DNA), p. 75
base (of RNA), p. 75
bond energy, p. 48
buffer, p. 52
carbohydrate, p. 55
carboxyl group, p. 68
cellulose, p. 59
chitin, p. 59
cholesterol, p. 65
competitive inhibitor, p. 73
complex carbohydrate, p. 58

covalent bond, p. 45
denaturation, p. 71
deoxyribonucleic acid (DNA), p. 75
disaccharide, p. 58
double bond, p. 46
double helix, p. 76
electron, p. 41
element, p. 40
enzyme, pp. 68, 72
fatty acid, p. 62
glycerol, p. 62
glycogen, p. 57
hydrogen bond, p. 46
hydrophilic, p. 62
hydrophobic, p. 62
inhibitor, p. 73
ion, p. 44
ionic bond, p. 46
ionic compound, p. 46
isotope, p. 42

lipid, p. 61
macromolecule, p. 55
mass, p. 41
molecule, p. 45
monosaccharide, p. 55
neutron, p. 41
noncompetitive inhibitor, p. 73
nucleic acid, p. 75
nucleotide, p. 75
nucleus, p. 41
peptide bond, p. 70
periodic table, p. 42
pH, p. 51
phospholipid, p. 66
polysaccharide, p. 58
primary structure, p. 70
protein, p. 67
proton, p. 41
quaternary structure, p. 70

radioactive, p. 42
ribonucleic acid (RNA), p. 75
saturated fat, p. 62
secondary structure, p. 70
simple sugar, p. 55
starch, p. 58

sterol, p. 65
substrate, p. 72
tertiary structure, p. 70
trans fat, p. 64
triglyceride, p. 52
unsaturated fat, p. 64
wax, p. 66

ABOUT THE CHAPTER OPENING PHOTO

Drops of dew bead on a dragonfly, magnifying parts of its compound eyes. The insects remain mostly immobile during the night and collect water droplets all over their bodies.

2·1–2·3 Atoms form molecules through bonding.

An atom is the smallest unit into which material can be divided without losing its essential properties. Molecules are atoms linked together.

ATOM STRUCTURE

All atoms are made up of protons and neutrons in the nucleus, and of electrons, which circle around the nucleus.

CARBON ATOM

(+) 6 Protons
● 6 Neutrons
• 6 Electrons

? 1. When we consider the mass of an atom, why do we ignore the weight of the electrons?

ELECTRON SHELLS AND ATOM STABILITY

The chemical characteristics of an atom depend on the number of electrons in its outermost shell. When atoms have electron vacancies in their outermost shell, they are more likely to interact with other atoms.

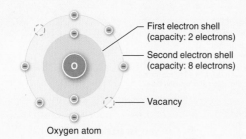

First electron shell (capacity: 2 electrons)

Second electron shell (capacity: 8 electrons)

Vacancy

Oxygen atom

UNSTABLE ATOM

H — Vacancy

Hydrogen atom

STABLE ATOM

He

Helium atom

? 2. Why are atoms with complete outer shells not likely to bond with another atom?

ATOM BONDING

Atoms can be bound together in three ways: covalent bonds, ionic bonds, and hydrogen bonds.

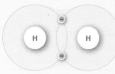

COVALENT BOND
A bond formed when atoms share electrons in order to become more stable, forming a molecule

H₂ molecule

IONIC BOND
An attraction between two oppositely charged ions, forming a compound

NaCl compound

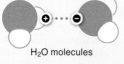

HYDROGEN BOND
An attraction between the slightly positively charged hydrogen atom of one molecule and the slightly negatively charged atom of another

H₂O molecules

? 3. Why is a hydrogen atom bonded to another hydrogen atom (to make an H₂ molecule) more stable than a hydrogen atom on its own?

? 4. Why are hydrogen bonds so much weaker than covalent or ionic bonds?

2·4–2·7 Water has features that enable it to support all life.

Water molecules easily form hydrogen bonds, giving water great cohesiveness and the ability to resist temperature changes, and making it a versatile solvent.

HYDROGEN BONDS

The hydrogen bonds between water molecules give water several of its most important characteristics that enable it to support all life.

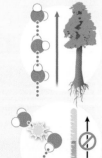

COHESIVENESS
Hydrogen bonds cause water molecules to "stick" together, allowing trees to transport the molecules from the soil to their leaves.

HIGH HEAT CAPACITY
Water resists heating because heat energy from the sun is used up breaking and re-forming hydrogen bonds.

LOW DENSITY AS A SOLID
When frozen, water becomes less dense due to the arrangement of molecules into a crystalline lattice.

GOOD SOLVENT
Water pries apart ionic bonds, dissolving ionic compounds.

? 5. Why is water such a good solvent?

? *Check Your Knowledge*

1. The atomic number of carbon is 6. Its nucleus must contain:

a) 6 neutrons and 6 protons.
b) 3 protons and 3 neutrons.
c) 6 neutrons and no electrons.
d) 6 protons and no electrons.
e) 6 protons and 6 electrons.

0 —— 42 —— 100
EASY   HARD

2. The second orbital shell of an atom can hold ____ electrons.

a) 2
b) 3
c) 4
d) 6
e) 8

0 — 20 —— 100
EASY   HARD

3. A covalent bond is formed when:

a) two nonpolar molecules associate with each other in a polar environment.
b) a positively charged particle is attracted to a negatively charged particle.
c) one atom gives up electrons to another atom.
d) two atoms share electrons.
e) two polar molecules associate with each other in a nonpolar environment.

0 — 11 —————— 100
EASY   HARD

4. Which of the following phenomena is most likely due to the high cohesiveness of water?

a) Lakes and rivers freeze from the top down, not the bottom up.
b) The fishing spider can walk across the surface of liquid water.
c) Adding salt to snow makes it melt.
d) The temperature of Santa Monica Bay, off the coast of Los Angeles, fluctuates less than the air temperature throughout the year.
e) All of the above are due to the cohesiveness of water.

0 ———— 53 —— 100
EASY   HARD

The pH of a fluid is a measure of how acidic or basic a solution is and depends on the concentration of H^+ ions present.

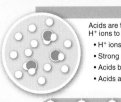

ACIDS

Acids are fluids that have a greater proportion of H^+ ions to OH^- ions.

• H^+ ions are very reactive.
• Strong acids are corrosive to metals.
• Acids break down food in your digestive tract.
• Acids are generally sour in taste.

BASES

Bases are fluids that have a greater proportion of OH^- ions to H^+ ions.

• OH^- ions bind with H^+ ions, neutralizing acids.
• Strong bases are caustic to your skin.
• Bases can be found in many household cleaners.
• Bases are generally bitter in taste and slippery.

0 1 2 3 4 5 6 7 8 9 10 11 12 13 14

? 6. Why does vomit taste so sour?

2·8—2·11 Carbohydrates are fuel for living machines.

Carbohydrates, made up of carbon, oxygen, and hydrogen, are the primary fuel for running all cellular machinery and also form much of the structure of cells.

SIMPLE CARBOHYDRATES

The simplest carbohydrates, including glucose—the most important carbohydrate to living organisms—are monosaccharides or simple sugars. They contain from three to seven carbon atoms.

GLUCOSE
$C_6H_{12}O_6$

FRUCTOSE
$C_6H_{12}O_6$

? 7. Why do carbohydrates function so well as a fuel for living organisms?

GLUCOSE IN YOUR BLOOD: 3 FATES

Depending upon whether energy is needed, glucose can be released for immediate use or stored on a short-term or long-term basis.

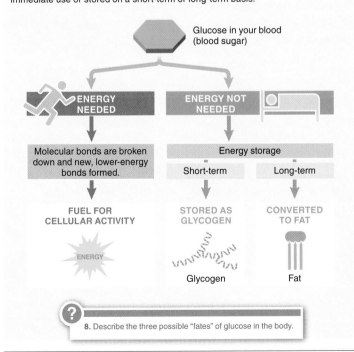

Glucose in your blood
(blood sugar)

ENERGY NEEDED	ENERGY NOT NEEDED

Molecular bonds are broken down and new, lower-energy bonds formed.

Energy storage

Short-term | Long-term

FUEL FOR CELLULAR ACTIVITY

ENERGY

STORED AS GLYCOGEN

Glycogen

CONVERTED TO FAT

Fat

? 8. Describe the three possible "fates" of glucose in the body.

COMPLEX CARBOHYDRATES

Multiple simple carbohydrates are sometimes linked together into more complex carbohydrates.

GLYCOGEN
Primary form of energy storage in animals

STARCH
Primary form of energy storage in plants

CHITIN
Forms the rigid outer skeleton of most insects and crustaceans

CELLULOSE
Structural material found in plants that is indigestible by humans

? 9. Why is it that if you take a piece of potato and leave it on your tongue for a while, it starts to taste sweet?

? 10. Why does cellulose pass through the digestive system unused?

?

5. Water can absorb and store a large amount of heat while increasing only a few degrees in temperature. Why?

a) The heat must first be used to break the hydrogen bonds rather than raise the temperature.

b) The heat must first be used to break the ionic bonds rather than raise the temperature.

c) The heat must first be used to break the covalent bonds rather than raise the temperature.

d) An increase in temperature causes an increase in adhesion of the water.

e) An increase in temperature causes an increase in cohesion of the water.

0 — 24 — 100
EASY — HARD

6. A chemical compound that releases H^+ into a solution is called:

a) a proton.

b) a base.

c) an acid.

d) a hydroxide ion.

e) a hydrogen ion.

0 — 32 — 100
EASY — HARD

7. Which of the following foods is not a significant source of complex carbohydrates?

a) fresh fruit

b) rice

c) pasta

d) oatmeal

e) All of the above are significant sources of complex carbohydrates.

0 — 40 — 100
EASY — HARD

8. Sucrose (table sugar) and lactose (the sugar found in milk) are examples of:

a) naturally occurring enzymes.

b) simple sugars.

c) monosaccharides.

d) disaccharides.

e) complex carbohydrates.

0 — 45 — 100
EASY — HARD

9. Which of the following statements about starch is incorrect?

a) Starch is the primary form of energy storage in plants.

b) Starch consists of a hundred or more glucose molecules joined together in a line.

c) Starch tastes sweet because it is made from glucose.

d) Starch is a polysaccharide.

e) All of the above statements about starch are correct.

0 — 66 — 100
EASY — HARD

10. Which of the following statements about fiber is incorrect?

a) Dietary fiber reduces the risk of colon cancer.

b) Fiber in the diet slows the passage of food through the intestines.

c) Humans cannot extract energy from fiber.

d) The cellulose of celery stalks and lettuce leaves is fiber.

e) Fiber scrapes the wall of the digestive tract, stimulating mucus secretion and aiding in the digestion of other molecules.

```
0                           100
EASY        47         HARD
```

11. A dietary fatty acid is liquid at room temperature (i.e., it has a low melting point) and contains carbon-carbon double bonds. It is most likely from:

a) a plant

b) a cow

c) a pig

d) a chicken

e) a lamb

```
0                           100
EASY     36            HARD
```

12. In an unsaturated fatty acid:

a) carbon-carbon double bonds are present in the hydrocarbon chain.

b) the hydrocarbon chain has an odd number of carbons.

c) the hydrocarbon chain has an even number of carbons.

d) no carbon-carbon double bonds are present in the hydrocarbon chain.

e) not all carbons in the hydrocarbon chain are bonded to hydrogen.

```
0                           100
EASY        51         HARD
```

13. Which statement about phospholipids is incorrect?

a) They are used as organisms' chief form of short-term energy.

b) They are hydrophobic at one end.

c) They are hydrophilic at one end.

d) They are a major constituent of cell membranes.

e) They contain glycerol linked to fatty acids.

```
0                           100
EASY       45          HARD
```

14. Proteins are an essential component of a healthy diet for humans (and other animals). Their most common purpose is to serve as:

a) raw material for growth.

b) fuel for running the body.

c) organic precursors for enzyme construction.

d) long-term energy storage.

e) inorganic precursors for enzyme construction.

```
0                           100
EASY        58         HARD
```

2·12—2·14 Lipids store energy for a rainy day.

Lipids are macromolecules—made up primarily from carbon, hydrogen, and oxygen—that are insoluble in water. Lipids are important in energy storage, as hormones, and in membrane structure. The breakdown of dietary fats releases more energy per gram than other macromolecules.

FATS (TRIGLYCERIDES)

Lipids composed of a head region and three long tails. They provide long-term energy storage and insulation. When broken down, they release significantly more energy per gram than other macromolecules.

Head: Glycerol

Tails: Fatty acids

STEROLS

Lipids composed of four interlinked rings of carbon atoms. They are important regulators of growth and development.

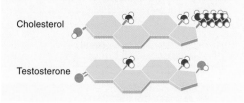

Cholesterol

Testosterone

PHOSPHOLIPIDS

The major component of the cell membrane that surrounds the contents of a cell and controls the flow of chemicals into and out of the cell.

Hydrophilic head (attracted to water)

Hydrophobic tails (not attracted to water)

11. Why do lipids contain so much more stored energy than carbohydrates?

12. Why do humans generally have a preference for fats in their diets?

13. Which types of lipids are important components of most cell membranes?

2·15—2·19 Proteins are versatile molecules that serve as building blocks.

Cells and tissues are primarily built from proteins, sequences of amino acids that fold into complex three-dimensional shapes. The atoms, especially nitrogen, present in the plant and animal proteins that an organism eats are essential to the organism's growth and repair.

PROTEIN STRUCTURE

The amino acid sequence of a protein determines how it folds into a particular three-dimensional shape. This shape determines many of the protein's features, such as which molecules it will interact with.

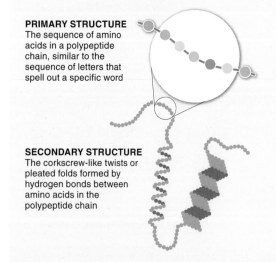

PRIMARY STRUCTURE
The sequence of amino acids in a polypeptide chain, similar to the sequence of letters that spell out a specific word

SECONDARY STRUCTURE
The corkscrew-like twists or pleated folds formed by hydrogen bonds between amino acids in the polypeptide chain

TERTIARY STRUCTURE
The complex three-dimensional shape formed by multiple twists and bends in the polypeptide chain, based on interactions between the amino acid side chains

QUATERNARY STRUCTURE
Two or more polypeptide chains bonded together

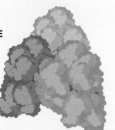

14. Describe two functions served by proteins in the body.

15. What does it mean to say that an amino acid is an "essential amino acid"?

16. How does the amino acid sequence of a protein affect its function?

ENZYMES

Enzymes are proteins that help initiate and speed up chemical reactions. They aren't permanently altered in the process, but rather can be used again and again.

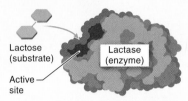

Lactose (substrate)

Lactase (enzyme)

Active site

1 Each enzyme has an active site that is a perfect fit for its substrate.

2 Like a key in a lock, lactose fits in the active site in lactase. The bond between the simple sugars is then broken.

Galactose

Glucose

3 The two simple sugars making up lactose are then released.

? 17. Why is protein shape so important for enzymes?

2·20–2·21 Nucleic acids store information on how to build and run a body.

The nucleic acids DNA and RNA are macromolecules that store information by having unique sequences of nucleotides. Both play central roles in directing protein production in organisms. RNA acts as a translator of the genetic code into proteins. It reads DNA sequences and directs the production of a sequence of amino acids.

DEOXYRIBONUCLEIC ACID (DNA)

Two sugar-phosphate backbones spiral around each other, forming a double helix. The backbones are connected to each other by nucleotide base pairs. The sequence of nucleotides contains the information about how to produce a particular protein.

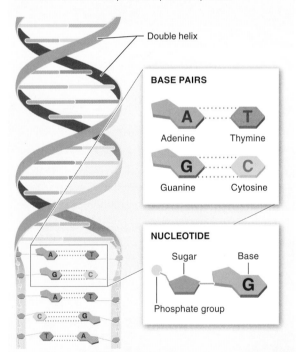

Double helix

BASE PAIRS

A — T
Adenine — Thymine

G — C
Guanine — Cytosine

NUCLEOTIDE

Sugar — Base

G

Phosphate group

? 18. How is information stored in the sequence of bases of a nucleic acid?

? 19. What determines the information in a segment of DNA? Why do researchers working on the Human Genome Project describe the sequence of nucleotides in only one strand when presenting a DNA sequence?

RIBONUCLEIC ACID (RNA)

There are three important structural differences between RNA and DNA.

Uracil

The sugar molecule in the RNA backbone contains an extra oxygen.

RNA has only one sugar-phosphate backbone, while DNA has two.

Instead of thymine, RNA has a similar base called uracil.

RNA acts as a middleman molecule—taking the instructions for protein production from DNA to another part of the cell where, in accordance with the RNA instructions, amino acids are pieced together into proteins.

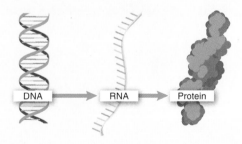

DNA → RNA → Protein

? 20. Describe the three ways in which RNA differs from DNA.

---

15. Dietary proteins:
   a) are considered "complete" only if they contain all of the amino acids required by humans.
   b) consist of all 20 amino acids required in the human body.
   c) are considered "complete" only if they contain the 12 non-essential amino acids required by humans.
   d) are nutritionally identical, since all are broken down into their constituent amino acids in the digestive system.
   e) can be obtained from animal sources but not plant sources.

0 — 52 — 100
EASY — HARD

16. The primary structure of proteins is often described as amino acids connected like beads on a string. In this same vein, which of the following images best describes a protein's quaternary structure?
   a) threads in a cloth
   b) needles in a haystack
   c) rungs on a ladder
   d) links on a chain
   e) coils in a spring

0 — 66 — 100
EASY — HARD

17. Which of the following statements about enzymes is incorrect?
   a) Enzymes can initiate chemical reactions.
   b) Enzymes speed up chemical reactions.
   c) Enzymes are proteins.
   d) Enzymes contain an active site for binding of particular substrates.
   e) Enzymes undergo a permanent change during the reactions they promote.

0 — 47 — 100
EASY — HARD

18. Which of the following nucleotide bases are present in equal amounts in DNA?
   a) adenine and cytosine
   b) thymine and guanine
   c) adenine and guanine
   d) thymine and cytosine
   e) adenine and thymine

0 — 23 — 100
EASY — HARD

19. Which type of macromolecule contains an organism's genetic information?
   a) polysaccharide
   b) monosaccharide
   c) fatty acid
   d) DNA
   e) phospholipid

0 — 6 — 100
EASY — HARD

3 | Cells

THE SMALLEST PART OF YOU

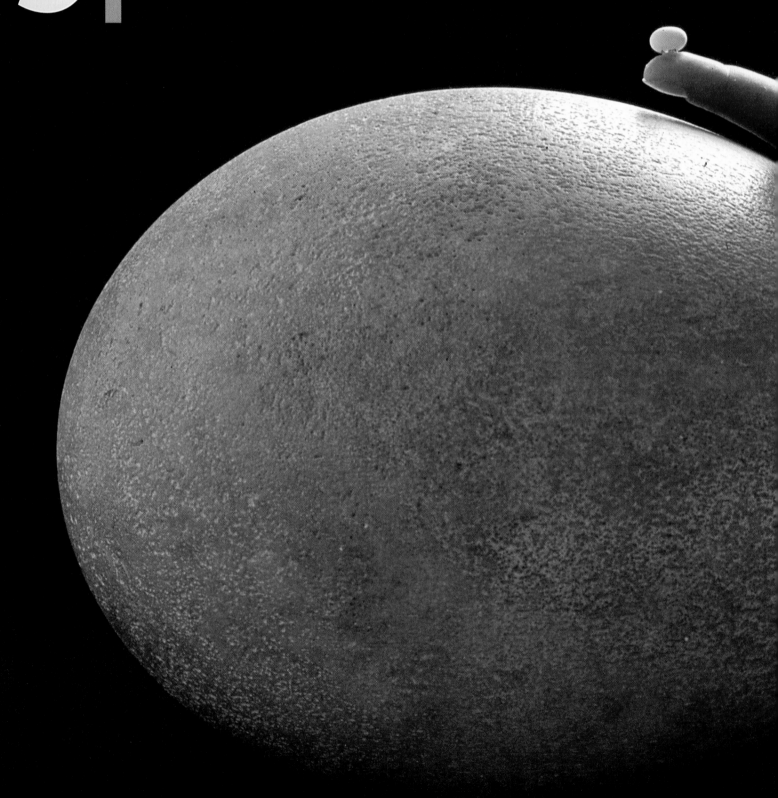

What is a cell?

Cell membranes are gatekeepers.

Molecules move across membranes in several ways.

Cells are connected and communicate with each other.

Nine important landmarks distinguish eukaryotic cells.

3·1–3·3
What is a cell?

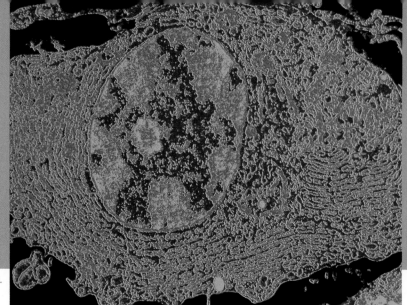

Human cell, packed with organelles.

3·1 All organisms are made of cells.

Where do we begin if we want to understand how organisms work? Given their complexity, this task can be daunting. Whether we are studying a creature as small as a flea or as large as an elephant or giant sequoia, all organisms are made of smaller units that are more easily studied and understood (**FIGURE 3-1**). The most basic unit of any organism is the **cell,** the smallest unit of life that can function independently and perform all the necessary functions of life. Understanding cell structure and function is the basis for our understanding of how complex organisms are organized.

The term "cell" was first used in the mid-1600s by Robert Hooke, an English scientist also known for his

contributions to philosophy, physics, and architecture. When he was made Curator of Experiments for the Royal Society of London, Hooke suddenly had access to many of the first microscopes available, and he began to examine everything he could get his hands on. Because Hooke thought the close-up views of a very thin piece of cork resembled a mass of small, empty rooms, he named these compartments *cellulae,* Latin for "small rooms."

After sufficient improvements were made to early microscopes in the 19th century, the central role of the cell in biology could be understood. By the 1830s, scientists realized that all plants and animals were made entirely from

Pixie's parasol mushroom

Ring-tailed lemurs

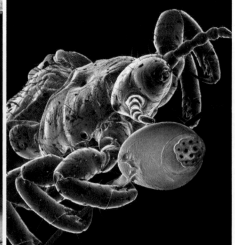

Head louse

FIGURE 3-1 **What do these diverse organisms have in common?** Cells.

cells. Subsequent studies revealed that every cell seemed to arise from the division of another cell. You, for example, are made up of at least 60 trillion cells, all of which came from just one cell: the single fertilized egg produced when an egg cell from your mother was fertilized by a sperm cell from your father.

The facts that (1) all living organisms are made up of one or more cells and (2) all cells arise from other, preexisting living cells are the foundations of **cell theory,** one of the unifying theories in biology, and one that is universally accepted by all biologists (**FIGURE 3-2**). As we see in Chapter 10, the origin of life on earth was a one-time deviation from cell theory: the first cells on earth probably originated from free-floating molecules in the oceans early in the earth's history (about 3.5 billion years ago). Since that time, however, all cells and thus all life have been produced as a continuous line of cells, originating from these initial cells.

Today, we know that the cell is a three-dimensional structure, like a fluid-filled balloon, in which many of the essential chemical reactions of life take place (such as the breakdown of carbohydrates for energy and the translation of the genetic code for protein production). Generally, these reactions involve transporting raw materials and fuel into the cell and exporting finished materials and waste products out of the cell. In addition, most, but not all, cell types contain DNA (deoxyribonucleic acid), a molecule that contains the information that directs the formation of various cellular products within the cell, the chemical reactions in the cell, and the cell's ability to reproduce itself. We explore all of these features of cell functioning in this and the next three chapters.

Q Aristotle wrote that "living things form quickly whenever… air and…heat are enclosed in anything." Why is he wrong?

To see a cell, you don't have to work in a lab or use a microscope. Just open your refrigerator. Chances are you've got a dozen or so visible cells in there: eggs. Although most cells are too small to see with the naked eye, there are a few exceptions, including hens' eggs from the supermarket. As long as they are unfertilized, which most store-bought hens' eggs are, each egg tends to contain just one cell. The ostrich egg, weighing about three pounds, contains the largest of all animal cells. (We should note, however, that by the time the ostrich lays a fertilized egg, the embryo inside has already gone through multiple divisions.)

In addition to being among the largest cells around, eggs are also the most valuable. Almas caviar, eggs from the beluga sturgeon, sells for nearly $700 per ounce. This value is exceeded only by that of human eggs, which currently

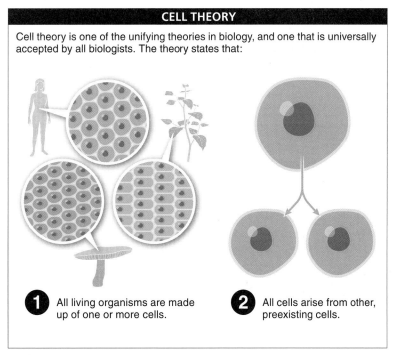

CELL THEORY

Cell theory is one of the unifying theories in biology, and one that is universally accepted by all biologists. The theory states that:

1 All living organisms are made up of one or more cells.

2 All cells arise from other, preexisting cells.

FIGURE 3-2 **The cell is the basic unit of life.**

Ostrich eggs weigh more than 3 pounds each

Beluga sturgeon eggs: $700 per ounce

EGG DONORS NEEDED
Healthy females.
Ages 18 - 30 donate to
infertile couples some of
the many eggs your
body disposes monthly.
COMPENSATION
$5000 - $8000
Call now
Reproductive Solutions
818/832-1494

10 μm

Human eggs: Thousands of dollars per egg

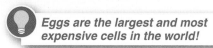
Eggs are the largest and most expensive cells in the world!

FIGURE 3-3 Not all cells are tiny. And some cells are extremely valuable!

fetch as much as $25,000 for a dozen or so eggs on the open market (**FIGURE 3-3**). (Human sperm cells command only about a penny per 20,000 cells!)

Most cells are much smaller than hens' eggs and ostrich eggs. Consider that, at this very moment, there are probably more than seven billion bacteria in your mouth—even if you just brushed your teeth! This is more than the number of people on earth. It is possible to squeeze so many bacteria in there because most cells (not just bacteria) are really, really tiny—so tiny that 2,000 red blood cells, lined up end to end, would just extend across a dime.

In this chapter, we investigate the two different kinds of cells that make up all of the organisms on earth, the processes by which cells control how materials move into and out of the cell, and how cells communicate with each other. We also explore some of the important structures found in many cells and the specialized roles these structures play in a variety of cellular functions. Along the way, we learn about some of the health consequences when cells malfunction.

TAKE-HOME MESSAGE 3·1

The most basic unit of any organism is the cell, the smallest unit of life that can function independently and perform all of the necessary functions of life, including reproducing itself. All living organisms are made up of one or more cells, and all cells arise from other, preexisting cells.

3·2 Prokaryotic cells are structurally simple but extremely diverse.

Although millions of diverse species live on earth and many of those organisms contain trillions of different cells, every cell falls into one of two basic categories.

A **eukaryotic cell** has a central control structure called a nucleus, which contains the cell's DNA. Organisms composed of eukaryotic cells are called **eukaryotes.** (The term "eukaryote" comes from the Greek words for "good" and "kernel," referring to the nucleus.)

A **prokaryotic cell** does not have a nucleus; its DNA simply resides in the cytoplasm. An organism consisting of a prokaryotic cell is called a **prokaryote.**

The first cells on earth were prokaryotes, making their appearance about 3.5 billion years ago (the term "prokaryote" comes from the Greek for "before" and "kernel," referring to the evolutionary origin of these cells before the eukaryotes). For a long time (1.5 billion years), the prokaryotes had the planet to themselves. All prokaryotes are one-celled organisms and are invisible to the naked eye.

Just two groups of prokaryotes exist: *bacteria* and *archaea.* All other organisms are eukaryotes. The largest group of prokaryotes is the bacteria. Bacteria are involved in many critical biological processes and affect human health in important ways. Those such as *Escherichia coli* (*E. coli*) live in your intestine and help your body make some essential vitamins. But some other strains of *E. coli* as well as some other bacteria species are responsible for illness. *Streptococcus pyogenes,* for example, causes strep throat. Less familiar to you may be the archaea; these microorganisms inhabit some of the harshest environments on earth, thriving in extremes of temperature, salinity, and pH.

Prokaryotes have four basic structural features (**FIGURE 3-4**).

1. A **plasma membrane** encompasses the cell (and sometimes is simply called the "cell membrane"). Anything inside the plasma membrane is referred to as "intracellular," and everything outside the plasma membrane is "extracellular."

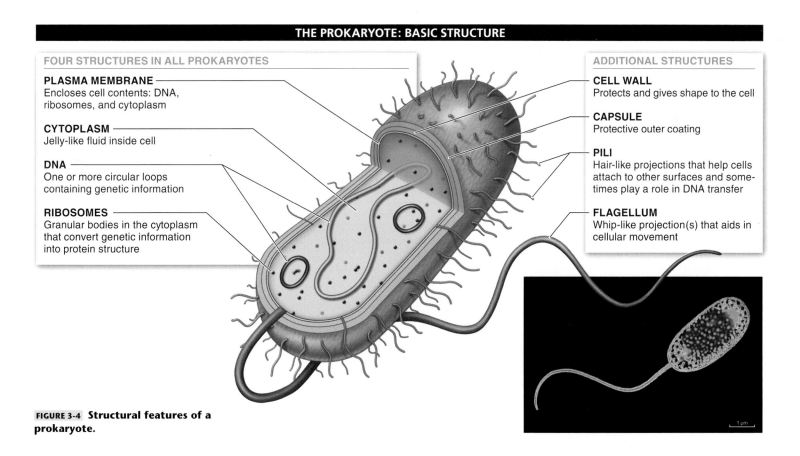

THE PROKARYOTE: BASIC STRUCTURE

FOUR STRUCTURES IN ALL PROKARYOTES

PLASMA MEMBRANE
Encloses cell contents: DNA, ribosomes, and cytoplasm

CYTOPLASM
Jelly-like fluid inside cell

DNA
One or more circular loops containing genetic information

RIBOSOMES
Granular bodies in the cytoplasm that convert genetic information into protein structure

ADDITIONAL STRUCTURES

CELL WALL
Protects and gives shape to the cell

CAPSULE
Protective outer coating

PILI
Hair-like projections that help cells attach to other surfaces and sometimes play a role in DNA transfer

FLAGELLUM
Whip-like projection(s) that aids in cellular movement

1 µm

FIGURE 3-4 Structural features of a prokaryote.

2. The **cytoplasm** refers to the cell's contents contained within the plasma membrane. This includes the jelly-like fluid, called the **cytosol,** and the cell's genome.

3. **Ribosomes** are little granular bodies where proteins are made; thousands of them are scattered throughout the cytoplasm.

4. Each prokaryote has one or more circular loops or linear strands of DNA.

Some prokaryotes have additional structures. For example, many have a rigid **cell wall** that protects and gives shape to the cell. Some have a sticky, sugary capsule as their outermost layer. This sticky outer coat provides protection and enhances the prokaryotes' ability to anchor themselves in place when necessary.

Many prokaryotes have a **flagellum** (*pl.* **flagella**), a long, thin, whip-like projection of the plasma membrane that rotates like a propeller and moves the cell through the medium in which it lives. Other appendages include **pili** (*sing.* **pilus**), much thinner, hair-like projections that help

prokaryotes attach to surfaces and can serve as "tubes" through which they exchange DNA.

Although prokaryotes are smaller, evolutionarily older, and structurally more simple than eukaryotes, they are fantastically diverse metabolically (i.e., in the way they break down and build up molecules). Among many other energy-usage "innovations" occurring in bacteria, some can fuel their activities in the presence or absence of oxygen and, depending on the type of bacteria, can use an extremely wide range of molecules as an energy source, such as the sulfur in deep-sea hydrothermal vents, or hydrogen gas, or light from the sun.

TAKE-HOME MESSAGE 3·2

Every cell on earth is either a eukaryotic or a prokaryotic cell. Prokaryotes, which have no nucleus, were the first cells on earth. They are single-celled organisms. Prokaryotes include the bacteria and archaea and, as a group, are characterized by tremendous metabolic diversity.

3·3 Eukaryotic cells have compartments with specialized functions.

In the two billion years that eukaryotes have been on earth, they have evolved into some of the most dramatic and interesting creatures, such as platypuses, dolphins, giant sequoias, and the Venus flytrap. Not all eukaryotes are multicellular, however. Many fungi are unicellular, and the Protista (or protists) are a huge group of eukaryotes, nearly

Heliconia *plant*

Cape porcupine

Yellow-fuzz cone slime mold

FIGURE 3-5 **Diversity of the eukaryotes.** Every organism that we can see without magnification is a eukaryotic organism.

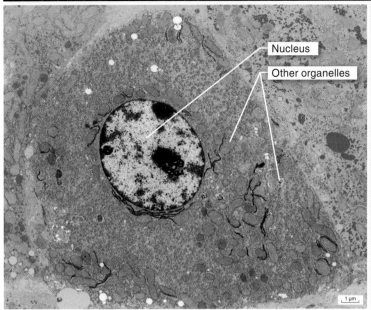

Nucleus

Other organelles

1 μm

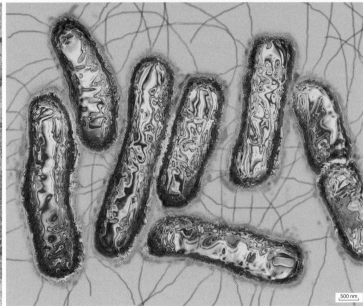

500 nm

TYPICAL EUKARYOTIC CELL FEATURES
- DNA contained in nucleus.
- Larger than prokaryotes—usually at least 10 times bigger.
- Cytoplasm contains specialized structures called organelles.

FIGURE 3-6 Comparison of eukaryotic and prokaryotic cells.

TYPICAL PROKARYOTIC CELL FEATURES
- No nucleus—DNA is in the cytoplasm.
- Internal structures mostly not organized into compartments.
- Much smaller than eukaryotes.

all of which are single-celled organisms visible only with a microscope. Nonetheless, because all prokaryotes are single-celled and thus invisible to the naked eye, every organism that we see around us is a eukaryotic organism, including all plants and animals (**FIGURE 3-5**).

Eukaryotic cells are about 10,000 times larger than prokaryotic cells in volume. They possess numerous structural features that make it easy to distinguish eukaryotes from prokaryotes under a microscope (**FIGURE 3-6**). Chief among the distinguishing features of eukaryotic cells is the presence of a **nucleus,** a membrane-enclosed structure that contains linear strands of DNA. In addition to a nucleus, eukaryotic cells usually contain in their cytoplasm several other specialized structures. Many of these structures, called **organelles,** are enclosed separately by their own lipid membranes.

The physical separation of compartments within a eukaryotic cell means that the cell has distinct areas in which different chemical reactions can occur simultaneously. In the mostly non-compartmentalized interior of a prokaryotic cell, random molecular movements quickly blend the chemicals throughout the cell, reducing the ease with which different reactions can occur simultaneously.

FIGURE 3-7 illustrates a generalized animal cell and a generalized plant cell. Because they share a common, eukaryotic ancestor, they have much in common. Both can have a plasma membrane, nucleus, cytoskeleton, and a host of organelles, including rough and smooth endoplasmic membranes, Golgi apparatus, and mitochondria. Animal cells have centrioles, which are not present in most plant cells. Plant cells have a rigid cell wall (as do fungi and many protists) and chloroplasts (also found in some protists). Plants also have a vacuole, a large central chamber (only occasionally found in animal cells). We explore each of these animal and plant organelles in detail later in this chapter.

When you compare a complex eukaryotic cell with the structurally simple prokaryotic cell, it's hard not to wonder about the origin of eukaryotic cells. We can't go back two billion years to watch the initial evolution of eukaryotic cells, but there is considerable evidence for some of what occurred. In particular, the **endosymbiosis theory** provides the best explanation for the presence of two organelles in eukaryotes: chloroplasts in plants and algae, and mitochondria in plants and animals. Chloroplasts enable plants and algae to convert sunlight into a more usable form of energy. Mitochondria help plants and

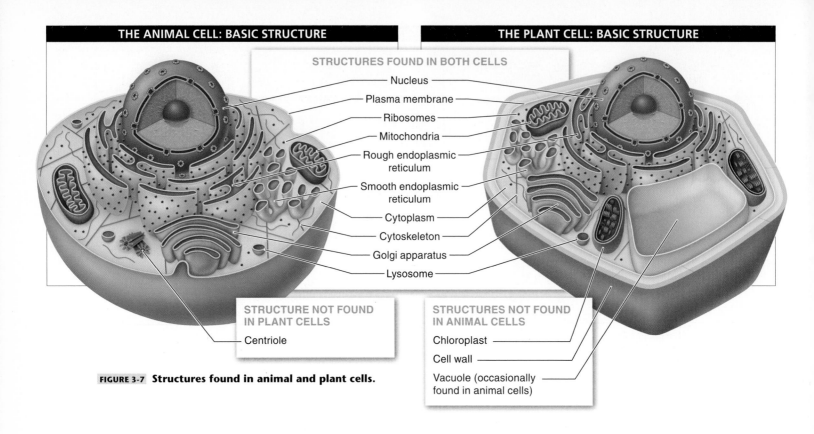

STRUCTURES FOUND IN BOTH CELLS

Nucleus

Plasma membrane

Ribosomes

Mitochondria

Rough endoplasmic reticulum

Smooth endoplasmic reticulum

Cytoplasm

Cytoskeleton

Golgi apparatus

Lysosome

STRUCTURE NOT FOUND IN PLANT CELLS

Centriole

FIGURE 3-7 **Structures found in animal and plant cells.**

STRUCTURES NOT FOUND IN ANIMAL CELLS

Chloroplast

Cell wall

Vacuole (occasionally found in animal cells)

animals harness the energy stored in food molecules. (Chapter 4, on energy, explains the details of both of these processes.)

Q Humans—at a microscopic level—may be part bacteria. How can that be?

According to the theory of endosymbiosis, two different types of prokaryotes may have set up close partnerships with each other. For example, some small prokaryotes capable of performing photosynthesis (the process by which plant cells capture light energy from the sun and transform it into the chemical energy stored in food molecules) may have come to live inside a larger "host" prokaryote. The photosynthetic "boarder" may have made some of the energy that it captured in photosynthesis available for use by the host.

After a long while, the two cells may have become more and more dependent on each other, until neither cell could live without the other (they became "symbiotic") and they became a single, more complex organism. Eventually, the photosynthetic prokaryote evolved into a **chloroplast,** the organelle in plant and eukaryotic algae cells in which

photosynthesis occurs. A similar scenario might explain how another large host prokaryote engulfed a smaller prokaryote unusually efficient at converting food and oxygen into easily usable energy and this smaller prokaryote evolved into a **mitochondrion,** the organelle in plant and animal cells that converts the energy stored in food into a form usable by the cell (**FIGURE 3-8**).

The idea of the role of endosymbiosis in the evolution of eukaryotes is supported by several observations. .

1. Chloroplasts and mitochondria are similar in size to prokaryotic cells and divide by splitting (fission), just like prokaryotes.

2. Chloroplasts and mitochondria have ribosomes, similar to those found in bacteria, that allow them to synthesize some of their own proteins; this ability is not found in other organelles, which rely on proteins made by cytoplasmic ribosomes.

3. Chloroplasts and mitochondria have small amounts of circular DNA, similar to the circular DNA in prokaryotes and in contrast to the linear DNA strands found in a eukaryote's nucleus.

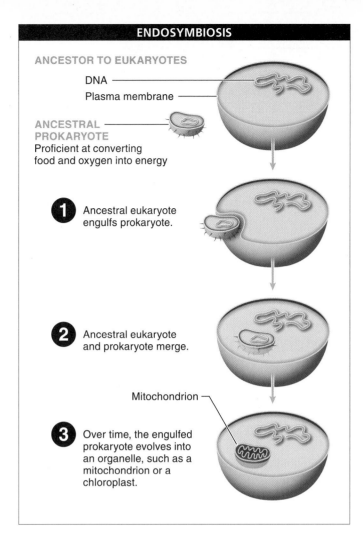

ENDOSYMBIOSIS

ANCESTOR TO EUKARYOTES

DNA

Plasma membrane

ANCESTRAL PROKARYOTE
Proficient at converting food and oxygen into energy

1 Ancestral eukaryote engulfs prokaryote.

2 Ancestral eukaryote and prokaryote merge.

Mitochondrion

3 Over time, the engulfed prokaryote evolves into an organelle, such as a mitochondrion or a chloroplast.

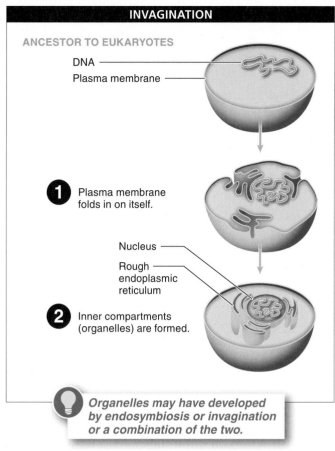

INVAGINATION

ANCESTOR TO EUKARYOTES

DNA

Plasma membrane

1 Plasma membrane folds in on itself.

Nucleus

Rough endoplasmic reticulum

2 Inner compartments (organelles) are formed.

Organelles may have developed by endosymbiosis or invagination or a combination of the two.

FIGURE 3-8 **How did eukaryotic cells become so structurally complex?** Two theories.

4. Analysis of chloroplast and mitochondrial DNA has revealed that it is highly related to bacterial DNA, much more closely than it is related to eukaryotic DNA.

The best current theory about the origin of the other organelles in eukaryotes is a process called **invagination.** The idea is that the plasma membrane around the cell may have folded in on itself to form the inner compartments, which subsequently became modified and specialized (see Figure 3-8).

TAKE-HOME MESSAGE 3·3

Eukaryotes are single-celled or multicellular organisms consisting of cells with a nucleus that contains linear strands of genetic material. The cells also commonly have organelles throughout their cytoplasm; these organelles may have originated evolutionarily through endosymbiosis or invagination, or both.

3·4 – 3·7
Cell membranes are gatekeepers.

Like gatekeepers, cell membranes control the movement of material into and out of the cell.

3·4 Every cell is bordered by a plasma membrane.

Just as skin covers our bodies, every cell of every living thing on earth is enclosed by a plasma membrane, a two-layered membrane that holds the contents of a cell in place and regulates what enters and leaves the cell. Plasma membranes are thin (a stack of a thousand would be only as thick as a single hair) and flexible, and in photos or diagrams the membranes often resemble simple plastic bags, holding the cell contents in place. This image is a gross oversimplification,

however. Membranes are indeed thin and flexible, but they are far from simple: a close look at a plasma membrane will reveal that its surface is filled with pores, outcroppings, channels, and complex molecules floating around within the two layers of the membrane itself (**FIGURE 3-9**).

Cells are perpetually interacting with their external environment. And in these interactions, the plasma

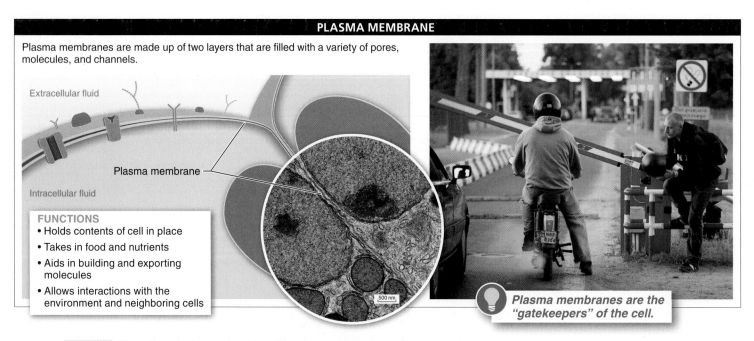

PLASMA MEMBRANE

Plasma membranes are made up of two layers that are filled with a variety of pores, molecules, and channels.

Extracellular fluid

Plasma membrane

Intracellular fluid

FUNCTIONS
- Holds contents of cell in place
- Takes in food and nutrients
- Aids in building and exporting molecules
- Allows interactions with the environment and neighboring cells

500 nm

Plasma membranes are the "gatekeepers" of the cell.

FIGURE 3-9 More than just an outer layer. The plasma membrane performs several critical functions beyond simply enclosing a cell's interior contents.

membranes must perform several critical functions beyond simply holding a cell's interior contents. The plasma membrane must function in ways that accomplish the following.

- The membrane enables the cell to take in food and nutrients and dispose of waste products.

- It allows the cell to take in water.

- It allows the cell to build and export molecules needed elsewhere in the body.

- It mediates communications with the external environment and other cells and adhesion to other cells or surfaces.

- Like a border control checkpoint, it controls the flow of molecules into and out of the cell.

The foundation of all plasma membranes is a layer of lipid molecules all packed together. These are a special type of lipid, called **phospholipids,** which, as you'll recall from Section 2-13, have what appear to be a head and two long tails. The head consists of a molecule of **glycerol** linked to a molecule containing phosphorus (**FIGURE 3-10**). This head region is said to be **polar,** because the electrons are not shared equally among the atoms, leading to regions of partial positive and partial negative charge. As you learned in Chapter 2, water is also a polar molecule and, for this reason, other polar molecules mix easily with water. Molecules that can mix with water are described as

hydrophilic ("water-loving") molecules. The two tails of the phospholipid are long chains of carbon and hydrogen atoms. Because the electrons in these bonds are shared equally, the carbon-hydrogen chains are **nonpolar.** And because they are nonpolar, these tails do not mix with water and are said to be **hydrophobic** ("water-fearing"). The chemical structure of phospholipids gives them a sort of split personality: their hydrophilic head region mixes easily with water, while their hydrophobic tail region does not mix with water.

The split personality of phospholipids makes them good membrane material. Once a large number of phospholipids are packed together with all of their heads facing one way and their tails the other, we have a sheet with one side that is hydrophilic and one that is hydrophobic. In the cell's plasma membrane, two of these sheets of phospholipids are arranged so that the hydrophobic tails are all in contact with one another and the hydrophilic heads are in contact with the watery solution outside and inside the cell (see Figure 3-10). This arrangement gives us another way to describe the structure of the plasma membrane: as a **phospholipid bilayer.**

The phospholipids are not locked in place in the plasma membrane; they just float around their side of the bilayer. They cannot pop out of the membrane or flop from one side to the other, because their hydrophobic tails always line up away from any watery solution. Just as similarly charged sides of two magnets push away from each other, so do the

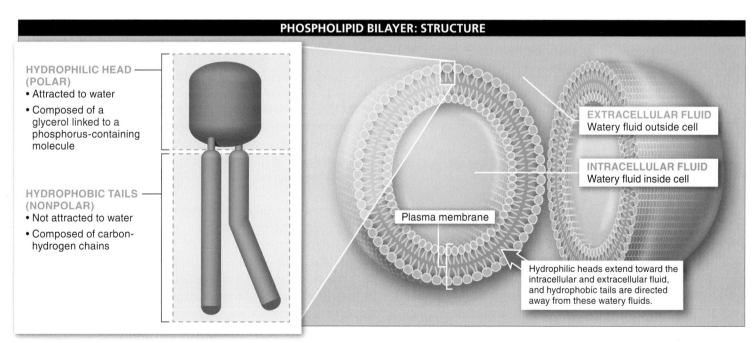

PHOSPHOLIPID BILAYER: STRUCTURE

HYDROPHILIC HEAD (POLAR)
- Attracted to water
- Composed of a glycerol linked to a phosphorus-containing molecule

HYDROPHOBIC TAILS (NONPOLAR)
- Not attracted to water
- Composed of carbon-hydrogen chains

EXTRACELLULAR FLUID
Watery fluid outside cell

INTRACELLULAR FLUID
Watery fluid inside cell

Plasma membrane

Hydrophilic heads extend toward the intracellular and extracellular fluid, and hydrophobic tails are directed away from these watery fluids.

FIGURE 3-10 **Good membrane material.** The phospholipid bilayer of the plasma membrane prevents fluid from leaking out of the cell.

hydrophobic tails in the center of the membrane push away from and avoid coming into contact with water molecules. Because the center part of the bilayer membrane is made up of hydrophobic lipids, the solution on one side of the membrane cannot leak across into the solution on the other side. In this way, the plasma membrane forms a boundary around the cell's contents.

TAKE-HOME MESSAGE 3·4

Every cell of every living organism is enclosed by a plasma membrane, a two-layered membrane that holds the contents of a cell in place and regulates what enters and leaves the cell.

3·5 Molecules embedded in the plasma membrane help it perform its functions.

The many different functions of the plasma membrane are accomplished with the help of different types of protein, carbohydrate, and lipid molecules embedded within or attached to the phospholipid bilayer. Many of these molecules float around, held in a proper orientation by hydrophobic and hydrophilic forces, but not always anchored in place. For these reasons, the plasma membrane is often described as a **fluid mosaic** (**FIGURE 3-11**).

For every 50–100 phospholipids in the membrane, there is one protein molecule. Some of these proteins, called **transmembrane proteins,** penetrate right through the lipid bilayer, from one side to the other. Others, called **surface**

proteins or peripheral proteins, reside primarily on the inner or outer surface of the membrane.

What determines whether a protein resides on the surface or extends through the bilayer? Its tertiary structure. Remember from Chapter 2 that all the amino acids that make up each protein have side chains that differ from one another chemically. Some of these side chains are hydrophobic, others are hydrophilic, and as a protein is assembled into its final shape, these side chains can cause parts of the protein to be attracted to hydrophobic or hydrophilic regions of the molecule. Because a transmembrane protein has both hydrophobic and hydrophilic regions, part of the protein can be positioned in the hydrophobic region in the center of the membrane while the other parts are in the hydrophilic regions. Surface membrane proteins, on the other hand, have an entirely hydrophilic structure and reside on the membrane surface, bound only to the head regions of the phospholipids. As a consequence, they can be positioned on either the outer or the inner surface of the membrane.

Once membrane proteins are in place, the hydrophobic and hydrophilic forces keep them properly oriented. Because all of the components of the plasma membrane are held in the membrane in this manner, they can float around without ever popping out.

There are several primary types of membrane proteins, each of which performs a different function (**FIGURE 3-12**).

1. Receptor proteins are surface or transmembrane proteins that bind to chemicals in the cell's external environment. In doing so, receptor proteins can attach a cell to the extracellular matrix or convey information from the outside to the inside of the cell, regulating certain processes within the cell. This receptor-mediated regulation is a process called signal transduction.

MOLECULES WITHIN THE PLASMA MEMBRANE

Extracellular fluid

Hydrophilic region
Hydrophobic region

Carbohydrates

Plasma membrane

Transmembrane protein

Lipids

Surface proteins

Intracellular fluid

Hydrophobic and hydrophilic forces determine the orientation of proteins in the plasma membrane.

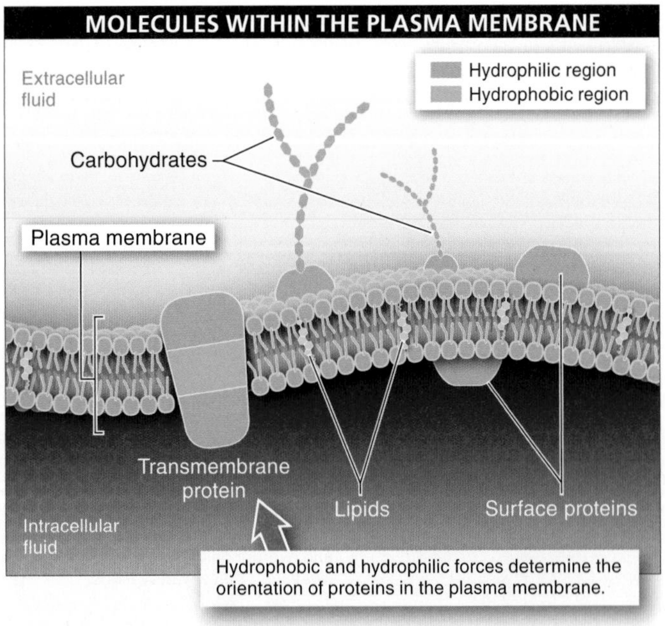

FIGURE 3-11 A fluid mosaic. Protein, carbohydrate, and lipid molecules are embedded in the plasma membrane, but many can float around because they are not always anchored in place.

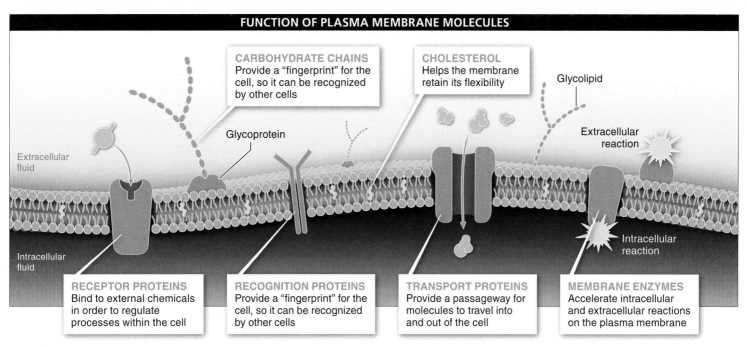

FUNCTION OF PLASMA MEMBRANE MOLECULES

CARBOHYDRATE CHAINS
Provide a "fingerprint" for the cell, so it can be recognized by other cells

CHOLESTEROL
Helps the membrane retain its flexibility

Glycolipid

Glycoprotein

Extracellular fluid

Extracellular reaction

Intracellular fluid

Intracellular reaction

RECEPTOR PROTEINS
Bind to external chemicals in order to regulate processes within the cell

RECOGNITION PROTEINS
Provide a "fingerprint" for the cell, so it can be recognized by other cells

TRANSPORT PROTEINS
Provide a passageway for molecules to travel into and out of the cell

MEMBRANE ENZYMES
Accelerate intracellular and extracellular reactions on the plasma membrane

FIGURE 3-12 **Plasma membrane molecules serve diverse roles.**

Cells in the heart, for example, have receptor proteins that bind to adrenaline, a chemical released into the bloodstream in times of extreme stress or fright. When adrenaline binds to these heart cells, the cells increase the heart's rate of contraction to pump blood through the body more quickly. You have experienced this reaction if you've ever been startled and felt your heart start to pound.

2. Recognition proteins are surface or transmembrane proteins that give each cell a "fingerprint" that makes it possible for the body's immune system (which fights off infections) to distinguish the cells that belong inside your body from those that are invaders and need to be attacked. Carbohydrates also play a role in recognition. Short glycoproteins on the outside of the cell membrane serve as part of the membrane's fingerprint. Like receptor proteins, recognition proteins can also help cells bind to or adhere to other cells or molecules within the extracellular matrix.

3. Transport proteins are transmembrane proteins that help polar or charged substances pass through the plasma membrane. Transport proteins come in a variety of shapes and sizes, making it possible for a wide variety of molecules to be transported.

4. Membrane enzymes are surface or transmembrane proteins that accelerate chemical reactions on the plasma membrane's surface. (A variety of membrane-bound **enzymatic proteins** exist, with some accelerating reactions

on the inside of the plasma membrane and others accelerating reactions on the outside of the plasma membrane.)

In addition to the various kinds of membrane proteins and the carbohydrate chains of membrane-bound glycoproteins, the lipid cholesterol can also be incorporated in a cell's plasma membrane. **Cholesterol** helps the membrane maintain its flexibility, preventing the membrane from becoming too fluid or floppy at moderate temperatures and acting as a sort of antifreeze, preventing the membrane from becoming too rigid at freezing temperatures. The membranes of some cells are about 25% cholesterol; other plasma membranes, such as those of most bacteria and plants, have no cholesterol at all.

TAKE-HOME MESSAGE 3·5

The plasma membrane is a fluid mosaic of proteins, lipids, and carbohydrates. Proteins found in the plasma membrane enable it to carry out most of its gatekeeping functions. The proteins act as receptors, help molecules enter and leave the cell, and catalyze reactions on the inner and outer cell surfaces. In conjunction with carbohydrates, some plasma membrane proteins identify the cell to other cells. And, in addition to the phospholipids that make up most of the plasma membrane, cholesterol is an important lipid in some membranes, influencing fluidity.

Faulty membranes can cause diseases.

The single most common fatal inherited disease in the United States is cystic fibrosis, a disease that results from an improperly functioning membrane. At any given time, about 30,000 people in the United States have cystic fibrosis.

Cystic fibrosis occurs when an individual inherits from both parents incorrect genetic instructions for producing one type of transmembrane protein that allows chloride ions to get into and out of cells. This transport protein occurs primarily in the membranes of cells in the lungs and digestive tract.

These genetic instructions can be defective in more than a thousand different ways, but the result is the same: malfunction of chloride passageways in a cell's membrane that causes gradual accumulation of chloride ions within cells. In nearly all cases of cystic fibrosis, two primary effects occur: an improper salt balance in the cells and a buildup of thick, sticky mucus—particularly in the lungs. Normal mucus helps to protect the lungs by trapping dust and bacteria. This mucus is then moved out of the lungs (helped along by coughing). The mucus produced by someone with cystic fibrosis, however, is too thick and sticky to be moved out of the lungs, so it collects there, where it impairs lung function and increases the risk of bacterial infection. Because of the improper cellular salt balance, one way to

test for cystic fibrosis is to measure the concentration of salt in the sweat—abnormally high concentrations indicate that the person may have the disease

Although many high-tech treatments have been promised for the sufferers of cystic fibrosis and a great deal of research is being done on this disease, one of the most common treatments is decidedly low-tech. Parents help their children with cystic fibrosis clear the mucus out of their lungs by holding them on a steep slant, almost upside down, and vigorously patting or thumping their chest and back. This shakes loose the mucus in their lungs and moves it to a place where they can cough it up (**FIGURE 3-13**). With careful treatment, the life expectancy of someone with cystic fibrosis can be 35–40 years or longer.

Faulty membranes also play a role in many other diseases, including heart disease (familial hypercholesterolemia), diabetes, and hormonal disorders (such as Graves disease). In most of these cases, the membrane component that is improperly functioning is commonly a receptor protein or a transport protein.

It shouldn't be surprising, then, to discover that much pharmaceutical research focuses on altering cell membrane functioning, One group of drugs that alter membrane function—called "beta-blockers"—is extremely effective at

TREATING CYSTIC FIBROSIS

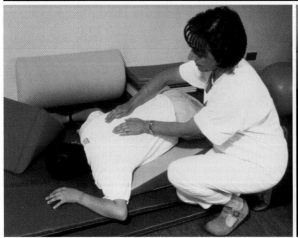

"Thumping" on the chest and back can loosen the mucus.

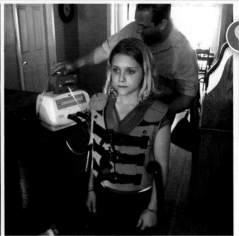

The vest, by inflating and deflating rapidly, can have a similar effect in the course of a 20-minute session.

The thick and sticky mucus produced by someone with cystic fibrosis collects in the lungs, impairing lung function and increasing the risk of bacterial infection.

FIGURE 3-13 **Moving mucus manually or with the use of an inhalation vest.**

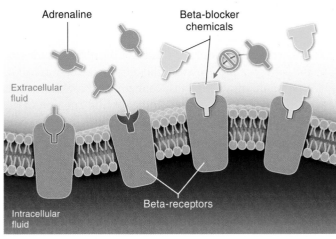
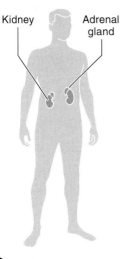

1 In stressful situations, the adrenal glands pump out adrenaline.

2 Adrenaline binds with beta-receptors on cells, causing a faster heartbeat and increased blood pressure.

3 Beta-blocker chemicals bind to receptors and prevent adrenaline from binding to the cell.

Kidney Adrenal gland Adrenaline Beta-blocker chemicals

Extracellular fluid

Intracellular fluid Beta-receptors

By binding to adrenaline receptors, beta-blockers reduce anxiety symptoms.

FIGURE 3-14 **Drugs can alter cell membrane function.** Adrenaline, the fight-or-flight chemical, cannot enter the cell when beta-blocking medications bind to the cell's beta-receptors.

reducing anxiety. This effect was discovered almost accidentally. The drugs were actually developed as a treatment for high blood pressure—the only use for which they are approved by the FDA. But as a result of several decades of clinical research studies documenting the effectiveness of beta-blockers in treating anxiety, many doctors now prescribe beta-blockers (in a legal practice called "off-label" prescribing) to control the symptoms of anxiety.

Q *Why do "beta-blockers" reduce anxiety?*

Here's how beta-blockers work. Many cells in your body, particularly the cells of the heart, have receptor proteins on their plasma membranes that can bind to adrenaline, which helps your body cope with stressful situations. These receptor proteins are called "beta-adrenergic receptors," or beta-receptors. In stressful situations, your adrenal glands pump out adrenaline (**FIGURE 3-14**). On reaching cells in your heart (among other locations in your body) and binding to beta-receptors, the adrenaline promptly causes your heart to beat faster and more forcefully, increasing your blood pressure in the process. This reaction is fine in a short-term fight-or-flight situation, but it is not healthy over the long run because the increased pressure can damage blood vessels. Depending on its severity, this reaction can also be problematic if you are giving a presentation or taking a test, or if you're in any other anxiety-producing situation.

When you take a beta-blocker pill, the pill dissolves and the chemicals travel throughout your body until they encounter the beta-receptors. They bind to the receptors, hold on, and block the adrenaline from doing its job. This outcome slows your heart rate, causes a reduction in blood pressure, and can bring great relief to those suffering from the sweating and trembling associated with anxiety.

TAKE-HOME MESSAGE 3·6

Normal cell functioning can be disrupted when cell membranes—particularly the proteins embedded in them—do not function properly. Such malfunctions can cause health problems, such as cystic fibrosis. But intentional disruption of normal cell membrane function can have beneficial, therapeutic effects, such as in the treatment of high blood pressure and anxiety.

3·7 Membrane surfaces have a "fingerprint" that identifies the cell.

Every cell in your body has a "fingerprint" made from a variety of molecules on the outside-facing surface of the cell membrane. Some of these membrane molecules differ from cell to cell, depending on the specific function of the cell. Others are common to all of your cells and tell your immune system, "I belong here." Cells with an improper fingerprint are recognized as foreign and are attacked by your body's defenses.

Throughout our evolutionary history, this system has been tremendously valuable in helping our bodies fight infection. In some cases, however, this vigilance is a problem, like a car alarm that goes off even when you don't want it to. Suppose you receive a liver (or any other organ) transplant. Even if the donor is a close relative, the molecular fingerprint on the cells of the donated liver is not identical to your own. Consequently, your body sees the new organ as a foreign object and puts up a fight against it (**FIGURE 3-15**). Because your body will naturally try to reject the new organ, doctors must administer drugs that suppress your immune system. Immune suppression helps you tolerate the new liver, but, as you can imagine, it leaves you without some of the defenses to fight off other foreign invaders, such as bacteria that may cause infection.

Q Why is it extremely unlikely that a person will catch HIV from casual contact—such as shaking hands—with an infected individual?

The AIDS-causing virus, HIV, uses the molecular fingerprints on plasma membranes to infect an individual's cells. These same molecular markers are also the reason that it's extremely unlikely that you can catch an HIV infection from casual contact with an infected individual, such as shaking his or her hand. The specific molecular markers involved in infection by HIV belong to a group of identifying markers called "clusters of differentiation." Abbreviated as "CD markers" and having names such as CD1, CD2, and CD3, these marker molecules are proteins embedded in the plasma membrane that enable the cell to bind to outside molecules and, sometimes, transport them into the cell.

One CD marker, called the CD4 marker, is found only on cells deep within the body and in the bloodstream, such as immune system cells and some nerve cells. It is the CD4 marker, in conjunction with another receptor, that is targeted by HIV. If the virus can find a cell with a CD4 marker, it can

infect that cell, and because the CD4 markers never occur on the surface of skin cells, casual contact such as touching is very unlikely to transmit the virus (**FIGURE 3-16**). Even if millions of HIV particles are present on a person's hands, they just can't gain access into any of the other person's surface cells.

MOLECULAR FINGERPRINTS AND ORGAN TRANSPLANTS

The molecular fingerprint on the cells of a donated liver is not identical to the molecular fingerprint of the recipient, even if the donor is a close relative. As a result, the recipient's immune system sees the new organ as a foreign object and puts up a fight against it.

Liver transplant recipient

Molecular fingerprint (on membrane surfaces of recipient)

Donor liver

Molecular fingerprint (on membrane surfaces of donor liver)

POSSIBLE OUTCOMES

The recipient's body rejects the new organ.

The liver is accepted after drugs are administered to suppress the immune system.

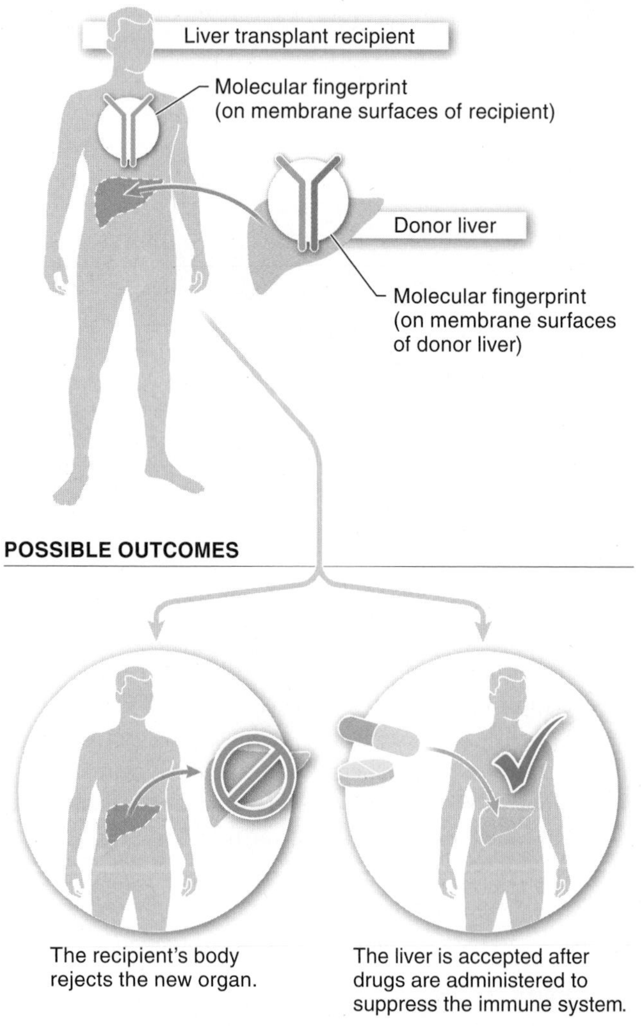

FIGURE 3-15 Mismatched molecular fingerprints can cause difficulty in organ transplantation.

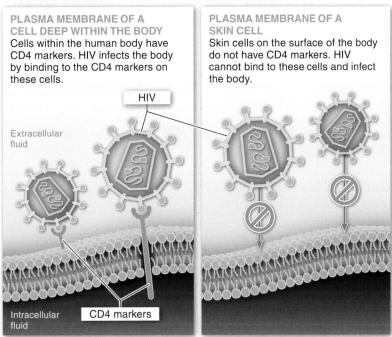

HIV is not spread through casual contact such as hugging, shaking hands, or sharing a drinking glass.

PLASMA MEMBRANE OF A CELL DEEP WITHIN THE BODY
Cells within the human body have CD4 markers. HIV infects the body by binding to the CD4 markers on these cells.

PLASMA MEMBRANE OF A SKIN CELL
Skin cells on the surface of the body do not have CD4 markers. HIV cannot bind to these cells and infect the body.

HIV

Extracellular fluid

Intracellular fluid

CD4 markers

FIGURE 3-16 HIV requires CD4 markers—not found on skin cells—to infect the body.

The most common ways, by far, of transmitting HIV involve the transfer of blood, semen, vaginal fluid, or breast milk from one individual to another. Particles of the virus as well as cells infected by the virus are present in these fluids. Consequently, the chief routes of transmission are from an infected mother to her child in breast milk, from an infected mother to her baby at birth, the use of contaminated needles, and unprotected sexual intercourse. Because an open sore or cut might expose some of the cells in your bloodstream to the outside world, however, it is not impossible for casual transmission of HIV to occur.

TAKE-HOME MESSAGE 3·7

Every cell in your body has a "fingerprint" made from a variety of molecules on the outside-facing surface of the cell membrane. This molecular fingerprint is key to the function of your immune system.

3·8 – 3·11
Molecules move across membranes in several ways.

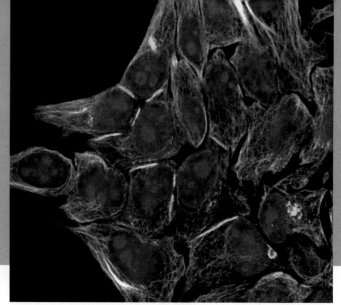

Passage of proteins between cells can be tracked (junctions between these human cancer cells are stained red and green).

3·8 Passive transport is the spontaneous diffusion of molecules across a membrane.

To function properly, cells must take in food and/or other necessary materials and must move out both metabolic waste and molecules produced for use elsewhere in the body. In some cases, this movement of molecules requires energy and is called active transport. (We cover active transport later in this chapter.) In other cases, the molecular movement occurs spontaneously, without the input of energy, and is called **passive transport.** There are two types of passive transport: diffusion and osmosis. (Osmosis is discussed in Section 3-9.)

Diffusion is passive transport in which a particle, called a **solute,** is dissolved in a gas or liquid (a **solvent**) and moves from an area of high solute concentration to an area of lower concentration (**FIGURE 3-17**). A difference in the concentration of solutes in two areas is called a *concentration gradient*—and the larger the difference in the concentration of the solutes in the two areas, the greater the concentration gradient is.

We say that molecules tend to move "down" their concentration gradient. This movement occurs because molecules move randomly and are equally likely to move in any direction (in the absence of any other forces). And so when certain molecules are highly concentrated, they keep bumping into each other and eventually end up evenly distributed. For a simple illustration, drop a tiny bit of food coloring into a bowl of water and wait for a few minutes. The molecules of dye are initially clustered together in a very high concentration. Gradually, they disperse down

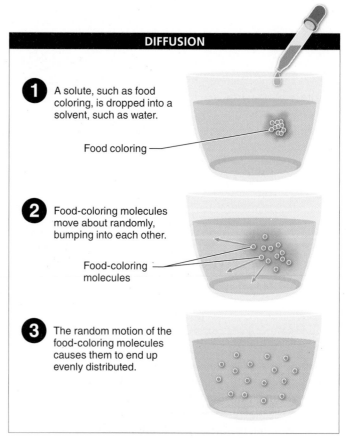

DIFFUSION

1. A solute, such as food coloring, is dropped into a solvent, such as water.

   Food coloring

2. Food-coloring molecules move about randomly, bumping into each other.

   Food-coloring molecules

3. The random motion of the food-coloring molecules causes them to end up evenly distributed.

FIGURE 3-17 **Diffusion: a form of passive transport that results in an even distribution of molecules.**

their concentration gradient until the color is equally spread throughout the bowl.

In cells, molecules such as oxygen (O_2) and carbon dioxide (CO_2) that are small and carry no charge can pass directly through the phospholipid bilayer of the membrane without the assistance of any other molecules, in a process called **simple diffusion.** Each time you take a breath, for example, there is a higher concentration of O_2 molecules in the air you pull into your lungs than in your blood. And so that oxygen diffuses across the plasma membranes of the lung cells and into your bloodstream, where red blood cells pick it up and deliver it to parts of your body where it is needed. Similarly, because CO_2 in your bloodstream is at a higher concentration than in the air in your lungs, it diffuses from your blood into the cells of your lungs and is released to the atmosphere when you exhale (**FIGURE 3-18**).

Most molecules, however, can't get through plasma membranes on their own. Polar molecules (electrically charged) are repelled by the hydrophobic middle region of the phospholipid bilayer. Or the molecules may be too big to squeeze through the membrane. These molecules may still be able to diffuse across the membrane, down their concentration gradient, with the help of a transport protein. Often, this transport protein spans the membrane and functions like a revolving door, allowing movement of molecules in either direction, depending on their concentration gradient. When spontaneous diffusion across a plasma membrane requires a transport protein, it is called **facilitated diffusion** (see Figure 3-18).

Defects in transport proteins can reduce facilitated diffusion or even bring it to a complete stop, with serious health consequences. Many genetic diseases are the result of inheriting incorrect genetic instructions for building transport proteins. In the disease cystinuria, incorrect genetic instructions result in a malformed transport protein in the plasma membrane. When structured and functioning properly, this transport protein facilitates the diffusion of some amino acids (including cysteine, from which the disease gets its name) out of the kidneys into the urine. When the protein is malformed, the transporters cannot facilitate this diffusion and these amino acids build up in the kidneys, forming painful and dangerous kidney stones.

Diffusion across membranes doesn't occur just in animals—we see it in all organisms. As we'll see in Chapter 4, one of the most important biological processes on earth is the diffusion of CO_2 from the atmosphere (an area of relatively high concentration) into the leaf cells of plants (areas of

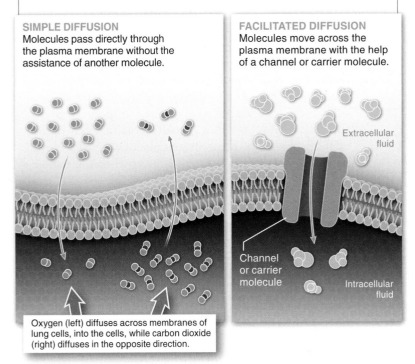

PASSIVE TRANSPORT

Passive transport occurs when molecules move across a membrane without energy input. Molecules move down their concentration gradients.

SIMPLE DIFFUSION
Molecules pass directly through the plasma membrane without the assistance of another molecule.

FACILITATED DIFFUSION
Molecules move across the plasma membrane with the help of a channel or carrier molecule.

Extracellular fluid

Channel or carrier molecule

Intracellular fluid

Oxygen (left) diffuses across membranes of lung cells, into the cells, while carbon dioxide (right) diffuses in the opposite direction.

FIGURE 3-18 **Simple and facilitated diffusion: no energy required.**

relatively low concentration), where it can be attached to other molecules, forming sugars through photosynthesis. At the same time, O_2 diffuses out of the leaves and into the atmosphere.

TAKE-HOME MESSAGE 3·8

For proper functioning, cells must acquire food molecules and/or other necessary materials from outside the cell. Similarly, metabolic waste molecules and molecules produced for use elsewhere in the body must move out of the cell. In passive transport—which includes simple and facilitated diffusion and osmosis—the molecular movement occurs spontaneously, without the input of energy. This generally takes place as molecules move down their concentration gradient.

Just as solute molecules will passively diffuse down their concentration gradients, water molecules will also move from areas of high concentration to areas of low concentration to equalize the concentration of water inside and outside the cell. The diffusion of water across a membrane is a special type of passive transport called **osmosis** (**FIGURE 3-19**). Just as solute molecules may diffuse across a plasma membrane, molecules of water also move across the membrane, equalizing the water concentration

inside and outside the cell. Although some water can pass through the lipid bilayer, the hydrophobic region severely limits this flow. Most of the rapid movement of water in and out of cells occurs through "water channels," called *aquaporins,* which are transmembrane proteins with hydrophilic channels through which the water molecules pass in single file.

Osmosis can have some dramatic effects on cells. As we saw above, many molecules just can't move across a cell membrane. But while the molecule can't move out of the cell and down its concentration gradient, water can move into the cell down *its* concentration gradient. And as water diffuses into the cell, the cell will get larger.

When a cell is in a fluid environment (referred to as a solution), the amount of dissolved substances (solutes) in that solution may be (1) equal to, (2) less than, or (3) greater than the concentration of dissolved substances in the cell. This relationship between the concentrations of solutes inside the cell and solutes outside the cell is referred to as **tonicity** (see Figure 3-19).

1. If the concentration of solutes outside the cell is *equal* to the concentration inside the cell, the outside solution is **isotonic.** Water still moves between the solution and the cell, but because it moves at the same rate in both directions, there is no net change in the amount of water inside versus outside the cell

2. If the concentration of solutes outside the cell is *lower* than the concentration inside the cell, the outside solution is **hypotonic.** In this situation, if the cell membrane is not permeable to the solutes, more water will move into the cell than out of it, and the cell will swell.

3. If the concentration of solutes outside the cell is *higher* than the concentration inside the cell, the outside solution is **hypertonic.** In this situation, if the cell membrane is not permeable to the solutes, more water will move out of the cell than into it, and the cell will shrivel.

You can see osmosis in action in your own kitchen (**FIGURE 3-20**). Take a stalk of celery and leave it on the counter for a couple of hours. As water evaporates from the cells of the celery stalk, it will shrink and become limp.

Q Drinking seawater can be deadly. Why?

OSMOSIS

Osmosis is a type of passive transport by which water diffuses across a membrane, in order to equalize the concentration of water inside and outside the cell. The direction of osmosis is determined by the total amount of solutes on either side of the membrane.

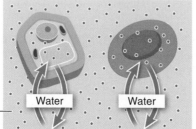

PLANT CELL | ANIMAL CELL (RED BLOOD CELL)

ISOTONIC SOLUTION
- Solute concentrations are balanced.
- Water movement is balanced.

Extracellular fluid

Water | Water

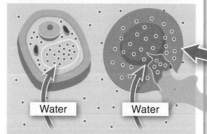

PLANT CELL | ANIMAL CELL

HYPOTONIC SOLUTION
- Solute concentrations are lower in the extracellular fluid.
- Water diffuses into cells.

Water | Water

Unlike plant cells, animal cells may explode in hypotonic solutions because they don't have a cell wall to limit cellular expansion.

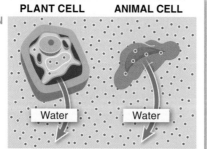

PLANT CELL | ANIMAL CELL

HYPERTONIC SOLUTION
- Solute concentrations are higher in the extracellular fluid.
- Water diffuses out of cells.

Water | Water

 Water will always move toward a region having a greater concentration of solutes.

FIGURE 3-19 **Osmosis overview.**

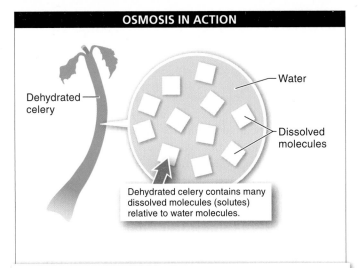

OSMOSIS IN ACTION

Dehydrated celery

Water

Dissolved molecules

Dehydrated celery contains many dissolved molecules (solutes) relative to water molecules.

WHEN PLACED IN DISTILLED WATER

Distilled water contains fewer dissolved molecules than the celery cells. Water molecules diffuse into the celery, equalizing the water concentration inside and outside the cells. The celery becomes crisp.

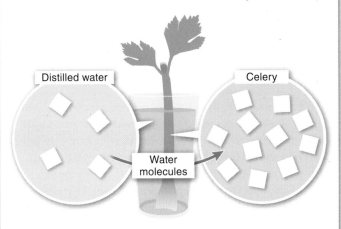

Distilled water

Celery

Water molecules

WHEN PLACED IN SALT WATER

Salt water contains more dissolved molecules than the celery cells. Water molecules diffuse out of the celery. The celery becomes even more shriveled.

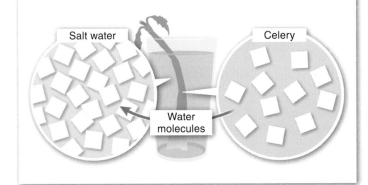

Salt water

Celery

Water molecules

FIGURE 3-20 Osmosis in your kitchen.

You can make it crisp again by placing it in a solution of distilled water. Why distilled water? Because it contains very few dissolved molecules—it is a hypotonic solution. The

cells in the celery stalk, on the other hand, contain many dissolved molecules, such as salt. Because those dissolved molecules can't easily pass across the plasma membrane of the celery stalk cells, the water molecules move down their concentration gradient and into celery cells. Would the celery regain its crispness if you placed it in concentrated salt water? No, because there are more dissolved solute molecules (and fewer water molecules) in the hypertonic salt solution than inside the celery cells. What little water remains in the celery cells would actually move, by osmosis, out of the celery, causing it to shrivel even more.

Q How do laxatives relieve constipation? A more practical use of osmosis is seen with the laxative Milk of Magnesia. This product contains magnesium salts, which are poorly absorbed from the digestive tract. As a consequence, after a dose of the laxative, water moves by osmosis from the surrounding cells into the intestines. The water softens the feces in the intestines and increases the fecal volume, thereby relieving constipation.

It's important to note that the direction of osmosis is determined only by a difference in the *total concentration* of all the molecules dissolved in the water: it does not matter what the solutes are, only how many molecules of solutes there are. To determine which way the water molecules will move, you need to determine the total amount of "dissolved stuff" on either side of the membrane. The water will move toward the side with the greater concentration of solute. For this reason, even small increases in the salinity of lakes can have disastrous consequences for the organisms living there, from fish to bacteria. Conversely, putting animal cells—such as red blood cells—in distilled water causes them to explode, because water will diffuse into the cell (which contains more solutes) and the cell will swell and burst. This does not generally happen to plant cells because (as we see later in this chapter) the plant cell plasma membrane is surrounded by a rigid cell wall that limits the amount of cell expansion possible when water moves in by osmosis.

> ### TAKE-HOME MESSAGE 3·9
> The diffusion of water across a membrane is a special type of passive transport called osmosis. Water moves from an area with a lower concentration of solutes to an area with a higher concentration of solutes. Water molecules move across the membrane until the concentration of water inside and outside the cell is equalized.

In active transport, cells use energy to move small molecules into and out of the cell.

Molecules can't always move spontaneously and effortlessly in and out of cells. Sometimes their transport needs energy, in which case the process is called **active transport.** Such energy expenditures may be necessary if the molecules or ions to be moved are being moved against their concentration gradient. In all active transport, proteins embedded in the membrane act like motorized revolving doors, pushing molecules across cell membranes regardless of the concentration of those molecules on either side of the membrane. The two distinct types of active transport, primary and secondary, differ only in the source of the fuel that keeps the revolving doors spinning.

The process of digestion provides an example of **primary active transport,** the type of active transport that occurs when energy from ATP is used to fuel the transport of molecules. To help break down food into more digestible bits, the cells lining your stomach create an acidic environment by pumping large numbers of H^+ ions (also called protons) into the stomach contents, against their concentration gradient (**FIGURE 3-21**). As a result, H^+ ions are more numerous in the stomach contents than inside the cells lining the stomach. All of this H^+ pumping increases your ability to digest the food but comes at a great energetic cost—in the form of usage of the high-energy molecule ATP—because the protons would not normally flow into a region against their concentration gradient. (We explore ATP in more detail in Chapter 4.)

Many transport proteins use an indirect method of fueling their activities rather than using energy released directly from ATP. In the process of **secondary active transport,** the transport protein simultaneously moves one molecule against its concentration gradient while letting another flow down its concentration gradient. Although no ATP is used directly in this process, at some other time and in some other location, energy from ATP was used to pump one of the molecules involved in the secondary transport process against its concentration gradient. The process is a bit like using energy to pump water up to the top of a high water tower. Later, the water can be allowed to run out of the tower over a water wheel, which can, in turn, power a process such as grinding wheat into flour. Our bodies frequently use the energy from one reaction that occurs spontaneously to fuel another reaction that requires energy.

TAKE-HOME MESSAGE 3·10

In active transport, movement of molecules across a membrane requires energy. Active transport is necessary if the molecules to be moved are very large or if they are being moved against their concentration gradient. Proteins embedded in the plasma membrane act like motorized revolving doors to actively transport (pump) the molecules.

ACTIVE TRANSPORT

Active transport occurs when the movement of molecules into and out of a cell requires the input of energy. For example, in response to eating, the cells lining your stomach use ATP to pump large numbers of H^+ ions into the stomach.

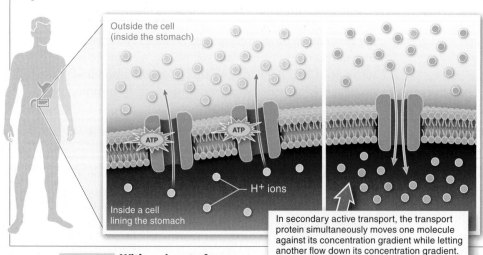

Outside the cell
(inside the stomach)

H^+ ions

Inside a cell
lining the stomach

In secondary active transport, the transport protein simultaneously moves one molecule against its concentration gradient while letting another flow down its concentration gradient. No ATP is used directly in the process.

Active transport in the stomach increases your ability to digest food.

FIGURE 3-21 **With an input of energy, molecules can be moved against concentration gradients.**

3·11 Endocytosis and exocytosis are used for bulk transport of particles.

Many substances are just too big to get into or out of a cell by passive or active transport. To absorb large particles, such as bacterial invaders, cells engulf them with their plasma membrane in a process called **endocytosis.** To export large particles, such as digestive enzymes manufactured for use elsewhere in the body, they often use the process of **exocytosis.**

There are three types of endocytosis: phagocytosis, pinocytosis, and receptor-mediated endocytosis. All three involve the basic process of the plasma membrane oozing around an object outside the cell, surrounding it, forming a little pocket called a *vesicle,* and then pinching off the vesicle so that it is inside the cell but separated from the rest of the cell contents.

Phagocytosis and Pinocytosis Relatively large particles are engulfed by cells in a process called **phagocytosis** (**FIGURE 3-22**). Amoebas and other unicellular protists, as well as white blood cells, use phagocytosis to consume entire organisms, either as food or as their way of defending against pathogens (disease-causing organisms or substances). Whereas "phagocytosis" comes from the Greek for "eat" and "container," the term **pinocytosis** comes from the Greek for "drink" and "container" and describes the process of cells taking in dissolved particles and liquids. The two processes are largely the same, except that the vesicles formed during pinocytosis are generally much smaller than those formed during phagocytosis.

Receptor-Mediated Endocytosis The third type of endocytosis, **receptor-mediated endocytosis,** is much more specific than either phagocytosis or pinocytosis. Receptor molecules on the surface of a cell recognize and bind to one specific type of molecule. For one receptor it might be insulin, for another it might be cholesterol. Many receptors of the same type are often clustered together in a cell's plasma membrane. When the appropriate molecule binds to each of the receptor proteins, the membrane begins to fold inward, first forming a little pit and then completely engulfing the molecules, which are still attached to their receptors.

One of the most important examples of receptor-mediated endocytosis involves cholesterol (**FIGURE 3-23**). Most cholesterol that circulates in the bloodstream is in the form of particles called low-density lipoproteins, or LDL. Each

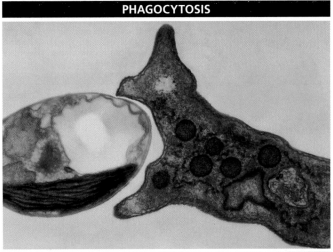

PHAGOCYTOSIS

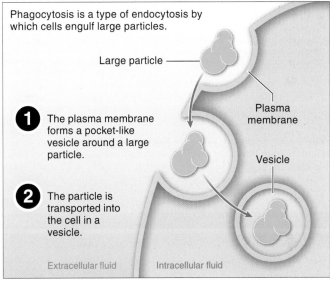

Phagocytosis is a type of endocytosis by which cells engulf large particles.

Large particle

Plasma membrane

Vesicle

1 The plasma membrane forms a pocket-like vesicle around a large particle.

2 The particle is transported into the cell in a vesicle.

Extracellular fluid   Intracellular fluid

FIGURE 3-22 **Through phagocytosis, amoebas and other unicellular protists, as well as white blood cells, consume other organisms for food or for defense.**

molecule of LDL is a cholesterol globule coated by phospholipids. Proteins embedded within the LDL's phospholipid coat are recognized by receptor proteins built into the plasma membranes of liver cells. Once bound to the receptors, the LDL molecule is consumed by the cell by endocytosis. Inside the cell, the cholesterol is broken down and used to make a variety of other useful molecules, such as the hormones estrogen and testosterone.

Circulating cholesterol often builds up on the walls of arteries, reducing blood flow and causing the artery to

RECEPTOR-MEDIATED ENDOCYTOSIS

Receptor-mediated endocytosis is a type of endocytosis by which cells engulf specific particles.

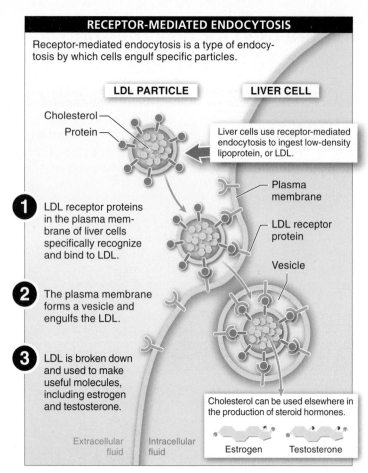

LDL PARTICLE

Cholesterol

Protein

LIVER CELL

Liver cells use receptor-mediated endocytosis to ingest low-density lipoprotein, or LDL.

1 LDL receptor proteins in the plasma membrane of liver cells specifically recognize and bind to LDL.

Plasma membrane

LDL receptor protein

Vesicle

2 The plasma membrane forms a vesicle and engulfs the LDL.

3 LDL is broken down and used to make useful molecules, including estrogen and testosterone.

Extracellular fluid

Intracellular fluid

Cholesterol can be used elsewhere in the production of steroid hormones.

Estrogen   Testosterone

FIGURE 3-23 **Receptor proteins aid in endocytosis.** This process is important in the ability of your liver to remove cholesterol from your bloodstream.

harden. Too much circulating cholesterol in LDL molecules can lead to cardiovascular disease and death (**FIGURE 3-24**). Individuals lucky enough to have large numbers of LDL receptors on the plasma membranes of their liver cells have a significantly lower risk of cardiovascular disease.

Q *Faulty cell membranes are a primary cause of cardiovascular disease. What modification to these membranes might be an effective treatment?*

Conversely, some individuals are at risk of early onset of cardiovascular disease because they consume food laden with too much cholesterol (such as egg yolks, cheese, and sausages) or have the misfortune of inheriting genes that code for faulty liver cell membranes that have few LDL receptors—a disorder called familial hypercholesterolemia. In the extreme case where an individual is born with no LDL receptors, circulating cholesterol accumulates in the arteries so rapidly that cardiovascular disease begins to develop even before puberty, and death from a heart attack can occur before the age of 30.

Exocytosis The movement of molecules out of cells takes place throughout the body. For example, cells in the pancreas produce a chemical called insulin that moves throughout the circulatory system. Insulin informs body cells that there is glucose in the bloodstream that ought to

HEART DISEASE

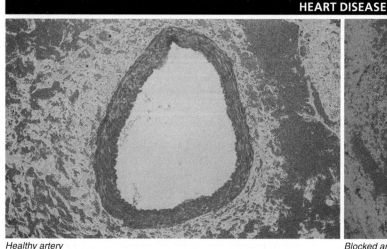

Healthy artery

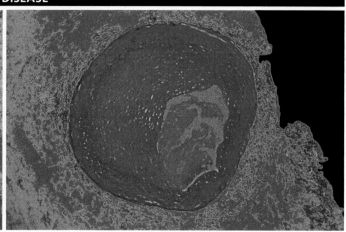

Blocked artery

FIGURE 3-24 **Comparison of a healthy artery and an artery choked by the buildup of cholesterol.**

When circulating cholesterol levels exceed the LDL receptors' capacity for binding and removing the lipid, blood flow can become reduced or blocked and the risk of cardiovascular disease increases.

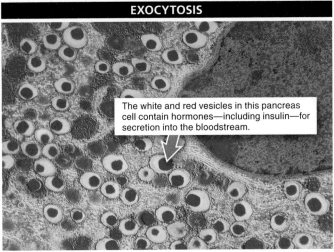

The white and red vesicles in this pancreas cell contain hormones—including insulin—for secretion into the bloodstream.

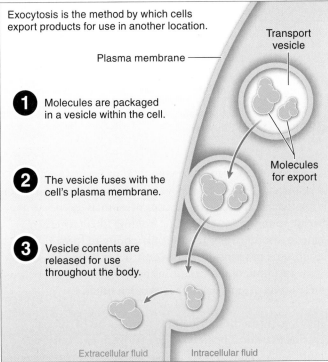

Exocytosis is the method by which cells export products for use in another location.

Plasma membrane

Transport vesicle

1 Molecules are packaged in a vesicle within the cell.

Molecules for export

2 The vesicle fuses with the cell's plasma membrane.

3 Vesicle contents are released for use throughout the body.

Extracellular fluid   Intracellular fluid

FIGURE 3-25 **Exocytosis moves molecules out of the cell.** This is how insulin is exported from the cells where it is synthesized.

be taken in and used for energy. The insulin molecule is much too large to pass out through the plasma membranes of the cells where it is manufactured. As a result, after molecules of insulin are produced, they are coated with a phospholipid membrane to form a vesicle. The insulin-carrying vesicle then moves through the cytoplasm to the inner surface of the cell's plasma membrane. Once there, the phospholipid membrane surrounding the insulin and the phospholipid membrane of the cell fuse together, dumping the contents of the vesicle out of the cell, where it can enter the bloodstream (**FIGURE 3-25**).

Exocytosis is not restricted to large molecules. In the brain and other parts of the nervous system, for example, communication between cells occurs as one cell releases large numbers of very small molecules, called neurotransmitters, by exocytosis.

TAKE-HOME MESSAGE 3·11

When materials cannot get into a cell by diffusion or through a pump (for example, when the molecules are too big), cells can engulf the molecules or particles with their plasma membrane in a process called endocytosis. Similarly, molecules can be moved out of a cell by exocytosis. In both processes, the plasma membrane moves to surround the molecules or particles and forms a little vesicle that is pinched off inside the cell (endocytosis) or fuses with the plasma membrane and dumps its contents outside the cell (exocytosis).

3·12
Cells are connected and communicate with each other.

Cells reach out and connect; no telephone lines required.

3·12 Connections between cells hold them in place and enable them to communicate with each other.

So far in this chapter we have examined the cell as a free-living and independent entity. The majority of cells in any multicellular organism, however, are connected to other cells. Cells adhere to other cells in a variety of ways, involving numerous types of protein and glycoprotein adhesion molecules. We examine three primary types of connections between animal cells: (1) tight junctions, (2) desmosomes, and (3) gap junctions (**FIGURE 3-26**).

Tight junctions form continuous, water-tight seals around cells and also anchor cells in place. Much like the caulking around a tub or sink that keeps water from leaking into the surrounding walls, tight junctions prevent fluid flow between cells. Tight junctions are particularly important in the small intestine, where digestion occurs. Cells lining the small intestine absorb nutrients from the watery fluid moving through your gut. If the fluid—and the resident bacteria—inside the intestine were to leak between the cells and into your body cavity, you would not be able to extract sufficient energy and nutrients from your food, and the bacteria would make you sick. The tight junctions instead force fluid to pass *into* the cells that line the intestine, where the nutrients can be used.

Desmosomes are like spot welds or rivets that fasten cells together into strong sheets. They occur at irregular intervals and function like fastened Velcro: they hold cells together but are not water-tight, allowing fluid to pass around them. Desmosomes and other similar junctions are found in much of the tissue that lines the cavities of animals' bodies. They also are found in muscle tissue, holding fibers together. Genetic disorders that reduce cells' ability to form desmosome proteins or lead to destruction of desmosomes by the immune system result in the formation of blisters, where layers of skin separate from each other.

Finally, **gap junctions** are pores surrounded by special proteins that form open channels between two cells (see Figure 3-26). Functioning like secret passageways, these junctions are large enough for salts, sugars, amino acids, and the chemicals that carry electrical signals to pass through, but are too small for the passage of organelles or very large molecules such as proteins and nucleic acids. Gap junctions are an important mechanism for cell-to-cell communication. In the heart, for example, the electrical signal telling muscle cells to contract is passed from cell to cell through gap junctions. Gap junctions are also important in allowing a cell to recognize that it has bumped up against another cell; chemicals flowing from one cell to the next can signal the body to stop producing cells of a particular type.

Compared with normal cells, cancer cells have fewer gap junctions. And research suggests that the resulting

TIGHT JUNCTIONS
Form a water-tight seal between cells, like caulking around a tub

DESMOSOMES
Act like Velcro and fasten cells together

GAP JUNCTIONS
Act like secret passageways and allow materials to pass between cells

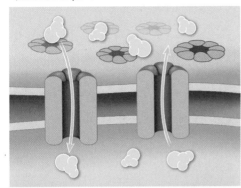

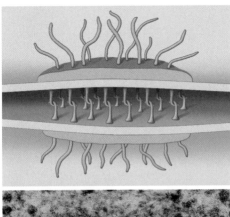

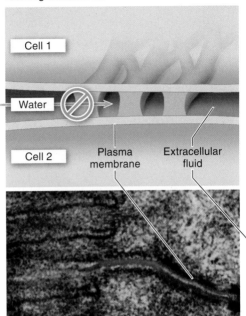

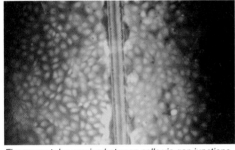

Cell 1

Water

Cell 2

Plasma membrane

Extracellular fluid

Tight junction between small intestine epithelial cells

Desmosome between heart muscle cells

Fluorescent dye moving between cells via gap junctions

FIGURE 3-26 **Cell connections: tight junctions, desmosomes, and gap junctions.**

Q Is a breakdown of cell-to-cell communication related to cancer?

reduction in intercellular communication among cancer cells may be important in the formation of masses of cells, called tumors, and the spread of cancer cells throughout the body. (Interestingly, treatments that cause an increase in the number of gap junctions tend to reduce tumor growth.)

TAKE-HOME MESSAGE 3·12

In multicellular organisms, most cells are connected to other cells. The connections can form a water-tight seal between the cells (tight junctions), can hold sheets of cells together while allowing fluid to pass between neighboring cells (desmosomes), or can function like secret passageways, allowing the movement of cytoplasm, molecules, and other signals between cells (gap junctions).

3·13 – 3·22
Nine important landmarks distinguish eukaryotic cells.

Cell of the voodoo lily. How many different organelles can you spot in this plant cell?

3·13 The nucleus is the cell's genetic control center.

The nucleus is the largest and most prominent organelle in most eukaryotic cells. In fact, the nucleus is generally larger than any prokaryotic cell. If a cell were the size of a large lecture hall or movie theater, the nucleus would be the size of a big-rig 18-wheeler truck parked in the front rows. The nucleus has two primary functions, both related to the fact that most of the cell's DNA resides in the nucleus. First, the nucleus is the genetic control center, directing most cellular activities by controlling which molecules are produced, and in what quantity. Second, the nucleus is the storehouse for hereditary information (**FIGURE 3-27**).

Three important structural components stand out in the nucleus (**FIGURE 3-28**). First is the **nuclear membrane,** sometimes called the nuclear envelope, which surrounds the nucleus and separates it from other parts of the cytoplasm. Unlike most plasma membranes, however, the nuclear membrane consists of two bilayers, one on top of the other, much like the double-bagging of groceries at the market. The nuclear membrane is not a bag, though. It is perforated, covered with tiny pores made from multiple proteins embedded in the phospholipid membranes and spanning both bilayers. These pores enable large molecules to pass from the nucleus to the cytosol and from the cytosol to the nucleus.

Nuclear pores are large enough to permit free diffusion of small molecules, such as water, sugars, and ions. The

FIGURE 3-27 The nucleus holds the genetic information that makes it possible to build organisms. That's why identical twins, carrying the same genetic information, look so similar.

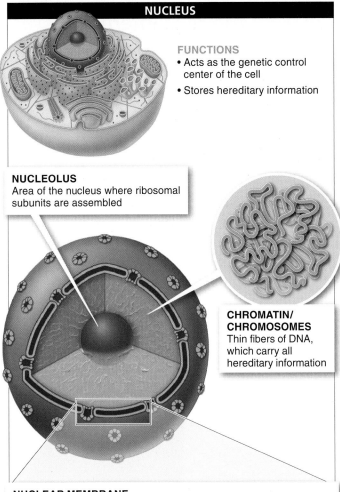

NUCLEUS

FUNCTIONS
- Acts as the genetic control center of the cell
- Stores hereditary information

NUCLEOLUS
Area of the nucleus where ribosomal subunits are assembled

CHROMATIN/ CHROMOSOMES
Thin fibers of DNA, which carry all hereditary information

NUCLEAR MEMBRANE
Two bilayers, covered in pores, that surround the nucleus

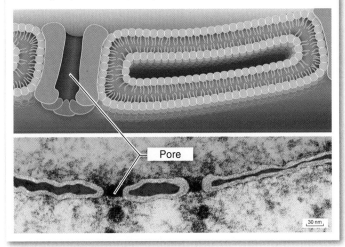

Pore

30 nm

FIGURE 3-28 **The nucleus: the cell's genetic control center.**

nuclear pore protein complexes, however, are able to regulate the transport of many other molecules across the nuclear envelope in both directions. Two types of transport, in particular, are important. These include 1) the movement from the cytosol into the nucleus of proteins that interact with DNA and catalyze nuclear activities, and 2) the movement from the nucleus to the surrounding cytosol of RNA and RNA-protein complexes.

The second prominent structure in the nucleus is the **chromatin,** a mass of long, thin fibers consisting of DNA with some proteins attached that keep the DNA from getting impossibly tangled. Most of the time, as the DNA directs cellular activities, the chromatin resembles a plate of spaghetti. When it's time for cell division (a process described in detail in Chapter 6), the chromatin coils up and the threads become shorter and thicker until they become visible as chromosomes, the compacted, linear DNA molecules that carry hereditary information.

A third structure in the nucleus is the **nucleolus,** an area near the center of the nucleus where subunits of the ribosomes, a critical part of the cellular machinery, are assembled. Ribosomes are like little factories in which the information stored in the DNA is used to construct proteins, including enzymes and the proteins that make up tissues such as the bark of trees or the bone of vertebrates. The ribosomes are built in the nucleolus but pass through the nuclear pores and into the cytosol before starting their protein-production work.

TAKE-HOME MESSAGE 3·13

The nucleus is usually the largest and most prominent organelle in the eukaryotic cell. It directs most cellular activities by controlling which molecules are produced and in what quantity. The nucleus is the storehouse for hereditary information.

3·14 Cytoplasm and the cytoskeleton form the cell's internal environment, provide its physical support, and can generate movement.

If you imagine yourself inside a cell the size of a big lecture hall, it might come as a surprise that you can barely see that big rig of a nucleus parked in the front rows. Visibility is almost zero, not only because the room is filled with jelly-like cytoplasm, but also because there is a dense web of thick and thin, straight and branched ropes, strings, and scaffolding, running every which way throughout the room.

This inner scaffolding of the cell, which is made from proteins, is the **cytoskeleton** (FIGURE 3-29). It has three chief purposes. First, it gives animal cells shape and support—making red blood cells look like little round doughnuts (without the hole in the middle) and giving neurons their very long, thread-like appearance. Plant cells are shaped primarily by their cell wall (a structure we discuss later in the chapter), but they also have a cytoskeleton. Second, the cytoskeleton controls the intracellular traffic flow, serving as a series of tracks on which a variety of organelles and molecules are guided across and around the inside of the cell. And third, because the elaborate scaffolding of the cytoskeleton is dynamic and can generate force, it gives all cells some ability to control their movement.

Three types of protein fibers make up the cytoskeleton. **Microtubules,** the thickest, are linear polymers of a protein and look like rigid, hollow tubes. Like intracellular conveyor belts, they are the tracks to which molecules and organelles within the cell can become attached and moved along. Microtubules also help to pull chromosomes apart during cell division. Continuously built, disassembled, and rebuilt, microtubules rarely last more than about 10 minutes in a cell. **Intermediate filaments,** a second type of cytoskeleton fiber, are durable, rope-like systems of numerous different overlapping proteins. They give cells great strength. **Microfilaments** are the thinnest elements in the cytoskeleton. Long, solid, rod-like fibers, microfilaments help generate forces, including those important in cell contraction and cell division.

CYTOSKELETON

FUNCTIONS
- Acts as the inner scaffolding of the cell
- Provides shape and support
- Controls intracellular traffic flow
- Enables movement

THREE TYPES OF PROTEIN FIBERS IN THE CYTOSKELETON

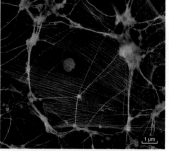

MICROTUBULES
- Thick, hollow tubes
- The tracks to which molecules and organelles within the cell may attach and be moved along

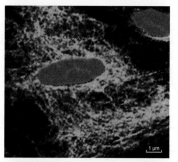

INTERMEDIATE FILAMENTS
- Durable, rope-like systems of numerous overlapping proteins
- Give cells great strength

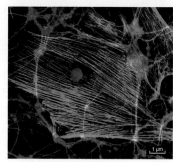

MICROFILAMENTS
- Long, solid rod-like fibers
- Help with cell contraction and cell division

FIGURE 3-29 **The cytoskeleton: the cell's inner scaffolding.**

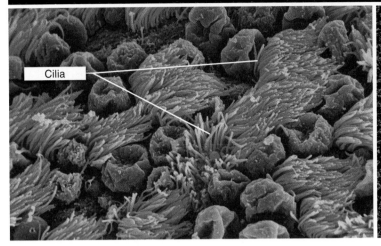

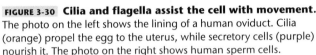

With gentle beating, cilia can move fluid past cells, whereas flagella, with whip-like motions, can move the cells themselves.

FIGURE 3-30 **Cilia and flagella assist the cell with movement.** The photo on the left shows the lining of a human oviduct. Cilia (orange) propel the egg to the uterus, while secretory cells (purple) nourish it. The photo on the right shows human sperm cells.

A couple of microtubule-based structures are sometimes present in cells and can help to move the cell through its environment (or in stationary cells, can help move the environment past the cell). **Cilia** (*sing.* **cilium**) are short projections often found in large numbers on a single cell (**FIGURE 3-30**). Cilia beat swiftly, often in unison and in ways that resemble blades of grass in a field, blowing in the wind. Cilia can move fluid along and past a cell. This movement can accomplish many important tasks, including sweeping the airways to our lungs to clear them of debris (such as dust) in the air we breathe.

Flagella are much longer than cilia. They occur in many prokaryotes and single-celled eukaryotes, and many algae and plants have cells with one or more flagella. But in animals, cell types with one or more flagella are very rare. One of these cell types, however, has a critical role in every animal species: sperm cells. With a flagellum for a tail, sperm are among the most mobile of all animal cells. Some spermicidal birth control methods prevent conception by disabling the flagellum and immobilizing the sperm cells.

TAKE-HOME MESSAGE 3·14

The inner scaffolding of the cell, which is made from proteins, is the cytoskeleton. Consisting of three types of protein fibers—microtubules, intermediate filaments, and microfilaments—the cytoskeleton gives animal cells their shape and support, gives cells some ability to control their movement, and serves as a series of tracks on which organelles and molecules are guided across and around the inside of the cell.

3·15 Mitochondria are the cell's energy converters.

Cars have it easy: we generally put only a single type of fuel in them, and it is exactly the same fuel every time. With human bodies it's a different story. During some meals we put in meat and potatoes. Other times, fruit or bread or vegetables, popcorn and gummy bears, pizza and ice cream. Yet no matter what we eat, we expect our body to utilize the energy in these various foods to power all the reactions that make it possible for us to breathe, move, and think. The

mitochondria are the organelles that make this possible (**FIGURE 3-31**).

Mitochondria (*sing.* **mitochondrion**) are the cell's all-purpose energy converters, and they are present in nearly all plant cells, animal cells, and every other eukaryotic cell. Our mitochondria allow us to convert the energy contained in the chemical bonds of carbohydrates, fats, and proteins into

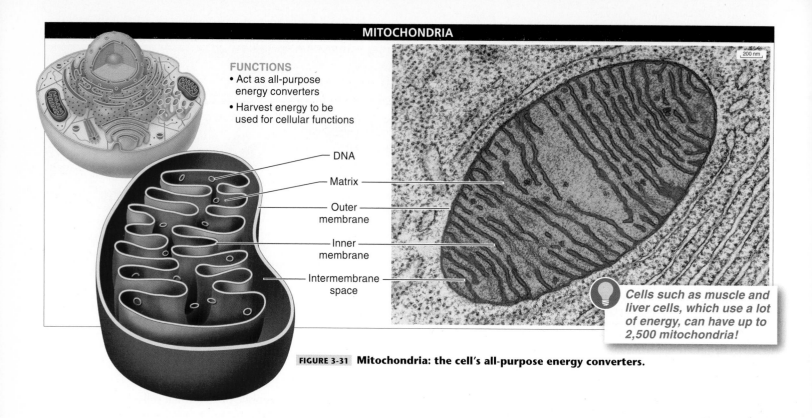

FUNCTIONS
- Act as all-purpose energy converters
- Harvest energy to be used for cellular functions

DNA

Matrix

Outer membrane

Inner membrane

Intermembrane space

200 nm

Cells such as muscle and liver cells, which use a lot of energy, can have up to 2,500 mitochondria!

FIGURE 3-31 Mitochondria: the cell's all-purpose energy converters.

carbon dioxide, water, and ATP. Cells use high-energy ATP molecules to fuel all their functions and activities. (ATP and how it works are described in detail in Chapter 4.) Because this energy conversion requires a significant amount of oxygen, mitochondria consume most of the oxygen used by each cell. In humans, for example, our mitochondria consume as much as 80% of the oxygen we breathe. Mitochondria give a significant return on this investment by producing about 90% of the energy our cells need to function.

As with so many aspects of our bodies, form follows function. Cells that are not very metabolically active, such as some fat storage cells in humans, have very few mitochondria. Other cells, such as muscle, liver, and sperm cells in animals and fast-growing root cells in plants—all of which have large energy requirements—are packed densely with mitochondria (**FIGURE 3-32**). Some of these cells have as many as 2,500 mitochondria! (See Section 3-16 for a description of an experimental approach to documenting how a cell's structure can change when the cell's function must change.)

To visualize the structure of a mitochondrion, imagine a plastic sandwich bag. Now take another,

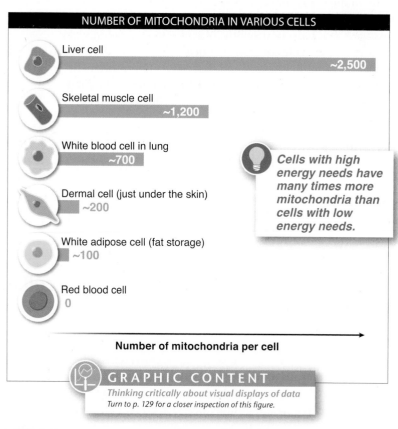

NUMBER OF MITOCHONDRIA IN VARIOUS CELLS

Liver cell — ~2,500

Skeletal muscle cell — ~1,200

White blood cell in lung — ~700

Dermal cell (just under the skin) — ~200

White adipose cell (fat storage) — ~100

Red blood cell — 0

Number of mitochondria per cell

Cells with high energy needs have many times more mitochondria than cells with low energy needs.

GRAPHIC CONTENT
Thinking critically about visual displays of data
Turn to p. 129 for a closer inspection of this figure.

FIGURE 3-32 How does the number of mitochondria vary among different types of cells?

bigger plastic bag and stuff it inside the sandwich bag. That's the structure of mitochondria: there is a smooth outer membrane and a scrunched-up inner membrane. This construction forms two separate compartments within the mitochondrion: a region outside the inner plastic bag (called the **intermembrane space**) and another region, called the **mitochondrial matrix,** inside the inner plastic bag. This bag-within-a-bag structure has important implications for energy conversion. And having a heavily folded inner membrane that is much larger than the outer membrane provides a huge amount of surface area on which to conduct chemical reactions. (We discuss the details of mitochondrial energy conversions and the role of mitochondrial structure in Chapter 4.)

As we learned in our earlier discussion of endosymbiosis, mitochondria may very well have existed, billions of years ago, as separate, single-celled, bacteria-like organisms. They are similar to bacteria in size and shape, and may have originated when symbiotic bacteria took up permanent residence within other cells. Perhaps the strongest evidence for this is that mitochondria have their own DNA (see Figure 3-31). Anywhere from 2 to 10 copies of its own little ring-shaped DNA are mixed in among the approximately 3,000 proteins in each mitochondrion. This DNA carries the instructions for making 13 important mitochondrial proteins necessary for metabolism and energy production.

We're always taught that our mothers and fathers contribute equally to our genetic composition, but this isn't quite true. The mitochondria in every one of your cells (and the DNA that comes with them) come from the mitochondria that were initially present in your mother's egg, which, when fertilized by your father's sperm, developed into you. In other words, all of your mitochondria are descended from your mother's mitochondria. The tiny sperm contributes DNA, but no cytoplasm and, hence, no mitochondria.

Consequently, mitochondrial DNA is something that we inherit exclusively from our mothers. This is true not only in humans but in most multicellular eukaryotes. As the fertilized egg develops into a two-celled, then four-celled, embryo, the mitochondria split themselves by a process called fission, the same process of division and DNA duplication used by bacteria—so there are always a sufficient number of mitochondria for the newly produced cells. This similarity between mitochondria and bacteria is another characteristic that supports the theory that mitochondria were originally symbiotic bacteria.

Given the central role of mitochondria in converting the energy in food molecules into a form that is usable by cells, you won't be surprised to learn that mitochondrial malfunctions can have serious consequences. Recent research has focused on possible links between defective mitochondria and diseases characterized by fatigue and muscle pain. It seems, for instance, that many cases of "exercise intolerance"—extreme fatigue or cramps after only slight exertion—may be related to defective mitochondrial DNA.

> **Q** We all have more DNA from one parent than the other. Who is the bigger contributor: mom or dad? Why?

TAKE-HOME MESSAGE 3·15

In mitochondria, which are found in nearly all eukaryotic cells, the energy contained in the chemical bonds of carbohydrate, fat, and protein molecules is converted into carbon dioxide, water, and ATP, the energy source for all cellular functions and activities. Mitochondria may have their evolutionary origins as symbiotic bacteria living inside other cells.

? **3·16 THIS IS HOW WE DO IT**

Can cells change their composition to adapt to their environment?

Figuring out how to approach a question in biology can be one of the most important steps in trying to understand how things work. In most cases, the best strategy is to take the simplest approach. Researchers wondered, for example, how an animal's cells might

respond to exposure to a much colder environment. Here's what they did.

They used 15 cats from five litters. From each litter, one cat served as a control, while the others were exposed to extreme cold—a temperature of −30° C—for two periods of 1 hour per day. The rest of the time, all of the cats were kept at 20° C.

| **Why did the researchers use multiple cats from each litter?**

After one week of this treatment, the team collected samples of fat tissue from all of the cats, which they examined microscopically. In the fat tissue, they measured the number and size of mitochondria, the number of tiny blood vessels (capillaries) per cell, and the size of the cells.

| **What was the purpose of counting the capillaries?**

The results were clear and dramatic, as shown here (note that a micrometer, μm, also called a micron, is one-millionth of a meter).

	Control	Cold-stressed
Number of mitochondria (per μm³)	1.48 ± 0.11	1.74 ± 0.10
Size of mitochondria (μm³)	0.13 ± 0.04	0.48 ± 0.13
Number of capillaries (per cell)	0.34 ± 0.12	0.71 ± 0.2
Cell size (μm)	75 ± 2	18 ± 2

The experimental approach was elegantly simple. The researchers used just a single manipulation—exposure to cold—and then observed the consequences. This made it straightforward for drawing conclusions. Cold stress caused a change in fat cells. In the animals exposed to cold, the mitochondria became more numerous, with each tripling in size. The number of blood vessels doubled. And the fat cell size was reduced to less than a quarter of the size found in the cells of the control group animals.

| **Why do you think the fat cells in the cats exposed to cold shrank so much?**

Essentially, it appears that cells in the fat tissue of cold-exposed animals upgraded their heat-generating system. And by making use of stored lipids to fuel the intensive heat production, they depleted a significant portion of their energy reserves.

| **What change in the study would increase your confidence in the conclusions?**

TAKE-HOME MESSAGE 3·16

Form follows function in an organism's cells and reflects their environment. When cells must perform intensive heat production, for example, they significantly increase the number and size of their mitochondria. They also increase the blood supply to the tissue and make use of existing stores of energy.

3·17 Lysosomes are the cell's garbage disposals.

Garbage. What does a cell do with all the garbage it generates? Mitochondria wear out after about 10 days of intensive activity, for starters. And white blood cells constantly track down and consume bacterial invaders, which they then have to dispose of. Similarly, the thousands of ongoing reactions of cellular metabolism produce many waste macromolecules that cells must digest and recycle. Many eukaryotic cells deal with this garbage by maintaining hundreds of versatile floating "garbage disposals" called lysosomes (**FIGURE 3-33**).

Lysosomes are round, membrane-enclosed, acid-filled vesicles that dispose of garbage. They are filled with about 50 different digestive enzymes and a super-acidic fluid, a corrosive broth so powerful that if a lysosome were to burst, it would almost immediately kill the cell by rapidly digesting all of its component parts. The selection of enzymes in the lysosome represents a broad spectrum of chemicals designed for dismantling macromolecules that are no longer needed by the cell or are generated as by-products of cellular metabolism.

Some of the enzymes break down lipids, others carbohydrates, others proteins, and still others nucleic acids. Consequently, when a cell consumes a particle of food or even an invading bacterium via phagocytosis, the cell directs

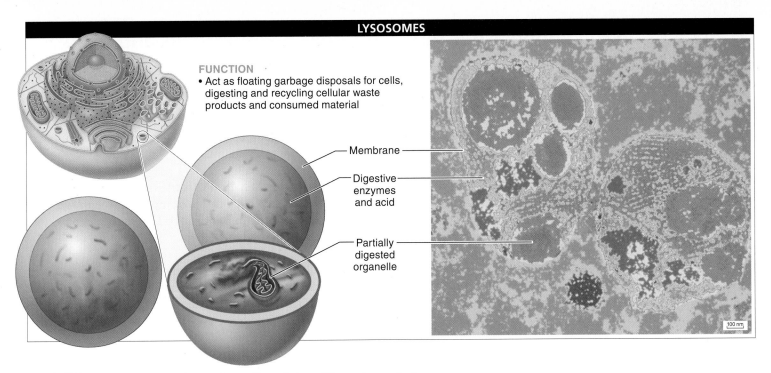

FUNCTION
• Act as floating garbage disposals for cells, digesting and recycling cellular waste products and consumed material

Membrane

Digestive enzymes and acid

Partially digested organelle

100 nm

FIGURE 3-33 **Lysosomes: digestion and recycling of the cell's waste products.**

the material to lysosomes for dismantling. And, ever the efficient system, the cell releases most of the component parts of molecules that are digested, such as the amino acids from proteins, back into its cytoplasm, where they can be reused by the cell as raw materials.

Some immune system cells tend to have particularly large numbers of lysosomes, most likely because of their great need for disposing of the by-products of disease-causing bacteria.

With 50 different enzymes necessary for lysosomes to carry out their metabolic salvaging act, malfunctions sometimes occur. In a common genetic disorder called Tay-Sachs disease, an individual inherits an inability to produce a critical lipid-digesting enzyme. Even though the lysosomes cannot digest certain lipids, the cells continue to send lipids to the lysosomes, where they accumulate, undigested. The

lysosome swells until it bursts and digests the whole cell or until it chokes the cell to death. This process occurs in large numbers of cells within the first few years of life, and eventually leads to the child's death.

Although the matter has long been debated among biologists, it does seem that plant cells also contain lysosomes, compartments with similar digestive broths that carry out the same digestive processes as in animals.

TAKE-HOME MESSAGE 3·17

Lysosomes are round, membrane-enclosed, acid-filled organelles that function as a cell's garbage disposals. They are filled with about 50 different digestive enzymes and enable a cell to dismantle macromolecules, including disease-causing bacteria.

3·18 In the endoplasmic reticulum, cells build proteins and lipids and disarm toxins.

The information for how to construct the molecules essential to a cell's smooth functioning and survival is stored in the DNA found in the cell's nucleus. The energy used to construct these molecules and to run cellular functions comes primarily from the mitochondria. The actual production and modification of biological molecules, however, occurs in a system of organelles called the **endomembrane system** (FIGURE 3-34). This mass of interrelated membranes spreads out from and surrounds the nucleus, forming chambers within the cell that contain their own mixtures of chemicals. The endomembrane system takes up as much as one-fifth of the cell's volume and is responsible for many of the fundamental functions of the cell.

Polypeptide chains are assembled into functional proteins and lipids are produced within the membrane-enclosed compartments of the endoplasmic reticulum. Additionally, products for export to other parts of the body are modified and packaged here, and many of the toxic chemicals that find their way into our bodies—from recreational drugs to antibiotics—are broken down and neutralized in the endomembrane system.

Because the process of protein production follows a somewhat linear path, we explore the endomembrane system sequentially, beginning just outside the nucleus with the endoplasmic reticulum.

Rough Endoplasmic Reticulum Perhaps the organelle with the most cumbersome name, the **rough endoplasmic reticulum,** or **rough ER** ("endoplasmic reticulum" is derived from the Greek for "within" and "anything molded" and the Latin for "small net"), is a large series of interconnected, flattened sacs that look like a stack of pancakes. These sacs are connected directly to the nuclear envelope. In most eukaryotic cells, the rough ER almost completely surrounds the nucleus (FIGURE 3-35). It is called "rough" because its surface is studded with little bumps. These bumps are ribosomes, the cell's protein-making machines, and cells with high rates of protein production generally have large numbers of ribosomes. We cover the details of ribosome structure and protein production in Chapter 5.

The primary function of the rough ER is to fold and package proteins that will be shipped to other locations in the endomembrane system, on the cell surface, or outside the cell. Poisonous frogs, for example, package their poison in the rough ER of the cells where it is produced before transporting it to the poison glands on their skin. Proteins that are used within the cell itself are generally produced on free-floating ribosomes in the cytoplasm.

Smooth Endoplasmic Reticulum As its name advertises, the **smooth endoplasmic reticulum,** or **smooth ER,** is part of the endomembrane network that is smooth, because there are no ribosomes bound to it (FIGURE 3-36). Although it is connected to the rough ER, it is farther from the nucleus and, besides lacking ribosomes, differs slightly in appearance. Whereas the rough ER looks like stacks of pancakes, the smooth ER sometimes looks like a collection of branched tubes.

The smooth surface gives us the first hint that smooth ER has a different job than rough ER. Because ribosomes are absolutely essential for

OVERVIEW OF THE ENDOMEMBRANE SYSTEM

FUNCTIONS
• Produces and modifies molecules to be exported to other parts of the organism
• Breaks down toxic chemicals and cellular by-products

Rough endoplasmic reticulum

Smooth endoplasmic reticulum

Golgi apparatus

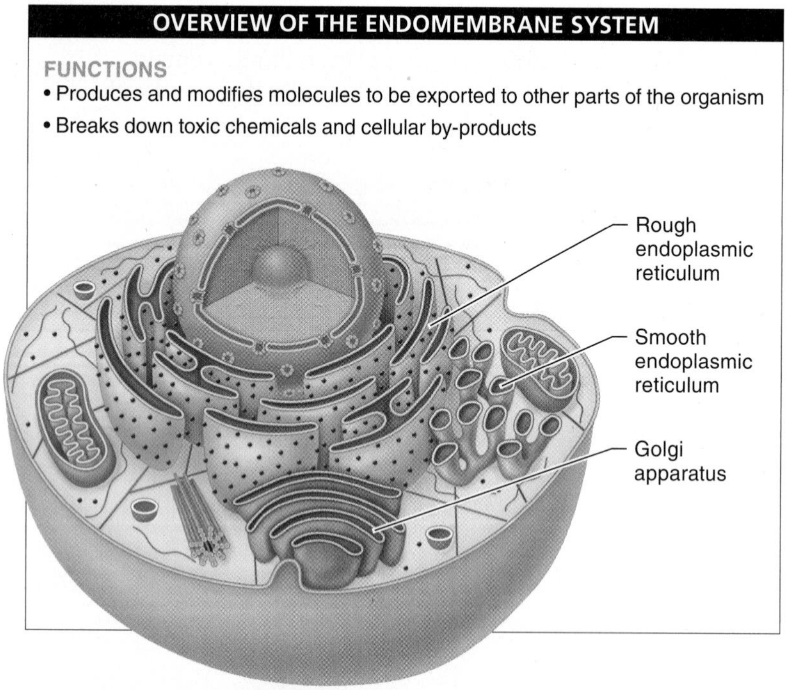

FIGURE 3-34 The endomembrane system: the rough endoplasmic reticulum, smooth endoplasmic reticulum, and Golgi apparatus.

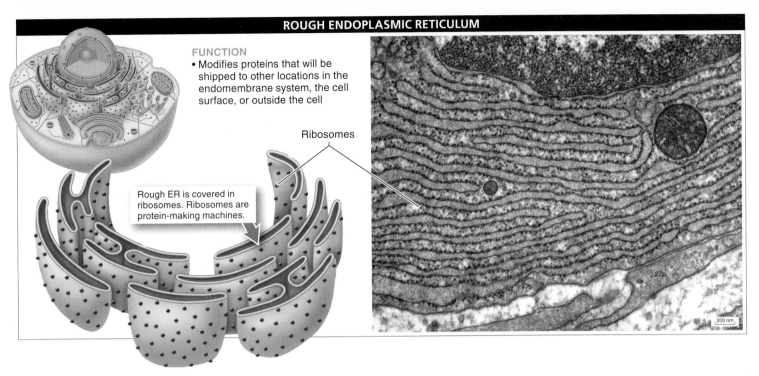

ROUGH ENDOPLASMIC RETICULUM

FUNCTION
- Modifies proteins that will be shipped to other locations in the endomembrane system, the cell surface, or outside the cell

Ribosomes

Rough ER is covered in ribosomes. Ribosomes are protein-making machines.

200 nm

FIGURE 3-35 **The rough endoplasmic reticulum is studded with ribosomes.**

protein production, we can guess that, with few ribosomes, the smooth ER is not extensively involved in folding or packaging proteins. It isn't. Instead, it primarily takes part in the synthesis of lipids such as fatty acids, phospholipids, and steroids, as well as carbohydrates. Exactly which lipids are produced varies throughout the organism and across plant

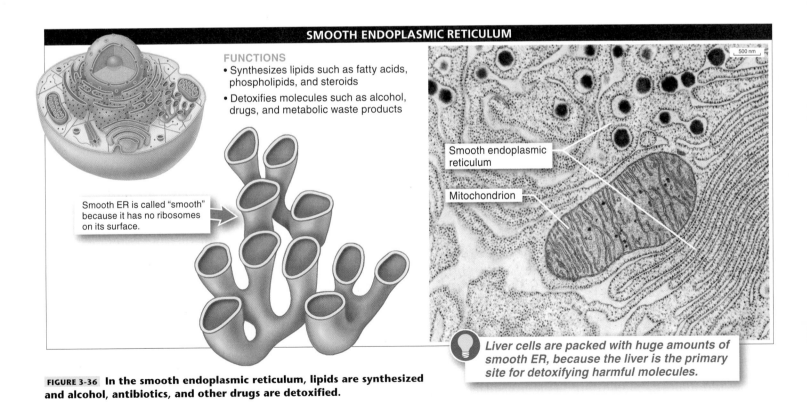

SMOOTH ENDOPLASMIC RETICULUM

FUNCTIONS
- Synthesizes lipids such as fatty acids, phospholipids, and steroids
- Detoxifies molecules such as alcohol, drugs, and metabolic waste products

500 nm

Smooth endoplasmic reticulum

Mitochondrion

Smooth ER is called "smooth" because it has no ribosomes on its surface.

Liver cells are packed with huge amounts of smooth ER, because the liver is the primary site for detoxifying harmful molecules.

FIGURE 3-36 **In the smooth endoplasmic reticulum, lipids are synthesized and alcohol, antibiotics, and other drugs are detoxified.**

and animal species. Inside the smooth ER of mammalian ovaries and testes, for example, the hormones estrogen and testosterone are produced. Inside the smooth ER of liver and fat cells, other lipids are produced. Following the same packaging process that occurs for proteins in the rough ER, lipids produced by the smooth ER are packaged in transport vesicles and then sent to other parts of the cell or to the plasma membrane for export.

Another critical responsibility of the smooth ER—particularly the smooth ER in human liver cells—is to help protect us from the many dangerous molecules that get into our bodies. Alcohol, antibiotics, barbiturates, amphetamines, or other stimulants that we may consume, along with many toxic metabolic waste products formed in our bodies, are made less harmful by the detoxifying enzymes in the smooth ER. And just as a bodybuilder's muscles get bigger and bigger in response to weightlifting, our smooth ER proliferates in cells that are exposed to large amounts of particular drugs.

As we see over and over in living organisms, form follows function. Not surprisingly, then, just as cells with high rates of protein production have large numbers of ribosomes, we find huge amounts of smooth ER in liver cells because they are the primary sites of molecular detoxification. Other cells that are packed with ER (both rough and smooth) include plasma cells in the blood that produce immune system

Q How can long-term use of one drug increase your resistance to another, different drug that you have never encountered?

proteins for export and pancreas cells that secrete large amounts of digestive enzymes.

Chronic exposure to many drugs (from antibiotics to heroin) can induce a proliferation of smooth ER, particularly in the liver, and of the smooth ER's associated detoxification enzymes. This proliferation in turn increases tolerance to the drugs, necessitating higher doses to achieve the same effect. This increased detoxification capacity of the cells often enables them to better detoxify other compounds, even if the person has never been exposed to these chemicals. This increased detoxification capacity can lead to problems. An individual who has been exposed to large amounts of certain drugs, for example, may end up responding less well to antibiotics.

TAKE-HOME MESSAGE 3·18

The production and modification of biological molecules in eukaryotic cells occurs in a system of organelles called the endomembrane system, which includes, among other organelles, the rough and smooth endoplasmic reticulum. In rough ER, proteins that will be shipped elsewhere in the body are folded and packaged. In the smooth ER, lipids and carbohydrates are synthesized and alcohol, antibiotics, and other drugs are detoxified.

3·19 The Golgi apparatus processes products for delivery throughout the body.

Moving farther outward from the nucleus, we encounter another organelle within the endomembrane system: the Golgi (GOHL-jee) apparatus (**FIGURE 3-37**). The **Golgi apparatus** processes molecules synthesized in the cell—primarily proteins and lipids—and packages those that are destined for use elsewhere in the body. The Golgi apparatus is also a site of carbohydrate synthesis, including the complex polysaccharides found in many plasma membranes. The Golgi apparatus, which is not connected to the endoplasmic reticulum, is a flattened stack of membranes (each of which is called a Golgi body) that are *not* interconnected.

After transport vesicles bud from the endoplasmic reticulum, they move through the cytoplasm until they

reach the Golgi apparatus. Here, the vesicles fuse with the Golgi apparatus membrane and dump their contents into a Golgi body. The molecules pass through about four successive Golgi body chambers. In each Golgi body, enzymes make slight modifications to the molecules (such as the addition or removal of phosphate groups or sugars). The processing that occurs in the Golgi apparatus often involves tagging molecules (much like adding a postal address or tracking number) to direct them to some other part of the organism. After they are processed, the molecules bud off from the Golgi apparatus in a vesicle, which then moves into the cytosol. If the molecules are destined for delivery and use elsewhere in the body, the transport vesicle eventually fuses with the cell's plasma

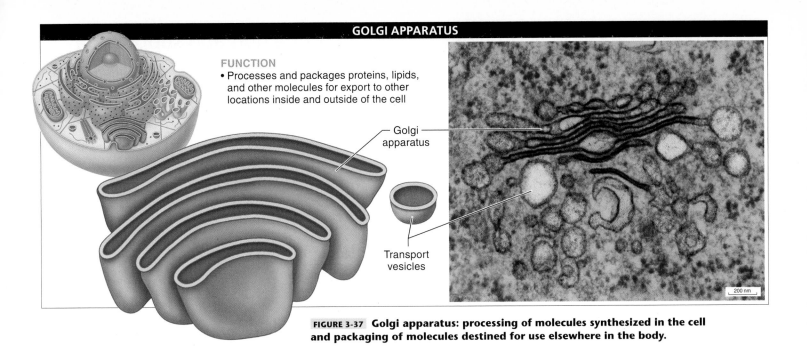

FUNCTION
• Processes and packages proteins, lipids, and other molecules for export to other locations inside and outside of the cell

Golgi apparatus

Transport vesicles

200 nm

FIGURE 3-37 **Golgi apparatus: processing of molecules synthesized in the cell and packaging of molecules destined for use elsewhere in the body.**

membrane and dumps the molecules into the bloodstream via exocytosis.

FIGURE 3-38 shows how the various parts of the endomembrane system work to produce, modify, and package molecules within a cell.

TAKE-HOME MESSAGE 3·19

The Golgi apparatus—another organelle within the endomembrane system—processes molecules synthesized in the cell and packages those that are destined for use elsewhere in the body.

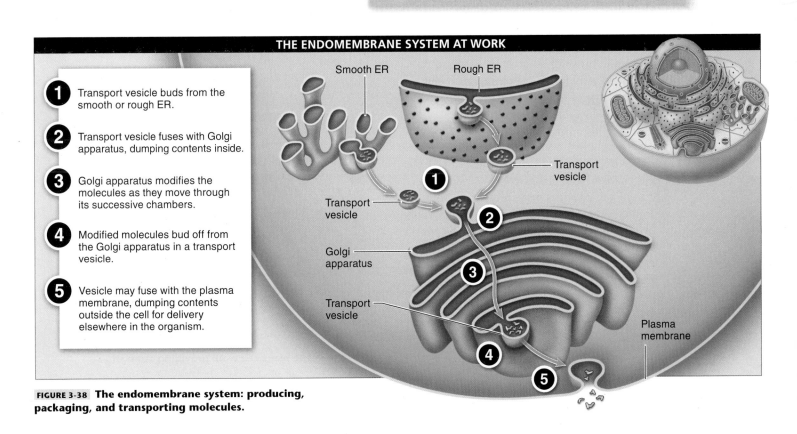

THE ENDOMEMBRANE SYSTEM AT WORK

Smooth ER   Rough ER

1 Transport vesicle buds from the smooth or rough ER.

2 Transport vesicle fuses with Golgi apparatus, dumping contents inside.

3 Golgi apparatus modifies the molecules as they move through its successive chambers.

4 Modified molecules bud off from the Golgi apparatus in a transport vesicle.

5 Vesicle may fuse with the plasma membrane, dumping contents outside the cell for delivery elsewhere in the organism.

Transport vesicle

Transport vesicle

Golgi apparatus

Transport vesicle

Plasma membrane

FIGURE 3-38 **The endomembrane system: producing, packaging, and transporting molecules.**

3·20 The cell wall provides additional protection and support for plant cells.

Until this point, we've discussed organelles and structures that are common to both plants and animals. Now we're going to look at several structures that are not found in all eukaryotic cells. Because plants are rooted in the ground, unable to move, they have several special needs beyond those of animals. In particular, they can't outrun predators or outmaneuver their competitors to get more food. If they need more light, they have few options beyond simply growing larger or taller. And to resist plant-eating animals, they must simply reduce their edibility.

One structure that helps plants achieve these goals is the cell wall, which surrounds the plasma membrane. The cell wall is made largely from long fibers of a polysaccharide called cellulose. Note that although animal cells do not have cell walls, some non-plants such as archaea, bacteria, protists, and fungi also have cell walls; their cell walls are made of polysaccharides other than cellulose.

In plants, cell wall production can be a multistage process. Initially, when the cell is still growing, a primary cell wall is laid down, along with some glue that helps adjacent cells adhere to each other. Sometime later in life, a secondary cell wall is laid down. This secondary cell wall usually contains a complex molecule called lignin, which gives the wall much of its strength and rigidity. Oddly, this strength remains even after the plant cell dies—hence wood's great value to humans as a construction material, among other things.

The cell wall is nearly 100 times thicker than the plasma membrane. The tremendous structural strength it confers on plant cells enables some plants to grow several hundred feet tall and provides some protection from insects and other animals that might eat the plant. Cell walls also help to make plants more water-tight—an important feature in organisms that cannot move out of the hot sun to reduce water loss through evaporation (**FIGURE 3-39**).

Surprisingly, despite its great strength, the cell wall does not completely seal off plant cells from one another. Rather, it is porous, allowing water and solutes to reach the plasma membrane. Additionally, the cells in most plants have anywhere from

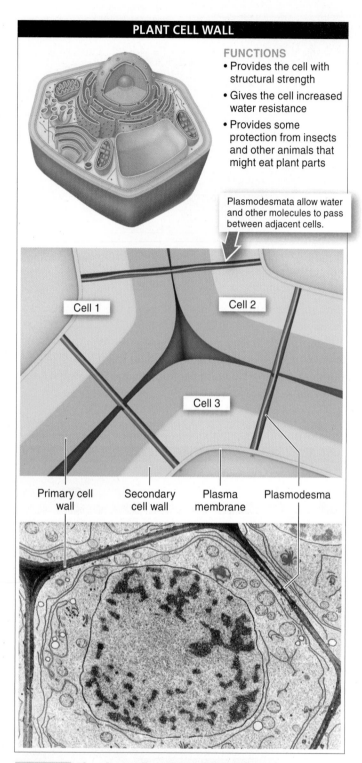

PLANT CELL WALL

FUNCTIONS
- Provides the cell with structural strength
- Gives the cell increased water resistance
- Provides some protection from insects and other animals that might eat plant parts

Plasmodesmata allow water and other molecules to pass between adjacent cells.

Cell 1   Cell 2   Cell 3

Primary cell wall   Secondary cell wall   Plasma membrane   Plasmodesma

FIGURE 3-39 **The plant cell wall: providing strength.**

1,000 to 100,000 microscopic tube-like channels, called **plasmodesmata** (*sing.* **plasmodesma**) (see Figure 3-39), connecting the cells to each other and enabling communication and transport between them. In fact, because so many of the cells are connected to one another, sharing cytoplasm and other molecules, some biologists have wondered whether we should consider a plant as just one big cell. How would you respond to that idea?

TAKE-HOME MESSAGE 3·20

The cell wall is an organelle found in plants (and in some other non-animal organisms). It is made primarily from the carbohydrate cellulose, and it surrounds the plasma membrane of the cell. The cell wall confers tremendous structural strength on plant cells, gives plants increased resistance to water loss, and provides some protection from insects and other animals that might eat the plant. In plants, plasmodesmata connect cells and enable communication and transport between them.

3·21 Vacuoles are multipurpose storage sacs for cells.

If you look at a mature plant cell through a microscope, one organelle, the **central vacuole,** usually stands out more than all the others because it is so huge and appears empty (**FIGURE 3-40**). Although it may look like an empty sac, the central vacuole is anything but. Surrounded by a membrane, filled with fluid, and occupying from 50% to 90% of a plant cell's interior space, the central vacuole can play an important role in five different areas of plant life. (Vacuoles are also found in some other eukaryotes, including some protists, fungi, and animals, but they tend to be particularly prominent in plant cells.)

1. **Nutrient storage.** The vacuole stores hundreds of dissolved substances, including amino acids, sugars, and ions.

2. **Waste management.** The vacuole retains waste products and degrades them with digestive enzymes, much like the lysosome in animal cells.

3. **Predator deterrence.** The poisonous, nasty-tasting materials that accumulate inside the vacuoles of some plants make a powerful deterrent to animals that might try to eat parts of the plant.

4. **Sexual reproduction.** The vacuole may contain pigments that give some flowers their red, blue, purple, or other colors, enabling them to attract birds and insects that help the plant reproduce by transferring pollen.

5. **Physical support.** High concentrations of dissolved substances in the vacuole can cause water to rush into the cells through the process of osmosis. The increased fluid pressure inside the vacuole can cause the cell to

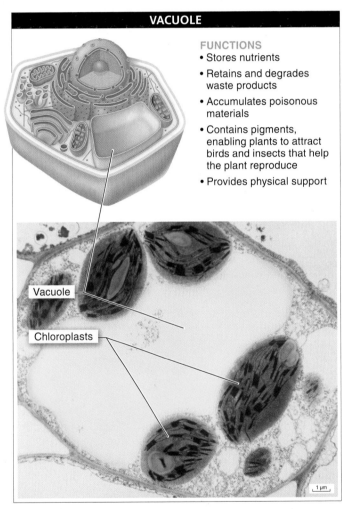

VACUOLE

FUNCTIONS
- Stores nutrients
- Retains and degrades waste products
- Accumulates poisonous materials
- Contains pigments, enabling plants to attract birds and insects that help the plant reproduce
- Provides physical support

Vacuole

Chloroplasts

1 μm

FIGURE 3-40 **The vacuole: multipurpose storage.**

enlarge a bit and push out the cell wall. This process is responsible for the pressure (called **turgor pressure**) that allows stems, flowers, and other plant parts to stand upright. The ability of non-woody plants to stand upright is due primarily to turgor pressure. Wilting is the result of a loss of turgor pressure.

TAKE-HOME MESSAGE 3·21

In plants, vacuoles can occupy most of the interior space of the cell. Vacuoles are also present in some other eukaryotic species. In plants, they function as storage spaces and play a role in nutrition, waste management, predator deterrence, reproduction, and physical support.

3·22 Chloroplasts are the plant cell's solar power plant.

It would be hard to choose the "most important organelle" in a cell, but if we had to, the chloroplast would be a top contender. The chloroplast, an organelle found in all plants and eukaryotic algae, is the site of photosynthesis—the conversion of light energy into the chemical energy of food molecules, with oxygen as a by-product. Because all photosynthesis in plants and algae takes place in chloroplasts, these organelles are directly or indirectly responsible for everything we eat and for the oxygen we breathe. Life on earth would be vastly different without the chloroplast (**FIGURE 3-41**).

In a green leaf, each cell has about 40–50 chloroplasts. This means there are about 500,000 chloroplasts per square millimeter of leaf surface (that's more than 200 million chloroplasts in an area the size of a small postage stamp). Chloroplasts are oval, somewhat flattened objects. On the outside, the chloroplast is encircled by two distinct layers of membranes (similar to the structure of mitochondria). The fluid in the inside compartment, called the **stroma,** contains some DNA and much protein-making machinery.

With a simple light microscope, it is possible to see little spots of green within the chloroplasts. Up close, these spots look like stacks of pancakes. Each stack consists of numerous interconnected little flattened sacs called **thylakoids,** and it is on the membranes of the thylakoids that the light-collecting for photosynthesis occurs. (In Chapter 4, we discuss the details of how chloroplasts convert the energy in sunlight into the chemical energy stored in sugar molecules.)

A peculiar feature of chloroplasts, mentioned in Section 3-3, is that they resemble photosynthetic bacteria, particularly with their circular DNA (which contains

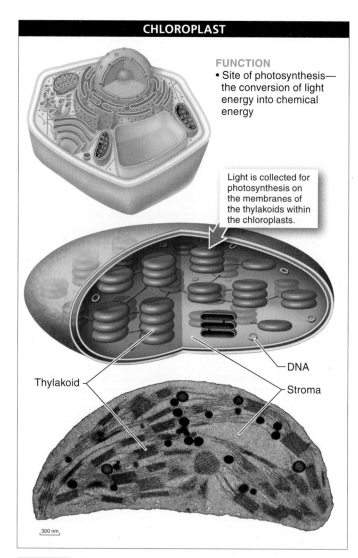

CHLOROPLAST

FUNCTION
• Site of photosynthesis— the conversion of light energy into chemical energy

Light is collected for photosynthesis on the membranes of the thylakoids within the chloroplasts.

DNA

Thylakoid

Stroma

300 nm

FIGURE 3-41 **The chloroplast: location of photosynthesis.**

REVIEW OF CELL STRUCTURES

STRUCTURE	ANIMALS	PLANTS	FUNCTION
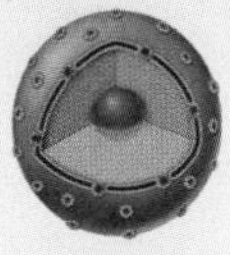 Nucleus	✓	✓	Directs cellular activity and stores hereditary information
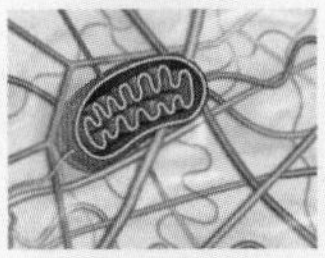 Cytoskeleton	✓	✓	Provides structural shape and support and enables cellular movement
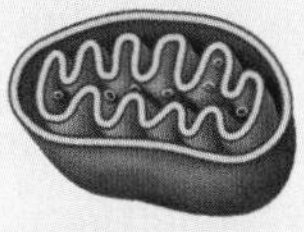 Mitochondrion	✓	✓	Harvests energy for cellular functions
 Lysosome	✓	✓	Digests and recycles cellular waste products and consumed material
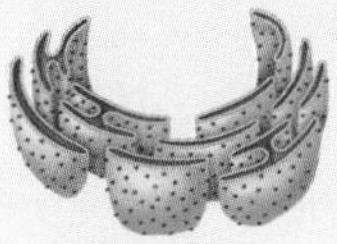 Rough ER	✓	✓	Modifies proteins that will be shipped elsewhere in the organism
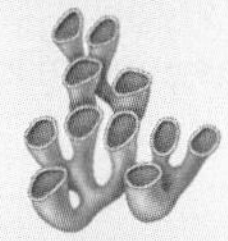 Smooth ER	✓	✓	Synthesizes lipids and detoxifies molecules
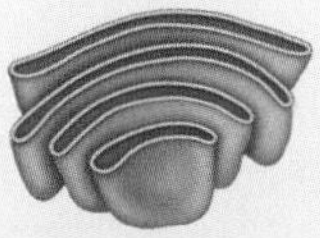 Golgi apparatus	✓	✓	Processes and packages proteins, lipids, and other molecules
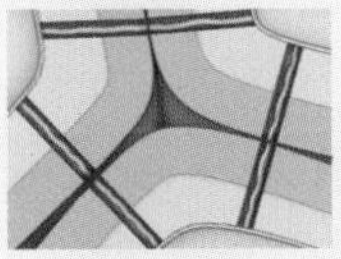 Cell wall	🚫	✓	Provides structural strength, protection, and increased resistance to water loss
 Vacuole	Sometimes	✓	Stores nutrients, degrades waste products, provides pigments and structural support
 Chloroplast	🚫	✓	Performs photosynthesis

FIGURE 3-42 **Review of major cellular structures and their functions.**

many of the genes essential for photosynthesis). Also, the dual outer membrane of the chloroplast is consistent with the idea that, long ago, a cell engulfed a photosynthetic bacterium, enveloping it with its plasma membrane in endocytosis. These features have given rise to the belief that chloroplasts might originally have been bacteria that were engulfed by a predatory cell. According to the endosymbiosis theory (see Section 3-3), the bacteria remained alive and, rather than becoming a meal, became the cell's meal ticket, providing food for the cell in exchange for protection.

FIGURE 3-42 summarizes the major parts of a cell, in animals and plants, and their functions.

TAKE-HOME MESSAGE 3·22

The chloroplast is the organelle in plants and algae that is the site of photosynthesis—the conversion of light energy into chemical energy, with oxygen as a by-product. Chloroplasts may originally have been bacteria that were engulfed by a predatory cell by endosymbiosis.

Drinking Too Much Water Can Be Dangerous!

People have long understood the risks of becoming dehydrated, particularly during activities such as marathon running in which a great deal of fluid can be lost through sweating. Only recently, however, has it been recognized that too much water—called "water intoxication"—can also be dangerous. In 2007, a 28-year-old woman competing in a water-drinking contest died just a few hours after consuming about two gallons of water without urinating. Similarly, the death of a runner during the 2002 Boston marathon was attributed to water intoxication. Numerous other cases of serious injury and death from water intoxication have been reported, resulting from water-drinking games, hazing, and extreme training programs.

Q: **When people sweat, what do they lose?** Sweat consists of water and some dissolved solids, primarily salt. As the sweat evaporates, the body is cooled.

Q: **How do people respond to nausea and dizziness, the warning signs of dehydration?** By drinking water. Because the water they drink lacks salt and other solutes, sodium imbalances can occur, particularly in the bloodstream and the fluid around cells (extracellular fluid).

Q: **If you drink lots (two gallons or more) of water—which has few or no solutes—where will it move once it gets into your bloodstream and extracellular fluid?** It moves by osmosis into your cells, where there is a higher concentration of solutes (primarily sodium) and a lower concentration of water molecules.

Q: **What happens to your cells if they absorb too much water?** They swell, sometimes to the point of rupturing. Such swelling and rupturing in the enclosed brain cavity can cause vomiting and confusion. Disastrously, these symptoms are often mistaken for the symptoms of dehydration and so the consumption of water is recommended, making the situation worse. Further swelling can lead to seizures, coma, and even death.

Q: **What should you do?** Because water is considered the least toxic chemical compound, many people mistakenly assume there are no dangers associated with consuming it. This includes runners. As a consequence, many marathon organizers have been reducing the number of water stops they offer runners during the race and testing symptomatic runners' sodium levels in medical tents along the course. Consuming salt tablets and salty snacks helps keep blood sodium levels in a healthy range.

Check Your Knowledge

Thinking critically about visual displays of data

1. What are the variables in this graph?

2. What additional information would make this figure more helpful? Why?

3. What can you conclude from this figure?

4. Is "number of mitochondria per cell" the best measure of a cell's "energy-generating capacity"? Can you think of a reason why this might not be a perfect measure? (Hint: muscle cells can be much, much larger than liver cells.)

5. Based on these data, can you make any of your own predictions?

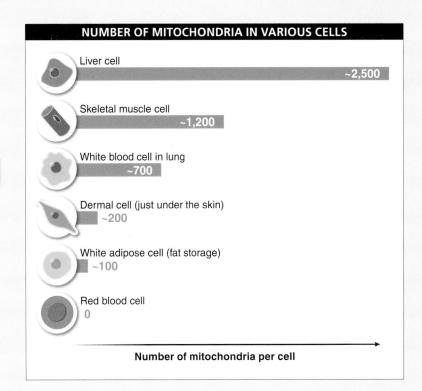

NUMBER OF MITOCHONDRIA IN VARIOUS CELLS

Liver cell — ~2,500

Skeletal muscle cell — ~1,200

White blood cell in lung — ~700

Dermal cell (just under the skin) — ~200

White adipose cell (fat storage) — ~100

Red blood cell — 0

Number of mitochondria per cell

See answers at the back of the book.

Key Terms in Cells

active transport, p. 106
cell, p. 86
cell theory, p. 87
cell wall, p. 90
central vacuole, p. 125
chloroplast, p. 92
cholesterol, p. 97
chromatin, p. 113
cilium (*pl.* cilia), p. 115
cytoplasm, p. 90
cytoskeleton, p. 114
cytosol, p. 90
desmosome, p. 110
diffusion, p. 102
endocytosis, p. 107
endomembrane system, p. 120
endosymbiosis theory, p. 91
enzymatic protein, p. 97
eukaryote, p. 89
eukaryotic cell, p. 89
exocytosis, p. 107
facilitated diffusion, p. 103
flagellum (*pl.* flagella), pp. 90, 115
fluid mosaic, p. 96
gap junction, p. 110
glycerol, p. 95

Golgi apparatus, p. 122
hydrophilic, p. 95
hydrophobic, p. 95
hypertonic, p. 104
hypotonic, p. 104
intermediate filaments, p. 114
intermembrane space, p. 117
invagination, p. 93
isotonic, p. 104
lysosome, p. 118
membrane enzymes, p. 97
microfilaments, p. 114
microtubules, p. 114
mitochondrial matrix, p. 117
mitochondrion (*pl.* mitochondria), pp. 92, 115
nonpolar, p. 95
nuclear membrane, p. 112
nucleolus, p. 113
nucleus, p. 91
organelle, p. 91
osmosis, p. 104
passive transport, p. 102
phagocytosis, p. 107
phospholipid, p. 95
phospholipid bilayer, p. 95
pilus (*pl.* pili), p. 90
pinocytosis, p. 107

plasma membrane, p. 89
plasmodesma (*pl.* plasmodesmata), p. 125
polar, p. 95
primary active transport, p. 106
prokaryote, p. 89
prokaryotic cell, p. 89
receptor protein, p. 96
receptor-mediated endocytosis, p. 107

recognition protein, p. 97
ribosome, p. 90
rough endoplasmic reticulum (rough ER), p. 120
secondary active transport, p. 106
simple diffusion, p. 103
smooth endoplasmic reticulum (smooth ER), p. 120

solute, p. 102
solvent, p. 102
stroma, p. 126
surface protein, p. 96
thylakoid, p. 126
tight junction, p. 110
tonicity, p. 104
transmembrane protein, p. 96
transport protein, p. 97
turgor pressure, p. 126
vacuole (central), p. 125

ABOUT THE CHAPTER OPENING PHOTO

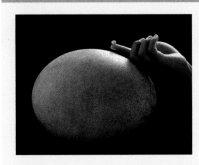

The hummingbird egg weighs just 0.02 ounce, while the egg of the elephant bird (extinct) has a volume of 2 gallons.

? Check Your Knowledge

1. Which of the following statements about the cell theory is correct?

   a) All living organisms are made up of one or more cells.

   b) All cells arise from other, preexisting cells.

   c) All eukaryotic cells contain symbiotic prokaryotes.

   d) All prokaryotic cells contain symbiotic eukaryotes.

   e) Both a) and b) are correct.

0 ——(11)—————— 100
EASY   HARD

2. Which of the following statements about prokaryotes is incorrect?

   a) Prokaryotes appeared on earth before eukaryotes.

   b) Prokaryotes have circular pieces of DNA within their nuclei.

   c) Prokaryotes contain cytoplasm.

   d) Prokaryotes contain ribosomes.

   e) Some prokaryotes can conduct photosynthesis.

0 —————————(51)—— 100
EASY   HARD

3. Which of the following facts supports the claim that mitochondria developed from bacteria that, long ago, were incorporated into eukaryotic cells by the process of phagocytosis?

   a) Mitochondria have flagella for motion.

   b) Mitochondria have proteins for the synthesis of ATP.

   c) Mitochondria are the "powerhouses" of the cell.

   d) Mitochondria are small and easily transported across cell membranes.

   e) Mitochondria have their own DNA.

0 ————————(37)———— 100
EASY   HARD

3·1–3·3 What is a cell?

The cell is the smallest unit of life that can perform all of the necessary activities of life. All living organisms are made up of one or more cells.

CELL THEORY

Cell theory is one of the unifying theories in biology, and one that is universally accepted by all biologists. The theory states that:

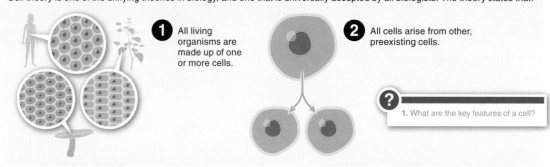

1 All living organisms are made up of one or more cells.

2 All cells arise from other, preexisting cells.

> ? 1. What are the key features of a cell?

PROKARYOTIC CELLS

TYPICAL PROKARYOTIC CELL FEATURES
- No nucleus—DNA is in the cytoplasm.
- Internal structures mostly not organized into compartments.
- Much smaller than eukaryotes.

Bacterial cell

> ? 2. Describe four structural features of prokaryotes.

EUKARYOTIC CELLS

TYPICAL EUKARYOTIC CELL FEATURES
- DNA contained in nucleus.
- Larger than prokaryotes—usually at least 10 times bigger.
- Cytoplasm contains specialized structures called organelles.

Animal cell

Plant cell

> ? 3. Describe two evolutionary mechanisms by which organelles may have originated.

3·4–3·7 Cell membranes are gatekeepers.

Every cell is enclosed by a plasma membrane, a two-layered structure that holds the contents of a cell in place and regulates what enters and leaves the cell.

THE PLASMA MEMBRANE

The plasma membrane is a fluid mosaic of proteins, lipids, and carbohydrates.

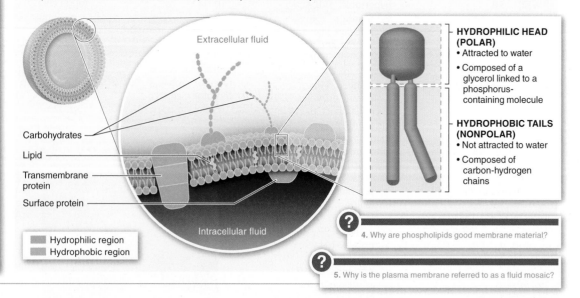

Extracellular fluid

Carbohydrates

Lipid

Transmembrane protein

Surface protein

Intracellular fluid

☐ Hydrophilic region
☐ Hydrophobic region

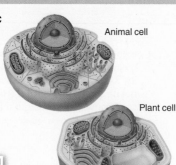

HYDROPHILIC HEAD (POLAR)
- Attracted to water
- Composed of a glycerol linked to a phosphorus-containing molecule

HYDROPHOBIC TAILS (NONPOLAR)
- Not attracted to water
- Composed of carbon-hydrogen chains

> ? 4. Why are phospholipids good membrane material?

> ? 5. Why is the plasma membrane referred to as a fluid mosaic?

FUNCTIONS OF PLASMA MEMBRANE MOLECULES

Plasma membrane molecules serve diverse roles.

RECEPTOR PROTEINS
Bind to external chemicals in order to regulate processes within the cell

RECOGNITION PROTEINS
Provide a "fingerprint" for the cell, so it can be recognized by other cells

TRANSPORT PROTEINS
Provide a passageway for molecules to travel into and out of the cell

MEMBRANE ENZYMES
Accelerate intracellular and extracellular reactions on the plasma membrane

CARBOHYDRATE CHAINS
Provide a "fingerprint" for the cell, so it can be recognized by other cells

CHOLESTEROL
Helps the membrane retain its flexibility

? 6. Why do chloride ions accumulate within some cells of individuals with cystic fibrosis?

? 7. Describe two ways in which HIV is commonly transmitted.

3·8–3·11 Molecules move across membranes in several ways.

Cells must import food molecules and other necessary materials from outside the cell and export metabolic waste and molecules produced for use elsewhere.

PASSIVE TRANSPORT

In passive transport, the molecular movement occurs spontaneously, without the input of energy. This generally occurs as molecules move down their concentration gradient.

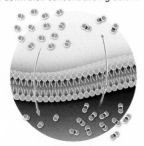

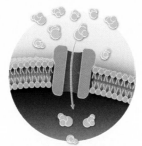

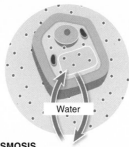

Water

SIMPLE DIFFUSION
Molecules pass directly through the plasma membrane without the assistance of another molecule.

FACILITATED DIFFUSION
Molecules move across the plasma membrane with the help of a carrier molecule.

OSMOSIS
Water molecules diffuse across a membrane until the concentration of water inside and outside the cell is equalized.

? 8. Why are osmosis and simple and facilitated diffusion classified as "passive" transport?

? 9. Describe the movement of water between a cell and a surrounding isotonic solution.

ACTIVE TRANSPORT

Active transport is necessary if the molecules to be moved are very large or if they are being moved against their concentration gradient, which requires energy in the form of ATP. Proteins embedded in the plasma membrane actively transport (pump) the molecules.

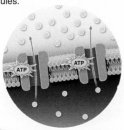

ATP   ATP

? 10. Why is energy required for active transport to take place?

ENDOCYTOSIS

The plasma membrane surrounds an object that is outside the cell, forming a little pocket called a vesicle.

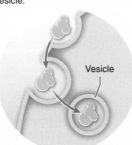

Vesicle

? 11. Describe the role of the plasma membrane in endocytosis.

EXOCYTOSIS

A vesicle within a cell fuses with the cell's plasma membrane. Vesicle contents are released outside the cell.

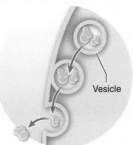

Vesicle

?

4. Hydrophobic molecules can pass freely through the plasma membrane, but ions and polar molecules are impeded by the hydrophobic core. For this reason, plasma membranes can be considered:
   a) partially permeable.
   b) impermeable.
   c) hydrophobic.
   d) hydrophilic.
   e) None of these terms properly describe plasma membranes.

0 — **15** — 100
EASY — HARD

5. Drugs called beta-blockers do all of the following except:
   a) reduce high blood pressure.
   b) block signaling through adrenaline receptors.
   c) reduce outward symptoms of anxiety.
   d) bind to the cytoplasmic side of a receptor protein.
   e) reduce the effects of adrenaline on the heart.

0 — **44** — 100
EASY — HARD

6. Cellular "fingerprints":
   a) are exposed on the cytoplasmic side of the membrane.
   b) are made from cholesterol.
   c) are "erased" by HIV.
   d) can help the immune system distinguish "self" from "non-self."
   e) All of the above are correct.

0 — **42** — 100
EASY — HARD

7. The movement of molecules across a membrane from an area of high concentration to one of low concentration is best described as:
   a) active transport.
   b) inactivated transport.
   c) passive transport.
   d) channel-mediated diffusion.
   e) electron transport.

0 — **31** — 100
EASY — HARD

8. The transport of water across a membrane from a solution of lower solute concentration to a solution of higher solute concentration is best described as:
   a) osmosis.
   b) facilitated diffusion.
   c) receptor-mediated transport.
   d) active transport.
   e) general diffusion.

0 — **38** — 100
EASY — HARD

131

9. In an experiment, you measure the concentration of a polar molecule inside and outside a cell. You find that the concentration is high and gradually increasing inside the cell. What is your best hypothesis for the process you are observing?

a) facilitated diffusion

b) passive transport

c) simple diffusion

d) active transport

e) endocytosis

0 — 44 — 100
EASY — HARD

10. Which of the following structures allow the passage of small molecules between animal cells?

a) nucleoli

b) tight junctions

c) desmosomes

d) gap junctions

e) black holes

0 — 38 — 100
EASY — HARD

11. The largest structure in a eukaryotic cell is the _____ and it is surrounded by _____ membrane(s).

a) nucleus; one

b) nucleus; two

c) Golgi apparatus; one

d) mitochondrion; two

e) mitochondrion; one

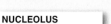

0 — 39 — 100
EASY — HARD

12. The cytoskeleton:

a) is a viscous fluid found in all cells.

b) fills a cell's nucleus but not the other organelles.

c) gives an animal cell shape and support, but cannot control movement.

d) helps to coordinate intracellular movement of organelles and molecules.

e) All of the above are correct.

0 — 55 — 100
EASY — HARD

13. Which of the following statements about mitochondria is correct?

a) Mitochondria are found in both eukaryotes and prokaryotes.

b) There tend to be more mitochondria in fat cells than in liver cells.

c) Most plant cells contain mitochondria.

d) Mitochondria may have originated evolutionarily as photosynthetic bacteria.

e) All of the above are correct.

0 — 70 — 100
EASY — HARD

3·12 Cells are connected and communicate with each other.

In multicellular organisms, most cells are connected to other cells by specialized structures that hold them in place and enable them to communicate with each other.

THREE DIFFERENT CONNECTIONS BETWEEN ANIMAL CELLS

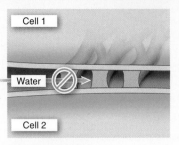

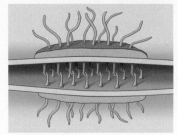

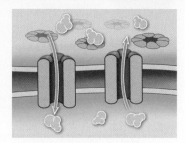

TIGHT JUNCTIONS
Form a water-tight seal between cells, like caulking around a tub

DESMOSOMES
Act like Velcro and fasten cells together

GAP JUNCTIONS
Act like secret passageways and allow materials to pass between cells

? 12. Where in the body would you expect to find cells with tight junctions? Why?

3·13–3·22 Nine important landmarks distinguish eukaryotic cells.

Specialized structures in cells perform specific life-sustaining functions.

NUCLEUS

The nucleus is the genetic control center of eukaryotic cells. It directs protein production and is the storehouse for all hereditary information.

NUCLEOLUS
Area of the nucleus where ribosomal subunits are assembled

CHROMATIN/ CHROMOSOMES
Thin fibers of DNA, which carry all hereditary information

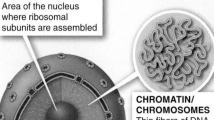

NUCLEAR MEMBRANE
Two bilayers, covered in pores, that surround the nucleus

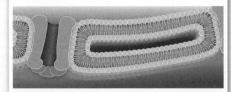

? 13. Describe the two primary functions of the nucleus.

CYTOSKELETON

The cytoskeleton is the inner scaffolding of a cell, giving it shape and support and serving as a series of tracks on which organelles and molecules are guided around the inside of the cell.

MICROTUBULES
• Thick, hollow tubes
• The tracks to which molecules and organelles within the cell may attach and be moved along

INTERMEDIATE FILAMENTS
• Durable, rope-like systems of numerous overlapping proteins
• Give cells great strength

MICROFILAMENTS
• Long, solid rod-like fibers
• Help with cell contraction and cell division

? 14. Describe the three types of protein fibers that make up the cytoskeleton.

MITOCHONDRIA

Mitochondria are found in virtually all eukaryotic cells and act as all-purpose energy converters, harvesting energy to be used for cellular functions.

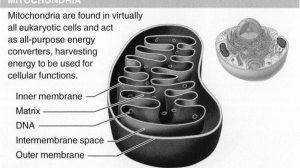

Inner membrane

Matrix

DNA

Intermembrane space

Outer membrane

? 15. Why do mitochondria consume most of the oxygen used by each cell?

LYSOSOMES

Lysosomes are acid-filled organelles that function as cellular garbage disposals.

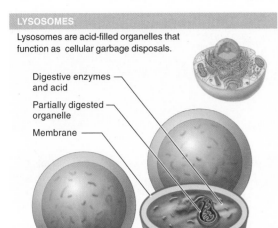

Digestive enzymes and acid

Partially digested organelle

Membrane

?
16. Describe how lysosomes function as a cell's "garbage disposal."

ENDOPLASMIC RETICULUM

The production and modification of biological molecules in eukaryotic cells occurs in a system of organelles called the endomembrane system, which includes the rough and smooth endoplasmic reticulum.

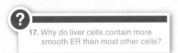

 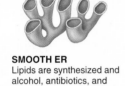

ROUGH ER
Folds and packages proteins to be shipped elsewhere in the body.

SMOOTH ER
Lipids are synthesized and alcohol, antibiotics, and other drugs are detoxified.

?
17. Why do liver cells contain more smooth ER than most other cells?

GOLGI APPARATUS

Another organelle in the endomembrane system, the Golgi apparatus processes and packages molecules destined for use elsewhere in the body.

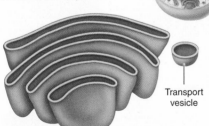

Transport vesicle

?
18. Name two molecules that are synthesized in the cell and packaged by the Golgi apparatus.

PLANT CELL WALL

The cell wall provides structural strength, increases resistance to water loss, and provides some protection from insects and other animals that might plant parts.

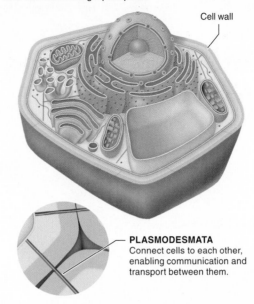

Cell wall

PLASMODESMATA
Connect cells to each other, enabling communication and transport between them.

VACUOLES

Vacuoles are storage spaces found in several eukaryotic taxa—and particularly prominent in plant cells—that play a role in nutrition, waste management, predator deterrence, reproduction, and physical support.

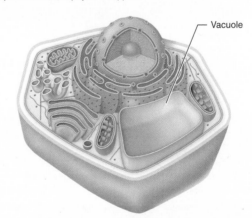

Vacuole

CHLOROPLASTS

The chloroplast is the organelle in plants and algae that is the site of photosynthesis—the conversion of light energy into chemical energy, with oxygen as a by-product.

Thylakoid

?

14. Which of the following organelles is not present in animal cells?
a) lysosome
b) Golgi apparatus
c) rough endoplasmic reticulum
d) mitochondrion
e) chloroplast

0 — **14** — 100
EASY — HARD

15. Given that a cell's structure reflects its function, what function would you predict for a cell with a large Golgi apparatus?
a) movement
b) secretion of digestive enzymes
c) transport of chemical signals
d) rapid replication of genetic material and coordination of cell division
e) attachment to bone tissue

0 — **59** — 100
EASY — HARD

16. Cell walls:
a) occur only in plant cells.
b) are not completely solid, having many small pores.
c) confer less structural support than the plasma membrane.
d) dissolve when a plant dies.
e) are made primarily from phospholipids.

0 — **66** — 100
EASY — HARD

17. Which of the following organelles is not found in both plant and animal cells?
a) nucleus
b) rough endoplasmic reticulum
c) mitochondrion
d) smooth endoplasmic reticulum
e) central vacuole

0 — **41** — 100
EASY — HARD

18. In plant cells, chloroplasts:
a) serve the same purpose that mitochondria serve in animal cells.
b) are the site of conversion of light energy into chemical energy.
c) play an important role in the breakdown of plant toxins.
d) have their own linear strands of DNA.
e) Both b) and d) are correct.

0 — **40** — 100
EASY — HARD

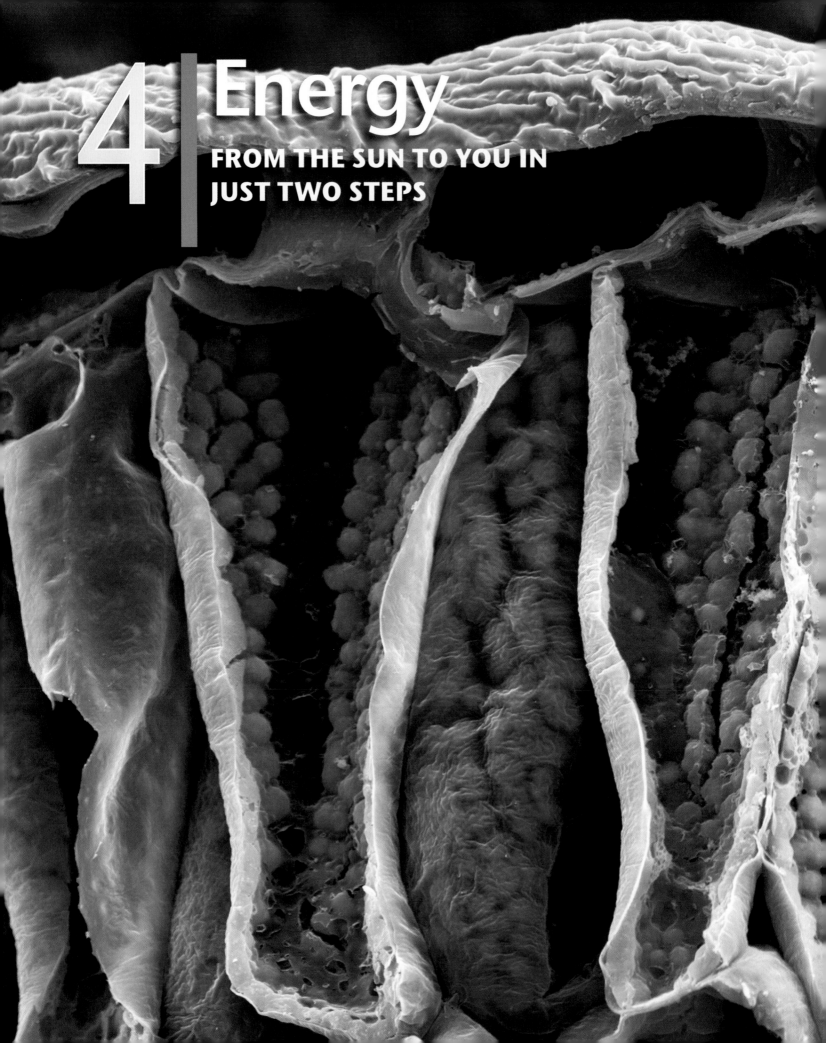

4 | Energy

FROM THE SUN TO YOU IN JUST TWO STEPS

Energy flows from the sun and through all life on earth.

Photosynthesis uses energy from sunlight to make food.

Cellular respiration converts food molecules into ATP, a universal source of energy for living organisms.

There are alternative pathways to energy acquisition.

4·1 – 4·4
Energy flows from the sun and through all life on earth.

Sunlight shines through the redwoods.

4·1 Cars that run on french fry oil? Organisms and machines need energy to work.

Imagine that you are on a long road trip. Your car's fuel gauge is nearing empty, so you pull off the highway. Instead of driving into a gas station, however, you head to the back of a fast-food restaurant and fill the fuel tank with used cooking grease. You head back to the highway, ready to drive several hundred miles before needing another pit stop.

Q *Humans can get energy from food. Can machines?*

Fast food for your car? Yes! The idea isn't as far-fetched as it sounds. In fact, on the roads of America today, many vehicles run on **biofuels,** fuels produced from plant and animal products (**FIGURE 4-1**). Most vehicles, however, run on **fossil fuels** such as gasoline. These fuels (which also include oil, natural gas, and coal) are produced from the decayed remains of plants and animals modified over millions of years by h·· t, pressure, and bacteria.

It turns out that biofuels, fossil fuels, and the food fuels that supply energy to most living organisms are chemically similar. This fact is not surprising because energy from the sun is the source of the energy stored in the chemical bonds between the atoms in all these fuels. Let's investigate how fuels provide energy.

When we burn gasoline, long chains of carbon and hydrogen atoms are broken down. As the bonds linking those atoms are broken and molecules with lower-energy bonds are formed, carbon dioxide (CO_2) and water are produced, and a lot of energy that was stored in the

chemical bonds holding each gasoline molecule together is released. (An interesting fact: the energy released in burning one gallon of gas is equivalent to the caloric content of 15 large cheese pizzas.) In an automobile engine, some of this released energy is harnessed to push pistons, spin a crankshaft, turn wheels, and move the car.

Animal fats and the oils in many plants—such as those used to cook french fries—share an important chemical feature with gasoline. Like gasoline, these fats and oils contain chains of carbon and hydrogen atoms bound together, and

FIGURE 4-1 **Biofuel technology.** Biofuels are chemically similar to fossil fuels.

just as with gasoline, breaking these bonds and forming new, lower-energy bonds releases large amounts of energy (and water and CO_2). If this released energy can be captured efficiently, it, too, can be used to push pistons and turn car wheels.

Cars that run on biofuels are more than just a technological trick. The production of biofuels requires only the plant or animal source, sunlight, air, water, and a relatively short amount of time—a few months or years, depending on the source. On the other hand, the production of fossil fuels such as coal or crude oil requires plant and animal remains and millions of years. This difference gives biofuels an important advantage over fossil fuels: they are a renewable resource. For this reason, biofuels point the way toward a future of reduced dependence on fossil fuels, the supplies of which are dwindling and the combustion of which has many harmful consequences, such as increasing global warming and releasing cancer-causing particles into the atmosphere.

Are we at the point yet where all our cars can run on biofuels? Not quite. There are several significant drawbacks to the increased use of biofuels, chief among them the destruction of forests, wetlands, and other ecologically important habitats resulting from the increased use of land to grow these fuels. (And fossil fuel and water are used in the process.) This is why the search for better fuels continues.

In this chapter we explore how plant, animal, and other living "machines" run on energy stored in chemical bonds. Nearly all life depends on capturing energy from the sun and converting it into forms that living organisms can use. This energy capture and conversion occurs in two important processes that mirror each other: (1) **photosynthesis,** the process by which plants capture energy from the sun and store it in the chemical bonds of sugars and (2) **cellular respiration,** the process by which all living organisms release the energy stored in the chemical bonds of food molecules and use it to fuel their lives (**FIGURE 4-2**). The sun to you in just two steps!

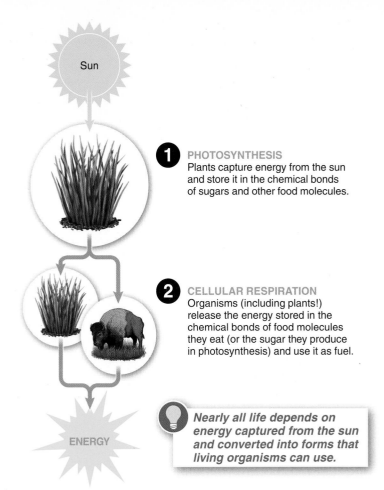

1 PHOTOSYNTHESIS
Plants capture energy from the sun and store it in the chemical bonds of sugars and other food molecules.

2 CELLULAR RESPIRATION
Organisms (including plants!) release the energy stored in the chemical bonds of food molecules they eat (or the sugar they produce in photosynthesis) and use it as fuel.

Nearly all life depends on energy captured from the sun and converted into forms that living organisms can use.

FIGURE 4-2 **Photosynthesis and cellular respiration.** The sun to you in just two steps!

TAKE-HOME MESSAGE 4·1

The sun is the source of the energy that powers most living organisms and other "machines." The energy from sunlight is stored in the chemical bonds of molecules. When these bonds are broken, energy is released, regardless of whether the bond is in a molecule of food, of a fossil fuel, or of a biofuel such as the oil in which french fries are cooked.

4·2 Energy has two forms: kinetic and potential.

"Batteries not included." For a child, those are pretty depressing words. We know that many of the toys and electronic gadgets that make our lives fun or useful (or both) need energy—usually in the form of batteries. Generating ringtones, lights, and movement requires energy. The same is true for humans, plants, and all other

living organisms: they need energy for their activities, from moving to reproducing to thinking.

Energy is the capacity to do work. And work is anything that involves moving matter against an opposing force. In the study of living things, we encounter two types of

KINETIC ENERGY

POTENTIAL ENERGY

FIGURE 4-3 **Two forms of energy.** Kinetic energy is the energy of motion; potential energy is energy stored in an object, such as water trapped behind a dam, or a skier poised at the top of a hill.

energy: kinetic and potential. **Kinetic energy** is the energy of motion. The kinetic energy of an object is the energy that it has due to its motion. Legs pushing bike pedals and birds flapping wings are examples of kinetic energy (FIGURE 4-3). Heat, which results from lots of molecules moving rapidly, is another form of kinetic energy. Because it comes from the movement of high-energy particles, light is also a form of kinetic energy—probably the most important form of kinetic energy on earth. (When we look at photosynthesis later in this chapter, we explore how sunlight is harnessed for producing food molecules.)

An object does not have to be moving to have the capacity to do work; it may have **potential energy,** which is stored energy that results from an object's location or position. Water behind a dam, for example, has potential energy. If a hole is opened in the dam, the water can flow through, and perhaps spin a waterwheel or turbine. A concentration gradient, which we discussed in Chapter 3, also has potential energy: if the molecules in an area of high concentration move toward an area of lower concentration, the potential energy of the gradient is converted to the kinetic energy of molecular movement, and this kinetic energy can do work. **Chemical energy,** the storage of energy in chemical bonds, is also a type of potential energy.

Because potential energy doesn't involve movement, it is a less obvious form of energy than kinetic energy. An apple has potential energy, as does any other type of food (FIGURE 4-4).

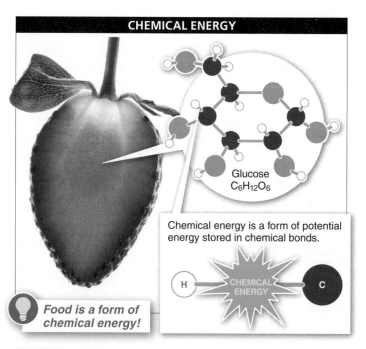

CHEMICAL ENERGY

Glucose
$C_6H_{12}O_6$

Chemical energy is a form of potential energy stored in chemical bonds.

H — CHEMICAL ENERGY — C

💡 *Food is a form of chemical energy!*

FIGURE 4-4 **The energy of chemical bonds.** The chemical bonds in food molecules are a form of potential energy.

Why? Because, during cellular respiration, the chemical energy stored in the chemical bonds making up the food can be released (when those bonds are broken and lower-energy bonds are re-formed), enabling you to run, play, and work. We explore cellular respiration, the energy-releasing breakdown of molecules, later in this chapter. But first we need to know more about the nature of energy.

TAKE-HOME MESSAGE 4·2

Energy, the capacity to do work, comes in two forms. Kinetic energy is the energy of moving objects, while potential energy, such as chemical energy, is *stored* energy that results from the position or location of an object.

4·3 As energy is captured and converted, the amount of energy available to do work decreases.

Every minute of every day—even on cloudy ones—the sun shines brightly, releasing tremendous amounts of energy. Organisms on earth cannot capture every bit of the sun's energy that makes it to the surface of the earth; indeed, most plants capture only a tiny fraction of the available energy. What happens to the rest? This unused energy does not simply disappear. Accountants would love to monitor the flow of energy because, as in a good accounting ledger,

all of the energy numbers add up perfectly. All energy from the sun can be accounted for. Some (probably less than 1%) is captured and transformed into usable chemical energy by organisms through photosynthesis (FIGURE 4-5). The rest of the energy from the sun is reflected back into space (probably about 30%) or is absorbed by land, the oceans, and the atmosphere (about 70%) and mostly transformed into heat. Heat is not easily harnessed to do work, however,

ENERGY TRANSFORMATIONS

KINETIC ENERGY TO POTENTIAL ENERGY

Light energy from the sun → Energy transformed into heat → Chemical energy stored in plants

POTENTIAL ENERGY TO KINETIC ENERGY

Chemical energy stored in muscles and liver → Energy transformed into heat → Kinetic energy of forward motion

FIGURE 4-5 **As energy is converted to do work, some energy is released as heat.**

THE NATURE OF ENERGY   PHOTOSYNTHESIS   CELLULAR RESPIRATION   ALTERNATIVE PATHWAYS

FIGURE 4-6 **Inefficient conversion.** Much of the energy used to fuel a combustion engine is converted to heat rather than forward motion.

and is therefore a much less useful form of energy than the energy transformed into chemical energy in plants (and stored as carbohydrates).

The same accounting also exists on a smaller level. If you eat a bowl of rice, some portion of the chemical energy stored in the bonds of the molecules that make up the rice grains is transformed into usable energy that can fuel your cells' activities. All the rest is transformed into heat and is ultimately lost into the atmosphere.

The fact that energy can change form but never disappear is an important feature of energy in the universe, whether we are looking at the sun and the earth or a human and her rice bowl. Just as energy can never disappear or be destroyed, energy can never be created. All the energy now present in the universe has been here since the universe began, and everything that has happened since then has occurred by the transformation of one form of energy into another. In all our eating and growing, driving and sleeping, we are simply transforming energy. The study of the transformation of energy from one type to another, such as from potential energy to kinetic energy, is called **thermodynamics,** and the **first law of thermodynamics** states that energy can never be created or destroyed. It can only change from one form to another.

Because plants capture less than 1% of the sun's energy, it might seem like they are particularly inefficient. But we humans are also rather inefficient at extracting the chemical energy of plants when we eat them. These inefficiencies occur because every time energy is converted

from one form to another, some of the energy is converted to heat. When a human converts the chemical energy in a plate of spaghetti into the kinetic energy of running a marathon, or when a car transforms the chemical energy of gasoline into the kinetic energy of forward motion, some energy is converted to heat, the least usable form of kinetic energy. In automobiles, for example, about three-quarters of the energy in gasoline is lost as heat (**FIGURE 4-6**).

The **second law of thermodynamics** states that every conversion of energy is not perfectly efficient and invariably includes the transformation of some energy into heat. Although heat is certainly a form of energy, it is almost completely useless to living organisms for fueling their cellular activity because it is not easily harnessed to do work. Put another way, the second law of thermodynamics tells us that although the quantity of energy in the universe is not changing, its quality is. Little by little, the amount of energy that is available to do work decreases. Now that we understand that organisms on earth cannot capture every single bit of energy released by the sun—and that energy conversions are inefficient—we can look at the chief energy currency of the cell: ATP.

TAKE-HOME MESSAGE 4·3

Energy is neither created nor destroyed but can change form. Each conversion of energy is inefficient, and some of the usable energy is converted to less useful heat energy.

4·4 ATP molecules are like free-floating rechargeable batteries in all living cells.

Much of the work that cells do requires energy. But even though light from the sun carries energy, as do molecules of sugar, fat, and protein, none of this light energy can be used directly to fuel chemical reactions in organisms' cells. First it must be captured in the bonds of a molecule called **adenosine triphosphate (ATP),** a free-floating molecule found in cells that acts like a rechargeable battery, temporarily storing energy that can then be used for cellular work in plants, animals, bacteria, and all the other organisms on earth. The use of ATP solves an important timing and coordination problem for living cells: a supply of ATP guarantees that the energy required for energy-consuming reactions will be available when it's needed.

ATP is a simple molecule with three components (**FIGURE 4-7**). At the center of the ATP molecule are two of these components: a small sugar molecule attached to a molecule called adenine. But it is the third component that makes ATP so effective in carrying and storing energy for a short time: attached to the sugar and adenine is a chain of three negatively charged phosphate groups (hence the "tri" in "triphosphate"). Because the bonds between these three phosphate groups must hold the groups together in the face of the three negative electrical charges that all repel one another, each of these bonds contains a large amount of

energy and is stressed and unstable. The instability of these high-energy bonds makes the three phosphate groups like a tightly coiled spring or a twig that is bent almost to the point of breaking. With the slightest push, one of the phosphate groups will pop off, displaced by water. And in the process, a little burst of energy is released that the cell can use.

It is precisely because each molecule of ATP is always on the brink of ejecting one of its phosphate groups that ATP is such an effective energy source inside a cell. As long as plenty of ATP molecules are around, they can energize the chemical reactions that make it possible for the cell to carry out the processes that require work, such as building muscle, repairing wounds, or growing roots. Each time a cell expends one of its ATP molecules to pay for an energetically expensive reaction, a phosphate is broken off and energy is released. What is left is a molecule with two phosphates, called ADP (adenosine *di*phosphate), and a separate phosphate group (labeled P_i).

An organism can then use ADP, a free-floating phosphate, and an input of kinetic energy to rebuild its ATP stocks (**FIGURE 4-8**). The kinetic energy is converted to potential energy when a free phosphate group attaches to the ADP molecule and makes ATP. In this manner, ATP functions

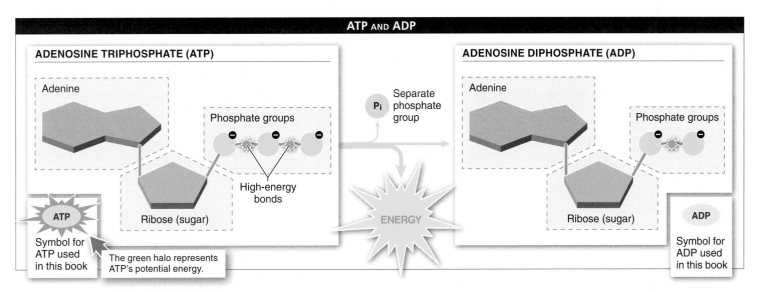

FIGURE 4-7 **The structure of ATP and ADP.** When ATP ejects one of its phosphate groups, energy is released as the ATP becomes ADP.

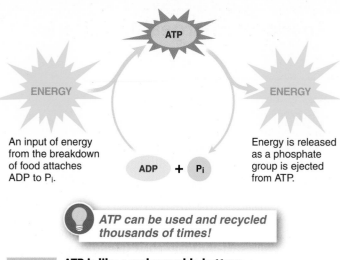

An input of energy from the breakdown of food attaches ADP to P$_i$.

ADP + P$_i$

Energy is released as a phosphate group is ejected from ATP.

💡 **ATP can be used and recycled thousands of times!**

FIGURE 4-8 **ATP is like a rechargeable battery.**

like a rechargeable battery. Where does the input of energy for recharging ATP come from? When we discuss photosynthesis, we'll see that plants, algae, and some bacteria can directly use light energy from the sun to make ATP from ADP and free-floating phosphate groups. These photosynthetic organisms can also use the energy generated from breaking down sugar and other molecules. That's the energy source that animals use, generating ATP from the energy contained in the bonds of their food molecules.

Whether it comes from the sun or from the breakdown of molecules such as sugar, the energy is used to re-create the unstable bond in the triphosphate chain of ATP. When energy is needed, the organism can again release it by breaking the bond holding the phosphate group to the rest of the molecule. Our bodies recycle ATP molecules in this way tens of thousands of times a day.

Here's the ATP story in a nutshell. Breaking down a molecule of sugar—in a glass of orange juice, for example—leads to a miniature burst of energy in one of your cells. The energy from the mini-explosion is put to work building the unstable high-energy bonds that attach phosphate groups to ADP molecules, creating new molecules of ATP. Later—perhaps only a fraction of a second later—when an energy-consuming reaction is needed, your cells can release the energy stored in the new ATP molecules.

TAKE-HOME MESSAGE 4·4

Cells temporarily store energy in the bonds of ATP molecules. This potential energy can be converted to kinetic energy and used to fuel life-sustaining chemical reactions. At other times, inputs of kinetic energy are converted to the potential energy of the energy-rich but unstable bonds in the ATP molecule.

4·5–4·11
Photosynthesis uses energy from sunlight to make food.

Lotus water lilies reach toward the sky.

4·5 Where does plant matter come from? Photosynthesis: the big picture.

Watching a plant grow over the course of a few years can seem like watching a miracle, or at least a very subtle magic trick. Of course it's neither, but the process is nonetheless amazing. Consider that in five years a tree can increase its weight by 150 pounds (68 kg) (**FIGURE 4-9**). Where does that 150 pounds of new tree come from?

Q When humans grow, the new tissue comes from food we eat. When plants grow, where does the new tissue come from?

Our first guess might be the soil. Could that be it? It's easy enough to weigh the soil in a pot when first planting a tree, and then weigh it again 5 years later. After 5 years, though, we find that the soil in our planter has lost less than a pound, nowhere near enough to explain the massive increase in the amount of plant material. Perhaps the new growth comes from the water? Wrong again. Although the older and much larger tree holds more water in its many cells, the water provided to the plant does not come close to accounting for the increase in the dry weight of the plant.

The amazing truth is that most of the new material comes from an invisible gas in the air. In the process of photosynthesis, plants capture carbon dioxide gas (CO_2) from the air. Using energy they get from sunlight, along with water and small amounts of chemicals usually found in soil, they produce solid, visible (and often tasty) sugars and other organic molecules that are used to make plant structures such as leaves, roots, stems, flowers, fruits, and seeds. In the process, the plants give off oxygen (O_2), a by-product that happens to be necessary for much of the life on earth—including all animal life!

FIGURE 4-9 **When plants grow, where does the new tissue come from?** From the dirt? From thin air?

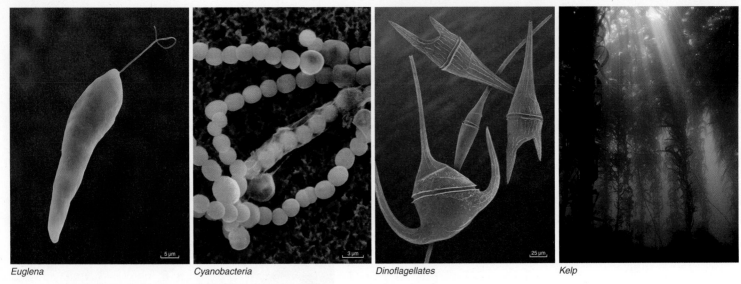

Euglena Cyanobacteria Dinoflagellates Kelp

FIGURE 4-10 **Plants aren't the only photosynthesizers.** Some bacteria and other unicellular organisms, along with kelp and other multicellular algae, are capable of photosynthesis.

Although plants are generally the most visible organisms that can capture light energy and convert it to organic matter, they are not the only organisms capable of photosynthesis. Some bacteria and many other unicellular organisms, along with kelp and other multicellular algae, are also capable of using the energy in sunlight to produce organic materials (FIGURE 4-10).

There are three inputs to the process of photosynthesis (FIGURE 4-11): light energy (from the sun), carbon dioxide (from the atmosphere), and water (from the ground). From these three inputs, the plant produces sugar and oxygen. As we'll see, photosynthesis is best understood as two separate events: a "photo" segment, during which light is captured, and a "synthesis" segment, during which sugar is built. In the "photo" reactions, light energy is captured and temporarily saved in energy-storage molecules. During this process, water molecules split and produce oxygen. In the "synthesis" reactions, the energy in the energy-storage molecules is used to assemble sugar molecules from carbon dioxide from the air.

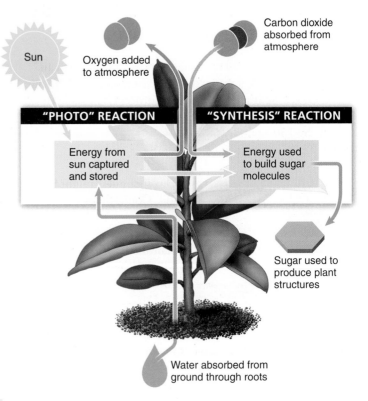

TAKE-HOME MESSAGE 4·5

Through photosynthesis, plants use water, the energy of sunlight, and carbon dioxide gas from the air to produce sugars and other organic materials. In the process, photosynthesizing organisms also produce oxygen, which makes all animal life possible.

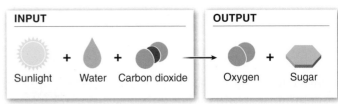

FIGURE 4-11 **Photosynthesis: the big picture.**

4·6 Photosynthesis takes place in the chloroplasts.

If a plant part is green, then you know it is photosynthetic. Leaves are green because the cells near the surface are packed full of **chloroplasts,** light-harvesting organelles, which make it possible for the plant to use the energy from sunlight to make sugars (their food) and other plant tissue (much of which animals use for food) (**FIGURE 4-12**). Other plant parts, such as stems, may also contain chloroplasts (in which case they, too, are capable of photosynthesis), but most chloroplasts are located within the cells in a plant's leaves.

Let's take a closer look at chloroplasts (**FIGURE 4-13**). The sac-shaped organelle is filled with a fluid called the **stroma.** Floating in the stroma is an elaborate system of interconnected membranous structures called **thylakoids,** which often look like stacks of pancakes. Once inside the chloroplast, you can be in one of two places: in the stroma or inside the thylakoids. The conversion of light energy to chemical energy—the "photo" part of photosynthesis—occurs inside the thylakoids. The production of sugars—taking place in the "synthesis" part of photosynthesis—occurs within the stroma.

We examine both the "photo" and the "synthesis" processes in greater detail later in this chapter. First, however, we examine the nature of light energy and of **chlorophyll,** the special molecule found in chloroplasts that makes the capture of light energy possible.

TAKE-HOME MESSAGE 4·6

In plants, photosynthesis occurs in chloroplasts, green organelles packed in cells near the plants' surfaces, especially in the leaves.

FIGURE 4-12 **Photosynthesis factories.** Cells near the leaf's surface are packed with photosynthetic chloroplasts.

Top edge of leaf

Photosynthetic cells packed with chloroplasts

Bottom edge of leaf

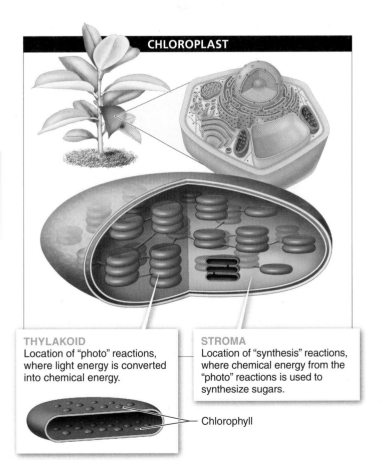

CHLOROPLAST

THYLAKOID
Location of "photo" reactions, where light energy is converted into chemical energy.

STROMA
Location of "synthesis" reactions, where chemical energy from the "photo" reactions is used to synthesize sugars.

Chlorophyll

FIGURE 4-13 **Chloroplast structure.** The chloroplast is where photosynthesis takes place in a plant.

Light energy travels in waves: plant pigments absorb specific wavelengths.

You can't eat sunlight. That's because sunlight is *light* energy rather than the chemical energy found in the bonds of food molecules. Photosynthesis is powered by **light energy,** a type of kinetic energy made up of little energy packets called **photons,** which are organized into waves. Photons can do work as they bombard surfaces such as your face (heating it) or a leaf (enabling it to build sugar from carbon dioxide and water).

Photons have various amounts of energy, and the length of the wave in which they travel corresponds to the amount of energy carried by the photon. The shorter the wavelength, the more energy the photon carries. Within a ray of light, there are super-high-energy photons (those with short wavelengths), relatively low-energy photons (those with longer wavelengths), and everything in between. This range, which is called the **electromagnetic spectrum,** extends from extremely short, high-energy gamma rays and X rays, with wavelengths as short as 1 nanometer (nm; a human

hair is about 50,000 nm in diameter), to very long, low-energy radio waves, with wavelengths as long as a mile (**FIGURE 4-14**).

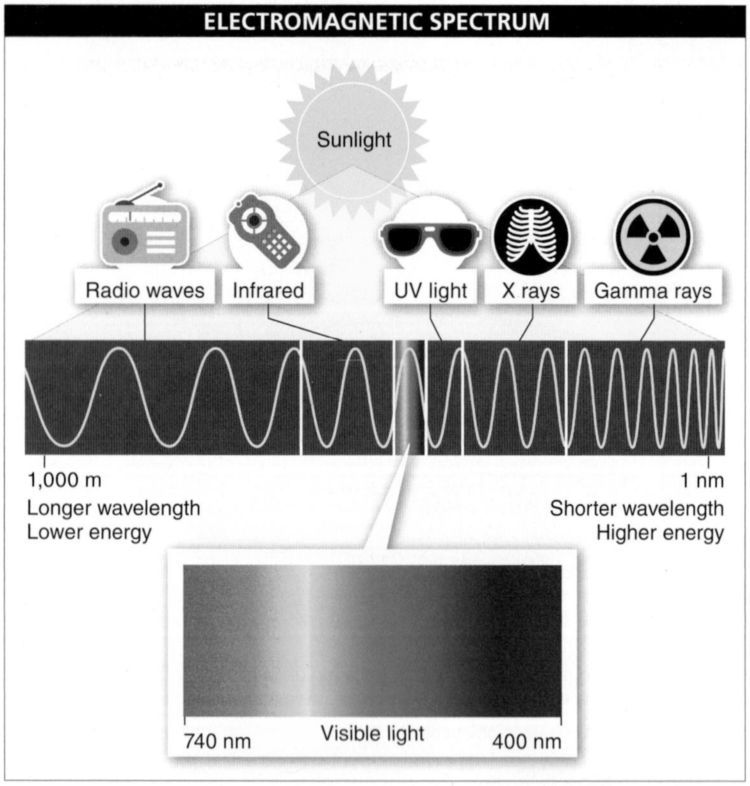

Just as we can't hear some super-high-pitched frequencies of sound (even though many dogs can), there are some wavelengths of light that are too short or too long for us to see. The light that we can see, visible light, spans all the colors of the rainbow. Humans (and some other animals) can see colors because our eyes contain light-absorbing molecules called **pigments.** These pigments absorb wavelengths of light within the visible range. The energy in these light waves excites electrons in the pigments, stimulating nerves in our eyes, which then transmit electrical signals to our brains. We perceive different wavelengths within the visible spectrum as different colors. The pigments in the human eye absorb many different wavelengths pretty well: that's why we can see so many colors. When plants use sunlight's energy to make sugar during photosynthesis, they also use the visible portion of the electromagnetic spectrum. Unlike the pigments in our eyes, however, plant pigments (the energy-capturing parts of a plant) absorb and use only a portion of visible light wavelengths.

Chlorophyll is the main pigment molecule in plants that absorbs light energy from the sun. Chlorophyll molecules are embedded in the thylakoid membranes of chloroplasts, which are found primarily in plants' leaves. Just as light energy excites electrons in the pigments responsible for color vision in humans, electrons in a plant's chlorophyll can become excited by certain wavelengths of light and can capture a bit of this light energy.

Plants produce several different light-absorbing pigments (**FIGURE 4-15**). The primary photosynthetic pigment, called **chlorophyll *a*,** absorbs red and blue-violet wavelengths of light. Every other wavelength generally travels through or bounces off this pigment. Chlorophyll *a* cannot efficiently absorb green light and instead reflects those wavelengths. We perceive the reflected light waves as green, and so the pigment and the leaves that contain it appear green. Another pigment, **chlorophyll *b*,** is similar in structure but absorbs blue and red-orange wavelengths and reflects yellow-green wavelengths. Some related pigments called **carotenoids** absorb blue-violet and blue-green wavelengths and reflect yellow, orange, and red wavelengths.

ELECTROMAGNETIC SPECTRUM

Sunlight

Radio waves | Infrared | UV light | X rays | Gamma rays

1,000 m
Longer wavelength
Lower energy

1 nm
Shorter wavelength
Higher energy

740 nm   Visible light   400 nm

FIGURE 4-14 **A spectrum of energy.** A ray of light emits high-energy photons, low-energy photons, and everything in between. Plants use only a fraction of the light's available energy.

PHOTOSYNTHETIC PIGMENTS

Plants produce several different light-absorbing pigments. Each photosynthetic pigment absorbs and reflects specific wavelengths.

Chlorophyll *a*
Chlorophyll *b*
Carotenoids

Energy absorption

Wavelength (nm)
400   500   600   700

Seasonal differences in the amount of pigment molecules present in leaves lead to the leaves changing color.

SPRING

Light from the sun

Light reflected
Light absorbed

Amount of pigment molecules present in leaves

Chlorophyll *a*   Chlorophyll *b*   Carotenoids

Photosynthetic pigments

FALL

Light reflected
Light absorbed

In the fall, chlorophyll *a* and *b* molecules are broken down and stored in branches.

Chlorophyll *a*   Chlorophyll *b*   Carotenoids

Photosynthetic pigments

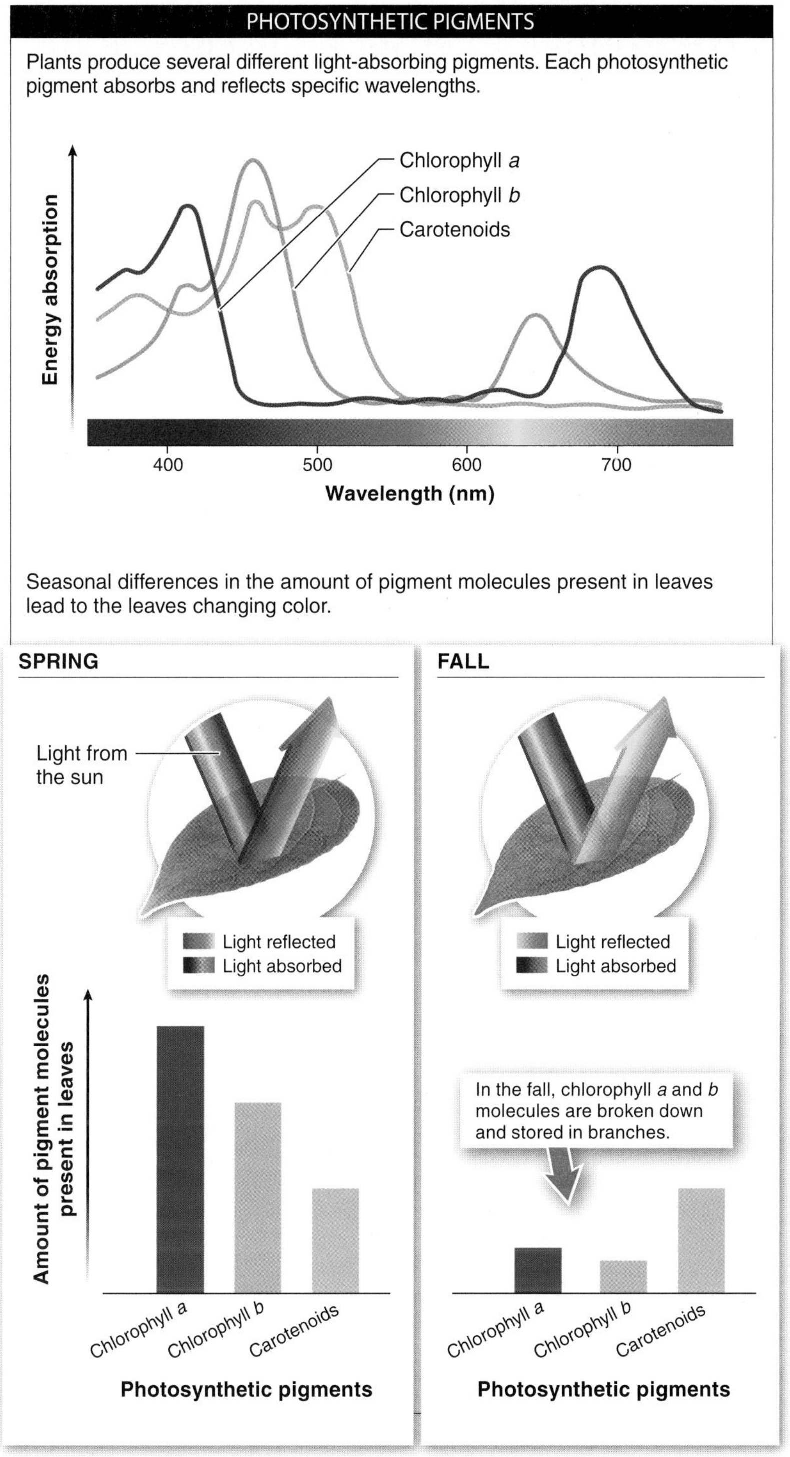

GRAPHIC CONTENT
Thinking critically about visual displays of data
Turn to p. 171 for a closer inspection of this figure.

FIGURE 4-15 **Plant pigments.** Each photosynthetic pigment absorbs and reflects specific wavelengths.

In the late summer, cooler temperatures cause some trees to prepare for the winter by shutting down chlorophyll production and reducing photosynthesis rates, much like an animal's hibernation. Gradually, the chlorophyll *a* and *b* molecules in the leaves are broken down and their chemical components are stored in the branches. As the amounts of chlorophyll *a* and *b* in the leaves decrease relative to the remaining carotenoids, the striking colors of the fall foliage are revealed (see Figure 4-15). During the rest of the year, chlorophylls *a* and *b* are so abundant in leaves that green masks the colors of the other pigments.

Q Why do the leaves of some trees turn beautiful colors each fall?

TAKE-HOME MESSAGE 4·7

Photosynthesis is powered by light energy, a type of kinetic energy made of energy packets called photons. Photons hit chlorophyll and other light-absorbing molecules in the chloroplasts of cells near the green surfaces of plants. These molecules capture some of the light energy and harness it to build sugar from carbon dioxide and water.

An organism can use energy from the sun only if it can convert the light energy of the sun into the chemical energy in the bonds between atoms. The most important molecule in this conversion is the pigment chlorophyll (**FIGURE 4-16**). When chlorophyll is hit by photons of certain wavelengths, the light energy bumps an electron (e⁻) in the chlorophyll molecule to a higher energy level, an *excited* state. Upon absorbing the photon, the electron briefly gains energy, and the potential energy in the chlorophyll molecule increases.

An electron in a photosynthetic pigment that is excited to a higher energy state generally has one of two fates. (1) The electron returns to its resting, unexcited state, releasing energy in the process, some of which may bump electrons in a nearby molecule to a higher energy state (while the rest of the energy is dissipated as heat). Or (2) the excited electron itself is passed to another molecule.

The passing of electrons from molecule to molecule is one of the chief ways that energy moves through cells. Many molecules carry or accept electrons during cellular activities. All that is required is that the acceptor must have a greater attraction for electrons than does the molecule from which it receives them. This receiver molecule, in turn, hands off electrons to another acceptor with an even greater attraction for them. A molecule that gains electrons always carries greater energy than it did before receiving the electron(s). For this reason, the passing of electrons from one molecule to another can be viewed as a passing of potential energy. In this way, energy moves through cells.

This transfer of electrons is one of the first steps of photosynthesis, the process that enables a plant to harness light energy from the sun and convert it to the more readily usable chemical energy. As we see later in this chapter, the dismantling of food molecules such as glucose to generate energy is also a story of breaking and rearranging chemical bonds as electrons pass from one atom or molecule to another.

Q Suppose a large meteor hit the earth. How could smoke and soot in the atmosphere wipe out life far beyond the area of direct impact?

Because particles in the atmosphere can block light from the sun and reduce the excitation of electrons in chlorophyll molecules, photosynthesis depends on a relatively clean atmosphere. Any

ENERGY MOVEMENT THROUGH CHLOROPHYLL

1 Light energy bumps an electron in the chlorophyll molecule to a higher, excited energy level.

Sun

e⁻ Higher energy state

Thylakoid Photons

Potential energy increases.

e⁻ Normal energy state

Chlorophyll

2 The excited electron generally has one of two different fates:

Some energy is transferred to a nearby molecule, where it excites another electron.

e⁻

e⁻

ENERGY

e⁻

e⁻

or

e⁻

The excited electron is transferred to a nearby molecule.

e⁻

e⁻

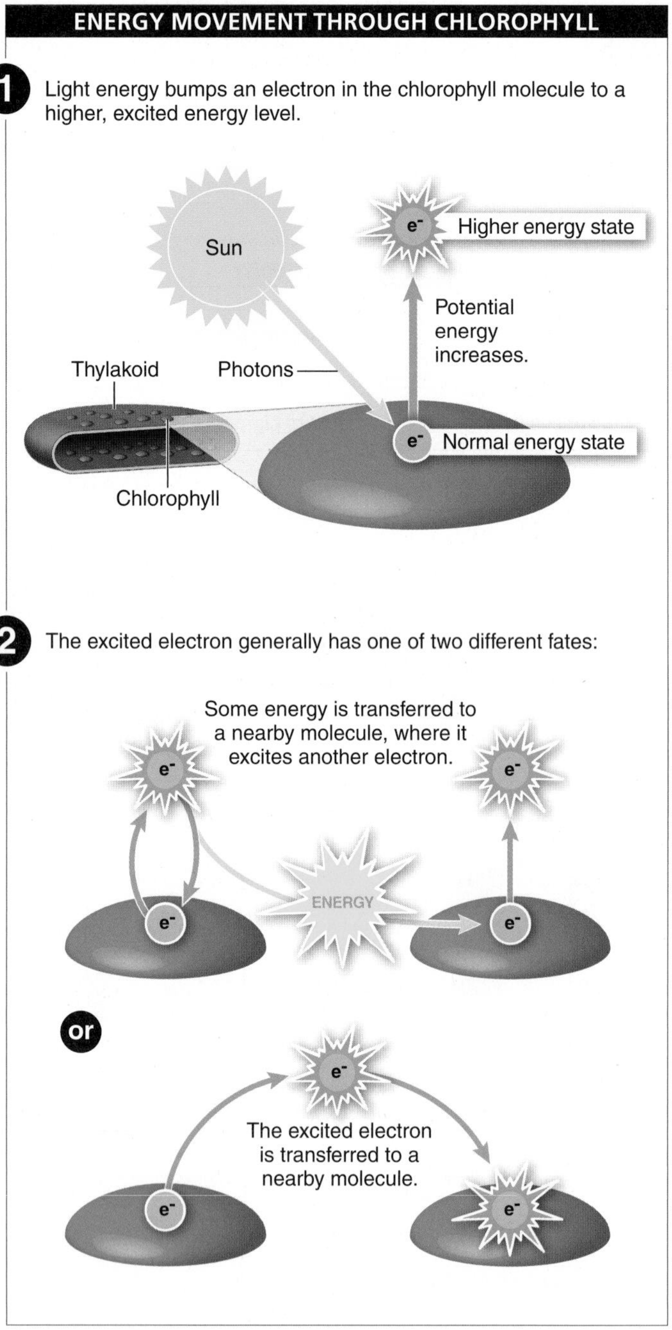

FIGURE 4-16 **Capturing light energy as excited electrons.** Chlorophyll electrons are excited to a higher energy state by light energy.

reduction in the available sunlight can have serious effects on plants. Scientists believe that if a large meteor hit the earth—as one did when the dinosaurs were wiped out

65 million years ago—smoke, soot, and dust in the atmosphere could block sunlight to such an extent that plants in the region, or possibly all of the plants on earth, could not conduct photosynthesis at high enough levels to survive. And when plants die off, all of the animals and other species that rely on plants for energy die as well. Nearly all life on earth is completely dependent on the continued excitation of electrons by sunlight.

TAKE-HOME MESSAGE 4·8

When chlorophyll is hit by photons, the light energy excites an electron in the chlorophyll molecule, increasing the chlorophyll's potential energy. The excited electrons can be passed to other molecules, moving the potential energy through the cell.

4·9 Photosynthesis in detail: the energy of sunlight is captured as chemical energy.

Photosynthesis is a complex process, but our understanding can be greatly aided by remembering one phrase: FOLLOW THE ELECTRONS. Where are they coming from? What are they passing through? Where are they going? What will happen to them when they get there?

In the first part of photosynthesis, the "photo" part, sunlight hits the chloroplasts of a plant's leaves and some of the energy in this sunlight is captured and stored in ATP and in another molecule, called NADPH, which stores energy by accepting high-energy electrons (**FIGURE 4-17**). After these transformations, the captured energy is ready to be used to make sugar molecules in the "synthesis" part of photosynthesis.

The energy-capturing process occurs in two **photosystems.** Embedded in the thylakoid membranes in the chloroplast, these photosystems are structures composed of light-catching pigments (including chlorophyll) and protein. As the pigments absorb photons, electrons in the pigments gain energy and become excited, and then return to their resting state, releasing energy. The released energy (but not the electrons) is transferred to neighboring pigment molecules. This process continues until the energy transferred among many pigment molecules makes its way to a chlorophyll *a* molecule at the center of the photosystem, and excites an electron there (**FIGURE 4-18**). This is where the electron journey begins.

The chlorophyll *a* molecule at the center of the photosystem is special, differing from the other pigment molecules in one key feature. When its electrons are boosted to an excited state, they do not return to their resting, unexcited state. Instead, the special chlorophyll *a* continually loses its excited electrons to a nearby molecule, called the **primary electron acceptor,** which acts like an electron vacuum.

The electrons taken away from the special chlorophyll *a* molecule must be replaced. The replacement electrons come from water. As long as photosynthesis is occurring, a constant supply of replacement electrons is required. Molecules of water, near the special chlorophyll *a* molecule in the thylakoid membrane, are continuously split. This split—in which four photons of light split two molecules of water into four electrons, four protons (H^+), and a molecule of O_2—provides the electrons necessary to replenish chlorophyll *a*'s electron

Q Why must plants get water for photosynthesis to occur?

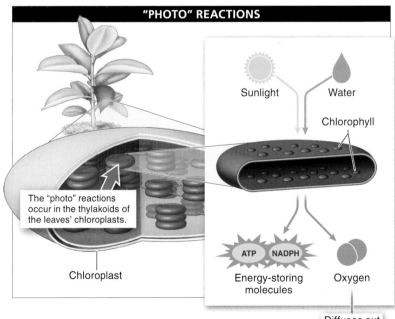

FIGURE 4-17 **Overview of the "photo" reactions.** Light energy is captured in the "photo" portion of photosynthesis. That energy is later used to power the building of sugar molecules.

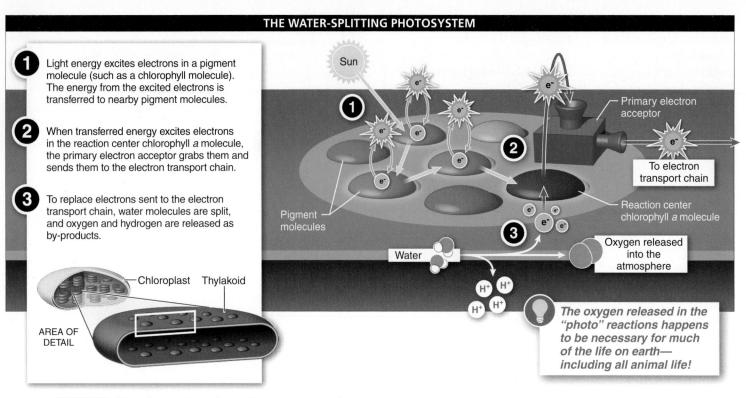

1 Light energy excites electrons in a pigment molecule (such as a chlorophyll molecule). The energy from the excited electrons is transferred to nearby pigment molecules.

2 When transferred energy excites electrons in the reaction center chlorophyll *a* molecule, the primary electron acceptor grabs them and sends them to the electron transport chain.

3 To replace electrons sent to the electron transport chain, water molecules are split, and oxygen and hydrogen are released as by-products.

Sun

Primary electron acceptor

To electron transport chain

Reaction center chlorophyll *a* molecule

Pigment molecules

Water

Oxygen released into the atmosphere

Chloroplast   Thylakoid

AREA OF DETAIL

The oxygen released in the "photo" reactions happens to be necessary for much of the life on earth— including all animal life!

FIGURE 4-18 **The photosystem that splits water molecules.** Splitting water provides electrons for photosynthesis.

supply. A convenient and life-sustaining by-product of the splitting of water in photosynthesis is the oxygen that is released from the cell, a by-product essential for much of the life on earth, including all animal life.

Once the primary electron acceptor gets hold of the high-energy electron from chlorophyll *a,* it passes it along like a hot potato to another molecule, which passes it to another, which passes it to yet another, in what is called an **electron transport chain** (**FIGURE 4-19**). The photosynthetic electron transport chain consists of two photosystems and several protein complexes that hand off electrons from one to the next. At each step in the electron transport chain's sequence of electron handoffs, the electrons fall to a lower energy state, and a little bit of energy is released. These bits of energy are harnessed to power pumps in the thylakoid membrane that move protons (which are also referred to as hydrogen ions or H^+ ions) from the stroma to the inside of the thylakoid. The pumps pack the protons inside the thylakoid sac at higher and higher concentrations. Think of a pump pushing water into an elevated tank, creating a store of potential energy that can gush out of the tank with great force and kinetic energy. Similarly, the protons eventually rush out of the thylakoid sacs with great force— and the force of the protons moving down their concentration gradient is harnessed to build energy-storing

ATP molecules, one of the two products of the "photo" portion of photosynthesis.

Recall that the energy-capturing and energy-transforming processes of the "photo" reactions occur in two photosystems (arrangements of chlorophyll and other light-catching molecules). The electron transport chain physically links the first photosystem to the second. As the traveling electrons continue their journey, they fill electron vacancies in the reaction center of the second photosystem, right next to the first photosystem (**FIGURE 4-20**). Like the first photosystem, the second photosystem also has numerous pigments that harness photons from the sun and pass the light energy to another special chlorophyll *a* molecule. The special chlorophyll *a* molecule at the center of this second photosystem has electron vacancies because, as in the first photosystem, when electrons in the special chlorophyll *a* molecule are boosted to an excited state, they are whisked away from the chlorophyll molecule by another primary electron acceptor. This electron acceptor then passes the electrons to a second electron transport chain. At the end of this second electron transport chain, the electrons are passed to a molecule called $NADP^+$, creating **NADPH,** a high-energy electron carrier. NADPH is the second important product of the "photo" portion of photosynthesis.

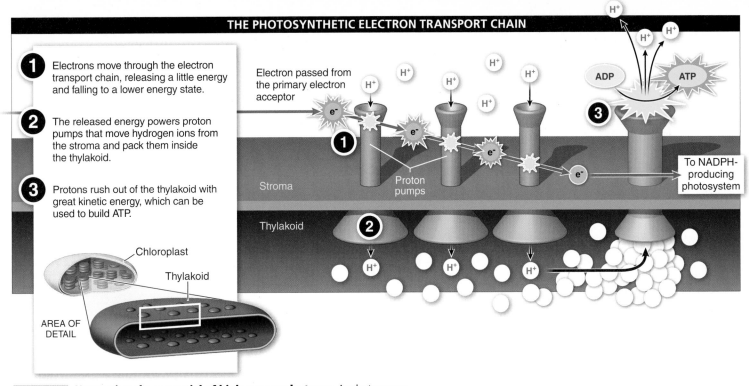

1. Electrons move through the electron transport chain, releasing a little energy and falling to a lower energy state.

2. The released energy powers proton pumps that move hydrogen ions from the stroma and pack them inside the thylakoid.

3. Protons rush out of the thylakoid with great kinetic energy, which can be used to build ATP.

FIGURE 4-19 **Harnessing the potential of high-energy electrons.** As electrons are passed from the primary electron acceptor to a chain of molecules embedded within the thylakoid membrane, called the electron transport chain, the released energy is used to create a proton gradient.

SUMMARY OF "PHOTO" REACTION COMPONENTS

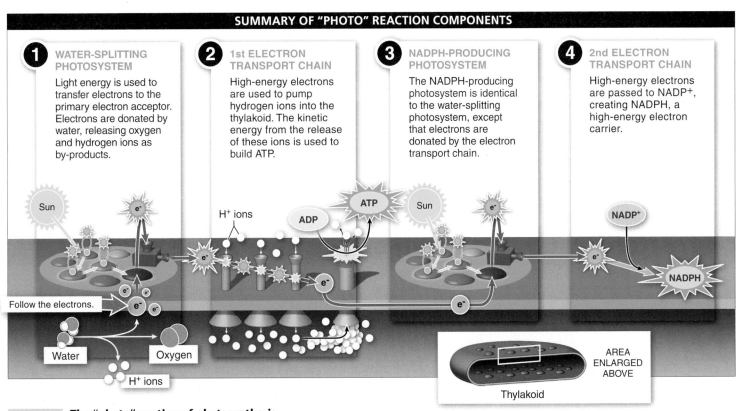

1. **WATER-SPLITTING PHOTOSYSTEM**

   Light energy is used to transfer electrons to the primary electron acceptor. Electrons are donated by water, releasing oxygen and hydrogen ions as by-products.

2. **1st ELECTRON TRANSPORT CHAIN**

   High-energy electrons are used to pump hydrogen ions into the thylakoid. The kinetic energy from the release of these ions is used to build ATP.

3. **NADPH-PRODUCING PHOTOSYSTEM**

   The NADPH-producing photosystem is identical to the water-splitting photosystem, except that electrons are donated by the electron transport chain.

4. **2nd ELECTRON TRANSPORT CHAIN**

   High-energy electrons are passed to NADP+, creating NADPH, a high-energy electron carrier.

FIGURE 4-20 **The "photo" portion of photosynthesis.**

With the electrons' passage through the second photosystem and arrival in NADPH, we now have the final products of the "photo" part of photosynthesis (which is also called the "light-dependent reactions"): we've captured light energy from the sun and converted it to the chemical energy of ATP and the high-energy electron carrier NADPH (see Figure 20). But we haven't made any food yet. In the next section, we cover the "synthesis" part of photosynthesis and see how plants use the energy in ATP and NADPH to produce sugar from carbon dioxide.

TAKE-HOME MESSAGE 4·9

There are two parts to photosynthesis. The first is the "photo" part, in which light energy is transformed into chemical energy, while splitting water molecules and producing oxygen. Sunlight's energy is first captured when an electron in chlorophyll is excited. As this electron is passed from one molecule to another, energy is released at each transfer, some of which is used to build the energy-storage molecules ATP and NADPH.

4·10 Photosynthesis in detail: the captured energy of sunlight is used to make food.

The "synthesis" part of photosynthesis takes place in a series of chemical reactions called the **Calvin cycle.** All the Calvin cycle reactions occur in the stroma of the leaves' chloroplasts, outside the thylakoids. Plants carry out these reactions using the energy stored in the ATP and NADPH molecules that are built in the "photo" portion of photosynthesis. This dependency links the light-gathering ("photo") reactions with the sugar-building ("synthesis") reactions (FIGURE 4-21).

If there is any part of photosynthesis that appears magical, it is the Calvin cycle. Just as a magician seems to make a rabbit appear from thin air, the Calvin cycle takes invisible molecules of CO_2 from the air and uses them to assemble visible—even edible—molecules of sugar. The processes in the Calvin cycle occur in three steps (FIGURE 4-22).

Fixation. First, using an enzyme called **rubisco,** plants pluck carbon from the air, where it occurs in the form of carbon dioxide (which has one carbon), and then attach, or "fix," it to a visible organic molecule (which has five carbon atoms) within the chloroplast. Not surprisingly, given its role as the critical chemical that enables plants to build food molecules, rubisco is the most abundant protein on earth.

1. **Sugar creation.** The newly built molecule is chemically modified: a phosphate from ATP is added, and the molecule receives some high-energy electrons from NADPH and is split in two. For every three carbon dioxide molecules added to the Calvin cycle, one sugar product is produced. This product, a three-carbon sugar, is called glyceraldehyde 3-phosphate (G3P).

Some of the G3P molecules are combined to make the six-carbon sugars glucose and fructose. These sugars can be used as fuel by the plant, enabling it to grow. They can also be used as fuel by animals that eat the plant.

2. **Regeneration.** Not all of the G3P molecules are used to produce sugars. In the third and final phase of the Calvin cycle, some G3P molecules are rearranged to regenerate the original five-carbon molecule in the

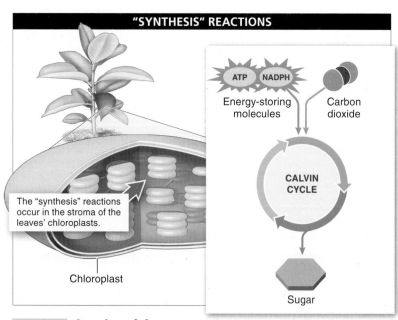

"SYNTHESIS" REACTIONS

The "synthesis" reactions occur in the stroma of the leaves' chloroplasts.

Chloroplast

Energy-storing molecules — ATP NADPH

Carbon dioxide

CALVIN CYCLE

Sugar

FIGURE 4-21 Overview of the "synthesis" reactions of photosynthesis.

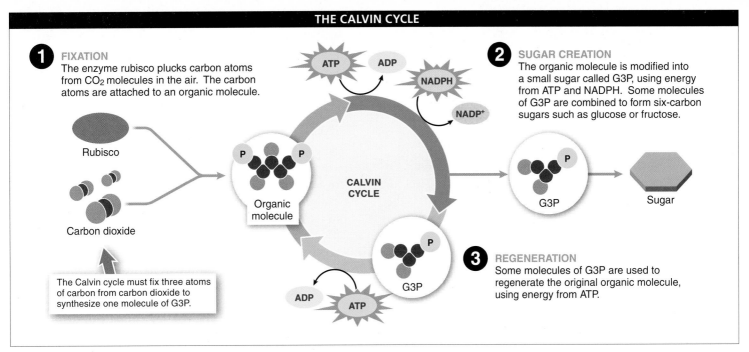

1 FIXATION
The enzyme rubisco plucks carbon atoms from CO₂ molecules in the air. The carbon atoms are attached to an organic molecule.

Rubisco

Carbon dioxide

The Calvin cycle must fix three atoms of carbon from carbon dioxide to synthesize one molecule of G3P.

Organic molecule

CALVIN CYCLE

ATP ADP

NADPH

NADP⁺

2 SUGAR CREATION
The organic molecule is modified into a small sugar called G3P, using energy from ATP and NADPH. Some molecules of G3P are combined to form six-carbon sugars such as glucose or fructose.

G3P

Sugar

3 REGENERATION
Some molecules of G3P are used to regenerate the original organic molecule, using energy from ATP.

G3P

ADP ATP

FIGURE 4-22 **Sugar synthesis.** In the Calvin cycle—the "synthesis" portion of photosynthesis—sugars are produced.

chloroplast to which the carbon from CO₂ is attached. Rearranging G3P to regenerate the starting molecule requires energy from ATP produced in the "photo" reactions of photosynthesis. With this regeneration, the Calvin cycle can continue to fix carbon and produce molecules of G3P.

Ultimately, to synthesize one molecule of G3P, the Calvin cycle requires three "turns" of the Calvin cycle and fixation of three atoms of carbon from carbon dioxide to the initial organic molecule; this process consumes nine molecules of ATP and six molecules of NADPH generated in the "photo" reactions of photosynthesis.

TAKE-HOME MESSAGE 4·10

The second part, or "synthesis" part, of photosynthesis is the Calvin cycle, which occurs in the stroma of chloroplasts. During this phase, carbon from CO₂ in the atmosphere is attached (fixed) to molecules in the chloroplasts, sugars are built, and molecules are regenerated to be used again in the Calvin cycle. The fixation, building, and regeneration processes consume energy from ATP and NADPH (the products of the "photo" part of photosynthesis).

4·11 The battle against world hunger can use plants adapted to water scarcity.

Sudan. Ethiopia. India. Somalia. Many of the world's regions with the highest rates of starvation are also places with the hottest, driest climates. This is not a coincidence. These climate conditions present difficult challenges for sustaining agriculture (FIGURE 4-23), and in the absence of stable crop yields, food production is unpredictable and the risk of starvation high. But evolutionary adaptations in some plants enable them to thrive in hot, dry conditions. Recent technological advances in agriculture use these innovative

evolutionary solutions to battle the problem of world hunger. In this section, we discuss some adaptations that allow plants to thrive when water is scarce. We also look at how humans use these adaptations to grow food in the dry, inhospitable climates where starvation rates are highest.

When it gets too hot and dry, animals can seek coolness in the shade. Plants, however, are anchored in place and do not have this option. Consequently, plants in hot, dry climates

FIGURE 4-23 **Nowhere to hide.** Plants that lose too much water can't always survive in extremely hot, dry weather.

can lose significant amounts of water through evaporation. Evaporation is a problem for plants because water is essential to photosynthesis, growth, and the transport of nutrients. Without water, plants cannot live long.

One method of combating water loss through evaporation is for plants to close their **stomata** (*sing.* stoma), small pores usually on the underside of leaves (**FIGURE 4-24**). These openings are the primary sites for gas exchange in plants: carbon dioxide for photosynthesis enters through these openings, and oxygen generated as a by-product in photosynthesis exits through

them. When open, the stomata also allow water to evaporate from the plant. Closing their stomata solves one problem for plants (too much water evaporation) but creates another: with the stomata shut, oxygen from the "photo" reactions of photosynthesis cannot be released from the chloroplasts, and carbon dioxide cannot enter. If there are no carbon dioxide molecules for sugar production, the Calvin cycle tries to fix carbon but instead finds only oxygen. Plant growth comes to a standstill and crops fail.

In some plants, including corn and sugarcane, a process has evolved that minimizes water loss but still enables the plants to make sugar when the weather is hot and dry. In the process called **C4 photosynthesis,** these plants add an extra set of steps to the usual process of photosynthesis, which is usually called C3 photosynthesis (**FIGURE 4-25**). In these steps, the plants produce an enzyme that functions like the ultimate "CO_2-sticky tape." This enzyme has a tremendously strong attraction for carbon dioxide; it can find and bind carbon even when CO_2 concentration is very low. As a consequence, the plant's stomata can be opened just a tiny bit, and let in just a little CO_2. Reducing the amount of stomata opening reduces evaporation and conserves water for the plant. (In contrast, rubisco, the usual enzyme that plants use to pluck carbon from the atmosphere, functions poorly when CO_2 is scarce, necessitating greater stomata opening.)

Q How might global warming be bad for agriculture?

This seems like such a good solution that we would expect all plants to use it. There is a catch, though. The extra steps in C4 photosynthesis require the plant to expend additional energy. Specifically, every time the plant generates a molecule of the "CO_2-sticky tape"

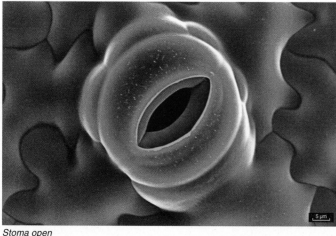
Stoma open

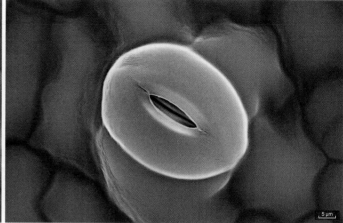
Stoma closed

FIGURE 4-24 **Plant stomata.** Carbon dioxide enters a plant through stomata, but water can be lost through the same openings.

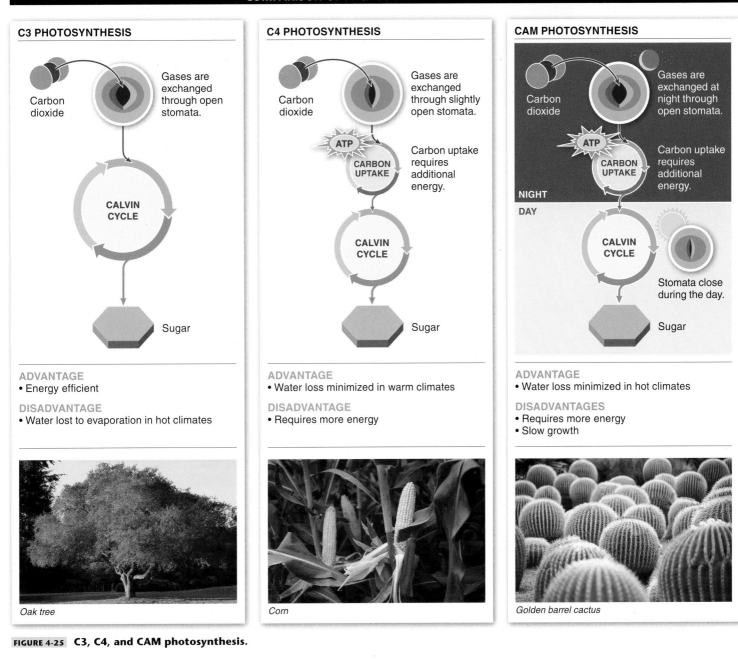

FIGURE 4-25 C3, C4, and CAM photosynthesis.

enzyme, it uses one molecule of ATP. It is acceptable to pay this energy cost only when the climate is so hot and dry that the plant would otherwise have to close its stomata and completely shut down all sugar production. If the climate is mild, however, plants conducting the more energetically expensive C4 photosynthesis would be out-competed by the more efficient plants conducting standard C3 photosynthesis. Not surprisingly, we see few C4 plants in the temperate regions of the world and little C4 photosynthesis among photosynthetic organisms living in

the oceans. In hot, dry regions, however, C4 plants are dominant and displace the C3 plants wherever both occur (**FIGURE 4-26**). With global warming, many scientists expect to see a gradual expansion of the ranges over which C4 plants grow, and believe that non-C4 plants will be pushed farther and farther away from the equator.

A third and similar method of carbon fixation, called **CAM** (for "crassulacean acid metabolism"), is also found in hot, dry areas. In this method, used by many cacti, pineapples,

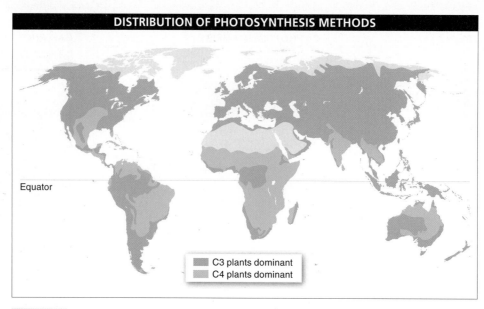

DISTRIBUTION OF PHOTOSYNTHESIS METHODS

Equator

- C3 plants dominant
- C4 plants dominant

FIGURE 4-26 **Global distribution of C3 and C4 plants.**

and other fleshy, juicy plants, the plants close their stomata during hot, dry days. At night, they open the stomata and let CO_2 into the leaves, where it binds temporarily to a holding molecule. During the day, when a carbon source is needed to make sugars in the Calvin cycle, the CO_2 is gradually released from the holding molecule, enabling photosynthesis to proceed while keeping the stomata closed to reduce water loss (see Figure 4-25). A disadvantage of CAM photosynthesis is that by completely closing their stomata during the day, CAM plants significantly reduce the total amount of CO_2 they can take in. As a consequence, they have much slower growth rates and cannot compete well with non-CAM plants under any conditions other than extreme dryness.

C4 and CAM photosynthesis originally evolved because they made it possible for plants to grow better in the world's hot and dry regions. Researchers are now experimenting with these adaptations as a way to fight world hunger. They have introduced into rice plants several genes from corn that code for the C4 photosynthesis enzymes. Once in the rice, these genes increase the rice plant's ability to photosynthesize, leading to higher growth rates and food yields. Whether the addition of C4 photosynthesis enzymes will make it possible to grow new crops on a large scale in previously inhospitable environments is not certain. Early results suggest, however, that this is a promising approach.

TAKE-HOME MESSAGE 4·11

C4 and CAM photosynthesis are evolutionary adaptations at the biochemical level that, although more energetically expensive than regular (C3) photosynthesis, allow plants in hot, dry climates to close their stomata and conserve water without shutting down photosynthesis.

A pit stop at the toddler refueling station.

4·12 How do living organisms fuel their actions? Cellular respiration: the big picture.

Food is fuel. And all the activities of life—growing, moving, reproducing—require fuel. Plants, most algae, and some bacteria obtain their fuel directly from the energy of sunlight, which they harness through photosynthesis. Less self-sufficient organisms, such as humans, alligators, and insects, must extract the energy they need from the food they eat. This energy comes from photosynthetic organisms either directly (from eating plants) or indirectly (from eating animals that eat plants) (**FIGURE 4-27**).

All living organisms—including plants (a fact that is often overlooked)—extract energy from the chemical bonds of molecules (which can be considered "food") through a process called cellular respiration (**FIGURE 4-28**). This process is a bit like photosynthesis in reverse. In photosynthesis, the energy of the sun is captured and used to build molecules of sugars, such as glucose. In cellular respiration, plants and animals break down the chemical bonds of sugar and other energy-rich food molecules (such as fats and proteins) to

FIGURE 4-27 **Living organisms require fuel (in one form or another).**

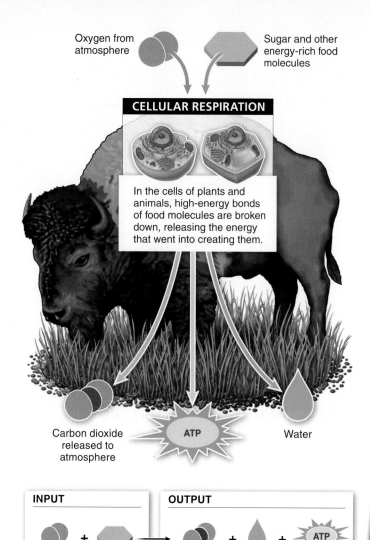

Oxygen from atmosphere

Sugar and other energy-rich food molecules

CELLULAR RESPIRATION

In the cells of plants and animals, high-energy bonds of food molecules are broken down, releasing the energy that went into creating them.

Carbon dioxide released to atmosphere

ATP

Water

INPUT		OUTPUT			
Oxygen + Sugar	→	Carbon dioxide + Water + ATP Energy

FIGURE 4-28 Cellular respiration: the big picture.

release the energy that went into creating them. (Don't confuse cellular respiration with the act of breathing, which is also called *respiration.*) As energy is released, cells capture and store it in the bonds of ATP molecules. This plentiful, readily available stored energy can then be tapped as needed to fuel the work of the life-sustaining activities and processes of all living organisms.

In humans and other animals, cellular respiration starts after we eat food, digest it, absorb the nutrient molecules into the bloodstream, and deliver them to the cells of our bodies. At this point, our cells begin to extract some of the energy stored in the bonds of the food molecules. We focus here on the breakdown of glucose, but later in this chapter we'll see that the process is similar for the breakdown of fats or lipids. Ultimately, when a food molecule has been completely processed, the cell has used the food molecule's stored energy (along with oxygen) to create a large number of high-energy-storing ATP molecules (which supply energy to power the cell's activities), water, and carbon dioxide (which is exhaled into the atmosphere).

TAKE-HOME MESSAGE 4·12

Living organisms extract energy through a process called cellular respiration, in which the high-energy bonds of sugar and other energy-rich molecules are broken, releasing the energy that went into creating them. The cell captures the food molecules' stored energy in the bonds of ATP molecules. This process requires fuel molecules and oxygen, and it yields ATP molecules, water, and carbon dioxide.

4·13 The first step of cellular respiration: glycolysis is the universal energy-releasing pathway.

To generate energy, fuels such as glucose and other carbohydrates, as well as proteins and fats, are broken down in three steps: (1) glycolysis, (2) the Krebs cycle, and (3) the electron transport chain. **Glycolysis** means the splitting (*lysis*) of sugar (*glyco-*), and it is the first step that all organisms on the planet take in breaking down food molecules; for many single-celled organisms, this one step is sufficient to provide all of the energy they need (**FIGURE 4-29**).

As **FIGURE 4-30** illustrates, glycolysis is a sequence of chemical reactions (there are 10 in all) through which

glucose is broken down, resulting in two molecules of a substance called **pyruvate.** Glycolysis has two distinct phases: an "uphill" preparatory phase and a "downhill" payoff phase.

Just as you sometimes have to spend money to make money, before any energy can be extracted from glucose, some energy must be added to the molecule. This addition occurs during the "uphill" phase. The additional energy (which comes from ATP) destabilizes the glucose molecule, making it ripe for chemical breakdown. Once the glucose can be broken down

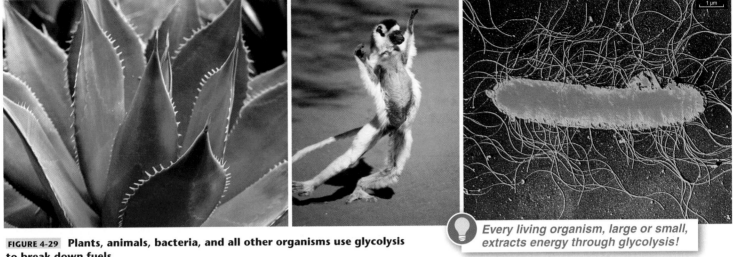

FIGURE 4-29 **Plants, animals, bacteria, and all other organisms use glycolysis to break down fuels.**

Every living organism, large or small, extracts energy through glycolysis!

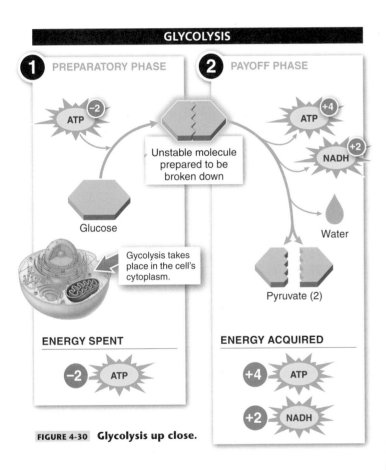

GLYCOLYSIS

1 PREPARATORY PHASE

2 PAYOFF PHASE

ATP −2

Unstable molecule prepared to be broken down

Glucose

Gycolysis takes place in the cell's cytoplasm.

ATP +4

NADH +2

Water

Pyruvate (2)

ENERGY SPENT

−2 ATP

ENERGY ACQUIRED

+4 ATP

+2 NADH

FIGURE 4-30 **Glycolysis up close.**

chemically, the "payoff phase" begins as energy stored in its bonds can be harnessed as the bonds are broken.

Three of the 10 steps in glycolysis yield energy. In two of these three steps, as bonds in the sugar are broken and other, lower-energy bonds are formed, the energy released is

quickly harnessed by the attachment of phosphate groups to molecules of ADP to create energy-rich ATP molecules. In the third energy-yielding step of glycolysis, electrons originally from the glucose are transferred to NAD^+ to become the high-energy electron carrier NADH. Later (in an electron transport chain in the mitochondria), this energy will be converted to ATP. The net result of glycolysis is that each glucose molecule is broken down into two molecules of pyruvate. During this breakdown, some of the released energy is captured in the production of energy-rich ATP molecules and molecules of the high-energy electron carrier NADH. Two molecules of water are also produced during glycolysis.

Glycolysis can proceed regardless of whether any oxygen is present. When oxygen is present, glycolysis isn't the end of the story. But in the absence of oxygen and in many yeasts and bacteria, glycolysis is the only game in town for fueling activity. Because single-celled organisms have much lower energy needs, they can function solely on the yields of glycolysis. For many organisms (including humans), however, glycolysis is a springboard to further energy extraction. The additional energy payoffs come from the Krebs cycle and the electron transport chain.

TAKE-HOME MESSAGE 4·13

Glycolysis is the initial phase in the process by which all living organisms harness energy from food molecules. Glycolysis occurs in a cell's cytoplasm and uses the energy released from breaking chemical bonds in food molecules to produce high-energy molecules, ATP and NADH.

4·14 The second step of cellular respiration: the Krebs cycle extracts energy from sugar.

Cells could stop extracting energy when glycolysis ends, but they rarely do so because that would be like leaving most of your meal on your plate. In glycolysis, only a small fraction of the energy stored in sugar molecules is recovered and converted to ATP and NADH. Cells get much more of an "energy bang" for their "food buck" in the steps following glycolysis, which occur in the mitochondria. This is why mitochondria are considered ATP "factories." In the mitochondria, the molecules produced from the breakdown of glucose during glycolysis are broken down further, during two steps that are dramatically more efficient at capturing energy: the Krebs cycle (the subject of this section) and the electron transport chain (the subject of the next section). In breaking down the products of glycolysis, the **Krebs cycle** produces some additional molecules of ATP and, more importantly, captures a huge amount of chemical energy by producing high-energy electron carriers.

Q Aerobic training can cause our bodies to produce more mitochondria in cells. Why is this beneficial?

Before the Krebs cycle can begin, however, the end products of glycolysis—two molecules of pyruvate for every molecule of glucose used— must be modified. First, the pyruvate molecules move from the cytoplasm into the mitochondria, where they undergo three quick modifications that prepare them to be broken down in the Krebs cycle (**FIGURE 4-31**).

Modification 1. Each pyruvate molecule passes a pair of its high-energy electrons (and a proton) to the electron-carrier molecule NAD$^+$, building two molecules of NADH.

Modification 2. Next, a carbon atom and two oxygen atoms are removed from each pyruvate molecule and released as carbon dioxide. The CO_2 molecules diffuse out of the cell and, eventually, out of the organism. In humans, for example, these CO_2 molecules pass into the bloodstream and are transported to the lungs, from which they are eventually exhaled.

Modification 3. In the final step in the preparation for the Krebs cycle, a giant compound known as coenzyme A attaches itself to the remains of each pyruvate molecule, producing two molecules called acetyl-CoA. Each acetyl-CoA molecule is now ready to enter the Krebs cycle.

There are eight separate steps in the Krebs cycle, but our emphasis is on its three general outcomes (**FIGURE 4-32**).

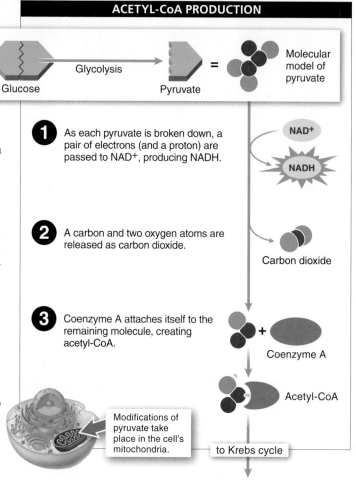

ACETYL-CoA PRODUCTION

Glycolysis: Glucose → Pyruvate = Molecular model of pyruvate

1 As each pyruvate is broken down, a pair of electrons (and a proton) are passed to NAD$^+$, producing NADH.

NAD$^+$
NADH

2 A carbon and two oxygen atoms are released as carbon dioxide.

Carbon dioxide

3 Coenzyme A attaches itself to the remaining molecule, creating acetyl-CoA.

Coenzyme A
Acetyl-CoA

Modifications of pyruvate take place in the cell's mitochondria.

to Krebs cycle

FIGURE 4-31 **Preparation of pyruvate.** In the mitochondria, pyruvate must be modified before it can be broken down in the Krebs cycle.

Outcome 1. *A new molecule is formed.* Acetyl-CoA adds its two-carbon acetyl group to a molecule of the starting material of the Krebs cycle, a four-carbon chemical called oxaloacetate, creating a six-carbon molecule.

Outcome 2. *High-energy electron carriers (NADH) are made and carbon dioxide is exhaled.* The six-carbon molecule then gives electrons to NAD$^+$ to make the high-energy electron carrier NADH. (Don't forget that the main purpose of the Krebs cycle is the capture of energy.) The six-carbon molecule also releases two carbon atoms along with four oxygen atoms to form two carbon dioxide molecules. In mammals (including humans), this CO_2 is carried by the bloodstream to the lungs, from where it is exhaled into the atmosphere.

Outcome 3. *The starting material of the Krebs cycle is re-formed, ATP is generated, and more high-energy electron carriers are formed.* After the CO_2 is released, the four-carbon molecule that remains from the original pyruvate-oxaloacetate molecule formed in Outcome 1 is modified and rearranged to once again form oxaloacetate, the starting material of the Krebs cycle. In the process of this reorganization, one ATP molecule is generated, and more electrons are passed to NAD⁺ and a molecule called FAD to form NADH and FADH₂, both of which are high-energy electron carriers. The formation of these high-energy electron carriers increases the energy yield of the Krebs cycle. One oxaloacetate is re-formed, and the cycle is ready to break down the second molecule of acetyl-CoA. Two turns of the cycle are necessary to completely dismantle our original molecule of glucose.

Now that we have seen the Krebs cycle in its entirety, let's trace the path of the original six carbons in the original glucose molecule. In a sense, the carbon atoms that were first plucked from the atmosphere to make sugar during photosynthesis have been exhaled back into the atmosphere as six molecules of carbon dioxide.

1. Glycolysis: the six-carbon starting point. Glucose is broken down into two molecules of pyruvate. No carbons are removed.

2. Preparation for the Krebs cycle: two carbons are released. Two pyruvate molecules are modified to enter the Krebs cycle, and they each lose a carbon atom in the form of two molecules of carbon dioxide.

3. Krebs cycle: the last four carbons are released. A total of four carbon atoms enter the Krebs cycle in the form of two molecules of acetyl-CoA, entering one at a time. For each turn of the Krebs cycle, two molecules of carbon dioxide are released. So the two final carbons are released into the atmosphere during the second turn of the wheel. Poof! The six carbon atoms that were originally present in our single molecule of glucose are no longer present.

So we've come full circle. In photosynthesis, carbon atoms from the air were used to build sugar molecules, which had energy stored in the bonds between their carbon, hydrogen, and oxygen atoms. In cellular respiration, the energy previously stored in the bonds of the sugar is converted to molecules of ATP, NADH, and FADH₂. Carbon atoms from sugar are exhaled back into the air as CO_2, and water is produced.

What happens to the high-energy electron carriers, NADH and FADH₂? They eventually give up their high-energy electrons to the final stage of cellular respiration, the electron transport chain. The energy released as those

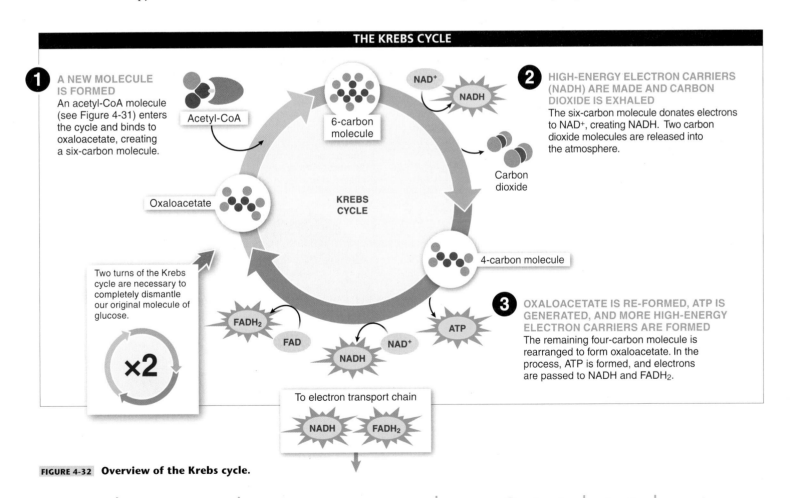

FIGURE 4-32 **Overview of the Krebs cycle.**

electrons pass through the transport chain is captured in the bonds of more ATP molecules. We explore that process in the next section.

Q Why might the malfunctioning of mitochondria play a role in lethargy or fatigue?

As we explore the important roles of the Krebs cycle and electron transport chain in generating usable energy for organisms, it's important to note that mitochondrial malfunctions have serious consequences for health. More than a hundred genetic mitochondrial disorders have been identified, all of which can lead to energy shortage, including muscle weakness, fatigue, and muscle pain. And, as noted in Section 3-15, it appears that many cases of extreme fatigue or cramps that occur after only slight exertion may be related to inherited mutations in the mitochondrial DNA. Damage to our cellular powerhouses, not surprisingly, can create a personal energy crisis.

TAKE-HOME MESSAGE 4·14

A huge amount of additional energy can be harvested by cells after glycolysis. First, the end product of glycolysis, pyruvate, is chemically modified. Then, in the Krebs cycle, the modified pyruvate is broken down step by step. This breakdown releases carbon into the atmosphere (as CO_2) as bonds are broken, and captures some of the released energy in two ATP molecules and numerous high-energy electron carriers for every glucose molecule.

4·15 The third step of cellular respiration: ATP is built in the electron transport chain.

How do we finally get a big payoff of usable energy from our glucose molecule? Glycolysis and the Krebs cycle produce a few molecules of ATP for each molecule of glucose broken down, but it is the energy held in the high-energy electron carriers NADH and $FADH_2$ that are formed in these processes that ultimately generates the largest amount of usable energy as ATP. In fact, almost 90% of the energy payoff from a molecule of glucose is harvested in the final step of cellular respiration, when the electrons from NADH and $FADH_2$ move along an electron transport chain. This process, like the Krebs cycle, takes place in the mitochondria.

In a manner similar to that seen in the chloroplast during photosynthesis, mitochondria convert kinetic energy (from electrons) into potential energy (a concentration gradient of protons). Two structural features of mitochondria are essential to their impressive ability to harness energy from molecules.

Feature 1. Mitochondria have a "bag-within-a-bag" structure that makes it possible for the regions inside and outside the "inner bag" to have different concentrations of molecules and makes it possible to harness the potential energy in the bonds of NADH and $FADH_2$ molecules to produce ATP (**FIGURE 4-33**).

Material inside the mitochondrion can lie in one of two spaces: (1) in the intermembrane space, which is outside the inner bag, or (2) in the **mitochondrial matrix,** which is inside the inner bag. With two distinct regions separated by

MITOCHONDRIA: A CLOSER LOOK AT STRUCTURE

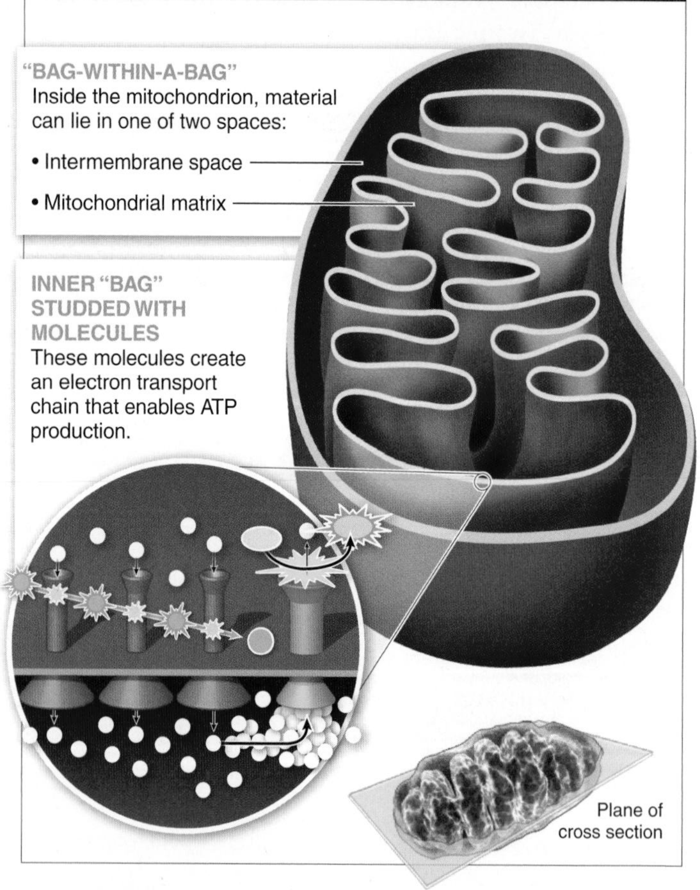

"BAG-WITHIN-A-BAG"
Inside the mitochondrion, material can lie in one of two spaces:
• Intermembrane space
• Mitochondrial matrix

INNER "BAG" STUDDED WITH MOLECULES
These molecules create an electron transport chain that enables ATP production.

Plane of cross section

FIGURE 4-33 "A bag-within-a-bag." The structure of mitochondria makes possible their impressive ability to harness energy from food molecules.

a membrane, the mitochondrion can create higher concentrations of molecules in one area or the other, creating a concentration gradient. And because a concentration gradient is a form of potential energy—molecules move from the high-concentration area to the low-concentration area the way water rushes down a hill—once a gradient is created, the energy released as the gradient dissipates can be used to do work. In the electron transport chain, this energy is used to build the energy-rich molecule ATP.

Feature 2. The inner bag of the mitochondrion is studded with molecules, mostly electron carriers, which are sequentially arranged as a "chain." This arrangement makes it possible for the molecules to hand off electrons in an orderly sequence.

Now let's explore how these features of mitochondria make it possible to harness energy from high-energy electron carriers (**FIGURE 4-34**).

Q Over-the-counter NADH pills provide energy to sufferers of chronic fatigue syndrome. Why might this be?

Step 1 of the electron transport chain begins with NADH and FADH$_2$ in the mitochondrial matrix (inside the inner bag) moving to the membrane. There, the high-energy electrons they carry are transferred to molecules embedded within the membrane. After they donate their electrons, the molecules that remain, NAD$^+$ and FAD, are recycled back to the Krebs cycle.

The membrane-embedded molecules pass the electrons to the next carrier, which passes the electrons to the next, and so on. At each handoff, a bit of energy is released. Thus, as electrons move from one carrier to another through the electron transport chain, they lose energy at each handoff.

At the end of the chain (step 2 in Figure 4-34), the lower-energy electrons are handed off to oxygen, which then combines with free H$^+$ ions in the mitochondrial fluid to form water.

As shown in step 3 of Figure 4-34, most of the energy released at each handoff from one electron carrier to another in the electron transport chain is used to pump protons (H$^+$ ions) from the mitochondrial matrix across the membrane and into the intermembrane space. As more and more protons are pumped across the membrane and packed into the intermembrane space, a concentration gradient is created. This gradient represents a significant source of potential energy.

If this description seems familiar, it is. In chloroplasts, during photosynthesis, great numbers of protons are pumped from the stroma outside the thylakoid sacs to the inside of the thylakoids. We likened this potential energy

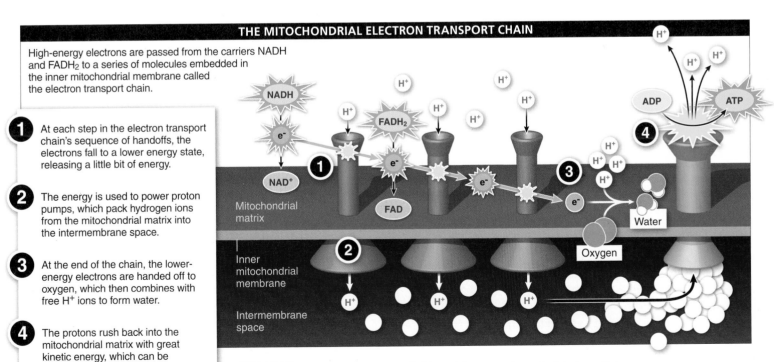

THE MITOCHONDRIAL ELECTRON TRANSPORT CHAIN

High-energy electrons are passed from the carriers NADH and FADH2 to a series of molecules embedded in the inner mitochondrial membrane called the electron transport chain.

1. At each step in the electron transport chain's sequence of handoffs, the electrons fall to a lower energy state, releasing a little bit of energy.

2. The energy is used to power proton pumps, which pack hydrogen ions from the mitochondrial matrix into the intermembrane space.

3. At the end of the chain, the lower-energy electrons are handed off to oxygen, which then combines with free H$^+$ ions to form water.

4. The protons rush back into the mitochondrial matrix with great kinetic energy, which can be used to build ATP.

FIGURE 4-34 **The big energy payoff.** Most of the energy harvested during cellular respiration is generated by the electron transport chain in the mitochondria.

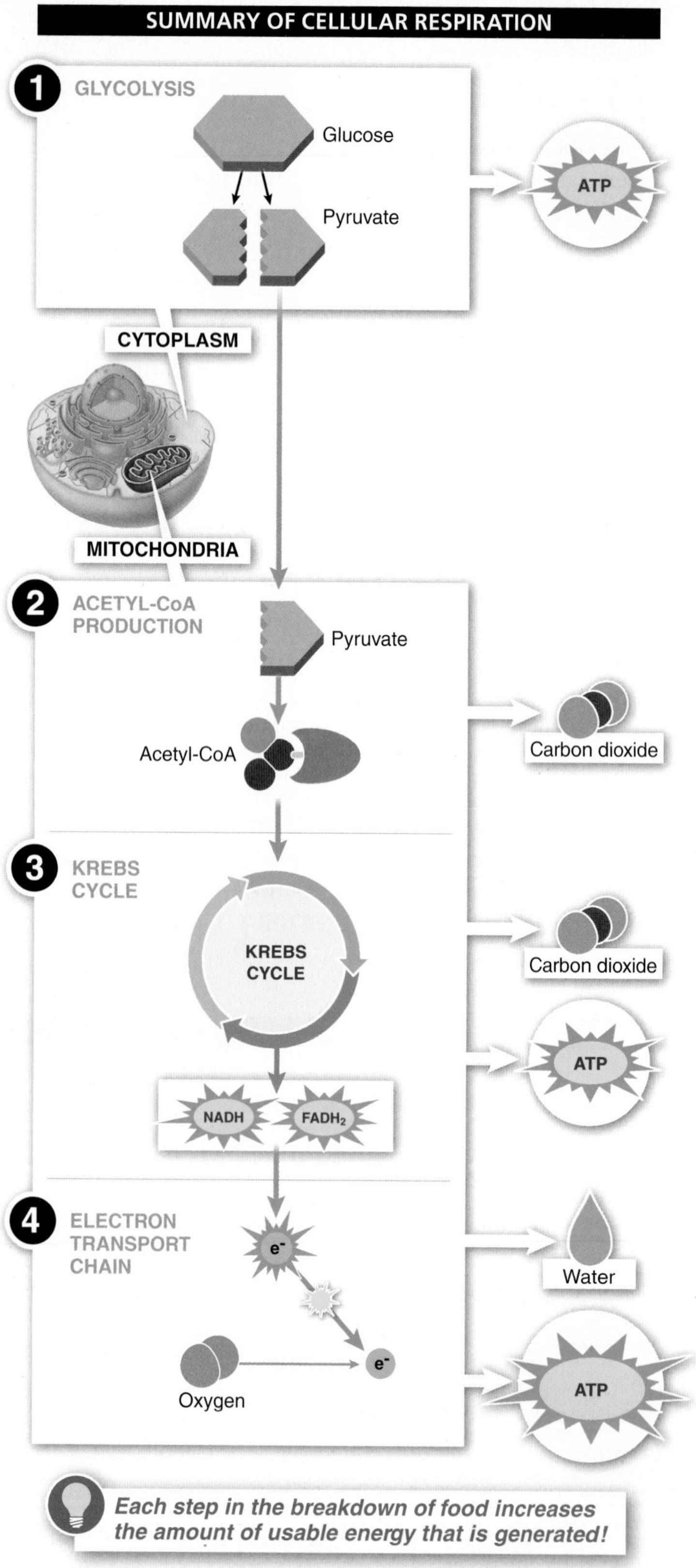

SUMMARY OF CELLULAR RESPIRATION

1 GLYCOLYSIS

Glucose

Pyruvate

ATP

CYTOPLASM

MITOCHONDRIA

2 ACETYL-CoA PRODUCTION

Pyruvate

Acetyl-CoA

Carbon dioxide

3 KREBS CYCLE

KREBS CYCLE

Carbon dioxide

ATP

NADH   FADH₂

4 ELECTRON TRANSPORT CHAIN

e⁻

e⁻

Oxygen

Water

ATP

Each step in the breakdown of food increases the amount of usable energy that is generated!

FIGURE 4-35 The steps of cellular respiration: from glucose to usable energy.

to the potential energy of water in an elevated tower, which can be released with great force. Similarly, in step 4 of the mitochondrial electron transport chain, the protons pumped into the intermembrane space rush back into the mitochondrial matrix through channels in the inner mitochondrial membrane. And as the protons pass through, the force of their flow fuels the attachment of free-floating phosphate groups to ADP to produce ATP.

In the end, the number of ATP molecules generated from the complete dismantling of one molecule of glucose is about 36, most of which are produced with the energy harnessed from high-energy electron carriers as they pass their electrons down the electron transport chain (**FIGURE 4-35**).

Given the central role of the electron transport chain in the generation of usable energy from the breakdown of food molecules, any interference in its functioning has dire consequences. And in fact, murder by cyanide poisoning, an old tradition in detective stories, is just such an interference. When cyanide gets into the mitochondria, it binds to a molecule in the electron transport chain, preventing it from accepting electrons. This halts the transfer of electrons and the pumping of protons across the mitochondrial membrane. As a consequence, the production of ATP that would occur when protons rushed back across the membrane down their concentration gradient ceases. Halting the production of ATP removes a cell's energy source, starving it very quickly. For this reason, cyanide poisoning can cause death within minutes.

Q Cyanide blocks the passage of electrons to oxygen in the electron transport chain. Why does this make it a toxic poison?

TAKE-HOME MESSAGE 4·15

The largest energy payoff of cellular respiration comes as electrons from the NADH and FADH₂ produced during glycolysis and the Krebs cycle move along the electron transport chain. The electrons are passed from one carrier to another and energy is released, pumping protons into the mitochondrial intermembrane space. As the protons rush back into the mitochondrial matrix, the force of their flow fuels the production of large amounts of ATP.

Can we combat the fatigue and reduced cognitive functioning of jet lag with NADH pills?

Often, scientific thinking is applied to questions about the natural world with the sole purpose of better understanding organisms and how they function. In other cases, research questions are formulated and investigated with the intention of applying the knowledge gained to address specific problems.

In many cases, the transition to applied science is a logical and natural outgrowth of basic research. Consider the question of treating the serious effects of jet lag.

What is jet lag and why is it of scientific interest?

Jet lag occurs when a person travels across several time zones and there is a mismatch between her body clock and the time of day or night at her destination. It is accompanied by a constellation of symptoms, including fatigue, gastrointestinal distress, memory loss, and reductions in cognitive performance.

Jet lag affects a large number of travelers including pilots and other air crew. It can have serious consequences because it hinders decision-making abilities, effective communication, and memory.

How is jet lag related to cellular respiration?

Researchers suspected that interventions targeting cellular respiration—with an eye toward increasing the rate at which cells generate ATP—might decrease the fatigue experienced in jet lag. In particular, they have focused on the high-energy electron carrier NADH. As we saw in the previous section, during cellular respiration energy is captured in the high-energy electron carriers NADH and FADH$_2$.

Why should NADH alleviate symptoms of jet lag?

If levels of NADH could be increased simply by taking the molecule in pill form, this might, given the central role of NADH as a source of potential energy, lead to increased production of usable energy through the electron transport chain. And so this hypothesis gave rise to a testable prediction: "Supplementing NADH should counteract some of the effects of jet lag, including reduced cognitive functioning and fatigue."

The experimental setup. The researchers used a randomized, controlled, double-blind experimental design. The participants were 36 volunteers, 35-55 years old, with at least 14 years of formal education and normal sleep schedules. During the study, the participants did not consume any caffeine, alcohol, or any medications known to affect nervous system functioning.

The volunteers were randomly assigned to one of two groups, placebo or NADH, and took a battery of tests to establish their baseline performance. They then took an overnight "red-eye" flight across four time zones, from California to Maryland. They arrived at 6 A.M., were given breakfast, and were then administered a pill containing either NADH or a placebo. At 9:30 A.M. and again at 12:30 P.M., they were given the same battery of tests.

Did NADH reduce the symptoms of jet lag?

Overall, the participants receiving NADH performed significantly better on four tests of cognitive functioning and reported less sleepiness. The results from the specific tests suggest that the improvements in cognitive functioning experienced by the NADH-taking subjects have real-world relevance.

1. Vigilance. Participants watched a computer monitor and responded each time they saw a particular symbol. They were scored on the number of errors they made, particularly "errors of omission" indicating lapses of attention. Here are the results for the two groups:

Placebo: 37% of subjects made omission errors.

NADH: 14% of subjects made omission errors.

2. Working memory. The participants were required to remember numbers and perform mental operations on them.

Placebo: Subjects answered 6.8 more problems per minute than in the baseline test.

NADH: Subjects answered 13.2 more problems per minute than in the baseline test.

3. Multi-tasking. Participants were required to shift between two different tasks that involved marking numbers on a spreadsheet.

> *Placebo:* Subjects increased performance by 19.2 points over baseline.
>
> Subjects' reaction time was *slower* than baseline by 0.44 seconds.
>
> *NADH:* Subjects increased performance by 77.5 points over baseline.
>
> Subjects' reaction time was *faster* than baseline by 0.15 seconds.

4. Visual perception. Participants viewed a 4 × 4 checkerboard pattern and, on the screen that followed, had to identify the matching pattern.

> *Placebo:* Subjects completed 1.4 more items per minute than at baseline.
>
> *NADH:* Subjects completed 5.4 more items per minute than at baseline.

5. Sleepiness. Participants self-reported their sleepiness on a 7-point scale.

> *Placebo:* 75% of subjects reported increased sleepiness.
>
> *NADH:* 25% of subjects reported increased sleepiness

What conclusions can we draw from these results?

The results reported in this well-designed study were clear and definitive. The placebo-receiving jet-lagged volunteers were more likely to make errors related to not paying attention, had greater difficulty with memory and concentration, and were less effective at multi-tasking. The researchers' conclusion, supported by the evidence, was that "NADH appears to be a suitable short-term countermeasure for the effects of jet lag on cognition and sleepiness."

What degree of confidence should we have that the question of whether NADH reduces jet lag is answered?

While the results do support the researchers' hypothesis, it still may be premature to consider the issue completely resolved. As they pointed out, the optimal doses of NADH still need to be investigated. Additionally, they only examined the subjects' response to NADH directly following the red-eye flight. It is not clear what the duration of the effect of NADH might be on cognition and sleepiness.

It is wise, also, to be aware of any biases—even unconscious biases—that might influence researchers. In this study, for example, the researchers reported that "Menuco Corporation funded the study." A quick search reveals that Menuco Corporation was founded by one of the study's authors and is a for-profit company that markets and sells NADH supplements. The company and author hold the patent for the manufacturing process of the NADH supplement used in the study.

These facts do not invalidate the results, however. The study was carefully controlled and well designed. And most important, the researchers described their methods in such detail that the research can be replicated. Given the small number of subjects in their study, as well as the increasing relevance of jet lag in today's world, replication of these findings would be an important factor in increasing our confidence in the generalizability of the conclusions.

TAKE-HOME MESSAGE 4·16

The symptoms of jet lag—including fatigue, memory loss, and reductions in cognitive performance—can impair the performance of people in many professions today. The results of a randomized, controlled, double-blind study support the hypothesis that an NADH supplement may be a suitable short-term countermeasure for these effects.

4·17 – 4·18
There are alternative pathways to energy acquisition.

Unlike cars, living organisms (such as this cheetah) use multiple types of fuel.

4·17 Beer, wine, and spirits are by-products of cellular metabolism in the absence of oxygen.

Every beer brewery, the entire wine industry, and all distilleries of whiskey, vodka, tequila, and other alcoholic beverages owe their existence to microscopic yeast cells scrambling to break down their food for energy under stressful conditions. To better understand how yeast metabolism produces alcohol, it helps to begin by investigating what happens when humans and other animals try to metabolize energy from sugar molecules under some stressful conditions.

If you run or swim as fast as you can, you soon feel a burning sensation in your muscles. Why? Your muscle cells are becoming very acidic. This acid buildup occurs when we demand of our bodies bursts of energy beyond what they can sustain (**FIGURE 4-36**). (The next-day muscle soreness is not due to acid buildup, which goes away in a matter of minutes or hours; this soreness is caused by damage to the muscle fibers. They break down a bit before growing stronger.)

With rapid, strenuous exertion, our bodies soon fall behind in delivering oxygen from the lungs to the bloodstream to the cells and finally to the mitochondria. Oxygen deficiency then limits the rate at which the mitochondria can break down fuel and produce ATP (**FIGURE 4-37**). This slowdown occurs because the electron transport chain requires oxygen as the final acceptor of all the electrons generated during glycolysis and the Krebs cycle. If oxygen is in short supply, the electrons from NADH (and $FADH_2$) have nowhere to go. Consequently, the regeneration of NAD^+ (and FAD) in the electron transport chain is halted, leaving no recipient molecules for the high-energy electrons harvested from the

breakdown of glucose and pyruvate, and the whole process of cellular respiration can grind to a stop. Organisms don't let this interruption last long, though; most have a backup method for breaking down sugar.

Among animals, there is an acceptor for the NADH electrons in the absence of oxygen: pyruvate, the end product of glycolysis. When pyruvate accepts the electrons, it forms lactic acid (see Figure 4-37). Once the NADH gives up its electrons, NAD^+ is regenerated, and glycolysis

During strenuous exertion, our muscles can require more oxygen than is available to them.

FIGURE 4-36 Energy production without oxygen. Organisms have a backup method for breaking down sugar when oxygen is not present.

can continue. But as lactic acid builds up, it causes a burning feeling in our muscles. It's not ideal, but to escape a predator or to exercise strenuously, the two ATP molecules generated from each glucose molecule during glycolysis—which can serve as an immediate energy source—are better than nothing.

Like humans, yeast normally use oxygen during their breakdown of food. And like humans, they have a backup method when oxygen is not available. But yeast make use of a different electron acceptor, and the resulting reaction leads to the production of all drinking alcohol.

After glycolysis in these single-celled organisms, pyruvate is usually converted to a molecule called *acetaldehyde,* releasing bubbles of CO_2 in the process (which, when yeast is used in baking, allows bread to rise). In the absence of oxygen, acetaldehyde accepts the electrons released from NADH, allowing glycolysis to resume. Acetaldehyde's acceptance of NADH's electrons results in the production of **ethanol,** the molecule that gives beer, wine, and spirits their kick. Ethanol can also be used as a fuel source because it can be combusted in much the same way as the biofuels generated from animal fats and plant oils described earlier in this chapter.

Fermentation is the process by which cells obtain energy in the absence of oxygen. It occurs when, following glycolysis, alternative molecules are used as electron acceptors.

Yeast is used in the production of beer.

FIGURE 4-38 **Just a by-product of metabolism under stressful conditions.** Beer, wine, and spirits are by-products of cellular metabolism in the absence of oxygen.

Interestingly, although ethanol is always the alcohol produced by fermentation, the flavor of the output of fermentation depends on the type of sugar metabolized by the yeast. Fruits, vegetables, and grains all give different results. If the sugar comes from grapes, wine is produced. If the sugar comes from a germinating barley plant, beer is produced. Potatoes, on the other hand, are the sugar source usually used to produce vodka.

Because yeast prefer the more efficient process of aerobic respiration, they produce alcohol only in the absence of oxygen. Fermentation tanks used in producing wine, beer, and other spirits are built to keep oxygen out so that yeast cells are forced to use their backup pathway of fermentation (**FIGURE 4-38**).

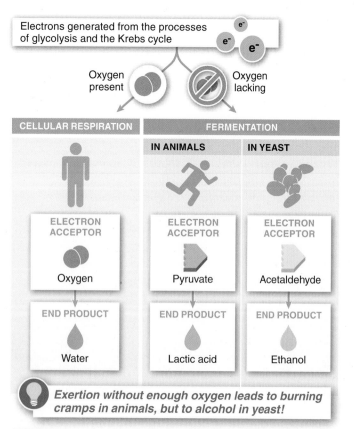

FIGURE 4-37 **Energy production with and without oxygen.**

Exertion without enough oxygen leads to burning cramps in animals, but to alcohol in yeast!

TAKE-HOME MESSAGE 4·17

Oxygen deficiency limits the breakdown of fuel because the electron transport chain requires oxygen as the final acceptor of electrons during the chemical reactions of glycolysis and the Krebs cycle. When oxygen is unavailable, yeast resort to fermentation, in which they use a different electron acceptor, acetaldehyde, and in the process generate ethanol, the alcohol in beer, wine, and spirits.

4·18 Eating a complete diet: cells can run on protein and fat as well as on glucose.

Most automobiles can run on only one type of fuel, gasoline. If you run out of fuel and cannot get to a gas station, you are out of luck. In this chapter, we have examined the steps by which plants, animals, and other organisms use glucose as fuel. But living organisms are more flexible than cars when it comes to fuel sources. They also have more complex needs because their fuel must provide not only energy but also the raw materials for growth.

Q If a person has a low-carb diet, what provides the fuel for his body?

Evolution has built humans and other organisms with the metabolic machinery that allows them to extract energy and other valuable chemicals from proteins, fats, and a variety of carbohydrates

(**FIGURE 4-39**). For that reason, we are able to consume and efficiently utilize meals comprising various combinations of molecules.

Sugars. In the case of dietary carbohydrates, many are polysaccharides—multiple simple sugars linked together—rather than solely the simple sugar glucose. Before they can be broken down by cellular respiration, the polysaccharides must first be separated by enzymes into glucose or related simple sugars.

Lipids. Dietary lipids are broken down into their two constituent parts: a glycerol molecule and fatty acids. The glycerol is chemically modified into one of the molecules produced during glycolysis. It enters the glycolysis pathway at that step and is broken down to yield energy. The fatty acids, meanwhile, are chemically modified into acetyl-CoA, at which point they enter the Krebs cycle.

Proteins. Proteins are chains of amino acids. After consumption, the chains are broken down chemically, then each amino acid is broken down into (1) an amino group that may be used in the production of tissue or excreted in the urine, and (2) a carbon compound that is converted to one of the intermediate compounds in glycolysis or the Krebs cycle, allowing the energy stored in its chemical bonds to be harnessed.

In the end, humans (indeed, all animals) are able to harvest energy from a variety of food sources beyond simple glucose.

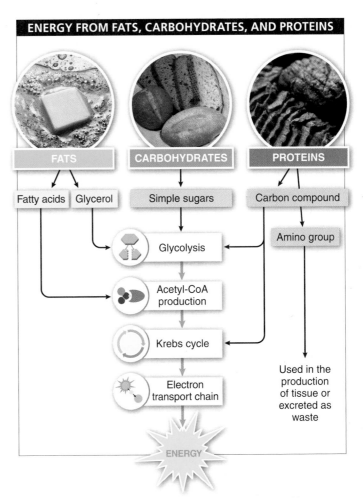

ENERGY FROM FATS, CARBOHYDRATES, AND PROTEINS

FATS → Fatty acids | Glycerol

CARBOHYDRATES → Simple sugars

PROTEINS → Carbon compound | Amino group

Glycolysis

Acetyl-CoA production

Krebs cycle

Electron transport chain

ENERGY

Used in the production of tissue or excreted as waste

FIGURE 4-39 *"More flexible than autos."* Animals are able to harvest energy from proteins, carbohydrates, and lipids.

Whether a meal contains carbohydrates, lipids, proteins, or some combination thereof, the nutrients are chemically modified in some preliminary steps and then fed into one of the intermediate steps in glycolysis or the Krebs cycle to furnish usable energy for the organism (see Figure 4-39).

TAKE-HOME MESSAGE 4·18

Humans and other organisms have metabolic machinery that allows them to extract energy and other valuable chemicals from proteins, fats, and carbohydrates in addition to the simple sugar glucose.

If you feed and protect your flowers in a vase, they'll last longer.

Like animals, plants require fuel for the cellular activities that keep them alive.

Q: **How do plants get the energy they need to stay alive?** Plants use photosynthesis to harness light energy, converting it to sugar molecules that serve as their food.

Q: **Can plants still photosynthesize once they're in a vase in your house?** Yes, but humans don't make it easy. Once cut, plants generally cannot produce sufficient sugar through photosynthesis. Light levels in houses tend to be a bit too low, and the loss of many or most plant leaves reduces the number of chloroplasts in which photosynthesis can occur. So they're starving to death.

Q: **Can you slow their demise?** Yes. Plants are able to take up sugar in the vase water and use it as an energy source for cell activities. So adding a bit of sugar to the vase is like putting fuel in their tank.

Q: **Is it that easy?** No. Putting sugar—a molecule with lots of energy stored within its chemical bonds—in the vase water is like offering a free lunch. And many, many organisms are looking for a free lunch. Unfortunately, when you add sugar, bacteria on the flower stems can grow rapidly, blocking the water-conducting tubes in the stems. This slows the flow not just of sugar, but of water as well.

Q: **What should you do?** With the addition to vase water of both sugar and an antibacterial chemical such as chlorine bleach, you can feed and protect your cut flowers, significantly increasing their longevity.

Conclusion: Most flowers will last longer if you cut their stems underwater and at a slant, to maximize water absorption. Then place the flowers in a flask with about 2 inches of warm water, which enhances the flow into the flower. At this point, add a spoonful of sugar and a drop or two of bleach. Then, after a few minutes, transfer the flowers to a vase.

Check Your Knowledge

Thinking critically about visual displays of data

1. What are the axes of these two graphs?

2. What variable(s) is presented? How was it measured? What do the colors represent?

3. Do you know the source of the information in the graph? Does that matter? Why or why not?

4. Why are there two graphs? What is the difference between them?

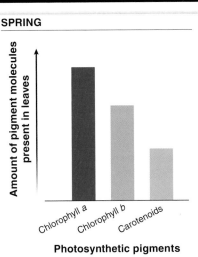
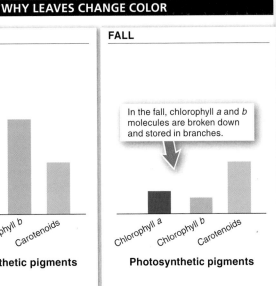

WHY LEAVES CHANGE COLOR

SPRING

Amount of pigment molecules present in leaves

Chlorophyll *a* | Chlorophyll *b* | Carotenoids

Photosynthetic pigments

FALL

In the fall, chlorophyll *a* and *b* molecules are broken down and stored in branches.

Chlorophyll *a* | Chlorophyll *b* | Carotenoids

Photosynthetic pigments

5. What can you conclude from this figure?

6. What additional information would make this figure more helpful? Why?

7. Do the data report experimental results? Is there a control group? An experimental group?

See answers at the back of the book.

Key Terms in Energy

adenosine triphosphate (ATP), p. 141
biofuel, p. 136
C4 photosynthesis, p. 154
Calvin cycle, p. 152
CAM (crassulacean acid metabolism), p. 155
carotenoid, p. 146
cellular respiration, p. 137
chemical energy, p. 138

chlorophyll, p. 145
chlorophyll *a*, p. 146
chlorophyll *b*, p. 146
chloroplast, p. 145
electromagnetic spectrum, p. 146
electron transport chain, p. 150
energy, p. 137
ethanol, p. 168
fermentation, p. 168

first law of thermodynamics, p. 140
fossil fuel, p. 136
glycolysis, p. 158
kinetic energy, p. 138
Krebs cycle, p. 160
light energy, p. 146
mitochondrial matrix, p. 162
NADPH, p. 151

photon, p. 146
photosynthesis, p. 137
photosystem, p. 149
pigment, p. 146
potential energy, p. 138
primary electron acceptor, p. 149
pyruvate, p. 158
rubisco, p. 152

second law of thermodynamics, p. 140
stoma (*pl.* stomata), p. 154
stroma, p. 145
thermodynamics, p. 140
thylakoid, p. 145

ABOUT THE CHAPTER OPENING PHOTO

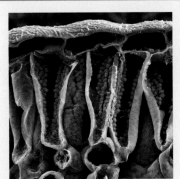

A scanning electron micrograph of a section through the leaf of the Christmas rose (*Helleborus niger*). In the body of the leaf (center) are numerous cells containing chloroplasts (green), the organelles that are the site of photosynthesis within the leaf.

4·1—4·4 Energy flows from the sun and through all life on earth.

Energy from the sun fuels all life on earth as the energy is converted to different forms.

Check Your Knowledge

1. Animal fats and plant oils are sometimes used as sources of fuel for automobile engines. How is energy harvested from these molecules?

   a) They contain long chains of hydrogen and carbon that, when broken, release the energy stored in the bonds linking the atoms together.

   b) They contain hydrogen and carbon tails linked by covalent bonds that, when broken, release chemical energy.

   c) They contain multiple phosphate groups that each release energy when "liberated" from the molecule chemically.

   d) They contain long hydrophobic regions that, when mixed with water, generate explosive resistances.

   e) They contain long carbon tails, and each atom has unpaired electrons that are released on exposure to extreme heat and pressure.

   0 — 30 — 100
   EASY   HARD

2. A cyclist rides her bike up a very steep hill. Which of the following statements properly describes this activity in energetic terms?

   a) Potential energy in food is converted to kinetic energy as the cyclist's muscles push her up the hill.

   b) Kinetic energy is highest when the cyclist is at the crest of the hill.

   c) The cyclist produces the most potential energy as she cruises down the hill's steep slope.

   d) Potential energy is greatest when the cyclist is at the top of the hill.

   e) Both a) and d) are correct.

   0 — 43 — 100
   EASY   HARD

3. Every time a source of energy is converted from one form to another:

   a) the potential energy of the system increases.

   b) heat is required.

   c) the second law of thermodynamics is violated.

   d) the total amount of energy in the universe is reduced by a tiny amount.

   e) some of the energy is converted to heat, which is not a very usable form of kinetic energy.

   0 — 28 — 100
   EASY   HARD

ENERGY CAPTURE AND CONVERSION

Plants capture energy from the sun and store it in the chemical bonds of sugars and other food molecules. Organisms release the energy stored in the chemical bonds of food molecules they eat and use it as fuel.

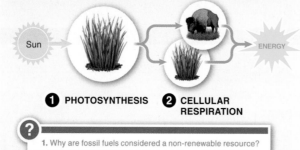

1 PHOTOSYNTHESIS **2** CELLULAR RESPIRATION

1. Why are fossil fuels considered a non-renewable resource?

KINETIC AND POTENTIAL ENERGY

Energy, the capacity to do work, comes in two forms.

KINETIC ENERGY   POTENTIAL ENERGY

KINETIC ENERGY
The energy of moving objects

POTENTIAL ENERGY
The energy that is stored in objects, including chemical bonds

2. In terms of energy, describe what happens when a ball at the top of a steep ramp is released.

ENERGY TRANSFORMATIONS

As energy is captured and converted, the amount of energy available to do work decreases. Some energy is released as heat.

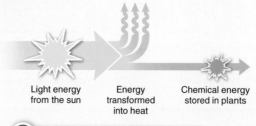

Light energy from the sun | Energy transformed into heat | Chemical energy stored in plants

3. Why is the conversion of energy from one form to another considered inefficient?

ADENOSINE TRIPHOSPHATE (ATP) AND ADENOSINE DIPHOSPHATE (ADP)

Cells temporarily store energy in the bonds of ATP molecules. This potential energy can be converted to kinetic energy and used to fuel life-sustaining chemical reactions.

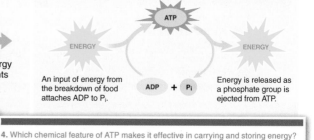

ATP

ENERGY   ENERGY

An input of energy from the breakdown of food attaches ADP to P_i.

$ADP + P_i$

Energy is released as a phosphate group is ejected from ATP.

4. Which chemical feature of ATP makes it effective in carrying and storing energy?

4·5—4·11 Photosynthesis uses energy from sunlight to make food.

In photosynthesis, plants transform light energy into the chemical energy of ATP and NADPH, while splitting water molecules and producing oxygen.

PHOTOSYNTHESIS: THE BIG PICTURE

Plants use water, the energy of sunlight, and carbon dioxide gas from the air to produce sugars and other organic materials. In the process, photosynthesizing organisms also produce oxygen, which makes all animal life possible.

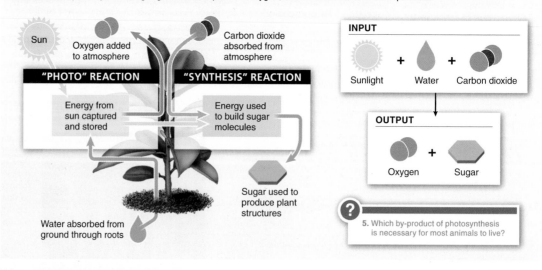

Sun

Oxygen added to atmosphere | Carbon dioxide absorbed from atmosphere

"PHOTO" REACTION | "SYNTHESIS" REACTION

Energy from sun captured and stored | Energy used to build sugar molecules

Sugar used to produce plant structures

Water absorbed from ground through roots

INPUT
Sunlight + Water + Carbon dioxide

OUTPUT
Oxygen + Sugar

5. Which by-product of photosynthesis is necessary for most animals to live?

"PHOTO" REACTIONS

In the "photo" part of photosynthesis, chloroplasts transform light energy into the chemical energy of ATP and NADPH, while splitting water molecules and producing oxygen.

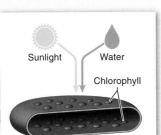
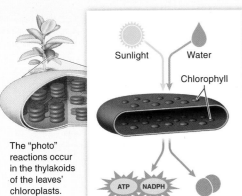

The "photo" reactions occur in the thylakoids of the leaves' chloroplasts.

"SYNTHESIS" REACTIONS

In the Calvin cycle, the "synthesis" part of photosynthesis, carbon from CO_2 in the atmosphere is attached to molecules in chloroplasts to build sugars. This production of sugars consumes ATP and NADPH generated in the "photo" part of photosynthesis.

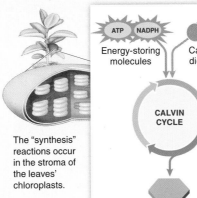
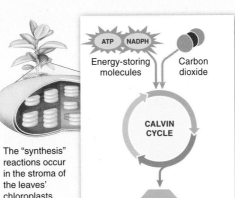

The "synthesis" reactions occur in the stroma of the leaves' chloroplasts.

---

6. In which parts of a plant are most of its chloroplasts typically found? Why?

7. Why do leaves usually appear green?

8. Describe one of the chief ways in which energy "moves" through cells within a plant.

9. As they are excited by light energy during photosynthesis, electrons sometimes move from one molecule to another. Other times they do not. Explain this difference.

10. Rubisco is the most abundant protein on earth. What does it do?

C3 PHOTOSYNTHESIS

The most common photosynthetic pathway among plant. For C3 photosynthesis to function efficiently, plants must open their stomata for access to carbon dioxide, making them vulnerable to water loss from evaporation.

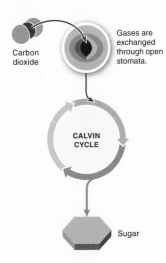

Carbon dioxide

Gases are exchanged through open stomata.

Sugar

C4 PHOTOSYNTHESIS

C4 photosynthesis can function efficiently even when a plant does not fully open its stomata, enabling the plant to reduce water loss from evaporation. It is energetically more costly than C3 photosynthesis and so is most common in dry, hot habitats where the benefits of reduced water loss are significant.

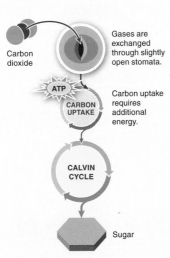

Carbon dioxide

Gases are exchanged through slightly open stomata.

Carbon uptake requires additional energy.

Sugar

CAM PHOTOSYNTHESIS

Plants using CAM photosynthesis open their stomata only at night. Because CAM plants close their stomata in the day, they reduce water loss, making this valuable in hot, dry habitats. But this reduces the overall amount of carbon dioxide available to the plants for photosynthesis, reducing their growth rates.

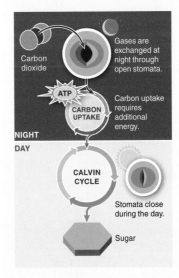

Carbon dioxide

Gases are exchanged at night through open stomata.

Carbon uptake requires additional energy.

NIGHT

DAY

Stomata close during the day.

Sugar

---

11. To reduce water loss via evaporation, plants may close their stomata. Although this action solves one problem for a plant, it creates another. Describe the new problem.

---

4. In your body, when energy is released during the breakdown of glucose:

a) adenosine monophosphate is created.

b) adenosine diphosphate is created.

c) some energy may be harnessed to build high-energy bonds that attach phosphate groups to ADP molecules.

d) molecules of ATP capture and absorb the heat from the reaction.

e) it forms adenosine-CoA.

5. A green plant can carry out photosynthesis if given nothing more than:

a) water, light, and carbon dioxide.

b) water, light, and oxygen.

c) carbon dioxide.

d) oxygen.

e) oxygen and carbon dioxide.

6. The actual production of sugars during photosynthesis takes place:

a) in the chloroplast outer membrane.

b) within the stroma, inside the thylakoids of the chloroplast.

c) within the stroma, outside the thylakoids of the chloroplast.

d) within the mitochondria.

e) within the thylakoid membranes.

7. The leaves of plants can be thought of as "eating" sunlight because:

a) light energy, like chemical energy released when the bonds of food molecules are broken, is a type of kinetic energy.

b) both light energy and food energy can be interconverted without heat loss.

c) the carbon-oxygen bonds within a photon of light release energy when broken by the enzymes in chloroplasts.

d) the carbon-hydrogen bonds within a photon of light release energy when broken by the enzymes in chloroplasts.

e) photons contain hydrocarbons.

8. A molecule of chlorophyll increases in potential energy:

a) when it binds to a photon.

b) when a photon strikes it, boosting electrons to a higher-energy excited state.

c) when it loses an electron.

d) only in the presence of oxygen.

e) None of the above. The potential energy of a molecule cannot change.

173

9. **Plants rely on water:**

a) to provide the protons necessary to produce chlorophyll.

b) to concentrate beams of sunlight on the reaction center.

c) to replenish oxygen molecules that are lost during photosynthesis.

d) to replace electrons that are excited by light energy and passed down an electron transport chain.

e) to serve as an energy source.

10. **During photosynthesis, which step is most responsible for a plant's acquisition of new organic material?**

a) the "building" of NADPH during the Calvin cycle

b) the excitation of chlorophyll molecules by photons of light

c) the "plucking" of carbon atoms from the air and fixing of the carbons to organic molecules within the chloroplast

d) the loss of water through evaporation

e) ATP made during the light reactions

0 ——————— 50 ——————— 100
EASY   HARD

11. **During C4 photosynthesis:**

a) plants use less ATP in making sugar.

b) plants can produce sugars even when they close their stomata to reduce water loss on hot days.

c) plants are able to generate water molecules to cool their leaves.

d) plants produce more rubisco.

e) plants are able to produce sugars without any input of carbon dioxide.

12. **During cellular respiration:**

a) oxygen is used to transport chemical energy throughout the body.

b) metabolic oxygen is produced.

c) light is converted to kinetic energy.

d) ATP is converted to water and sugar.

e) energy from the chemical bonds of food molecules is captured.

0 ——————— 50 ——————— 100
EASY   HARD

13. **During cellular respiration, most of the energy contained within the bonds of food molecules is captured in:**

a) the conversion of the kinetic energy of food to the potential energy of ATP.

b) the Krebs cycle and electron transport chain.

c) digestion.

d) glycolysis.

e) None of the above. Energy is lost, not gained, during cellular respiration.

4·12—4·16 Cellular respiration converts food molecules into ATP, a universal source of energy for living organisms.

Living organisms extract energy through a process called cellular respiration in which glucose and oxygen are converted to carbon dioxide, water, and energy.

CELLULAR RESPIRATION: THE BIG PICTURE

In cellular respiration, the high-energy bonds of sugar and other energy-rich molecules are broken, releasing the energy that went into creating them. This process requires fuel molecules and oxygen, and it yields ATP molecules, water, and carbon dioxide.

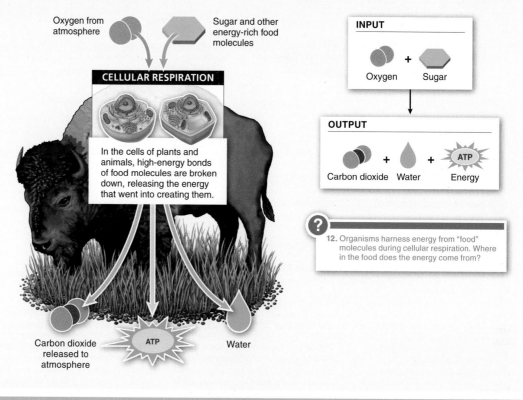

Oxygen from atmosphere

Sugar and other energy-rich food molecules

CELLULAR RESPIRATION

In the cells of plants and animals, high-energy bonds of food molecules are broken down, releasing the energy that went into creating them.

Carbon dioxide released to atmosphere

ATP

Water

INPUT

Oxygen + Sugar

OUTPUT

Carbon dioxide + Water + ATP Energy

12. Organisms harness energy from "food" molecules during cellular respiration. Where in the food does the energy come from?

MITOCHONDRIA: A CLOSER LOOK AT STRUCTURE

The structure of mitochondria makes possible their impressive ability to harness energy from food molecules.

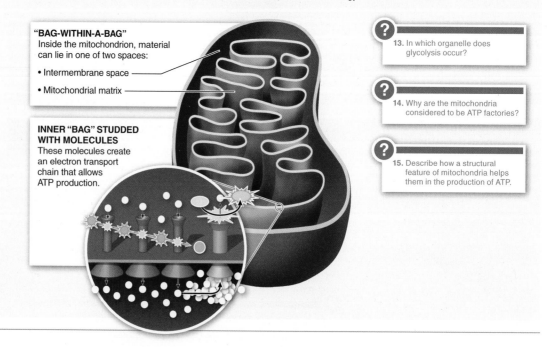

"BAG-WITHIN-A-BAG"
Inside the mitochondrion, material can lie in one of two spaces:

• Intermembrane space

• Mitochondrial matrix

INNER "BAG" STUDDED WITH MOLECULES
These molecules create an electron transport chain that allows ATP production.

13. In which organelle does glycolysis occur?

14. Why are the mitochondria considered to be ATP factories?

15. Describe how a structural feature of mitochondria helps them in the production of ATP.

The steps of cellular respiration: from glucose to usable energy.

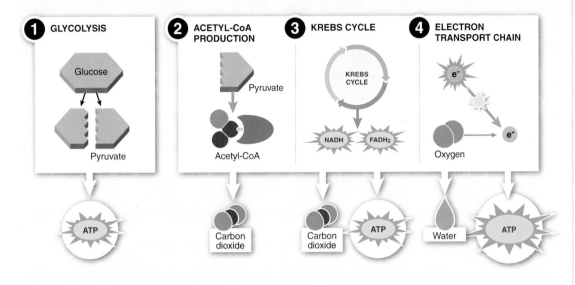

1 GLYCOLYSIS

Glucose → Pyruvate

ATP

2 ACETYL-CoA PRODUCTION

Pyruvate → Acetyl-CoA

Carbon dioxide

3 KREBS CYCLE

KREBS CYCLE

NADH   FADH₂

Carbon dioxide   ATP

4 ELECTRON TRANSPORT CHAIN

e⁻   →   e⁻

Oxygen

Water   ATP

4·17–4·18 There are alternative pathways to energy acquisition

Organisms can generate ATP when oxygen is lacking or when organic molecules other than glucose, such as lipids, proteins, and polysaccharides, are consumed.

ACQUIRING ENERGY WHEN OXYGEN IS NOT PRESENT

Oxygen deficiency limits the breakdown of fuel because the electron transport chain requires oxygen as the final acceptor of electrons during the chemical reactions of glycolysis and the Krebs cycle. When oxygen is unavailable, animals and yeast resort to fermentation, in which different electron acceptors are used. Animals use pyruvate, and in the process generate lactic acid. Yeast use acetaldehyde, and in the process generate ethanol, the alcohol in beer, wine, and spirits.

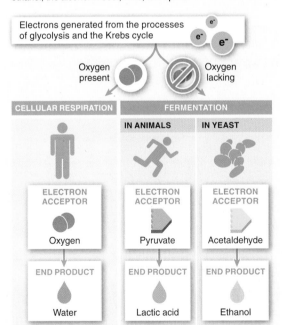

Electrons generated from the processes of glycolysis and the Krebs cycle

e⁻   e⁻   e⁻

Oxygen present   Oxygen lacking

CELLULAR RESPIRATION	FERMENTATION	
	IN ANIMALS	IN YEAST
ELECTRON ACCEPTOR	ELECTRON ACCEPTOR	ELECTRON ACCEPTOR
Oxygen	Pyruvate	Acetaldehyde
END PRODUCT	END PRODUCT	END PRODUCT
Water	Lactic acid	Ethanol

ACQUIRING ENERGY FROM SOURCES OTHER THAN GLUCOSE

Humans and other organisms have metabolic machinery that allows them to extract energy from sources other than glucose. Whether a meal contains carbohydrates, lipids, proteins, or some combination thereof, the nutrients are chemically modified in some preliminary steps and then fed into one of the intermediate steps in glycolysis or the Krebs cycle to furnish usable energy for the organism.

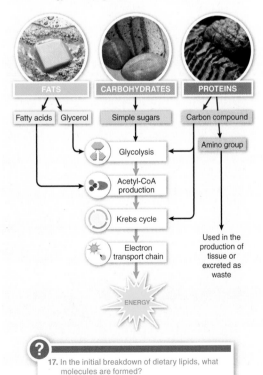

FATS   CARBOHYDRATES   PROTEINS

Fatty acids   Glycerol   Simple sugars   Carbon compound

Glycolysis

Acetyl-CoA production

Amino group

Krebs cycle

Electron transport chain

Used in the production of tissue or excreted as waste

ENERGY

16. A lack of oxygen limits the breakdown of fuel for most animals. Why?

17. In the initial breakdown of dietary lipids, what molecules are formed?

14. Which of the following energy-generating processes is the only one that occurs in all living organisms?
a) the Krebs cycle
b) glycolysis
c) combustion
d) photosynthesis
e) None of the above. There are no energy-generating processes that occur in all living organisms.

0 — EASY — 38 — HARD — 100

15. During the Krebs cycle:
a) the products of glycolysis are further broken down, generating additional ATP and the high-energy electron carrier NADH.
b) high-energy electron carriers pass their energy to molecules of sugar, which store the electrons as potential energy.
c) the products of glycolysis are broken down, generating ATP and ethanol.
d) cellular respiration can continue even in the absence of oxygen.
e) the products of glycolysis are converted to acetyl-CoA.

0 — EASY — 38 — HARD — 100

16. All alcoholic beverages are produced as the result of:
a) cellular respiration by bacteria that occurs in the absence of oxygen.
b) cellular respiration by bacteria that occurs in the absence of free electrons.
c) cellular respiration by yeast that occurs in the absence of free electrons.
d) cellular respiration that occurs in the absence of sugar.
e) cellular respiration by yeast that occurs in the absence of oxygen.

0 — EASY — 18 — HARD — 100

17. In harvesting the chemical energy of the molecules in food:
a) all macromolecules must first be converted to glucose.
b) all macromolecules must first be converted to hydrocarbon chains.
c) all macromolecules must first be converted to simple sugars.
d) organisms can use sugars, lipids, and proteins.
e) all macromolecules must first be converted to free-form amino acids.

0 — EASY — 59 — HARD — 100

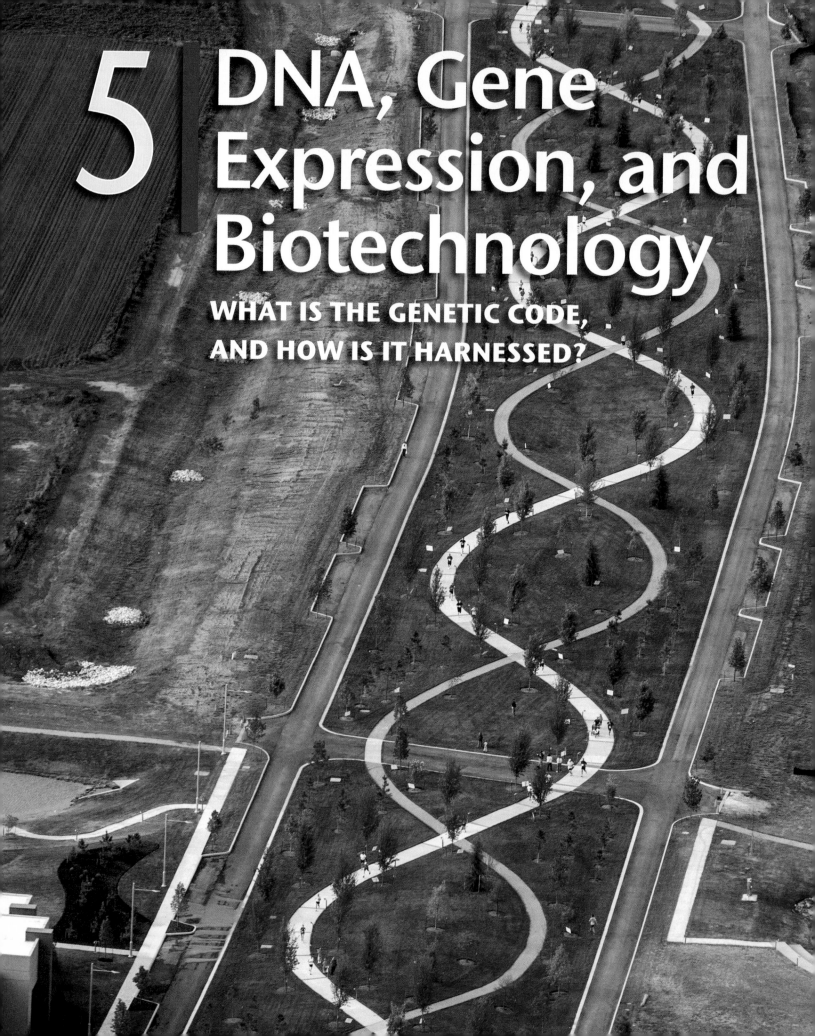

5 | DNA, Gene Expression, and Biotechnology

WHAT IS THE GENETIC CODE, AND HOW IS IT HARNESSED?

DNA: what is it, and what does it do?

Information in DNA directs the production of the molecules that make up an organism.

Damage to the genetic code has a variety of causes and effects.

Biotechnology is producing improvements in agriculture.

Biotechnology has the potential for improving human health (and criminal justice).

5·1–5·5
DNA: what is it, and what does it do?

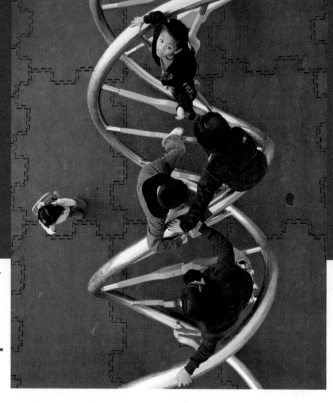

Children climb on a DNA sculpture.

5·1 Knowledge about DNA is increasing justice in the world.

In 1981, Julius Ruffin, a 27-year-old maintenance worker at Eastern Virginia Medical School, got onto an elevator and lost more than 20 years of his life. Several weeks earlier, a nursing student at the school had been raped by an attacker who broke into her apartment. When Ruffin got on the elevator, the student thought she recognized him as her attacker and called the police, who immediately arrested him. Ruffin's girlfriend testified that he was with her at the time of the attack, but on the basis of the victim's eyewitness testimony, a jury found Ruffin guilty and sentenced him to life in prison. Ruffin maintained his innocence—to no avail, until 2003. At that time, the state's Division of Forensic Science performed a DNA analysis on a swab of evidence that remained from the investigation. The analysis revealed that Ruffin was not the attacker. Rather, the analysis showed that the DNA perfectly matched that of a man who, in 2003, was already in a Virginia prison, serving time for rape. Ruffin was freed, but only after having served more than two decades in prison (**FIGURE 5-1**).

Q What is the most common reason that DNA analyses overturn incorrect criminal convictions?

Julius Ruffin's case is tragic, but it is not unusual. He is one of a group of 316 people in the United States (as of early 2014) who have been freed from prison as a result of DNA analyses. These unjustly imprisoned people spent an average of 13.5 years behind bars. Eighty percent had been convicted of sexual assault; 28% had been convicted of murder. In three-quarters of the cases, inaccurate eyewitness testimony played an important role in the guilty verdict.

(Recall, from Chapter 1, the experiments that revealed the unreliability of eyewitness identification.)

In this chapter, we take a close look at DNA, the molecule responsible for Julius Ruffin's exoneration and the deferred justice served to the 315 other people. All living organisms—people, plants, animals, bacteria, and otherwise—carry DNA in almost every cell in their body (with just a few exceptions). Like a social security number, every person's DNA is unique (with the exception of identical twins). In addition to being contained in our living cells, our DNA exists in what we leave behind. It is in our

FIGURE 5-1 **Vindicated by DNA evidence.** Julius Ruffin was released from prison after two decades of wrongful incarceration.

saliva, hair, blood, and even the dead skin cells that fall from our bodies. This is why DNA can serve as an individual identifier; it is a trail we all leave that is increasingly being used to ensure greater justice in our society.

The importance of DNA goes far beyond its function as an individual identifier, however. The information carried within this molecule, which is organized into individual units called genes, is among the most important of all biological knowledge. It contains instructions for the function of every enzyme and cell in our bodies and carries a record of the evolutionary history of lineages of cells and organisms. And, as witnessed by the following excerpts, it is often in the news.

> **❝**Selfish dictators may owe their behaviour partly to their genes, according to a study that claims to have found a genetic link to ruthlessness.**❞**
>
> — *Nature,* April 2008
>
> **❝***Too Many One-Night Stands? Blame Your Genes . . .* according to a new study, it may be fair to say that while you jolly well could help cheating, your particular genes did make things more difficult.**❞**
>
> — *Time* magazine, December 2010

In fact, it's nearly impossible to open a newspaper or watch a news report these days without being informed that yet another complex human characteristic or trait has been linked to a newly discovered gene (**FIGURE 5-2**). But what does that actually mean? What is this molecule that apparently exerts influence over everything from our fidelity to our ruthlessness? We explore these issues and more in this chapter, beginning with a look at the structure of DNA itself and how it contains the information for producing organisms. In later sections of the chapter, we learn how modern manipulations of DNA are

FIGURE 5-2 Genetics issues are in the news.

having far-reaching implications not just for human health but for agriculture as well.

> ## TAKE-HOME MESSAGE 5·1
>
> DNA is a molecule that all living organisms carry in almost every cell in their body. It contains instructions for the functions of every cell. Because every person's DNA is unique and because we leave a trail of DNA behind us as we go about our lives, DNA can serve as an individual identifier.

5·2 The DNA molecule contains instructions for the development and functioning of all living organisms.

Beginning in the 1900s and continuing through the early 1950s, a series of experiments revealed two important features of DNA. First, molecules of DNA are passed down from parent to offspring. Second, the instructions on how to create a body and control its growth, development, and behavior are encoded in the DNA molecule.

Given that DNA must be able to hold the instructions for how to produce every possible type of structure in every

living organism, scientists were in a frenzy to learn all they could about it. There was a spirited race to determine the chemical structure of DNA and to understand how the molecule was assembled and shaped so that it could hold and transmit so much information.

Many formidable scientists rose to the challenge of determining the structure of DNA. American Linus Pauling had already won a Nobel prize in chemistry for his work on

James Watson (left) and Francis Crick (right) figured out the exact structure of DNA

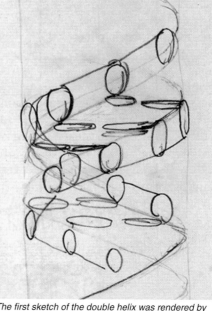

The first sketch of the double helix was rendered by Crick's wife, Odile

FIGURE 5-3 **Watson and Crick.**

elucidating the structure of molecules when he began investigating the structure of DNA. Simultaneously, Maurice Wilkins and Rosalind Franklin in England devoted their research to this task as well, and produced X-ray pictures of DNA that were critical to decoding its shape. But it was Englishman Francis Crick and American James Watson, working in Cambridge, England, who happened to put all the pieces together and deduce the exact structure of DNA (FIGURE 5-3).

As we'll see later in this chapter, their discovery was more than just a description of a molecule. As soon as they figured out DNA's structure, the answers to several other thorny problems in biology—such as how DNA might be able to duplicate itself—became apparent. We explore that process later, in the next chapter, but first we need to examine the structure of DNA in more detail.

DNA (deoxyribonucleic acid) is a **nucleic acid,** a macromolecule that stores information. It consists of individual units called **nucleotides,** which have three components: a molecule of sugar, a phosphate group (containing four oxygen atoms bound to a phosphorus atom), and a nitrogen-containing molecule called a **base.** The physical structure of DNA is frequently described as a "double helix." But what exactly is a double helix? Picture a long ladder twisted around like a spiral staircase and you'll have a good idea of what a DNA molecule looks like (FIGURE 5-4). The molecule has two distinct strands, like the vertical sides of a ladder. These are the "backbones" of the

DNA molecule, and each is made from two alternating molecules: a sugar, then a phosphate group, then a sugar, then a phosphate group, and so on. The sugar is always deoxyribose and the phosphate group is always the same, too. It is the shapes of the molecules in the backbone that cause the DNA "ladder" to twist.

The alternating sugars and phosphates hold everything in place, but they play only a supporting role. The rungs of the ladder are where things get interesting. Attached to each sugar, and protruding inward like half of a rung on the ladder, is one of four nitrogen-containing bases: adenine, thymine, cytosine, and guanine. When discussing DNA, these bases are usually referred to by their first letter: A, T, C, and G.

Both backbones of the ladder have a base protruding from each sugar. The base on one side of the ladder binds, via hydrogen bonds, to a base on the other side, and together these **base pairs** form the rungs of the ladder. They don't just pair up at random, though. Every time a C protrudes from one side, it forms hydrogen bonds with a G on the other side (and vice versa: a G always bonds to a C). Similarly, every time a T protrudes from one side, it forms hydrogen bonds with an A on the other side (and vice versa). For this reason, each DNA molecule always has the same number of Gs as Cs, and the same number of As as Ts. Because of these base-pairing rules, it also is true that if we know the base sequence for one of the strands in a DNA molecule, we know the sequence in the other. For this

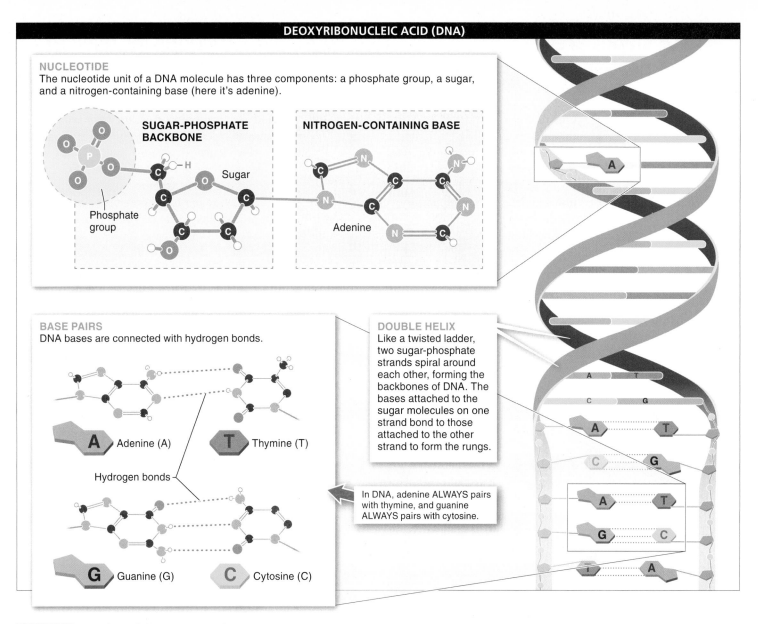

NUCLEOTIDE
The nucleotide unit of a DNA molecule has three components: a phosphate group, a sugar, and a nitrogen-containing base (here it's adenine).

SUGAR-PHOSPHATE BACKBONE

Phosphate group

Sugar

NITROGEN-CONTAINING BASE

Adenine

BASE PAIRS
DNA bases are connected with hydrogen bonds.

A Adenine (A)

T Thymine (T)

Hydrogen bonds

G Guanine (G)

C Cytosine (C)

DOUBLE HELIX
Like a twisted ladder, two sugar-phosphate strands spiral around each other, forming the backbones of DNA. The bases attached to the sugar molecules on one strand bond to those attached to the other strand to form the rungs.

In DNA, adenine ALWAYS pairs with thymine, and guanine ALWAYS pairs with cytosine.

FIGURE 5-4 **Overview of the structure of DNA.**

reason, a DNA sequence is described by writing the sequence of bases on only one of the strands.

If a human DNA molecule were really a twisted ladder, it would be a very, very long one. One molecule of DNA can have as many as 200 million base pairs, or rungs. How does such a molecule fit into a cell? The rungs are small, and the twisting of the molecule—twists upon twists—shortens it considerably: think about how, if you repeatedly twist your shoelace around and around, it becomes shorter and shorter. That's what happens with DNA. Let's now investigate how DNA's structure—the rungs of the ladder, in particular—enables it to carry information.

TAKE-HOME MESSAGE 5·2

DNA is a nucleic acid, a macromolecule that stores information. It consists of individual units called nucleotides, which consist of a sugar, a phosphate group, and a nitrogen-containing base. DNA's structure resembles a twisted ladder, with the sugar and phosphate groups serving as the backbones of the molecule and base pairs serving as the rungs. The sequence of bases on one side of the ladder-like DNA molecule complements that of the bases on the other side.

5·3 Genes are sections of DNA that contain instructions for making proteins.

Q Why is DNA considered the universal code for all life on earth?

One of DNA's most amazing features is that it embodies the instructions for building the cells and structures for almost every single living organism on earth (**FIGURE 5-5**). Thus, DNA is like a universal language, the letters of which are the bases A, T, C, and G. (Note that the sugar-phosphate backbone serves only to hold the bases in sequence, like the binding of a book. It does not convey genetic information.)

We've seen how the *structure* of DNA is like a spiral staircase. Another analogy may help you understand the *information-containing* aspect of DNA. You can think of an organism's DNA as a cookbook. Just as a cookbook contains detailed instructions on how to make a variety of foods (such as french toast, macaroni and cheese, or chocolate chip cookies), an organism's DNA carries the detailed instructions to build

an organism and keep it running. And just as a book can be viewed as a sequence of letters, with the book's meaning determined by which letters are strung together and in what order, a molecule of DNA can be viewed as a sequence of bases. Letters don't have much meaning on their own, of course, but when they are put together into words and sentences, their order holds a great deal of information. Similarly, the sequence in which bases appear in a molecule of an organism's DNA makes up a **code** that holds the detailed instructions for the building of the organism—chiefly in the form of instructions for the amino acid sequences of polypeptides. Upon processing and folding, these polypeptides become the numerous and varied protein molecules that make up an organism—whether it is a one-celled amoeba, a giant oak tree, or a biology student.

The full set of DNA present in an individual organism is called its **genome** (**FIGURE 5-6**). In prokaryotes, including all

Human (Homo sapiens)

Onion (Allium cepa)

Fruit fly (Drosophila melanogaster)

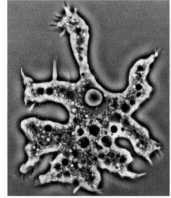
Amoeba (Hartmannella *sp.*)

Salamander (Amphiuma means)

 DNA provides the instructions for building virtually every organism on earth!

FIGURE 5-5 **DNA is the universal code for all life on earth.**

FIGURE 5-6 The genome unpacked.

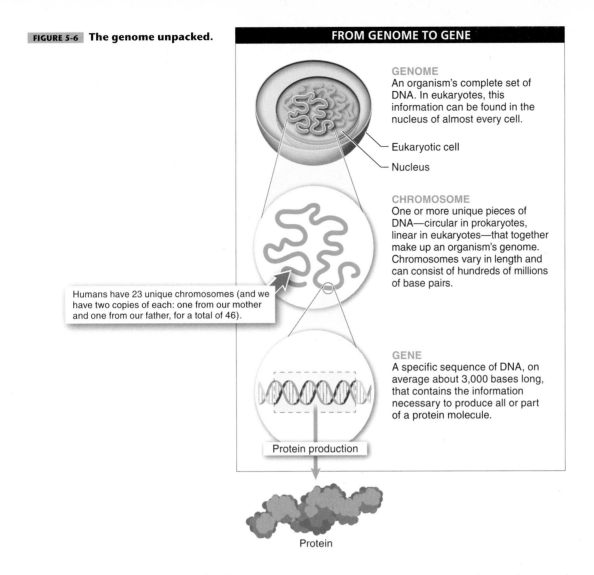

FROM GENOME TO GENE

GENOME
An organism's complete set of DNA. In eukaryotes, this information can be found in the nucleus of almost every cell.

Eukaryotic cell

Nucleus

CHROMOSOME
One or more unique pieces of DNA—circular in prokaryotes, linear in eukaryotes—that together make up an organism's genome. Chromosomes vary in length and can consist of hundreds of millions of base pairs.

Humans have 23 unique chromosomes (and we have two copies of each: one from our mother and one from our father, for a total of 46).

GENE
A specific sequence of DNA, on average about 3,000 bases long, that contains the information necessary to produce all or part of a protein molecule.

Protein production

Protein

bacteria, the information is contained within circular pieces of DNA. In eukaryotes, including humans, this information is laid out in long linear strands of DNA in the nucleus. Rather than being one super-long DNA strand, eukaryotic DNA exists as many smaller, more manageable pieces, called **chromosomes.** Humans, for example, have three billion base pairs, divided into 23 unique pieces of DNA. Because we have two copies of each (one from our mother, one from our father), we have 46 chromosomes.

Within the long sequences of bases in a cell's DNA molecules are relatively short sequences, on average about 3,000 bases long, that are the actual genes. The location or position of a gene on a chromosome is called a **locus** (*pl.* **loci**). "Gene" may seem like an impossibly nebulous concept, because the word is often used casually in the media as if it were some magical, irresistible, and mysterious force that controls our bodies and behavior. Beyond these vague descriptions, though, the word has a literal meaning. A **gene** is a sequence of bases (or, more precisely, base pairs) in a DNA molecule that carries the

information necessary for producing a functional product, usually a polypeptide or RNA molecule. Nothing more, nothing less.

Remember that a DNA molecule is like a ladder in which half of each rung is any one of the four bases—A, T, G, or C. Strung together, a segment might be read as "AAAGGCTAGGC . . ." continuing on for another 3,000 or so bases. Returning to our cookbook analogy, just as a particular sequence of letters in a cookbook may be read and understood as the directions for baking chocolate chip cookies, the sequence of bases in DNA also carries information. Perhaps the sequence spells out part of the instructions for producing a red blood cell, or for constructing the keratin that will form part of a curly strand of hair, or for assembling a chemical that alters your brain chemistry so that you exhibit a mood disorder or suicidal behavior.

Each gene is the instruction set for producing one particular molecule, usually a protein. For example, there is a gene in

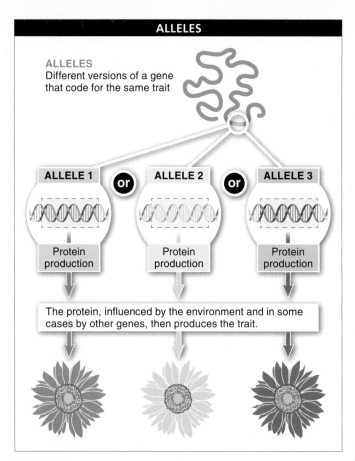

ALLELES

ALLELES
Different versions of a gene that code for the same trait

| ALLELE 1 | or | ALLELE 2 | or | ALLELE 3 |

Protein production

Protein production

Protein production

The protein, influenced by the environment and in some cases by other genes, then produces the trait.

FIGURE 5-7 *"Different versions of the same thing."* Alleles are alternative versions of a gene.

Within a species, individuals sometimes have slightly different instruction sets for a given protein, and these instructions can result in a different version of the same characteristic. These alternative versions of a gene that code for the same feature are called **alleles** (**FIGURE 5-7**)—and function like alternative recipes for chocolate chip cookies. Any single characteristic or feature of an organism is referred to as a **trait.** A simple hypothetical example will clarify the meaning of these terms. The color of a daisy's petals is a trait. The instructions for producing this trait are found in a gene that controls petal color. This gene may have many different alleles; one allele may specify the trait of red petals, another may specify white petals, and yet another may specify yellow petals (see Chapter 7). Similarly, one allele for eye color in fruit flies may carry the instructions for producing a red eye, while another, slightly different allele may have instructions for brown eyes. (Ultimately, though, the trait may be influenced not just by the genes an individual carries but by the way those genes interact with the environment, too.)

> ## TAKE-HOME MESSAGE 5·3
>
> DNA is a universal language that provides the instructions for building all the structures in all living organisms. The full set of DNA that an organism carries is called its genome. In prokaryotes, the DNA occurs in circular pieces. In eukaryotes, the genome is divided among smaller, linear strands of DNA. An organism's DNA pieces are generally called chromosomes. A gene is a sequence of bases in a DNA molecule that carries the information necessary for producing a functional product, usually a polypeptide or RNA molecule.

silk moths that codes for fibroin, the chief component of silk. And, there is a gene in humans that codes for triglyceride lipase, an enzyme that breaks down dietary fat.

5·4 Not all DNA contains instructions for making proteins.

Q An onion has five times as much DNA as a human. Why doesn't that make onions more complex than humans?

It is debatable whether humans are the most complex species on the planet, but surely we must be more complex than an onion. "Complexity" is somewhat subjective and can be assessed in a variety of ways, such as by counting the number of different cell types in the organism. But if we measured complexity simply as the amount of DNA an organism has, we'd have to say an onion

is more complex—it has more than five times as much DNA as a human (**FIGURE 5-8**)! We don't fare any better when compared with some other seemingly simple organisms, either. The salamander species *Amphiuma means,* for example, has about 25 times as much DNA as we do, and one species of amoeba—a single-celled organism—has almost 200 times as much!

Comparing the amount of DNA present in various species, in terms of both numbers of chromosomes and numbers of base pairs, reveals a paradox: there does not seem to be any

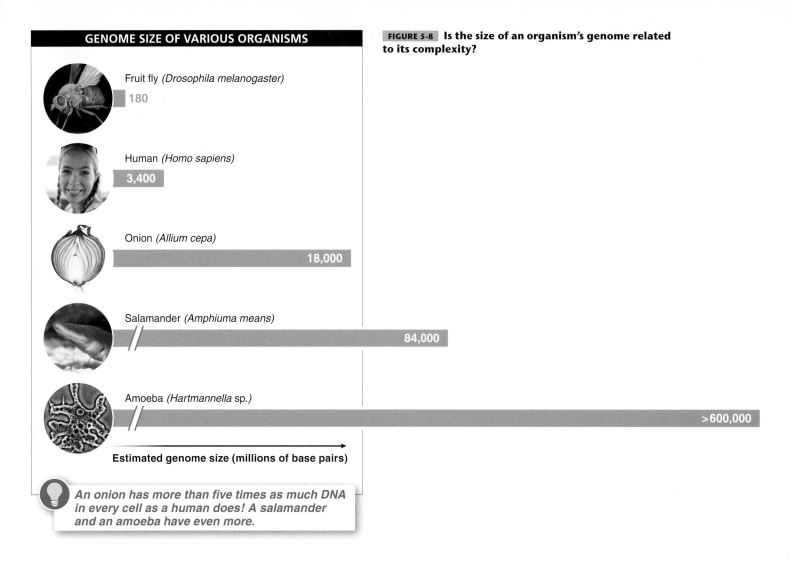

GENOME SIZE OF VARIOUS ORGANISMS

Fruit fly (*Drosophila melanogaster*)
180

Human (*Homo sapiens*)
3,400

Onion (*Allium cepa*)
18,000

Salamander (*Amphiuma means*)
84,000

Amoeba (*Hartmannella* sp.)
>600,000

Estimated genome size (millions of base pairs)

An onion has more than five times as much DNA in every cell as a human does! A salamander and an amoeba have even more.

FIGURE 5-8 **Is the size of an organism's genome related to its complexity?**

relationship between the size of an organism's genome and the organism's complexity.

The description earlier in this chapter about what DNA is and how genes code for proteins is logical and tidy, but it doesn't completely explain what we observe in cells. In humans, for example, genes make up only about 2% of the DNA (**FIGURE 5-9**). In many species, the proportion of the DNA that consists of genes is even smaller. In almost all eukaryotic species, the amount of DNA present far exceeds the amount necessary to code for all of the proteins in the organism. The fact is, a huge proportion of the base sequences in DNA do not code for proteins and have no known purpose. When it was first observed, some biologists even referred to this non-coding DNA as "junk DNA."

In what types of organisms do we find the most "junk DNA"? Bacteria and viruses tend to have very little

non-coding DNA; genes make up 90% or more of their DNA. It is in the eukaryotes (with the exception of yeasts) that we see an explosion in the amount of non-coding DNA (**FIGURE 5-10**). Non-coding regions of DNA often take the form of sequences that are repeated, sometimes thousands (or even hundreds of thousands) of times; often these repeats exist because some DNA sequences can make copies of themselves and move throughout the genome. Occasionally, the non-coding DNA consists of gene fragments, duplicate versions of genes, and pseudogenes (sequences that evolved from actual genes but accumulated mutations that made them lose their protein-coding ability). About 25% of the non-coding regions occur *within* genes—in which case they are called **introns.** About 75% of the non-coding regions occur *between genes* (see Figure 5-10).

In the end, the presence of this non-coding DNA is still not completely understood. Recent evidence has revealed, however, that some of the "non-coding" DNA actually does

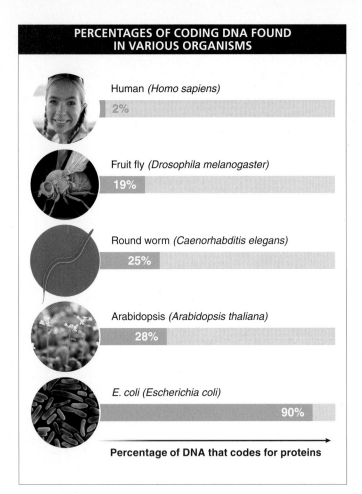

PERCENTAGES OF CODING DNA FOUND IN VARIOUS ORGANISMS

Human (*Homo sapiens*)
2%

Fruit fly (*Drosophila melanogaster*)
19%

Round worm (*Caenorhabditis elegans*)
25%

Arabidopsis (*Arabidopsis thaliana*)
28%

E. coli (*Escherichia coli*)
90%

Percentage of DNA that codes for proteins

FIGURE 5-9 **"Junk DNA."** The proportion of DNA that codes for proteins or RNA varies greatly among species. Of the species listed, which has the least "junk DNA"?

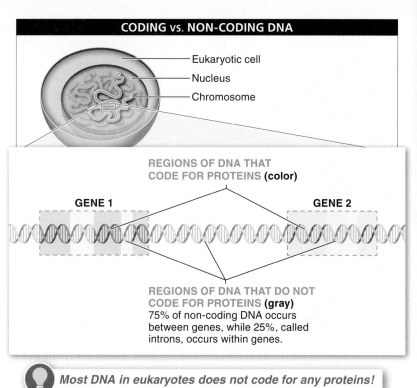

CODING vs. NON-CODING DNA

Eukaryotic cell
Nucleus
Chromosome

REGIONS OF DNA THAT CODE FOR PROTEINS (color)

GENE 1   GENE 2

REGIONS OF DNA THAT DO NOT CODE FOR PROTEINS (gray)
75% of non-coding DNA occurs between genes, while 25%, called introns, occurs within genes.

💡 *Most DNA in eukaryotes does not code for any proteins!*

FIGURE 5-10 **Non-coding regions of DNA.** These regions are found both between and within genes.

a reservoir of potentially useful sequences. In any case, the label "junk DNA" is a bad description.

encode extremely short RNA molecules (~20 nucleotides long) that function in gene regulation; they act as a "switch" that regulates when genes are turned on or off, or as a "volume knob" that influences the rate at which gene products are produced. Non-coding DNA may also serve as

TAKE-HOME MESSAGE 5·4

Only a small fraction of the DNA in eukaryotic species is in genes that code for proteins; the function of much of the rest is still poorly understood, although at least some of it plays important roles in the cell, such as gene regulation.

5·5 How do genes work? An overview.

Just as having a recipe for chocolate chip cookies is not the same thing as having the actual cookies, having a wealth of hereditary information about how to build muscle cells or leaf cells is not sufficient to produce an organism. Think about it: an organism's every cell contains all of the information needed to manufacture every protein in its body. This means that the skin cells on your arm contain the genes

for producing proteins found only in liver cells or red blood cells or muscle tissue—but they don't produce them. Having the instructions is not the same as having the products.

The genes in strands of DNA are a storehouse of information, an instruction book, but they are only one part of the process by which an organism is built. If the genes that

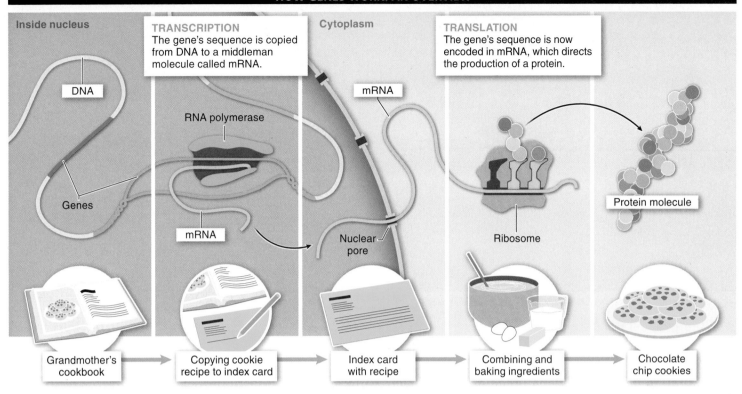

Inside nucleus

DNA

Genes

TRANSCRIPTION
The gene's sequence is copied from DNA to a middleman molecule called mRNA.

RNA polymerase

mRNA

Cytoplasm

mRNA

Nuclear pore

TRANSLATION
The gene's sequence is now encoded in mRNA, which directs the production of a protein.

Protein molecule

Ribosome

Grandmother's cookbook → Copying cookie recipe to index card → Index card with recipe → Combining and baking ingredients → Chocolate chip cookies

FIGURE 5-11 **Overview of the steps from gene to protein.**

an organism carries for a particular trait—its **genotype**—are like a recipe in a cookbook, the physical manifestation of the instructions—the organism's **phenotype**—is the cookie, or any of the other foods described by the recipes. And, just as you have to assemble the ingredients, mix them, then bake the dough to get a cookie, there are several steps in the production of the molecules, tissues, and even behaviors that make up a phenotype.

How does a gene (a sequence of bases in a section of DNA) affect a flower's color or the shape of a nose or the texture of a dog's fur (the phenotype)? The process occurs in two main steps: **transcription,** in which a copy of a gene's base sequence is made, and **translation,** in which that copy is used to direct the production of a polypeptide, which then, in response to a variety of factors, including the cellular environment, folds into a functional protein.

FIGURE 5-11 presents an overview of the processes of transcription and translation. In transcription, which in

eukaryotes occurs in the nucleus, the gene's base sequence, or code, is copied into a middleman molecule called **messenger RNA (mRNA).** (Because prokaryotes don't have a nucleus, transcription occurs in the cytoplasm.) This is like copying the information for the chocolate chip cookie recipe out of the cookbook and onto an index card. The mRNA then moves out of the nucleus into the cytoplasm, where translation allows the messages encoded in the mRNA to be used to build proteins.

TAKE-HOME MESSAGE 5·5

The genes in strands of DNA are a storehouse of information, an instruction book. The process by which this information is used to build an organism occurs in two main steps: transcription, in which a copy of a gene's base sequence is made, and translation, in which that copy is used to direct the production of a polypeptide.

Information in DNA directs the production of the molecules that make up an organism.

A display of the expression levels of many genes simultaneously, based on DNA microarray technology.

5·6 In transcription, the information coded in DNA is copied into mRNA.

If DNA is like a cookbook filled with recipes, transcription and translation are like cooking. In cooking, you use information about how to make chocolate chip cookies to actually produce the cookies. In an organism, the information about putting together proteins is used to build the proteins that the organism needs to function. In this section, we examine transcription, the first step in the two-step process by which DNA regulates a cell's activity and its synthesis of proteins (see Figure 5-11). In transcription, a copy is made of one specific gene within the DNA. Continuing our cookbook analogy, transcription is like copying a single recipe from the cookbook onto an index card. Transcription happens in four steps (**FIGURE 5-12**).

Step 1. Recognize and bind. To start the transcription process, the enzyme RNA polymerase recognizes a **promoter site,** a sequence in a gene that indicates the start of the gene and, in effect, tells the RNA polymerase, "Start here." RNA polymerase binds to the DNA molecule at the promoter site and unwinds it just a bit, so that only one strand of the DNA can be read.

Step 2. Transcribe. As the DNA strand is processed through the RNA polymerase, the RNA polymerase builds a copy—called a "transcript"—of the gene from the DNA molecule,

just as (to use another analogy) a court reporter transcribes and creates a record of everything that is said in a courtroom. This copy is called messenger RNA (mRNA) because, once this copy of the gene is created, it can move elsewhere in the cell and its message can be translated into a protein. Throughout transcription, DNA is unwound ahead of the RNA polymerase so that a single strand of the DNA can be read, and it is rewound after the polymerase passes.

The mRNA transcript is constructed from four different nucleotides, which have a structure similar to that of DNA nucleotides but with a different kind of sugar. Each pairs up with an exposed base on the unwound and separated DNA strand, following these rules:

If the DNA strand has a thymine (T), an adenine (A) is added to the mRNA.

If the DNA strand has an adenine (A), a uracil (U) is added to the mRNA.

If the DNA strand has a guanine (G), a cytosine (C) is added to the mRNA.

If the DNA strand has a cytosine (C), a guanine (G) is added to the mRNA.

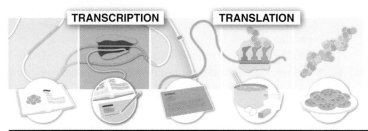
TRANSCRIPTION

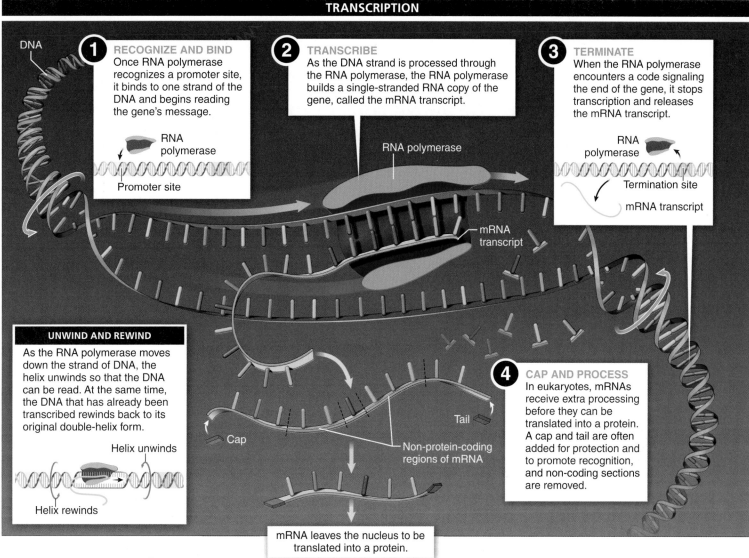

DNA

1 RECOGNIZE AND BIND
Once RNA polymerase recognizes a promoter site, it binds to one strand of the DNA and begins reading the gene's message.

RNA polymerase

Promoter site

2 TRANSCRIBE
As the DNA strand is processed through the RNA polymerase, the RNA polymerase builds a single-stranded RNA copy of the gene, called the mRNA transcript.

RNA polymerase

mRNA transcript

3 TERMINATE
When the RNA polymerase encounters a code signaling the end of the gene, it stops transcription and releases the mRNA transcript.

RNA polymerase

Termination site

mRNA transcript

UNWIND AND REWIND
As the RNA polymerase moves down the strand of DNA, the helix unwinds so that the DNA can be read. At the same time, the DNA that has already been transcribed rewinds back to its original double-helix form.

Helix unwinds

Helix rewinds

Cap

Non-protein-coding regions of mRNA

Tail

4 CAP AND PROCESS
In eukaryotes, mRNAs receive extra processing before they can be translated into a protein. A cap and tail are often added for protection and to promote recognition, and non-coding sections are removed.

mRNA leaves the nucleus to be translated into a protein.

FIGURE 5-12 **Transcription: copying the base sequence of a gene.** The first step in a two-step process by which DNA regulates a cell's activity and synthesis of proteins. (Shown here is eukaryotic transcription.)

Step 3. Terminate. When the RNA polymerase encounters a sequence of bases on the DNA at the end of the gene (called a termination sequence), it stops creating the transcript and detaches from the DNA molecule. After termination, the mRNA molecule is released as a free-floating, single-strand copy of the gene.

Step 4. Capping and editing. In prokaryotic cells, once an mRNA transcript is produced and begins to separate from the DNA it is ready to be translated into a protein (it doesn't have a nuclear membrane to cross). In eukaryotes, mRNAs receive extra processing before they can be translated into a protein. First, a cap and a tail may be added at the beginning and end of the transcript. Like the front and back covers of a book, these serve to protect the mRNA from damage and help the protein-making machinery recognize the mRNA. Second, because (as we saw earlier in the chapter) there may be some non-coding bits of the DNA that are transcribed into mRNA, those sections—the introns—are snipped out. Once the mRNA transcript has been edited, it is ready to leave the nucleus for the cytoplasm, where it will be translated into a polypeptide that will fold into a protein.

TAKE-HOME MESSAGE 5·6

Transcription is the first step in the two-step process of producing proteins based on instructions contained in DNA. In transcription (which occurs in the nucleus in eukaryotic cells), a single copy of one specific gene in the DNA is made in the form of a molecule of mRNA. When the mRNA copy of a gene is completed and processed, it moves to the cytoplasm, where it can be translated into a polypeptide.

5·7 In translation, the mRNA copy of the information from DNA is used to build functional molecules.

Once the mRNA molecule has moved out of the cell's nucleus and into the cytoplasm, the translation process begins. In translation, the information carried by the mRNA is read, and ingredients present in the cytoplasm are used to produce a protein (see Figure 5-11). The process of translation is like combining and baking the ingredients listed in our chocolate chip cookie recipe to produce a cookie.

Several ingredients must be present in the cytoplasm for translation to occur. First, there must be large numbers of free amino acids floating around. Recall from Chapter 2

that amino acids are the raw materials for building proteins and an essential component of our diet. Second, there must be **ribosomal subunits,** which are components of ribosomes, the protein-production factories where amino acids are linked together in the proper order to produce the protein. And there must also be molecules that can read the mRNA code and translate that message from a sequence of bases into a protein.

To read an encrypted message, you need to know the secret code by which to translate each of the encrypted characters into a readable character in the message. Similarly, to

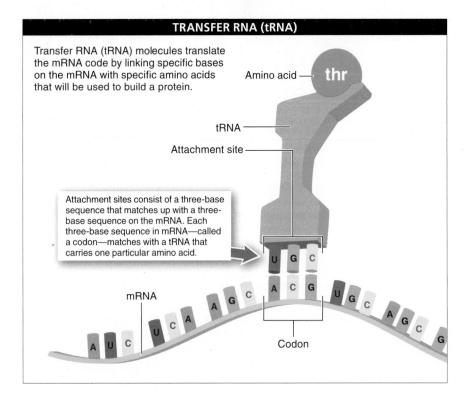

TRANSFER RNA (tRNA)

Transfer RNA (tRNA) molecules translate the mRNA code by linking specific bases on the mRNA with specific amino acids that will be used to build a protein.

Amino acid — thr

tRNA —

Attachment site —

Attachment sites consist of a three-base sequence that matches up with a three-base sequence on the mRNA. Each three-base sequence in mRNA—called a codon—matches with a tRNA that carries one particular amino acid.

U G C

mRNA

A U C · U C A · A G C · A C G · U G C · A G C · G

Codon

FIGURE 5-13 Each transfer RNA attaches a particular amino acid to the mRNA.

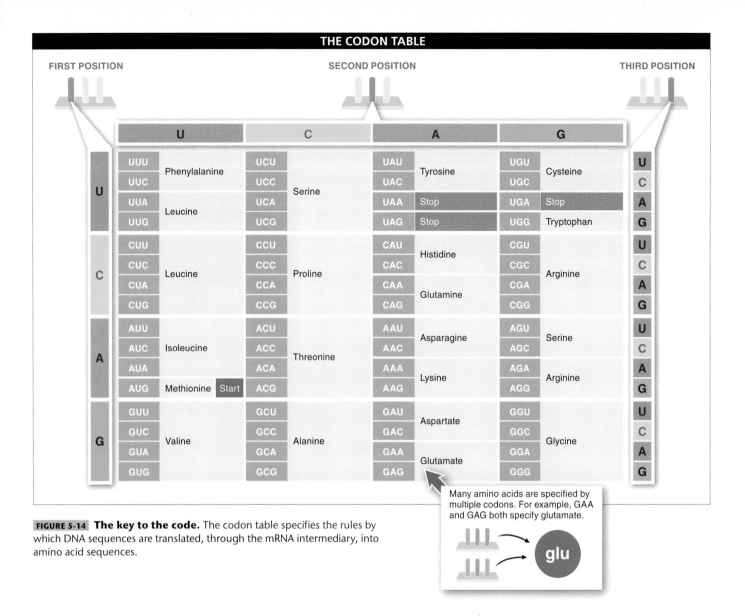

THE CODON TABLE

FIRST POSITION	SECOND POSITION: U	SECOND POSITION: C	SECOND POSITION: A	SECOND POSITION: G	THIRD POSITION
U	UUU Phenylalanine	UCU Serine	UAU Tyrosine	UGU Cysteine	U
U	UUC Phenylalanine	UCC Serine	UAC Tyrosine	UGC Cysteine	C
U	UUA Leucine	UCA Serine	UAA Stop	UGA Stop	A
U	UUG Leucine	UCG Serine	UAG Stop	UGG Tryptophan	G
C	CUU Leucine	CCU Proline	CAU Histidine	CGU Arginine	U
C	CUC Leucine	CCC Proline	CAC Histidine	CGC Arginine	C
C	CUA Leucine	CCA Proline	CAA Glutamine	CGA Arginine	A
C	CUG Leucine	CCG Proline	CAG Glutamine	CGG Arginine	G
A	AUU Isoleucine	ACU Threonine	AAU Asparagine	AGU Serine	U
A	AUC Isoleucine	ACC Threonine	AAC Asparagine	AGC Serine	C
A	AUA Isoleucine	ACA Threonine	AAA Lysine	AGA Arginine	A
A	AUG Methionine Start	ACG Threonine	AAG Lysine	AGG Arginine	G
G	GUU Valine	GCU Alanine	GAU Aspartate	GGU Glycine	U
G	GUC Valine	GCC Alanine	GAC Aspartate	GGC Glycine	C
G	GUA Valine	GCA Alanine	GAA Glutamate	GGA Glycine	A
G	GUG Valine	GCG Alanine	GAG Glutamate	GGG Glycine	G

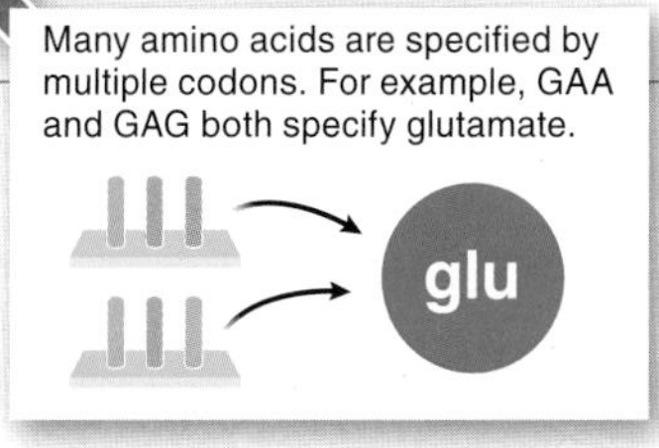

Many amino acids are specified by multiple codons. For example, GAA and GAG both specify glutamate.
glu

FIGURE 5-14 **The key to the code.** The codon table specifies the rules by which DNA sequences are translated, through the mRNA intermediary, into amino acid sequences.

understand which sequence of amino acids in a polypeptide is specified by a sequence of nucleotides in a DNA molecule—and in the mRNA molecule transcribed from the DNA—you need to know a secret code of sorts.

A special type of RNA molecules holds the secret code. These molecules, called **transfer RNA (tRNA),** interpret the mRNA code, translating the language of DNA—coded in the linear sequence of bases—into the language of proteins, coded in the linear sequences of amino acids.

Picture a molecule with two distinct ends (FIGURE 5-13). On one end of the tRNA molecule is an attachment site consisting of a three-base sequence that matches up with a three-base sequence on the mRNA transcript. This matchup enables the tRNA molecule to attach to the mRNA. Attached to the other end of the tRNA molecule is an amino acid. Each three-base sequence in mRNA—called a **codon**—matches with a tRNA molecule that carries a particular amino acid. The codon ACG, for example, is recognized by the tRNA molecule that carries the amino acid threonine. And the codon CAG is recognized by the tRNA molecule that carries glutamine. For every possible codon, there is one type of tRNA molecule that will recognize and bind to the mRNA at that point, and it will always carry the same amino acid.

The codon table (FIGURE 5-14), also called the "genetic code," describes which tRNA, and therefore which amino acid, is specified by each codon. With four possible bases in each of the three positions of a codon, 64 different codons are possible. Sixty-one of these codons specify amino acids, and 3 are "stop" sequences, indicating the end of translation. Because there are only 20 amino acids, many amino acids

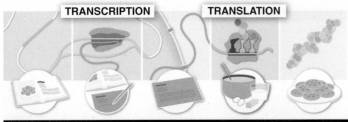

TRANSLATION

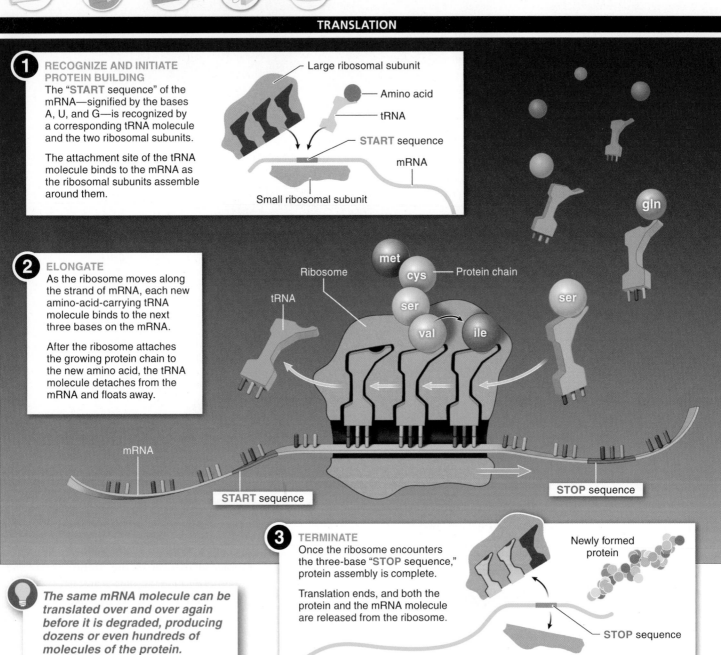

1 RECOGNIZE AND INITIATE PROTEIN BUILDING
The "START sequence" of the mRNA—signified by the bases A, U, and G—is recognized by a corresponding tRNA molecule and the two ribosomal subunits.

The attachment site of the tRNA molecule binds to the mRNA as the ribosomal subunits assemble around them.

Large ribosomal subunit
Amino acid
tRNA
START sequence
mRNA
Small ribosomal subunit

2 ELONGATE
As the ribosome moves along the strand of mRNA, each new amino-acid-carrying tRNA molecule binds to the next three bases on the mRNA.

After the ribosome attaches the growing protein chain to the new amino acid, the tRNA molecule detaches from the mRNA and floats away.

Ribosome
Protein chain
tRNA
met
cys
ser
val
ile
ser
gln
mRNA
START sequence
STOP sequence

The same mRNA molecule can be translated over and over again before it is degraded, producing dozens or even hundreds of molecules of the protein.

3 TERMINATE
Once the ribosome encounters the three-base "STOP sequence," protein assembly is complete.

Translation ends, and both the protein and the mRNA molecule are released from the ribosome.

Newly formed protein
STOP sequence

FIGURE 5-15 Translation: reading a sequence of nucleotides and producing protein. The second step in a two-step process by which DNA regulates a cell's activity and synthesis of proteins.

are specified by multiple codons. UGU and UGC, for example, both specify cysteine. With just a few small exceptions, the genetic code is the same in every organism on earth.

The translation of an mRNA molecule into a sequence of amino acids (that will then fold into the complex three-dimensional shape of a protein) occurs in three steps (**FIGURE 5-15**).

Step 1. Recognize and initiate protein building.
Translation begins in the cell's cytoplasm when the subunits of a ribosome, essentially a two-piece protein-building factory, recognize and assemble around a codon on the mRNA transcript called the start sequence. This start sequence is always the codon AUG. As the ribosomal subunits assemble themselves into a ribosome, the attachment site of a particular tRNA molecule also recognizes the start sequence on the mRNA and binds to it. This initiator tRNA carries the amino acid methionine (met). Thus, methionine is always the first amino acid in any protein that is produced. (Occasionally, in eukaryotes, this initial methionine is "edited out" later in the protein-building process.)

Step 2. Elongate. After the mRNA start sequence (AUG), the next three bases on the mRNA specify which amino acid–carrying tRNA molecule should bind to the mRNA. If the next three bases on the mRNA transcript are GUU, for example, a tRNA molecule that recognizes this sequence, and carrying its particular amino acid (valine, or val), will attach to the mRNA at that point. The ribosome then facilitates the connection of this second amino acid to the first. After the amino acid carried by a tRNA molecule is attached to the new amino acid, the tRNA molecule detaches from the mRNA and floats away.

As this process continues, the amino acid chain grows. The next three bases on the mRNA specify the next amino acid to be added to the first two. And the three bases after that specify the fourth amino acid, and so on. This process of progressively linking together the amino acids specified by

an mRNA strand is called **protein synthesis,** because all proteins are chains of amino acids, like beads on a string. Precision in this process is essential. If the tRNAs were to misread or skip or double-read any bases, they could alter the amino acid sequence (and possibly the normal functioning) of the protein specified by the mRNA.

Step 3. Terminate. Eventually, the ribosome arrives at the codon on the mRNA that signals the end of translation. Once the ribosome encounters this stop sequence, the assembly of the amino acid chain is complete. Translation ends, and the amino acid chain and mRNA molecule are released from the ribosome. As it is being produced, the amino acid chain folds and bends, based on the chemical features in the amino acid side chains (as we saw in Chapter 2). Through the folding and bending, the protein acquires its three-dimensional structure. The completed protein—such as a membrane protein or insulin or a digestive enzyme—may be used within the cell or packaged for delivery via the bloodstream to somewhere else in the body where it is needed.

Following the completion of translation, the mRNA strand may remain in the cytoplasm to serve as the template for producing another molecule of the same protein. In bacteria, an mRNA strand may last from a few seconds to more than an hour; in mammals, an mRNA may last several days. Depending on how long it lasts, the same mRNA strand may be translated hundreds of times. Eventually, it is broken down by enzymes in the cytoplasm.

TAKE-HOME MESSAGE 5·7

Translation is the second step in the two-step process by which information carried in DNA directs the synthesis of proteins. In translation, the information from a gene that has been encoded in the nucleotide sequence of an mRNA is read, and ingredients present in the cell's cytoplasm are used to produce a protein.

5·8 Genes are regulated in several ways.

Why do some people get sick more than others? Why is one person taller or shorter than another? More generally, what makes us look and act the way we do? We say that our

health, growth, and development are the result of interactions between our genes and our environment. But how do our genes and our environment actually interact?

It matters which genes a person has, but carrying a gene is not the only thing that matters when it comes to **gene expression**—the production of the protein that the gene's sequence codes for. A person's traits also depend on **gene regulation,** whether a gene is turned on—producing its protein product—or turned off. And this, commonly, hinges on a person's environment.

A powerful tool in the study of gene regulation is the microarray, a small chip that looks like a microscope slide that can be used to monitor the expression levels of thousands of genes simultaneously. Taking cells from one part of the body, researchers examine gene expression in that tissue. Microarrays are particularly useful in exploring how gene expression differs in response to an illness, or the treatment of an illness, or in response to aging.

For example, in a five-year medical study exploring the health consequences of loneliness, microarray analyses of gene activity in immune system cells revealed that 209 genes were expressed differently in two groups of participants. One group included individuals who were experiencing chronic social isolation and loneliness, while individuals in the other group were not.

Individuals in the "lonely" group had impaired transcription (decreased activity) of some genes associated with an effective stress response. They also had increased

Lonely, isolated individuals have very different patterns of immune system gene expression.

FIGURE 5-16 **Social isolation.**

activity in genes associated with inflammatory processes that are implicated in a variety of diseases (**FIGURE 5-16**). Related studies have shown a similarly reduced immune function among stressed-out individuals—including one study of students during a stressful exam period!

Controlling gene expression Cells have a wide variety of mechanisms by which they can control when

THE OPERON: AN OVERVIEW

An operon is a group of several genes, along with the elements that control their expression as a unit, all within one section of DNA. (Sizes of these elements are not shown to scale. The promoter and operator regions are typically much shorter than the genes.)

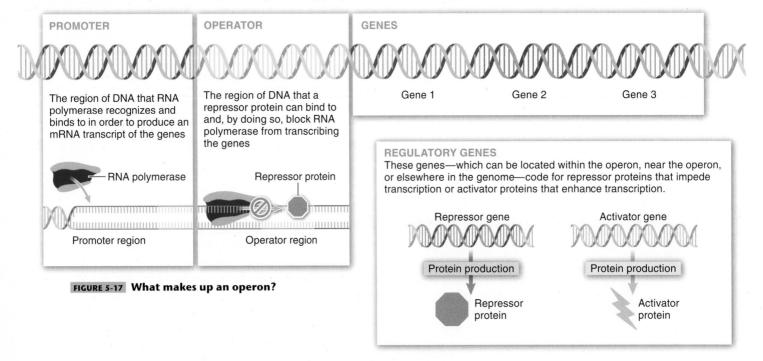

PROMOTER

The region of DNA that RNA polymerase recognizes and binds to in order to produce an mRNA transcript of the genes

—RNA polymerase

Promoter region

OPERATOR

The region of DNA that a repressor protein can bind to and, by doing so, block RNA polymerase from transcribing the genes

Repressor protein

Operator region

GENES

Gene 1   Gene 2   Gene 3

REGULATORY GENES

These genes—which can be located within the operon, near the operon, or elsewhere in the genome—code for repressor proteins that impede transcription or activator proteins that enhance transcription.

Repressor gene   Activator gene

Protein production   Protein production

Repressor protein   Activator protein

FIGURE 5-17 **What makes up an operon?**

individual genes are expressed. One of the main ways expression is controlled is by proteins called transcription factors, which bind to specific sequences in the DNA called regulatory sites, which are often located in front of genes. The regulation may be "positive control," in which binding of the regulatory protein initiates or speeds up gene expression, or "negative control," in which the protein slows or blocks gene expression.

Regulatory proteins can be produced by the same cell whose DNA they are regulating, or they may come from nearby cells. During the initial development of an organism, for example, gene regulation is extremely important in directing cells to differentiate into one type of tissue—perhaps muscle, rather than bone—and proteins produced by one cell commonly influence gene expression in neighboring cells.

Prokaryotic gene control and the lac operon

Regulation may also be influenced by the nutrients an organism takes in. One well-studied example concerns the case of how the bacterium *E. coli* adjusts its diet depending on which sugars are available in its environment. More specifically, although its preference is for glucose, in the absence of glucose, it will utilize lactose as an energy source. (*E. coli* cells, many of which live in your intestine, are likely to encounter lactose, which is the chief sugar in milk.)

The absorption and digestion of lactose, however, requires the expression of three genes, and bacteria do not waste their energy transcribing these genes, translating the proteins, folding them, and sending them to the correct destinations unless lactose is actually present. That regulation is accomplished with a set of regulatory sequences called an **operon,** which is a group of several genes and the elements that control their expression as a unit, all within one continuous segment of the cell's DNA (FIGURE 5-17).

Promoter. For a gene to be transcribed, RNA polymerase recognizes and binds to the promoter region, the specific sequence of nucleotides in the DNA that signals the beginning of the gene.

Operator. A molecule called a repressor protein can bind to the operator, the regulatory portion of the DNA, and by doing so it blocks RNA polymerase from transcribing the genes necessary for lactose metabolism.

Regulatory gene. The regulatory gene codes for the repressor protein that, when bound to the operator region, does the blocking of RNA polymerase's binding to the promoter site. The regulatory gene may be located in the operon it regulates or elsewhere in the genome.

THE LAC OPERON IN ACTION

IF LACTOSE IS NOT PRESENT: OPERON OFF

A repressor protein binds to the operator, preventing RNA polymerase from binding to the promoter and transcribing the genes necessary for lactose metabolism.

RNA polymerase

Repressor protein

Promoter   Operator   Gene 1   Gene 2   Gene 3

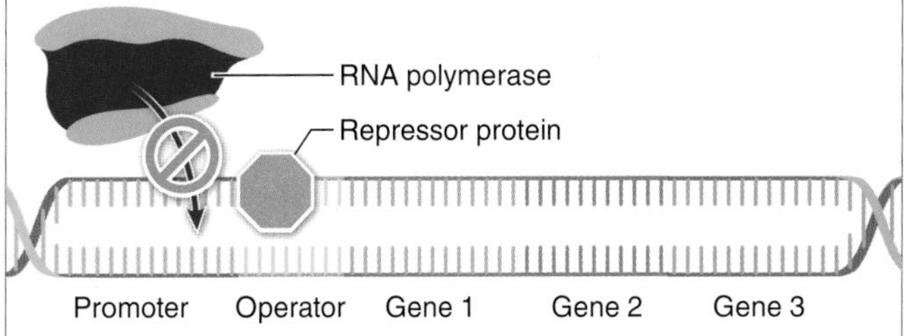

IF LACTOSE IS PRESENT: OPERON ON

Lactose (slightly modified) binds to the repressor protein, preventing it from binding to the operator. RNA polymerase can then bind to the promoter and transcribe the genes necessary for lactose metabolism.

Repressor protein

Lactose

RNA polymerase

Promoter   Operator   Gene 1   Gene 2   Gene 3

mRNA transcript

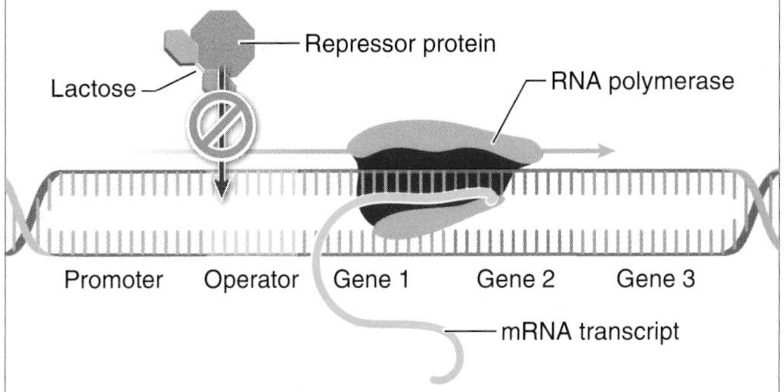

FIGURE 5-18 **Operon in action.** In *E. coli,* environmental conditions control the expression of the genes necessary for the metabolism of lactose.

In the absence of glucose (more on this below), here's how this operon (called the "lac operon") functions. If lactose is not present: (1) The regulatory gene produces the repressor protein. (2) The repressor protein binds to the operator region. And (3) RNA polymerase is blocked from transcribing the lactose metabolism genes (FIGURE 5-18).

If, on the other hand, lactose *is* present: (1) The regulatory gene produces the repressor protein. (2) Lactose binds to the repressor protein, altering its shape so that it can no longer bind to the operator. And (3) with no repressor bound to the DNA, RNA polymerase binds to the promoter and transcribes the genes necessary for lactose metabolism.

Note that the lac operon can be switched on only in the absence of glucose. If no glucose is present, a regulatory

MECHANISMS OF REGULATING GENE EXPRESSION

TRANSCRIPTION REGULATION

POSITIVE CONTROL
- Activators can initiate or speed up gene expression
- Enhancer sequences can speed RNA polymerase binding and gene transcription

NEGATIVE CONTROL
- Repressors can block or slow down gene expression
- Chemicals can bind to DNA and block gene transcription

POST-TRANSCRIPTION REGULATION

- mRNA processing, transport, and enzymatic breakdown can be blocked or accelerated
- mRNA translation and protein processing rates can be altered

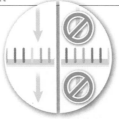

FIGURE 5-19 **Regulating gene expression.** Cells can increase or decrease gene expression in multiple ways.

protein called an *activator* helps the DNA unwind in the vicinity of the lac operon, enabling RNA polymerase to bind and transcribe genes. When glucose is present, however, the activator protein cannot bind to the DNA, and this prevents RNA polymerase from binding to and transcribing the lactose metabolism genes.

Eukaryotic gene control There are many other ways that genes can be regulated, besides operons. Many eukaryotes, for example, have enhancer sequences that increase transcription rates (positive control). Often far from the gene they regulate, these sequences, after binding a regulatory protein, bind transcription factors associated with RNA polymerase and enhance the rate at which the RNA polymerase binds to a promoter and transcribes a gene.

Additional mechanisms of gene control include the binding of chemicals to DNA that can block genes from being transcribed (negative control) and the regulation of mRNA. Messenger RNA can be regulated by altering its processing and transport, or by altering the rate at which mRNA is broken down in the cytoplasm. Translation of mRNAs can be regulated, too, as can the processing of proteins following translation (**FIGURE 5-19**).

TAKE-HOME MESSAGE 5·8

Environmental signals influence the turning on and turning off of genes. By binding to DNA, regulatory proteins can block or facilitate the binding of RNA polymerase and the subsequent transcription of genes. Regulation of gene expression also can occur in a variety of other ways that enhance or impede transcription, or alter mRNA's longevity and rate of degradation, or influence translation or protein processing.

5·9–5·11
Damage to the genetic code has a variety of causes and effects.

Damage to the genetic code can interfere with normal development.

5·9   What causes a mutation, and what are its effects?

Through the two-step process of transcription and translation, an organism converts the information held in its genes into the proteins necessary for life. But the process is only as good as the organism's underlying genetic information. Sometimes, something occurs to alter the sequence of bases in an organism's DNA. Such an alteration is called a **mutation,** and it can lead to changes in the structure and function of the proteins produced. Mutations can have a range of effects. Sometimes they result in a serious, even deadly, problem for an organism. Sometimes they have little or no detrimental effect. And occasionally—but very rarely—they may even turn out to be beneficial to the organism.

As an example of how mutations can affect organisms, consider the case of breast cancer in humans. When two human genes, called BRCA1 and BRCA2, are functioning properly, they help to prevent breast cancer by helping to repair DNA damage, preventing cells from accumulating the changes that lead to cancer. If the DNA sequence of either of these genes is altered through mutation and the gene's normal function is lost, the person carrying the gene has a significantly increased risk of developing breast cancer. (Because a variety of other factors, including environmental variables, are involved in development of cancer, it's impossible to know for certain whether these individuals will develop breast cancer.) Currently, more than 200 different mutations in the DNA sequences of these genes have been detected, each of which results in an increased risk of developing breast cancer.

Given the havoc they can cause for an organism, it's not surprising that mutations have a bad reputation. After all,

because they can change the protein produced, they can disrupt normal processes and harm the individual (**FIGURE 5-20**). But there are a couple of reasons why mutations' bad reputation may not be fully deserved. First, it turns out that many—perhaps even *most*—mutations are neutral, having neither a positive nor a negative effect on an organism's phenotype. This may be the case when a mutation occurs in a non-coding region of DNA, or when a change in DNA within a gene doesn't alter the function of the protein produced. Based on a recent study, researchers estimate that the rate of mutations in cells involved in reproduction is approximately 10^{-8} per base pair per generation.

A second reason that mutations' bad reputation may be undeserved is the paradoxical fact that mutations are essential to evolution. Those mutations that don't kill an organism, or reduce its ability to survive and reproduce, can be beneficial. Every genetic feature in every organism was, initially, the result of a mutation. (In Chapter 8, we explore the relationship between mutation and evolution.) Ultimately, most mutations you inherit from your parents will have no effect. And all of you are carrying mutations that you will never know about!

It's important to note that mutations can occur in an organism's gamete-producing cells (that is, cells that produce sperm or eggs) as well as in its non-sex cells (such as skin cells or cells in the lungs). Mutations in non-sex cells can have bad health consequences for the person carrying them. Many forms of cancer, such as lung cancer and skin cancer, result from such mutations. On the upside, non-sex-cell mutations are not passed on to your children.

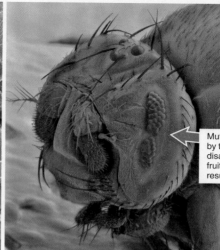

Mutations can change the protein produced by the altered gene, sometimes with disastrous consequences. Shown here: in fruit flies, a mutation to a gene for eye shape results in almost complete loss of the eye.

FIGURE 5-20 **Wreaking havoc.**

Mutations in the sex cells (gametes), on the other hand, do not have any adverse health effects on the person carrying them, but these mutations can be passed on to offspring. Individuals inheriting certain mutations from a parent—because the mutation occurred in the parent's sex cells, or the parent inherited the mutation from his or her parent—can be at increased risk for certain diseases such as breast cancer or cardiovascular disease. Inherited mutations can also have an effect before birth, sometimes causing miscarriages or the occurrence of birth defects.

In terms of how they physically affect DNA, mutations generally fall into two types: point mutations and chromosomal aberrations (**FIGURE 5-21**). In point mutations, one base pair is changed. In chromosomal aberrations, entire sections of a chromosome are altered.

Point mutations occur when one base pair in the DNA is substituted for another, or when a base pair is inserted or deleted. Insertions and deletions can be much more harmful than substitutions, because the amino acid sequence of a protein is affected. If a single base is added or removed, the three-base groupings in an mRNA get thrown off, and the entire sequence of amino acids stipulated "downstream" from that point will be wrong—the reading frame is shifted. It's almost like putting your hands on a computer keyboard, but offset by one key to the left or right, and then typing what should be a normal sentence. It comes out as gibberish.

Chromosomal aberrations are changes to the overall organization of the genes on a chromosome. Chromosomal aberrations are like the manipulation of large chunks of text when you are editing a term paper. The aberrations can involve the complete deletion of an entire section of DNA, the relocation of a gene from one part of a chromosome to

elsewhere on the same chromosome or even to a different chromosome, or the duplication of a gene, with the new copy inserted elsewhere on the chromosome or on a different chromosome. Whatever the type of aberration, a gene's expression—the production of the protein that its sequence codes for—can be altered, as well as the expression of the genes around it.

Given the potentially hazardous health consequences of mutations, it is advisable to minimize their occurrence. Can this be done? Yes and no. There are three chief causes of mutation and, although one of them is beyond our control, the risk of occurrence of the other two can be significantly reduced (**FIGURE 5-22**).

1. Spontaneous mutations. Some mutations arise by accident as long strands of DNA are duplicating themselves—at the rate of more than a thousand bases a minute in humans—when cells are dividing (you'll read more details on this process in Chapter 6). Most errors are corrected by DNA repair enzymes, but some slip through and there's not much we can do about them.

2. Radiation-induced mutations. Ionizing radiation, such as X rays, is radiation with enough energy to disrupt atomic structure—even break apart chromosomes—by removing tightly bound electrons. Because ionizing rays cannot pass through lead, the lead apron a doctor or dentist puts over you when you get an X ray protects your body from the ionizing radiation. However, even non-ionizing lower-energy radiation (which is not able to remove electrons) can damage DNA. Ultraviolet (UV) rays from the sun, for example, can be absorbed by certain bases in DNA and cause them to

Q Why do dentists put a heavy apron over you when they X-ray your teeth?

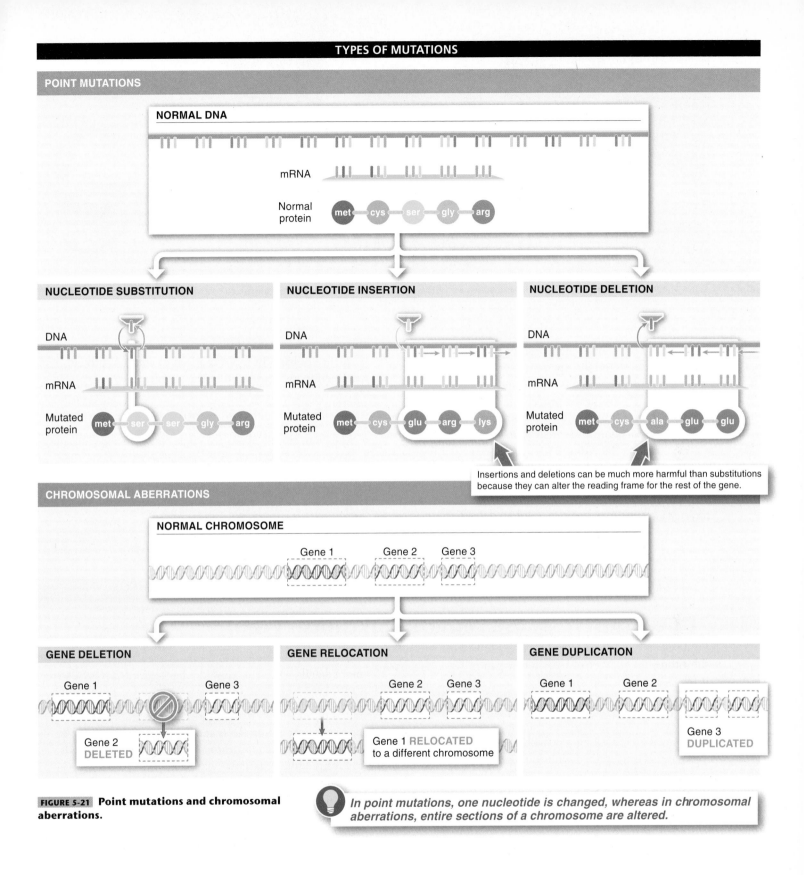

POINT MUTATIONS

NORMAL DNA

mRNA

Normal protein: met – cys – ser – gly – arg

NUCLEOTIDE SUBSTITUTION

DNA

mRNA

Mutated protein: met – ser – ser – gly – arg

NUCLEOTIDE INSERTION

DNA

mRNA

Mutated protein: met – cys – glu – arg – lys

NUCLEOTIDE DELETION

DNA

mRNA

Mutated protein: met – cys – ala – glu – glu

Insertions and deletions can be much more harmful than substitutions because they can alter the reading frame for the rest of the gene.

CHROMOSOMAL ABERRATIONS

NORMAL CHROMOSOME

Gene 1   Gene 2   Gene 3

GENE DELETION

Gene 1   Gene 3

Gene 2 DELETED

GENE RELOCATION

Gene 2   Gene 3

Gene 1 RELOCATED to a different chromosome

GENE DUPLICATION

Gene 1   Gene 2

Gene 3 DUPLICATED

FIGURE 5-21 **Point mutations and chromosomal aberrations.**

In point mutations, one nucleotide is changed, whereas in chromosomal aberrations, entire sections of a chromosome are altered.

rearrange bonds. This can prevent them from pairing correctly with the complementary DNA strand and can transform a cell into a cancer cell. This is why long-term sun exposure can contribute to the development of skin cancer.

Another source of dangerous radiation is found in the core of nuclear power plants, where radioactive atoms are used and produced in energy-generating reactions. The high energy of the radioactivity that fuels the production of

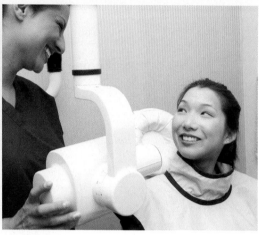

FIGURE 5-22 Gambling with mutation-inducing activities. You can increase or decrease your risk of mutations with your behavior: smoking, tanning, and radiation (but notice the protective lead apron).

Q Why is it dangerous to be near the core of a nuclear power plant?

usable energy can pass through your body and disrupt your DNA, causing point mutations and chromosomal aberrations. With the proper safety precautions, however, nuclear power plant workers can minimize their exposure to harmful radiation.

3. Chemical-induced mutations. Many chemicals, such as those found in cigarette smoke and in the exhaust from internal combustion engines, can also react with the atoms in DNA molecules and induce mutations.

In Section 5.11, we examine how even tiny changes in the sequence of bases in DNA can lead to errors in protein production and profound health problems.

TAKE-HOME MESSAGE 5·9

Mutations are alterations in a single base or changes in large segments of DNA that include several genes or more. They are rare, but when they do occur, they may disrupt normal functioning of the body (although many mutations are neutral). Extremely rarely, mutations may have a beneficial effect. Mutations play an important role in evolution.

? **5·10 THIS IS HOW WE DO IT**

Does sunscreen use reduce skin cancer risk?

Wear sunscreen.

It's a simple piece of advice, and it's endorsed by the American Medical Association. Exposure to the sun is implicated in many skin cancers. And the dramatic effect of sunscreen in making it possible to spend time in the sun without getting burned is obvious the first time you use it.

So it might come as a surprise that, according to many studies, sunscreen does not protect people from skin cancer. And even more surprising is that many studies have shown that sunscreen use is even associated with an *increased* risk of the most dangerous of all skin cancers, melanoma.

Let's look at what happens when you're exposed to ultraviolet radiation from the sun.

What happens when our skin cells are exposed to the sun?

When skin cells are exposed to a type of ultraviolet radiation called UVB radiation, the cells and their contents—including RNA and DNA molecules—can be damaged. This signals nearby cells to produce chemicals that cause inflammation—which you know as sunburn—as part of the process by which the damaged cells are cleared away.

The relationship between sun exposure and skin cancer has been noted and studied for more than 70 years:

• People with lighter-colored skin are much more sensitive to the sun than those with darker skin, burn more easily, and have more than twice the rate of melanoma When melanoma occurs in people with darker skin, it primarily affects areas with lighter skin color, such as the palms and soles.

• There is a strong relationship between the amount of time a person spends in the sun and the risk of melanoma.

Sunscreens contain chemicals that absorb or reflect some of the UV radiation. In doing so, they protect against many of the harmful effects of the sun, including sunburn. Nearly everyone assumed that the same wavelengths of sunlight that caused sunburn were also responsible for causing skin cancers. And so it followed that sunscreen would provide protection from skin cancer as well as sunburn.

Why would anyone question whether sunscreen has a positive effect?

Sunscreens first became widely available in the 1960s and 1970s. Since then, there has been an increase in the incidence of melanoma. Strangely, in countries where sunscreen is most recommended and most often used—Canada, the United States, Australia, and Scandinavia—the increase in melanoma has been greatest. Deaths from melanoma in the United States more than doubled between 1950 and 1990.

In a study in Sweden, researchers compared sunscreen use between two groups: 571 patients diagnosed with melanoma and 913 healthy controls. Contrary to the researchers' expectations, sunscreen users had a significantly higher risk for developing melanoma than non-users—almost twice as high! (This was just one of eight studies published between 1979 and 1995 that found a positive association between sunscreen use and melanoma.)

Sunscreen users were at greater risk for cancer! How can that be?

People using sunscreen may have a false sense of security. This may lead them to spend more time in the sun and to be less likely to wear hats or other clothing that offers protection. Further complicating matters, until the mid-1990s, sunscreens with full UVA and UVB protection (called "broad spectrum" sunscreens) were not available. And even in 2010, only one-third of sunscreens available in the United States offered protection from all the wavelengths of UV radiation in sunlight. For these reasons, because the sunscreens reduced sunburn while allowing more time in the sun, they may have caused people to unknowingly overexpose themselves to those wavelengths for which the sunscreen gave no protection.

Before throwing away your sunscreen, though, it's important to be sure that it really is of no benefit. This is an important element of scientific thinking and the guiding principle taken by researchers in a more recently published study about melanoma and sunscreen.

How can you figure out whether sunscreen users are actually protected from cancer?

All of the studies purporting to show a positive relationship between sunscreen use and melanoma risk were "case-controlled" studies. This is a study design in which individuals with a particular outcome (such as developing melanoma) are identified and then compared with a group of individuals who do not have that outcome (do not have melanoma). The groups are analyzed to see whether they differ in some significant way—such as sunscreen use. Such a study design has some limitations. It can identify factors influencing the outcome of interest, but the validity of any associations depends on how similar the two groups actually are. In this case, for example, an assumption was made that the healthy subjects and the subjects with melanoma experienced similar sun exposure.

The results from such studies are undermined when the groups are heterogeneous—for example, if those in the melanoma group had more sun exposure than those in the healthy group. In fact, subsequent investigations suggested that the case-controlled studies were hindered by "confounding variables" that caused comparison groups to differ in significant ways. (Can you think of ways the two groups might differ?)

"Randomized controlled trials." Why are they better than case-controlled studies?

An alternative—and usually more powerful—study design is a randomized controlled study. In 2011, researchers reported on just such a study in Queensland, Australia, which randomly assigned 1,621 adults to one of two groups: regular sunscreen use and discretionary sunscreen use. (Would it have been a better study if subjects in the second group had not been allowed to use sunscreen at all? Why do you think the researchers didn't do that?)

For 5 years, those in the sunscreen group received unlimited sunscreen and were asked to apply it every morning (and to reapply it after sweating, bathing, or long sun exposure). The discretionary sunscreen users were allowed to use sunscreen at their usual frequency (which included no use at all for some people). At the end of the 5-year treatment period, and continuing for 10 additional years (until 2006), the researchers noted the incidence of melanomas. (This was made possible because in Queensland, all melanomas must be reported to the Queensland Cancer Registry.) They also obtained, from questionnaires filled out by study participants, information about time spent outdoors and sunscreen usage.

The results were dramatic. Among the 812 subjects in the sunscreen-use group, during the 10-year follow-up period, 11 new melanomas were identified. Among the 809 subjects in the discretionary-use group, twice as many new melanomas were identified. Invasive melanomas, the most severe type—occurred almost four times more frequently among people in the discretionary group. Based on the questionnaires, there were no differences between the groups for any known risk factors, or for the amount of time they had spent in the sun during the trial.

Because this is the first randomized controlled study of sunscreen use and melanoma, cancer experts have described it as a potential "game changer." They view the results as a clear indication that melanoma-prevention strategies should incorporate efforts at increasing regular use of sunscreen.

> **If randomized controlled studies are so much better, why would anyone bother doing case-controlled studies?**

These studies reveal just how important well-controlled experimental design is to demonstrating causal relationships. The case-controlled studies were important, but their inherent limitations led to results that may not hold up under closer scrutiny. Randomized controlled trials, though, are difficult to conduct and relatively expensive—and it can be difficult or even impossible to ensure full compliance with treatments (or with non-treatment by individuals in control groups). But we have much greater confidence in the evidence from such studies.

TAKE-HOME MESSAGE 5·10

The relationship between sunscreen use and skin cancer is important but murky. Numerous case-controlled studies suggested that sunscreen use increased the incidence of melanoma, the most deadly type of skin cancer. But a more powerful, randomized controlled approach demonstrated that regular sunscreen use significantly reduces the risk of melanoma.

5·11 Faulty genes, coding for faulty enzymes, can lead to sickness.

Isabella joins her friends in sipping wine during a dinner party. As the meal progresses, her companions become tipsy. Their conversations turn racy, their moods relaxed. They refill their glasses, reveling in a little buzz. Not so for Isabella. Before her first glass is empty, she experiences a "fast-flush" response: her face turns crimson, her heart begins to race, and her head starts to pound. Worse still, she soon feels the need to vomit.

How can people respond so differently to alcohol? It comes down to a difference in a single base pair in their DNA, a single difference that can influence dramatically a person's behavior, digestion, respiration, and general ability to function. The base-pair change leads to the production of a non-functional enzyme, and the lack of a functional version of this enzyme leads to physical illness. Let's look at the details.

When we consume alcohol, our bodies start a two-step process to convert the alcohol molecules from their intoxicating form into

Q *Why do many Asians have unpleasant experiences associated with alcohol consumption?*

innocuous molecules. Each of the two steps is made possible by a different enzyme, whose assembly instructions are, for most people, coded in their DNA.

"Fast-flushers" such as Isabella complete the first step of breaking down alcohol, but not the second, because they carry defective genetic instructions for making aldehyde dehydrogenase, the enzyme that makes possible the second step of the process. A poisonous substance subsequently accumulates, and the symptoms of the fast-flush reaction are due to this substance's toxic effects in the body.

Approximately half of the people living in Asia carry a non-functional form of the gene for aldehyde dehydrogenase, a mutation that may confer a greater benefit than harm. In a study of 1,300 alcoholics in Japan, not a single one was a fast-flusher, even though half of all Japanese people are fast-flushers. The minor change in the genetic code that makes alcohol consumption an unpleasant experience may be responsible for the lower incidence of alcoholism among Japanese and other Asian people.

In many other cases, the link between a particular defective DNA sequence and physical illness is equally direct. Recall from Chapter 3 the case of Tay-Sachs disease. In Tay-Sachs disease, an individual inherits genes with a mutation that causes an inability to produce a critical lipid-digesting enzyme in their lysosomes, the cellular garbage disposals. Because these organelles cannot digest certain lipids, the lipids accumulate, undigested. The lysosomes swell until they eventually choke the cell to death. This occurs in numerous cells in the first few years of life, and ultimately leads to the child's death.

Although the details differ from case to case, the overall picture is the same for many, if not most, inherited diseases. The pathway from mutation to illness includes just four short steps (**FIGURE 5-23**).

1. A mutated gene codes for a non-functioning protein, commonly an enzyme.

2. The non-functioning enzyme can't catalyze a particular reaction as it normally would, bringing the reaction to a halt.

3. The molecule with which the enzyme would have reacted (to convert it to another substance) accumulates, just as half-made products would pile up on a blocked assembly line.

4. The accumulating chemical causes sickness and/or death.

The fact that many genetic diseases involve illnesses brought about by faulty enzymes suggests some strategies for treatment. These include administering medications that contain the normal-functioning version of the enzyme. For instance, lactose-intolerant individuals can consume the enzyme lactase, which for a short while gives them the ability to digest lactose. Alternatively, lactose-intolerant individuals can reduce their consumption of lactose-containing foods to keep the chemical from accumulating, thus reducing the problems that come from lactose overabundance.

TAKE-HOME MESSAGE 5·11

Many genetic diseases result from mutations that cause a gene to produce a non-functioning enzyme, which in turn blocks the functioning of a metabolic pathway.

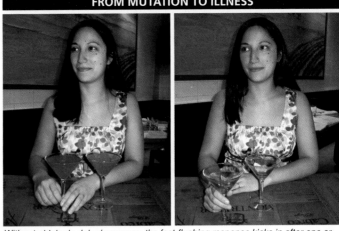

FROM MUTATION TO ILLNESS

Without aldehyde dehydrogenase, the fast-flushing response kicks in after one or two alcoholic drinks.

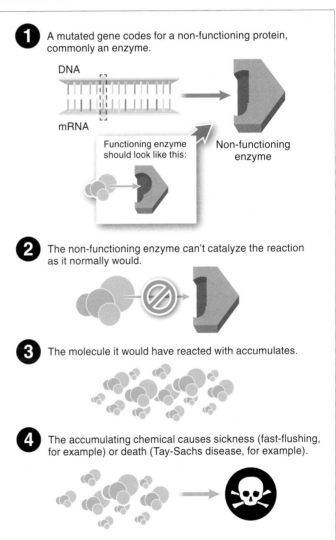

1 A mutated gene codes for a non-functioning protein, commonly an enzyme.

DNA

mRNA

Functioning enzyme should look like this:

Non-functioning enzyme

2 The non-functioning enzyme can't catalyze the reaction as it normally would.

3 The molecule it would have reacted with accumulates.

4 The accumulating chemical causes sickness (fast-flushing, for example) or death (Tay-Sachs disease, for example).

FIGURE 5-23 **Faulty enzymes can interfere with metabolism.**

5·12–5·14
Biotechnology is producing improvements in agriculture.

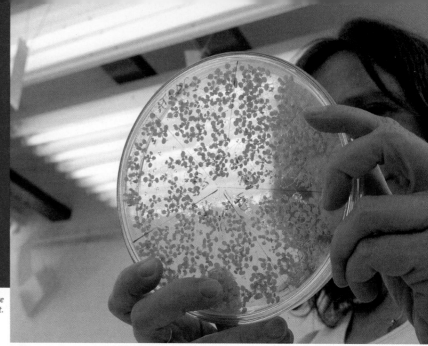

These genetically modified plants (Arabidopsis) are resistant to salinity and drought.

5·12 What is biotechnology?

Understanding a phenomenon (such as how DNA is transcribed and translated) is nice, but controlling it can take the satisfaction to an entirely new level. This may explain why there is so much excitement surrounding the field of **biotechnology,** in which organisms, cells, and their molecules are modified to achieve practical benefits. The modern emphasis in biotechnology is on **genetic engineering,** the manipulation of organisms' genetic material by adding, deleting, or transplanting genes from one organism to another.

In the remainder of this chapter, we explore the two primary areas in which biotechnology is applied: agriculture and human health. We begin by introducing the most important tools and techniques used in biotechnology.

How would you create a plant resistant to being eaten by insects? Or a colony of bacteria that can produce human insulin? Although there are many different uses of biotechnology, it employs a surprisingly small number of processes and tools that enable researchers to do the following (**FIGURE 5-24**):

1. **Chop** up the DNA from a donor organism that exhibits the trait of interest.

2. **Amplify** the small amount of DNA into larger quantities.

3. **Insert** pieces of DNA into bacterial cells or viruses.

FIVE TOOLS OF BIOTECHNOLOGY

CHOP up DNA from a donor species that exhibits a trait of interest.

AMPLIFY small samples of DNA into more useful quantities.

INSERT pieces of DNA into bacterial cells or viruses.

GROW separate colonies of bacteria or viruses, each containing some donor DNA.

IDENTIFY colonies of bacteria or viruses that have DNA for a trait of interest.

 Not all of the tools are used in all biotechnology applications—some use only one or a few of these techniques.

FIGURE 5-24 **Five important tools and techniques of most biotechnology procedures.**

4. **Grow** separate colonies of the bacteria or viruses, each of which contains a different inserted piece of donor DNA.

5. **Identify** colonies of bacteria or viruses that have DNA for the trait of interest (and propagate them).

To be sure, it's easier said than done, but these tools capture the essence of most modern biotechnology. Not all of the techniques are used in all biotech applications; nonetheless, they all are mainstays of today's biotechnology world. Let's explore each technique in a bit more detail.

Tool 1. Chopping up DNA from a donor organism. To begin, researchers select an organism that has a desirable trait. For example, they might want to produce human growth hormone in large amounts. Their first step would be to obtain a sample of human DNA and cut it into smaller pieces. Cutting DNA into small pieces requires the use of **restriction enzymes** (**FIGURE 5-25**).

Restriction enzymes have a single function: when they encounter DNA, they cut it into small pieces. These enzymes evolved to protect bacteria from attack by viruses. Upon encountering DNA from an invading virus, a restriction enzyme recognizes and binds to a particular sequence of four to eight bases on the invader's DNA and cuts there, thus making it impossible for the virus to reproduce within the bacterial cell. Since all DNA has the same basic structure, these enzymes can cut DNA from any source as long as the specific four- to eight-base sequence is present (bacteria protect their own DNA by modifying it, or it would be cut up also). Dozens of different restriction enzymes exist, each of which recognizes and cuts a different sequence of bases in DNA.

Tool 2. Amplifying DNA pieces into larger quantities. Oftentimes in genetic engineering, only a small sample of donor DNA is taken and chopped up. And in many other situations, only a small amount of DNA is available for analysis or for some other biotechnology use. The **polymerase chain reaction (PCR)** is a laboratory technique that allows a tiny piece of DNA—a piece from the first step of genetic engineering, or, in another biotech use, a piece recovered from a crime scene—to be duplicated repeatedly (**FIGURE 5-26**).

The process involves heating up a solution with the DNA of interest, causing the two strands to separate. Then, in the presence of certain enzymes and plenty of nucleotides floating free in the solution, the DNA is cooled. As it cools, an enzyme matches free nucleotides with their complementary bases on each of the single strands and covalently links these nucleotides together to form a new, complementary, single strand. The result is two complete

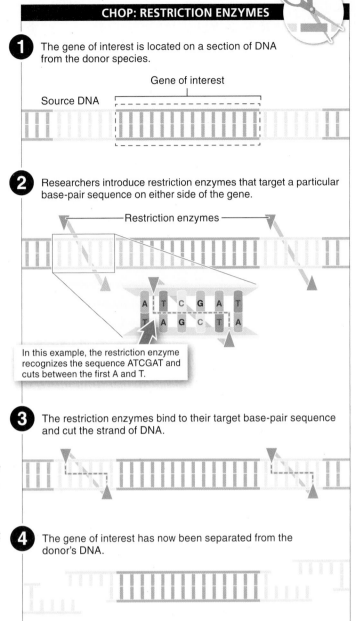

CHOP: RESTRICTION ENZYMES

1 The gene of interest is located on a section of DNA from the donor species.

Gene of interest

Source DNA

2 Researchers introduce restriction enzymes that target a particular base-pair sequence on either side of the gene.

Restriction enzymes

A T C G A T
T A G C T A

In this example, the restriction enzyme recognizes the sequence ATCGAT and cuts between the first A and T.

3 The restriction enzymes bind to their target base-pair sequence and cut the strand of DNA.

4 The gene of interest has now been separated from the donor's DNA.

FIGURE 5-25 **Restriction enzymes are used to isolate a gene of interest.**

double-stranded copies of the DNA of interest. This process of heating and cooling can be repeated again and again until there are billions of identical copies of the target sequence.

Tool 3. Inserting foreign DNA into the target organism. In the human growth hormone example mentioned above, the researchers might want to transfer the human growth hormone gene into the bacterium *E. coli,* creating **transgenic organisms.**

To create a transgenic organism (that is, an organism with DNA inserted from a different species), researchers must

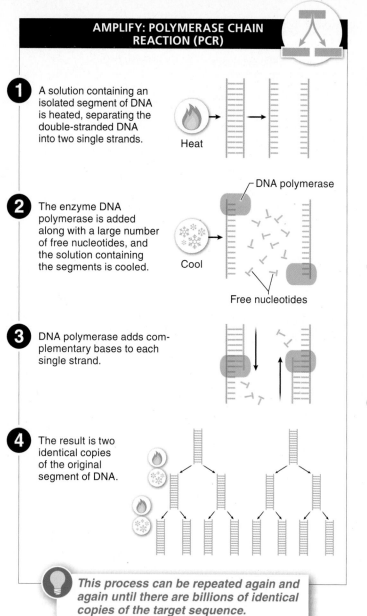

1 A solution containing an isolated segment of DNA is heated, separating the double-stranded DNA into two single strands.

Heat

2 The enzyme DNA polymerase is added along with a large number of free nucleotides, and the solution containing the segments is cooled.

Cool

DNA polymerase

Free nucleotides

3 DNA polymerase adds complementary bases to each single strand.

4 The result is two identical copies of the original segment of DNA.

This process can be repeated again and again until there are billions of identical copies of the target sequence.

FIGURE 5-26 **The polymerase chain reaction can duplicate a small strand of DNA repeatedly to form billions of copies.**

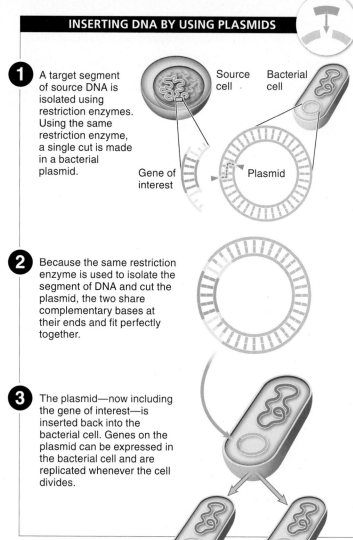

1 A target segment of source DNA is isolated using restriction enzymes. Using the same restriction enzyme, a single cut is made in a bacterial plasmid.

Source cell

Bacterial cell

Gene of interest

Plasmid

2 Because the same restriction enzyme is used to isolate the segment of DNA and cut the plasmid, the two share complementary bases at their ends and fit perfectly together.

3 The plasmid—now including the gene of interest—is inserted back into the bacterial cell. Genes on the plasmid can be expressed in the bacterial cell and are replicated whenever the cell divides.

FIGURE 5-27 **A gene chauffeur.** Plasmids can transfer DNA from one species to another.

physically deliver the DNA from a donor species into the recipient organism. This delivery often is accomplished using **plasmids,** circular pieces of DNA that can be incorporated into a bacterium's genome (**FIGURE 5-27**). Genes on the plasmid can then be expressed in the bacterial cell and are replicated whenever the cell divides, so both of the new cells contain the plasmid. In some cases, for the delivery process, genes are incorporated into viruses instead of plasmids. The viruses can then be used to infect organisms and transfer the genes of interest into those organisms.

Tool 4. Growing bacterial colonies that carry the DNA of interest: cloning. Once a piece of foreign DNA has been transferred to a bacterial cell, every time the bacterium divides, it creates a **clone,** a genetically identical cell that contains that inserted DNA. The term **cloning** describes the production of genetically identical cells, organisms, or DNA molecules, a process that occurs each time a bacterium divides. With numerous rounds of cell division, it is possible to produce a huge number of clones, all of which transcribe and translate the gene of interest.

In a typical recombinant DNA experiment—that is, one in which DNA from multiple sources is brought together—a large amount of DNA may be chopped up with restriction enzymes, incorporated into plasmids, and introduced into bacterial cells. The bacteria are then allowed to divide repeatedly, with each bacterial cell producing a clone of the foreign DNA fragment it carries. Together, all of the different cells containing all of the different fragments of the original DNA are called a **clone library** or a **gene library** (FIGURE 5-28).

Tool 5. Identifying bacterial colonies that have received the gene of interest (and propagating them). Chopping up human DNA and inserting the pieces (each of which carries different genes) into the genomes of bacteria, which divide repeatedly, leads to a large population of bacterial cells, with many carrying useful genetic information. But the information is in no particular order, much like a bookstore in which all the books are uncatalogued and in complete disarray. How can a researcher interested in working with bacterial cells that contain just the one human gene capable of producing human growth hormone identify and separate those bacteria from the other bacteria in the population?

Researchers have developed a way to make the bacteria of interest identify themselves (FIGURE 5-29). First, a chemical is added to the entire population of bacterial cells, separating all the double-stranded DNA into single strands. Next, a short sequence of single-stranded DNA is washed over the bacteria. Called a **DNA probe,** this DNA contains a sequence complementary to part of the gene of interest, and it has also been modified so that it is radioactive. Bacteria with the gene of interest bind to this probe and glow with

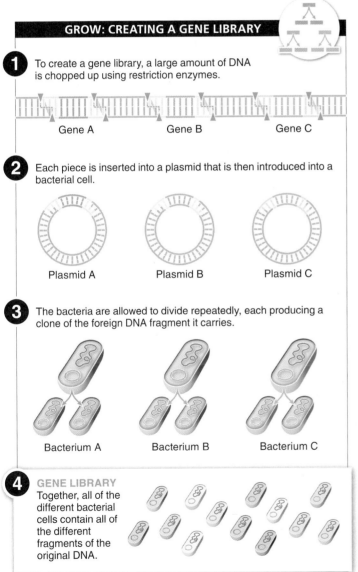

GROW: CREATING A GENE LIBRARY

1 To create a gene library, a large amount of DNA is chopped up using restriction enzymes.

Gene A   Gene B   Gene C

2 Each piece is inserted into a plasmid that is then introduced into a bacterial cell.

Plasmid A   Plasmid B   Plasmid C

3 The bacteria are allowed to divide repeatedly, each producing a clone of the foreign DNA fragment it carries.

Bacterium A   Bacterium B   Bacterium C

4 GENE LIBRARY
Together, all of the different bacterial cells contain all of the different fragments of the original DNA.

IDENTIFYING A GENE BY USING A DNA PROBE

GENE LIBRARY

1 To locate a gene of interest among the millions of clones in a gene library, the bacteria are washed with a chemical that breaks down the DNA, making it single-stranded.

Gene of interest

2 A radioactive probe is then washed over the single-stranded DNA.

DNA probe

The probe is a short length of DNA that contains a sequence of bases complementary to the gene of interest.

3 The DNA probe binds to the complementary base sequence found in the gene of interest and glows with radioactivity, allowing the gene to be easily identified.

G T A C T
C A T G A

T A G C T

A T C C G

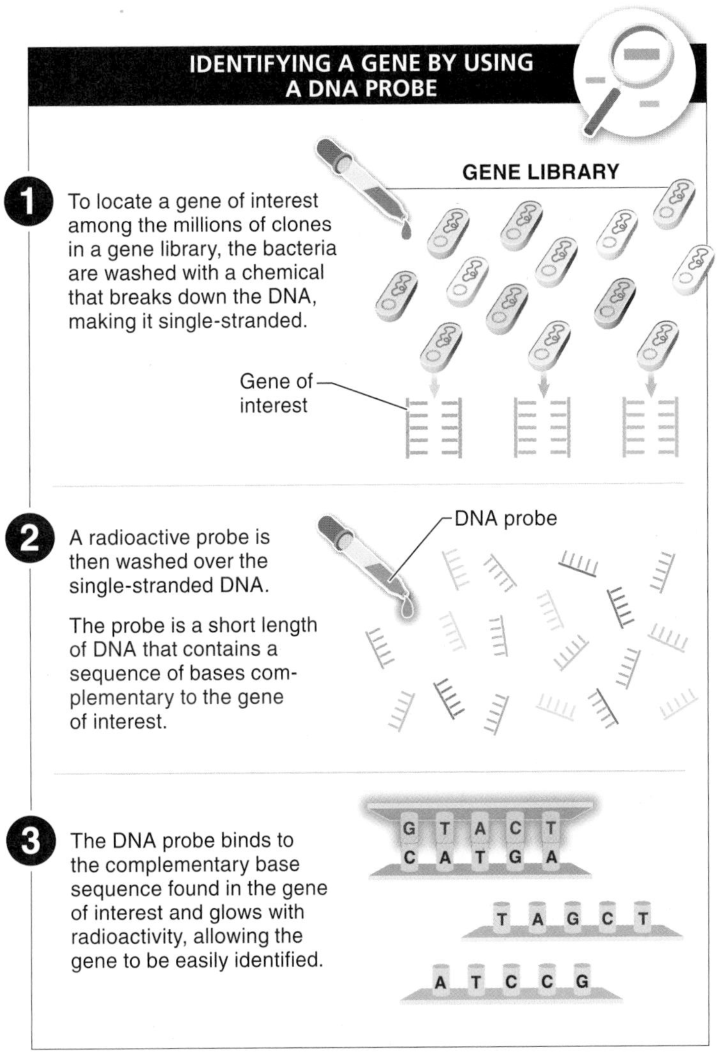

FIGURE 5-29 **Finding the needle in a haystack.** A DNA probe is used to locate the desired clone in a gene library.

FIGURE 5-28 **A DNA archive.** A gene library (or clone library) is a collection of cloned DNA fragments.

radioactivity. These cells can then be separated out and grown in large numbers—for example, vats of *E. coli* that produce human growth hormone. (Note that because the "correct" folding of a protein depends not just on the amino acid sequence specified by a gene but also on the appropriate cellular conditions, genes transplanted from one species to another don't always produce proteins with the exact same shape. Such folding problems can lead to proteins that do not function as expected.)

We turn next to how these tools and techniques are used to develop products. Keep in mind, however, that the field is still in its infancy.

TAKE-HOME MESSAGE 5·12

Biotechnology is the use of technology to modify organisms, cells, and their molecules to achieve practical benefits. Modern molecular methods make it possible to cut and copy DNA from one organism and deliver it into another. The methods include the use of naturally occurring restriction enzymes for cutting DNA, the polymerase chain reaction for amplifying small amounts of DNA, insertion of the DNA into bacterial or viral vectors, and the cloning and identification of cells with the transferred DNA of interest.

5·13 Biotechnology can improve food nutrition and make farming more efficient and eco-friendly.

Your breakfast cereal is probably fortified with vitamins and minerals. And for snacking you may eat protein bars that have as much protein as a full chicken breast. It shouldn't come as a surprise, then, that farmers have begun using biotechnology to improve on the natural levels of vitamins, minerals, and other nutrients in the fruits, vegetables, and livestock they produce.

For thousands of years, humans have been practicing a relatively crude and slow form of **genetic engineering**—the manipulation of a species' genome in ways that do not normally occur in nature. In its simplest form, genetic engineering is the careful selection of the plants or animals

to be used as the breeders for a crop or animal population. Through this process, farmers and ranchers have produced meatier turkeys, seedless watermelons, and big, juicy corn kernels (**FIGURE 5-30**). But what used to take many generations of breeding can now be accomplished in a fraction of the time, using **recombinant DNA technology**, the combination of DNA from two or more sources into a product.

In crop plants, for example, the process begins with the identification of a new characteristic, such as larger size or faster ripening time, that farmers would like in a particular crop. Traditionally, breeders would then search for an

FIGURE 5-30 Ancestral corn and modern corn—selected for larger, juicier kernels.

organism within the same species that had the desirable trait, breed it with their crop organisms, and hope that the offspring would express the trait in the desired way. With recombinant DNA technology, the desired trait can come from *any* species, so the pool of organisms from which the trait can be taken becomes much larger. Organisms produced with recombinant DNA technology are referred to as genetically modified organisms, or GMOs.

Although the rewards are potentially huge, in practice, the process of creating transgenic species that are more nutritious or have other desirable traits turns out to be difficult. Nonetheless, the results so far hint at a fruitful marriage of agriculture and technology.

Q How might a genetically modified plant help 500 million malnourished people?

Nutrient-rich "golden rice" Almost 10% of the world's population suffers from vitamin A deficiency, which causes blindness in a quarter-million children each year and a host of other illnesses in people of all ages. These nutritional problems are especially severe in southern Asia and sub-Saharan Africa, where rice is a staple of most diets. Addressing this global health issue, researchers have developed what may be the model for solving problems with biotechnology. It involves the creation of a new crop called "golden rice."

Mammals generally make vitamin A from beta-carotene, a substance found in abundance in most plants (it's what makes carrots orange), but not in the edible part of rice grains. Researchers set out to change this by inserting into the rice genome, from other species, three genes that code for the enzymes used in the production of beta-carotene: two genes from the daffodil plant and one bacterial gene (**FIGURE 5-31**). It's clear that the transplanted genes are working, because the normally white rice grain takes on a golden color from the accumulated beta-carotene. And since golden rice was first developed in 1999, new lines have been produced that use only two transplanted genes, yet produce almost 25 times the vitamin A found in the original strains. Field tests of golden rice are still under way, but it is viewed as one of the most promising applications yet of biotechnology.

> **❝**Italians come to ruin most generally in three ways, women, gambling, and farming. My family chose the slowest one.**❞**
> — POPE JOHN XXIII

While the development of golden rice demonstrates that biotechnology, in the near future, may help us produce food that is more nutritious, biotechnology has already had a much more profound impact than this on agricultural practices in the United States. It is not a stretch to say that we are in the midst of a revolution—a green revolution— and that few people are aware of it. This revolution is the extent to which biotechnology has reduced the costs, both environmental and financial, of producing the plants and animals we eat.

Currently, more than 170 million acres worldwide are planted with genetically modified crops, most containing built-in insecticides and herbicide resistance, representing more than a forty-fold increase over the past 10 years. The financial benefits to farmers—at least in the short run—are

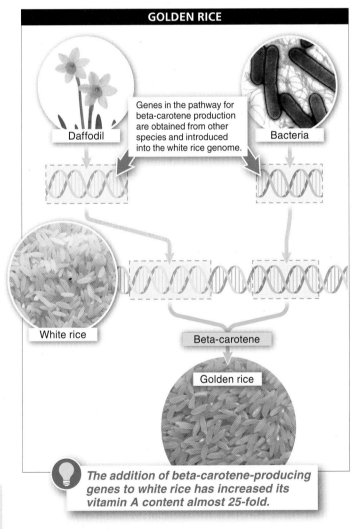

GOLDEN RICE

Daffodil

Genes in the pathway for beta-carotene production are obtained from other species and introduced into the white rice genome.

Bacteria

White rice

Beta-carotene

Golden rice

💡 *The addition of beta-carotene-producing genes to white rice has increased its vitamin A content almost 25-fold.*

FIGURE 5-31 **The potential to prevent blindness in 250,000 people each year.** Engineering rice to prevent blindness by increasing its vitamin A content.

14%—
86%
Corn

7%—
93%
Cotton

7%—
93%
Soybeans

■ Proportion of crops that are not genetically modified
■ Proportion of crops that are genetically modified

GRAPHIC CONTENT
Thinking critically about visual displays of data
Turn to p. 225 for a closer inspection of this figure.

FIGURE 5-32 **A significant portion of crops grown in the United States are genetically modified.**

so great that more and more of them are embracing the genetically modified crops.

Q **What genetically modified foods do most people in the United States consume (usually without knowing it)?**

The numbers are surprising: 86% of all corn grown in the United States is genetically modified; 93% of all cotton grown is genetically modified; and 93% of all soybeans grown are genetically modified (**FIGURE 5-32**). Two factors explain much of the extensive adoption of genetically modified plants in U.S. agriculture. (1) Many plants have had insecticides engineered into them, which can reduce the amounts of insecticides used in agriculture. (2) Many plants also have herbicide-resistance genes engineered into them. Such herbicide-resistant plants can reduce the amount of plowing required around crops to remove weeds. As a consequence, then, the use of genetically modified plants can reduce both the costs of producing food and the loss of topsoil to erosion.

Here are some of the biggest successes in the application of recombinant DNA technology to agriculture.

Insect and herbicide resistance Insect pests have a field day on agricultural crops. Crops planted at high

densities and nurtured with ample water and fertilizer represent a huge potential food resource for insects. Every year, about 40 million tons of corn are unmarketable as a consequence of insect damage. Increasingly, however, farmers have been enjoying greater success in their battles against insect pests.

Farmers owe much of this success to soil-dwelling bacteria of the species *Bacillus thuringiensis* (in brief, Bt). These bacteria produce spores containing crystals that are poisonous to insects but harmless to the crop plants and to people. In insects, within an hour of ingestion, Bt crystals cause pores to develop throughout the digestive system, paralyzing the insects' gut and making them unable to feed. Within a few days, the insects die from a combination of tissue damage and starvation.

Beginning in 1961, the toxic Bt crystals were included in the pesticides sprayed on crop plants. Then in 1995, the gene coding for the production of the Bt crystals was inserted directly into the DNA of many different crop plants, including corn, cotton, and potatoes. As a consequence, farmers no longer need to apply huge amounts of Bt-containing pesticides (**FIGURE 5-33**);

Q **How can genetically modified plants lead to reduced pesticide use by farmers?**

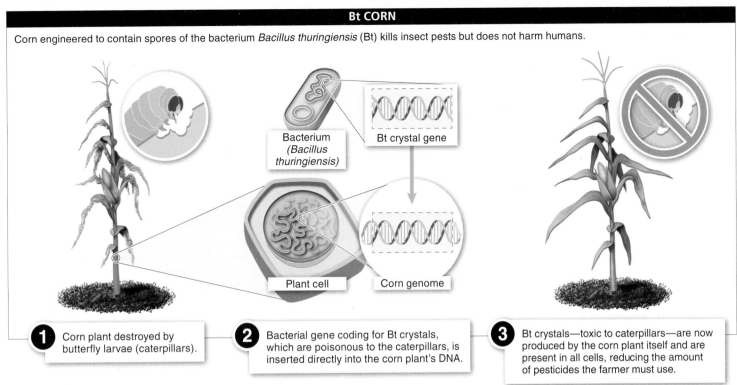

Corn engineered to contain spores of the bacterium *Bacillus thuringiensis* (Bt) kills insect pests but does not harm humans.

Bacterium
(*Bacillus
thuringiensis*)

Bt crystal gene

Plant cell

Corn genome

1 Corn plant destroyed by butterfly larvae (caterpillars).

2 Bacterial gene coding for Bt crystals, which are poisonous to the caterpillars, is inserted directly into the corn plant's DNA.

3 Bt crystals—toxic to caterpillars—are now produced by the corn plant itself and are present in all cells, reducing the amount of pesticides the farmer must use.

FIGURE 5-33 **Help from bacteria in growing disease-resistant corn.**

instead, the plants themselves produce insect-killing Bt crystals. There is no evidence that humans are harmed by the Bt crystals, even when they are exposed to very high levels.

Bacteria have come to the aid of farmers in their fight against pests in another way, too. In the 1990s, researchers discovered a bacterial gene that gives the bacteria resistance to herbicides, and they introduced the gene into crop plants. Integration of this gene into the plants' DNA gives them resistance to herbicides, allowing farmers to kill weeds with herbicides but leave the crop plants unharmed, greatly increasing yields (**FIGURE 5-34**).

Faster growth and bigger
bodies Agriculture includes the cultivation not just of plants but also of animals. And for the first time, the U.S. Food and Drug Administration is close to approving for human consumption a genetically modified animal. The animal in question is a transgenic Atlantic salmon that carries a growth hormone gene from another species (Chinook salmon), along with a region of DNA from a third species (ocean pout) that acts as an "on" switch, facilitating transcription of the growth hormone gene. The transgenic fish, which is

reported by its creators to taste the same as regular Atlantic salmon, grows much more quickly and reaches market size within 18 months rather than the usual three years (**FIGURE 5-35**).

The FDA has reported that the transgenic salmon "is as safe to eat as food from other Atlantic salmon." Numerous fisheries experts, food safety experts, environmental groups,

FIGURE 5-34 **Crop duster.** Herbicides like the one applied by this crop duster must kill weeds while leaving the crop unharmed.

The larger salmon carries a growth hormone gene that keeps it growing year-round rather than in the summer only.

FIGURE 5-35 Bigger salmon.

and consumer groups, however, continue to express concerns about a wide variety of safety and environmental issues and the process by which the safety and environmental impacts of the transgenic species have been evaluated. Health concerns include the possibility that consuming the salmon will cause increased rates of allergic

reactions, as well as unknown effects that may stem from potentially higher levels of hormones present in the fish.

Most troubling to environmental groups is the risk that the larger, faster-growing transgenic fish will escape from their enclosed breeding facilities into their natural habitat—something that many experts believe is inevitable. If this occurs, environmentalists fear that the fish might harm wild salmon populations, many of which are listed as endangered, because the transgenic salmon can consume more resources and may grow too large to be consumed by its natural predators. It is unclear what the outcome would be.

TAKE-HOME MESSAGE 5·13

Biotechnology has led to important improvements in agriculture by using transgenic plants and animals to produce more nutritious food. Even more significant is the extent to which biotechnology has reduced the environmental and financial costs of producing food, through the creation of herbicide-resistant and insect-resistant crops. The ecological and health risks of such widespread use of transgenic species are not fully understood and are potentially great.

5·14 Fears and risks: are genetically modified foods safe?

Chickens without feathers look ridiculous (**FIGURE 5-36**). But such a genetically modified breed was developed with a valuable purpose in mind: "naked" birds are easier and less expensive to prepare for market, benefiting farmers by lowering their costs and benefiting consumers by lowering prices. These chickens, however, have turned out to be unusually vulnerable to mosquito attacks, parasites, and disease, and ultra-sensitive to sunlight. They also have difficulty mating, because the males are unable to flap their wings. Researchers currently are working to address these problems.

Naked chickens teach us an important lesson about genetically modified plants and animals. Although the new breed of featherless chickens was produced by relatively low-tech genetic engineering—the traditional animal husbandry method of crossbreeding two different types of chickens—rather than by more modern recombinant DNA

technology, the new breed ended up having not just the desired trait of no feathers but also some unintended and undesirable traits. Now, as more genetically modified foods are created using modern methods of recombinant DNA technology, the same risks of also creating unintended and potentially harmful traits must be weighed. For these and other reasons—some legitimate and rational, others irrational—many people have concerns about the production and consumption of genetically modified foods (**FIGURE 5-37**). These concerns are outlined in the following discussion.

Organisms that we want to kill may become invincible. Pesticide-resistant canola plants were cultivated in Canada, making it possible for farmers to apply herbicides freely to kill the weeds but not the canola crop. But the pesticide-resistant canola plants accidentally spread to neighboring farms and grew out of control, because traditional herbicides could not kill them. Similarly, there is the

Featherless birds are cheaper for farmers and consumers. But there are unintended consequences, including vulnerability to mosquitoes and other parasites.

FIGURE 5-36 "Naked" birds.

possibility that insect pests will develop resistance to the Bt produced by genetically modified crops, which will also make these pests resistant to Bt pesticides applied to crops that are not genetically modified.

Organisms that we don't want to kill may be killed inadvertently. Monarch butterflies feed on milkweed plants. Recent research has demonstrated that if pollen from plants genetically modified to contain the insect-killing Bt genes accidentally lands on milkweed plants and is consumed by monarch butterflies, the butterflies can be killed, which may significantly reduce their populations. Although such an incident has not occurred outside experimental fields, it illustrates a risk that may be hard to control.

Genetically modified crops are not tested or regulated adequately. It is impossible to really know whether a new technology has been tested adequately. Still, scientists and lawmakers have been working toward an organized and responsible set of policies designed to ensure that sufficient safety testing is done. For example, laboratory procedures for working with recombinant DNA have been established, and researchers

have developed techniques that make it impossible for most genetically engineered organisms to survive outside the specific conditions for which they are developed.

As an example of the degree of testing of genetically engineered foods, the Monsanto Company has had its strain of herbicide-resistant soybeans evaluated and approved by 31 different regulatory agencies in 17 different countries, including, in the United States, the Department of Agriculture, the Food and Drug Administration, and the Environmental Protection Agency. In a recent report on genetically modified animals, however, an expert committee of the U.S. National Academy of Sciences warned that GMOs still pose risks that the government is unable to evaluate. Technology is moving so fast that it is difficult to even know what the new risks might be.

> ❝And he gave it for his opinion, 'that whoever could make two ears of corn, or two blades of grass, to grow upon a spot of ground where only one grew before, would deserve better of mankind, and do more essential service to his country, than the whole race of politicians put together.'❞
>
> — JONATHAN SWIFT, *Gulliver's Travels,* 1726

Eating genetically modified foods is dangerous. In the 1990s, a gene from Brazil nuts was used to improve the nutritional content of soybeans. The genetically modified soybeans had better nutritional content, but they also acquired some

FIGURE 5-37 **Consumer fears.** Protesters voice opposition to the use of genetically modified organisms (GMOs).

allergy-causing chemicals previously present in the Brazil nuts but not in soybeans. This outcome illustrates the risk that some unwanted features might be passed from species to species in the creation of transgenic organisms. In this case, all of the genetically modified soybeans were destroyed and this research program was suspended. To date, no evidence has appeared to suggest that consumption of any genetically modified foods is dangerous.

Loss of genetic diversity among crop plants is risky. As increasing numbers of farmers stop using non-GMO crops in favor of one or a few genetically modified strains of crops, the genetic diversity of the crops declines. This can make them more vulnerable to environmental changes or pests. The Irish Potato Famine is an example of the value of genetic diversity in crops. In the mid-1800s, much of the population of Ireland depended on a diet of potatoes. Because most of the potato crops had been propagated from cuttings from the same plant, they were all genetically the same. When the crops were infected by a rot-causing mold, all of the potato plants were susceptible and most were wiped out, causing a famine responsible for the deaths of more than a million people.

Hidden costs may reduce the financial advantages of genetically modified crops. When seed companies create genetically modified seeds with crop traits desirable to farmers, the companies also engineer sterility into the seeds. As a consequence, the farmers must purchase new seeds for each generation of their crops. Such increases in long-term costs and dependency on seed companies must be factored in by the farmers.

Another argument sometimes made in opposition to genetically modified foods is that such organisms are not "natural" and, for that reason, must be harmful. This is one argument that is flawed and should not be a cause for concern. Smallpox, HIV, poison ivy, and cyanide, after all, are natural. The smallpox vaccine, on the other hand, is unnatural. Innumerable other valuable technological developments are equally unnatural. There simply is no value in knowing whether something is natural or unnatural when evaluating whether or not it is good and desirable.

In the end, we must compare the risks of producing genetically modified foods with the benefits. The cost-benefit analyses will have to include the potential to reduce food costs and the ability to reduce environmental degradation by agriculture. For example, with genetically modified, pest-resistant crops, farmworkers will greatly benefit from spending less time applying pesticides. These benefits of reduced pesticide exposure will be significantly greater for workers in the less-developed countries, where safety regulations for pesticide use are more frequently ignored. Unfortunately, establishing the risks of genetically modified foods—as well as the ecological magnitude of those risks—is difficult, and this important issue must be evaluated extremely closely and regulated carefully.

TAKE-HOME MESSAGE 5·14

More and more genetically modified foods are being created using modern methods of recombinant DNA technology. Among the public, however, numerous legitimate fears remain about the potentially catastrophic risks of these foods, given that their development relies on such new technology, and about the long-term financial advantages they offer.

5·15 – 5·18
Biotechnology has the potential for improving human health (and criminal justice).

In 1996, Dr. Ian Wilmut created a sheep named Dolly, the first mammal to be cloned from an adult somatic cell.

5·15 The treatment of diseases and the production of medicines are improved with biotechnology.

You can't always get what you want. In the best of all worlds, biotechnology would *prevent* debilitating human diseases. Next best would be to *cure* diseases once and for all. But these noble goals are not always possible, so biotechnology often is directed at the more practical goal of *treating* diseases, usually by producing medicines more efficiently and more effectively than they can be produced with traditional methods. Biotechnology has had some notable successes in achieving this goal. The treatment of diabetes is one such success story.

Type 1 diabetes, often called juvenile diabetes, is a chronic disease in which the body cannot produce insulin, a chemical that allows cells to take up and break down sugar from the blood. Type 2 diabetes, which accounts for 90% of all cases of diabetes, is a metabolic disorder in which blood sugar levels rise higher than normal as a result of insulin resistance and insufficient insulin production. Complications from both types of diabetes can include vascular disease, kidney damage, and nerve damage. Approximately one-third of all people with diabetes (including the vast majority of those with type 1 diabetes) treat their condition with one or more daily injections of insulin. As recently as 1980, the insulin that most diabetics used was extracted from the pancreas of cattle or pigs that had been killed for meat. For most people, these insulin injections kept the disease under control. But the traditional process of collecting insulin this way was difficult and costly.

Everything changed in 1982, when a 29-year-old entrepreneur, Bob Swanson, joined scientist Herbert Boyer to transform the potential of recombinant DNA technology. In doing so, they started the biotech revolution. Working with the scientist Stanley Cohen, Swanson and Boyer used restriction enzymes to snip out the human DNA sequence that codes for the production of insulin. They then inserted this sequence into the bacterium *E. coli,* creating a transgenic organism. After cloning the new, transgenic bacteria, the team was able to grow vats and vats of the bacterial cells, all of which churned out human insulin (**FIGURE 5-38**). The drug could be produced efficiently in huge quantities and made available for patients with diabetes. This was the first genetically engineered drug approved by the U.S. Food and Drug Administration and it continues to help millions of people every day.

Q *Why do some bacteria produce human insulin?*

Perhaps even more significant than providing a better source of insulin, Swanson, Boyer, and Cohen's application of biotechnology revealed a generalized process for genetic engineering. It instantly opened the door to a more effective method of producing many different medicines to treat diseases. Today, more than 1,500 companies work in the recombinant DNA technology industry, and their products generate more than $40 billion in revenues each year.

Several important achievements followed the development of insulin-producing bacteria. Here are just two examples.

1. Human growth hormone (HGH). Produced by the pituitary gland, human growth hormone has dramatic effects throughout the body. It stimulates protein synthesis, increases the utilization of body fat for energy to fuel metabolism, and

FIGURE 5·38 **Lifesaving insulin.** Human insulin is engineered through recombinant DNA technology.

stimulates the growth of virtually every part of the body (FIGURE 5·39). Insufficient growth hormone production, usually due to pituitary malfunctioning, leads to dwarfism.

When treated with supplemental HGH, individuals with dwarfism experience additional growth. Until 1994, however, HGH treatment was prohibitively expensive because the growth hormone could be produced only by extracting and purifying it from the pituitary glands of human cadavers. Through the creation of transgenic bacteria, using a technique

Sylvester Stallone—age 61 in this photo—used human growth hormone (along with strength training) to develop larger muscles for a film role.

FIGURE 5·39 **Bulking up with a little (illegal) help.**

similar to that used in the creation of insulin-producing bacteria, HGH can now be produced in virtually unlimited supplies and made available to more people who need it.

The availability of HGH, which can increase strength and endurance, may be irresistibly tempting to some people (who don't need it for medical reasons)—even at $7,500 for a month's supply. Recent sporting scandals suggest that the illegal use of HGH occurs frequently among elite swimmers, cyclists, and other athletes.

2. Erythropoietin. Produced primarily by the kidneys, erythropoietin (also known as EPO) is a hormone that regulates the production of red blood cells. Numerous clinical conditions (nutritional deficiencies and lung disease, among others) and treatments (such as chemotherapy) can lead to anemia, a lower than normal number of red blood cells, which reduces an individual's ability to transport oxygen to tissues and cells. This lack of sufficient oxygen, in turn, can cause a variety of symptoms, including weakness, fatigue, and shortness of breath.

First cloned in 1985, recombinant human erythropoietin (rhu-EPO) is now produced in large amounts in hamster ovaries. It is used to treat many forms of anemia. Worldwide sales of EPO are in the billions of dollars.

EPO has been at the center of several "blood doping" scandals in professional cycling. This hormone increases the oxygen-carrying capacity of the blood, so some otherwise healthy athletes have used EPO to improve their athletic performance. It can be very dangerous, though. By increasing the number of red blood cells, the blood can become much thicker, and this can increase the risk of heart attack.

Q What is "blood doping"? How does it improve some athletes' performance?

Beyond these and other medicines currently produced by transgenic organisms, plans are under way to create a variety of other useful products for treating disease—including potatoes that produce antibodies enabling a more effective response to illness. In the next section we examine the strategies for preventing genetic diseases, and the much less successful attempts to cure diseases through biotechnology.

TAKE-HOME MESSAGE 5·15

Biotechnology has led to some notable successes in treating diseases, usually by producing medicines more efficiently and effectively than they can be produced with traditional methods.

5·16 Gene therapy: biotechnology can help diagnose and prevent genetic diseases, but has had limited success in curing them.

Would you want to know? Once, this was just a hypothetical question: if you carried a gene that meant you were likely to develop a particular disease later in your life, would you want to know about it? Or another question: would you want to know if there's a good chance that your future children will be born with a genetic disease? Now, for better or for worse, these are becoming real-life questions that we all must address. And there is more at stake than simply peace of mind. As biotechnology develops the tools to identify some of the genetic time-bombs that many of us carry, it also carries the danger that such information may become the basis for greater discrimination than we have ever known.

Intervening to prevent diseases through biotechnology focuses on answering certain questions posed at three different points in time.

1. Is a given set of parents likely to produce a baby with a genetic disease? Many genetic diseases occur only if an individual inherits two copies of the disease-causing gene, one from each parent. This is true for Tay-Sachs disease, cystic fibrosis, and sickle-cell anemia, among others. Individuals with only a single copy of the disease-causing gene never fully show the disease, but they may pass on the disease gene to their children. Consequently, two healthy parents (that is, having no disease symptoms) may produce a child with the disease. In these cases, it can be beneficial for the parents to be screened to determine whether they carry a disease-causing copy of the gene. Such screening, combined with genetic counseling and testing of embryos following fertilization, can reduce the incidence of a genetic disease dramatically. Since screening began in 1969, the incidence of Tay-Sachs disease, for example, has fallen by more than 75% (**FIGURE 5-40**).

2. Will a baby be born with a genetic disease? Once fertilization has occurred, it is possible to test an embryo or developing fetus for numerous genetic problems. Prenatal genetic screening can detect cystic fibrosis, sickle-cell anemia, Down syndrome, and a rapidly growing list of other disorders.

To screen the fetus, doctors must examine some of the fetal cells and/or the amniotic fluid (which surrounds the fetus in the uterus and carries many chemicals produced by the developing embryo). Cells and fluid are usually collected by

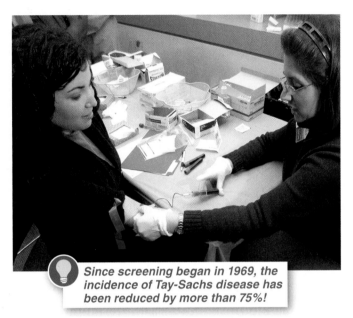

Since screening began in 1969, the incidence of Tay-Sachs disease has been reduced by more than 75%!

FIGURE 5-40 Genetic screening can determine the presence of the Tay-Sachs gene.

amniocentesis or chorionic villus sampling (CVS), techniques that we explore in detail in Chapter 6.

3. Is an individual likely to develop a genetic disease later in life? DNA technology can also be used to detect disease-causing genes in individuals who are currently healthy but are at increased risk of developing an illness later. Early detection of many diseases, such as breast cancer, prostate cancer, and skin cancer, greatly enhances the ability to treat the disease and reduce the risk of more severe illness or death.

These potential benefits of genetic technology come with significant potential costs. People who have a gene that puts them at increased risk of developing a particular disease, for example, might be discriminated against, even though they are not currently sick and may never suffer from the particular disease. Although a federal Genetic Information Nondiscrimination Act was signed into law in 2008, the law does not cover life insurance, disability insurance, and long-term care insurance. Insurance companies have already denied such coverage on discovering that an individual carries a gene that puts him or her at increased risk of disease. Additionally, parents who discover that their developing fetus will develop a painful, debilitating, or fatal disease soon after birth are confronted with the difficult question of how to proceed.

When it comes to curing a disease by using biotechnology, there is good news and bad news. The good news is that, in the 1990s, a handful of humans with a usually fatal genetic disease called severe combined immunodeficiency disease (SCID) were completely cured through the application of biotechnology. The bad news is that it has not been possible to apply these promising techniques to other diseases.

It's not for a lack of trying. There have been more than 500 other clinical trials for **gene therapies** designed to treat or cure a variety of diseases by inserting a functional gene into an individual's cells to replace a defective version of the gene. But no clear successes. Not one.

Let's examine the case of SCID, which has served as a model for gene therapy. SCID is a condition in which a baby is born with an immune system unable to properly produce a type of white blood cell. This leaves the infant vulnerable to most infections and usually leads to death within the first year of life (**FIGURE 5-41**). In gene therapy for SCID, researchers removed from an affected baby's bone marrow some **stem cells,** cells that have the ability to develop into any type of cell in the body. In bone marrow, they normally produce white blood cells, but in individuals with SCID, a malfunctioning gene disrupts normal white blood cell production.

Next, in a test tube, the bone marrow stem cells were infected with a transgenic virus carrying the functioning gene. In cases where the technique worked, the virus inserted the good gene into the DNA of the stem cells, which were then injected back into the baby's bone marrow. There, the cells could produce normal white blood cells, permanently curing the disease. Although this strategy worked to cure several cases of SCID, treatment has been suspended indefinitely following the recent deaths of two patients from illness related to their treatment.

Q Why has gene therapy had such a poor record of success in curing diseases?

Difficulties with gene therapy have been encountered in several different areas, usually related to the organism used to transfer the normal-functioning gene into the cells of a person with a genetic disease. They include:

Living in a protective bubble made it possible for this child with severe combined immunodeficiency to survive for 12 years.

FIGURE 5-41 **Protected by a bubble.**

1. Difficulty getting the working gene into the specific cells where it is needed.

2. Difficulty getting the working gene into enough cells and at the right rate to produce a physiological effect.

3. Difficulty arising from the transfer organism getting into unintended cells.

4. Difficulty regulating gene expression.

Beyond these technical problems, for most diseases the malfunctioning gene has not been identified or the disease is caused by more than one malfunctioning gene. And finally, it is important to keep in mind that gene therapy targets cells in the body other than sperm and eggs. Consequently, while a disease might, in theory, be cured in an individual, he or she can still pass on the disease-causing gene(s) to offspring. It's not clear what the future holds for gene therapy, but a great deal of research is still in progress.

TAKE-HOME MESSAGE 5·16

Biotechnology tools have been developed to reduce suffering and the incidence of diseases, but come with significant potential costs. Gene therapy has had a poor record of success in curing human diseases, primarily because of technical difficulties in transferring normal-functioning genes into the cells of a person with a genetic disease.

Cloning—ranging from genes to organs to individuals—offers both promise and perils.

Cloning. Perhaps no scientific word more readily conjures horrifying images of the intersection of curiosity and scientific achievement. But is fear the appropriate emotion to feel about this burgeoning technology? Perhaps not.

For starters, let's clarify what the word means. "Cloning" actually refers to a variety of different techniques. To be sure, cloning can refer to the creation of new individuals that have exactly the same genome as the donor individual—a process called "whole organism cloning." That is, a clone is like an identical twin, except that it may differ in age by years or even decades. It is also possible to clone tissues (such as skin) and entire organs from an individual's cells. And, as we saw in Section 5-12, it is possible to clone genes.

Cloning took center stage in the public imagination in 1997, when Ian Wilmut, a British scientist, and his colleagues first reported that during the previous year they had cloned a sheep—which they named Dolly. Their research was based on ideas that went back to 1938, when Hans Spemann first proposed the experiment of removing the nucleus from an unfertilized egg and replacing it with the nucleus from the cell of a different individual. Although the process used by Wilmut and his research group was difficult and inefficient, it was surprisingly simple in concept (FIGURE 5-42). They removed a cell from the mammary gland of a grown sheep, put its nucleus into another sheep's egg from which the nucleus had been removed, induced the egg to divide as if it were a naturally fertilized egg, and transplanted it into the uterus of a surrogate mother sheep. Out of 272 tries, they achieved just one success. But that was enough to show that the cloning of an adult animal was possible.

Shortly after news of Dolly's birth, teams set about cloning a variety of other species, including mice, cows, pigs, and cats (FIGURE 5-43). Not all of this work was driven by simple curiosity. For farmers, cloning could have real value. It can take a long time to produce animals with desirable traits from an agricultural perspective—such as increased milk production in cows. And with each successive generation of breeding, it can be difficult to maintain these traits in the population. But with transgenic techniques and whole animal cloning, large numbers of valuable animals with such traits can be produced and maintained.

Medical researchers, too, see much to gain from cloning. In particular, transgenic animals containing human genes—

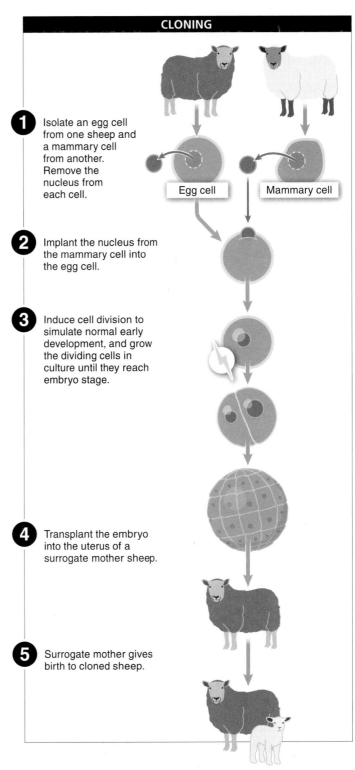

CLONING

1 Isolate an egg cell from one sheep and a mammary cell from another. Remove the nucleus from each cell.

Egg cell   Mammary cell

2 Implant the nucleus from the mammary cell into the egg cell.

3 Induce cell division to simulate normal early development, and grow the dividing cells in culture until they reach embryo stage.

4 Transplant the embryo into the uterus of a surrogate mother sheep.

5 Surrogate mother gives birth to cloned sheep.

FIGURE 5-42 **No longer science fiction.** The steps used in the cloning of Dolly the sheep.

FIGURE 5-43 **Genetically identical cloned animals.** The cloning of animals can maintain desirable traits from generation to generation.

Q Are there any medical justifications for cloning? such as the hamsters producing rhu-EPO, discussed earlier—can be very valuable. But can a human be cloned? At this point, it is almost certain that the cloning of a human will be possible. Many people wonder, though, whether such an endeavor should be pursued. There is near unanimity among scientists that human cloning to produce children should not be attempted. Some of the reasons cited relate to problems of safety for the mother and the child, legal and philosophical issues relating to the inability of cloned individuals to give consent, problems of the exploitation of women, and concerns regarding identity and individuality. Governments are struggling to develop wise regulations for this new world.

TAKE-HOME MESSAGE 5·17

Cloning of individuals has potential benefits in agriculture and medicine, but ethical questions linger.

5·18 DNA is an individual identifier: the uses and abuses of DNA fingerprinting.

In another time, Colin Pitchfork, a murderer and rapist, would have walked free. But in 1987, he was captured and convicted, betrayed by his DNA, and is now serving two life sentences in prison. Pitchfork's downfall began when he raped and murdered two 15-year-old high school girls in a small village in England in the 1980s. The police thought they had their perpetrator when a man confessed, but only to the second murder. He denied any involvement with the first murder, which perplexed the police because the details of the two crimes strongly suggested that the same person committed both.

At the time, British biologist Alec Jeffreys made the important discovery that there are small pieces of DNA in human chromosomes that are tremendously variable in their base sequences. In much the same way that each person has a driver's license number or social security number that differs from everyone else's, these DNA fragments are so variable that it is extremely unlikely that two people would have identical sequences at these locations. Thus, a comparison between these regions in a person's DNA sample and in DNA-containing evidence left at a crime scene would enable police to determine whether the evidence sample came from that person.

Jeffreys analyzed DNA left by the murderer-rapist on the two victims and found that it did indeed come from a single person, and that person was *not* the man who had confessed to one of the crimes. The original suspect was released and has the distinction of being the first person cleared of a crime through DNA fingerprinting. To track down the criminal, police then requested blood samples from all men in the area of the crimes who were between 18 and 35 years old—a practice that many viewed as an invasion of privacy. The police collected and analyzed more than 5,000 blood samples. This procedure led them to Colin Pitchfork, whose DNA matched perfectly the DNA left on both victims, and ultimately was the evidence responsible for his conviction (**FIGURE 5-44**). (He almost slipped through, having persuaded a friend to give a blood

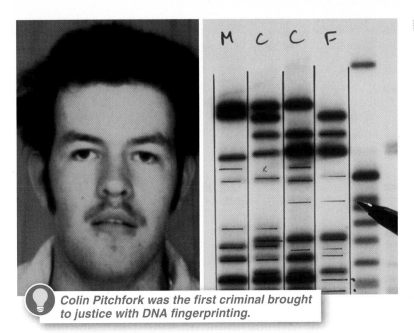

FIGURE 5-44 **Betrayed by his DNA.**

Colin Pitchfork was the first criminal brought to justice with DNA fingerprinting.

sample in his name. But when the friend was overheard telling the story in a pub, police tracked down Pitchfork to get a blood sample.)

DNA fingerprinting is now used extensively in forensic investigations, in much the same way that regular fingerprints have been used for the past 100 years. But regular fingerprinting is limited in its usefulness for many crimes because, often, no fingerprints are left behind. DNA samples are frequently left behind, though, usually in the form of semen, blood, hair, skin, or other tissue. As a consequence, this technology has been directly responsible for bringing thousands of criminals to justice and, as we saw at the beginning of the chapter, for establishing the innocence of more than 300 people wrongly convicted of murder and other capital crimes. Let's examine how DNA fingerprinting is done, why it is such a powerful forensic tool, and why it is not foolproof.

The DNA from different humans is almost completely identical. More than 99.9% of the DNA sequences of two individuals are the same, because we're all of the same species and thus share a common evolutionary history. Even so, in comparing two individuals' genomes of three billion base pairs each, a one-tenth of a percent difference still translates to about three million base-pair differences. These differences are responsible for the fact that all individuals have their own unique genome. (The lone exception? Identical twins, whose DNA is exactly the same.) Thus, when we are trying to evaluate whether the DNA from a crime scene matches that from a suspect, the analysis focuses on the parts of our DNA

that differ. There are thousands of these highly variable regions in the human genome.

Among these thousands of variable regions, one particular type is used for the determination of a person's genetic fingerprint. These regions are called STRs (for short tandem repeats) and are characterized by a short sequence (commonly four or five nucleotides) that repeats over and over within the region. The number of repeats is what varies among individuals.

An individual—we'll call her Individual A—has two copies of each chromosome, one from her mother and one from her father. At one STR region (on chromosome 2, for example), the number of times the sequence repeats is likely to differ on the maternal and the paternal copies of that chromosome in Individual A. The sequence may repeat 3 times on the maternal copy of her chromosome 2, and it may repeat 14 times on the paternal copy. In such a case, Individual A is said to have two different alleles for this STR region: 3 and 14. In contrast, in Individual B, for the same STR region on chromosome 2, the sequence may repeat 5 and 11 times (FIGURE 5-45).

For an STR region within the human genome, there typically are about 10 different alleles that occur within a population, and each is shared by about 10% of all individuals in a population. So the likelihood that two individuals carry the same two alleles is about 1 in 100. This is unlikely, but given enough people, many are likely to carry the same alleles. If two different STR regions are analyzed, the likelihood that

Short tandem repeats, or STRs, are sequences of DNA (commonly 4 or 5 nucleotides) that repeat over and over again. They occur in some of the most highly variable regions of an individual's DNA.

INDIVIDUAL A

Short tandem repeat (STR)

NUMBER OF REPEATS

Chromosome from mother

3

Chromosome from father

14

STR region

3 / 14

Individual A's alleles for this STR region

INDIVIDUAL B

Chromosome from mother

5

Chromosome from father

11

For a given region, the STR sequence is the same, but its number of repeats differs among individuals (and usually differs on the maternal and paternal copies of a chromosome that an individual carries).

5 / 11

Individual B's alleles for this STR region

FIGURE 5-45 **Biotechnology in forensics.** Forensic scientists can use highly variable regions of DNA to genetically link a person to DNA-containing evidence left at a crime scene.

two individuals have the same four alleles is $1/10 \times 1/10 \times 1/10 \times 1/10$, or 1/10,000. This is much rarer, but still would lead to multiple individuals within the same large city carrying the same four alleles. The real power of DNA fingerprinting comes from simultaneously determining the alleles an individual carries (that is, their genotype) not simply at one or two STR locations but at *13 different* STR locations. This is the number used by the FBI in constructing DNA fingerprints in the United States, and it makes the probability that two individuals would have exactly the same genotype extremely low (see Figure 5-45).

Q What is a DNA fingerprint?

For each STR locus analyzed, an individual's genotype is determined by using PCR to amplify that

A DNA fingerprint is created by determining which alleles an individual carries for 13 different STR regions.

1 **AMPLIFY THE STR REGION**
For each of the 13 STR regions used to construct an individual's DNA fingerprint, the DNA fragment containing each STR region is amplified using PCR, resulting in huge numbers of those fragments.

The amplified DNA fragments differ in size depending on how many times the repeating unit of that STR is repeated.

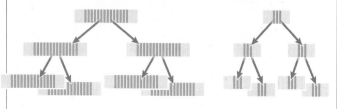

2 **SORT THE FRAGMENTS BY SIZE**
Amplified DNA fragments are poured into an electrophoresis gel and an electrical charge is applied.

Because DNA is a negatively charged molecule, the fragments move toward the positively charged electrode. Smaller pieces (having fewer repeats) move across the gel more quickly than larger pieces.

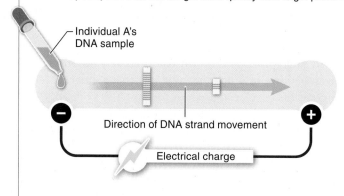

Individual A's DNA sample

Direction of DNA strand movement

Electrical charge

3 **IDENTIFY THE GENOTYPE**
The number of repeats within an STR region (indicating an individual's genotype) is determined by comparing the length of the fragments containing that STR region with DNA fragments of known lengths.

FIGURE 5-46 **Your unique identifier: a DNA fingerprint.** DNA fingerprints are being used to match a suspect's DNA to DNA found at the scene of a crime.

region, then measuring the length of the STR region using a technique called electrophoresis, which allows researchers to see how long a stretch of DNA is. The length of the region can then be used to determine the number of times the STR is repeated. For a single STR region, an individual's genotype is expressed by two numbers, reflecting the number of STR repeats in the copies inherited from the mother and from the father. And a person's full DNA fingerprint is a string of 26 numbers that includes the two numbers for each of 13 STRs.

In court, a suspect's genotype might be compared with the DNA fingerprint obtained from evidence found at the crime scene. DNA samples from different people will produce different 26-number "fingerprints," whereas different samples of DNA from one person will have exactly the same genotype (**FIGURE 5-46**).

In the end, despite universally accepted methods, DNA fingerprinting still is not foolproof. Numerous incidences of human error—accidental as well as intentional—have been documented, ranging from mislabeled test tubes to tissue from a suspect being added to evidence from a crime scene. So we should not blindly draw conclusions solely from this one type of evidence. Nonetheless, DNA fingerprinting is an increasingly valuable tool for law enforcement. The FBI, for example, has reported that nearly one-third of their suspects are cleared immediately by DNA testing, and that many more criminals, because of DNA fingerprinting, now plead guilty to the crimes they have committed.

TAKE-HOME MESSAGE 5·18

Comparisons of highly variable DNA regions can be used to identify tissue specimens and determine the individual from whom they came.

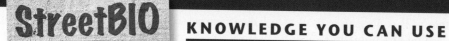

Mixing aspirin and alcohol can lead to metabolic interference and unexpected inebriation.

Ethanol is the form of alcohol found in cocktails, beer, and wine. When you consume alcohol, the first step in your body's metabolic breakdown of the ethanol is carried out by the enzyme alcohol dehydrogenase.

Q: What happens to ethanol molecules that are *not* broken down? Any ethanol molecules in the bloodstream that are not broken down by alcohol dehydrogenase make their way to the brain. Once there, they cause you to feel a bit inebriated—or drunk, if the amount of ethanol is sufficiently large. Aspirin has the unintended side effect of disabling alcohol dehydrogenase, thus interfering with its normal functioning.

Q: What happens if someone takes two aspirins and then drinks several alcoholic beverages? By blocking the activity of alcohol dehydrogenase, aspirin interferes with a person's ability to break down ethanol. As a consequence, the ethanol remains in the body longer and has a more pronounced effect on the brain, producing greater inebriation (26% greater, according to one study).

Q: What can you conclude? Medications, even over-the-counter products, can have unexpected (and unintended) physiological consequences, particularly when mixed with alcohol consumption. Exercise great caution when taking them.

Check Your Knowledge

Thinking critically about visual displays of data

1. What proportion of corn grown in the United States is genetically modified?

2. From the figure, can you determine whether there is more genetically modified corn or genetically modified cotton produced in the United States? How do you do this? And if it is not possible, why isn't it?

3. What is the "take-home message" from this figure? Does it influence your thoughts on genetically modified crops? Why or why not?

4. Why are data given for proportions of genetically modified crops rather than absolute amounts? Does this alter your interpretation of the graph?

5. Worldwide, the proportion of corn grown that is genetically modified is just 26%, while the proportion for soybeans is 77% and for cotton is 49%. How would this information add value to the figure?

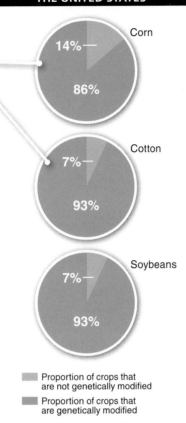

GENETICALLY MODIFIED CROPS IN THE UNITED STATES

Corn
14%
86%

Cotton
7%
93%

Soybeans
7%
93%

Proportion of crops that are not genetically modified

Proportion of crops that are genetically modified

6. The proportion of corn grown in the United States that is genetically modified has changed since 2001 as follows:

Year	Proportion
2001	26%
2002	34%
2003	40%
2004	47%
2005	52%
2006	61%
2007	73%
2008	80%
2009	85%
2010	86% (shown in the graph)

Show two different ways that you could display these data. What is the clearest conclusion someone would draw from these figures?

See answers at the back of the book.

Key Terms in DNA, Gene Expression, and Biotechnology

ABOUT THE CHAPTER OPENING PHOTO

The double-helix sidewalks of McMillian Park in Huntsville, Alabama, will link businesses and a nonprofit educational and research center.

REVIEW & REHEARSE

5
DNA, GENE EXPRESSION, and BIOTECHNOLOGY

5·1–5·5 DNA: what is it, and what does it do?

The DNA molecule contains instructions for the development and functioning of all living organisms.

DEOXYRIBONUCLEIC ACID (DNA)

DNA, the genetic material for all living organisms, is a double-helical molecule consisting of two "backbone" chains of sugar and phosphate molecules, with the chains linked by pairs of nucleotide bases.

? Check Your Knowledge

1. A person's DNA is carried in:
a) muscle cells.
b) hair.
c) saliva.
d) skin cells.
e) All of the above contain a person's DNA.

0 — 23 — 100
EASY · HARD

2. Which of the following are always the same in every DNA molecule?
a) the sugar
b) the base
c) the phosphate group
d) Only a) and b) are always the same.
e) Only a) and c) are always the same.

0 — 33 — 100
EASY · HARD

3. The full set of an individual organism's DNA is called its:
a) complement.
b) genome.
c) nucleosome.
d) nucleotide.
e) chromosome.

0 — 15 — 100
EASY · HARD

4. In humans, genes make up _____ of the DNA.
a) about 75%
b) 100%
c) less than 5%
d) about 10%
e) about 50%

0 — 38 — 100
EASY · HARD

NUCLEOTIDE
The nucleotide unit of a DNA molecule has three components: a phosphate group, a sugar, and a nitrogen-containing base (here it's adenine).

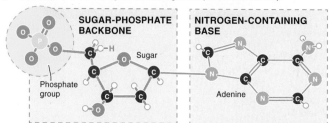

SUGAR-PHOSPHATE BACKBONE

Phosphate group · Sugar

NITROGEN-CONTAINING BASE

Adenine

BASE PAIRS
DNA bases are connected with hydrogen bonds.

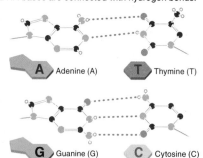

A — Adenine (A)
T — Thymine (T)
G — Guanine (G)
C — Cytosine (C)

DOUBLE HELIX
Like a twisted ladder, two sugar-phosphate strands spiral around each other, forming the backbones of DNA. The bases attached to the sugar molecules on one strand bond to those attached to the other strand to form the rungs.

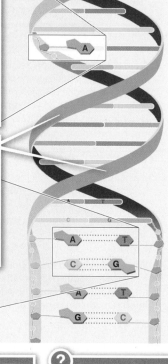

? 1. How and why is DNA helpful in ensuring greater justice in our society?

? 2. After sequencing a molecule of DNA, you discover that 20% of the bases are cytosine. What percentage of the bases would you expect to be thymine? Why?

? 3. What function does the sugar-phosphate backbone of a DNA molecule serve?

FROM GENOME TO GENE

GENOME
An organism's complete set of DNA. In eukaryotes, this information can be found in the nucleus of almost every cell.

CHROMOSOME
One or more unique pieces of DNA that together make up an organism's genome. Chromosomes vary in length and can consist of hundreds of millions of base pairs.

GENE
A specific sequence of DNA, on average about 3,000 bases long, that contains the information necessary to produce all or part of a protein molecule.

Protein production

Protein

ALLELES
Alleles are different versions of a gene that code for the same trait.

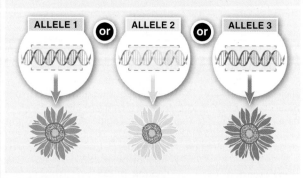

ALLELE 1 or **ALLELE 2** or **ALLELE 3**

CODING vs. NON-CODING DNA
Only a small fraction of the DNA in eukaryotic species is in genes that code for proteins; the function of much of the rest is still poorly understood, although at least some of it plays important roles in the cell, such as gene regulation.

? 4. Only a small fraction of the DNA in eukaryotic species codes for proteins. The remainder is sometimes referred to as "junk DNA." Why is this a poor description of the non-coding DNA?

5·6—5·8 Information in DNA directs the production of the molecules that make up an organism.

Information in the genes is transcribed into the sequences of mRNA molecules. These mRNA sequences are then translated as they direct the construction of proteins.

TRANSCRIPTION

A single copy of one specific gene in the DNA is made in the form of a molecule of mRNA. When the mRNA copy of a gene is completed, it moves to the cytoplasm, where it can be translated into a polypeptide.

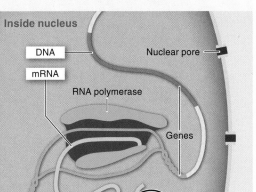

Inside nucleus

- DNA
- mRNA
- Nuclear pore
- RNA polymerase
- Genes

5. What is the role of transcription? Why is this step necessary? Why isn't translation the first step in using DNA information to build a new organism?

TRANSLATION

The information from a gene that has been encoded in the nucleotide sequence of an mRNA is read, and ingredients present in the cell's cytoplasm are used to produce a protein.

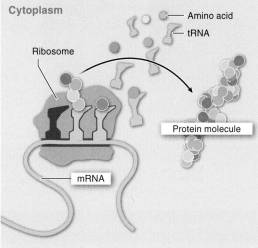

Cytoplasm

- Amino acid
- tRNA
- Ribosome
- Protein molecule
- mRNA

6. When the mRNA copy of a gene is completed in a eukaryotic cell, why does the mRNA move to the cytoplasm?

TRANSFER RNA (tRNA)

Transfer RNA (tRNA) molecules translate the mRNA code by linking specific bases on the mRNA with specific amino acids that will be used to build a protein.

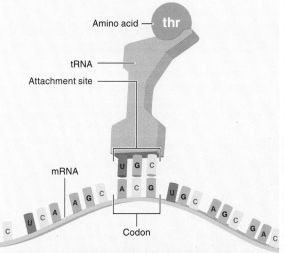

- Amino acid — thr
- tRNA
- Attachment site
- mRNA
- Codon

Attachment sites consist of a three-base sequence that matches up with a three-base sequence on the mRNA. Each three-base sequence in mRNA—called a codon—matches with a tRNA that carries one particular amino acid.

7. What is the function of tRNA in translation?

MECHANISMS OF REGULATING GENE EXPRESSION

Regulation can occur in ways that enhance or impede transcription, or alter the longevity and degradation of mRNA, or influence translation or protein processing.

TRANSCRIPTION REGULATION

POSITIVE CONTROL
- Activators can initiate or speed up gene expression
- Enhancer sequences can speed RNA polymerase binding and gene transcription

NEGATIVE CONTROL
- Repressors can block or slow down gene expression
- Chemicals can bind to DNA and block gene transcription

POST-TRANSCRIPTION REGULATION

- mRNA processing, transport, and enzymatic breakdown can be blocked or accelerated
- mRNA translation and protein processing rates can be altered

8. *E. coli* prefers glucose as an energy source, but if lactose, and not glucose, is present, the *E. coli* cells can survive. How is *E. coli* able to respond to this environmental change?

---

5. Genotype is to phenotype as:
a) cookie is to oven.
b) cookie is to recipe.
c) cookbook is to cookie.
d) recipe is to cookie.
e) oven is to cookie.

0 — 21 — 100
EASY ——— HARD

6. To start the transcription process, a large molecule, _____, recognizes a _____.
a) RNA polymerase; messenger RNA
b) DNA polymerase; termination site
c) DNA polymerase; promoter site
d) RNA polymerase; promoter site
e) DNA polymerase; messenger RNA

0 — 31 — 100
EASY ——— HARD

7. During transcription, at the point where the DNA strand being copied has an adenine, a(n) _____ is added to the _____.
a) thymine; tRNA
b) cytosine; DNA
c) uracil; DNA
d) adenine; mRNA
e) uracil; mRNA

0 — 36 — 100
EASY ——— HARD

8. There are different _____ molecules for each of the 20 different amino acids that are used in building proteins.
a) ribosomal subunit
b) tRNA
c) mRNA
d) DNA
e) elongation

0 — 52 — 100
EASY ——— HARD

9. Deletions and substitutions are two types of mutations. Which type is more likely to cause mistranslations of proteins?
a) Substitutions, because they shift the reading frame and cause downstream amino acids to be changed.
b) Substitutions, because one protein is substituted for another protein.
c) Deletions, because they shift the reading frame and cause downstream amino acids to be changed.
d) Deletions, because one protein is deleted.
e) None of the above are correct.

0 — 37 — 100
EASY ——— HARD

10. Which of the following statements about mutations is incorrect?

a) Mutations are very rare.

b) A mutation in DNA always leads to changes in the structure and function of the protein produced.

c) When one nucleotide base pair is replaced by another, this constitutes a point mutation.

d) One general type of mutation is a change to the overall organization of chromosomal genes.

e) X rays can cause mutations.

0 — 63 — 100
EASY   HARD

11. Which of the following statements about the metabolism of ethanol (which is present in alcoholic beverages) is incorrect?

a) Individuals who produce non-functioning aldehyde dehydrogenase exhibit "fast-flushing."

b) The process requires two enzymes: alcohol dehydrogenase and isopropyl dehydrogenase.

c) Individuals who are "fast-flushers" are less likely to become alcoholics.

d) Aspirin interferes with the action of alcohol dehydrogenase.

e) None. All of the statements above are correct.

0 — 56 — 100
EASY   HARD

12. The polymerase chain reaction (PCR):

a) makes it possible to create huge numbers of copies of tiny pieces of DNA.

b) enables researchers to determine the sequence of a complementary strand of DNA when they have only single-stranded DNA.

c) utilizes RNA polymerase to build strands of DNA.

d) can create messenger RNA molecules from small pieces of DNA.

e) All of the above are correct.

0 — 60 — 100
EASY   HARD

13. Which of the following is not a difficulty that medicine has encountered in its attempts to cure human diseases through gene therapy?

a) The transfer organism—usually a virus—may get into unintended cells and cause disease.

b) It is difficult to get the working gene into the specific cells where it is needed.

c) It is difficult to get the working gene into enough cells at the right rate to have a physiological effect.

d) For many diseases, a malfunctioning gene has not been identified.

e) All of the above are difficulties encountered in attempts to cure human diseases through gene therapy.

0 — 24 — 100
EASY   HARD

5·9—5·11 Damage to the genetic code has a variety of causes and effects.

Mutations are alterations to the sequence of bases in an organism's DNA and can lead to changes in the structure and function of the proteins produced.

MUTATIONS CAN HAVE A RANGE OF EFFECTS

Mutations are rare, but when they do occur, they may disrupt normal functioning of the body (although many mutations are neutral). Extremely rarely, mutations may have a beneficial effect. Mutations play an important role in evolution.

Normal fruit fly   Mutant fruit fly

TYPES OF MUTATIONS

Mutations generally fall into two types: point mutations and chromosomal aberrations.

POINT MUTATIONS
One base pair is changed. A base pair in the DNA can be substituted for another, or a base pair can be inserted or deleted.

CHROMOSOMAL ABERRATIONS
Entire sections of a chromosome are altered. Aberrations can involve the complete deletion of an entire section of DNA, the relocation of a gene, or the duplication of a gene.

9. Mutations often are considered negative events with bad consequences. Why is this view of mutations somewhat of a paradox?

10. Which step in a metabolic pathway is disrupted in Tay-Sachs disease?

5·12—5·14 Biotechnology is producing improvements in agriculture.

Biotechnology is the use of technology to modify organisms, cells, and their molecules to achieve practical benefits.

FIVE IMPORTANT TOOLS AND TECHNIQUES OF MOST BIOTECHNOLOGY PROCEDURES

With modern molecular methods, researchers can cut and copy DNA and deliver it to new organisms, not necessarily of the same species.

 CHOP up DNA from a donor species that exhibits a trait of interest.

 AMPLIFY small samples of DNA into more useful quantities.

 INSERT pieces of DNA into bacterial cells or viruses.

 GROW separate colonies of bacteria or viruses, each containing some donor DNA.

 IDENTIFY colonies of bacteria or viruses that have DNA for a trait of interest.

THE BENEFITS OF BIOTECHNOLOGY

Biotechnology has led to important advances in agriculture by using transgenic plants and animals to produce more nutritious food. Biotechnology has also reduced the environmental and financial costs of producing food through the creation of herbicide-resistant and insect-resistant crops. The ecological and health risks of such widespread use of transgenic species are not fully understood and are potentially great.

11. Describe two ways in which advances in biotechnology have helped make farming more efficient.

FEARS AND RISKS ASSOCIATED WITH BIOTECHNOLOGY

 More and more genetically modified foods are being created using modern methods of recombinant DNA technology. Among the public, however, numerous legitimate fears remain about the potentially catastrophic risks of these foods, given that their development relies on such new technology, and about the long-term financial advantages they offer.

12. The creation of genetically modified foods raises many concerns, including that the loss of genetic diversity may have bad consequences. Why is this a concern?

5·15–5·18 Biotechnology has the potential for improving human health (and criminal justice).

Advances in biotechnology have led to some successes in treating diseases, automated methods for analyzing DNA sequences, and the potential benefits of cloning.

USING BIOTECHNOLOGY TO TREAT DISEASE

Biotechnology has had some successes in treating diseases, usually by producing medicines more efficiently and effectively than with traditional methods. For example, the human insulin that is used to treat diabetes is engineered through recombinant DNA technology.

? **13.** Describe three major applications of biotechnology and their impact on human health.

GENE THERAPY

Biotechnology tools have been developed to reduce suffering and the incidence of diseases, but come with significant potential costs. Gene therapy has had a poor record of success in curing human diseases, primarily because of technical difficulties in transferring normal-functioning genes into the cells of a person with a genetic disease.

? **14.** How has the advent of genetic screening for prospective parents led to decreases in the incidence of fatal genetic diseases?

CLONING

Cloning—which includes the production of genetically identical cells, organisms, or DNA molecules of individuals—has potential benefits in medicine and agriculture (the cloning of animals can maintain desirable traits from generation to generation), but ethical questions linger.

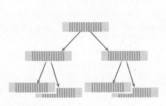

? **15.** What is a potential benefit to farmers of cloning technology?

SHORT TANDEM REPEATS (STRs)

Short tandem repeats, or STRs, are sequences of DNA (commonly 4 or 5 nucleotides) that repeat over and over again. They occur in some of the most highly variable regions of an individual's DNA.

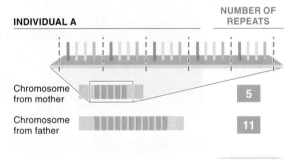

INDIVIDUAL A — **NUMBER OF REPEATS**

Chromosome from mother — **5**

Chromosome from father — **11**

For a given region, the STR sequence is the same, but its number of repeats differs among individuals (and usually differs on the maternal and paternal copies of a chromosome that an individual carries).

5 / 11 Individual A's alleles for this STR region

CREATING A DNA FINGERPRINT

A DNA fingerprint is created by determining which alleles an individual carries for 13 different STR regions.

1 **AMPLIFY THE STR REGION**
For each of the 13 STR regions used to construct an individual's DNA fingerprint, the DNA fragment containing each STR region is amplified using PCR, resulting in huge numbers of those fragments.

The amplified DNA fragments differ in size depending on how many times the repeating unit of that STR is repeated.

2 **SORT THE FRAGMENTS BY SIZE**
Amplified DNA fragments are poured into an electrophoresis gel and an electrical charge is applied.

Because DNA is a negatively charged molecule, the fragments move toward the positively charged electrode. Smaller pieces (having fewer repeats) move across the gel more quickly than larger pieces.

3 **IDENTIFY THE GENOTYPE**
The number of repeats within an STR region (indicating an individual's genotype) is determined by comparing the length of the fragments containing that STR region with DNA fragments of known lengths.

?

14. Golden rice:

a) grows without a husk, thereby reducing the processing required before it can be consumed.

b) can make vitamin A without beta-carotene.

c) could help prevent blindness due to vitamin A deficiency in 250,000 children each year.

d) supplies more vitamin A in one serving than an individual needs in a full week.

e) is one of the most recent developments in organic farming.

0 — EASY ——32—— 100 HARD

15. Which of the following statements about Bt crystals is correct?

a) They are produced by soil-dwelling bacteria of the species *Bacillus thuringiensis*.

b) The gene coding for the production of Bt crystals has been genetically engineered into the genome of dairy cows, increasing their milk production sixfold.

c) They are produced by the polymerase chain reaction (PCR).

d) They are produced by most weedy species of plants.

e) All of the above are correct.

0 — EASY ———62—— 100 HARD

16. Recombinant human _____ is/are at the center of the "blood doping" scandals in professional cycling.

a) erythropoietin (EPO)

b) growth hormone (HGH)

c) insulin

d) red blood cells

e) stem cells

0 — EASY ——47——— 100 HARD

17. Short tandem repeat sequences of DNA:

a) are characteristic of genes that code for biochemical traits rather than structural traits.

b) are used in biotechnology when creating a clone.

c) are produced when a mutation occurs in a non-sex cell.

d) can be used to produce a DNA fingerprint.

e) are produced when a mutation occurs in a sperm-producing or egg-producing cell.

0 — EASY ———64—— 100 HARD

6 | Chromosomes and Cell Division

CONTINUITY AND VARIETY

There are different types of cell division.

Mitosis replaces worn-out old cells with fresh new duplicates.

Meiosis generates sperm and eggs and a great deal of variation.

There are sex differences in the chromosomes.

Deviations from the normal chromosome number lead to problems.

6·1–6·4
There are different types of cell division.

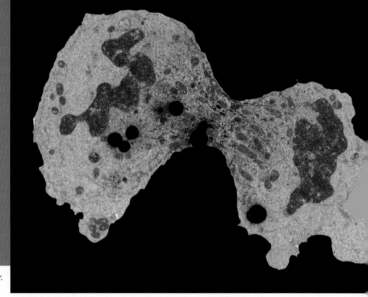

Cell division in a Ewing sarcoma bone tumor.

6·1 Immortal cells can spell trouble: cell division in sickness and in health.

Once you are fully grown, do you have just one complete set of cells that live as long as you do? The answer is *no.* Although some of the cells in your body may last for many decades, throughout most of your body your cells are continually dying off, and the ones that remain divide and replace the cells you've lost, in an ongoing process. But can this cell replacement go on forever? And does a cell even know how old it is?

Actually, a cell does have a feature that provides an approximate measure of how old it is. Just as a car comes with an odometer, which keeps track of how far the car has been driven, animal cells have a sort of counter that reflects how many times the cell has divided. This "counter" is a section of non-coding, repetitive DNA, called a **telomere,** that serves as a protective cap and is located at each tip of every chromosome, right next to the genes that direct the processes that keep the organism alive (**FIGURE 6-1**).

Every time a cell divides, making an exact copy of itself, its DNA divides as well. However, each time the DNA divides, the process by which chromosomes are duplicated causes the telomere at each end of every chromosome to get a bit shorter. After many cell divisions, the telomeres can become so short that additional cell divisions cause the loss of functional, essential DNA, and that means almost certain death for the cell.

Telomeres in humans are repeats of the nucleotide sequence TTAGGG. It's the identical sequence in many vertebrate

species and very similar in most other eukaryotes. The number of times the sequence repeats varies across species; in humans, the number is about 2,000. At birth, the telomeres in most human cells are long enough to support 80–90 cell divisions. In a 70-year-old person, the telomeres are long enough for only about 20–30 divisions.

Occasionally, individuals are born with telomeres that impair functioning of a protein that helps maintain the nucleus of cells. This interferes with normal cell functioning in numerous ways, including damaging the cell's telomeres, causing them to be much shorter than normal. In these people, the normal functioning of many genes is disrupted, and their cells and tissues begin to appear aged very soon after birth (**FIGURE 6-2**). As a consequence, children born with this disorder rarely live beyond the age of 13.

This observation, in conjunction with the observation that telomeres get shorter with each cell division, might seem to suggest that continually rebuilding the telomere after each round of division would be an effective strategy for allowing a cell and its descendants to function for a longer time than normal. Such a line of cells would never die. By extension, it's tempting to imagine that constantly rebuilding telomeres might act like a fountain of youth. Unfortunately, it does not.

We know this because there are some cells that do acquire this feature. These cells rebuild their telomeres after each cell division, restoring the chromosomes' protective caps. For single-celled organisms and for the cells that divide to

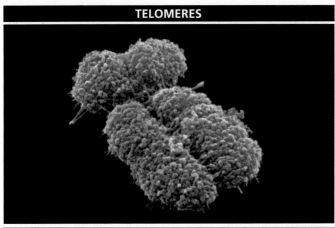

TELOMERES

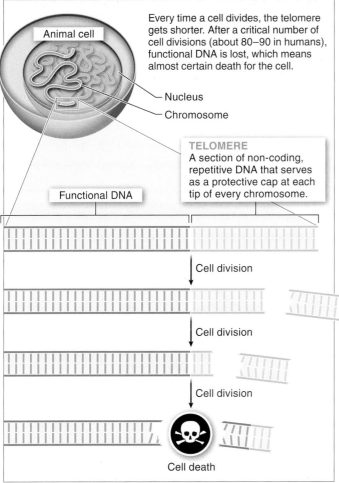

Every time a cell divides, the telomere gets shorter. After a critical number of cell divisions (about 80–90 in humans), functional DNA is lost, which means almost certain death for the cell.

Animal cell

— Nucleus

— Chromosome

TELOMERE
A section of non-coding, repetitive DNA that serves as a protective cap at each tip of every chromosome.

Functional DNA

Cell division

Cell division

Cell division

Cell death

FIGURE 6-1 **A cellular "odometer."** Telomeres limit cells to a fixed number of divisions.

produce sperm and eggs in multicellular organisms (cells that must go through many thousands of rounds of cell division), this telomere rebuilding is essential.

Unfortunately, for most of the other types of cells that rebuild their telomeres with each cell division, the telomere rebuilding

presents a big problem: the cells are unable to stop dividing. Such cells commonly go by another name: cancer. Because telomere rebuilding occurs in (and possibly is necessary to) many human cancers, researchers have hopes that inhibiting it might help fight cancer. This process is already the target of some anti-cancer drugs now being researched. In any case, because one of the defining features of cancer is runaway cell division, discovering a cure for cancer will necessarily involve a deep understanding of cell division.

In this chapter, we investigate the processes that enable cells to divide and create new cells. Prokaryotes exhibit one method of cell division (called binary fission), and it serves all of their cell division needs. Eukaryotes exhibit two methods of cell division, mitosis and meiosis, each of which has a specific purpose in a eukaryote's life cycle. We'll explore normal cell division and discuss what happens when cell division does not proceed in the normal way.

TAKE-HOME MESSAGE 6·1

Cell division is an ongoing process in most organisms and their tissues; disruptions to normal cell division can have serious consequences. In eukaryotic cells, a protective section of DNA called the telomere, at each end of every chromosome, plays a role in keeping track of cell division, getting shorter every time the cell divides. If telomeres become too short, additional cell divisions cause the loss of essential DNA and cell death. Cells that rebuild their telomeres with each division can become cancerous.

FIGURE 6-2 **Only children.** These siblings were born with a genetic condition known as Hutchinson-Gilford progeria syndrome. Born with shorter-than-normal telomeres, they appear to age at a rapid rate.

As a method for storing genetic information, DNA has complete market saturation. All life on earth uses it. This is pretty remarkable considering the tremendous diversity of life that exists on our planet—from single-celled bacteria to multicellular plants and animals. One way in which different organisms' DNA varies is in how it is organized into chromosomes.

The most important part of a eukaryotic chromosome may be the DNA molecule, which carries information about how to accomplish the processes needed to support the life of the organism. But eukaryotic chromosomes (and some prokaryotic chromosomes) are made of more than just DNA. The eukaryotic **chromosome** is composed of **chromatin,** a linear DNA strand bound to and wrapped tightly around proteins called **histones,** which keep the DNA from getting tangled and enable it to be tightly and efficiently packed in an orderly manner inside the nucleus. Plants and animals usually have between 10 and 50 chromosomes (although there are species with as few as 2 and others with more than a hundred).

Most prokaryotes—the bacteria and archaea—have less DNA than eukaryotes. They carry their genetic information in a single, circular chromosome, a strand of DNA that is attached at one site to the cell membrane (**FIGURE 6-3**). And when it is time for them to reproduce, they use a method called **binary fission,** which means "division in two" (**FIGURE 6-4**). This process begins with **replication,** the method by which a cell creates an exact duplicate of each chromosome.

Replication in prokaryotes begins as the double-stranded DNA molecule unwinds from its coiled-up configuration. Once the strands are uncoiled, they split apart like a zipper, with bases exposed on each of the two separated, single-stranded circular molecules of DNA. As the double-stranded molecule unzips, enzymes bind to the

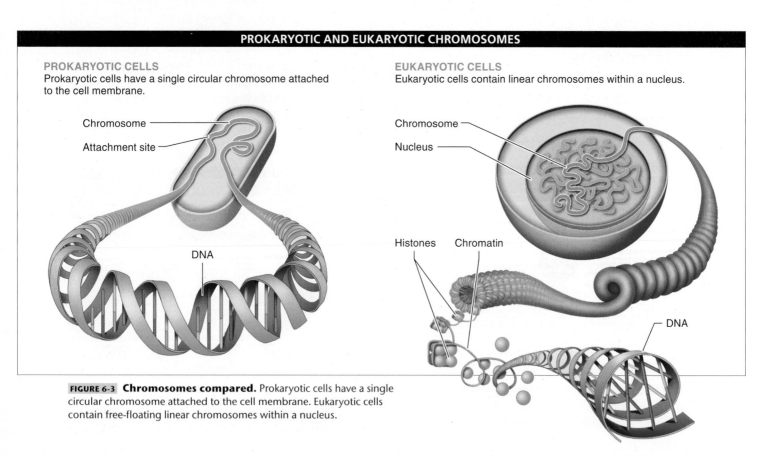

PROKARYOTIC AND EUKARYOTIC CHROMOSOMES

PROKARYOTIC CELLS
Prokaryotic cells have a single circular chromosome attached to the cell membrane.

Chromosome
Attachment site
DNA

EUKARYOTIC CELLS
Eukaryotic cells contain linear chromosomes within a nucleus.

Chromosome
Nucleus
Histones   Chromatin
DNA

FIGURE 6-3 **Chromosomes compared.** Prokaryotic cells have a single circular chromosome attached to the cell membrane. Eukaryotic cells contain free-floating linear chromosomes within a nucleus.

PROKARYOTIC CELL DIVISION: BINARY FISSION

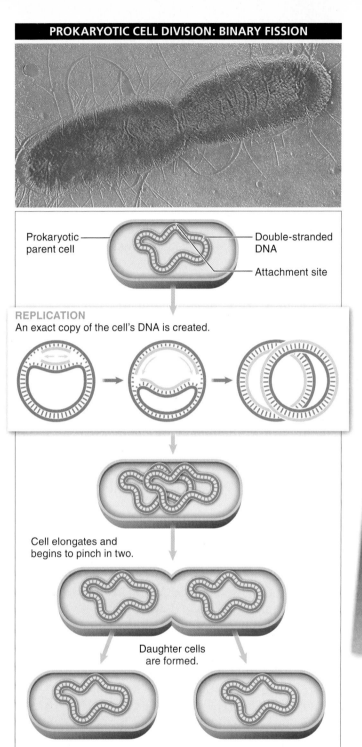

Prokaryotic parent cell

Double-stranded DNA

Attachment site

REPLICATION
An exact copy of the cell's DNA is created.

Cell elongates and begins to pinch in two.

Daughter cells are formed.

Binary fission results in two genetically identical daughter cells.

DNA and attach free-floating nucleotides to the growing DNA backbone, matching A to T and G to C, thus creating two identical double-stranded DNA molecules.

The two newly created circular chromosomes attach to the inside of the plasma membrane at different spots from each other. The original cell, called the **parent cell,** then pinches in until it divides into two new cells, called **daughter cells.** Each of the daughter cells has an identical two-stranded copy of the original two-stranded circular chromosome.

In some prokaryotes, such as the bacteria called *E. coli* that live in our digestive system, the complete process of binary fission can occur very quickly—often in as little as 20 minutes. Binary fission is considered **asexual reproduction,** because the daughter cells inherit their DNA from a single parent cell and thus are genetically identical to the parent.

TAKE-HOME MESSAGE 6·2

In most bacteria and archaea, the genetic information is carried in a single, circular chromosome, a strand of DNA that is attached at one site to the cell membrane. Eukaryotes have much more DNA than do bacteria and organize it into linear chromosomes within the nucleus. Bacteria divide by a type of asexual reproduction called binary fission: first, the circular chromosome duplicates itself, then the parent cell splits into two new, genetically identical daughter cells.

There is a time for everything in the eukaryotic cell cycle.

In life, we often go through phases that are defined by the primary focus of our interests and activities. For many years we go to school. For a while we may tend to our career. Then, for a spell, our personal life may take center stage. Most eukaryotic cells go through phases, too. They spend long periods of time occupied with activities relating solely to their growth, and then may suspend those activities as they segue into a period devoted exclusively to reproducing themselves. This alternation of activities between processes related to growth and processes related to cell division is called the **cell cycle** (**FIGURE 6-5**).

Before we go any farther in discussing the cell cycle, we need to note an important distinction among the cells of the body. All the cells of a multicellular eukaryotic organism can be divided into two types: **somatic cells** are the cells forming the body of the organism; **reproductive cells** are the sex cells—that is, the **gametes** (sperm and eggs)—and the cells that give rise to them.

Somatic cells and reproductive cells use different methods of producing new cells. In this section, we focus on cell division as it occurs in somatic cells. Later in the chapter, we examine cell division that leads to the production of sex cells.

The cell cycle describes the series of phases in somatic cell division. There are two main phases in the cell cycle: **interphase,** during which the cell grows and prepares to divide, and the **mitotic phase** (or **M phase**), during which first the nucleus and genetic material within the cell

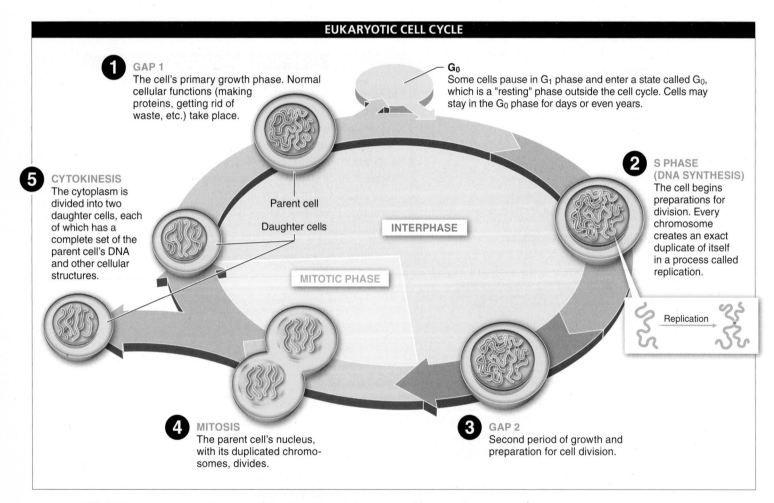

EUKARYOTIC CELL CYCLE

1 GAP 1
The cell's primary growth phase. Normal cellular functions (making proteins, getting rid of waste, etc.) take place.

G_0
Some cells pause in G_1 phase and enter a state called G_0, which is a "resting" phase outside the cell cycle. Cells may stay in the G_0 phase for days or even years.

5 CYTOKINESIS
The cytoplasm is divided into two daughter cells, each of which has a complete set of the parent cell's DNA and other cellular structures.

Parent cell

Daughter cells

INTERPHASE

MITOTIC PHASE

2 S PHASE
(DNA SYNTHESIS)
The cell begins preparations for division. Every chromosome creates an exact duplicate of itself in a process called replication.

Replication

4 MITOSIS
The parent cell's nucleus, with its duplicated chromosomes, divides.

3 GAP 2
Second period of growth and preparation for cell division.

FIGURE 6-5 **The cycle of cellular activity.** The eukaryotic cell cycle is the series of events that results in the division of somatic cells (cells that make up the body of an organism).

divide, then the rest of the cellular contents divide. Interphase is further divided into three distinct sub-phases, described below.

Interphase The three sub-phases of interphase can be summarized as follows.

Gap 1 (G_1). During this period, a cell may grow and develop as well as performing its various cellular functions (making proteins, getting rid of waste, and so on). Most cells spend most of their time in the Gap 1 phase. Some cells pause in the Gap 1 phase and enter a state called G_0, which is a quiescent or "resting" phase outside the cell cycle in which no cell division occurs. Cells may stay in the G_0 phase for days or even years—or permanently in some cases, such as most neurons and heart muscle cells—before resuming cell division.

DNA synthesis (S phase). During this phase, the cell begins to prepare for cell division, first by creating an exact duplicate of each chromosome by replication. Before replication, each chromosome's DNA consists of a single long, linear molecule. After replication, each chromosome's DNA has become a pair of identical long, linear pieces, held together near the center; the region where the two pieces are in contact is called the **centromere.** (As we'll see, the centromere also serves as the point to which spindle fibers can attach.)

Gap 2 (G_2). This phase, which is generally much shorter than Gap 1, is usually characterized by significant growth, as well as high rates of protein synthesis in preparation for division. This phase differs from Gap 1, though, because the genetic material has now been duplicated.

Mitotic Phase This period begins with **mitosis,** a process in which the parent cell's nucleus, including its chromosomes, divides. Mitosis is generally followed by **cytokinesis** (which may begin prior to the end of mitosis), during which the cytoplasm is divided into two daughter cells, each of which has a complete set of the parent cell's DNA and other cellular structures. The mitotic phase is usually the shortest period in the eukaryotic cell cycle.

The cell cycle describes the series of phases in somatic cell division, but it's not a one-size-fits-all cycle, with all cells moving from growth phases through cell division phases perpetually and inevitably. There is great variation in how cells move through the cycle. Animal embryos, at one end of a continuum, may move through the cell cycle so quickly that they spend almost no time at all in the G_1 and G_2 phases. Heart muscle cells and brain neurons, at the other end of the continuum, may never pass through the cell division phase, remaining in a non-dividing state for decades.

Variation in the rates of cell division is regulated by a **cell-cycle control system,** a group of molecules, mostly proteins, within a cell that coordinates the events of the cell cycle. This control system functions through a system of **checkpoints,** critical points in the cell cycle at which progress is blocked—and cells are prevented from dividing—until specific signals trigger continuation of the process. Checkpoints in the cell cycle make it possible for cells to (a) reduce the likelihood of completing cell division when errors have occurred in the process, and (b) respond to feedback conveying information about the cell's internal and external environment.

The signals that trigger transitions to subsequent phases in the cell cycle most commonly consist of **growth factors,** which provide feedback about the cell's environment and can signal that division is appropriate.

There are three primary checkpoints that regulate the cell cycle in eukaryotes (**FIGURE 6-6**).

1. G_1/S checkpoint: assessing DNA damage and cell growth. Occurring near the end of the G_1 phase, this is the

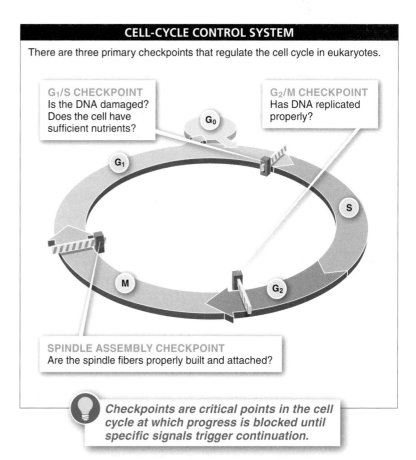

CELL-CYCLE CONTROL SYSTEM

There are three primary checkpoints that regulate the cell cycle in eukaryotes.

G_1/S CHECKPOINT
Is the DNA damaged?
Does the cell have sufficient nutrients?

G_2/M CHECKPOINT
Has DNA replicated properly?

G_0

G_1

S

M

G_2

SPINDLE ASSEMBLY CHECKPOINT
Are the spindle fibers properly built and attached?

Checkpoints are critical points in the cell cycle at which progress is blocked until specific signals trigger continuation.

FIGURE 6-6 **Control of the cell cycle.** Cells use a system of checkpoints to ensure proper conditions have been achieved before the cell divides.

point when a cell "decides" whether it will proceed to the S phase and complete cell division, or delay cell division, or enter into an extended "resting" phase, G_0.

Although often described as an extended "resting" phase, G_0 is not actually a time of rest for a cell; in fact, it may be characterized by great metabolic activity that is critical to an organism's proper functioning. Rather, G_0 is simply a non-dividing, non-growing phase. In humans, liver cells spend almost all of their time in G_0 phase, dividing only once every year or two. As noted above, many neurons and muscle cells never leave G_0, living and functioning for many decades without dividing.

Several situations can prevent a cell from passing this G_1/S restriction point. Cells cultured in the lab, for example, will delay division at this point if the medium in which they are grown does not contain sufficient nutrients. If a cell's DNA has been damaged, too, the cell may be blocked from entering the S phase (and DNA repair mechanisms may be triggered). A malfunction in the G_1/S checkpoint can prevent cells with damaged DNA from blocking cell division, leading to uncontrolled cell division and cancer.

2. G_2/M checkpoint: assessing DNA synthesis. Just before beginning mitosis, a cell reaches the G_2/M checkpoint. This checkpoint serves as a "mitosis-readiness"

assessment. If it is passed, indicating that no DNA damage is detected, the cell initiates the complex process of mitosis. If not, the cell will typically undergo repair of damaged DNA.

3. Spindle assembly checkpoint: assessing anaphase readiness during mitosis. This important cell-cycle checkpoint occurs during mitosis. At this point, cell-cycle control mechanisms assess whether the chromosomes have aligned properly at the metaphase plate and whether there is appropriate tension (pull) on them. If this checkpoint is passed, the cell completes cell division.

TAKE-HOME MESSAGE 6·3

Eukaryotic somatic cells alternate in a cycle between cell division and other cell activities. The cell division portion of the cycle is called the mitotic phase. The remainder of the cell cycle, called interphase, consists of two gap phases (during which cell growth and other metabolic activities occur) separated by a DNA synthesis phase, during which the genetic material is replicated. A cell-cycle control system functions through a series of checkpoints, critical points in the cell cycle at which progress is blocked—and cells are prevented from dividing—until specific signals trigger continuation of the process.

6·4 Cell division is preceded by chromosome replication.

In one of the great understatements in the scientific literature, James Watson and Francis Crick wrote the following sentence in their paper describing the structure of DNA: "It has not escaped our notice that the specific pairing we have postulated immediately suggests a possible copying mechanism for the genetic material." The researchers' ability to explain how DNA copied itself was a critical feature of their description of DNA's structure. After all, every single time any cell in any organism's body divides, that cell's DNA must first duplicate itself so that each of the two new daughter cells has all of the genetic material of the original parent cell. The process of DNA duplication, as we've seen, is called replication.

What is it about DNA's structure that makes duplication so straightforward? Watson and Crick were referring to the feature of DNA called **complementarity,** meaning that in the double-stranded DNA molecule, the base on one strand always has the same pairing partner (called the

complementary base) on the other strand: A pairs with T (and vice versa), and G pairs with C (and vice versa).

With this consistent pattern of pairing, when the DNA molecule separates into two strands, it is possible to perfectly reconstruct, for each strand, all the information on the missing half, because one strand carries all the information needed to construct its complementary strand. Just before cells divide, the DNA molecule unwinds and "unzips," and each half of the unzipped molecule serves as a template on which the missing half is reconstructed. At the end of the process of reconstructing the missing halves, there are two DNA molecules—each identical to the original DNA molecule—one for each of the two new cells (**FIGURE 6-7**).

Before we describe the process by which DNA is duplicated, it is useful to revisit the structure of DNA because it dictates the directions along the linear molecule in which replication occurs. Recall from Chapter 5 that the individual units that

DNA COMPLEMENTARITY

Complementary base pairing makes it possible to produce two identical strands by separating the parent molecule and using each strand as a template to build a new complementary strand.

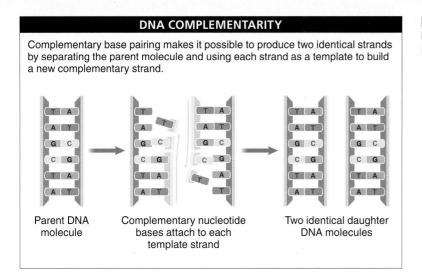

Parent DNA molecule

Complementary nucleotide bases attach to each template strand

Two identical daughter DNA molecules

FIGURE 6-7 **Meant for each other.** The nucleotide base on one strand always has the same pairing partner on the other strand.

make up DNA are nucleotides, which have three components: a nitrogen-containing molecule called a base, a phosphate group, and a molecule of a five-carbon sugar. Each of the five carbon atoms in the sugar molecules is given a number. The nitrogenous base is attached to the 1′ (pronounced "one prime") carbon. An –OH group is attached to the 3′ carbon. And the phosphate group is attached to the sugar's 5′ carbon atom (**FIGURE 6-8**).

The process of DNA replication occurs in two steps: (1) unwinding and separation of the two strands, and (2) reconstruction and elongation of new complementary strands (**FIGURE 6-9**). A feature of DNA that has important consequences for the process of replication is that its two strands run in opposite directions. So the backbone of the DNA molecule is a long alternation of sugar, phosphate, sugar, phosphate, sugar, phosphate, and so on, going from top to bottom in a 5′ to 3′ pattern on one side and going from bottom to top in a 5′ to 3′ pattern on the opposite side.

1. Unwinding and separation. Replication begins at a specific site, called the **origin of replication,** where the coiled, double-stranded DNA molecule unwinds and separates into two strands, like a zipper unzipping. In prokaryotes, there is a single origin of replication, while eukaryotes have multiple origin sites on each chromosome. At the origin of replication, a complex of proteins binds to the DNA. One of the proteins, an enzyme called **DNA helicase,** unwinds the coiled DNA and separates the two complementary strands. The unwinding and separating of the two DNA strands creates what is called a **replication fork.**

2. Reconstruction and elongation. In the reconstruction and elongation process, each of the single strands becomes

a double strand as the appropriate complementary base of a nucleotide pairs with the exposed base, and enzymes link the nucleotides of the new strand together.

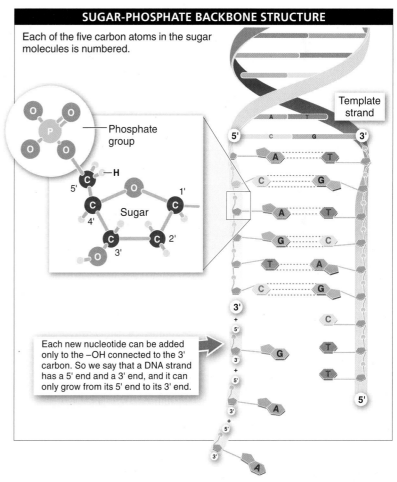

SUGAR-PHOSPHATE BACKBONE STRUCTURE

Each of the five carbon atoms in the sugar molecules is numbered.

Template strand

Phosphate group

Sugar

Each new nucleotide can be added only to the –OH connected to the 3′ carbon. So we say that a DNA strand has a 5′ end and a 3′ end, and it can only grow from its 5′ end to its 3′ end.

FIGURE 6-8 **A look at the structure of DNA.** The DNA molecule has two strands of nucleotides held together by hydrogen bonds between bases. The two strands wind into a double helix.

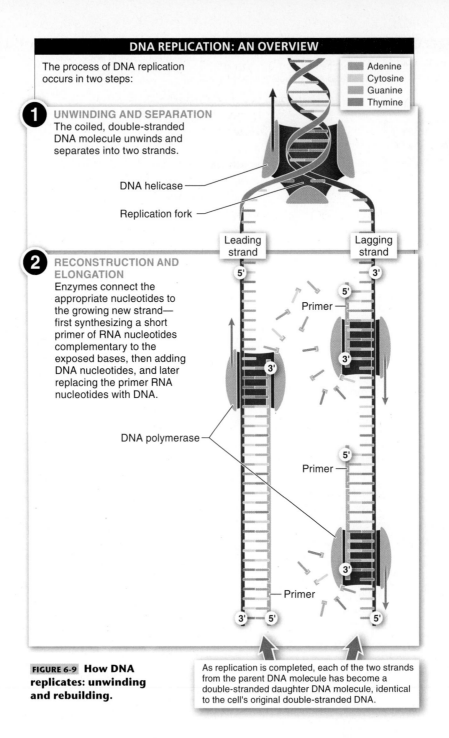

DNA REPLICATION: AN OVERVIEW

The process of DNA replication occurs in two steps:

	Adenine
	Cytosine
	Guanine
	Thymine

1 **UNWINDING AND SEPARATION**
The coiled, double-stranded DNA molecule unwinds and separates into two strands.

DNA helicase

Replication fork

Leading strand

Lagging strand

5'

3'

2 **RECONSTRUCTION AND ELONGATION**
Enzymes connect the appropriate nucleotides to the growing new strand—first synthesizing a short primer of RNA nucleotides complementary to the exposed bases, then adding DNA nucleotides, and later replacing the primer RNA nucleotides with DNA.

Primer

5'

3'

3'

DNA polymerase

Primer

5'

Primer

3'

3'   5'   5'

FIGURE 6-9 **How DNA replicates: unwinding and rebuilding.**

As replication is completed, each of the two strands from the parent DNA molecule has become a double-stranded daughter DNA molecule, identical to the cell's original double-stranded DNA.

As we described above, the two strands of the original DNA molecule run in opposite directions (5′ to 3′ on one side and 3′ to 5′ on the other). This fact, along with the fact that new nucleotides can be added only to the 3′ end of a growing strand, means the synthesis of DNA on the two template strands must occur in opposite directions.

At the replication fork, a group of several proteins, called a replication complex, binds to each of the exposed strands. This complex includes the enzymes **primase** and **DNA**

polymerase. Bound to a template strand, primase synthesizes a short sequence of RNA nucleotides, called a primer, that is complementary to the template strand at that point. DNA polymerase then adds new DNA nucleotides to the 3′ end of the primer as it builds a strand complementary to the template strand.

For one of the template strands, called the **leading strand,** the DNA is unwound and separated and the complementary strand is built as the replication complex simply adds nucleotides in the 5′ to 3′ direction, *toward* the replication fork as it moves along the strand. The new double-stranded DNA can then rewind.

The other template strand, which is called the **lagging strand,** requires a more complicated, discontinuous process. At the replication fork, primase adds an RNA primer to the lagging strand. Building the complementary strand in the 5′ to 3′ direction, however, leads to the addition of nucleotides in the direction opposite to the movement of the replication fork. DNA polymerase then continues the synthesis of the complementary strand, *away from* the replication fork, until it encounters a previously synthesized section (see Figure 6-9).

While this occurs, however, the replication fork has continued to move in the opposite direction, exposing additional single-stranded DNA on the lagging strand. And so again, primase adds an RNA primer, after which DNA polymerase continues the synthesis away from the replication fork and toward the point at which the previous fragment began. This process of **discontinuous replication** produces short stretches of newly synthesized fragments. On each fragment, the RNA primer is then replaced by DNA nucleotides, as on the leading strand, a nd another enzyme then joins the ends of the newly synthesized fragments. That portion of the newly synthesized DNA molecule can then wind into a double helix.

DNA replication occurs in all types of cells, somatic and reproductive, before cell division. The end result of replication is two double-stranded DNA molecules that carry virtually the same genetic information. This makes it possible for the somatic cells of the body that are produced by cell division to be virtually identical genetically. Note, though, that the daughter molecules are not completely identical. Replication is accompanied by a very low—but non-zero—rate of errors that are not corrected.

Q Errors sometimes occur when DNA duplicates itself. Why might that be a good thing?

DNA Proofreading and Error Correction A variety of mutations (such as mismatched bases and added or deleted segments) can occur during replication (or when DNA is damaged by some external source, such as X rays). Because DNA polymerases perform proofreading functions as well as excision and repair functions, many of these errors are caught and repaired during or after replication or whenever they occur.

If an error remains, however, the sequences in a replicated DNA molecule (including the genes) can be different from those in the parent molecule. A changed sequence may then produce a different mRNA sequence, which may in turn produce a different amino acid sequence. As a result, the changed sequence may lead to the production of a different protein. The error rate in eukaryotes is very low—about one error per several billion base pairs—but it is not zero. And when an error is introduced into the DNA of a gamete-producing cell (i.e., one that divides to make sperm or eggs), a new gene can sometimes enter a population and be acted on by evolution. We explore the relationship between mutation and evolution in more detail in Chapter 8.

> **TAKE-HOME MESSAGE 6·4**
>
> Every time a cell divides, that cell's DNA must first duplicate itself so that each of the two new daughter cells has all the genetic material of the original parent cell. The process of DNA duplication, called replication, is catalyzed by several important enzymes and occurs in two steps: unwinding and separation of the two strands, and reconstruction and elongation of the new complementary strands. The end result is two double-stranded DNA molecules that carry virtually the same genetic information as the parent DNA. Although enzymes proofread and repair DNA during and after replication, some errors may remain.

6·5–6·8
Mitosis replaces worn-out old cells with fresh new duplicates.

For most of the cell cycle, chromosomes resemble a plate of spaghetti more than the familiar condensed, replicated "X" shape seen in photos.

6·5 Most cells are not immortal: mitosis generates replacements.

Look around your room. Dust is everywhere. What is it? It is primarily dead skin cells. In fact, you (and your friends and family) slough off millions of dead skin cells each day—yet your skin doesn't disappear. Why not? Because your body replaces the sloughed-off cells. When the cells wear out, your body creates replacements through the process of mitosis.

Q What is dust? Why is it your fault?

Mitosis has just one purpose: to enable existing cells to generate new, genetically identical cells. There are two different reasons for this need (**FIGURE 6-10**).

1. Growth. During development, organisms get bigger. Growth happens in part through the creation of new cells. In fact, if you want to see cell division in action, one surefire place to find it is at the tip of a plant root: at a growth rate of about half an inch per day, the root is one of the fastest-growing parts of a plant.

2. Replacement. Cells must be replaced when they die. The wear and tear that come from living can physically damage cells. The daily act of shaving, for example, damages thousands of cells on a man's face. It's nothing to

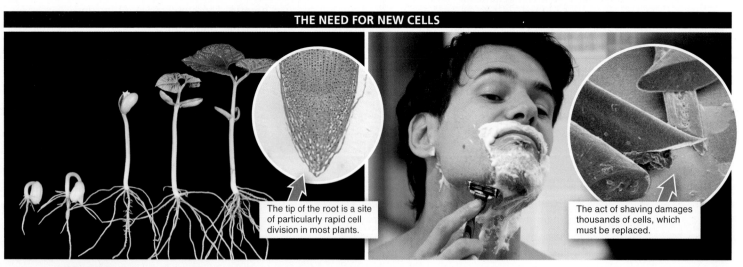

THE NEED FOR NEW CELLS

The tip of the root is a site of particularly rapid cell division in most plants.

The act of shaving damages thousands of cells, which must be replaced.

FIGURE 6-10 **Reasons for mitosis.** Mitosis is important in an organism's growth and the replacement of cells.

worry about, though. Microscopic views of human skin reveal several distinct layers, with the outermost layers—the layers under assault during shaving—made up primarily of dead cells. These cells help protect us from infection and also reduce the rate at which the underlying living cells dry out. The living cells that exist just below the layers of dead cells are produced at a high rate by mitosis.

Some other cells that must be replaced actually die on purpose, in a planned process of cell suicide called **apoptosis** (A-pop-TOE-siss). This seemingly counterproductive strategy is employed in parts of the body where the cells are likely to accumulate significant genetic damage over time and are therefore at high risk of becoming cancer cells (a process described later in this chapter). Cells targeted for apoptosis include many of the cells lining the digestive tract as well as those in the liver, two locations where cells are almost constantly in contact with harmful substances.

Every day, a huge number of cells in an individual must be replaced by mitosis. In humans this number is in the billions. Nearly all the somatic cells of the body—that is, all cells except sperm- and egg-producing cells—undergo mitosis. There are a few notable exceptions, as we've already seen. Heart muscle cells and most neurons, in particular, do not seem to divide, or, if they do divide, they do so at very, very slow rates. (We don't know why this is so.)

The rate at which mitosis occurs in animals varies dramatically for different types of cells. The most rapid cell division takes place in the bone marrow (as red blood cells

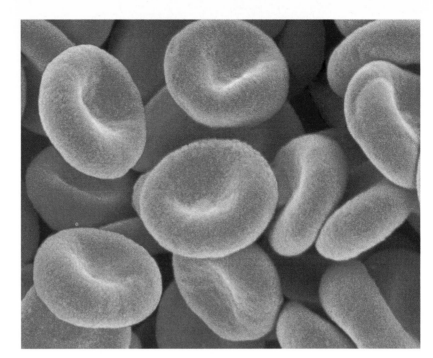

FIGURE 6-11 **Short-lived cells.** Red blood cells in your body are replaced about every 120 days.

are produced) and in the cells lining various tissues and organs. The average red blood cell, for example, is in circulation for only about two to four months and then must be replaced (**FIGURE 6-11**). The cells lining the intestines are replaced about every three weeks. Hair follicles contain some of the most rapidly dividing cells.

TAKE-HOME MESSAGE 6·5

Mitosis enables existing cells to generate new, genetically identical cells. This makes it possible for organisms to grow and to replace cells that die.

6·6 Overview: mitosis leads to duplicate cells.

For mitosis to begin, the parent cell replicates its DNA, creating a duplicate copy of each chromosome. Once this task is completed, mitosis can take place. Mitosis occurs in just four steps, in which the now-duplicated chromosomes are separated into identical sets in two separate nuclei, after which the cytoplasm and the rest of the cell are divided into two cells that

pinch apart (**FIGURE 6-12**). Where once there was one parent cell, now there are two identical daughter cells (**FIGURE 6-13**).

Before we explore the process of mitosis in more detail, it is helpful to clarify some of the terminology used to describe the important structures and processes.

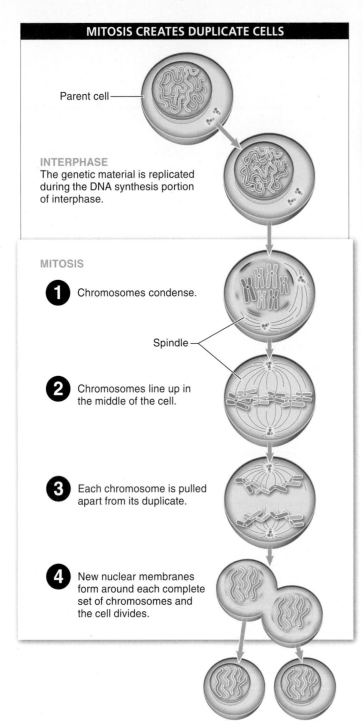

MITOSIS CREATES DUPLICATE CELLS

Parent cell

INTERPHASE
The genetic material is replicated during the DNA synthesis portion of interphase.

MITOSIS

1 Chromosomes condense.

Spindle

2 Chromosomes line up in the middle of the cell.

3 Each chromosome is pulled apart from its duplicate.

4 New nuclear membranes form around each complete set of chromosomes and the cell divides.

FIGURE 6-12 **A simplified introduction to mitosis.**

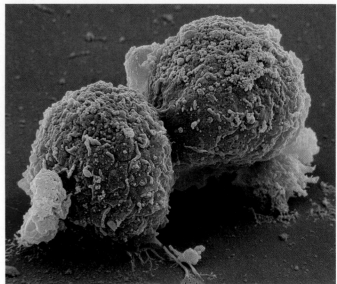

FIGURE 6-13 **These two human embryonic stem cells have just completed mitosis.**

Before mitosis begins, two important events occur.

1. The chromosomes replicate, becoming two identical linear DNA molecules. The two DNA molecules are held together at a region called the **centromere.** Throughout mitosis, until the centromere splits, each of the identical DNA molecules is called a **chromatid;** together, the two are called **sister chromatids** (**FIGURE 6-14**).

2. The sister chromatids begin the process of **condensation,** in which they coil tightly and become compact—in contrast to the uncondensed and tangled state of the chromosomes prior to replication, during most of interphase.

When the sister chromatids condense, they look like the letter X. This appearance can be confusing. As we saw earlier, chromosomes are not X-shaped. They are *linear.* The reason the genetic material appears X-shaped in most photos is that the only time it is coiled tightly and is thus thick enough to be seen (and photographed) is after it has condensed in preparation for cell division. But this occurs only after replication.

Q Animal chromosomes are linear. Why do they look like the letter X in pictures?

Because a lot of room is required for the sister chromatids to separate, the membrane around the nucleus is dismantled and disappears early in mitosis. At the same time, a structure

Generally, when you look at a cell through a microscope, you won't see any chromosomes because the cells you're looking at are in some part of their interphase. During all of the cell cycle except mitosis, chromosomes are uncoiled and spread out in a diffuse way, like a mass of spaghetti. And because they are so stretched out, they are not dense enough to be visible.

THE GENETIC MATERIAL DURING CELL DIVISION

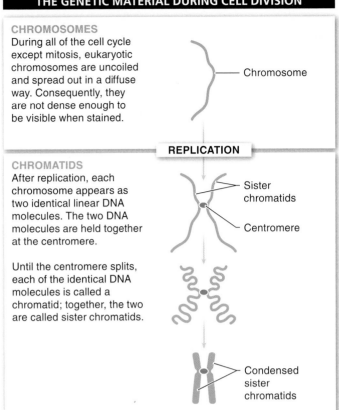

CHROMOSOMES
During all of the cell cycle except mitosis, eukaryotic chromosomes are uncoiled and spread out in a diffuse way. Consequently, they are not dense enough to be visible when stained.

Chromosome

REPLICATION

CHROMATIDS
After replication, each chromosome appears as two identical linear DNA molecules. The two DNA molecules are held together at the centromere.

Until the centromere splits, each of the identical DNA molecules is called a chromatid; together, the two are called sister chromatids.

Sister chromatids

Centromere

Condensed sister chromatids

FIGURE 6-14 **From one chromosome: two sister chromatids.**

called the **spindle,** composed of proteins—mostly hollow tubes called **microtubules**—is assembled. The spindle microtubules, part of the cell's cytoskeleton, can be thought of as a group of parallel threads stretching across the cell between its two ends, or poles. In animal cells, the threads connect at each pole to a structure called the **centriole.** These threads (known as **spindle fibers**) attach to the centromeres and pull the sister chromatids to the middle of the cell. During mitosis, they'll eventually pull the chromatids apart as cell division proceeds.

> **TAKE-HOME MESSAGE 6·6**
>
> Mitosis is the process by which cells duplicate themselves. Mitosis follows chromosome replication and leads to the production of two daughter cells from one parent cell.

6·7 The details: mitosis is a four-step process.

Let's look at the process of mitosis in a bit more detail, keeping in mind that the ultimate result is the production of two cells with identical chromosomes.

Interphase: In Preparation for Mitosis, the Chromosomes Replicate Processes essential to cell division take place even before the mitotic phase of the cell cycle begins. During the DNA synthesis part of interphase, sister chromatids are formed as every chromosome replicates itself. Each pair of sister chromatids is held together at a centromere.

Mitosis The actual process of cell division occurs in four steps (**FIGURE 6-15**).

1. Prophase: following replication, the sister chromatids condense. Prophase begins when the sister chromatids have condensed sufficiently to be seen with a light microscope. At this point, the spindle forms and the nuclear envelope breaks down.

2. Metaphase: the chromatids congregate at the cell center. After condensing, the pairs of sister chromatids seem to move aimlessly around the cell, but eventually they line up at the cell's center, pulled by spindle fibers attached to a disk-like group of proteins, called a **kinetochore,** that develops on the centromere of each pair. At the end of metaphase, all the chromatid pairs are lined up in an orderly fashion, straddling the center in a "single-file" congregation that is called the metaphase plate. The chromatids are at their most condensed during this part of mitosis.

3. Anaphase: the chromatids separate and move in opposite directions. In anaphase, the spindle microtubules attached to the centromeres begin pulling each chromatid in the sister chromatid pairs toward opposite poles of the cell. From each pair of sister chromatids, the centromere splits as one DNA molecule is pulled in one direction and the other, identical DNA molecule is pulled in the opposite direction. At the end of anaphase, one full set of chromosomes is at one end of the cell and another identical

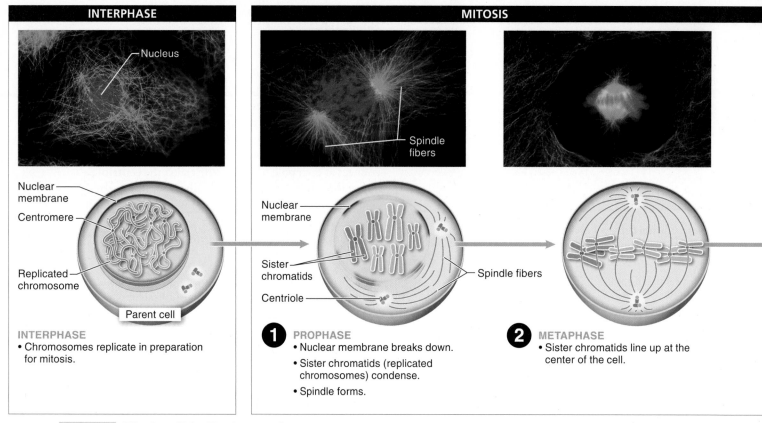

Nucleus

Spindle fibers

Nuclear membrane

Centromere

Replicated chromosome

Parent cell

Nuclear membrane

Sister chromatids

Centriole

Spindle fibers

Spindle fibers

INTERPHASE
• Chromosomes replicate in preparation for mitosis.

1 PROPHASE
• Nuclear membrane breaks down.
• Sister chromatids (replicated chromosomes) condense.
• Spindle forms.

2 METAPHASE
• Sister chromatids line up at the center of the cell.

FIGURE 6-15 Mitosis: cell duplication, step by step.

full set is at the other end. These chromosome sets will eventually occupy the nucleus of each of the two new daughter cells that result from the cell division.

4. Telophase: new nuclear membranes form around the two complete chromosome sets. With two full, identical sets of chromosomes collected at either end of the cell, the parent cell is prepared to divide into two genetically identical cells. In this last step, called telophase, the chromosomes begin to uncoil and fade from view, nuclear membranes reassemble, and the cell begins to divide into two.

The process of mitosis, the division of a cell's nucleus into two nuclei, each of which includes an identical set of chromosomes, is generally accompanied by cytokinesis. During cytokinesis, the cell's cytoplasm is also divided into

approximately equal parts, with some of the organelles going to each of the two new cells. When cytokinesis is complete, the two new daughter cells, each with an identical nucleus containing identical genetic material, enter interphase and begin the business of being cells.

Usually, mitosis occurs without errors and only when needed. In rare cases, however, something goes wrong that can cause cell division to proceed unchecked. In such cases, cancer can arise.

TAKE-HOME MESSAGE 6·7

The ultimate result of mitosis and cytokinesis is the production of two genetically identical cells.

6·8 Cell division out of control may result in cancer.

Too much of a good thing can be bad. This is especially true when cell division runs amok, a situation that can lead to cancer. **Cancer** is defined as unrestrained cell growth and

division that can damage adjacent tissues. Some cancers can metastasize, or spread to other locations in the body. Cancer can cause serious health problems and is the second leading

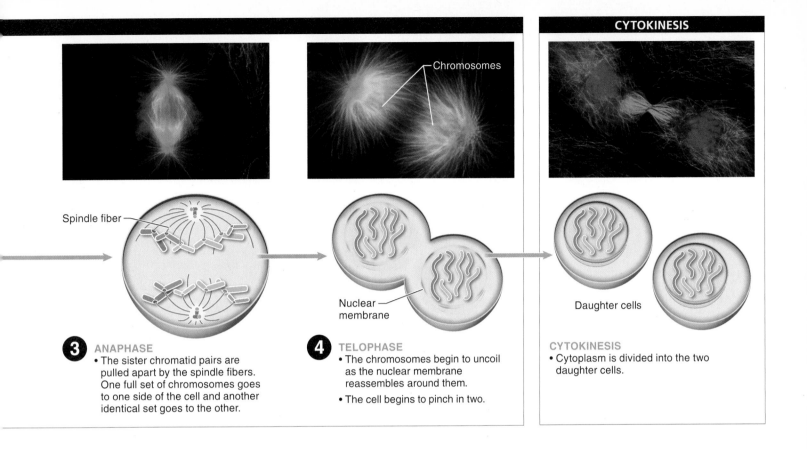

Chromosomes

Spindle fiber

Nuclear membrane

Daughter cells

3 ANAPHASE
• The sister chromatid pairs are pulled apart by the spindle fibers. One full set of chromosomes goes to one side of the cell and another identical set goes to the other.

4 TELOPHASE
• The chromosomes begin to uncoil as the nuclear membrane reassembles around them.
• The cell begins to pinch in two.

CYTOKINESIS
• Cytoplasm is divided into the two daughter cells.

cause of death in the United States, responsible for more than 20% of all deaths. Only heart disease causes more deaths.

Cancer occurs when some disruption of the DNA in a normal cell interferes with the cell's ability to regulate cell division. DNA disruption can be caused by chemicals that mutate DNA or by sources of high energy such as X rays, the sun, or nuclear radiation. Cancer can even be caused by some viruses. However it begins, once a cell loses control over its cell cycle, cell division can proceed unrestrained (**FIGURE 6-16**).

> **Cancer cells are those which have forgotten how to die.**
>
> — HAROLD PINTER, playwright, from the poem "Cancer Cells," 2002

Cancer cells have several features that distinguish them from normal cells; the three most significant differences are the following:

1. Cancer cells lose their "contact inhibition." Most normal cells divide until they come into contact with other cells or collections of cells (tissues). At that point, they stop dividing. Cancer cells, however, ignore the signal that they are at high density and continue to divide.

2. Cancer cells can divide indefinitely. As we saw in Section 6-1, most normal human cells can divide 80 to 90 times. After that point, the cell may continue living but it loses the ability to divide. Cancer cells, on the other hand, never lose their ability to divide and continue to do so indefinitely, even in the presence of conditions that would normally halt the cell cycle before cell division. (Cancer cells can divide indefinitely because they are able to rebuild their telomeres following each cell division, as we saw in Section 6-1.)

3. Cancer cells have reduced "stickiness." Cells are normally held together by adhesion molecules, proteins within cell membranes. And cancer cells, too, usually group together, forming a tumor. But the membranes of cancer cells tend to have reduced adhesiveness, causing them to stick to each other less than do non-cancerous cells.

Tumors caused by excessive cell growth and division are of two very different types: benign and malignant. Benign tumors, such as many moles, are just masses of normal cells that do not spread. They can usually be removed safely without any lasting consequences. Malignant tumors, on

CANCER CELLS

Cancer cells have several features that distinguish them from normal cells.

CANCER CELLS HAVE NO CONTACT INHIBITION

Normal cells divide until they bump up against other cells or tissues, at which point they stop dividing.

Cancer cells, however, ignore the signal and continue to divide, piling on top of one another.

Normal cells

Cancer cells

CANCER CELLS DIVIDE INDEFINITELY

Normal somatic cells can divide a limited number of times. Cancer cells never lose their ability to divide, and continue to do so indefinitely.

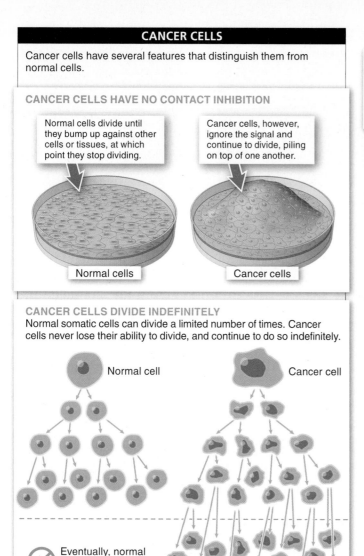

Normal cell

Cancer cell

Eventually, normal cells are no longer able to divide.

Cancer cells continue to divide indefinitely.

CANCER CELLS HAVE REDUCED "STICKINESS"

The membranes of cancer cells tend to have reduced adhesiveness, causing them to stick to each other less than do non-cancerous cells.

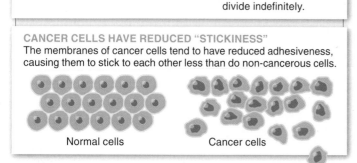

Normal cells

Cancer cells

FIGURE 6-16 **Problem cells.** Cancer occurs when there is a disruption in cells' ability to regulate cell division and a reduced adhesion among the cells.

the other hand, are the result of unrestrained growth of cancerous cells. They grow continuously and shed cells (**FIGURE 6-17**). The shedding of cancer cells from malignant tumors is how cancer spreads, a process called metastasis (meh-TASS-tuh-siss). In this process, cancer cells separate

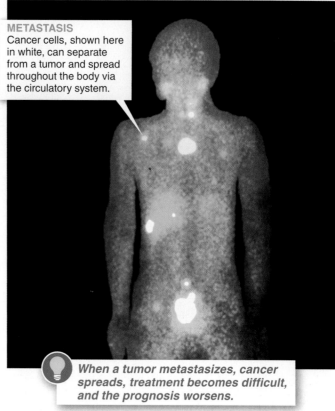

METASTASIS
Cancer cells, shown here in white, can separate from a tumor and spread throughout the body via the circulatory system.

When a tumor metastasizes, cancer spreads, treatment becomes difficult, and the prognosis worsens.

FIGURE 6-17 **Multiple tumors.** Metastasis spreads cancer cells throughout the body.

from a tumor and invade the circulatory system, which spreads them to different parts of the body where they can cause the growth of additional tumors.

How does cancer actually kill the organism? Somewhat surprisingly, it's not because of some chemical or genetic property of the cancer cells themselves. It's simpler than that. As a tumor gets larger, it takes up more and more space and presses against neighboring cells and tissues (**FIGURE 6-18**). Eventually, the tumor may block other cells and tissues from carrying out their normal functions and even kill them. This cell dysfunction or cell death can have disastrous consequences when the affected normal tissue controls processes critical to life, such as breathing, heart function, or the detoxification processes in the liver.

Q What is cancer? How does it usually cause death?

To treat cancer, the rapidly dividing cells must be removed surgically or killed, or their division at least slowed down. Currently, the killing and slowing down are done in two ways: by chemotherapy and by radiation.

In chemotherapy, drugs that interfere with cell division are administered, slowing down the growth of tumors. Because

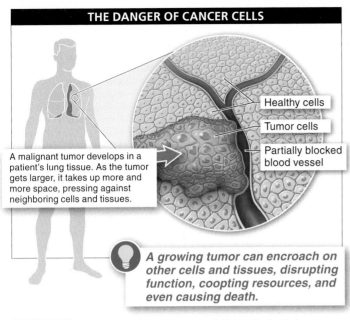

THE DANGER OF CANCER CELLS

A malignant tumor develops in a patient's lung tissue. As the tumor gets larger, it takes up more and more space, pressing against neighboring cells and tissues.

Healthy cells

Tumor cells

Partially blocked blood vessel

💡 *A growing tumor can encroach on other cells and tissues, disrupting function, coopting resources, and even causing death.*

FIGURE 6-18 **Cancer's effects.** Cancer cells harm the body by crowding and disrupting normal cells.

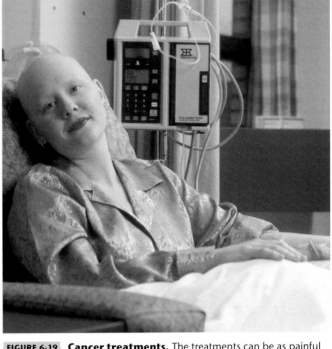

FIGURE 6-19 **Cancer treatments.** The treatments can be as painful and difficult as the disease.

these drugs interfere with rapidly dividing cells throughout the body (not just the rapidly dividing cancer cells), they can have very unpleasant side effects. In particular, chemotherapy drugs disrupt normal systems that rely on the rapid and constant production of new cells. For instance, chemotherapy often causes extreme fatigue and shortness of breath because it reduces the rate at which red blood cells are produced, thus limiting the amount of oxygen that can be transported throughout the body. By interfering with the division of bone marrow stem cells, chemotherapy also reduces the production of platelets and white blood cells and thus increases bruising and bleeding, as well as increasing susceptibility to infection. Another location of rapidly dividing cells commonly affected by chemotherapy and radiation is the hair follicle. As a consequence, many people lose their hair when undergoing chemotherapy. It usually grows back, however, when treatments stop.

Like chemotherapy, radiation works by disrupting cell division. Unlike the drugs used in chemotherapy, however, which circulate throughout the entire body, radiation therapy directs high-energy radiation only at the part of the body where a tumor is located. As with chemotherapy, the radiation process is not perfect, and nearby tissue is often harmed as well. The significant adverse effects of chemotherapy and radiation treatment on normal tissue and the suffering this leads to have caused some patients to comment that the treatment for cancer can initially feel as

bad as the disease itself—especially because patients may be asymptomatic at the time of treatment (**FIGURE 6-19**).

What causes cancer? Researchers have made significant progress toward answering this question at the cellular level. Most cancer is caused by mutations in a cell's DNA that disrupt the normal processes that control and regulate the cell cycle. More specifically, the cancer-causing mutations seem to affect two different types of genes: those that stimulate cell growth and those that restrain it.

Is there hope for a complete cure soon? Not yet. Nonetheless, many potentially successful therapies for cancer treatment and prevention are on the horizon. Extensive research is being conducted on the mechanisms by which genes controlling the cell cycle are damaged in cancer cells, for example, and how such damage might be prevented or reversed.

TAKE-HOME MESSAGE 6·8

Cancer is unrestrained cell growth and cell division, which lead to large masses of cells that may cause serious health problems. It often results from mutations to genes important in controlling the cell cycle, thus reducing the effectiveness of the checkpoints. Treatment focuses on killing or slowing down the fast-growing and dividing cells, usually using chemotherapy and/or radiation.

6·9–6·13
Meiosis generates sperm and eggs and a great deal of variation.

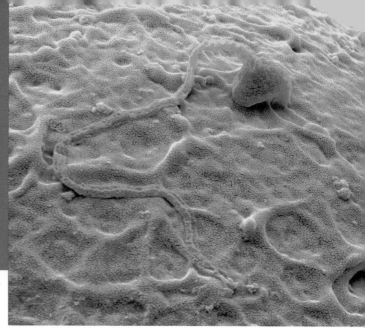

A human sperm cell fertilizing an egg.

6·9 Overview: sexual reproduction requires special cells made by meiosis.

There are two ways in which an organism can reproduce. Many organisms, including bacteria, fungi, and even some plants and animals, undergo **asexual reproduction,** in which a single parent produces identical offspring. Other organisms, including most animals and plants, undergo **sexual reproduction,** in which offspring are produced by the fusion of two reproductive cells in the process of **fertilization.** And some species, particularly among plants, can use both methods. In sexual reproduction, because a combination of DNA from two separate individuals is passed on to offspring, the resulting offspring are genetically different from their parents and, for reasons explained later, from one another.

What would happen if sexually reproducing organisms, humans included, produced reproductive cells through mitosis? Both parents would contribute a full set of genes— that is, 23 *pairs* of chromosomes in humans—to create a new individual, and the new offspring would inherit 46 pairs of chromosomes in all. And when that individual reproduced, if he or she contributed 46 pairs of chromosomes and his or her mate also contributed 46 pairs, their offspring would have 92 pairs of chromosomes. Where would it end? The genome would double in size every generation. That wouldn't work at all.

In sexually reproducing organisms, a way has evolved to avoid such chromosome overload. The solution is **meiosis,** a process that enables organisms to make special reproductive cells, the gametes, which have only half as many chromosomes as the rest of the cells in the organism's body (the somatic cells). In other words, in anticipation of

combining one individual's genome with another's during fertilization, meiosis reduces each individual's genome by half in the gametes. In humans, for example, each gamete cell has only one set of 23 chromosomes, rather than two sets.

In genetics, the term **diploid** refers to cells that have two copies of each chromosome (in humans, two sets of 23 chromosomes, for 46 chromosomes in total), and the term **haploid** refers to cells that have one copy of each chromosome. Thus, somatic cells are diploid, and gametes, the cells produced in meiosis, are haploid. At fertilization, two haploid cells, each with one set of 23 chromosomes, merge and create a new individual with the proper diploid human genome of 46 chromosomes. And when the time comes to reproduce, this new individual, through meiosis, also will produce haploid gametes that have only a single set of 23 chromosomes. With sexual reproduction, then, diploid organisms produce haploid gametes that fuse at fertilization to restore the diploid state (**FIGURE 6-20**). In this way, meiosis maintains a stable genome size in a species.

Although there are some variations on this pattern of alternation between the haploid and diploid states, in most cases, multicellular animals produce simple haploid cells for reproduction. And after two of those gametes come together to form a diploid fertilized egg, multiple cell divisions by mitosis again produce a diploid, multicellular animal.

Meiosis achieves more than just a reduction in the amount of genetic material in gametes. As a diploid individual, you

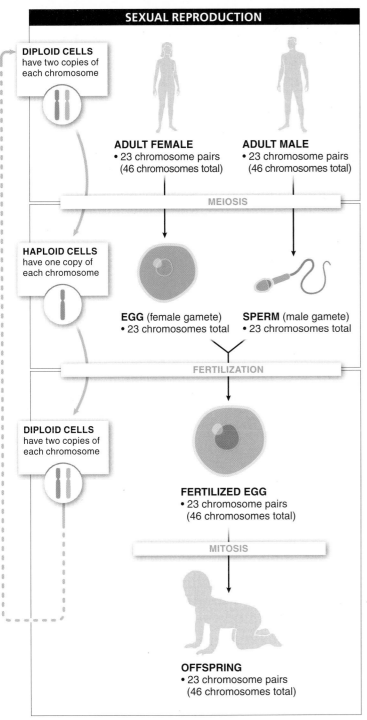

SEXUAL REPRODUCTION

DIPLOID CELLS
have two copies of
each chromosome

ADULT FEMALE
• 23 chromosome pairs
(46 chromosomes total)

ADULT MALE
• 23 chromosome pairs
(46 chromosomes total)

MEIOSIS

HAPLOID CELLS
have one copy of
each chromosome

EGG (female gamete)
• 23 chromosomes total

SPERM (male gamete)
• 23 chromosomes total

FERTILIZATION

DIPLOID CELLS
have two copies of
each chromosome

FERTILIZED EGG
• 23 chromosome pairs
(46 chromosomes total)

MITOSIS

OFFSPRING
• 23 chromosome pairs
(46 chromosomes total)

FIGURE 6-20   Sexual reproduction. Diploid organisms produce haploid gametes that fuse at fertilization and return to the diploid state.

have two copies (i.e., two alleles) of every gene in every somatic cell: one from your mother and one from your father. When making haploid cells from cells that are diploid, an individual creates cells that have one allele for each trait rather than two. Which of these two alleles is included in each gamete cannot be predicted.

FIGURE 6-21   One mother, one father: varied offspring. Variation generated in the production of sperm and eggs causes each sibling to carry a unique set of alleles.

Each egg or sperm is produced with a varied combination of maternal and paternal alleles, so each of the gametes an individual makes carries a unique set of alleles. As a consequence, each offspring inherits a slightly different assemblage of each parent's alleles. All offspring resemble their parents, yet none resembles them in exactly the same way. This variation among offspring can be seen among the five Jackson brothers in the photo that opens this chapter (and among the Kennedy brothers in **FIGURE 6-21**).

In all, then, meiosis has two important outcomes:

1. It reduces the amount of genetic material in gametes.

2. It produces gametes that differ from one another with respect to the combinations of alleles they carry.

In the next sections, we examine the exact steps that lead to these outcomes and investigate the consequences of all the variation that meiosis and sexual reproduction can generate.

TAKE-HOME MESSAGE 6·9

Meiosis is the process by which reproductive cells are produced in sexually reproducing organisms. It results in gametes that have only half as much genetic material as the parent cell and that differ from one another in the combinations of alleles they carry.

Sperm and egg are produced by meiosis: the details, step by step.

Mitosis is an all-purpose process for cell division. It occurs all over the body, all the time. Meiosis, on the other hand, is a special-purpose process. In sexually reproducing animals, it takes place in just a single place: the **gonads** (the ovaries and testes). And it takes place for just a single reason: the production of gametes (reproductive cells). To better understand meiosis, let's examine in detail the steps by which a diploid animal cell creates haploid cells, each of which is different from the other.

Meiosis, then, starts with a diploid cell—not just any diploid cell, but one of the specialized diploid cells, found in the gonads, that is capable of undergoing meiosis. Thus, for humans, meiosis starts with a cell that has 46 chromosomes. These include a maternal copy and a paternal copy of each of 22 chromosomes—each of these pairs is called a **homologous pair,** or homologues—along with two additional chromosomes (one from each parent), called *sex chromosomes* (**FIGURE 6-22**).

Before meiosis can occur (and just as we saw with mitosis), each of these 46 chromosomes is duplicated during the cell's interphase. This means that when meiosis begins, we have 92 (2 × 46) strands of DNA, after replication.

Unlike mitosis, which has only one cell division, cells undergoing meiosis divide twice. In the first division, the homologues separate. In other words, for each of the 23 chromosome pairs, the maternal sister chromatid pairs

MEIOSIS: REDUCING THE GENOME BY HALF

INTERPHASE
Each chromosome in a homologous pair replicates to form a sister chromatid.

(Chromosomes shown condensed here for diagrammatic purposes.)

Diploid parent cell

Homologues
Maternal chromosome
Paternal chromosome

Sister chromatids

MEIOSIS I
In the first division of meiosis, the homologous pairs separate.

Haploid daughter cells (gametes)

MEIOSIS II
In the second division of meiosis, the sister chromatids separate. This results in four haploid cells, each containing just one copy of each chromosome, rather than a homologous pair.

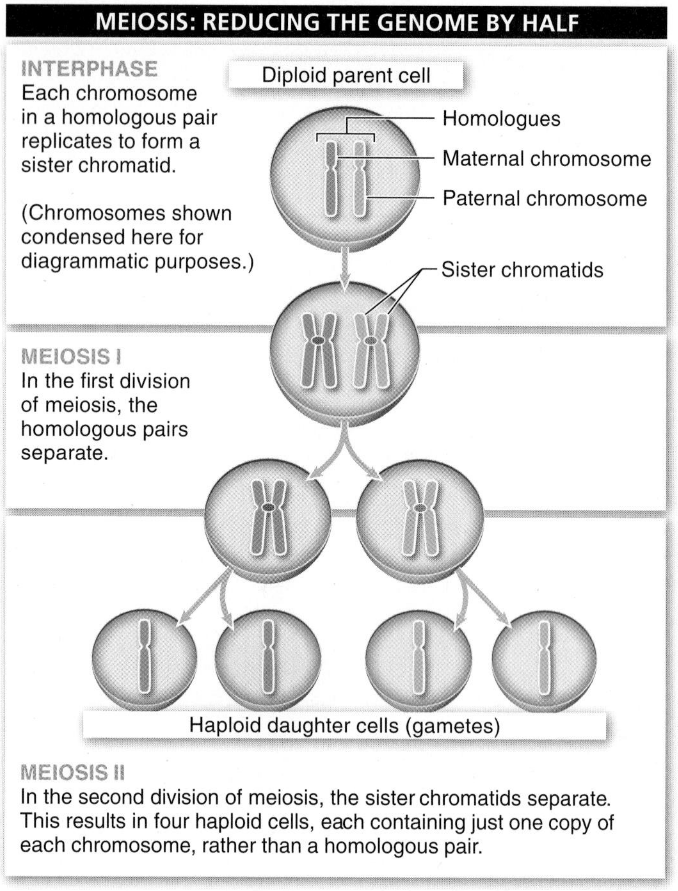

FIGURE 6-23   **Meiosis reduces the genome by half in anticipation of combining it with another genome.**

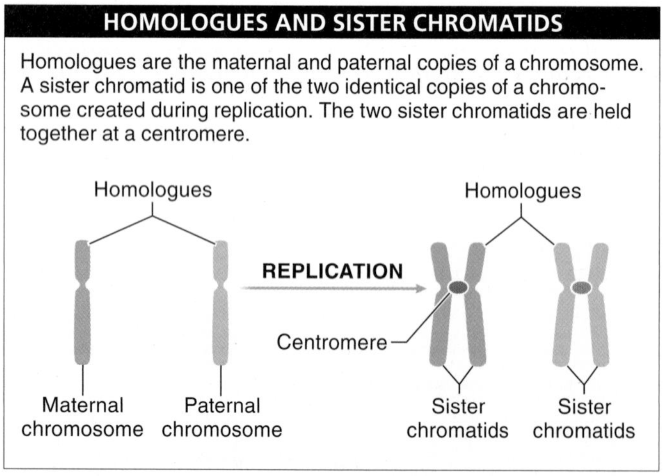

HOMOLOGUES AND SISTER CHROMATIDS

Homologues are the maternal and paternal copies of a chromosome. A sister chromatid is one of the two identical copies of a chromosome created during replication. The two sister chromatids are held together at a centromere.

Homologues

Homologues

REPLICATION

Centromere

Maternal chromosome
Paternal chromosome
Sister chromatids
Sister chromatids

FIGURE 6-22   **Chromosome vocabulary: homologues and sister chromatids.**

and the paternal sister chromatid pairs separate into two new cells. In the second division, each of the two new cells divides again, so that each of the four daughter cells contains a single chromosome from the homologous pair. At the end of meiosis, there are four new cells, each of which has 23 strands of DNA—that is, 23 chromosomes (**FIGURE 6-23**). Note that, in animals, (1) none of these four cells will undergo any further cell division—they do not become parent cells for a new cycle of cell division—and (2) only specialized types of diploid cells located in the gonads can undergo meiosis.

Interphase: In Preparation for Meiosis, the Chromosomes Replicate   Before meiosis begins, preparations for cell division occur that are similar to those taking place in the interphase of mitosis in the somatic cell

cycle. In particular, every chromosome creates an exact duplicate of itself by replication. The chromosomes that were each a single, long, linear piece of genetic material become a pair of identical long, linear pieces, held together at the centromere.

Meiosis begins following this replication step during interphase; it proceeds through eight steps (**FIGURE 6-24**).

Meiosis Division I: The Homologues Separate
The first meiotic division takes place in four stages.

1. Prophase I: chromosomes condense and crossing over occurs. This is by far the most complex of all the phases of meiosis. As in mitosis, it begins with all of the replicated genetic material condensing.

As the sister chromatids become shorter and thicker, the homologous chromosomes come together. This is where the process diverges from mitosis. The homologous chromosomes (each of which has become a pair of identical chromatids) line up, touching. Under a microscope, the two homologous pairs of sister chromatids appear as pairs of X's lying on top of each other.

At this point, the sister chromatids that are next to each other do something that makes every sperm or egg cell genetically unique: they swap little segments of DNA. In other words, when this is happening inside you, some of the genes that you inherited from your mother get swapped onto the strand of DNA you inherited from your father, and the corresponding bits from your father are inserted into the DNA strand from your mother. This event is called **genetic recombination** (or, more often, just **recombination**) or **crossing over,** and it can take place at several spots (up to dozens) on each chromatid. As a result of crossing over, every sister chromatid ends up having a unique mixture of your genetic material. Note that crossing over occurs only during the production of gametes, in meiosis. It does not occur during mitosis. Following crossing over, the nuclear membrane disintegrates.

We explore crossing over in more detail in Section 6-12. The remaining steps of meiosis are relatively straightforward.

2. Metaphase I: all chromosomes line up along the center of the cell. After crossing over, each pair of homologous chromosomes (that is, the pairs of X's lying on top of each other) moves to the center of the cell, pulled by the spindle fibers to form the arrangement called the metaphase plate. Remember that each pair of homologous

chromosomes includes the maternal and paternal versions of the chromosome (with crossed-over segments) *and* the replicated copy of each—four strands in all.

3. Anaphase I: homologues are pulled to either side of the cell. This phase is the beginning of the first cell division that occurs during meiosis. In anaphase, the homologues are pulled apart toward opposite poles of the cell. One of the homologues (consisting of two sister chromatids) goes to one pole, the other to the opposite pole. At this point, something else occurs that, along with crossing over, contributes to making all the products of meiosis genetically unique. The maternal and paternal sister chromatid pairs are pulled to the ends of the cell in a random fashion called **random assortment.** As a result of random assortment, the pairs of sister chromatids gathered at each pole are a mix of maternal and paternal sister chromatids.

Imagine all the different combinations that can occur in a species with 23 pairs of chromosomes. For chromosome pair 1, perhaps the maternal homologue goes to the "top" pole and the paternal to the "bottom" pole. And for pair 2, perhaps the maternal homologue also goes to the top and the paternal to the bottom. But maybe for pair 3, it is the paternal homologue that goes to the top and the maternal homologue to the bottom. For each of the 23 pairs, you cannot predict which will go to the top and which will go to the bottom. There is a *huge* number of possible combinations. (Can you figure out how many? Here's a hint: if there were 2 chromosome pairs, the answer would be 4; if there were 3 chromosome pairs, the answer would be 8.)

4. Telophase I and cytokinesis: nuclear membranes reassemble around sets of sister chromatid pairs, and two daughter cells form. After the pairs of chromatids arrive at the two poles of the cell, nuclear membranes re-form, then cytokinesis occurs: the cytoplasm divides, and the cell membrane pinches the cell into two daughter cells. Each daughter cell has its own nucleus that contains the genetic material—two sister chromatids for each of the 23 chromosomes in humans.

Meiosis Division II: Separating the Sister Chromatids
There is a brief interphase after the first division of meiosis. In some organisms, the DNA molecules (now in the form of chromatid pairs) briefly uncoil and fade from view. In others, the second part of meiosis begins immediately. *It is important to note that in the brief interphase before prophase II, there is no replication of any of the chromosomes.* The second part of meiosis, like the first division of meiosis, is a four-phase process.

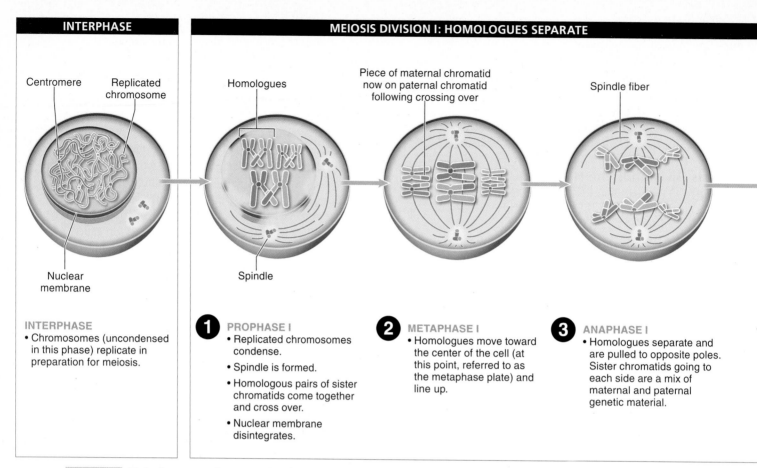

Centromere   Replicated chromosome

Homologues

Piece of maternal chromatid now on paternal chromatid following crossing over

Spindle fiber

Nuclear membrane

Spindle

INTERPHASE
• Chromosomes (uncondensed in this phase) replicate in preparation for meiosis.

1 PROPHASE I
• Replicated chromosomes condense.
• Spindle is formed.
• Homologous pairs of sister chromatids come together and cross over.
• Nuclear membrane disintegrates.

2 METAPHASE I
• Homologues move toward the center of the cell (at this point, referred to as the metaphase plate) and line up.

3 ANAPHASE I
• Homologues separate and are pulled to opposite poles. Sister chromatids going to each side are a mix of maternal and paternal genetic material.

FIGURE 6-24   Meiosis: generating reproductive cells, step by step.

5. Prophase II: chromosomes re-condense. The second division of meiosis begins with prophase II. In this phase, the genetic material in each of the two daughter cells once again coils tightly, making the pairs of chromatids visible under the microscope. (Unlike prophase I, no crossing over occurs during prophase II.)

6. Metaphase II: sister chromatid pairs line up at the center of the cell. In each of the two daughter cells, the sister chromatid pairs (each pair appearing as an X) move to the center of the cell, pulled by spindle fibers attached to the centromere, where the sister chromatids are held together. The congregation of all the genetic material in the center of each daughter cell is visible as a flat metaphase plate.

7. Anaphase II: sister chromatids are pulled to opposite sides of the cell. This phase starts with 46 pieces of DNA in each cell created by meiosis I: that is, 23 sister chromatid pairs. During this anaphase, the fibers attached to the centromere begin pulling each chromatid in the 23 sister chromatid pairs toward opposite poles of the daughter cell. When anaphase II is finished, each of what will become the four daughter cells (you could think of them as granddaughter cells!) has one single copy of each of the 23 chromosomes.

8. Telophase II and cytokinesis: nuclear membranes reassemble and the two daughter cells pinch into four haploid gametes. After the sister chromatids for all 23 chromosomes have been pulled to opposite poles, the cytoplasm divides. The cell membrane pinches each cell into two new daughter cells, nuclear membranes begin to re-form, and the process comes to a close.

In humans, the outcome of one diploid cell undergoing meiosis is the creation of four haploid daughter cells, each with just one set of 23 individual chromosomes. These chromosomes contain a combination of traits from the individual's diploid set of chromosomes.

TAKE-HOME MESSAGE 6·10

Meiosis in animals occurs only in gamete-producing cells. It is preceded by DNA replication and consists of two rounds of cellular division, one in which homologous pairs of sister chromatids separate and a second in which the sister chromatids separate. The final product of meiosis in a diploid organism is four haploid gametes.

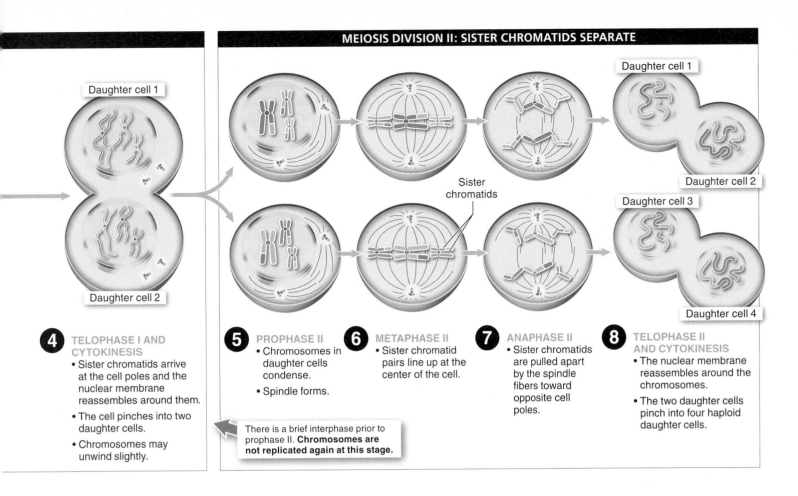

Daughter cell 1

Daughter cell 2

Sister chromatids

Daughter cell 1

Daughter cell 2

Daughter cell 3

Daughter cell 4

4 TELOPHASE I AND CYTOKINESIS
• Sister chromatids arrive at the cell poles and the nuclear membrane reassembles around them.
• The cell pinches into two daughter cells.
• Chromosomes may unwind slightly.

5 PROPHASE II
• Chromosomes in daughter cells condense.
• Spindle forms.

There is a brief interphase prior to prophase II. **Chromosomes are not replicated again at this stage.**

6 METAPHASE II
• Sister chromatid pairs line up at the center of the cell.

7 ANAPHASE II
• Sister chromatids are pulled apart by the spindle fibers toward opposite cell poles.

8 TELOPHASE II AND CYTOKINESIS
• The nuclear membrane reassembles around the chromosomes.
• The two daughter cells pinch into four haploid daughter cells.

6·11 Male and female gametes are produced in slightly different ways.

No matter how small or large they are, no matter whether they're plants or animals, for all sexually reproducing organisms there is just one way to distinguish males from females. Regardless of the species, the defining feature is always the same (and if you're thinking of anything visible to the naked eye, you're wrong). When there are two sexes—as there are in nearly every sexually reproducing animal species—the females are the sex that produces the larger gamete, and the males produce the smaller, more motile, gamete (**FIGURE 6-25**). Because all gametes in animals are produced through meiosis, it is a slight variation in how meiosis works in males and females that leads to this difference.

The female gamete is larger than the male gamete because it has more cytoplasm. During the production of sperm, the two divisions occur just as described in Section 6-10, resulting in four evenly sized cells that become sperm. During the production of eggs, things are a bit different.

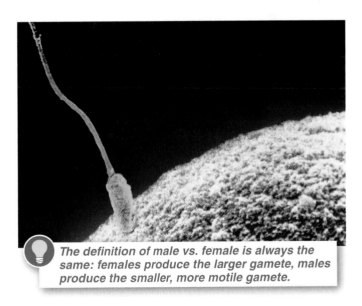

The definition of male vs. female is always the same: females produce the larger gamete, males produce the smaller, more motile gamete.

FIGURE 6-25 **Egg and sperm.** In sexually reproducing organisms, the female produces the larger gamete.

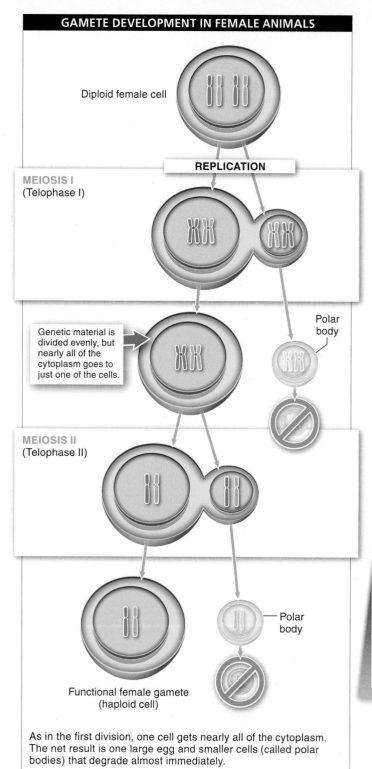

GAMETE DEVELOPMENT IN FEMALE ANIMALS

Diploid female cell

REPLICATION

MEIOSIS I
(Telophase I)

Genetic material is divided evenly, but nearly all of the cytoplasm goes to just one of the cells.

Polar body

MEIOSIS II
(Telophase II)

Polar body

Functional female gamete (haploid cell)

As in the first division, one cell gets nearly all of the cytoplasm. The net result is one large egg and smaller cells (called polar bodies) that degrade almost immediately.

FIGURE 6-26 Unequal distribution of cytoplasm results in one large egg.

The first division occurs just as described in Section 6-10, except that in telophase I, as the cell divides, the genetic material is evenly divided, but nearly all of the cytoplasm goes to one of the cells and almost none goes to the other. The smaller cell is called a **polar body** and degrades almost immediately in most animals (although it sometimes goes through a second division, after which, in most animals, both of those polar bodies degrade). Then, in the second meiotic division of the larger cell, there is again an unequal division of cytoplasm.

As in the first division, one of the new cells gets nearly all of the cytoplasm and the other gets almost none, forming another polar body. The net result of meiosis in the production of eggs is one large egg with lots of cytoplasm and two or three small polar bodies with very little cytoplasm that degrade and never function as gametes (**FIGURE 6-26**).

Ultimately, whether eggs or sperm are produced, each gamete ends up with just one copy of each chromosome. That way, the fertilized egg that results from the fusion of sperm and egg carries two complete sets of chromosomes, so the developing individual will be diploid. The extra cytoplasm carried by the egg contains a large supply of nutrients and other chemical resources to help with the initial development of the organism following fertilization.

TAKE-HOME MESSAGE 6·11

In species with two sexes—including nearly every sexually reproducing plant and animal species—females are the sex that produces the larger gamete, and males produce the smaller gamete. Whether it is male or female gametes that are being produced, each gamete ends up with just one copy of each chromosome.

6·12 Crossing over and meiosis are important sources of variation.

Genetically speaking, there are two ways to create unique individuals. The obvious way is for an organism to carry an allele that is not present in any other individuals. Alternatively—and equally successful in creating uniqueness—an individual can carry a *collection* of alleles, no single one of which is unique, that has never before occurred in another individual. Both types of novelty introduce important variation into a population of organisms. The process of crossing over, or genetic recombination (**FIGURE 6-27**), which occurs during prophase I in meiosis, creates a significant amount of the second type of variation.

As we saw in Section 6-10, during the first prophase of meiosis, the sister chromatids of homologous chromosomes all come together. Let's review exactly what it means. Let's look at just one of your homologous pairs of chromosomes, the homologous pair of chromosome 15. When we refer to the "homologous pair" of chromosome 15, remember that this pair includes two copies of chromosome 15: one copy from your mother (which you inherited from the egg that was fertilized to create you) and one copy from your father (which you inherited from the sperm that fertilized the egg). Each chromosome in the pair carries the same genes, but because they came from different people, they don't necessarily have the same alleles.

Once the sister chromatids of the homologous chromosome pairs line up (so that there are now four chromatids in two pairs lying very close together), regions that are close together can swap segments. A piece of one of the maternal chromatids—perhaps including the first 100 genes on the strand of DNA—may swap places with the same segment in a paternal chromatid. Elsewhere, a stretch of 20 genes in the middle may be swapped from the other maternal chromatid with one of the paternal chromatids. The points at which chromatids exchange genetic material during recombination are called **chiasmata** (*sing.* **chiasma**). Every time a swap of DNA segments occurs, an identical amount of genetic material is exchanged, so all four chromatids still contain the complete set of genes that make up the chromosome. The *combination* of alleles on each chromatid, though, is now different.

Suppose there are genes relating to eye color, hair color, and height on a particular chromosome. After crossing over, the linear strands still have instructions for all of those traits. But whereas one chromatid previously may have had instructions for brown eyes, brown hair, and short height, it may now have some differences—perhaps brown eyes, blond hair, and tall height. All of the alleles from your parents are still carried on one DNA molecule or another. But the *combination of traits* that are linked together on a single chromatid is new. And when a gamete, let's say it's an egg, carrying a new combination of alleles is fertilized by a sperm, the developing individual will carry a completely novel set of alleles. And so without creating new versions of any traits (such as yellow eyes or purple hair), crossing over still creates gametes with collections of alleles that may never have existed together. In Chapter 8, we'll see that this variation is tremendously important for evolution.

TAKE-HOME MESSAGE 6·12

Although it doesn't create new versions of any traits, crossing over during the first prophase of meiosis creates gametes with combinations of traits that may never have existed before; this variation is important for evolution.

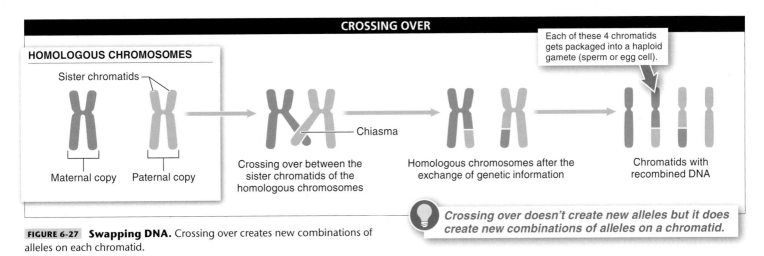

CROSSING OVER

HOMOLOGOUS CHROMOSOMES

Sister chromatids

Maternal copy   Paternal copy

Crossing over between the sister chromatids of the homologous chromosomes

Chiasma

Homologous chromosomes after the exchange of genetic information

Each of these 4 chromatids gets packaged into a haploid gamete (sperm or egg cell).

Chromatids with recombined DNA

Crossing over doesn't create new alleles but it does create new combinations of alleles on a chromatid.

FIGURE 6-27   Swapping DNA. Crossing over creates new combinations of alleles on each chromatid.

As we've seen, there are two fundamentally different ways in which cells and organisms can reproduce. On the one hand, there is mitosis and asexual reproduction through binary fission, fragmentation, or vegetative processes; on the other hand, there is meiosis and sexual reproduction. Is one method better than the other? It depends. In fact, the more appropriate question is, what are the advantages and disadvantages of each method and under what conditions do the benefits outweigh the costs?

What Are the Advantages of Sexual Reproduction? Sexual reproduction leads to offspring that are genetically different from one another and from either parent, through three different routes (**FIGURE 6-28**).

1. **Combining alleles from two parents at fertilization.** First and foremost, with sexual reproduction, a new individual comes from the fusion of gametes from two different individuals. Each of these parents comes with his or her own unique set of genetic material.

2. **Crossing over during the production of gametes.** As we saw, crossing over during prophase I of meiosis causes every chromosome in a gamete to carry a mixture of an individual's maternal and paternal genetic material.

3. **Shuffling and reassortment of homologues during meiosis.** Recall that as the homologues for each chromosome are pulled to opposite poles of the cell during the first division of meiosis, maternal and paternal homologues are randomly pulled to each pole. This means that there is an extremely large number of different combinations of maternal and paternal homologues that could end up in each gamete.

Q Bacteria reproduce asexually, whereas most plants and animals reproduce sexually. Is either method better than the other?

The variability among the offspring produced by sexual reproduction enables populations of organisms to cope better with changes in their environment. After all, if the environment is gradually changing from one generation to the next, individuals producing many genetically different offspring increase the likelihood that one of their offspring will carry a set of genes particularly suited to the new environment. Over time, populations of sexually reproducing organisms can quickly adapt to changing environments. It's like buying different lottery tickets—the more tickets you buy, the more likely it is that one of them will be a winner.

SOURCES OF GENETIC VARIATION

There are multiple reasons why offspring are genetically different from their parents and one another.

ALLELES COME FROM TWO PARENTS
Each parent donates his or her own set of genetic material.

CROSSING OVER
Crossing over during meiosis produces a mixture of maternal and paternal genetic material on each chromatid.

REASSORTMENT OF HOMOLOGUES
The homologues and sister chromatids distributed to each daughter cell during meiosis are a random mix of maternal and paternal genetic material.

FIGURE 6-28 **Creating many different combinations of alleles.** As a result of crossing over and meiosis, each individual developing from a separate fertilized egg is genetically unique.

What Are the Disadvantages of Sexual Reproduction? In the yellow dung fly (*Scathophaga stercoraria*), males sometimes wrestle each other for mating access to a female. The female awaits the outcome of the battle in a pile of dung. Occasionally, females drown in the dung pile as they wait. Males, too, can be at risk during reproduction. This is seen most dramatically in species such as praying mantises and black widow spiders, in which the female will attempt to eat the male (and often succeed in doing so) during or after mating, gaining not just gametes but a nutritious meal as well (**FIGURE 6-29**)!

Dangers associated with mating are just one of the downsides to sexual reproduction. There are several others. First, when an individual reproduces, only half of its offspring's alleles will come from that organism. The other half will come from the other parent. With asexual reproduction, an individual produces nearly identical offspring, so there is a very efficient transfer of genetic information from one generation to the next. Second, with sexual reproduction it takes time and energy to find a partner. This is energy that asexual organisms can devote to additional reproduction. Third, as we see with the dung flies, sex can be a risky proposition because organisms make themselves vulnerable to predation, disease, and other calamities during mating and reproduction. And finally, as we'll see later in this chapter, the complex cellular division required for sexual reproduction offers opportunities for mistakes, sometimes leading to chromosomal disorders.

What Are the Advantages of Asexual Reproduction?
With asexual reproduction, the advantages and disadvantages are more or less reversed. Because asexual reproduction involves only a single individual, it can be fast and easy. Some bacteria can divide, forming a new generation, every 20 minutes (**FIGURE 6-30**). And for organisms in isolated habitats or when establishing new populations, asexual reproduction can be advantageous as well. Asexual reproduction is efficient, too. Offspring carry all of the genes that their parent carried—they are genetically identical. If the environment is stable, it is beneficial for organisms to produce offspring as similar to themselves as possible.

What Are the Disadvantages of Asexual Reproduction?
The downside to asexual reproduction is that the more closely an offspring's genome resembles its

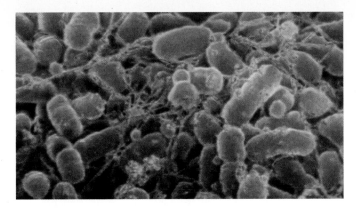

BACTERIAL POPULATION GROWTH

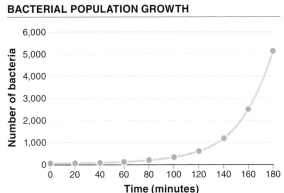

 Asexual reproduction can be fast! Some bacteria can divide every 20 minutes.

FIGURE 6-30 **Pluses and minuses.** This bacterial colony can reproduce quickly and efficiently. But because the cell division is asexual, it doesn't produce genetic variety to enable adaptation should the environment change.

parent's, the less likely it is that the offspring will be suited to the environment when it changes.

In the end, we still see large numbers of species that reproduce asexually and large numbers that reproduce sexually. It seems that conditions favoring each occur in the great diversity of habitats in the world. That we see both sexual and asexual reproduction also highlights the recurring theme in biology that there often is more than one way to solve a problem.

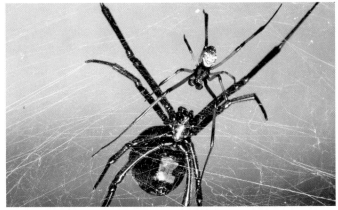

FIGURE 6-29 **Sexual reproduction has its risks.** Female black widow spiders attempt to eat the male during or after mating.

TAKE-HOME MESSAGE 6·13

There are two fundamentally different ways that cells and organisms can reproduce: (1) mitosis and asexual reproduction, and (2) meiosis and sexual reproduction. Asexual reproduction can be fast and efficient, but it leads to genetically identical offspring that carry all of the genes that their parent carried, which could be disadvantageous in a changing environment. Sexual reproduction leads to offspring that are genetically different from one another and from either parent, but it takes more time and energy and can be risky.

6·14–6·15
There are sex
differences in the
chromosomes.

XX and XY: only one of the 23 chromosome pairs determines sex in humans.

6·14 How is sex determined in humans?

In humans, the sex of a baby is determined by its father. The sequence of events involved in sex determination is instigated by one special pair of chromosomes, the sex chromosomes, which carry information that directs a growing fetus to develop as a male or as a female.

Let's take a closer look at the human sex chromosomes. We noted that there are 23 pairs of chromosomes in every somatic cell. These can be divided into two different types: 1 pair of sex chromosomes and 22 pairs of non-sex chromosomes. The human sex chromosomes are called the **X and Y chromosomes** (**FIGURE 6-31**).

How do the X and Y chromosomes differ from the other chromosomes? All of the genetic information is stored on the chromosomes in all the cells of an organism's body. But most of this information is not sex specific—that is, if you are building an eye or a neuron or a skin cell or a digestive enzyme, it doesn't matter whether it is for a male or for a female; the instructions are the same for both sexes. Some genetic information, however, instructs the body to develop into one sex or the other. That information is found on the sex chromosomes.

An individual has two copies of all the non-sex chromosomes (called autosomes). One copy is inherited from the mother, one from the father. Individuals also have two copies of the sex chromosomes, but not always two copies of the same kind. Males have one copy of the X chromosome and one copy of the Y chromosome.

Q Which parent determines a baby's sex? Why?

Females, on the other hand, don't have a Y chromosome but instead have two copies of the X chromosome.

So how does the father determine the sex of the baby? During meiosis in females, the gametes that are produced

FIGURE 6-31 X and Y: the human sex chromosomes.

carry only one copy of each chromosome. This is true for the sex chromosomes, too. So from the two copies of the X chromosome carried by females, half of the gametes end up with a copy of one of those X chromosomes while the rest of the gametes inherit the other X chromosome. Thus, every egg has an X for its one sex chromosome. During meiosis in males, the sperm that are produced also carry one copy of each chromosome, including the sex chromosomes, but in this case, half of the sperm produced inherit the X chromosome and the other half inherit the Y chromosome. At fertilization, an egg bearing a single X chromosome (and one copy of all the non-sex chromosomes) is fertilized by a sperm bearing one copy of all the non-sex chromosomes and *either* an X chromosome or a Y chromosome. When the sperm carries an X, the baby will have two X chromosomes and will develop as a female. When the sperm carries a Y, the baby will have an X and a Y and so will develop as a male (FIGURE 6-32).

Q We know that no genetic information on the Y chromosome is essential for producing a normally functioning human. Why?

Given the distribution of X and Y chromosomes into males and females, it must be true that no essential genetic information is carried on the Y chromosome. Why? Because females don't have a Y chromosome in any of their cells, yet they are able to develop and live normal, healthy lives. For this reason, we know that nothing on the Y chromosome is absolutely necessary for the development of a normally functioning human.

Physically, the X and Y chromosomes look very different from each other. The X chromosome is relatively large and carries a great deal of genetic information relating to a large number of non-sex-related traits. The Y chromosome is tiny and carries genetic information about only a very small number of traits. The genetic instructions on the Y chromosome instruct the fetal gonads to develop as testes rather than ovaries. Once this is done, very little additional genetic input from the Y chromosome is necessary. Instead, the hormones produced by the testes or ovaries generally direct the rest of the body to develop as a male or female. In the next sections we investigate some of the ramifications of males having two different sex chromosomes and look at some of the systems for sex determination that have evolved in other groups of organisms.

HOW SEX IS DETERMINED IN HUMANS

Individuals have two copies of the sex chromosomes in every cell.

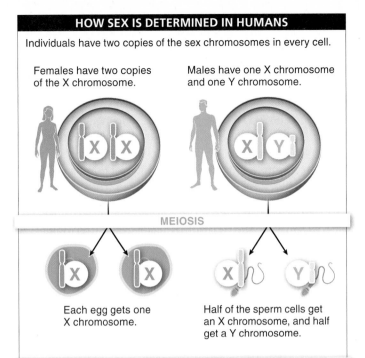

Females have two copies of the X chromosome.

Males have one X chromosome and one Y chromosome.

MEIOSIS

Each egg gets one X chromosome.

Half of the sperm cells get an X chromosome, and half get a Y chromosome.

FERTILIZATION

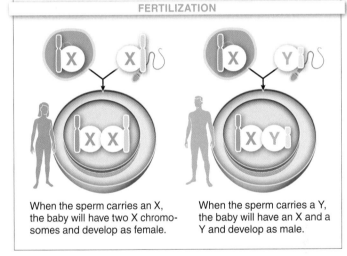

When the sperm carries an X, the baby will have two X chromosomes and develop as female.

When the sperm carries a Y, the baby will have an X and a Y and develop as male.

FIGURE 6-32 **Sex determination in humans.** The sex of offspring is determined by the father.

TAKE-HOME MESSAGE 6·14

In humans, the sex chromosomes carry information that directs a growing fetus to develop as either a male (if a Y chromosome is present) or a female (if no Y chromosome is present). Sex determination depends on the sex chromosome inherited from the father.

6·15 The sex of offspring is determined in a variety of ways in non-human species.

For something as fundamental to a species as sex determination, you might imagine that one method evolved and all species use it. The world is more diverse than that. In most plants there aren't even distinct males and females. In corn, for example, every individual produces both male and female gametes. All earthworms and garden snails are also capable of producing both male and female gametes. Such organisms are called **hermaphrodites,** because male and female gametes are produced by a single individual. But even among the species with separate males and females there are several different methods of sex determination. Humans that have both an X and a Y chromosome are male, and those that have two X chromosomes are female. This sex determination pattern is common among eukaryotes and is seen in all mammals. Sex determination differs in other organisms, including some birds, insects, and reptiles (**FIGURE 6-33**).

Birds In birds (as well as in some fish and butterfly species), the mother determines the sex in much the same way that fathers determine sex in humans. Females have one copy of two different sex chromosomes, called the Z

and the W chromosomes, while males have only one type, carrying two copies of the Z chromosome. Consequently, the sex of bird offspring is determined by the female rather than the male.

Ants, Bees, and Wasps In these insects, sex is determined by the number of chromosome sets an individual possesses. Males are haploid, having a single set of chromosomes, and females are diploid, carrying two sets of chromosomes. In this method, females produce haploid eggs by meiosis. They mate with males and store the sperm in a sac. As each egg is produced, the female can fertilize it with sperm she has stored, in which case the offspring will be diploid and female. Alternatively, she can lay the unfertilized egg, which develops into a haploid, male individual. Just think, in these species, males don't have a father, yet they do have a grandfather!

Turtles In some species, sex determination is controlled by the environment rather than by the number or types of chromosomes an organism has. In most turtles, for

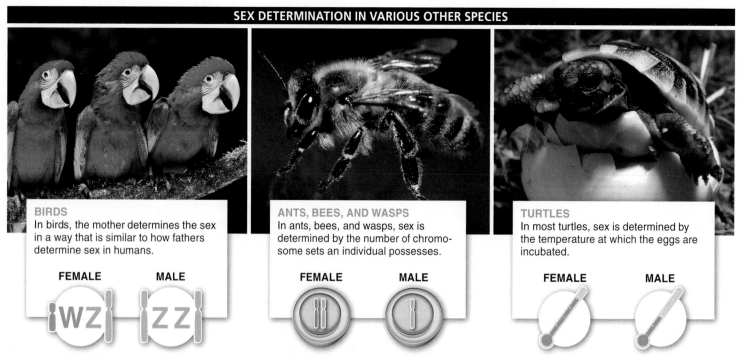

SEX DETERMINATION IN VARIOUS OTHER SPECIES

BIRDS
In birds, the mother determines the sex in a way that is similar to how fathers determine sex in humans.

FEMALE MALE

WZ ZZ

ANTS, BEES, AND WASPS
In ants, bees, and wasps, sex is determined by the number of chromosome sets an individual possesses.

FEMALE MALE

TURTLES
In most turtles, sex is determined by the temperature at which the eggs are incubated.

FEMALE MALE

FIGURE 6-33 **Other methods of sex determination.** In some species, sex is determined by the type of chromosomes, the number of chromosomes, or incubation temperature of the eggs.

example, offspring's sex is determined by the temperature at which the eggs are kept. Eggs kept relatively hot during incubation become females, while eggs incubated at cooler temperatures become males. The sex of some lizards and crocodiles is also determined this way.

TAKE-HOME MESSAGE 6·15

A variety of methods are used for sex determination across the world of plant and animal species. These include the presence or absence of sex chromosomes, the number of chromosome sets, and environmental factors such as incubation temperature.

? 6·16 THIS IS HOW WE DO IT

Can the environment determine the sex of a turtle's offspring?

Unexpected observations may be a sign that our ideas about how the world works are not quite right. As such, they provide a great opportunity for scientists to solve a problem. Figuring out how to approach the problem, however, can be like solving a puzzle, and the solution can have far-reaching implications and unexpected importance.

In 1966, Madeline Charnier reported a surprising observation in an obscure publication from West Africa. For a lizard species she observed, it seemed that the sex ratio of the offspring produced was influenced by the environment. In warmer temperatures, most of the eggs that hatched contained females. And in cooler temperatures, most of the hatched eggs contained males.

As members of a species in which our sex is determined by the *chromosomes* we inherit from our parents, it isn't surprising that we'd assume that sex determination would be the same for all animals. Charnier's observation, however, which spurred biologists to notice similar patterns of sex determination in other reptiles and some amphibian species, suggested that our understanding of sex determination wasn't quite right.

Can you propose how to test the "incubation temperature determines sex" hypothesis?

To test whether incubation temperature really could influence offspring sex, researchers conducted a laboratory experiment. They set up incubators at two different temperatures: one cool (25° C) and one warm (30.5° C). They then collected turtle eggs. From each clutch of eggs, they put half of the eggs in the cool

incubator and half in the warm incubator. As the eggs hatched, the results were as clear as could be. At 25° C, all 210 offspring produced were males. At 30.5° C, all 211 offspring produced were females. The researchers concluded that—at least in the lab—the temperature could influence the sex of the offspring.

When it comes to temperature, how does a lab differ from turtles' natural environment?

In the initial lab experiment, the temperature was constant. In the turtles' natural habitat, it fluctuates every day: it is cool at night and warmer in the day. To better approximate natural conditions, the researchers conducted two additional experiments. First, in the laboratory incubators they created fluctuating temperatures. In one, the temperature fluctuated from 20° to 30° C and back again each day. In the other, the temperature fluctuated similarly, but between 23° and 33° C. The results were identical to the first experiment. In the cooler incubator, 100% of the offspring hatching were male. In the warmer incubator, 100% were female.

In the final study, the eggs collected from each clutch were again divided evenly, but this time they were incubated in the turtles' natural habitat. Half were buried at a shaded nesting site that received little sun exposure (and rarely exceeded 30° C). The other half were buried at a nesting site that was exposed to the sun (and often exceeded 30° C). Of the 100 eggs hatching at the shaded site, all were males. Of the 127 eggs hatching at the exposed site, 123 were females. Based on these results, would you be confident that incubation temperature does indeed influence the sex of the offspring? Can you think of any alternative explanations?

> **Is temperature-dependent sex determination just a curiosity of nature or are there any important implications for humans?**

Based on models of climate change, it is likely that the average temperature in North America will increase by about 4° C over the next 100 years. Scientists have predicted that such a rapid change could have a significant impact on biological systems. There has been relatively little direct empirical evidence for such impacts, however. One researcher, though, decided to test this prediction using temperature-dependent sex determination in turtles.

> **How could we evaluate whether climate change might have significant biological consequences?**

Frederic Janzen monitored 390 nests of painted turtles over a five-year period. In each year, he identified every nest on an island in the Mississippi River. He also noted the average temperature during the month that the eggs were incubating. After the eggs hatched, he brought the turtles into the lab, where he could determine their sex.

Janzen discovered a strong linear relationship between the average temperature and the offspring sex ratio.

Average Temperature (° C)	Sex Ratio (% males)
21.0	100
23.0	90
23.6	33
24	40
25.2	0

He concluded that an increase of even a few degrees could result in the extinction of the painted turtle as a result of producing no males. He further suggested that this was just one possible dramatic impact of rapid climate change and that it might be a bellwether for broader biological consequences.

TAKE-HOME MESSAGE 6·16

Observations of some lizard and turtle species reveal that sex determination is influenced by the temperature during incubation of the eggs. Both in the lab and in natural habitats, at cooler temperatures more males develop, and at warmer temperatures more females develop. Climate change is predicted to bring an increase of about 4° C in North America in the next 100 years. This increase is likely to have adverse effects on turtles with temperature-dependent sex determination and may signal far greater impacts of climate change on biological systems.

6·17–6·18
Deviations from the normal chromosome number lead to problems.

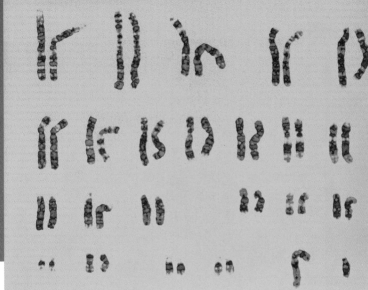

Karyotype displaying the chromosome pairs of a normal human male.

6·17 Down syndrome can be detected before birth: karyotypes reveal an individual's entire chromosome set.

Reproduction becomes riskier for women as they become older. Increasingly, their gametes contain incorrect numbers of chromosomes or chromosomes that have been damaged (**FIGURE 6-34**). These problems can lead to effects on the offspring that range from minor to fatal.

To find out whether their offspring may have a disorder associated with incorrect numbers of chromosomes, parents can request a quick test for some common genetic problems even before their baby is born. This information is available from an analysis of an individual's **karyotype,** a visual display

of the complete set of chromosomes. A karyotype can be made for adults or children, but it is most commonly done for a fetus. A karyotype is a useful diagnostic tool because it can be prepared very early in the fetus's development to assess whether it has an abnormality in the number of chromosomes or in their structure. And because the test shows all of the chromosomes, even the sex chromosomes, it also reveals the sex of the fetus.

Preparing a karyotype takes five steps. (1) Some cells must be obtained from the individual. (2) The cells are then encouraged to divide by culturing them in a test tube with nutrients. (3) After a few days to two weeks of cell division, the cells are treated with a chemical that stops them exactly midway through cell division—a time when the chromosomes are coiled thickly and are more visible than usual. (4) The cells are then placed on a microscope slide, and a stain is added that binds to the chromosomes, making them more easily visible. (5) Finally, the chromosomes are arranged by size and shape and displayed on a monitor (**FIGURE 6-35**).

The first step—obtaining cells to prepare the karyotype—is relatively easy in an adult or child. Usually, cells are collected from a small blood sample (red blood cells do not have a nucleus or chromosomes, but white blood cells do). Collecting cells from a fetus, on the other hand, poses a special problem. Two different methods for collecting cells have been developed.

1. Amniocentesis. This procedure can be done approximately three to four months into a pregnancy (**FIGURE 6-36**). In one quick motion (and without the use of anesthetic), a 3- to 4-inch needle is pushed through the abdomen, through the

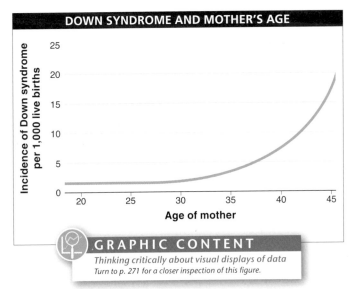

DOWN SYNDROME AND MOTHER'S AGE

Incidence of Down syndrome per 1,000 live births

25
20
15
10
5
0

20   25   30   35   40   45

Age of mother

GRAPHIC CONTENT
Thinking critically about visual displays of data
Turn to p. 271 for a closer inspection of this figure.

FIGURE 6-34   Reproduction risks and maternal age. Down syndrome incidence increases as the age of the mother increases.

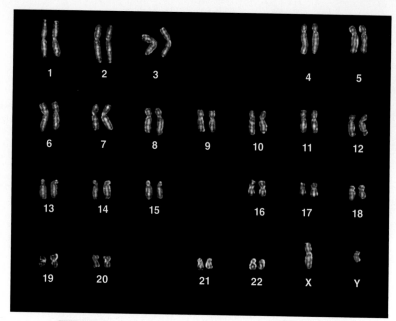

FIGURE 6-35 **A human karyotype.** A visual display of a complete set of chromosomes.

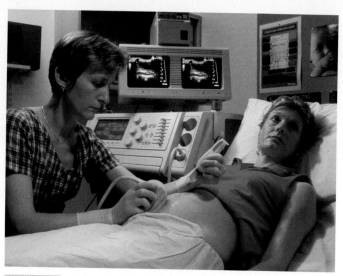

FIGURE 6-36 **Amniocentesis.** Fluid surrounding a fetus (and containing some of its cells) is extracted and analyzed to determine the genetic composition of the developing fetus.

amniotic sac, and into the amniotic fluid that surrounds and protects the fetus. Using ultrasound for guidance, the doctor aims for a small pocket of fluid as far as possible from the fetus and withdraws about 2 tablespoons of fluid. This fluid contains many cells from the fetus, which can then be used for karyotype analysis. Any chromosomal abnormalities in the fetus will be present in these cells—and many of them can be seen in the karyotype.

2. Chorionic villus sampling (CVS). In this procedure, rather than sampling cells from the amniotic fluid, a small bit of tissue is removed from the **placenta,** the temporary organ that allows the transfer of gases, nutrients, and waste products between a mother and fetus. A needle is inserted

either through the abdomen or through the vagina and cervix, again using ultrasound for guidance. Then a small piece of the finger-like projections from the placenta is removed by the syringe. Because much of the placenta develops from the fertilized egg, much of the placenta consists of cells with the same genetic composition as the embryo. The chief advantage of CVS over amniocentesis is that it can be done several weeks earlier in the pregnancy, usually between the 10th and 12th weeks.

The resulting karyotype, whether from amniocentesis or CVS, will reveal whether the fetus carries an extra copy of any of the chromosomes or lacks a copy of one or more chromosomes. Of all the chromosomal disorders detected by karyotyping, Down syndrome is the most commonly observed. Named after John Langdon Down, the doctor who first described it, in 1866, the syndrome is revealed by the presence of an extra copy of chromosome 21 (FIGURE 6-37). (For this reason, the condition that causes Down syndrome is also called "trisomy 21.") Striking about 1 in every 1,000 children born, Down syndrome is characterized by a suite

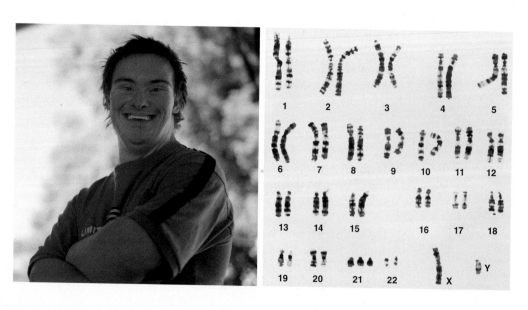

FIGURE 6-37 **Trisomy 21.** An extra copy of chromosome 21 causes Down syndrome.

of physical and mental characteristics that includes learning disabilities, a flat facial profile, heart defects, and increased susceptibility to respiratory difficulties.

Down syndrome and other disorders caused by a missing chromosome or an extra copy of a chromosome are a consequence of **nondisjunction,** the unequal distribution of chromosomes during meiosis. Nondisjunction can occur at two different points in meiosis: homologues can fail to separate during meiosis I, or sister chromatids can fail to separate during meiosis II (**FIGURE 6-38**). In both cases, nondisjunction results in an egg or sperm with zero or two copies of a chromosome rather than a single copy. Any of the chromosomes can fail to separate during cell division, but the ramifications of trisomy (having an extra copy of one chromosome in every cell) are greater for chromosomes with larger numbers of genes.

When trisomy occurs for chromosomes with greater numbers of genes, the likelihood that the developing embryo will survive to birth is reduced. Consequently, we tend to see cases of trisomy that involve only the chromosomes with the fewest genes, such as chromosomes 13, 15, 18, 21, and 22. In fact, observations show that trisomy 1 is never seen (all such fertilized eggs die before implantation in the uterus), trisomy 13 occurs in 1 in 20,000 newborns (and most die soon after birth), while trisomy 21, as we've seen, occurs in 1 in 1,000 newborns (many of whom live long lives).

Q Why do older women have more babies with Down syndrome?

As we mentioned at the beginning of this section, as women become older there are increased problems associated with reproduction. Older women, for example, have more babies with Down syndrome. Why does the risk of having a baby with Down syndrome or some other disorder that results from trisomy increase with increasing age for women, but significantly less so for men? As women age, their gametes tend to have more errors. The reason is that those eggs began meiosis near the time the woman was born, and they may not complete it until she is 40 or more years old. During those decades, the cells may develop problems—including a reduction in the size and stability of the spindle fibers and reduced cohesion between sister chromatids—that interfere with normal cell division. In men, the cells that undergo meiosis are relatively young because new sperm-producing cells are produced every couple of weeks after puberty.

Whereas lacking a non-sex chromosome or having an extra chromosome usually has serious consequences for health, we'll see in the next section that lacking or having an extra X or Y chromosome has consequences that are much less severe.

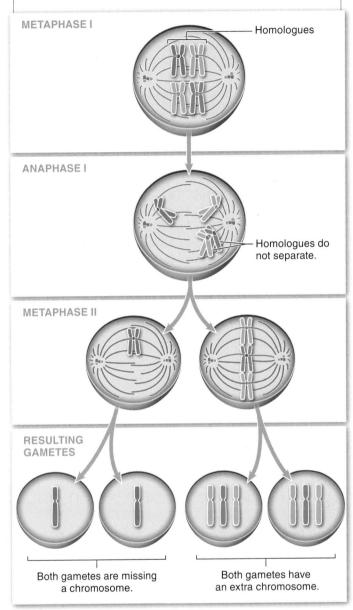

NONDISJUNCTION

Nondisjunction is the unequal distribution of chromosomes during meiosis. The resulting gametes have zero or two copies of a chromosome, rather than a single copy.

METAPHASE I — Homologues

ANAPHASE I — Homologues do not separate.

METAPHASE II

RESULTING GAMETES

Both gametes are missing a chromosome.

Both gametes have an extra chromosome.

FIGURE 6-38   **An error during meiosis produces gametes with too many or too few chromosomes.**

TAKE-HOME MESSAGE 6·17

A karyotype is a visual display of a complete set of chromosomes. It is a useful diagnostic tool because it can be prepared early in fetal development to assess whether there is an abnormality in the number of chromosomes or in their structure, such as in Down syndrome. Down syndrome is caused by having an extra copy of chromosome 21.

Life is possible with too many or too few sex chromosomes.

It is usually fatal to have one too many or one too few of the non-sex chromosomes. When it comes to the sex chromosomes, however, the situation is less extreme. Many individuals are born lacking one of the sex chromosomes or with an additional X or Y chromosome, and they usually survive. We can get a glimpse into the role of the sex chromosomes by looking at the physical and mental consequences of having an abnormal number of these chromosomes. In each of the cases below, the condition is caused by nondisjunction of the sex chromosomes during the production of sperm or eggs (**FIGURE 6-39**).

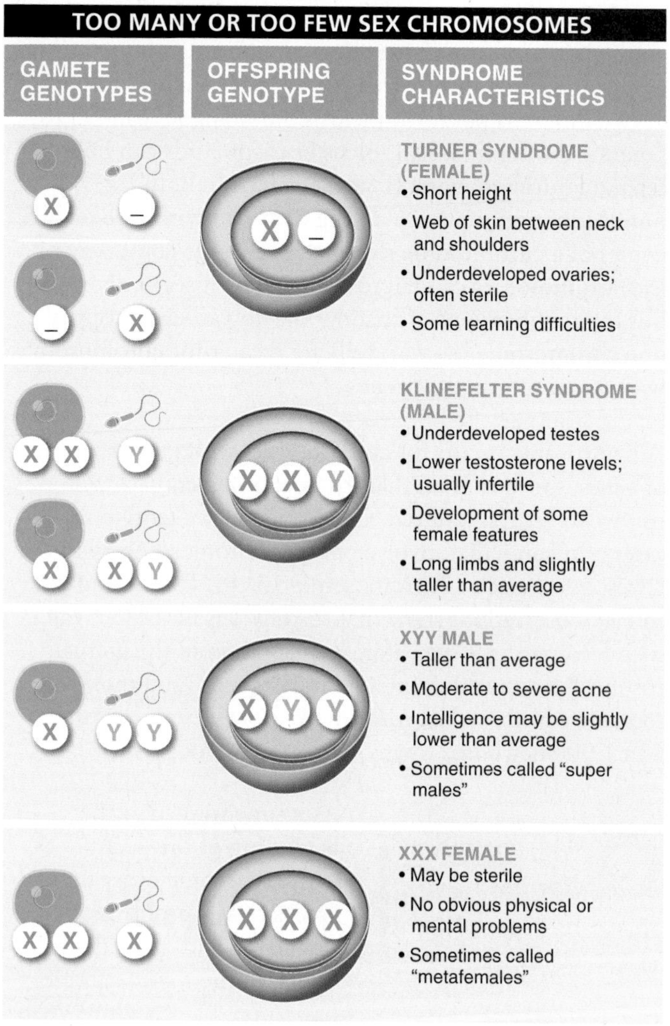

Turner Syndrome: X Approximately 1 in 5,000 females carry only one X chromosome (and no Y chromosome), exhibiting a condition called Turner syndrome, denoted as X_ (or sometimes XO). This is the only condition in humans in which a person can survive without one of a pair of chromosomes. Instead of having 46 chromosomes in every cell, these individuals have only 45. As common as Turner syndrome is, in 98% of the fertilized eggs in which this condition occurs, the egg is spontaneously aborted long before a fetus can come to term.

There are both physical and mental consequences of the absence of a second sex chromosome.

- Women with Turner syndrome are usually relatively short, averaging 4 feet 8 inches in height.

- They develop a web of skin between the neck and shoulders.

- The ovaries never fully mature, so the women are almost always sterile.

- The breasts and other secondary sex characteristics develop incompletely.

- Intelligence is usually normal, but some learning difficulties are common.

Klinefelter Syndrome: XXY An individual who has two X chromosomes is female; an individual with an X and a Y is male. But what happens when both of those conditions exist? An individual who carries two X chromosomes and a Y chromosome, a condition known as Klinefelter syndrome, develops as a male. This is because, as we saw earlier, the Y

TOO MANY OR TOO FEW SEX CHROMOSOMES

GAMETE GENOTYPES	OFFSPRING GENOTYPE	SYNDROME CHARACTERISTICS
X / _	X _	**TURNER SYNDROME (FEMALE)** • Short height • Web of skin between neck and shoulders • Underdeveloped ovaries; often sterile • Some learning difficulties
XX / Y X / XY	XXY	**KLINEFELTER SYNDROME (MALE)** • Underdeveloped testes • Lower testosterone levels; usually infertile • Development of some female features • Long limbs and slightly taller than average
X / YY	XYY	**XYY MALE** • Taller than average • Moderate to severe acne • Intelligence may be slightly lower than average • Sometimes called "super males"
XX / X	XXX	**XXX FEMALE** • May be sterile • No obvious physical or mental problems • Sometimes called "metafemales"

FIGURE 6-39 **Characteristics of individuals with too many or too few sex chromosomes.**

chromosome carries genetic instructions that cause fetal gonads to develop as testes; if these instructions are absent, the fetal gonads develop as ovaries. The extra X chromosome, however, does cause Klinefelter males to be somewhat feminized, although this effect can be reduced through treatments such as testosterone supplementation. Approximately 1 in 1,000 males have the genotype XXY, making this one of the most common genetic abnormalities in humans.

Klinefelter syndrome has some physical and mental consequences.

> **Q** If a person has two X chromosomes but also has a Y chromosome, is the individual male or female?

- Men with this syndrome have testes, but they are smaller than average. Because the testes are small, levels of testosterone are low and the men are almost always infertile.

- They develop some female features, including reduced facial and chest hair and some breast development.

- They have long limbs and are slightly taller than average (about 6 feet on average).

- They learn to speak at a later age than average and tend to have language impairments.

- Some individuals have further additional X chromosomes, with the karyotypes XXXY or even XXXXY. These males also exhibit Klinefelter syndrome but more frequently have mental retardation.

A hermaphrodite is an individual with functioning male and female reproductive organs capable of producing both male and female gametes. Often it is mistakenly assumed that a person with Klinefelter syndrome must be a hermaphrodite because he has both two X chromosomes (which would usually make an individual a female) and an X and a Y chromosome (usually making the individual a male). Hermaphroditism is common among invertebrates and occurs in some fish and other vertebrates, but contrary to urban legends, human hermaphrodites do not exist. Some men with Klinefelter syndrome may have some features of the opposite sex; however, they do not produce female gametes and so are not hermaphrodites. Individuals with sex characteristics that preclude definitive identification as male or female are called "intersex."

Q Do human hermaphrodites exist?

XYY Males There is no official name for the condition in which an individual has one X chromosome and two Y chromosomes, although such individuals are sometimes referred to as "super males." This chromosomal abnormality occurs in approximately 1 in 1,000 males (and in about 1 in 325 males who are 6 feet or taller). There are no distinguishing features at birth to indicate that an individual carries an extra Y chromosome, and the vast majority live their lives normally without knowing they have an extra chromosome in every cell.

Several consequences of the XYY condition have been well documented.

- XYY males are relatively tall (about 6 feet 2 inches on average).

- They tend to have moderate to severe acne.

- Their intelligence usually falls within the same range as that of XY males, although the average may be slightly lower.

Because, in the United States, about three to five times as many XYY males are found in prisons (as a percentage of the prison population) as in the population as a whole, there has been tremendous controversy over whether the extra Y chromosome predisposes individuals to criminal behavior. These observations were not part of randomized, controlled, double-blind experiments, however, so they are open to numerous alternative interpretations.

In one study, for example, researchers identified all male Danish citizens born in Copenhagen between 1944 and 1947 and obtained sex chromosome determinations for the 4,139 of these men who were taller than 6 feet 0.4 inches (184 cm). From this sample, they found that the level of intellectual functioning (as indicated by scores on an army selection test taken by the men) was significantly lower among the XYY males than among XY males. Based on this finding, they suggested that the increased representation of XYY males in prison was more likely to be due to the fact that males of lower intelligence are more likely to be apprehended, regardless of which chromosomes they carry. The controversy remains unresolved, and the vast majority of XYY males are not in prison.

Although an error during meiosis in either the father or the mother can cause Klinefelter or Turner syndrome, it is only an error in meiosis in a male that can give rise to an XYY child. Why? Because only males carry Y chromosomes. Consequently, an XYY child must receive both of his Y chromosomes from his father. This means that the father's sperm cell must have contained two copies of the Y, a result of nondisjunction during meiosis in the father.

XXX Females Sometimes called "metafemales," individuals with three X chromosomes occur at a frequency of about 1 in 1,000 women. Very few studies of this condition have been completed, although initial observations suggest that some XXX females are sterile but otherwise have no obvious physical or mental problems.

TAKE-HOME MESSAGE 6·18

Although it is usually fatal to have one too many or one too few of the non-sex chromosomes, individuals born with only a single sex chromosome that is an X, or with an additional X or Y chromosome, usually survive—though often with physical and/or mental problems.

Can You Select the Sex of Your Baby?
(Would You Want To?)

Q: **Given our knowledge of the behavior of the X and the Y chromosomes, is it more likely that these sex-selection techniques involve manipulations of sperm or of eggs? Why?** Because all eggs have X chromosomes, the X chromosome doesn't play a role in determining the sex of a baby. Instead, the fertility clinics must somehow sort and separate sperm cells based on whether they carry an X or a Y chromosome (men produce approximately equal numbers of each).

Q: **What feature of sperm must be used in separating them?** This process must be based on a way of distinguishing between sperm carrying an X chromosome and sperm carrying a Y chromosome. After the two types of sperm are distinguished, they must then be separated. Sperm with the desired sex chromosome are then used to fertilize the woman's egg (which may be done within her body, using artificial insemination, or in a Petri dish, after which a fertilized embryo is transferred into the woman's reproductive tract).

Q: **What techniques might make it possible to separate sperm with an X chromosome from those**
with a Y chromosome? One method involves determining, by weighing sperm, which cells have more DNA. The heavier sperm must be carrying the X rather than the Y chromosome because the X is so much larger. Another method is to add a fluorescent dye to sperm that temporarily attaches to DNA. Because sperm with an X chromosome have more DNA, more of the dye attaches to them and they are more fluorescent. A machine then sorts the sperm one by one, and at ovulation, insemination is performed using the sperm with the desired sex chromosome.

Q: **Does it work?** The technique is still fairly new, but initial results suggest that the procedures have a success rate of 70% to 90%.

Q: **Concerns for you to ponder.** What are the ethical concerns raised by selecting the sex of one's baby? Are there some circumstances in which such a selection would be more acceptable than in others? Would it be acceptable to select for a baby with one eye color versus another? Why would that process be more difficult than selecting for sex?

Check Your Knowledge

Thinking critically about visual displays of data

1. According to this graph, what is the probability (i.e., 1 in 800, or 1 in 256, etc.) that a baby born to a 35-year-old woman will have Down syndrome?

2. Is a woman of age 45 who gives birth more likely than not to have a baby with Down syndrome? What is the exact probability that her baby will not have Down syndrome?

3. From this graph, can you conclude that most babies with Down syndrome are born to women older than 35?

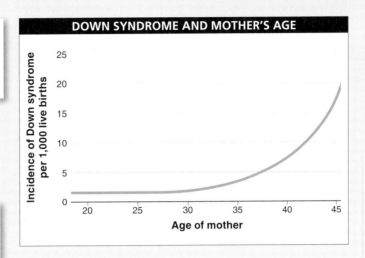

DOWN SYNDROME AND MOTHER'S AGE

(Graph: Incidence of Down syndrome per 1,000 live births vs. Age of mother)

4. Can you determine from the graph what the chances of having a baby born with Down syndrome is after age 45? What factors might have been responsible for the graph not showing data for women older than 45?

5. Among woman aged 35 and older only, the incidence of fetuses produced that have an extra copy of chromosome 21 may be higher than the graph indicates. Why?

See answers at the back of the book.

Key Terms in Chromosomes and Cell Division

anaphase, p. 245
apoptosis, p. 243
asexual reproduction, pp. 235, 250
binary fission, p. 234
cancer, p. 246
cell cycle, p. 236
cell-cycle control system, p. 237
centriole, p. 245
centromere, pp. 237, 244
checkpoints, p. 237
chiasmata (*sing.* chiasma), p. 257
chromatid, p. 244
chromatin, p. 234

chromosome, p. 234
complementarity, p. 238
complementary base, p. 238
condensation, p. 244
crossing over, p. 253
cytokinesis, p. 237
daughter cell, p. 235
diploid, p. 250
discontinuous replication, p. 241
DNA helicase, p. 239
DNA polymerase, p. 240
DNA synthesis (S phase), p. 237
fertilization, p. 250
gamete, p. 236

G_0, p. 237
Gap 1 (G_1), p. 237
Gap 2 (G_2), p. 237
genetic recombination, p. 253
gonad, p. 252
growth factors, p. 237
haploid, p. 250
hermaphrodite, p. 262
histone, p. 234
homologous pair (homologues), p. 252
interphase, p. 236
karyotype, p. 265
kinetochore, p. 245
lagging strand, p. 241

leading strand, p. 240
meiosis, p. 250
metaphase, p. 245
microtubules, p. 245
mitosis, p. 237
mitotic phase (M phase), p. 236
nondisjunction, p. 267
origin of replication, p. 239
parent cell, p. 235
placenta, p. 266
polar body, p. 256
primase, p. 240
prophase, p. 244
random assortment, p. 253

recombination, p. 253
replication, p. 234
replication fork, p. 239
reproductive cell, p. 236
sexual reproduction, p. 250
sister chromatids, p. 244
somatic cell, p. 236
spindle, p. 245
spindle fiber, p. 245
telomere, p. 232
telophase, p. 244
X and Y chromosomes, p. 260

ABOUT THE CHAPTER OPENING PHOTO

Sibling similarities—such as those seen among the Jackson brothers (shown here performing in 1973)—illustrate the phenotypic consequences of genetic similarity; but the subtle differences also reveal variability among offspring in sexually reproducing species. The phenotypic variation stems from a variety of sources, including crossing over, the shuffling and reassortment of homologues during meiosis, and the contributions from two parents.

6·1–6·4 There are different types of cell division.

Cell division is an ongoing process in most organisms and their tissues; disruptions to normal cell division can have serious consequences.

? Check Your Knowledge

1. Which of the following statements about telomeres is incorrect?

   a) They function like a counter, keeping track of how many times a cell has divided.

   b) At birth, they are long enough to permit approximately 80–90 cell divisions in most cells.

   c) They are slightly shorter in prokaryotic cells than in eukaryotic cells.

   d) They function like a protective cap on chromosomes.

   e) They contain no critical genes.

   0 ——————62———— 100
   EASY   HARD

2. Prokaryotic cells can divide by:

   a) mitosis.

   b) binary fission.

   c) meiosis.

   d) both mitosis and binary fission.

   e) none of the above.

   0 ——31——————— 100
   EASY   HARD

3. In multicellular organisms, cells that undergo mitotic division but not meiotic division are called _____ cells.

   a) somosis

   b) skin

   c) interphase

   d) somatic

   e) germ

   0 ——22——————— 100
   EASY   HARD

4. DNA replication is facilitated by the fact that the base on one strand of the double helix always has the same partner on the other strand. This feature of DNA is called:

   a) complementarity.

   b) duplication.

   c) cytokinesis.

   d) transcription.

   e) redundancy.

   0 ———35————— 100
   EASY   HARD

TELOMERES

In eukaryotic cells, a protective section of DNA called the telomere, at each end of every chromosome, plays a role in keeping track of cell division, getting shorter every time the cell divides.

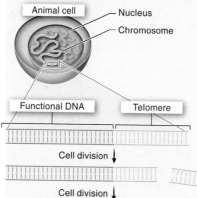

Animal cell — Nucleus
— Chromosome

Functional DNA   Telomere

Cell division ↓

Cell division ↓

Cell division ↓

Cell death

If telomeres become too short, additional cell divisions cause the loss of essential DNA and cell death. Cells that rebuild their telomeres with each division can become cancerous.

? 1. Each time a cell divides, the telomere gets a little shorter. If it gets too short, this can interfere with genes and lead to cell death. Why would a therapy that helped cells rebuild telomeres not be the answer to the quest for "eternal youth"?

PROKARYOTIC AND EUKARYOTIC CHROMOSOMES

In most prokaryotes, genetic information is carried in a single, circular DNA strand (chromosome). Eukaryotes have much more DNA, organized into linear chromosomes in the nucleus.

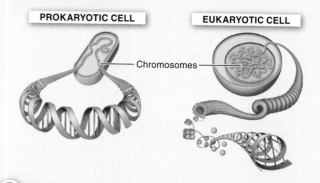

PROKARYOTIC CELL   EUKARYOTIC CELL

— Chromosomes —

? 2. Compare and contrast the genetic information found in prokaryotes and eukaryotes.

PROKARYOTIC CELL DIVISION: BINARY FISSION

Bacteria divide by a type of asexual reproduction called binary fission.

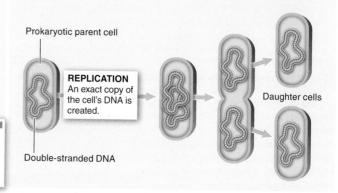

Prokaryotic parent cell

REPLICATION
An exact copy of the cell's DNA is created.

Daughter cells

Double-stranded DNA

EUKARYOTIC CELL CYCLE

In the cell cycle, eukaryotic somatic cells alternate between division, called the mitotic phase, and other cell activities, called interphase, which consists of two gap phases separated by a DNA synthesis phase.

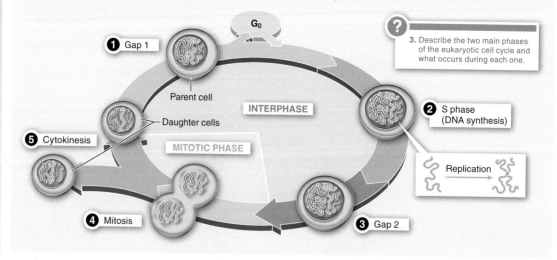

G₀

① Gap 1

Parent cell

INTERPHASE

Daughter cells

⑤ Cytokinesis

MITOTIC PHASE

④ Mitosis

③ Gap 2

② S phase (DNA synthesis)

Replication

? 3. Describe the two main phases of the eukaryotic cell cycle and what occurs during each one.

DNA COMPLEMENTARITY

Complementary base pairing makes it possible to produce two identical strands by separating the parent molecule and using each strand as a template to build a new complementary strand.

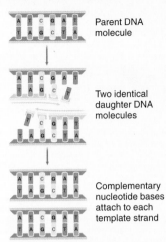

Parent DNA molecule

Two identical daughter DNA molecules

Complementary nucleotide bases attach to each template strand

DNA REPLICATION

Every time a cell divides, its DNA must first replicate so that each of the two new cells has all the genetic material of the parent cell.

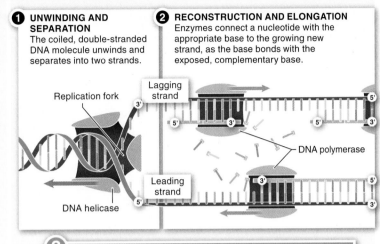

1 UNWINDING AND SEPARATION
The coiled, double-stranded DNA molecule unwinds and separates into two strands.

2 RECONSTRUCTION AND ELONGATION
Enzymes connect a nucleotide with the appropriate base to the growing new strand, as the base bonds with the exposed, complementary base.

Replication fork

Lagging strand

DNA polymerase

Leading strand

DNA helicase

4. Every time a cell divides, why must the DNA in the cell make a copy of itself?

---

5. Mitosis results in:

a) daughter cells with twice as much genetic material as the parent cell and a unique collection of alleles.

b) eight daughter cells.

c) daughter cells with the same number and composition of chromosomes.

d) gametes.

e) genetically varied offspring.

0 EASY — 45 — 100 HARD

6. Using a light microscope, it is easiest to see chromosomes:

a) during mitosis and meiosis, because the condensed chromosomes are thicker and therefore more prominent.

b) during interphase, when they are concentrated in the nucleus.

c) in the mitochondria, because the chromosomes are circular.

d) during asexual reproduction.

e) during interphase, because they are uncoiled and more linear.

0 EASY — 40 — 100 HARD

7. The division of the cytoplasm during cell division is called:

a) cytoplasm splicing.

b) cytokinesis.

c) vegetative growth.

d) cytodivision.

e) hybridization.

0 — 9 — 100
EASY HARD

8. Which of the following statements about tumors is incorrect?

a) Benign tumors pose less of a health risk than malignant tumors.

b) Malignant tumors shed cancer cells that can spread in the body.

c) Tumors are caused by excessive cell growth and division.

d) Malignant tumor cells can travel to other parts of the body in a process called metastasis.

e) Tumors contain cells with abnormally high contact inhibition.

0 EASY — 48 — 100 HARD

6·5–6·8 Mitosis replaces worn-out old cells with fresh new duplicates.

In mitosis, cells generate new, genetically identical cells, enabling organisms to grow and replace cells.

MITOSIS

Mitosis occurs in four steps, following replication of the chromosomes, to produce two genetically identical daughter cells from one parent cell.

5. What is the main purpose of mitosis?

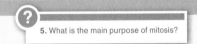

Parent cell

MITOSIS

Daughter cells

INTERPHASE
Chromosomes replicate in preparation for mitosis.

PROPHASE
The nuclear membrane breaks down and sister chromatids condense.

METAPHASE
Sister chromatids line up at the center of the cell.

ANAPHASE
Sister chromatid pairs are pulled apart by the spindle fibers.

TELOPHASE
Chromosomes uncoil and the cell begins to pinch in two.

CYTOKINESIS
Cytoplasm is divided into two identical daughter cells.

6. Describe the two events that must occur before mitosis begins.

7. Describe the phase of the cell cycle during which the events make it possible to end up with sufficient genetic material for two identical daughter cells.

CANCER

Cancer is unrestrained cell growth and division, leading to large masses of cells (tumors). Treatment focuses on killing the cells or slowing down cell growth and division. Cancer cells have several features that distinguish them from normal cells.

Cancer cells have no contact inhibition.

Cancer cells divide indefinitely.

Cancer cells have reduced "stickiness."

8. Describe three ways in which cancer cells differ from "typical" cells. Why is it that when someone has cancer, the properties of the cancer cells are not a direct cause of death?

9. During meiosis but not during mitosis:

a) haploid gametes are produced that are identical in their allelic composition.

b) the cytoplasm divides.

c) chromosomes line up in the center of the cell during metaphase.

d) genetic variation among the daughter cells is increased.

e) two identical daughter cells are produced.

10. Sister chromatids are:

a) the result of crossing over.

b) identical molecules of DNA resulting from replication.

c) homologous chromosomes.

d) produced in meiosis but not in mitosis.

e) single-stranded.

11. Which of the following is the best way to distinguish male from female?

a) Males are larger.

b) Males are more brightly colored.

c) Males produce motile gametes.

d) Males are more aggressive.

e) All of the above statements are correct.

12. A potential disadvantage of asexual reproduction is that:

a) it increases the time required to find a mate.

b) it allows perpetuation of a population even when the members are isolated.

c) it allows population size to increase rapidly.

d) it produces genetically uniform populations.

e) it requires additional rounds of meiosis, which is energetically much more costly than mitosis.

13. A karyotype of one of your skin cells would reveal a total of 46 chromosomes. How many of these are maternally inherited non-sex chromosomes?

a) 23

b) 20

c) 46

d) 22

e) 24

6.9–6.13 Meiosis generates sperm and eggs and a great deal of variation.

Meiosis occurs only in gamete-producing cells of sexually reproducing organisms.

SEXUAL REPRODUCTION

Diploid organisms produce haploid gametes that fuse at fertilization and return to the diploid state.

9. Why do the gametes produced by meiosis have only half the genetic material of the parent cell?

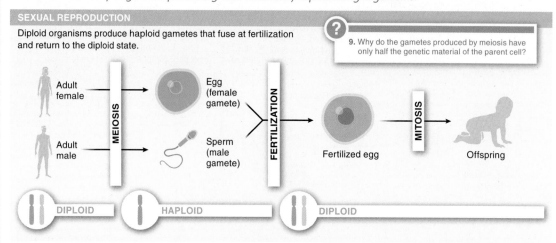

MEIOSIS

Meiosis consists of two rounds of cell division: separation of homologous pairs of sister chromatids, then separation of the sister chromatids. The final products are haploid gametes.

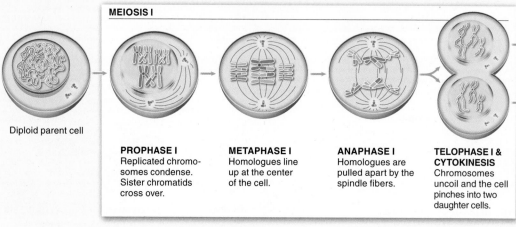

MEIOSIS I

Diploid parent cell

PROPHASE I
Replicated chromosomes condense. Sister chromatids cross over.

METAPHASE I
Homologues line up at the center of the cell.

ANAPHASE I
Homologues are pulled apart by the spindle fibers.

TELOPHASE I & CYTOKINESIS
Chromosomes uncoil and the cell pinches into two daughter cells.

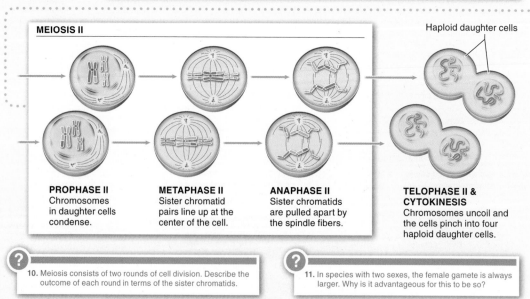

MEIOSIS II

Haploid daughter cells

PROPHASE II
Chromosomes in daughter cells condense.

METAPHASE II
Sister chromatid pairs line up at the center of the cell.

ANAPHASE II
Sister chromatids are pulled apart by the spindle fibers.

TELOPHASE II & CYTOKINESIS
Chromosomes uncoil and the cells pinch into four haploid daughter cells.

10. Meiosis consists of two rounds of cell division. Describe the outcome of each round in terms of the sister chromatids.

11. In species with two sexes, the female gamete is always larger. Why is it advantageous for this to be so?

SOURCES OF GENETIC VARIATION

There are multiple reasons why offspring are genetically different from their parents and one another.

ALLELES COME FROM TWO PARENTS
Each parent donates his or her own set of genetic material (a copy of one of the alleles they carry for each gene).

CROSSING OVER
During gamete production, crossing over during meiosis produces a mixture of maternal and paternal genetic material on each chromatid.

REASSORTMENT OF HOMOLOGUES
During gamete production, the homologues and sister chromatids distributed to each daughter cell during meiosis are a random mix of their maternal and paternal genetic material.

12. One important type of genetic variation does not require the creation of a new allele. How can that be?

13. Why is crossing over considered such an important event with respect to evolution?

14. Seventy to ninety percent of the genetic material in a sperm or egg cell made in your body could be inherited from your mother. Why isn't there always a 50/50 split (half from your mother and half from your father)? Is the reverse situation (more genetic material from your father than from your mother) possible?

6·14—6·16 There are sex differences in the chromosomes.

Sex chromosomes carry information that directs a growing fetus to develop as a male or as a female.

SEX DETERMINATION IN HUMANS

When the sperm carries an X, the baby will have two X chromosomes and develop as female.

When the sperm carries a Y, the baby will have an X and a Y and develop as male.

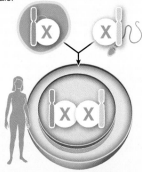

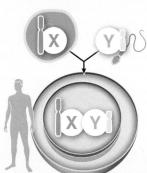

15. What are the differences between sex chromosomes and non-sex chromosomes?

16. Within a species, if one sex (i.e., males or females) has two different sex chromosomes while the other has two of the same chromosomes, which of these sexes will determine the sex of any offspring they have together? Why?

6·17—6·18 Deviations from the normal chromosome number lead to problems.

An error during meiosis can produce gametes with too many or too few chromosomes, usually leading to serious consequences for health.

NONDISJUNCTION

Nondisjunction is the unequal distribution of chromosomes during meiosis. The resulting gametes have zero or two copies of a chromosome, rather than a single copy.

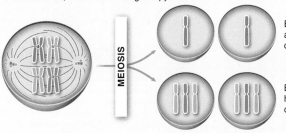

MEIOSIS

Both gametes are missing a chromosome.

Both gametes have an extra chromosome.

17. Healthy individuals may have just one sex chromosome, as long as it is an X chromosome. Why can't a person survive with a Y chromosome and no X chromosome?

---

14. Which of the following is not a method of chromosomal sex determination that occurs in nature?

a) In birds, the mother's sex chromosomes determine the sex of the offspring.

b) In sea turtles, eggs laid in hot sand become females and eggs laid in cooler sand become males.

c) In humans, the presence of a Y chromosome makes an individual male, even if he also possesses two X chromosomes.

d) In bees, the eggs that the queen allows to be fertilized become females and the eggs not fertilized become males.

e) All of the above are naturally occurring examples of sex determination.

```
0                           100
EASY        38         HARD
```

15. A karyotype reveals:

a) the shape of the spindle.

b) 23 pairs of chromosomes.

c) the number, shapes, and sizes of chromosomes in an individual cell.

d) the sex chromosomes but not the non-sex chromosomes.

e) the non-sex chromosomes but not the sex chromosomes.

```
0                           100
EASY     20            HARD
```

16. Nondisjunction:

a) is the unequal division of the genetic material during cell division.

b) occurs during meiosis but not mitosis.

c) occurs in males but not females among mammals, and in females but not males among birds.

d) leads to a missing chromosome or extra chromosome.

e) Both a) and d) are correct.

```
0                           100
EASY          36       HARD
```

17. No karyotype has ever shown a person to have one Y chromosome and no X chromosome. Why?

a) No meiotic event could produce an egg without an X chromosome.

b) Individuals are probably born with this karyotype, but have such slight phenotypic abnormalities that they live out their lives normally and are unaware of this genetic anomaly.

c) Genes on the Y chromosome are expressed only when an X chromosome is present.

d) The X chromosome contains genes that are essential for life.

e) A karyotype would not reveal such an abnormality.

```
0                           100
EASY              65   HARD
```

7 | Genes and Inheritance

FAMILY RESEMBLANCE: HOW TRAITS ARE INHERITED

Why do offspring resemble their parents?

Probability and chance play central roles in genetics.

How are genotypes translated into phenotypes?

Some genes are linked together.

7·1–7·5
Why do offspring resemble their parents?

Upon maturation, the baby Emperor Penguin will closely resemble its parents.

7·1 Family resemblance: your father and mother each contribute to your genetic makeup.

Q How can a single bad gene make you smell like a rotten fish?

Can a gene be cruel? Of course not. But if one could, consider this candidate: in humans, there is a gene for an enzyme called FMO3 (flavin-containing mono-oxygenase-3), which breaks down a chemical in our bodies that smells like rotting fish. Some unfortunate individuals inherit a defective FMO3 gene and can't break down the noxious chemical. Instead, their urine, sweat, and breath excrete it, causing them to smell like rotting fish. Worst of all, because the odor comes from within their bodies, they cannot wash it away no matter how hard they try. Called "fish odor syndrome," this disorder often causes those afflicted by it to suffer ridicule, social isolation, and depression.

About 200 cases of fish odor syndrome have been reported, but researchers suspect that it may be under-reported because many people with the symptoms do not seek help. Currently, there is no cure for the disorder, although there are a few ways of reducing the odor such as reducing consumption of eggs, fish, and some meats that contain certain chemicals, including sulfur.

For individuals born with this malady, beyond their own suffering there looms a scary question: "Will I pass this condition on to my children?" They might also wonder how they came to have the disorder, particularly if neither of their parents suffered from it.

Fortunately, most of us do not have to worry about fish odor syndrome, but we do wonder about many other inherited traits that we may or may not pass down to our children. Where do we begin and our parents end (**FIGURE 7-1**)?

In many cases, the answers to questions about heredity are simple. Recall from Chapter 6 that sexually reproducing organisms inherit one copy of each chromosome from each parent so that they carry two copies of every chromosome in

FIGURE 7-1 Where do we begin and our parents end? Family resemblances reveal that many traits are inherited.

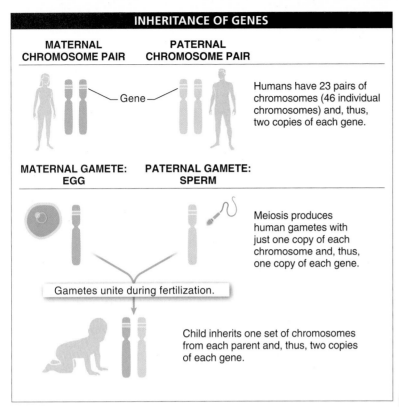

INHERITANCE OF GENES

MATERNAL CHROMOSOME PAIR   **PATERNAL CHROMOSOME PAIR**

Gene

Humans have 23 pairs of chromosomes (46 individual chromosomes) and, thus, two copies of each gene.

MATERNAL GAMETE: EGG   **PATERNAL GAMETE: SPERM**

Meiosis produces human gametes with just one copy of each chromosome and, thus, one copy of each gene.

Gametes unite during fertilization.

Child inherits one set of chromosomes from each parent and, thus, two copies of each gene.

FIGURE 7-2 **How we inherit our genes.** Each human offspring inherits one maternal set of 23 chromosomes and one paternal set.

fish odor allele, and if, when they have children, it comes together with a defective FMO3 allele from another person, the fish odor trait will be expressed. In this way, some alleles can exist in a population without always revealing themselves.

In this chapter, we explore how heredity works. Inheritance follows some simple rules that allow us to make sense of patterns of family resemblance such as facial features or hair texture and even to predict the likelihood that an offspring will inherit a trait. We'll also examine why the behavior of some traits is easy to predict while many other traits have less straightforward patterns of inheritance, yet still can be studied experimentally and yield predictable patterns.

TAKE-HOME MESSAGE 7·1

Offspring resemble their parents because they inherit genes—instructions for biochemical, physical, and behavioral traits, some of which are responsible for diseases—from their parents.

every cell (except their sex cells). Humans, for example, have 23 pairs of chromosomes (46 individual chromosomes). At each location—referred to as a locus (*pl.* loci)—on the two chromosomes of a pair is the same gene: one copy from the mother and one from the father (**FIGURE 7-2**). As we noted in Chapter 5, each of the two copies of the gene is called an **allele.**

The gene for FMO3 is on chromosome 1. There is a normal version—or allele—of the gene for FMO3, which most people carry, and there is a rare, defective version that is responsible for fish odor syndrome. As long as a person has at least one normal version of the FMO3 gene, he or she will produce enough of the enzyme to break down the fishy chemical. But if a person inherits two copies of the defective version of the fish odor gene, one from each parent, that person will inherit the disorder (**FIGURE 7-3**).

When it comes to having children, there is a bright side, at least. Although individuals with fish odor syndrome carry two defective copies of the allele and will pass on one set of the bad instructions in every sperm or egg cell that they make, their children won't necessarily inherit the disease. As long as the other parent supplies a normal version of the FMO3 gene, the child will not have fish odor syndrome. The unaffected children, though, will carry a silent copy of the

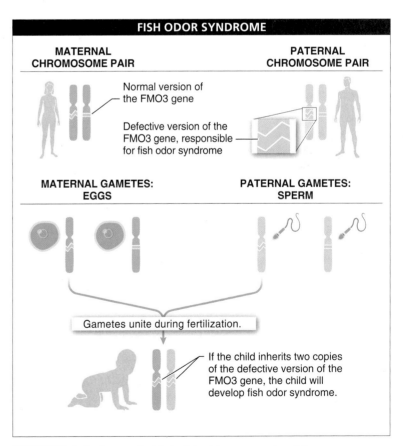

FISH ODOR SYNDROME

MATERNAL CHROMOSOME PAIR   **PATERNAL CHROMOSOME PAIR**

Normal version of the FMO3 gene

Defective version of the FMO3 gene, responsible for fish odor syndrome

MATERNAL GAMETES: EGGS   **PATERNAL GAMETES: SPERM**

Gametes unite during fertilization.

If the child inherits two copies of the defective version of the FMO3 gene, the child will develop fish odor syndrome.

FIGURE 7-3 **Unlucky catch.** A baby inheriting two copies of the defective FMO3 gene will develop fish odor syndrome.

FAMILY RESEMBLANCE   GENETICS AND CHANCE   PHENOTYPES FROM GENOTYPES   GENE LINKAGE

7·2 Some traits are controlled by a single gene.

Like father, like son. No one will ever seek a paternity test to prove that Jaden Smith is the son of Will Smith. All it takes is a quick look at the facial features shared by the two actors and we are sure they are related (FIGURE 7-4). We see all around us that offspring generally resemble their parents more than they resemble other random individuals in the population—a consequence of the passing of characteristics from parents to offspring through their genes. This is **heredity.**

Observing heredity is easy. Elucidating how it works is not. For thousands of years before the mechanisms of heredity were discovered and understood, plant and animal breeders recognized that there is a connection from parents to offspring across generations. In ancient Greece, for example, the poet Homer extolled the tremendous benefits to society that came from the skillful breeding of horses. The awareness of beneficial breeding practices enabled farmers to systematically create strains of crops, livestock, and even pets with desirable traits (FIGURE 7-5).

Once breeders recognized the existence of heredity, they began selecting individual plants or animals with the desired traits to breed with each other, in the hope that

> "HEREDITY
> I AM the family face;
> Flesh perishes, I live on,
> Projecting trait and trace
> Through time to times anon,
> And leaping from place to place
> Over oblivion.
>
> The years-heired feature that can
> In curve and voice and eye
> Despise the human span
> Of durance—that is I;
> The eternal thing in man,
> That heeds no call to die."
>
> — THOMAS HARDY, *in Moments of Vision and Miscellaneous Verses,* 1917

the offspring would also have those desirable traits. Since then, it has been possible to create a rich world of sweeter corn, loyal dogs, docile livestock, beautiful flowers, and more. And these breeders did it all without ever really understanding how heredity works.

FIGURE 7-4 Father and son: Will and Jaden Smith.

Once breeders recognized the existence of heredity, selective breeding—such as for reduced size in horses—became possible.

FIGURE 7-5 **Honey, I shrunk the horse!** Selective breeding was used to produce this tiny horse.

which single-gene traits pass from parent to offspring is the easiest pattern of inheritance to decipher, we first explore this process. Then we'll expand our model of heritability to account for the passing on of more complex traits.

TAKE-HOME MESSAGE 7·2

Some traits are determined by instructions that an individual carries on a single gene, and these traits exhibit straightforward patterns of inheritance.

Practical successes were many, but people didn't understand exactly how the outcomes were achieved. They knew the result of selective breeding, but why and how it worked, or didn't always work, was a mystery. Patterns of similarity among related individuals are impossible to make sense of without an underlying understanding of how heredity works. This understanding is the province of the field of genetics.

Mechanisms of heredity become a bit easier to grasp when we focus on visible traits with well-established inheritance patterns. Coat color and fur length in cats fits this bill well. Virtually every short-haired cat, for instance, has at least one short-haired parent. And virtually every completely white-haired cat has at least one white-haired parent. In humans, many genetic diseases have similarly simple patterns of inheritance. The neurodegenerative disorder known as Huntington's disease, for example, follows such a pattern: individuals who develop the disease have a parent who developed the disease. Traits that are determined by the instructions a person carries on one gene are called **single-gene traits** (**FIGURE 7-6**).

It is important to note here that most human characteristics are influenced by multiple genes, as well as by the environment. However, because the mechanism by

EXAMPLES OF SINGLE-GENE TRAITS

FUR LENGTH IN CATS

Long-haired cat   Short-haired cat

COAT COLOR IN CATS

White-haired cat   Colored-haired cat

FIGURE 7-6 **Fur length and coat color in cats: single-gene traits.** Some traits are determined by the instructions an organism carries on one gene.

Mendel learned about heredity by conducting experiments.

What do parents "give" their offspring that confers similarity? It wasn't until the mid-1800s that any real headway was made on answering this question. At that time, Gregor Mendel, a monk living in what is now the Czech Republic, conducted studies that not only shed light on this question but practically answered it completely. Mendel understood the essence of the genetics puzzle and set out to piece it all together.

When Mendel turned to questions about heredity, there were no obvious answers. One idea that had been popular since the late 1600s suggested that an entire pre-made human—albeit a tiny one—was contained in every sperm cell (**FIGURE 7-7**). Though imaginative, this idea could not explain why children resemble both their mothers and their fathers, not just their fathers. Another popular idea suggested that offspring reflect a simple blending of their two parents' traits by blending of the blood. But how then, could brown-eyed parents give birth to blue-eyed children?

> **Q** Individuals who share common ancestry are called "blood relatives," yet they don't actually share any blood. How might the phrase "blood relatives" be a reflection of early conceptions of inheritance?

Sometimes scientific breakthroughs are made because a new technique is invented or a lucky observation is made by chance. Neither played a role in Mendel's success. He didn't do anything radically new, but simply applied the tried-and-true process of methodical experimentation and scientific thinking (as described in Chapter 1). In particular, three features of Mendel's research were critical to its success (**FIGURE 7-8**).

1. He chose a good organism to study: the garden pea. It's not that Mendel had a particular fondness for vegetables. Moreover, his goals were to understand inheritance in all organisms, not just plants. But cats and dogs or even mice wouldn't have served his purposes very well, because they would have been too hard to take care of in the large numbers he required—thousands and thousands of individuals. Humans, too, would have made a terrible study organism. We take too long to breed (and won't produce offspring on command). Pea plants, on the other hand, were simply the right tool. They are relatively easy to fertilize manually by "pollen dusting." A single **cross**—the process in which male pollen (carrying sperm) is used to fertilize eggs—produces numerous offspring. In addition, pea plants reproduce quickly, so Mendel could conduct experiments that included multiple generations.

2. Mendel chose to focus on easily categorized traits like shape and color. For instance, all peas of the variety that Mendel studied are either round or wrinkled in shape, with nothing in between. In addition, all peas are either yellow or green in color, never any intermediate shade. In all, Mendel looked at seven different traits of the pea plants, but for each trait only two variants ever appeared. Mendel and his research assistants could easily observe and unambiguously identify the trait variants.

3. Mendel began his studies by first repeatedly breeding together similar plants until he had many distinct populations, each of which was unvarying for a particular trait. He described these plants as **true-breeding** for that trait because they always produced offspring with the same variant of the trait as the parents. For example, when true-breeding round-pea plants were crossed together, they always produced plants with round peas. True-breeding purple-flowered

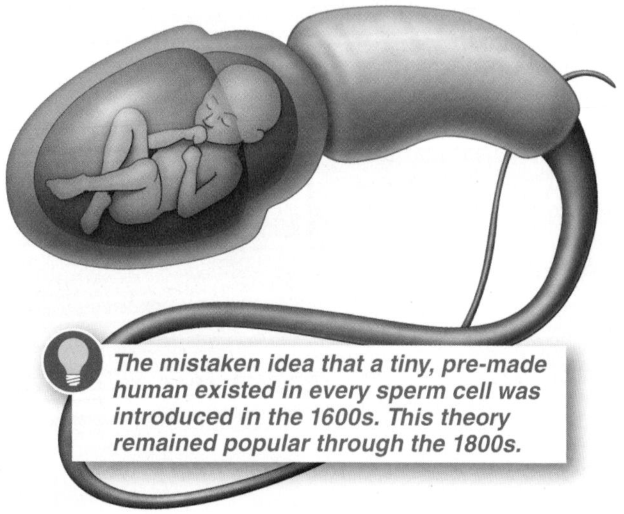

💡 *The mistaken idea that a tiny, pre-made human existed in every sperm cell was introduced in the 1600s. This theory remained popular through the 1800s.*

FIGURE 7-7 **The "pre-made human."** A tiny human in every sperm cell? Why would children have any resemblance to their mothers?

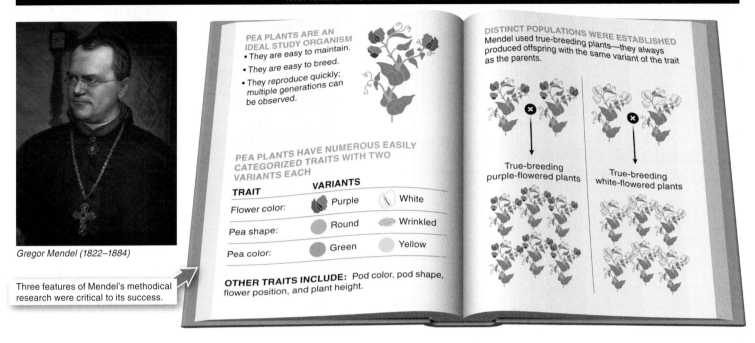

Gregor Mendel (1822–1884)

Three features of Mendel's methodical research were critical to its success.

PEA PLANTS ARE AN IDEAL STUDY ORGANISM
• They are easy to maintain.
• They are easy to breed.
• They reproduce quickly; multiple generations can be observed.

PEA PLANTS HAVE NUMEROUS EASILY CATEGORIZED TRAITS WITH TWO VARIANTS EACH

TRAIT	VARIANTS	
Flower color:	Purple	White
Pea shape:	Round	Wrinkled
Pea color:	Green	Yellow

OTHER TRAITS INCLUDE: Pod color, pod shape, flower position, and plant height.

DISTINCT POPULATIONS WERE ESTABLISHED
Mendel used true-breeding plants—they always produced offspring with the same variant of the trait as the parents.

True-breeding purple-flowered plants

True-breeding white-flowered plants

FIGURE 7-8 Mendel's pea plant experiments. Careful planning and a well-selected study organism contributed to the success of Mendel's experiments.

pea plants always produced purple-flowered offspring, while true-breeding white-flowered pea plants always produced white-flowered offspring. It took a lot of prep work to establish these populations, but once Mendel had them, he was in a position to set up all of the different crosses that enabled him to piece together the genetics puzzle.

Once he had obtained his groups of true-breeding plants, Mendel began a straightforward process of experimentation. He crossed plants with different traits—for example, he crossed plants with green peas and plants with yellow peas—and observed large numbers of their offspring, counting the number of plants that produced green peas and the number of plants that produced yellow peas. (This cross, written as "green × yellow cross," always resulted in plants that produced only yellow peas.) Mendel devised a hypothesis that would explain his observations and generate predictions about the outcomes of further crosses. Then he conducted those crosses to see whether the predictions generated by his hypothesis were borne out.

There was a simple elegance to Mendel's work: from his predictions he articulated simple questions that could be answered by his experiments. For example, "If I cross a plant that produces wrinkled peas with a plant that produces smooth peas, will the offspring plants produce wrinkled or smooth peas?" His crosses would come out either one way or the other. Everything was black and white. This process of performing well-thought-out, rigorous experiments was a radical innovation for biology. We explore his results in the next section and see that they allowed Mendel to clearly and definitively describe how the traits he studied were inherited.

TAKE-HOME MESSAGE 7·3

In the mid-1800s, Gregor Mendel conducted studies that help us understand heredity. He focused on easily observed and categorized traits in garden peas and applied methodical experimentation and rigorous hypothesis testing to determine how traits are inherited.

7·4 Segregation: you've got two copies of each gene but put only one copy in each sperm or egg.

One odd and recurring result spurred Mendel to figure out the mechanism by which traits could be passed from parent to offspring. Just as brown-eyed parents can have blue-eyed children, sometimes traits that weren't present in either parent pea plant would show up in their offspring. When plants with purple flowers were fertilized by pollen from other plants with purple flowers, they produced mostly purple-flowered offspring, but sometimes they produced plants with white flowers.

How was it possible to produce white flowers from a purple cross? Where did the whiteness come from? Here's where Mendel's meticulous and methodical experiments paid off. Let's follow his process. First, he started with some true-breeding white-flowered plants. Then he got some true-breeding purple-flowered plants. He wondered: which color wins out when a white-flowered plant is crossed with a purple-flowered plant? The answer was definitive: purple wins (**FIGURE 7-9**).

All of the offspring from these crosses were purple, every time. For this reason, Mendel called the purple-flower trait **dominant**, and he considered the white-flower trait to be the **recessive** trait. In general, a dominant trait masks the effect of a recessive trait when an individual carries both the dominant and the recessive versions of the instructions for the trait.

Things got a bit more interesting when Mendel took the purple-flowered plants that came from the cross between true-breeding purple- and white-flowered plants and bred these purple offspring with each other. He found that these mixed-parentage plants were no longer true-breeding. Occasionally, they would produce white-flowered offspring. (To be exact, of the 929 offspring plants Mendel examined, 705 had purple flowers, and 224 had white flowers.) Apparently, the directions for building white flowers—last seen in one of their grandparents—were still lurking inside their purple-flowered parent plants. The existence of traits that could disappear for a generation and then show up again was quite perplexing.

Mendel devised a simple hypothesis to explain this pattern of inheritance. It incorporated three ideas that helped him (and now help us) make predictions about crosses (**FIGURE 7-10**).

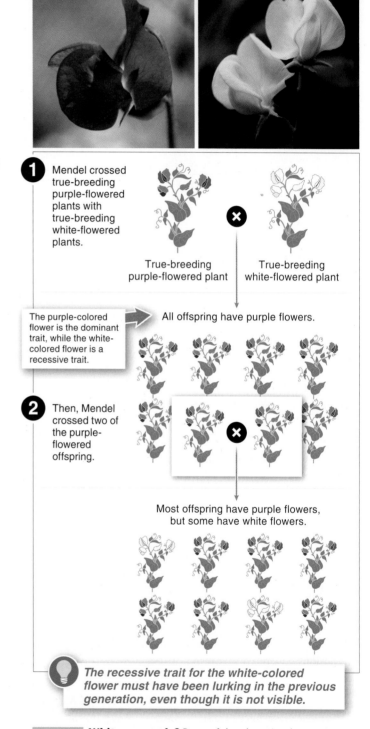

DOMINANT AND RECESSIVE TRAITS

1 Mendel crossed true-breeding purple-flowered plants with true-breeding white-flowered plants.

True-breeding purple-flowered plant   True-breeding white-flowered plant

The purple-colored flower is the dominant trait, while the white-colored flower is a recessive trait.

All offspring have purple flowers.

2 Then, Mendel crossed two of the purple-flowered offspring.

Most offspring have purple flowers, but some have white flowers.

The recessive trait for the white-colored flower must have been lurking in the previous generation, even though it is not visible.

FIGURE 7-9 **White or purple?** By careful and repeated crosses among pea plants, Mendel determined that there were "dominant" and "recessive" traits.

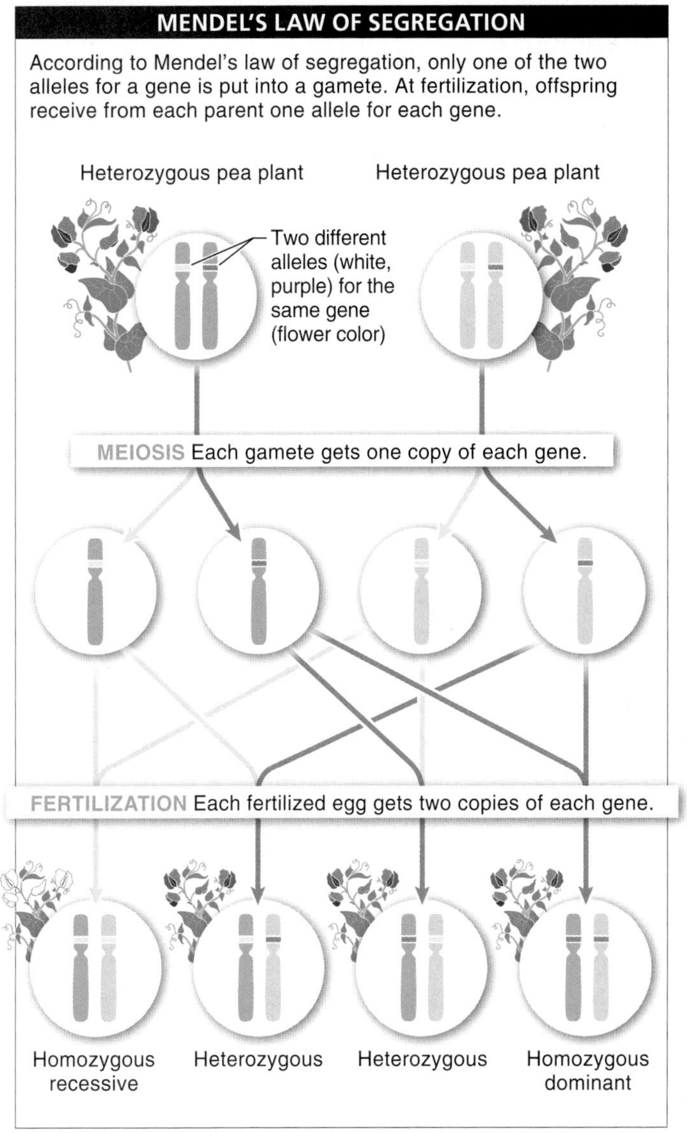

According to Mendel's law of segregation, only one of the two alleles for a gene is put into a gamete. At fertilization, offspring receive from each parent one allele for each gene.

Heterozygous pea plant   Heterozygous pea plant

Two different alleles (white, purple) for the same gene (flower color)

MEIOSIS Each gamete gets one copy of each gene.

FERTILIZATION Each fertilized egg gets two copies of each gene.

Homozygous recessive   Heterozygous   Heterozygous   Homozygous dominant

FIGURE 7-10 **Segregation of alleles in meiosis.** Organisms have two copies of each gene but place only one copy into each gamete during the process of meiosis (as described in Chapter 6).

1. Rather than passing on the trait itself, **each parent puts into every sperm or egg it makes a single set of instructions for building the trait.** Today, we call that instruction set a gene.

2. **Offspring receive two copies of the instructions for any trait.** Often, both sets of instructions are identical, and the offspring produce the traits according to those instructions. Other times, though, each parent contributes a slightly different set of instructions—that is, a different allele—for that trait. So, for example, pea plants have two alleles for flower color: a purple-flower allele and a white-

flower allele. Each is a different allele, but because both specify instructions for producing color in flowers, they are both genes for flower color. Same gene (flower-color gene), different alleles (purple and white).

3. **The trait observed in an individual depends on the *two* copies of the gene it inherits from its parents.** When an individual inherits the same two alleles for this gene, the individual's genotype for that gene is said to be **homozygous** and the individual shows the trait specified by the instructions embodied in those alleles. When an individual inherits a different allele from each parent, the individual's genotype for that gene is said to be **heterozygous.** Dominant and recessive alleles are defined by their action when they are in the heterozygous state: if there is a phenotypic effect of one of the alleles but not the other allele, the first is called the dominant allele and is said to "mask" the effect of the other allele, which is called the recessive allele. Note that this is the sole defining feature. "Dominant" does not mean that an allele is advantageous or more common than a recessive allele.

> **❝There were no questions.❞**
> —Entry in meeting notes following Mendel's first public presentation of his ideas on how heredity works. (Tragically, no one in the audience of scientists had any idea what he was talking about. Not until about 40 years later did the world understand his discoveries.)

When an individual reproduces, it contributes just one of its two copies of a gene to its offspring. The other parent will contribute the other allele. So, when sperm and eggs are made, each sex cell gets only one copy of a gene—as opposed to the two copies present in every other cell in the body. For a male who is heterozygous for a particular gene, for example, it means that half of the sperm he produces will have one of the alleles and half will have the other. The idea that, of the two copies of each gene everyone carries, only one of the two alleles gets put into each gamete is so significant and important that it is called **Mendel's law of segregation.**

TAKE-HOME MESSAGE 7·4

Each parent puts a single set of instructions for building a particular trait into every sperm or egg he or she makes. This instruction set is called a gene. The trait observed in an individual depends on the two copies (alleles) of the gene it inherits from its parents.

Observing an individual's phenotype is not sufficient for determining its genotype.

Things are not always as they appear. Take skin coloration, for example. Humans and many other animals have a gene that contains the information for producing melanin, one of the chemicals responsible for giving our skin its coloring (**FIGURE 7-11**). This gene is one of many that influence skin color. Unfortunately, there is also a defective, non-functioning version of the melanin gene that is passed along through some families. An individual who inherits two copies of the defective version of the gene cannot produce pigment and has a condition known as albinism, a disorder characterized by little or no pigment in the eyes, hair, and skin. But it is impossible to tell whether a normally pigmented individual carries one of these defective alleles just by looking at his or her appearance— we would need to get a genetic analysis done to discover this information.

The inability to deduce an individual's genetic makeup through simple observation is a general problem in genetics: physical appearances don't necessarily reflect the underlying genes. A normally pigmented individual may carry two copies of the pigment-producing allele or may have just one. In either case, the individual will look the same. The outward appearance of an individual is called its **phenotype.** A phenotype includes features visible to the naked eye such as flashy coloration, height, or the presence of antlers. A phenotype also includes less easily visible

characteristics such as the chemicals an individual produces to clot blood or digest lactose. An individual's phenotype even includes the behaviors it exhibits.

Underlying the phenotype is the **genotype.** This is an organism's genetic composition. We usually speak of an individual's genotype in reference to a particular trait. For example, an individual's genotype might be described as "homozygous for the recessive allele for albinism," meaning that the person has two recessive alleles of the gene. Another individual's genotype for the melanin gene might be described as "heterozygous" (a recessive allele and a dominant allele.) Occasionally, the word "genotype" is also used as a way of referring to *all* of the genes an individual carries.

When an organism exhibits a recessive trait, such as albinism, we know with certainty what its genotype is for the melanin gene. When it shows the dominant trait, on the other hand, it's impossible to distinguish whether the organism carries two copies of the dominant allele (homozygous dominant) or carries one copy of the dominant allele and one of the recessive (heterozygous). Because it's not possible to discern the genotypes of two individuals with the same phenotype just by looking at them, much of genetic analysis must make use of clever experiments and careful record-keeping.

To analyze and predict the outcome of crosses, first we must assign symbols to represent the different variants of a gene. Generally we use an uppercase letter for the dominant allele and lowercase letter for the recessive allele. In the case of pigmentation/albinism, we use the letter "*m*" because the trait is caused by the gene carrying instructions for production of the pigment melanin. We represent the genotype of the albino giraffe as *mm,* because that individual must carry two copies of the recessive allele, *m.* A giraffe that is pigmented must have the genotype *MM* or *Mm.* If we don't know which of the two possible genotypes the pigmented individual has, we can write *M_* where _ is a placeholder for the unknown second allele.

We can trace the possible outcomes of a cross between two individuals using a handy tool called the **Punnett square.** In **FIGURE 7-12** we illustrate the cross between a true-breeding pigmented individual, *MM,* and an albino, *mm.* Along the top of the square we list, individually, the two alleles that one of the parents produces, and along the left

PHENOTYPE: Little or no pigment in the eyes, hair, and skin

GENOTYPE: Homozygous for the recessive allele for albinism

FIGURE 7-11 One gene, much pigment. Carrying two non-functioning versions of the gene carrying instructions for production of the pigment melanin causes albinism.

PUNNETT SQUARE: ALBINISM

A Punnett square is a useful tool for determining the possible outcomes of a cross between two individuals.

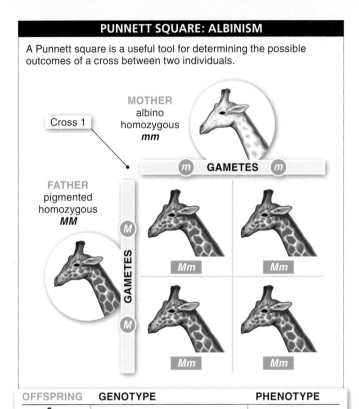

Cross 1

MOTHER
albino
homozygous
mm

GAMETES

FATHER
pigmented
homozygous
MM

GAMETES

OFFSPRING	GENOTYPE	PHENOTYPE
	All heterozygous *Mm*	All pigmented

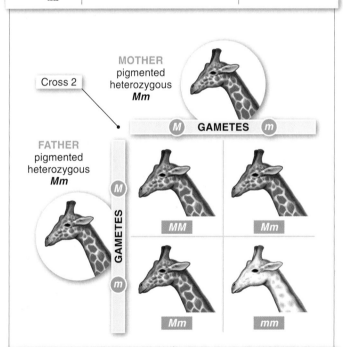

Cross 2

MOTHER
pigmented
heterozygous
Mm

GAMETES

FATHER
pigmented
heterozygous
Mm

GAMETES

OFFSPRING	GENOTYPE	PHENOTYPE
	1/4 homozygous dominant *MM*	3/4 pigmented
	2/4 heterozygous *Mm*	
	1/4 homozygous recessive *mm*	1/4 albino

FIGURE 7-12 **Predicting the outcome of crosses.** A Punnett square shows us the likelihood of the albino genotype occurring in offspring.

side of the square we list the two alleles that the other parent produces. We split up an individual's two alleles in this way because, although the individual carries two alleles, only one of the alleles is contained in each sperm or egg cell that it produces. The two gametes that come together at fertilization produce the genotype of the offspring.

In the four cells of the Punnett square, we enter the genotypes of all the possible offspring resulting from our cross. Each cell contains one allele given at the head of the column and one allele at the left of the row. In Cross 1 illustrated in Figure 7-12, every possible offspring would be heterozygous and would be normally pigmented, because it receives a dominant allele from the pigmented parent and a recessive allele from the albino parent.

In the bottom half of Figure 7-12 (Cross 2), we trace the cross between two heterozygous individuals. Note that each parent produces two kinds of gametes, one with the dominant allele and one with the recessive allele. This cross has three possible outcomes: one-quarter of the time the offspring will be homozygous dominant (*MM*), one-quarter of the time the offspring will be homozygous recessive (*mm*), and the remaining half of the time the offspring will be heterozygous (*Mm*). Phenotypically, three-quarters of the offspring will be normally pigmented (*MM* or *Mm*) and one-quarter will be albino (*mm*).

TAKE-HOME MESSAGE 7·5

It is not always possible to determine an individual's genetic makeup, known as its genotype, by observation of the organism's outward appearance, known as its phenotype. For a particular trait, an individual may carry a recessive allele whose phenotypic effect is masked by the presence of a dominant allele. Much genetic analysis makes use of clever experiments and careful record-keeping, often using Punnett squares, to determine organisms' genotypes.

Probability and chance play central roles in genetics.

Because of the role of chance in genetics, we cannot always predict the exact outcome of a genetic cross.

7·6 Chance is important in genetics.

Sometimes genetics is a bit like gambling. Even with perfect information, it can still be impossible to know the genetic outcome with certainty. It's like flipping a coin: you can know every last detail about the coin, but you still can't know whether the coin will land on heads or tails. The best you can do is define the probability of each possible outcome.

The rules of probability (the same ones that govern coin tosses and the rolling of dice) have a central role in genetics, for two reasons. The first is a consequence of segregation, the process Mendel described, in which each gamete an individual produces receives only one of the two copies of each gene that the individual carries in most of its other cells. As a result, it is equally likely that the haploid gamete—the sperm or egg—will include one allele or the other allele of the two that the individual carries. It is impossible to know which allele it will be. The second reason probability plays a central role in genetics is that fertilization, too, is a chance event. All of the sperm or eggs produced by an individual are different from one another, and any one of those gametes may be the gamete involved in fertilization. Thus, knowing everything about the alleles a parent carries is not always enough to be able to determine with certainty which alleles his or her offspring will carry.

Let's explore how we can make predictions in games of chance, as well as in matings, based on probabilities. If an individual is homozygous for a trait, 100% of his or her gametes will carry that allele. Any gamete produced by an individual heterozygous for a trait has a 50% probability of carrying the dominant allele and a 50% probability of carrying the recessive allele. Let's use these probabilities in an example.

If a male is heterozygous for albinism (*Mm*) and a female is albino (*mm*), what is the probability that an offspring of

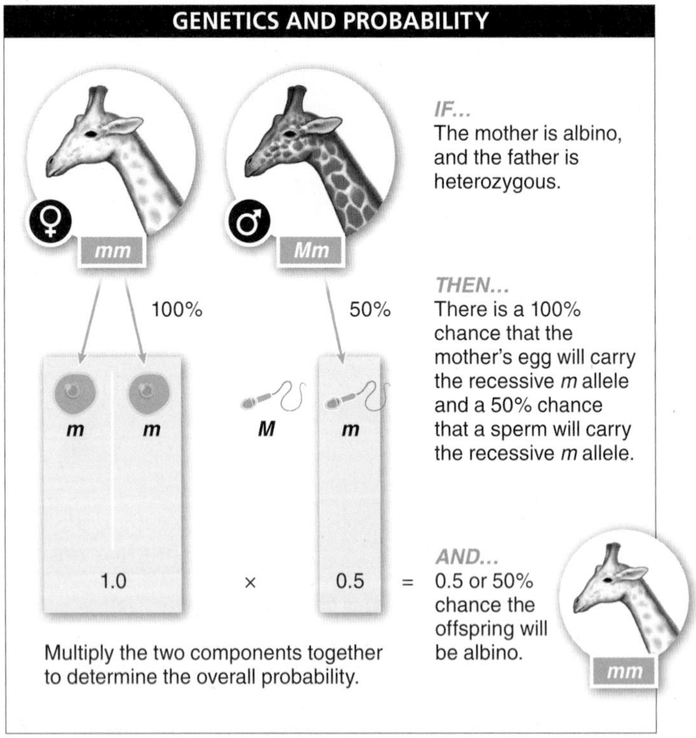

GENETICS AND PROBABILITY

IF...
The mother is albino, and the father is heterozygous.

THEN...
There is a 100% chance that the mother's egg will carry the recessive *m* allele and a 50% chance that a sperm will carry the recessive *m* allele.

AND...
0.5 or 50% chance the offspring will be albino.

100%   50%

m   *m*   *M*   *m*

1.0   ×   0.5   =   0.5 or 50% chance the offspring will be albino.

Multiply the two components together to determine the overall probability.

FIGURE 7-13 Using probability to determine the chance of inheriting albinism.

theirs will be homozygous for albinism (*mm*)? To get the *mm* outcome, two events must occur. First, the father's gamete must carry the recessive allele (*m*), and second, the mother's gamete must carry the recessive allele (*m*). In this case, the probability of a homozygous recessive offspring is 0.5 (the probability that the father's gamete carries *m*) times 1.0 (the probability that the mother's gamete carries *m*), for a probability of 0.5. This is a general rule when determining the likelihood of a complex event occurring: if you know the probability of each component that must occur, you multiply all the probabilities together to get the overall probability of that complex event occurring (**FIGURE 7-13**).

Consider another example, involving Tay-Sachs disease. As described in Chapter 3, Tay-Sachs is caused by malfunctioning lysosomes that do not digest cellular waste properly. It leads to death in early childhood. Tay-Sachs occurs if a child inherits two recessive alleles for the Tay-Sachs gene. If each parent is heterozygous for the Tay-Sachs gene, what is the probability that their child will have Tay-Sachs disease?

To solve this, break down the event of the child having Tay-Sachs into two separate events: first, it must inherit a recessive allele, *t*, from its heterozygous father. Because half the father's sperm will have the dominant *T* allele and half will have the *t* allele, the probability of the child inheriting the *t* allele is 0.5. Second, the child must also inherit a recessive allele from its heterozygous mother. Because half of the mother's eggs will have the *T* allele, while the other half have the *t* allele, the probability of the child inheriting the *t* allele from its mother is also 0.5. So the overall probability of the child having Tay-Sachs disease is 0.5 × 0.5 = 0.25, or 1 in 4 (**FIGURE 7-14**).

Of course, if the couple has only one child, we can't predict *with certainty* whether the child will have Tay-Sachs. The mathematical probability tells us, however, that if the couple had an infinite number of children, we could expect that one-fourth of them would have Tay-Sachs disease. This fact

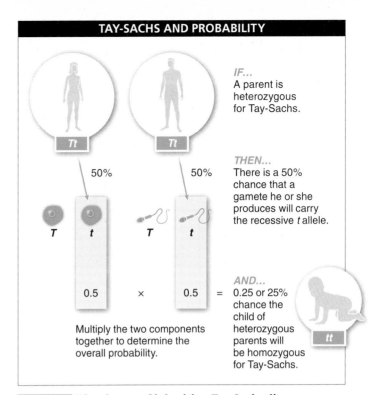

TAY-SACHS AND PROBABILITY

IF... A parent is heterozygous for Tay-Sachs.

50%   50%

THEN... There is a 50% chance that a gamete he or she produces will carry the recessive *t* allele.

0.5 × 0.5 = 0.25 or 25% chance the child of heterozygous parents will be homozygous for Tay-Sachs.

AND...

Multiply the two components together to determine the overall probability.

FIGURE 7-14 **The chance of inheriting Tay-Sachs disease.** The probability that a child will inherit Tay-Sachs disease from two heterozygous parents is 25%.

may not be much help for heterozygous couples trying to decide whether to have a child or not.

TAKE-HOME MESSAGE 7·6

Probability plays a central role in genetics. In segregation, each gamete that an individual produces receives only one of the two copies of each gene the individual carries in its other cells, but it is impossible to know which allele goes into the gamete. Chance plays a role in fertilization, too: all of the sperm or eggs produced by an individual are different from one another, and any one of those gametes may be the gamete involved in fertilization.

7·7 A test-cross enables us to figure out which alleles an individual carries.

How can you see something that is invisible to the naked eye? Genes are too small to be seen, and so determining an individual's genotype requires indirect methods. Suppose you are in charge of the alligators at a zoo. Some of your alligators come from a population in which white, albino

alligators have occasionally occurred, although none of your alligators is white. Because white alligators—those having two recessive pigmentation alleles, *mm* ("*m*" for melanin)—are popular with zoo visitors, you would like to produce some at your zoo through a mating program.

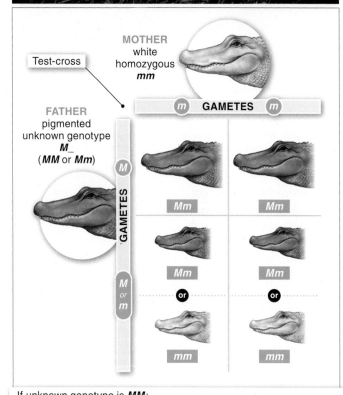

TEST-CROSS: WHITE ALLIGATORS

Test-cross

MOTHER
white
homozygous
mm

FATHER
pigmented
unknown genotype
M_
(***MM*** or ***Mm***)

GAMETES

m   *m*

GAMETES

M

Mm   *Mm*

*M
or
m*   *Mm*   *Mm*

or   or

mm   *mm*

If unknown genotype is **MM**:

OFFSPRING	GENOTYPE	PHENOTYPE
	All heterozygous **Mm**	All pigmented

If unknown genotype is **Mm**:

OFFSPRING	GENOTYPE	PHENOTYPE
	2/4 heterozygous **Mm**	2/4 pigmented
	2/4 homozygous recessive **mm**	2/4 white

FIGURE 7-15 A test-cross can reveal an unknown genotype. In this test-cross, a homozygous white female alligator is bred with a normally colored male of unknown genotype. The color of their offspring will help identify whether the male is homozygous dominant or heterozygous.

The problem is that you cannot be certain of the genotype of your alligators. They might be homozygous dominant, *MM*, or they might be heterozygous, *Mm*. In either case, their phenotype is normal coloration. How can you figure out the genotype of a particular alligator? This is a challenge to animal breeders, but not an insurmountable one. Genes may be invisible, but their identity can be revealed by a simple tool called the **test-cross.**

In the test-cross, you cross (i.e., mate) an individual exhibiting a dominant trait but whose genotype is unknown with an individual that is homozygous recessive. Then you examine the phenotypes of their offspring. In the case of your breeding program for albino alligators, you could borrow an albino alligator from another zoo and breed your unknown-genotype alligator (genotype: *M_*) with that albino alligator (genotype: *mm*). There are two possible outcomes, and they will reveal the genotype of your unknown-genotype alligator (**FIGURE 7-15**). If your alligator is homozygous dominant (*MM*), it will contribute a dominant allele, *M*, to every offspring. Even though the albino alligator will contribute the recessive allele, *m,* to all its offspring, all the offspring of this cross will be heterozygous, *Mm,* and none of them will be albino.

If, on the other hand, your unknown-genotype alligator is heterozygous, *Mm,* half of the time it will contribute a recessive allele, *m,* to the offspring. In every one of those cases, the offspring will be homozygous recessive and thus albino.

So the cleverness of the test-cross is that when you cross your unknown-genotype organism with an individual showing the recessive trait (and so having the known genotype of *mm*), the offspring will reveal the previously unknown makeup of the parent. To be confident in concluding that the unknown-genotype alligator has the *MM* genotype, though, you'd have to observe many, many offspring. After all, even if its genotype is *Mm,* quite a few offspring in a row might be normally pigmented, with the genotype *Mm.* Eventually, however, a heterozygous individual is likely to produce an offspring with the homozygous recessive genotype, *mm,* and the albino phenotype.

TAKE-HOME MESSAGE 7·7

In a test-cross, an individual that exhibits a dominant trait but has an unknown genotype is mated with an individual that is homozygous recessive. The phenotypes of the offspring reveal whether the unknown-genotype individual is homozygous dominant (all of the offspring exhibit the dominant trait) or heterozygous (half of the offspring show the dominant trait and half show the recessive trait).

7·8 We use pedigrees to decipher and predict the inheritance patterns of genes.

People want to know things about the future, such as: What is the likelihood that I will have a child with a particular genetic disease, say hemophilia? Or what is my own risk of developing a genetic disease, such as Huntington's disease, later in my life? Geneticists who study these and other diseases want to know other things: How is a particular disease or trait inherited? Is it recessive or dominant? Is it carried on the sex chromosomes or on one of the other chromosomes? A **pedigree** is a type of family tree that can help answer these questions.

Dog and horse breeders often speak of "pedigreed" animals, a feature that adds tremendous value to the animals. This is because, with knowledge of an animal's family tree, or complete lineage, it is much less likely that any genetic surprises will occur as the animal develops and reproduces.

In a pedigree, information is gathered from as many related individuals as possible across multiple generations (**FIGURE 7-16**). Starting from the bottom, each row in the chart represents a generation, listing all of the children in their order of birth, their sex, and whether or not they have a particular trait. Working up the pedigree, the children's parents are indicated and, above them, their parents' parents, for as far back as data are available. Squares

Q Why do breeders value "pedigreed" horses and dogs so much?

represent males and circles represent females, and these shapes are shaded to indicate that an individual exhibits the trait of interest. Sometimes the genotype (as much of it as is known) is also listed for each individual.

By analyzing which individuals manifest the trait and which do not, it may be possible to deduce a pattern of inheritance for the trait—or, at least, rule out certain patterns. For example, for dominant traits, all affected individuals must have at least one parent who exhibits the trait. In contrast, for recessive traits, an individual can exhibit a recessive trait even if both parents are unaffected (i.e., do not exhibit the trait). In this case, the individual's parents must be heterozygous for that trait, each carrying one dominant and one recessive allele.

The pedigree can also help to determine whether a trait is carried on the sex chromosomes (X or Y, which we discuss in Section 7-13) or on one of the non-sex chromosomes (also called "autosomes"). Traits that are controlled by genes on the sex chromosomes are called **sex-linked traits.** Recessive sex-linked traits, for example, appear more frequently in males than in females, whereas dominant sex-linked traits appear more frequently in females. These patterns may become obvious only on inspection of a large pedigree.

An example of how pedigrees can help determine how traits are inherited is given in **FIGURE 7-17** . Anury is a condition seen in dogs and some other animals in which the animal has no tail. The pedigree reveals that anury is inherited as a recessive trait, because unaffected parents can have offspring with the disorder.

Can you determine the genotype of the individual labeled "1" in Figure 7-17? Why must that female be heterozygous (*Aa*) for anury? (By the way, an individual that carries one allele for a recessive trait, and so does not exhibit the trait but can have offspring that do, is referred to as a **carrier** of the trait.) Must her mate have the same genotype?

On the same pedigree, note that two individuals in the second row of the pedigree have a puppy (indicated by "?"). What is the probability that this puppy has anury? Examine the pedigree carefully and see whether you can come up with the answer; the next paragraph will guide you through if you get stuck.

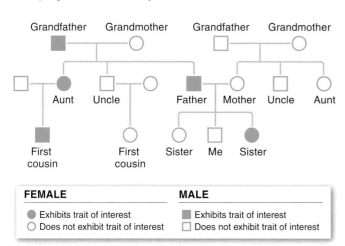

PEDIGREE

A pedigree is a useful tool to document a trait of interest across multiple generations of family members.

Grandfather Grandmother   Grandfather Grandmother

Aunt Uncle   Father Mother Uncle Aunt

First cousin   First cousin   Sister Me Sister

FEMALE	MALE
● Exhibits trait of interest	■ Exhibits trait of interest
○ Does not exhibit trait of interest	□ Does not exhibit trait of interest

FIGURE 7-16 Family tree. A pedigree maps the occurrence of a trait in a family.

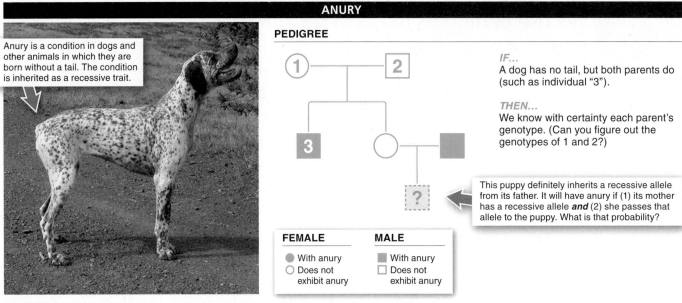

PEDIGREE

Anury is a condition in dogs and other animals in which they are born without a tail. The condition is inherited as a recessive trait.

IF...
A dog has no tail, but both parents do (such as individual "3").

THEN...
We know with certainty each parent's genotype. (Can you figure out the genotypes of 1 and 2?)

This puppy definitely inherits a recessive allele from its father. It will have anury if (1) its mother has a recessive allele *and* (2) she passes that allele to the puppy. What is that probability?

FEMALE
● With anury
○ Does not exhibit anury

MALE
■ With anury
□ Does not exhibit anury

FIGURE 7-17 **Pedigree puzzle: tracing the occurrence of anury in a family of dogs.**

In the cross producing the mystery puppy, the father has anury and so must be homozygous recessive; he will definitely pass on one *a* to the son. The mother's genotype can be *AA, Aa,* or *aA,* given that both of her parents are heterozygous. Consequently, she has a ⅔ (2 in 3) probability of being a carrier of the *a* allele. If she is, then there is a ½ (50%) chance that she will pass the allele on to her offspring, and he will have anury. The probability, therefore, is ⅔ × ½ = ⅓, or a 1 in 3 chance that the puppy will have anury.

Q Researchers often use plants and small insects to study inheritance patterns— even when they're more interested in how a trait works in humans. Why?

As we'll discuss in later sections of this chapter, some traits may not show complete dominance and many traits are also influenced by the environment, so it is not always completely obvious what a trait's mode of inheritance is. In such cases, the more individuals we can include in the pedigree, the

more accurate the analysis. Some human pedigrees contain thousands of individuals and stretch back six or more generations. With some other species it is possible to analyze tens of thousands of individuals per generation for a dozen or more generations. This is why plants and small insects (among other organisms) are excellent for studying inheritance patterns.

Once we have an idea about how a trait is inherited, we can identify individuals who are carriers of a recessive trait and make informed predictions about a couple's risk of having a child with a particular disorder. This knowledge can be useful in conjunction with prenatal testing and treatment.

TAKE-HOME MESSAGE 7·8

Pedigrees help scientists, doctors, animal and plant breeders, and prospective parents determine the genes that individuals carry and the likelihood that the offspring of two individuals will exhibit a given trait.

7.9–7.15
How are genotypes translated into phenotypes?

ROBERT EISENSTAEDT
B.S. Public Communications

DAVID EISER
B.S. Biology/Psychology

DAVID ELLIOTT
B.A. Arts and Sciences

WILLIAM ELLIOTT
B.S. Management

JUDY ELLNER
B.S. Human Development

NANCY ELLWANGER
B.S. Human Development

LINTON EMORY
B.S. Engineering

DAVID ENG
B.S. Management

Phenotype diversity—which is all around us—has multiple sources.

7·9 Incomplete dominance and codominance: the effects of both alleles in a genotype can show up in the phenotype.

As Mendel saw it, the world of genetics was straightforward and simple. We should be so lucky. Each of the traits he studied were coded for by a single gene with two alleles—one completely dominant and one recessive—and with no environmental effects. This, however, doesn't capture the complexity of the world beyond Mendel's pea plants. So, in this and the following sections we build up a more complex model of how genes influence the building of bodies.

We begin with the observation that the phenotype of heterozygous individuals sometimes differs from that of either of the homozygotes, and instead reflects the influence of both alleles rather than a clearly dominant allele.

One situation in which complete dominance is not observed is called **incomplete dominance,** in which the phenotype of a heterozygote is intermediate between the phenotypes of the two homozygotes. An example of incomplete dominance we can easily observe is the flower color of snapdragons (**FIGURE 7-18**).

We can obtain true-breeding (homozygous) lines of snapdragons with red flowers and true-breeding (homozygous) lines that produce only white flowers. When plants from these two populations are crossed, we would expect—if one allele were dominant over the other—either all red or all white flowers. Instead, such crosses always produce plants with pink flowers. Interestingly, when we cross two plants with pink flowers, we get ¼ red-flowered plants, ½ pink-flowered plants, and ¼ white-flowered plants.

How can we interpret this cross? It seems that the plants with white flowers have the genotype $C^W C^W$ and produce no pigment. At the other extreme, the plants with red flowers have the genotype $C^R C^R$ and produce a great deal of pigment. The letter "C" refers to the fact that the gene codes for color, and the superscript "W" or "R" refers to an allele producing no pigment (**w**hite) or **r**ed pigment. We use these designations for the genotypes because it isn't clear that either white or red is dominant over the other, and so neither should be represented by uppercase or lowercase. The pink flowers receive one of the pigment-producing C^R alleles and one of the no-pigment-producing C^W alleles, and so produce an intermediate amount of pigment. Ultimately, the intensity of pigmentation just depends on the amount of pigment chemical that is made by the flower-color genes.

An example of incomplete dominance in humans can be seen in the processing of cholesterol in the bloodstream. There is a plasma membrane receptor that allows cells (chiefly those in the liver) to remove cholesterol from the bloodstream (see Section 3-11), and the gene that codes for this receptor exhibits incomplete dominance. Individuals who carry two copies of a mutant allele (called FH) for this gene produce few (or even no) LDL receptors. In these individuals, circulating cholesterol levels are high and cardiovascular disease develops at a very young age. Individuals carrying one FH allele and one allele that codes for normal-functioning LDL receptors have significantly reduced levels of circulating cholesterol and a correspondingly lower risk of cardiovascular disease. Individuals carrying two copies of the allele for normal-functioning LDL receptors produce twice as many LDL

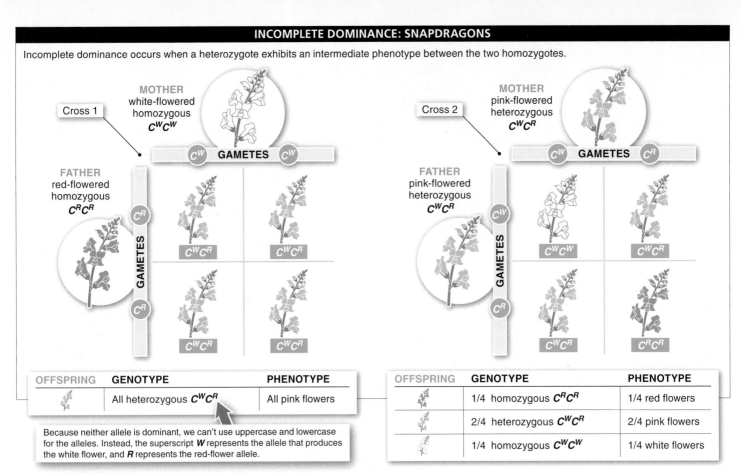

INCOMPLETE DOMINANCE: SNAPDRAGONS

Incomplete dominance occurs when a heterozygote exhibits an intermediate phenotype between the two homozygotes.

Cross 1

MOTHER
white-flowered homozygous
$C^W C^W$

FATHER
red-flowered homozygous
$C^R C^R$

GAMETES C^W C^W

GAMETES C^R C^R

$C^W C^R$ $C^W C^R$

$C^W C^R$ $C^W C^R$

OFFSPRING	GENOTYPE	PHENOTYPE
	All heterozygous $C^W C^R$	All pink flowers

Because neither allele is dominant, we can't use uppercase and lowercase for the alleles. Instead, the superscript **W** represents the allele that produces the white flower, and **R** represents the red-flower allele.

Cross 2

MOTHER
pink-flowered heterozygous
$C^W C^R$

FATHER
pink-flowered heterozygous
$C^W C^R$

GAMETES C^W C^R

GAMETES C^W C^R

$C^W C^W$ $C^W C^R$

$C^W C^R$ $C^R C^R$

OFFSPRING	GENOTYPE	PHENOTYPE
	1/4 homozygous $C^R C^R$	1/4 red flowers
	2/4 heterozygous $C^W C^R$	2/4 pink flowers
	1/4 homozygous $C^W C^W$	1/4 white flowers

FIGURE 7-18 Pink snapdragons demonstrate incomplete dominance. When true-breeding white and red snapdragons are crossed, offspring have pink flowers.

CODOMINANCE

FIGURE 7-19 With codominance, a heterozygous individual shows features of both alleles.

receptors as heterozygotes and experience significantly lower levels of circulating cholesterol and a much lower risk of cardiovascular disease.

A second situation in which complete dominance is not observed is called **codominance,** in which the heterozygote displays characteristics of both homozygotes, playfully represented in **FIGURE 7-19** (although "shirt phenotype," of course, has no genetic basis). In codominance, neither allele masks the effect of the other. An actual example of codominance occurs with feather color in chickens. When white chickens are crossed with black chickens, all the offspring have both white and black feathers.

TAKE-HOME MESSAGE 7·9

Sometimes the effects of both alleles in a heterozygous genotype are evident in the phenotype. With incomplete dominance, the phenotype of a heterozygote appears to be an intermediate blend of the phenotypes of the two homozygotes. With codominance, a heterozygote has a phenotype that exhibits characteristics of both homozygotes.

7·10 What's your blood type? Some genes have more than two alleles.

Do you know your blood type? It can be O, A, B, or AB. Each of these different blood types (also called blood groups) indicates something about the physical characteristics of your red blood cells and has implications for blood transfusions—both giving and receiving blood. The blood groups are also interesting from a genetic perspective, because they illustrate a case of **multiple allelism,** in which a single gene has more than two alleles. Each individual still carries only two alleles—one from the mother and one from the father. But if you survey all of the alleles present in the *population,* you will find more than just two alleles.

Inheritance of the ABO blood groups provides the simplest example of multiple allelism, because there are only three alleles. We can call these alleles I^A, I^B, and i. The I^A and I^B alleles are both completely dominant to i, so individuals are considered to have blood type A whether they have the genotype $I^A I^A$ or $I^A i$ (**FIGURE 7-20**). Similarly, an individual with the genotype $I^B I^B$ or $I^B i$ is considered to have blood type B. If you carry two copies of the i allele, you have blood type O. The I^A and I^B alleles are codominant with each other, so the genotype $I^A I^B$ gives rise to blood type AB. Consequently, with these three alleles in the population, individuals can be one of four different blood types: A, B, AB, or O.

What are the phenotypes of these alleles? An individual's blood-type alleles carry instructions that direct construction of a specific set of chemicals, called antigens, that protrude from every red blood cell. These antigens are molecules (chiefly carbohydrates bound to protein) that jut from the surface of a cell and can "turn on" a body's defenses against foreign invaders. The I^A allele directs the production of A antigens all over the surface of red blood cells. Similarly, the I^B allele directs the production of B antigens on all red blood cells. The i allele does not code for the A antigen *or* the B antigen.

This means that individuals with blood type AB have red blood cells with both A and B antigens, while individuals with blood type O have red blood cells that have neither A nor B antigens on their surface (**FIGURE 7-21**).

Antigens on red blood cells play a role in the body's disease-fighting immune system. Antigens are like signposts, telling

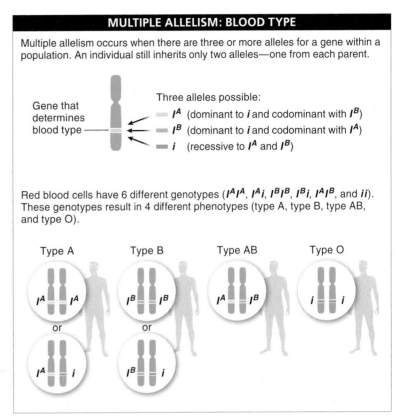

MULTIPLE ALLELISM: BLOOD TYPE

Multiple allelism occurs when there are three or more alleles for a gene within a population. An individual still inherits only two alleles—one from each parent.

Gene that determines blood type

Three alleles possible:
I^A (dominant to i and codominant with I^B)
I^B (dominant to i and codominant with I^A)
i (recessive to I^A and I^B)

Red blood cells have 6 different genotypes ($I^A I^A$, $I^A i$, $I^B I^B$, $I^B i$, $I^A I^B$, and ii). These genotypes result in 4 different phenotypes (type A, type B, type AB, and type O).

Type A

I^A I^A

or

I^A i

Type B

I^B I^B

or

I^B i

Type AB

I^A I^B

Type O

i i

FIGURE 7-20 Multiple allelism. There are three different alleles—I^A, I^B, and i—for blood type in humans.

the immune system whether a cell belongs in the body or not. If a red blood cell with the wrong antigens enters your bloodstream, your immune system recognizes it as a foreign invader and destroys it. Such an attack is initiated by molecules in the bloodstream called *antibodies,* which attack only foreign antigens. Individuals with only A antigens on their red blood cells produce antibodies that attack B antigens. If these cells encounter a red blood cell with B antigens, they attack it. Such an immune response can lead to destruction of red blood cells, low blood pressure, and even death. Under normal circumstances, antibodies do not encounter a red blood cell with foreign antigens. Such an event may occur, however, if red blood cells with foreign antigens are accidentally injected into the person's bloodstream in a transfusion.

Individuals with only B antigens on their red blood cells produce antibodies that attack A antigens. Individuals with

BLOOD TYPE, ANTIGENS, AND ANTIBODIES

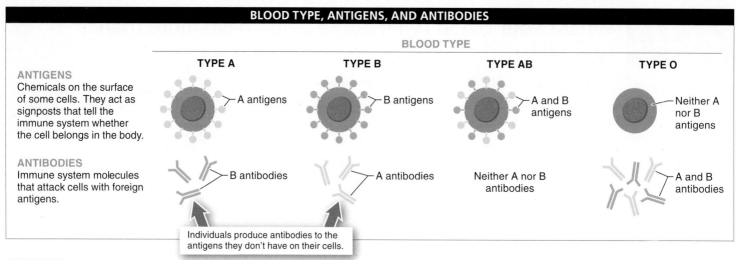

BLOOD TYPE

ANTIGENS
Chemicals on the surface of some cells. They act as signposts that tell the immune system whether the cell belongs in the body.

TYPE A	TYPE B	TYPE AB	TYPE O
A antigens	B antigens	A and B antigens	Neither A nor B antigens

ANTIBODIES
Immune system molecules that attack cells with foreign antigens.

B antibodies — A antibodies — Neither A nor B antibodies — A and B antibodies

Individuals produce antibodies to the antigens they don't have on their cells.

FIGURE 7-21 Friend or foe? Antigens are signposts that tell the immune system whether or not a cell belongs in the body.

Q *Why are people with type O blood considered "universal donors"? Why are those with type AB considered "universal recipients"?*

blood type O, who have neither A nor B antigens on their red blood cells, produce antibodies that attack both A and B antigens. Individuals with blood type AB don't produce either type of antibody (or else they would have antibodies that attacked their own blood cells). From this information, we can deduce which blood types can be used in transfusions. Individuals with blood type O are universal donors, because their red blood cells have no A or B antigens and so do not trigger a reaction from either type of antibody. And individuals with blood type AB are universal recipients, because they do not produce antibodies to either the A or B antigen. This is shown in **FIGURE 7-22**.

THE SCIENCE BEHIND BLOOD DONATION

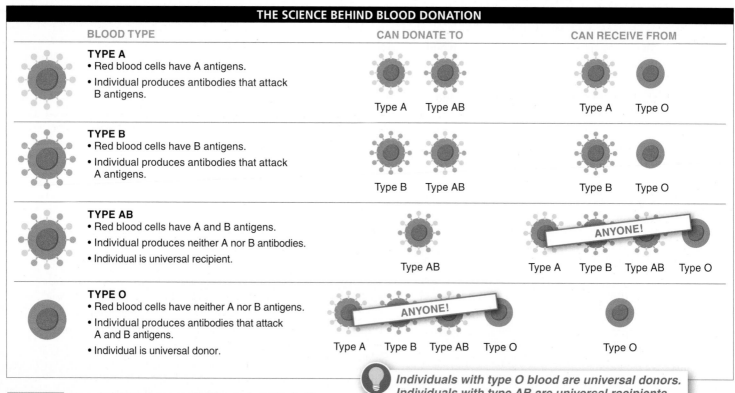

BLOOD TYPE	CAN DONATE TO	CAN RECEIVE FROM
TYPE A • Red blood cells have A antigens. • Individual produces antibodies that attack B antigens.	Type A Type AB	Type A Type O
TYPE B • Red blood cells have B antigens. • Individual produces antibodies that attack A antigens.	Type B Type AB	Type B Type O
TYPE AB • Red blood cells have A and B antigens. • Individual produces neither A nor B antibodies. • Individual is universal recipient.	Type AB	ANYONE! Type A Type B Type AB Type O
TYPE O • Red blood cells have neither A nor B antigens. • Individual produces antibodies that attack A and B antigens. • Individual is universal donor.	ANYONE! Type A Type B Type AB Type O	Type O

Individuals with type O blood are universal donors. Individuals with type AB are universal recipients.

FIGURE 7-22 Mapping blood compatibility. An individual will mount an immune response to red blood cells if he or she produces an antibody that attacks an antigen present on the donated cells.

Another marker on the surface of red blood cells is the Rh blood group marker. (Note that the Rh blood group is not an example of multiple allelism. A single gene with just two alleles determines the presence of the Rh marker. However, like the ABO blood groups, the Rh marker restricts the type of blood a person can receive in a transfusion.) Individuals who possess red blood cells that carry the Rh cell surface marker have one or two copies of the dominant Rh marker allele, and they are said to be "Rh-positive." This "positive" (or "+") is noted along with their ABO blood type, as in "O-positive" or "B-negative." Individuals who have two copies of the recessive allele for this gene do not have any Rh markers, and they are described as "negative," as in "O-negative" or "A-negative." If, during a blood transfusion, individuals who are Rh-negative are exposed to Rh-positive blood, their

immune system attacks the Rh antigens as foreign invaders—an immune response that can vary from mild, which passes unnoticed, to severe, which can lead to death.

Beyond the ABO marker groups, there are many, many genes with multiple alleles—a dozen or even more alleles in some cases. In fact, one gene for eye color in fruit flies has more than 1,000 different alleles!

> ### TAKE-HOME MESSAGE 7·10
>
> In multiple allelism, a single gene has more than two alleles. Each individual still carries only two alleles, but in the population, more than just two alleles exist. This is the case for the ABO blood groups in humans.

7·11 Multigene traits: how are continuously varying traits such as height influenced by genes?

When babies are born, the parents are often curious about how tall their child will grow to be. Old wives' tales suggest a couple of ways for predicting height. If the baby is a boy, they say to add 5 inches to the mother's height and average that with the father's height. If it is a girl, subtract 5 inches from the father's height and average that with the mother's height. Alternatively, the lore says, just take the child's height at two years of age and double it.

These methods can be surprisingly accurate at making predictions, but they don't really help us understand the underlying reason *why* height can be predicted so well. The reason these methods work is that genes play a strong role in influencing height, and so offspring do resemble their parents in measurable ways. But unlike the simple case of Mendel's pea plants, where a single gene with two alleles determines the height of the plant, for humans and most animals, adult height—a continuously varying trait—is influenced by many different genes. Such a trait is said to be **polygenic.**

Recent research has identified at least 180 heritable loci that influence adult height. For each of these loci, the alleles that a person carries play a role in determining his or her height. Individuals with "tall" variants for more of the genes tend to be taller than those with the "tall" alleles for fewer of the genes. The term **additive effects** describes what happens when the effects of the alleles of multiple genes all contribute to the ultimate phenotype.

Of the genes influencing height, a large number play roles in skeletal growth and hormone pathways. One of these

height genes, for example, is on chromosome 15 and codes for an enzyme that converts testosterone to estrogen. This enzyme influences height because estrogen helps bones fuse at their ends and thus stop growing. The variety of heights seen among humans—from very short to very tall, with every height in between—reflects the fact that height is a trait influenced by contributions from multiple genes, as well as the environment (**FIGURE 7-23**).

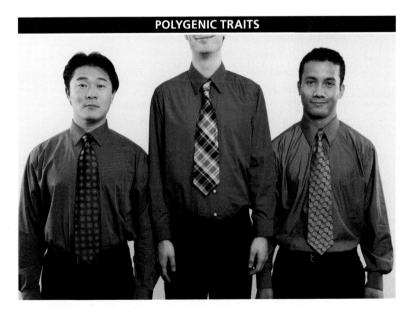

POLYGENIC TRAITS

FIGURE 7-23 From many genes, one trait. Height and skin color are multigene traits.

Many other physical traits are influenced by multiple genes, including eye color in humans. Eye color was long believed to be controlled by a single gene with a dominant brown-eye allele and a recessive blue-eye allele, but it now seems that eye color is the result of the interactions between at least two genes and possibly more—a situation that makes more sense, given the significant continuous variation in eye color seen among adults.

Many behavioral traits are influenced by multiple genes. The developmental disorder known as autism, for example, seems to be the result of alterations in numerous—perhaps as many as 10 or even 20—different genes. Individuals with autism have difficulty interacting with others, particularly in making emotional connections. Autistic individuals also tend to have narrowly focused and repetitive interests. Autism is notoriously difficult to study, though, because its symptoms are varied, as is their intensity. It turns out that these variations in the disorder may result from the different combinations of the many genes involved in autism, much as different alleles for the many height genes may work together to produce a variety of heights.

Interestingly, the genes responsible for autism may also influence some desirable characteristics. This idea—called the "geek theory of autism"—was first suspected when an upsurge in cases of autism in children was noticed in areas with large numbers of high-tech workers, including Silicon Valley in California and near Cambridge, England, in the heart of the United Kingdom's high-technology industry. Subsequently, researchers have published findings showing an over-representation of autistic children among parents working in the fields of engineering, physics, computer science, and math (although the researchers point out that autistic children are born to parents across all professions and socio-economic backgrounds). Perhaps alleles for genes that contribute to making individuals good at computer programming and solving complex technology problems can produce autism when someone carries too many of them.

> **TAKE-HOME MESSAGE 7·11**
>
> Many traits, including continuously varying traits such as height and eye color, are influenced by multiple genes.

7·12 Sometimes one gene influences multiple traits.

Just as multiple genes can influence one trait, some individual genes can influence multiple, unrelated traits, a phenomenon called **pleiotropy.** In fact, this may be true of nearly all genes. Consider sickle-cell disease (also called sickle-cell anemia), a potentially fatal condition in which individuals produce defective red blood cells that change their shape, becoming sickle-shaped, when they lose the oxygen they carry. The defective blood cells can't effectively transport oxygen to tissues, and they accumulate in blood vessels, causing extreme pain. Individuals with sickle-cell disease suffer shortness of breath and numerous other problems that lead to a significantly reduced life span. The gene responsible for sickle-cell disease encodes the molecule hemoglobin, the oxygen-carrying molecule in red blood cells: Hb^A is the allele for normal hemoglobin, and Hb^S is the abnormal, "sickle-cell" allele. Sickle-cell anemia occurs in individuals homozygous for the sickle-cell allele, $Hb^S Hb^S$, but not in individuals carrying at least one copy of the normal allele, Hb^A (although heterozygous individuals produce both normal and sickling red blood cells, just not enough to cause sickle-cell anemia).

Q What is the benefit of "almost" having sickle-cell disease?

This hemoglobin gene is pleiotropic because, although it is just one gene, it causes a cascade of different phenotypic effects. Individuals homozygous or heterozygous for the sickle-cell allele, $Hb^S Hb^S$ or $Hb^A Hb^S$, have abnormal hemoglobin, red blood cell deformation, and circulatory problems. Moreover, individuals with these genotypes also are resistant to the parasite that causes malaria. The resistance is due to the fact that the malarial parasite—which lives in red blood cells—cannot survive well in cells that carry the defective version of the hemoglobin gene. And because even individuals who are heterozygous for the sickle-cell allele have a significant number of sickling red blood cells, their bloodstream is just not a hospitable environment for the malarial parasite (**FIGURE 7-24**). Consequently, this one gene influences multiple traits.

Another example of pleiotropy is the SRY gene. Named for "**s**ex-determining **r**egion on the **Y** chromosome," this gene causes fetal gonads to develop as testes shortly after fertilization. Following the testes' secretion of testosterone, a cascade of other developmental changes then takes place, such as development of the internal and external male

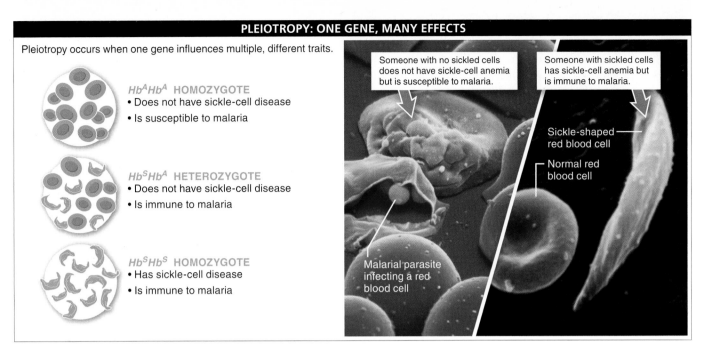

PLEIOTROPY: ONE GENE, MANY EFFECTS

Pleiotropy occurs when one gene influences multiple, different traits.

$Hb^A Hb^A$ **HOMOZYGOTE**
• Does not have sickle-cell disease
• Is susceptible to malaria

$Hb^S Hb^A$ **HETEROZYGOTE**
• Does not have sickle-cell disease
• Is immune to malaria

$Hb^S Hb^S$ **HOMOZYGOTE**
• Has sickle-cell disease
• Is immune to malaria

Someone with no sickled cells does not have sickle-cell anemia but is susceptible to malaria.

Someone with sickled cells has sickle-cell anemia but is immune to malaria.

Sickle-shaped red blood cell

Normal red blood cell

Malarial parasite infecting a red blood cell

FIGURE 7-24 **From one gene, multiple traits.** The allele for sickle-cell disease is pleiotropic: it causes red blood cells to form an unusual, sickled shape, and it also provides resistance to malaria.

reproductive structures (including the prostate gland, seminal vesicles, vas deferens, penis, and scrotum). Ultimately, the SRY gene is responsible for numerous behavioral characteristics as well. These characteristics are described further in Chapter 9.

> **TAKE-HOME MESSAGE 7·12**
>
> In pleiotropy, one gene influences multiple, unrelated traits. Most, if not all, genes may be pleiotropic.

7·13 Why are more men than women color-blind? Sex-linked traits differ in their patterns of expression in males and females.

The patterns of inheritance of most traits do not differ between males and females. When a gene is on an autosome (one of the non-sex chromosomes), both males and females inherit two copies of the gene, one from their mother and one from their father. The likelihood that an individual inherits one particular genotype rather than another does not differ between males and females.

Traits coded for by the sex chromosomes, on the other hand, have different patterns of expression in males and females. One of the most easily observed examples of this phenomenon is red-green color-blindness. On the X chromosome in humans, there is a gene that carries the instructions for producing light-sensitive proteins in the eye that make it possible to distinguish between the colors red

and green. As long as an individual has at least one functioning copy of this gene, he or she produces sufficient amounts of the protein to have normal color vision.

There is a rare allele for this gene, however, that produces a non-functioning version of the light-sensitive protein. Having some of this non-functioning protein is not a problem as long as the person also carries another, normal version of the gene and produces some of the functioning protein.

Here's the problem: men get only one chance to inherit the normal version of the gene that codes for red-green color vision. The gene is on the X chromosome, and men inherit this chromosome only from their mother. Women get two

FAMILY RESEMBLANCE GENETICS AND CHANCE PHENOTYPES FROM GENOTYPES GENE LINKAGE

299

Q If a man is color-blind, did he inherit this condition from his mother, his father, or both parents?

chances. Although a woman may inherit the defective allele from one parent, she can still inherit the normal allele from the other parent. As long as she inherits the normal gene from one parent, she will have normal color vision. (If she inherits the defective allele from both parents, she will be red-green color-blind.) As we would predict, then, the frequency of red-green color-blindness is significantly greater in males than in females (**FIGURE 7-25**). Approximately 7% to 10% of men exhibit red-green color-blindness, while fewer than 1% of women are red-green color-blind.

Although males exhibit sex-linked recessive traits more frequently than do women, the situation is reversed for sex-linked dominant traits. In these cases, because females have two chances to inherit the allele that causes the trait, they are more likely to have the allele and thus exhibit the trait than are males, who have only one chance to inherit the allele.

TAKE-HOME MESSAGE 7·13

The patterns of inheritance of most traits do not differ between males and females. However, when a trait is coded for by a gene on a sex chromosome, such as color vision on the X chromosome, the pattern of expression differs for males and females.

SEX-LINKED TRAITS: COLOR-BLINDNESS

A sex-linked trait is carried on the X chromosome. Women carry two copies of the X chromosome, while men carry an X chromosome and a Y chromosome.

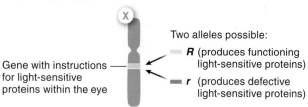

Gene with instructions for light-sensitive proteins within the eye

Two alleles possible:
- **R** (produces functioning light-sensitive proteins)
- **r** (produces defective light-sensitive proteins)

TO BE COLOR-BLIND...

Male must inherit color-blindness allele (**r**) from his mother.

Female must inherit color-blindness allele (**r**) from both parents.

TO HAVE NORMAL VISION...

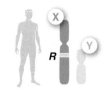 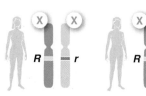

Male must inherit normal color-vision allele (**R**) from his mother.

Female can inherit normal color-vision allele (**R**) from either her mother or her father.

FIGURE 7-25 **Sex-linked traits such as color-blindness are not expressed equally among males and females.**

? ## 7·14 THIS IS HOW WE DO IT

What is the cause of male-pattern baldness?

It is a rare scientist who is interested only in finding a particular answer. Most are drawn to scientific thinking because it illuminates productive ways of *approaching* a problem. The greatest advances in understanding often come about by exploring new approaches to old problems. And here we explore one of the oldest.

Why do men lose their hair? Conventional wisdom has long suggested that baldness in men is a trait they inherit from their mother. This hypothesis goes back to 1916, when Dorothy Osborn published the first scientific study putting forth heredity as a cause of baldness. In her

paper, she took aim squarely at one of the most widely held hypotheses of the time: that the wearing of hats caused baldness due to pressure they put on blood vessels that nourish the scalp.

| Can you propose how to test the "hats cause baldness" hypothesis?

Falsifiable hypotheses have the potential to be rejected. It was easy for Osborn to demonstrate that, as she put it, "the hat is not to blame." In spite of definitive evidence against them, however, numerous equally wrong hypotheses persist today: "The hair follicles are clogged from shampoo and too much washing." "Not enough blood is circulating around the scalp." "Hair gels and other products are toxic."

> **Are these hypotheses falsifiable? Propose how to test each of them.**

What Osborn found was that it's actually pretty difficult to study the inheritance patterns of baldness carefully. For starters, it's very common—about 50% of men experience some balding by the age of 50, and 70% by age 70.

> **Why do these observations make conventional pedigree analysis difficult?**

Also, because balding increases with age, it's hard to know whether younger, non-bald men will go bald later or never at all. And it's unclear whether all patterns of balding have the same underlying causes or should be considered different phenomena.

A modern approach. In 2005, researchers took a very modern approach to studying the inheritance of baldness. They studied in detail the DNA of 391 men—including 201 balding men—from 95 families in which at least two brothers exhibited early-onset male-pattern baldness. For comparison, they also examined the DNA of additional, unrelated men who were either under the age of 40 with male-pattern baldness or were over the age of 60 and unaffected by baldness.

The most common pattern the researchers found was that the men with male-pattern baldness were significantly more likely to share one particular stretch of DNA on their X chromosome. The unaffected men were significantly more likely to share a different sequence of DNA.

> **Balding men commonly shared a DNA sequence on their X chromosome. What does that tell us about whom they inherited that DNA from?**

The finding that the DNA region implicated in baldness was on the X chromosome confirmed the long-standing observation that male-pattern baldness is a trait passed down to men from their mothers. Men receive their sole X chromosome from their mother and their Y chromosome from their father.

Future directions? This research resolved one long-standing debate. In the researchers' words, "the average phenotypic resemblance should be greater between affected males and their maternal grandfathers than between affected males and their fathers." More

importantly, perhaps, it illuminated an avenue for research on the *treatment* of male-pattern baldness. The DNA sequence associated with male-pattern baldness, it turned out, is located within the region of a single gene—a gene already suspected to play a role in triggering baldness. The gene carries instructions for the production of androgen receptors. Androgens are male sex hormones, such as testosterone.

The DNA sequence that was found so much more frequently among individuals exhibiting male-pattern baldness was associated with higher activity of the androgen receptor gene. And although the researchers were not able to determine the exact relationship between the androgen receptor gene and male-pattern baldness, they suspect that the DNA sequence somehow increases the impact of androgens, thereby leading to hair loss. This is consistent, they noted in their paper, with the finding that castrated males—who produce almost no androgens at all—don't go bald.

> **What strategies for treating male-pattern baldness do these observations suggest?**

Stopping short of the one-gene = one trait conclusion. The researchers concluded that this maternally inherited allele does play a central role in the inheritance of male-pattern baldness. But they stopped short of saying that it is the only contributor to male-pattern baldness. And in a later study, they identified some autosomal genes (on chromosome 20) that contribute to a smaller, but still important, similarity between fathers and sons in patterns of male-pattern baldness.

> **This research increases our confidence that baldness in men is a trait they inherit from their mother. No one would say, however, that one or several genes *cause* male-pattern baldness. Discuss the several reasons why such a definitive assertion should not be made.**

TAKE-HOME MESSAGE 7·14

Observations of male-pattern baldness within families and comparisons with unrelated individuals suggest that the baldness is caused by a sex-linked gene that codes for an androgen receptor. Males inheriting an allele—always from their mother—for higher activity of the androgen receptor gene are more likely to have male-pattern baldness than males inheriting an alternative allele.

It is a very serious warning, in boldfaced capital letters: **"PHENYLKETONURICS: CONTAINS PHENYLALANINE."** But it's not next to a skull and crossbones on a glass bottle in a chemistry lab, it's on cans of diet soda intended for human consumption (**FIGURE 7-26**). Most of us either don't notice the warning or ignore it. Still, it's always there, and to some people it's a matter of life and death. What does it mean?

At the most fundamental level, this warning is on products that contain phenylalanine because an organism's phenotype is a product of its genes in combination with its environment. In this case, specifically, the warning is for people who have a particular genotype that, in the presence of the amino acid phenylalanine, can be deadly.

Sometimes our bodies use phenylalanine directly to build proteins, adding it to a growing amino acid chain. At other times, phenylalanine is chemically converted into another amino acid, called tyrosine. The body may then use tyrosine as one of the building blocks as it constructs proteins, and in a variety of other functions.

The problem is this: at birth, some people carry two copies of a mutant version of the gene that is supposed to produce the enzyme that converts phenylalanine into tyrosine. The mutant gene produces a malfunctioning enzyme, and none

of the body's phenylalanine is converted into tyrosine. Little by little, as these individuals consume phenylalanine, it builds up in their bodies because none of it is converted into tyrosine. If this continues, babies usually begin to show symptoms within 3 to 6 months. Within a few years, so much accumulates that it reaches toxic levels and poisons the brain, leading to mental retardation and other serious health problems. The disease is called phenylketonuria, or PKU.

Here's where the warning label comes in. By limiting the amount of phenylalanine in their diet (from diet soda and other sources), individuals with the two mutant alleles for processing phenylalanine can avoid the toxic buildup of the amino acid in their brain. In essence, they modify their environment (their diet) so that it contains only a tiny, carefully monitored amount of phenylalanine. Because all newborn babies in the United States are screened for PKU at birth, they can be treated with the appropriate diet immediately. And in an environment free of excess phenylalanine, the PKU mutant alleles are harmless. This example highlights the fact that genes, by themselves, do not "code" for physical characteristics. Rather, genes interact with the environment to produce physical characteristics. Unless you have information about both the genes and the environment, it is not usually possible to know what the phenotype will be.

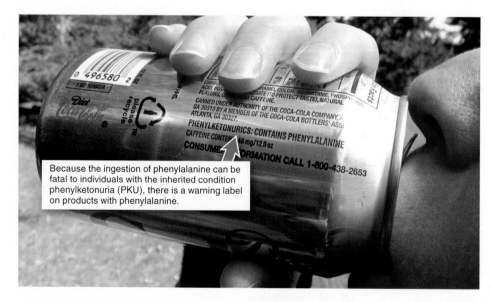

Because the ingestion of phenylalanine can be fatal to individuals with the inherited condition phenylketonuria (PKU), there is a warning label on products with phenylalanine.

FIGURE 7-26 Change the environment, "cure" the disease.

Q Could you create a temporarily spotted Siamese cat with an ice pack? Why?

A non-life-threatening illustration of the interaction of genes with the environment can be seen at a pet shop. Siamese cats (as well as Himalayan rabbits) carry genes that produce dark pigmentation. These genes interact strongly with the environment and are heat-sensitive. Dark pigment is produced only in relatively cold areas of the animal's body, while warm areas remain very light in color. This is why the fur on the coldest parts of the body—the ears, paws, tail, and tip of the face—becomes the darkest, while the fur on the rest of the body remains cream-colored or white (**FIGURE 7-27**). For Siamese cats living in cold climates and spending a lot of time outside, it's interesting to notice that they become significantly darker in color during the winter months. Those that lounge indoors all winter remain lighter in color.

There are thousands of other cases in which genes' interactions with the environment influence their ultimate effects in the body. The very fact that identical twins—who inherit exactly the same set of alleles at the time of fertilization—don't die on the same day at the same moment and from the same cause reveals that environmental variation influences the expression of genes. The scope of environmental influences ranges from traits with large and obvious environmental effects, such as body weight and its relationship with caloric intake, to traits such as eye color that are barely influenced by environmental effects, to traits with complex and subtle interactions with the environment, such as intelligence or personality.

Q James Watson, the co-discoverer of DNA, once wrote that when we completed the Human Genome Project, we would have "the complete genetic blueprint of man." Why might that be a poor metaphor?

Because of the role that environmental factors play in influencing phenotypes, DNA is not like a blueprint for a house. There is nearly always significant interaction between the genotype and the environment that influences the exact phenotype produced. The use of this metaphor is problematic to the extent that it suggests the phenotype is determined *solely* by the genotype. If that were the case, there would be no reason to invest in better schools, physical fitness

Some pigment genes produce dark pigment only under cold conditions—such as on the tail, nose, ears, and feet of these animals.

FIGURE 7-27 Heat-sensitive fur color. Some pigment-producing genes produce the dark pigment of fur only under cold conditions. That's why these animals have darker patches of fur on their extremities.

regimens, nutritional monitoring, self-help efforts, or any other process by which individuals or societies try to improve people's lives (i.e., alter phenotypes) by enriching their environment.

TAKE-HOME MESSAGE 7·15

Genotypes are not like blueprints that specify phenotypes. Phenotypes are generally a product of the genotype in combination with the environment.

7·16–7·17
Some genes are linked together.

Different genes influence red hair and freckles, so why are they often inherited together?

7·16 Most traits are passed on as independent features: Mendel's law of independent assortment.

Sometimes you can be right about something for the wrong reason. This happened to Gregor Mendel. He didn't know that genes were carried on chromosomes. He believed that the units of heredity were just free-floating entities within cells. Given this perspective, it made sense to him that the inheritance pattern of one trait wouldn't influence the inheritance of any other trait. He believed that all genes behaved independently.

> **"** ... the relation of each pair of different characters in hybrid union is independent of the other differences in the two original parental stocks. **"**
>
> — GREGOR MENDEL, clearly articulating the idea of independent assortment in his publication "Experiments in Plant Hybridization" (1865)

It helps to consider an example. Earlier in this chapter we saw that all cats with completely white fur have at least one parent that also has completely white fur. This is because white fur is caused by a single dominant gene.

Imagine that you had a true-breeding population of cats with white fur (remember, "true-breeding" for a trait

means that all offspring always manifest the trait; in the case of white fur, all of the individuals have the genotype WW). Now suppose that an individual in this population mated with a cat from a true-breeding population of cats that all had some colored fur (all individuals have the genotype ww). All of their offspring would have white fur, but they would be heterozygous (Ww), getting a dominant allele from their white-furred parent and a recessive allele from the other parent. If two heterozygotes had offspring together, though, they would produce three-quarters white-furred and one-quarter non-white-furred offspring, with genotypes in the ratio of $\frac{1}{4}WW$, $\frac{1}{2}Ww$, and $\frac{1}{4}ww$. That is just what Mendel observed for traits in pea plants.

But what if we concurrently observed another characteristic of these cats? Suppose the original true-breeding population of white-furred cats (WW) was also true-breeding for long hair (all ll), a condition caused by carrying two recessive alleles for a single gene. And suppose that individuals in the colored-fur population (ww) were also true-breeding for short hair (all LL). The question is, do the alleles an individual inherits for the white-fur trait influence which alleles that individual inherits for fur length? And the

MENDEL'S LAW OF INDEPENDENT ASSORTMENT

Mendel's law of independent assortment states that one trait does not influence the inheritance of another trait.

IF...
Parents are both heterozygous for both traits (i.e., "doubly heterozygous"). Four different types of gametes are produced by each: **LW**, **Lw**, **lW**, and **lw**.

MOTHER
short hair
heterozygous **Ll**
white fur
heterozygous **Ww**

FATHER
short hair
heterozygous **Ll**
white fur
heterozygous **Ww**

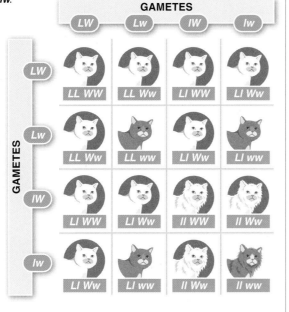

GAMETES: LW, Lw, lW, lw

	LW	Lw	lW	lw
LW	LL WW	LL Ww	Ll WW	Ll Ww
Lw	LL Ww	LL ww	Ll Ww	Ll ww
lW	Ll WW	Ll Ww	ll WW	ll Ww
lw	Ll Ww	Ll ww	ll Ww	ll ww

THEN...
The genotype proportions for **L/l** are still ¼ **LL**, ½ **Ll**, and ¼ **ll**. And the genotype proportions for **W/w** are still ¼ **WW**, ½ **Ww**, and ¼ **ww**.

OFFSPRING	GENOTYPE	PHENOTYPE
	1/4 homozygous dominant **LL**	3/4 short-haired
	2/4 heterozygous **Ll**	
	1/4 homozygous recessive **ll**	1/4 long-haired
	1/4 homozygous dominant **WW**	3/4 white-furred
	2/4 heterozygous **Ww**	
	1/4 homozygous recessive **ww**	1/4 non-white-furred

 In this example, having white fur does not affect which alleles are inherited for fur length.

FIGURE 7-28 Independent assortment of genes.

answer is that they do not (**FIGURE 7-28**). Rather, the first cross of a long-haired, all-white cat with a cat having short, colored fur would result in offspring heterozygous for both traits—referred to as **dihybrid**—and expressing each of the dominant traits.

Phenotypically, all of the offspring from this cross would have short, completely white fur. In a mating between two of these doubly heterozygous individuals—referred to as a **dihybrid cross**—three-quarters of the offspring would have the dominant trait and one-quarter would have the recessive trait, regardless of which trait you are tallying. In other words, neither trait influences the inheritance pattern for the other trait; all traits are inherited independently of each other. This is known as **Mendel's law of independent assortment.**

In the next section we'll see that, despite Mendel's correct understanding that separate traits are inherited independently, his belief that this happened because all genes just float freely around in the cell was not correct. The genes, as we now know, are carried on chromosomes. And this sometimes leads to situations in which independent assortment does *not* occur.

TAKE-HOME MESSAGE 7·16

Genes tend to behave independently, such that the inheritance pattern of one trait doesn't usually influence the inheritance of any other trait.

7·17 Red hair and freckles: genes on the same chromosome are sometimes inherited together.

Most redheads have pale skin and freckles. This simple observation is problematic for the law of independent assortment as Mendel imagined it. After all, in his law, he asserted that the inheritance of one trait does not influence the inheritance of another. But clearly having a trait, such as red hair, seems to influence the presence of another trait, pale skin (**FIGURE 7-29**). Strictly speaking, Mendel's second law is not true for *every* pair of traits. Sometimes the alleles for two genes are inherited together and expressed almost as a package.

Q Why do most redheads have pale skin?

In the case of humans, for example, there are about 21,000 genes in our genome. Yet we have only 23 unique chromosomes (two copies of each). Thus, genes influencing different traits must be on the same chromosome, maybe even right next to each other. When they are close together, we say that they are **linked genes.** One 2008 study, for example, demonstrated a link between human genes that influence hair color and skin pigmentation (including freckles).

Why are linked genes inherited together? To answer this, we must revisit the behavior of chromosomes during the production of gametes, discussed in Chapter 6. When you made that sperm or egg by meiosis, only one of the two copies of each of your chromosomes ended up in the gamete. It may

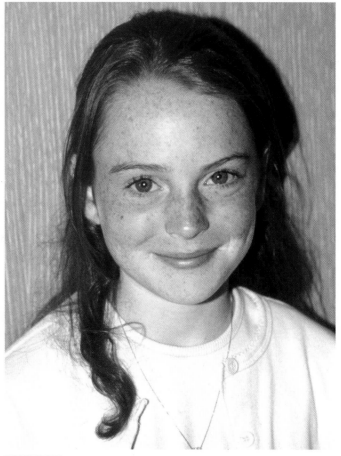

FIGURE 7-29 **Violating the law of independent assortment.** Red hair and freckles are often inherited as a package deal.

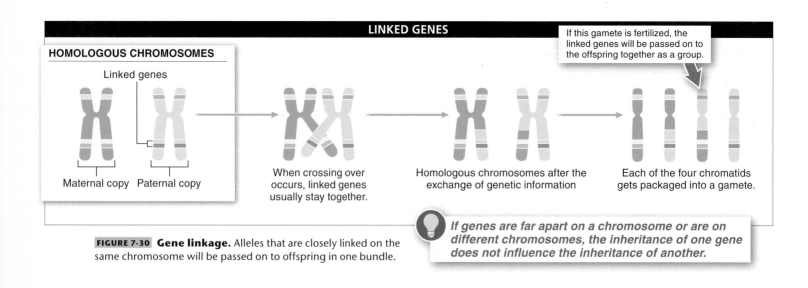

LINKED GENES

HOMOLOGOUS CHROMOSOMES

Linked genes

Maternal copy   Paternal copy

When crossing over occurs, linked genes usually stay together.

Homologous chromosomes after the exchange of genetic information

Each of the four chromatids gets packaged into a gamete.

If this gamete is fertilized, the linked genes will be passed on to the offspring together as a group.

FIGURE 7-30 **Gene linkage.** Alleles that are closely linked on the same chromosome will be passed on to offspring in one bundle.

💡 *If genes are far apart on a chromosome or are on different chromosomes, the inheritance of one gene does not influence the inheritance of another.*

have been the one from your mother or it may have been the one from your father. In either case, all of the alleles that were on the chromosome from that one parent were passed on as a group to the child that resulted from the fertilization involving that gamete. This process continues generation after generation. The linked alleles never get split up unless, during meiosis, recombination occurs between them, moving one or more to the other chromosome in the pair so that they now become linked with the alleles on that chromosome (**FIGURE 7-30**).

When alleles are linked closely on the same chromosome, Mendel's second law doesn't hold true. It is very surprising—and was fortunate for Mendel—that of the seven pea plant traits he chose for his studies, none of them were close together on the same chromosome. For this reason, they all behaved as if they weren't linked.

TAKE-HOME MESSAGE 7·17

Sometimes, having one trait influences the presence of another trait. This is because the alleles for two genes are inherited and expressed almost as a package deal when the genes are located close together on the same chromosome.

Can a Gene Nudge Us toward Novelty-Seeking (and Spicy Foods)?

On human chromosome 11 there is a gene called DRD4. This gene carries the instructions for building a receptor for the brain chemical dopamine within the membranes of many of your brain cells. All humans have the dopamine receptor gene DRD4 which helps control dopamine, a chemical messenger that alters the activity of the brain's pleasure centers and influences initiative and motivation. A short sequence near the beginning of the DRD4 gene, called "promoter polymorphism–521C/T," has two different variants, alleles C and T. Dopamine activity is increased in individuals with one or two copies of the T allele, and decreased in individuals with the CC genotype.

Q: Can a single gene influence your personality? In 2000, researchers reported that individuals carrying two copies of the C allele for the DRD4 gene, as opposed to zero or one copy of the C allele, were more likely to exhibit certain personality traits, particularly novelty-seeking. The subjects in the study responded to a commonly used questionnaire, called the temperament and character inventory. They ranked how well certain statements described them, such as "I have sometimes done things just for kicks or thrills" or "I think things through before coming to a decision." The subjects were then evaluated for seven dimensions of personality traits. When it came to their score for "novelty-seeking," there was a small but statistically significant difference in how the subjects scored, depending on their genotype for this DRD4 gene.

By 2008, 11 studies had been published on the relationship between the CC genotype and novelty-seeking. In a "meta-study" that evaluated all the published findings together, researchers concluded that these studies demonstrated a significant association between novelty-seeking and the –521C/T genotype. They summarized the results graphically.

Researchers have subsequently reported a variety of behavioral differences among people with higher novelty-seeking scores,

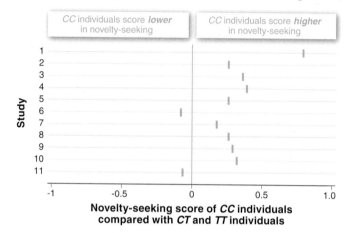

including increased propensities for engaging in high-risk sports, alcohol and drug consumption, and attraction to high-risk vocations. These people have even been reported to show a preference for spicy foods!

Q: Do you want your employers or insurance companies to know which alleles you carry for the DRD4 gene? Information about your genetic predisposition for diseases such as breast cancer or colon cancer can put you at risk for discrimination. Might a person be at risk for discrimination based on his or her genotype for the DRD4 gene? Should it be taken into account that the personality differences influenced by DRD4 are small, that they are affected by other genes, and that there is tremendous variation in personality traits that is unrelated to the DRD4 gene? Would you want to know what your genotype is? Why?

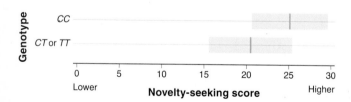

GRAPHIC CONTENT

Thinking critically about visual displays of data

See p. 309 for a closer inspection of this figure.

Check Your Knowledge

GRAPHIC CONTENT

Thinking critically about visual displays of data

1. The top graph displays the novelty-seeking score from subjects who responded to a questionnaire on character and temperament. Based on this information, what was the average novelty-seeking score for individuals with the CC genotype?

2. What does the CC genotype refer to? How does it differ from the CT or TT genotypes?

3. What information is conveyed by the light blue bars? Why is this information helpful?

4. What does it mean if a blue line is at 0?

5. The 11 studies shown all evaluate the same thing, yet their results are not identical. Why might that be?

6. What conclusion can you draw from this figure? How certain are you about that conclusion?

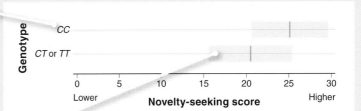

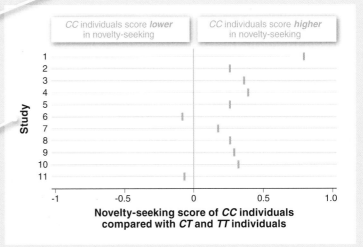

See answers at the back of the book.

Key Terms in Genes and Inheritance

additive effects, p. 297
allele, p. 279
carrier, p. 291
codominance, p. 294
cross, p. 282
dihybrid, p. 305

dihybrid cross, p. 305
dominant, p. 284
genotype, p. 286
heredity, p. 280
heterozygous, p. 285

homozygous, p. 285
incomplete dominance, p. 293
linked gene, p. 306
Mendel's law of independent assortment, p. 305

Mendel's law of segregation, p. 285
multiple allelism, p. 295
pedigree, p. 291
phenotype, p. 286
pleiotropy, p. 298

polygenic, p. 297
Punnett square, p. 286
recessive, p. 284
sex-linked trait, p. 291
single-gene trait, p. 281
test-cross, p. 290
true-breeding, p. 282

ABOUT THE CHAPTER OPENING PHOTO

Identical twins inherit the same genetic material. Shown here are sets of twins posing for a group picture at a gathering of more than five hundred sets of twins in Germany.

7·1–7·5 Why do offspring resemble their parents?

Offspring inherit genes—instruction sets for biochemical, physical, and behavioral traits, some of which are responsible for diseases—from their parents.

? Check Your Knowledge

1. Most genes come in alternative forms called:

a) alleles.

b) heterozygotes.

c) gametes.

d) chromosomes.

e) homozygotes.

0 — **23** — 100
EASY — HARD

2. Traits that are determined by a single gene:

a) occur in single-celled organisms, but not in humans.

b) are common in humans.

c) must occur on the X chromosome.

d) include eye color and skin color.

e) can have only two alleles.

0 — **69** — 100
EASY — HARD

3. Pea plants were well suited for Mendel's breeding experiments for all of the following reasons except:

a) Peas exhibit variations in a number of observable characteristics, such as flower color and seed shape.

b) Mendel could control the pollination between different pea plants.

c) It is easy to obtain large numbers of offspring from any given cross.

d) Many of the characteristics that vary in pea plants are not linked closely on the same chromosome.

e) Peas have a particularly long generation time.

0 — **31** — 100
EASY — HARD

4. The law of segregation states that:

a) the transmission of genetic diseases within families is always recessive.

b) an allele on one chromosome will always segregate from an allele on a different chromosome.

c) gametes cannot be separate and equal.

d) the number of chromosomes in a cell is always divisible by 2.

e) the two alleles for a given trait segregate into different gametes.

0 — **30** — 100
EASY — HARD

INHERITANCE OF GENES

Each human offspring inherits one maternal set of 23 chromosomes and one paternal set.

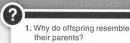

1. Why do offspring resemble their parents?

MATERNAL CHROMOSOME PAIR   **PATERNAL CHROMOSOME PAIR**

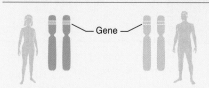

Gene

Humans have 23 pairs of chromosomes (46 individual chromosomes) and, thus, two copies of each gene.

MATERNAL GAMETE: EGG   **PATERNAL GAMETE: SPERM**

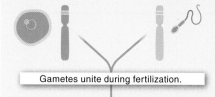

Each human gamete has just one copy of each chromosome and, thus, one copy of each gene.

Gametes unite during fertilization.

Child inherits one set of chromosomes from each parent and, thus, two copies of each gene.

GREGOR MENDEL

In the mid-1800s, Gregor Mendel conducted studies that help us understand heredity. He focused on easily observed and categorized traits in garden peas and applied methodical experimentation and rigorous hypothesis testing to determine how traits are inherited.

2. Why is it simpler to describe heredity using single-gene traits?

3. Describe the main focus of Mendel's research on pea plants.

MENDEL'S LAW OF SEGREGATION

According to Mendel's law of segregation, each parent puts a single set of instructions for building a particular trait into every sperm or egg it makes. This instruction set is called a gene. The trait observed in an individual depends on the two copies (alleles) of the gene it inherits from its parents.

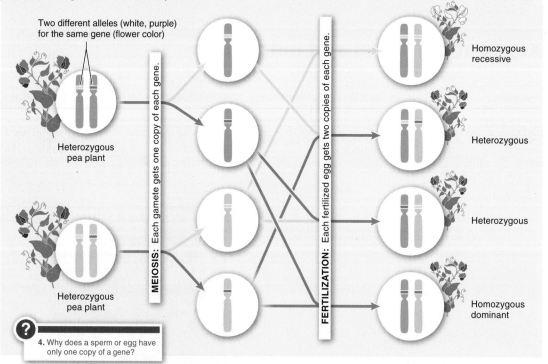

Two different alleles (white, purple) for the same gene (flower color)

Heterozygous pea plant

Heterozygous pea plant

MEIOSIS: Each gamete gets one copy of each gene.

FERTILIZATION: Each fertilized egg gets two copies of each gene.

Homozygous recessive

Heterozygous

Heterozygous

Homozygous dominant

4. Why does a sperm or egg have only one copy of a gene?

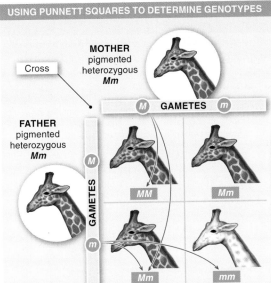

MOTHER
pigmented
heterozygous
Mm

Cross

GAMETES

FATHER
pigmented
heterozygous
Mm

GAMETES

MM   Mm

Mm   mm

It is not always possible to determine an individual's genetic makeup, known as its genotype, by observation of the organism's outward appearance, known as its phenotype. For a particular trait, an individual may carry a recessive allele whose phenotypic effect is masked by the presence of a dominant allele. We can trace the possible outcomes of a cross between 2 individuals using a tool called a Punnett square.

OFFSPRING	GENOTYPE	PHENOTYPE
	1/4 homozygous dominant *MM*	3/4 pigmented
	2/4 heterozygous *Mm*	
	1/4 homozygous recessive *mm*	1/4 albino

? 5. Compare and contrast the terms "phenotype" and "genotype."

7·6—7·8 Probability and chance play central roles in genetics.

Sometimes genetics is a bit like gambling. Even with perfect information, it can still be impossible to know the genetic outcome with certainty.

GENETICS AND PROBABILITY

It is possible to determine the probability of a complex genetic event occurring if you know the probability of each component. You multiply all the probabilities together to get the overall probability of the event.

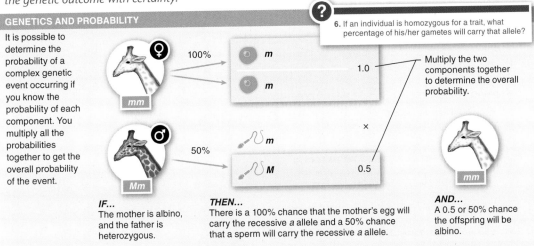

? 6. If an individual is homozygous for a trait, what percentage of his/her gametes will carry that allele?

100%   *m*   1.0

mm

♂   50%   *m*   Multiply the two components together to determine the overall probability.

Mm   *M*   0.5

×

mm

IF...
The mother is albino, and the father is heterozygous.

THEN...
There is a 100% chance that the mother's egg will carry the recessive *a* allele and a 50% chance that a sperm will carry the recessive *a* allele.

AND...
A 0.5 or 50% chance the offspring will be albino.

PEDIGREES

Pedigrees help scientists, doctors, animal and plant breeders, and prospective parents determine the genes that individuals carry and the likelihood that the offspring of two individuals will exhibit a given trait.

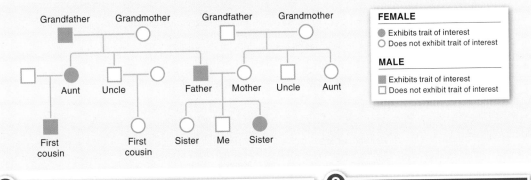

Grandfather   Grandmother   Grandfather   Grandmother

Aunt   Uncle   Father   Mother   Uncle   Aunt

First cousin   First cousin   Sister   Me   Sister

FEMALE
● Exhibits trait of interest
○ Does not exhibit trait of interest

MALE
■ Exhibits trait of interest
□ Does not exhibit trait of interest

? 7. In what situation is a test-cross helpful in determining an individual's genotype?

? 8. What does it mean to be the "carrier of a trait"?

---

5. In pea plants, purple flower color is dominant to white flower color. If two pea plants that are true-breeding for purple flowers are crossed, in the offspring:

a) all of the flowers will be purple.

b) three-quarters of the flowers will be purple and one-quarter will be white.

c) half of the flowers will be purple and one-quarter will be white.

d) one-quarter of the flowers will be purple and three-quarters will be white.

e) all of the flowers will be white.

0 — 23 — 100
EASY — HARD

6. Pea flowers may be purple (*P*) or white (*p*). Pea seeds may be round (*R*) or wrinkled (*r*). What proportion of the offspring from a cross between purple-flowered, round-seeded individuals (heterozygous for both traits) will have both white flowers and round seeds?

a) 1/16

b) 1/2

c) 3/16

d) 3/4

e) 9/16

0 — 57 — 100
EASY — HARD

7. The test-cross:

a) makes it possible to determine the genotype of an individual of unknown genotype that exhibits the dominant version of a trait.

b) is a cross between an individual whose genotype for a trait is not known and an individual homozygous recessive for the trait.

c) sometimes requires the production of multiple offspring to reveal the genotype of an individual whose genotype is unknown (but who exhibits the dominant phenotype).

d) Only a) and b) are correct.

e) Choices a), b), and c) are correct.

0 — 43 — 100
EASY — HARD

8. Which of the following statements is correct regarding pedigree analysis?

a) Darkened squares or circles always represent individuals with the trait being traced.

b) White squares or circles always represent heterozygous individuals.

c) Horizontal lines connect siblings.

d) The length of the vertical lines is dependent on the relatedness between two individuals.

e) Squares represent females, and circles represent males.

0 — 33 — 100
EASY — HARD

9. All of the offspring of a black hen and a white rooster are gray. The simplest explanation for this pattern of inheritance is:

a) multiple alleles.

b) codominance.

c) incomplete dominance.

d) incomplete heterozygosity.

e) sex linkage.

0 — EASY ——(44)—— 100 HARD

10. A woman with type B blood and a man with type A blood could have children with which of the following phenotypes?

a) AB only

b) AB or O only

c) A, B, or O only

d) A or B only

e) A, B, AB, or O

0 — EASY —(27)——— 100 HARD

11. Which of the following traits shows a polygenic method of inheritance?

a) flower color in snapdragons

b) blood type in humans

c) seed color in peas

d) sickle-cell disease in humans

e) skin color in humans

0 — EASY ——(44)—— 100 HARD

12. The impact of a single gene on more than one characteristic is called:

a) incomplete dominance.

b) environmentalism.

c) balanced polymorphism.

d) pleiotropy.

e) codominance.

0 — EASY —(31)——— 100 HARD

13. A rare, X-linked dominant condition in humans, congenital generalized hypertrichosis, is marked by excessive hair growth all over a person's body. Which of the following statements about this condition is incorrect?

a) The son of a woman with this disease has just slightly more than a 50% chance of having this condition.

b) All daughters of a man with this condition will have the condition.

c) The daughter of a woman with this disease has just slightly more than a 50% chance of having this condition.

d) Every son of a woman with this disorder will have this condition.

e) The son of a man with this condition is no more likely to have the condition than the son of a man who doesn't have this condition.

0 — EASY ————(69)— 100 HARD

7·9–7·15 How are genotypes translated into phenotypes?

The world in which each trait is coded for by a single gene with two alleles—one completely dominant and one recessive—and with no environmental effects at all doesn't quite capture the complexity of the world beyond Mendel's pea plants.

INCOMPLETE DOMINANCE

Sometimes the effects of both alleles in a heterozygous genotype are evident in the phenotype. With incomplete dominance, the phenotype of a heterozygote appears to be an intermediate blend of the phenotypes of the two homozygotes. For example, when true-breeding white and red snapdragons are crossed, offspring have pink flowers.

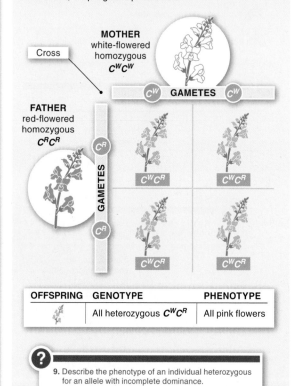

Cross

MOTHER
white-flowered
homozygous
C^WC^W

GAMETES C^W C^W

FATHER
red-flowered
homozygous
C^RC^R

GAMETES C^R C^R

C^WC^R C^WC^R

C^WC^R C^WC^R

OFFSPRING	GENOTYPE	PHENOTYPE
	All heterozygous C^WC^R	All pink flowers

9. Describe the phenotype of an individual heterozygous for an allele with incomplete dominance.

CODOMINANCE

In codominance, neither allele masks the effect of the other and the heterozygote displays characteristics of both homozygotes. For example, when white chickens are crossed with black chickens, the offspring all have both white and black feathers.

POLYGENIC TRAITS

Many traits, including continuously varying traits such as height, eye color, and skin color, are influenced by multiple genes.

11. What does it mean for a trait to be polygenic?

MULTIPLE ALLELISM

Multiple allelism occurs when there are three or more alleles for a gene within a population. An individual still inherits only two alleles—one from each parent.

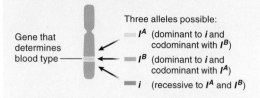

Gene that determines blood type

Three alleles possible:

I^A (dominant to i and codominant with I^B)

I^B (dominant to i and codominant with I^A)

i (recessive to I^A and I^B)

Red blood cells have 6 different genotypes (I^AI^A, I^Ai, I^BI^B, I^Bi, I^AI^B, and ii). These genotypes result in 4 different phenotypes (type A, type B, type AB, and type O).

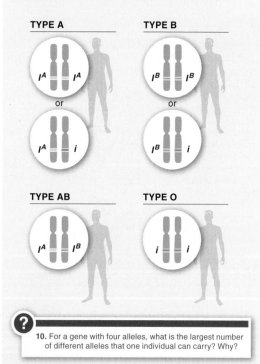

TYPE A
I^A I^A
or
I^A i

TYPE B
I^B I^B
or
I^B i

TYPE AB
I^A I^B

TYPE O
i i

10. For a gene with four alleles, what is the largest number of different alleles that one individual can carry? Why?

PLEIOTROPY

Pleiotropy occurs when one gene influences multiple, unrelated traits. The allele for sickle-cell disease is pleiotropic: it causes red blood cells to form an unusual, sickled shape, and it also provides resistance to malaria.

Hb^AHb^A HOMOZYGOTE
• Does not have sickle-cell disease
• Is susceptible to malaria

Hb^SHb^A HETEROZYGOTE
• Does not have sickle-cell disease
• Is immune to malaria

Hb^SHb^S HOMOZYGOTE
• Has sickle-cell disease
• Is immune to malaria

12. Describe how being heterozygous for the sickle-cell allele can help provide resistance to malaria.

SEX-LINKED TRAITS

The patterns of inheritance of most traits do not differ between males and females. However, when a trait is coded for by a gene on a sex chromosome, such as color vision on the X chromosome, the pattern of expression differs for males and females.

13. Why are sex-linked recessive traits more commonly observed in males than females?

ENVIRONMENTAL FACTORS

Genotypes are not like blueprints that specify phenotypes. Rather, phenotypes are a product of the genotype in combination with the environment.

14. How does the case of PKU support the notion that our phenotype is a product of our genotype in combination with the effect of our environment?

7·16—7·17 Some genes are linked together.

Most traits are passed on as independent features; however, alleles that are closely linked on the same chromosome can be passed on to offspring in one bundle.

MENDEL'S LAW OF INDEPENDENT ASSORTMENT

Many genes tend to behave independently, such that the inheritance pattern of one trait doesn't usually influence the inheritance of any other trait.

IF...
Parents are both heterozygous for both traits (i.e., "doubly heterozygous"). Four different types of gametes are produced by each:
LW, Lw, lW, and *lw.*

MOTHER
short hair
heterozygous *Ll*
white fur
heterozygous *Ww*

FATHER
short hair
heterozygous *Ll*
white fur
heterozygous *Ww*

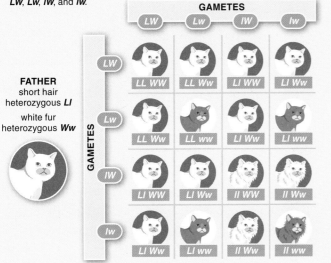

THEN...
The genotype proportions for *L/l* are still ¼ *LL*, ½ *Ll*, and ¼ *ll*. And the genotype proportions for *W/w* are still ¼ *WW*, ½ *Ww*, and ¼ *ww*.

OFFSPRING	GENOTYPE	PHENOTYPE
	1/4 homozygous dominant *LL*	3/4 short-haired
	2/4 heterozygous *Ll*	
	1/4 homozygous recessive *ll*	1/4 long-haired
	1/4 homozygous dominant *WW*	3/4 white-furred
	2/4 heterozygous *Ww*	
	1/4 homozygous recessive *ww*	1/4 non-white-furred

15. What is Mendel's second law? Why doesn't it apply for all traits?

LINKED GENES

When genes are located close together on the same chromosome, the alleles for genes are inherited and expressed almost as a package deal.

HOMOLOGOUS CHROMOSOMES

Linked genes

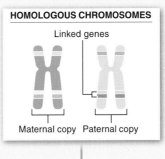

Maternal copy   Paternal copy

When crossing over occurs, linked genes usually stay together.

Homologous chromosomes after the exchange of genetic information

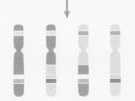

Each of the four chromatids gets packaged into a gamete.

---

14. Individuals carrying two non-functioning alleles for the gene producing the enzyme that converts phenylalanine into tyrosine:
a) will develop phenylketonuria (PKU).
b) may or may not develop PKU, because diet also plays a role in determining their phenotype.
c) will develop both PKU and Tay-Sachs disease, because the genes causing the traits are linked.
d) will develop both PKU and Tay-Sachs disease, because one pleiotropic gene influences both traits.
e) do not need to worry about their phenylalanine consumption, because PKU depends on the environment.

0 ——————— 58 ——————— 100
EASY   HARD

15. Because of Mendel's law of independent assortment:
a) individuals with red hair are more likely to have freckles.
b) skin color and hair texture tend to be inherited together.
c) we know that genes cannot exist as free-floating entities within a cell but must be carried on chromosomes.
d) the alleles coding for one trait do not usually influence the inheritance pattern for another trait.
e) Both a) and b) are correct.

0 ——————— 51 ——————— 100
EASY   HARD

16. Thousands (or even tens of thousands) of different traits make up an individual. For this reason:
a) in a species with 23 different chromosomes, some traits must be coded for by genes on the same chromosome.
b) the environment must influence more than half of our traits.
c) all genes must be pleiotropic.
d) knowing the person's phenotype is insufficient for determining their genotype.
e) All of the above are correct.

0 ——————— 76 ——————— 100
EASY   HARD

8 | Evolution and Natural Selection

DARWIN'S DANGEROUS IDEA

Evolution is an ongoing process.

Darwin journeyed to a new idea.

Four mechanisms can give rise to evolution.

Through natural selection, populations of organisms can become adapted to their environments.

The evidence for evolution is overwhelming.

8·1
Evolution is an ongoing process.

What big eyes you have! The glasswing butterfly, Cithaerias aurorina, from Peru.

8·1 We can see evolution occurring right before our eyes.

What's the longest that you've ever gone without food? Twenty-four hours? Thirty-six hours? If you've ever gone hungry that long, you probably felt as if you were going to die of starvation, but humans can survive days, even weeks, without food. In 1981, for example, 27-year-old Bobby Sands went on a hunger strike. Forsaking all food and consuming only water, he gradually deteriorated and ultimately died—after 66 days without food.

Each hour without food is even more dangerous when you're tiny. A fruit fly, for example, can only last just under a day—20 hours, give or take a few (**FIGURE 8-1**). Fruit flies can't live long without food because their tiny bodies don't hold very large caloric reserves. But could you breed fruit flies that could live longer than 20 hours, on average? Yes.

First, start with a population of 5,000 ordinary fruit flies. In biology, the term **population** means a group of organisms of the same species living in a particular geographic region. The region can be a small area such as a test tube or a large area such as a lake. In your experiment, your population occupies a cage in your laboratory.

From these 5,000 fruit flies, you will choose only the 20% of flies that can survive the longest without food:

1. Remove the food from the cage with the population of 5,000 flies.

2. Wait until 80% of the flies have starved to death, then put a container of food into the cage.

3. After the surviving flies eat, they'll have the energy to reproduce. When they do, collect and transfer the eggs to a new cage.

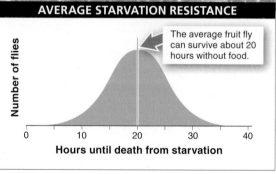

AVERAGE STARVATION RESISTANCE

The average fruit fly can survive about 20 hours without food.

Number of flies

Hours until death from starvation

FIGURE 8-1 **How long can a fruit fly survive without food?**

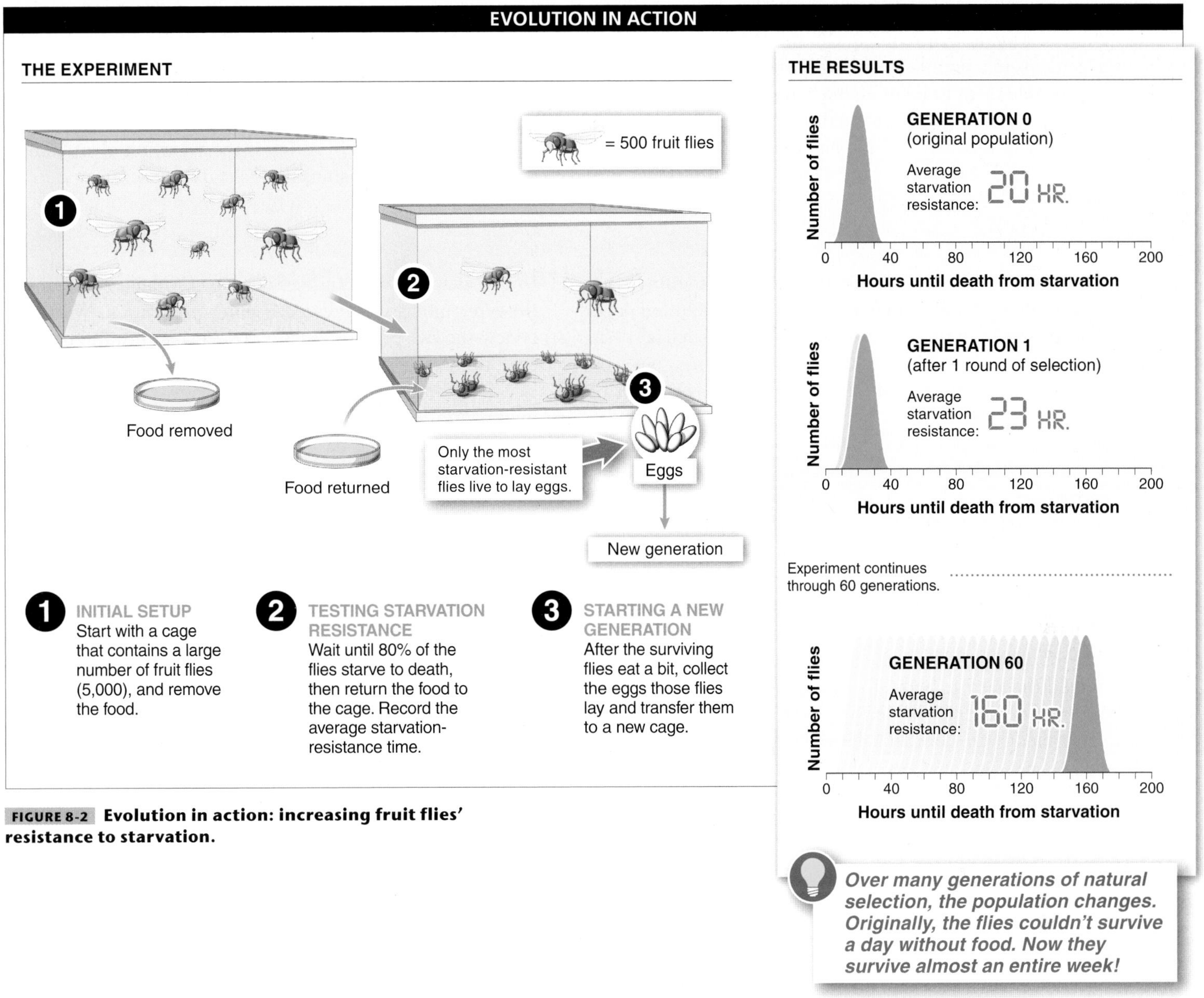

THE EXPERIMENT

= 500 fruit flies

1 Food removed

2

3

Food returned

Only the most
starvation-resistant
flies live to lay eggs.

Eggs

New generation

1 **INITIAL SETUP**
Start with a cage
that contains a large
number of fruit flies
(5,000), and remove
the food.

2 **TESTING STARVATION
RESISTANCE**
Wait until 80% of the
flies starve to death,
then return the food to
the cage. Record the
average starvation-
resistance time.

3 **STARTING A NEW
GENERATION**
After the surviving
flies eat a bit, collect
the eggs those flies
lay and transfer them
to a new cage.

FIGURE 8-2 **Evolution in action: increasing fruit flies'
resistance to starvation.**

THE RESULTS

GENERATION 0
(original population)

Average
starvation
resistance: **20 HR.**

Hours until death from starvation

GENERATION 1
(after 1 round of selection)

Average
starvation
resistance: **23 HR.**

Hours until death from starvation

Experiment continues
through 60 generations.

GENERATION 60

Average
starvation
resistance: **160 HR.**

Hours until death from starvation

*Over many generations of natural
selection, the population changes.
Originally, the flies couldn't survive
a day without food. Now they
survive almost an entire week!*

You can now start a new generation consisting only of the
offspring of those fruit flies able to survive without food
for the longest amount of time. When these eggs hatch,
the flies show increased resistance to starvation. The
average fly in the new generation can live for about
23 hours without food. And again, some of the new flies
survive for more than 23 hours, some for less (**FIGURE 8-2**).

What if you kept repeating these three steps? Start each
new generation using eggs only from the fruit flies in the
top 20% of starvation resistance. After five generations,
would the average starvation-resistance time for a fly in
your latest population be even higher? In fact, it would.
With each new generation, you would see a slight
increase, and with the fifth generation, the population's
average survival time would be about 32 hours.

After 60 generations of allowing only the flies that are best
at surviving without food to reproduce, how has the
population changed? Amazingly, the *average* fly in the
resulting population can survive for more than 160 hours
without food (see Figure 8-2). In 60 generations, the flies
have gone from an average starvation resistance of less than
a day to one of nearly a week! At the point where the food
is removed, the flies in generation 60 are noticeably fatter
than the flies in generation 0.

What happened? In a word: **evolution.** That is, there was a
genetic change in the population of fruit flies living in the
cage. In the generation 60 population, even the fly with the
worst starvation resistance is more than seven times better
at resisting starvation than the best fly in the original
population. Later in this chapter we'll discuss the

mechanisms by which evolution can occur, but, in short, in this situation it is the consequence of certain individual organisms in a population being born with characteristics that enable them to be more likely to survive and to reproduce than other individuals in the population. In this experiment, the 20% of fruit flies that were the most starvation resistant had a huge reproductive advantage over less starvation-resistant flies because they were the *only* flies in the population that survived to reproduce.

This experiment answers a question that is sometimes perceived as complex or controversial: does evolution occur? The answer is an unambiguous *yes*. We can watch it happen in the lab whenever we want. Recall from our discussion of scientific thinking (see Chapter 1) that for us to have confidence in our hypotheses about anything, they must be testable and reproducible. To be certain of the results in the fruit fly study, researchers carried out the starvation-resistance experiment described above five separate times. The results were the same every time.

In this experiment we changed starvation resistance in fruit flies, but what if we used dogs instead of flies and, instead of allowing the most starvation-resistant individuals to reproduce, we allowed only the smallest dogs to reproduce? Would the average body size of individual dogs in subsequent generations decrease over time? Yes. What if the experiment were done in a natural habitat, in the absence of human intervention, on rabbits, and only the fastest rabbits escaped death from predation by foxes? Would the average running speed in rabbits increase over time? Yes. We know these answers because each of these experiments or observations has already been done. An unfortunate example of evolution in action is the evolution of resistance to antibiotics in numerous illness-causing bacteria, including some bacteria responsible for pneumonia and tuberculosis.

In this chapter, we examine how evolution takes place and the types of changes it can cause in a population. We also review the five primary lines of evidence indicating that evolution and natural selection are processes that help us clarify all other ideas and facts in biology. Let's begin our investigation with a look at how the idea of evolution by natural selection was developed. This knowledge will help us better understand why evolution by natural selection is regarded as one of the most important ideas in human history, why it is sometimes considered a dangerous idea, and why it generates emotional debate.

TAKE-HOME MESSAGE 8·1

The characteristics of the individuals present in a population can change over time. We can observe such change in nature and can even cause it to occur.

8·2–8·4
Darwin journeyed to a new idea.

The Galápagos land iguana, Conolophus subcristatus.

8·2 Before Darwin, many people believed that all species had been created separately and were unchanging.

Charles Darwin grew up in an orderly world. In the Victorian British society he inhabited, many beliefs about humans and our place in the world had changed little over the previous two centuries. Biblical explanations were sufficient for most natural phenomena: most people thought the earth was about 6,000 years old. And with the occasional exception of a flood or earthquake or volcanic eruption, the earth was believed to be mostly unchanging. People recognized that organisms existed in groups called species or kinds. (In Chapter 10, we discuss in more detail what a species is; for now, we'll just say that individual organisms in a given species can interbreed with each other but not with members of another species.) People in Darwin's society also generally believed that all species, including humans, had been created at the same time and that, once created, they never changed and never died out.

Darwin threw into question some long- and dearly held beliefs about the natural world, and he forever changed our perspective on the origins of humans and our relationship to all other species. He didn't smash his society's worldview to pieces all at once, though, and he didn't do it by himself (**FIGURE 8-3**).

DARWIN'S INFLUENCES

GEORGES-LOUIS LECLERC, COMTE DE BUFFON
(1707–1788)
Suggested that the earth was much older than previously believed.

GEORGES CUVIER
(1769–1832)
By documenting fossil discoveries, showed that extinction had occurred.

JEAN-BAPTISTE LAMARCK
(1744–1829)
Suggested that living species might change over time.

CHARLES LYELL
(1797–1875)
Argued that geological forces had gradually shaped the earth and continue to do so.

The young Charles Darwin

FIGURE 8-3 Scientists who shaped Darwin's thinking.

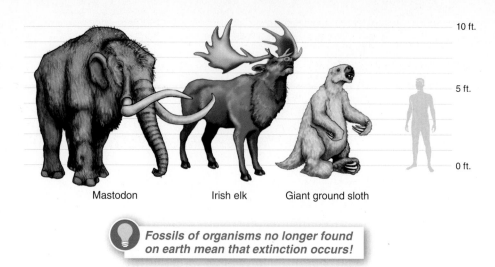

Mastodon   Irish elk   Giant ground sloth

Fossils of organisms no longer found on earth mean that extinction occurs!

FIGURE 8-4 **Extinction occurs.** Deep in coal mines, Cuvier discovered the fossilized remains of very large animals no longer found on earth.

In the 1700s and 1800s, scientific thought was advancing at a rapid pace. In 1778, the respected French naturalist Georges-Louis Leclerc, Comte de Buffon shook things up by suggesting that the earth must be about 75,000 years old. He arrived at this age by estimating that 75,000 years was the minimum time required for the planet to cool from a molten state. In the 1790s, Georges Cuvier began to explore the bottoms of coal and slate mines and found fossil remains that were unlike any living species. Many interpreted the Bible in such a way that made Cuvier's discoveries unthinkable, believing that biblical accounts did not allow for species to be wiped out. Cuvier's publications documented giant fossils (including the Irish elk, the mastodon, and the giant ground sloth) that bore no resemblance to any currently living animals (**FIGURE 8-4**). And although Cuvier was not a proponent of the idea of evolution, the fossils he discovered allowed only one explanation: extinction was a fact. Troubling as this observation was for the prevailing Western worldview, it was only the beginning.

Not only was it starting to seem that species could disappear from the face of the earth, but several scientists, including Darwin's own grandfather, Erasmus Darwin, began to suggest that living species might change over time. And in the early 1800s, the biologist Jean-Baptiste Lamarck

championed a popular idea about the mechanisms by which this change might occur. The idea—that change came about chiefly through organisms' use or disuse of particular features—was wrong, but the increased willingness among scientists to question previously sacred "truths" contributed to an atmosphere of unfettered scientific thought in which it was possible to challenge convention.

Perhaps the revolutionary ideas that most inspired Darwin were those of the geologist Charles Lyell. In his 1830 book *Principles of Geology,* Lyell argued that geological forces had shaped the earth and were continuing to do so, producing mountains and valleys, cliffs and canyons, through gradual but relentless change. This idea that the physical features of the earth were constantly changing would most closely parallel Darwin's idea that the living species of the earth, too, were gradually—but constantly—changing.

TAKE-HOME MESSAGE 8·2

In the 18th and 19th centuries, scientists began to overturn many commonly held beliefs in the Western world, including that the earth was only about 6,000 years old and that all species had been created separately and were unchanging. These gradual changes in scientists' beliefs helped shape Charles Darwin's thinking.

8·3 A job on a 'round-the-world survey ship allowed Darwin to indulge his love of nature and make observations that enabled him to develop a theory of evolution.

Charles Darwin was born into a wealthy family in England in 1809. Never at the top of his class, Charles professed to hate schoolwork. Nonetheless, at 16, he went to the University of Edinburgh to study medicine, following his father's footsteps. He was bored, though, and left at the end of his second year, when the prospect of watching gruesome surgeries (in the days before anesthesia) was more than he could bear.

At his father's urging, Darwin then pursued the ministry, studying theology at the University of Cambridge. Although he never felt great inspiration in his theology studies, Darwin was in heaven at Cambridge, where he could pursue his real love, the study of nature.

Shortly after graduation in 1831, Darwin landed his dream job, a position as a "gentleman companion" for the captain of HMS *Beagle,* on a five-year, 'round-the-world surveying expedition. This job came as a huge relief to the young Darwin, who was not really interested in becoming a minister.

Once on the *Beagle,* Darwin found that he didn't actually like sea travel. He was seasick for much of his time on board, writing to his cousin: "I hate every wave of the ocean with a fervor which you . . . can never understand." To avoid nausea, Charles spent as much time as possible on shore. It was there that he found that fieldwork was his true calling. At each stop, he would eagerly investigate the new worlds he found. In Brazil, he was enthralled by tropical forests. In Patagonia, he explored beaches and cliffs, finding spectacular fossils from huge extinct mammals. Elsewhere, he explored coral reefs and barnacles, always packing up specimens for museums and recording his observations for later use. He was like a schoolboy on permanent summer vacation (**FIGURE 8-5**).

The only book Darwin took with him on the *Beagle* was Lyell's *Principles of Geology,* which he read again and again. Intrigued by the book's premise that the earth is constantly changing, Darwin's mind was ripe for fresh ideas—particularly as he began observing seemingly inexplicable things. How could he explain marine fossils high in the Andes, hundreds of miles from the nearest ocean? The idea that the earth was an ever-changing planet would serve Darwin well when, a few years into its journey, the *Beagle* stopped at the Galápagos Islands, off the northwest coast of South America.

This group of volcanic islands was home to many unusual species, from giant tortoises to extremely docile lizards, which

DARWIN'S 'ROUND-THE-WORLD VOYAGE

London, England
Atlantic Ocean
Galápagos Islands
Pacific Ocean
Indian Ocean

Galápagos Islands

FIGURE 8-5 **Like a schoolboy on permanent summer vacation.** In 1831, Darwin set out on HMS *Beagle* on a five-year surveying expedition that took him around the globe.

made so little effort to run away that Darwin had to avoid stepping on them. Darwin was particularly intrigued by the wide variety of birds, especially the finches, which seemed dramatically more variable than those he had seen in other locations.

Darwin noticed two important and unexpected patterns on his voyage that would be central to his discovery of a mechanism for evolution. The first involved the finches he collected and donated to the Zoological Society of London. Darwin had assumed that their differences were equivalent to differences between tall and short, curly-haired and straight-haired people. That is, he thought that all the finches were of the same species but with different physical characteristics or **traits,** such as body size, beak shape, or

RESEMBLANCE OF ISLAND FINCH SPECIES TO MAINLAND FINCH SPECIES
Although the finches found on the Galápagos Islands had slightly different physical characteristics, they still resembled the single mainland finch.

600 mi

Galápagos Islands

Ecuador (mainland)

Mainland finch

GALÁPAGOS FINCHES

Ground finch

Tree finch

Warbler finch

RESEMBLANCE OF EXTINCT SPECIES TO LIVING SPECIES IN SAME AREA
Around the world, there is a striking similarity between the fossils of extinct species and the living species in that same area.

10 ft

Glyptodont (extinct)

Hairy armadillo (alive today)

FIGURE 8-6 Darwin observed unexpected patterns.

feather color. The staff of the Zoological Society, however, could see from the birds' physical differences that the birds were not reproductively compatible and that there were 13 unique lineages—a different species for every one of the Galápagos Islands that Darwin had visited. Moreover, although the birds were different species, they all resembled very closely the single species of finch living on the closest mainland, in Ecuador (**FIGURE 8-6**).

This resemblance seemed a suspicious coincidence to Darwin. Perhaps the island finches resembled the mainland species because they used to be part of the same mainland population. Over time they may have separated and diverged from the original population and gradually formed new—but similar—species. Darwin's logic was reasonable, but his idea flew in the face of all the scientific thinking of the day.

The second important but unexpected pattern Darwin noted was that, throughout his voyage, at every location there was a striking similarity between the fossils of extinct species and the living species in that same area. In Argentina, for instance, he found some giant fossils from a group of organisms called "glyptodonts." The extinct glyptodonts looked just like armadillos, a species that still flourished in the same area. Or rather, they looked like armadillos on steroids: the average

armadillo today is about the size of a house cat and weighs 10 pounds, while the glyptodonts were giants at 10 feet long and more than 4,000 pounds (see Figure 8-6).

If glyptodonts had lived in South America in the past, and armadillos were currently living there, why was only one of the species still alive? And why were the glyptodont fossils found only in the same places where modern armadillos lived? Darwin deduced that glyptodonts resembled armadillos because they were their ancient relatives. Again, it was a logical deduction, but it contradicted the scientific dogma of the day that species were unchanging and extinction did not occur.

TAKE-HOME MESSAGE 8·3

Charles Darwin was able to focus on studying the natural world when, in 1831, he got a job on a ship conducting a five-year, 'round-the-world survey. Darwin noted unexpected patterns among fossils and living organisms. Fossils resembled but were not identical to living organisms in the same area. And finch species on each of the Galápagos Islands differed from each other in small but significant ways. These observations helped Darwin develop his theory of how species might change over time.

Darwin didn't have a great "Eureka!" moment about evolution while on the *Beagle.* The wheels were turning in his head after his voyage, however, and inspiration struck when he was reading "for amusement" *Essay on the Principle of Population,* by the economist Thomas Malthus. Malthus prophesied doom and gloom for populations, including humans, based on his calculations that populations had the potential to grow much faster than food supplies could. Darwin speculated that, rather than the future holding certain catastrophe for all, maybe the best individuals would "win" in the ensuing struggle for existence, and the worst would "lose." If so, he suddenly saw that "favourable variations would tend to be preserved, and unfavourable ones to be destroyed."

In 1842, Darwin prepared a draft of his ideas in a 35-page paper, written in pencil, and he fleshed it out over the next couple of years. He knew that his idea was important—so important that he wrote a letter to his wife instructing her, in the case of his sudden death (an odd request, given that he was only 35 years old), to give his "sketch" to a competent person, along with about $1,000 and all of his books, so that this person could complete it for him.

He also had an inkling that his ideas would rock the world. In a letter to a close friend, he wrote: "At last gleams of light have come, and I am almost convinced (quite contrary to the opinion I started with) that species are not (it is like confessing to murder) immutable." But "confessing to murder" was apparently more than Darwin was ready for. Inexplicably, he put his sketch into a drawer, where it remained for 14 years—even as his friends, including Charles Lyell, warned him that he should publish his ideas in case someone else came up with the same ideas independently. In 1858, Darwin found that his friends' warnings had "come true with a vengeance."

In a letter to Darwin, Alfred Russel Wallace (**FIGURE 8-7**), a young British biologist in the throes of malaria-induced mania in Malaysia, laid out a clear description of the process of evolution by natural selection, after having read Malthus's book. He asked Darwin to "publish it if you think it is worthy." Although he may have been inspired by his fever to put his thoughts in writing, they were careful and precise, and resulted from many years of thought and exposure to the writings of many of the same scientists who influenced Darwin. Crushed at having been scooped,

FIGURE 8-7 **Alfred Russel Wallace.** Darwin and Wallace independently identified the process of evolution by natural selection.

Darwin wrote that "all my originality will be smashed." He had numerous friends among the most prominent scientists of the day, however, and they arranged for a joint presentation of Wallace's and Darwin's work to the Linnaean Society of London. As a result, both Darwin and Wallace are credited for the first description of evolution by natural selection.

Darwin then sprang into action, rapidly putting together his thoughts and observations and completing, 16 months later, a full book. In 1859, he published *The Origin of Species* (its full title is *On the Origin of Species by Means of Natural Selection, or the Preservation of Favoured Races in the Struggle for Life*).

EVOLVING BELIEFS IN DARWIN'S WORLD

BEFORE:

• All organisms were put on earth by a creator at the same time.

• Organisms are fixed:
  no additions, no subtractions.

• Earth is about 6,000 years old.

• Earth is mostly unchanging.

First published in 1859

AFTER:

• Organisms change over time.

• Some organisms have gone extinct.

• Earth is more than 6,000 years old.

• The geology of earth is not constant, but always changing.

 The ideas Darwin described in The Origin of Species *reflected a new dynamic view of life on earth.*

FIGURE 8-8 **Reconsidering the world.** Ideas commonly held by the general public before and after Darwin.

The book was an instant hit, selling out on its first day, provoking public discussion and debate and ultimately causing a wholesale change in the scientific understanding of natural selection and many other important evolutionary ideas. Where once the worldview was of a young earth, populated by unchanging species all created at one time, with no additions or extinctions, now there was a new dynamic view of life on earth: descent with modification—species could and did change over time, and as some species split into new species, others became extinct (**FIGURE 8-8**).

Darwin's theory has proved to be among the most important and enduring contributions in all of science. It has stimulated an unprecedented diversity of theoretical and applied research, and it has withstood repeated experimental and observational testing. With the background of Darwin's elegant idea in hand, we can now examine its details.

TAKE-HOME MESSAGE 8·4

After putting off publishing his thoughts on natural selection for more than 15 years, Darwin did so only after Alfred Russel Wallace independently came up with the same idea. The two men published a joint presentation on their ideas in 1858, and Darwin published a much more detailed treatment in *The Origin of Species* in 1859, sparking wide debate and discussion about the processes and patterns of evolution.

8·5–8·10
Four mechanisms can give rise to evolution.

Close-up of the forewing (elytron) of the green tiger beetle, (Cicindela campestris). The coloration may help deter predators by resembling a species of wasp with a harmful sting.

8·5 Evolution occurs when the allele frequencies in a population change.

Suppose you were put in charge of a large population of tigers in a zoo. Almost all of them are orange and brown with black stripes. Occasionally, though, unusual all-white tigers are born. This white phenotype is the result of a rare pair of alleles that suppresses the tiger's production of most fur pigment (**FIGURE 8-9**). (Recall from Section 7-1 that alleles are variants of a gene and that most mammals inherit one allele from their mother and one from their father for each gene.) Because visitors flock to zoos to see white tigers, you want to increase the proportion of your tiger population that is white. How would you do so?

One possibility is that you could alter the population by trying to produce more animals with the white phenotype. In this case, your best strategy would be to try to breed the white tigers with each other. Over time, this would lead to more and more white tigers. And as the generations go by (this will take a while, though, since the generation time for tigers is about eight years), your population will include a higher proportion of white tigers. When this happens, you will have witnessed evolution, a change in the proportion of alleles for the pigment-suppression gene in the population.

Another way you might increase the proportion of white tigers in your population would be to acquire some white tigers from another zoo. By directly adding white tigers to your population, the population would include a higher proportion of white tigers (you might even trade some of your orange tigers, which would further increase the proportion of your population that is white). By adding

white tigers or removing orange tigers, you will again have witnessed evolution in your population.

As this example illustrates, evolution doesn't involve changing the genetics or physical features of *individuals.* Individuals do not evolve. Rather, you change the proportions of the alleles in the *population.* An allele's frequency is the proportion in which it is present in a population relative to the other alleles of the same gene, much like its "market share." In the tiger example, the white-fur alleles initially had little market share, perhaps as small as 1%. Over time, though, they came to make up a larger and larger proportion of the total fur-pigment alleles in the population. And as this happened, evolution occurred.

Any time an allele's market share changes, evolution is taking place. Perhaps the most dramatic way we see this is in acts of predation. When one organism kills another, the dead animal's alleles will no longer be passed on in the population. And if certain alleles make it more likely that one animal will be killed by another—maybe they alter its coloration, or speed, or some other defense mechanism—those alleles are likely to lose market share. The converse is equally true: evolution occurs when alleles increase in frequency.

Darwin demonstrated that natural selection could be an efficient mechanism of evolution and a powerful process that results in adapting populations to their environments. Evolution and natural selection, however, are not the same thing. Natural selection is one way that evolution can occur,

ALLELE FREQUENCIES

● Proportion of orange fur-pigment alleles in the population

● Proportion of white fur-pigment alleles in the population

Tiger population

Evolution is a change in the allele frequencies of a population over time. For example, a change in the proportion of pigment alleles in the population of tigers means that evolution has occurred.

FIGURE 8-9 **Evolution defined.** Evolution is a change in allele frequencies in a population.

but it is not the only mechanism of evolutionary change. It is simply one of them, and the only one that leads to adaptation (**FIGURE 8-10**). The four evolutionary mechanisms are:

MUTATION
An alteration of the base-pair sequence in the DNA of an individual's gamete-producing cells that changes an allele's frequency.

GENETIC DRIFT
A random change in allele frequencies, unrelated to any allele's influence on reproductive success.

MIGRATION
A change in allele frequencies caused by individuals moving into or out of a population.

NATURAL SELECTION
A change in allele frequencies that occurs when individuals with one version of a heritable trait have greater reproductive success than individuals with a different version of the trait.

FIGURE 8-10 **Four ways in which evolution can occur.**

1. Mutation	3. Migration
2. Genetic drift	4. Natural selection

Keeping in mind that evolution is genetic change in a population, we'll now explore each of these four processes that lead to such genetic changes.

TAKE-HOME MESSAGE 8·5

Evolution is a change in allele frequencies within a population. It often occurs by four different mechanisms: mutation, genetic drift, migration, and natural selection.

8·6 Mutation—a direct change in the DNA of an individual—is the ultimate source of all genetic variation.

In describing the first of four evolutionary mechanisms, it is helpful to keep in mind our precise definition of

evolution: a change in the allele frequencies found in a population. **Mutation** is an alteration of the base-pair sequence of an individual's DNA, and when this alteration occurs in the DNA that codes for a particular gene, the change in the DNA sequence may change that allele. (See Section 5-9 for a detailed discussion of the mutation process.) Say, for example, that a mutation changes one of a person's two blue-eye alleles into a brown-eye allele. If this mutation occurs in the sperm- or egg-producing cells, it can be passed on to the next generation; the offspring may carry the brown-eye allele. When this happens, the proportion of blue-eye alleles in the population is slightly reduced, and the proportion of brown-eye alleles is slightly increased. Evolution has occurred.

When considering mutation as a mechanism of evolution, it is important to remember that the cells of a multicellular eukaryotic organism can be divided into two types: somatic cells and reproductive cells. Somatic cells—the cells forming the body of an organism—are not passed from parent to offspring. Only reproductive cells are. For that reason, only mutations that affect reproductive cells can be inherited, and so we consider only those mutations as altering the allele frequencies within a population. The causes of mutations, however, are the same for somatic and reproductive cells.

Mutations can occur spontaneously during the complex process of cell division and, as we saw in Chapter 5, can be induced by a variety of environmental factors—including radiation, such as X rays, and some chemicals. But although these factors can induce mutations and influence the rate at which mutations occur, they do not generally influence exactly *which* mutations occur. Thus, we tend to say that "mutations are random." What this means, more precisely, is that (1) we cannot predict ahead of time which individuals will have which mutations, and (2) we cannot predict whether the consequences of a mutation will be benign, harmful, or useful. This doesn't mean, however, that all mutations have an equal probability of occurring. They do not. Some mutations are far more likely than others.

Treatment of an agricultural pest with harmful chemicals, for example, might increase the number of mutations occurring in that pest population, but it does not increase the likelihood that a particular mutation is beneficial or detrimental.

In addition to being a mechanism of evolutionary change, mutation has another, more important, role relevant to all mechanisms of evolution, including natural selection: mutation is the ultimate source of genetic variation in a population. We saw that a mutation may lead to the conversion of one allele to another that is already found within the population, as in our blue eye and brown eye example. More importantly, though, a mutation may create a completely novel allele that codes for the production of a new protein (**FIGURE 8-11**). That is, a change in the base-

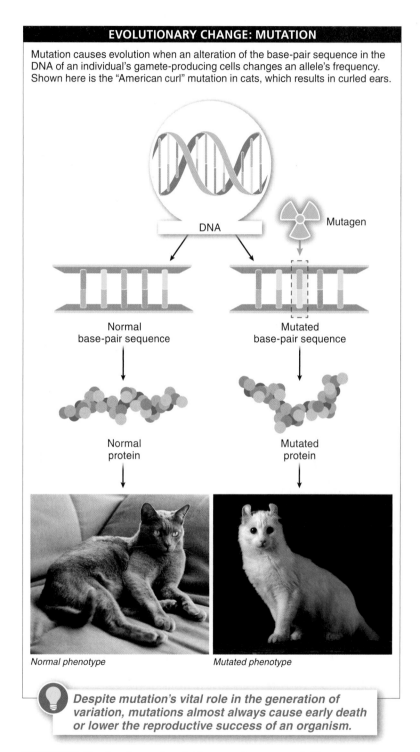

EVOLUTIONARY CHANGE: MUTATION

Mutation causes evolution when an alteration of the base-pair sequence in the DNA of an individual's gamete-producing cells changes an allele's frequency. Shown here is the "American curl" mutation in cats, which results in curled ears.

DNA

Mutagen

Normal base-pair sequence

Mutated base-pair sequence

Normal protein

Mutated protein

Normal phenotype

Mutated phenotype

Despite mutation's vital role in the generation of variation, mutations almost always cause early death or lower the reproductive success of an organism.

FIGURE 8-11 **Mechanisms of evolutionary change: mutation.**

pair sequence of a person's DNA may cause the production of a gene product that has never existed before in the human population: instead of blue or brown eyes, the mutated gene might code for yellow or red eyes. If such a new allele occurs in the sperm- or egg-producing cells, and if it does not significantly reduce an individual's "reproductive fitness" (which we'll consider below), the new allele can be passed on to offspring and remain in the population. At some future time, the mutation might even confer higher fitness, in which case natural selection could lead to an increase in frequency of this allele in the population. For this reason, mutation is critical to natural selection: all variation—the raw material for natural selection—must initially come from mutation.

Despite this vital role in the generation of variation, however, nearly all mutations have either no impact on an organism's fitness or a negative impact by causing early death or reducing the organism's reproductive success. Mutations to a normally functioning allele that codes for a normally functioning protein typically result in either an allele that

codes for the same protein or a new allele that codes for a non-functioning protein. The latter case almost inevitably reduces an organism's fitness. For this reason, our bodies protect our sperm- or egg-producing DNA with a variety of built-in error-correction mechanisms. As a result, mutations are rare (in humans, on the order of 1 mutation in every 30 million base pairs in each generation).

TAKE-HOME MESSAGE 8·6

Mutation is an alteration of the base-pair sequence in an individual's DNA. If such an alteration changes an allele in an individual's gamete-producing cells, the frequency of alleles has changed and this constitutes evolution within the population. Mutations can be caused by high-energy radiation or chemicals in the environment and also can appear spontaneously. Mutation is the only way that new alleles can be created within a population, and thus it generates the variation on which natural selection can act.

8·7 Genetic drift is a random change in allele frequencies in a population.

Another evolutionary mechanism is **genetic drift,** a random change in allele frequencies in a population. For example, consider two alleles for a particular trait, let's say a cleft chin. We'll assume that it's a single-gene trait, with two alleles, and that cleft is dominant over smooth chin. (Note: although research on cleft chins suggests a strong genetic influence on the trait, the mode of inheritance hasn't been determined with certainty.) In this case, individuals with either one or two copies of the dominant allele (*CC* or *Cc*) exhibit the cleft chin (**FIGURE 8-12**). Now suppose that two heterozygous (*Cc*) people have one child. Which combination of alleles will that child receive? This is impossible to predict because it depends completely on which sperm fertilizes which egg—the luck of the draw. If the couple's sole child inherits a recessive allele from each parent, will the *population's* allele frequencies change? Yes. The new individual's two recessive alleles (*cc*) increase the proportion of recessive alleles in the population. It is equally likely that this couple's only child will receive two dominant alleles (*CC*). In either case, because a change in allele frequencies has occurred, evolution has happened.

This impact of genetic drift is much greater in small populations than in large populations. In a large

population, any one couple having a child with two recessive alleles is likely to be offset by another couple having one child with two dominant alleles. When that happens, the overall allele frequencies in the population do not change, and evolution has not occurred.

One of the most important consequences of genetic drift is that it can lead to **fixation** for one allele of a gene in a population (see Figure 8-12). Fixation results when an allele's frequency in a population reaches 100% (and the frequency of all other alleles of that gene becomes 0%). If this happens, there is no more variability in the population for this gene; all individuals will always produce offspring carrying only that allele (until new alleles arise through mutation). For this reason, genetic drift reduces the genetic variation in a population.

The important factor that distinguishes genetic drift from natural selection is that the change in allele frequencies is not related to the alleles' influence on reproductive success. This doesn't mean that genetic drift is confined to alleles that have no effect on fitness. Rather, it just refers to the cause of an allele's change in frequency. Multiple evolutionary mechanisms can occur simultaneously—an allele favored by natural selection,

GENETIC DRIFT

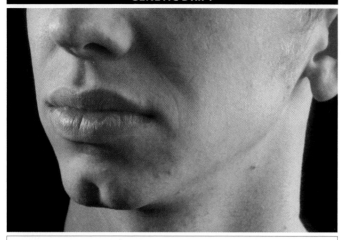

Genetic drift causes evolution when an allele's frequency changes for any of several reasons that are unrelated to the allele's influence on reproductive success.

POPULATION BEFORE GENETIC DRIFT
Allele frequencies:
- cleft chin (dominant)
- smooth chin (recessive)

Neither allele is related to reproductive success. Inheritance is based solely on chance.

REPRODUCTION
In this example, a heterozygous couple (*Cc*) could have two children that are homozygous recessive (*cc*), causing an increase in the proportion of recessive alleles in the population.

POPULATION AFTER GENETIC DRIFT
There are now more recessive alleles in the population than before.

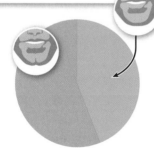

FIXATION
Genetic drift leads to fixation when an allele's frequency becomes 100% in a population. If this occurs, there is no longer genetic variation for the gene.

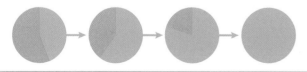

FIGURE 8-12 Mechanisms of evolutionary change: genetic drift. Genetic drift has the greatest impact in small populations.

for example, may simultaneously increase or decrease in frequency due to genetic drift.

Two special cases of genetic drift, the founder effect and population bottleneck effect, are important in the evolution of many populations.

Founder Effect A small number of individuals may leave a population and become the founding members of a new, isolated population. The founder population may have different allele frequencies than the original, "source" population, particularly if the founders are a small group. If this new population does have different allele frequencies, evolution has occurred. Because the founding members of the new population will give rise to all subsequent individuals, the new population will be dominated by the genetic features that happened to be present in that group of founding fathers and mothers. This type of genetic drift is called the **founder effect.**

The Amish population in the United States is thought to have been established by a small number of founders, some of whom happened to carry the allele for *polydactyly*— the condition of having

Q Why are Amish people more likely than other people to have extra fingers and toes?

extra fingers and toes. As a consequence, this trait, while rare, now occurs much more frequently among the Amish than it does worldwide (**FIGURE 8-13**).

Population Bottleneck Effect Occasionally, a famine, disease, or rapid environmental change causes the deaths of a large proportion of individuals in a population. Because the population is quickly reduced to a small fraction of its original size, this reduction is called a **bottleneck.** If the catastrophe is equally likely to strike any member of the population, the remaining members are essentially a random, small sample of the original population. For this reason, the remaining population may not possess the same allele frequencies as the original population. Thus, the consequence of such a population bottleneck would be evolution through genetic drift (**FIGURE 8-14**).

Just such a population bottleneck occurred with cheetahs near the end of the last ice age, about 10,000 years ago. Although the cause is unknown— possibly environmental cataclysm or human hunting pressures—it appears that nearly all cheetahs died. And although the population rebounded, all cheetahs living

Genetic drift occurs via the founder effect when the founding members of a new population have different allele frequencies than the original source population.

SOURCE POPULATION
Allele frequencies:
● 5 digits per hand (recessive)
● >5 digits per hand (dominant)

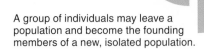

A group of individuals may leave a population and become the founding members of a new, isolated population.

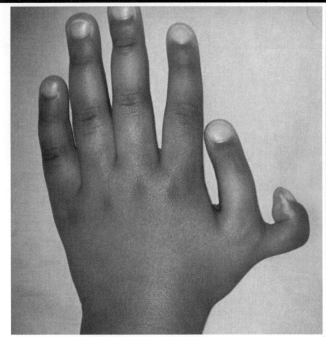

NEWLY FOUNDED POPULATION
If allele frequencies for a specific gene differ among a population's founders relative to the source population, a trait (such as polydactyly) may be disproportionately frequent in the new population.

FIGURE 8-13 One way that genetic drift occurs: the founder effect.

Genetic drift occurs via the bottleneck effect when famine, disease, or rapid environmental change causes the deaths of a large, random proportion of the population, and the surviving individuals have different allele frequencies than the original population.

SOURCE POPULATION

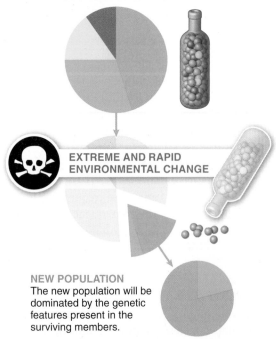

EXTREME AND RAPID ENVIRONMENTAL CHANGE

NEW POPULATION
The new population will be dominated by the genetic features present in the surviving members.

All cheetahs living today can trace their ancestry back to a dozen or so individuals that happened to survive a population bottleneck about 10,000 years ago!

FIGURE 8-14 Another way that genetic drift occurs: the bottleneck effect.

today can trace their ancestry back to a dozen or so lucky individuals that survived the bottleneck. As a result of this past instance of evolution by genetic drift, there is almost no genetic variation in the current population of cheetahs. (And, in fact, a cheetah is able to accept a skin graft from any other cheetah, much as identical twins can from each other.)

TAKE-HOME MESSAGE 8·7

Genetic drift is a random change in allele frequencies within a population, unrelated to the alleles' influence on reproductive success. Genetic drift is a significant mechanism of evolutionary change, primarily in small populations.

8·8 Migration into or out of a population may change allele frequencies.

The third mechanism of evolutionary change is **migration.** Migration, also called **gene flow,** is the movement of some individuals of a species from one population to another (**FIGURE 8-15**) This movement from population to population within a species distinguishes migration from the founder effect, in which individuals migrate to a new habitat previously unpopulated by that species. If migrating individuals survive and reproduce in the new population, and if they carry a different proportion of alleles than the individuals in their new home, then the recipient population experiences a change in allele frequencies and, consequently, experiences evolution. And because alleles are simultaneously lost from the population that the migrants left behind, that population, too, will experience a change in its allele frequencies and, thus, will evolve.

Gene flow between two populations is influenced by the mobility of the organisms and by barriers, such as mountains or rivers. And as we'll see in Chapter 16, human activities, too, can dramatically influence the migratory potential of organisms, as we sometimes transport species. This can happen intentionally, as when people purchase exotic pets or non-native species for their gardens, or unintentionally, such as when marine species get trapped in the ballast water tanks of cargo ships, later to be released in ports on the other side of the world when the tanks are flushed.

MIGRATION (GENE FLOW)

Gene flow causes evolution if individuals move from one population to another, causing a change in allele frequencies in either population.

1 BEFORE MIGRATION
Two populations of the same species exist in separate locations. In this example, they are separated by a mountain range.

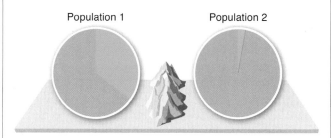

2 MIGRATION
A group of individuals from Population 1 migrates over the mountain range.

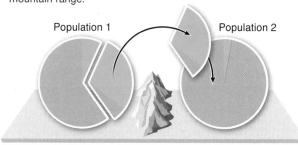

3 AFTER MIGRATION
The migrating individuals are able to survive and reproduce in the new population.

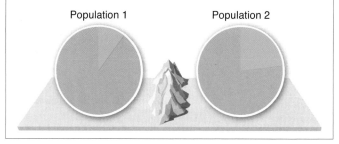

TAKE-HOME MESSAGE 8·8

Migration, or gene flow, leads to a change in allele frequencies in a population as individuals move into or out of the population.

FIGURE 8-15 Mechanisms of evolutionary change: migration (gene flow).

8·9 When three simple conditions are satisfied, evolution by natural selection is occurring.

The fourth mechanism of evolutionary change is **natural selection.** This is the mechanism that Darwin identified in *The Origin of Species,* in which he noted that three conditions are necessary for natural selection to occur:

1. There must be variation for the trait within a population.

2. That variation must be heritable (that is, capable of being passed from parents to offspring).

3. Individuals with one version of the trait must produce more offspring than those with a different version of the trait.

Let's examine these conditions more closely.

Condition 1: Variation for a Trait Close your eyes and imagine a dog. What does the dog look like? If 50 people were asked this question, we would probably get 50 different dog descriptions (FIGURE 8-16). Some are big, some are small. Some have short hair, some long. They vary in just about every way you can imagine. Likewise, if 50 people were to imagine a human face, a similarly broad

range of images would pop into their heads. Variation is all around us. Beyond making the world an interesting place to live in, variation serves another purpose: it is the raw material on which evolution works.

Variation is not limited to physical features such as fur color or face shape. Organisms vary in physiological and biochemical ways, too. Some people can quickly and efficiently metabolize alcohol, for example. Others find themselves violently ill soon after sipping a glass of wine. Similarly, we vary in our susceptibility to poison ivy or diseases such as malaria. Behavioral variation—from temperament to learning abilities to interpersonal skills—is dramatic and widespread, too. So impressed was Darwin with the variation he observed throughout the world that he devoted the first two chapters of *The Origin of Species* to a discussion of variation in nature and among domesticated animals.

Condition 2: Heritability The second condition that Darwin identified as necessary for natural selection was a no more complex discovery than the first: for natural selection to happen, offspring must inherit the trait from

FIGURE 8-16 **The first condition for natural selection is variation for a trait.**

their parents. Although inheritance was poorly understood in Darwin's time, it was not hard to see that, for many traits, offspring look more like their parents than like some other, random individual in the population (FIGURE 8-17). Animal breeders had long known that the fastest horses generally give birth to the fastest horses. Farmers, too, understood that the plants with the highest productivity generally produce seeds from which highly productive plants grow. And everyone knew that children resemble their parents. It was enough to know that this similarity between offspring and parents exists—it was not necessary to understand how it occurs or to be able to quantify it. We call the transmission of traits from parents to their children through genetic information **inheritance** or **heritability.**

Condition 3: Differential Reproductive Success

It would be nice to say that Darwin made a stunning and insightful discovery for the third of the three conditions necessary for natural selection, but he didn't. Rather, he derived the third condition for natural selection from three fairly simple observations. First, more organisms are born than can survive. Second, organisms are continually struggling for existence. Lastly, some organisms are more likely than others to survive and reproduce. In a world of limited resources, finding food or shelter is a zero-sum game: if one organism is feasting, another is likely to be starving.

FIGURE 8-17 **The second condition for natural selection is heritability.** Goldie Hawn and daughter Kate Hudson resemble each other.

This three-part observation led Darwin to his third condition for natural selection, which is called **differential reproductive success:** from all the variation existing in a population, individuals with traits most suited to survival and reproduction in their environment generally leave more offspring than do individuals with other traits (FIGURE 8-18).

The tiniest dog in a litter has reduced differential reproductive success. Its more robust siblings prevent access to the food it needs to grow and thrive.

FIGURE 8-18 **The third condition for natural selection is differential reproductive success.**

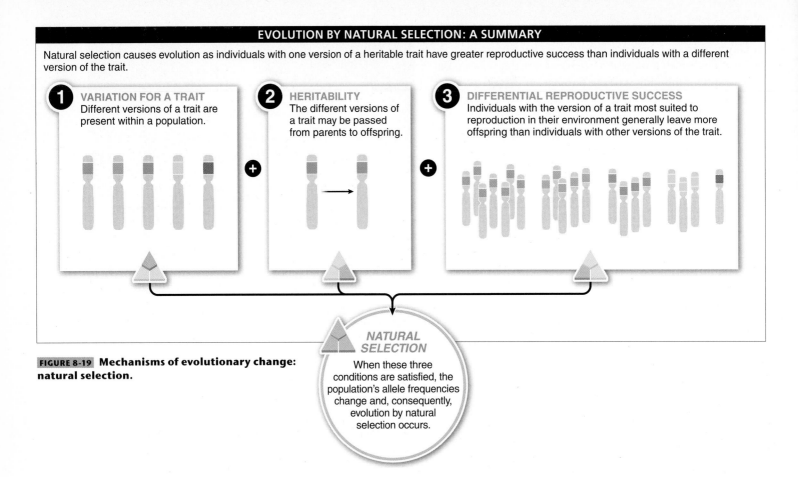

Natural selection causes evolution as individuals with one version of a heritable trait have greater reproductive success than individuals with a different version of the trait.

1 VARIATION FOR A TRAIT
Different versions of a trait are present within a population.

+

2 HERITABILITY
The different versions of a trait may be passed from parents to offspring.

+

3 DIFFERENTIAL REPRODUCTIVE SUCCESS
Individuals with the version of a trait most suited to reproduction in their environment generally leave more offspring than individuals with other versions of the trait.

NATURAL SELECTION
When these three conditions are satisfied, the population's allele frequencies change and, consequently, evolution by natural selection occurs.

FIGURE 8-19 Mechanisms of evolutionary change: natural selection.

For example, in the food experiments with fruit flies, we saw that the flies inheriting the ability to pack on fat when food is available end up leaving more offspring than those inheriting a poor ability to pad their little fruit fly frames with fat deposits. The portly fruit flies have greater reproductive success than other individuals in the population.

That's it. Natural selection—one of the most influential and far-reaching ideas in the history of science—takes place when three basic conditions are met (**FIGURE 8-19**):

1. Variation for a trait

2. Heritability of that trait

3. Differential reproductive success based on that trait

When these three conditions are satisfied, evolution by natural selection is occurring. It's nothing more and nothing less. Over time, the traits that lead some organisms to have greater reproductive success than others will increase in frequency in a population, while traits that reduce reproductive success will become less and less common.

When it comes to some traits, the reason they specifically confer greater reproductive success is that they make the individual more attractive to the opposite sex. Such traits—including the brightly colored feathers of male peacocks, the large antlers of male red deer, and a variety of other "ornaments" that increase an individual's status or appeal—increase in frequency because they satisfy the three conditions for natural selection. This natural selection for mating success is called **sexual selection.**

Q In most agricultural pests treated with pesticides, a resistance to the pesticides evolves. How does this happen?

Another way of looking at natural selection is to focus not on the winners (the individuals who are producing more offspring) but on the losers. Natural selection can be viewed as the elimination of some heritable traits from a population. If you carry a trait that makes you a slower-running rabbit, for example, you are more likely to be eaten by a fox (**FIGURE 8-20**). If running speed is a heritable trait (and it is), the next generation in a population contains fewer slow rabbits. Over time, the population is changed by natural selection. It evolves.

NATURAL SELECTION IN NATURE

1 VARIATION FOR A TRAIT
Running speed in rabbits can vary from one individual to the next.

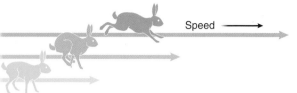

Speed ⟶

2 HERITABILITY
The trait of running speed is passed on from parents to their offspring.

3 DIFFERENTIAL REPRODUCTIVE SUCCESS
In a population, rabbits with slower running speeds are eaten by the fox, and their traits are not passed on to the next generation.

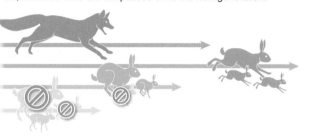

FIGURE 8-20 Removing the losers. Natural selection can be thought of as the elimination from a population of traits that confer poor reproductive success.

One of Darwin's contemporaries, Thomas Huxley, supposedly cursed himself when he first read *The Origin of Species,* saying that he couldn't believe he didn't figure it out on his own. Each of the three basic conditions is indeed simple and obvious. The brilliant deduction, though, was to put the three together and appreciate the consequences.

TAKE-HOME MESSAGE 8·9

Natural selection is a mechanism of evolution that occurs when there is heritable variation for a trait, and individuals with one version of the trait have greater reproductive success than do individuals with a different version of the trait. Natural selection can also be thought of as the elimination of alleles that reduce the reproductive rate of individuals carrying those alleles, relative to the reproductive rate of individuals who do not.

8·10 A trait does not decrease in frequency simply because it is recessive.

Do recessive traits gradually become less common in a population? In the early 1900s, there was much discussion among biologists about this question. They wondered: "If the allele for brown eyes is dominant over the allele for blue eyes, why doesn't a population eventually become all brown-eyed?"

R. C. Punnett (originator of the Punnett square, described in Chapter 7) posed this question to G. H. Hardy, a mathematician with whom Punnett played cricket. Hardy replied the next day, and he published his answer in *Science* in 1908. It turns out that Wilhelm Weinberg answered the question independently six months earlier. But because he published his findings in an obscure journal, scientists didn't appreciate his result for another 35 years.

Their result underlies all evolutionary genetics and is referred to as the Hardy-Weinberg Law. It reveals several things, including that the answer to the question of whether recessive traits become rarer in a population is *no.* A trait does not decrease in frequency simply because it is recessive. Here's how Hardy and Weinberg demonstrated it.

First, we refer to the frequency of the dominant allele in the population, A, as "p" and the frequency of the recessive allele, a, as "q." Since every allele in the population has to be either A or a, we can say that $p + q = 1$. The percentage of the two alleles must total 100%. If we know the frequency of either allele in the population, we can subtract it from 1 to calculate the frequency of the other allele.

Next, we can predict how common each genotype in the population will be. The frequency of *AA* is just the probability that an individual gets two copies of allele *A,* which is $p \times p$, or p^2. Applying the same math, the frequency of *aa* individuals in the population is simply $q \times q$, or q^2. Predicting the frequency of heterozygous individuals, *Aa*, is slightly more complicated. Because the dominant allele may come from either the mother or the father, the frequency of the *Aa* genotype is $2 \times p \times q$, or $2pq$. In other words, it's actually the frequency of the *Aa* genotype (i.e., getting the dominant allele from the mother) and the *aA* genotype (i.e., getting the dominant allele from the father), which is $p \times q$ plus $q \times p$, which simplified to $2pq$.

Consider an example (**FIGURE 8-21**). Suppose a population of 1,000 kangaroo rats has the following phenotype and genotype frequencies: 16 are dark brown (*BB*), 222 are spotted (*Bb*), and 762 are light brown (*bb*). The trait shows incomplete dominance, with the allele for dark brown color, *B,* dominant over the allele for light brown color, *b,* and the heterozygote having a spotted phenotype. Because every individual in the population has two alleles for the coat-color gene (one from each parent), there are twice as many alleles as members of the population. So a population of 1,000 kangaroo rats has 2,000 alleles of the gene for coat color.

Now let's look at the allele frequencies. Each *BB* individual has two copies of *B*, and each *Bb* individual has one copy:

Frequency of $B = [(2 \times 16) + 222]/2{,}000 = 0.127 = p$

Frequency of $b = [(2 \times 762) + 222]/2{,}000 = 0.873 = q$

Since we know the allele frequencies, *p* (0.127) and *q* (0.873), we can determine the genotype frequencies of the offspring produced in this population. According to the equations above, they should be:

Frequency of $BB = p^2 = (0.127)^2 = 0.016$

Frequency of $Bb = 2pq = 2(0.127)(0.873) = 0.222$

Frequency of $bb = q^2 = (0.873)^2 = 0.762$.

If 1,000 kangaroo rats are produced, we expect to see genotype frequencies that are the same as in the parent generation: 16 *BB*, 222 *Bb,* and 762 *bb*. And from these genotype frequencies, we expect the following allele frequencies among the offspring they produce:

Frequency of $B = (2 \times 16 + 222)/2{,}000 = 0.127 = p$

Frequency of $b = (2 \times 762 + 222)/2{,}000 = 0.873 = q$

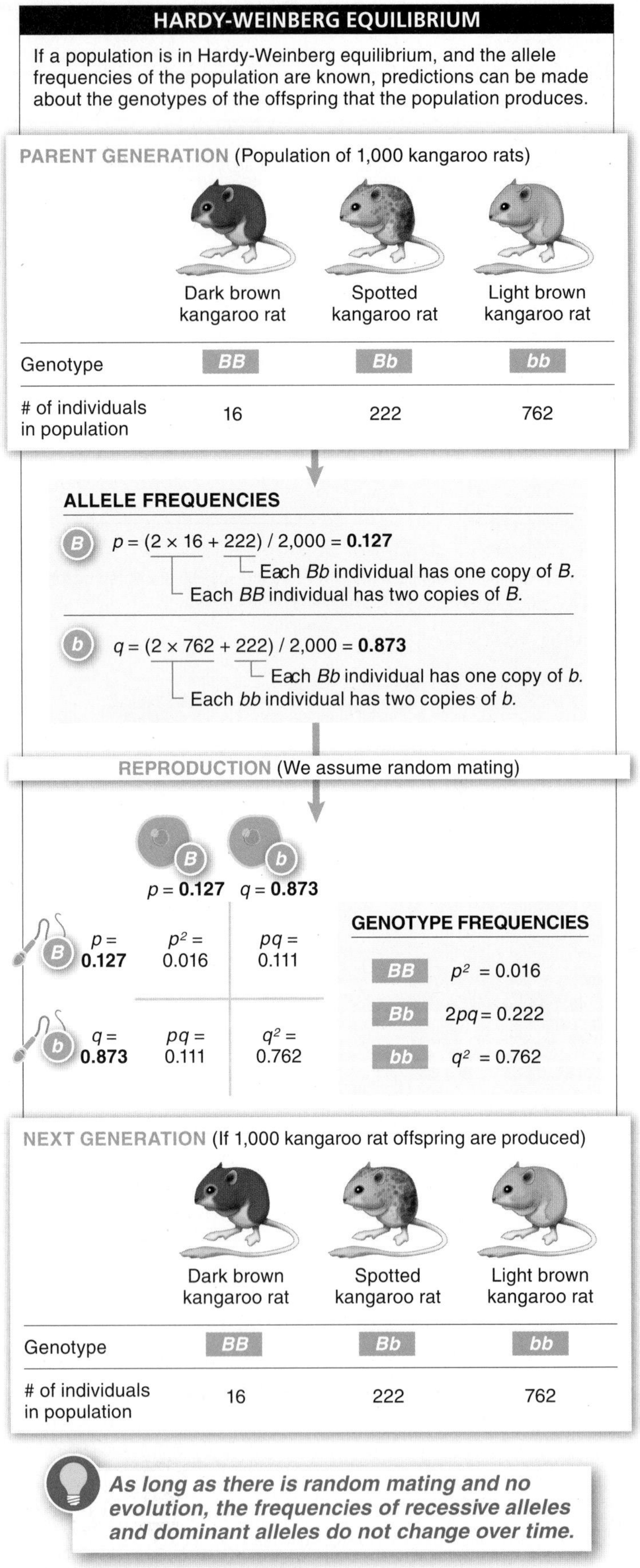

FIGURE 8-21 Recessive alleles don't necessarily disappear from populations.

HARDY-WEINBERG EQUILIBRIUM

If a population is in Hardy-Weinberg equilibrium, and the allele frequencies of the population are known, predictions can be made about the genotypes of the offspring that the population produces.

PARENT GENERATION (Population of 1,000 kangaroo rats)

	Dark brown kangaroo rat	Spotted kangaroo rat	Light brown kangaroo rat
Genotype	BB	Bb	bb
# of individuals in population	16	222	762

ALLELE FREQUENCIES

B $p = (2 \times 16 + 222) / 2{,}000 = \textbf{0.127}$

     Each *Bb* individual has one copy of *B*.
     Each *BB* individual has two copies of *B*.

b $q = (2 \times 762 + 222) / 2{,}000 = \textbf{0.873}$

     Each *Bb* individual has one copy of *b*.
     Each *bb* individual has two copies of *b*.

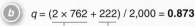

REPRODUCTION (We assume random mating)

$p = 0.127$ $q = 0.873$

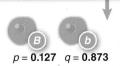

GENOTYPE FREQUENCIES

	$p =$ 0.127	$p^2 =$ 0.016	$pq =$ 0.111
	$q =$ 0.873	$pq =$ 0.111	$q^2 =$ 0.762

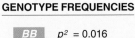

BB	$p^2 = 0.016$
Bb	$2pq = 0.222$
bb	$q^2 = 0.762$

NEXT GENERATION (If 1,000 kangaroo rat offspring are produced)

	Dark brown kangaroo rat	Spotted kangaroo rat	Light brown kangaroo rat
Genotype	BB	Bb	bb
# of individuals in population	16	222	762

 As long as there is random mating and no evolution, the frequencies of recessive alleles and dominant alleles do not change over time.

Notice that the frequencies are unchanged. The allele frequencies 0.127 and 0.873 will always produce the same genotype frequencies—0.16, 0.222, and 0.762—which, in turn, will always have the same allele frequencies. Put another way: the recessive allele *doesn't* decrease over time! It doesn't change at all.

This is true as long as individuals are not dying off specifically because they carry the recessive allele (or because they carry the dominant), in which case the allele frequencies would be changing due to natural selection. It also holds true as long as any of the other mechanisms of evolution that we discussed earlier are not acting on the population—that is, only if mutations, migration, or genetic drift aren't altering the allele frequencies. In each of these exceptions, the allele frequencies would be changing through evolution. The Hardy-Weinberg conclusion also assumes random mating, in which the alleles are randomly coming together in all possible genotypes.

As long as these assumptions hold true—that is, random mating and no evolution—allele frequencies will not change over time, and the Hardy-Weinberg equations allow us to predict the genotype frequencies we should see.

What if we examine a population and find that the genotype frequencies we observe are *not* those predicted by the Hardy-Weinberg equations? There may be, for example, fewer heterozygotes than predicted. If this occurs, we say that the population is not in *Hardy-Weinberg equilibrium,* and we know that either evolution or non-random mating is occurring, or both. If there are fewer heterozygotes than we expected, for example, it may be that more heterozygotes than homozygotes are dying, for some reason—perhaps predation. Or maybe individuals are preferentially mating with individuals having a similar phenotype. In either case, our calculations help us better understand the forces influencing the population and suggest further lines of investigation.

TAKE-HOME MESSAGE 8·10

If we know the frequency of each allele in a population, we can predict the genotypes and phenotypes we should see in that population. If the phenotypic frequencies in a population are not those predicted from the allele frequencies, the population is not in Hardy-Weinberg equilibrium, because an assumption of the equations has been violated. Either non-random mating or evolution is occurring. But as long as the Hardy-Weinberg assumptions are not violated, recessive alleles and dominant alleles do not change their frequencies over time.

A harp seal pup's white coat helps it hide from polar bears.

8·11 Traits causing some individuals to have more offspring than others become more prevalent in the population.

"Survival of the fittest." This is perhaps one of the most famous but most misunderstood phrases. Taken literally, it seems to be circular: those organisms that survive *must* be the fittest. But this is true only if being fit is defined as the ability to survive. As we'll see, in evolution, the word *fitness* is not defined by an organism's ability to survive or its physical strength or its health. Rather, fitness has everything to do with an organism's reproductive success.

Here's an interesting side note: the phrase "survival of the fittest" was coined not by Darwin but by Herbert Spencer, an influential sociologist and philosopher. Moreover, the phrase did not appear in *The Origin of Species* when Darwin first published it. It wasn't until the fifth edition, 10 years later, that he used the phrase in describing natural selection.

Before we see how fitness affects natural selection and a population's adaptation to its environment, let's define fitness. **Fitness** is a measure of the relative amount of reproduction of an individual with a particular phenotype compared with the reproductive output of individuals of the same species with alternative phenotypes.

Suppose there are two fruit flies. One fly carries the genes for a version of a trait that allows it to survive a long time without food. The other has the genes for a different version of the trait that allows it to survive only a short

while without food. Which fly has the greater fitness? If the environment is one in which there are long periods of time without food, such as in the experiment described at the beginning of the chapter, the fly that can live a long time without food is likely to produce more offspring than the other fly, and so over the course of its life it has greater fitness. The alleles carried by an individual with high fitness will increase their proportion in a population over time, and the population will evolve.

There are three important elements to an organism's fitness.

1. An individual's fitness is measured relative to other genotypes or phenotypes in the population. Those traits that confer the highest fitness will generally increase in frequency in a population, and their increase will always come at the expense of alternative traits that confer lower fitness.

2. Fitness depends on the specific environment in which the organism lives. The fitness value of having one trait versus another depends on the environment. A sand-colored mouse living in a beach habitat will be more fit than a chocolate-colored mouse. But that same sand-colored mouse will practically call out to potential predators if it lives in the darker brush away from the beach. An organism's fitness, although genetically based, is not fixed in stone and unchanging—it can change over time and across habitats (**FIGURE 8-22**).

3. Fitness depends on an organism's reproductive success compared with other organisms in the population. If you carry an allele that gives you the trait of surviving for 200 years, but that also causes you to be sterile, your fitness is zero; that allele will never be passed down to future generations, and its frequency in the population will soon be zero. On the other hand, if you inherit an allele that gives you a trait that causes you to die at half the age of everyone else, but also causes you to have twice as many offspring as the average individual, your fitness is increased. It is reproductive success that is all-important in determining whether particular traits increase in frequency in a population.

Q "Survival of the fittest" is a misnomer. Why?

"Survival of the fittest" is a misleading phrase because it is the individuals with the greatest *reproductive output* that are the most fit in any population. It becomes a more meaningful phrase if we consider it as a description of the fact that those alleles that increase an individual's fitness will "survive" in a population more than those that decrease an individual's fitness.

TAKE-HOME MESSAGE 8·11

Fitness is a measure of the relative amount of reproduction by an individual with a particular phenotype, compared with the reproductive output of individuals with alternative phenotypes. An individual's fitness can vary, depending on the environment in which it lives.

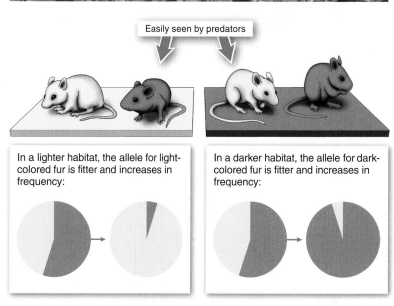

FITNESS AND HABITAT

Easily seen by predators

In a lighter habitat, the allele for light-colored fur is fitter and increases in frequency:

In a darker habitat, the allele for dark-colored fur is fitter and increases in frequency:

FIGURE 8-22 An organism's fitness depends on the environment in which it lives.

8·12 Organisms in a population can become better matched to their environment through natural selection.

If you put a group of humans on the moon, will they flourish? If you take a shark from the ocean and put it in your swimming pool, will it survive? In both cases, the answer is *no*. Organisms are rarely successful when put into novel environments. And the stranger the new environment, the less likely it is that the transplanted organism will survive. Why is that?

As Darwin noted over and over during his travels, the organisms that possess traits that allow them to better exploit the environment in which they live will tend to produce more offspring than the organisms with alternative traits. With passing generations, a population will be made up of more and more of these fitter organisms. And, as a consequence, populations of organisms will tend to be increasingly well matched or adapted to their environment.

Adaptation refers both to the process by which organisms become better matched to their environment *and* to the specific features that make an organism more fit. Examples of

adaptations abound. Bats have an extremely accurate type of hearing (called echolocation) for navigating and finding food, even in complete darkness. Porcupine quills make porcupines almost impervious to predation (**FIGURE 8-23**). Mosquitoes produce strong chemicals that prevent blood from clotting, so that they can extract blood from other animals.

FIGURE 8-23 **Adaptations increase fitness.** Quills are an adaptation that reduces (but doesn't eliminate) the risk of predation for porcupines.

TAKE-HOME MESSAGE 8·12

Adaptation, which refers both to the process by which organisms become better matched to their environment and to the specific traits that make an organism more fit, occurs as a result of natural selection.

8·13 Natural selection does not lead to perfect organisms.

In Lewis Carroll's *Through the Looking-Glass,* the Red Queen tells Alice that "in this place it takes all the running you can do, to keep in the same place." She might have been speaking about the process of evolution by natural selection. After all, if the least fit individuals are continuously weeded out of the population, we might logically conclude that, eventually, fitness will reach a maximum, and all organisms in all populations will be perfectly adapted to their environment. But this never happens. That is where the Red Queen's wisdom comes in.

Consider one of the many clearly documented cases of evolution in nature: the beak size of Galápagos finches. Over the course of a multidecade study, biologists Rosemary Grant and Peter Grant closely monitored the average size of the finches' beaks. They found that the average beak size within a population fluctuated according to the food supply. During dry years—when the finches had to eat large, hard seeds—birds with bigger, stronger beaks were more successful and multiplied. During wet years, smaller-beaked birds were more successful, because there was a surplus of small, soft seeds.

The ever-changing "average" finch beak illustrates that adaptation does not simply march toward some optimal endpoint (**FIGURE 8-24**). Evolution in general, and natural selection specifically, does not guide organisms toward

"betterness" or perfection. Natural selection is simply a process by which, in each generation, the alleles that cause organisms to have the traits that make them most fit in that environment tend to increase in frequency. If the environment changes, the alleles that are most favored may change, too.

> **❝** I was taught that the human brain was the crowning glory of evolution so far, but I think it's a very poor scheme for survival.**❞**
>
> — KURT VONNEGUT, **American writer**

Q *Why doesn't natural selection lead to perfect organisms?*

In the next to last paragraph of *The Origin of Species,* Darwin wrote: "as natural selection works solely by and for the good of each being, all corporeal [physical] and mental endowments will tend to progress towards perfection." In this passage, he overlooks several factors that prevent populations from progressing inevitably toward perfection:

1. Environments change quickly. Natural selection may be too slow to adapt the organisms in a population to such a constantly moving target.

2. Variation is needed as the raw material of selection— remember, it is the first necessary condition for natural selection to occur. If a mutation creating a

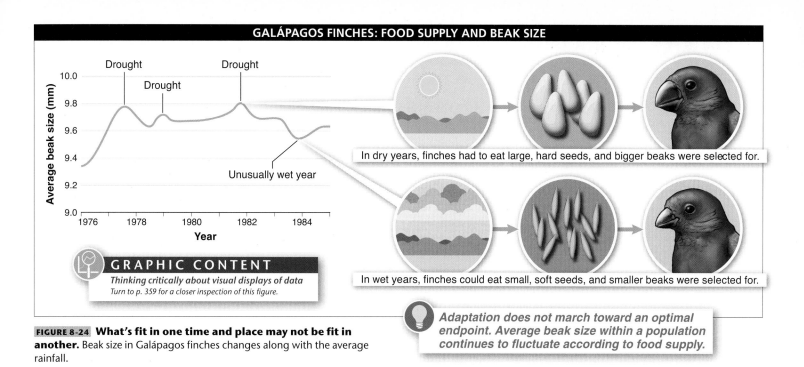

GALÁPAGOS FINCHES: FOOD SUPPLY AND BEAK SIZE
Average beak size (mm)
10.0
9.8
9.6
9.4
9.2
9.0
1976
1978
1980
1982
1984
Year
Drought
Drought
Drought
Unusually wet year
In dry years, finches had to eat large, hard seeds, and bigger beaks were selected for.
In wet years, finches could eat small, soft seeds, and smaller beaks were selected for.
GRAPHIC CONTENT
Thinking critically about visual displays of data
Turn to p. 359 for a closer inspection of this figure.
Adaptation does not march toward an optimal endpoint. Average beak size within a population continues to fluctuate according to food supply.

FIGURE 8-24 What's fit in one time and place may not be fit in another. Beak size in Galápagos finches changes along with the average rainfall.

new, "perfect" version of a gene never arises, the individuals within a population will never be perfectly adapted.

3. There may be different alleles for a trait, each causing individuals to have the same fitness. In this case, each allele represents an equally fit "solution" to the environmental challenges.

TAKE-HOME MESSAGE 8·13

Natural selection does not lead to organisms perfectly adapted to their environment, because (1) environments can change more quickly than natural selection can adapt organisms; (2) mutation does not produce all possible alleles; and (3) there is not always a single, optimum adaptation for a given environment.

8·14 Artificial selection is a special case of natural selection.

In practice, plant and animal breeders understood natural selection before Darwin did; they just didn't know that they understood it. Farmers bred crops for maximum yield, and dog, horse, and pigeon fanciers selectively bred the animals with their favorite traits to produce more and more of the offspring with more and more exaggerated versions of those traits.

The process used by animal breeders and farmers is called **artificial selection,** but the underlying genetic process does not differ from natural selection, because the three conditions are satisfied. Artificial selection is considered a special case of natural selection, however, because the differential reproductive success is being determined by humans rather than by nature. Apple growers, for example, use artificial selection to produce the wide variety of apples now available. What is important is that it is still differential reproductive success, and the results are no different. It was a stroke of genius for Darwin to recognize that the same process farmers were using to develop new and better crop varieties is occurring naturally in every population on earth, and always has been.

TAKE-HOME MESSAGE 8·14

Animal breeders and farmers are making use of natural selection when they modify their animals and crops through selective breeding, because the three conditions for natural selection are satisfied. Since the differential reproductive success is determined by humans rather than by nature, this type of natural selection is also called artificial selection.

Natural selection can change the traits in a population in several ways.

Certain traits are easily categorized—blue eyes and brown eyes, or white tiger fur and orange tiger fur, for example. Other traits, such as height, are influenced by many genes and environmental factors, so a continuous range of phenotypes occurs (**FIGURE 8-25**) Whether a given trait is influenced by one gene or by a complex interaction of many genes and the environment, it can be subject to natural selection and be changed in any of several ways.

Directional Selection In **directional selection,** individuals with one extreme of the range of variation in the population have higher fitness. Milk production in cows is an example. There is a lot of variation in milk production from cow to cow. As you might expect, farmers select for breeding those cows with the highest milk production and have done so for many decades. The result of such selection is not surprising: between 1920 and 1945, average milk production increased by about 50% in the United States (**FIGURE 8-26**).

Those at the other end of the range, the cows that produce the smallest amounts of milk, have reduced fitness, since the farmers do not allow them to reproduce, limiting the

FIGURE 8-26 **More milk?** Cows have been selected for their ability to produce more and more milk.

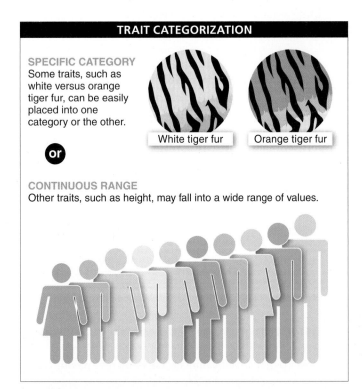

TRAIT CATEGORIZATION

SPECIFIC CATEGORY
Some traits, such as white versus orange tiger fur, can be easily placed into one category or the other.

White tiger fur   Orange tiger fur

or

CONTINUOUS RANGE
Other traits, such as height, may fall into a wide range of values.

FIGURE 8-25 **Some traits fall into clear categories, others range continuously.**

number of "less milk" alleles in the next generation. In subsequent generations, the average value for the milk-production trait increases. Note, however, that the variation for the trait in the cow population decreases a bit, too, because the alleles contributing to one of the phenotypic extremes are eventually removed from the population.

Hundreds of experiments in nature, in laboratories, and on farms demonstrate the power of directional selection. In fact, it is one of the marvels of laboratory selection and animal breeding that nearly any trait chosen to be exaggerated through directional selection responds dramatically— sometimes with absurd results (**FIGURE 8-27**). Turkeys, for instance, have been selected for increased size of their breast muscles, which makes them more profitable for farmers. Directional selection has been so successful that the birds' breast muscles are now so large that it has become impossible for them to mate. All turkeys on large-scale,

Q Turkeys on poultry farms have such large breast muscles that they can't get close enough to each other to mate. How can such a trait evolve?

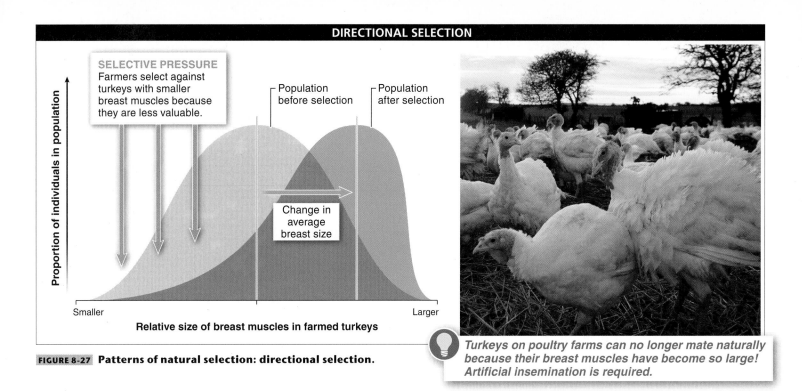

DIRECTIONAL SELECTION

SELECTIVE PRESSURE
Farmers select against turkeys with smaller breast muscles because they are less valuable.

Population before selection

Population after selection

Change in average breast size

Proportion of individuals in population

Smaller — Relative size of breast muscles in farmed turkeys — Larger

FIGURE 8-27 Patterns of natural selection: directional selection.

Turkeys on poultry farms can no longer mate naturally because their breast muscles have become so large! Artificial insemination is required.

industrial poultry farms must now reproduce through artificial insemination!

Stabilizing Selection How much did you weigh at birth? Was it more than 10 pounds? More than 11? Was it close to the 22 ½ pounds that Carmelina Fedele's child weighed when born in Italy in 1955? Or was it 5 pounds or less? Unlike directional selection, for which there is increased

fitness at one extreme and reduced fitness at the other, **stabilizing selection** is said to take place when individuals with intermediate phenotypes are the most fit. The death rate among babies, for example, is lowest between 7 and 8 pounds, and is higher for both lighter and heavier babies (**FIGURE 8-28**). The variation in birth weight has decreased as stabilizing selection has reduced the frequencies of genes associated with high and low birth weights.

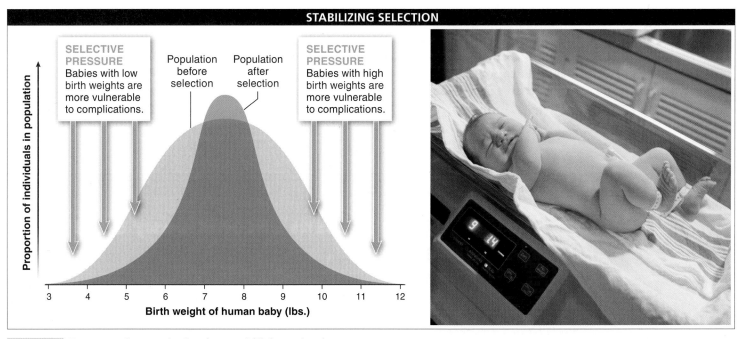

STABILIZING SELECTION

SELECTIVE PRESSURE
Babies with low birth weights are more vulnerable to complications.

Population before selection

Population after selection

SELECTIVE PRESSURE
Babies with high birth weights are more vulnerable to complications.

Proportion of individuals in population

3 4 5 6 7 8 9 10 11 12
Birth weight of human baby (lbs.)

FIGURE 8-28 Patterns of natural selection: stabilizing selection.

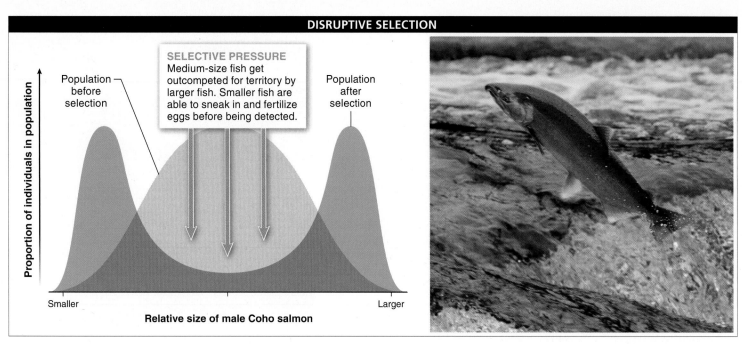

SELECTIVE PRESSURE
Medium-size fish get outcompeted for territory by larger fish. Smaller fish are able to sneak in and fertilize eggs before being detected.

Population before selection

Population after selection

Proportion of individuals in population

Smaller

Larger

Relative size of male Coho salmon

FIGURE 8-29 Patterns of natural selection: disruptive selection.

Modern technologies, including Caesarean deliveries and premature-birth wards, allow many babies to live who would not have survived without such technology. This has the unintended consequence of reducing the selection against any alleles causing those "extreme" traits, reducing the rate of their removal from the population.

Disruptive Selection There is a third kind of natural selection, in which individuals with extreme phenotypes experience the highest fitness, and those with intermediate phenotypes have the lowest. Although examples of this type of selection, called **disruptive selection,** are rare in nature, the results are not surprising. Among some species of fish—the coho salmon, for instance—only the largest males acquire good territories. They generally enjoy relatively high reproductive success. While the intermediate-size fish regularly get run out of the good territories, some tiny males

are able to sneak in and fertilize eggs before the territory owner detects their presence. Consequently, we see an increase in the frequency of small and large fish, with a reduction in the frequency of medium-size fish (**FIGURE 8-29**).

TAKE-HOME MESSAGE 8·15

Acting on traits for which populations show a large range of phenotypes, natural selection can change populations in several ways. These include directional selection, in which the average value for the trait increases or decreases; stabilizing selection, in which the average value of a trait remains the same while extreme versions are selected against; and disruptive selection, in which individuals with extreme phenotypes have the highest fitness.

? 8·16 THIS IS HOW WE DO IT

By picking taller plants, do humans unconsciously drive the evolution of smaller plants?

There are almost as many ways to approach problems scientifically as there are problems. One consistent element, however, is a clear and concise statement of your

hypothesis. Articulating a hypothesis as concisely and precisely as possible often sheds light on how you might make the observations or collect the necessary data to test it. This step can help you take a big idea or issue and break it down to a more manageable, answerable question. For example, let's consider a problem that conservation biologists wonder (and worry) about: the impacts of human activities on plant populations.

What's an example of an overly broad research question?

Can humans unwittingly cause evolutionary changes in plants just by making use of them? Researchers addressed this big question with a simple but powerful investigation of a medicinal plant, the Tibetan snow lotus, *Saussurea laniceps*—a rare plant that lives in rocky habitats high in the eastern Himalayas.

In traditional Chinese and Tibetan medicine, snow lotus flowers are used to treat headaches and high blood pressure. The flowers are also valued as souvenirs. As a consequence, snow lotus flowers are highly sought after. But they're not picked at random: larger flowers (found on taller plants), which are believed to be more potent, are strongly preferred by the human harvesters. It's this fact that made it possible for the researchers to tackle the research question noted above.

What's a more manageable version of the research question?

Is human harvesting of the snow lotus causing evolutionary change to the plants?

How might we recast the question as a testable hypothesis?

Because humans like to pick the tallest of the wild-growing snow lotus plants, they are causing evolution. Or we could get even more precise:

Human harvesting of the tallest snow lotus plants is causing the evolution of smaller plants.

How could the researchers determine whether the plants are becoming shorter over time?

This was their first step: they simply measured the snow lotus plants from recent collections sold in China.

Is there a "control group" with which to compare today's plants?

Here's where the researchers got clever. They came up with two different, but equally valuable, control groups. For the first, they headed to herbaria. These are the repositories of specimens acquired by explorers and collectors that serve as a tool for cataloging plant biodiversity around the world. The researchers measured snow lotus plants from eight

herbaria and from numerous field collections, noting the date that each plant had been collected. What did their comparison reveal?

Prior to 1920, the average snow lotus plant height was *22 cm;* the average snow lotus plant height collected after 2000 was just *14 cm.* Without question, the more recently the plant had been collected, the smaller it was. (When they statistically evaluated the relationship between the year of collection and the plant height, they found a significant negative correlation: $r^2 = 0.44$; $p < 0.0001$; $N = 218$.) From these collections, the researchers also measured snow lotus plants from a closely related species that is not generally picked. Over the course of 120 years, there was no decline in the average plant height for this relative. Snow lotus plants are getting smaller only for the species that is harvested by humans.

Could another comparison be made?

To further test their hypothesis, the researchers were able to make another, possibly even better, comparison. They compared the snow lotus size at a heavily harvested site with plants growing in protected, sacred Tibetan areas, where flower harvesting is not permitted. Here's what they found:

	Protected Areas	Heavily Harvested Areas
Plant height (average ± SE)	22.7 ± 2.0 cm	13.4 ± 0.4 cm

The plants were about 40% shorter in the heavily harvested sites than in the protected areas!

Did the researchers prove their point? And does it matter?

The evidence strongly supports the researchers' hypothesis. Humans have inadvertently imposed directional selection on the snow lotus. The researchers noted, too, that the intensity of the selection seems to have increased since the 1970s, as better roads have increased access to the habitat of the snow lotus and greater interest in alternative medicines has created a larger market for them.

Because smaller snow lotus plants produce fewer seeds, the fitness of the heavily harvested populations may be reduced. If this is the case, the conservation status of the plant may be threatened. Moreover, the case of the snow lotus may be just one example. If the phenomenon of humans unconsciously acting as a selective force when harvesting desirable species is more general, the implications for the conservation of species could be great.

Can you think of other naturally occurring species that humans value and may harvest in ways that lead to inadvertent evolution?

TAKE-HOME MESSAGE 8·16

Humans value the Tibetan snow lotus flower for medicinal and other purposes. Because people prefer to harvest the largest snow lotus plants, there has been selection for smaller plants. Data from herbaria collections and from snow lotus populations in protected areas reveal a significant negative trend in height over the past hundred years.

8·17 Natural selection can cause the evolution of complex traits and behaviors.

We have seen that natural selection can change allele frequencies and modify the frequency with which simple traits, such as fur color or turkey-breast size or plant height, appear in a population. But what about complex traits, such as behaviors, that involve numerous physiological and neurological systems? For instance, can natural selection improve maze-running ability in rats?

The short answer is *yes*. Remember, evolution by natural selection is occurring, changing the allele frequencies for traits, whenever (1) there is variation for the trait, (2) that variation is heritable, and (3) there is differential reproductive success based on that trait. These conditions can easily be satisfied for complex traits, including behaviors.

> ❝All things must change
> To something new, to something strange.❞
>
> — HENRY WADSWORTH LONGFELLOW,
> **American poet**

In 1954, to address this question, William Thompson trained a group of rats to run through a maze for a food reward (**FIGURE 8-30**). He found a huge amount of variation in the rats' abilities: some rats learned much more quickly than others how to run the maze. Thompson then selectively bred the fast learners with each other and the slow learners with each other. Over several generations, he developed two separate populations: rats descended from a line of fast

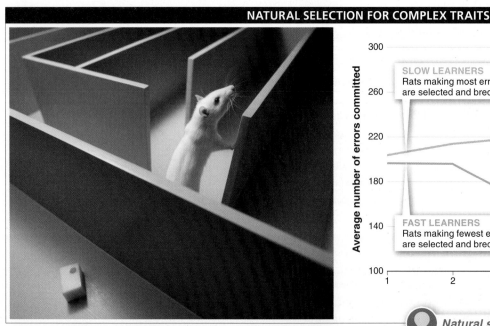

NATURAL SELECTION FOR COMPLEX TRAITS

SLOW LEARNERS
Rats making most errors in the maze are selected and bred with each other.

After 6 generations of selection, the rats selected for fast learning made only half as many errors as the slow learners.

FAST LEARNERS
Rats making fewest errors in the maze are selected and bred with each other.

Average number of errors committed (y-axis: 100, 140, 180, 220, 260, 300)

Generation (x-axis: 1, 2, 3, 4, 5, 6)

💡 *Natural selection can produce and alter complex traits, including behaviors.*

FIGURE 8-30 **"Not too complex . . ."** Natural selection can alter maze-running behavior in rodents.

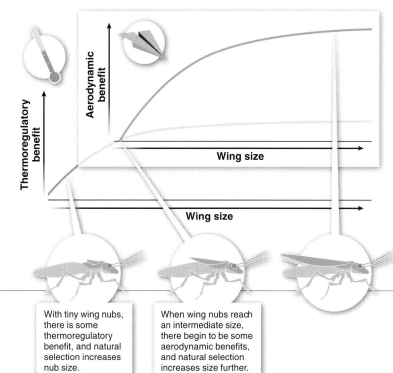

💡 *Traits selected for one function may be co-opted by natural selection for a new function.*

FIGURE 8-31 **How could natural selection produce an insect wing, when the primitive wing nub had no aerodynamic benefits?**

With tiny wing nubs, there is some thermoregulatory benefit, and natural selection increases nub size.

When wing nubs reach an intermediate size, there begin to be some aerodynamic benefits, and natural selection increases size further.

maze-learners and rats descended from a line of slow maze-learners. After only six generations, the slow learners made twice as many errors as the fast learners before mastering the maze, while the fast learners were adept at solving complex mazes that would give many humans some difficulty. Sixty years later, it's still unclear which particular genes are responsible for maze-running behavior, yet the selection experiment still demonstrates a strong genetic component to the behavior.

Q How can a wing evolve if 1% of a wing doesn't help an organism fly or glide at all?

Natural selection can also produce complex traits in unexpected, roundabout ways. One vexing case involves the question of how natural selection could produce an organ as complex as a fly wing, when 1% or 2% of a wing—that is, an incomplete structure—doesn't help an insect to fly. In other words, while it is clear how natural selection can preserve and increase the frequencies of fitness-increasing traits in populations, how do these soon-to-be-useful traits increase during the early stages if they don't increase the organism's fitness (**FIGURE 8-31**)?

The key to answering this question is that 1% of a wing doesn't actually need to function as a wing to increase an individual's fitness. Often, structures are enhanced or elaborated on by natural selection because they enhance fitness by serving some other purpose. Experiments using models of insects demonstrated that, as expected, a small percentage of a wing, in

the form of a nub or "almost wing," confers no benefit at all when it comes to flying. (The nubs don't even help flies keep their orientation during a "controlled fall.") The incipient wings *do* help the insects address a completely different problem, though. They allow much more efficient temperature control, so that an insect can gain heat from the environment when the insect is cold and dissipate heat when the insect is hot. Experiments on heat-control efficiency, in fact, show that as small nubs become more and more pronounced, they are more and more effective, probably conferring increased fitness on the individual fly—but only up to a point.

Eventually, the thermoregulatory benefit stops increasing, even if the nub length continues to increase. But it is right around this point that the proto-wing starts to confer some aerodynamic benefits (see Figure 8-31). Consequently, natural selection may continue to increase the length of this "almost wing," but now the fitness increase is due to a wholly different effect. Such functional shifts explain the evolution of numerous complex structures that we see today, and these shifts may be common in the evolutionary process.

TAKE-HOME MESSAGE 8·17

Natural selection can change allele frequencies for genes involved in complex physiological processes and behaviors. Sometimes a trait that has been selected for one function is later modified to serve a completely different function.

8·18–8·22
The evidence for evolution is overwhelming.

The Denise's Pygmy seahorse (Hippocampus denise) hides itself in a sea fan.

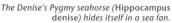

8·18   The fossil record documents the process of natural selection.

<blockquote>
❝ It is indeed remarkable that this theory [evolution] has been progressively accepted by researchers, following a series of discoveries in various fields of knowledge. The convergence, neither sought nor fabricated, of the results of work that was conducted independently is in itself a significant argument in favor of this theory. ❞

— POPE JOHN PAUL II, 1996
</blockquote>

In the 150 or so years since Darwin first published *The Origin of Species,* thousands of studies have been conducted on his theory of evolution by natural selection, both in the laboratory and in natural habitats. A wide range of modern methodologies has also been developed—such as genome sequencing and algorithms for comparing genome similarities and constructing the most likely phylogenies—all contributing to a much deeper understanding of the process of evolution. This ongoing accumulation of evidence overwhelmingly supports the basic premise that Darwin put forward, while filling in many of the gaps that frustrated him.

In the remainder of the chapter we review the five primary lines of evidence demonstrating the occurrence of evolution:

1. **The fossil record**—physical evidence of organisms that lived in the past

2. **Biogeography**—patterns in the geographic distribution of living organisms

3. **Comparative anatomy and embryology**—growth, development, and body structures of major groups of organisms

4. **Molecular biology**—examination of life at the level of individual molecules

5. **Laboratory and field studies**—implementation of the scientific method to observe and study evolutionary mechanisms

The first of the five lines of evidence is the fossil record. Although it has been central to much documentation of the occurrence of evolution, it is a very incomplete record. After all, the soft parts of an organism tend to decay rapidly and completely after death. And there are only a few environments (such as tree resin, tar pits, the bottom of deep lakes, and continental shelves) in which the processes of erosion and decomposition are so reduced that an organism's hard parts, including bones, teeth, and shells, can be preserved for thousands or even millions of years. These remains, called **fossils,** can be used to reconstruct what organisms must have looked like long ago. Such reconstructions often provide a clear record of evolutionary change, including unique insights into the patterns and processes of history (**FIGURE 8-32**).

The use of **radiometric dating** helps in painting a clearer picture of organisms' evolutionary history by telling us the age of the rock in which a fossil is found (**FIGURE 8-33**). In Darwin's time, it was assumed that the deeper in the earth a fossil was found, the older it was. Radiometric dating goes a step farther, making it possible to determine not just the *relative* age of fossils but also their *absolute, numerical* age. This is accomplished by evaluating the amounts of certain radioactive isotopes present in the fossil-containing rocks or the layers above and below them. Radioactive isotopes in a rock begin breaking down into more stable

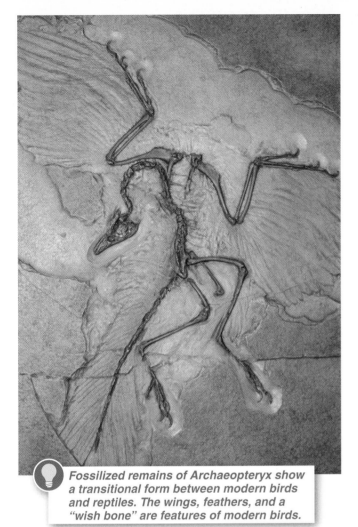

Fossilized remains of Archaeopteryx show a transitional form between modern birds and reptiles. The wings, feathers, and a "wish bone" are features of modern birds.

FIGURE 8-32 **Evidence for evolution: the fossil record.** Fossils can be used to reconstruct the appearance of organisms that lived long ago.

Paleontologists must deal with the fact, however, that fossilization is an exceedingly unlikely event, and when it does occur, it represents only those organisms that (1) happened to live in that particular area, (2) could be preserved under certain chemical conditions, and (3) had physical structures that can leave fossils. For this reason, the fossil record is unavoidably incomplete. Entire groups of organisms have left no fossil record at all, and for others the record has numerous gaps. Still, for what there is, the fossil record can be incredibly detailed and interesting, and fossils have been found that link all of the major groups of vertebrates.

The evolutionary history of horses is among the most well-preserved in the fossil record. First appearing in North America about 55 million years ago, horses radiated around the world, with more recent fossils appearing in Eurasia and Africa. These fossils exhibit distinct adaptations to those differing environments. Later, about 1.5 million years ago, much of the horse diversity—including all North American horse species—disappeared, leaving only a single remaining genus, or group of species, called *Equus*. Because there is now only one horse genus on earth, it is tempting to imagine a simple linear path

compounds as soon as the rock is formed, and they do so at a constant rate. Nothing can alter this. By measuring the relative amounts of the radioactive isotope and its leftover decay product in the rock where a fossil is found, the age of the rock, and thus of the fossil, can be calculated.

Radiometric dating confirms that the earth is very old. Rocks more than 3.8 billion years old have been found on all of the earth's continents, with the oldest so far found in northwestern Canada. By using the radioactive isotope uranium-238, with a half-life of 4.5 billion years, researchers have determined that the earth is about 4.6 billion years old and that the earliest organisms appeared at least 3.5 billion years ago. Radiometric dating also makes it possible to put the fossil record in chronological order. By dating all the fossils discovered in one locale, paleontologists can better evaluate whether the organisms are related to each other and how groups of organisms changed over time.

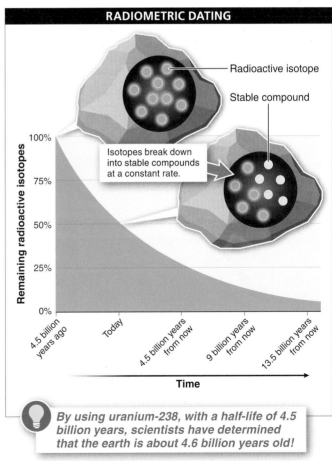

RADIOMETRIC DATING

By using uranium-238, with a half-life of 4.5 billion years, scientists have determined that the earth is about 4.6 billion years old!

FIGURE 8-33 **How old is that fossil?** Radiometric dating using uranium-238 helps to determine the age of rocks and the fossils in them.

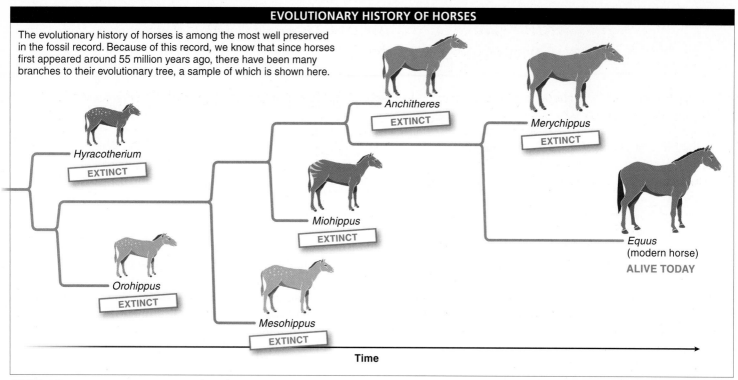

EVOLUTIONARY HISTORY OF HORSES

The evolutionary history of horses is among the most well preserved in the fossil record. Because of this record, we know that since horses first appeared around 55 million years ago, there have been many branches to their evolutionary tree, a sample of which is shown here.

Anchitheres
EXTINCT

Merychippus
EXTINCT

Hyracotherium
EXTINCT

Miohippus
EXTINCT

Equus
(modern horse)
ALIVE TODAY

Orohippus
EXTINCT

Mesohippus
EXTINCT

Time

FIGURE 8-34 **An evolutionary family tree.** The branching evolutionary tree of the horse.

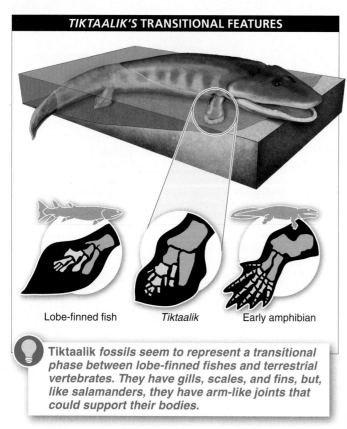

TIKTAALIK'S TRANSITIONAL FEATURES

Lobe-finned fish   Tiktaalik   Early amphibian

💡 *Tiktaalik fossils seem to represent a transitional phase between lobe-finned fishes and terrestrial vertebrates. They have gills, scales, and fins, but, like salamanders, they have arm-like joints that could support their bodies.*

FIGURE 8-35 **A missing link?**

from modern horses straight back through their 55-million-year evolutionary history. But that's just not how evolution works. In reality, there have been numerous branches of horses that have split off over evolutionary time, flourished for millions of years, and only recently gone extinct. What we see living today is just a single branch of a greatly branched evolutionary tree (**FIGURE 8-34**).

The fossil record provides another valuable piece of evidence for evolution, in the form of fossils with transitional features. These are fossils that demonstrate a link between groups of species thought to have shared a common ancestor. One such fossil is *Tiktaalik* (**FIGURE 8-35**). First found in northern Canada and estimated to be 375 million years old, *Tiktaalik* fossils seem to represent a transitional phase between lobe-finned fishes and terrestrial vertebrates. These creatures, like lobe-finned fishes, had gills, scales, and fins, but they also had arm-like joints in their fins and could support their bodies with their limbs in much the same way that salamanders can.

TAKE-HOME MESSAGE 8·18

Radiometric dating confirms that the earth is very old and allows scientists to determine the age of fossils. By analyzing fossil remains, paleontologists can reconstruct what organisms looked like long ago, learn how organisms were related to each other, and understand how groups of organisms evolved over time.

8·19 Geographic patterns of species distributions reflect species' evolutionary histories.

When it comes to species distributions, history matters. Species were not designed from scratch to fill a particular niche. Rather, whatever arrived in a geographic region first—usually a nearby species—took up numerous different lifestyles in numerous different habitats, and the populations ultimately adapted to and evolved in each environment. In Hawaii, it seems that a finch-like descendant of the honeycreepers arrived 4–5 million years ago and rapidly evolved into a large number of diverse species. The same process occurred and continues to occur in all locales, not just on islands.

The study of the distribution patterns of living organisms around the world is called **biogeography.** This is the second line of evidence that helps us to see that evolution takes place and to better understand the process. The patterns of biogeography that Darwin and many subsequent researchers noticed provide strong evidence that evolutionary forces are responsible for these patterns. Species often more closely resemble other species that live less than a hundred miles away but in radically different habitats than they resemble species living thousands of miles away in nearly identical habitats. In Hawaii, for example, nearly every bird is some sort of modified honeycreeper, from seed-eating honeycreepers to curved-bill nectar-feeding honeycreepers (**FIGURE 8-36**).

Large, isolated habitats also have interesting biogeographic patterns. Australia and Madagascar are filled with unique organisms that are clearly not closely related to organisms elsewhere. In Australia, for example, marsupial species, rather than placental mammals, fill all of the usual roles. There are marsupial "wolves," marsupial "mice," marsupial "squirrels," and marsupial "anteaters" (**FIGURE 8-37**).

The marsupials of Australia physically resemble their placental counterparts for most traits, but molecular analysis shows that they are actually more closely related to one another, sharing a common marsupial ancestor. Their relatedness to each other is also revealed by similarities in their reproduction: females give birth to offspring at a relatively early stage of development, and the offspring finish their development in a pouch. The presence of marsupials in Australia does not simply mean that marsupials are better adapted than placentals to Australian habitats. When placental organisms are transplanted to Australia they do just fine, often thriving to the point of endangering the native species. Instead, it appears that the terrestrial placental mammals in Australia

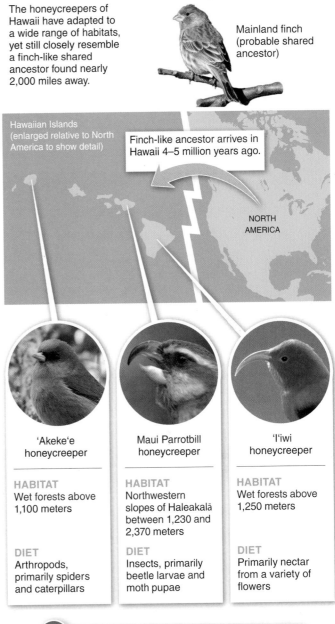

BIOGEOGRAPHY: HAWAIIAN HONEYCREEPERS

The honeycreepers of Hawaii have adapted to a wide range of habitats, yet still closely resemble a finch-like shared ancestor found nearly 2,000 miles away.

Mainland finch (probable shared ancestor)

Hawaiian Islands (enlarged relative to North America to show detail)

Finch-like ancestor arrives in Hawaii 4–5 million years ago.

NORTH AMERICA

'Akeke'e honeycreeper

HABITAT
Wet forests above 1,100 meters

DIET
Arthropods, primarily spiders and caterpillars

Maui Parrotbill honeycreeper

HABITAT
Northwestern slopes of Haleakalā between 1,230 and 2,370 meters

DIET
Insects, primarily beetle larvae and moth pupae

'I'iwi honeycreeper

HABITAT
Wet forests above 1,250 meters

DIET
Primarily nectar from a variety of flowers

 The Hawaiian honeycreepers resemble a common ancestor from mainland North America, but all have unique features.

FIGURE 8-36 **Evidence for evolution: biogeography.**

disappeared about 55 million years ago—although the reasons are not clear—giving the marsupials the chance to flourish and diversify.

AUSTRALIAN MARSUPIALS

Sugar glider

Numbat

Tasmanian wolf

PLACENTAL COUNTERPARTS

Flying squirrel

Giant anteater

Gray wolf

FIGURE 8-37 **Evidence for evolution: biogeography.** Many Australian marsupials resemble placental counterparts, though they are not closely related.

 Though less related to each other than you are to a shrew, these marsupials and their placental counterparts have come to resemble each other as natural selection has adapted them to similar habitats.

Biogeographic patterns such as those seen in honeycreepers and in the marsupials of Australia illustrate that evolution doesn't necessarily lead to the same "solutions" each time a particular set of environmental conditions occurs. Rather, the traits in populations that happen to be in a particular location gradually change, and those species become better adapted to the habitats they occupy.

TAKE-HOME MESSAGE 8·19

Observing geographic patterns of species distributions—noting similarities and differences among species living close together but in very different habitats and among species living in similar habitats but located far from one another—helps us understand the evolutionary histories of populations.

8·20 Comparative anatomy and embryology reveal common evolutionary origins.

If you observe any vertebrate embryo while it's developing, you will see that it passes through a stage in which it has little gill pouches on the sides of the neck. It will also pass through a stage in which it has a long bony tail. This is true whether it is a human embryo or that of a turtle or a chicken or a shark. The gill pouches disappear before birth in all but the fishes. Similarly, we don't find humans with tails. Why do these features exist during an embryo's development? Such common embryological stages indicate that the organisms share a common ancestor, from which all have been modified (**FIGURE 8-38**). Study of these developmental stages and the adult body forms of organisms provides our third line of evidence for the occurrence of evolution.

Among adult animals, several features of anatomy reveal the ghost of evolution in action. We find, for example, that many related organisms show unusual similarities that can be explained only through evolutionary relatedness. The forelimbs of mammals are used for a variety of very different functions in bats, porpoises, horses, and humans (**FIGURE 8-39**). If each had been designed specifically for the uses necessary to that species—flying, swimming, running, grasping—we would expect dramatically different designs. And yet, in each of these species we see the same bones—modified extensively—betraying the fact that they share a common ancestor. These features are called **homologous structures.**

Q Why do all vampire bats grow teeth specialized for grinding solid food, when the bats have a completely liquid diet?

At the extreme, homologous structures sometimes come to have little or no function at all. Such evolutionary leftovers, called **vestigial structures,** exist because they had value in an ancestor species. Some vestigial structures in mammals include the molars that continue to grow in vampire

HOMOLOGOUS STRUCTURES: EMBRYO STAGE

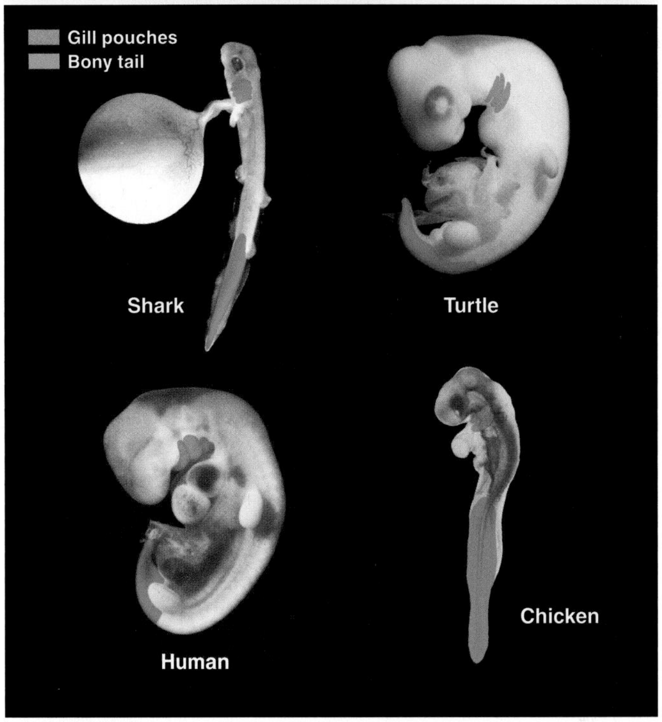

- Gill pouches
- Bony tail

Shark

Turtle

Human

Chicken

FIGURE 8-38 **Evidence for evolution: embryology.** Structures derived from common ancestry can be seen in embryos.

bats, even though these bats consume a completely liquid diet (**FIGURE 8-40**); eye sockets (with no eyes) in some populations of cave-dwelling fishes; and in whales, pelvic bones that are attached to nothing (but in nearly all other

HOMOLOGOUS BONE STRUCTURES

Humerus

Radius
Ulna

Phalanges
Metacarpals
Carpals

Human

Horse

Bat

Porpoise

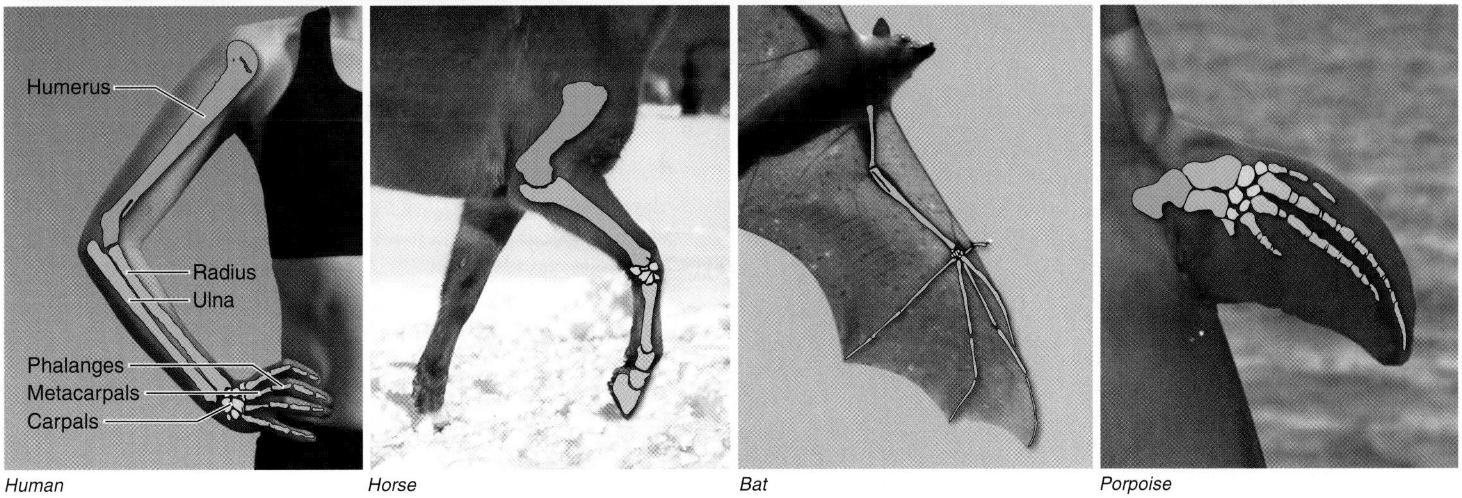

FIGURE 8-39 **Evidence for evolution: comparative anatomy.** Homologous bone structures among some mammals.

The similarities in the bone structure of the forelimbs of mammals demonstrate common ancestry.

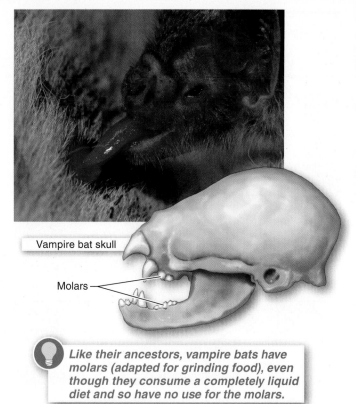

Vampire bat skull

Molars

Like their ancestors, vampire bats have molars (adapted for grinding food), even though they consume a completely liquid diet and so have no use for the molars.

FIGURE 8-40 Evidence for evolution: vestigial structures.

mammals serve as an important attachment point for leg bones). Even humans may have a vestigial organ—the appendix. It is greatly enlarged in our relatives the great apes, acting as host to cellulose-digesting bacteria that aid in breaking down the plants in the apes' diet, but in humans it seems to serve no purpose (although recent studies suggest it provides some immune system support).

Not all organisms with similar-looking adaptations actually share recent ancestors. We see flying mammals (bats) and flying insects (locusts) (FIGURE 8-41). Likewise, dolphins and penguins live in similar habitats and have flippers that help them swim. In both examples, however, the analogous structures developed from different original structures. Natural selection, in a process called **convergent evolution,** may act on different starting materials (such as a flipper or a forelimb) and modify them until they serve similar purposes—much as we saw in the marsupial and placental mammals in Figure 8-37.

Different starting materials come to perform the same function through convergent evolution.

FIGURE 8-41 Evidence for evolution: convergent evolution and analogous structures.

TAKE-HOME MESSAGE 8·20

Similarities in the anatomy and development of different groups of organisms and in their physical appearance can reveal common evolutionary origins.

New technologies for deciphering and comparing the genetic code provide our fourth line of evidence for evolution. From microscopic bacteria to flowering plants to insects and primates, the molecular instructions for building organisms and transmitting hereditary information are the same: four simple bases, arranged in an almost unlimited variety of sequences. All living organisms share the same genetic code.

When we examine the similarity of DNA among related individuals within a species, we find that they share a greater proportion of their DNA than do unrelated individuals. This is not unexpected; you and your siblings got all of your DNA from the same two parents, while you and your cousins got half of your DNA from the same two grandparents. The more distantly you and another individual are related, the more your DNA differs.

We can measure the similarity between the DNAs of two species by comparing their DNA sequences for individual genes. For example, let's look at the gene that codes for a polypeptide used to build hemoglobin. In vertebrate animals, hemoglobin is found inside red blood cells, where its function is to carry oxygen throughout the body. It is made up of two types of chains of amino acids, the alpha chain and the beta chain. In humans, the beta chain has 146 amino acids. In rhesus monkeys, this beta chain is nearly identical: of the 146 amino acids, 138 are the same as those found in human hemoglobin, and only 8 are different. In dogs, the sequence is still somewhat similar, with 114 of the same amino acids. When we look at non-mammals such as birds, we find about 101 amino acid similarities with humans. And between humans and lamprey eels (still vertebrates, but with more than 500 million years since our last common ancestor), there are 21 amino acid similarities.

The differences in the amino acid sequence of the beta hemoglobin chain (and thus the alleles that code for this structure) seem to indicate that humans have more recently shared a common ancestor with rhesus monkeys than with dogs. And that we have more recently shared an ancestor with dogs than with birds or lampreys. These findings are just as we would expect, based on estimates of evolutionary

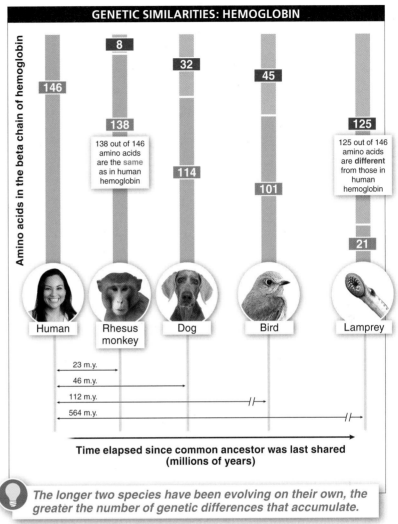

GENETIC SIMILARITIES: HEMOGLOBIN

Amino acids in the beta chain of hemoglobin

- Human: 146
- Rhesus monkey: 8 / 138 — 138 out of 146 amino acids are the same as in human hemoglobin
- Dog: 32 / 114
- Bird: 45 / 101
- Lamprey: 125 / 21 — 125 out of 146 amino acids are different from those in human hemoglobin

23 m.y.
46 m.y.
112 m.y.
564 m.y.

Time elapsed since common ancestor was last shared (millions of years)

The longer two species have been evolving on their own, the greater the number of genetic differences that accumulate.

FIGURE 8-42 An evolutionary clock? Genetic similarities (and differences) demonstrate species relatedness.

relatedness made from comparative anatomy and embryology, as well as from the fossil record. It is as if a molecular clock is ticking. The longer two species have been evolving separately, the greater the number of changes in amino acid sequences—or "ticks of the clock" (**FIGURE 8-42**).

TAKE-HOME MESSAGE 8·21

All living organisms share the same genetic code. The degree of similarity in the DNA of different species can reveal how closely related they are and the amount of time that has passed since their last common ancestor.

8·22 Laboratory and field experiments enable us to watch evolution in progress.

A fifth line of evidence for the occurrence of evolution comes from multigeneration experiments and observations. People once thought that evolution was too slow a process to be observed in action. However, by choosing the right species—preferably organisms with very short life spans—and designing careful experiments, we can observe and measure evolution as it is happening.

In one clever study, researchers looked at populations of grass on golf courses—a habitat where lawn-mower blades represent a significant source of mortality. All the grass on the golf courses was of the same species, but on the putting greens it was cut very frequently, on the fairways it was cut only occasionally, and in the rough it was almost never cut at all (**FIGURE 8-43**). Over just a few years, significant changes took place in these different grass populations. The grass plants on the greens came to be short lived, with rapid development to reproductive age and a very high seed output. For the plants in these populations, life was short and reproduction came quickly. (For those in which it did not come quickly, it did not come at all; hence their lack of representation in the population.) Plants on the fairways had slightly slower development and a reduced seed output, while those in the rough had the slowest development and the lowest seed output of all. When plants from each of the habitats were collected and grown in greenhouses under identical conditions, the dramatic differences in growth, development, and life span remained, confirming that the frequencies of the various alleles controlling the traits of life span and reproductive output in the three populations had changed. In other words, evolution had occurred.

One particularly disturbing line of evidence for the occurrence of evolution in nature comes from the evolution of antibiotic-resistant strains of bacteria that cause illness in humans. In the 1940s, when penicillin was first used as a treatment for bacterial infections, it was uniformly effective in killing *Staphylococcus aureus*. However, some strains of *Staphylococcus* contain alleles that make them resistant to penicillin. Because penicillin has become such a pervasive toxin in the environment of *Staphylococcus,* natural selection has led to an increase in the frequency of alleles that make these strains resistant to the drug. As a consequence, humans are increasingly at risk for becoming infected with *Staphylococcus* and getting diseases such as pneumonia and meningitis. Today, more than 90% of

isolated *S. aureus* strains are resistant to penicillin (**FIGURE 8-44**). In the 1960s, the antibiotics methicillin and oxacillin had nearly complete effectiveness against *Staphylococcus*. Today, though, nearly a third of staph infections are resistant to these antibiotics as well. The meaning of such unintentional natural selection "experiments" is clear and consistent: evolution is occurring all around us.

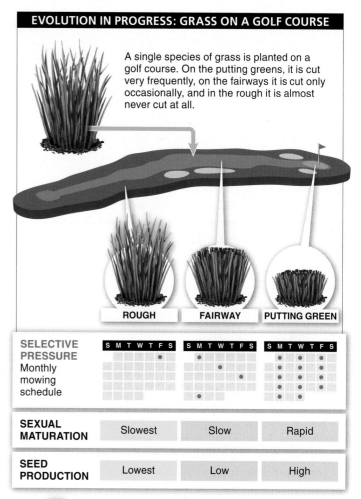

EVOLUTION IN PROGRESS: GRASS ON A GOLF COURSE

A single species of grass is planted on a golf course. On the putting greens, it is cut very frequently, on the fairways it is cut only occasionally, and in the rough it is almost never cut at all.

	ROUGH	FAIRWAY	PUTTING GREEN
SELECTIVE PRESSURE Monthly mowing schedule	S M T W T F S	S M T W T F S	S M T W T F S
SEXUAL MATURATION	Slowest	Slow	Rapid
SEED PRODUCTION	Lowest	Low	High

 Over the course of only a few years, grass plants from the same stock had developed into three distinct populations as a result of the frequency at which they were cut.

FIGURE 8-43 Evolution in progress: grasses. Different mowing patterns can cause evolution in the grass on a golf course.

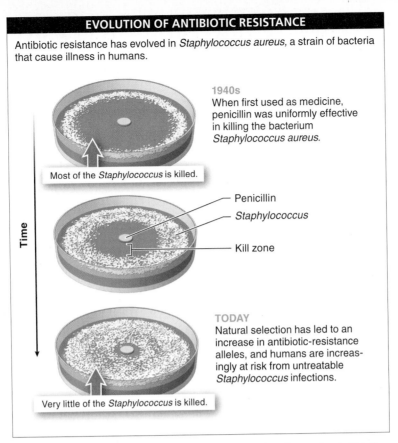

EVOLUTION OF ANTIBIOTIC RESISTANCE

Antibiotic resistance has evolved in *Staphylococcus aureus*, a strain of bacteria that cause illness in humans.

1940s
When first used as medicine, penicillin was uniformly effective in killing the bacterium *Staphylococcus aureus*.

Most of the *Staphylococcus* is killed.

Time

Penicillin

Staphylococcus

Kill zone

TODAY
Natural selection has led to an increase in antibiotic-resistance alleles, and humans are increasingly at risk from untreatable *Staphylococcus* infections.

Very little of the *Staphylococcus* is killed.

FIGURE 8-44 **Evolution in progress: disease-causing bacteria.** Antibiotic resistance has evolved in *Staphylococcus*.

Finally, let's return to the starvation-resistant fruit flies introduced at the beginning of this chapter. They are an unambiguous demonstration that natural selection can produce dramatic changes in a population, and that it can bring about these changes very quickly. And, perhaps most important, we saw that replicating the same evolutionary process over and over again produced the same predictable results. In the laboratory, on the farm, in the doctor's office, and in deserts, streams, and forests, evolution is occurring.

Reflecting on both the process and the products of evolution and natural selection, in the final paragraph of

The Origin of Species Darwin eloquently wrote: "There is grandeur in this view of life . . . and that . . . from so simple a beginning endless forms most beautiful and most wonderful have been, and are being, evolved."

TAKE-HOME MESSAGE 8·22

Replicated, controlled laboratory selection experiments and long-term field studies of natural populations allow us to watch and measure evolution as it occurs.

Evolution
What It Is and What It Is Not . . .

Misconceptions about evolution and how it occurs can arise from misunderstandings about subtleties in the process, from incomplete or erroneous accounts of evolution in schools and the media, and from purposeful attempts to interfere with the understanding of evolution.

Misconception: **Evolution is a theory about the origin of life.** *Clarification:* Although scientists are interested in (and are investigating) how life began, evolutionary biology is focused on the processes that occurred (and still occur) *after* life began, including those responsible for the diversification and branching of species and those adapting populations to their environments.

Misconception: **Evolution is not observable or testable.** *Clarification:* Evolution is both observable and testable. Long-term, controlled, and replicated laboratory experiments on many species have demonstrated evolution (see an example in Section 8-1) and speciation (formation of new species). Numerous studies of natural populations have also documented evolution taking place (see Section 8-13). And, as in astronomy and geology, sophisticated hypothesis testing can be done using observations in the natural world.

Misconception: **Evolution is only a theory and hasn't been proved.** *Clarification:* This misconception stems from confusion between the colloquial definition of a theory as a "guess" or "hunch" and the scientific definition of a theory as an explanation, based on many lines of evidence, that generates and supports predictions and is tested in many ways (see Section 1-10).

Misconception: **Individuals evolve (and are "trying" to adapt).** *Clarification:* Natural selection can lead to populations of organisms adapted to their environment (see Sections 8-12 and 8-16). But the process does not involve "trying" to adapt—it is simply a consequence of some individuals passing on their genes at a higher rate than individuals carrying alternative versions of those genes, and it is the population rather than any one individual that changes over time (see Section 8-11).

Misconception: **Evolution is goal-oriented and leads to progress and "optimal" solutions.** *Clarification:* Alleles that make it impossible for the individuals carrying them to live in a particular environment are weeded out of the population. But this is not the same as creating perfect or optimal organisms. Rather, the traits and features of organisms that increase in frequency within a population are simply those that cause the organisms carrying them to leave more descendants than are left by individuals with alternative traits (see Sections 8-9 and 8-16). As the environment changes, so do the traits that are most fit. Several factors can prevent populations from becoming better adapted to a particular environment (see Section 8-13).

Misconception: **There are no transitional fossils; the gaps in the fossil record disprove evolution.** *Clarification:* Because fossils are formed only under a narrow range of environmental conditions, biologists do not expect to find all transitional forms, so the gaps do not disprove evolution. Numerous transitional fossils *have* been found, including those between dinosaurs and modern birds, and those between whales and their terrestrial mammalian ancestors (see Section 8-18).

Misconception: **The theory of evolution says that life originated, and evolution proceeds, randomly, by chance.** *Clarification:* Although chance is one element of evolution—there is a random element to the generation of mutations, for example— many non-random factors are central to how evolution proceeds. For example, the differential reproductive success of some variants— such as bacteria with antibiotic-resistance genes—that leads to natural selection is not random (see Section 8-12).

Misconception: **Most biologists have rejected evolution.** *Clarification:* All modern theories of science are continually tested, verified, and modified as new and deeper understandings of the natural world emerge. That is how science works. The presidents of the National Academy of Sciences, the American Association for the Advancement of Science, and the National Science Teachers Association recently wrote a joint statement saying that "evolutionary theory has stood the test of time in serving as the most comprehensive scientific explanation for the diversity of life on Earth and it is accepted by the overwhelming majority of scientists." There have been no credible challenges to the basic principles of evolution and how it proceeds.

Check Your Knowledge

1. What does the curvy line represent?

2. Finch beaks don't grow or shrink much once the bird reaches maturity. What is responsible for the average beak size changing from year to year?

3. Three droughts and one very wet year are highlighted. Why is that information included here?

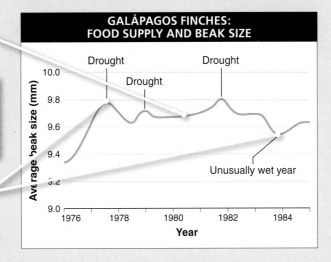

GALÁPAGOS FINCHES: FOOD SUPPLY AND BEAK SIZE

Drought   Drought   Drought

Unusually wet year

Average beak size (mm)

Year

4. If you wanted to make predictions about finches' beak size in the future, what information would be helpful?

5. Is there an "optimum" beak size for a finch? How does this figure help you answer that question?

6. What do you think is the "Take-Home Message" of this graph?

See answers at the back of the book.

Key Terms in Evolution and Natural Selection

adaptation, p. 339
artificial selection, p. 341
biogeography, p. 351
bottleneck effect, p. 329
convergent evolution, p. 354

differential reproductive success, p. 333
directional selection, p. 342
disruptive selection, p. 344
evolution, p. 317
fitness, p. 338

fixation, p. 328
fossil, p. 348
founder effect, p. 329
gene flow, p. 331
genetic drift, p. 328
heritability, p. 333

homologous structure, p. 353
inheritance, p. 333
migration, p. 331
mutation, p. 326
natural selection, p. 332

population, p. 316
radiometric dating, p. 348
sexual selection, p. 334
stabilizing selection, p. 343
trait, p. 321
vestigial structure, p. 353

ABOUT THE CHAPTER OPENING PHOTO

A barn owl swoops to seize a mouse.

8·1 Evolution is an ongoing process.

Characteristics of the individuals that make up a population can change over time. We can observe such change in nature and can even cause such change to occur.

EVOLUTION IN ACTION

In an experiment that tests how long fruit flies can survive without food, evolution occurs right before our eyes. In the experiment, only the fruit flies that survive the longest go on to reproduce and populate the next generation.

? 1. Describe one benefit of conducting research on evolution in a short-lived species such as fruit flies.

The average starvation resistance time is recorded for each generation. Over many generations of natural selection, the population changes. The fruit flies of generation 60 can survive without food much longer than the first generation.

GENERATION 0
Number of flies
Average starvation resistance:
20 HR.
Hours until death from starvation

GENERATION 1
Number of flies
Average starvation resistance:
23 HR.
Hours until death from starvation

GENERATION 60
Number of flies
Average starvation resistance:
160 HR.
Hours until death from starvation

? *Check Your Knowledge*

1. **Selecting for increased starvation resistance in fruit flies:**

   a) has no effect, because starvation resistance is not a trait that influences fruit flies' fitness.

   b) has little effect; ongoing mutation counteracts any benefits from selection.

   c) cannot increase their survival time, because there is no genetic variation for this trait.

   d) has no effect; starvation resistance is dependent on the effects of too many genes.

   e) can produce populations in which the average time to death from starvation is 160 hours.

   0 ——————— 36 ——————— 100
   EASY · · · · · · · · · · · HARD

2. **Georges Cuvier's discovery of fossils of Irish elk and giant ground sloths:**

   a) supports scientific evidence for extinction.

   b) was possible because of Buffon's determination that the earth was more than 6,000 years old.

   c) was possible only following Darwin's publication of *The Origin of Species.*

   d) was made in deep ocean trenches.

   e) suggested that species are immutable.

   0 ——————— 31 ——————— 100
   EASY · · · · · · · · · · · HARD

3. **While on the *Beagle*, Darwin was intrigued by glyptodonts because:**

   a) glyptodont fossils resembled a species that currently existed but were much, much larger.

   b) glyptodont populations thrived on islands but were always wiped out by predators when they were brought to the mainland.

   c) glyptodont fossils varied in size across the different islands in the Galápagos archipelago.

   d) armadillos occurred in Africa but not in South America; glyptodont fossils in South America revealed the continents had once been joined.

   e) their dental structures indicated they consumed seeds larger than those produced by any plants currently found on earth.

   0 ——————— 26 ——————— 100
   EASY · · · · · · · · · · · HARD

8·2–8·4 Darwin journeyed to a new idea.

Charles Darwin developed a theory of evolution by natural selection that explained how populations of species can change over time.

DARWIN'S INFLUENCES

In the 18th and 19th centuries, gradual changes in scientists' beliefs helped shape Charles Darwin's thinking.

 GEORGES-LOUIS LECLERC, COMTE DE BUFFON (1707–1788) Suggested that the earth was much older than previously believed.

 GEORGES CUVIER (1769–1832) By documenting fossil discoveries, showed that extinction had occurred.

 JEAN-BAPTISTE LAMARCK (1744–1829) Suggested that living species might change over time.

 CHARLES LYELL (1797–1875) Argued that geological forces had gradually shaped the earth and continue to do so.

? 2. Describe three commonly held Western beliefs about the natural world that were overturned in the 18th and 19th centuries.

DARWIN'S OBSERVATIONS

Darwin noted unexpected patterns among living organisms he observed and fossils he discovered while on the voyage of the *Beagle*.

FINCHES
The 13 species of finches found on the Galápagos islands all had slightly different physical characteristics, but they still resembled the single mainland finch.

Glyptodont (extinct)   Hairy armadillo (alive today)

FOSSILS
Around the world, there is a striking similarity between the fossils of extinct species found in an area and the living species in that same area.

? 3. How did Darwin's time on the *Beagle* help him develop his ideas on evolution?

? 4. How did the study of fossils give Darwin insight into natural selection?

? 5. Who came up with the theory of evolution by natural selection, independent of Charles Darwin? Why might he be less well-known today?

8.5—8.10 Four mechanisms can give rise to evolution.

Evolution—a change in the allele frequencies in a population—can occur via mutation, genetic drift, migration, or natural selection.

THE EVOLUTION OF POPULATIONS

Evolution is a change in the allele frequencies of a population over time. For example, a change in the proportion of pigment alleles in the population of tigers means that evolution has occurred.

ALLELE FREQUENCIES

● Proportion of orange fur-pigment alleles in the population

○ Proportion of white fur-pigment alleles in the population

? 6. What is evolution?

MUTATION

Mutation is an alteration of the base-pair sequence in an individual's DNA. Such an alteration constitutes evolution if it changes an allele that the individual carries.

Mutated base-pair sequence

Mutated protein

? 7. Describe three causes of mutation.

GENETIC DRIFT

Genetic drift is a random change in allele frequencies within a population, unrelated to the alleles' influence on reproductive success.

Population before genetic drift

Population after genetic drift

Two special cases of genetic drift:

FOUNDER EFFECT
The founding members of a new population can have different allele frequencies than the original source population.

BOTTLENECK EFFECT
The surviving members of a catastrophic event can have different allele frequencies than the source population.

? 8. What is genetic drift? Why is it a more potent agent of evolution in small populations than in large populations?

? 9. What does it mean when fixation for an allele occurs in a population?

MIGRATION

Migration, or gene flow, leads to a change in allele frequencies in a population as individuals move into or out of the population.

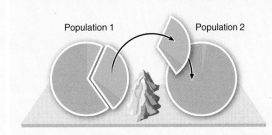

Population 1   Population 2

NATURAL SELECTION

Natural selection is a mechanism of evolution that occurs when the following three conditions are satisfied:

VARIATION FOR A TRAIT
Different traits are present in individuals of the same species.

HERITABILITY
Traits are passed on from parents to their children.

DIFFERENTIAL REPRODUCTIVE SUCCESS
In a population, individuals with traits most suited to reproduction in their environment generally leave more offspring than individuals with alternative traits.

? 10. Distinguish between evolution and natural selection. (Restrict your answer to 30 or fewer words.)

? 11. Do recessive alleles tend to decrease in frequency in a population? Why or why not?

?

4. Evolution occurs:
   a) only when the environment is changing.
   b) only through natural selection.
   c) almost entirely because of directional selection.
   d) only through natural selection, genetic drift, migration, or mutation.
   e) by altering physical traits but not behavioral traits.

   0 — **19** — 100
   EASY   HARD

5. Which of the following statements about mutations is incorrect?
   a) Mutations are almost always random with respect to the needs of the organism.
   b) A mutation is any change in an organism's DNA.
   c) Most mutations are harmful or neutral for the organism in which they occur.
   d) The origin of genetic variation is mutation.
   e) All of these statements are correct.

   0 — **34** — 100
   EASY   HARD

6. In a fish population in a shallow stream, the genotypic frequencies of yellowish-brown fish and greenish-brown fish changed significantly after a flash flood randomly swept away individuals from that stream. This change in genotypic frequency was most likely attributable to:
   a) gene flow.
   b) disruptive selection.
   c) directional selection.
   d) stabilizing selection.
   e) genetic drift.

   0 — **65** — 100
   EASY   HARD

7. When a group of individuals colonizes a new habitat, the event is likely to be an evolutionary event, because:
   a) members of a small population have reduced rates of mating.
   b) gene flow increases.
   c) mutations are more common in novel environments.
   d) new environments tend to be inhospitable, reducing survival there.
   e) small founding populations are rarely genetically representative of the initial population.

   0 — **52** — 100
   EASY   HARD

8. "Survival of the fittest" may be a misleading phrase to describe the process of evolution by natural selection, because:

a) it is impossible to determine the fittest individuals in nature.

b) survival matters less to natural selection than does reproductive success.

c) natural variation in a population is generally too great to be influenced by differential survival.

d) during population bottlenecks, it is the least fit individuals that have the greatest survival.

e) reproductive success, on its own, does not necessarily guarantee evolution.

0 ——————54—————— 100
EASY   HARD

9. Evolutionary adaptation:

a) refers both to the process by which populations become better matched to their environment and to the features of an organism that make it more fit than other individuals.

b) cannot occur in environments influenced by humans.

c) is possible only when there is no mutation.

d) is responsible for the fact that porcupines are at an unusually high risk of predation.

e) occurs for physical traits but not behaviors.

0 ——————14—————— 100
EASY   HARD

10. Adaptations shaped by natural selection:

a) are magnified and enhanced through genetic drift.

b) are unlikely to be present in humans living in industrial societies.

c) may be out of date, having been shaped in the past under conditions that differed from those in the present.

d) represent perfect solutions to the problems posed by nature.

e) are continuously modified so that they are always matched to the environment in which a population lives.

0 ——————77—————— 100
EASY   HARD

11. In a population in which a trait is exposed to stabilizing selection:

a) neither the average value nor the variation for the trait changes.

b) both the average value and the variation for the trait increase.

c) the average value increases or decreases, and the variation for the trait decreases.

d) the average value for the trait stays approximately the same, and the variation for the trait decreases.

e) the average value for the trait stays approximately the same, and the variation for the trait increases.

0 ——————48—————— 100
EASY   HARD

8·11—8·17 Through natural selection, populations of organisms can become adapted to their environments.

When there is variation for a trait, and the variation is heritable, and there is differential reproductive success based on that trait, evolution by natural selection is occurring.

FITNESS

Fitness is a measure of the relative amount of reproduction of an individual with a particular phenotype, as compared with the reproductive output of individuals with alternative phenotypes. An individual's fitness can vary, depending on the environment in which the individual lives.

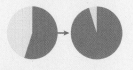

In a lighter habitat, the allele for light-colored fur is fitter and increases in frequency:

In a darker habitat, the allele for dark-colored fur is fitter and increases in frequency:

12. Describe three important components to an organism's evolutionary fitness.

ADAPTATION

Adaptation—meaning both the process by which the organisms in a population become better matched to their environment and the specific features that make an organism more fit—occurs as a result of natural selection.

However, adaptation does not lead to perfect organisms. For example, the average beak size in Galápagos finches fluctuates according to average rainfall and food supply.

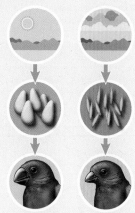

TRAIT CATEGORIZATION

Some traits fall into clear categories, others range continuously.

SPECIFIC CATEGORY
Some traits, such as white versus orange tiger fur, can be easily placed into one category or the other.

CONTINUOUS RANGE
Other traits, such as height, may fall into a wide range of values.

13. How does the increasing frequency of antibiotic-resistant strains of bacteria represent an example of the occurrence of evolution?

14. In *The Origin of Species*, Charles Darwin wrote: "We may look with some confidence to a secure future of great length. And as natural selection works solely by and for the good of each being, all corporeal and mental endowments will tend to progress towards perfection." Give three reasons why he was wrong.

PATTERNS OF NATURAL SELECTION

Natural selection can change populations in several ways.

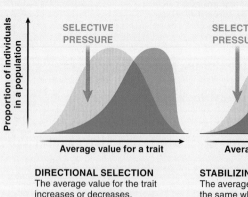 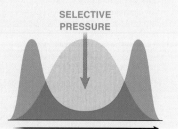

Proportion of individuals in a population

SELECTIVE PRESSURE

Average value for a trait

DIRECTIONAL SELECTION
The average value for the trait increases or decreases.

SELECTIVE PRESSURE   SELECTIVE PRESSURE

Average value for a trait

STABILIZING SELECTION
The average value of a trait remains the same while extreme versions of the trait are selected against.

SELECTIVE PRESSURE

Average value for a trait

DISRUPTIVE SELECTION
Individuals with extreme phenotypes have the highest fitness.

15. Describe two examples of artificial selection in agriculture and/or animal breeding.

16. How does modern medicine alter the selective pressures on birth weight in humans?

17. Describe an example of a trait selected for one function that later was modified to serve a different function.

8·18–8·22 The evidence for evolution is overwhelming.

Many overwhelming lines of evidence document the occurrence of evolution and point to the central and unifying role of evolution by natural selection in helping us to better understand all other ideas and facts in biology.

THE FOSSIL RECORD

Analysis of fossil remains enables biologists to reconstruct what organisms looked like long ago, learn how organisms were related to each other, and understand how groups of organisms evolved over time.

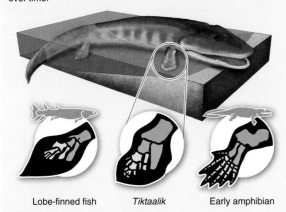

Lobe-finned fish   *Tiktaalik*   Early amphibian

? 18. Describe two limitations of the fossil record in documenting evolution.

BIOGEOGRAPHY

Observing geographic patterns of species distributions helps us to understand the evolutionary histories of populations. For example, Hawaiian honeycreepers have adapted to a wide range of habitats, yet still closely resemble a finch-like shared ancestor found nearly 2,000 miles away.

Hawaiian Islands

Mainland finch (probable shared ancestor)

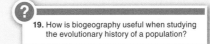

'Akeke'e honeycreeper   Maui Parrotbill honeycreeper   'I'iwi honeycreeper

? 19. How is biogeography useful when studying the evolutionary history of a population?

COMPARATIVE ANATOMY AND EMBRYOLOGY

Similarities in the anatomy and development of different groups of organisms and in their physical appearance can reveal common evolutionary origins.

■ Gill pouches   ■ Bony tail

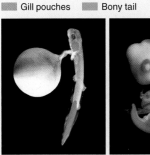
Shark embryo

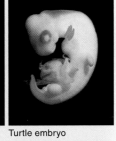
Turtle embryo

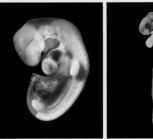
Human embryo

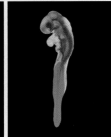
Chicken embryo

? 20. Giving an example of each, compare and contrast homologous and analogous structures.

COMMON GENETIC SEQUENCES

All living organisms share the same genetic code. The degree of similarity in the DNA sequences of different species can reveal how closely related they are and the amount of time that has passed since they last shared a common ancestor.

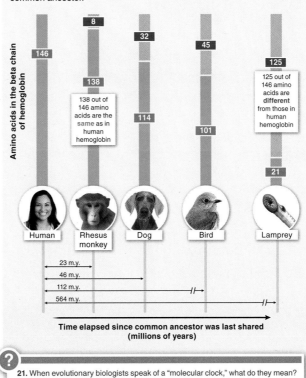

Amino acids in the beta chain of hemoglobin

8
32
45
146
138
125
114
101
21

125 out of 146 amino acids are **different** from those in human hemoglobin

138 out of 146 amino acids are the **same** as in human hemoglobin

Human   Rhesus monkey   Dog   Bird   Lamprey

23 m.y.
46 m.y.
112 m.y.
564 m.y.

Time elapsed since common ancestor was last shared (millions of years)

? 21. When evolutionary biologists speak of a "molecular clock," what do they mean?

?

12. Maze-running behavior in rats:
a) is too complex a trait to be influenced by natural selection.
b) is a heritable trait.
c) is not influenced by natural selection, because it does not occur in rats' natural environment.
d) shows no variation.
e) is influenced primarily by mutation.

0 — EASY — 52 — 100 — HARD

13. A fossil is defined most broadly as:
a) the preserved pieces of hard parts (e.g., shell) of extinct animals.
b) any preserved remnant or trace of an organism from the past.
c) the preserved bones of vertebrates.
d) a piece of an organism that has turned into rock.
e) the product of preservation of intact animal bodies.

0 — EASY — 41 — 100 — HARD

14. Which of the following statements about marsupials is correct?
a) They fill many niches in Australia that are occupied by placental mammals elsewhere.
b) They are less fit than placental mammals.
c) They have become extinct as a result of the greater fitness of placental mammals.
d) They are more closely related to each other than to placental mammals.
e) Both a) and d) are correct.

0 — EASY — 46 — 100 — HARD

15. What can be concluded from comparing differences in molecular biology between different species?
a) Extremely different species are fundamentally unrelated in any way.
b) Only DNA sequences can be used to compare species' relatedness.
c) Birds are more closely related to humans than dogs are.
d) Genetic similarities and differences demonstrate species relatedness.
e) The longer two species have been evolving on their separate paths, the fewer the genetic differences between them.

0 — EASY — 47 — 100 — HARD

16. Evolution:
a) occurs too slowly to be observed.
b) can occur in the wild but not in the laboratory.
c) is responsible for the increased occurrence of antibiotic-resistant bacteria.
d) does not occur in human-occupied habitats.
e) None of these statements are correct.

0 — EASY — 48 — 100 — HARD

9 Evolution and Behavior

COMMUNICATION, COOPERATION, AND CONFLICT IN THE ANIMAL WORLD

Behaviors are traits that can evolve.

Cooperation, selfishness, and altruism can be better understood with an evolutionary approach.

Sexual conflict can result from disparities in reproductive investment by males and females.

Communication and the design of signals evolve.

9.1–9.4
Behaviors are traits
that can evolve.

In Borneo, a macaque hangs upside down on a
branch to take a sip of water.

9·1 Behavior has adaptive value, just like other traits.

What are your favorite things to eat? Perhaps donuts or hot fudge sundaes top your list. Or maybe bananas, or cheeseburgers and french fries. One thing is almost certain about the food preferences of any human: the list is filled with calorie-rich substances. Put another way, when it comes to eating, humans show an almost complete aversion to dirt or pebbles or other substances from which no energy can be extracted (FIGURE 9-1).

In controlled studies, humans demonstrate a clear and consistent preference for sweet and fatty foods. Specifically,

FIGURE 9-1 **Mud pie for dinner?** Humans show an aversion to eating dirt (and other substances from which we cannot extract nutrition).

when people are presented with different mixtures, their preferences increase as the sugar concentration in the food increases—up to a point, after which adding more sugar makes the food less preferable. But the preference becomes stronger and stronger as the food's fat content increases. In short, we prefer sweet (but not too sweet) foods that are packed with as much fat as possible (FIGURE 9-2).

Watch birds in a field and you'll see that they, too, have preferences. A starling, for example, does not eat everything in its path. Rather, it walks through the grass, probing the ground and only occasionally picking up something to eat, such as a beetle larva. Shore crabs, too, have definite preferences. When presented with mussels of different sizes, shore crabs prefer the mussels that provide the greatest amount of energy relative to the amount of energy a crab must expend to open the mussel. The largest mussels take too long to open and so are less preferred; the smallest mussels are easily opened but don't provide enough caloric content to make the effort worthwhile (FIGURE 9-3). Thus, the research shows, shore crabs prefer medium-size mussels.

Humans, starlings, shore crabs, and all other animals have taste preferences for the same reason: animals (including humans), unlike plants, cannot create their own food, so they must consume materials from which they can extract the most energy and acquire essential nutrients. If they don't, they die. We may believe that some things simply taste good while others do not, but these preferences are a direct result of an interaction between the materials we eat and molecules within our taste buds. And those molecules

At this point, adding more sugar makes food less preferable, while the preference continues to increase as the fat content increases.

Human preference for food

— Sugar content
— Fat content

Sugar and fat content in food

 In controlled experiments, humans exhibit preferences for sugary foods, high in fat— and the more fat, the stronger the preference.

FIGURE 9-2 Fatty and sweet. Across cultures, humans prefer sweet, but not too sweet, foods that are packed with fat.

• Conflict, aggression, and territoriality

• Cooperation, alliance-building, and sociality

• Competing for food and avoiding predators

• Migration and navigation

• Behavioral control of body temperature

• Courtship and mate choice

• Pair bonding and fidelity

• Breeding and parental behavior

• Communication

• Learning and tool use

In Chapter 8, we learned about the evolutionary origins of many *physical traits* and the general process by which natural selection can lead to the evolution of organisms adapted to their environments. Similarly, in this chapter, we explore the evolution of *behavioral traits.* We'll concentrate on three broad areas of animal behavior—cooperation and conflict, mating and parenting, and communication—focusing on why those behaviors evolved and answering this question: how do these behaviors contribute to an organism's fitness, survival, and reproduction?

Our approach to the study of evolution and behavior parallels our approach to the study of the evolution of non-behavioral

are produced by a large number of genes that have been shaped by natural selection.

Taste preferences directly influence the evolutionary fitness of organisms; the feeding choices of animals affect the number of offspring they can produce. As a consequence, natural selection can shape feeding behaviors just as it can bring about changes in physical structures.

Before we continue, let's define exactly what we mean by "behavior." **Behavior** encompasses any and all of the actions performed by an organism, often in response to its environment or to the actions of another organism. Feeding behavior is only one of many behaviors influenced by natural selection.

The scope of animal behavior is vast and varied. Even a brief listing of well-studied topics in animal behavior is impressively long:

ENERGY-EFFICIENT FEEDING BEHAVIOR HAS EVOLVED

A shore crab feeds on a mussel

 Shore crabs preferentially choose mussels that provide the most energy relative to the effort it takes to open the shell.

FIGURE 9-3 Efficient eater.

traits. Recall from the discussion of natural selection in the previous chapter that when a heritable trait increases an individual's reproductive success relative to that of other individuals, that trait tends to increase in frequency in the population. This is Darwin's mechanism for evolution by natural selection and, as we saw in Chapter 8, its effects are evident all around us, from the fancy ornamentation of male peacocks' feathers produced by sexual selection to the cryptic coloration that camouflages so many organisms. In this chapter, we see that *behavior* is just as much a part of an organism's phenotype as is its anatomical structures, and that behavior is produced and shaped by natural selection.

TAKE-HOME MESSAGE 9·1

Behavior encompasses any and all of the actions performed by an organism. When a heritable trait increases an individual's reproductive success relative to that of other individuals, that trait tends to increase in frequency in the population. Behavior is a part of an organism's phenotype, and as such it can be produced and shaped by natural selection.

9·2 Some behaviors are innate.

In his study of pea plants, Gregor Mendel described a single gene for plant height that caused a plant to be either tall or short. The production of a trait such as plant height, however, is not completely genetically determined. Certain environmental conditions, such as the type of soil and the availability of water, nutrients, and sunlight, also have a role to play. Nearly all physical traits of all organisms are the products not only of genes but also of environmental conditions. When it comes to the production of behaviors, the environment also plays an important role.

The degree to which a behavior depends on the environment for expression, however, varies a great deal. At one extreme are

behaviors—called **instincts** or **innate behaviors**—that don't require any environmental input to develop. Innate behaviors are present in all individuals in a population and do not vary much from one individual to another or over an individual's life span. An example of innate behavior is a **fixed action pattern.** Triggered in response to a specific signal called a **sign stimulus,** a fixed action pattern is a sequence of behaviors that requires no learning, does not vary among individuals, and, once started, runs to completion. Here are just two examples of fixed action patterns.

1. **Egg retrieval in geese.** When a goose spots an egg outside its nest (a sign stimulus), a fixed action pattern is

SOME BEHAVIORS ARE INNATE

In geese, the sight of an egg outside the nest triggers a fixed action pattern: the goose uses a side-to-side egg-retrieval movement all the way back to the nest, even if the egg is taken away during the process.

Male stickleback during non-breeding season

Male stickleback during breeding season

During the breeding season, a red belly on any other male stickleback fish triggers an aggressive response.

FIGURE 9-4 **No learning required.**

triggered. The goose gets out of the nest and rolls the egg back, using a side-to-side motion, keeping the egg tucked underneath its bill. Once started, a goose continues the behavior to completion, even if the egg is taken away on the way to the nest (**FIGURE 9-4**).

2. Aggressive displays and attacks by stickleback fish. The bellies of male stickleback fish turn bright red when the breeding season arrives. During this time, a male stickleback reacts aggressively to the sight of a red belly (a sign stimulus) on any other male stickleback. In fact, the males become antagonistic at the sight of anything remotely resembling a red belly. One researcher even noticed that his sticklebacks performed aggressive displays every day when a red mail truck drove by his window.

> **TAKE-HOME MESSAGE 9·2**
>
> Like any physical trait, behavior can depend on the environment for expression, though the degree of that dependence varies. Instincts, or innate behaviors, develop without any environmental input. They are behaviors that are present in all individuals in a population and do not vary much from one individual to another or over an individual's life span. A fixed action pattern, a type of innate behavior, is a sequence of behaviors that requires no learning, does not vary, and runs to completion once started.

9·3 Some behaviors must be learned (and some are learned more easily than others).

In contrast to innate behaviors are those behaviors that are influenced to a much greater degree by the individual's environment. These behaviors are acquired, altered, and modified over time in response to past experiences. That is, they require **learning.** There can be tremendous variation among behaviors that require learning: some come relatively easily and are learned by most individuals in a population, while other behaviors are less easily learned.

Consider a trait common to most primates, including humans: fear of snakes. In the wild, rhesus monkeys are afraid of snakes. The sight of a snake causes the monkeys to engage in fear-related responses, including making alarm calls and rapidly moving away from the snake.

Observations of monkeys in captivity, however, reveal that they aren't born with a fear of snakes. Captive rhesus monkeys will reach over a plastic model of a snake to get a peanut. Experiments also show that monkeys learn to fear snakes if they see another monkey terrified at the sight of a snake, even if they only see the reaction on television (**FIGURE 9-5**). They will no longer reach over a snake model to get a peanut, even when they are very hungry. Instead, they scream and move as far away from the artificial snake as possible. Studies on humans show an equally easily learned fear of snakes. Rather than shrieking and cowering in the corner of a cage, however, we respond with sweaty palms and an increased heart rate.

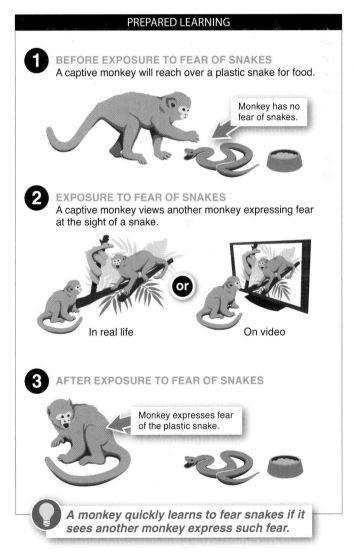

PREPARED LEARNING

1 **BEFORE EXPOSURE TO FEAR OF SNAKES**
A captive monkey will reach over a plastic snake for food.

Monkey has no fear of snakes.

2 **EXPOSURE TO FEAR OF SNAKES**
A captive monkey views another monkey expressing fear at the sight of a snake.

In real life   On video

3 **AFTER EXPOSURE TO FEAR OF SNAKES**

Monkey expresses fear of the plastic snake.

A monkey quickly learns to fear snakes if it sees another monkey express such fear.

FIGURE 9-5 **Easily learned.** The monkey can learn to fear snakes by observing fear in other monkeys.

Q Why is it so much easier for an infant to learn a complex language than for a college student to learn biology?

When behaviors are learned easily and by all (or nearly all) individuals, this is called **prepared learning.** In addition to the snake-fearing behavior of monkeys, examples of prepared learning abound. The acquisition of language in humans is a dramatic example. Most children don't talk until they are a year old, but by the age of three they understand most rules of sentence construction, and the average six-year-old who is a native English-speaker already has a vocabulary of about 13,000 words. These skills are impressive, particularly given that, at this age, children are generally not competent in reading, writing, and fine motor coordination.

Examples of prepared learning also underscore the fact that organisms don't learn everything with equal ease. For example, researchers demonstrated how easily monkeys learned to fear snakes, but in a related experiment, they found that the monkeys were not prepared to learn other behaviors as easily. The researchers altered the videotape that previously showed a monkey expressing fear on encountering a snake to show a monkey having the same fear reaction in response to a flower or a toy rabbit. The captive monkeys—which had never seen flowers or rabbits or snakes—did not learn to fear flowers or rabbits.

Q Human babies quickly and easily develop a fear of snakes. Yet they don't easily develop a fear of guns. Why?

These observations point to an evolutionary basis for the acquisition of certain behaviors. It seems that organisms are well-prepared to learn behaviors that were important to their ancestors' survival and reproductive success over the course of their evolutionary history, and are less prepared to learn behaviors irrelevant to their evolutionary success. Consider, for example, that although human babies quickly and easily develop a fear of snakes, they don't easily develop a fear of guns. Given that more than 30,000 people are killed by guns in the United States each year, while fewer than three dozen people are killed by snakes, it would seem that we ought to be very afraid of guns and relatively unconcerned about snakes. But we are built in just the opposite way (**FIGURE 9-6**). The likely explanation is that evolution can be slow in producing populations that are adapted to their environments—that is, evolutionary change cannot always keep up with a rapidly (in evolutionary terms) changing environment.

Like all genes, any genes involved in behaviors have been handed down to us from our ancestors. Snakes caused many human deaths over the course of our evolutionary history. In contrast, guns didn't kill a single person until very recently on the evolutionary time scale. Accordingly, we still fear our ancient enemy, the snake, and have no instinctual response to novel threats, regardless of how deadly.

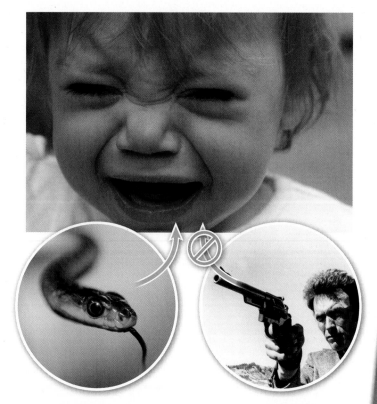

FIGURE 9-6 **Instincts out of date.** A human infant is born with a predisposition for developing a fear of snakes.

TAKE-HOME MESSAGE 9·3

In contrast to innate behaviors are those behaviors that are influenced more by the individual's environment, requiring some learning, and are often modified over time in response to past experiences. Organisms are well-prepared to learn behaviors that were important to the reproductive success of their ancestors, and less prepared to learn behaviors irrelevant to their evolutionary success.

9.4

Complex-appearing behaviors don't require complex thought in order to evolve.

Why do humans and other animals have sex? Is it because they are thinking, robot-like: "Must maximize reproductive success. Must maximize reproductive success"? Of course not. Yet they go about their lives behaving as if they were.

Q Do animals consciously act in order to improve their reproductive success?

In actuality, natural selection produces organisms that exhibit relatively simple behaviors in response to certain situations or environmental conditions— behaviors that to us may seem complex and sophisticated. Let's look at an example. An animal that experiences pleasure when it has sex has the incentive to seek out additional opportunities to experience that pleasure. As long as reproductive success is enhanced in the process, it is not necessary (from the reproductive-success standpoint) for the animal to deliberately seek that outcome. In other words, natural selection doesn't have to produce animals consciously trying to maximize reproductive success. It only needs to produce animals that behave in a way that actually results in reproductive success, and this outcome will occur if organisms experience a pleasurable sensation by having sex. It's like an evolutionary shortcut. Behaviors that lead to a specific outcome that increases the animal's relative reproductive success will be favored by natural selection.

We can investigate experimentally whether natural selection sometimes results in such evolutionary short cuts—which we do by trickery. Consider the example of egg retrieval in geese. When an egg falls out of the nest, a goose retrieves it in exactly the same way every time. Is a goose able to do this because it vigilantly keeps tabs on all of its eggs? This seemed to be the case, until researchers decided to trick some geese. The researchers learned that putting *any* object that remotely resembles an egg near a goose's nest triggers the retrieval process. Geese will retrieve beer cans, door knobs, and a variety of other objects that will not increase the animal's reproductive success. Moreover, if multiple egg-like items are just outside the nest, the goose will retrieve the largest item first.

To test the limits of the retrieval behavior, researchers began putting larger and larger models of eggs outside the nests. When given a choice between an actual goose egg and an artificial egg the size of a basketball, the goose always tried to retrieve the giant egg. Rather than keeping tabs on their eggs, geese seem to be following a rule of thumb to retrieve any nearby egg-like objects, preferentially retrieving the larger objects first (**FIGURE 9-7**). Thus, natural selection can produce organisms that exhibit silly

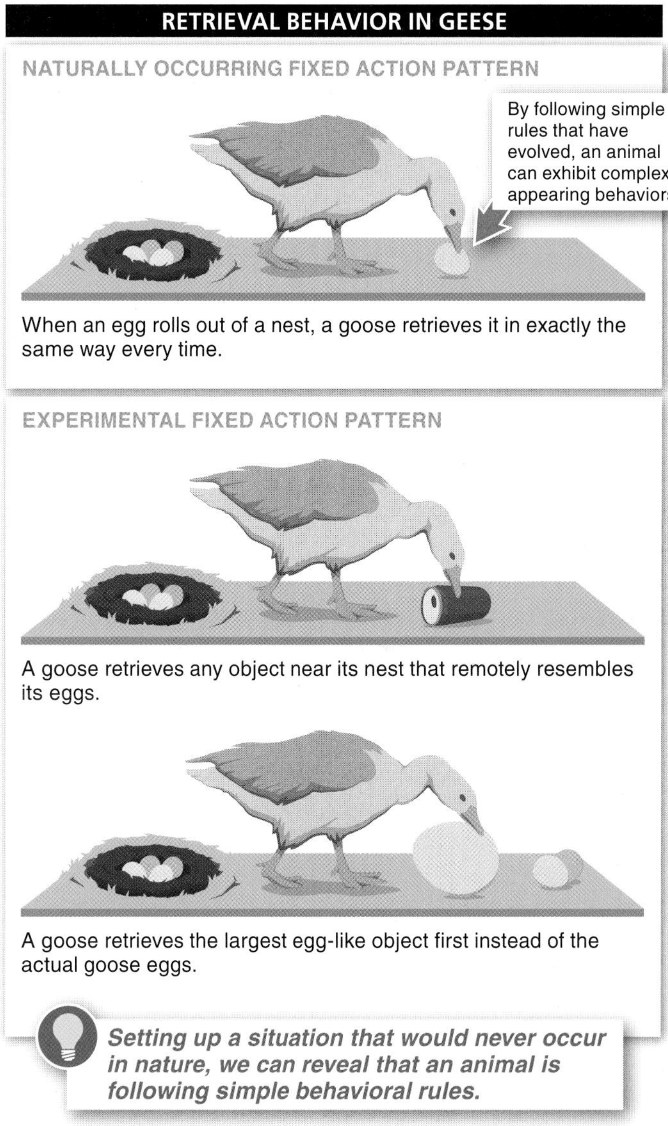

RETRIEVAL BEHAVIOR IN GEESE

NATURALLY OCCURRING FIXED ACTION PATTERN

By following simple rules that have evolved, an animal can exhibit complex-appearing behaviors.

When an egg rolls out of a nest, a goose retrieves it in exactly the same way every time.

EXPERIMENTAL FIXED ACTION PATTERN

A goose retrieves any object near its nest that remotely resembles its eggs.

A goose retrieves the largest egg-like object first instead of the actual goose eggs.

Setting up a situation that would never occur in nature, we can reveal that an animal is following simple behavioral rules.

FIGURE 9-7 **Programmed to retrieve.** The goose will retrieve any egg-like object outside its nest.

and clearly maladaptive (i.e., fitness-decreasing) behaviors in experiments—the goose trying to retrieve a basketball-sized egg—if, in nature, individuals' performance of these simple actions nearly always leads to fitness-increasing behaviors—keeping eggs in the nest.

Taking an evolutionary approach to the study of the behavior of animals, including humans, is not new. Charles Darwin wrote in *On the Origin of Species:* "In the distant future I see open fields for far more important researches. Psychology will be based on a new foundation." And in the past several decades, Darwin's prediction has increasingly proved true. In a synthesis of evolutionary biology and psychology, researchers taking an approach called "evolutionary psychology" have begun to view the human brain and human behaviors, including emotions, as traits produced by natural selection, selected as a result of their positive effects on survival and reproduction. So, too, has the evolutionary approach to studying behavior begun to influence the field of economics. The 2002 Nobel prize for economics was awarded to researchers who used insights from evolutionary biology and psychology in their analyses of human decision making.

In the rest of this chapter, working from the understanding that organisms' traits, including their behaviors, have evolved by natural selection, we examine the question of why organisms behave as they do. We begin with an exploration of selfishness and cooperation in animals—in particular, how natural selection has produced organisms that engage in apparent acts of altruism.

TAKE-HOME MESSAGE 9·4

If certain behavior in natural situations usually increases an animal's relative reproductive success, the behavior will be favored by natural selection. The natural selection of such behaviors does not require the organism to consciously try to maximize its reproductive success.

Cooperation, selfishness, and altruism can be better understood with an evolutionary approach.

An arctic ground squirrel in Alaska.

9.5 "Kindness" can be explained.

If we look closely, we see many behaviors in the animal world that *appear* to be **altruistic behaviors**—that is, they seem to come at a cost to the individual performing them, while benefiting a recipient. When discussing altruism, we define costs and benefits in terms of their contribution to an individual's fitness.

Take the case of the Australian social spider (*Diaea ergandros*). After giving birth to about 50 hungry spiderlings, the mother's body slowly liquefies into a nutritious fluid that the newborn spiders consume. Over the course of about 40 days, her offspring literally eat her alive; they ultimately kill their mother, but start their lives well-nourished. The cost to the mother is huge—she is unable to produce any more offspring—but there is a clear benefit to her many offspring (**FIGURE 9-8**).

Such altruistic-appearing behavior is so common in the natural world that it puzzled Darwin. Natural selection, he believed, generally works to produce selfish behavior. The alleles that cause the individual carrying them to have the greatest reproductive success should become more common in any given generation, relative to other alleles for that

💡 *This mother spider is actually being selfish—not altruistic—when she lets her offspring eat her alive.*

FIGURE 9-8 Parental care to the extreme. The female *Diaea ergandros* spider feeds her offspring with her own body.

same gene. At the same time, alleles that cause the individual carrying them to help increase other individuals' reproductive success at the expense of their own success should decrease their relative frequency in a population. Put another way, natural selection should never produce altruistic behavior. Darwin worried that if the apparent instances of altruism were indeed truly altruistic, they would prove fatal to his theory.

As it turns out, Darwin's theory is safe. Virtually all of the apparent acts of altruism in the animal kingdom prove, on closer inspection, to be not truly altruistic; instead, they have evolved as a consequence of either **kin selection** or **reciprocal altruism.**

Kindness Toward Close Relatives: Kin Selection

Kin selection is a strategy by which one individual assisting another can compensate for its own decrease in fitness if it is helping a close relative in a way that increases the relative's fitness. Kin selection can lead to the evolution of apparently-altruistic behavior toward close relatives. Suppose that, for one gene, you carry allele *K,* which causes you to behave in a way that increases the fitness of a close relative, while decreasing your own fitness. Because you and your relatives tend to share many of the same alleles, including allele *K,* your altruism may increase the frequency of *K* in the population. Thus the increased fitness of a relative you help might compensate for your own reduced fitness, because allele *K* will, overall, increase its frequency in the population when you help your relatives increase their reproductive output.

Kindness Toward Unrelated Individuals: Reciprocal Altruism
Reciprocal altruism can lead to the evolution of apparently-altruistic behavior toward unrelated individuals. Suppose that, for another gene, you carry allele *R,* which causes you to help individuals to whom you are not related. This helps to increase their fitness and decreases

yours in the process. Allele *R* might still increase its market share in the population, if the individuals whom you help become more likely to return the favor and help you in the future, increasing your fitness.

Seen in this light, both kin selection and reciprocal altruism can lead to the evolution of behaviors that are *apparently* altruistic but, in actuality, are beneficial—from an evolutionary perspective—to the individuals engaging in the behaviors.

In the next two sections, we explore kin selection and reciprocal altruism in more detail. We also investigate some of the many testable predictions about when acts of apparent kindness should occur, whom they should occur between, and how we could increase or decrease the frequency of their occurrence through a variety of modifications to the environment.

Occasionally, we see individuals engaging in behavior that *is* genuinely altruistic. We'll examine how and why certain environmental situations cause individuals to behave in a way that decreases their fitness and, as such, is evolutionarily maladaptive. It is important to note here that even as genes play a central role in cooperation and conflict, people's ability to override impulses toward selfishness is responsible for some of the rich diversity in human behavior that we see around us.

TAKE-HOME MESSAGE 9·5

Many behaviors in the animal world appear to be altruistic. In almost all cases, the apparent acts of altruism are not truly altruistic; they have evolved as a consequence of either kin selection or reciprocal altruism and, from an evolutionary perspective, are beneficial to the individual engaging in the behavior.

9·6 Apparent altruism toward relatives can evolve through kin selection.

The grasslands of the western United States are home to large colonies of Belding's ground squirrels, living in underground burrows. Because the colonies are as large as hundreds or thousands of individuals, they attract many birds of prey, which succeed in killing a squirrel in about 10% of attempts. Squirrels have a system for reducing predation risk that resembles a neighborhood watch program. When an aerial predator approaches, it is common for one squirrel, standing on top of a burrow, to

produce a loud whistle-like alarm call, warning the colony of the impending danger. On hearing an alarm call, squirrels quickly take cover in their burrows, reducing their risk of death. Making an alarm call is a very dangerous activity: about half the time that an alarm call is made, the caller is killed by the predator (**FIGURE 9-9**).

Some squirrels see predators and make alarm calls, while other squirrels see predators and keep their mouths shut. Why

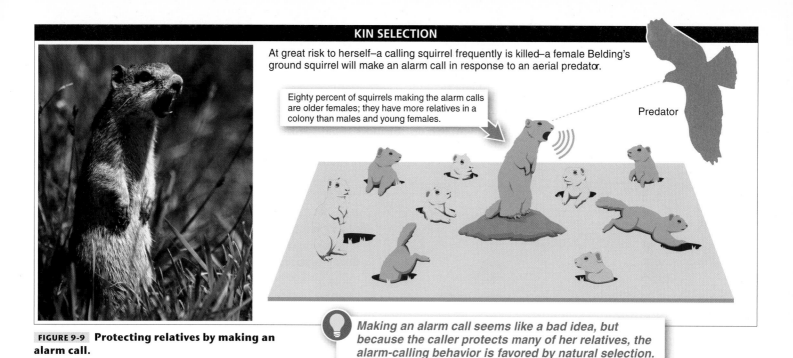

At great risk to herself–a calling squirrel frequently is killed–a female Belding's ground squirrel will make an alarm call in response to an aerial predator.

Eighty percent of squirrels making the alarm calls are older females; they have more relatives in a colony than males and young females.

Predator

FIGURE 9-9 **Protecting relatives by making an alarm call.**

Making an alarm call seems like a bad idea, but because the caller protects many of her relatives, the alarm-calling behavior is favored by natural selection.

would any squirrel make an alarm call? And which squirrels are most likely to engage in this altruistic-appearing behavior? It certainly isn't random: about 80% of the squirrels making alarm calls are female. Moreover, older females are five times more likely than young females to make alarm calls.

These differences reveal that alarm calling is about protecting relatives. The more kin an individual is likely to have, the more likely that individual is to sound the alarm. Males travel long distances to live in new colonies shortly after reaching maturity, so most adult males don't live near their parents, siblings, or any other relatives except their own offspring. Females, on the other hand, remain near the area where they were born and are likely to have many close relatives nearby. Older females, then, are likely to have the largest number of relatives.

The biologist W. D. Hamilton expressed the idea of "kin selection," that one individual assisting another could compensate for its own decrease in fitness if it helped a close relative in a way that increased the relative's fitness. After all, he realized, the recipient of the aid is likely to have at least some genes in common with the altruistic individual. And the more genes they share (i.e., the more closely the individuals are related), the more likely it is that the alleles passed down by the recipient of the altruistic-like behavior to its offspring are the same alleles found in the altruistic individual.

Thus, altruistic-appearing behavior will most likely occur when the benefits to close relatives are greater than the cost

to the individual performing the behavior. That is, the more closely related two individuals are, the more likely they are to act altruistically toward each other. It seems that individuals are acting selflessly/altruistically, but they are really acting in their genes' best interests.

Belding's ground squirrels probably don't know exactly how closely they are related to the squirrels around them, so it is likely that natural selection has led to an evolutionary shortcut that allows them to behave as if they did know. In a clever experiment that tested this idea, researchers trapped adult female squirrels and relocated them to distant ground squirrel colonies in which they had no close relatives. If the transplanted females were able to determine exactly how closely they were related to the squirrels around them, they would not make alarm calls. To do so would put the caller at risk without benefiting her genes. What the researchers found, though, was that transplanted females were just as likely to sound the alarm in the colony of strangers as they were in their home territories, surrounded by close relatives (**FIGURE 9-10**). It seems that female ground squirrels have evolved to follow a simple rule that says, "If I am an older female, I behave as if I have many close relatives around me."

Based on the idea of kin selection, it is necessary to redefine an individual's fitness. An individual's fitness is not measured just by his or her total reproductive output. Fitness also includes the reproductive output that individuals bring about through their seemingly altruistic

INAPPROPRIATE ALARM CALLING IN UNNATURAL SITUATIONS

Researchers transplanted an older female Belding's ground squirrel to a new colony. She had no relatives there but still made alarm calls, revealing that animals may sometimes follow simple behavioral rules.

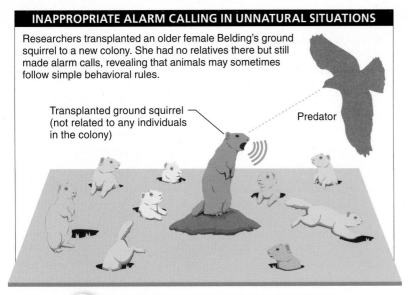

Transplanted ground squirrel (not related to any individuals in the colony)

Predator

A transplanted female Belding's ground squirrel that is not related to the squirrel colony behaves as though she is.

FIGURE 9-10 **Following simple behavioral rules.**

behaviors toward their close kin. This redefined measure of fitness is called **inclusive fitness.**

Q James Joyce wrote: "Whatever else is unsure in this stinking dunghill of a world, a mother's love is not." According to the concept of kin selection, he may not be completely right. Why?

No two individuals—with the exception of identical twins—are genetically identical. And because different individuals do not share all of the same alleles, we expect that they should experience some conflict. In other words, because their genetic interests are not always aligned, they sometimes will differ in the outcome that is best for them when they interact. One disturbing example of such conflict takes place between a pregnant woman and her developing fetus.

A mother's and a fetus's interests differ when it comes to the question of how much food—doled out as nutrients in the blood flow across the placenta—the fetus ought to get. There is a point at which it is in the mother's best interests to reduce the amount of glucose and other nutrients given to the fetus. Her genes gain if she saves some of her resources for future fetuses, to whom she will be related just as closely as she is to the present fetus.

Now consider the fetus's "point of view." Future siblings will carry some but not all of the same genes as the fetus. Put

another way, a fetus is always more closely related to itself than to its siblings. Consequently, the fetus does not necessarily benefit from sacrificing nutritional intake for the sake of future siblings. In a sense, it is a sibling rivalry that starts before the sibling is even conceived! The conflict results in a physiological battle throughout pregnancy. The fetus produces chemicals that increase the diameter of the mother's blood vessels, thus increasing the amount of sugar delivered to the fetus. In response, the mother produces more insulin, a chemical that has exactly the opposite effect, reducing the amount of sugar in the bloodstream that is available to the fetus. In some mothers this conflict causes gestational diabetes, the pregnant woman's inability to properly regulate her blood sugar levels—a condition that disappears as soon as the baby is born. In all pregnancies, though, the conflict escalates until the mother is producing a thousand times the normal amount of insulin.

CONFLICT AND KIN SELECTION

A fetus produces chemicals that increase the diameter of the mother's blood vessels–increasing the amount of food that the fetus gets. In response, the mother produces insulin, reducing the amount of sugar in the bloodstream. Gestational diabetes can result when this conflict escalates.

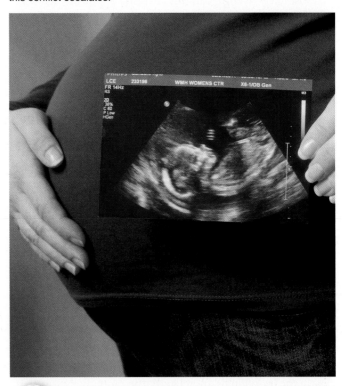

Because a mother's and a fetus's genetic interests are not perfectly aligned, there is conflict about how much food the fetus should get.

FIGURE 9-11 **Competing for nutrients.**

Gestational diabetes might also be the result of conflict between the fetus's alleles that come from its father and those that come from its mother. The parents may have different interests concerning how much to invest in the current offspring versus how much to save for future offspring (**FIGURE 9-11**). (Paternally inherited genes may benefit from greater resource uptake by the fetus, while maternally inherited genes may benefit from reduced resource uptake.)

Just as close relatives are more likely to help each other than are strangers, the converse is also true: the less closely related two individuals are, the more likely they are to experience conflict. Male lions, for example, on taking over a pride usually kill unrelated cubs (although females defend their young aggressively), causing the females to become reproductively ready sooner than they would if they continued nursing their cubs.

Q Is a child living with one or more stepparents at greater risk of abuse than a child of the same age living with his or her biological parents?

The idea of kin selection gives rise to another prediction about conflict among humans. Specifically, it implies that child abuse, when it occurs, is more likely to be abuse of a child by his or her stepparent than abuse of a child by his or her biological parent. Is this the case?

Numerous studies across many different cultures, including an evaluation of 20,000 reports from the American Humane Association in the United States, support what has been called the "Cinderella syndrome." These findings included, for example, an estimate that the probability that a preschooler will be abused is about 1 in 3,000 for a child living with two biological parents (with whom he or she shares considerable genetic relatedness) and 40 in 3,000 for a child living with a stepparent (with whom he or she has no genetic relatedness). This difference in risk for children living in a home with a stepparent versus those living with their biological parents remains even when socioeconomic factors are taken into account. And the same effect has been noted in multiple other cultures. It is important to note, however, that in the overwhelming majority of stepfamilies no abuse occurs.

> ### TAKE-HOME MESSAGE 9·6
> Kin selection is apparently-altruistic behavior in which an individual that assists a genetic relative compensates for its own decrease in direct fitness by helping increase the relative's fitness and, consequently, its own inclusive fitness.

9·7 Apparent altruism toward unrelated individuals can evolve through reciprocal altruism.

It is ironic that studies of altruism among animals reveal that natural selection has primarily produced selfish behavior. And, as we saw in the previous section, when behavior appears to be altruistic, this is frequently because individuals are helping kin and, by doing so, are promoting the reproduction of copies of the genes (i.e., the alleles) that they share with close relatives. Does any apparently-altruistic behavior toward non-relatives occur in the animal world? Yes, but there's not much of it. Here we'll explore the conditions that give rise to reciprocal altruism, how it may have arisen, and why it is so common among humans but so rare among most other animal species.

We start by examining one species with well-documented altruistic-appearing behavior: the vampire bat. Vampire bats live in social groups of 8–12 mostly unrelated adults,

roosting primarily in caves and hollow trees. They feed by landing on large mammals such as cattle, horses, and pigs, piercing the skin with their razor-sharp teeth, and drinking the blood that flows from the wound.

Because of their small body size (about the size of your thumb) and their very high metabolic rate, vampire bats must consume almost their entire body weight in blood each night. If they go for more than about 60 hours without finding a meal, they are likely to die from starvation. Here's where the apparent altruism comes in: a bat that has not found food and is close to death will beg food from a bat that has recently eaten. In many cases, the bat that has just eaten will regurgitate some of the blood it has consumed into the mouth of the hungry bat, saving it from starvation. This act obviously has very high benefit

A vampire bat may vomit up a blood meal for an unrelated individual—saving that individual's life. Bats only do this for individuals that will return the favor, so their own survival is improved in the long run and the behavior is favored by natural selection.

FIGURE 9-12 **Returning the favor.**

for the recipient of the blood, but it comes at a cost to the sharing bat, which loses some of the caloric content of a meal it has just obtained (**FIGURE 9-12**).

Kin selection is responsible for some of the blood sharing (females often regurgitate blood for their own offspring), but, in many cases, bats give blood to unrelated individuals. How might this behavior have arisen? To answer the question, it is important to note three other features of vampire bats. First, they are able to recognize more than a hundred distinct individual bats. Second, bats that receive blood donations from non-relatives reciprocate significantly more than average with the bats that have shared blood with them. And third, bats that are not familiar with each other (and do not have a history of helping each other) generally do not regurgitate for each other.

One method proposed to explain the evolution of this apparent altruism is that the bats giving blood to other bats in need are repaid the favor when they are in need of blood. In other words, the act only seems to be selfless, when in actuality it is selfish. With such reciprocal altruism, both individuals (at different encounters) give up something of

relatively low value in exchange for getting something of great value at a later time when they need it most. In other words, they are storing goodwill in another individual, in much the same way that a person might put money in a bank for a rainy day. In both cases, individuals are protected from some of the world's uncertainties.

Taken together, studies of bats and other mammals show that reciprocal altruism can evolve if the following three conditions are met:

1. Repeated interactions among individuals, with opportunities to be both the donor and the recipient of altruistic-appearing acts

2. Benefits to the recipient that are significantly greater than the costs to the donor

3. The ability to recognize and punish cheaters, individuals that are recipients of altruistic-appearing acts but do not return the favor

In the absence of these conditions, selfishness is expected to be the norm among unrelated individuals. But in the presence of all three conditions, reciprocity masquerading as altruism is likely to occur, as it does with blood sharing among the vampire bats. The three conditions required for the evolution of reciprocal altruism are not satisfied in many animal species, which may be why altruistic-appearing behavior among unrelated individuals is rare.

As rare as it is in other species, reciprocal altruism is very common among humans. Friendship, among other human relationships, is built on reciprocity and is almost universal. The opportunity for friendship may be enhanced by our long life span and our ability to recognize thousands of faces and keep track of cheaters. These features are essential in individuals engaging in reciprocal altruism, because an individual becomes very vulnerable when he or she acts in an altruistic manner toward an unrelated individual. The risk is that the altruism will not be repaid, in which case the cheater enjoys greater fitness than the altruist.

Q Why are humans among the few species to have friendships?

While cooperators, those who repay altruistic-appearing behavior, have evolutionary advantages over loners, this advantage disappears if the cooperators are the givers all or most of the time, never or rarely getting anything in return. The importance of keeping track

Q Why is it easier to remember gossip than physics equations?

FIGURE 9-13 **Keeping track of valuable social information.**

The importance of keeping track of cheaters and of identifying good potential reciprocity partners (i.e., other cooperators) may be why humans are so interested in social information and seemingly trivial gossip.

about health, generosity, social status, and reproduction. Researchers have even hypothesized that the evolution of maintaining an interest in social information, while valuable in most contexts, may now have some maladaptive manifestations. We may find ourselves interested in and distracted by the social lives of Angelina Jolie, Brad Pitt, or other individuals whom we'll most likely never meet.

of cheaters and of identifying good potential reciprocity partners (i.e., other cooperators) may be why humans are so interested in social information and seemingly trivial gossip. As with many of the complex behaviors we have explored, humans probably use some rules of thumb when making decisions about reciprocal altruism. This might include keeping track of *all* available social information to best identify any promising individuals with whom to engage in reciprocal altruism (**FIGURE 9-13**). This includes information

TAKE-HOME MESSAGE 9·7

In reciprocal altruism, an individual engages in an altruistic-appearing act toward another individual. Although giving up something of value, the actor does so only when likely to get something of value at a later time. Reciprocal altruism occurs only if individuals have repeated interactions and can recognize and punish cheaters, conditions satisfied in humans but in few other species.

9·8 In an "alien" environment, behaviors produced by natural selection may no longer be adaptive.

The world in which alarm calling evolved did not include biologists with pickup trucks who trapped female squirrels, drove them long distances, and released them into colonies of unrelated individuals. So, when a female squirrel suddenly finds herself in this alien environment, her evolved behaviors can no longer be expected to be adaptive—no more than a human could be expected to survive on the moon. And individuals' genes and all the behaviors they influence cannot change overnight. Adaptation to a new environment takes time, and the more quickly an environment changes, the more likely it is that the evolved behaviors of a population will no longer be appropriate.

Let's consider another human example: donating money to refugees on another continent. To understand how natural

selection could lead to this behavior, we need to understand a bit about human evolutionary history. We know from archaeological deposits that for more than two million years, our ancestors lived as hunter-gatherers in small groups of a few hundred people, at most. Their success depended on joint efforts against predators and in killing prey; being "nice" paid off when the prospect of hunting alone or sleeping outside the camp meant almost certain death. The loners died, so we are descended from those who could work well with others.

It is only recently, the blink of an eye in evolutionary time, that humans invented agriculture, industrialization, and the means of food production and distribution. As a result of these changes, most of us have easy access to unlimited

FIGURE 9-14 Helping those you may never meet.

How You Can Help › Syria Refugee Crisis

Syria Refugee Crisis

INTERNATIONAL
RESCUE
COMMITTEE

From Harm to Home

DONATE NOW.
Support our work ›

TAKE ACTION.
Advocate for change ›

SIGN UP
Get IRC News.

<your e-mail address> ›

Charitable acts can give us pleasure—but in today's world can be evolutionarily maladaptive.

amounts of food and our typical group size has increased dramatically; on a given day, you may see 10 or even 100 *times* as many people as a hunter-gatherer ancestor might have seen. Additionally, if you were a hunter-gatherer, all of the people in your group would be a regular part of your life. You would have many opportunities to help them and, in turn, to be helped by them. Reciprocity paid. And we evolved so that "altruistic" acts gave us pleasure, stimulating parts of our brain in ways that made us want to repeat such actions.

It is with this brain that you approach the issue of refugees halfway across the world, or a homeless family somewhere in the United States. It is likely—perhaps almost certain—that you will not have repeated interactions with any of those people. And you will probably never be in a position to be helped by them. From an evolutionary perspective, your action will almost certainly not increase your fitness. But in the world in which humans evolved, such altruistic-appearing behavior would most likely have been reciprocated at some future time—and thus your instincts guide you to, and reward you for, your kindness (**FIGURE 9-14**).

From the weight-control difficulties that come from easy access to food, to charitable contributions to people in

faraway places, to alarm calling in transplanted squirrels, when organisms of any species find themselves in a situation where there is a **mismatch** between the environment they are in and the environment to which they are evolutionarily adapted, we expect (and see) behaviors that appear to be (and are) not evolutionarily adaptive. Still, understanding the process of natural selection can help us make sense of these behaviors (which in some cases can even seem nonsensical).

Can you think of other ways in which the modern industrial environment differs from the environment to which humans are adapted? How do these differences lead to situations in which our behavior, like the behavior of the transplanted Belding's ground squirrel, may not be adaptive from an evolutionary perspective?

> ### TAKE-HOME MESSAGE 9·8
> When there is a mismatch between the environment organisms are in and the environment to which they are adapted, the behaviors they exhibit are not necessarily evolutionarily adaptive.

9.9 Selfish genes win out over group selection.

Kin selection and reciprocal altruism can evolve in a population of animals, as we have seen, and examples of kin selection abound in nature. As a consequence, casual observers of nature frequently see individuals acting in ways that appear altruistic, even though these individuals are truly acting—from the perspective of evolutionary fitness—in their own selfish interests. This nearly universal selfishness raises the question of whether evolution ever leads to behaviors that are good for the species or population but detrimental to the individual exhibiting the behavior, a process called **group selection.**

It might seem that evolution would favor individuals that behave in a manner that benefits the group, even if it comes at a cost to the individual's own inclusive fitness. But this does not happen. Behaviors that reduce an individual's fitness (relative to that of other individuals in the population) are not likely to evolve.

Let's look at an example. Imagine that a new allele appears in a population (perhaps through mutation) that causes the individual carrying the allele to double its reproductive output, even though this might spell doom for the species as individuals overuse their resources. An individual carrying this "selfish" allele will pass on more copies of the allele to its offspring than an individual carrying the alternative allele, coding for the production of fewer offspring, will pass on to its offspring—and this will lead to excessive, "selfish" consumption of the species' resources. The selfish offspring, in turn, will pass on the selfish allele at a higher rate than the alternative allele is passed on. This scenario may lead to extinction of the species, yet it still occurs. With its market share perpetually increasing, the selfish allele eventually is favored by evolution and will predominate (**FIGURE 9-15**).

Now consider what happens when a new allele appears that causes an individual to reduce its reproductive output below what is best for that individual's fitness. Just because a behavior (determined by an "unselfish" allele) leads to a better outcome for the group, this doesn't mean that natural selection will favor that behavior. Even if it reduces the likelihood that, in the long run, the population goes extinct, such an allele does not increase its market share

relative to the alternative "normal" or "selfish" allele. Instead, natural selection generally causes increases in the frequencies of alleles that benefit the *individual* carrying them, even when this comes at the expense of the group. In some special situations, it is possible for natural selection to lead to group selection. But the stringent conditions necessary for this to occur are so rarely found in nature that we almost never see it.

DOES GROUP SELECTION OCCUR?

Group selection describes the evolution of a trait that is beneficial for the species or population while decreasing the fitness of the individual exhibiting the trait.

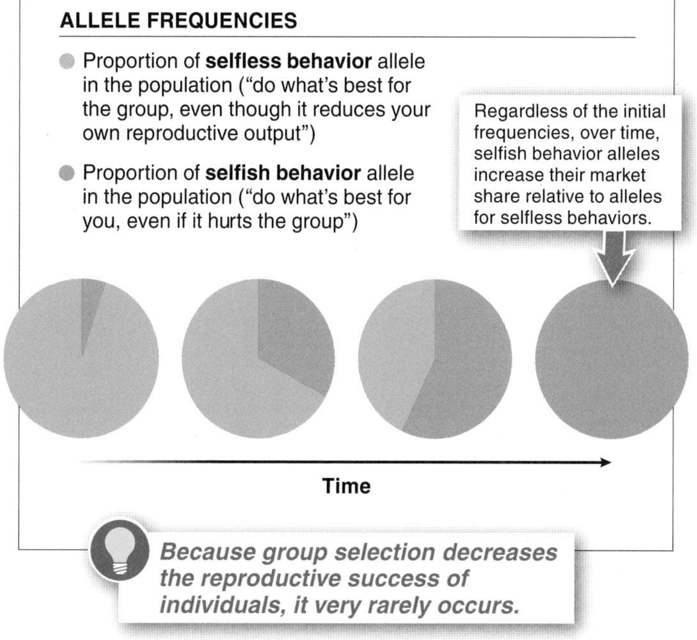

ALLELE FREQUENCIES

- Proportion of **selfless behavior** allele in the population ("do what's best for the group, even though it reduces your own reproductive output")
- Proportion of **selfish behavior** allele in the population ("do what's best for you, even if it hurts the group")

Regardless of the initial frequencies, over time, selfish behavior alleles increase their market share relative to alleles for selfless behaviors.

Time

Because group selection decreases the reproductive success of individuals, it very rarely occurs.

FIGURE 9-15 Can a "selfless" gene increase in frequency in a population?

TAKE-HOME MESSAGE 9·9

Behaviors that are good for the species or population but detrimental to the fitness of the individual exhibiting such behaviors are not generally produced in a population under natural conditions.

9·10–9·16
Sexual conflict can result from disparities in reproductive investment by males and females.

A red-crowned crane doing a mating dance.

9·10 There are big differences in how much males and females must invest in reproduction.

As we've seen, many behaviors have evolved that influence the ways in which animals interact with each other, reducing conflict and sometimes leading to cooperation. One aspect of life that necessarily involves interaction—for sexually reproducing species—is reproduction, from courtship and mating to parental investment and the forming and breaking of ties. In this and the next few sections, we'll explore these behaviors and the factors that influence them.

How many babies can a woman produce over her lifetime? The number is probably higher than you would guess: a Russian woman had 69 children (in 27 pregnancies). But even this high number is greatly exceeded by the 888 offspring produced by one man (Emperor of Morocco from 1672 to 1727). The large difference between the number of offspring that can be produced by females and males is less surprising. Among other species of mammals, the pattern is consistent: a male elephant seal can produce 100 offspring, while the maximum produced by a female over her lifetime is 8; a male red deer can produce 24 offspring, while the maximum produced by a female is 14. Here we examine the physical differences between males and females and how they lead to differences in sexual behavior.

The very definition of "male" and "female" hinges on a physical difference between the sexes. Recall from Chapter 6 that in species with two distinct sexes, a **female** is defined as the sex that produces the larger gamete, while a **male** produces the smaller gamete (**FIGURE 9-16**). At conception, the mother's material and energetic contribution to the offspring—her **reproductive investment**—exceeds the

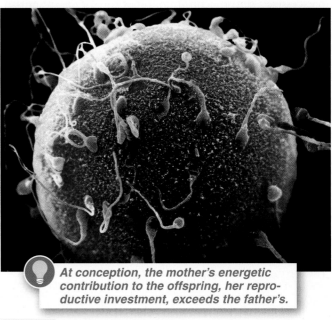

At conception, the mother's energetic contribution to the offspring, her reproductive investment, exceeds the father's.

FIGURE 9-16 **Small sperm, big egg.**

father's. This is true for all animals, whether mammals, birds, insects, or sharks. (It is also true for plants.) Not only are female gametes larger, they tend to be relatively immobile and produced in smaller numbers. Male gametes, on the other hand, though smaller, are more plentiful and very motile.

This discrepancy in size and quantity between the sperm produced by a male and the eggs produced by a female sets the stage for evolutionary developments that magnify this initial difference in reproductive investment. For starters, the difference in the number of gametes that males and females can produce means that males have the potential to produce many, many more offspring than females. Put another way, a male's **total reproductive output,** the lifetime number of offspring he can produce, tends to increase as the number of females he is able to fertilize increases. A female, on the other hand, does not generally increase her reproductive success by mating with additional males beyond the first (**FIGURE 9-17**).

Q *Why do males usually compete for females rather than the opposite?* Because additional matings usually lead to greater increases in reproduction (and fitness) for males, selective pressure has resulted in the evolution of some differences in male and female reproductive behavior. For males, the most effective way to maximize their reproductive success often is to find and gain access to mating opportunities with additional females. For females, an effective way to maximize their reproductive success often is to put more effort into parenting and less effort into mating.

When it comes to putting more effort into parenting, two physical differences that can exist between males and females are particularly important. First, in species with internal fertilization, which includes most mammals, fertilization takes place in the female. The offspring also grow and develop within the female's body. The amount of energy females invest in reproduction is therefore much greater than males' investment. Females' reproductive investment also limits their reproductive output; they can be pregnant only once at a time.

A second important physical difference between females and males occurs in the mammals: lactation takes place in females and not in males (with a very small number of exceptions). In these species, then, nurturing during both pregnancy and lactation can be accomplished only by the female. This difference in reproductive investment has led to the evolution of some very different reproductive behaviors in males and females.

Turn to p. 399 for a closer inspection of this figure.

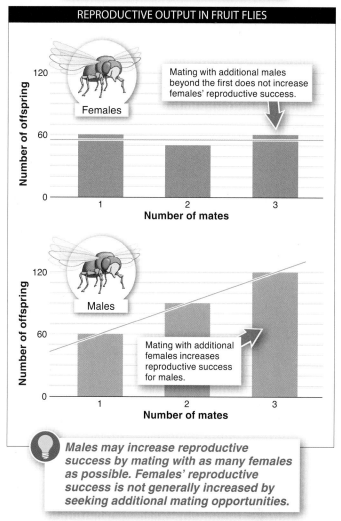

Males may increase reproductive success by mating with as many females as possible. Females' reproductive success is not generally increased by seeking additional mating opportunities.

FIGURE 9-17 **Maximizing reproductive success in fruit flies.**

Although the gamete size difference is consistent across all animals (i.e., the egg is always bigger and energetically more costly to make than the sperm), the physical differences between the sexes in the early nurturing (i.e., gestation and feeding) of offspring can vary considerably across animal species. In some cases, male and female investments become more nearly equal after fertilization. In birds, for example, before emergence of the chicks, much of the development of the fertilized egg is external: the female lays an egg, but either the male or the female can protect and incubate the developing embryo. Further, birds do not lactate. Once hatched, the chicks must be fed—a task that can be done by both parents.

Accordingly, in many bird species, the maximum lifetime reproductive output of males and females is similar. In

The female mammal invests much energy in the production and care of offspring.

Amphibian eggs are left to develop on their own; there is very little maternal or paternal reproductive investment after fertilization.

Among most bird species, males and females have an equal reproductive investment.

FIGURE 9-18 **Male and female reproductive investment differs across species.** Female investment is greatest when offspring develop internally and are fed by lactation after birth. In animals whose offspring develop externally, males or females (or neither parent) can provide the care.

the kittiwake gull, for example, the largest number of offspring produced by a male is 26, while for a female the number is 28. External fertilization in fish and amphibian species further reduces the reproductive investment of the female—she does not spend any energy as the fertilized eggs begin developing into embryos (**FIGURE 9-18**).

Another profound consequence of internal fertilization is that a male cannot be 100% certain that any offspring a female produces are his progeny. Because it is possible for a female to mate with multiple males, any of whom could be the father, male mammals and birds will always have some degree of **paternity uncertainty.**

In the next few sections, we'll continue to explore how the physical differences between males and females in

reproductive investment, along with paternity uncertainty, have led to the evolution of differences in male and female reproductive behavior.

TAKE-HOME MESSAGE 9·10

In mammals and many other animals, there are important physical differences between males and females relating to reproduction. Fertilization usually takes place in the female. Lactation takes place only in the female. And in species in which fertilization occurs inside the female, males cannot be certain that offspring are their progeny. These physical differences have led to the evolution of differences in male and female reproductive behavior.

9·11 Males and females are vulnerable at different stages of the reproductive exchange.

Suppose you are on your college campus and a person of the opposite sex comes up to you and says, "Hi. I have been noticing you around campus. I find you very attractive. Would you go out with me tonight?" What percentage of men would answer *yes?* And women? In a study conducted at Florida State University in 1978 and 1982, the

percentage answering *yes* was 50% for both males and females.

Now imagine the question was "Would you have sex with me tonight?" Among the men, 75% said *yes;* among the women, not a single one said *yes.* It is tempting to interpret

FIGURE 9-19 **The choosier sex.** Males and females differ in their attitude toward mating opportunities. For the female more than the male, the choice of the wrong mate could have expensive consequences.

such results as a consequence solely of the culture of Western society, characterized by significant differences in the expectations and tolerances of male and female sexual conduct. However, cultural expectations don't adequately explain the consistency of the pattern found across Western and non-Western societies: men and women differ in their approach to sexuality.

Studies of male-female differences in selectiveness about sexual partners, drawing on a wide range of human cultures—including many as far removed from Western influence as possible, such as the Trobriand Islanders of Melanesia in the South Pacific—reveal a consistent difference in men's and women's willingness to have sex. Why might this difference exist?

Humans, like nearly all mammals, are characterized by a greater initial reproductive investment by females. For females, the cost of a poor mating choice can have significant consequences—pregnancy and lactation, with offspring from a low-quality male or a male who does not provide any parental investment or access to valuable resources (**FIGURE 9-19**). For a male, the consequences of a poor choice are less dire—little beyond the time and energy involved in mating.

Two differences in the sexual behavior of males and females across the animal kingdom have evolved:

1. The sex with the greater energetic investment in reproduction is more discriminating about mating.

2. Members of the sex with the lower energetic investment in reproduction compete among themselves for access to the higher-investing sex.

A dramatic illustration of how a high reproductive investment leads to the evolution of choosiness in mating behavior comes from the insect world. When bush crickets mate, the male loses about a quarter of its body weight in contributing a massive ejaculate (the equivalent of nearly 50 pounds of semen in a human), which the female then uses for energy (**FIGURE 9-20**). It can represent up to one-tenth of her lifetime caloric intake.

When bush crickets mate, the male loses about a quarter of his body weight contributing a massive ejaculate that the female uses for energy. This makes males very choosy.

FIGURE 9-20 **A costly decision.** Because of the investment the male bush cricket is required to make when fertilizing a female, he chooses a mate very carefully.

Not surprisingly, male crickets are very choosy when selecting a mate. They reject small females that would produce relatively few offspring. Females, you won't be surprised to learn, spend a great deal of effort courting males.

The sex with the greater reproductive investment must be choosy: a poor choice of mate could be disastrous. The sex with the lower initial reproductive investment (usually males), on the other hand, is not made vulnerable through its mating choices, because the matings are of little energetic consequence.

The point of greatest vulnerability for males comes when they provide parental care to offspring. Due to paternity uncertainty, there is some chance that the male may be investing in offspring that are not his own. This has significant evolutionary costs: rather than increasing his own fitness, he is increasing the fitness of another male. Females, conversely, are less vulnerable at the point of providing parental care, because a female can be completely certain that the offspring she gives birth to are her own.

In the next two sections, we explore the evolution of reproductive behaviors that help males and females reduce their vulnerability and maximize their reproductive success. As we explore some general patterns of reproductive behavior among males and females, it is important to keep two critical points in mind.

1. There is tremendous variability across species in male and female behaviors. For example, the use of DNA fingerprinting to identify the parents of bird offspring has led to some surprising revelations. In significantly more cases than researchers originally predicted, offspring have different fathers than observers assumed. It seems that things are not always as they appear to observing biologists. As a consequence, much of the new research calls into question some long-held assumptions about the behavioral consequences of physical differences in reproductive investment among birds versus mammals. There are general behavioral patterns among animals, but as is becoming increasingly clear, these patterns are not universal features of their biology, and a reasonable skepticism is important when identifying and interpreting broad trends across large groups of species.

2. Throughout history, there have been many cases of people using observations and scientific findings to justify a wide variety of discriminatory thoughts and behaviors—for example, that if male mammals "naturally" tend to be less faithful to their mates, this reduces individual responsibility for behaviors among humans. Such thinking ignores both the tremendous variation in behavior among species and the power of cultural norms and socialization to influence and shape human behavior, while encouraging the inappropriate assumption that biology can supply meaningful insights into morality or ethical decision making.

TAKE-HOME MESSAGE 9·11

Differing patterns of investment in reproduction make males and females vulnerable at different stages of the reproductive process. This has contributed to the evolution of differences in their sexual behavior. The sex with greater energetic investment in reproduction is more discriminating about mates, and members of the sex with a lower energetic investment in reproduction compete among themselves for access to the higher-investing sex.

9·12 Tactics for getting a mate: competition and courtship can help males and females secure reproductive success.

Female choosiness (and the male-male competition it leads to) tends to increase the likelihood that a female will select only those males that have plentiful resources or relatively high-quality genes, either of which is beneficial to the female, allowing her to produce more or better offspring—where "better" may mean increased disease resistance or physical traits that will be found attractive by future mates. Female choosiness is manifested by one or more of four general rules.

1. Mate only after subjecting a male to courtship rituals. In many bird species, the female requires the male to perform an elaborate and time-consuming courtship dance before she will mate with him. For the western grebe, for example, this courtship dance involves fancy dives into water, graceful hovering, various head movements, and flamboyant twists and turns. The courtship process can go on for several days. But if the male passes the time-consuming audition, he can generally be counted on to stick around to see a brood through hatching and early care (**FIGURE 9-21**).

2. Mate only with a male who controls valuable resources. Territorial defense is a common means by which males compete for access to females. Among arctic ground squirrels, for example, a female chooses a mate based, in

FACTORS IN MATE SELECTION

COURTSHIP RITUALS
A female grebe requires the male to perform a courtship dance before she will mate with him.

CONTROL OF VALUABLE RESOURCES
Female yellow-bellied marmots prefer rock outcroppings that provide retreats for escape from predators and for hibernation (and are controlled by dominant males).

GIFTS UP FRONT
A female hanging fly will not mate with a male unless he brings her a large offering of food.

GOOD LOOKS
A female peacock is attracted to a male with the most beautiful tail feathers.

FIGURE 9-21 Four factors that influence a female's choice of mate.

> "It is a truth universally acknowledged, that a single man in possession of a good fortune must be in want of a wife."
>
> — JANE AUSTEN, *Pride and Prejudice*, 1813

3. **Mate only with a male who contributes a large parental investment up front.** Better than a believable pledge to commit resources to future offspring is an actual exchange in which a female requires a male to give her his parental investment up front, in the form of resources that will help her maximize her reproductive success. In the hanging fly, for example, a female will not mate with a male unless he brings her a big piece of food, called a **nuptial gift**—usually a dead insect. The larger the food item, the longer she will mate; and the more she eats, the larger the number of eggs she will lay. After about 20 minutes of mating, though, when a male has transferred all of the sperm that he can, he is likely to break off the mating and take back whatever remains of the "gift," which he may use to try to attract another mate. Nuptial feeding is common among birds and insects.

4. **Mate only with a male that has a valuable physical attribute.** Male-male competition for the chance to mate with females can also take a more literal form: actual physical contests. Across the animal kingdom, from dung beetles to hippopotamuses, male-male contests determine the dominance rankings of males. Females then mate primarily with the highest-ranking males.

In a similar process, rather than choosing the best-fighting or largest males, females sometimes base their choice on some other physical attribute, such as antler size in red deer, the bright red chest feathers of frigate birds, or the elaborate tail feathers of the male peacock (see Figure 9-21). In each case, for the female, the physical feature serves as an indicator of the relative quality of the male, possibly because the feature is correlated with the male's health.

With so many examples of male-male fighting as part of the courtship rituals that have evolved for attracting a mate, it is reasonable to ask: why is it so rare for females to fight? And why do females generally not have to advertise their health with flashy feathers or other ornamentation? The answer is that as long as females are making the greater investment in reproduction, nearly any male will mate with them. Consequently, there is nothing more to be gained by trying to outcompete other females or otherwise attract the attention of males.

Q Why do so few women get into barroom brawls?

part, on the territory he defends, which is where she will reside after mating. With greater quality and quantity of resources in his territory, a male is better able to attract females, whose reproductive success can be increased if the territory is rich in resources.

Moreover, among humans, social and cultural values have powerful influences over mating behavior, complicating interpretation. We are not lumbering robots, destined to follow some genetic program. Researchers have noted some subtle manifestations of female health and fertility, including waist-to-hip ratios and patterns of facial and body symmetry. There is a rich and complex world of mating tactics, many of which we do not fully understand.

TAKE-HOME MESSAGE 9·12

As a consequence of male-female differences in initial reproductive investment, males tend to increase their reproductive success by mating with many females and have evolved to compete among themselves to get the opportunity to mate.

9·13 Tactics for keeping a mate: mate guarding can protect a male's reproductive investment.

If a male simply abandons a female after mating and searches for other mating opportunities, rather than making any investment in the potential offspring, he has no risk of investing further energy in offspring that are not his. This is one way to minimize the potential costs associated with paternity uncertainty. If you don't play, you can't lose.

This strategy, however, is not necessarily the most effective way for a male to maximize his reproductive success. If a male has a hundred or even a thousand matings, but no offspring survive, that behavior is not evolutionarily successful. Consequently, in species for which offspring's survival can be enhanced with greater parental investment—including those species not well-developed at birth—there is an incentive for a male to provide some additional investment in offspring, even though such behavior makes him vulnerable to paternity uncertainty.

In situations where males provide parental care, it is common for the male to reduce his vulnerability through some form of **mate guarding.** In contrast to a female, who can be certain that any offspring emerging from her body is hers, a male inhabits a "danger zone" that lasts as long as the female is fertile. If she mates with any other males during this time, the offspring she produces may not be his. If he is going to make any investments of time or energy that will benefit the offspring, he benefits by minimizing his risk in the danger zone. It is during this period that mate guarding is particularly common.

Q Why do so few females guard their mates as aggressively as males do?

Mate-guarding methods range from the simple to the macabre. If a male wants to ensure that a female does not mate with another male, why bother to stop mating at all? In many species, males take this approach to reduce their risk in the danger zone. Among house flies, even though the male has completed the transfer of sperm to the female in 10 minutes

of copulation, he does not separate from her for a full hour. Moths go even further and continue to mate for a full 24 hours. And in the extreme case of this strategy, certain frog species continue individual bouts of mating for several months (**FIGURE 9-22**). In humans, this would be equivalent to almost 10 years for a single round of intercourse.

In a slightly subtler form of mate guarding that occurs in reptiles, insects, and many mammalian species, after copulation, males block the passage of additional sperm into the female by producing a copulatory plug. Formed in the female reproductive tract from coagulated sperm and mucus, copulatory plugs can be very effective. Male garter snakes that encounter a female snake with a copulatory plug, for example, do not attempt to court or mate with her, treating her instead as if she were not available.

💡 *In some frog species, individual bouts of mating continue for several months. This is the equivalent in humans of ten-year-long episodes of intercourse!*

FIGURE 9-22 Preventing paternity uncertainty. A prolonged period of mating prevents the female from accessing other males, assuring the male of reproductive success.

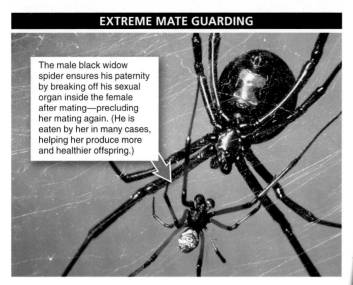

EXTREME MATE GUARDING

The male black widow spider ensures his paternity by breaking off his sexual organ inside the female after mating—precluding her mating again. (He is eaten by her in many cases, helping her produce more and healthier offspring.)

FIGURE 9-23 A reasonable trade-off? The male black widow spider ensures his paternity by an extreme form of mate guarding.

A much more extreme form of mate guarding occurs in the black widow spider: the male breaks off his sexual organ inside the female, preventing her from ever mating again. Interestingly, when the act is completed, the female usually kills and eats the male (**FIGURE 9-23**). In sealing his mate's reproductive tract, the male assures himself of fathering the offspring, and in consuming her mate's nutrient-filled body, the female gets resources that help her produce the offspring.

> ## TAKE-HOME MESSAGE 9·13
>
> Mate guarding can, in general, increase reproductive success by reducing additional mating opportunities for a partner, and can improve a male's reproductive success by increasing his paternity certainty and thus reducing his vulnerability when he makes investment in offspring.

❓ 9·14 THIS IS HOW WE DO IT

When paternity uncertainty seems greater, is paternal care reduced?

Evolutionary theory generates numerous predictions on how organisms will make decisions about parental investment. In the previous section we saw that there should be a relationship between a male's certainty of paternity and his investment in the offspring. When paternity certainty is low, males should benefit by reducing their parental investment and, instead, seeking additional mating opportunities.

Often, however, it can be difficult to test such predictions experimentally. The results of a manipulation can be difficult to interpret—is an observed decrease in paternal investment really a response to a perceived decrease in paternity certainty, or is it simply a consequence of the experimenter's presence or interference?

A powerful strategy to address this challenge is to use an experimental manipulation that makes contrasting predictions in two different situations. That way, a response to the manipulation in one situation provides evidence in favor of the prediction, while a response in the second situation provides evidence for a *lack* of response to the manipulation. This serves as a sort of internal control in the experiment. Here's how a researcher used this approach in two clever experiments.

The system Bluegill sunfish live in lakes and rivers in North America. Most males reach maturity at age 7 years. During the breeding season, males use their tail to create a depression in the sandy bottom and chase away almost everything that approaches this nest. Females come and lay eggs—which the male fertilizes—and leave shortly afterward. The male remains to guard the eggs and the small offspring after they hatch. Typically, he doesn't even take a break to forage during this period of parental investment.

About 20% of the bluegill males in a population mature at age 2 years, at a much smaller size. These males, called "cuckold males," hide near the nests of other males and attempt to sneak into the nest, fertilize the eggs, and escape without being detected by the nest "owner."

Male bluegills are unable to distinguish between eggs they have fertilized and eggs fertilized by another male. They can, however, tell whether just-hatched offspring are their biological offspring (from a chemical cue in the offspring's urine).

Experiment 1 The researcher randomly chose 34 nests. Around each nest he placed two glass containers, each containing two small cuckold males, and left them there for the duration of the egg-laying. The cuckold males couldn't fertilize the eggs, but they could be seen by the male at the nest. As a control, the researcher placed two empty glass containers around each of 20 other nests.

A day after the eggs were laid, the researcher placed a glass container with a predator fish (that eats eggs and just-hatched fish) near each nest, and then evaluated parental care by measuring how vigorously the nest owner defended his eggs over the course of two 30-second periods. Parental care was measured again after the offspring hatched.

In each case, the researcher allotted a "parental care score," reflecting the intensity of the male's guarding of the eggs and defense of the hatched offspring.

> **Would you expect the presence of cuckold males to influence a nest owner's perception of paternity certainty?**

Prediction a: The presence of the cuckold males should reduce the nest owner's paternity certainty and therefore reduce his egg-guarding efforts.

> **Should the presence of cuckold males influence a nest owner's perception of paternity certainty *after* the offspring hatch? Why or why not?**

Prediction b: After the offspring hatch, the nest owner can determine whether they are his genetic offspring, so he should not exhibit any reduction in parental care relative to males in the control group.

Results of Experiment 1 Manipulation: Cuckold males nearby, but all eggs fertilized by nest owner

Parental Care Score

	Egg Guarding	Offspring Guarding
Prediction	Reduced	Unchanged
Actual results:		
No rivals (control)	80 ± 10	90 ± 10
Rivals present	52 ± 7	95 ± 10
Change in care	Reduced	Unchanged

> **How much did the presence of cuckold males reduce egg guarding? How much did it alter offspring guarding?**

Experiment 2 The researcher randomly chose 20 new nests (none from Experiment 1) and, the day after egg-laying and fertilization, he removed one-third of the eggs from each nest and replaced them with unrelated fertilized eggs from another male's nest. Then, as in the first experiment, the researcher placed a glass container with a predator fish near the nest and evaluated parental care.

> **Should a nest owner show reduced parental care of eggs that were swapped in from another nest? Why or why not?**

Prediction a: Prior to hatching of the eggs, the nest owner should exhibit the same egg-guarding efforts regardless of whether or not the eggs were swapped.

> **Should a nest owner show reduced parental care of hatched offspring after eggs were swapped? Why or why not?**

Prediction b: After the offspring hatch, because the nest owner can determine whether they are his genetic offspring, he should exhibit reduced parental care relative to the control males.

Results of Experiment 2 Manipulation: Eggs swapped with those fertilized by a different male.

Parental Care Score

	Egg Guarding	Offspring Guarding
Prediction	Unchanged	Reduced
Actual results:		
Eggs not swapped (control)	90 ± 10	73 ± 9
Eggs swapped	95 ± 10	50 ± 8
Change in care	Unchanged	Reduced

> **How much was egg guarding reduced when unrelated eggs were swapped into the nest? How much was offspring guarding changed?**

> **What conclusions can you draw from these results?**

In each of the two experiments, males decreased their parental care relative to males in the control group in response to signs that the offspring were less likely to be their own genetic offspring. The experiments provide strong evidence that genetic relatedness to offspring plays an important role in parental care by a male bluegill sunfish.

TAKE-HOME MESSAGE 9·14

Experimental manipulations of the cues of paternity certainty can increase or decrease a male's parental investment in accordance with the prediction that decision making about parental investment reflects perceptions of genetic relatedness.

Monogamy versus polygamy: mating behaviors can vary across human and animal cultures.

As we continue our tour of animal mating behavior, we turn again to the elephant seal. In December of each year, the males appear on islands off the coast of northern California, where they compete with each other for possession of the beach. Through bloody fights they establish dominance hierarchies, with the biggest males—which are 13 feet long and weigh more than 2 tons—generally winning.

In mid-January, the females arrive and are ready to mate. They congregate in large groups on just a few prime beaches. Because the females stick close together, the biggest males, who control the prime beaches, gain access to almost all of the females when they dominate the other males in the competition for sexual access to females (**FIGURE 9-24**). In one study that observed 115 males, the 5 highest-ranking males fathered 85% of the offspring. While nearly every female will mate and produce offspring, the majority of males never get the chance to mate during the 10–20 years of their lives.

The elephant seals' mating pattern exemplifies **polygamy,** a system in which some individuals attract multiple mates while other individuals attract none. Polygamous mating

systems can be subdivided into **polygyny,** in which individual males mate with multiple females, and **polyandry,** in which individual females mate with multiple males. Polygamy can be contrasted with **monogamy,** in which most individuals mate and remain with just one other individual. Polygamy and monogamy are two types of **mating systems,** which describe the patterns of mating behavior in a species. In this section, we explore the features of environments and species that influence mating systems and survey the range of mating systems observed in nature.

As we have seen, throughout the animal world, as a result of their relative parental investment, females are choosy about which males they mate with, and males compete for access to mating opportunities. Not surprisingly, multiple females often end up selecting the same male—usually a male on a territory rich in resources or a male with unusually pronounced physical features, such as antler size. Such female selection is the reason for the fancy ornamentation we see in male peacocks. Although the male peacock's tail is like a giant bull's-eye to predators, the number of eyespots on the tail is directly related to how well he can attract a mate: below 140 eyespots, he

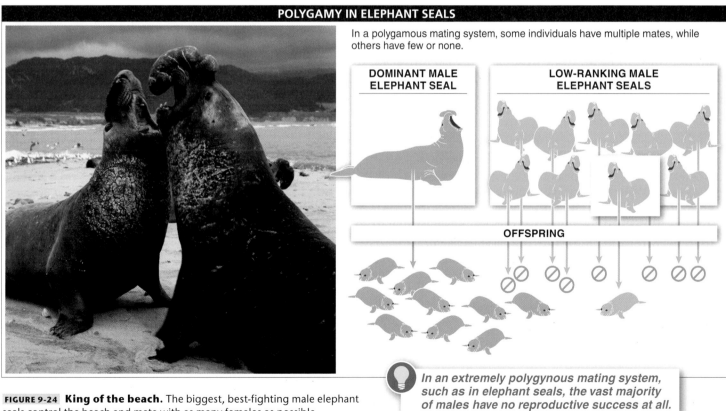

POLYGAMY IN ELEPHANT SEALS

In a polygamous mating system, some individuals have multiple mates, while others have few or none.

DOMINANT MALE ELEPHANT SEAL

LOW-RANKING MALE ELEPHANT SEALS

OFFSPRING

FIGURE 9-24 King of the beach. The biggest, best-fighting male elephant seals control the beach and mate with as many females as possible.

In an extremely polygynous mating system, such as in elephant seals, the vast majority of males have no reproductive success at all.

gets no mates; at 150, he gets two to three mates on average; with 160 eyespots, he gets six or more mates (see Figure 9-21).

Identifying a population's mating systems is not as easy as the elephant seal example might lead us to believe. Three issues, in particular, complicate the task. First, there are often differences between animals' mating behavior and their bonding behavior. That is, it may seem that a male and female have formed a **pair bond**—in which they spend a high proportion of their time together, often over many years, sharing a nest or other "home" and contributing equally to parental care of offspring in what appears to be a monogamous relationship. Closer inspection (often including DNA analysis of the offspring), however, sometimes reveals that the male and/or the female may be mating with other individuals in the population, and perhaps the mating system is better described as a variation on polygamy. A second difficulty in defining a species' mating system arises because the mating system may vary within the species. That is, some individuals may be monogamous, while others are polygamous. And the mating system may even change over the course of an individual's life. A third difficulty is that males and females often differ in their mating behavior. In the elephant seals described above, it could be said that the females are all mating monogamously, while the males are polygamous.

Examination of birds and mammals in general, however, reveals one sharp split. As we discussed in Section 9-10, the vast majority of female mammals make a greater parental investment than male mammals. In birds, females and males have a more equal parental investment. Does the difference in parental investment patterns in birds and mammals lead to different mating systems? Yes. In mammals, polygyny is the most common mating system across all large groups, from rodents to primates. Polygyny is a consequence of the significant female investment and lesser male investment. Males can generally benefit more by seeking additional mating opportunities, leading to male-male competition. In birds, the relatively equal parental investment by males and females has led to much less polygyny. More than 90% of the approximately 10,000 bird species we know about appear to be monogamous (although, in most cases, it is serial monogamy) (**FIGURE 9-25**).

And what of humans? Across a variety of cultures, males consistently have greater variance in reproductive success than females—that is, some males have very high reproductive success and many others have little or none. **FIGURE 9-26** , for

MONOGAMOUS BIRDS

💡 *Of the approximately 10,000 species of birds, more than 90% appear to be monogamous.*

FIGURE 9-25 **Let's stay together.** Parental investment that is roughly equal often leads to monogamous mating behavior in birds.

example, presents data from one of the first investigations of this phenomenon. It is from a study of the Xavante Indians of Brazil. The Xavante were selected because they were a pre-industrial population subsisting primarily as hunter-gatherers, with no access to reliable birth control and with almost no contact with Western cultures. The *average*

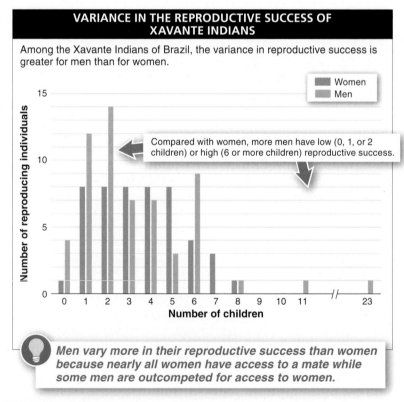

VARIANCE IN THE REPRODUCTIVE SUCCESS OF XAVANTE INDIANS

Among the Xavante Indians of Brazil, the variance in reproductive success is greater for men than for women.

Compared with women, more men have low (0, 1, or 2 children) or high (6 or more children) reproductive success.

(y-axis: Number of reproducing individuals; x-axis: Number of children)

💡 *Men vary more in their reproductive success than women because nearly all women have access to a mate while some men are outcompeted for access to women.*

FIGURE 9-26 **Larger male variance in reproductive success.**

Q Are humans monogamous or polygamous? number of children (3.6) does not differ between men and women, but the *range* does: some men have very large numbers of offspring (as many as 23!), and many have none. There is significantly less variability in reproductive success from one woman to another. Interestingly, a 2010 study using data from 7,710 women and men living in the contemporary United States similarly found significantly greater variance in reproductive success among men than among women.

Some differences in the variance in reproductive success among males and females, which can be significantly influenced by socialization and other cultural forces, are generally associated with polygynous mating systems.

However, the difference between the sexes found in the Brazilian study is quite small when compared with many other mammalian species, such as elephant seals, and is close to that seen in populations with a monogamous mating system. Humans, consequently, seem to have a mating system close to, but not completely, monogamous.

> **TAKE-HOME MESSAGE 9·15**
>
> Mating systems—monogamy, polygyny, and polyandry—describe the variation in number of mates and the reproductive success of males and females. They are influenced by the relative amounts of males' and females' parental investment.

9·16 Sexual dimorphism is an indicator of a population's mating behavior.

Male elephant seals are three to four times the size of female elephant seals. In contrast, the males and females of most bird species are the same size. In these species, even expert bird-watchers often cannot distinguish between the sexes, except when the female bird is carrying eggs (**FIGURE 9-27**). When the sexes of a species do differ in size

SEXUAL DIMORPHISM

With sexual dimorphism, the sexes differ in size or appearance.

SEXUAL MONOMORPHISM

With sexual monomorphism, the sexes are indistinguishable.

BEHAVIORS ASSOCIATED WITH SEXUAL DIMORPHISM
- One parent invests more in caring for the offspring.
- Mating system tends toward polygamy.
- One sex (usually females) is choosier when selecting a mate.
- One sex (usually males) competes for access to mating opportunities with the other sex.

BEHAVIORS ASSOCIATED WITH SEXUAL MONOMORPHISM
- Both parents invest (approximately) equally in caring for the offspring.
- Mating system tends toward monogamy.
- Both sexes are equally choosy when selecting a mate.

FIGURE 9-27 The same or not the same? In monogamous species, the males and females are similar in appearance and behavior. In polygamous species, they tend to differ.

BEHAVIORS EVOLVE   COOPERATION AND ALTRUISM   SEXUAL CONFLICT   COMMUNICATION

393

or appearance, this is called **sexual dimorphism.** Why do species have such dramatic differences in the degree to which males and females resemble each other?

Q It's almost impossible to distinguish males from females in most bird species. Why does this tell us they are monogamous?

Body size is an important clue to behavior. We have seen how male elephant seals have a winner-take-all tournament for control of the beach, and as the winner, a male has access to mating opportunities. Because the largest individual will have the most offspring, there is selection for larger and larger body size. There is no selection for larger body size among female elephant seals, because females of any size can mate. Hence the dramatic size difference between the sexes.

In addition to body size, coloration can also be a clue to behavior. Because females of some polygynous species choose the males with the brightest or flashiest coloration rather than the largest males, male-male competition sometimes results in differences in physical appearance between the sexes other than (or in addition to) size.

What happens when there is little male-male competition for mates? Among bird species, males can provide significant investment in offspring. Eggs have to be incubated, and because many chicks emerge in a very poorly developed condition, it takes two parents to raise them. So a pair tends to stay together for a season or longer.

If each female can pair up with only one male, there really isn't much competition for mates. And without male-male competition, there is little selection for increased size. As a consequence, little sexual size dimorphism occurs in most bird species.

In turn, we can predict a bit about the parental practices of a species just by looking at a picture of a male and a female and examining the ratio of body sizes. If the two are dramatically different, as in elephant seals, we can hypothesize that the smaller sex is doing most of the care of the offspring and that the species is more likely to be polygamous than monogamous.

Q Men are bigger than women. What does that suggest about our evolutionary history of monogamy versus polygamy?

TAKE-HOME MESSAGE 9·16

Differences in the level of competition among individuals of each sex for access to mating opportunities can lead to the evolution of male-female differences in body size and other aspects of appearance. In polygynous species, this results in larger males that are easily distinguished visually from females. In monogamous species, there are few such differences between males and females.

9·17–9·18
Communication and the design of signals evolve.

A mona monkey (Cercopithecus mona) *in western Africa bares its teeth.*

9·17 Animal communication and language abilities evolve.

The challenge of communication is central to the evolution of animal behaviors, whether they relate to cooperation, conflict, or the many interactions inherent to selecting a mate. An animal must be able to convey information about its status and intentions to other individuals. And equally important, animals must be able to interpret and assess the information they get from other individuals so as to reduce the likelihood of being tricked or taken advantage of. In this section and the next, we explore the evolution of a variety of methods of animal communication and the ways in which communication can influence fitness and the evolution of other behaviors.

Not all animals need to open their mouths when they have something important to communicate. When the female silkworm moth is ready to mate, for example, she releases a potent chemical called bombykol into the air. As molecules of the chemical come in contact with the bushy antennae of a male silkworm moth—who might be as far as three miles away—he recognizes the mate attractant and responds by flying upwind, looking for the source of the chemical, until he zeroes in on the female and they mate.

This is just one example—a chemical one—of **communication,** an action or signal on the part of one individual that informs or alters the behavior of another individual (**FIGURE 9-28**). A huge range of animal behaviors requires communication, including defending territory, alerting individuals to threats, announcing one's readiness for mating, establishing one's position in a dominance hierarchy, and caring for offspring.

Animals communicate with many different types of signals. Three types are most common.

1. Chemical. Molecules released by an individual into the environment that trigger behavioral responses in other individuals are called **pheromones.** They include the airborne mate attractant of the silkworm moth, territory markers such as those found in dog urine, and trail markers used by ants. They are even used by humans: some chemicals that influence the length of the menstrual cycle in women are responsible for the synchronization of menstrual cycles that occurs in women living in group housing situations, such as in dormitories and prisons.

2. Acoustical. Sounds that trigger behavioral responses are abundant in the natural world. These include the alarm calls of Belding's ground squirrels described earlier, the complex songs of birds, whales, frogs, and crickets vying for mates, and the territorial howling of wolves.

3. Visual. Individuals often display visual signals of threat, dominance, or health and vigor. Examples include the male baboon's baring of his teeth and the vivid tail feathers of the male peacock.

With increasingly complex forms of communication, increasingly complex information can be conveyed. One extreme case is the honeybee **waggle dance.** When a honeybee scout returns to the hive after successfully locating a source of food, she performs a set of maneuvers on a honeycomb that resemble a figure 8, while wiggling her abdomen. Based on the specific angle at which the scout dances (relative to the sun's position in the sky) and

BEHAVIORS EVOLVE   COOPERATION AND ALTRUISM   SEXUAL CONFLICT   COMMUNICATION

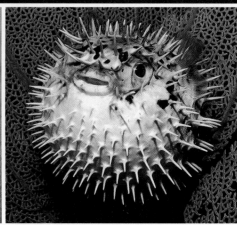

CHEMICAL COMMUNICATION
Pheromones released by one individual and detected by another, using its antennae, can trigger behavioral responses.

AUDITORY COMMUNICATION
Sounds, such as the roar of a lion, are a common method of triggering behavioral responses.

VISUAL COMMUNICATION
Organisms can convey information, such as threat or receptivity, with visual displays. Here, a balloon-fish puffs up its body in response to a predator.

FIGURE 9-28 Chemical, auditory, and visual communication. Many animal behaviors require communication: the ability to convey and receive information from other animals.

the duration of her dance, the bees that observe the dance are able to leave the hive and quickly locate the food source.

The complexity and power of the waggle dance raises an important question: at what point does communication become language? **Language** is a very specific type of communication in which arbitrary symbols represent concepts, and a system of rules, called grammar, dictates the way the symbols can be manipulated to communicate and express ideas (**FIGURE 9-29**).

It's not always easy to identify language. Consider one type of communication in vervet monkeys. Vervet monkeys live in groups, and when an individual sees a predator coming, it makes an alarm call to warn the others. Unlike in most other alarm-calling species, the vervet monkey's alarm call differs depending on whether the predator is a hawk, a snake, or a leopard. And the response of other vervet monkeys to the call is appropriate to the type of predator approaching. The snake alarm causes individuals to stand on their toes and look down. The leopard alarm causes individuals to quickly climb the nearest tree. And the hawk alarm causes vervet monkeys to look to the sky.

Does alarm calling in vervet monkeys or the honeybee waggle dance constitute language? This is debatable. Is the form of American Sign Language taught to chimpanzees and gorillas a language (see Figure 9-29)? There are opinions on both sides. Interestingly, the great apes have shown a capacity for creating new signs, expressing abstract ideas, and referring to things that are distant in time or space—important features of human language.

COMPLEX FORMS OF ANIMAL COMMUNICATION

Honeybee waggle dance

Orangutan and trainer giving hand signals

Children listening to a teacher

FIGURE 9-29 Dancing, signing, and speaking. A variety of ways have evolved in animals to convey complex information to others.

One thing that is certain, however, is that human language is near the most complex end of the communication continuum.

Language influences many of the other behaviors discussed throughout this chapter. The evolution of reciprocal altruism, for example, may be influenced by language, as this makes it easier for individuals to convey their needs and resources to other individuals. Similarly, in courtship and the maintenance of reproductive relationships, language gives individuals a tool for conveying complex information relating to the resources and value they may bring to the interaction.

TAKE-HOME MESSAGE 9·17

Methods of communication—chemical, acoustic, and visual—have evolved among animal species, enabling them to convey information about their condition and situation. These abilities influence fitness and the evolution of almost all other behaviors.

9·18 Honest signals reduce deception.

Among Natterjack toads, found in northern Europe, males produce booming calls to attract females. Because females desire large males and size determines the volume of the calls, females will push their way through murky swamps until they reach the loudest croaker. They are drawn by calls that can be heard a mile away and are often louder than the legal noise limit for a car engine (**FIGURE 9-30**).

The Natterjack's call is an **honest signal,** a signal that cannot be faked and is given when both the individual making the signal and the individual responding to it have the same interests. An honest signal is one that carries the most accurate information about an individual or situation, and animals that respond to signals of any sort have evolved to value most highly those signals that cannot be faked. A loud Natterjack's call, for example, cannot be faked by a small Natterjack toad. And so there is evolutionary pressure for animals to both produce and respond to honest signals, because these behaviors enable them to maximize their fitness by selecting the best possible mate.

When one animal can increase its fitness by deceiving another, deception can also be expected to evolve. An example might be baby birds begging for food from their parent. If an allele that causes a chick to exaggerate its need for food leads to faster growth and better health for that individual, the allele is likely to increase in frequency in the population.

Communication and signaling, therefore, are features of populations that are continually evolving. We would expect this evolutionary "arms race" between honest signals and deception to lead, over time, to both increasingly unambiguous signals and ever more sophisticated patterns of deception.

In this chapter, we've seen how our understanding of the behavior of animals, including humans, has been greatly expanded by applying scientific thinking, an experimental approach, and careful consideration of the process by which the evolutionary mechanism of natural selection shapes populations. We return to an important point highlighted at the beginning of this chapter: an organism's phenotype is not limited to its physical traits but includes its behaviors as well. Consequently, behaviors respond to selective pressures and evolve.

HONEST SIGNAL

The volume of a male Natterjack toad's call affects females' mating decisions. Loud calls can only be made by large males—smaller males can't fake it—so it is an honest signal, carrying accurate information.

FIGURE 9-30 **Loud and true.**

TAKE-HOME MESSAGE 9·18

Animals have evolved to rely primarily on signals that cannot easily be faked, in order to gain the maximum amount of information.

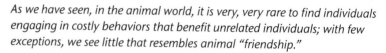
How to win friends and influence people

As we have seen, in the animal world, it is very, very rare to find individuals engaging in costly behaviors that benefit unrelated individuals; with few exceptions, we see little that resembles animal "friendship."

Humans are a rare exception—every day, in myriad ways, we see friendly behaviors. Just as we put money in the bank for a rainy day, we buffer ourselves from the world's uncertainties by storing goodwill in our neighbors. Cooperators have evolutionary advantages over loners.

Q: **Why is "kindness" a risky behavior evolutionarily?** From an evolutionary perspective, we remain vulnerable. With each costly act we perform for unrelated individuals, strangers and friends alike, there is the risk that our efforts will be in vain, our energies lost. For that reason, in our evolution as reciprocal altruists, we acquired a hesitancy to stick our necks out.

Q: **How can we help others feel less vulnerable and more willing to cooperate?** Our ability to override the impulses that often push us toward selfishness is one of the hallmarks of being human. Still, in each interaction, there may be an individual on the other side who is also feeling vulnerable. By taking steps, often absurdly simple, to address the unconscious vulnerabilities that others feel, we can increase the likelihood that they choose cooperation, reciprocity, and friendship.

- Learn and use other people's names. It tells them you recognize them specifically and, consequently, that you understand exactly who it is that you "owe." Smile and make eye contact—even when attempting to change lanes in traffic. These gestures feel like the beginning of a relationship, thus stimulating favor-granting instincts.

- Embrace etiquette. Acknowledge your debts to others. Be effusive and public in your thanks for kindnesses done to you. Send thank-you cards.

- Take the first step in reciprocity. Human cooperation is so tied to reciprocal exchange that even tiny gestures of good faith can play an important role in building relationships. Whenever possible, give gifts. Even small ones. Even when they are not required. (Especially when they are not required.)

- Develop a good reputation. Become known for generosity, for loyalty, for remembering and acknowledging kindnesses done to you.

Can you think of other steps?

Check Your Knowledge

Thinking critically about visual displays of data

1. According to the top bar graph, how many offspring does a female with one mate produce? How does this compare with the number of offspring produced by a male with one mate?

2. What is the number of offspring produced by females with two mates? Why is this lower than the number for females mating with just one male? What additional data would help you answer this question?

3. How many offspring do you think a female with four mates would have? What about a male with four mates?

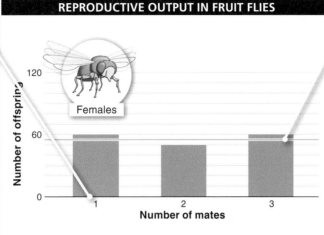

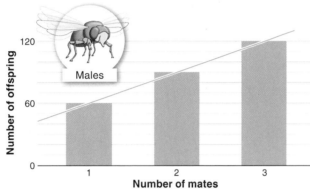

REPRODUCTIVE OUTPUT IN FRUIT FLIES

Females — Number of offspring vs. Number of mates

Males — Number of offspring vs. Number of mates

4. According to this figure, what limits the reproductive output of a female fruit fly? What limits a male's reproductive output?

5. Why is it important to have the same axes on both graphs?

See answers at the back of the book.

Key Terms in Evolution and Behavior

altruistic behavior, p. 373
behavior, p. 367
communication, p. 395
female, p. 382
fixed action pattern, p. 368
group selection, p. 381
honest signal, p. 397
inclusive fitness, p. 376
innate behavior, p. 368
instinct, p. 368

kin selection, p. 374
language, p. 396
learning, p. 369
male, p. 382
mate guarding, p. 388
mating system, p. 391
mismatch, p. 380
monogamy, p. 391
nuptial gift, p. 387
pair bond, p. 392
paternity uncertainty, p. 384
pheromone, p. 395

polyandry, p. 391
polygamy, p. 391
polygyny, p. 391
prepared learning, p. 370
reciprocal altruism, p. 374
reproductive investment, p. 382
sexual dimorphism, p. 394
sign stimulus, p. 368
total reproductive output, p. 383
waggle dance, p. 395

ABOUT THE CHAPTER OPENING PHOTO

Wolves communicate and bond through howling.

9.1–9.4 Behaviors are traits that can evolve.

As long as they satisfy the three necessary conditions for natural selection (variation, heritability, and differential reproductive success), behaviors can evolve just as physical traits can.

BEHAVIOR

Behavior encompasses any and all of the actions performed by an organism. Behavior is as much a part of an organism's phenotype as is an anatomical structure, and as such it can be produced and shaped by natural selection.

 1. What would you expect to happen to a trait that increases an individual's reproductive success relative to that of other individuals?

INNATE BEHAVIORS

Innate behaviors are present in all individuals in a population and do not vary much from one individual to another or over an individual's life span.

In geese, the sight of an egg outside the nest triggers a fixed action pattern: the goose uses a side-to-side egg-retrieval movement all the way back to the nest, even if the egg is taken away during the process.

The bellies of male stickleback fish turn bright red when the breeding season arrives. During this time, a male stickleback reacts aggressively to the sight of a red belly on any other male stickleback.

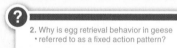 **2.** Why is egg retrieval behavior in geese referred to as a fixed action pattern?

COMPLEX-APPEARING BEHAVIORS

Complex-appearing behaviors don't necessarily require complex thought to evolve. The natural selection of such behaviors does not require the organism to consciously try to maximize its reproductive success.

LEARNED BEHAVIORS

Learned behaviors are those behaviors that are influenced more by the individual's environment, requiring some learning, and are often altered and modified over time in response to past experiences. Organisms are well-prepared to learn behaviors that have been important to the reproductive success of their ancestors, and less prepared to learn behaviors irrelevant to their evolutionary success. For example, a monkey quickly learns to fear snakes if it sees another monkey express such fear.

1 **BEFORE EXPOSURE TO FEAR OF SNAKES**
A captive monkey will reach over a plastic snake for food.

2 **EXPOSURE TO FEAR OF SNAKES**
A captive monkey views another monkey expressing fear at the sight of a snake.

In real life   or   On video

3 **AFTER EXPOSURE TO FEAR OF SNAKES**
A captive monkey expresses fear of the plastic snake.

 **3.** Which type of behaviors are we most well prepared to learn?

4. If an over-sized egg—too big to be a goose egg—is placed next to a goose's nest, what will the goose do?

Check Your Knowledge

1. An animal will preferentially feed on:

a) the largest prey it can find.

b) the prey that provides the most energy relative to effort.

c) the smallest prey it can find.

d) the prey that provides the most calories relative to the prey size.

e) the prey that provides the most calories relative to the predator size.

0 ——30—— 100
EASY   HARD

2. From an evolutionary perspective, behavior can best be viewed as:

a) a trait that arises by learning, not by natural selection.

b) non-heritable.

c) a trait subject to drift and mutation, but not natural selection.

d) part of the phenotype.

e) All of the above are correct.

0 ——61—— 100
EASY   HARD

3. During the breeding season, the sight of a red belly on any other stickleback triggers aggressive behavior in a male stickleback. This is called:

a) a fixed action pattern.

b) prepared learning.

c) aggressive conditioning.

d) sexual selection.

e) male-male competition.

0 ——51—— 100
EASY   HARD

4. Babies in the United States easily develop a fear of snakes. Yet they don't easily develop a fear of guns. Why?

a) Humans cannot develop fears of inanimate objects.

b) Evolution can be slow in producing populations that are adapted to their environments.

c) Babies are more likely to encounter snakes than guns in the United States.

d) Fewer individuals are killed by guns than by snakes in the United States each year.

e) All of the above are correct.

0 ——22—— 100
EASY   HARD

9.5–9.9 Cooperation, selfishness, and altruism can be better understood with an evolutionary approach.

Many behaviors in the animal world that appear to be altruistic behaviors can be explained.

APPARENT ALTRUISM

Many behaviors in the animal world appear altruistic. In almost all cases, such acts are not truly altruistic and have evolved as a consequence of kin selection or reciprocal altruism, and from an evolutionary perspective are beneficial to the individual engaging in the behavior.

5. Virtually all of the apparent acts of altruism observed in the animal kingdom prove to have evolved as a consequence of what factor?

KIN SELECTION

Kin selection describes apparently altruistic behavior in which an individual that assists a genetic relative compensates for its own decrease in direct fitness by helping increase the relative's fitness and, consequently, its own inclusive fitness.

At great risk to herself, a female Belding's ground squirrel will make an alarm call in response to a predator, which is likely to save individuals to whom she is related. Because she protects her genetic relatives, the alarm-calling behavior is favored by natural selection.

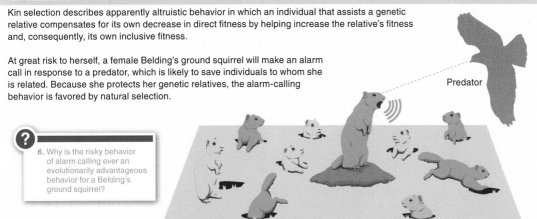

Predator

6. Why is the risky behavior of alarm calling ever an evolutionarily advantageous behavior for a Belding's ground squirrel?

LEARNED BEHAVIOR

Reciprocal altruism describes altruistic-appearing acts in which the individual performing the act is likely to get something of value from the recipient at a later time. The evolution of reciprocal altruism requires that individuals have repeated interactions and can recognize and punish cheaters, conditions satisfied in humans but in few other species.

A vampire bat may vomit up a blood meal for an unrelated individual—saving that individual's life. Bats preferentially do this for individuals that have done so for them previously (and will return the favor), improving their own survival in the long run.

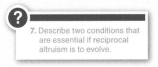
7. Describe two conditions that are essential if reciprocal altruism is to evolve.

GROUP SELECTION

Group selection describes the evolution of a trait that is beneficial for the species or population while decreasing the fitness of the individual exhibiting the trait. Behaviors that are good for the species or population but detrimental to the individual exhibiting such behaviors are not generally produced in a population under natural conditions.

ALLELE FREQUENCIES

● Proportion of **selfless behavior** allele in the population ("do what's best for the group, even though it reduces your own reproductive output")

● Proportion of **selfish behavior** allele in the population ("do what's best for you, even if it hurts the group")

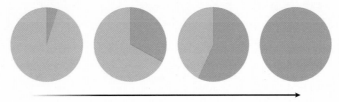

Time

8. An organism in an environment that differs from the environment to which it is evolutionarily adapted may exhibit maladaptive behaviors. Using an example, explain why.

9. Will natural selection favor a behavior that leads to a better outcome for the population but not for the individual? Explain why or why not.

---

5. In Belding's ground squirrels, why are females much more likely than males to make alarm calls?

a) Belding's ground squirrels have a sex ratio that is biased toward females.

b) Females invest more in food storage, so they are more likely to lose their lives or their food if a predator attacks.

c) Belding's ground squirrels have a sex ratio that is biased toward males.

d) Females tend to remain in the area where they were born, so the females that call are warning their own kin.

e) Males forage alone, so their alarm calls would be useless.

0 — 18 — 100
EASY   HARD

6. The genetic contribution of an individual to subsequent generations through its own offspring and through it influence on the survival of other relatives is:

a) individual fitness.

b) inclusive fitness.

c) altruism.

d) inclusive altruism.

e) the protector effect.

0 — 50 — 100
EASY   HARD

7. Vampire bats:

a) sometimes regurgitate blood into the mouth of another bat that is close to starving, but the likelihood of this is a function of whether the individuals are genetically related.

b) are one of the few animal species that exhibit kin selection.

c) sometimes regurgitate blood into the mouth of an unrelated bat that is close to starving.

d) exhibit reciprocal altruism but not kin selection.

e) There are no such things as vampire bats; they're found only in a Dracula novel.

0 — 37 — 100
EASY   HARD

8. Altruistic behavior in animals may result from kin selection, a process in which:

a) genes promote the survival of copies of themselves when behaviors by animals possessing those genes assist other animals that share those genes.

b) aggression within sexes increases the survival and reproduction of the fittest individuals.

c) companionship is advantageous to animals because, in the future, they can recognize and help individuals that have helped them.

d) aggression between the sexes increases the survival and reproduction of the fittest individuals.

e) companionship is advantageous to animals because, in the future, they can recognize individuals that have helped them and request help once again.

0 — 25 — 100
EASY   HARD

9. All of the following conditions are necessary for reciprocal altruism to evolve in a species except:

a) the ability to recognize different individuals.

b) the ability to punish cheaters who do not reciprocate.

c) repeated interactions with the same individuals.

d) in at least one of the sexes, individuals always remaining to live near their kin.

e) None of the above are necessary for the evolution of reciprocal altruism.

0   **31**   100
EASY   HARD

10. In a situation where males guard eggs and care for the young without help from the female, which of the following statements would most likely be correct?

a) The males are larger and more brightly colored in order to attract the very best females.

b) The males and females are equally brightly colored, but males court females aggressively.

c) The population is monogamous with no sexual dimorphism.

d) A single male controls a harem of females to which he has exclusive reproductive access.

e) The females are more brightly colored than males and court males aggressively.

0   **44**   100
EASY   HARD

11. In mammals, as well as many other animals, males generally compete for females. The best explanation for this phenomenon is:

a) males are more aggressive.

b) males, on average, have higher fitness.

c) females have a higher parental investment.

d) males are choosy.

e) females are better looking.

0   **26**   100
EASY   HARD

12. Mate guarding is a reproductive tactic that functions to:

a) reduce paternity uncertainty.

b) increase the female's investment in offspring.

c) reduce the male's reproductive investment.

d) reduce the female's fitness.

e) increase the number of mates to which a male has access.

0   **32**   100
EASY   HARD

9·10–9·16 Sexual conflict can result from disparities in reproductive investment by males and females.

There are differences between males and females in the patterns of investment in reproduction.

DIFFERENCES IN MALE AND FEMALE REPRODUCTIVE BEHAVIORS

In mammals and many other animals, there are important physical differences between males and females relating to reproduction. These differences have led to the evolution of differences in male and female reproductive behaviors.

Males tend to increase their reproductive success by mating with many females and have evolved to compete among themselves to get the opportunity to mate. Females do not increase their reproductive success through extra matings, but rather by caring for their offspring and being choosy when selecting a mate.

10. Describe two ways in which, in mammals, males and females initially make unequal energetic investments in reproduction.

11. Compare and contrast the stages of the mammalian reproductive process at which males and females are most vulnerable.

FACTORS IN MATE SELECTION

Female choosiness (and the male-male competition it leads to) tends to increase the likelihood that a female will select only those males that have plentiful resources or relatively high-quality genes. Female choosiness is manifested by four general rules.

COURTSHIP RITUALS
A female grebe requires the male to perform a courtship dance before she will mate with him.

GIFTS UP FRONT
A female hanging fly will not mate with a male unless he brings her a large offering of food.

CONTROL OF VALUABLE RESOURCES
Female yellow-bellied marmots prefer rock outcroppings that provide retreats for escape from predators and for hibernation (and are controlled by dominant males).

GOOD LOOKS
A female peacock is attracted to a male with the most beautiful tail feathers.

12. Female deer tend to choose as a mate a male deer possessing large, prominent antlers. What does the size of the antlers signify to the female?

MATE GUARDING

Mate guarding reduces additional mating opportunities for a partner and can improve a male's reproductive success by increasing his paternity certainty, reducing his vulnerability when investing in offspring.

Mating systems describe the variation in number of mates and in the reproductive success of males and females. They are influenced by the relative amounts of males' and females' parental investment.

POLYGAMOUS MATING SYSTEMS
Some individuals attract multiple mates while other individuals attract none. Polygamous mating systems can be subdivided into polygyny, in which individual males mate with multiple females (such as in elephant seals), and polyandry, in which individual females mate with multiple males.

MONOGAMOUS MATING SYSTEMS
Individuals mate and remain with just one other individual.

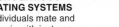

13. In a polygynous mating system, many males may produce no offspring. Why?

14. Describe the three types of mating systems.

Differences in the level of competition among individuals of each sex for access to mating opportunities can lead to the evolution of male-female differences in body size and other aspects of appearance.

SEXUAL DIMORPHISM
• The sexes differ in size or appearance.

• One parent invests more in caring for the offspring.

• Mating system tends toward polygamy.

• One sex (usually males) competes for access to mating opportunities with the other sex.

SEXUAL MONOMORPHISM
• The sexes are indistinguishable.

• Both parents invest (approximately) equally in caring for the offspring.

• Mating system tends toward monogamy.

15. Female moorhens are larger and more aggressive than males. They also compete among themselves for access to the smaller, fatter males. Which sex do you think provides more parental care? Explain your answer.

9·17–9·18 Communication and the design of signals evolve.

The challenge of communication is central to the evolution of animal behaviors.

Methods of communication have evolved in animal species, enabling individuals to convey information about their condition and situation. These abilities influence fitness and the evolution of almost all other behaviors.

CHEMICAL COMMUNICATION
Pheromones released by one individual and detected by another, using its antennae, can trigger behavioral responses.

AUDITORY COMMUNICATION
Sounds, such as the roar of a lion, are a common method of triggering behavioral responses.

VISUAL COMMUNICATION
Organisms can convey information, such as threat or receptivity, with visual displays. Here, a balloonfish puffs up its body in response to a predator.

16. Why are "honest signals" so named?

13. Relative to birds, more mammalian species are:
a) polygynous.
b) monogamous.
c) polyandrous.
d) hermaphroditic.
e) sexually monomorphic.

0 — EASY ——— **59** ——— 100 HARD

14. In a species such as pigeons, in which males are almost indistinguishable in appearance from females, the most likely mating system is:
a) monomorphism.
b) monogamy.
c) polygyny.
d) polyandry.
e) It is impossible to predict the mating system with only this information.

0 — EASY ——— **37** ——— 100 HARD

15. If you find a species of fish in which males are much more brightly colored and larger than females, what might you infer about their mating system?
a) The degree of sexual dimorphism does not give any information about the mating system.
b) They are simultaneous hermaphrodites.
c) They exhibit parallel monogamy.
d) They are serially monogamous.
e) They are polygynous.

0 — EASY ——— **39** ——— 100 HARD

16. Polygynous species:
a) usually employ external fertilization.
b) are usually sexually dimorphic, with males larger and more highly ornamented.
c) are usually sexually dimorphic, with females larger and more highly ornamented.
d) usually have males and females that are physically indistinguishable.
e) are more commonly found among birds than among mammals.

0 — EASY ——— **29** ——— 100 HARD

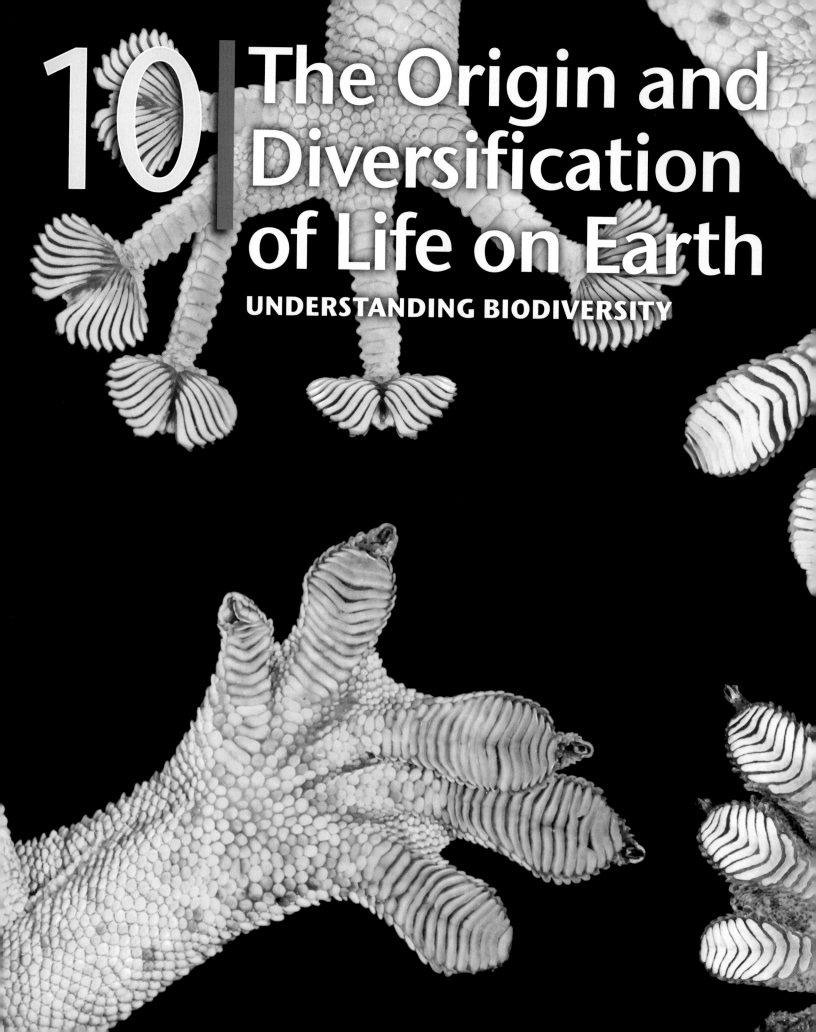

10 The Origin and Diversification of Life on Earth

UNDERSTANDING BIODIVERSITY

10·1–10·3
Life on earth most likely originated from non-living materials.

Life on earth arose in a very different environment than we experience today. (Shown here: the formation of volcanic rock in Hawaii.)

10·1 Complex organic molecules arise in non-living environments.

In the beginning, there was nothing. Now there is something. That, in a nutshell, describes one of the most important, yet difficult to resolve, questions in science: how did life on earth begin? We open our investigation by describing a few things that we do know and exploring how they help us speculate about the things we do not yet know. First, let's give a basic definition of what we mean by "life." **Life** is defined by the ability to replicate and by the presence of some sort of metabolic activity (the chemical processes by which molecules are acquired and used and energy is transformed in controlled reactions). In the next section we expand on this definition and explore the transition from non-living to living.

Earth formed about 4.5 billion years ago from clouds of dust and gases, and as it very gradually cooled, a crust formed at the surface and condensing water formed the oceans. This probably took several hundred million years. The oldest rocks, found in Canada, are about 3.8 billion years old. And the earliest life forms appeared not long after these first rocks formed: fossilized bacteria-like cells have been found in rocks that are 3.4 billion years old.

From that initial point in earth's formation, tremendous **biodiversity** arose—variety and variability among all genes, species, and ecosystems. In this chapter we explore how this biodiversity might have come to be, how we name groups of organisms, and how we determine the relatedness of these groups to one another. We begin by returning to the question of the origin of life on earth.

How did these first organisms arise? Some have suggested that life may have originated elsewhere in the universe and traveled to earth, possibly on a meteor. It is hotly debated, however, whether microbes could even survive the multi-million-year trip to earth in the cold vacuum of space with no protection from ultraviolet and other forms of radiation. Experimental data have been unable to answer this question definitively. The vast majority of scientists

CONDITIONS ON EARTH AT THE TIME LIFE BEGAN

Earth's early atmosphere had almost no oxygen and was filled with many compounds released during volcanic eruptions.

Small organic molecules eventually formed, providing the building blocks of life.

FIGURE 10-1 Darwin's "warm little pond." The first life on earth tolerated an atmosphere without oxygen.

believe, instead, that life originated on earth, probably in several distinct phases.

Phase 1: The formation of small molecules containing carbon and hydrogen

The conditions on earth around the time of the origin of life were very different from those of today. In particular, chemical analyses of old rocks reveal that no oxygen gas was present. The atmosphere included large amounts of carbon dioxide, nitrogen, methane, ammonia, hydrogen, and hydrogen sulfide. Most of these molecules were produced by volcanic eruptions. It was this environment that probably served as the cradle of life, or what Darwin called the "warm little pond" (FIGURE 10-1).

Critical to the origin of life was the formation of small molecules containing carbon and hydrogen. As we saw in Chapter 2, carbon readily bonds with hydrogen and with other carbon atoms, creating molecules with a huge variety of forms that interact with each other in reactions that would eventually become the processes of life. There are several plausible scenarios for how these small organic molecules might have formed. The most likely comes from some simple but revealing four-step experiments done in 1953 by a 23-year-old student named Stanley Miller and his advisor, Harold Urey (FIGURE 10-2).

1. They created a model of the "warm little pond" and their best estimate of earth's early atmosphere: a flask of water with H_2, CH_4 (methane), and NH_3 (ammonia).

2. They subjected this mini-world to sparks, to simulate lightning.

3. They cooled the atmosphere so that any compounds formed in it would rain back down into the water.

4. They waited, and then they examined the contents of the water to see what happened.

They didn't have to wait long to get exciting results. Within a matter of days—not millions of years or even a few months—they discovered many organic molecules, including five different amino acids, in their primordial sea. (Using more sensitive equipment, recent re-analyses reveal that all 20 amino acids present in living organisms were produced during the course of these experiments.) This was the first demonstration that complex organic molecules could have arisen in earth's early environment. However, questions remain, such as whether it is reasonable to think that the environment Urey and Miller assumed to exist on the early earth is actually likely to have existed. Still, the experiments are a promising first step, suggesting that complex organic compounds—including amino acids, the primary constituents of proteins and an essential

component of living systems—could have been produced from inorganic chemicals and lightning energy in the primitive environment of earth.

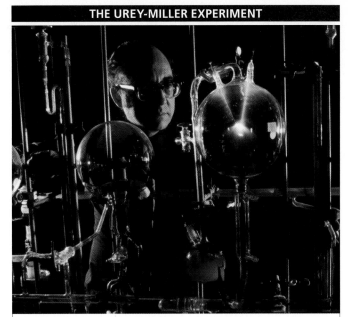

THE UREY-MILLER EXPERIMENT

Stanley Miller (shown) and Harold Urey developed a simple four-step experiment that demonstrated how complex organic molecules could have arisen in earth's early environment.

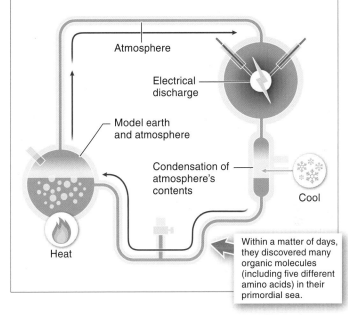

Atmosphere

Electrical discharge

Model earth and atmosphere

Condensation of atmosphere's contents

Cool

Heat

Within a matter of days, they discovered many organic molecules (including five different amino acids) in their primordial sea.

FIGURE 10-2 **A promising first step.** The Urey-Miller experiment generated organic molecules from hydrogen, methane, and ammonia.

TAKE-HOME MESSAGE 10·1

Under conditions similar to those thought to exist on early earth, small organic molecules can form, and these molecules have some chemical properties of life.

10·2 Cells and self-replicating systems evolved together to create the first life.

After the generation of numerous organic molecules such as amino acids, the second phase in the generation of life from non-life was probably the assembly of these building block molecules into self-replicating, information-containing molecules. This is where things get a bit more speculative. It's complicated enough to generate a complex organic molecule, but it's a whole lot more complicated to generate an organic molecule that can replicate itself. Researchers believe that to get to the replication phase, enzymes, or something with the catalytic activity of enzymes, were required.

Phase 2: The formation of self-replicating, information-containing molecules Recently, researchers discovered that the nucleic acid RNA can do what proteins do—catalyze reactions necessary for replication—meaning that this single, relatively simple molecule could have been a self-replicating (i.e., able to make copies of itself) system and a precursor to cellular life.

These self-replicating molecules raise an important question: When exactly was the threshold between living and non-living crossed? It has been proposed that in the early world, self-replicating nucleic acid molecules carried the information on how to replicate *and* served as the machinery to actually carry out the replication. Is that enough? At this point, it is reasonable to ask, again: what exactly is "life"?

Fossils of 3.4-billion-year-old cells have been found in rocks in South Africa and Australia. These cells appear to be prokaryotic cells, similar to living bacterial cells, with no nucleus, no organelles, and a circular strand of genetic information (**FIGURE 10-3**). Some even look as if they are in the process of dividing. Although, superficially, the fossilized cells look like just a couple of circles, several lines of evidence support the idea that they are indeed remnants of cells. These include: (1) the age of the rocks themselves has been reliably determined, (2) the size of the circles is similar to that of modern-day prokaryotes, and (3) the ratio of two carbon isotopes ($^{12}C/^{13}C$) is more characteristic of fossilized organisms than of typical rocks that do not contain fossils. But many questions remain, including: Were these cells the first living organisms on earth? And were they descendants of earlier, self-replicating molecules of RNA? Because no earlier fossils have been found, it is difficult to answer these questions.

Earlier, we mentioned that there are characteristics shared by all living organisms and living systems and that these

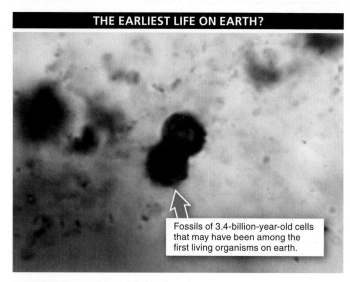

THE EARLIEST LIFE ON EARTH?

Fossils of 3.4-billion-year-old cells that may have been among the first living organisms on earth.

FIGURE 10-3 **Ancient prokaryotic cells.**

include the ability to replicate and the ability to carry out some sort of metabolism. So the self-replicating RNA molecules were right on the border; they were able to replicate, but not able to carry out metabolism. We now explore the critical third phase in the generation of life from non-life: the development of a membrane that separated these self-replicating small molecules from their surroundings, thus forming cells and facilitating metabolic activity.

Phase 3: The development of a membrane, enabling metabolism and creating the first cells One way that self-replicating molecules could have acquired chemicals and used them in the controlled reactions of metabolism was by packaging the molecules within membranes. As we saw in Chapter 3, cell membranes are semi-permeable barriers that separate the inside of cells from their external environment. Membranes make numerous aspects of metabolism possible. In particular, they make it possible for chemicals inside the cell to be at higher concentrations than they are outside the cell. Differences in chemical concentrations inside and outside a cell are essential to most life-supporting reactions.

So, if we could combine a self-replicating molecule and some metabolic chemicals into a unit, surrounded by a membrane, life would be possible. But how could this have happened initially? Some evidence suggests that the first

THE ORIGIN OF LIFE?

How did the first organisms on earth arise? Most scientists believe that life originated on earth, probably in several distinct phases.

1 FORMATION OF SMALL MOLECULES CONTAINING CARBON AND HYDROGEN
Because of carbon's chemical structure, it can form a huge variety of molecules with widely varying functions.

2 FORMATION OF SELF-REPLICATING, INFORMATION-CONTAINING MOLECULES
The nucleic acid RNA can both carry information and catalyze the reactions necessary for replication. Thus, this single molecule could have been a self-replicating system and precursor to cellular life.

3 DEVELOPMENT OF A MEMBRANE
Membranes that compartmentalized the self-replicating molecules from their surroundings facilitated metabolic activity.

FIGURE 10-4 How did life arise?

cells may simply have formed spontaneously. Specifically, researchers have found that mixtures of phospholipids placed in water or salt solutions tend to spontaneously form small spherical units that resemble living cells. These units may even "sprout" new buds at their surface, appearing to divide. Because these cell-like units don't have any genetic material, however, they cannot be considered to be alive. But if, at some point, units like these incorporated some self-replicating molecules inside, maybe by forming around them, such **microspheres** might have been important in the third phase in the generation of life from non-life: the compartmentalization of self-replicating, information-containing molecules into cells. If this did occur, the final step in the creation of something from nothing would be complete (**FIGURE 10-4**).

The exact process by which life on earth originated is still uncertain, and research continues. What is clear, however, is that life now abounds in great diversity. In this chapter we explore that diversity and how it is generated. In doing so, we'll investigate what a species is and consider how individual species split and create additional species.

TAKE-HOME MESSAGE 10·2

The earliest life on earth, which resembled bacteria, appeared about 3.5 billion years ago, not long after the earth was formed. Evidence supports the idea that self-replicating molecules, possibly RNAs, may have formed in earth's early environment and later acquired or developed membranes, enabling them to replicate and making metabolism possible—the two conditions that define life.

? **10·3 THIS IS HOW WE DO IT**

Could life have originated in ice, rather than in a "warm little pond"?

In science, when do we need to think outside the box?

Phenomena in the natural world don't always give up their secrets easily. When trying to better understand some process in nature, applying a new experimental technique occasionally allows a breakthrough. Other times, it may be the discovery of new or unexpected evidence that provides the breakthrough. But sometimes progress requires a more radical approach and thinking outside the box.

Researchers have been questioning some of the most basic assumptions about life's origin on earth, as described in the first two sections of this chapter. For example, the most widely held view is that life on earth emerged from a particular type of environment, one that was warm or hot and was wet— something along the lines of the "warm little pond" Darwin had speculated about in *The Origin of Species*. Evidence from the experiments of Miller and Urey, as well as other research, supports this view.

But what if icy baths, not warm ponds, were the "incubator" of life?

That's thinking outside the box. And the researchers suggesting this idea have provided some intriguing evidence and proposed some clever ideas to support it. Their ideas build upon the broad consensus about the most important physical and chemical requirements for the initial generation of self-replicating molecules.

Chemical requirement 1 Precursor molecules need to last a while and need to come in contact with each other.

Conventional assumption: In living cells today, compartments make this duration and closeness of contact possible. But before life appeared on earth, under warm, wet conditions, some sort of chambers or microspheres may have spontaneously formed and served this purpose.

Novel approach: It turns out that, as water freezes, tiny compartments form within the ice. On early earth, low temperature may have slowed the degradation of any precursor molecules—including RNA—that formed. And the tiny compartments may have held those precursor molecules close together, making it possible for them to react with each other.

Chemical requirement 2 Precursor molecules need to exhibit catalytic properties.

Conventional assumption: At warm or hot temperatures—but not at colder temperatures—molecules move quickly and collide frequently, enhancing reaction rates.

Novel approach: Although reactions usually slow down as the temperature drops, some actually speed up. As water freezes, the ice crystals form only from pure water. If there are any impurities present—such as salt or cyanide—they are excluded from the crystals and concentrated in small chambers of liquid water within the ice. At these higher concentrations, the molecules collide more frequently, even as the temperature drops. As a consequence, certain reactions can occur *more* rapidly—possibly including the creation and elongation of the first RNA chains.

| Is it even feasible that ice was present on early earth and precursor molecules could have formed in it?

Intriguing observations and evidence Researchers have carried out experiments in which they prepare—and then freeze—tubes containing seawater and the building blocks of RNA. After thawing out the tubes, they find numerous RNA molecules, some long enough to be able to act as enzymes.

And recent evaluations of glacier-encased land north of the Arctic Circle suggest that 4 billion years ago, the sun may have been dimmer, and the earth may have cooled so much that ice covered the oceans. Some scientists have even gone so far as to describe earth as a "giant snowball" at that time.

| Has exploration of the plausibility of ice as the initial medium of RNA replication answered the questions about how life on earth originated?

There is still plenty of skepticism about the idea that the primordial soup was a cold soup. Many researchers suspect that reported evidence of RNA chains forming under freezing temperatures may reflect accidental contamination, or that the chains actually formed during the thawing-out process.

As a case study of scientific thinking in action, though, this example illuminates the importance of evaluating the assumptions underlying our hypotheses. And we get a glimpse of how a fresh perspective can remove constraints that might limit our ability to see solutions to problems.

| Is there any value to false starts (and even dead ends) encountered in research investigations?

Keeping an open mind is more important than rigidly holding onto an idea. Observations and evidence must take the central role in guiding our interpretations and understanding of natural processes.

TAKE-HOME MESSAGE 10·3

As researchers investigate how life on earth might have originated, some are questioning the long-held assumptions that self-replicating molecules with catalytic properties are most likely to have formed in a warm, wet environment. They've proposed that the laws of chemistry and the properties of water as it freezes may actually favor ice as the initial incubator of life. The answer is unclear, but the process of scientific thinking is guiding investigators to develop and test their hypotheses.

10·4–10·7
Species are the basic units of biodiversity.

A world of beauty: There are more than 280,000 species of plants.

10·4 What is a species?

A cat and a mouse are different things. Oprah Winfrey and Bill Gates, on the other hand, are different *versions* of the same thing. That is, although they are different individuals, both are humans. We generally distinguish between different kinds of living organisms: a rose versus a daisy, a wasp versus a fly, a snake versus a frog. Similarly, we lump some organisms into the same group and recognize them as different versions of the same thing: red roses and yellow roses, Chihuahuas and Dalmatians. How do we know when to classify two individuals as members of different groups and when to classify them as members of the same group (**FIGURE 10-5**)?

Biologists use the word **species** to label different kinds of organisms. According to the **biological species concept,** species are populations of organisms that interbreed, or

could possibly interbreed, with each other under natural conditions, and that cannot interbreed with organisms from other such groups (**FIGURE 10-6**).

Notice that the biological species concept completely ignores physical appearance when defining a species and instead emphasizes **reproductive isolation,** the inability of individuals from two populations to produce fertile offspring with each other, thereby making it impossible for gene exchange between the populations to occur.

Let's clarify two important features of the biological species concept. First, it says that members of a species are actually interbreeding or *could possibly* interbreed. This emphasis means that just because two individuals are physically separated, they aren't necessarily of different species. A person

THE SAME
We easily recognize Cristiano Ronaldo of Real Madrid and Beyoncé Knowles as members of the same species.

DIFFERENT
The Persian buttercup and the pink rose or the Bornean orangutan and the golden lion tamarin look similar but are not from the same species.

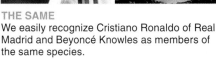

FIGURE 10-5 **Appearance is not an accurate identifier.**

WHAT MAKES A SPECIES?

Herd of African elephants in Namibia

SPECIES ARE...
• Populations of organisms that interbreed with each other
• Or could possibly breed, under natural conditions
• And are reproductively isolated from other such groups

FIGURE 10-6 **An interbreeding population.**

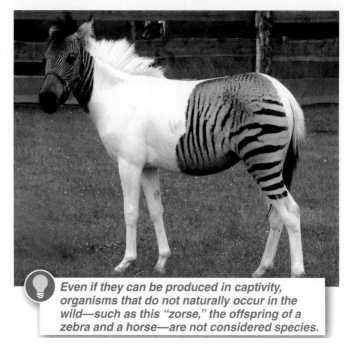

Even if they can be produced in captivity, organisms that do not naturally occur in the wild—such as this "zorse," the offspring of a zebra and a horse—are not considered species.

FIGURE 10-7 **A species? Interbreeding is not enough.**

living in the United States and a person living in Iceland, for example, may not be able to mate because of the distance, but if they were brought to the same location, they could. So, although the people are not in the same population, we do not consider them to be reproductively isolated. Second, our definition refers to "natural" conditions. This distinction is important because occasionally, in captivity, individuals may

interbreed that would not interbreed in the wild, such as zebras and horses (**FIGURE 10-7**).

There are two types of barriers that prevent individuals of different species from reproducing: prezygotic and postzygotic barriers (**FIGURE 10-8**). (Remember, an egg that has been fertilized by a sperm cell is a *zygote.*)

REPRODUCTIVE ISOLATION: KEEPING SPECIES SEPARATE

PREZYGOTIC BARRIERS
• Individuals are physically unable to mate with each other.
  OR
• If individuals are able to mate, the male's reproductive cell is unable to fertilize the female's reproductive cell.

POSTZYGOTIC BARRIERS
• Matings produce hybrid individuals that do not survive long after fertilization.
  OR
• If hybrid offspring survive (such as the mules above), they are infertile or have reduced fertility.

FIGURE 10-8 **Barriers to reproduction.**

Prezygotic barriers make it impossible for individuals to mate with each other or, if they can mate, make it impossible for the male's reproductive cell to fertilize the female's reproductive cell. These barriers include situations in which the members of the two species have different courtship rituals or have physical differences that prevent mating or fertilization.

Postzygotic barriers occur after fertilization and generally prevent the production of fertile offspring from individuals of two different species. These barriers are responsible for the production of **hybrid** individuals that either do not survive long after fertilization or, if they do, are infertile or have reduced fertility. Mules, for example, are the hybrid offspring of horses and donkeys, and although they can survive, they cannot produce their own offspring.

At first glance, the biological species concept seems to make it possible to determine unambiguously whether individuals belong to the same or different species. For plants and animals this is usually true. For many other organisms, however, it is impossible to apply the biological species concept. We examine those species in Section 10-6 and explore some alternative methods of identifying species. Even though it is not perfect, however, the biological species concept gives us an important tool for conceptualizing and categorizing the tremendous diversity of organisms found on earth.

TAKE-HOME MESSAGE 10·4

Species are generally defined as populations of individuals that interbreed with each other, or could possibly interbreed, and that cannot interbreed with organisms from other such groups. This concept can be applied easily to most plants and animals, but is not applicable for many other types of organisms.

10·5 How do we name species?

Keeping track of a large group of anything requires an organizational system. Many libraries, for example, catalogue their books by using the Dewey decimal system, which organizes books into 10 different classes, further subdivides each class into 10 divisions, and each division into 10 sections.

With the huge number of species on earth, a classification system is particularly important. Biologists use the system developed by the Swedish biologist Carolus Linnaeus in the mid-1700s. Here's how it works (**FIGURE 10-9**). Every species is given a scientific name that consists of two parts, a **genus** (*pl.* genera) and a **specific epithet.** Linnaeus named humans *Homo sapiens,* meaning "wise man." *Homo* is the genus and *sapiens* is the specific epithet, and *Homo sapiens* is the species name. (The genus is capitalized, and both genus and specific epithet are italicized.) The redwood tree has the name *Sequoia sempervirens.* The strength of Linnaeus's system is that it is hierarchical—that is, each element of the system falls under a single element in the level just above it.

In the Linnaean system, the species is the narrowest classification for an organism. The name for every species within a genus is unique, but many different species may be in the same genus. Similarly, many genera are grouped within a **family.** And many families are grouped within an **order.** Orders are grouped within a **class.** Classes are grouped within a **phylum.** And, as Linnaeus originally set it up, all phyla were classified under one of three **kingdoms:** the animal kingdom, the plant kingdom, or the "mineral kingdom."

Today, many of the species classifications that Linnaeus described have been revised, and some of his designations, such as the "mineral kingdom," have been left out, while two other kingdoms (fungi and protists) have been added. Also, all of the kingdoms are now classified under an even higher order of classification, the **domain.** But Linnaeus's basic hierarchical structure remains, and all life on earth is still named using this system, with all organisms belonging to one of the three domains: the bacteria, the archaea, and the eukarya.

When new species are discovered, they are given names based on the Linnaean system. The higher-level groups into which a new species should be classified are generally clear—that is, it's usually obvious whether something is in the animal or plant kingdom, or whether it is in the mammalian or amphibian class. But when scientists are assigning a specific epithet, they frequently have a little fun. In recent years, for example, an amphibian fossil was named *Eucritta melanolimnetes,* which translates roughly into "creature from the black lagoon." And in honor of Elvis Presley, a wasp species was named *Preseucoila imallshookupis.* A variety of other celebrities also have had species named after them, including the Beatles, Mick Jagger, Kate Winslet, and Steven Spielberg.

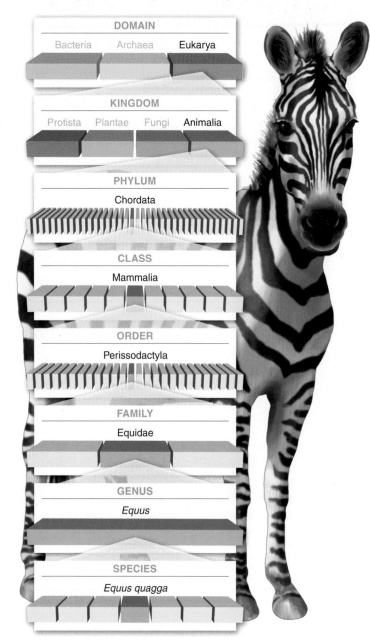

THE ORGANIZATION OF LIFE

Each species on earth, like *Equus quagga*, is given a scientific name and is categorized according to hierarchical groups.

DOMAIN

Bacteria   Archaea   **Eukarya**

KINGDOM

Protista   Plantae   Fungi   **Animalia**

PHYLUM

Chordata

CLASS

Mammalia

ORDER

Perissodactyla

FAMILY

Equidae

GENUS

Equus

SPECIES

Equus quagga

FIGURE 10-9 **Name that zebra.**

Plants and animals often are referred to by common names, which are different from their "official" names in the Linnaean system and are based on similarities in appearance. These common names, however, can cause confusion. "Fish" for example is used as part of the names of jellyfish, crayfish, and silverfish, none of which is closely related to the vertebrate group of fishes that includes salmon and tuna. Later in this chapter we'll see how an important goal of modern classification and the naming of organisms is to link an organism's classification more closely with its evolutionary history than with its physical characteristics.

TAKE-HOME MESSAGE 10·5

Each species on earth is given a unique name, using a hierarchical system of classification. Every species falls into one of three domains.

10·6 Species are not always easily defined.

Biologists, like all humans, can be biased. When investigating the natural world, for example, they often focus on plants and animals, to the exclusion of the rest of the earth's rich biodiversity. This gets them into trouble when it comes to an idea such as the biological species concept. While the biological species concept is remarkably useful when describing most plants and animals, it falls short of representing a universal and definitive way of distinguishing many life forms (**FIGURE 10-10**).

Difficulties in classifying asexual species The biological species concept defines species as populations of interbreeding individuals. But this is a useless distinction for the asexual species of the world. Recall from Chapter 6 that asexual reproduction, common among single-celled organisms (including all bacteria), many plants, and some animals, is a form of reproduction that doesn't involve fertilization or even two individuals. Rather, the cell (or cells) of an individual simply divides, creating new individuals. Because asexual reproduction does not involve partners or interbreeding, the concept of reproductive isolation is not meaningful, and it might seem that every individual should be considered a separate species. Clearly, that's not a helpful rule to follow.

Difficulties in classifying fossil species When classifying fossil species, differences in the size and shape of fossil bones from different individuals can never definitively reveal whether there was reproductive isolation between those individuals. This makes it impossible to apply the biological species concept.

Difficulties in determining when one species has changed into another Based on fossils, it seems that modern-day humans, *Homo sapiens,* probably evolved from a related species called *Homo heidelbergensis* about 250,000–400,000 years ago. This seems reasonable, until you consider that your parents—who are in the species *H. sapiens*—were born to your *H. sapiens* grandparents, who were born to your *H. sapiens* great-grandparents, and so on. If humans evolved from *H. heidelbergensis,* at what *exact* point did *H. heidelbergensis* turn into *H. sapiens?* It may not be possible to identify the exact point at which this change occurred.

Q Chihuahuas and Great Danes generally can't mate. Does that mean they are different species?

Difficulties in classifying ring species Living in central Asia are some small, insect-eating songbirds called greenish warblers. Unable to live at higher elevations of the Tibetan mountain range, the warblers live in a ring around it. At the southern end of the mountain range, in northwest India, the warblers interbreed with each other. Along either side of the range, the warbler population is split, and warblers on one side do not interbreed with those on the other, because the mountain range separates them. Where the two "side" populations meet up again at the northernmost end of the mountain range, in the forests of Siberia, they can no longer interbreed.

What happened? Gradual changes in the warblers on each side of the mountain range accumulated so that the two

populations that meet up in Siberia are sufficiently different, physically and behaviorally, that they have become reproductively incompatible. But because the two non-interbreeding populations in the north are connected by gene flow through other populations farther south, there is no exact point at which one species stops and the other begins. So where do you draw the line? The greenish

---

THE BIOLOGICAL SPECIES CONCEPT DOESN'T ALWAYS WORK

The biological species concept is remarkably useful when describing most plants and animals, but it doesn't work for distinguishing all life forms.

CLASSIFYING ASEXUAL SPECIES
Asexual reproduction does not involve interbreeding, so the concept of reproductive isolation is no longer meaningful.

CLASSIFYING FOSSIL SPECIES
Differences in size and shape of fossil bones cannot reveal whether there was reproductive isolation between the individuals from which the bones came.

DETERMINING WHEN ONE SPECIES HAS CHANGED INTO ANOTHER
There is rarely a definitive moment marking the transition from one species to another.

CLASSIFYING RING SPECIES
Two non-interbreeding populations may be connected to each other by gene flow through another population, so there is no exact point where one species stops and the other begins.

CLASSIFYING HYBRIDIZING SPECIES
Hybridization—the interbreeding of closely related species—sometimes occurs and produces fertile offspring, suggesting that the borders between the species are not clear cut.

FIGURE 10-10 A useful concept that can't always be easily applied.

warblers are just one example of the more than 20 such **ring species** that have been described.

Difficulties in classifying hybridizing species

Increasingly, **hybridization**—the interbreeding of closely related species—has been observed among plant and animal species. This phenomenon fits with the biological species concept as long as postzygotic barriers have evolved, so that the hybrids are unable to reproduce. But in some cases, such as among butterflies in the genus *Heliconius,* the hybrids have high survival rates *and* are fertile, whether interbreeding with one another or with individuals of either of the two parental species. This suggests that the borders between the species are not clear-cut.

All of these shortcomings have prompted the development of several alternative approaches to defining what a species is. These alternatives tend to focus on aspects of organisms other than reproductive isolation as defining features. The most commonly used alternative is the **morphological species concept,** which characterizes species based on physical features such as body size and shape. Although the choice of which features to use is subjective, an important aspect of the morphological species concept is that it can be used effectively to classify asexual species. And because it

doesn't require knowledge of whether individuals can actually interbreed, the morphological species concept is a bit easier to use when observing organisms in the wild.

When it comes to species definitions, although the biological species concept is the most widely used and can be applied without difficulty to most plants and animals, we should not expect one size to fit all, and there will probably never be a universally applicable definition of what a species is. From asexual species to ring species to hybridizing species, the diversity of the natural world is simply too great to fit into neat, completely defined and distinct little boxes. Nonetheless, scientists can generally use a species definition that is satisfactory for a particular situation.

> ### TAKE-HOME MESSAGE 10·6
>
> The biological species concept is useful when describing most plants and animals, but it falls short of representing a universal and definitive way of distinguishing many life forms. Difficulties arise when trying to classify asexual species, fossil species, species arising over long periods of time, ring species, and hybridizing species. In these cases, alternative approaches to defining species can be used.

10·7   How do new species arise?

Biologists don't really have a clue about exactly how many species there are on earth. Estimates vary tremendously, from 5 million to 100 million. Biologists do know, however, the process by which all species arose.

The process of **speciation,** in which one species splits into two distinct species, occurs in two phases and requires more than just evolutionary change in a population. The first phase of speciation is *reproductive isolation,* through which two populations are separated from one another and so come to have independent evolutionary fates. The second phase is *genetic divergence,* in which two populations evolving separately accumulate physical and behavioral differences over time. These differences may arise as random neutral mutations, or through each population adapting differently to features of their separate environments, including to different predators and different types of food. When differences have accumulated that prevent members of the two populations from interbreeding, we say that speciation has occurred.

Often, the initial reproductive isolation necessary for speciation comes about when two populations are geographically separated. Although this is an effective and common way for speciation to occur, speciation can also occur without it.

Speciation with geographic isolation: allopatric speciation   Suppose one population of squirrels is split into two separate populations because the local climate grows wetter and a river forms that splits the habitat in two. Because the squirrels cannot cross the river, the populations on either side are reproductively isolated from each other. Over time, the two populations have different evolutionary paths as they accumulate different mutations and adapt to particular features of their separate habitats, which may differ. Eventually, the two populations might genetically diverge so much that even if the river separating them disappeared and the populations came back into contact, squirrels from the two groups could no longer interbreed. In fact, two species of antelope ground squirrels formed on the north and south rims of the Grand Canyon as a result of this type of speciation. Speciation that occurs as a result of geographic isolation is known as **allopatric speciation** (FIGURE 10-11).

ALLOPATRIC SPECIATION

Allopatric speciation occurs when a geographic barrier causes one group of individuals in a population to be reproductively isolated from another group.

1 INITIAL POPULATION

2 REPRODUCTIVE ISOLATION
Suppose a river forms through the squirrels' habitat, separating the population. Because the squirrels cannot cross the river, the two groups are now reproductively isolated.

3 GENETIC DIVERGENCE
Over time, the populations on either side diverge enough genetically that they are no longer able to interbreed.

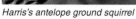
Harris's antelope ground squirrel

White-tailed antelope ground squirrel

FIGURE 10-11 Geographic isolation can result in genetic divergence. Two species of antelope ground squirrel that have evolved on the north and south rims of the Grand Canyon.

Another example of allopatric speciation is seen in the various finch species of the Galápagos Islands, the same finches that Darwin observed and collected while on his 'round-the-world trip on HMS *Beagle*. Individual finches from the nearest mainland, now Ecuador, originally colonized one or more of the islands. (And later, additional islands may have been colonized by birds from the mainland or from previously colonized islands.) But because the islands are far apart, the finches tended not to travel between them, and the populations remained reproductively isolated from one another. Consequently, 14 different finch species have evolved in the Galápagos Islands, each species adapting to the predominant food source on its particular island and now having features that allow it to specialize in eating certain of the wide range of insects, buds, and seeds found on the islands (FIGURE 10-12). Only one species of finch is found in mainland Ecuador.

The barrier doesn't have to be an expanse of water. A forming glacier could split a population into two or more isolated populations. Or a drop in the water level of a lake might expose strips of land that divide the lake into detached, smaller bodies of water, separating one large population of fish into two distinct populations. In each case, the result is the same: geographic isolation that enforces reproductive isolation.

Researchers can easily create new species in the lab, using an analogous strategy. In one experiment, a single population of fruit flies (*Drosophila pseudoobscura*) was divided into two. The two populations were then maintained on different diets—one on the sugar maltose and the other on a starch-based food. After only eight generations of enforced reproductive isolation and adaptation to their differing nutritional environments, the populations had diverged sufficiently to form separate species; when the populations were mixed, the fruit flies from one population would no longer interbreed with flies from the other population.

Q Could you create a new species in the laboratory? How?

Speciation without geographic isolation: sympatric speciation Speciation can also occur among populations that overlap geographically. This is called **sympatric speciation.** Among vertebrates, it is rare for populations of the same animal to become reproductively isolated when they coexist in the same area, so this method of speciation is relatively uncommon. But it *is* common among plants, and it occurs in one of two ways.

During cell division in plants (both in reproductive cells and in other cells of the plant body), an error sometimes occurs in which the chromosomes are duplicated but a cell does not

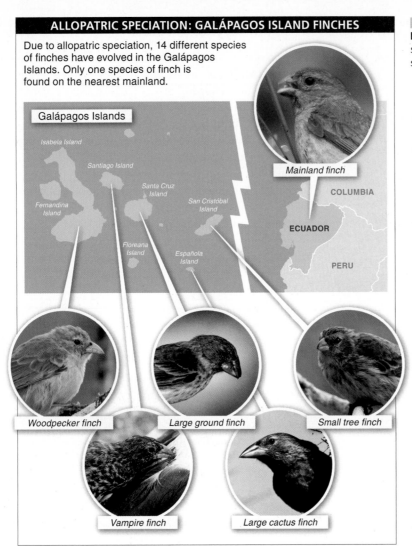

ALLOPATRIC SPECIATION: GALÁPAGOS ISLAND FINCHES

Due to allopatric speciation, 14 different species of finches have evolved in the Galápagos Islands. Only one species of finch is found on the nearest mainland.

Galápagos Islands

Isabela Island

Santiago Island

Santa Cruz Island

Fernandina Island

San Cristóbal Island

Floreana Island

Española Island

Mainland finch

COLUMBIA

ECUADOR

PERU

Woodpecker finch

Large ground finch

Small tree finch

Vampire finch

Large cactus finch

FIGURE 10-12 **Five of the 14 Galápagos Island finch species.** Each species of finch specializes in eating the insects, buds, and/or seeds found in its island habitat.

divide. This creates a new cell that can then grow into an individual with twice as many sets of chromosomes as the parent from which it came. The new individual may have four sets of chromosomes, for example, while the original individual had two. This doubling of the number of sets of chromosomes is called **polyploidy** (**FIGURE 10-13**).

The individual with four sets of chromosomes can no longer interbreed with individuals having only two sets, because their offspring would have three sets (two sets from the parent that had four, and one set from the parent that had two), which could not divide evenly during cell division. The individual with four sets can, however, propagate through self-fertilization or by mating with other individuals that have four sets. As a consequence, the individuals with four sets of chromosomes have achieved instant reproductive isolation from the original population and are therefore considered a new species.

Although rare in animals, speciation by polyploidy has occurred several times among some species of tree frogs.

A much more common method of sympatric speciation occurs when plants of different (but closely related) species interbreed, forming a hybrid. The hybrid may not be able to interbreed with either of the parental species, but it may be able to propagate asexually—as many plants can. And meiosis often is disrupted in hybrids, causing a chromosome doubling (polyploidy). When this occurs, the hybrids with a doubled number of chromosome sets can interbreed with each other (see Figure 10-13). This method of speciation has led to the production of a large number of important crop plants, including wheat, bananas, potatoes, and coffee.

Whether populations are separated from each other allopatrically or sympatrically, speciation is not considered complete until sufficient differences have evolved in the two

SYMPATRIC SPECIATION

Sympatric speciation results in the reproductive isolation of populations that coexist in the same area. Two scenarios lead to this method of speciation.

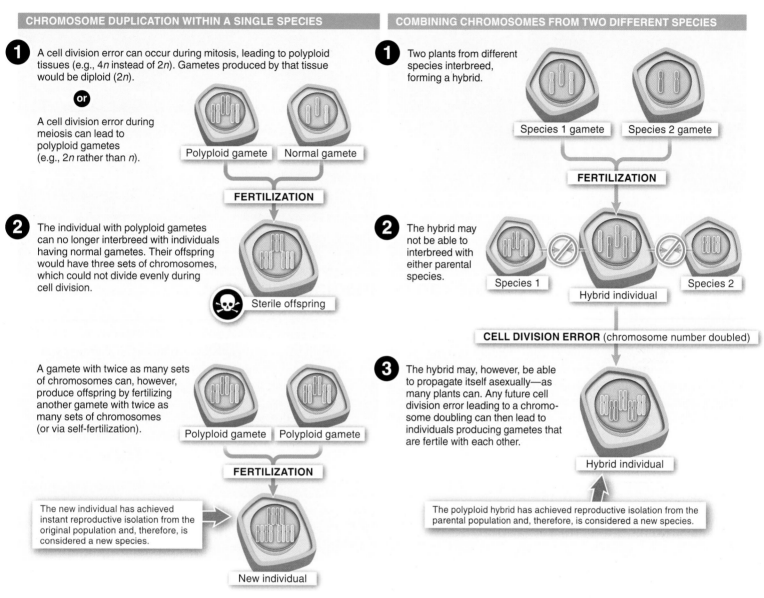

CHROMOSOME DUPLICATION WITHIN A SINGLE SPECIES

1 A cell division error can occur during mitosis, leading to polyploid tissues (e.g., 4*n* instead of 2*n*). Gametes produced by that tissue would be diploid (2*n*).

or

A cell division error during meiosis can lead to polyploid gametes (e.g., 2*n* rather than *n*).

Polyploid gamete   Normal gamete

FERTILIZATION

2 The individual with polyploid gametes can no longer interbreed with individuals having normal gametes. Their offspring would have three sets of chromosomes, which could not divide evenly during cell division.

Sterile offspring

A gamete with twice as many sets of chromosomes can, however, produce offspring by fertilizing another gamete with twice as many sets of chromosomes (or via self-fertilization).

Polyploid gamete   Polyploid gamete

FERTILIZATION

The new individual has achieved instant reproductive isolation from the original population and, therefore, is considered a new species.

New individual

COMBINING CHROMOSOMES FROM TWO DIFFERENT SPECIES

1 Two plants from different species interbreed, forming a hybrid.

Species 1 gamete   Species 2 gamete

FERTILIZATION

2 The hybrid may not be able to interbreed with either parental species.

Species 1   Hybrid individual   Species 2

CELL DIVISION ERROR (chromosome number doubled)

3 The hybrid may, however, be able to propagate itself asexually—as many plants can. Any future cell division error leading to a chromosome doubling can then lead to individuals producing gametes that are fertile with each other.

Hybrid individual

The polyploid hybrid has achieved reproductive isolation from the parental population and, therefore, is considered a new species.

FIGURE 10-13 **Speciation without geographic isolation.** Plants may genetically diverge through sympatric speciation.

populations that they could no longer interbreed even if they did come in contact.

Separation of two populations of a species can occur relatively quickly—as long as it takes for a new river (or freeway) to divide one large population into two—but the genetic divergence that causes true reproductive isolation can take a very long time, sometimes thousands of years. For this reason, speciation can be difficult to study and observe.

TAKE-HOME MESSAGE 10·7

Speciation is the process by which one species splits into two distinct species that are reproductively isolated. It can occur by polyploidy or by a combination of reproductive isolation and genetic divergence.

10·8–10·10
Evolutionary trees help us conceptualize and categorize biodiversity.

The diversity of life on earth can be thought of as branching like a tree.

10·8 The history of life can be imagined as a tree.

Although Carolus Linnaeus's method for naming species was an important step in categorizing and cataloguing earth's biodiversity, his underlying assumption about where species came from was wrong. He believed, as did all biologists of his day, that all species had been created at the same time and were unchanging. For this reason, his classification of organisms into hierarchical groups was based on nothing more than his own evaluation of how physically similar various organisms appeared to be.

A hundred years later, however, Charles Darwin proposed and documented that species could, in fact, change and give

HOW TO READ AN EVOLUTIONARY TREE

Evolutionary trees reveal the evolutionary history of species and the sequence of speciation events that gave rise to them.

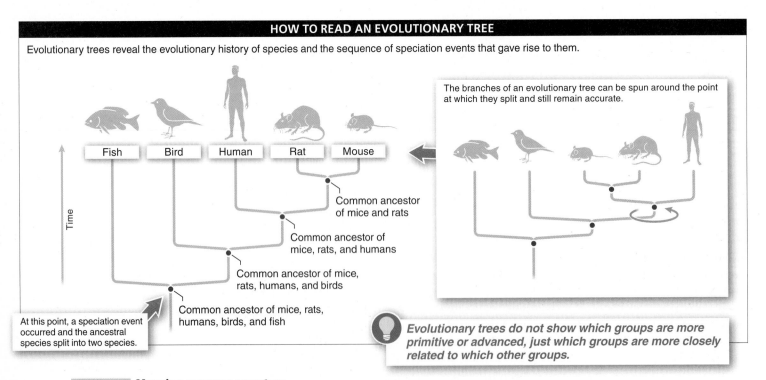

The branches of an evolutionary tree can be spun around the point at which they split and still remain accurate.

Fish | Bird | Human | Rat | Mouse

Time

Common ancestor of mice and rats

Common ancestor of mice, rats, and humans

Common ancestor of mice, rats, humans, and birds

Common ancestor of mice, rats, humans, birds, and fish

At this point, a speciation event occurred and the ancestral species split into two species.

Evolutionary trees do not show which groups are more primitive or advanced, just which groups are more closely related to which other groups.

FIGURE 10-14 Mapping common ancestors.

rise to new species. With Darwin, the classification of species acquired a new goal and a more important function. In *The Origin of Species* Darwin wrote: "Our classifications will come to be, as far as they can be so made, genealogies." That is, Darwin proposed that the classifications of organisms would resemble family trees that link generations of parents and offspring over long periods of time. With these words, Darwin was the first to link classification with evolution.

Linnaean classification involved placing organisms within groups as a function of their apparent similarity with each other. The modern incarnation of Darwin's vision of classification is called **systematics** and has the broader goal of reconstructing the **phylogeny,** or evolutionary history, of organisms. That is, through systematics, all species—even extinct species—are named and arranged in a manner that indicates the common ancestors they share and the points at which the species diverged. A complete phylogeny of all organisms is like a family tree for all species, past and present.

A phylogenetic tree not only shows the relationships among organisms but also presents a hypothesis about evolutionary history (**FIGURE 10-14**). At the beginning of life on earth there was the first living organism, one that could replicate itself. Then a **speciation event** occurred, after which the population of the first living organism split into independent evolutionary lineages. The phylogenetic tree had its first branch, and there was biodiversity. Over

hundreds of millions of years, speciation events continued to occur, and today the tree has branches with millions (or possibly tens of millions) of tips that represent all the species on earth. Moving up a branch of any evolutionary tree from its trunk toward its tips, we can see when groups split, with each branching point representing a speciation event. And it doesn't stop here—speciation continues to add new branches all the time.

It is important to remember that phylogenies are hypotheses and, like any hypotheses, they are subject to revision and modification. In the next section, we explore how phylogenetic trees are constructed and look at the rich information they convey about the history of life on earth. We also investigate the modern methods of molecular systematics and see how, with new molecular and DNA evidence, biologists are revising many earlier phylogenies that were based on organisms' physical features.

TAKE-HOME MESSAGE 10·8

The history of life can be visualized as a tree; by tracing from the branches back toward the trunk, we can follow the pathway back from descendants to their ancestors. The tree reveals the evolutionary history of all species and the sequence of speciation events that gave rise to them.

10·9 Evolutionary trees show ancestor-descendant relationships.

Many human traditions have a sacred "Tree of Life." The bodhi tree that the Buddha sat beneath to achieve the ultimate truth is the Tree of Enlightenment; the apple tree in the Garden of Eden is the Tree of Knowledge. The history of the relationship between all organisms that constitute life on earth is another tree of life.

The trunk and branches of an evolutionary tree represent ancestor-descendant relationships that link living organisms with all life that has ever existed on earth. The evolutionary tree of life can be thought of as one giant tree, but as a practical matter, biologists often study only particular branches. These branches can be illustrated as big trees or little trees and can be expanded to include whatever organisms the biologist is studying: the tree might, for example, include all animals or just the rodents.

In evolutionary trees, it does not matter on which side of the tree you put a particular group of organisms. Any branches can be spun around the **nodes** at which they split (see Figure 10-14). This pinwheel effect means that you cannot assume that rats are more evolutionarily advanced than mice, or vice versa. They are equally advanced in the sense that both groups derived from the same speciation event.

Evolutionary trees tell us many things, but one thing they do not tell us is which groups are most "primitive" and which are most "advanced." This property of phylogenetic trees can serve to undermine some of our most sacred beliefs, such as the

Q Are humans more advanced, evolutionarily, than cockroaches? Can bacteria be considered "lower" organisms?

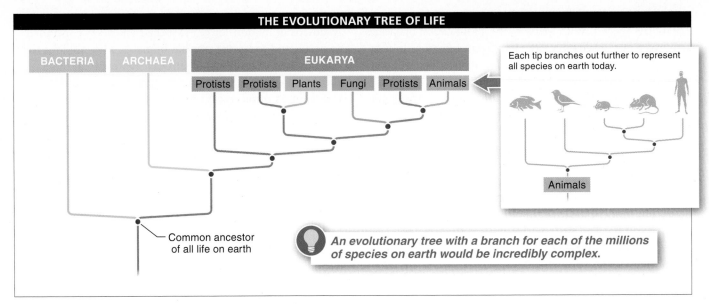

THE EVOLUTIONARY TREE OF LIFE

BACTERIA   ARCHAEA   EUKARYA

Protists   Protists   Plants   Fungi   Protists   Animals

Each tip branches out further to represent all species on earth today.

Animals

Common ancestor of all life on earth

An evolutionary tree with a branch for each of the millions of species on earth would be incredibly complex.

FIGURE 10-15 A growing tree. The evolutionary tree of life has branches with millions of tips representing all species on earth. As speciation events occur, new branches are added to the tree.

one that humans are the pinnacle of evolution. Many trees can be drawn to support this idea, including Figure 10-14. But notice that if you rotate this tree around any one of its nodes, you can get a number of different trees.

What trees do tell us is which groups are most closely related to which other groups. One of the most interesting revelations of "tree thinking" is that—despite appearances—fungi, such as mushrooms, yeasts, and molds, are more

closely related to animals than to plants (**FIGURE 10-15**). We investigate this surprising fact in Chapter 11.

Biologists use the term **monophyletic** to describe any group in which all of the individuals are more closely related to each other than to any individuals outside that group. Monophyletic groups are determined by looking at the nodes of the trees. For example, birds and crocodiles, taken together, compose a monophyletic group because they share a more recent common ancestor (designated by node A in **FIGURE 10-16**) than either group shares with lizards or mammals. Lizards and crocodiles, taken together, do not compose a monophyletic group, because their common ancestor (at node B) is also shared by birds. But birds, crocodiles, and lizards, all taken together, *do* compose a monophyletic group, by virtue of all three sharing the common ancestor at node B. Similarly, birds, crocodiles, lizards, and mammals also compose a monophyletic group. Notice also that in Figure 10-15, the **protists** (or "Protista"), a kingdom of mostly single-celled eukaryotes, are not actually a monophyletic group, contrary to original expectations.

Constructing evolutionary trees by comparing similarities and differences among organisms

Reading an evolutionary tree reveals which groups are most closely related and approximately how long ago they shared a common ancestor. But how are evolutionary trees—which might hypothesize historical events, such as the cat-dog split, that happened 60 million years ago—constructed in the first place?

Until recently, these trees were assembled by looking carefully at numerous physical features of species and generating tables

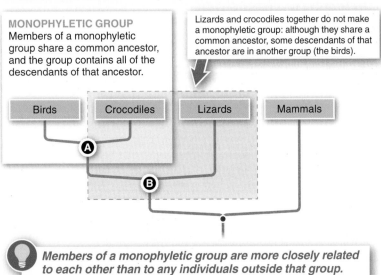

MONOPHYLETIC GROUPS

MONOPHYLETIC GROUP
Members of a monophyletic group share a common ancestor, and the group contains all of the descendants of that ancestor.

Lizards and crocodiles together do not make a monophyletic group: although they share a common ancestor, some descendants of that ancestor are in another group (the birds).

Birds   Crocodiles   Lizards   Mammals

A

B

Members of a monophyletic group are more closely related to each other than to any individuals outside that group.

FIGURE 10-16 Similar appearance doesn't necessarily mean groups are monophyletic.

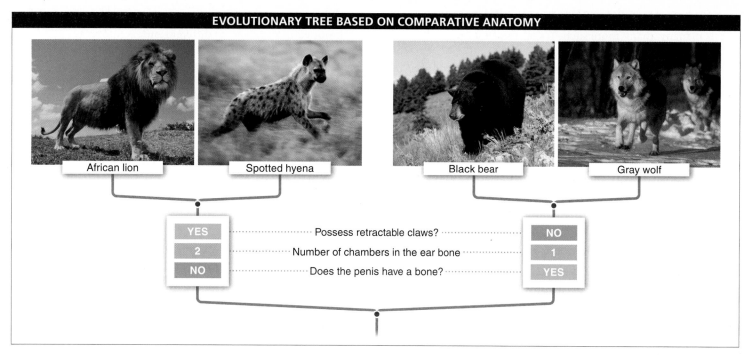

EVOLUTIONARY TREE BASED ON COMPARATIVE ANATOMY

African lion | Spotted hyena | Black bear | Gray wolf

YES	NO
2	1
NO	YES

Possess retractable claws?

Number of chambers in the ear bone

Does the penis have a bone?

FIGURE 10-17 **Looking for clues in body structures.** Before DNA sequencing became available, physical features of species were used to determine evolutionary relatedness.

that compared these features across the species. **FIGURE 10-17** is a simple example of such a table, showing a clear split between the characteristics of the lion and the hyena, on the one hand, and those of the wolf and the bear, on the other. For most of the 20th century, biologists classifying organisms would often use 50 or more traits to generate a tree.

> ❝We all should know that diversity makes for a rich tapestry, and we must understand that all the threads of the tapestry are equal in value no matter what their color.❞
>
> — MAYA ANGELOU, in *Wouldn't Take Nothing for My Journey Now*, 1993

Then, beginning in the 1980s, biologists began using molecular sequences rather than physical traits to generate evolutionary trees. The rationale for this approach is that organisms inherit DNA from their ancestors and so, as species diverge, their DNA sequences also diverge, becoming increasingly different. As more time passes following the splitting of one species into two, the differences in their DNA sequences become greater. By comparing how similar the DNA sequences are between two groups, it is possible to estimate how long ago they shared a common ancestor (**FIGURE 10-18**).

In theory, the construction of evolutionary trees using DNA sequences is not really different from using physical traits.

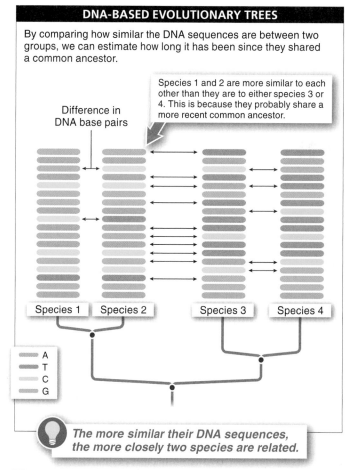

DNA-BASED EVOLUTIONARY TREES

By comparing how similar the DNA sequences are between two groups, we can estimate how long it has been since they shared a common ancestor.

Species 1 and 2 are more similar to each other than they are to either species 3 or 4. This is because they probably share a more recent common ancestor.

Difference in DNA base pairs

A
T
C
G

Species 1 | Species 2 | Species 3 | Species 4

The more similar their DNA sequences, the more closely two species are related.

FIGURE 10-18 **DNA sequences reveal evolutionary relatedness.**

Each DNA base pair can be thought of as a "trait," and many, many more such "traits" can be compared.

Although most evolutionary trees are now produced by using molecular sequence comparisons, when it comes to actually naming species we still use Linnaeus's system, assigning a species name and fitting the new species within the other categories such as kingdom, class, and family. This can be difficult, because there is no objective way to assign groups above the level of species—for example, deciding whether multiple genera should be grouped in the same or different families, or whether multiple families should be grouped in the same or different orders.

TAKE-HOME MESSAGE 10·9

Evolutionary trees constructed by biologists are hypotheses about the ancestor-descendant relationships among species. The trees represent an attempt to describe which groups are most closely related to which other groups.

10·10 Similar structures don't always reveal common ancestry.

Whales and sharks have fins; bats and many insects have wings. Do bats have wings because they are most closely related to insects and inherited their wings from a common, winged ancestor, or did insects' and bats' wings evolve independently? As the evolutionary tree in **FIGURE 10-19** makes clear, insects' wings and bats' wings evolved independently—an adaptation that arose separately on more than one occasion.

The mapping of species' characteristics onto phylogenetic trees provides us with the story of evolution. Before the 1980s, biologists had tried to decode the process of evolution by comparing physical features of organisms. Then, with the advent of methods for comparing DNA sequences, tracing the history of life on earth became a more rigorous science. Let's look at a case that illuminates why the original methods were weaker (**FIGURE 10-20**).

Initially, biologists thought that African golden moles belonged in the order known as insectivores, which includes shrews, hedgehogs, and other moles. This belief seemed reasonable because the moles have many characteristics in common with these other animals: they are small, they have long, narrow snouts, their eyes are tiny, and they live in underground burrows. Biologists thought that this group of characteristics evolved just once, and that all species in the insectivore order possessed these characteristics because they inherited them from a common ancestor.

Recent evidence from DNA sequencing, however, surprised everyone who studies these animals. The DNA evidence reveals that African golden moles are more closely related to elephants than to the insectivores, including all of the other mole species! Why do they look so similar to the insectivores? Because of a phenomenon called **convergent evolution,**

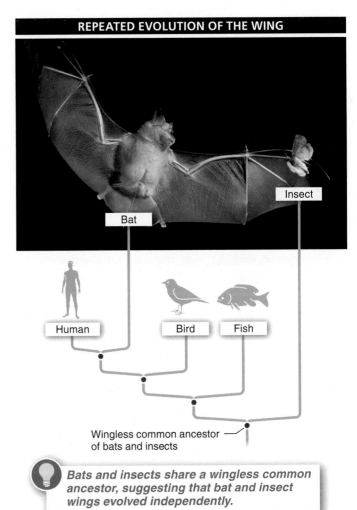

REPEATED EVOLUTION OF THE WING

Insect

Bat

Human   Bird   Fish

Wingless common ancestor of bats and insects

Bats and insects share a wingless common ancestor, suggesting that bat and insect wings evolved independently.

FIGURE 10-19 **Separate paths to flight.** Because wings are an adaptation that has evolved independently many times in populations living in similar environments under similar selective pressures, the presence of wings does not necessarily indicate evolutionary relatedness.

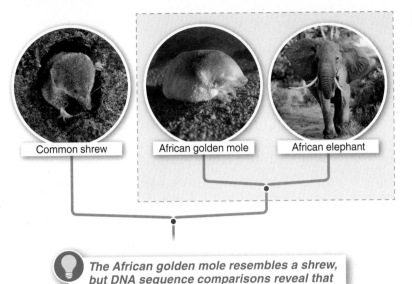

Common shrew

African golden mole

African elephant

💡 *The African golden mole resembles a shrew, but DNA sequence comparisons reveal that it is more closely related to an elephant.*

Q Humans have several fused vertebrae at the bottom of their spinal cord that look like a tiny, internal "tail." Are we evolving a tail or losing a tail?

which occurs when populations of different organisms live in similar environments and so experience similar selective forces. **Analogous traits** are characteristics (such as bat wings and insect wings) that are similar because they were produced by convergent evolution, not because they descended from a common structure in a shared ancestor. Features that are inherited from a common ancestor are called **homologous features.** All mammals have hair, for example, because they inherited this trait from a common ancestor.

Analogous features are problematic when biologists are constructing evolutionary trees, because they are the result of natural selection rather than relatedness. As a consequence, these traits should not be considered a sign of common ancestry when constructing an evolutionary tree. But how do we know whether traits are homologous or analogous? Through DNA analysis. DNA sequences do not become similar during convergent evolution, so molecular phylogenies cannot be fooled. This is an important reason why molecular-based evolutionary trees are preferable to trees based on physical features.

Although analogous traits and the process of convergent evolution are not helpful in constructing evolutionary trees, that doesn't mean they are useless to biologists. Far from it. Convergence provides some of the best evidence for the power of natural selection. For example, from the 19th to the 20th centuries in industrialized parts of Europe, a large number of distinct butterfly and moth species became darker and darker. This change was not the result of their sharing a common ancestor, but instead came about because industrialization had caused an accumulation of dark soot on the bark of many trees. This soot accumulation caused light-colored moths and butterflies to stand out against the background when they landed on trees, making them an easier target for predators than dark-colored individuals. As a consequence, natural selection led to a rapid evolution of body color, with many species converging on a darker coloration at the same time.

TAKE-HOME MESSAGE 10·10

Evolutionary trees are best constructed by comparing organisms' DNA sequences rather than physical similarities, because convergent evolution can cause distantly related organisms to appear closely related, but it doesn't increase their DNA sequence similarity.

More than 400 Anolis lizard species exist. (Shown here: A blue-throated anole, Anolis chrysolepsis, from Ecuador.)

10·11   Macroevolution is evolution above the species level.

When water runs over rocks, it wears them away. The process is simple and slow, yet powerful enough to have created the Grand Canyon. To be sure, water running over rocks does not always make a Grand Canyon. Nonetheless, no additional physical processes are necessary. The process of evolution has a lot in common with a stream of water running over rocks: in the short term, it produces small changes in a population, yet the accumulation of these changes over the long term can be "canyonesque" (**FIGURE 10-21**). Let's consider some examples.

The production of 200-ton dinosaurs from rabbit-size reptile ancestors. The diversification from a single species of flowering plant into more than 230,000 species. These large-scale examples, which are the products of evolutionary change involving the origins of entirely new groups of organisms, are referred to as **macroevolution.** These examples can be contrasted with phenomena that involve changes in allele frequencies in a population—referred to as **microevolution**—such as the increase in milk production in cows during the first half of the 20th century, or the gradual change in the average beak size of birds with changing patterns of rainfall (**FIGURE 10-22**).

No additional processes necessary! Just as a stream of water running over rocks can create the Grand Canyon, accumulated microevolutionary changes can lead to dramatic macroevolutionary changes.

FIGURE 10-21 **From a trickle of water to the Grand Canyon.**

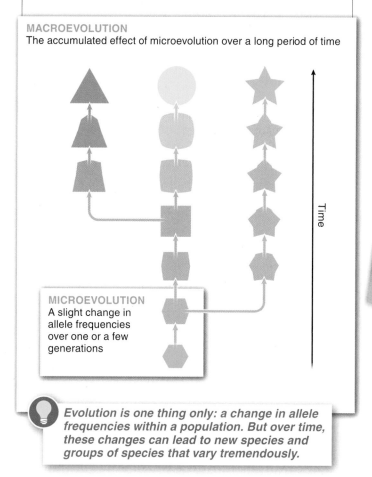

EVOLUTION: MICRO vs. MACRO

Microevolutionary changes in allele frequencies in a population over time can lead to macroevolution, changes on a grand scale, including vast diversification of species.

MACROEVOLUTION
The accumulated effect of microevolution over a long period of time

Time

MICROEVOLUTION
A slight change in allele frequencies over one or a few generations

Evolution is one thing only: a change in allele frequencies within a population. But over time, these changes can lead to new species and groups of species that vary tremendously.

FIGURE 10-22 The scale of evolution.

These micro and macro events might seem like two very different processes, but they are not. Evolution, whether at the micro or macro level, is one thing only: a change in allele frequencies over time. In the short term, over one or a few generations, evolution can appear as a slight and gradual change within a species. But over a longer period of time, the accumulated effects, acting continuously and combined with reproductive isolation of populations, can lead to the dramatic phenomena described as macroevolution. Just as a trickle of water, given enough time, produced the Grand Canyon, so evolution has created the endless forms of diversity on earth. In a sense, microevolution is the process and macroevolution is the result.

TAKE-HOME MESSAGE 10·11

The process of evolution—changes in allele frequencies within a population—in conjunction with reproductive isolation is sufficient to produce speciation and the rich diversity of life on earth.

10·12 The pace of evolution is not constant.

If you listened in on some scientists debating their research, you might hear one side describing "evolution by jerks" and the other side making light of "evolution by creeps." They're not gearing up for a fight out in the schoolyard. They're just debating the pace of evolution.

The traditional model of how evolutionary change occurred was that populations changed slowly but surely, gradually accumulating sufficient genetic differences for speciation—hence the phrase "evolution by creeps." Spurred on by findings from the fossil record that do not always support this view, however, researchers have come to believe that evolution may often occur in a different way: brief periods of

rapid evolutionary change immediately after speciation, followed by long periods with relatively little change—hence "evolution by jerks." This newer view of the pace of evolution, in which long periods of relatively little evolutionary change are punctuated by bursts of rapid change, is called **punctuated equilibrium** (FIGURE 10-23).

In nature, we can find examples of both gradual change and the irregular pattern of punctuated equilibrium. For many groups of organisms, such as many mollusks and mammals, we see in the fossil record a long period in which particular species seem to go through very little change, followed by the appearance of a large number of newer, but clearly related,

THE TEMPO OF EVOLUTION

The pace of evolution varies for different species. Some species have evolved gradually over time, while others spend vast amounts of time with little change.

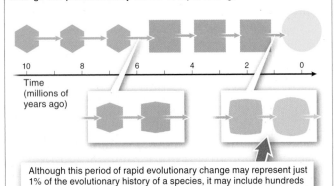

GRADUAL CHANGE
Evolution by creeps: The pace of evolution occurs gradually in incremental steps.

10   8   6   4   2   0
Time (millions of years ago)

PUNCTUATED EQUILIBRIUM
Evolution by jerks: Long periods of relatively little evolutionary change are punctuated by bursts of rapid change.

10   8   6   4   2   0
Time (millions of years ago)

Although this period of rapid evolutionary change may represent just 1% of the evolutionary history of a species, it may include hundreds or thousands of generations—over thousands of years in the case of primates, or months for bacteria.

FIGURE 10-23 **The pace of evolution can be rapid or slow.**

species, without seeing any fossils representing the gradual transition between the newer and older fossils. For other groups, such transitional fossils do exist. For example, very complete sets of fossils reveal the transitional sequence from fox-like ancient terrestrial mammals to modern whales.

In the end, there is no single rate at which evolution occurs across all species. Some species have spent vast periods of time with little change at all. The coelacanth species of fish, for example, seems to have undergone almost no physical changes over more than 300 million years. Other species have clearly changed gradually, over time, for their entire evolutionary history. For any organism, the rate of evolutionary change depends on the selective forces acting on the population. Strong and directional forces may act over long periods of time and lead to rapid change, while stabilizing selection may lead to very little change.

Darwin foreshadowed the discussion about the pace of evolution in *The Origin of Species,* writing that "the periods during which species have undergone modification, though long as measured in years, have probably been short in comparison with the periods during which they retain the same form."

TAKE-HOME MESSAGE 10·12

The pace at which evolution occurs can be rapid or very slow. In some cases, the fossil record reveals long periods with little evolutionary change punctuated by rapid periods of change. In other cases, species may change at a more gradual but consistent pace.

10·13 Adaptive radiations are times of extreme diversification.

Even the greatest of success stories often owe something to a bit of luck. As mammals, it might seem that we owe a little of our success to luck. Flash back to 65 million years ago. Our mammalian ancestors had their place among the organisms on earth, but these rodent-size, insect-eating, nocturnal creatures didn't remotely possess the dominant position we hold today. It was the dinosaurs' time, and the giant reptiles dominated the earth.

All that changed in an instant when the earth was struck by an asteroid, about 6 miles (10 km) in diameter, near what is now the eastern part of Mexico. This caused the

environmental conditions on earth to change quickly and drastically. We'll explore the details of this catastrophic event in the next section, but one outcome was extreme: almost all of the dinosaur species were wiped out. Our mammalian ancestors, though, survived and found themselves living on a planet where most of their competitors had suddenly disappeared. We were in the right place at the right time.

What followed was an explosive expansion of mammalian species. In a brief period of time, a small number of species diversified into a much larger number of species, able to live in a wide diversity of habitats. Called an **adaptive radiation,** such a large and rapid diversification has occurred many times throughout history (**FIGURE 10-24**).

Three different phenomena tend to trigger adaptive radiations. After one of these events, surviving species find themselves in locations where they suddenly have access to plentiful new resources.

1. Mass extinction events. With the near-total disappearance of the dinosaurs, a world of "opportunities" opened up for the mammals. Where previously the dinosaurs had prevented mammals from utilizing resources, mammals suddenly had few competitors. Not surprisingly, the number of mammalian species increased from perhaps just a few hundred to more than 4,000 species in about 130 genera. This happened over about 10 million years, barely the blink of an eye by geological standards. Following other large-scale extinctions, numerous other groups that suddenly lost most of their competitors experienced similar adaptive radiations.

It is interesting to note that although, from time to time, sudden and extreme events lead to mass extinctions that wipe out a large proportion of the species on earth, in every case these mass extinctions are followed by a time of explosive speciation in the groups that survive

2. Colonization events. In a rare event, one or a few birds or small insects will fly off from a mainland and end up on a distant island group, such as Hawaii or the Galápagos Islands. Once there, they tend to find a large number of opportunities for adaptation and diversification. In the Galápagos, as we learned in Section 10-7, 14 finch species evolved from a single species found on the nearest mainland, 600 miles away. In Hawaii, there are several hundred species of fruit flies, all believed to have evolved from one species that colonized the islands—perhaps blown there by a storm, or carried there stuck in the feathers of a bird—and experienced an adaptive radiation.

3. Evolutionary innovations. In the world of computers, software developers are always looking for the "killer app"— the new application so useful that it immediately leads to huge success, opening up a large new niche in the software market or greatly expanding an already-existing niche. The first spreadsheet, email program, and web browser were all killer apps. In nature, evolution sometimes produces killer apps, too. These are innovations such as the wings and rigid

ADAPTIVE RADIATION

Adaptive radiation occurs when a small number of species diversify into a larger number of species. Three phenomena tend to trigger adaptive radiation.

MASS EXTINCTION EVENTS
With their competition suddenly eliminated, remaining species can rapidly diversify.

COLONIZATION EVENTS
Moving to a new location with new resources (and possibly fewer competitors), colonizers can rapidly diversify.

EVOLUTIONARY INNOVATIONS
With the evolution of an innovative feature that increases fitness, a species can rapidly diversify.

FIGURE 10-24 A rapid diversification of species.

outer skeleton that appeared in insects and helped them diversify into the most successful group of animals, with more than 800,000 species today (more than a hundred times the number of mammalian species). The flower is another innovation that propelled an explosion of diversity and ensured the evolutionary success of flowering plants relative to the non-flowering plants, such as ferns and pine trees. Today, about 9 out of 10 plant species are flowering plants.

TAKE-HOME MESSAGE 10·13

Adaptive radiations—brief periods of time during which a small number of species diversify into a much larger number of species—tend to be triggered by mass extinctions of potentially competing species, colonizations of new habitats, or the appearance of evolutionary innovations.

There have been several mass extinctions on earth.

> *Forests keep disappearing, rivers dry up, wild life's become extinct, the climate's ruined and the land grows poorer and uglier every day.*
> — ANTON CHEKHOV, *Uncle Vanya*, 1899

If the past is a guide to the future, we know this: no species lasts forever. Speciation is always producing new species, but **extinction,** the complete loss of all individuals in a species population, takes them away. Extinction is always occurring and is faced by all species.

For any given time in earth's history, it is possible to estimate the rate of extinctions. This rate can be expressed in several difference ways, such as the number of species that go extinct over a given period of time. And the evidence reveals that these rates are far from constant (**FIGURE 10-25**). Although the details differ in most cases, extinctions generally fall into one of two categories: "background" extinctions or mass extinctions.

While extinctions are occurring all the time, **background extinctions** are those that occur at lower rates, during periods other than times of mass extinctions. Background extinctions occur mostly as the result of natural selection. Competition with other species, for example, may reduce a species' size or the range over which it can roam or grow. Or a species might be too slow to adapt to gradually changing environmental conditions and becomes extinct as its individuals die off.

Mass extinctions are periods during which a large number of species on earth become extinct over a relatively short period. There have been at least five mass extinctions on earth and, during each of these extinctions, 50% or more of the animal species living at the time became extinct.

There is a fundamental difference between background and mass extinctions that goes beyond differences in rates. They have different causes. Mass extinctions are due to extraordinary and sudden changes to the environment (such as an asteroid impact). As a consequence, nothing more than bad luck is responsible for the extinction of species during mass extinctions; fit and unfit individuals alike perish.

Of the five mass extinctions during the past 500 million years, the most recent is also the best understood: the asteroid crash, described in the previous section, that cleared the way

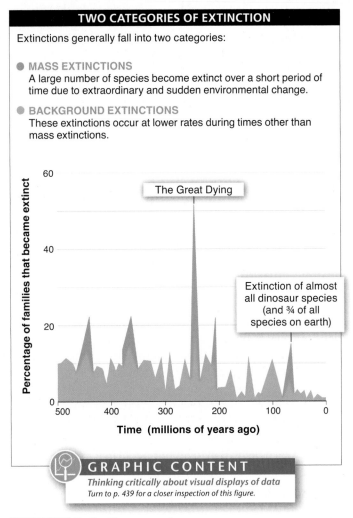

TWO CATEGORIES OF EXTINCTION

Extinctions generally fall into two categories:

● **MASS EXTINCTIONS**
A large number of species become extinct over a short period of time due to extraordinary and sudden environmental change.

● **BACKGROUND EXTINCTIONS**
These extinctions occur at lower rates during times other than mass extinctions.

The Great Dying

Extinction of almost all dinosaur species (and ¾ of all species on earth)

Percentage of families that became extinct

Time (millions of years ago)

GRAPHIC CONTENT
Thinking critically about visual displays of data
Turn to p. 439 for a closer inspection of this figure.

FIGURE 10-25 **Extinction never sleeps.**

for our mammalian ancestors to diversify. This massive asteroid smashed into the Caribbean near the Yucatán Peninsula of Mexico 65 million years ago (**FIGURE 10-26**). The impact left a crater more than 100 miles wide and probably created an enormous fireball that caused fires worldwide, followed by a cloud of dust and debris that blocked all sunlight from the earth and disturbed the global climate for months. In the aftermath of this catastrophe, about 75% of all species on earth were wiped out, including almost all dinosaurs, a tremendously successful group that had been thriving for 150 million years. (Among the rare surviving dinosaurs were those that would become the birds.)

As bad as the dinosaur-devastating asteroid event was, from a biodiversity perspective it is not the worst catastrophe in

Global climate change! Worldwide mass extinction! Catastrophic events set the stage for explosive speciation of the remaining groups, including the mammals.

FIGURE 10-26 **Crashing into Mexico's Yucatán Peninsula.** Sixty-five million years ago, an asteroid 6 miles in diameter wiped out three-fourths of all species on earth, including almost all the dinosaurs.

earth's history. That distinction falls to a mass extinction that took place 250 million years ago, called the Great Dying (see Figure 10-25). Although the cause is not clear—hypotheses include an asteroid impact, continental drift, a supernova (a star explosion), or extreme volcanic eruptions—more than 95% of all marine life became extinct, along with almost 75% of all terrestrial vertebrates. The causes of the three other mass extinctions also are poorly understood, but the evidence in the fossil record of the tremendous upheavals is clear.

An important question now being debated is whether we are currently in the midst of a human-caused mass extinction event. We'll explore whether this is true—and why—in Chapter 16.

TAKE-HOME MESSAGE 10·14

As new species are being created, others are lost through extinction, which may be a consequence of natural selection or large, sudden changes in the environment. Mass extinctions are periods during which a large number of species on earth become extinct over a short period of time.

10·15–10·18
An overview of the diversity of life on earth: organisms are divided into three domains.

The flower hat jellyfish (Olindias formosa), a rare species that lives in the oceans off Brazil, Argentina, and southern Japan.

10·15 All living organisms are classified into one of three groups.

Biological diversity can be humbling. At the very least, it makes it harder to believe that humans are particularly special. We are not at the center of earth's "family tree." Nor are we at its peak. We are simply one branch.

When Linnaeus first put together his system of classification, he saw a clear and obvious split: all living organisms were either plant or animal. Plants could not move and could make their own food. Animals could move but could not make their own food. So in Linnaeus's original classification, all organisms were put in either the animal kingdom or the plant kingdom (his third, "mineral" kingdom, now abandoned, included only non-living matter).

With the refinement of microscopes and subsequent discovery of the rich world of **microbes**—microscopic organisms—the two-kingdom system was inadequate. Where did the microbes belong? Some could move, but many of those could also make their own food, seeming to put them somewhere between plants and animals. And the problems didn't stop with the microbes: mushrooms and molds, among other organisms originally categorized as plants, didn't move but they didn't make their own food either—they digested the decaying plant and animal material around them.

The two-kingdom system gave way in the 1960s to a five-kingdom system. At its core,

the new system was a division based on the distinction between *prokaryotic* cells (those without nuclei) and *eukaryotic* cells (those with nuclei). The prokaryotes were put in one kingdom, where the only residents were the bacteria: single-celled organisms with no nucleus, no organelles, and genetic material in the form of a circular strand of DNA. The eukaryotes—having a nucleus, compartmentalized organelles, and individual, linear pieces of DNA—were divided into four separate kingdoms: plants, animals, fungi, and protists.

The classification of organisms took a huge leap forward in the 1970s and 1980s, and the five-kingdom system had to be discarded. Until that point, organisms had been classified primarily based on their appearance. But because the ultimate goal of classification had changed to reconstructing phylogenetic trees that reflected the evolutionary history of earth's diversity, Carl Woese, an American biologist, and his colleagues began classifying organisms by their nucleotide sequences.

Woese assumed that the more similar the genetic sequences were between two species, the more closely related they were, and he built phylogenetic trees accordingly. The only way Woese could compare the evolutionary relatedness of all the organisms present on earth today was by examining one molecule that was

Q How can one molecule support the claim that all living organisms probably evolved from a single common ancestor?

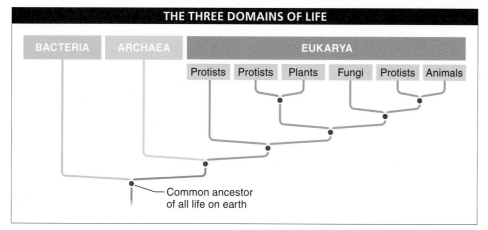

THE THREE DOMAINS OF LIFE

BACTERIA   ARCHAEA   EUKARYA

Protists   Protists   Plants   Fungi   Protists   Animals

Common ancestor
of all life on earth

FIGURE 10-27 All living organisms are classified into one of three groups.

found in *all* living organisms and looking at the degree to which it differed from species to species. He discovered a perfect candidate for this role: a molecule called ribosomal RNA, which helps translate genes into proteins (see Chapter 5). Ribosomal RNA has the same function in all organisms on earth, almost certainly because it comes from a common ancestor. Over time, however, its genetic sequence (i.e., the DNA that codes for it) has changed a bit. Tracking these changes makes it possible to reconstruct the process of diversification and change that has taken place.

The trees that Woese's genetic sequence data generated had some big surprises. First and foremost, the sequences revealed that the biggest division in the diversity of life on earth was not between plants and animals. It wasn't even between prokaryotes and eukaryotes. The new trees revealed instead that the diversity among microbes was much, much greater than ever imagined—particularly because of the discovery of a completely new group of prokaryotes called **archaea** (*sing.* archaeon), which thrive in some of the most extreme environments on earth and differ greatly from bacteria. The tree of life was revised to show three primary branches, called domains: the bacteria, the archaea, and the eukarya (**FIGURE 10-27**). The domain names are often capitalized: Bacteria, Archaea, and Eukarya.

Woese put the domains above the kingdom level in the Linnaean system. In Woese's new system, which is the most widely accepted classification scheme today, the bacteria and archaea domains each have one kingdom and the eukarya domain has four. Because both bacteria and archaea are microscopic, it can be hard to believe that the two domains are as different from each other as either domain is from the eukarya. However, each of the three domains is monophyletic, meaning that each contains species that share a common ancestor and includes all descendants of that ancestor. Close inspection even reveals that the archaea

are more closely related to the eukarya than they are to the bacteria.

The three-domain, six-kingdom approach is not perfect and is still subject to revision. As we saw earlier, for example, within the eukarya, the single-celled protists have turned out to be much more diverse than initially thought and, problematically, are not a monophyletic group. Increasingly, it is recognized that they should be split into multiple kingdoms. Also problematic is that bacteria sometimes engage in **horizontal gene transfer,** which means that, rather than passing genes simply from "parent" to "offspring," they transfer genetic material directly into another species. This process complicates the attempt to determine phylogenies based on sequence data, because it creates situations in which two organisms might have a similar genetic sequence not because they share a common ancestor, but as a result of a direct transfer of the sequence from one species to another.

Additionally, a fourth group of incredibly diverse and important biological entities, the **viruses,** is not even included in the tree of life, because they are not considered to be living organisms. Viruses can replicate, but they can have metabolic activity only by taking over the metabolic processes of another organism. Their lack of metabolic activity puts viruses just outside the definition of life that we use in this book, but some scientists do view viruses as living.

}Q Are viruses alive?

The most commonly accepted tree of life suggests that, after the origin of life, the following sequence of events occurred (**FIGURE 10-28**):

1. The bacteria arose from the first self-replicating, metabolizing cells.

2. There was a split between the bacteria and a line that gave rise to the archaea and eukarya.

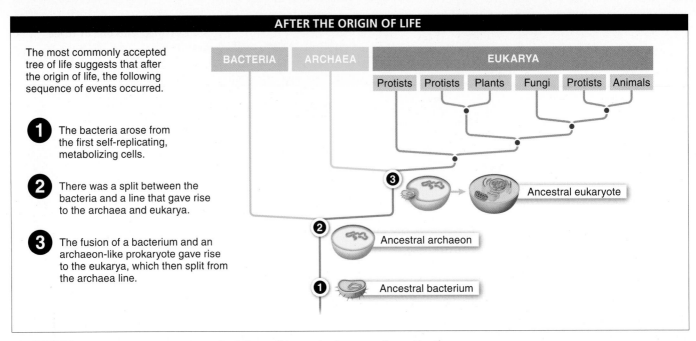

The most commonly accepted tree of life suggests that after the origin of life, the following sequence of events occurred.

BACTERIA **ARCHAEA** **EUKARYA**

Protists Protists Plants Fungi Protists Animals

1 The bacteria arose from the first self-replicating, metabolizing cells.

2 There was a split between the bacteria and a line that gave rise to the archaea and eukarya.

3 The fusion of a bacterium and an archaeon-like prokaryote gave rise to the eukarya, which then split from the archaea line.

Ancestral eukaryote

3

2 Ancestral archaeon

1 Ancestral bacterium

FIGURE 10-28 **From self-replicating, metabolizing cells to complex organisms.** The biggest branches on the tree of life separate the three domains.

3. The fusion of a bacterium and an archaeon-like prokaryote gave rise to the eukarya, which then split from the archaea line.

The next three sections introduce each of the three domains, surveying the broad diversity that has evolved within each domain and the common features that link all the members of each. Chapters 11, 12, and 13 cover the domains in greater detail.

TAKE-HOME MESSAGE 10·15

All life on earth can be divided into three domains—bacteria, archaea, and eukarya—which reflect species' evolutionary relatedness to each other. Plants and animals are just two of the four kingdoms in the eukarya domain, encompassing only a small fraction of the domain's diversity.

10·16 The bacteria domain has tremendous biological diversity.

Morning breath is stinky. And learning the cause of the offensive smell may make the situation even worse. You see, when you wake up, your mouth contains huge amounts of bacterial waste products. Perhaps the only consolation is that it gives us a glimpse into just how diverse and resourceful bacteria are.

Q Why is morning breath so stinky?

At any given time, there are several hundred species of bacteria in your mouth—mostly on your tongue—all competing for the resources you put there (**FIGURE 10-29**). Some of the bacteria are aerobic, requiring oxygen for their metabolism, and others are anaerobic. At night, because the flow of saliva slows down and the oxygen content of your mouth

decreases, the anaerobic bacteria get the upper hand in terms of growth and reproduction. These bacteria metabolize food bits in your mouth, plaque on your teeth and gums, and dead cells from the lining of your mouth, breaking down proteins in these materials to use as their energy source. Because proteins are made from amino acids, some of which contain the smelly chemical sulfur, their breakdown leads to the odor in the accumulating waste products.

Once you wake up, you breathe more and produce more saliva, both of which increase the oxygen level in your mouth. This tips the battle for space and food back in favor of the aerobic bacteria. Because aerobic bacteria prefer carbohydrates as their energy source and because carbohydrates don't contain sulfur, the sulfur smell goes away as the aerobic bacteria start to outcompete the anaerobic bacteria. The aerobic bacteria are, of course, also filling your mouth with waste products, it's just that their waste products don't smell as bad.

On a small scale, your mouth reveals some of the tremendous biological versatility of the bacteria: hundreds of species can live in a tiny area—a teaspoon of soil, for example, is home to more than a billion bacteria—they can thrive in a variety of unexpected habitats, they can utilize a variety of food sources, and they can survive and thrive with or without oxygen. Looking around the world, we find the clear dominance of bacteria. By any measure, this is their planet: the biomass of bacteria (if they were all collected, dried out, and weighed) exceeds that of all the plants and animals on earth (**FIGURE 10-30**). Bacteria live in soil, air, water, arctic ice, and volcanic vents. Many can even make their own food, utilizing light from the sun or harnessing energy from chemicals such as ammonia.

While the various species differ in many ways, the bacteria are a monophyletic group, sharing a common ancestor. For this reason, they all have a few features in common. All bacteria are single-celled organisms with no nucleus or organelles, with one or more circular molecules of DNA as their genetic material, and using several methods of exchanging genetic information. Because they are asexual—they reproduce without a partner, just by dividing—the biological species concept cannot be applied to bacteria when classifying them into narrower categories. As a consequence, bacteria are classified on the basis of physical appearance or, preferably, genetic sequences.

Many people think of bacteria as illness-causing organisms. However, although bacteria are responsible for many diseases, including strep throat, cholera, syphilis,

💡 *Bacteria—shown here on a toothbrush bristle—thrive in a huge range of habitats and have a biomass greater than that of all the plants and animals on earth.*

FIGURE 10-30 We're outnumbered.

pneumonia, botulism, anthrax, leprosy, and tuberculosis, disease-causing bacteria are only a small fraction of the domain, and bacteria seem to get less credit for their many positive effects on our lives. Consider that bacteria (*E. coli*) living in your gut help your body digest the food you eat and, in the process, make certain vitamins your body needs. Other bacteria produce antibiotics such as streptomycin. Still others live symbiotically with plants as small fertilizer factories, converting nitrogen into a form that is usable by the plant. Bacteria also give taste to many foods, from sour cream to cheese, yogurt, and sourdough bread. Increasingly, bacteria are used in biotechnology—from those that can metabolize crude oil and help in the cleanup of spills to transgenic bacteria used in the production of insulin and other medical products, as described in Chapter 5.

We explore the great diversity of the bacteria in more detail in Chapter 13.

FIGURE 10-29 Bacteria thrive on your tongue and in your mouth.

TAKE-HOME MESSAGE 10·16

All bacteria share a common ancestor and have a few features in common. They are prokaryotic, asexual, single-celled organisms with no nucleus or organelles, with one or more circular molecules of DNA as their genetic material, and using several methods of exchanging genetic information. Bacteria have a much broader diversity of metabolic and reproductive abilities than do the eukarya.

In a bubbling hot spring in Yellowstone National Park, the temperature ranges from boiling water (212° F, or 100° C) down to a relatively cool 165° F (74° C) at the surface. It would seem a most inhospitable place for life. Yet researchers have found 38 different species of archaea thriving there. Perhaps even more surprising, the genetic differences among these species are more than double the genetic differences between plants and animals.

In the freezing waters of Antarctica, too, archaea abound. More than a third of the organisms in the Antarctic surface waters are archaea. In swamps, completely devoid of oxygen, and in the extremely salty water of the Dead Sea, the story is the same: where once it was assumed that no life could survive, the archaea not only exist but thrive and diversify (**FIGURE 10-31**).

Q *Will life be found elsewhere in the universe? How does the discovery of archaea alter that likelihood?*

If biologists ever find themselves thinking that we have a handle on the breadth of life that is possible on earth, they would be wise to remember the archaea. Until relatively recently, our perception of life on earth was that there were bacteria and there were eukaryotes, such as the plants and animals. But in the past several decades, as researchers have explored some of the most unlikely of habitats, they have found archaea thriving. There are entire worlds of life on earth that we never even imagined.

Champagne pool, Waiotapu, New Zealand

Geogemma barossii, Strain 121, from a 121° C hydrothermal vent in the Pacific Ocean

FIGURE 10-31 **Archaea can thrive in a diversity of environments, including the most inhospitable-seeming places.**

Analyses of genetic sequences indicate that the archaea and the bacteria diverged about 3 billion years ago and that the eukarya split off from the archaea approximately 2.5 billion years ago. The archaea are grouped in one kingdom within the domain archaea, but we have no idea how many species exist. Given that they are the dominant microbes in the deep seas, it may very well be that archaea—of which we were completely ignorant until recent decades—are the most common organisms on earth. It is still too early to tell.

We do know that, like bacteria, all archaea are single-celled prokaryotes. For that reason, under a microscope they look very similar to bacteria. Several physical features distinguish them from the bacteria, however. Specifically, the archaeal cell walls contain polysaccharides not found in either bacteria or eukaryotes. The archaea also have cell membranes, ribosomes, and some enzymes similar to those found in the eukarya.

The archaea exhibit tremendous diversity and are often divided into five groups based on their physiological features.

1. Thermophiles ("heat lovers"), which live in very hot places

2. Halophiles ("salt lovers"), which live in very salty places

3. High- and low-pH-tolerant archaea

4. High-pressure-tolerant archaea, found as deep as 4,000 meters (about 2.5 miles) below the ocean surface, where the pressure is almost 6,000 pounds per square inch (compared with an air pressure of less than 15 pounds per square inch at sea level)

5. Methanogens, which are anaerobic and produce methane

There also seem to be large numbers of archaea living in relatively moderate environments that are also commonly home to bacteria. We explore the great diversity of the domain archaea in more detail in Chapter 13.

TAKE-HOME MESSAGE 10·17

Archaea, many of which are adapted to life in extreme environments, physically resemble bacteria but are more closely related to eukarya. Because they thrive in many habitats that humans have not yet studied well, including the deepest seas and oceans, they may turn out to be much more common than currently believed.

10·18 The eukarya domain consists of four kingdoms: plants, animals, fungi, and protists.

After exploring the archaea and bacteria domains, turning to the eukarya feels like coming home. We are most familiar with this domain: all of the living organisms that we can see are eukarya. Plants, mushrooms, slime molds, insects, fish, birds, and, of course, mammals, including humans. There are three kingdoms of eukarya that can be seen with the naked eye: plants, animals, and fungi (**FIGURE 10-32**). All are made up from eukaryotic cells—they have a membrane-enclosed nucleus—and each kingdom is almost entirely multicellular.

A fourth kingdom contains the protists, which are often too small to be seen by the naked eye, and is a sort of grab bag that includes a wide range of mostly single-celled eukaryotic organisms, including amoebas, paramecia, and algae. Discovery of new species of protists, as with the other microscopic organisms on earth (the bacteria and archaea), continues at a very high rate. As we noted earlier, biologists now know that the protists are not a monophyletic group and are increasingly splitting them into multiple kingdoms within the eukarya.

Because they are so much easier to see than bacteria and archaea, a disproportionate number of the named species on earth are in the domain eukarya. In fact, of the 1.5 million named species, the majority are eukarya, with about half being insects. This is more a result of the interests and biases of biologists than a reflection of the relative numbers of actual species in the world. We explore the great diversity of the eukarya in Chapters 11 (on animals) and 12 (on plants and fungi).

TAKE-HOME MESSAGE 10·18

All living organisms that we can see with the naked eye (and many that are too small to be seen) are eukarya, including all plants, animals, and fungi. The eukarya are unique among the three domains in having cells with organelles.

THE EUKARYA

Slime mold

Pink calla lily

Mushrooms

Blackbuck antelope

FIGURE 10-32 **All shapes and sizes.** There is a huge diversity of eukaryotic organisms.

Do Racial Differences Exist on a Genetic Level?

In a fraction of a second, we are aware of someone's size, gender, approximate age—and race. But what is race? Does race have biological meaning? And can humans be categorized into groups that can scientifically be called races?

Q: What is race, biologically speaking? The biological species concept hinges on a straightforward question: can two individuals interbreed under natural conditions? The concept of race, conversely, is not straightforward. Some biologists have proposed that races are groups of interbreeding populations that have different and distinct traits but are reproductively compatible. But it is important to note that, in humans, beyond phylogenetic distinctions, religion, culture, ethnicity, and geography are common aspects of many legal, social, and political conceptions of race. So do such social constructions of "race" reflect meaningful divisions?

Q: What do human racial groupings reflect? Are blacks and whites genetically different? The answer is obviously *yes;* black-skin alleles differ from white-skin alleles. Moreover, the prevalence of some genetic diseases varies by race and ethnicity. Sickle-cell anemia, for example, is relatively common among Africans and Southeast Asians because the genes that cause it also improve a person's resistance to malaria, a disease prevalent in Africa and Southeast Asia.

Yet Africans from the southernmost part of the continent have no greater risk of sickle-cell anemia than do the Japanese, because malaria is similarly rare in both of their homelands. Should we use this trait rather than skin color when determining a person's race? If we did, southern Africans would be grouped with the Japanese and northern Africans with southern Asians (who share their genetic resistance to malaria and the higher risk of sickle-cell anemia).

This observation reveals an important problem with racial groupings, one that DNA sequence comparisons also show: *similarity in skin color between two people doesn't tell you anything about overall genetic similarity.* There is, for example, a huge variation in blood type among Africans: some are type O, some AB, others A or B. But the same goes for any population in the world, Asians and Turks, Russians and Spaniards. Yet we do not group people according to blood type or any other physiological and biochemical variations.

Q: What is the future of race? Racial groupings such as "Hispanic" or "black" or "Asian" do not reflect consistent genetic differences. Still, race remains an important part of individual, social, cultural, and ethnic identity for many, and its use is mandated by federal agencies. For these reasons, it will remain a powerful social construct, even as scientific consensus emerges that racial groupings do not reflect meaningful genetic distinctions.

Check Your Knowledge

Thinking critically about visual displays of data

1. What is the approximate rate of background extinctions over the time period shown here?

2. What distinguishes the mass extinctions from the background extinctions in this figure?

3. Why do you think the graph records the percentage of families becoming extinct rather than the percentage of species becoming extinct?

4. Why isn't the green line indicating the rate of background extinctions a straight line?

5. Rates of extinction are generally estimated based on the fossil record. Why might this not be an accurate measure of the magnitude of a mass extinction?

6. Conservation biologists believe it is important that we keep a close watch on the current rate of extinctions. Why might that be?

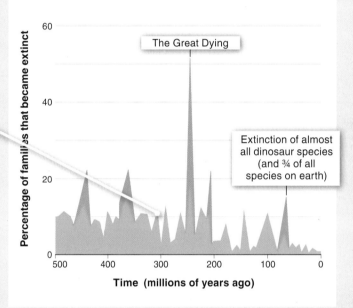

TWO CATEGORIES OF EXTINCTION

Extinctions generally fall into two categories:

● MASS EXTINCTIONS
A large number of species become extinct over a short period of time due to extraordinary and sudden environmental change.

● BACKGROUND EXTINCTIONS
These extinctions occur at lower rates during times other than mass extinctions.

The Great Dying

Extinction of almost all dinosaur species (and ¾ of all species on earth)

Percentage of families that became extinct (y-axis: 0, 20, 40, 60)

Time (millions of years ago) (x-axis: 500, 400, 300, 200, 100, 0)

See answers at the back of the book.

Key Terms in The Origin and Diversification of Life on Earth

ABOUT THE CHAPTER OPENING PHOTO

Millions of hairs on each gecko foot produce an electrical attraction force that enables them to adhere to surfaces, walk up walls, and even run upside down across ceilings. Shown here: several gecko feet adhering to polished glass.

REVIEW & REHEARSE

10

THE ORIGIN and DIVERSIFICATION of LIFE on EARTH

 Check Your Knowledge

1. In a set of classic experiments performed in the early 1950s, Urey and Miller subjected an experimental system composed of H_2, CH_4 (methane), and NH_3 (ammonia) to electrical sparks. A few days later, they found _____ in their system.

a) amino acids
b) DNA
c) microspheres
d) cells
e) RNA

EASY **13** HARD

2. What kind of molecule is thought to be the most likely to have been the first genetic material?

a) protein
b) DNA
c) carbohydrate
d) RNA
e) microsphere

EASY **25** HARD

3. According to the biological species concept, species are natural populations of organisms that have the potential to interbreed and are _____ isolated from other such populations.

a) behaviorally
b) prezygotically
c) postzygotically
d) geographically
e) reproductively

EASY **31** HARD

4. The classification rank that includes genera but not orders is:

a) species.
b) domain.
c) class.
d) kingdom.
e) family.

0 ———————— 100
EASY **50** HARD

10·1–10·3 Life on earth most likely originated from non-living materials.

The earliest life on earth, which resembled bacteria, appeared about 3.5 billion years ago, not long after the earth was formed.

WHAT IS LIFE?

Life is defined by the ability to replicate and by the presence of some sort of metabolic activity (the chemical processes by which molecules are acquired and used and energy is transformed in controlled reactions).

? **1.** Which groups of organisms currently living on earth most resemble the earliest life forms?

THE UREY-MILLER EXPERIMENT

In 1953, Stanley Miller and Harold Urey developed a simple four-step experiment that demonstrated how complex organic molecules could have arisen in earth's early environment.

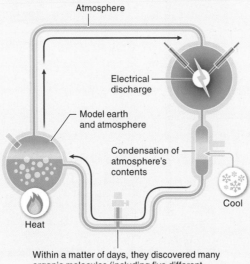

Atmosphere

Electrical discharge

Model earth and atmosphere

Condensation of atmosphere's contents

 Cool

Heat

Within a matter of days, they discovered many organic molecules (including five different amino acids) in their primordial sea.

THE ORIGIN OF LIFE?

How did the first organisms on earth arise? Most scientists believe that life originated on earth, probably in several distinct phases.

FORMATION OF SMALL MOLECULES CONTAINING CARBON AND HYDROGEN
Because of carbon's chemical structure, it can form a huge variety of molecules with widely varying functions.

FORMATION OF SELF-REPLICATING, INFORMATION-CONTAINING MOLECULES
The nucleic acid RNA can both carry information and catalyze the reactions necessary for replication. Thus, this single molecule could have been a self-replicating system and precursor to cellular life.

DEVELOPMENT OF A MEMBRANE
Membranes that compartmentalized the self-replicating molecules from their surroundings facilitated metabolic activity.

? **2.** Why is the development of a membrane system an important phase in the origin of life?

10·4–10·7 Species are the basic units of biodiversity.

Species are distinct biological entities, named using a hierarchical system of classification.

WHAT IS A SPECIES?

According to the biological species concept, species are populations of organisms that interbreed, or could possibly interbreed, with each other under natural conditions, and that cannot interbreed with organisms from other such groups.

? **3.** Using the biological species concept, define a species.

REPRODUCTIVE ISOLATION

Reproductive isolation is the inability of individuals from two populations to produce fertile offspring with each other, thereby making it impossible for gene exchange between the populations to occur.

THE ORGANIZATION OF LIFE

Each species on earth, like *Equus quagga*, is given a scientific name and is categorized according to hierarchical groups.

DOMAIN
Bacteria   Archaea   Eukarya

KINGDOM
Protista   Plantae   Fungi   Animalia

PHYLUM
Chordata

CLASS
Mammalia

ORDER
Perissodactyla

FAMILY
Equidae

GENUS
Equus

SPECIES
Equus quagga

4. In the Linnaean system, which is the narrowest classification for an organism?

5. Explain the difference between defining a species according to the biological species concept and according to the morphological species concept. Include an example of why one method might be more appropriate than the other for certain situations.

THE BIOLOGICAL SPECIES CONCEPT DOESN'T ALWAYS WORK

The biological species concept is very useful when describing most plants and animals, but it doesn't work for distinguishing all life forms.

CLASSIFYING ASEXUAL SPECIES
Asexual reproduction does not involve interbreeding, so the concept of reproductive isolation is no longer meaningful.

CLASSIFYING FOSSIL SPECIES
Differences in size and shape of fossil bones cannot reveal whether there was reproductive isolation between the individuals from which the bones came.

DETERMINING WHEN ONE SPECIES HAS CHANGED INTO ANOTHER
There is rarely a definitive moment marking the transition from one species to another.

CLASSIFYING RING SPECIES
Two non-interbreeding populations may be connected to each other by gene flow through another population, so there is no exact point where one species stops and the other begins.

CLASSIFYING HYBRIDIZING SPECIES
Hybridization—the interbreeding of closely related species—sometimes occurs and produces fertile offspring, suggesting that the borders between the species are not clear cut.

ALLOPATRIC SPECIATION

Allopatric speciation occurs when a geographic barrier causes one group of individuals in a population to be reproductively isolated from another group.

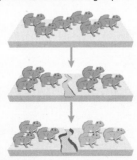

6. Why must the first phase of speciation involve reproductive isolation?

SYMPATRIC SPECIATION

Sympatric speciation results in the reproductive isolation of populations that coexist in the same area. Two scenarios lead to this method of speciation.

CHROMOSOME DUPLICATION WITHIN A SINGLE SPECIES

A cell division error during meiosis can lead to polyploid gametes. An individual with polyploid gametes can no longer interbreed with individuals having normal gametes. That individual can, however, interbreed with another individual that has polyploid gametes (or via self-fertilization). The new individual has achieved instant reproductive isolation from the original population and, therefore, is considered a new species.

COMBINING CHROMOSOMES FROM TWO DIFFERENT SPECIES

Two plants from different species interbreed, forming a hybrid. The hybrid may not be able to interbreed with either parental species. The hybrid may, however, be able to propagate itself asexually. Any future cell division error leading to a chromosome doubling can then lead to individuals producing gametes that are fertile with each other. The polyploid hybrid has achieved reproductive isolation from the parental population and, therefore, is considered a new species.

5. Populations of *Larus* gulls around the North Pole show an unusual pattern of reproductive isolation: each population is able to interbreed with its neighboring populations, but populations separated by larger geographic distances are not able to interbreed. *Larus* gulls are an example of a(n) _____ species.

  a) ring
  b) polyploid
  c) circular
  d) Escher
  e) linked

0 ——————— 62 ——————— 100
EASY   HARD

6. In animals, it is believed that the most common mode of speciation is:

  a) autopolyploidy.
  b) chromosomal.
  c) directional.
  d) allopatric.
  e) sympatric.

0 ——————— 54 ——————— 100
EASY   HARD

7. Polyploidy:

  a) arises only when an error in meiosis results in diploid gametes instead of haploid gametes.
  b) is a common method of sympatric speciation for animals.
  c) arises when allopatric speciation causes plants to have fewer sets of chromosomes than their parent plants.
  d) is an increased number of sets of chromosomes.
  e) always results in allopatric speciation.

0 ——————— 46 ——————— 100
EASY   HARD

8. In the plant kingdom, all of the species are descended from a single common ancestor. In terms of phylogeny, what type of tree of life is this?

  a) monophyletic
  b) uniphyletic
  c) punctuated
  d) sympatric
  e) allopatric

0 ——————— 27 ——————— 100
EASY   HARD

9. Phylogenetic trees should be viewed as:

a) true genealogical relationships among species.

b) the result of vertical, but never horizontal, gene transfer.

c) intellectual exercises, not to be interpreted literally.

d) representations of allopatric speciation events.

e) hypotheses regarding evolutionary relationships among groups of organisms.

0 — 34 — 100
EASY — HARD

10. The difference between microevolution and macroevolution is that:

a) macroevolution takes place over long periods of time, whereas microevolution takes place over relatively short periods.

b) macroevolution occurs with physical structures, whereas microevolution occurs with physiological traits.

c) microevolution occurs with physical structures, whereas macroevolution occurs with physiological traits.

d) microevolution has been proven, whereas macroevolution is very speculative.

e) microevolution occurs in prokaryotes, whereas macroevolution takes place among eukaryotes.

0 — 47 — 100
EASY — HARD

11. The idea of punctuated equilibrium challenges which component of Darwin's theory of evolution?

a) steady change

b) gradualism

c) species stasis

d) Both a) and b) are correct.

e) None of the above are correct.

0 — 45 — 100
EASY — HARD

12. Which of the following scenarios would best facilitate adaptive radiation?

a) A population of birds native to an island archipelago is forced to relocate to the mainland by a storm.

b) A population of cheetahs goes through an event in which all genetic diversity in the population is wiped out.

c) Darker-colored moths have a selective advantage over lighter-colored moths due to industrial soot on trees.

d) A population of birds becomes stranded on an island archipelago.

e) All of the above are equally likely to facilitate adaptive radiation.

0 — 78 — 100
EASY — HARD

10·8–10·10 Evolutionary trees help us conceptualize and categorize biodiversity.

Evolutionary trees reveal the evolutionary history of species and the sequence of speciation events that gave rise to them.

THE EVOLUTIONARY TREE OF LIFE

The evolutionary tree of life has branches with millions of tips representing all species on earth. As speciation events occur, new branches are added to the tree.

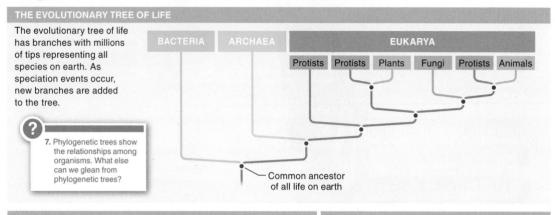

7. Phylogenetic trees show the relationships among organisms. What else can we glean from phylogenetic trees?

Common ancestor of all life on earth

MONOPHYLETIC GROUPS

Members of a monophyletic group share a common ancestor, and the group contains all of the descendants of that ancestor.

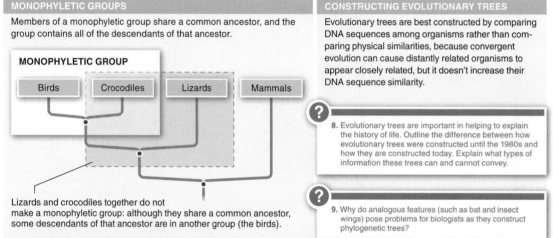

MONOPHYLETIC GROUP

Birds — Crocodiles — Lizards — Mammals

Lizards and crocodiles together do not make a monophyletic group: although they share a common ancestor, some descendants of that ancestor are in another group (the birds).

CONSTRUCTING EVOLUTIONARY TREES

Evolutionary trees are best constructed by comparing DNA sequences among organisms rather than comparing physical similarities, because convergent evolution can cause distantly related organisms to appear closely related, but it doesn't increase their DNA sequence similarity.

8. Evolutionary trees are important in helping to explain the history of life. Outline the difference between how evolutionary trees were constructed until the 1980s and how they are constructed today. Explain what types of information these trees can and cannot convey.

9. Why do analogous features (such as bat and insect wings) pose problems for biologists as they construct phylogenetic trees?

10·11–10·14 Macroevolution gives rise to great diversity.

The process of evolution in conjunction with reproductive isolation is sufficient to produce speciation and the rich diversity of life on earth.

EVOLUTION: MICRO vs. MACRO

Evolution is one thing only: a change in allele frequencies within a population. But over time, these changes can lead to new species and groups of species that vary tremendously.

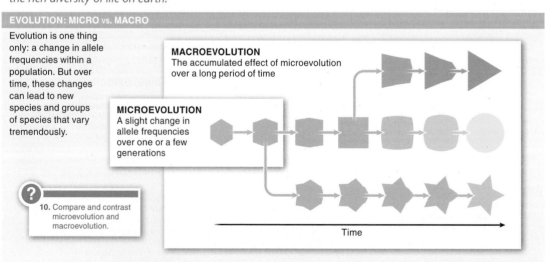

MACROEVOLUTION
The accumulated effect of microevolution over a long period of time

MICROEVOLUTION
A slight change in allele frequencies over one or a few generations

Time

10. Compare and contrast microevolution and macroevolution.

THE TEMPO OF EVOLUTION

The pace of evolution varies for different species. Some species have evolved gradually over time, while others spend vast amounts of time with little change.

GRADUAL CHANGE
Evolution by creeps: The pace of evolution occurs gradually in incremental steps.

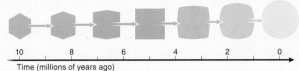

Time (millions of years ago)

PUNCTUATED EQUILIBRIUM
Evolution by jerks: Long periods of relatively little evolutionary change are punctuated by bursts of rapid change.

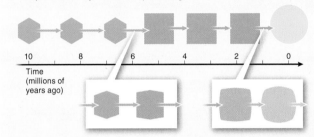

Time (millions of years ago)

EXTINCTION

Over time, species can be lost through extinction, which may be a consequence of natural selection or large, sudden changes in the environment. Mass extinctions are periods during which a large number of species on earth become extinct over a short period of time. These events are usually followed by periods of unusually rapid adaptive radiation and diversification of the remaining species.

ADAPTIVE RADIATION

Adaptive radiations—brief periods of time during which a small number of species diversify into a much larger number of species—tend to be triggered by mass extinctions of potentially competing species, colonizations of new habitats, or the appearance of evolutionary innovations.

11. In mammals, the shape of the skull, as well as the shape and size of the teeth, often reflects the method of feeding. A skull with a more rounded jaw and shorter, narrower teeth indicates a mammal that is well adapted to pulling bunches of leaves off trees and grinding them up. A skull with a more square jaw indicates a mammal adapted to grazing and cropping vegetation at the ground level. Explain how the adaptations of long, flat teeth and a more square jaw or muzzle could have triggered adaptive radiation in horses.

12. What sort of evidence do we use to help develop models of evolutionary change?

13. What is the fundamental difference between background extinctions and mass extinctions?

10·15–10·18 An overview of the diversity of life on earth: organisms are divided into three domains.

All life on earth can be divided into three domains—bacteria, archaea, and eukarya—which reflect species' evolutionary relatedness to each other.

14. Why did the species classification system created in the 1960s have to be revised in the 1970s and 1980s?

BACTERIA

All bacteria share a common ancestor and have a few features in common. They are prokaryotic, asexual, single-celled organisms with no nucleus or organelles, with one or more circular molecules of DNA as their genetic material, and using several methods of exchanging genetic information. Bacteria have a much broader diversity of metabolic and reproductive abilities than do the eukarya.

15. Why is it impossible to apply the biological species concept to the classification of bacteria?

ARCHAEA

Archaea, many of which are adapted to life in extreme environments, physically resemble bacteria (because both domains include only prokaryotes) but are more closely related to eukarya. Because they thrive in many habitats that humans have not yet studied well, including the deepest seas and oceans, they may turn out to be much more common than currently believed.

16. Archaea are single-celled prokaryotes, and under a microscope they look very similar to bacteria. Which physical features distinguish them from the bacteria?

EUKARYA

All living organisms that we can see with the naked eye (and many that are too small to be seen) are eukarya, including all plants, animals, and fungi. The eukarya are unique among the three domains in having cells with organelles.

17. Which features make the domain eukarya unique?

---

13. The mass extinction that occurred on earth 65 million years ago was immediately followed by:
a) the rise of archaea.
b) the emergence of the first non-photosynthetic organisms.
c) the rise of the reptiles, including the dinosaurs.
d) an increase in atmospheric oxygen levels.
e) the rapid divergence and radiation of modern mammals.

0 — 46 — 100
EASY — HARD

14. Prokaryotes are classified into _____ domain(s).
a) 1
b) 2
c) 3
d) 4
e) 5

0 — 42 — 100
EASY — HARD

15. All protists are alike in that all are:
a) bacterial.
b) photosynthetic.
c) eukaryotic.
d) prokaryotic.
e) archaeal.

0 — 42 — 100
EASY — HARD

16. Which of the following pairs of domains share the most recent common ancestor?
a) archaea and eukarya
b) bacteria and eukarya
c) archaea and bacteria
d) None of the above; all three domains evolved from different ancestors.
e) None of the above; all three domains are equally related to each other.

0 — 57 — 100
EASY — HARD

17. Which of the following groups would be placed nearest the fungi in an evolutionary tree based on DNA sequences?
a) plants
b) bacteria
c) animals
d) archaea
e) protists

0 — 41 — 100
EASY — HARD

11 | Animal Diversification

VISIBILITY IN MOTION

Animals are just one branch of the Eukarya domain.

Invertebrates—animals without a backbone—are the most diverse group of animals.

The phylum Chordata includes vertebrates, animals with a backbone.

All terrestrial vertebrates are tetrapods.

11.1–11.3
Animals are just one branch of the Eukarya domain.

A scarlet macaw (Ara macao) in flight,
Tambopata National Reserve, Peru.

11·1 What is an animal?

Looking at a kangaroo, it's obvious that it is an animal. The same goes for a mosquito, squid, or earthworm. It gets a bit more difficult, though, when you look at a sponge. Is it an animal? Or is it a plant? Or is it something else? This is when it is helpful to have some guidelines for identifying and organizing the biodiversity we can see on earth.

For the sponge, the answer is that it is classified as an animal. Here's why. When defining what is or is not an animal, we look for certain features that are common to all animals. Surprisingly, considering the enormous differences

among even the few animals mentioned above, just three characteristics are generally sufficient to define **animals** (FIGURE 11-1).

1. All animals eat other organisms. That is, rather than manufacturing their own food, as plants and some bacteria do through photosynthesis, animals consume plants, bacteria, other animals, fungi, or some combination of these. And because the best way to catch other organisms is to move toward them or chase after them, that brings us to the second characteristic that sets animals apart from plants.

WHAT IS AN ANIMAL?

Rock Agama (Agama agama) male eating grasshopper

ANIMALS EAT OTHER ORGANISMS
All animals acquire energy by consuming other organisms.

Eastern gray kangaroos (Macropus giganteus)

ANIMALS MOVE
All animals have the ability to move—at least at some stage of their life cycle.

Nembrotha kubaryana nudibranch, a type of sea slug

ANIMALS ARE MULTICELLULAR
Animals consist of multiple cells and have body parts that are specialized for different activities.

FIGURE 11-1 **Characteristics of an animal.**

2. All animals move—*at least, at some stage of their life cycle.* That qualifying phrase is important because some animals move only during a period called the "larval stage" that comes early in their lives, and when they become adults they fasten themselves to a surface and stay there. If you visit a rocky seashore, you can find two of these animals—mussels and barnacles—living on rocks in the tidal zone. Organisms that are fastened in place, such as adult mussels and barnacles, are said to be **sessile,** but even animals that are sessile as adults moved when they were larvae.

3. All animals are multicellular, and they generally have body parts that are specialized for different activities. For example, most animals possess sensory organs (eyes, ears, nose, tongue, antennae, whiskers, and so on) that enable them to acquire information about their environment and, among other things, detect potential prey and predators.

There are a few other general features of animals—most reproduce sexually, for example—but these three are the easiest criteria to use in determining whether or not you're looking at an animal.

> **TAKE-HOME MESSAGE 11·1**
>
> Animals are organisms that share three characteristics: all of them eat other organisms, all can move during at least one stage of their development, and all are multicellular.

11·2 There are no "higher" or "lower" species.

Humans often reveal their biases when speaking of other organisms. There are two possible states for a species. It can be *extant,* meaning that it currently exists. Or it can be *extinct.* From an evolutionary perspective, references to extant species as "lower" or "primitive" does not make sense. Rather, these labels usually reflect nothing more than how similar a species is to humans. Evolution can lead to greater adaptation between an organism and its environment, but it does not necessarily lead to greater complexity or to some other feature that can be identified as "higher" or more advanced. The fact that a species exists means it can do the things all organisms must do: find food, escape predators, and reproduce (**FIGURE 11-2**).

Darwin wrote a note to himself in the margin of a book he was reading that proposed a theory of cosmic and biological evolution: "Never use the word higher or lower." It was an important insight, but one that can be difficult to adhere to—particularly as we turn our attention to the animals.

Throughout his writings, for the many species he described, Darwin emphasized the process of adaptation and the relationship between populations of organisms and the environments in which they lived. One species was never better or worse than another. Rather, each species was differentiated from the others, with specializations that adapted individuals of that species to the particular niche in which they lived. Different environments posed different challenges. All extant species, as evidenced by their existence and persistence, are able to overcome those challenges.

Our tour through the animal kingdom begins with animals having a simple body plan, such as sponges, jellyfishes, and worms. It continues through the mollusks and arthropods—complex animals that do not have a backbone. And it culminates with the vertebrates: mammals (including humans), amphibians, reptiles (including birds), and fishes. Taking this path allows us to explore the major evolutionary transitions in the chronological order in which they occurred. This order does not provide any information

💡 *From an evolutionary perspective, every living species is successful.*

FIGURE 11-2 **Two equally successful organisms: the earthworm and the tiger.**

about the relative success of any species. Nor does it imply that humans are the "crown of creation."

> **((I have been studying the traits and dispositions of the lower animals (so-called), and contrasting them with the traits and dispositions of man. I find the result humiliating to me.))**
> — MARK TWAIN, in *The Lowest Animal* (written ~1897, first published 1962)

Although each extant species represents some measure of evolutionary success, certain adaptations have led to the rapid and extensive diversification of particular *groups* of species. We investigate nine of the phyla (there are approximately 36 animal phyla in all) with the most species (shown in Figure 11-3) and the evolutionary innovations that contributed to their success. These groups represent only a quarter of all animal phyla, but they account for about 99% of the animal species on earth today.

TAKE-HOME MESSAGE 11·2

From an evolutionary perspective, it is inappropriate to view any species as "higher" or "lower." Certain adaptations have led to the rapid and extensive diversification of particular groups of species. Of the approximately 36 animal phyla, 9 phyla account for more than 99% of all described animal species.

11·3 Four key distinctions divide the animals.

More than two-thirds of the almost two million species on earth that have been identified are animals. All of these animals share a common ancestor, an ancestral protist. More specifically, this last common ancestor to all of the animals is thought to have been a free-living unicellular organism resembling a sperm in size and shape. With such a large number of animals to organize and identify, it is useful to divide them into a few large categories, based on their evolutionary relatedness.

Analyses of DNA and RNA sequences have helped biologists identify when (and the extent to which) species or groups of species share a common ancestor. These analyses have made it possible, for example, to better classify the animals into major monophyletic groupings. Each of these groupings hinges on a particular adaptation and on the question of whether or not the animals of a species have descended from an ancestor with that adaptation (**FIGURE 11-3**). We can find the appropriate place for an animal in a phylogenetic tree based on the answers to a series of questions.

At any level of classification, of course, there are additional characteristics that can be used to further subdivide and classify the animals. We'll explore many of those characteristics throughout this chapter, but first let's look at four of the key distinctions.

1. Does the animal have specialized cells that form defined tissues? If an animal has no tissues—groups of cells with a common structure and function—then we do not need to ask any further questions. It belongs to the phylum Porifera, better known as the sponges. Sponges are aggregations of similar cells, with little coordination of activities among the cells. All other animals form a monophyletic group of animals with clearly defined tissues. Humans, for example, have highly specialized cells such as skin cells, muscle cells, and sensory cells. For all non-sponge animals, we then can ask a second question.

2. Does the animal develop with radial symmetry or bilateral symmetry? All animals with defined tissues develop in a shape that has some sort of symmetry: radial or bilateral. **Radial symmetry** describes animals with a body structured like a pie, such as jellyfishes, corals, and sea anemones. Slow-moving or free-floating, radially symmetrical animals have no front or back ends, and it is possible to make multiple slices, all going through the center, that divide the organism into identical pieces.

Organisms with defined tissues that do not develop with radial symmetry develop with **bilateral symmetry.** These organisms (such as humans, cows, and scorpions) have left and right sides that are mirror images. The animals can move adeptly, searching for food and avoiding predators. As we'll see, in some animals, such as starfish, embryonic development is bilaterally symmetrical, but then the animal becomes radially symmetrical as an adult. Where these animals fall in the classification by body symmetry depends on the kind of symmetry they show as embryos, not as adults.

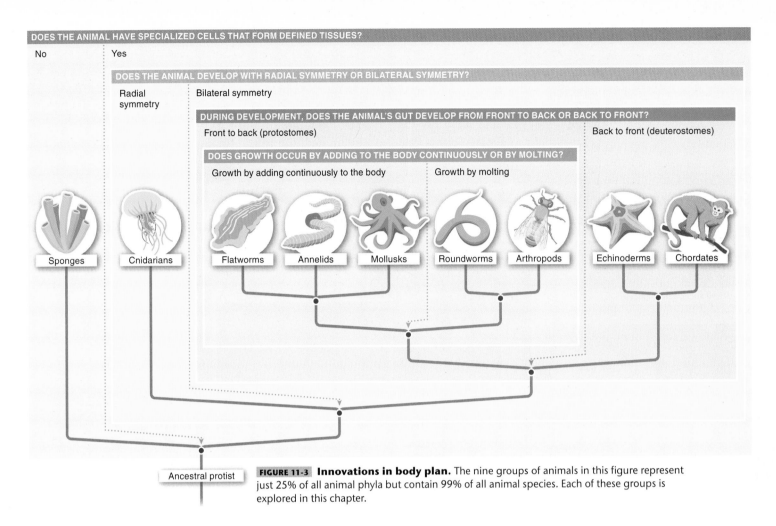

No   Yes

DOES THE ANIMAL DEVELOP WITH RADIAL SYMMETRY OR BILATERAL SYMMETRY?

Radial   Bilateral symmetry
symmetry

DURING DEVELOPMENT, DOES THE ANIMAL'S GUT DEVELOP FROM FRONT TO BACK OR BACK TO FRONT?

Front to back (protostomes)   Back to front (deuterostomes)

DOES GROWTH OCCUR BY ADDING TO THE BODY CONTINUOUSLY OR BY MOLTING?

Growth by adding continuously to the body   Growth by molting

Sponges   Cnidarians   Flatworms   Annelids   Mollusks   Roundworms   Arthropods   Echinoderms   Chordates

Ancestral protist

FIGURE 11-3 **Innovations in body plan.** The nine groups of animals in this figure represent just 25% of all animal phyla but contain 99% of all animal species. Each of these groups is explored in this chapter.

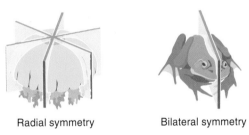

Radial symmetry   Bilateral symmetry

Among the monophyletic group of animals with defined tissues that develop with bilateral symmetry, we can make a further phylogenetic division based on the answer to a third question.

3. During gut development, does the mouth or anus form first? Early in the evolution of animals—about 630 million years ago—a major split occurred that allows us to separate the bilaterally symmetrical animals with defined tissues into two distinct monophyletic lineages: the **protostomes** (which can be translated as "mouth first") and the **deuterostomes** ("mouth second").

These names refer to the way the gut develops—from front to back or from back to front. In protostomes, the gut

develops from front to back, so the first opening that forms becomes the mouth of the adult animal, and the last opening becomes the anus. In deuterostomes, the gut develops from back to front. The anus is the first opening to form, and the mouth is the second. Although it's something you can see only when closely observing animals' embryonic development, the protostome-deuterostome split is one of the most basic divisions of animals, in terms of revealing their evolutionary relatedness to one other.

In building an overall phylogeny of the animals, we can further subdivide into monophyletic groups the animals that (1) have defined tissues, (2) develop with bilateral symmetry, and (3) are protostomes. To do this, we need to ask a fourth question.

4. Does growth occur by molting or by adding to the animal's body in a continuous manner? Among the bilaterally symmetrical animals that are protostomes, there is an important distinction between animals that molt and those that do not. Animals that molt (such as lobsters and insects) shed their exoskeleton (a hard outer layer) and replace it with a larger one at regular intervals

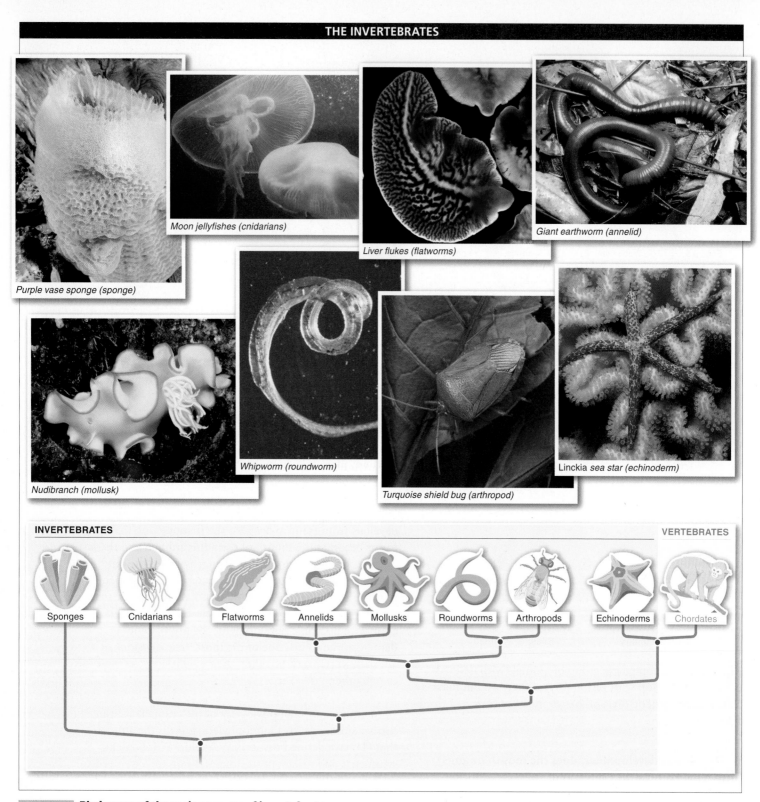

Purple vase sponge (sponge)

Moon jellyfishes (cnidarians)

Liver flukes (flatworms)

Giant earthworm (annelid)

Nudibranch (mollusk)

Whipworm (roundworm)

Turquoise shield bug (arthropod)

Linckia *sea star (echinoderm)*

INVERTEBRATES

VERTEBRATES

Sponges · Cnidarians · Flatworms · Annelids · Mollusks · Roundworms · Arthropods · Echinoderms · Chordates

FIGURE 11-4 Phylogeny of the major groups of invertebrates.

during development. Other animals (such as clams and octopuses) grow by adding to the size of their body in a continuous manner.

All animals share a common ancestor, the ancestral protist, but these four key distinctions help us identify and organize all of the approximately 36 phyla of living animals into monophyletic groups. This is important in helping us to understand the evolutionary history of animals because within any of the groups, we know that all of the individuals are more closely related to one another than to any individuals outside that group.

Additional evolutionary transitions have occurred within the groups created by these four distinctions, allowing further categorization of animals into monophyletic groups. These categories help us understand the sequence of evolutionary events that have resulted in the tremendous animal diversity we see in the world today.

In spite of the evolutionary rationale of organizing animal diversity within monophyletic groups, there is a distinction that may be more commonly used: whether the animal has a backbone or not. **Invertebrates**—including sponges—are animals that do not have a backbone; **vertebrates** are animals that do have a backbone.

Invertebrates are the largest and most diverse group of animals, comprising more than 95% of all living animal species (**FIGURE 11-4**). And as "animals without backbones,"

the grouping has a convenient and easy-to-apply definition. But the invertebrates are not a monophyletic group. It's a grouping that reminds us that—until the recent advent of molecular methods for building accurate phylogenetic trees—we have long been fooled by convergent evolution into believing that organisms that *appear* similar also share close evolutionary relatedness.

In this case, all invertebrates are protostomes except for one large group, whose members are all deuterostomes. Called echinoderms, and including sea stars, sea urchins, and sand dollars, this animal phylum shows the same from-back-to-front gut development as the vertebrates, and these animals are, in fact, the vertebrates' closest relatives.

The invertebrates include eight separate evolutionary lineages with an enormous diversity of body forms, body sizes, habitats, and behaviors. We examine them in detail in the following sections.

TAKE-HOME MESSAGE 11·3

The animals probably originated from an ancestral protist. Four key distinctions divide the extant animals into monophyletic groups: (1) tissues or not, (2) radial or bilateral symmetry, (3) protostome or deuterostome development, and (4) growth through molting or through continuous addition to the body.

11·4–11·12
Invertebrates—animals without a backbone—are the most diverse group of animals.

The most diverse animals on earth? More than 25% of all described species are beetles. Shown here is a wood-boring beetle (Madecassia rothschildi) *from Madagascar.*

11·4 Sponges are animals that lack tissues and organs.

In the natural world, we see many things that are hard to believe. Consider this: sponges (phylum Porifera) are animals. They are multicellular, they eat other organisms, and (for part of their life) they are mobile (**FIGURE 11-5**). We'll begin our investigation of animal diversity with this group that may be the least "animal-like," and the simplest, of all the animals.

Sponges have no specialized tissues or organs, and most lack any symmetry. Despite their simplicity, however, sponges are remarkably efficient at obtaining food. Living as sessile suspension-feeders, sponges strain suspended matter and food particles from water. You can think of the anatomy of an adult sponge as a cylinder with pores in its walls. Although they have no tissues, sponges do have several different cell types. Epidermal cells—flattened, skin-like cells, connected to each other—cover the outside of the sponge. Collar cells, each of which has a long, whip-like flagellum, cover the inside. The middle layer is a gel, with embedded fibers that stiffen the sponge's body (**FIGURE 11-6**).

The beating of each collar cell's flagellum creates a current that carries water upward and out through the opening at the top of the sponge. That water is replaced by an inward flow of water through the pores on the sides of the sponge. The incoming water carries bacteria and algae, other microscopic organisms, and small particles of organic material into the sponge's body. In the "collars" of the collar cells, sticky mucus traps some of this material, and some particles are taken up into the cells by endocytosis

FIGURE 11-5 An overview of the sponges.

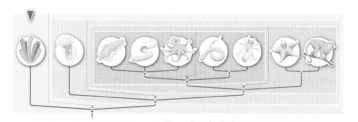

THE SPONGES

Brown tube sponge

Purple sponge

COMMON CHARACTERISTICS
- No tissues or organs
- Body consists of a hollow tube with pores in its wall
- Feed by pumping in water, along with bacteria, algae, and small particles of organic material, through their pores
- Free-swimming larvae
- Sessile as adults

MEMBERS INCLUDE
- About 5,000 species

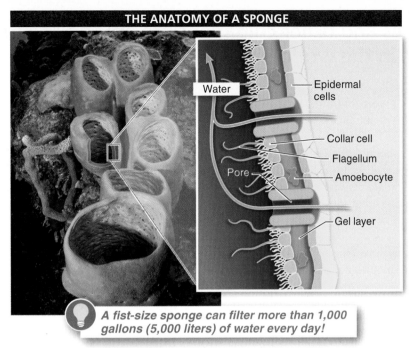

THE ANATOMY OF A SPONGE

Water

Epidermal cells

Collar cell

Flagellum

Amoebocyte

Pore

Gel layer

A fist-size sponge can filter more than 1,000 gallons (5,000 liters) of water every day!

FIGURE 11-6 **Sponge structure.**

(see Section 3-11). The volume of water moving through its body daily can be as much as 20,000 times the volume of the sponge.

Sponges are hermaphrodites. That is, each individual contains both male and female reproductive organs, but it produces only one kind of gamete (eggs or sperm) at a time. Sponges that are acting as males release a cloud of sperm that swim to other sponges, which are acting as females, and fertilize their eggs. The eggs develop into larvae, which are released into the water and then drift for a few days before settling on a rock or coral outcrop and developing into a sessile adult sponge.

Sponges also reproduce asexually: small buds form on the outside of the sponge and eventually break off, settle to the bottom, and grow into new sponges. Even a piece accidentally broken off by wave action or a collision with a passing fish will grow into a new sponge. Most remarkable is the ability of sponge cells to reassemble. Picture this experiment. You put a living sponge into a food blender and puree it, then strain the liquid through a fine sieve to remove all the chunks, leaving a suspension of individual cells, which you then dump into an aquarium. Within a day you'll see small clumps of sponge forming as the individual cells move around and attach to each other when they meet, and within a week a new sponge will form. For an even more dramatic demonstration, you can repeat the experiment with two sponge species of different colors. Not only do the sponges

Q Is your kitchen sponge an animal? Is it alive?

reassemble, but the cells from each species assemble only with other cells of the same species, so you'll have two sponges, each with only the cells of its own species.

Because the dried skeletons of some sponges are soft and absorbent, humans have long made use of them. Sponges have been used as padding in helmets and as a tool for painting, for example. And in ancient Rome, a sponge on a stick was commonly used as toilet paper. But all of these sponges are far from living. They are just a matrix of soft, elastic fibers of a protein called spongin, isolated from dried sponges. Because over-fishing has dramatically reduced the populations of these sponges, they are not widely available. Rather, the "sponges" used in most kitchens and bathrooms today are nearly always produced from synthetic materials. Loofah sponges, contrary to their name, are made from plant material.

TAKE-HOME MESSAGE 11·4

Sponges are among the simplest of the animal lineages. A sponge consists of a hollow tube with pores in its wall; it has no tissues or organs. Sponges reproduce sexually (by producing eggs and sperm) and asexually (by budding). The fertilized eggs grow into free-swimming larvae that settle and develop into sessile, filter-feeding adult sponges.

11·5 Jellyfishes and other cnidarians are among the most poisonous animals in the world.

If you feel a sting when you are swimming in the ocean, you have met a jellyfish. If you are lucky, it will be one of the thousands of species that feed on tiny floating organisms called plankton, because these species have only mild stings. If it is a species that feeds on larger prey such as fish or shrimp, however, you are definitely unlucky. These jellyfishes have powerful stings that can cause extreme pain or even death.

The jellyfishes, along with sea anemones and corals, belong to the phylum Cnidaria (pronounced nigh-DARE-ee-ah). As with all animals except the sponges, the 11,000 species in this phylum have defined tissues. The cnidarians also all have a simple body plan, with radial symmetry (FIGURE 11-7).

There are two types of cnidarian bodies: a sessile polyp and a free-floating medusa, which is the form most people picture when they think of a jellyfish. In some species, individuals spend part of their life cycle as a polyp and part as a medusa. Other species exist only as medusas or, as in corals and sea anemones, only as polyps.

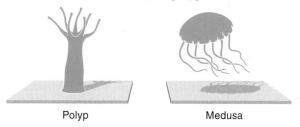

Polyp   Medusa

Cnidarians are carnivores and use their tentacles to capture and feed on a wide variety of marine organisms, from protists to fish and shellfish. Their method of capturing prey relies on a sort of weapon that is unique to the cnidarians—a stinging cell called a cnidocyte. All cnidarians have tentacles, located near their mouth, that are armed with rows of stinging cnidocytes. Each cnidocyte has a coiled thread with barbs inside and a "trigger" on the outside. When something comes in contact with the trigger, perhaps prey (or perhaps your leg as you swim in the ocean), the coiled thread is ejected and, like a harpoon, can penetrate the prey, often injecting a toxin (FIGURE 11-8). The Portuguese man-o'-war is one of the dangerously poisonous species of cnidarians. Being stung by a Portuguese man-o'-war is painful (extremely painful if the unlucky swimmer becomes entangled in the tentacles and receives hundreds of stings), but these encounters are rarely fatal.

We explore here some of the great diversity among three familiar groups of cnidarians.

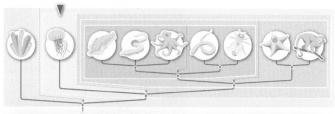

THE CNIDARIANS

Mosaic jellyfish

Dahlia anemone

Sun coral

COMMON CHARACTERISTICS
- Radially symmetrical
- Tentacles armed with rows of stinging cells, used to paralyze prey

MEMBERS INCLUDE
- Jellyfishes
- Sea anemones
- Corals

FIGURE 11-7   An overview of the cnidarians.

Corals   The corals live as small (just a few millimeters long) polyps in large colonial groups, familiar as the beautiful structures found in coral reefs. Corals sting prey and use their tentacles to catch plankton (and even, on

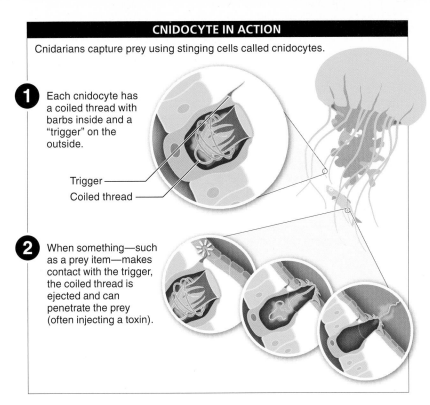

CNIDOCYTE IN ACTION

Cnidarians capture prey using stinging cells called cnidocytes.

1 Each cnidocyte has a coiled thread with barbs inside and a "trigger" on the outside.

Trigger
Coiled thread

2 When something—such as a prey item—makes contact with the trigger, the coiled thread is ejected and can penetrate the prey (often injecting a toxin).

FIGURE 11-8 Stinging cells are used to capture prey.

the polyps to conduct photosynthesis, and the polyps gain oxygen and nutrients produced by the algae. Strangely, when coral polyps get too hot they expel their zooxanthellae. Because the zooxanthellae are responsible for the colors of many corals, as well as much of their nutrition, when coral polyps expel their zooxanthellae, the coral appears white—a phenomenon called coral bleaching. The zooxanthellae may return to the polyps when the water cools, but repeated bleaching events often lead to the death of the coral polyps. The world's oceans are already warming, and coral bleaching events are becoming more and more frequent (**FIGURE 11-9**).

Q How is global warming affecting the coral reefs of the world?

occasion, small fish), which are directed into the mouth by the tentacles, then digested in the stomach. Corals most commonly reproduce sexually, with both external fertilization (releasing sperm and eggs into the water, where they can fuse and form larvae) and internal fertilization (in which only sperm are released and can fertilize eggs within female corals). Corals can also reproduce asexually.

Corals grow in a wide variety of shapes—from free-standing spherical or branched forms to crusts growing on rocks or other corals. All of this is possible because corals secrete calcium carbonate to create hard shells on which a multitude of individual coral polyps can live as a colony. Such assemblies of giant calcium carbonate skeletons and corals are called **coral reefs.** Coral reefs provide an environment that is home to a greater diversity of species than any other marine habitat. The Great Barrier Reef, located in the Coral Sea, off the coast of Queensland, Australia, extends more than 1,500 miles (about 2,600 km) from north to south. A truly awesome display of biological productivity, this reef is the largest biological structure in the world and is easily visible from the International Space Station.

Coral reefs are among the first casualties of global warming. Although coral polyps can catch prey using their cnidocytes, they obtain most of their nutrition from algae, called zooxanthellae, that live symbiotically within the polyps. Each partner contributes and receives something of value in the relationship. The algae use the carbon dioxide produced by

Healthy coral

Bleached coral

With warming water, corals eject their photosynthetic symbionts, causing coral "bleaching." Without the colored algae, the corals appear white and may die, endangering the coral reef and the diversity it supports.

FIGURE 11-9 Coral bleaching.

Sea Anemones Sea anemones resemble flowers, and some of the colorful species are popular additions to saltwater aquaria. The polyp form of a sea anemone looks a bit like an upside-down medusa form of a jellyfish—the tentacles with their stinging cells are at the top of the anemone's body, surrounding its mouth. Sea anemones have a larval stage that swims freely, then settles on a rock and metamorphoses into the adult form. Even as adults, many sea anemones can crawl a few inches a day, and wandering sea anemones are often found in seaweed.

Jellyfishes The jellyfishes (which are not fish!) range tremendously in size. The Asian giant jellyfish is more than 6 feet (about 2 m) across and weighs nearly 500 pounds (more than 200 kg). When swarms of this giant jellyfish appear, they clog the nets of fishing boats with a sticky mass of toxic stingers. At the opposite end of the size scale is the Irukandji jellyfish, about the size of a hen's egg. This species is so deadly that, in 2007, the sighting of just five Irukandji in Hervey Bay, Queensland, completely halted the filming of a major Hollywood movie (*Fool's Gold*), which had to be completed in the safety of a studio.

> ### TAKE-HOME MESSAGE 11·5
> Corals, sea anemones, and jellyfishes are radially symmetrical animals with defined tissues, in the phylum Cnidaria. All cnidarians are carnivores and use specialized stinging cells located in their tentacles to capture prey.

11·6 Flatworms, roundworms, and segmented worms come in all shapes and sizes.

It's a good thing that biologists don't run bait shops—if they did, a simple trip to buy worms for fishing would be way too complicated. The name "worm" is commonly applied to a long, skinny, slimy animal without a backbone, but you can find animals fitting that description in eight different phyla. You will never encounter most of them, and that's a good thing, because many are parasites that cause really unpleasant diseases. We consider here just three phyla—**flatworms** (phylum Platyhelminthes), **roundworms** (phylum Nematoda), and **segmented worms,** or **annelids** (phylum Annelida)—that illustrate most of the diversity of the animals we call worms.

All of the worms described here have defined tissues and are protostomes—the gut develops from front to back. Unlike the cnidarians, however, the worms (along with all of the remaining animals we discuss in this chapter) develop with a body plan characterized by bilateral symmetry, which adapts them for forward movement.

It is important to note, however, that worms do not make up a monophyletic group. Their general body plan—that is, a long, cylindrical, legless, soft-bodied animal without a backbone—has evolved several times independently, and their evolutionary relationships are not always obvious. Before their evolutionary relationships were understood, these groups of organisms were all classified by Linnaeus as worms, simply because of these similarities in their appearance.

We know today, however, that convergent evolution is responsible for much of the resemblance and that the segmented worms (annelids), for example, are probably more closely related to mollusks (including snails, clams, and octopuses) than to the roundworms. And the roundworms are believed to be more closely related to the arthropods (insects) than to either the flatworms or the segmented worms (see Figure 11-3). From an evolutionary perspective, "worm" is now an obsolete—and potentially misleading—label.

Flatworms Unlike the roundworms and segmented worms, flatworms have no body cavity. That is, the space between the body wall and the digestive tract is not a fluid-filled cavity. Possibly as a consequence of this, the flatworms have no specialized respiratory or circulatory organs and so have a reduced capacity for diffusion of gases and nutrients.

Like the segmented worms, but unlike the roundworms, flatworms grow by adding body mass rather than by molting (**FIGURE 11-10**). Flatworms have well-defined head and tail regions, with clusters of light-sensitive cells called eyespots that help them orient themselves within their environment. Most flatworms are hermaphroditic, each individual producing both male and female gametes, and they engage in both sexual and asexual reproduction. Flatworms include more than 20,000 species and can be parasites (the flukes and the tapeworms) or brilliantly colored, free-living aquatic creatures.

Many flatworms have a digestive system, though among those that do, their gut has only one opening, requiring them to both consume food and eliminate undigested waste through the same opening. The digestive systems of roundworms and segmented worms, conversely, have two

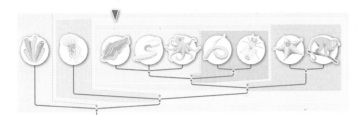

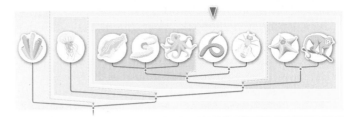

THE FLATWORMS

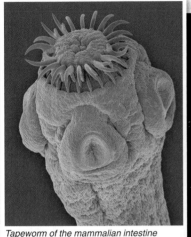

Tapeworm of the mammalian intestine

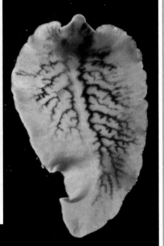

Liver fluke

COMMON CHARACTERISTICS
- Well-defined head and tail regions
- Hermaphroditic and can engage in both sexual and asexual reproduction
- Some have a single opening in the body, which serves as a mouth and an anus

MEMBERS INCLUDE
- Tapeworms
- Flukes

FIGURE 11-10 An overview of the flatworms.

openings. One large group of flatworms that lack a digestive system is the tapeworms. These parasitic worms live in their host's gut and absorb nutrients from the host directly through their body wall.

All 5,000 species of tapeworms are parasites, and most have a two-stage life cycle that is split between two different host species. For example, the common tapeworm that infects dogs spends the other half of its life cycle in fleas, while the human pork tapeworm splits its life cycle between pigs and humans. Tapeworms have long, flat bodies made up of repeated segments, each of which is a reproductive unit. Mature tapeworms spread by breaking off segments, which are then shed in the feces of an infected individual.

THE ROUNDWORMS

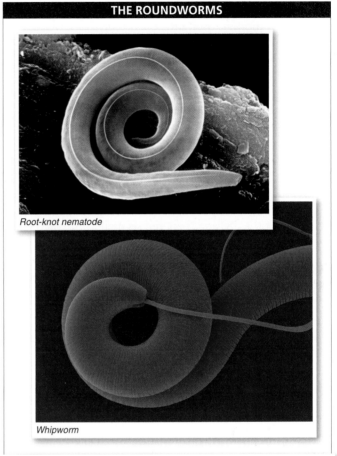

Root-knot nematode

Whipworm

COMMON CHARACTERISTICS
- Long, narrow unsegmented body
- Bilaterally symmetrical
- Surrounded by a strong, flexible cuticle
- Must molt in order to grow larger

MEMBERS INCLUDE
- More than 90,000 species (scientists estimate there may be five times as many species that have not yet been identified)

FIGURE 11-11 An overview of roundworm diversity.

Roundworms The roundworms, also called nematodes, are probably the most numerically abundant animals on earth—a spoonful of garden soil contains several thousand individuals, and some species produce more than 200,000 eggs every day. Unlike segmented worms and flatworms, they grow by molting. More than 90,000 roundworm species have been named, and there may be five times as many species that have not yet been identified (**FIGURE 11-11**).

Many soil-dwelling roundworms live in the roots of plants, damaging or even killing the plant. About 15,000 species of roundworms are parasites of other animals, and roundworms are responsible for a large number of human diseases. Most roundworms are transmitted by fecal contamination of soil or food.

Tiny parasitic roundworms called filariae are responsible for several tropical diseases, including some that have an enormous social and economic impact in certain parts of the world. Elephantiasis, for example, is a disease in which filariae transmitted by the bite of a mosquito block the lymph ducts so that fluid accumulates in the limbs or scrotum, causing grotesque swelling. These filariae are found in India, Africa, South Asia, Pacifica, and tropical regions of the Americas.

Segmented Worms (Annelids) The segmented worms, or annelids, number about 13,000 species and are easy to recognize by the grooves running around the body that mark the divisions between segments—if you've seen an earthworm, you are familiar with these grooves. The segmented worms are protostomes with defined tissues, and they do not molt. They are organized into three different groups: marine polychaetes (pronounced POL-lee-keets), terrestrial oligochaetes, and leeches (**FIGURE 11-12**).

Polychaetes are marine worms, living on the seafloor. *Polychaete* means "many bristles," and the combination of segments and bristles makes polychaetes easy to recognize. Some species burrow through the mud and extract organic material from it. Tube worms use sand grains or limestone to make a tube in which they live, with just their waving tentacles exposed. Small particles of food are trapped in mucus on the tentacles and transported to the worm's mouth.

Earthworms, which belong to the group called *oligochaetes* ("few bristles"), are the annelid worms you are most likely to have seen. The night crawler is a typical earthworm. It gets its name from its habit of emerging from its burrow on rainy nights to crawl across the surface of the ground. More than 4,000 earthworm species have been named, ranging in size from less than half an inch (about a centimeter) to the enormous length of the giant Gippsland earthworm, 7–10 feet (2–3 m)!

Earthworms are bulk-feeders—that is, as an earthworm burrows, it consumes particles of soil and organic material. The organic material is digested as it passes through the worm's one-way gut, and the fecal material plus the inorganic part of the soil is excreted as feces, called castings. Earthworm castings are valued by gardeners as soil supplements, and an

THE ANNELIDS

Polychaete worm

Earthworm

Leech

COMMON CHARACTERISTICS
• Segmented body

MEMBERS INCLUDE
• Marine polychaetes (about 9,000 species)
• Earthworms (more than 4,000 species)
• Leeches (about 500 species)

FIGURE 11-12 **An overview of annelid diversity.**

earthworm can produce its own weight in castings every day. Earthworms' activities mix the soil components, create a more uniform mixture of nutrients, and expedite the breakdown of organic materials in the soil, thus making the nutrients available for plants. Because they perform all of these tasks, earthworms have enormous economic value to agriculture. A government study published in 2008 estimated the economic value of earthworms in Ireland alone to be more than $1 billion per year.

Have you ever waded in a pond or swamp and emerged to find a long, dark brown worm clinging to your leg? If so, you have met the third major group of annelids, the *leeches*. Probably you pulled the leech off (not an easy thing to do, because leeches are both slippery and stretchy) and threw it as far as you could. If you had looked at it carefully, though, you'd have noticed that, like earthworms, leeches have segmented bodies.

Not all leeches are blood-suckers—in fact, more than half the leech species are predators. The horse leech, for example, which reaches a length of 8 inches (about 20 cm), feeds on smaller annelid worms, snails, and aquatic insect larvae.

TAKE-HOME MESSAGE 11·6

Worms are found in several different phyla and are not a monophyletic group. All are bilaterally symmetrical protostomes with defined tissues. The flatworms and segmented worms (annelids) do not molt; the roundworms do. Flatworms include parasitic flukes and tapeworms, many of which infect humans. Many roundworms are parasites of plants or animals and are responsible for several widespread human diseases. Earthworms are annelids that play an important role in recycling dead plant material.

11·7 Most mollusks live in shells.

Just because they are related, that doesn't mean organisms will have much of a physical resemblance. Consider these animals. A colossal predatory squid, more than 40 feet long, with eyes the size of beach balls and a razor-sharp beak. A small snail—*escargot* in French—one of the more than 700 million consumed in France each year. An oyster, with its nearly 2,000-year-old reputation as an aphrodisiac (unsupported by any scientific proof, it turns out). All of these are mollusks, among the most diverse groups of animals and occupying an impressive and impressively varied position within the human imagination.

More than 100,000 mollusk species have been named (and many more have not yet been named). The members of this large phylum (Mollusca) live in the ocean, in fresh water, and on land and include many familiar creatures in addition to those mentioned above, including clams, scallops, mussels, and octopuses. They are so diverse that it is difficult to describe any single defining characteristic. The position of mollusks within the animal phylogeny, however, reflects several important characteristics that they all share: they have defined tissues, are bilaterally symmetrical, and are protostomes. Further, all mollusks grow by adding tissue rather than by molting.

Several additional features are also common to many mollusks. Some groups of mollusks have a shell that protects the soft body, a mantle (the tissue that secretes calcium

carbonate to form the shell), and a sandpaper-like tongue structure, called the radula, that is used during feeding.

Here, we examine three of the major groups of mollusks. The animals in these groups share most of the features common to mollusks, but with very different body plans: **gastropods, bivalve mollusks,** and **cephalopods** (FIGURE 11-13).

Gastropods Snails and slugs are gastropod mollusks. *Gastropod* means "belly foot," and these mollusks get their name from the expanded foot on the bottom of their body, which allows them to climb a vertical surface as easily as they glide across a horizontal one. Snails have a one-piece, curled shell, and slugs are snails that have just a tiny remnant of a shell that is not even visible because it is covered by the mantle. Found in both aquatic and terrestrial environments, snails and slugs account for three-quarters of all mollusks.

The snail's shell is its primary protection against predators, but for terrestrial slugs and sea slugs, which have very little shell material, other defense mechanisms are necessary. A terrestrial slug relies on slime for defense: when a slug is attacked, it secretes slime that sticks to the predator. Worse still, anything that touches the slime coating the predator sticks to it, so a bird that attacks a slug quickly finds that its beak and face are covered by pieces of dead leaves, clods of soil, and twigs. Sea slugs have other methods of defense.

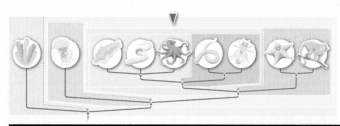

THE MOLLUSKS

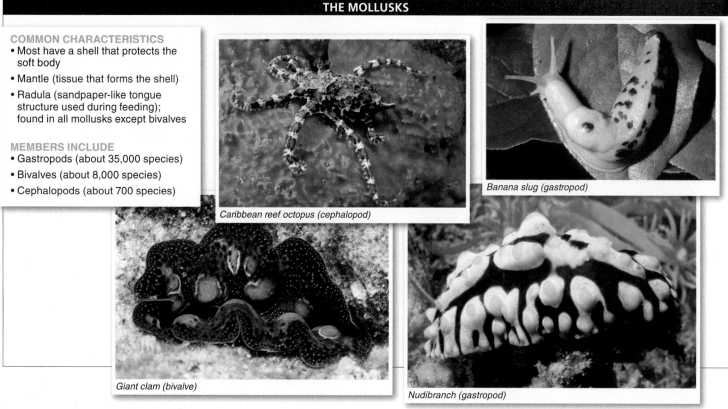

COMMON CHARACTERISTICS
- Most have a shell that protects the soft body
- Mantle (tissue that forms the shell)
- Radula (sandpaper-like tongue structure used during feeding); found in all mollusks except bivalves

MEMBERS INCLUDE
- Gastropods (about 35,000 species)
- Bivalves (about 8,000 species)
- Cephalopods (about 700 species)

Caribbean reef octopus (cephalopod)

Banana slug (gastropod)

Giant clam (bivalve)

Nudibranch (gastropod)

FIGURE 11-13 **An overview of mollusks.**

Some species synthesize toxic chemicals, while those that feed on sponges can sometimes recycle sponge toxins into their own slime. One of the most remarkable uses of another animal's defense equipment is found among sea slugs that eat sea anemones. Anemones are protected by stinging cells, and when eaten, some of these cells are transferred to the sea slug's own skin, enabling it to sting other creatures.

Bivalve Mollusks The bivalve mollusks are soft-bodied animals protected by a pair of shells hinged together by a ligament. Clams, scallops, oysters, and mussels are examples of bivalves. There are about 8,000 bivalve species and most of them live in the ocean, although there are some freshwater species. Clams spend their lives buried in mud or sand, scallops live on the seafloor, and oysters and mussels fasten themselves to underwater objects, such as rocks, the pilings that support piers, and the hulls of boats. All bivalves are filter-feeders that draw a current of water in

through a tube called the "incurrent siphon," across their gills, where tiny food particles are captured, and out through the "excurrent siphon."

When a grain of sand is trapped in the shell of a bivalve, the mantle may secrete layer after layer of a shell-like protein material that covers the sand grain and forms the iridescent gem called a pearl. Oysters are the best known source of pearls, but clams and other bivalve mollusks also form pearls.

Cephalopods The third major group of mollusks is the cephalopods (**FIGURE 11-14**). It includes 6 nautilus species and more than 600 species of squids and octopuses. The nautilus has an external shell, squids have very small shells that are covered by the mantle, and octopuses have lost the shell entirely. In all these species, the tentacles appear to grow directly from the head, which explains the name: *cephalopod* translates as "head-footed."

SQUIDS
- 8 short tentacles and 2 long sucker-bearing tentacles
- Free-swimming
- About 300 species

OCTOPUSES
- 8 short tentacles called arms
- Bottom-dwellers, living in coral reefs and on rocky coasts
- About 300 species

NAUTILUSES
- Chambered shell used for protection
- Free-swimming
- 6 species

FIGURE 11-14 **An overview of cephalopod diversity.**

The most obvious feature of cephalopods may be their tentacles, which they use to walk and swim and to capture prey. Squids—many of which are ferocious predators—have eight short tentacles called arms and two long, sucker-bearing tentacles that are used to capture prey.

After a slow approach to its unsuspecting prey, a squid propels its tentacles forward with astonishing speed—accelerations of more than 800 feet (about 250 m) per second per second have been measured, the equivalent of your car accelerating from 0 to 6,000 mph in 10 seconds! The suckers on the tentacles adhere to the prey and draw it back toward the squid, where the arms take over, turning and manipulating the prey as it is bitten by the squid's sharp beak and pulled into the mouth by the tongue-like radula.

Squids are often more than 3 feet (about 1 m) long, and members of some species may exceed 40 feet (about 13 m). When a squid is in a hurry, it swims tail-end first, using jet propulsion. Water is expelled through the siphon at the tentacle end of the body, shooting the animal backward.

Octopuses, which can have tentacles that spread 12 feet (about 4 m), are bottom-dwellers that live in coral reefs and on rocky coasts. Because they have no shell at all, octopuses can squeeze through astonishingly small openings, and their ability to escape from even carefully covered tanks is a perpetual challenge for the keepers of zoos and aquaria. Keepers are accustomed to arriving in the morning and finding an octopus wandering around the room.

In the next section we further explore the cephalopods and consider the question of whether their skills of manual dexterity make them the smartest of the invertebrates.

TAKE-HOME MESSAGE 11·7

Mollusks are protostome invertebrates that do not molt. They are the second most diverse phylum of animals and include snails and slugs, clams and oysters, and squids and octopuses. Most mollusks have a shell for protection, a mantle of tissue that wraps around their body, and a specialized tongue called a radula.

Are some animals smarter than others?

Squids and octopuses are wide-ranging predators that must locate, capture, and subdue other animals so they can eat them. This is a far more active lifestyle than that of bivalves or snails, and it requires quick movements and rapid responses to stimuli. All of these features are conspicuous elements of the behavior of an octopus as it moves through a coral reef, inserting a tentacle into every opening. In fact, as an expert multi-tasker, an octopus can capture a prey item with one of its tentacles, use a second tentacle to hold something else, and still keep the remaining six tentacles in motion, searching for more prey. In addition, octopuses are very good at manipulating things with their tentacles. It doesn't take a captive octopus long to learn how to unscrew the top on a glass jar to reach a fish that's swimming inside—after all, that's just a minor variation on twisting a clam to open it (**FIGURE 11-15**).

Because of their manipulative skills and easily observed exploratory behavior, octopuses are sometimes featured on TV shows and in magazine articles, where they are often described as the smartest invertebrates and their intelligence is compared to that of mammals. But this is the sort of claim that we must view skeptically.

Q Are octopuses smart?

What is intelligence, after all? The concept of intelligence loses its meaning when we are comparing animals that live in completely different worlds and respond to entirely different stimuli. We can say that octopuses are very good at doing the things that octopuses need to do—searching for prey, capturing it, and manipulating it, all within their marine environment, for example. But is following a complex sequence of actions the same thing as intelligence? Spiders, for example, construct elaborate webs that are strategically placed in the flight paths of insects; they detect the impact of an insect on the web and rush out to wrap the insect in silk before it can escape. Does that require more or less intelligence than opening a glass jar to eat a fish? And what about the red squirrels of northern Europe? Every fall they hide more than 3,000 acorns, cones, and nuts, then recover more than 80% of their hidden nuggets of food over the course of the winter. A human probably couldn't manage that.

If intelligence is defined, in part, as the ability to do some problem-solving task valued by humans, the concept of "intelligence" may actually be relevant only to humans. Applying the question of intelligence to non-human species may just be an assessment of how well they can accomplish tasks that are essential to their survival. For this reason, questions of comparative animal intelligence are not generally useful. Rather, animals' abilities should be considered as evolutionary responses to particular selective pressures imposed by their environment, and the fact that a species currently survives should be taken as an indication of some mastery of its particular niche. As stated earlier in this chapter, from an evolutionary perspective, any species alive today must be considered a success.

TAKE-HOME MESSAGE 11·8

The predatory behavior of octopuses involves exploration and manipulation, behaviors that humans often consider to be intelligent. But the concept of intelligence cannot be applied objectively to other species, which have evolved in response to the selective forces at work in their own particular niches.

FIGURE 11-15 **An octopus attempts to open a jar.**

When it comes to species diversity, the **arthropods** take the cake, outnumbering all other forms of life. In all, the phylum contains almost one million recognized species, or approximately 75% of all animal species and 60% of *all species* on earth (FIGURE 11-16). Taken together, there are more than a billion billion (1 followed by 18 zeroes) arthropods alive at any given time! And about 80% of these are insects. Put another way, on earth, right now, there are more than 150 million insects for every single human. And there are about six times as many species of beetles as there are species of all birds, mammals, fishes, amphibians, and reptiles combined.

The phylum of arthropods (Arthropoda) contains bilaterally symmetrical, protostome invertebrates that have a distinct head, midsection (called the thorax), and abdomen, a rigid external covering (called an **exoskeleton**) made of a stiff carbohydrate known as chitin, and legs with joints (think of all the joints in a lobster's claw) (FIGURE 11-17). The four major lineages of arthropods are:

- Millipedes and centipedes

- Chelicerates (including horseshoe crabs, spiders, mites, ticks, and scorpions)

- Crustaceans (including lobsters, crabs, shrimp, and barnacles)

- Insects

The arthropod phylum is so large that, even if we excluded all the insects—the most diverse arthropod group—it would still contain more species than any other phylum.

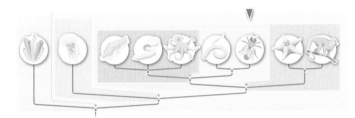

THE ARTHROPODS

COMMON CHARACTERISTICS
- Segmented body with a distinct head, thorax, and abdomen
- Exoskeleton made of chitin
- Jointed appendages

MEMBERS INCLUDE
- Insects (more than 800,000 species)
- Arachnids (about 60,000 species)
- Crustaceans (about 52,000 species)
- Millipedes and centipedes (about 10,000 species)

Elephant hawk moth (insect)

Blue millipede

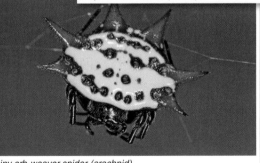

Spiny orb-weaver spider (arachnid)

Spider crab (crustacean)

 Arthropods make up about 75% of the animal species on earth.

FIGURE 11-16 **Arthropod diversity: insects, millipedes and centipedes, arachnids, and crustaceans.**

Cecropia ant   *Millipede*   *Jumping spider*   *Sally lightfoot crab*

INSECTS
- Three pairs of walking legs
- Legs are located on the thorax
- Life cycle consists of separate life stages

MILLIPEDES AND CENTIPEDES
- Many pairs of legs (two pairs per segment in millipedes; one pair per segment in centipedes)
- Long, segmented body

ARACHNIDS
- Usually four pairs of walking legs
- Legs are located on the thorax
- Specialized mouthparts
- Predators

CRUSTACEANS
- Many pairs of legs
- Usually five pairs of appendages that extend from the head
- Mostly aquatic

 FIGURE 11-17 **Exploring the arthropods.**

Millipedes and Centipedes The names *millipede* and *centipede* translate to "a thousand feet" and "a hundred feet" and refer to the enormous numbers of legs on these long, skinny animals. They don't really have a thousand—or even a hundred—feet, but millipedes do have more legs than centipedes, and that is the easiest way to tell them apart. Millipedes and centipedes have long, segmented bodies that seem almost worm-like, but their jointed legs and hard exoskeletons are characteristics of arthropods. Both live among fallen leaves, especially in forests where the leaf litter layer is usually cool and moist. Millipedes feed on decaying plant material, while centipedes are predators that use fangs equipped with venom to kill insects and even snakes and small mammals. Centipedes use their jaws to tear prey into pieces that are small enough to swallow.

Chelicerates There are about 60,000 species of chelicerates, a group that includes four horseshoe crab species and the arachnids, land-dwelling arthropods that include spiders, scorpions, mites, and ticks. Most have four pairs of walking legs and a specialized feeding apparatus. Arachnids are distinguished from other arthropods in that they have only two body regions instead of three.

Arachnids are predators. Many spiders construct webs by spinning silk fibers (which consist of interconnected protein molecules). Orb-weaving spiders construct intricate spiral-shaped webs that trap (and sometimes even lure) flying insects. When a spider captures prey, it uses its fangs to inject venom that contains two types of enzymes. The first set of enzymes disrupts the prey's nervous system and paralyzes it, then other enzymes dissolve its internal organs. When the prey's innards have liquefied, the spider sucks out the contents.

The venom of most spiders is usually harmless to animals as large as humans, but two North American spiders are dangerous. The black widow spider is easily identified by a red hourglass marking on the bottom surface of the female. The brown recluse spider lacks distinctive markings, so it is harder to identify—but it is more dangerous. Both species are likely to move indoors, constructing webs in little-used spaces such as closets and cabinets and behind furniture. Brown recluse bites can create open wounds that can last for months or years and sometimes require repeated skin grafts to close.

Scorpions look dangerous to humans, but their sting is no worse than a bee's. Scorpions are lethal to insects, hunting them at night. They seize prey with their claws and use their stinger only if the prey struggles. Some of the scorpion's mouth structures are covered with spines that grind the prey, while glands secrete enzymes that liquefy it, and the scorpion sucks up the soup.

Crustaceans If arachnids are the most feared arthropods, the aquatic crustaceans—lobsters, crayfish, crabs, and shrimp—may be the most desirable arthropods, at least at meal time. In the United States, the top three fishery crops, by annual market value, are shrimp, crabs, and lobsters.

There are about 52,000 species of crustaceans, and although crustacean species can look very different—from sessile barnacles to mobile, tiny krill—they have one feature in common: five pairs of appendages extending from the head. Three of these appendages are used for feeding, and two are antennae that sense the environment.

Crustaceans have many pairs of legs, and their legs are modified for many purposes. Shrimp and barnacles (which are closely related to shrimp, but spend their adult lives within a shell, fastened to a rock) have legs with comb-like projections that they use to capture plankton. Free-living crustaceans (shrimp, crabs, lobsters) have modified limbs in the abdominal region that are used to hold eggs or newly hatched young.

Although most crustaceans are aquatic, the inconspicuous wood lice (also known as pill bugs, sow bugs, or roly-polies) are terrestrial forms that you can find when you turn over a rock in your garden. Wood lice are herbivores and play an important role in recycling dead plant material.

Insects Mostly terrestrial, insects have three pairs of walking legs, and most also have one or two pairs of wings extending from the thorax. Insects abound in nearly all terrestrial habitats and are, by far, the most diverse group of arthropods. Every species of plant, for example, is eaten by at least one insect species. Wings and the ability to fly, as well as the developmental process of metamorphosis, have played important roles in the great success and diversification of insects. We explore these features in more detail in Section 11-11.

TAKE-HOME MESSAGE 11·9

The arthropods are protostome invertebrates, and with nearly one million species (and probably at least as many more yet to be identified), they outnumber all other forms of life in species diversity. Centipedes are predators with fangs that inject venom, and millipedes are herbivores that feed on dead plant material. Spiders and scorpions are predatory arthropods that eat insects and, occasionally, small vertebrates. Lobsters, crabs, shrimp, and barnacles are predatory marine crustaceans. Insects, with adaptations that include the ability to fly and metamorphosis, are the most diverse group of arthropods.

❓ 11·10 THIS IS HOW WE DO IT

How many species are there on earth?

When contemplating a question in biology, the most useful approach often resembles a sort of "back-of-the-envelope" calculation. This typically requires gross simplification, but it can illuminate the information needed to come to a more accurate answer, while revealing assumptions that may have a large influence on the answer obtained.

As we consider the diversity of life on earth, one seemingly straightforward question comes to mind.

How many distinct species are there on our planet?

It might come as a surprise to learn that we don't really know the answer. And it's even a bit embarrassing that we aren't very confident about making even a ballpark estimate. Some scientists put the number at around 3 million, while others believe it could be greater than 100 million.

In a famous paper published in 1982—a paper actually short enough to fit on the back of an envelope—the biologist Terry Erwin laid out a simple approach to estimating the answer. His strategy was two-pronged. It began by posing a question that he could answer with a high degree of accuracy.

How many species of beetles are there in one species of tree?

More specifically, Erwin set out to get a definitive count of all the beetle species living in the canopy of a single species of tree (*Luehea seemannii*) in a tropical forest in Panama. His methods were not elegant. Over the course of three seasons, he sprayed a fog of pesticide into the canopy of 19 trees of this species, then collected and identified all of the beetles that fell to the ground.

Here's what he found: 1,143 different species of beetles fell from the trees. Because other researchers had reported collecting additional beetle species from the same species of tree, he rounded off the number to 1,200 species.

The second prong of Erwin's strategy was to estimate how this number was likely to translate into the numbers of other insect species living in the diverse tree species in the tropics. In essence, from data he had high confidence in, he made an extrapolation.

> **How is the number of beetle species in one tree species related to the number of all insect species in all types of trees?**

This required several educated guesses (Erwin is, after all, an expert on beetles), along with a bit of speculation. His back-of-the-envelope calculations included estimates of the proportion of beetle species that are host specific (13.5%), the number of unique trees per hectare (~70), the proportion of all arthropod species that are beetles (40%), the proportion of arthropods that are in the forest canopy as opposed to elsewhere (67%), and the number of unique species of tropical trees (50,000). When he multiplied everything together, his estimates led him to conclude that there were almost 30 million species of arthropods in tropical forests. This number was remarkable because it was about 10 times higher than most previous estimates of the total number of species on earth!

Understanding the self-correcting and collaborative nature of scientific thinking, Erwin did not claim this was the true and final answer. He said, "I would hope someone will challenge these figures with more data." Erwin had, after all, made numerous assumptions. (Which of these seem likely to have had a significant influence on his final answer?)

Taking Erwin's ideas as a starting point, numerous researchers have noted (and tried to improve on) his numerous assumptions. In 2010, researchers using more data and more realistic assumptions came up with a better estimate. Rather than making precise estimates for each parameter, they used probability distributions (such as for the proportion of beetle species specific to each species of tree).

This enabled the researchers to come up with a whole range of estimates, each having a likelihood associated with it, and their estimates were much lower than Erwin's 30 million. One of their models generated an estimate of the median number of tropical arthropod species as 3.7 million, with a 90% confidence interval that the true number was between 2.0 and 7.4 million.

These estimates and the process by which they've been generated give us greater confidence that we know the answer. Unfortunately, this isn't an answer to our original question. It's an answer to an easier question: how many *arthropod* species are there in the *tropics*?

The reason for limiting the scope of the question is that it is much easier to define and identify beetle or other arthropod species. But does our answer bring us closer to knowing how many species, in total, there are on earth? That is difficult to know. Consider another question.

> **What types of species are the most difficult to estimate (or count)?**

Given the impossibility of applying the biological species concept to asexual species, for example, how should we define and identify species of prokaryotes and viruses? And, are the numbers of marine species proportional to the numbers of arthropod species.

The back-of-the-envelope approach described here can be a useful starting point for these new calculations as well.

TAKE-HOME MESSAGE 11·10

Determining how many distinct species there are on earth has long been a challenge for biologists. By counting the exact number of beetle species in one species of tropical tree and estimating several parameters that can be used to extrapolate this number to the number of species of all arthropods across the entire tropics, it is possible to make headway on this challenge. Initial estimates using this method suggest there may be many more species on earth than previously believed.

Flight and metamorphosis produced the greatest adaptive radiation ever.

Two characteristics of insects that have been central to their success are flight and the way they cope with the change in body size as they grow. Wings and the ability to fly are adaptations that first appeared in insects. With flight, insects are able to avoid many predators and can efficiently search for food and for mates. Flight also enables organisms to more quickly leave one location and disperse to another that is potentially more hospitable.

As arthropods, insects have a modular body plan and a rugged exoskeleton, to which their muscles are attached. Like a suit of armor, the exoskeleton provides excellent protection from injury and predators. It also helps individuals conserve water and resist drying out. One challenge associated with having an exoskeleton, however, is that it prohibits growth. Nearly all insects (and a small number of other arthropod species) overcome this constraint with a process of growth

Q Mammals get bigger and bigger the more they eat. Why don't insects?

(shared by the roundworms, of phylum Nematoda) that is very different from the pattern seen in other animals, such as mammals. After hatching, their life is divided into three completely different stages (**FIGURE 11-18**).

1. Larva. An egg hatches into a larva (*pl.* larvae), which looks completely different from the adult form. A caterpillar is the larval stage of a butterfly or moth, and a grub is the larval stage of a beetle. A larva's job is to eat and grow large enough to enter the next life stage, the pupa. Larvae are able to grow by passing through numerous stages (usually 4–8) in which they shed their exoskeleton, increase in size, and develop a new, slightly larger exoskeleton, in the process of molting.

2. Pupa. Eventually, a larva completely covers itself with a casing. Covered by the case—which has different names (such as cocoon or chrysalis) in different insect groups—the

---

METAMORPHOSIS

COMPLETE METAMORPHOSIS
Complete metamorphosis is the division of an organism's life history into three completely different stages (occurs in 83% of insect species).

INCOMPLETE METAMORPHOSIS
Incomplete metamorphosis is the pattern of growth and development in which an organism does not pass through separate, dramatically different life stages (occurs in 17% of insect species).

LARVA
An egg hatches into a larva, which eats (and eats), growing as it molts several times until large enough to enter the pupal stage.

PUPA
The larva encloses itself in a case, body structures are broken down, and new structures are assembled into the adult form.

ADULT
The adult emerges from the pupa and no longer grows. Its primary function is to reproduce.

NYMPH
An egg hatches into a nymph, resembling a small version of the adult (but without wings or reproductive organs), growing as it molts several times.

ADULT
Upon reaching adult size, the nymph stops molting and the adult forms.

 The separation of life stages has contributed to the enormous ecological diversity of insects.

FIGURE 11-18 **Metamorphosis.** Some insects have a life history divided into unique stages, each with a distinct purpose.

insect is now called a pupa. In the pupa, genes are activated that code for the adult body form, and the proteins that made up the larva are broken down to amino acids, which are recycled to synthesize adult proteins and reassembled into the adult form. This rebuilding process is called **metamorphosis,** from two Greek words meaning "change of form." The entire process requires several days, and by the end of the pupal stage, the adult insect is curled tightly inside the pupal case.

3. Adult. When the **adult** form hatches from the pupal case, it is at its full size. Insects do not grow after they emerge from their pupae—for this reason, some adults do not even eat. Rather than eating and growing, the job of an adult insect is reproduction.

The developmental process called **complete metamorphosis,** which includes these three stages, occurs only in insects and allows the larva and adult to act as if they were animals from different species, each optimized to perform very specialized tasks. Caterpillars eat leaves, for example, and spend their entire larval period on a single plant. In contrast, butterflies feed on nectar that they collect by flying from flower to flower, saving up enough resources to lay or fertilize eggs. As a larva, the animal feeds and grows. As an adult, it reproduces. The genes that control the larval body form are different from the genes that determine the adult form, so natural selection can act on the larval and adult stages independently.

About 83% of insect species go through complete metamorphosis, and this separation of the life stages has been an important factor in helping insects diversify into nearly 20 times as many named species as vertebrates. Among those insect species that do not undergo the dramatic changes of complete metamorphosis, typically the eggs hatch as nymphs (the juvenile form), which resemble a smaller version of the adult. These juveniles do not have wings or reproductive organs, but they live in the same habitats as adults and eat the same foods. They then undergo several molts as they grow, and stop molting when they reach adult size. This pattern of growth and development is called **incomplete metamorphosis,** and it occurs in grasshoppers and cockroaches, among many other species.

TAKE-HOME MESSAGE 11·11

The ability to fly and the developmental process of metamorphosis as a means of overcoming the constraints of having a rugged exoskeleton have contributed to the enormous ecological diversity of insects. The life cycle of most insects includes a larval stage that is devoted to feeding and growth, a pupal stage during which metamorphosis occurs, and an adult stage in which the insect reproduces.

11·12 Echinoderms are vertebrates' closest invertebrate relatives and include sea stars, sea urchins, and sand dollars.

Appearances can be deceiving. Sea stars and other members of the phylum Echinodermata live in the oceans, are radially symmetrical (as adults), and are covered by spiny plates. And yet, because they are deuterostomes (their gut forms from back to front), it turns out that they are more closely related to humans than to any other invertebrate group. Echinoderms stand as an example of how classifying organisms based on evolutionary relatedness—rather than on a comparison of observable physical traits—can reveal a great deal about the force of natural selection and its dramatic effects on body form in adapting populations of organisms to their environment.

The echinoderms include about 6,000 species of marine animals, most of which are enclosed by a hard skeleton of spiny plates (**FIGURE 11-19**). Adult sea stars (starfishes and

brittle stars), probably the most recognizable echinoderms, have five or more appendages evenly distributed around their body circumference. They move with equal ease in any direction, and their sensory organs are distributed around their circumference. Adult sea urchins and sand dollars don't have projecting arms, but they also are radially symmetrical.

Although radial symmetry is characteristic of adult echinoderms, their larvae are bilaterally symmetrical. Bilaterally symmetrical larvae are evidence that echinoderms evolved from bilaterally symmetrical ancestors, and their radial symmetry as adults is an adaptation associated with their mode of locomotion and feeding specializations.

Echinoderms do not have a brain. They have a nervous system that consists of a central ring of nerves, with

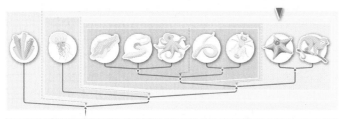

THE ECHINODERMS

Red-and-yellow necklace sea star

Sand dollar

Spines of a fire urchin

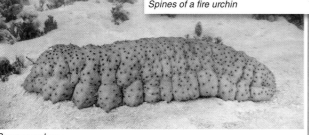

Sea cucumber

branches that extend into each of their appendages and help them gain information about and respond to their environment. Echinoderms creep on little tube feet that are extensions of an internal system of water-filled canals radiating throughout the body. Tube feet carpet the undersides of sea stars' arms and extend up and around the bodies of sea urchins and sand dollars, extending and contracting in waves to grasp and release the surface on which they move as the animal glides along.

Although each tube foot is tiny, there are thousands of them, and their combined force allows a sea star to pull the two shells of a clam or mussel apart. Once it has the shells apart, an extraordinary and unexpected thing happens. The sea star pushes its stomach out through its mouth and inserts it into the opened shells. The stomach then secretes digestive enzymes that break down the clam or mussel tissue, and the sea star absorbs the soup that results. Sea urchins feed on algae, scraping them loose from rocks with sharp tooth-like surfaces made of calcium. Sand dollars capture floating particles of algae and other organic matter by trapping them in streams of mucus, which are moved toward their mouth by beating cilia.

TAKE-HOME MESSAGE 11·12

Because they are deuterostomes (as are vertebrates), echinoderms are the invertebrates that are the closest evolutionary relatives to the vertebrates (and other chordates). Their aquatic larvae are bilaterally symmetrical and share some anatomical features with chordates, but adult echinoderms are radially symmetrical.

COMMON CHARACTERISTICS
• Enclosed by a hard skeleton of spiny plates
• Larvae are bilaterally symmetrical and share some anatomical features with chordates
• Adults are radially symmetrical
• Undersides are covered with tube feet that aid in locomotion and grasping

MEMBERS INCLUDE
• Sea stars (about 1,600 species)
• Sea urchins and sand dollars (about 940 species)
• Sea cucumbers (about 1,100 species)

FIGURE 11-19 **The echinoderms.**

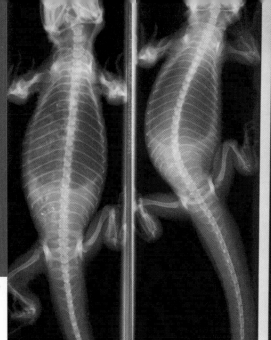
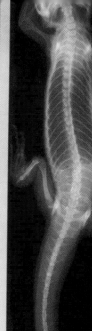

11·13 – 11·15
The phylum Chordata includes vertebrates, animals with a backbone.

Animals in motion: these X-ray images show a tuatara (a New Zealand reptile) walking.

11·13 All vertebrates are members of the phylum Chordata.

Vertebrates are deuterostome animals, having defined tissues and bilateral symmetry. Vertebrates are part of a phylum called Chordata (the chordates) that includes three major groups in total; the two other groups are the tunicates and the lancelets (**FIGURE 11-20** 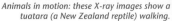). For at least a portion of their life cycle, members of all three of the major groups of chordates possess four distinct chordate body structures that are found in no other animal groups (**FIGURE 11-21**).

1. The **notochord,** a rod of tissue extending from head to tail, is the structure that gives chordates their name (from the Latin for "cord"). The notochord stiffens the body when muscles contract during movement. The simplest chordates—those most distantly related to the vertebrates—retain the notochord throughout life. In the more complex chordates such as vertebrates, however, the notochord is present only in the early embryo and is replaced by the backbone (vertebral column) as the embryo develops.

2. A **dorsal hollow nerve cord** extends along the animal's back (its dorsal side) from its head to its tail. In vertebrates, this nerve cord eventually forms the central nervous system, which consists of the spinal cord and the brain. Other kinds of animals (worms, insects, and so on) also have a nerve cord, but it lies in the lower portion of the front (ventral) part of the body and is solid instead of hollow.

3. **Pharyngeal slits** are present in the embryos of all chordates, but in many chordates (including humans) the slits disappear as the animal develops. The earliest chordates were aquatic, and to breathe and feed they passed water through slits in the pharyngeal region (the area between the

back of the mouth and the top of the throat). In humans, pharyngeal slits are present only in the embryo; our gills were lost far back in evolutionary time.

4. A **post-anal tail** is another chordate characteristic. The posterior (back) end of the digestive system is the anus. The region of the body extending beyond the opening of the anus is referred to as "post-anal." All vertebrates have a tail in this location, but some, including humans, have a tail only for a brief period, during embryonic development.

Although all chordates share these four characteristic structures, the chordates exhibit tremendous physical and ecological diversity.

Tunicates (Urochordata; about 2,000 species) are invertebrate marine animals that have defined tissues, bilateral symmetry, and deuterostome development, in which the gut develops from back to front. The adults are about the size of your thumb and look like balls of brownish green jelly. You can find them attached to docks and the mooring lines of boats. It is the free-swimming larvae of tunicates that reveal their chordate characteristics. Tunicate larvae have a distinct notochord, dorsal hollow nerve cord, and tail, all of which disappear during development. The sessile adult tunicates only vaguely resemble their larvae. Adult tunicates are filter-feeders: cilia draw a current of water through the mouth into the pharynx and out through the pharyngeal slits. Microscopic food items are trapped by a layer of mucus stretched across the slits.

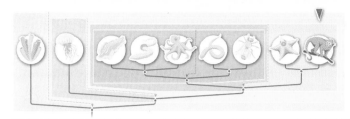

THE CHORDATES

COMMON CHARACTERISTICS
All chordates possess four common body structures, although in many chordates, these structures are only present during specific life stages.

• Notochord
• Dorsal hollow nerve cord
• Pharyngeal slits
• Post-anal tail

MEMBERS INCLUDE
• Tunicates (about 2,000 species)
• Lancelets (about 20 species)
• Vertebrates (about 56,000 species)

Golden lion tamarin (vertebrate)

Compound ascidians (tunicates)

Schooling bannerfish (vertebrates)

Lancelet

Thorny devil lizard (vertebrate)

FIGURE 11-20 **An overview of chordates.**

THE DISTINCT BODY STRUCTURES OF CHORDATES

The members of Chordata are defined by four distinct body structures present in all chordates but in no other animal groups.

NOTOCHORD
A rod of tissue extending from the head to the tail
• Simpler chordates retain the notochord throughout life
• In more complex chordates, the notochord in early embryos is replaced by a backbone

PHARYNGEAL SLITS
Slits through which water is passed in order to breathe and feed
• In many chordates (including humans), the slits disappear as the animal develops

DORSAL HOLLOW NERVE CORD
Nerve cord that extends along the animal's back (its dorsal side)
• In vertebrates, the nerve cord eventually forms the spinal cord and brain

POST-ANAL TAIL
Tail that extends beyond the posterior (back) end of the digestive system
• Some vertebrates (including humans) have a tail only briefly, during embryonic development

FIGURE 11-21 **Characteristic chordate body structures.**

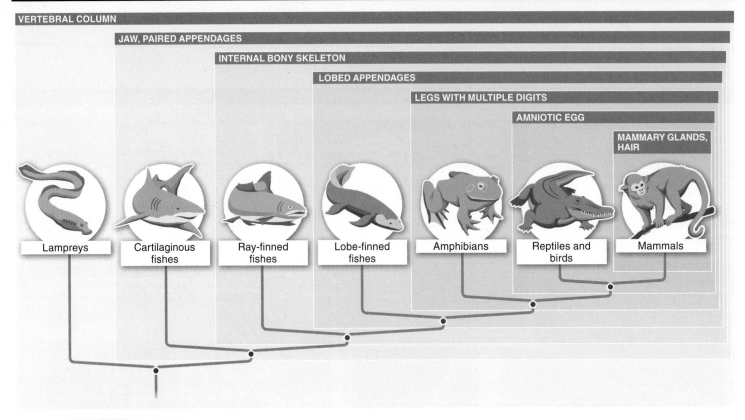

VERTEBRAL COLUMN

JAW, PAIRED APPENDAGES

INTERNAL BONY SKELETON

LOBED APPENDAGES

LEGS WITH MULTIPLE DIGITS

AMNIOTIC EGG

MAMMARY GLANDS, HAIR

Lampreys | Cartilaginous fishes | Ray-finned fishes | Lobe-finned fishes | Amphibians | Reptiles and birds | Mammals

FIGURE 11-22 **Vertebrate phylogeny.**

Lancelets (Cephalochordata; about 20 species), also bilaterally symmetrical deuterostomes, are slender, eel-like invertebrate animals, about the length of your little finger, that live in coastal waters. Unlike tunicates, both embryonic and adult lancelets have all the chordate characteristics. Lancelets are also filter-feeders.

Vertebrates (Vertebrata; about 56,000 species) are the most diverse group of chordates (**FIGURE 11-22**). Because they are deuterostomes, they are bilaterally symmetrical, with defined tissue and from-back-to-front gut development. Vertebrate differ from the other chordates in two important ways.

1. They have a backbone, formed when a column made from hollow bones (or cartilage in some organisms), called vertebrae, forms around the notochord. This backbone surrounds and protects the dorsal hollow nerve cord.

2. They have a head, at the front (anterior) end of the organism, containing a skull, a brain, and sensory organs.

The size range of vertebrates is huge. The smallest is a frog species from New Guinea that is just over a quarter-inch (6.4 mm) long. And some tiny hummingbirds and shrews weigh only a tenth of an ounce (about 3 grams). Blue whales, weighing in at more than 130 tons (120,000 kg), are the largest. Vertebrates move by swimming, burrowing, crawling, walking, running, and flying, and they are present in all habitats on earth. But despite all these differences, all vertebrates have the four chordate characteristics at some stage of their lives. In the remaining sections we focus on the vertebrates, which are by far the most diverse and widely dispersed group of chordates.

TAKE-HOME MESSAGE 11·13

All chordates have four characteristic structures: a notochord, a dorsal hollow nerve cord, pharyngeal slits, and a post-anal tail. The three subphyla of chordates, though superficially very different, are united by possessing these four structures at some stage of their life cycle.

11·14 The evolution of jaws and fins gave rise to the vast diversity of vertebrate species.

The earliest vertebrates were fish-like animals that lived more than 500 million years ago and did not have jaws. Two kinds of jawless vertebrates still exist: the lampreys (with 41 species) and the hagfishes (with 43 species). Most lamprey species do not feed as adults, instead living off energy stores built up by filter feeding in their larval stage. Some lamprey species, however, are parasitic and attach to other fishes by creating suction with a circular oral disk that is studded with sharp spines. First, the spines scrape an opening through the scales and skin. Then, a protein produced by the lamprey keeps the wound from closing, and the lamprey feeds on the blood and body fluids that seep out. Hagfishes feed on dead animals, using two spiny dental plates that they embed in their prey.

Lampreys and hagfishes are about as basic as a vertebrate can be. They have a simple, tube-like shape with no fins or limbs. And because they do not have jaws, they cannot seize prey or chew food before swallowing it. The evolution of fins and jaws is what set the stage for the explosion of diversity in the vertebrates.

Q Why are fins and jaws needed?

Fishes swim the same way that lampreys and hagfishes swim—that is, by bending the body from side to side to create an S-shaped wave that moves from head to tail. A typical fish has seven fins serving different purposes: to drive the fish forward, to minimize rolling from side to side, and for steering and stopping. The combined effects of the seven fins allow the fish to swim rapidly in a straight line when it is in open water or to weave its way through a dense stand of plants or corals.

The evolution of fins was paralleled by the evolution of jaws, because the two structures work together. Fins get you to the organism you are going to eat, and jaws capture and kill it (**FIGURE 11-23**).

The development of jaws and fins allowed great diversification in vertebrate body shapes and composition. There are three categories of jawed fish. **Cartilaginous fishes,** with about 880 species, include the sharks and rays (**FIGURE 11-24**). Cartilaginous fishes are characterized by a skeleton made entirely of cartilage, a solid but slightly flexible connective tissue—the same tissue that gives your nose and ears their shape.

Ray-finned fishes have a rigid skeleton made from bone, which, like cartilage, is a solid connective tissue consisting of specialized cells and an extracellular material (matrix) that the cells secrete. Bone, however, is much less flexible

THE EVOLUTION OF JAWS AND FINS

FISHES WITHOUT JAWS OR FINS
• Tail propels organism through water
• Feed by attaching oral disk to prey

Oral disk

FISHES WITH JAWS AND FINS
• Fins provide controlled movement through water
• Jaws lined with sharp teeth allow for seizing and chewing prey

💡 *The development of fins and jaws set the stage for the evolutionary explosion of vertebrate diversity.*

FIGURE 11-23 The lamprey and the shark.

Stingray

Coral grouper

Ray fin

Australian lungfish

Lobe fin

CARTILAGINOUS FISHES
• Characterized by a skeleton made completely from cartilage
• About 880 species, including sharks and rays

RAY-FINNED FISHES
• Characterized by rigid bones and fins lined with hardened rays
• Possess a swim bladder, which aids in flotation
• About 27,000 species, including almost everything you think of as "fish"

LOBE-FINNED FISHES
• Characterized by two pairs of sturdy fins on the underside of their body
• 6 species of lungfish and 2 species of coelacanths

FIGURE 11-24 **Fishes with jaws: cartilaginous, ray-finned, and lobe-finned fishes.**

than cartilage because its extracellular matrix is mineralized by crystals of calcium phosphate. Ray-finned fishes have a mouth at the narrow tip, or apex, of the body, and their fins, made from webs of skin, are supported with hardened rays made of bone.

Ray-finned fishes are the largest group of jawed fishes and the most diverse group of vertebrates, with about 27,000 species. This group includes almost everything you think of as "fish," from salmon to goldfish.

In addition to a bony skeleton, an important evolutionary development in the ray-finned fishes was the swim bladder, a gas-filled organ that keeps the fish from sinking. This bladder is believed to be homologous to the lungs of land-dwelling (terrestrial) vertebrates, with both having evolved from simple air sacs connected to the gut.

Cartilaginous fishes, which have no swim bladder, must constantly move through the water or they will sink.

The third group of jawed fishes, the **lobe-finned fishes,** are represented by just eight species. These fishes have sturdy pelvic and pectoral fins on the underside of their body (see Figure 11-25), which, unlike other fins, have a central appendage containing numerous bones and muscles that connect the fins to the body. As we see in the next section, lobe-fins were useful in initiating the move onto land.

TAKE-HOME MESSAGE 11·14

The development of two structures in fishes—fins and jaws—set the stage for the enormous diversity of modern vertebrates.

11·15 The movement onto land required lungs, a rigid backbone, four legs, and eggs that resist drying.

Movement from water onto land required four major evolutionary adaptations: lungs, a backbone, four legs, and eggs that won't dry out. All four of these characteristics are necessary for a land animal, and all four evolved in predatory fishes that lived in shallow water, the immediate ancestors of terrestrial vertebrates.

As we've noted, lungs probably evolved from the air sacs that were also evolutionary precursors to the swim bladder found in ray-finned fishes. All of the terrestrial vertebrates—amphibians, reptiles (including birds), and

Where did legs and lungs come from?

mammals—are descendants of the lobe-finned fishes that lived during the Devonian period, some 400 million years ago. At this time, the lobe-finned fishes had lungs and the rudiments of limbs, but they were still fully aquatic and living near the shore. The four sturdy fins on the underside of their body helped them move through shallow water, and lungs allowed them to breathe air when the oxygen concentration in the warm, stagnant water was low. Thus, these fishes already had some of the basic characteristics of terrestrial animals. The jointed bones in the fins of lobe-finned fishes don't look much like your arms and legs, but they are homologous structures.

To move onto land, a vertebrate needed more than just legs and lungs. It needed structural support to resist the pull of gravity. Each vertebra of a terrestrial vertebrate has projections that interlock with projections from the vertebra ahead of it and the vertebra behind it. These interlocking projections prevent the backbone from sagging under the pull of gravity, and the body weight is transmitted through the limbs to the ground.

The last innovation necessary to move onto land was an egg that resists drying out (**FIGURE 11-25**). When eggs are deposited on land, they are exposed to air and lose water by evaporation. The eggs of terrestrial animals need a waterproof covering—a membrane and a shell—to prevent them from drying out before they hatch. The appearance of eggs with a shell (about 380 million years ago) further facilitated the evolution of entirely terrestrial vertebrates, the groups that evolved into mammals and reptiles (including birds). But these initial egg coverings were soft and likely still required a watery environment. The eggs of terrestrial vertebrates have a water-tight membrane that keeps the embryo surrounded by a bath of fluid called amniotic fluid.

TAKE-HOME MESSAGE 11·15

Four adaptations were important in the transition of life from water to land. Fins were modified into limbs. Vertebrae were modified to transmit body weight through the limbs to the ground. The site of gas exchange was transferred from gills and swim bladders to lungs. And terrestrial vertebrate eggs with membranes and a shell resisted drying out.

FROM WATER TO LAND

The transition of vertebrates from life in water to life on land required overcoming three main obstacles. Four major evolutionary innovations allowed for this transition.

PROBLEM: RESPIRATION

Aquatic animals use gills to acquire dissolved oxygen from water. The transition onto land required the ability to breathe air.

SOLUTION: LUNGS
Gas exchange was transferred from the gills to lungs, which evolved from the air sacs (that were also the evolutionary precursors to the swim bladder found in ray-finned fishes).

Ray-finned fish

Early terrestrial vertebrate

PROBLEM: GRAVITY

Water is buoyant, so aquatic animals don't need much structural support. The transition onto land required structural support to resist the pull of gravity.

SOLUTIONS: LIMBS and MODIFIED VERTEBRAE

Limbs evolved from the jointed fins found on the underside of lobe-finned fishes.

Vertebrae were modified to transmit the body weight through the limbs to the ground.

Lobe-finned fish

Early terrestrial vertebrate

PROBLEM: EGG DESICCATION

The transition onto land required an egg that resisted drying out when exposed to air.

SOLUTION: AMNIOTIC EGG
Terrestrial animals developed a waterproof eggshell, which prevents eggs from drying out before they hatch.

FIGURE 11-25 How did vertebrates make the transition from life in water to life on land?

11·16 – 11·20
All terrestrial vertebrates are tetrapods.

Capybara (Hydrochoerus hydrochaeris) rodents found in South America. They can reach more than 200 pounds (91kg)!

11·16 Amphibians live a double life.

The ancestors of all vertebrates that live on land had lungs and four legs. Some modern vertebrates, such as whales and snakes, have evolved so that their limbs are reduced to a few shrunken and unused bones. Whales and some snakes have returned to living in water, and some salamanders have lost lungs and breathe through their skin, but these are recent changes, occurring in species whose ancestors had lungs and four legs. Taxonomically, all terrestrial vertebrates are **tetrapods** (*tetra* = four; *poda* = feet).

Terrestrial vertebrates are divided into two main groups: (1) animals (called **non-amniotes**) such as amphibians that reproduce in water and do not have desiccation-proof amniotic eggs; and (2) animals (called **amniotes**) such as reptiles, birds, and mammals that have amniotic eggs. We discuss amphibians here, then reptiles and mammals in the next two sections.

The first terrestrial vertebrates were **amphibians,** from the Greek word *amphibios,* meaning "living a double life." There are two stages in the life of most amphibians: a water-breathing juvenile form and an air-breathing adult form. The adults of many amphibian species lay their eggs in water (**FIGURE 11-26**). Amphibian eggs lack the waterproof

---

THE AMPHIBIAN LIFE CYCLE

Amphibians, such as frogs, toads, and salamanders, are terrestrial vertebrates (tetrapods) with non-amniotic eggs. Most species live on land as adults, but develop in water.

EGGS
Amphibians have non-amniotic eggs, which must be laid in water to prevent desiccation.

JUVENILES
Amphibians spend their juvenile stage underwater and undergo metamorphosis to develop legs and lungs.

ADULTS
Only the adults are true land animals; however, most of the species in this group stay close to water to lay their eggs.

FIGURE 11-26 **From egg to larva to adult: the amphibian life cycle.**

layers that allow amniotes to lay eggs on land, so most of the 6,000 species in this group are still significantly tied to life in the water—because they must always be near water to lay their eggs, which are simple structures, not unlike fish eggs. Frogs and toads make up the vast majority of amphibians (about 5,400 species). Other amphibians include salamanders (550 species) and caecilians, a group of legless, burrowing animals (170 species).

The juvenile stage of frogs, called a tadpole, lives in water, lacks legs, and eats algae. A few weeks (or up to many months in some species) after hatching, metamorphosis occurs, and the tadpole develops legs, lungs, and a digestive system fit for its adult life as a carnivore. Adult frogs have thin, moist skin through which gas (oxygen and carbon dioxide) exchange with the air can take place. When a frog

is underwater, for example, oxygen diffuses directly from the water into the blood.

The past two decades have seen a stunning decline in amphibian populations throughout the world, and almost a third of all amphibian species are endangered or threatened. This situation is attributable to a combination of causes: climate change, habitat degradation, fungal diseases, and increased pollution.

TAKE-HOME MESSAGE 11·16

Amphibians are terrestrial vertebrates, but the adults of most species still lay eggs in water. The eggs hatch into aquatic juveniles.

11·17 Birds are reptiles in which feathers evolved.

Soon after amniotic vertebrates appeared, two different evolutionary lineages began to diverge (**FIGURE 11-27**). One of these lineages is the mammals, the group to which humans belong. The other lineage is the one referred to in this chapter as "reptiles (including birds)." That is an awkward term, but the evolutionary tree shows why it is necessary. Surprisingly, birds (about 9,700 species) are one branch of the reptile lineage that also includes snakes and

lizards (about 8,000 species), turtles (about 300 species), crocodiles and alligators (23 species), the New Zealand tuatara (2 species)—and the dinosaurs.

The characteristics that hold the bird-crocodile-dinosaur group together are mostly similarities in bones (especially the bones of the skull and legs) and DNA sequences. Reptiles are amniotes and thus have amniotic eggs. You are

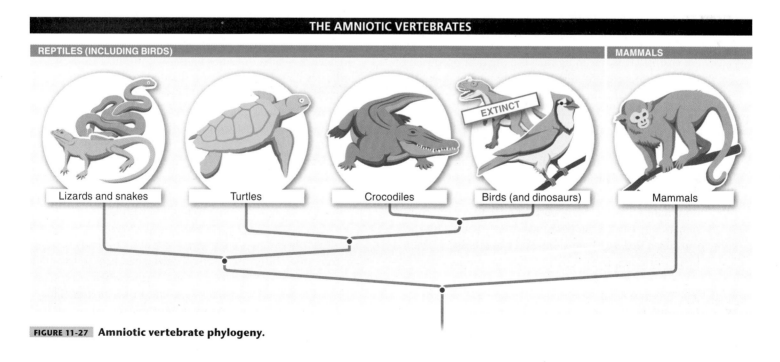

FIGURE 11-27 Amniotic vertebrate phylogeny.

REPTILES
- Skin is covered in scales
- Body temperature is controlled by external conditions, such as air temperature (exothermic)
- Include snakes and lizards (about 8,000 species), turtles (about 300 species), crocodiles and alligators (23 species), and tuatara (2 species)

BIRDS
- Have feathers and wings, providing insulation and enabling flight
- Body temperature is maintained by heat generated from cellular respiration (endothermic)
- Include about 9,700 species

FIGURE 11-28 **Comparison of reptiles and birds.**

familiar with the hard-shelled eggs of birds, and some lizard, snake, and turtle eggs also have hard shells. Other species in these groups have eggs with a paper-like shell.

Most people find it confusing that birds are grouped with turtles, lizards and snakes, and alligators and crocodiles, and that confusion is easy to understand (**FIGURE 11-28**). After all, birds are covered by feathers and can fly, whereas the other reptiles have bare skin and do not fly. And there is an important physiological difference as well: birds are **endotherms,** meaning they use the heat produced by cellular respiration to raise their body temperature above air temperature, whereas other reptiles are **ectotherms**—they bask in the sun to raise their body temperature and seek the shade when the air is too warm.

How can animals as different as crocodiles and hawks be closely related? During the Mesozoic era, from about 250 million to 65 million years ago, dinosaurs were the dominant terrestrial vertebrates. The fossil record shows a remarkably clear series of animal forms that bridge the transition between bare-skinned dinosaurs and feathered birds. Birds are, in fact, just one group of dinosaurs.

The most apparent difference between reptiles (including most dinosaurs) and birds is the presence of feathers in birds. The fossil record

Q Birds are reptiles?

reveals that feathers evolved before birds. Fossil evidence also suggests that the initial evolution of feathers probably had nothing to do with flight.

Among reptiles, many different species had feathers. A small, agile dinosaur called *Sinosauropteryx,* for example, living about 120 million years ago, had feathers that were simple spiky filaments on its neck, back, and tail, along with shorter filaments covering its body (**FIGURE 11-29**). The 2010 discovery of fossilized pigment-packed organelles—called melanosomes—within these feathers indicates that the feathers and filaments were probably brightly colored. Researchers believe that they may have been used by male *Sinosauropteryx* for courtship displays to females, as well as for aggressive displays to other males. *Sinosauropteryx* and many of the other feathered dinosaurs were latecomers, however. The pigeon-sized *Archaeopteryx,* which was among the first bird species, was already, by 147 million years ago, using its feathers to fly; by 140 million years ago, the skies were filled with birds.

Another big difference between birds and reptiles is that, as noted earlier, birds use internally generated heat to maintain a high body temperature, whereas reptiles rely on the sun to heat their bodies. This evolutionary change may have been related to feathers. As feathers evolved for display

Q What were the first feathers used for?

and flight, they may also have provided some insulation. This insulation may then have enabled the evolution of high rates of cellular respiration and the maintenance of a high and constant body temperature.

Colorful feathers may have originally been used for behavioral displays. In modern birds, feathers provide insulation and aid in flight.

FIGURE 11-29 *Sinosauropteryx:* **an early feathered, flightless dinosaur.**

> ## TAKE-HOME MESSAGE 11·17
>
> Birds are a branch of the reptile lineage but, unlike other reptiles, possess feathers and can generate body heat. The complex anatomical and physiological systems that we see in extant animals, such as feathers and endothermy in birds, are the products of hundreds of millions of years of step-by-step changes that began with simple structures. Feathers were originally colorful structures, possibly used for behavioral displays; additional functions such as insulation and flight evolved later.

11·18 Mammals are animals that have hair and produce milk.

As we saw in the previous section, two different evolutionary lineages began to diverge after amniotic vertebrates appeared. One gave rise to the reptiles, and the other gave rise to the mammals. Originally small, nocturnal insect-eaters, the early mammals remained small. And, judging from the scarce fossil record, they were never in great abundance while the dinosaurs were around. Early mammals had several features in common that are still present in all mammals today. Two important features are **hair**—dead cells filled with the protein keratin—which serves as an insulator, and **mammary glands** in female mammals, which enable them to produce nutritious, calorie-rich milk and nurse their young.

As the earliest mammals evolved, there was a gradual transition from bodies with short legs that projected out horizontally from the trunk, like the legs of an alligator, to bodies with long legs held vertically beneath the trunk, like those of a dog (**FIGURE 11-30**). These anatomical changes would have allowed the earliest mammals to run faster and

FIGURE 11-30 **How did endothermy evolve in early mammals?**

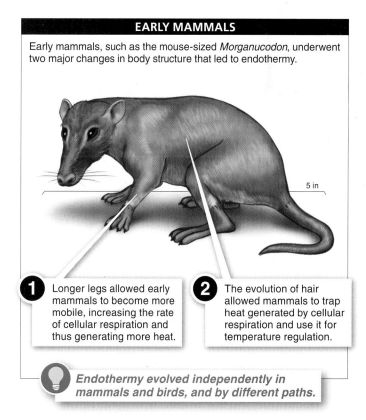

EARLY MAMMALS

Early mammals, such as the mouse-sized *Morganucodon*, underwent two major changes in body structure that led to endothermy.

5 in

1 Longer legs allowed early mammals to become more mobile, increasing the rate of cellular respiration and thus generating more heat.

2 The evolution of hair allowed mammals to trap heat generated by cellular respiration and use it for temperature regulation.

Endothermy evolved independently in mammals and birds, and by different paths.

Mammals are endothermic vertebrates that have hair and produce milk for their young.

Platypus

Kangaroo

African elephant

MONOTREMES
- Females lay eggs
- Females produce milk, but do not have nipples—babies suck milk from the hairs on their mother's chest
- Only 5 species survive—the platypus and 4 species of spiny animals called echidnas

MARSUPIALS
- Females give birth after a short period of development
- Females of most species have a pouch where the young complete their development
- About 300 species, including kangaroos, koalas, wallabies, and possums

PLACENTAL MAMMALS
- Females have a placenta that provides oxygen and nutrients to embryos in the uterus
- About 4,500 species

 FIGURE 11-31 **Three groups of mammals: monotremes, marsupials, and placentals.**

farther to capture prey. But an increase in muscle activity for running had to be accompanied by an increase in the rate of cellular respiration to provide the energy to keep the leg muscles working. With this increase in metabolism, along with the evolution of hair, the early mammals could trap the heat produced and use it for temperature regulation, setting the stage for endothermy (but through a different sequence of events than in the birds).

One feature common to most mammals is **viviparity,** or giving birth to babies ("live birth") rather than laying eggs, but it is not a defining mammalian characteristic—it isn't common to all mammals. **Monotremes** retain the ancestral condition of laying eggs. Monotremes do produce milk, but they do not have nipples—their mammary glands open directly onto the skin, so babies can lap up milk from the skin and hair.

Today, only five species of monotremes survive: the platypus and four species of spiny animals called echidnas. The platypus lives in streams and rivers in eastern Australia. It uses its broad, leathery, electrosensitive bill to probe under rocks and sunken logs for prey such as aquatic insect larvae and crayfish, which it finds by sensing the electrical activity of their muscles as they try to hide. The echidnas—three species in Australia and one in New Guinea—also have a leathery electrosensitive snout.

The remaining two lineages of mammals, the **marsupials** and the **placentals,** are viviparous, but the newborn young of marsupials and placentals are quite different (**FIGURE 11-31**). Marsupials are called the "pouched mammals" because the female of most marsupial species has a pouch on her abdomen in which the young complete their development, following a short period of embryonic life in the uterus.

Placental mammals (including humans) take their name from the *placenta,* which is the structure responsible for the transfer of nutrients, respiratory gases, and metabolic waste products between the mother and the developing fetus. Dense capillary beds in the placenta send finger-like projections into the wall of the uterus, where they are surrounded by blood from the mother's circulation.

TAKE-HOME MESSAGE 11·18

Hair and mammary glands are defining characteristics of mammals. Monotremes are egg-laying mammals. Marsupial mammals give birth after a short period of development in the uterus, and the newborn completes its development in the mother's pouch. Placental mammals have a placenta that provides oxygen and nutrients to the fetus as it undergoes a longer development in the uterus.

The primates, the evolutionary lineage to which humans belong, originated about 55 million years ago. We can get an idea of what the ancestral primate looked like from the modern species of tree-living (arboreal) mammals commonly called prosimians—a group that includes the lemurs. This is because many of the anatomical characteristics of humans and the other primates can be traced to our arboreal origin. Our forward-directed eyes and binocular vision that allow us to judge distances accurately, our shoulder and elbow joints that allow our arms to rotate, and our fingers and opposable thumbs, and our toes, that allow us to grasp objects—all these are traits we inherited from our arboreal ancestors (FIGURE 11-32).

Humans are part of the primate lineage that includes the groups commonly referred to as the New World and Old World monkeys (both of which have tails) and the apes (which lack tails). Among the apes, gibbons and orangutans live in pairs or alone, while gorillas and chimpanzees live in social groups that consist of one or more adult males and several females—the ancestral social structure for human societies. Within the apes, genetic and anatomical characteristics show that chimpanzees are our closest living relatives (FIGURE 11-33). Human and chimpanzee genes are

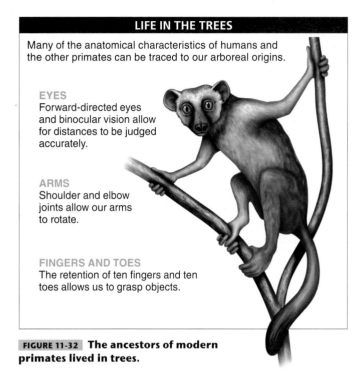

LIFE IN THE TREES

Many of the anatomical characteristics of humans and the other primates can be traced to our arboreal origins.

EYES
Forward-directed eyes and binocular vision allow for distances to be judged accurately.

ARMS
Shoulder and elbow joints allow our arms to rotate.

FINGERS AND TOES
The retention of ten fingers and ten toes allows us to grasp objects.

FIGURE 11-32 **The ancestors of modern primates lived in trees.**

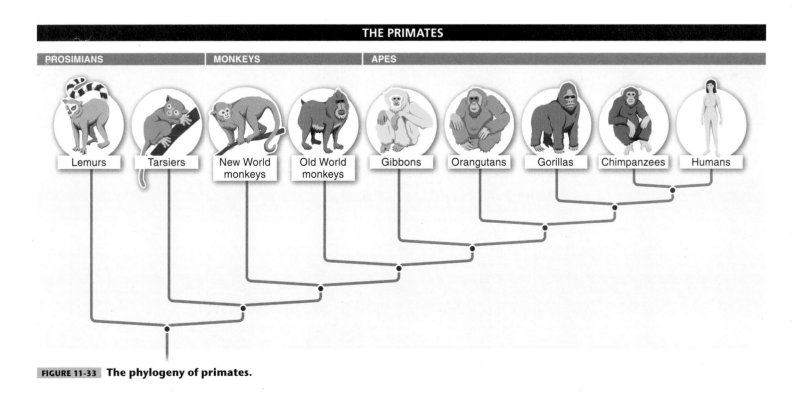

THE PRIMATES

PROSIMIANS | MONKEYS | APES

Lemurs · Tarsiers · New World monkeys · Old World monkeys · Gibbons · Orangutans · Gorillas · Chimpanzees · Humans

FIGURE 11-33 **The phylogeny of primates.**

very similar: their base sequences differ only by about 1%, and one-third of human and chimpanzee genes are identical. The amount of genetic difference between humans and chimpanzees indicates that the chimpanzee and modern human lineages separated only five or six million years ago (see Section 8-21).

Humans differ from chimpanzees in three major anatomical characteristics: humans are bipedal (we normally walk on two legs, whereas chimpanzees usually walk on four legs), humans are bigger than chimpanzees, and the human brain is about three times the size of the chimpanzee brain.

When we trace the appearance of these human characteristics through the fossil record, we find that humans did not become bipedal, big, and brainy all at once. Instead, the three characteristics evolved one by one. Bipedality evolved first, then brain volume increased, and finally body size increased, accompanied by a further increase in brain size.

What are the advantages of walking on two feet rather than four? Shifting from walking on four legs to walking on two legs required changes in several parts of the skeleton. The evolution of bipedalism was not a simple process, but it was one of the first important changes in the human lineage, and it seems to have set the stage for all the changes that followed. The primary advantage of bipedal locomotion for early humans was probably energetic efficiency, though the issue is still debated as research continues. Bipedal locomotion at walking speed uses less energy than quadrupedal locomotion and frees the hands for carrying and for tool use.

GRAPHIC CONTENT
Thinking critically about visual displays of data
Turn to p. 487 for a closer inspection of this figure.

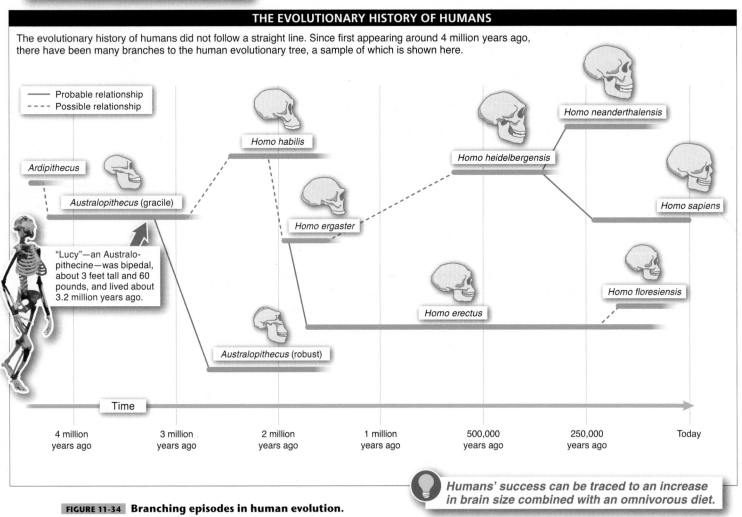

THE EVOLUTIONARY HISTORY OF HUMANS

The evolutionary history of humans did not follow a straight line. Since first appearing around 4 million years ago, there have been many branches to the human evolutionary tree, a sample of which is shown here.

——— Probable relationship
- - - Possible relationship

Ardipithecus

Australopithecus (gracile)

Homo habilis

Homo ergaster

Homo heidelbergensis

Homo neanderthalensis

Homo sapiens

"Lucy"—an Australopithecine—was bipedal, about 3 feet tall and 60 pounds, and lived about 3.2 million years ago.

Homo erectus

Homo floresiensis

Australopithecus (robust)

Time

| 4 million years ago | 3 million years ago | 2 million years ago | 1 million years ago | 500,000 years ago | 250,000 years ago | Today |

Humans' success can be traced to an increase in brain size combined with an omnivorous diet.

FIGURE 11-34 **Branching episodes in human evolution.**

About 3.5 or 4 million years ago, several bipedal groups, the australopithecines, appeared (FIGURE 11-34). They were no larger than chimpanzees and had the same brain volume, just a bit larger than a pint jar (350–400 cc). The fossil known as Lucy was the first of many australopithecines to be discovered. An adult female, Lucy was only 3 feet (about 1 m) tall and probably weighed no more than 60 pounds (a bit less than 30 kg). The group of australopithecines to which Lucy belonged lived in grassy habitats with scattered trees; their hands and feet retained the curved finger and toe bones that are characteristic of arboreal animals.

Lucy and the other members of the closely related species of australopithecines probably foraged on the ground for food and climbed into trees to escape predators and perhaps, at night, to sleep. With jaws and teeth much like the jaws and teeth of chimpanzees, they had a diet probably much like that of chimpanzees, a mixture of leaves, soft fruits, and nuts.

Following the appearance of bipedal australopithecines, a second branching episode in human evolution produced the earliest species in our own genus, *Homo* (from the Latin word for "human"). These earliest humans had brain volumes about twice those of chimpanzees, but little increase in body size. The species in this radiation had markedly smaller teeth than their australopithecine ancestors, a change that might indicate they had started to use tools instead of their teeth for the initial preparation of food. Stone tools are found in the same deposits as the

fossils of *Homo habilis,* and this species may have been the first to use tools. Up to this time, all of human evolution had taken place in the southern and eastern parts of Africa, but studies suggest that about two million years ago, one branch, represented by *Homo erectus,* migrated out of Africa and spread through eastern Europe and Asia, while a second branch, represented by *Homo ergaster,* remained in Africa.

Additional branching episodes in human evolution (see Figure 11-34) gave rise to several species of humans, including Neandertals (*Homo neanderthalensis*), "Flores man" (*Homo floresiensis,* described in the next section), and our own species (*Homo sapiens*). This radiation coincided with an increase in body size to approximately the height and weight of modern humans and an increase in brain volume to nearly twice that of earlier ancestors. In addition, the body form of the species that evolved during this branching episode looked like that of modern humans, with shorter arms and longer legs.

TAKE-HOME MESSAGE 11·19

Humans' forward-looking eyes, hands and feet with 10 fingers and 10 toes, and shoulder and elbow joints that allow the arms to rotate are characteristics retained from our arboreal ancestors. The early ancestors of humans left the trees and took up life on the ground, where they walked on two legs.

11·20 How did we get here? The past 200,000 years of human evolution.

About 200,000 years ago, a new human species branched off the *Homo ergaster* lineage in Africa—the first modern *Homo sapiens.* Two hundred thousand years is a ballpark figure based on a combination of molecular and fossil evidence. Then, about 60,000 years ago, a small group— probably no more than 100—of these modern humans left Africa and ultimately spread across the earth.

Mitochondrial DNA shows that the initial human migration out of Africa followed three major pathways (FIGURE 11-35). One path turned west after leaving Africa and spread into Europe. A second path turned southeast and spread into southern Asia and through the Indo-Australian Archipelago to Australia.

Q *Were two species of humans ever alive at the same time? If so, what happened?*

The third migration went northeast, populating northern Asia. This is the group that crossed the Bering Straits bridge about 15,000 years ago and spread southward through the Americas. Even as some groups migrated from Africa, many individuals remained.

There was something about the world that those *H. sapiens* groups moved into as they left Africa that we would find bizarre—*they were not alone.* Populations of at least three other species of humans (i.e., species in the genus *Homo*) were already present in some of the areas that modern humans were moving into. That situation is totally different from the world we know today, in which we are the

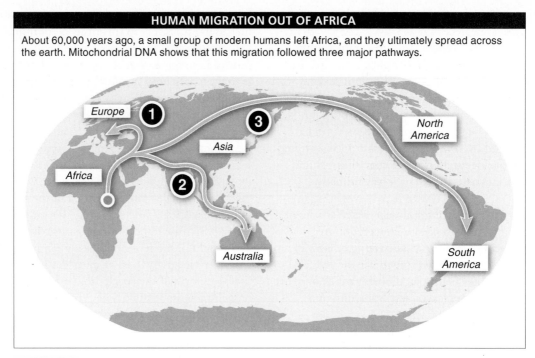

HUMAN MIGRATION OUT OF AFRICA

About 60,000 years ago, a small group of modern humans left Africa, and they ultimately spread across the earth. Mitochondrial DNA shows that this migration followed three major pathways.

FIGURE 11-35 *Homo sapiens* **on the move: three migratory paths out of Africa.**

only human species. What would it be like to live with species that were as similar to us as dogs are to wolves, coyotes, and jackals? How would you interact with an individual who was clearly human, but not human in exactly the same way that you are?

This is the situation that the first modern *H. sapiens* encountered when they left Africa. Neandertals (*H. neanderthalensis*) had already spread across Europe and the Middle East. Neandertals were about the same size as modern humans, but more robust and muscular. Fossils of Neandertals often include bones that had been broken and healed. These injuries provide two types of information about Neandertals.

1. Neandertals must have lived in organized groups that included a social support system, because these fossils reveal serious injuries. The injured individuals would have been incapacitated for days or even weeks until they healed, and family or clan members must have cared for them during this period. Ritual burials of Neandertals have also been found, and these provide additional support for the hypothesis that they lived in organized social groups.

2. The pattern of injuries found in Neandertal skeletons is distinctive—the only modern counterpart is found in

professional rodeo bull and bronco riders, people who come in very close contact with large, angry animals. Neandertals probably hunted large mammals (bison, mammoths, wooly rhinoceroses) with short spears that were used for close-up jabbing instead of being thrown from a safe distance.

The *H. sapiens* who migrated through the Indo-Australian Archipelago encountered two species of humans: *H. erectus* and *H. floresiensis*. *Homo erectus* had migrated to Asia at least a million years earlier and was still present on the island of Java when modern humans arrived, about 50,000 years ago. Although *H. erectus* was the same size as modern humans, it had a smaller brain—an average brain volume of just over a quart (about 1,000 cc) for *H. erectus* compared with a quart and a half (about 1,400 cc) for *H. sapiens*. Nonetheless, *H. erectus* had technological skills. These may have been the first humans to use fire (the evidence on this is not definitive), and they almost certainly built boats that allowed them to move along coasts and from island to island.

In 2003, researchers discovered stone tools and fossils of a species of human only 3 feet (about 1 m) tall and with a brain volume of just over a pint (about 350 cc) on Flores Island, which lies east of Java. The paleontologists who discovered this species gave it the scientific name *Homo floresiensis* ("Flores man"), but the world press promptly called it the Hobbit because of its tiny size. When the dwarf human *H. floresiensis* lived, Flores Island was also home to a dwarf elephant, which *H. floresiensis* may have eaten, and a giant monitor lizard, which may have eaten *H. floresiensis*.

After modern *H. sapiens* groups spread into the areas occupied by the three other human species, these three other species disappeared. Neandertals became extinct about 30,000 years ago, *H. erectus* about 27,000 years ago, and *H. floresiensis* about 12,000 years ago. Why did these three species vanish? We're not sure. Previously, researchers thought that modern *H. sapiens* must have exterminated the other species, either by monopolizing access to food and living space or by killing them in battles for food and space. But recent results from DNA sequencing suggest that *H. sapiens* actually encountered and interbred with Neandertals, just as they emerged from Africa. This conclusion is based on the finding that at least 1% to 4% of the genetic makeup of most modern humans includes Neandertal DNA.

Now we are the only extant human species, a situation that is unique in the evolutionary history of humans. Ever since the first branching event, about four million years ago, multiple, closely related species of humans had coexisted. We are the first human species to be alone.

TAKE-HOME MESSAGE 11·20

Modern humans (*Homo sapiens*) evolved in Africa about 200,000 years ago, and all living humans are descended from that evolutionary radiation. About 60,000 years ago, a small group of modern humans moved out of Africa, and the descendants of this group ultimately populated Europe, Asia, and the Americas. Three other species of humans were living at this time, all of which became extinct between 30,000 and 12,000 years ago, after modern humans had spread into the areas where they were living.

Where Are You From?
"Recreational Genomics" and the Search for Clues to Your Ancestry in Your DNA

How much do you know about your ancestral heritage? Is it enough to know that you are Italian and Asian? More than two dozen companies now offer a promise to help you discover your genetic ancestry—usually for $100 to $1,000. Can they deliver what they promise?

Q: **Are clues to your ancestry hidden within your DNA?** The short answer is yes. Researchers use comparisons of DNA sequences to explore personal ancestries *within* our species, even preparing pie charts that purport to identify a person's ancestral proportions (for example, 10% West African, 70% Middle Eastern/North African, and 20% European).

Q: **How is DNA used to build a family tree?** Most tests are based on one of three types of analysis. (1) Mitochondrial DNA tests evaluate highly variable sequences in maternally inherited mitochondrial DNA. (2) Y-chromosome tests look at variation in the paternally inherited Y chromosome. (3) Autosomal DNA tests examine highly variable sections of DNA scattered throughout the chromosomes. In each case, a sample of cells is collected (usually by swabbing the inside of the cheeks), then technicians determine the base sequences at a number of locations (a few dozen to several hundred) and compare those sequences with sequences from thousands of individuals in populations around the world. The matching of your sequences with those of other samples is then interpreted as potential evidence of shared ancestry.

Q: **Genetic ancestry testing may be good fun, but is it good science?** Unfortunately, the limitations of commercially available tests of genetic ancestry are significant. These are among the most important limitations:

• High genetic variation among individuals within most populations makes it difficult to identify specific sequences that can reliably indicate membership in a population. As an analogy: if Native Americans are highly variable for a trait (e.g., height), we can't use that trait to prove someone's Native American ancestry.

• High rates of gene flow between populations reduce the reliability with which any sequence can demonstrate membership in one particular population. As an analogy: dark hair is extremely common in Asia, but having dark hair doesn't necessarily indicate Asian ancestry.

• Evaluating too few genetic loci, of which just a small number happen to be similar, can lead to the conclusion that individuals are much more genetically similar than they actually are.

• A DNA match between two individuals living today is *not* a match with an ancestor. Rather, it suggests that the two people may have inherited the DNA sequence from a common ancestor.

Q: **What can you conclude?** Until we have a much more comprehensive catalog of genetic variation—including precise estimates of how much variation exists within populations and how much exists between populations—the results of genetic ancestry testing remain overly speculative and insufficiently reliable. Conclusions may be incorrect and misleading. But, because the conceptual underpinnings are solid, the future of genetic ancestry testing is promising.

Check Your Knowledge

Thinking critically about visual displays of data

THE EVOLUTIONARY HISTORY OF HUMANS

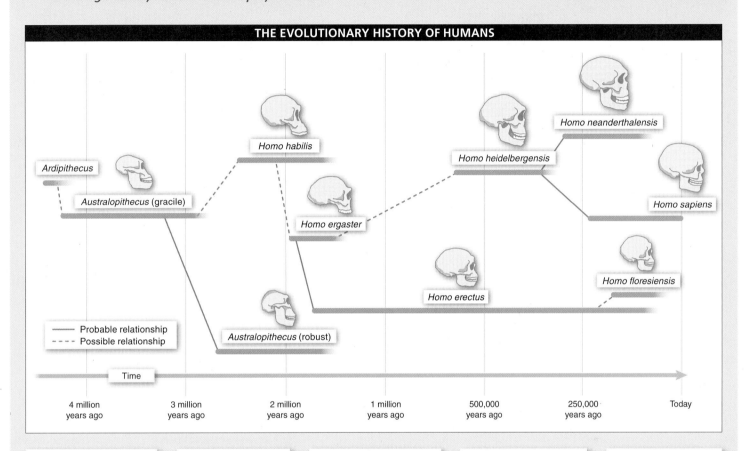

Ardipithecus

Australopithecus (gracile)

Homo habilis

Homo ergaster

Homo heidelbergensis

Homo neanderthalensis

Homo sapiens

Homo erectus

Homo floresiensis

Australopithecus (robust)

——— Probable relationship
- - - - Possible relationship

Time

| 4 million years ago | 3 million years ago | 2 million years ago | 1 million years ago | 500,000 years ago | 250,000 years ago | Today |

1. Moving back in time from today, which species are the most likely to be the direct ancestors of modern humans? Which are not? Why?

2. When did the genus *Homo* first appear? How did it differ from earlier species in this evolutionary tree?

3. What is the difference between the dotted purple lines and the solid purple lines? Why might that information change over time?

4. What is the largest number of different species on the human "family tree" that were alive at the same time? When was that?

5. How long ago did *Homo sapiens* become the only living representative of the genus *Homo*?

See answers at the back of the book.

Key Terms in Animal Diversification

ABOUT THE CHAPTER OPENING PHOTO

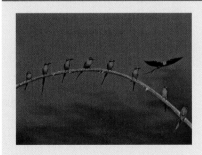

A group of the highly sociable southern carmine bee-eaters (*Merops nubicoides*) in the Luangwa Valley, Zambia.

REVIEW & REHEARSE

11

ANIMAL DIVERSIFICATION

? Check Your Knowledge

1. Which of the following is not a characteristic of all animals?

a) They are able to move at some point in their life.

b) They get their energy by eating other organisms.

c) They are multicellular.

d) They are sexually reproducing at some point in their life cycle.

e) All of the above are characteristics of all animals.

0 —————————— 45 —————————— 100
EASY   HARD

2. The type of adaptation most likely to occur over evolutionary time is one that:

a) increases the physiological complexity and sophistication of the individuals in a population.

b) increases the physiological simplicity of the individuals in a population.

c) causes the individuals in a population to have greater intelligence.

d) causes the individuals in a population to more closely resemble humans.

e) causes the individuals in a population to more efficiently reproduce in the environment in which they live.

0 —————————— 45 —————————— 100
EASY   HARD

3. The lineage that first separated from the common ancestor of all animals, and retains many of those primitive features to this day, includes which of the following modern organisms?

a) the grasshopper

b) the jellyfish

c) the sponge

d) the earthworm

e) the sea turtle

0 —————————— 38 —————————— 100
EASY   HARD

11·1–11·3 Animals are just one branch of the Eukarya domain.

Animals are multicellular organisms that feed on other organisms and can move during at least one stage of their life.

WHAT IS AN ANIMAL?

Three characteristics are common to all animals.

ANIMALS EAT OTHER ORGANISMS
All animals acquire energy by consuming other organisms.

ANIMALS MOVE
All animals have the ability to move—at least at some stage of their life cycle.

ANIMALS ARE MULTICELLULAR
Animals consist of multiple cells and have body parts that are specialized for different activities.

? 1. Describe the three characteristics that define animals.

FOUR KEY DISTINCTIONS DIVIDE THE ANIMALS

The 9 groups of animals in this figure represent just 25% of all animal phyla but contain 99% of all animal species.

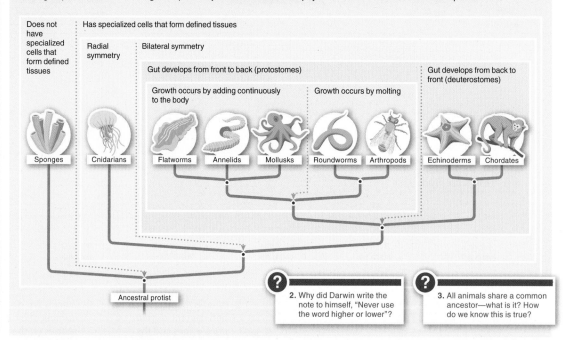

Does not have specialized cells that form defined tissues

Has specialized cells that form defined tissues

Radial symmetry

Bilateral symmetry

Gut develops from front to back (protostomes)

Gut develops from back to front (deuterostomes)

Growth occurs by adding continuously to the body

Growth occurs by molting

Sponges | Cnidarians | Flatworms | Annelids | Mollusks | Roundworms | Arthropods | Echinoderms | Chordates

Ancestral protist

? 2. Why did Darwin write the note to himself, "Never use the word higher or lower"?

? 3. All animals share a common ancestor—what is it? How do we know this is true?

11·4–11·12 Invertebrates—animals without a backbone—are the most diverse group of animals.

Invertebrates, defined as animals without a backbone, are the largest and most diverse group of animals, comprising 96% of all living animal species. The invertebrates are not a monophyletic group, however, and include protostomes and some (but not all) of the deuterostomes.

THE SPONGES

COMMON CHARACTERISTICS
• No tissues or organs
• Body consists of a hollow tube with pores in its wall
• Feed by pumping in water, along with bacteria, algae, and small particles of organic material, through their pores
• Free-swimming larvae
• Sessile as adults

MEMBERS INCLUDE
• About 5,000 species

? 4. Describe the two ways in which sponges can reproduce.

THE CNIDARIANS

COMMON CHARACTERISTICS
• Radially symmetrical
• Tentacles armed with rows of stinging cells, used to paralyze prey

MEMBERS INCLUDE
• Jellyfishes
• Sea anemones
• Corals

? 5. Compare and contrast the two types of cnidarian bodies.

THE FLATWORMS

COMMON CHARACTERISTICS
- Well-defined head and tail regions
- Hermaphroditic and can engage in both sexual and asexual reproduction
- Some have a single opening in the body, which serves as a mouth and an anus

MEMBERS INCLUDE
- Tapeworms
- Flukes

THE ROUNDWORMS

COMMON CHARACTERISTICS
- Long, narrow unsegmented body
- Bilaterally symmetrical
- Surrounded by a strong, flexible cuticle
- Must molt in order to grow larger

MEMBERS INCLUDE
- More than 90,000 species (scientists estimate there may be five times as many species that have not yet been identified)

 6. What important ecological role do roundworms play?

THE ARTHROPODS

COMMON CHARACTERISTICS
- Segmented body with a distinct head, thorax, and abdomen
- Exoskeleton made of chitin
- Jointed appendages

MEMBERS INCLUDE
- Insects (more than 800,000 species)
- Arachnids (about 60,000 species)
- Crustaceans (about 52,000 species)
- Millipedes and centipedes (about 10,000 species)

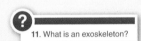 11. What is an exoskeleton?

 12. Describe how you could estimate the number of species of beetles in a temperate region.

13. Compare and contrast the benefits and costs of having an exoskeleton.

 14. Which two features of insects contributed to their great adaptive radiation?

THE ANNELIDS

COMMON CHARACTERISTICS
- Segmented body

MEMBERS INCLUDE
- Marine polychaetes (about 9,000 species)
- Earthworms (more than 4,000 species)
- Leeches (about 500 species)

 7. You have a friend who confuses a terrestrial slug with a segmented worm (annelid). Describe the characteristics of the slug that you could point out to him that are not found in any annelids.

THE MOLLUSKS

COMMON CHARACTERISTICS
- Most have a shell that protects the soft body
- Mantle (tissue that forms the shell)
- Radula (sandpaper-like tongue structure used during feeding); found in all mollusks except bivalves

MEMBERS INCLUDE
- Gastropods (about 35,000 species)
- Bivalves (about 8,000 species)
- Cephalopods (about 700 species)

 8. What is the specialized tongue found in mollusks called? How does it function?

 9. Why are questions of comparative animal intelligence not generally useful?

 10. What evidence is typically used to show that octopuses are "smart"?

THE ECHINODERMS

COMMON CHARACTERISTICS
- Enclosed by a hard skeleton of spiny plates
- Larvae are bilaterally symmetrical and share some anatomical features with chordates
- Adults are radially symmetrical
- Undersides are covered with tube feet that aid in locomotion and grasping

MEMBERS INCLUDE
- Sea stars (about 1,600 species)
- Sea urchins and sand dollars (about 940 species)
- Sea cucumbers (about 1,100 species)

 15. Why are echinoderms viewed as an example of how classifying organisms based on evolutionary relatedness can reveal a great deal about the force of natural selection?

---

4. Sponges are sessile, meaning that they:

   a) reproduce asexually.
   b) are parasitic and depend on their host for a constant supply of nutrients.
   c) have exoskeletons that they must shed as they grow.
   d) live within shells they find on the ocean floor.
   e) live attached to a solid structure and do not move around.

   0 ——————[13]—————————— 100
   EASY   HARD

5. In cnidarians, cnidocytes are primarily used for:

   a) creation of water flow across the body wall.
   b) formation of free-living medusas.
   c) secretion of digestive enzymes.
   d) prey capture and defense.
   e) muscular contraction during movement.

   0 ———————[25]————————— 100
   EASY   HARD

6. The mollusk's mantle is used primarily for:

   a) feeding.
   b) gas exchange.
   c) producing the shell.
   d) excretion.
   e) reproduction.

   0 —————————[36]——————— 100
   EASY   HARD

7. The phylum Arthropoda includes all of the following kinds of animals except:

   a) snails.
   b) crabs.
   c) crayfish.
   d) butterflies.
   e) scorpions.

   0 ——————————————[54]—— 100
   EASY   HARD

8. For which of the following groups of organisms is it most difficult to estimate species numbers?

   a) bees, wasps, and ants
   b) bacteria
   c) tropical trees
   d) beetles
   e) mammals

   0 ————————[29]———————— 100
   EASY   HARD

9. Which of the following echinoderms has radial symmetry during its larval stage?

a) sea star

b) brittle star

c) sea urchin

d) sand dollar

e) None of the above have radial symmetry during the larval stage.

0 — 45 — 100
EASY   HARD

10. Which one of the following characteristics distinguishes all chordates from all other animals?

a) a vertebral column

b) a dorsal hollow nerve cord

c) collar cells

d) bilateral symmetry during embryonic or larval development

e) an amniotic egg

0 — 50 — 100
EASY   HARD

11. Which of the following are chordates?

a) fishes

b) humans

c) frogs

d) All of the above are chordates.

e) Only a) and c) are chordates.

0 — 36 — 100
EASY   HARD

12. Which of these animals is a tetrapod that does not produce amniotic eggs?

a) salamander

b) human

c) monkey

d) elephant

e) python

0 — 48 — 100
EASY   HARD

13. Why is the amniotic egg considered a key evolutionary innovation?

a) It prohibits external fertilization, thereby facilitating the evolutionary innovation of internal fertilization.

b) It has an unbreakable shell.

c) It greatly increases the likelihood of survival of the eggs in a terrestrial environment.

d) It enables eggs to float in an aquatic medium.

e) It extends the time of embryonic development.

0 — 37 — 100
EASY   HARD

11·13–11·15 The phylum Chordata includes vertebrates, animals with a backbone.

All chordates have four characteristic structures: a notochord, a dorsal hollow nerve cord, pharyngeal slits, and a post-anal tail.

THE CHORDATES

COMMON CHARACTERISTICS

All chordates possess four common body structures, although in many chordates, these structures are only present during specific life stages.

- Notochord
- Dorsal hollow nerve cord
- Pharyngeal slits
- Post-anal tail

MEMBERS INCLUDE

- Tunicates (about 2,000 species)
- Lancelets (about 20 species)
- Vertebrates (about 56,000 species)

16. Describe the four distinct chordate body structures.

THE DISTINCT BODY STRUCTURES OF CHORDATES

NOTOCHORD
A rod of tissue extending from the head to the tail

- Simpler chordates retain the notochord throughout life
- In more complex chordates, the notochord in early embryos is replaced by a backbone

DORSAL HOLLOW NERVE CORD
Nerve cord that extends along the animal's back (its dorsal side)

- In vertebrates, the nerve cord eventually forms the spinal cord and brain

PHARYNGEAL SLITS
Slits through which water is passed in order to breathe and feed

- In many chordates (including humans), the slits disappear as the animal develops

POST-ANAL TAIL
Tail that extends beyond the posterior (back) end of the digestive system

- Some vertebrates (including humans) have a tail only briefly, during embryonic development

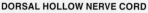

17. Which two evolutionary innovations in vertebrates resulted in their eventual domination among the large animals?

FROM WATER TO LAND

The transition of vertebrates from life in water to life on land required overcoming three main obstacles. Four major evolutionary innovations allowed for this transition.

PROBLEM: RESPIRATION

Aquatic animals use gills to acquire dissolved oxygen from water. The transition onto land required the ability to breathe air.

SOLUTION: LUNGS
Gas exchange was transferred from the gills to lungs, which evolved from the air sacs (that were also the evolutionary precursors to the swim bladder found in ray-finned fishes).

PROBLEM: GRAVITY

Water is buoyant, so aquatic animals don't need much structural support. The transition onto land required structural support to resist the pull of gravity.

SOLUTIONS: LIMBS and MODIFIED VERTEBRAE
Limbs evolved from the jointed fins found on the underside of lobe-finned fishes. Vertebrae were modified to transmit the body weight through the limbs to the ground.

PROBLEM: EGG DESICCATION

The transition onto land required an egg that resisted drying out when exposed to air.

SOLUTION: AMNIOTIC EGG
Terrestrial animals developed a waterproof eggshell, which prevents eggs from drying out before they hatch.

18. What is the most likely evolutionary precursor to the vertebrate lungs?

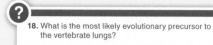

19. Why is it more important for an amphibian than a reptile to live in close proximity to water?

11·16–11·20 All terrestrial vertebrates are tetrapods.

Terrestrial vertebrates include amphibians, reptiles (including birds), and mammals.

THE TERRESTRIAL VERTEBRATES

AMPHIBIANS
- Most species live on land as adults, but develop in water.
- Body temperature is controlled by external conditions, such as air temperature (exothermic)
- Include frogs and toads (about 5,400 species), salamanders (about 550 species), and caecilians (170 species)

REPTILES
- Skin is covered in scales
- Body temperature is controlled by external conditions, such as air temperature (exothermic)
- Include snakes and lizards (about 8,000 species), turtles (about 300 species), crocodiles and alligators (23 species), and tuatara (2 species)

BIRDS
- Have feathers and wings, providing insulation and enabling flight
- Body temperature is maintained by heat generated from cellular respiration (endothermic)
- Include about 9,700 species

MAMMALS
- Skin is covered in hair
- Body temperature is maintained by heat generated from cellular respiration (endothermic)
- Produce milk for their young
- Include placental mammals (about 4,500 species), marsupials (about 300 species), and monotremes (5 species)

20. Why does an ectotherm, such as a snake, require only an occasional meal to survive, while an endotherm, such as a mouse, requires far more frequent meals?

21. In mammals, what is the main function of the placenta?

EARLY MAMMALS

Early mammals, such as the mouse-sized Morganucodon, underwent two major changes in body structure that led to endothermy.

1 Longer legs allowed early mammals to become more mobile, increasing the rate of cellular respiration and thus generating more heat.

2 The evolution of hair allowed mammals to trap heat generated by cellular respiration and use it for temperature regulation.

LIFE IN THE TREES

Many of the anatomical characteristics of humans and the other primates can be traced to our arboreal origins.

EYES
Forward-directed eyes and binocular vision allow for distances to be judged accurately.

ARMS
Shoulder and elbow joints allow our arms to rotate.

FINGERS AND TOES
The retention of ten fingers and ten toes allows us to grasp objects.

HUMAN EVOLUTION

Modern humans evolved in Africa between 200,000 and 100,000 years ago, and all living humans are descended from that evolutionary radiation. The evolutionary success of humans can be traced to an increase in brain size, possibly in conjunction with an omnivorous diet and increased caloric intake.

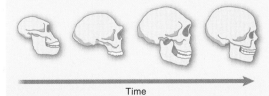

Time

HUMAN MIGRATION OUT OF AFRICA

About 60,000 years ago, a small group of modern humans left Africa, and they ultimately spread across the earth. Mitochondrial DNA shows that this migration followed three major pathways.

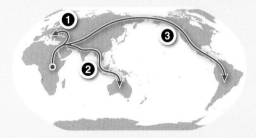

22. The fossil record of early humans reveals that they had much smaller teeth than their ancestors. What does this suggest about their diet? Why?

23. Describe the three major anatomical differences between humans and chimpanzees.

24. The fossil record contains evidence of healed broken bones among the Neandertals. What does this reveal about their social organization and ways of acquiring food? Why?

14. Which came first, the chicken or the egg?
a) The chicken, because the amniotic egg did not evolve until the first chicken appeared.
b) The egg, because the amniotic egg evolved well before the first birds.
c) The chicken, because, during speciation, the adult stage always precedes the juvenile stage.
d) The egg, because the chicken is not a real species.
e) It is impossible to determine, because eggs leave no fossils.

0 — 42 — 100
EASY   HARD

15. Marsupials and which of the following groups combine to make a monophyletic group?
a) birds
b) carnivores
c) primates
d) monotremes
e) placental mammals

0 — 43 — 100
EASY   HARD

16. According to the fossil record, the first modern humans—*Homo sapiens*—appeared approximately _____ years ago.
a) 6 million
b) 6,000
c) 190 million
d) 200,000
e) 1 million

0 — 55 — 100
EASY   HARD

12 | Plant and Fungi Diversification

WHERE DID ALL THE PLANTS AND FUNGI COME FROM?

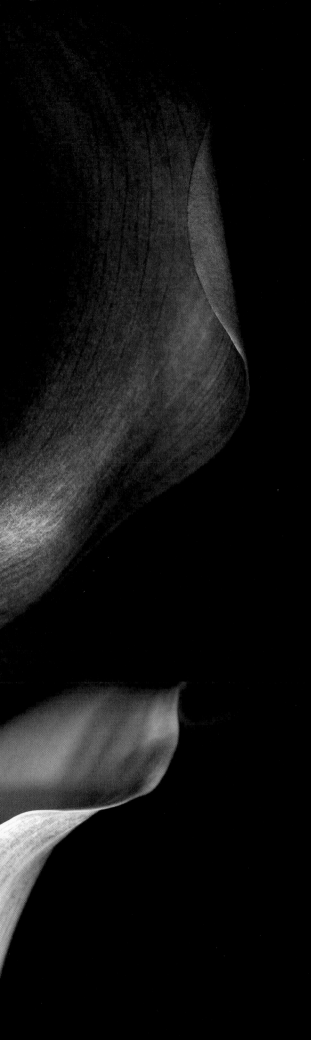

Plants are just one branch of the eukarya.

The first plants had neither roots nor seeds.

The advent of the seed opened new worlds to plants.

Flowering plants are the most diverse and successful plants.

Plants and animals have a love-hate relationship.

Fungi and plants are partners but not close relatives.

12·1
Plants are just one branch of the eukarya.

The land plant Psilotum nudum, *growing in Hawaii Volcanoes National Park.*

12·1 What makes a plant?

An animal can move from place to place to get food or water, to avoid being eaten, or to find a mate. It's harder for a plant. It must do all the things an animal does—obtain food and water, protect itself from predators, and reproduce—*but* can't move a centimeter in the process. Wherever a seed puts down roots, that's where the plant stays anchored for its entire life.

So, plants are organisms that are fixed in place, but that's only one aspect of being a plant. A **plant** is a multicellular eukaryote that produces its own food by carrying out photosynthesis, using energy from sunlight to convert carbon dioxide and water to sugar—and has an embryo that develops within the protected environment of the female

parent. Plants occur almost exclusively on land, and they vary in size from less than 0.04 inch (1 millimeter) to 380 feet (116 meters) tall (FIGURE 12-1). (Note that although there are a small number of exceptions to our definition of a plant, it applies to the vast majority of the more than 280,000 species of plants on earth today.)

There are other multicellular and photosynthetic eukaryotes on earth, including many species of algae, such as seaweed, as well as aquatic species that are the closest relatives of land plants. They differ from plants, however, in that they live only in water or on very moist land surfaces. This is in sharp contrast to most plants, which can live even in deserts.

WHAT IS A PLANT?

PLANTS CREATE THEIR OWN FOOD
Almost all plants carry out photosynthesis, using energy from sunlight to convert carbon dioxide and water into sugar.

PLANTS ARE SESSILE AND (MOSTLY) TERRESTRIAL
Plants are anchored in place at their bases and occur almost exclusively on land.

PLANTS ARE MULTICELLULAR
Plants consist of multiple cells and have structures that are specialized for different functions.

FIGURE 12-1 **The defining characteristics of a plant.**

Over evolutionary time, dodder has lost its chlorophyll and cannot conduct photosynthesis. It steals nutrients from the plant on which it grows.

FIGURE 12-2 Dodder: a parasitic, non-photosynthetic plant.

Moving the light source every few days, growers can make "lucky bamboo" grow in spirals.

OBTAINING FOOD
Because plants can't move to reach sunlight, they bend in place and grow toward light.

Diploid ferns

Haploid ferns

FINDING A MATE
Male and female plants can't meet to reproduce, so they have developed ways of getting the male gamete to the female gamete, including alternating haploid and diploid life stages, and using other organisms to transport the male gametes.

As we have said, most plants make their own food by the process of photosynthesis. However, a plant can't live on carbohydrates—the product of photosynthesis—alone. A plant needs nitrogen to build proteins, phosphorus to make ATP, and salts to create concentration gradients between the inside and outside of cells. Plants use **roots,** the part of a plant below ground, to obtain these needed substances from the soil. Above ground, plants have a **shoot** that consists of a stem and leaves. The stem is the structure that supports the main photosynthetic organ of a plant: its leaves.

When we think of plants, we often think of them as "chlorophyll-containing," but some plants have no chlorophyll—their ancestors had chlorophyll (which is why they are classified as plants), but they have lost almost all of it over evolutionary time—and they can't carry out photosynthesis. Instead, they live as parasites that steal nutrients from other plants. Dodder is an example of a parasitic plant that almost completely lacks chlorophyll and gets its sugar from the host plant it grows on (**FIGURE 12-2**). You can probably find dodder growing on plants in a nearby vacant lot or at the roadside.

RESISTING PREDATION
Plants can't run from predators, so they have developed adaptations such as thorns to defend themselves.

FIGURE 12-3 Plants must overcome the constraints of being immobile.

Q If a plant growing in the shade can't move to a new, sunnier location, how can it reach the available sunlight?

Plants that carry out photosynthesis need sunlight, and that creates a challenge for some plants. Think about plants growing on the floor of a forest. It's mostly shady under the tree canopy, with just a few sunny spots where light penetrates the tree branches overhead (**FIGURE 12-3**). If a seed sprouts (germinates) in the shade, the young plant needs to somehow get to the nearest sunny

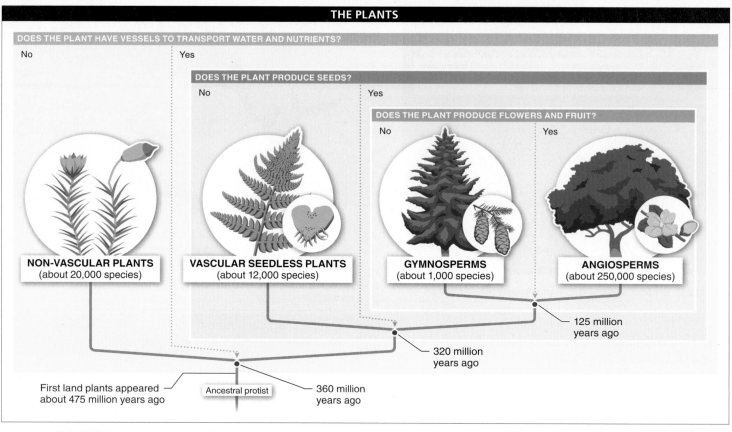

DOES THE PLANT HAVE VESSELS TO TRANSPORT WATER AND NUTRIENTS?

No   Yes

DOES THE PLANT PRODUCE SEEDS?

No   Yes

DOES THE PLANT PRODUCE FLOWERS AND FRUIT?

No   Yes

NON-VASCULAR PLANTS
(about 20,000 species)

VASCULAR SEEDLESS PLANTS
(about 12,000 species)

GYMNOSPERMS
(about 1,000 species)

ANGIOSPERMS
(about 250,000 species)

125 million years ago

320 million years ago

First land plants appeared about 475 million years ago

Ancestral protist

360 million years ago

FIGURE 12-4   **Phylogeny of the plants.**

spot. Plants can do that, of course, but not by moving. Instead, they grow toward the light. You may have seen a plant called a "lucky bamboo" (it's not really bamboo), which can grow in spirals or loops. The growers have produced these shapes by moving the light source every few days, forcing the plant to bend to follow the light. If you buy lucky bamboo, you must continue rotating the light source or it will start to grow straight, like any other plant.

Sex can also be a challenge for an organism that is anchored in place by its roots. Not surprisingly, plants have a sex life very different from that of animals. Consider humans as a typical animal. From the moment an egg is fertilized by a sperm, a human has the diploid number of chromosomes—that is, two sets of chromosomes, one set from each parent. Only the egg and sperm are haploid—an egg or a sperm has only one set of chromosomes. In plants, the haploid stage allows the male and female to reproduce, even though the plants will never meet because they can't move. Some plants enlist the help of animals to carry the male gamete to the female gamete.

Resisting predators is another challenge for organisms that cannot move. Running away is how most animals react to a predator, but plants have found other ways to defend themselves. Thorns, for example, are an anatomical defense that plants use to avoid predation (see Figure 12-3). Plants

also use chemicals to deter predators, as many people learn when they develop an itchy rash from poison ivy.

The earliest land plants were the first multicellular organisms to live on land. They were **non-vascular,** meaning that they had no tube-like vessels to transport water and nutrients (FIGURE 12-4). The subsequent evolution of land plants was a series of radiations of forms with characteristics that made them increasingly independent of water. The evolutionary tree of plants shows these stages clearly: first the development of vessels to conduct water from the soil through the plant (the vascular plants), then seeds that provide nutrients to get the next generation off to a good start (the gymnosperms), and finally flowers that allow plants to entice or trick insects and birds into spreading the plant's male gametes (the angiosperms). We'll look at each of these plant adaptations in this chapter. We also explore the ecologically important kingdom of fungi. Although fungi are not plants, they are very closely associated with plants.

TAKE-HOME MESSAGE 12·1

Plants are multicellular organisms that spend most of their lives anchored in one place by their roots. Characteristics evolved that made it possible for plants to successfully obtain food, reproduce, and protect themselves from predation on land, despite their inability to move.

Uluhe ferns (Dicranopteris linearis) *growing in Hawaii Volcanoes National Park.*

12·2 Colonizing land brings new opportunities and new challenges.

The aquatic ancestors of land plants were green algae, which are classified by many taxonomists as part of the plant kingdom. Like the land plants, green algae are multicellular, photosynthetic eukaryotes, but they live only in water or on very moist land surfaces. As water dwellers, green algae do not require specialized structures to obtain water and nutrients; water simply enters their cells by osmosis, and the nutrients they require are in solution in the water that surrounds them. Some green algae, such as sea lettuce and stoneworts, even look like land plants. But there is enough uncertainty among taxonomists about the best classification of green algae that some classify them within the protist kingdom, and they consider some green algae that look like slime on rocks—organisms called coleochaetes (pronounced кон-lee-oh-keets)—to be the closest relatives of plants (**FIGURE 12-5**). An individual coleochaete is about the size of a pinhead and is only one cell-layer thick. Coleochaetes can withstand exposure to air, so they survive when the water level in a lake falls and leaves them high and dry. This resistance to drying out was the first evolutionary step that plants took as they moved from water to land.

The first land plants appeared about 472 million years ago. They weren't impressive-looking, just some patches of low-growing green stems at the water's edge. They did not have any of the structures we associate with plants today—no roots, leaves, or flowers. But from an evolutionary perspective, those early plants were enormously important because, until terrestrial plants evolved, there was nothing on land for other land organisms to eat. Thus, the first

land plants not only set the stage for the tremendous diversity of plant life that we know today but also paved the way for the evolution and diversification of land animals.

As land plants emerged, they faced the same two challenges that were to confront the first terrestrial animals, some 25 million years later: supporting themselves against the

Coleochaetes, a type of green alga, are the closest relative to plants. They can withstand exposure to air when water levels fall.

Ancestors of plants began the transition from water to land with the evolution of resistance to drying.

FIGURE 12-5 **Plants' closest relative: a green alga called a coleochaete.**

PLANTS ARE A BRANCH OF EUKARYA

PLANTS WITHOUT ROOTS OR SEEDS

THE ADVENT OF THE SEED

FLOWERING PLANTS

PLANT AND ANIMAL RELATIONSHIPS

FUNGI

497

When plants emerged onto land, they faced the same two challenges that terrestrial animals faced 25 million years later.

PROBLEM: GRAVITY

ADAPTATION: The earliest plants grew very close to the ground, as mosses do today, in order to resist the pull of gravity.

FIGURE 12-6 **Leaving the water.**

PROBLEM: DESICCATION

ADAPTATION: Plants developed an outer waxy layer called a cuticle that covers their entire surface.

pull of gravity and reducing evaporation so they didn't dry out. The second of these problems was the more urgent. The earliest plants did not have to grow upward—they could creep along the ground—but they *did* have to avoid drying out (**FIGURE 12-6**). The material that protects all land plants from drying is a shiny, waxy layer on the stem and leaves called the **cuticle.**

TAKE-HOME MESSAGE 12·2

The first land plants were small, had no leaves, roots, or flowers, and could grow only at the water's edge. Nonetheless, they set the stage for the enormous diversity of terrestrial plants and animals on earth today.

12·3 Mosses and other non-vascular plants lack vessels for transporting nutrients and water.

The earliest land plants were low-growing for a reason: they had no structures that could transport water and nutrients from the soil upward into the plant. The only way that these substances could move was by diffusing from one cell into an adjacent cell, and so on. Diffusion is a slow process, and plants that rely on it can grow only a few centimeters tall.

Despite the limitations of diffusion, three groups of plants—liverworts, hornworts, and mosses—all known as **bryophytes,** still use diffusion to move substances through their bodies, rather than having any sort of "circulatory system." More than 12,000 species of mosses grow in habitats extending from arctic and alpine regions to the tropics. Liverworts and hornworts are small (less than an inch in height), simple plants that grow in moist and shady places and resemble flattened moss. These three types of bryophyte

plants are referred to as "non-vascular" because they do not have vessels to transport water and food. Water and nutrients are absorbed into the outermost layer of cells by projections that penetrate a few micrometers into the soil. Because these projections are so short, non-vascular plants must either live in places where the soil is always moist or become dormant when the soil surface dries out (**FIGURE 12-7**).

To adapt to land, the non-vascular plants had to develop a method of reproduction that protected the plant embryo from drying out and provided it with a source of nutrients. What made this possible was a life cycle of alternating haploid and diploid generations, which is radically different from the life cycle of humans and most other animals. In animals, the haploid gametes, at fertilization, produce a new, diploid cell that becomes a multicellular organism,

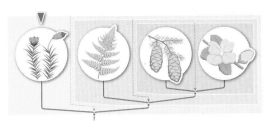

THE NON-VASCULAR PLANTS

Moss

Liverwort

Hornwort

COMMON CHARACTERISTICS
• Distribute water and nutrients throughout the plant by diffusion
• Release haploid spores, which grow and produce gametes
• Life cycle with multicellular haploid and diploid phases

MEMBERS INCLUDE
• Mosses (about 12,000 species)
• Liverworts (about 8,000 species)
• Hornworts (about 100 species)

FIGURE 12-7 **Overview of the non-vascular plants.**

MOSS LIFE CYCLE

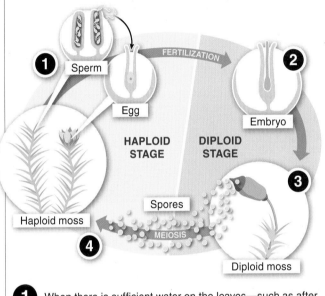

The spongy part of the moss is the haploid form. Hovering above, on stalks, are the spore capsules, the diploid form.

1. When there is sufficient water on the leaves—such as after rainfall—sperm swim from male reproductive structures to female reproductive structures, where they fertilize the egg.

2. A diploid embryo forms and develops into an adult diploid moss.

3. The diploid moss develops a capsule, which bursts and releases haploid spores.

4. A spore lands on moist soil and grows into an adult haploid moss.

FIGURE 12-8 **Alternation of generations in mosses.**

which, in the adult stage, starts the process all over again. In plants, however, there is an "alternation of generations."

All plants exhibit the alternation of generations, but the life cycle of a moss provides a useful example of how the alternation of generations works. When you look at a spongy mass of moss (or any other non-vascular plant),

what you see is the *haploid* (or **gametophyte**) part of the life cycle, rather than the *diploid* (or **sporophyte**) part of the life cycle (**FIGURE 12-8**). For this reason, mosses are described as having a dominant gametophyte. The adult moss plants are haploid plants—that is, all the cells have only one set of chromosomes. There are male and female moss plants that have male and female reproductive

Peat can be dried and then burned as fuel to heat homes.

In traditional Scotch whisky production, the burning of peat during the distillation process provides the smoky flavor.

 Non-vascular plants such as peat moss have many important economic and ecological uses, from flood control to gardening to the production of Scotch whisky.

FIGURE 12-9 **Uses of peat.**

structures located among the feathery leaves at the tips of the stems. Water collects here during rainstorms, allowing the sperm to "swim" (or be splashed) from the male structures to fertilize eggs in the female structures. Once the egg is fertilized in the female structure, a diploid zygote is formed and divides to become an embryo.

The embryo is sheltered within the female reproductive structure, which provides water and nutrients for the growing embryo. Eventually, the female structure elongates to such a degree that it breaks in two and forms a capsule extending over the top of the plant like a raised fist. Inside the capsule are haploid **spores**—single cells, containing DNA, RNA, and a few proteins. When a capsule ruptures, it releases hundreds of spores. The spores that land in moist, sheltered spots grow into new (haploid) male or female moss plants.

Mosses are important components of many soils. Without them, wind erosion can strip away unprotected soil, leaving tree trunks supported in mid-air by their roots.

FIGURE 12-10 **Non-vascular plants reduce erosion.** Without mosses, soils are more fragile. Shown here: a eucalyptus tree with eroded roots in New Zealand.

Some non-vascular plants are economically or ecologically important. "Peat moss" (*Sphagnum* moss), found in many parts of the world, is sold as a soil enhancer for gardening. In areas where trees are scarce, peat is dried and burned as fuel. In places with monsoon seasons, such as Malaysia and Indonesia, peat bogs—consisting of partially decayed moss—are important in flood control, because they can absorb enormous amounts of water and release it over a period of months. Mounds of burning peat are also used to dry the barley used in producing Scotch whisky, giving Scotch its distinctive smoky taste (**FIGURE 12-9**).

Unlikely as it seems, non-vascular plants can grow in deserts, and mosses (along with lichens and cyanobacteria) are important components of the biological crust that holds desert soils in place. The crust cements the soil particles together and allows the soil to resist wind erosion, but it is extremely fragile and very slow to regenerate. When people or cattle walk over the crust, they break it into small pieces that cannot resist wind, and erosion can then strip a meter or more of soil in a few decades, leaving tree trunks supported in mid-air by their roots (**FIGURE 12-10**).

TAKE-HOME MESSAGE 12·3

Non-vascular plants—mosses, liverworts, and hornworts—have scarcely evolved beyond the stage of the earliest land plants. They lack roots and vessels to move water and nutrients from the soil into the plant, and they reproduce with spores formed when a sperm from a male reproductive structure "swims" through a drop of rainwater to fertilize the egg in a female reproductive structure.

12·4 The evolution of vascular tissue made large plants possible.

What a difference some tubes make. That's all vascular tissue is—an infrastructure of tubes that begins in a plant's roots and extends up its stem and out to the tips of its leaves. The evolution of vascular tissue allowed early land plants to transport water and nutrients faster and more effectively than the cell-to-cell diffusion that non-vascular plants must rely on. Vascular plants' roots penetrate far enough into the soil to reach moisture even when the soil surface is dry. Roots that reach deep into the soil also provide the support that a plant needs to grow upward without falling over. As a consequence, **vascular plants** can grow taller than non-vascular plants and are more successful in areas where the surface of the ground dries out between rains (FIGURE 12-11).

Ferns are the most familiar of the primitive vascular plants. During the Carboniferous period, which extended from 360 to 300 million years ago, ferns were a major component of the huge swamp forests that eventually formed the coal deposits we now mine. A related group of plants, the horsetails, like their Carboniferous ancestors, grow in wet habitats where the oxygen concentration around the roots is low because the spaces between soil particles are filled with water (rather than air). Horsetails developed hollow stems that allow oxygen from the air to diffuse down to the roots. The thin leaves of horsetails have a single nutrient- and water-carrying vessel extending from the base to the tip.

Simply having vessels, as horsetails do, is an important evolutionary innovation, but ferns have a further one. In addition to the central vessel in each leaflet, the leaves of ferns have vessels branching from the central vessel to the edges of the leaflet. This arrangement places a channel for the movement of water and nutrients close to each cell in the leaf.

Like non-vascular plants, horsetails and most ferns reproduce with spores. In horsetails, the central hollow stem is topped by a conical structure where the haploid spores are produced and released. Many ferns have **sporangia** (*sing.* **sporangium**) on the undersides of the leaves where spores are produced (FIGURE 12-12). Because the haploid gametophyte is much smaller and simpler than the diploid sporophyte, ferns are described as having a dominant sporophyte.

Horsetails and ferns are taller than non-vascular plants, so their spores can be blown by the wind when they are

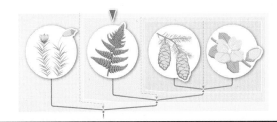

THE VASCULAR SEEDLESS PLANTS

Katote tree ferns

Common horsetail

Canary Island hare's foot fern

COMMON CHARACTERISTICS
- Distribute water and nutrients throughout the plant with a "circulatory system" of vascular tissue
- Release haploid spores, dispersed by the wind, which grow and produce gametes
- Life cycle (unlike in animals) with multicellular haploid and diploid phases

MEMBERS INCLUDE
- Ferns (about 12,000 species)
- Horsetails (about 15 species)

FIGURE 12-11 **Snapshot of the vascular seedless plants: ferns and horsetails.**

FERN LIFE CYCLE

Diploid fern

Haploid fern

released, and they may settle some distance from the parent plant. This increased dispersal ability was an important adaptation—the non-vascular plants are so low-growing that wind can't play much of a role in moving their spores.

A spore that lands on moist soil grows into a tiny heart-shaped structure called a **prothallus,** which is the free-living haploid stage of a fern (see Figure 12-12). The prothallus produces the haploid gametes: some cells produce eggs and others produce sperm. A sperm "swims" through drops of rainwater to fertilize an egg, and the fertilized egg (a diploid zygote with two sets of chromosomes) grows into an adult fern.

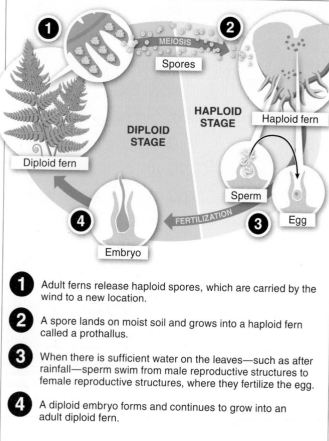

1 Adult ferns release haploid spores, which are carried by the wind to a new location.

2 A spore lands on moist soil and grows into a haploid fern called a prothallus.

3 When there is sufficient water on the leaves—such as after rainfall—sperm swim from male reproductive structures to female reproductive structures, where they fertilize the egg.

4 A diploid embryo forms and continues to grow into an adult diploid fern.

FIGURE 12-12 Haploid and diploid life stages in ferns. The photograph shows the sporangia, the spore-producing bodies, located on the underside of most fern leaves, on the diploid plant. The diploid and haploid plants in the fern life cycle are very different in appearance.

TAKE-HOME MESSAGE 12·4

Vessels are an effective "circulatory system" to carry water and nutrients up from the soil to a plant's leaves. The first vascular plants—including the earliest ferns and horsetails—were able to grow much taller than their non-vascular predecessors.

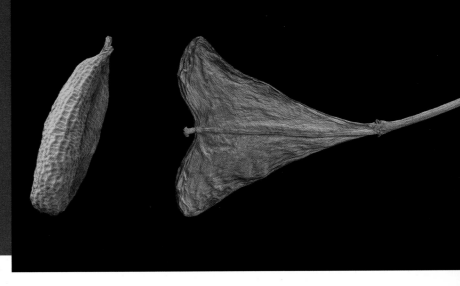

12·5–12·7
The advent of the seed opened new worlds to plants.

Shepherd's purse (Capsella rubella) seed and capsule.

12·5 What is a seed?

As we've seen, the evolutionary development of vascular tissue allowed plants to grow larger. Vascular plants could colonize areas where the soil at the surface of the ground was not always wet, because their roots could penetrate the soil to reach water and nutrients. The next big innovation in plant evolution was the **seed,** an embryonic plant with its own supply of water and nutrients encased within a protective coating (**FIGURE 12-13**).

Unlike spores, which are single cells that contain only DNA, RNA, and a few proteins, seeds contain both a multicellular embryo and a store of nutrients, mostly starch. This nutritive tissue—called **endosperm** in flowering plants and trees—can fuel the seed's initial growth. A seedling draws energy from the endosperm while it extends its leaves upward to begin photosynthesis and its roots downward to reach water and nutrients in the soil. There are two modern groups of seed-

---

SEEDS: STRUCTURE AND GROWTH

Seeds for sale at a market in Jordan

STRUCTURE
Fertilization produces a diploid seed, which contains a multicellular embryo and a store of carbohydrate (endosperm) to fuel its initial growth.

Protective coating
Endosperm
Embryo

GROWTH
A seedling draws energy from the endosperm while it extends its leaves upward to begin photosynthesis and its roots downward into the soil to reach water and nutrients.

FIGURE 12-13 **What is a seed?** A seed is a package that contains a multicellular embryo and a store of carbohydrate, the endosperm, to fuel its initial growth. This makes many seeds desirable as food sources.

PLANTS ARE A BRANCH OF EUKARYA · PLANTS WITHOUT ROOTS OR SEEDS · **THE ADVENT OF THE SEED** · FLOWERING PLANTS · PLANT AND ANIMAL RELATIONSHIPS · FUNGI

producing plants: **gymnosperms** (including pines, firs, and redwoods) and **angiosperms** (all flowering plants and trees).

Q Do plants have sex? How are seeds formed?

Like plants that do not produce seeds, seed plants have a life stage, called the gametophyte, which produces haploid gametes (sperm and egg). This is analogous to humans' production of haploid gametes. Because the gametophyte is much smaller than the sporophyte in seed plants, these plants are considered sporophyte dominant. **Pollen grains** and **ovules** are the male and female gametophytes, respectively, of seed plants. In brief (we expand on this later in the chapter), a haploid female gamete (egg) forms inside the ovule. When a pollen grain lands near the ovule, it produces a tube that grows into the ovule. Sperm from the pollen grain move through the pollen tube into the ovule and fertilize the egg. The external layer of the ovule then forms the seed coat.

Another of the challenges that plants face is dispersing their seeds. The seed stage is the only opportunity most plants have to scatter their offspring, and seeds and seed pods have many ways to do this. These range from the forceful send-off of exploding seed pods, to seeds that hitch a ride in or on passing animals, to those that are so small and light that they can float in water or air (such as dandelion seeds)—not to mention the use of fruits in seed dispersal. As we'll see later in the chapter, many methods for dispersing seeds have evolved.

TAKE-HOME MESSAGE 12·5

Seeds are a way that plants can give their offspring a good start in life. A seed contains a multicellular embryo plus a store of carbohydrate and other nutrients. Seeds are distributed by wind, animals, or water.

12·6 With the evolution of the seed, gymnosperms became the dominant plants on earth.

Gymnosperms include four major groups: the conifers, cycads, gnetophytes, and ginkgo (just one species) (FIGURE 12-14). All of the 900 or more species of gymnosperms are seed-bearing plants that produce ovules on the edge of a cone-like structure. During the middle of the Mesozoic era, 160 million years ago, the forests that dinosaurs walked through consisted entirely of gymnosperms—pine trees, redwoods, cycads, and their relatives. Flowering plants (angiosperms) did not appear for another 35 million years. Indeed, when it comes to population size and habitat range, pine trees and their relatives are among the most evolutionarily successful groups of plants on earth, growing on every continent except Antarctica and extending from sea level to the tree line on mountains (FIGURE 12-15).

Pines, spruces, firs, redwoods, and their relatives are familiar to residents of the temperate regions: many of these conifers have needle-like leaves—but that is not true of all gymnosperms. The ginkgo has distinctive fan-shaped leaves, and cycads have palm-like fronds with many small leaflets (see Figure 12-15). Both are nearly identical to fossils from the Mesozoic era.

The reproductive structures of gymnosperms—the cones—are male or female. The pine cones you are probably familiar

with are the female cones, which produce the ovules and, eventually, the seeds (FIGURE 12-16). The male cones are smaller and release pollen that is blown by the wind, and some of it reaches the ovules, which lie beneath

Q Why does so much tree pollen coat the windshields of cars parked outside in the spring?

the protruding scales of the female cones. The quantity of pollen released by conifers is beyond imagination—pollen-heavy air in a pine forest becomes hazy, and every surface is coated with yellow. Wind dispersal is clearly an inefficient method of **pollination** (getting the pollen to the vicinity of the ovule): billions of pollen grains are wasted for each grain that lands on a female cone and produces sperm to fertilize an egg in an ovule. Nonetheless, this "brute force" method of ensuring fertilization works: for more than 200 million years, pines have been successfully pollinated by wind.

When the pollen arrives at the female cone, a pollen tube forms and transports the haploid sperm to the ovule, where fertilization occurs and a diploid embryo begins to grow. The embryo develops slowly within the female cone over the course of many months, until the scales of the cone open and release the seed, ready to sprout into a new plant

THE GYMNOSPERMS

COMMON CHARACTERISTICS
- Distribute water and nutrients throughout the plant with a "circulatory system" of vascular tissue
- Reproductive structures called cones produce the gametes
- Fertilization produces seeds

MEMBERS INCLUDE
- Conifers (about 600 species)
- Cycads (about 300 species)
- Gnetophytes (about 65 species)
- Ginkgo (1 species)

Ginkgo

Douglas fir

Cycad

Welwitschia

GROUPS OF GYMNOSPERMS

CONIFERS
- Most commonly found in colder temperate and sometimes drier regions of the world
- Commonly have needle-shaped leaves
- Important source of timber
- Include pines, spruces, firs, cedars, hemlocks, yews, larches, and cypresses

CYCADS
- Slow-growing gymnosperms of tropical and subtropical regions
- Most resemble palm trees
- Several species are facing extinction in the wild

GNETOPHYTES
- Composed of 3 groups: *Gnetum, Ephedra,* and *Welwitschia*
- Most gnetophyte species are *Ephedra,* a shrub-like plant sometimes used as a herbal remedy for respiratory ailments

GINKGO
- *Ginkgo biloba* is the only remaining species
- Distinctive fan-shaped leaves
- The outer covering of the seeds emits a foul odor

FIGURE 12-15 Major groups of gymnosperms: conifers, cycads, gnetophytes, and ginkgo.

MALE CONE
The male cone releases pollen grains that require wind to reach a female cone.

FEMALE CONE
The female cone has ovules on the protruding scales. They produce seeds when fertilized by pollen.

 Although billions of pollen grains are wasted for each one that fertilizes an egg, gymnosperms have successfully used wind pollination for more than 200 million years.

FIGURE 12-16 **Cones are the reproductive structures of gymnosperms.**

(**FIGURE 12-17**). With the evolution of seeds, gymnosperms developed a life cycle with no free-living haploid stage such as we see in non-vascular plants.

This evolutionary change in life history probably reflects the different degrees of evolutionary fitness of haploid versus diploid organisms. A haploid organism has just one copy of each chromosome, which means it has just one copy of each allele. If the haploid organism carries a defective allele for a critically important gene, the organism is doomed, because it has only that one defective version of the gene. In contrast, diploid organisms have two sets of chromosomes and thus two copies of each allele. If one copy is defective, the second copy can function as a backup, enabling the plant to produce the gene product. As a result, mutations are much less likely to be lethal for diploid organisms than for haploid organisms.

1 Male cones release pollen grains that are dispersed by the wind to ovules found beneath the scales of female cones.

2 Pollen grains release sperm that fertilize an egg within the ovule.

3 Fertilization creates a diploid embryo that matures into a seed.

4 Eventually, the seed is released from the female cone and grows into an adult tree.

The evolution of seeds in gymnosperms eliminated the free-living haploid life stage seen in mosses and ferns.

FIGURE 12-17 **Assisted by the wind: the life cycle of gymnosperms.**

TAKE-HOME MESSAGE 12·6

Gymnosperms (pine trees and their relatives) were the earliest plants to produce seeds. This mode of reproduction offered advantages over the spores of earlier plants and gave gymnosperms the boost they needed to become the dominant plants of the early and middle Mesozoic era. Gymnosperms depend on wind to carry their pollen. Conifers protect the developing seeds in the female cone.

12·7 Conifers include the tallest and longest-living trees.

In addition to having the widest geographic range and the greatest number of species among the gymnosperms, the cone-bearing trees—the conifers—include both the tallest and the oldest living organisms on earth (**FIGURE 12-18**). The four tallest trees in the world are conifers: a coast redwood that is 380 feet (116 m) tall, a Douglas fir and a Sitka spruce, each 318 feet (97 m), and a Sierra redwood at 311 feet (95 m). But not all conifers are big: there are also miniature species of conifers, such as the shore pine, which can be just 20 centimeters tall. The oldest tree trunk belongs to a Great Basin bristlecone pine, the Methuselah tree, which has lived for more than 4,800 years (a slightly older tree was cut down in 1964). There's more to a tree than its trunk, though, and trunks can be replaced. That is what a Norway spruce in Sweden has done: the current trunk is about 600 years old, but the roots are 9,550 years old. This tree has persisted by sending up a new trunk each time the old one died.

Trees can grow large and live to great ages because woody plants can be exceptionally strong and resistant to attack by herbivores. A cross section of a tree trunk shows the structural characteristics that allow a tree to stand erect and transport water to leaves more than 100 meters above the ground. Rigidity is provided primarily by the heartwood, which is a core of dead tissue that contains complexly cross-linked molecules.

Bark covers the outside of the tree trunk and branches. The outer layers of bark are dead tissue that can be shed without damage to the tree, protecting the living tissue from attack by plant-eating insects. Part of the success of conifers can also be traced to their ability to defend themselves by exuding a sticky pine pitch that can engulf and smother insects.

TALLEST AND OLDEST

Conifers such as this redwood can reach up to 380 feet tall.

Conifers such as this bristlecone pine can live for more than 4,800 years.

Conifers have grown taller and reached older ages than any other plants.

FIGURE 12-18 Towering and ancient gymnosperms.

TAKE-HOME MESSAGE 12·7

Conifers are the success stories among gymnosperms, with more species and a larger geographic range than the rest of the gymnosperms combined. Rigidity, an exterior layer of bark, and the ability to exude sticky pitch protect conifers, allowing them to grow taller and live longer than any other plants.

12·8–12·10
Flowering plants are the most diverse and successful plants.

The bleeding heart flower (Lamprocapnos spectabilis) is a popular garden plant.

12·8 Angiosperms are the dominant plants today.

The appearance of flowering plants (angiosperms) about 135 million years ago in the Cretaceous period set the stage for the botanical world we know today, with flowering trees, flowering bushes, and all the grasses and herbaceous (non-woody) plants we see around us. The vast majority of plants on earth are flowering plants in the angiosperm group (**FIGURE 12-19**). Many of the early flowering plants would look familiar to us, and angiosperms dominate the plant world now, with some 250,000 species, compared with approximately 1,000 species of gymnosperms.

Flowers come in a bewildering variety of sizes, shapes, and colors, but they all have similar structures: a supporting stem with modified leaves—the flashy petals and the sepals, which form a (usually) green wrapping that encloses the flower while it is in bud. Most angiosperms combine the male and female reproductive structures in the same flower. The male structure is the **stamen** and includes the **anther,** which produces the pollen, and its supporting stalk, the **filament.** The female reproductive structure is the **carpel.** It has an enclosed **ovary** at its base, which contains one or more ovules in which eggs develop; a stalk (the **style**), extending from the ovary; and a sticky tip (the **stigma**) (**FIGURE 12-20**).

Pollination, in angiosperms, is the transfer of pollen from the male reproductive structures to the female reproductive structures of a flower—sometimes the same flower, and sometimes a different flower (on the same plant or on a different plant). Pollination is a multi-step process in which a pollen grain sticks to the stigma, forms a tube that grows until it reaches the ovule, and thus provides a route for sperm to travel down the tube to fertilize the egg. In the next section, we explore the tremendously varied methods by which angiosperms get the male gametes to the female gametes.

Although most flowers contain both male and female structures, several thousand species of plants produce flowers that are only male or only female. Sometimes, male and female flowers are borne on different individual plants, but in quite a few angiosperm species, the same plant has some male flowers and some female flowers, often side by side. Maize (corn) is a familiar plant with male and female flowers in different places on the same plant, although you might not recognize the reproductive structures as flowers. The tassel at the top of the plant is the male flower, and the ear is formed by female flowers.

TAKE-HOME MESSAGE 12·8

Flowering plants appeared more than 100 million years ago and diversified rapidly to become the dominant plants in the modern world. A flower houses a plant's reproductive structures. Most flowers have both male and female reproductive structures, but some species have flowers with only male or only female structures.

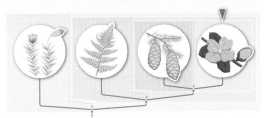

THE ANGIOSPERMS

COMMON CHARACTERISTICS
• Distribute water and nutrients throughout the plant with a "circulatory system" of vascular tissue
• Produce flowers, which produce gametes
• Seeds are enclosed within an ovary

MEMBERS INCLUDE
• Flowering trees, bushes, herbs, and grasses (about 250,000 species)

Apple tree

Water lily

French lavender

Victoria plums

FIGURE 12-19 Snapshot of the angiosperms.

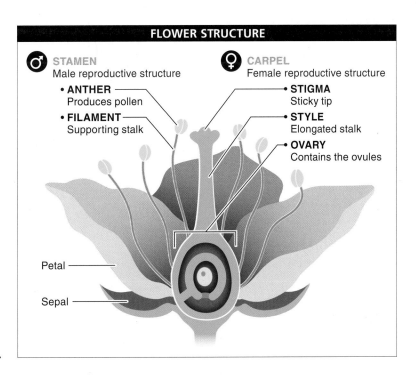

FLOWER STRUCTURE

♂ **STAMEN**
Male reproductive structure

• **ANTHER**
Produces pollen
• **FILAMENT**
Supporting stalk

♀ **CARPEL**
Female reproductive structure

• **STIGMA**
Sticky tip
• **STYLE**
Elongated stalk
• **OVARY**
Contains the ovules

Petal

Sepal

FIGURE 12-20   A flower houses a plant's reproductive structures.

12·9 A flower is nothing without a pollinator.

Picture a flower. What does it look like? Flowers differ hugely from one to another, and we all have our favorites. There are giant sunflowers, colorful roses, and some rotten-smelling orchids and lilies, to name just a few. The label "flower" would seem to indicate little beyond the fact that flowers are usually brightly colored, frequently have a prominent odor, and generally sit at the end of a plant's branches. Closer inspection, however, reveals that, with very few exceptions, all flowers have the same four fundamental structures: sepals, petals, stamens, and carpels. Here we examine the importance of flowers to sexual reproduction in plants.

Fertilization occurs when the male gamete merges with the female gamete. But that's a lot easier said than done: the male gamete has to get to where the female gamete is. The first step toward this is pollination—when a pollen grain makes the journey to the stigma.

A small number of angiosperms achieve pollination by releasing tremendous amounts of pollen into the wind, as do the gymnosperms, or into water—on the slim chance that some will land on the female reproductive organs of another plant of the same species. Most angiosperms, though, have a different way of moving pollen from the anthers of one flower to the stigma of another: they use animals to carry it (**FIGURE 12-21**).

To ensure that animals will visit a flower, picking up and delivering their pollen cargo, two strategies for achieving pollination have evolved among the flowering plants.

Q Why are some flowers so flashy?

1. Trickery. The plant deceives some animals into carrying its pollen from one plant to another. Some orchid species, for example, produce flowers that resemble female wasps. The mimicry is so good that male wasps mount the flower and attempt to mate with it. The male wasp twirls wildly on the flower, like a cowboy on a bucking bronco, repeatedly whacking his head against the strategically located anthers and getting pollen stuck all over his head and body. That is not enough for the plant to achieve pollination—but it's a start. If that male wasp gets fooled again by another orchid flower and mounts it in an attempt to mate, he will inadvertently deposit some of the pollen from his body onto the also strategically placed stigma of that flower. In the end, the wasp does not gain from his actions, but the orchids have an effective system of pollination.

2. Bribery. The plant bribes some animals to carry its pollen from one plant to another. Rather than just using trickery, the plant offers something of value to the animal. For this mutually beneficial method (a "mutualism"; see Section 15-14) to work, the plant must produce (a) a sticky pollen,

---

STRATEGIES FOR ATTRACTING POLLINATORS

TRICKERY
This isn't a bee! This orchid flower deceives the male bee into carrying its pollen. When the bee attempts to mate with the flower it picks up pollen which it delivers to other orchids.

FIGURE 12-21 Delivering precious cargo (inadvertently).

BRIBERY
Some plants offer something of value to an animal, bribing the animal to carry pollen from one plant to another. Here, a bee, covered in pollen, flies from flower to flower in search of nectar.

 Angiosperms have developed a way to transfer pollen efficiently from the anthers of one flower to the stigma of another: get an animal to carry it!

COLORS AND PATTERNS
- **WHITE:** Nocturnal pollinators, such as moths and bats
- **BRIGHT COLORS:** Visually oriented diurnal pollinators, such as birds, butterflies, and bees

FLOWER STRUCTURE
- **TUBE:** Pollinators with long tongues, such as moths
- **INTRICATE/CLOSED:** Pollinators such as bees

ODORS
- **SWEET:** Pollinators with a good sense of smell, such as moths, butterflies, and bees
- **STINKY:** Pollinators looking for rotten meat on which to lay eggs, such as flies
- **NO ODOR:** Pollinators with a poor sense of smell, such as birds

NECTAR
- **ABUNDANT:** Pollinators with high energy needs, such as bees, birds, and butterflies
- **NECTAR ABSENT:** Pollinators, such as flies, looking for a place to lay eggs, or beetles, looking for petals, pollen, and other parts to eat

FIGURE 12-22 Evolving together: plants and pollinators. You can often determine the type of animal that pollinates a flower just by examining the features of the flower.

(b) a flower that catches the attention of the pollinator, and, most important, (c) something of value to the pollinator. The payoff for the pollinator can be food, such as nutritious nectar rich in sugars and amino acids, or perhaps a safe, hospitable location for an insect to lay its eggs.

The variety of flower structures is tremendous. They differ in shape, color, smell, time of day they are open, whether or not they produce nectar, and whether or not their pollen is edible. Pollinators also vary widely and include birds (mostly hummingbirds), bees, flies, beetles, butterflies, moths, and even some mammals (mostly bats) (**FIGURE 12-22**). In each case, there has been strong coevolution between the plants and their pollinators: the plants become increasingly effective at attracting the pollinators and deterring other species from visiting the flower, while the pollinators become increasingly effective at exploiting the resources offered by the plants. Because of this strong coevolution between the plant and

animal species, for most flowers we can determine, just by examining the features of the flower, the type of animal that will pollinate it.

Pollination is just one step toward fertilization. Fertilization itself doesn't happen until the male and female sex cells meet and fuse. We explore this process in the next section.

TAKE-HOME MESSAGE 12·9

Angiosperms rely on animals to carry pollen from the anthers of one flower to the stigma of another. Flowers are conspicuous structures that advertise their presence with colors, shapes, patterns, and odors. Using flowers, plants are able to trick or bribe animals into transporting male gametes to female gametes, where fertilization can occur.

The process of fertilization—the fusion of two gametes to form a zygote—*begins* when a pollen grain lands on a stigma of a flower, but that is far from the end. In angiosperms, the process is called **double fertilization,** because there are two separate fusions of male nuclei (each carrying a complete set of the organism's genetic material) from the pollen grain with female nuclei in an ovule. As we'll see, double fertilization is a more efficient system than the type of fertilization in gymnosperms, because whenever an embryo is produced at fertilization (and only then), so, too, is a more substantial, ready-made food source.

When we left off in the previous section, a pollen grain—which will deliver the haploid male gamete—had just arrived on the female's stigma. The pollen grain forms a tube that extends downward through the stigma and style to the ovary, ultimately entering an ovule. As the pollen tube grows, two haploid sperm are formed within it by mitosis, and both move down the pollen tube (**FIGURE 12-23**).

While the pollen tube is growing, a cell within the ovule undergoes meiosis, forming four haploid cells, called spores. Three of these usually degenerate. The surviving haploid spore enlarges and undergoes mitosis, forming an embryo sac. A large, central cell in the embryo sac has two haploid nuclei. One of the other cells in the embryo sac is the haploid female gamete (the egg), which remains near the place where the pollen tube will be guided into the ovule by two other haploid cells within the embryo sac.

When the pollen tube reaches the ovule, one of the two sperm fuses with the egg to form a zygote. The other sperm fuses with the two nuclei in the middle of the embryo sac to form the endosperm, which has *three sets* of chromosomes (i.e., is triploid). The process is called double fertilization because two sperm enter the ovule and combine with female cells in two separate events, forming (1) a zygote (with two sets of chromosomes) and (2) an endosperm (with three sets of chromosomes).

The final steps in producing a seed occur as the diploid zygote cell undergoes multiple mitotic divisions to form an embryo, while the triploid cells multiply mitotically to produce the endosperm that will provide nutritional support for the seedling through its initial growth stages.

FIGURE 12-23 Double fertilization fortifies the seeds of angiosperms.

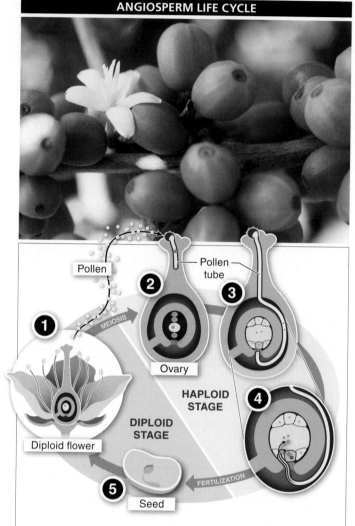

ANGIOSPERM LIFE CYCLE

Pollen

Pollen tube

Ovary

HAPLOID STAGE

DIPLOID STAGE

Diploid flower

Seed

MEIOSIS

FERTILIZATION

1 Pollen grains are released from the male anthers of a flower, delivering the male haploid gamete to the stigma—the sticky tip of the female reproductive structure.

2 The pollen grain produces a tube that extends through the stigma to the ovary. Meanwhile, a cell within the ovule undergoes meiosis, forming four haploid cells, called spores. Three of these usually degenerate.

3 The surviving haploid spore undergoes mitosis, forming an embryo sac that contains seven haploid cells. One cell becomes the egg, another cell—with two nuclei—will form the endosperm following fertilization by a sperm cell.

4 When the pollen tube reaches the ovule, two sperm are released. In a process called double fertilization, one of the two sperm fuses with the egg to form a zygote. The other sperm fuses with the two nuclei in the middle of the embryo sac to form the endosperm, which nourishes the embryo.

5 The zygote and the endosperm continue to develop within the ovule, forming a seed that will eventually be released and grow into a mature plant.

The outer layers of the ovule form the seed coat that will protect the seed until it sprouts. The enclosure of the seed within the ovary is a distinction between angiosperms and gymnosperms. The seeds of a gymnosperm are unenclosed and are sometimes referred to as "naked."

Q **What advantage does double fertilization give angiosperms?**

In gymnosperms, a single sperm fuses with an egg to form a zygote in an ovule that contains hundreds of other female cells, which are used as a source of nutrients for the resulting embryo. By contrast, there are two important advantages that double fertilization offers angiosperms.

1. Double fertilization initiates formation of endosperm only when an egg is fertilized. Waiting to be sure that an egg is fertilized is a good strategy, because making endosperm is a large energy investment for a plant. Gymnosperms invest that energy up front, and nutrients in ovules that are not fertilized are wasted because those "seeds" do not contain an embryo. In contrast, angiosperms do not waste energy forming endosperm in ovules that will not contain embryos.

2. Angiosperms can produce smaller gametes than gymnosperms, because the large energetic reserves will be produced only *after* fertilization occurs. The small size of the male and female gametes of angiosperms ensures that seeds are produced quickly. FIGURE 12-24 shows that as different reproductive strategies have evolved in plants, their gametes have become progressively smaller. Rapid production of seeds allows angiosperms to grow as annual plants (plants that complete their life cycle from sprouting to seed production in one growing season), which is something gymnosperms cannot do.

Q **Do flowers with both male and female structures fertilize themselves?**

Outbreeding, the combination of haploid cells from two different individuals, produces offspring with greater genetic diversity (that is, carrying a greater diversity of alleles) than offspring that result from inbreeding, the combination of a male and female gamete from the same individual. Angiosperms have a variety of ways to increase the chance that only sperm from another individual will fertilize the female gamete. Plants that have male and female reproductive structures in different flowers or even on different individuals, for example, increase the chance that a pollen grain landing on a stigma comes from a different plant. But many flowers with both male and female parts also have mechanisms to prevent self-fertilization. For

HAPLOID AND DIPLOID LIFE STAGES

HAPLOID STAGE	DIPLOID STAGE

NON-VASCULAR PLANTS
The majority of the life cycle is spent in the haploid stage.

VASCULAR SEEDLESS PLANTS
The haploid and diploid stages are both multicellular and physically independent of one another.

GYMNOSPERMS
The evolution of seeds almost completely eliminates the prominent haploid stage seen in mosses and ferns.

ANGIOSPERMS
Haploid gametes are further reduced in size, enabling more rapid seed production.

 As plants have developed different reproductive strategies, they have progressed from having a prominent haploid stage of life to simply having haploid gametes.

FIGURE 12-24 **Overview of the haploid and diploid stages of plant life cycles.**

example, the anthers may mature before the stigma, so the stigma is not ready to receive pollen when the anthers are active. And when the stigma becomes functional, the anthers are no longer producing pollen. Other plants use a molecular recognition system to prevent inbreeding: proteins on the surface of the stigma will not allow pollen from the same individual to form a pollen tube.

TAKE-HOME MESSAGE 12·10

Angiosperms undergo double fertilization, which ensures that a plant does not invest energy in forming endosperm for an ovule that has not been fertilized. Angiosperms have also developed methods to reduce the occurrence of self-fertilization and increase genetic variation among offspring.

12·11 – 12·12
Plants and animals have a love-hate relationship.

Fruits such as these peaches are structures that contain fertilized seeds.

12·11 Fleshy fruits are bribes that flowering plants pay animals to disperse their seeds.

Leaving home is an inevitable part of growing up. But as difficult as it may be, imagine you had to move away but didn't have a car or moving van—or even legs. How would you do it? This is yet another issue facing plants. Ingenious methods have evolved for transporting the male gametes to the female sex organs for fertilization, but that still results in all the fertilized eggs, and all the new offspring they develop into, living right on the female plant—or in nearly the same location, if they manage to fall off the parent plant. This is a

problem because parent and offspring would end up competing for the same light, space, soil nutrients, and other resources.

The solution is the fruit, a structure that aids in dispersing seeds—the reproductive packets made up of the embryo, some food reserves, and a hard coat. The fruit usually develops from the ovary (and sometimes from nearby tissue), right around the seeds, which develop from the

SEED DISPERSAL

HITCHING A RIDE
Some seed pods have spines or projections that attach them to passing animals.

FLYING AND FLOATING
The structure of some seeds allows them to be carried away from the parent plant by wind or water.

PROVIDING A FOOD SOURCE
Fleshy fruit is a form of bait that lures an animal to eat the seed and carry it far from the parent plant before eliminating it.

FIGURE 12-25 **Methods of dispersing seeds.**

ovules. This should cause you to look at a field of flowers differently. Given that the ovary is part of a flower, after pollination and fertilization, that flower will turn into a fruit. Another way of looking at it is that every fruit you eat was once a flower.

Pea pods, sunflower seeds, corn kernels, and hazelnuts are a few examples of fruits. These are "dry fruits," and the seeds they contain are transported by wind, water, or animals (**FIGURE 12-25**). Many angiosperms, though, make **fleshy fruits** that consist of the ovary plus some additional parts of the flower. For example, the core of an apple is the ovary and the flesh of the apple is derived from adjacent parts of the flower. Blueberries, watermelons, oranges, tomatoes, and peaches are other examples of fleshy fruits.

The fleshy part of a fruit is often larger than the ovary, and an angiosperm invests a lot of energy in producing fleshy fruits. The payoff comes when an animal eats the fruit and then defecates the seeds at a location far from the parent plant. In other words, fleshy fruit is the bait that some flowering plants use to get animals to disperse their seeds.

When you look at fruits, you can see several characteristics that help this system work.

- Fruits are colorful—red is the most common color, followed by yellow and orange. All of these colors contrast with green foliage and make fruits conspicuous.

- Fruits typically taste good—plants pour sugars into fruits, and their sweetness appeals to many animals.

- Fruit is good for animals, serving as a significant source of sugar, enabling animals to produce ATP for their needs. Additionally, fruit has other nutritional value; many birds incorporate the red and yellow colors of fruits into their own color patterns, and colorful male birds are appealing to female birds. Thus, male birds

that eat fruit are likely to produce a lot of offspring who also eat fruit—a good deal for the plant.

Of course, there's one more requirement to make fruit a successful way for a plant to disperse its seeds: the seeds must survive passage through the animal's digestive tract and be able to germinate when they come out. That's not a problem—seeds are fully viable when they emerge. It's easy enough to test this yourself. The next time you eat fresh tomatoes, check the toilet about 24 hours later. The seeds that have passed through your digestive system are still intact. If you want to test their viability, that's easy—recover a few, rinse them off, and put them on a wet paper towel in a sealed plastic bag. If you want to do a controlled experiment, you can take some seeds directly from a tomato and put them in another bag. Put both bags in a warm, dark place, and check them in a couple of days. You'll find that both sets of seeds have sprouted. (Or, you can just take our word for it.)

Seeds not only survive being eaten by animals—some seeds will not germinate *unless* they are eaten. For some plants, the pulp that surrounds the seeds in the fruit inhibits germination, and it must be digested away before the seed will sprout. The seeds of other plants have chemicals in the seed coats that must be removed by the acidity of an animal's stomach before the seeds will germinate. And there's another plus to this system: the seeds are deposited with a bit of manure that provides nutrients for the seedlings.

> **Q** Can seeds still sprout after being eaten by an animal?

TAKE-HOME MESSAGE 12·11

Following pollination and fertilization, plants often enlist animals to disperse their fruits, which contain the fertilized seeds, depositing them at a new location where the seedlings can grow. Fruits are made from the ovary and, occasionally, some surrounding tissue.

12·12 Unable to escape, plants must resist predation in other ways.

The interactions of plants and animals are not simple. On the one hand, plants depend on animals to pollinate their flowers and disperse their seeds. On the other hand, many kinds of animals eat plants, and plants are vulnerable because they can't run away.

Just because plants can't run away from plant-eating animals, however, they are not defenseless. A host of defensive

devices that have evolved in plants give them some protection against being eaten. These defenses fall into two categories: anatomical structures, such as thorns, and chemical compounds, including hallucinogens (**FIGURE 12-26**).

Spines, prickles, and thorns are a common way to discourage herbivores, and some of these structures are impressively large. The acacias, a group of plants that grow

ANATOMICAL STRUCTURES
Some plants have spines, spikes, and thorns that deter predators. Some trees have thick layers of bark that are shed in order to get rid of attacking insects.

STICKY TRAPS
Conifers exude pitch, a sticky substance that can engulf and smother attacking insects.

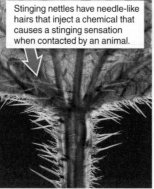

Locoweed contains substances that, when eaten, can cause cattle and horses to become lethargic and stop feeding.

Stinging nettles have needle-like hairs that inject a chemical that causes a stinging sensation when contacted by an animal.

CHEMICAL COMPOUNDS
Some plants synthesize chemicals that induce physiological and behavioral changes in the animals eating them.

FIGURE 12-26 **Spines, sticky traps, and toxic compounds.**
Plants have developed a range of defenses against predation.

as large bushes or small trees, are notoriously spiny. In these plants, the deterrent effects of spines are sometimes enhanced by the presence of ants that live in the swollen bases of the spines and fight off other animals—from insects to mammals, including humans—that would feed on their host acacia. Spines on palm fronds carry bacteria that infect the wounds that the spines create, providing a long-lasting reminder that it's not a good idea to try to eat a palm frond. In most plants, the surfaces of leaves and stems have microscopic hairs that protect the plant against insect herbivores by exuding a sticky fluid trap, but stinging plants such as nettles use larger hairs to inject a potent fluid into any animal that brushes against a leaf.

Many chemical defenses make plants bitter or otherwise unpalatable, but some plants play hardball by producing chemicals that affect the nervous or reproductive systems of their predators. Locoweeds, which occur all over western North America, contain alkaloids that can cause permanent damage to the nervous system. When cattle and horses eat locoweed, they tire quickly and stop eating. Other defensive chemicals are hallucinogens. Tetrahydrocannabinol, or THC, in the leaves and buds of marijuana plants disorients mammals and causes them to stop eating.

Many of the chemicals that plants produce to protect themselves have physiological effects in humans, and for thousands of years humans have used plants for medicinal purposes. The aboriginal peoples of North America and Eurasia, for example, knew that extracts of willow or aspen bark could relieve pain. Salicin is the compound in the bark that relieves pain, and we still use acetylsalicylic acid (aspirin) for pain relief. Opium poppies produce a more powerful pain reliever; foxglove flowers contain a compound that slows the heart rate; and a compound from the ipecacuanha plant can induce vomiting.

Plants with medicinal uses are sought after (a process called "bioprospecting"), and representatives of pharmaceutical companies trudge through rain forests and deserts to find them. The rewards from such a discovery can be great. For example, artemisin from Chinese sweet wormwood is effective in treating malarial infections that are resistant to conventional pharmaceutical compounds.

There's another role that chemicals can play in a plant's defense systems. Although plants can't move, they do send messages. They release volatile chemicals when they are attacked by insects, and these airborne chemicals can warn nearby plants of the impending threat or can call protective insects to their aid. When plants are nibbled by insects, they release the chemical methyl jasmonate (MJ), which is

Some plants living in nitrogen-deficient soil have turned the tables, becoming predators on insects.

A bug nymph trapped in the northern pitcher plant (Sarracenia purpurea) in an acidic bog in Michigan.

FIGURE 12-27 **Carnivorous plants feed on insects.**

carried with the breeze to nearby plants. When these plants detect the presence of MJ, they ramp up production of their own defensive chemicals so that they are prepared if the

insects move in to feast on them. Insects can also detect MJ, and insects that prey on other insects are attracted to plants emitting the chemical, because it signals the presence of prey. When corn and cotton plants are attacked by caterpillars, release of MJ summons parasitic wasps that deposit their eggs inside the caterpillars. When the eggs hatch, the wasp larvae eat the caterpillars from the inside out.

About 900 species of plants turn the tables—and eat insects. They use the protein in their insect meals to supplement the nitrogen compounds they get from the soil. Insectivorous plants are most common in boggy areas, because boggy soil often has low nitrogen concentrations (FIGURE 12-27). Pitcher plants are an example—their name comes from conspicuous pitcher-shaped structures that capture insects. It's easy for an insect to enter the open top of a pitcher, but hard to get out, because the inner walls are lined with hair-like structures that point downward, forcing the insect into a pool of liquid. Some pitcher plants merely absorb the nitrogen that is released when the trapped insects decay, but other species secrete digestive enzymes into the pool. These enzymes, which are basically the same as the enzymes in your stomach, digest the insects and make the nitrogen available more rapidly.

TAKE-HOME MESSAGE 12·12

Plants have a wide range of defenses against herbivorous animals, from physical defenses such as thorns to chemicals that have complex effects on animals' physiology. Plants respond to insect attack by synthesizing chemicals that make the plant less palatable. Some plants living in soil that is deficient in nitrogen have switched roles, preying on insects.

12·13–12·16
Fungi and plants
are partners but not
close relatives.

Cup fungi (Cookeina tricholoma) *from Borneo.*

12·13 Fungi are closer to animals than they are to plants.

Think about your toes for a minute. Have you ever had an itchy burning sensation between them? Almost all of us have athlete's foot at some time during our life. One study found that 15% of all people in the United Kingdom currently had this fungal infection.

Fungi make up their own monophyletic kingdom (see Chapter 10) within the eukarya domain. Most fungi are multicellular, sessile decomposers (**FIGURE 12-28**). Although they were originally thought to be plants lacking chlorophyll, it turns out that they have little in common with plants. In fact, DNA sequence comparisons reveal that

fungi are more closely related to animals than they are to plants (**FIGURE 12-29**). As eukaryotes, fungi have all the basic cellular components you would expect to find: nuclei, mitochondria, an endomembrane system, and a cytoskeleton. They also have cell walls, but the cell walls, instead of including cellulose, as in plants, are made of the carbohydrate chitin, a chemical also important in producing the exoskeleton of insects.

Fungi most likely arose from a unicellular, flagellated, aquatic protist more than 500 million years ago (and possibly as long as 1.3 billion years ago). Close to 100,000 species of fungi

WHAT IS A FUNGUS?

FUNGI ARE DECOMPOSERS OR SYMBIONTS
Fungi acquire energy by breaking down the tissues of dead organisms or by absorbing nutrients from living organisms.

FUNGI ARE SESSILE
Fungi are anchored to the organic material on which they feed.

FUNGI HAVE CELL WALLS MADE OF CHITIN
Fungi have cell walls containing chitin, a chemical that is also important in producing the exoskeleton of insects.

FIGURE 12-28 **The defining characteristics of fungi.**

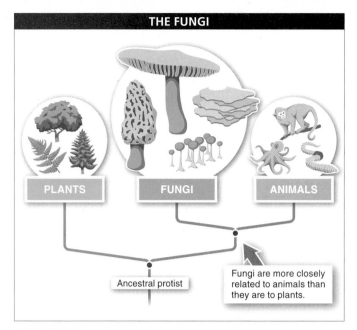

THE FUNGI

PLANTS

FUNGI

ANIMALS

Ancestral protist

Fungi are more closely related to animals than they are to plants.

FIGURE 12-29 **Where the fungi fit in.**

have been described, but the total number of species is estimated to be about 1.5 million. These species are divided into about seven phyla, but the overall phylogenetic classification is still very much in flux, with many of the relationships unresolved. Molecular evidence is, increasingly, guiding the process. About 98% of the described species belong to two monophyletic phyla, Ascomycota (about 64,000 species) and Basidiomycota (about 31,000 species). Besides these phyla, the two phyla with the largest numbers of described species are Microsporidia, with about 1,300 species, and Chytridiomycota, with about 700 species. These two groups, however, are not monophyletic.

The most commonly encountered types of fungi are (1) yeasts (the only single-celled fungi), truffles, and morels of the phylum Ascomycota; (2) mushrooms, of the phylum Basidiomycota; and (3) molds (which are not a phylogenetic grouping), such as *Penicillium*—the source of the antibiotic penicillin—of the phylum Ascomycota, and black bread mold, of the monophyletic phylum Zygomycota (**FIGURE 12-30**).

The fungus that causes athlete's foot is multicellular and consists of thread-like structures made up of long strings of cells called **hyphae** (pronounced HIGH-fee). Because of their thinness and length, hyphae have an enormous surface area. This means they are very good at taking up nutrients from your skin, but it also makes them very vulnerable to drying out. The fungus responsible for athlete's foot grows best in moist places—like on the skin between your toes.

Unless you have an active athlete's foot infection right now, however, your most recent encounter with a fungus was most likely when you last ate bread. Bread rises because yeast cells are mixed into the dough, which is then kept in a warm place while the yeast consumes sugar and produces carbon dioxide through fermentation. The carbon dioxide released by the yeast makes the dough swell into a loaf.

TAKE-HOME MESSAGE 12·13

Fungi are eukaryotes with the same internal cellular elements as other eukaryotes—and one distinctive feature: a cell wall formed from the carbohydrate chitin. Some fungi, the yeasts, live as individual cells; most other fungi are multicellular.

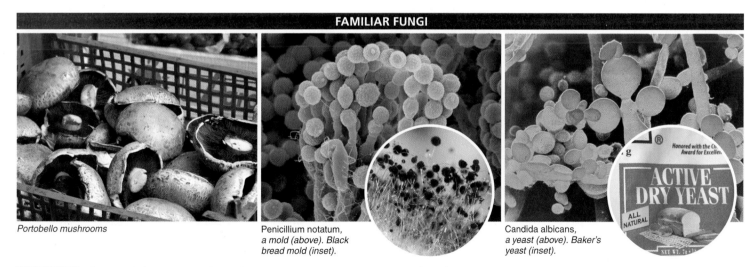

FAMILIAR FUNGI

Portobello mushrooms

Penicillium notatum, a mold (above). Black bread mold (inset).

Candida albicans, a yeast (above). Baker's yeast (inset).

ACTIVE DRY YEAST
ALL NATURAL
Honored with the Ch... Award for Excellen...
NET WT...

FIGURE 12-30 **The fungi that humans know best.** There are more than 1.5 million fungal species, usually divided among seven phyla. Those most recognizable to humans are mushrooms of the Basidiomycota (basidiomycetes), yeasts of the Ascomycota (ascomycetes), and molds (a term that does not refer to a monophyletic group), which occur in multiple phyla.

PLANTS ARE A BRANCH OF EUKARYA

PLANTS WITHOUT ROOTS OR SEEDS

THE ADVENT OF THE SEED

FLOWERING PLANTS

PLANT AND ANIMAL RELATIONSHIPS

FUNGI

519

12·14 Fungi have some structures in common, but exploit an enormous diversity of habitats.

As we've seen, most fungi are multicellular and are composed of long strings of barely visible, thread-like hyphal cells. The hyphae interconnect to form a mass of tissue called a **mycelium,** the form in which a fungus spends most of its time, usually underground or in a decaying tree or log. Unless you dug down to reach the mycelium, you would not know a fungus was there.

The structure that most people associate with fungi is the mushroom. But a mushroom is just a temporary reproductive structure (or "fruiting body"), part of a complex reproductive cycle that includes both sexual and asexual reproduction in some fungi (FIGURE 12-31). This cycle consists of several steps.

1. Underground, genetically distinct haploid hyphal cells join together. But their nuclei *don't* fuse, so instead of being diploid, they are "dikaryotic," meaning that each cell in the hypha has two nuclei.

2. The dikaryotic mycelium can grow and spread for years.

3. At some point, tightly packed dikaryotic hyphae may form a mushroom.

4. The haploid nuclei in the dikaryotic hyphae fuse in some cells, putting the mushroom into a diploid state.

5. In meiosis, the diploid cells produce huge numbers of haploid spores (up to a billion in a single mushroom!), which are dispersed by wind or water or on the bodies of animals.

6. After landing in a hospitable place, the spores grow as haploid hyphae, and the cycle begins anew.

Q What is the largest living organism in the world?

Fungi have an unusual and effective method of getting nutrition. Unlike humans, they digest their food outside their "body." While growing underground, hyphae secrete strong enzymes that break down the organic molecules around them, and the hyphae then absorb the nutrient-rich fluid. How effective are fungi at absorbing nutrients and growing? Extremely good. In fact, in eastern Oregon, a yellow honey mushroom fungus covers an area of about 4 square miles (nearly 10 square kilometers) (FIGURE 12-32). This fungus is estimated to be at least 2,400 years old, and it may be more than 8,000 years old, which would place it in the category of the oldest living organism. Moreover, by area, it is the largest.

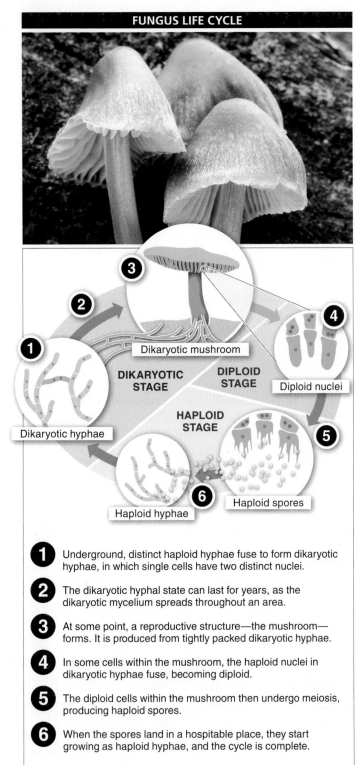

FUNGUS LIFE CYCLE

Dikaryotic mushroom

DIKARYOTIC STAGE

DIPLOID STAGE

Diploid nuclei

HAPLOID STAGE

Dikaryotic hyphae

Haploid hyphae

Haploid spores

1 Underground, distinct haploid hyphae fuse to form dikaryotic hyphae, in which single cells have two distinct nuclei.

2 The dikaryotic hyphal state can last for years, as the dikaryotic mycelium spreads throughout an area.

3 At some point, a reproductive structure—the mushroom—forms. It is produced from tightly packed dikaryotic hyphae.

4 In some cells within the mushroom, the haploid nuclei in dikaryotic hyphae fuse, becoming diploid.

5 The diploid cells within the mushroom then undergo meiosis, producing haploid spores.

6 When the spores land in a hospitable place, they start growing as haploid hyphae, and the cycle is complete.

FIGURE 12-31 **The three-stage reproductive cycle of a fungus.**

A yellow honey mushroom fungus in eastern Oregon is the largest organism by area on earth, covering four square miles, mostly underground. Just a small cluster of the fungus's temporary reproductive structures—what we call "mushrooms"—are shown here.

Area of largest individual yellow honey mushroom fungus (4 sq. mi.)

Relative size of football field for reference

Football field (360 ft. long)

Blue whale (90 ft. long)

African elephant (12 ft. tall)

Adult human (6 ft. tall)

FIGURE 12-32 **The largest organism in the world: a yellow honey mushroom fungus.**

organisms, you are a parasite. Just as fungi can live off dying trees, they can also grow in and on people. "Mycosis" is a general term for a disease that is caused by a fungus, such as athlete's foot, or related fungi that cause jock itch, beard itch, scalp itch, ringworm, and toenail fungus. These fungi get their nutrition by digesting some of the organic molecules of your body! And all of these fungal diseases are quite contagious, because they are spread by spores that can linger on moist surfaces and in clothing, so they tend to be a problem in places like dormitories, gymnasiums, and fitness centers.

Fungi can also thrive in poorly ventilated spaces in buildings (**FIGURE 12-33**). Molds are multicellular fungi that are responsible for many unpleasant effects. People living or working in a "sick building" can experience burning or watering eyes, a runny nose, and itchy skin—allergic reactions to the proteins in fungal spores. More severe effects, such as cancers and miscarriages, are probably produced by toxins released when the fruiting bodies of the fungi disintegrate. Curing a sick building can be so expensive that in some cases the entire building is destroyed and rebuilt.

It wouldn't be fair to mention so many undesirable effects of fungi without describing some of the many fungi that are beneficial to humans, particularly one that has benefited millions of humans: the fungus that produces penicillin. Alexander Fleming was a

But huge as it is, the fungus is nonetheless hard to detect, because the mycelium is underground. The fungus appears on the surface only as mushrooms (edible, but not particularly tasty) and as hyphae in dying and dead trees. Indeed, no one even knew that this enormous fungus existed until 2000, when Oregon foresters sought an explanation for an epidemic of dead and dying trees in the Malheur National Forest. (They discovered that the fungus was killing trees by causing their roots to rot.)

Fungi can grow in many different habitats because, as **decomposers,** all they need for their nutrient supply is some sort of organic material that they can break down. Fungi play an enormously important ecological role in speeding the decay of organic material in forests. They don't need light, so they can grow underground or inside dead trees and logs.

If you break down the tissues of dead organisms, you are a decomposer, and if you break down the tissues of living

Mold contaminations can cause a variety of irritating and potentially dangerous health problems.

FIGURE 12-33 **Cleaning out toxic mold.** A building overrun with mold spores may cause its occupants to become ill.

PLANTS ARE A BRANCH OF EUKARYA

PLANTS WITHOUT ROOTS OR SEEDS

THE ADVENT OF THE SEED

FLOWERING PLANTS

PLANT AND ANIMAL RELATIONSHIPS

FUNGI

521

researcher studying bacteria. In 1928, just prior to taking a month-long vacation, he stacked some of his used Petri dishes containing bacterial colonies in a corner of his lab. When he returned from his vacation, he found that some of the dishes had been contaminated by a fungus. Before discarding them, he noticed that near the fungal infection, the bacterial colonies were dead, while farther away from it, they continued to grow. Suspecting that the fungus might be producing something toxic to the bacteria, he cultured it and discovered that it did produce a substance that killed many different types of bacteria. He identified the fungus as *Penicillium.* Following the development by others of a method to purify the antibacterial compound produced by the fungus, penicillin became a "wonder drug," responsible for saving millions of lives by helping to treat bacterial infections.

Many mushrooms are gastronomic delicacies, and commercially grown portobello and shiitake mushrooms,

for example, command high prices. Truffles—the underground reproductive structures of a fungus called *Tuber*—can sell for as much as $3,500 per pound. Because truffles grow underground, there is nothing on the surface to indicate their presence, and truffle hunters use trained pigs or dogs to locate them by scent.

TAKE-HOME MESSAGE 12·14

Fungi are decomposers, and all they need to thrive is organic material to consume and a moist environment so their hyphae don't dry out. Fungi have complex life cycles, with both sexual and asexual phases, and the parts of a fungus that are most often visible are its temporary spore-producing bodies. Fungi can cause a variety of health problems, but also are responsible for antibacterial medicines such as penicillin.

12·15 Most plants have fungal symbionts.

If you examine the roots of a plant with a microscope, you will find round structures and fibers closely associated with the fine rootlets and root hairs. These are root fungi, or **mycorrhizae** (pronounced my-koh-RYE-zee). Some mycorrhizae have hyphae that press closely against the walls of root hair cells. Others send hyphae through the root cell walls and into the space between the cell wall and the plasma membrane (FIGURE 12-34).

Beneficial associations between roots and fungi are ancient—they have been found in 400-million-year-old fossils of early land plants—and nearly all species of modern plants have them. Lumber companies have found that they have greater success when replanting areas they have clear-cut if the conifer seedlings are first grown in soil that has been inoculated with mycorrhizal fungi. Home gardeners can buy potting soil with mycorrhizal spores or

MYCORRHIZAE

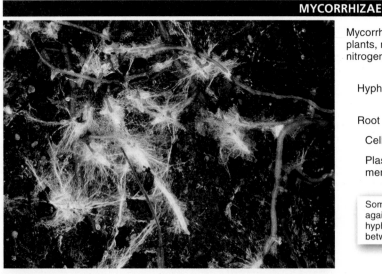

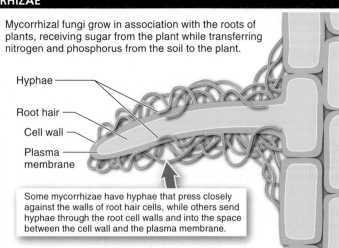

Mycorrhizal fungi grow in association with the roots of plants, receiving sugar from the plant while transferring nitrogen and phosphorus from the soil to the plant.

Hyphae

Root hair

Cell wall

Plasma membrane

Some mycorrhizae have hyphae that press closely against the walls of root hair cells, while others send hyphae through the root cell walls and into the space between the cell wall and the plasma membrane.

FIGURE 12-34 **The association between mycorrhizal fungi and plants is beneficial to both.**

SYMBIOTIC FUNGI ARE BENEFICIAL TO THEIR HOSTS

Researchers investigating the invasive velvetleaf plant evaluated how well offspring survived. They compared the survival of offspring from parents grown with and without symbiotic root fungi and reported the data below.

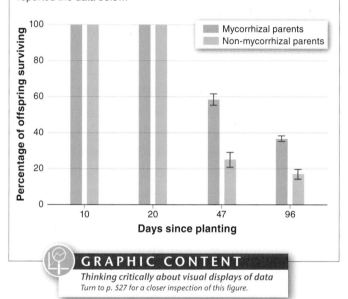

GRAPHIC CONTENT
Thinking critically about visual displays of data
Turn to p. 527 for a closer inspection of this figure.

FIGURE 12-35 **Plant partners.** Root fungi (mycorrhizae) can improve plant functioning.

buy spores compressed into tablets that can be placed in the soil when seedlings are transplanted. The association between mycorrhizal fungi and plants is beneficial to both partners. The mycorrhizal fungus benefits by drawing sugar from the plant, which it uses to support its own cellular respiration. The fungus extracts phosphorus and nitrogen from the soil and releases them inside the roots of the plant, allowing it to grow faster and larger. Research on the impact of root fungi on their plant hosts even demonstrated that the *offspring* of parental plants with mycorrhizae had significantly better survival than the offspring of parental plants without mycorrhizae (**FIGURE 12-35**). (In this case,

unfortunately, the fungi were enhancing the growth and survival of an invasive plant called velvetleaf, which causes hundreds of millions of dollars of damage each year to corn crops in the United States.)

Not all plants play fair, however. Some take nutrients from the fungus and give nothing in return. One of the best known of these mycorrhizal parasites is the ghost pipe (named for its ghostly white appearance), which grows in forests in the northern hemisphere (**FIGURE 12-36**). The ghost pipe is an angiosperm related to heaths and heathers. Its roots are associated with mycorrhizal fungi that are also connected to the roots of trees. The ghost pipe obtains all its nutrients from or through the mycorrhizae: nitrogen and phosphorus are provided by the mycorrhizal fungus, and sugar travels through the fungus from tree roots to the ghost pipe.

Some plants manipulate mycorrhizae to gain a competitive advantage over other plants. Garlic mustard, for example, was brought to North America by European colonists to add a tangy flavor to salads. Initially it was planted in kitchen gardens, but it escaped into the wild, where it has proved to be an extremely successful invasive species. Garlic mustard excretes compounds from its roots that interfere with the partnership between mycorrhizal fungi and local native plants. Because the native plants *do* need mycorrhizal fungi to prosper, the destruction of these fungi by garlic mustard weakens the plants. This chemical warfare is so successful that garlic mustard has wiped out entire populations of native woodland plants and is threatening many forests in the central United States.

FIGURE 12-36 **Not all plants play fair.** Ghost pipe is a plant that parasitizes mycorrhizae for nutrients.

Fungi also form mutually beneficial relationships with chlorophyll-containing bacteria and/or algae. These two-way or three-way partnerships are called **lichens,** and they grow on surfaces such as tree trunks and rocks. The fungus is fed by its photosynthetic partner and helps it absorb water and nutrients. In addition, the fungus provides nutrients for its partners by excreting enzymes that dissolve organic material and acid that dissolves rock.

TAKE-HOME MESSAGE 12·15

Plants and fungi have a close and mutually beneficial association in mycorrhizae. Mycorrhizal fungi grow in intimate association with the roots of most plants, receiving sugar from the plant and transferring nitrogen and phosphorus to the plant.

? 12·16 THIS IS HOW WE DO IT

Can beneficial fungi save our chocolate?

People love chocolate. It's a plant product at the heart of a huge economic industry. More than 3.5 million tons of cocoa beans are produced each year, and the global chocolate market is valued at close to $100 billion per year.

Chocolate is made from the seeds of the cacao tree (*Theobroma cacao*), a plant that is native to Central and South America. Twenty to 60 seeds are harvested from the large fruits, called cacao pods. The seeds are fermented, dried, roasted, cracked, and crushed, after which they may be combined with other ingredients, depending on the product.

Worldwide production of chocolate, unfortunately, is significantly reduced each year by fungal diseases that affect cacao plants. The fungi reduce the ability of trees to grow and to produce pods, as well as rotting the interior of the seed pods. In Peru, for example, the "frosty pod rot" fungus alone has reduced cacao production by one-third.

Because farmers would like alternatives to the high use of pesticides—which are expensive and not very effective—researchers are looking for other ways to reduce the impact of these fungal pathogens.

How could you approach the problem of finding ways to fight fungal pathogens?

Fungi often live within plants. In fact *every single plant species examined* has been found to harbor fungi inside its tissue. These fungi are extremely diverse and, in many cases, do not seem to harm their plant host. For this reason, researchers have taken the approach of searching for species of fungi that might be able to outcompete or

even harm the pathogenic fungi responsible for three of the most common fungal diseases.

To get started, the cacao researchers identified three distinct issues or questions requiring investigation.

1. Are there fungal species with anti-pathogen activity?

2. Can those species be introduced into cacao plants (without harming the plants)?

3. Within cacao plants, do these fungi actually limit damage by the pathogenic fungi and improve the plant's productivity?

To investigate the first question, researchers isolated 110 different types (called *morphospecies*) of fungi that can live in cacao trees and seem not to harm them. They then cultivated these fungi in Petri dishes, along with the pathogenic fungi. In each case, they evaluated which fungi "won" the interaction—which of two interacting species survived in the Petri dish.

Why are simple laboratory experiments a good place to start?

This approach made it possible to efficiently identify fungi with anti-pathogen activity to serve as "biocontrol agents." In fact, the researchers discovered 53 different fungi that showed antagonism toward the major pathogenic fungi. Of these, they identified three candidates that had both anti-pathogen activity and good colonizing ability.

The next step, to investigate the second question, was to attempt to deliver the helpful fungi into cacao seedlings. The researchers first germinated cacao seeds in sterile soil.

They then grew the seedlings in plastic shade houses that prevented the leaves from being exposed to any spores of airborne pathogenic fungi, and sprayed the plants with a suspension containing spores of a potentially beneficial fungus (while other plants were not sprayed, serving as controls).

❙ Why is it necessary to document an effective method of inoculating seedlings with the beneficial fungus?

Evaluating leaf tissue after 10, 25, and 33 days, the researchers identified fungal growth and estimated the percentage of fungal colonization. After 33 days, they found that colonization had occurred in 39%, 62%, and 92% of the cacao plants, depending on which of the three potentially beneficial fungi they had used for inoculation.

❙ Might beneficial fungi harm pathogenic fungi in Petri dishes in the lab, but not in trees in the field? Why?

The third, and possibly most important, issue examined was whether the beneficial species would reduce the impact of the pest species on cacao trees' productivity. Here, the results were a bit mixed.

In a greenhouse study, the researchers found 25% leaf mortality among trees infected with one of the pathogens. Among trees infected with both the pathogen and the potentially beneficial fungus, leaf mortality was just under 10%, a statistically significant reduction.

In the field, however, the results were less dramatic. Across four farms, researchers observed 320 experimental trees (inoculated with the potentially beneficial fungus) and 320 control trees (not inoculated with the potentially beneficial fungus). They found no reduction in the incidence of pod loss due to the pathogenic fungi. This may have been a consequence of an unusually low rate—less than 20%—of infection by the pathogenic fungi on these farms. The researchers did note, however, that in plants treated with the potentially beneficial fungus there was no reduction in plant growth *and* there was a 10% reduction in *lesions* caused by the pathogenic fungus.

❙ What can we conclude from these results?

It's clear that—at least using current methods—biocontrol isn't a magic bullet that wipes out the major fungal diseases of cacao trees. But researchers are optimistic.

❙ Which aspects of these experimental investigations are the most promising?

The researchers' optimism stems from several of their results. (1) They were able to identify potentially "good" fungi quickly in lab experiments. (2) They were able to inoculate cacao trees with "good" fungi. And (3) there seems to be at least some inhibition of pathogen activity in trees inoculated with "good" fungi.

The researchers are continuing their work, evaluating the impact of other factors on the anti-pathogen impact of the "good" fungi. These factors include (1) the timing of the spraying of seedlings to inoculate them, (2) the impact of using multiple potentially "good" fungal morphospecies simultaneously, (3) the amount of shade versus sun the trees are exposed to, and (4) the extent of pathogen infection.

TAKE-HOME MESSAGE 12·16

Pathogenic fungi significantly reduce the productivity of cacao plants—with a significantly adverse effect on the world chocolate market. Other fungi have been identified that are antagonistic to the pathogens when cultured in Petri dishes. Inoculation with these potentially "good" fungi reduces the pathogenic fungi's impact on plant growth in greenhouse experiments and shows promise when used in the field.

StreetBIO KNOWLEDGE YOU CAN USE

Yams: nature's fertility food?

Do you know any twins? Have you ever wondered how common twins are? In the United States and Europe, about 12 in every 1,000 births are twin births. In Asia, the number is slightly lower, at 8 sets of twins per 1,000 births. Elsewhere, the rates are similar—except for southwest Africa, particularly in Nigeria, which has been called "The Land of Twins." There, the rate of twin births among the Yoruba people is more than four times that in the United States, with about 50 pairs of twins per 1,000 births. (Triplet births are unusually common, too, occurring 16 times more frequently in Nigeria than in the United States.)

Q: **Why are so many twins born in Nigeria? Is it something in the water?** Nope. But that's not too far off. It's the diet of the Yoruba people. A staple in their diet is the white yam, a starchy vegetable that looks a bit like a potato; many people eat yams several times a day.

Q: **What's in the yams?** Some preliminary studies suggest that an estrogen-related compound in yams is responsible for stimulating the ovaries, increasing the likelihood, in any given month, that more than one egg will be released.

Q: **How can we be sure it's the yams?** At this point, the relevant data are still being collected. In one intriguing laboratory study, rats fed a diet rich in yams *doubled* their litter size. In another, the circulating levels of the follicle-stimulating hormone (which stimulates growth and maturity of follicles in the ovaries) and estradiol were both 32% higher in rats fed diets rich in yams than in rats fed a standard diet, while the levels of luteinizing hormone (which triggers ovulation) were 158% higher in the yam-fed rats.

Q: **So, will it work for you?** It's unclear. Anecdotes abound about women having twins after purposely increasing their yam consumption. But a randomized, controlled, double-blind study has yet to be conducted. How would you set up and analyze a study like that? (It is important to note that the yams consumed in Nigeria are not the same as American yams, which are sweet potatoes and do not contain the steroid levels of yams grown in Nigeria.)

Check Your Knowledge

GRAPHIC CONTENT

Thinking critically about visual displays of data

1. What can you conclude from this graph?

2. What percentage of the offspring from parents with mycorrhizal fungi were surviving after 47 days? What was the percentage surviving from parents without mycorrhizal fungi?

3. There are error bars for the percentage of offspring surviving at 47 days and at 96 days after planting. Why are there no error bars at 10 days and 20 days after planting?

4. What does it mean to say that the differences in offspring survival between mycorrhizal and non-mycorrhizal parents are "statistically significant"?

5. If you could have offspring survival data for one additional point in time, when would it be (i.e., how many days after planting)? Why?

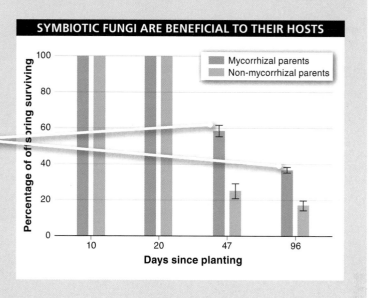

See answers at the back of the book.

Key Terms in Plant and Fungi Diversification

angiosperm, p. 504
anther, p. 508
bryophyte, p. 498
carpel, p. 508
cuticle, p. 498
decomposer, p. 521
double fertilization, p. 512

endosperm, p. 503
filament, p. 508
fleshy fruit, p. 515
flower, p. 508
gametophyte, p. 499
gymnosperm, p. 504
hyphae, p. 519

lichen, p. 524
mycelium, p. 520
mycorrhizae, p. 522
non-vascular plant, p. 496
ovary, p. 508
ovule, p. 504

plant, p. 494
pollen grain, p. 504
pollination, p. 504
prothallus, p. 502
root, p. 498
seed, p. 503
shoot, p. 495

sporangium, p. 501
spore, p. 500
sporophyte, p. 499
stamen, p. 508
stigma, p. 508
style, p. 508
vascular plant, p. 501

ABOUT THE CHAPTER OPENING PHOTO

Calla lily flowers.

REVIEW & REHEARSE
12
PLANT and FUNGI DIVERSIFICATION

? *Check Your Knowledge*

1. Within the plant kingdom, relative to the gymnosperms, the angiosperm species are:

a) monophyletic.

b) more evolutionarily advanced.

c) non-vascular.

d) seedless.

e) more multicellular.

0 ——— 31 ——— 100
EASY ——— HARD

2. Which of the following adaptations arose first in early plants?

a) vessels to transport water and food

b) seeds

c) roots

d) pollen

e) resistance to drying out

0 ——— 59 ——— 100
EASY ——— HARD

3. Which is the best brief description of the vascular system of the very first terrestrial plants?

a) The first plants developed specialized vessel cells that conducted water.

b) The first plants did not possess a vascular system.

c) The first plants had a very basic vascular system with a simple method of internal transport.

d) The first plants had only long, needle-like leaves, from which water could evaporate easily.

e) None of the above are correct.

0 ——— 56 ——— 100
EASY ——— HARD

4. Which of the following statements about ferns is incorrect?

a) Their sporophyte is dominant.

b) They require liquid water for fertilization.

c) Their seeds are dispersed by the wind.

d) They have vascular tissue for distributing water and nutrients throughout the plant.

e) Their spores are contained in sporangia.

0 ——— 64 ——— 100
EASY ——— HARD

12·1 Plants are just one branch of the eukarya.

Plants are photosynthetic, multicellular organisms that spend most of their lives anchored in one place by their roots.

WHAT IS A PLANT?

The defining characteristics of a plant:

PLANTS CREATE THEIR OWN FOOD
Almost all plants carry out photosynthesis, using energy from sunlight to convert carbon dioxide and water into sugar.

PLANTS ARE SESSILE AND (MOSTLY) TERRESTRIAL
Plants are anchored in place at their bases and occur almost exclusively on land.

PLANTS ARE MULTICELLULAR
Plants consist of multiple cells and have structures that are specialized for different functions.

? **1.** Describe two challenges facing plants as a consequence of their lack of mobility.

PHYLOGENY OF THE PLANTS

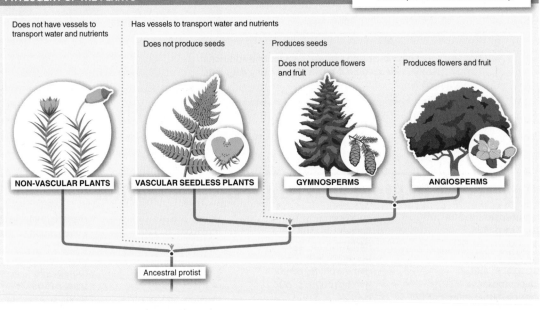

Does not have vessels to transport water and nutrients

Has vessels to transport water and nutrients

Does not produce seeds

Produces seeds

Does not produce flowers and fruit

Produces flowers and fruit

NON-VASCULAR PLANTS **VASCULAR SEEDLESS PLANTS** **GYMNOSPERMS** **ANGIOSPERMS**

Ancestral protist

12·2—12·4 The first plants had neither roots nor seeds.

The first land plants were small, had no leaves, roots, or flowers, and could grow only near water.

MOVING ONTO LAND PRESENTS CHALLENGES

When plants emerged onto land, they faced the same two challenges that terrestrial animals faced 25 million years later.

PROBLEM: GRAVITY
ADAPTATION: The earliest plants grew very close to the ground, as mosses do today, in order to resist the pull of gravity.

PROBLEM: DESICCATION
ADAPTATION: Plants developed an outer waxy layer called a cuticle that covers their entire surface.

? **2.** The earliest land plants grew very close to the ground. Why?

NON-VASCULAR PLANTS

COMMON CHARACTERISTICS
- Distribute water and nutrients throughout the plant by diffusion
- Release haploid spores, which grow and produce gametes
- Life cycle with multicellular haploid and diploid phases

MEMBERS INCLUDE
- Mosses (about 12,000 species)
- Liverworts (about 8,000 species)
- Hornworts (about 100 species)

3. How is fertilization in mosses dependent on water?

VASCULAR SEEDLESS PLANTS

COMMON CHARACTERISTICS
- Distribute water and nutrients throughout the plant with a "circulatory system" of vascular tissue
- Release haploid spores, dispersed by the wind, which grow and produce gametes
- Life cycle (unlike in animals) with multicellular haploid and diploid phases

MEMBERS INCLUDE
- Ferns (about 12,000 species)
- Horsetails (about 15 species)

4. What distinguishes a vascular plant from a non-vascular plant? How does this difference affect their geographic distribution?

5. Mosses and ferns differ from gymnosperms and angiosperms in their reproductive strategies in which of the following ways?

a) Mosses and ferns rely on liquid water for fertilization, whereas angiosperms and gymnosperms do not need liquid water for fertilization.

b) Mosses and ferns have much larger seeds than do angiosperms and gymnosperms.

c) Mosses and ferns use wind pollination, whereas angiosperms and gymno-sperms use insects for pollination.

d) Mosses and ferns are primarily diploid in their adult (reproductive) form, whereas gymnosperms and angio-sperms are primarily haploid.

e) Mosses and ferns are primarily haploid in their adult form, whereas gymnosperms and angiosperms are primarily diploid.

0 — 44 — 100
EASY — HARD

6. Which of the following is characteristic of gymnosperms?

a) They are more diverse than the angiosperms.

b) They are wind-pollinated.

c) The gametophyte generation is dominant.

d) They are water-pollinated.

e) All of the above are correct.

0 — 42 — 100
EASY — HARD

7. In terms of their adaptation to living on land, how are reptiles similar to the seed plants?

a) Both reptiles and seed plants became completely independent of water.

b) Reptiles eat plants.

c) Seed plants and reptiles have developed structures such as cuticles and impermeable skin to minimize desiccation.

d) Reptiles and seed plants have developed structures that house their gametes and protect them from the surrounding environment.

e) Both c) and d) are correct.

0 — 35 — 100
EASY — HARD

8. Which of the following terms includes all of the others in the list?

a) angiosperm
b) fern
c) vascular plant
d) gymnosperm
e) seed plant

0 — 39 — 100
EASY — HARD

12·5–12·7 The advent of the seed opened new worlds to plants.

A seed, which contains a multicellular embryo and a store of nutrients, is a way for plants to give their offspring a good start in life.

SEEDS: STRUCTURE AND GROWTH

STRUCTURE
Fertilization produces a diploid seed, which contains a multicellular embryo and a store of carbohydrate (endosperm) to fuel its initial growth.

GROWTH
A seedling draws energy from the endosperm while it extends its leaves upward to begin photosynthesis and its roots downward into the soil to reach water and nutrients.

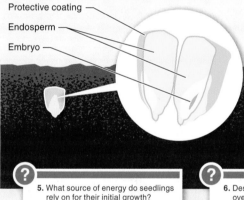

Protective coating
Endosperm
Embryo

5. What source of energy do seedlings rely on for their initial growth?

6. Describe an advantage that seeds have over the spores of earlier land plants.

GYMNOSPERMS

COMMON CHARACTERISTICS
- Distribute water and nutrients throughout the plant with a "circulatory system" of vascular tissue
- Reproductive structures called cones produce the gametes
- Fertilization produces seeds

MEMBERS INCLUDE
- Conifers (about 600 species)
- Cycads (about 300 species)
- Gnetophytes (about 65 species)
- Ginkgo (1 species)

7. Why are conifers considered a "success story" among gymnosperms?

CONES

Cones are the reproductive structures of gymnosperms.

MALE CONE
The male cone releases pollen grains that require wind to reach a female cone.

FEMALE CONE
The female cone has ovules on the protruding scales. They produce seeds when fertilized by pollen.

9. Unlike higher plants such as angiosperms and gymnosperms, all bryophytes lack:

a) stomata.

b) cuticles.

c) alternation of generations.

d) water transport mechanisms.

e) roots.

0 ————⦿69———— 100
EASY   HARD

10. Anthers and stigmas are found on:

a) bryophytes.

b) fungi.

c) angiosperms.

d) gymnosperms.

e) all of the above.

0 —⦿35————————— 100
EASY   HARD

11. Which of the following is not an example of a group of angiosperms?

a) cacti

b) cherry trees

c) pine trees

d) grasses

e) orchids

0 ——⦿43———————— 100
EASY   HARD

12. Over the evolutionary history of plants:

a) the sporophyte has become smaller, though more independent.

b) the gametophyte and sporophyte have grown increasingly independent of each other.

c) there has been a trend toward gametophyte dominance.

d) there has been a trend toward gametophyte dependence on the sporophyte.

e) the gametophyte has become larger, though more dependent.

0 ————————⦿73—— 100
EASY   HARD

13. Which of the following comparisons and contrasts between fungi and plants is incorrect?

a) Fungi cannot photosynthesize, but plants can.

b) Both fungi and plants use chitin as a structural stabilizer.

c) Fungi are heterotrophs (i.e., cannot fix carbon and so must use organic carbon for growth), but plants are not.

d) Both fungi and plants have cell walls.

e) Both fungi and plants have a sexual stage in their reproductive cycle.

0 ——————⦿62———— 100
EASY   HARD

12·8–12·10 Flowering plants are the most diverse and successful plants.

Flowering plants (angiosperms) appeared in the Cretaceous period, over 100 million years ago, and diversified rapidly to become the dominant plants in the modern world.

ANGIOSPERMS

COMMON CHARACTERISTICS

• Distribute water and nutrients throughout the plant with a "circulatory system" of vascular tissue

• Produce flowers, which produce gametes

• Seeds are enclosed within an ovary

MEMBERS INCLUDE

• Flowering trees, bushes, herbs, and grasses (about 250,000 species)

FLOWER STRUCTURE

A flower houses a plant's reproductive structures.

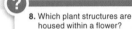

♂ **STAMEN** Male reproductive structure

• **ANTHER** Produces pollen

• **FILAMENT** Supporting stalk

♀ **CARPEL** Female reproductive structure

• **STIGMA** Sticky tip

• **STYLE** Elongated stalk

• **OVARY** Contains the ovules

Petal

Sepal

? 8. Which plant structures are housed within a flower?

? 9. If all animals became extinct, what changes would occur in plant communities? Why?

? 10. Why is double fertilization considered more efficient than the type of fertilization that occurs in gymnosperms?

STRATEGIES FOR ATTRACTING POLLINATORS

TRICKERY
Some plants deceive animals into carrying pollen from one plant to another. Shown here: This isn't a bee! It's actually an orchid flower. Pollen is picked up and delivered by male bees that attempt to mate with the flower.

BRIBERY
Some plants offer something of value to an animal, bribing the animal to carry pollen from one plant to another. Here, a bee, covered in pollen, flies from flower to flower in search of nectar.

12·11–12·12 Plants and animals have a love-hate relationship.

Overcoming a challenge of immobility, plants often use the assistance of animals to disperse their seeds, and they have a wide range of defenses against herbivorous animals.

SEED DISPERSAL

Plants have developed a range of seed dispersal methods.

HITCHING A RIDE
Some seed pods have spines or projections that attach them to passing animals.

FLYING AND FLOATING
The structure of some seed allow them to be carried away from the parent plant by wind or water.

PROVIDING A FOOD SOURCE
Fleshy fruit is a form of bait that lures an animal to eat the seed and carry it far from the parent plant before eliminating it.

? 11. How does development of a fleshy fruit benefit plants?

PLANT DEFENSES

Plants have developed a range of defenses against predation.

ANATOMICAL STRUCTURES
Some plants have spines, spikes, and thorns that deter predators. Some trees have thick layers of bark that are shed in order to get rid of attacking insects.

STICKY TRAPS
Conifers exude pitch, a sticky substance that can engulf and smother attacking insects.

CHEMICAL COMPOUNDS
Some plants synthesize chemicals that induce physiological and behavioral changes in the animals eating them.

? 12. Describe two plant structures that provide protection from predators.

12·13–12·16 Fungi and plants are partners but not close relatives.

Most fungi are multicellular, sessile decomposers.

WHAT IS A FUNGUS?

The defining characteristics of fungi:

FUNGI ARE DECOMPOSERS OR SYMBIONTS
Fungi acquire energy by breaking down the tissues of dead organisms or by absorbing nutrients from living organisms.

FUNGI ARE SESSILE
Fungi are anchored to the organic material on which they feed.

FUNGI HAVE CELL WALLS MADE OF CHITIN
Fungi have cell walls containing chitin, a chemical that is also important in producing the exoskeleton of insects.

FUNGI CLASSIFICATION

DNA sequence analyses reveal that fungi are more closely related to animals than they are to plants.

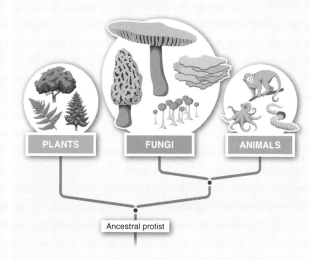

PLANTS FUNGI ANIMALS

Ancestral protist

? 13. Describe two features that distinguish fungi from other eukarya.

? 14. Describe two facets of fungi that provide value to humans.

LIFE HISTORY OF FUNGI

Fungi are decomposers, and all they need to thrive is organic material to consume and a moist environment so their hyphae don't dry out. Fungi can grow almost anywhere that is moist, and they can attain enormous sizes. Fungi have complex life cycles, with both sexual and asexual phases, and the parts of a fungus that are most often visible are its temporary spore-producing bodies.

MYCORRHIZAE

Mycorrhizal fungi grow in association with the roots of plants, receiving sugar from the plant while transferring nitrogen and phosphorus from the soil to the plant.

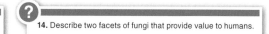

Hyphae
Root hair
Cell wall
Plasma membrane

? 15. How do fungi benefit plants? What benefit do the fungi derive from this relationship?

? 16. How do mycorrhizal fungi benefit from growing in association with plant roots?

?

14. Dispersal of fungal spores is typically done by:
   a) movement of cilia.
   b) insects.
   c) hummingbirds.
   d) wind.
   e) movement of flagella.

0 — 25 — 100
EASY HARD

15. Just before the first diploid cell is formed in fungal life cycles, the cell is said to be:
   a) haploid.
   b) a gamete.
   c) a mycelium.
   d) a hypha.
   e) dikaryotic.

0 — 62 — 100
EASY HARD

16. You are taking a hike down a forest trail and see the familiar sight of a mushroom on the ground. This visible portion of a fungal body is the structure also referred to as a:
   a) hypha.
   b) fruiting body.
   c) thallus.
   d) spore sac.
   e) mycelium.

0 — 36 — 100
EASY HARD

17. In most cases, the relationship between roots and fungi in mycorrhizae can best be described as:
   a) mycelium.
   b) competition.
   c) trickery.
   d) symbiosis.
   e) parasitism.

0 — 39 — 100
EASY HARD

531

13 | Evolution and Diversity Among the Microbes

BACTERIA, ARCHAEA, PROTISTS, AND VIRUSES: THE UNSEEN WORLD

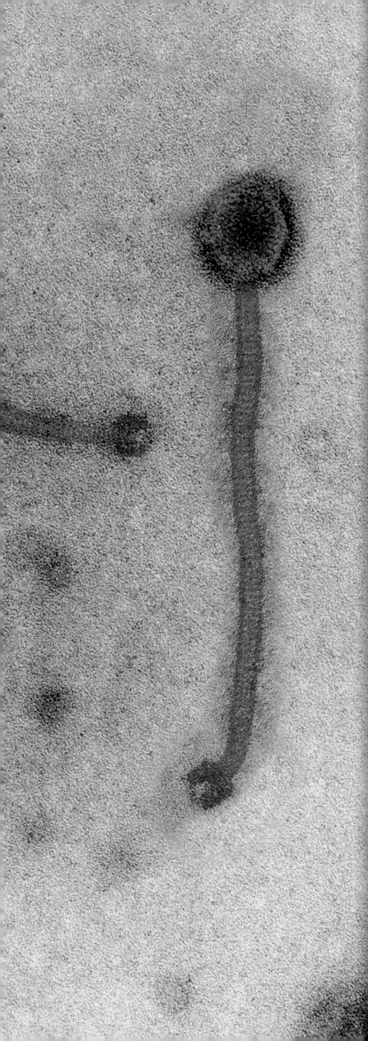

There are microbes in all three domains.

Bacteria may be the most diverse of all organisms.

In humans, bacteria can have harmful or beneficial health effects.

Archaea exploit some of the most extreme habitats.

Most protists are single-celled eukaryotes.

Viruses are at the border between living and non-living.

There are microbes in all three domains.

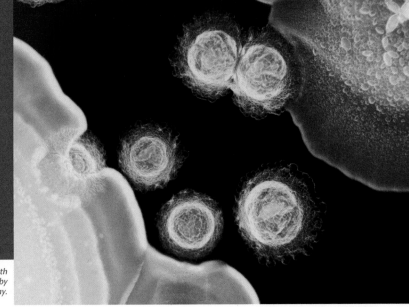

Dental plaque, a film that forms on teeth, is caused by the growth of bacterial colonies (shown here). It can generally be removed by brushing, but, if left, can lead to tooth decay.

13·1 Not all microbes are closely related evolutionarily.

Microbe is an appropriately descriptive name, but it's sloppy. It is simply a combination of Greek words that means "small life." We could use "microbe" to point to any one of many different types of organisms too small to see without magnification. In Chapters 3 and 10 we introduced two very different kinds of microbes: bacteria and viruses.

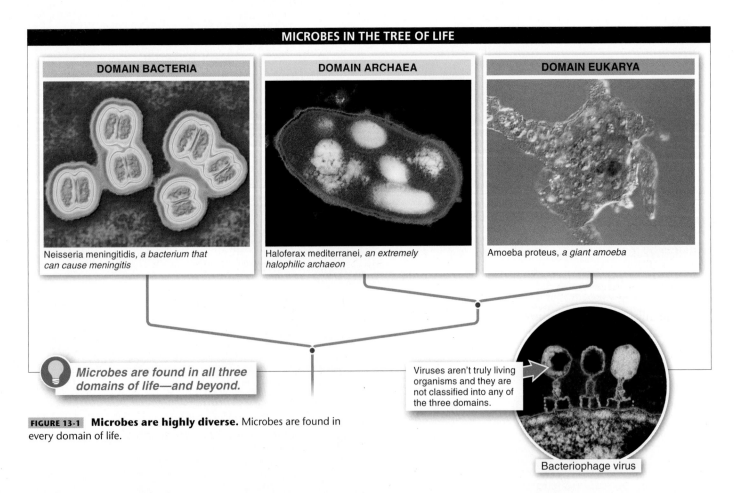

MICROBES IN THE TREE OF LIFE

DOMAIN BACTERIA	DOMAIN ARCHAEA	DOMAIN EUKARYA

Neisseria meningitidis, a bacterium that can cause meningitis

Haloferax mediterranei, an extremely halophilic archaeon

Amoeba proteus, a giant amoeba

💡 **Microbes are found in all three domains of life—and beyond.**

Viruses aren't truly living organisms and they are not classified into any of the three domains.

FIGURE 13-1 **Microbes are highly diverse.** Microbes are found in every domain of life.

Bacteriophage virus

In this chapter we explore bacteria and viruses in more detail, and also include two other kinds of microbes—the protists and the archaea. The amoeba, a kind of protist, may be familiar to you, but the archaea are probably unfamiliar to you. That's not surprising, because archaea were only recently recognized as a distinct group of microbes. The group doesn't even have a common name (unless you like "the group formerly known as archaebacteria"). We'll just call them archaea.

It is hard to make generalizations about microbes, because they are grouped together simply because they are small, not because they all share a recent common ancestor. In fact, microbes occur in all three domains of life—bacteria, archaea, and eukarya—and so the various types of microbes could not be more widely separated (**FIGURE 13-1**). In this

chapter we focus on the tiny organisms from each of these domains. The microbes in the domains bacteria and archaea are prokaryotic, although archaea have some characteristics similar to those of bacteria and some similar to those of eukaryotes. Protists are the mostly microbial members of the domain eukarya. And viruses, another type of microbe, are not classified into any domain at all because they are only at the borderline of life.

TAKE-HOME MESSAGE 13·1

Microbes are grouped together only because they are small, not because of evolutionary relatedness. They occur in all three domains of life and also include the viruses, which are not included in any of the domains.

13·2 Microbes are the simplest but most successful organisms on earth.

Humans are large organisms, and being large comes with some "baggage." We need a skeletal system to support our weight, and a respiratory system to take in oxygen and get rid of carbon dioxide. We need a circulatory system to move oxygen, carbon dioxide, and other molecules around our bodies, and a digestive system to take in food and break it down. We even need a nervous system so that our brain knows what distant parts of our body are doing.

Q How can a microbe function when its body is just a single cell?

Microbes—the most abundant organisms on earth—don't have skeletal, respiratory, circulatory, digestive, or nervous systems, because they are too small to need them. Take an amoeba, for example—its volume is about a million billion (10^{15}) times smaller than a human. When you are that tiny, the force of gravity is trivial, so the amoeba needs no skeleton to support it. Plenty of oxygen diffuses inward across its cell membrane, and carbon dioxide diffuses outward, so an amoeba does not need a respiratory system; and because every part of its interior is close to the body surface, it doesn't need a circulatory system to transport gases. An amoeba doesn't need a digestive system either: it eats by enclosing food items in a piece of its cell membrane and digests the food with the same enzymes it uses to recycle its own proteins, lipids, and carbohydrates. And no part of an amoeba is far enough

from any other part to require a specialized nervous system for communication.

Most microbes are even smaller than an amoeba: a typical bacterium or archaeon is about one thousand million billion (10^{18}) times smaller than a human, and an influenza virus is about one thousand billion trillion (10^{24}) times smaller than you are (**FIGURE 13-2**).

They may be invisible, but microbes could never be considered unsuccessful just because of their (lack of) size. They are actually more successful than humans in almost every imaginable way.

Microbes are genetically diverse. More than 500,000 kinds of microbes have been identified by their unique nucleotide sequences, and further studies will almost certainly distinguish millions of additional microbial species.

Microbial species live in almost every habitat on earth; among them, they can eat almost anything. As you read these lines, more than 400 species of microbes are thriving in your intestinal tract (**FIGURE 13-3**), 500 more species thrive in your mouth, and nearly 200 species call your skin home. The microbes that live in and on you eat mostly what you eat—some of the bacteria in your mouth and intestine compete with you, trying to digest your food before you can, and others use the waste products you release after you

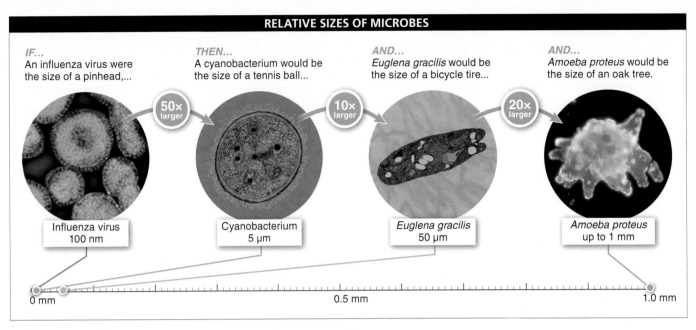

RELATIVE SIZES OF MICROBES

IF...
An influenza virus were the size of a pinhead,...

THEN...
A cyanobacterium would be the size of a tennis ball...

AND...
Euglena gracilis would be the size of a bicycle tire...

AND...
Amoeba proteus would be the size of an oak tree.

50× larger

10× larger

20× larger

Influenza virus
100 nm

Cyanobacterium
5 µm

Euglena gracilis
50 µm

Amoeba proteus
up to 1 mm

0 mm   0.5 mm   1.0 mm

FIGURE 13-2 The most abundant organisms on earth are too small to see.

have broken down the food. Others feed on the leftovers released by the breakdown of your cells during the normal process of cell renewal.

Living conditions in the human body are relatively moderate. Other microbes inhabit some of the toughest environments on earth—in the almost boiling water of hot springs, at depths a mile below the earth's surface, and more than a mile deep in the oceans, where hydrothermal vents emit water at 400° C (750° F).

Microbes are abundant. Surface seawater contains more than 100,000 bacterial cells per milliliter, and diatoms (protists in the eukarya domain) are as abundant there as are the

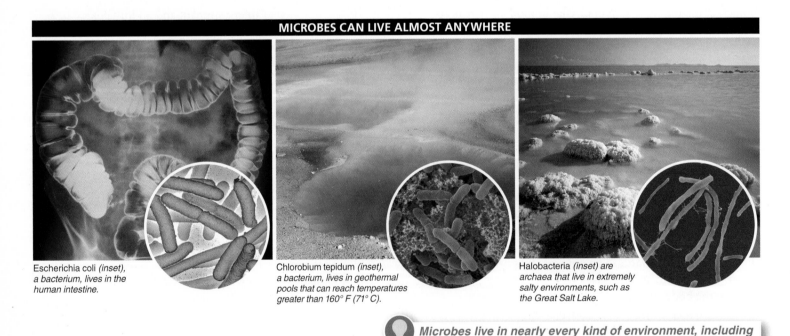

MICROBES CAN LIVE ALMOST ANYWHERE

Escherichia coli (inset), a bacterium, lives in the human intestine.

Chlorobium tepidum (inset), a bacterium, lives in geothermal pools that can reach temperatures greater than 160° F (71° C).

Halobacteria (inset) are archaea that live in extremely salty environments, such as the Great Salt Lake.

Microbes live in nearly every kind of environment, including water at temperatures of up to 750° F and as low as 5° F!

FIGURE 13-3 Microbes are everywhere on earth.

TOTAL NUMBER OF CELLS IN THE HUMAN BODY (TRILLIONS)

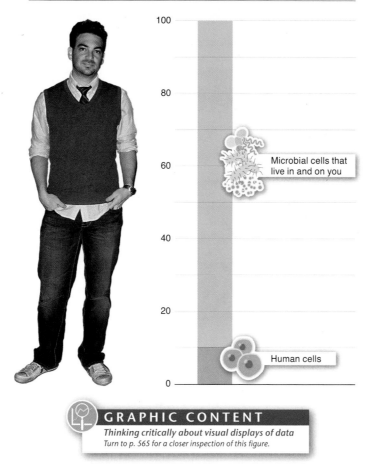

Microbial cells that live in and on you

Human cells

GRAPHIC CONTENT

Thinking critically about visual displays of data
Turn to p. 565 for a closer inspection of this figure.

FIGURE 13-4 **Microbe majority.** Your body has more microbial cells than human cells.

bacteria. These densities translate to about 8,000 million billion trillion (8×10^{30}) individuals of just these two kinds of microbes in the world's oceans. Your own body is a testament to the abundance of microbes: it contains about 100 trillion cells, but only one-tenth of those cells are actually human cells—the remaining 90 trillion cells are the microbes that live in and on you (**FIGURE 13-4**). You're a minority in your own body.

TAKE-HOME MESSAGE 13·2

Microbes are very small, simple organisms, but they do everything that larger, multicellular organisms do. They can live anywhere, from moderate to extreme environments. There are millions of different kinds of microbes on earth, in enormous numbers.

13·3–13·5
Bacteria may be the most diverse of all organisms.

The bacterium Staphylococcus epidermidis *is part of the normal flora on human skin and many mucous membranes.*

13·3 What are bacteria?

A bacterium is the simplest organism you can imagine—and in many ways, it is the most efficient. A bacterium has a cell envelope consisting of a plasma membrane and, in most cases, a cell wall—that's what it needs to maintain conditions inside that are different from conditions outside. Inside the cell envelope it has cytoplasm—the substance that fills all kinds of cells (including your own). Because bacteria are prokaryotes, they have no organelles. Proteins in the cytoplasm carry out essential functions, such as digesting molecules of food and transferring the energy gained to ATP. DNA in the cytoplasm carries the instructions for making those proteins. And messenger RNAs carry this information to ribosomes, where the proteins are synthesized. That's it—everything an organism needs, with no extra baggage.

Bacteria may be classified by their shape: some are spherical cells (known as the cocci), some are rod-shaped (the bacilli), and others are spiral-shaped (the spirilli) (**FIGURE 13-5**). Bacteria usually reproduce by binary fission, and in a few hours, a single cell can form a culture containing thousands of cells.

As a bacterial cell divides, the number of cells doubles every generation, producing a colony of cells, each of which is a clone of the original cell. Colonies of different species of bacteria look different. The familiar human intestinal bacterium *Escherichia coli,* for example, forms beige or gray colonies that have smooth margins and a shiny, mucus-like covering. Species of *Proteus,* which are often responsible for spoiling food because they can grow at refrigerator

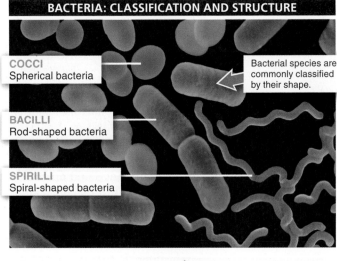

BACTERIA: CLASSIFICATION AND STRUCTURE

COCCI
Spherical bacteria

Bacterial species are commonly classified by their shape.

BACILLI
Rod-shaped bacteria

SPIRILLI
Spiral-shaped bacteria

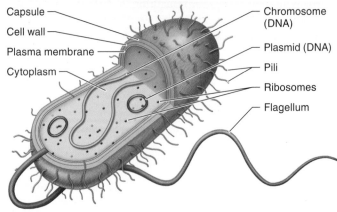

Capsule
Cell wall
Plasma membrane
Cytoplasm

Chromosome (DNA)
Plasmid (DNA)
Pili
Ribosomes
Flagellum

FIGURE 13-5 **Bacteria basics.** Bacteria are single-celled organisms that lack a nucleus.

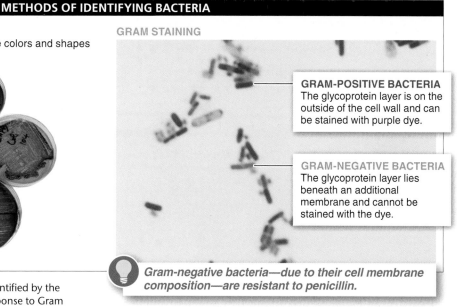

METHODS OF IDENTIFYING BACTERIA

APPEARANCE
Some bacteria can be identified by looking at the colors and shapes of their colonies.

GRAM STAINING

GRAM-POSITIVE BACTERIA
The glycoprotein layer is on the outside of the cell wall and can be stained with purple dye.

GRAM-NEGATIVE BACTERIA
The glycoprotein layer lies beneath an additional membrane and cannot be stained with the dye.

Gram-negative bacteria—due to their cell membrane composition—are resistant to penicillin.

FIGURE 13-6 **Bacterial IDs.** Bacteria are often identified by the appearance of an entire colony or by the cells' response to Gram staining.

temperatures, form colonies with a surface that looks like a contour map.

Microbiologists can often identify bacteria simply by looking at the colors and shapes of their colonies. They can get additional information by examining a single cell under a microscope, but living bacterial cells are transparent, so you can't see them with an ordinary microscope unless they have been dyed. In 1884, Hans Christian Gram, a Danish microbiologist, described a method of staining the cell walls of bacteria to make them visible under a microscope, and a **Gram stain** is still the first test microbiologists use when they are identifying an unknown bacterium (**FIGURE 13-6**).

Gram-positive bacteria are colored purple by the stain, because their cell wall has a thick layer of a glycoprotein called **peptidoglycan.** In *Gram-negative bacteria,* the layer of peptidoglycan is thinner and lies beneath an additional membrane, and it is not stained by the dye. Penicillin is effective in treating infections by Gram-positive bacteria because it interferes with the formation of peptidoglycan cross-links. Penicillin is less effective on Gram-negative

bacteria because it does not pass through the outer membrane that covers the peptidoglycan layer.

Being able to distinguish these two groups of bacteria is a big help to microbiologists trying to identify a bacterium, but the peptidoglycan is there because it is important for the bacterial cells—not to make life easier for microbiologists. The extensive interlocking bonds of the long peptidoglycan molecules provide strength to the cell wall. Many bacteria also have a **capsule** that lies outside the cell wall. This capsule can restrict the movement of water out of the cell and allow bacteria to live in dry places, such as on the surface of your skin. In other cases, the capsule is important in allowing the bacteria to bind to solid surfaces such as rocks or to attach to human cells.

TAKE-HOME MESSAGE 13·3

Bacteria are efficient single-celled organisms, with an envelope surrounding the cytoplasm, which contains the DNA (they have no nuclei and no intracellular organelles). A single bacterial cell can grow into a colony of cells.

13·4 Bacterial growth and reproduction is fast and efficient.

The time it takes for a bacterium to reproduce can be very short; most bacteria have generation times between 1 and 3 hours, and some are even shorter. *Escherichia coli,* for

example, has a generation time of 20 minutes in optimal conditions, so a single *E. coli* cell could give rise to a population of 20 billion cells in less than 12 hours.

Bacteria carry genetic information in two structures: the chromosome and plasmids. The genes that provide instructions for all of the cell's basic life processes are usually located in a circular DNA molecule, the bacterial chromosome. Most bacteria have just one chromosome, but some have more than one. A bacterial chromosome is organized more efficiently than a eukaryotic chromosome, in two ways. First, in bacteria, the genes that code for proteins with related functions—enzymes that play a role in a pathway that breaks down food for energy, for example—are often arranged right next to one another on the chromosome. This makes it possible to efficiently control the transcription of all the genes together. Second, almost all the DNA in a bacterial chromosome codes for proteins, so bacteria do not use time and energy transcribing mRNA that will not be translated. As we saw in Chapter 5, as much as 90% of the DNA in the chromosomes of eukaryotes does not code for genes, and after transcription, this is edited out of the mRNA before it is translated into protein.

A second type of structure that carries genetic information in bacteria is a circular DNA molecule called a **plasmid.** Plasmids carry genes for specific functions. For example, *metabolic plasmids* carry genes enabling bacteria to break down specific substances, such as toxic chemicals; *resistance plasmids* carry genes enabling bacteria to resist the effects of antibiotics; and *virulence plasmids* carry genes that control how sick an infectious bacterium makes its victim. Many bacteria have one or more (sometimes more than a hundred) plasmids. The strain of *E. coli* that has made news headlines after causing illness among many patrons of some fast-food restaurants carries a virulence plasmid that magnifies the effects of a gene for a sometimes-lethal toxin. (*E. coli* strains without this virulence plasmid are normal components of the bacterial community of the human intestine.)

Q What would be the benefit of being able to transfer genetic information directly from one adult human to another?

When a bacterium divides, it creates two new daughter cells, each "offspring" carrying the genetic information that was present in the chromosome of the mother cell. Thus, binary fission transmits genetic information from one bacterial generation to the next (**FIGURE 13-7**). However, bacteria can also transfer genetic information laterally—to other individuals *within* the same generation—through any of three different processes: conjugation, transduction, and transformation (**FIGURE 13-8**).

Conjugation is the process by which one bacterium transfers a copy of some of its genetic information to another bacterium—even when the two bacteria are

different species. It's very much like plugging your phone or iPod into a computer to transfer songs from the computer to the mobile device. This is not reproduction, because you start with one iPod and one computer, and that's what you have when you finish. But the iPod now contains songs that it did not have before. Plasmid transfer does exactly that for a bacterium—you still have just two bacterial cells, but the newly acquired plasmid has given the recipient bacterium genetic information that it did not have before. These genes could enable the bacterium to make an enzyme that allows it to metabolize a new chemical or to defend itself against a new antibiotic.

Transduction occurs when a kind of virus called a bacteriophage (one type is shown in Figure 13-1) infects a

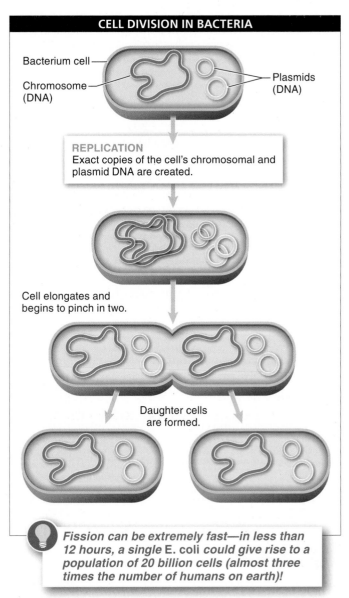

CELL DIVISION IN BACTERIA

Bacterium cell

Chromosome (DNA)

Plasmids (DNA)

REPLICATION
Exact copies of the cell's chromosomal and plasmid DNA are created.

Cell elongates and begins to pinch in two.

Daughter cells are formed.

Fission can be extremely fast—in less than 12 hours, a single E. coli *could give rise to a population of 20 billion cells (almost three times the number of humans on earth)!*

FIGURE 13-7 **Binary fission.** This asexual cell division method is used by prokaryotes.

METHODS OF GENETIC EXCHANGE IN BACTERIA WITHIN THE SAME GENERATION

CONJUGATION
A bacterium transfers a copy of some or all of its DNA to another bacterium, giving the second bacterium genetic information it did not have before.

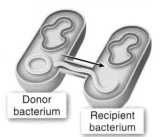

Donor bacterium

Recipient bacterium

TRANSDUCTION
A virus containing pieces of bacterial DNA inadvertently picked up from its previous host infects a new bacterium, and passes new bacterial genes to the bacterium.

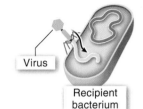

Virus

Recipient bacterium

TRANSFORMATION
A bacterium can take up DNA—potentially including alleles it did not carry—from its surroundings (usually from bacteria that have died).

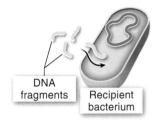

DNA fragments

Recipient bacterium

FIGURE 13-8 Lateral transfer of genetic information: conjugation, transduction, and transformation.

bacterial cell. The virus reproduces inside the bacterial cell, and sometimes, inadvertently, the new virus particles contain pieces of bacterial DNA in addition to or instead of the viral DNA. When these viruses are released and infect new bacterial cells, the bacterial DNA can be inserted into the host bacterium's chromosome, passing new bacterial genes to that bacterium.

Transformation is the process by which bacterial cells scavenge DNA from their environment. This DNA comes from other bacterial cells that have burst open, releasing their cellular contents. The circular chromosomes break into short lengths of DNA, which can then be taken up by living bacterial cells and inserted into their own chromosomes, potentially adding genes they did not originally have.

TAKE-HOME MESSAGE 13·4

Bacteria undergo binary fission. They grow rapidly, and their genes are efficiently organized in groups with related functions: virtually all the DNA codes for proteins. Bacteria sometimes carry genes for specialized traits on plasmids (small DNA molecules), which can be transferred from one bacterial cell to another by conjugation. DNA can also be transferred laterally between bacterial cells by transduction or transformation.

13·5 Metabolic diversity among the bacteria is extreme.

One important attribute that makes bacterial diversity possible is that bacteria can metabolize almost anything. (Not all bacteria can metabolize everything. Rather, there is a huge variety of bacteria, each type with its own particular set of metabolic specializations.) Some can even use energy from light to make their own food, just as plants do. Microbiologists place bacteria into different "trophic" (feeding) categories that reflect their metabolic specialization.

Chemical organic feeders (**chemoorganotrophs**) are bacteria that consume organic molecules, such as carbohydrates. You probably see the products of organic feeders every time you take a shower—they are responsible for the pink deposits on the shower curtain and other discolorations on shower tiles (**FIGURE 13-9**). Most of the

bacteria that live in and on your body are also organic feeders. Some compete with you to metabolize the food you eat. Others digest things you can't digest.

Chemical inorganic feeders (**chemolithotrophs,** meaning "rock feeders") are able to use a completely different type of food as their source of energy: inorganic molecules such as ammonia, hydrogen sulfide, hydrogen, and iron. The most common inorganic feeders are the iron bacteria responsible for the brown stains that form on plumbing fixtures in regions where tap water contains high levels of iron. Sulfur bacteria are associated with iron bacteria, and these are responsible for the slimy black deposits that you will probably find if you lift the stopper out of the drain in your bathroom sink.

CHEMOORGANOTROPHS
Feed on organic molecules

CHEMOLITHOTROPHS
Feed on inorganic molecules
(shown here: rusticles—created as bacteria
break down iron—on the wreck of the Titanic)

PHOTOAUTOTROPHS
Use energy from sunlight to produce glucose
via photosynthesis

FIGURE 13-9 **A bigger palate than yours.** Bacteria can
metabolize almost anything.

 *Bacteria are resourceful and can extract food from a huge
range of sources: the sun, inorganic molecules in your
drainage pipes, and even your shower curtain/door!*

On a larger scale, inorganic feeders are responsible for the acidic drainage that is a by-product of mining. The desirable ore makes up only a small part of the total amount of rock that is removed from a mine. The portion that does not contain ore is discarded on the ground surface in piles called "tailings." This material is often rich in minerals such as pyrites (iron sulfides). Inorganic feeders can gain energy by oxidizing these minerals and, in the process, they release compounds that combine with rainwater to produce strong acids, such as sulfuric acid. When this acidic water drains into streams, it can kill fish and aquatic plants and insects.

Bacteria that use the energy from sunlight (**photoautotrophs,** or "light self-feeders") contain chlorophyll and use light energy to convert carbon dioxide to glucose by photosynthesis. The floating mats of gooey green material that you see in roadside ditches are a type of photoautotroph called cyanobacteria.

The cyanobacteria living today closely resemble the first photosynthetic organisms that appeared on earth about 2.6 billion years ago. Cyanobacteria could use solar energy to build organic compounds from carbon dioxide, and in the process they broke down water molecules to release free oxygen. Before cyanobacteria, the earth's atmosphere contained no free oxygen—instead, air consisted almost entirely of nitrogen and carbon dioxide. The accumulation of oxygen released by cyanobacteria is called the **Oxygen Revolution.** Oxygen—which humans depend on—now makes up about 21% of the volume of air, and cyanobacteria still release important quantities of oxygen into the atmosphere. One of the common ways that bacteria are classified is as aerobic or anaerobic, depending on whether they require or do not require oxygen for growth—although some bacteria, called facultative anaerobes, utilize oxygen if it is present but can also switch to anaerobic respiration when oxygen is absent.

TAKE-HOME MESSAGE 13·5

Some bacteria eat organic molecules, some eat minerals, and still others carry out photosynthesis. About 2.6 billion years ago, the photosynthesizing bacteria were responsible for the first appearance of free oxygen in the earth's atmosphere.

13·6–13·10
In humans, bacteria can have harmful or beneficial health effects.

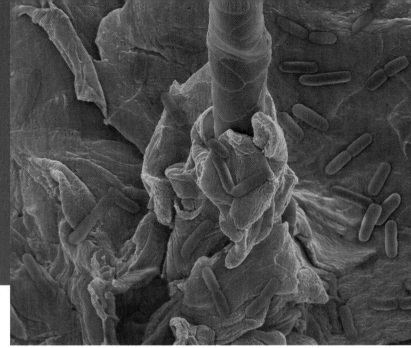

Escherichia coli (E. coli) *bacteria (pink) on the surface of human skin and hair follicle.*

13·6 Many bacteria are beneficial to humans.

Do you like yogurt? You can thank bacteria for it—*Lactobacillus acidophilus* and several other species of bacteria are added to milk to create yogurt. As the bacterial cells use the lactose (milk sugar) for energy, the by-product is lactic acid, which reacts with the milk proteins to produce the characteristic taste and texture of yogurt. If you buy a brand of yogurt that is labeled as containing "live cultures," you are consuming living bacterial cells as you eat the yogurt (**FIGURE 13-10**). Bacteria are also used to produce many other foods, such as cheeses, and they (along with yeasts) are used in the production of beer, wine, and vinegar. Industrial microbiology is a multi-billion-dollar industry.

But you can thank bacteria for more than just tasty snacks—you owe your life to them. Hundreds of species of bacteria grow in and on your body; these microbes are called your "normal flora." The normal flora take up every spot on your body that a disease-causing bacterium could adhere to, and they consume every potential source of nutrition, making it difficult for a disease-causing bacterium to gain a foothold. Thus, maintaining a robust population of these benign bacteria is your first line of defense against infection by harmful bacteria.

Probiotic therapy is a method of treating infections by deliberately introducing benign bacteria in numbers large enough to swamp the harmful forms. *Lactobacillus*

Many bacteria are beneficial. Those living in yogurt, for example, can take up residence in your digestive tract and improve your extraction of nutrients from food.

FIGURE 13-10 **Yogurt contains beneficial bacteria.**

acidophilus, which is a normal inhabitant of the human body, is used to treat gastrointestinal upsets, such as traveler's diarrhea, and urinary tract infections. In addition to replicating so vigorously that it crowds out harmful bacteria, *L. acidophilus* releases lactic acid, which interferes with the growth of other bacteria and prevents them from adhering to the walls of the urinary tract and bladder.

TAKE-HOME MESSAGE 13·6

A disease-causing bacterium must colonize your body before it can make you sick, and your body is already covered with harmless bacteria. If the population of harmless bacteria is dense enough, it will prevent invading bacteria from gaining a foothold.

? 13·7 THIS IS HOW WE DO IT

Are bacteria thriving in our offices, on our desks?

Bacteria can live in a huge variety of habitats. As humans create new, "artificial" environments, do bacteria thrive in these habitats as well? Some researchers wondered, in particular, about the extent to which bacteria might be thriving in a habitat increasingly important to many people in the industrialized world: the office building. They noted that in crowded offices, with shared bathrooms, meeting rooms, and other common areas, bacteria might be brought in and shed in large numbers. So they decided to investigate.

To evaluate the abundance of office bacteria, where could you take samples?

Because they wanted to draw some general conclusions, the researchers decided to collect samples in three different cities. They chose Tucson, Arizona (where one of the researchers lived), New York, and San Francisco. And to avoid being overly influenced by any one office—which might be unusually hospitable (or inhospitable) to bacteria—they selected 30 office buildings in each city. Within each office they sampled a small area on a desktop. They noted the gender of the occupant of each office where they collected a sample.

The samples were collected by wiping a cotton swab across a small (13 cm²) area of the desk. Each swab was stored in a sterile tube on ice and sent to the laboratory in Arizona. In the lab, the researchers swirled the swab in a sterile saline (saltwater) solution, and then rubbed it across the surface of a culture medium on a small Petri plate. The culture medium was a gel containing a nutrient broth in which bacteria could grow. They then kept the plate at 30° C (86° F) for five days, allowing any bacteria present to grow.

How could the researchers evaluate the abundance of bacteria in the different offices?

The researchers simply counted the number of colonies growing on each plate. After five days, colonies were visible to the naked eye. To evaluate any large-scale patterns, they calculated the average number of colonies that grew on the 30 plates from each city. They also compared the average number of colonies from women's desks and men's desks.

I What did they find?

The large number of samples they collected made it possible to find some clear and interesting results. First, they plotted the average number of colonies on the plates from each city:

They found a significant difference among the cities: San Francisco had the lowest number of colonies per sample (about 170), and New York had the highest (more than 275).

The researchers found an even more striking difference when they compared men's and women's desks:

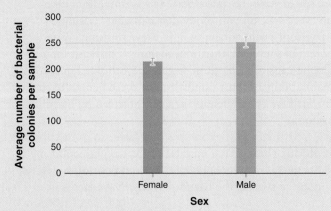

The samples from men's desks contained significantly more bacterial colonies (about 250, on average) than those from women's desks (about 215).

What can we conclude from these results?

When these research results were first published in 2012, many newspapers jumped at the opportunity to proclaim that "men are grubbier than women." This conclusion, after all, seemed to confirm previously published research reporting that men wash their hands and brush their teeth less frequently than women. But the researchers themselves were more conservative in the conclusions they drew.

Why might these results reflect something other than poorer hygiene in men?

The researchers pointed out that men are bigger than women, on average. So men have more skin surface and larger nasal and oral cavities. Consequently, they simply have more space for bacteria to colonize. And that, in turn, may lead to more bacteria being shed onto their desk surfaces.

Are there any other interesting or useful observations that you could make from this experimental approach?

This study was part of a larger project in which the researchers also looked at the variety of bacteria in each office. They found a broader diversity of bacterial types in Tucson than in the other two cities, but no difference in the diversity of bacteria living on women's versus men's desks.

Are there any practical consequences of these results?

An important result was that most of the bacteria they identified typically live on or in humans and are not harmful. The presence or absence of these bacteria probably has no health consequences.

Interestingly, the researchers noted that the Clorox Corporation provided some of the funding for this study. Although they clearly stated that the funders had no role in the study design or analysis, perhaps these methods will be used in the future to evaluate the effectiveness of various cleaning products or regimens.

TAKE-HOME MESSAGE 13·7

The abundance of bacteria in human habitats can be evaluated and compared, using simple swabbing and culturing methods. In office buildings, bacteria are extremely common, but with significant differences in abundance depending on the location of the office and the gender of its occupant.

13·8 Bacteria cause many human diseases.

The number of **pathogenic** (disease-causing) bacteria is very small compared with the total number of bacterial species, but some pathogens kill millions of people annually, despite advances in medicine and sanitation. Some bacteria are always pathogenic (such as those that cause cholera, plague, and tuberculosis), but others (the ones responsible for acne, strep throat, scarlet fever, and "flesh-eating" necrotizing fasciitis) are normal parts of the communities of bacteria that live in or on humans. These bacteria in our "normal flora" become pathogenic only under special circumstances.

The cholera epidemic that devastated London in 1854 became a milestone in epidemiology (the study of the occurrence of disease outbreaks), when Dr. John Snow identified the Broad Street pump as the source of infection by mapping the pattern of deaths (**FIGURE 13-11**). When Dr.

Snow persuaded the authorities to shut down the pump by removing the handle, new cases of cholera in the area dropped sharply.

Cholera epidemics are a continuing threat, especially in areas without effective sanitation. Epidemics of cholera are occurring now in parts of Iraq where the sewage systems have been destroyed and in Haiti, following the earthquake of 2010. The strains of cholera found in these areas are especially potent; as the bacteria multiply inside their hosts, they cause severe diarrhea with a massive loss of water. Victims of these severe strains of cholera are incapacitated and soon die, but before they die they release billions of cholera bacteria in diarrhea—an effective strategy for the bacteria, which contaminate the water used for drinking or bathing and rapidly infect new victims.

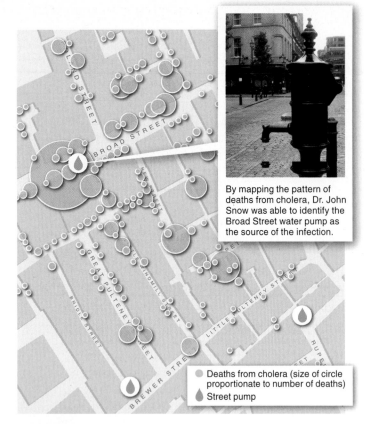

By mapping the pattern of deaths from cholera, Dr. John Snow was able to identify the Broad Street water pump as the source of the infection.

● Deaths from cholera (size of circle proportionate to number of deaths)
💧 Street pump

FIGURE 13-11 **Contaminated!** Water from the Broad Street pump in London was the source of a cholera outbreak that killed more than 500 people.

Streptococcus pyogenes is a normal part of the bacterial community of your nose and mouth. Normally it is harmless, and many people never experience any of the diseases it can cause. But when the population of *S. pyogenes* is not held in check by competition with other members of the bacterial community, it can become a pathogen. Strep throat (more formally known as "streptococcal pharyngitis") is the most common disease caused by *S. pyogenes;* it produces a severe sore throat and a distinctive rash of red abscesses with white pus at the top of the throat

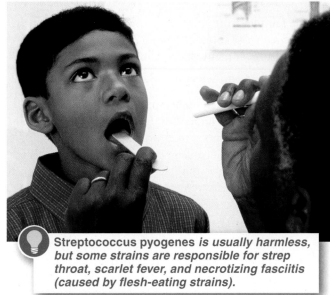

Streptococcus pyogenes is usually harmless, but some strains are responsible for strep throat, scarlet fever, and necrotizing fasciitis (caused by flesh-eating strains).

FIGURE 13-12 **Strep throat is the most common disease caused by *S. pyogenes*.**

(**FIGURE 13-12**). Some strains of *S. pyogenes* produce a toxin that is released into the bloodstream and produces a red rash that spreads across the skin—the disease called scarlet fever. Most threatening of all is an infection caused by strains of *S. pyogenes* that have a toxin that allows them to enter body tissues, where they produce necrotizing fasciitis. These strains of *S. pyogenes* are known by their well-earned name, "the flesh-eating bacteria."

TAKE-HOME MESSAGE 13·8

Some bacteria always cause disease, and others are harmful only under certain conditions. For example, *Streptococcus pyogenes* can be harmless, but under some conditions it releases toxins that are responsible for strep throat, scarlet fever, and necrotizing fasciitis.

13·9 Sexually transmitted diseases reveal battles between microbes and humans.

To many microbes, the human genitals and reproductive tract are desirable places to find shelter, nourishment, and opportunities for reproducing and dispersing. Unfortunately, these microbes can cause problems for humans in the form of **sexually transmitted diseases** (**STDs**). STDs produce symptoms of varying severity, from mild to extreme discomfort, to sterility, or even death. It is estimated that more than 300 million new cases of STDs occur each year worldwide.

STDs are caused by all types of microbes—bacteria, viruses, fungi, protists—and even by some arthropods (**FIGURE 13-13**). The organisms are passed from the mucous membranes (of the genitals, anus, and mouth) of one individual to those of

CAUSE	EXAMPLES	SYMPTOMS	TREATMENT
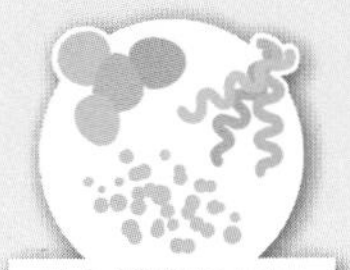 BACTERIUM	• Gonorrhea	Often none; sometimes painful urination, genital discharge, or irregular menstruation	Several antibiotics can successfully cure gonorrhea; however, drug-resistant strains are increasing.
	• Syphilis	Often no symptoms for years; eventual sores, skin rash, and if untreated, organ damage	Penicillin, an antibiotic, can cure a person in the early stages of syphilis.
	• Chlamydia	Often none; sometimes painful urination, genital discharge	Chlamydia can be easily treated and cured with antibiotics.
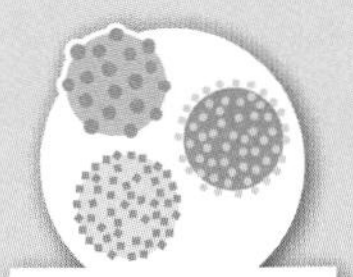 VIRUS	• HIV/AIDS	Initial symptoms range from none to flu-like; late stages involve severe infections and death	Currently no cure. Antiretroviral treatment can slow progression. Drug-resistant strains occur.
	• Genital herpes	Often none; outbreaks include sores on genitals, flu-like symptoms	Currently no cure. Antiviral medications can shorten and prevent outbreaks.
	• Human papilloma virus (HPV)	Often none; some types can lead to genital warts, others can cause cervical cancer	A vaccine prevents HPV, and is recommended for girls age 11–12. Warts and cancerous lesions can be removed.
 PROTIST	• Trichomoniasis	Painful urination and/or vaginal discharge in women; often no symptoms in men	Trichomoniasis can usually be cured with prescription drugs.
 FUNGUS	• Yeast infections	Genital itching or burning and/or vaginal discharge in women; genital itching in men	Yeast infections can usually be cured with antifungal suppositories or creams.
 ARTHROPOD	• Crab lice	Visible lice eggs or lice crawling or attached to pubic hair; itching in the pubic and groin area	Crab lice can be treated with over-the-counter lotions.

FIGURE 13-13 The most common STDs.

another during sexual contact; they are also sometimes transmitted by needles used for drug injections.

Although most STDs are curable with drugs, two characteristics of STDs make them nearly impossible to completely eradicate from a population: (1) their symptoms may be mild or absent, causing many people to unwittingly pass an infection to their partners, and (2) to prevent reinfection, both partners must be treated simultaneously. Furthermore, because most microbes have such high reproductive rates, populations of a microbe can evolve

quickly and become resistant to existing drugs. The treatment of STDs is one of the most pressing public health issues in the world today.

TAKE-HOME MESSAGE 13·9

Sexually transmitted diseases (STDs) are caused by a variety of organisms, including bacteria, viruses, protists, fungi, and arthropods. Worldwide, more than 300 million people are newly infected each year.

13·10 Bacteria's resistance to drugs can evolve quickly.

Penicillin was the first antibiotic to be manufactured and used widely against illness-inducing bacteria. It came into use during the Second World War and caused a revolution in the care of the wounded. But antibiotic-resistant infections soon appeared, and the number of resistant bacteria has increased rapidly ever since (**FIGURE 13-14**). Now, 70 years after the first use of antibiotics, many bacteria are resistant to many, or even to most, antibiotics. In the United States, more people now die of *Staphylococcus aureus* infections that are resistant to many different antibiotics (MRSA infections) than die of HIV/AIDS. A few years ago, antibiotic-resistant staph infections were acquired only in hospitals, but now these infections have spread to the community at large.

Microbes live everywhere they can, and compete for the best places to attach themselves and for the richest sources

of food to eat. This competition takes many forms: rapid growth to crowd competitors out of a living space, superior ability to take in nutrients and thus to starve competitors, and—this is the key to both the benefits of antibiotics and the problem of antibiotic resistance—production of chemicals that kill other microbes, or at least stop them from growing. Antibiotics are produced by microbes to help them compete with other microbes, and most of the antibiotics we use today are derived from microbes.

Bacteria and other microbes have developed a variety of ways to resist antibiotics. For example, some bacteria pump an antibiotic out of their cells as fast as it enters, so it never reaches a lethal concentration

Q Where do antibiotics come from, and why do they so quickly lose their effectiveness?

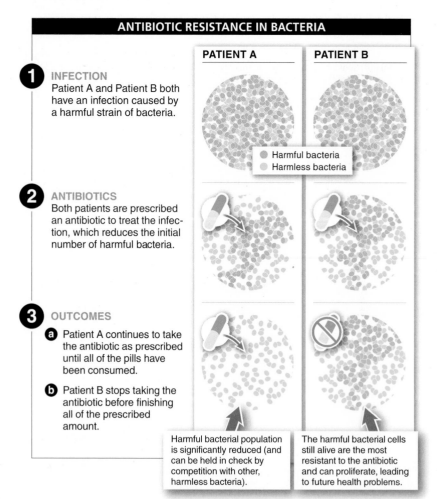

ANTIBIOTIC RESISTANCE IN BACTERIA

| | PATIENT A | PATIENT B |

1 INFECTION
Patient A and Patient B both have an infection caused by a harmful strain of bacteria.

● Harmful bacteria
● Harmless bacteria

2 ANTIBIOTICS
Both patients are prescribed an antibiotic to treat the infection, which reduces the initial number of harmful bacteria.

3 OUTCOMES

a Patient A continues to take the antibiotic as prescribed until all of the pills have been consumed.

b Patient B stops taking the antibiotic before finishing all of the prescribed amount.

Harmful bacterial population is significantly reduced (and can be held in check by competition with other, harmless bacteria).

The harmful bacterial cells still alive are the most resistant to the antibiotic and can proliferate, leading to future health problems.

FIGURE 13-14 Evolution of resistance. Seventy years after the appearance of antibiotics, many bacteria are resistant to many, or even most, antibiotics.

inside the bacterial cell. Other bacteria have proteins that bind to the antibiotic molecule and block its lethal effect. Still others carry that approach a step further—they have enzymes that break down the antibiotic molecules, which are then used as fuel to help the bacteria grow faster! Antibiotic resistance within a population of bacteria can spread quickly, because many of the genes that code for resistance are on plasmids. This means that a bacterium carrying a resistance gene can transmit the gene to other bacteria by conjugation; there's no need to wait for natural selection over multiple generations.

Q **Why is it essential to take every dose of an antibiotic prescribed by a doctor?**

When an antibiotic is taken as prescribed—that is, at the times specified on the label and until all the pills have been consumed—the population of target bacteria is greatly reduced. All of the target bacteria that remain are resistant to the antibiotic, but there are not very many of them. The growth of these resistant bacteria will be held in check by competition with other types of bacteria. But, if you stop taking the antibiotic before you have finished all of the prescription, many of the target bacterial cells will remain alive, including the ones that are most resistant to the antibiotic. These resistant cells will be the founders of a new population of bacteria in your body, so the next time you take that drug it will be ineffective. Even worse, taking antibiotics when they are not needed—to treat a viral infection, for example—selects for resistant bacteria without providing any benefit. Antibiotics have no effect on viruses.

The use of antibiotics in agriculture is another reason for the spread of antibiotic resistance. Low concentrations of antibiotics are routinely added to the feed for cattle, hogs, chickens, and turkeys. This can be beneficial in the short term, promoting growth and minimizing disease in the crowded conditions of commercial meat and milk production (**FIGURE 13-15**). But in the long run it can have

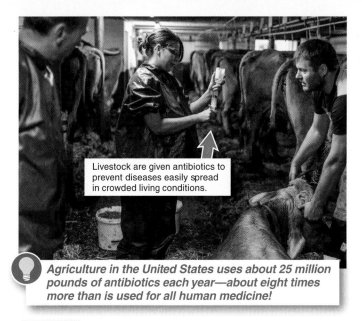

Livestock are given antibiotics to prevent diseases easily spread in crowded living conditions.

Agriculture in the United States uses about 25 million pounds of antibiotics each year—about eight times more than is used for all human medicine!

FIGURE 13-15 **Antibiotics are used in agriculture.**

disastrous consequences, as the practice can lead to selection for bacteria resistant to the antibiotics.

The antibiotics can also pass through the food chain to humans. Data gathered by the Union of Concerned Scientists indicate that agriculture in the United States uses about 25 million pounds of antibiotics each year—about eight times more than is used for all human medicine!

TAKE-HOME MESSAGE 13·10

Antibiotic resistance routinely evolves in microbes, and plasmid transfer allows an antibiotic-resistant bacterium to pass that resistance to other bacteria. Excessive use of antibiotics in medicine and agriculture has made several of the most important pathogenic bacteria resistant to every known antibiotic, and infections caused by these bacteria are nearly impossible to treat.

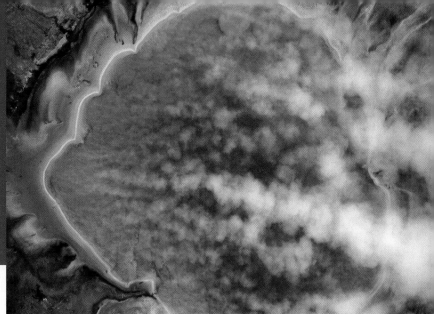

Grand Prismatic Spring hydrothermal vent in Yellowstone National Park, Wyoming. Hot springs are home to many microbes, including thermophilic archaea.

13·11   Archaea are profoundly different from bacteria.

When you look at the photograph of archaea in **FIGURE 13-16** , you may find it hard to believe that they are even a little bit different from bacteria, let alone *profoundly* different. Both live as single cells or colonies of cells, both are surrounded by a plasma membrane, and both have species with flagella that twirl like propellers. In fact, until the 1970s, biologists considered the archaea to be bacteria. Only then did comparisons of DNA reveal that archaea are as different from bacteria as humans are.

Those studies of archaeal nucleotide sequences stimulated other comparisons, which identified additional differences among bacteria, archaea, and eukarya. For example, the chemical compositions of the plasma membranes, cell walls, and flagella of archaea are qualitatively different from those of bacteria. And beyond the large DNA sequence differences and the differences in plasma membranes, cell walls, and flagella, a third difference reflects the phylogenetic positioning of archaea between bacteria and eukarya on the tree of life (see Figure 13-1): eukaryotes have a distinct cell nucleus that is separated from the cytoplasm by a nuclear membrane, whereas bacteria and archaea have neither a nucleus nor a nuclear membrane. The presence of a nucleus protects the chromosomes of a eukaryotic cell and allows the cell to control what molecules interact with its DNA.

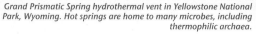

Archaea look very much like bacteria. But closer inspection—of their physiology, biochemistry, and DNA—reveals them to be profoundly different from all bacteria.

FIGURE 13-16 Appearances are deceiving.

> ### TAKE-HOME MESSAGE 13·11
>
> Archaea possess characteristics that place them between bacteria and eukaryotes on the tree of life. Archaea and bacteria may look similar, but they have significant differences in their DNA sequences, as well as in their plasma membranes, cell walls, and flagella. Furthermore, neither archaea nor bacteria resemble eukarya in one key way: only eukarya have a distinct cell nucleus and nuclear membrane.

Archaea are famous for their ability to live in places where life would seem to be impossible, such as in water around hydrothermal vents that emerge from the seafloor, which would be boiling if it were at sea level. At 2,000 meters below the surface, the water pressure reaches 200 atmospheres, or almost 3,000 pounds per square inch, and the 400° C water coming from the vent cannot boil. Archaea thrive there. Most organisms die at temperatures between 40° and 50° C, because their protein molecules are denatured, so the ability of archaea to survive at temperatures above 100° C is truly remarkable.

Equally impressive is the ability of archaea to live in water as acidic as pH 0 or as salty as a saturated NaCl solution, and to metabolize an extraordinary range of compounds—including sulfur, iron, and hydrogen gas—to obtain energy (FIGURE 13-17). Some bacteria also can tolerate extreme physical and chemical conditions, and organisms that can live in these conditions, both bacteria and archaea, are called **extremophiles** ("lovers of extreme conditions").

The extremophile nature of so many archaea leads to a practical problem that limits our knowledge of them. How do you grow an organism in your lab if it requires 200 atmospheres of pressure and a water temperature of 100° C or higher, and eats a gas that is toxic to humans? It's not easy, and consequently only about one-quarter of the identified archaeal species have been cultured in the lab. We know that the other species exist, because their transfer RNA has been identified in samples taken from the environment, and biologists are confident that thousands, perhaps millions, of species of archaea await discovery.

Extremophile archaea have important applications in bioengineering and environmental remediation. For example, *Thermus aquaticus,* an extremophile archaeon that lives in hot springs where the water is nearly boiling, produces an enzyme important in making PCR possible (see Chapter 5), and this enzyme is now sold for laboratory use under the name *Taq* polymerase. Far from being destroyed at high temperatures, *Taq* polymerase works best at 75° C, a temperature at which most human enzymes would simply fall apart.

> **❝** He was a killer, a thing that preyed, living on the things that lived, unaided, alone, by virtue of his own strength and prowess, surviving triumphantly in a hostile environment where only the strong survive.**❞**
> — JACK LONDON, *The Call of the Wild,* 1903

Bioengineers and biotechnologists believe that extremophile archaea and bacteria that thrive at high temperatures and pressures and metabolize toxic substances have enormous potential for industries that carry out activities under such conditions. Recent experiments have demonstrated the ability of some archaea to efficiently degrade hydrocarbons, making it possible for them to be used in the removal of sludge that accumulates in oil refinery tanks and, potentially, in the cleanup of contaminated environments such as oil slicks (FIGURE 13-18). Naturally occurring archaea

FIGURE 13-17 **Living in the harshest environments.** Archaea thrive in many extreme environments, such as this salt pond.

FIGURE 13-18 **Practical applications for archaea.** The metabolic activity of archaea has potential for use in the cleanup of oil spills.

Archaea in your intestine break down a chemical bond found in beans (a bond humans cannot break) ... but the process generates gas, which can cause discomfort as it tries to escape.

are helping to break down some of the more than 200 million gallons of oil released into the Gulf of Mexico following an oil rig explosion and subsequent massive leak in 2010. Other archaea show promise in clearing mineral deposits from pipes in the cooling systems of power plants.

But not all archaea are extremophiles. Archaea live everywhere that bacteria do, and many of them live in places you would find comfortable. In fact, you are home to many archaea and, on occasion, some of them make their presence all too obvious.

Beans are notorious for their tendency to produce gas in the intestine, but archaea are actually the culprits (FIGURE 13-19). Beans contain a couple of carbohydrates with chemical bonds that are not broken down very well by any human enzymes. Methane-producing archaea, on the other hand, have no such difficulties, producing an enzyme that targets

those bonds. As a result, it is the archaea in your intestine that digest most of these carbohydrates. But in the process of breaking these bonds, they produce gases that, as they escape the digestive system, can cause considerable distress.

TAKE-HOME MESSAGE 13·12

Archaea can tolerate extreme physical and chemical conditions that are impossible for most other living organisms. Archaea are hard to study, however, because many require extreme heat or pressure to grow, and these conditions are not easy to provide in a laboratory. Nonetheless, the ability of archaea to grow in such extreme conditions makes them potentially valuable for industrial and environmental purposes. Not all archaea are extremophiles; some live in moderate conditions and even in the human intestine.

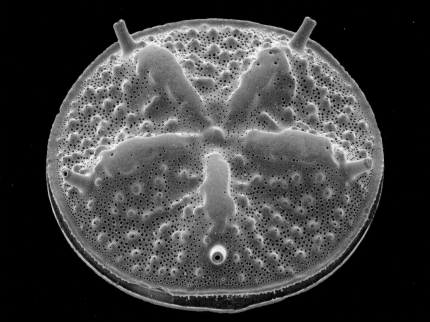

Diatoms, single-celled photosynthetic algae that form an important part of the plankton, have mineralized cell walls that provide protection and support.

13·13 The first eukaryotes were protists.

For the first 2 billion years of life on earth, organisms were extremely small—bacteria and archaea less than 10 micrometers across, one-eighth the diameter of a human hair. But in rocks about 1.9 billion years old, we find fossils of new kinds of organisms that are 10 times larger (**FIGURE 13-20**). These are a group of organisms called acritarchs (a name that can be translated as "confusing old things"). They were the first eukaryotes.

The larger size is the first thing you notice about these fossil cells, but the internal changes in the cells are even more significant: they were the basis for the success of the entire eukaryote lineage, including humans. For the first time in the history of life, cells had internal structures that carried out specific functions. These structures, the cellular organelles, perform the specialized activities that make eukaryotic cells more complex than prokaryotes.

The nucleus is an evolutionary innovation that appeared for the first time in protists. The nucleus is a region inside the cell that is enclosed by its own double-layered membrane. The nuclear membrane was probably formed by the fusion of folds of the plasma membrane that surrounded the cell. We know that many modern prokaryotes have complex infoldings of their plasma membranes—these increase the total surface area of the cell and allow better exchange of material between the cell and the external environment— and infoldings of this type may have played a role in

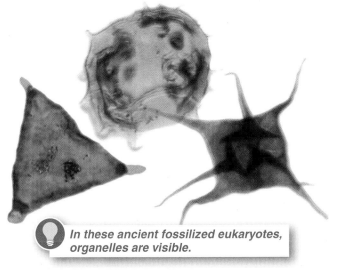

In these ancient fossilized eukaryotes, organelles are visible.

FIGURE 13-20 **Ancient protists.** These acritarch fossils represent the first eukaryotes on earth. The oldest fossils of this type were found in rock 1.9 billion years old.

endosymbiosis and the formation of the first organelles (see Figure 3-8). Infoldings probably fused around the DNA, for example, creating a membrane-enclosed compartment containing the cell's DNA. Thus, the cell nucleus was formed, separated by a double membrane from the cytoplasm of the cell.

Subsequently, these first nuclear membranes developed two specializations: they incorporated proteins that controlled the movement of molecules into and out of the nucleus, and they extended outward from the nucleus to form a folded membrane called the endoplasmic reticulum. The endoplasmic reticulum is the part of a eukaryotic cell where some proteins are assembled. Further development of internal membranes produced the Golgi apparatus, where newly synthesized proteins are given some final processing steps, and sac-like structures called lysosomes, which contain enzymes that break down damaged molecules. Finally, a lineage of protists took in a guest—a bacterial cell that subsequently became the mitochondrion, the organelle that produces most of the ATP synthesized by a eukaryotic cell.

TAKE-HOME MESSAGE 13·13

The nucleus is an evolutionary innovation that appeared for the first time in protists. Early protists took in a bacterial cell that subsequently became the mitochondrion, an organelle in eukaryotic cells that produces ATP.

13·14 There are animal-like protists, fungus-like protists, and plant-like protists.

Six major lineages of protists have been named, but even that classification does not capture the huge diversity of the group, and some of the best-known protists do not fit into any of the six lineages. Protists include forms that are very much like animals, others that seem a lot like fungi, and still others that look like plants (**FIGURE 13-21**).

Animal-like protists Some protists propel themselves quickly around their environment and appear to hunt for prey. These animal-like protists, which include *Paramecium,* are the ciliates, and get their name from the cilia (hair-like projections) that cover their body surfaces and propel the cells through water. *Paramecium* feeds by the process of **phagocytosis:** cilia in a funnel-shaped structure called the gullet create an inward flow of water that carries bacteria and other small particles of food with it. These particles accumulate at the inner end of the gullet, where a portion of the plasma membrane bulges inward and eventually breaks free, forming a food vacuole that drifts into the interior of the cell. Enzymes and hydrochloric acid enter the food vacuole from the cytoplasm, and in this acidic environment the engulfed food items are broken down into molecules that diffuse from the vacuole into the cytoplasm. When all of the digestible material has been consumed, the undigested contents of the vacuole are expelled.

Fungus-like protists Some protists resemble fungi: living as heterotrophs (organisms that are unable to fix carbon through photosynthesis, but require carbon for growth), establishing sheet-like colonies of cells on surfaces (such as the grout in shower stalls), using spores to reproduce, and sometimes producing fruiting bodies. These are the slime molds. They generally spread without any individual cells moving, but rather by adding new cells at the edges of the

DIVERSITY OF PROTISTS

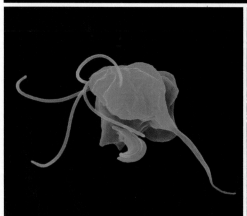

ANIMAL-LIKE PROTISTS
Some protists, such as *Trichomonas vaginalis* shown here, move around and hunt for prey like an animal.

FUNGI-LIKE PROTISTS
Some protists, such as the "multi-goblet" slime mold shown here, live as heterotrophs and form sheet-like colonies of cells like a fungus.

PLANT-LIKE PROTISTS
Some protists, such as the kelp forest shown here, are multicellular and carry out photosynthesis like a plant.

FIGURE 13-21 **Protists come in all shapes and sizes.**

colony. Some slime molds, however, called plasmodial slime molds, are oozing masses of gooey material that flow along a surface, engulfing bacteria, fungi, and small bits of organic material as they go. The streaming of a slime mold in its feeding phase is easy to observe with a microscope, and such a slime mold can flow around, over, or through almost anything—it can even flow through a window screen and reassemble itself on the other side! You may have seen an irregularly shaped blob of yellow material in a moist, shaded garden—that was a plasmodial slime mold.

Remarkably, a plasmodial slime mold is a very large, single cell (but has multiple nuclei). Slime molds divide by binary fission, just like other cells, but in this case a single cell may cover an area of several square centimeters! All of the nuclei in the cell undergo mitosis simultaneously.

Plant-like protists Other members of the protist kingdom grow in water and resemble plants. These include the protists referred to colloquially as algae and seaweeds. The term "seaweed" generally refers to the macroscopic, multicellular marine algae, many of which are used as a source of food for humans. And although they are all protists, seaweeds encompass several groups that do not share a common multicellular ancestor. These include some red algae and some green algae (from which the land plants most likely evolved), as well as some of the brown algae—such as the giant kelp that grows in water 30 meters deep. (The seaweeds, of course, represent parts of the protist kingdom that obviously are not "microbial.")

Brown algae cover large portions of the rocks in the intertidal zone. Giant kelp, growing in temperate regions of the North and South Pacific and off the Atlantic coast of South Africa, are among the fastest growing organisms on earth, with some growing as much as 60 meters in a single year. Kelp forests are an enormously diverse habitat; more than 1,000 species of fishes, crustaceans, snails, and mammals make their homes in these marine "forests." Sea otters wrap a kelp frond around their waist when they sleep, and gray whales hide in kelp forests to escape from killer whales.

Although many protists are unicellular, there are many exceptions. As we've seen, many green and brown algae are multicellular, composed of many different cells and cell types, performing different functions. Additionally, some other protists, such as spirogyra, are colonial, living as collections of cells, each of which can carry out all of its life processes independent of the other cells.

Also among the plant-like protists are the diatoms (**FIGURE 13-22**). Diatoms are unicellular organisms that live in ponds, lakes, and rivers as well as in the oceans. They are so small that, for some species, 30 individuals could be lined up

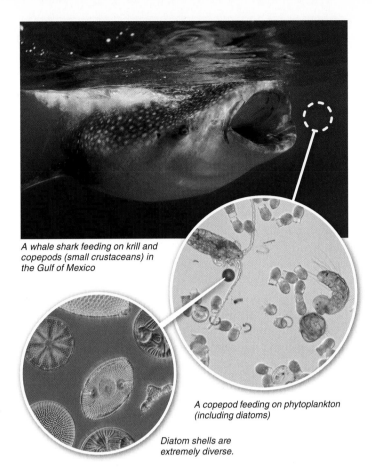

A whale shark feeding on krill and copepods (small crustaceans) in the Gulf of Mexico

A copepod feeding on phytoplankton (including diatoms)

Diatom shells are extremely diverse.

FIGURE 13-22 **Diatoms are aquatic photosynthetic protists.** Diatoms are food for copepods, which are an important source of food for many larger predators.

across the width of a human hair. A characteristic of the diatoms is that they are enclosed in a shell made of silica. Many species of diatoms float in the water, forming part of the phytoplankton—the collection of microscopic organisms that fix carbon dioxide and release oxygen. Phytoplankton can reach densities of hundreds of thousands of cells per liter, and it accounts for about one-quarter of the photosynthetic production of oxygen on earth, occupying a critical position in marine food chains. Small fishes and shrimp-like organisms called copepods feed on phytoplankton, and these small predators are eaten by larger predators, which are eaten by still larger predators. Thus the diversity of life in the marine habitat relies on diatoms and the other microbes that make up the phytoplankton.

TAKE-HOME MESSAGE 13·14

Protists are a diverse group of mostly unicellular eukaryotic organisms. The ciliates, such as *Paramecium,* are animal-like protists. Plasmodial slime molds are fungus-like protists. Colonial protists and multicellular protists such as the giant kelp can be enormous and are plant-like in appearance.

13·15 Some protists can make you very sick.

A **parasite** is an organism that lives in or on another organism, called a **host,** and damages it. A parasitic protist called *Plasmodium* that is transmitted by a mosquito is responsible for a worldwide epidemic disease—malaria. Malaria occurs in tropical parts of Africa, Asia, and Latin America, and it is common in the eastern Mediterranean as well. Between 350 million and 500 million people have clinical cases of malaria, and about 1 million people die of it each year. Malaria is the leading cause of death for children younger than 5 years old in sub-Saharan Africa; somewhere on the African continent, about every 30 seconds, a child dies of malaria.

Neither the incidence of malaria nor the rate of mortality has changed very much since *Plasmodium* was identified as the cause of malaria nearly a century ago. The reason for the lack of progress in combating malaria is the nature of *Plasmodium*—it has a complex development that makes it nearly invisible to the host's immune system (FIGURE 13-23).

The human immune system has a difficult time fighting a malarial infection because, to survive from one generation to the next, *Plasmodium* parasites go through a series of distinct developmental stages. And as the *Plasmodium* cells change from one stage to another, the parasite produces different cell surface proteins. Thus *Plasmodium* stays ahead of the human immune system by constantly changing the way it appears.

Although the immune system doesn't usually have much success fighting malarial infection, some individuals produce red blood cells that are inhospitable to *Plasmodium,* conferring a resistance to the malaria-causing protist. These individuals carry an allele, called Hb^S, which (as we learned in Chapter 7) codes for a variant of the normal adult hemoglobin that causes hemoglobin molecules to stick together in long chains that distort the red blood cell (see Figure 13-23), giving the cell a sickled shape. The sickled red blood cells leak substances that are essential to *Plasmodium,* making the cells inhospitable to the parasite. Although carrying a copy of the Hb^S allele confers resistance to malaria, individuals who carry two copies of the Hb^S allele suffer from sickle-cell anemia—a severely debilitating genetic disease. Individuals with a single Hb^S allele are said to have "sickle-cell trait."

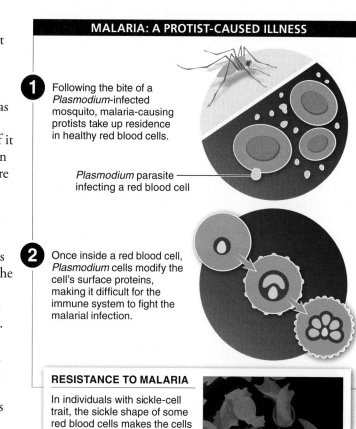

MALARIA: A PROTIST-CAUSED ILLNESS

1 Following the bite of a *Plasmodium*-infected mosquito, malaria-causing protists take up residence in healthy red blood cells.

Plasmodium parasite infecting a red blood cell

2 Once inside a red blood cell, *Plasmodium* cells modify the cell's surface proteins, making it difficult for the immune system to fight the malarial infection.

RESISTANCE TO MALARIA

In individuals with sickle-cell trait, the sickle shape of some red blood cells makes the cells inhospitable to the protist. This makes these individuals resistant to malaria.

FIGURE 13-23 **Malaria-infected blood cells and sickled blood cells.** The malaria parasite grows inside red blood cells and destroys them. Because sickle-cell trait changes the shape of some red blood cells, these cells resist infection by malaria parasites.

TAKE-HOME MESSAGE 13·15

Some protists cause debilitating diseases. *Plasmodium,* the protist responsible for malaria, is one of these. *Plasmodium* has characteristics that protect it from the human immune system, but some humans do have a defense against malaria: people with sickle-cell trait make red blood cells that are inhospitable to the parasite.

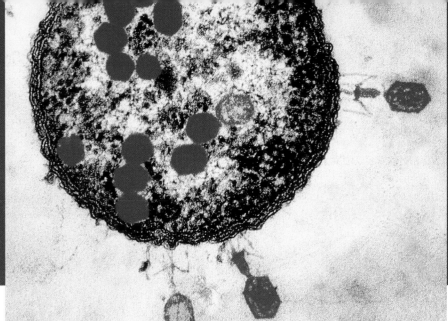

13·16–13·19
Viruses are at the border between living and non-living.

Colored transmission electron micrograph of T2 bacteriophage viruses (red) attacking an E. coli bacterium.

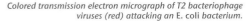

13·16 Viruses are not exactly living organisms.

A virus is not a cell, and that is why viruses do not fit into any of the three domains of life. A virus particle (called a virion) consists of genetic material inside a container made of protein. Some viruses also contain a few enzymes. That's all there is to a virus. The protein container is called the **capsid,** and the genetic material can be either DNA or RNA. Some viruses wrap themselves in a bit of the plasma membrane of the host cell as they are released. A virus of this type is called an enveloped virus—the flu virus is an example. Non-enveloped viruses are enclosed only by the protein container; the virus that causes the common cold is an example of a non-enveloped virus (**FIGURE 13-24**).

There is almost nothing inside the capsid of a virus except DNA or RNA. A virus particle does not carry out any metabolic processes, and it does not control the inward or outward movement of molecules to make conditions inside the virus particle different from conditions outside. Viruses just wait for a chance to insert their genetic material into a living cell.

BASIC STRUCTURE OF A VIRUS

All viruses have a container (the capsid) that holds their genetic material, and sometimes the capsid is wrapped in a membrane (an envelope).

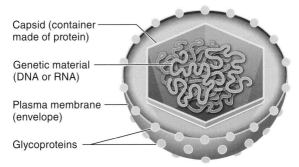

Capsid (container made of protein)

Genetic material (DNA or RNA)

Plasma membrane (envelope)

Glycoproteins

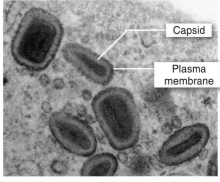

ENVELOPED VIRUS
Enveloped viruses wrap themselves in a bit of the plasma membrane of the host cell as they are released.

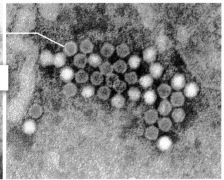

Capsid

Plasma membrane

NON-ENVELOPED VIRUS
Non-enveloped viruses are enclosed only by a capsid.

FIGURE 13-24 **The simple structure of viruses.**

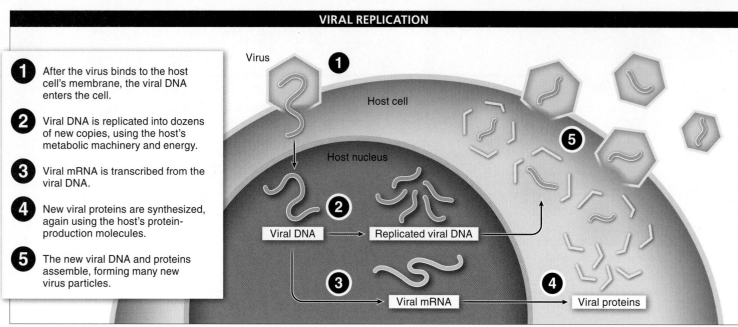

1. After the virus binds to the host cell's membrane, the viral DNA enters the cell.

2. Viral DNA is replicated into dozens of new copies, using the host's metabolic machinery and energy.

3. Viral mRNA is transcribed from the viral DNA.

4. New viral proteins are synthesized, again using the host's protein-production molecules.

5. The new viral DNA and proteins assemble, forming many new virus particles.

FIGURE 13-25 **Making more viruses.** A virus duplicates its own genetic material (DNA in the example shown here) by taking over the resources and metabolic machinery of a host cell.

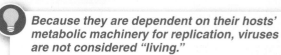

Because they are dependent on their hosts' metabolic machinery for replication, viruses are not considered "living."

We speak of "catching" a cold, but what actually happens is that a virus catches us. Viruses identify the cells that they can infect by recognizing the structure of glycoprotein molecules on the surfaces of cells. Every cell in your body has these molecules. They are embedded in the plasma membrane and extend outward from the cell. Your immune system uses these proteins to identify the cells as part of you, and it does not react to proteins that it recognizes as self rather than non-self.

Viruses have cracked the glycoprotein code, and when they find a cell with the appropriate glycoprotein on its surface, they bind to that cell's plasma membrane and insert their genetic material into the cell (**FIGURE 13-25**). Inside the host cell, the viral DNA or RNA takes over the cellular machinery and uses it to produce more viruses. Viruses carry out nearly all of their activities by hijacking materials and organelles in the host cell. Viral proteins are synthesized in exactly the same way as host cell proteins—mRNA binds to ribosomes, and tRNA matches the correct amino acid to each mRNA codon. The mRNA is produced from the virus's DNA or RNA, but all of the protein-building machinery comes from the host cell, as does the ATP required to synthesize the new viral protein.

There are also some other types of non-living infectious agents. Prions, for example, are misfolded proteins that form plaques and interfere with normal tissue. Usually acquired through ingestion of an infected animal (or its bodily fluids), prions can cause a variety of degenerative neurological diseases in humans. One of these, Creutzfeldt-Jakob disease, is sometimes referred to as a human form of mad cow disease: it is characterized by the development of holes in brain tissue, causing the tissue to appear sponge-like. It leads to dementia, memory loss, and disruption of balance and coordination. No cure or treatment is known, and few victims survive more than a year following the appearance of symptoms.

TAKE-HOME MESSAGE 13·16

A virus is not alive, but it can carry out some of the same functions as living organisms, provided that it can get inside a cell. A virus takes over the protein-making machinery of the host cell to produce more viral genetic material (RNA or DNA) and protein. The viral proteins and genetic material are assembled into new virus particles and released from the cell.

Many diseases are caused by viruses. Some viral diseases have been responsible for worldwide epidemics, called pandemics. The influenza pandemic of 1918–1919 killed at least 20 million people, and possibly as many as 50 million. In the current HIV/AIDS pandemic, more than 75 million people have been infected and about 35 million people have died of AIDS. Worldwide, about 1.6 million people died of AIDS-related illnesses in 2012.

Other viral diseases, such as the common cold, are not usually serious. Herpes is another common viral infection in humans, caused by two related viruses. Oral herpes is an infection near the mouth and can cause cold sores lasting two to three weeks. More than three-quarters of all adults in the United States are infected with the virus, with more than 50 million people experiencing outbreaks each year. Genital herpes, a sexually transmitted disease, affects about one in six people between the ages of 14 and 49 in the United States. The viruses causing herpes are present in and released from sores and spread from one person to another by skin-to-skin contact, but an infected person can also pass on the virus even when she or he has no symptoms. Although a variety of treatments are available for both types of herpes, including antiviral drugs that can reduce the severity and duration of symptoms, there is currently no cure.

DNA viruses have base-pair sequences that are stable over time, because the enzymes that replicate the DNA check for errors and correct them during replication. From a health perspective, the fact that DNA viruses do not change rapidly makes it much easier to fight and treat diseases caused by these viruses, using vaccines. Vaccination against a DNA virus such as smallpox, for example, provides years of protection.

RNA viruses, by contrast, change quickly, because the enzymes that carry out RNA replication do not have error-checking mechanisms. As a consequence, because the RNA-replicating enzymes make errors as they assemble new RNA molecules, RNA viruses are continuously mutating into new forms. The common influenza virus that causes outbreaks of flu every year, for example, is an RNA virus. And because flu viruses mutate so fast, the virus that causes health problems changes from one flu season to the next, necessitating a different flu vaccine every year.

Q Why do flu viruses change so quickly?

About every 50 years, a new variety of influenza causes a pandemic. The 20th century had three major influenza pandemics, and all of them originated in the same peculiar way. In each of these pandemics, a bird flu virus gained the ability to infect human cells by passing through pigs, and then the virus spread rapidly through the human population worldwide. The most famous of these pandemics was that of 1918–1919, mentioned above (**FIGURE 13-26**). It was called the Spanish influenza, because it

---

VIRAL OUTBREAKS

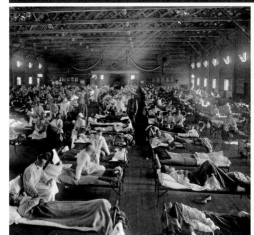

The influenza pandemic of 1918–1919 killed millions.

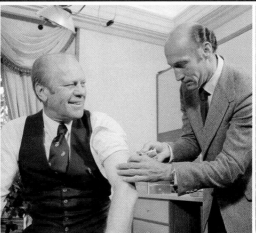

President Gerald Ford receives the swine flu vaccine in the White House in 1976.

In 2002–2003, the virus that causes severe acute respiratory syndrome (SARS) infected several thousand people around the world.

FIGURE 13-26 Viral outbreaks can threaten large populations.

seemed to enter Europe through Spain, but it originated in Asia. The Asian flu and Hong Kong flu pandemics also originated in Asia, as did smaller viral outbreaks, such as Korean flu (1947), swine fever (1976), and SARS (severe acute respiratory syndrome, 2002–2003), which did not become pandemics.

TAKE-HOME MESSAGE 13·17

Many diseases are caused by viruses. DNA viruses are relatively stable, because DNA-replicating enzymes check for errors and correct them during replication. RNA viruses change quickly, however, because RNA-replicating enzymes do not have error-checking mechanisms.

13·18 Viruses infect a wide range of organisms.

Viruses are nearly everywhere. There are viruses that infect animals, viruses that infect plants, even viruses that infect bacteria. Most viruses infect just one species, or only a few closely related species, and enter only one kind of cell in that species. But some viruses can infect a wide range of hosts—the rabies virus, for example, can infect any mammal. Pet dogs and cats are routinely immunized against it, but wild mammals, such as bats, skunks, and raccoons, can get rabies. That is why, if you are bitten by one of those wild animals, your treatment includes rabies shots as a precaution.

The glycoproteins on the surface of a virus determine which host species the virus can infect and which tissues of the host it can enter. Influenza A viruses (the ones that cause flu outbreaks every year), for example, have two types of glycoprotein that have different functions. One glycoprotein matches that of a host cell, and allows the virus to enter the cell. The other glycoprotein allows the virus to get back out of the cell, releasing new virus particles that can infect other cells (**FIGURE 13-27**).

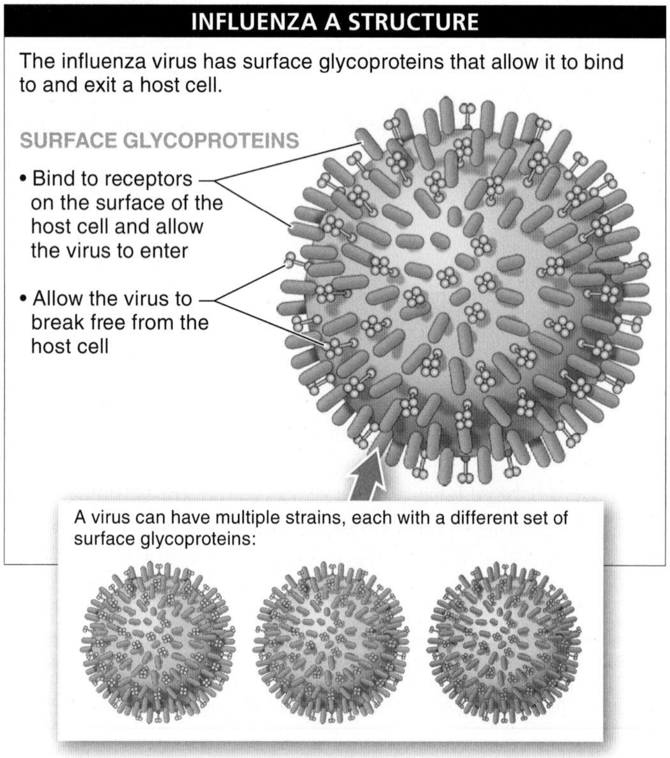

Influenza A is an example of a virus that can move from one species to another, and as we noted above, all of the influenza pandemics in the past century began with the transmission of a bird flu virus to humans. Here's how "species jumping" can occur.

INFLUENZA A STRUCTURE

The influenza virus has surface glycoproteins that allow it to bind to and exit a host cell.

SURFACE GLYCOPROTEINS

• Bind to receptors on the surface of the host cell and allow the virus to enter

• Allow the virus to break free from the host cell

A virus can have multiple strains, each with a different set of surface glycoproteins:

FIGURE 13-27 **A virus's surface proteins.** The influenza virus has surface glycoproteins that allow it to bind to and exit a host cell.

Q What role does a pig play in the transmission of virus from a bird to a human?

Viruses that infect birds don't bind well to glycoproteins in human cells, making it difficult for bird flu viruses to infect humans. However, the cells of pigs have glycoproteins that allow both human and bird flu viruses to bind to them. Thus, a pig cell can be infected by a human flu virus and a bird flu virus at the same time. But influenza viruses are not very careful when they incorporate newly synthesized RNA into new particles. If RNA strands from a human flu virus and a bird flu virus happen to be in a pig cell at the same time, some virus particles released from the cell might include RNA from both the bird and the human virus. These influenza viruses are a new strain, and they may be able to infect humans (**FIGURE 13-28**).

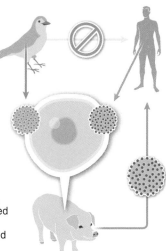

Viruses that infect birds don't bind well to glycoproteins in human cells, making it difficult for bird viruses to infect humans.

However, a virus that infects humans and a virus that infects birds can meet if both have infected a pig cell.

Because viral RNA replication is error-prone, the two viral RNA molecules can get packaged together into a new virus particle. The new virus can have features from the bird virus and be capable of infecting humans.

FIGURE 13-28 **Inside a pig, bird viruses can gain the ability to infect humans.**

Since 1997, one strain of bird flu has spread from Hong Kong through most of Asia and into Turkey, France, Germany, and England. This is the most recent avian influenza virus that can also infect humans (not needing pigs as intermediary). The virus readily infects birds, and tens of millions of chickens, turkeys, and ducks have been killed in attempts to eradicate the infection. Because the virus's host-entry glycoproteins do not bind well to human cells, the virus does not easily infect people and spread from human to human: so far, nearly all of the people who have been infected by the avian influenza virus have contracted it through close contact with infected flocks of birds or by eating birds that died of the viral infection. Only a half-

dozen cases are known in which the virus seems to have been transmitted from one person to another. However, avian flu can be deadly when it does infect humans—more than half of the 650 human cases reported as of 2014 have ended in death.

> ### TAKE-HOME MESSAGE 13·18
>
> Glycoproteins on the surfaces of viruses determine the types of cells a virus can invade. Most viruses infect just one species, or only a few closely related species, and enter only one kind of cell in that species.

13·19 HIV illustrates the difficulty of controlling infectious viruses.

New infectious diseases—some caused by bacteria and others by viruses—emerge quite frequently. Many of these new diseases originate in other species of animals and subsequently acquire the ability to infect humans. There are enough of these new diseases to fill the monthly issues of the journal *Emerging Infectious Diseases,* which is published by the Centers for Disease Control and Prevention. But even in that

crowd, **acquired immunodeficiency syndrome (AIDS)** stands out. AIDS is caused by the **human immunodeficiency virus (HIV),** which is derived from a strain of the simian immunodeficiency virus (SIV) that jumped from chimpanzees to humans in the early 1900s. HIV has all of the characteristics that make viral diseases hard to control—plus an additional characteristic of its own.

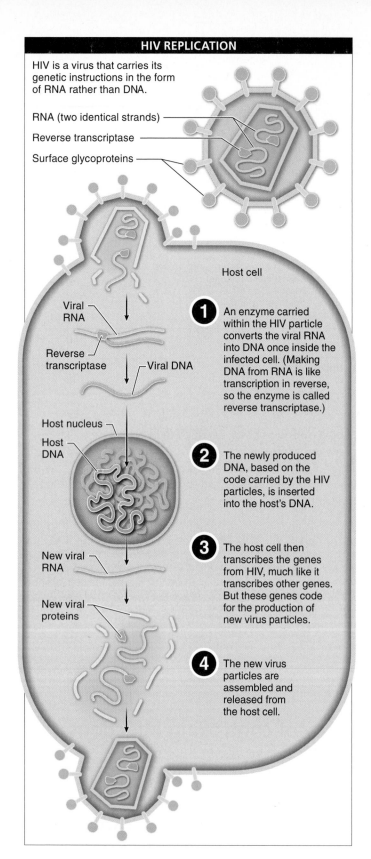

HIV REPLICATION

HIV is a virus that carries its genetic instructions in the form of RNA rather than DNA.

RNA (two identical strands)

Reverse transcriptase

Surface glycoproteins

Host cell

Viral RNA

Reverse transcriptase

Viral DNA

1 An enzyme carried within the HIV particle converts the viral RNA into DNA once inside the infected cell. (Making DNA from RNA is like transcription in reverse, so the enzyme is called reverse transcriptase.)

Host nucleus

Host DNA

2 The newly produced DNA, based on the code carried by the HIV particles, is inserted into the host's DNA.

New viral RNA

3 The host cell then transcribes the genes from HIV, much like it transcribes other genes. But these genes code for the production of new virus particles.

New viral proteins

4 The new virus particles are assembled and released from the host cell.

FIGURE 13-29 HIV infection. This RNA virus takes over the replicating machinery of a white blood cell to produce a new generation of HIV particles. HIV mutates rapidly during this process, because the reverse transcriptase enzyme is highly error-prone.

HIV mutates easily. HIV is a **retrovirus,** an RNA-containing virus that also contains a viral enzyme called reverse transcriptase that uses a strand of viral RNA as a template to synthesize a single strand of DNA. That DNA strand, in turn, is used as a template to make a complementary strand of DNA, and the resulting double-stranded DNA is integrated into the host cell's DNA. From here, the DNA is used by the host cell's machinery to make more viral RNA (**FIGURE 13-29**). Reverse transcriptase is so error-prone, however, that virtually every copy of HIV in an infected individual's body has a different mutation.

This enormous genetic variation in the HIV particles circulating in an infected person's body makes the infection hard to treat. Virus particles with different mutations can have different proteins on their surfaces, and these surface proteins change each time the virus replicates inside a host cell. Each new generation of HIV in the infected individual contains viruses with surface proteins that his or her immune system has never seen. Furthermore, some of the HIV mutations will confer resistance to the drugs that are being used to treat the patient, and new drugs must be used.

HIV attacks white blood cells. All of those problems would apply to any disease caused by a retrovirus, but HIV offers an additional challenge: it targets cells in the host's immune system, especially white blood cells, and particularly those that search for and attack invading bacteria or viruses. During the incubation period, which can last for many years, HIV infects white blood cells. New ones are produced to replace those killed by the virus, however, and the infected person has virtually no symptoms. Nonetheless, HIV is present in the individual's body fluids during the incubation period and can be transmitted to other individuals. HIV testing is used to detect infections during this stage.

Each time HIV infects another white blood cell, the reverse transcriptase makes errors in transcribing the RNA to DNA, and eventually one of the mutations allows the virus to bind to the glycoprotein on the

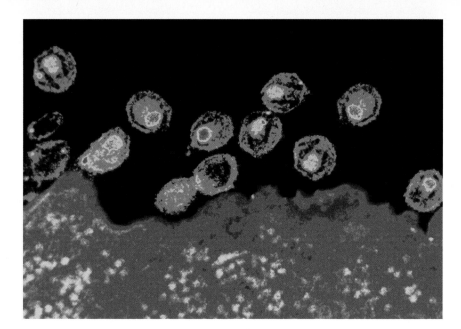

FIGURE 13-30 HIV attacks white blood cells essential for identifying foreign invaders.

surface of a specialized type of white blood cell—a bacteria- and virus-hunting white blood cell that is critically important in identifying disease-causing threats. A suitable mutation may occur in a couple of years, though more often it takes about 10 years or longer. But when it does happen, it signals a new stage in the HIV infection: the development of AIDS (**FIGURE 13-30**).

The immune system collapses. Normally, white blood cells all work together to identify and destroy cells that have been infected by a virus. When HIV begins to kill the cells that hunt for viruses and bacteria, the immune system begins to fail—it can no longer respond to HIV, or to any other

infectious agent. Patients with AIDS develop multiple infections, bacterial and viral, as well as cancers, because they have lost the immune system cells that would normally have marked infected and cancerous cells for destruction.

TAKE-HOME MESSAGE 13·19

HIV is especially difficult to control. Mutations change the properties of the retrovirus so that it is hard for the immune system to recognize it, and they produce variants that are resistant to the drugs used to treat the HIV infection.

 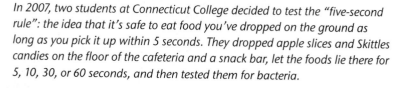
The five-second rule
How clean is that food you just dropped?

In 2007, two students at Connecticut College decided to test the "five-second rule": the idea that it's safe to eat food you've dropped on the ground as long as you pick it up within 5 seconds. They dropped apple slices and Skittles candies on the floor of the cafeteria and a snack bar, let the foods lie there for 5, 10, 30, or 60 seconds, and then tested them for bacteria.

Q: Is the five-second rule valid? The students found no bacteria on the food picked up within 30 seconds. After a minute, the apple slices had picked up some bacteria, but the Skittles had none (in a later experiment, they found bacteria on Skittles only after 5 minutes). The students concluded that the five-second rule should get an extension, proposing that you have 30 seconds to pick up moist foods and more than a minute to pick up dry foods without risk of bacterial contamination.

Q: Should we worry less about bacterial contamination of food? This question is a serious one and warrants more study. Each year in the United States there are more than 76 million cases of illness caused by contaminated food, including more than 5,000 that are fatal.

Q: Scientific method and drawing conclusions: can you generalize from a small number of observations to all possible situations? In the Connecticut College study, the students examined two food types, dropped in just two locations, and onto surfaces with unknown concentrations of bacteria. In another study, published in the *Journal of Applied Microbiology*, researchers focused on *Salmonella bacteria*. Because it has been documented that bacteria can survive on clothes, hands, sponges, cutting boards, and utensils for several days, the researchers decided to drop food on surfaces known to be covered in bacteria.

Q: Which factor is more important for dropped food: location or duration? Armed with slices of bologna and pieces of bread, the researchers found that when dropped on surfaces covered with *Salmonella* and left for a full minute, both types of food took up 1,500–80,000 bacteria. And although picking up the food quickly—within just 5 seconds—reduced by 90% the number of bacteria present, 150–8,000 bacteria still had time to hitch a ride on the food in that first 5 seconds. These experiments were replicated, with no significant differences, three times, on separate days.

A grubby conclusion: It's probably safe to say that if you're in the Connecticut College cafeteria, you can eat your dropped food as long as you pick it up quickly. But for all other situations, a bit more caution is advised—particularly, taking notice of the "drop zone." Unsanitary surfaces likely to have microbes will contaminate your food almost immediately. So hold on tight.

Check Your Knowledge

Thinking critically about visual displays of data

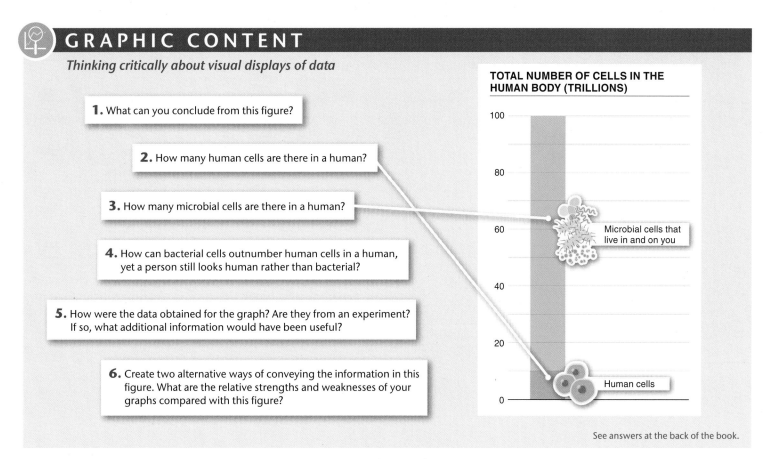

1. What can you conclude from this figure?

2. How many human cells are there in a human?

3. How many microbial cells are there in a human?

4. How can bacterial cells outnumber human cells in a human, yet a person still looks human rather than bacterial?

5. How were the data obtained for the graph? Are they from an experiment? If so, what additional information would have been useful?

6. Create two alternative ways of conveying the information in this figure. What are the relative strengths and weaknesses of your graphs compared with this figure?

TOTAL NUMBER OF CELLS IN THE HUMAN BODY (TRILLIONS)

Microbial cells that live in and on you

Human cells

See answers at the back of the book.

Key Terms in Evolution and Diversity Among the Microbes

acquired immunodeficiency syndrome (AIDS), p. 561
capsid, p. 557
capsule, p. 539
chemolithotroph, p. 541
chemoorganotroph, p. 541
conjugation, p. 540
extremophile, p. 551
Gram stain, p. 539
host, p. 556
human immunodeficiency virus (HIV), p. 561
microbe, p. 534

Oxygen Revolution, p. 542
parasite, p. 556
pathogenic, p. 545
peptidoglycan, p. 539
phagocytosis, p. 554
photoautotroph, p. 542
plasmid, p. 540
probiotic therapy, p. 543
retrovirus, p. 562
sexually transmitted disease (STD), p. 546
transduction, p. 540
transformation, p. 541

ABOUT THE CHAPTER OPENING PHOTO

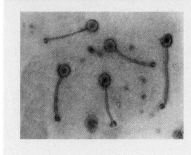

Colored transmission electron micrograph showing bacteriophage viruses that infect bacteria from the genus *Listeria*. Some food-borne species of *Listeria* are a serious threat to human health; this has prompted research into bacteriophages that might be used to help control such pathogens.

13·1–13·2 There are microbes in all three domains.

Microbes are the simplest but probably also the most successful organisms on earth.

MICROBES IN THE TREE OF LIFE

Microbes are grouped together only because they are small, not because of evolutionary relatedness. They occur in all three domains of life, and also include the viruses.

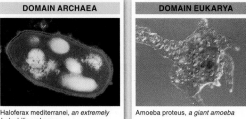

DOMAIN BACTERIA

Neisseria meningitidis, *a bacterium that can cause meningitis*

DOMAIN ARCHAEA

Haloferax mediterranei, *an extremely halophilic archaeon*

DOMAIN EUKARYA

Amoeba proteus, *a giant amoeba*

VIRUSES
Viruses aren't truly living organisms and they are not classified into any of the three domains.

? 1. Why doesn't an amoeba need a respiratory system?

? 2. It is difficult to make generalizations about the group of organisms sometimes referred to as "microbes." Why?

? **Check Your Knowledge**

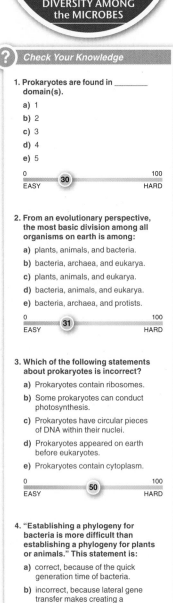

1. Prokaryotes are found in _____ domain(s).
   a) 1
   b) 2
   c) 3
   d) 4
   e) 5

   0 ——●30——— 100
   EASY   HARD

2. From an evolutionary perspective, the most basic division among all organisms on earth is among:
   a) plants, animals, and bacteria.
   b) bacteria, archaea, and eukarya.
   c) plants, animals, and eukarya.
   d) bacteria, animals, and eukarya.
   e) bacteria, archaea, and protists.

   0 ——●31——— 100
   EASY   HARD

3. Which of the following statements about prokaryotes is incorrect?
   a) Prokaryotes contain ribosomes.
   b) Some prokaryotes can conduct photosynthesis.
   c) Prokaryotes have circular pieces of DNA within their nuclei.
   d) Prokaryotes appeared on earth before eukaryotes.
   e) Prokaryotes contain cytoplasm.

   0 ————●50——— 100
   EASY   HARD

4. "Establishing a phylogeny for bacteria is more difficult than establishing a phylogeny for plants or animals." This statement is:
   a) correct, because of the quick generation time of bacteria.
   b) incorrect, because lateral gene transfer makes creating a phylogeny simpler.
   c) incorrect, because the metabolic diversity of bacteria makes classification impossible.
   d) correct, because it is difficult to obtain enough morphological data from bacteria.
   e) correct, because bacteria can engage in lateral gene transfer.

   0 —————●62—— 100
   EASY   HARD

13·3–13·5 Bacteria may be the most diverse of all organisms.

Different species of bacteria are found virtually everywhere and can eat virtually anything.

BACTERIA STRUCTURE

Bacteria are single-celled organisms. Their cytoplasm is surrounded by a membrane and contains the DNA (they have no nuclei and no intracellular organelles).

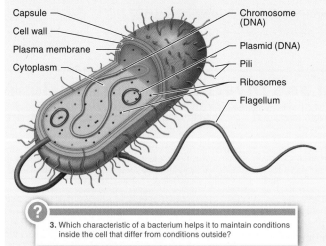

Capsule

Cell wall

Plasma membrane

Cytoplasm

Chromosome (DNA)

Plasmid (DNA)

Pili

Ribosomes

Flagellum

? 3. Which characteristic of a bacterium helps it to maintain conditions inside the cell that differ from conditions outside?

METHODS OF IDENTIFYING BACTERIA

Microbiologists can often identify bacteria based on the colors and shapes of their colonies. They can get additional information by examining whether they change color when stained by a dye.

GRAM-POSITIVE BACTERIA
The glycoprotein layer is on the outside of the cell wall and can be stained with purple dye.

GRAM-NEGATIVE BACTERIA
The glycoprotein layer lies beneath an additional membrane and cannot be stained with the dye.

CELL DIVISION IN BACTERIA

Bacteria divide by binary fission, creating two new daughter cells, each of which carries the genetic information that was present in the chromosome of the mother cell. Bacterial populations can grow rapidly—with some bacteria able to divide every 20 minutes.

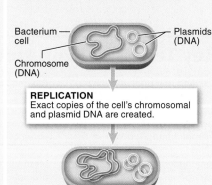

Bacterium cell

Plasmids (DNA)

Chromosome (DNA)

REPLICATION
Exact copies of the cell's chromosomal and plasmid DNA are created.

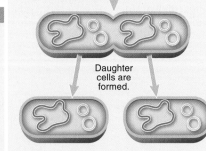

Cell elongates and begins to pinch in two.

Daughter cells are formed.

METHODS OF GENETIC EXCHANGE IN BACTERIA WITHIN THE SAME GENERATION

Bacteria are able to transfer genetic information laterally—that is, to other individuals—through any of three different processes: conjugation, transduction, and transformation.

CONJUGATION
A bacterium transfers a copy of some or all of its DNA to another bacterium, giving the second bacterium genetic information it did not have before.

TRANSDUCTION
A virus containing pieces of bacterial DNA inadvertently picked up from its previous host infects a new bacterium, and passes new bacterial genes to the bacterium.

TRANSFORMATION
A bacterium can take up DNA—potentially including alleles it did not carry—from its surroundings (usually from bacteria that have died).

? 4. Describe three types of lateral transfer of genetic information in bacteria.

METABOLIC DIVERSITY AMONG BACTERIA

Some bacteria eat organic molecules, some eat minerals, and still others carry out photosynthesis. Some require oxygen to live, others do not.

CHEMOORGANOTROPHS
Feed on organic molecules

CHEMOLITHOTROPHS
Feed on inorganic molecules

PHOTOAUTOTROPHS
Use energy from sunlight to produce glucose via photosynthesis

? 5. A common distinction among bacteria is between those that are aerobic and those that are anaerobic. What is the meaning of this distinction?

?

5. A population of *Escherichia coli* can double every _____ in an ideal laboratory culture.
   a) 20 seconds
   b) 2 minutes
   c) 2 days
   d) 20 minutes
   e) 2 hours

   0 —— **38** —— 100
   EASY   HARD

6. Plasmids containing genes for antibiotic resistance can be exchanged between bacterial cells in a culture by:
   a) conjugation.
   b) artificial exchange.
   c) cloning.
   d) transduction.
   e) conduction.

   0 —— **48** —— 100
   EASY   HARD

7. Which group of organisms utilizes the largest variety of energy sources?
   a) prokaryotes
   b) fungi
   c) protists
   d) invertebrate animals
   e) vertebrate animals

   0 —— **53** —— 100
   EASY   HARD

8. Which of the following statements about antibiotics is incorrect?
   a) Penicillin was the first antibiotic widely used to fight bacterial infections.
   b) Antibiotics help microbes compete with other microbes.
   c) Antibiotic-resistant microbes are selected for in humans who are taking antibiotics.
   d) Antibiotics, though effective against viruses, are not effective against bacteria.
   e) Antibiotics are used not just in human health care but also in agriculture.

   0 —— **44** —— 100
   EASY   HARD

9. Which of the following domains are the most closely related, in that they share a unique common ancestor?
   a) archaea and bacteria
   b) archaea and eukarya
   c) bacteria and eukarya
   d) None of the above; all three domains evolved from different ancestors.
   e) None of the above; all three domains are equally related to one another.

   0 —— **60** —— 100
   EASY   HARD

13·6–13·10 In humans, bacteria can have harmful or beneficial health effects

Bacterial effects on human health vary widely. Many are beneficial and many are neutral, while some can be harmful.

BENEFICIAL BACTERIA

Many bacteria are beneficial. Those living in yogurt, for example, can take up residence in your digestive tract and improve your extraction of nutrients from food while also producing vitamins. Bacteria are also used in the production of many other foods, including cheeses, and (along with yeasts) of beer, wine, and vinegar.

HARMFUL BACTERIA

Some bacteria always cause disease, and others do no harm except under certain conditions. For example, *Streptococcus pyogenes* can be harmless, but under some conditions it releases toxins that are responsible for strep throat, scarlet fever, and necrotizing fasciitis (caused by the flesh-eating strains).

? 6. What produces the characteristic taste and texture of yogurt?

? 7. In the mid-1800s, how was the spread of cholera slowed in London?

ANTIBIOTIC RESISTANCE

Antibiotic resistance routinely evolves in microbes. Excessive use of antibiotics in medicine and agriculture has made several of the most important pathogenic bacteria resistant to every known antibiotic, and infections caused by these bacteria are nearly impossible to treat.

If a patient stops taking an antibiotic before finishing all of the prescribed amount, many of the target bacterial cells may remain (and multiply), including those most resistant to the antibiotic.

? 8. Why do more people die from *Staphylococcus aureus* infections than from HIV/AIDS?

SEXUALLY TRANSMITTED DISEASES

Sexually transmitted diseases (STDs) are caused by a variety of organisms, including bacteria, viruses, protists, fungi, and arthropods. Worldwide, more than 300 million people are newly infected each year.

BACTERIUM	VIRUS	PROTIST	FUNGUS	ARTHROPOD
• Gonorrhea	• HIV/AIDS	• Trichomoniasis	• Yeast infections	• Crab lice
• Syphilis	• Genital herpes			
• Chlamydia	• Human papilloma virus (HPV)			

? 9. Describe two characteristics of STDs that make it nearly impossible to completely eliminate these diseases.

10. Unlike other microbes, archaea are able to thrive:

   a) in very extreme environments (with respect to temperature, salinity, pressure, etc.).

   b) only at the typical body temperature of their mammalian hosts.

   c) inside eukaryotic cell organelles.

   d) in lakes and rivers (i.e., in fresh water).

   e) in well-aerated soils.

0 — 17 — 100
EASY — HARD

11. Most archaeal species:

   a) have yet to be discovered.

   b) are pathogenic.

   c) are bacteriophages.

   d) are animal parasites.

   e) are beneficial to humans.

0 — 35 — 100
EASY — HARD

12. All protists are alike in that all are:

   a) photosynthetic.

   b) eukaryotic.

   c) parasitic.

   d) colonial.

   e) multicellular.

0 — 34 — 100
EASY — HARD

13. Which of the following statements about protists is correct?

   a) Protists were the first group of organisms to have a nucleus.

   b) Protists were the first group of bacteria.

   c) Protists are the only group of photosynthetic archaea.

   d) Protists are unique among the eukaryote kingdoms in that they are sessile.

   e) All protists use cilia for feeding.

0 — 31 — 100
EASY — HARD

14. Which of the following statements about *Plasmodium* is correct?

   a) It causes sickle-cell anemia.

   b) Because it changes its surface proteins frequently, it is largely invisible to the human immune system.

   c) It is transmitted by *Paramecium*.

   d) It confers resistance to malaria.

   e) It is most common in North America.

0 — 57 — 100
EASY — HARD

13·11–13·12 Archaea exploit some of the most extreme habitats.

Although archaea resemble bacteria, evolutionarily they are more closely related to the eukarya.

ARCHAEA STRUCTURE

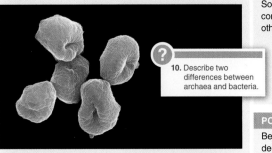

Although archaea (shown here) and bacteria are similar in appearance (and both domains are prokaryotes), they have significant differences in their DNA sequences, as well as differences in their plasma membranes, cell walls, and flagella.

? 10. Describe two differences between archaea and bacteria.

ARCHAEA THRIVE IN EXTREME CONDITIONS

Some archaea can tolerate extreme physical and chemical conditions that are impossible for most other living organisms, while others live in moderate conditions and even in the human intestine.

? 11. Some bacteria and archaea are called extremophiles. What does "extreme" refer to here?

POTENTIAL USES OF ARCHAEA

Because some archaea are able to efficiently degrade hydrocarbons, it may someday be possible to utilize them in the removal of sludge that accumulates in oil refinery tanks and in the cleanup of contaminated environments such as oil slicks.

? 12. Why might scientists eager to learn more about archaea have a hard time doing so?

13·13–13·15 Most protists are single-celled eukaryotes.

The nucleus is an evolutionary innovation that appeared for the first time in protists.

EARLY PROTISTS

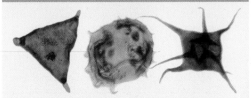

The earliest eukaryotic organisms—having internal structures, called cellular organelles, that performed specialized activities—were protists. Protist-resembling fossils have been found in rock 1.9 billion years old.

? 13. Describe the cellular structure that first appeared in the protists.

? 14. Which group of protists is referred to as the "animal-like protists"? Why?

MALARIA: A PROTIST-CAUSED ILLNESS

1 Following the bite of a *Plasmodium*-infected mosquito, malaria-causing protists take up residence in healthy red blood cells.

2 Once inside a red blood cell, *Plasmodium* cells modify the cell's surface proteins, making it difficult for the immune system to fight the malarial infection.

? 15. Which feature of *Plasmodium*, the protist responsible for malaria, helps it stay ahead of the human immune system?

DIVERSITY OF PROTISTS

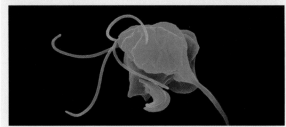

ANIMAL-LIKE PROTISTS

Some protists, such as *Trichomonas vaginalis* shown here, move around and hunt for prey like an animal.

FUNGI-LIKE PROTISTS

Some protists, such as the "multi-goblet" slime mold shown here, live as heterotrophs and form sheet-like colonies of cells like a fungus.

PLANT-LIKE PROTISTS

Some protists, such as the kelp forest shown here, are multicellular and carry out photosynthesis like a plant.

A virus is not alive, but it can carry out some of the same functions as living organisms, provided that it can get inside a cell.

BASIC STRUCTURE OF A VIRUS

All viruses have a container (the capsid) that holds their genetic material, and sometimes the capsid is wrapped in a membrane (an envelope).

Capsid (container made of protein)

Genetic material (DNA or RNA)

Plasma membrane (envelope)

Glycoproteins

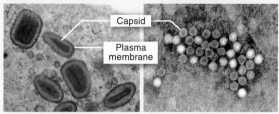

Capsid

Plasma membrane

ENVELOPED VIRUS
Enveloped viruses wrap themselves in a bit of the plasma membrane of the host cell as they are released.

NON-ENVELOPED VIRUS
Non-enveloped viruses are enclosed only by a capsid.

16. Why are viruses not considered "living"?

VIRAL REPLICATION

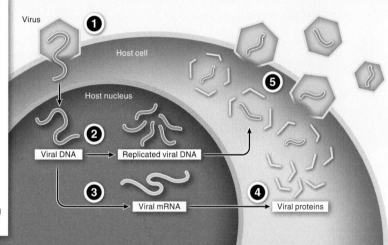

1 After the virus binds to the host cell's membrane, the viral DNA enters the cell.

2 Viral DNA is replicated into dozens of new copies, using the host's metabolic machinery and energy.

3 Viral mRNA is transcribed from the viral DNA.

4 New viral proteins are synthesized, again using the host's protein-production molecules.

5 The new viral DNA and proteins assemble, forming many new virus particles.

Virus

Host cell

Host nucleus

Viral DNA → Replicated viral DNA

Viral mRNA

Viral proteins

VIRUSES AND HEALTH

Many diseases are caused by viruses. Some viral diseases, such as the common cold, are not usually serious, but others have been responsible for worldwide epidemics, called pandemics. The influenza pandemic of 1918–1919 killed at least 20 million people, and possibly as many as 50 million. In the current HIV/AIDS pandemic, nearly 40 million people have been infected, with an annual mortality of about 3 million.

17. Why might host organisms find DNA viruses easier to fight than RNA viruses?

VIRAL TRANSMISSION THROUGH "SPECIES JUMPING"

Viruses that infect birds don't bind well to glycoproteins in human cells, making it difficult for bird viruses to infect humans.

However, a virus that infects humans and a virus that infects birds can meet if both have infected a pig cell.

Because viral RNA replication is error-prone, the two viral RNA molecules can get packaged together into a new virus particle. The new virus can have features from the bird virus and be capable of infecting humans.

18. Which feature of a virus determines which host species it can infect?

HIV

Infection by the retrovirus HIV is especially difficult to control. High rates of mutation continually change the properties of the virus, hindering recognition by the immune system. Additionally, the high rate of mutation commonly causes the production of variants that are resistant to the drugs used to treat the HIV infection.

15. Many biologists do not consider viruses to be alive. Which of the following characteristics of viruses leads to this conclusion?

  a) Viruses lack a metabolic system.

  b) Viruses do not respond to external stimuli.

  c) Viruses are unable to reproduce on their own.

  d) All of the above are correct.

  e) Only a) and c) are correct.

0 — 44 — 100
EASY   HARD

16. The genetic information in all viruses is:

  a) DNA.

  b) RNA.

  c) protein.

  d) a polymerase enzyme.

  e) DNA or RNA.

0 — 30 — 100
EASY   HARD

17. Viruses are able to infect:

  a) humans.

  b) plants.

  c) bacteria.

  d) birds.

  e) All of the above.

0 — 7 — 100
EASY   HARD

18. Which of the following statements about HIV is incorrect?

  a) It is a virus.

  b) It attacks red blood cells.

  c) It mutates frequently.

  d) It is derived from a simian immunodeficiency virus.

  e) It contains RNA but not DNA.

0 — 42 — 100
EASY   HARD

14 Population Ecology

PLANET AT CAPACITY:
PATTERNS OF POPULATION GROWTH

Population ecology is the study of how populations interact with their environments.

A life history is like a species summary.

Ecology influences the evolution of aging in a population.

The human population is growing rapidly.

14·1–14·6
Population ecology is the study of how populations interact with their environments.

African elephants (Loxodonta africana) *at the Chobe River in Botswana.*

14·1 What is ecology?

Lobster tastes delicious. Many people consider it one of the finest delicacies. It's not surprising, then, that catching and selling lobsters is big business. Generating almost $200 million a year for the State of Maine alone, it is among the most economically important businesses in New England. Now imagine for a minute that you were in charge of the lobster industry. The 6,000 lobstermen in Maine depend on your managing the industry so that not only can they catch and sell enough lobsters to survive, but sufficient numbers of lobsters are left to ensure there will be lobsters to catch in all the years to come. How would you do it? You would be faced with some tough questions.

- How many lobsters should you allow each lobsterman to take each day? Each year?

- Should you require that certain lobsters be thrown back? If so, should they be the biggest or the smallest? Males or females?

- Is it better to increase the size of the lobster population or maintain it at current levels?

- But wait a minute. Forget about these obviously complex questions and start with a seemingly simpler one: how many lobsters currently live in the waters off the coast of Maine? Is this even possible to determine (FIGURE 14-1)?

Welcome to ecology.

These difficult questions, and many others like them, are all part of **ecology,** a subdiscipline of biology defined as the study of the interactions between organisms and their environments. But don't be fooled by the simple definition. Ecology encompasses a very large range of interactions and units of observation and is studied at different levels (FIGURE 14-2). These include:

Ecologists have a challenging task managing natural populations when it's difficult to count the number of individuals present or determine how many the environment can support.

FIGURE 14-1 Managing valuable resources. Population growth models can help.

Ecology is the study of interactions between organisms and their environments. It can be studied at many levels, including:

INDIVIDUALS
Individual organisms

POPULATIONS
Groups of individual organisms that interbreed with each other

COMMUNITIES
Populations of different species that interact with each other within a locale

ECOSYSTEMS
All living organisms, as well as non-living elements, that interact in a particular area

FIGURE 14-2 **From individuals to ecosystems.** Ecologists study living organisms and their relationship to the environment.

- *Individuals:* How do individual organisms respond biochemically, physiologically, and behaviorally to their environments?

- *Populations:* How do groups of interbreeding individuals change over time, in terms of growth rates, distributions, and genetic compositions?

- *Communities:* How do populations of different species in a locale interact?

- *Ecosystems:* How do the living and non-living elements interact in a particular area, such as a forest, desert, or wetland?

We explore ecology at all of these levels throughout this and the next two chapters. We focus in this chapter on **population ecology,** a subfield of ecology that focuses on populations of organisms of a species and how they interact with the environment. We also examine the special case of human population growth and its impact on the environment, as well as the ways in which ecological knowledge can contribute to effective conservation policies.

TAKE-HOME MESSAGE 14·1

Population ecology is the study of the interactions between populations of organisms and their environments, particularly their patterns of growth and how they are influenced by other species and by environmental factors.

14·2 A population perspective is necessary in ecology.

As we begin our study of population ecology, a subtle but critical shift in perspective is required. It is a shift from the individual to the population—a group of organisms of the same species living in a particular geographic region—as the primary focus. Most ecological processes cannot be observed or studied within an individual. Rather, they emerge when considering the entire group of individuals that regularly exchange genes in a particular locale (**FIGURE 14-3**).

As an example of how a population perspective is needed, consider the case of adaptations produced by natural selection. Genetic changes don't occur within an individual. That is, a longer neck did not evolve in one individual giraffe. Instead, the genetic changes occurred within a population of giraffes over time. As a consequence of differential reproductive success among the individuals of a population with different neck lengths, there came to be

Population ecology examines features that cannot be studied on an individual organism, such as population size.

FIGURE 14-3 A change in perspective. Ecology requires a focus on populations rather than on isolated individuals.

more individuals with longer necks over time. Birth rates, death rates, and immigration and emigration rates are also features not of individuals but of populations.

In the next sections, we explore how populations grow, a question important both to the management of consumable natural resources—as in the lobster fisheries described above—and to the conservation of rare and endangered species.

TAKE-HOME MESSAGE 14·2

Most ecological processes cannot be observed or studied within an individual. Rather, we need to consider the entire group of individuals that regularly exchange genes in a particular locale.

14·3 Populations can grow quickly for a while, but not forever.

In a stable population, how many of the five million eggs laid by a female cod over the course of her life survive and grow to adulthood? How many of an elephant's babies survive to adulthood and reproduce? These may seem like difficult questions to answer, requiring complex mathematical models of population growth and special knowledge about cod or elephants. But actually, the most fundamental fact about population growth is ridiculously simple.

How many of the female cod's eggs and the female elephant's babies will, on average, make it to adulthood? Just two.

Q Who leaves more surviving offspring, a pair of elephants or a pair of rabbits?

If a population is not growing, each individual is ultimately replacing itself with a new individual. This is true for all species. Two of those cod eggs will make it. And two of those baby elephants will make it. For a stable population, on average two survive, rather than one, because both a male and a female contributed to each offspring. If each pair produced only one surviving offspring, the population would get smaller and smaller. But what happens to the rest of the cod eggs

and elephant offspring? Along the way from gamete to adult, they may not survive. For example, an egg may not get fertilized. Or a fertilized egg may not survive to give rise to an offspring. Predation and other factors may prevent a youngster's surviving to adulthood. Or an adult may not find a mate. There are too many risks to name them all.

What would happen if more than two cod or two elephants survived? If there were more than one offspring per individual, the population would grow until the earth was covered with cod, elephants, and every other species. But that hasn't happened, and it can't. Let's investigate why.

> **"** There is no exception to the rule that every organic being naturally increases at so high a rate that, if not destroyed, the Earth would soon be covered by the progeny of a single pair. **"**
> — CHARLES DARWIN, *The Origin of Species,* 1859

We start by figuring out how a population grows if each individual does more than just "replace" itself. To calculate

EXPONENTIAL GROWTH

Exponential growth occurs when each individual produces more than the single offspring necessary to replace itself. According to realistic (and moderate) estimates of birth and death rates, a population of just 500 elk would grow to more than a billion individuals within 80 years and eventually would cover the earth.

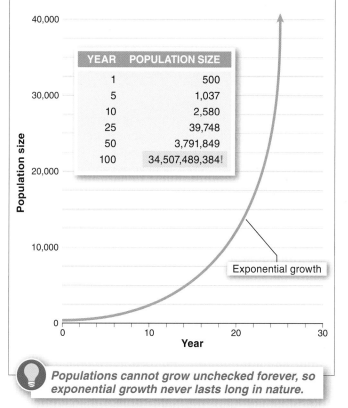

YEAR	POPULATION SIZE
1	500
5	1,037
10	2,580
25	39,748
50	3,791,849
100	34,507,489,384!

Exponential growth

Populations cannot grow unchecked forever, so exponential growth never lasts long in nature.

FIGURE 14-4 **Unchecked growth.** Any animal or plant species that grew exponentially would eventually cover the planet.

the **growth rate** of a population—the change in the number of individuals in the population in some unit of time, such as a year—we need two pieces of information: the number of individuals in the population now, N, and the per capita growth rate, r, which is simply the birth rate minus the death rate. The growth of the population in a year is the growth rate times the number of individuals present to start with: $r \times N$.

If the birth rate is greater than the death rate, the population gets bigger. If the death rate is bigger than the birth rate, the population gets smaller. For example, if there are 500 individuals in a population ($N = 500$) and, over the course of a year, 125 offspring are born, the birth rate is 125/500, or 0.25 births per individual. And if 25 of the 500 individuals die during the same year, the death rate is 25/500, or 0.05 deaths per individual. In this population, the growth rate, r, is 0.25 − 0.05, or 0.20 individuals per individual per year. But how many individuals is that?

Since there were 500 individuals to start with, the population would increase by 0.20×500, or 100 individuals, during a year (the time period observed). After a second year, if it grew at the same rate, the population growth would be 0.20×600, or 120 new individuals. This would mean the population now has 720 individuals.

With the same calculations, we could predict the population size for the next 10, 50, or even 100 years (**FIGURE 14-4**). When a population's size increases at a rate proportional to its current size—in other words, the bigger the population, the faster it grows—the growth is called **exponential growth.** As the J-shaped graph reveals, the size of a population growing exponentially becomes astronomical very quickly. In 10 years, the population of 500 elk would grow to 2,580. In 50 years, it would reach almost four million, and after 80 years it would surpass a billion individuals. It's clear that unchecked exponential population growth ends badly.

But the world isn't overrun with cod or elephants or elk, so, clearly, exponential growth doesn't occur for long. In the next section, we explore why populations cannot keep growing unchecked forever.

TAKE-HOME MESSAGE 14·3

Populations tend to grow exponentially, but this growth is eventually limited.

14·4 A population's growth is limited by its environment.

Life gets harder for individuals when it's crowded. Whether you are an insect, a plant, a small mammal, or a human, difficulties arise in conjunction with increasing competition for limited resources. In particular, as population size increases, organisms experience:

- Reduced food supplies, due to competition

- Diminished access to places to live and breed, also due to competition

- Increased incidence of parasites and disease, which spread more easily when their hosts live at higher density

- Increased predation risk, as predator populations grow in response to the increased availability and visibility of prey

Limitations such as these on a population's growth are a consequence of **population density**—the number of individuals in a given area—and are called **density-dependent factors.** They cause more than discomfort: with increased density, a population's growth is reduced as limited resources slow it down. This ceiling on growth is the **carrying capacity, K,** of the environment. As a population approaches the carrying capacity, death rate increases, emigration rate increases (as individuals seek more hospitable places to live), and a reduction in birth rate usually occurs, as low food supplies result in poor nutrition, which, in turn, reduces fertility (**FIGURE 14-5**).

Here's how the carrying capacity of an environment influences a population's growth. We start with our exponential growth equation:

$$\text{Population growth} = r \times N$$

Then we multiply by a term that can slow down exponential growth:

$$\text{Population growth} = (r \times N) \times [(K - N)/K]$$

When N, the population size, is small relative to the carrying capacity, K, it means there are plenty of resources to go around. In this situation, the term $[(K - N)/K]$ is close to 1. To illustrate this, let's choose a large number for K, say, 100,000, and a small number for N, say, 1,000. This would mean $[(K - N)/K]$ equals $[(100,000 - 1,000)/100,000]$, or 99,000)/100,000, which equals 0.99.

 With increased density, a population's growth can be reduced by limited resources.

FIGURE 14-5 **Fighting over scarce resources.**

So in this situation, we are multiplying normal exponential growth by 0.99. This is almost the same as multiplying it by 1, which doesn't change it. As a consequence, the population is growing exponentially (or very close to exponentially).

On the other hand, if the population size, N, is close to the value of K, it means that the environment is nearly full to capacity, and the term $[(K - N)/K]$ is close to 0. Let's plug in some numbers again. We can use 100,000 again for K. But let's use 99,000 for N. In this case, $[(K - N)/K]$ equals $[(100,000 - 99,000)/100,000]$, or 1,000/100,000, which equals 0.01. In this situation, we multiply normal exponential growth by 0.01, which reduces it to almost zero. In other words, population growth slows more and more as the population size nears the environment's carrying capacity.

When a population grows exponentially at first, but then slows as the population size approaches the carrying capacity, the growth is called **logistic growth** and makes an S-shaped curve (**FIGURE 14-6**).

Density-independent factors can also knock a population down, acting like "bad luck" limits to population growth.

Logistic growth describes population growth that is gradually reduced as the population nears the environment's carrying capacity.

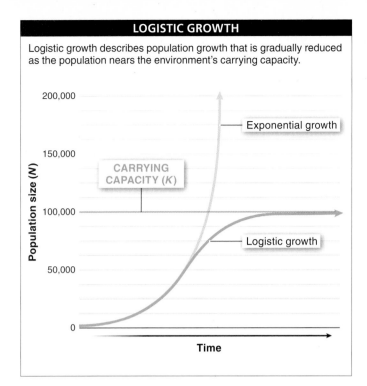

FIGURE 14-6 **Lack of resources limits growth.** Logistic growth is population growth that has stabilized because of limited resources.

These forces strike populations without regard for population size, by increasing the death rate or decreasing the number and rate of offspring produced. They are mostly weather- or geology-based, including calamities such as floods, earthquakes, and fires. They also include habitat destruction by other species, such as humans. The population hit by the disaster may be at the environment's carrying capacity, or it may be in the initial stages of exponential growth, before its growth is slowed by the carrying capacity. In either case, the density-independent force simply knocks down the size. The population then resumes logistic growth.

In an environment where these "bad luck" events repeatedly occur, a population might never have time to reach the carrying capacity. Instead, as the series of jagged curves in **FIGURE 14-7** show, the population might perpetually be in the exponential growth portion of the logistic growth curve, with periodic massive mortality events.

The growth of populations doesn't always appear as a smooth S-shaped logistic growth curve. For some populations, particularly humans, the carrying capacity of an environment is not set in stone. Consider that, in 1883, an acre of farmland in the United States produced an

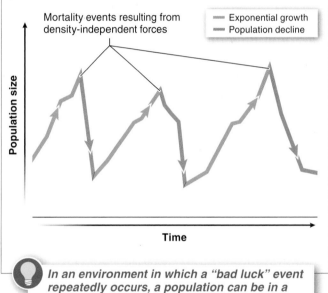

Density-independent forces—including fires, floods, and earthquakes—can dramatically reduce population size.

In an environment in which a "bad luck" event repeatedly occurs, a population can be in a perpetual state of exponential growth with periodic massive mortality events.

FIGURE 14-7 **Events can limit population growth.**

average of 11.5 bushels of corn. By 1933, this was up to 19.5 bushels per acre. By 1992, it had increased to 95 bushels per acre. And by 2013, it had increased to 159 bushels per acre! How did this happen? The development of agricultural technologies—including the use of vigorous hybrid varieties of corn, rich fertilizers, crop rotation, and effective pest management—has allowed farmers to produce more and more food from the same

FIGURE 14-8 **Efficient crop production.** Advances in agriculture make it possible to feed the world's growing population.

amount of land (**FIGURE 14-8**). Of course, over this same time, the carrying capacity has most likely been *decreased* for many other species trying to live in the same environment as humans and their crops.

> ### TAKE-HOME MESSAGE 14·4
>
> A population's growth can be constrained by density-dependent factors: as density increases, a population reaches the carrying capacity of its environment, and limited resources put a ceiling on growth. It can also be reduced by density-independent factors such as natural or human-caused environmental calamities.

Development of agricultural technologies is one example of how carrying capacity can be increased.

14·5 Some populations cycle between large and small.

Nature is not always tidy. Exponential and logistic growth equations, for instance, help us understand the concept of population growth, but real populations don't always show such "textbook" growth patterns. Sometimes they vary greatly in their rates and patterns of growth.

Sometimes populations don't grow logistically, as occasionally happens for desert locusts.

FIGURE 14-9 **Population explosion!**

Locust swarms of biblical proportions certainly attest to the unpredictability of population growth (**FIGURE 14-9**). In northwest Africa, the desert locusts (migrating grasshoppers) normally live as solitary individuals in relatively small, scattered populations. In 2004, however, the population size increased rapidly, probably due to unusually good rains and mild temperatures. As the rainy season ended and green areas gradually shrank, the locusts became progressively concentrated in smaller and smaller areas. At this point, for reasons related to overcrowding but not completely understood, the locusts behaved like a mob, rather than solitary individuals. Giant swarms—some including tens of millions of insects—began flying across huge expanses of land in search of food. They completely consumed the crops on giant swaths of farmland, causing more than $100 million in damage.

There is more than one way for population growth to deviate from the standard pattern. The explosive locust population growth, for example, occurs at unpredictable intervals. Another unusual pattern is the oscillations of the lynx and snowshoe hare populations of Canada. As seen in **FIGURE 14-10** , rather than undergoing smooth logistic growth, the populations of both the snowshoe hare and its predators, the lynx, cycle regularly between increases to very large numbers and crashes to much smaller numbers.

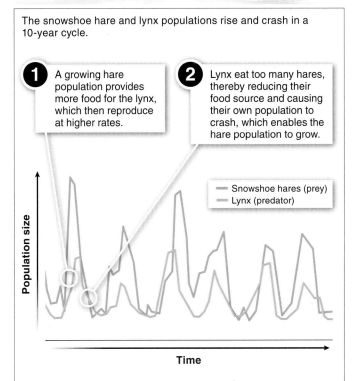

POPULATION OSCILLATIONS

The snowshoe hare and lynx populations rise and crash in a 10-year cycle.

1 A growing hare population provides more food for the lynx, which then reproduce at higher rates.

2 Lynx eat too many hares, thereby reducing their food source and causing their own population to crash, which enables the hare population to grow.

— Snowshoe hares (prey)
— Lynx (predator)

Population size

Time

FIGURE 14-10 Predator and prey. Population numbers—based on the number of pelts sold to the Hudson Bay Company by trappers—reveal cycles.

Thanks to the Hudson's Bay Company, which kept detailed records for decades on the number of pelts it purchased from fur trappers, this population cycling is well documented. Although its cause is not fully certain, to some extent, the predator and the prey cause their own cycling:

1. The hare population size grows,

2. providing more food for the lynx,

3. which then reproduce at a higher rate,

4. causing them to eat great numbers of the hares,

5. thereby reducing the hare population (and the lynx's food source),

6. causing the lynx population to crash,

7. enabling the hare population to grow, and the cycle to begin anew.

In discussing these examples of unusual population growth, we must not lose sight of the fact that, for the most part, regularly cycling populations and populations with periodic outbreaks of huge numbers are more the exception than the rule. In general, the logistic growth pattern describes populations better than any other model.

One population-growth myth that demands debunking is the story of lemmings and their supposed suicides. Do lemmings jump off cliffs when their population becomes too large? In a word: *no.* Lemming populations do occasionally experience large increases in size. Then, just as locust behavior changes when population density gets too great, lemming behavior changes, and many individuals migrate in search of less crowded habitats and more food. As they enter unfamiliar territory, some lemmings may suffer increased rates of death. But these deaths are not suicides, and they do not occur in large groups.

Q Do lemmings commit suicide by jumping off cliffs when their populations get too big?

How did the myth of lemming mass suicides become so prevalent? In the filming of a 1958 Disney documentary, *White Wilderness,* many lemmings were placed on a giant snow-covered turntable so that it would appear they were migrating. In a later scene, lemmings were filmed on a cliff overlooking a river and were then chased into the water as if in a mad rush. None of these scenes were of actual lemming migrations, and in nature, lemmings never behave like this.

There are practical reasons for predicting population sizes and growth rates, as we'll see. But it can be difficult to translate simple growth models into feasible management practices.

TAKE-HOME MESSAGE 14·5

Although the logistic growth pattern is better than any other model for describing the general growth pattern of populations, some populations cycle between periods of rapid growth and rapid shrinkage.

Suppose you worked for a logging company and were responsible for selecting which trees in an eastern hardwood forest to cut down. Your responsibilities would be similar to those you'd have if you were in charge of the lobster fisheries discussed at the beginning of this chapter. In the case of the hardwood forest, how many trees would you cut down for maximum wood yield?

The maximum yield on any given day would be obtained by cutting *everything* down (known as clear-cutting). No other strategy could yield more right now. But such a harvest can be done only once, so it isn't really a management strategy. A more sensible strategy would include harvesting some individuals and leaving others for the future. With that in mind, which trees (or lobsters) would you harvest so as to minimally affect the population's growth? Ideally, you would harvest those individuals that are no longer contributing to the population growth. But there may not be many post-reproductive individuals in the population, or it may be too hard to identify them.

What's the better solution? For long-term management, it is better to harvest some but leave others still growing and reproducing for harvest at a later time, so the population can persist indefinitely. The special case in which as many individuals as possible are removed from the population without impairing its growth is called the **maximum sustainable yield** (**FIGURE 14-11**). The value in such a harvest is that it can be carried out repeatedly, clearly yielding more in the long run than the shortsighted strategy of a complete, one-time harvest.

Your first step as manager is to determine the maximum sustainable yield for the resource, calculated as the point at which the population is growing at its fastest rate. In Figure 14-6, we can see that in logistic growth, the population is getting larger at the fastest rate when its size is equal to half the carrying capacity. At this midpoint, scarcity of mates is not a problem, as it can be at low population levels, nor is competition a problem, as it can be near the carrying capacity—one of the reasons that the population doesn't grow at all when it reaches the environment's carrying capacity.

Maximum sustainable yield is a useful concept, applicable not just to tree or lobster harvesting but also to managing nearly all natural resources. In fact, there are 31 official U.S.

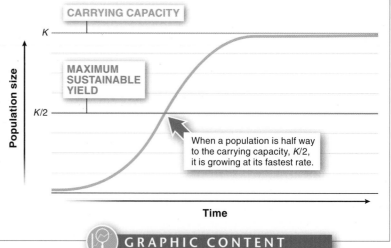

MANAGING NATURAL RESOURCES

Effective and sustainable management of natural resources requires the determination of a population's maximum sustainable yield, the point at which the maximum number of individuals are being added to the population (and so can be harvested or utilized).

CARRYING CAPACITY

K

MAXIMUM SUSTAINABLE YIELD

$K/2$

Population size

When a population is half way to the carrying capacity, $K/2$, it is growing at its fastest rate.

Time

FIGURE 14-11 **Tree harvest.** Wood is a renewable resource, but how many trees can we cut down (and which ones?) without irreversibly depleting the supply?

GRAPHIC CONTENT

Thinking critically about visual displays of data
Turn to p. 603 for a closer inspection of this figure.

government agencies mandated to utilize the maximum sustainable yield concept in determining harvest levels. This is, however, nearly always an impossible task. Why is that so?

Q Almost all natural resource managers working for the U.S. government fail to do their job exactly as mandated. Why?

There are numerous reasons. For starters, if maximum sustainable yield occurs when a population is at half its carrying capacity, do we first wait until the population stabilizes at its carrying capacity to determine what half of that carrying capacity would be? But then isn't it inefficient to sit around waiting for the population to reach its carrying capacity, when we want to maintain it at half that size? Or can we just estimate carrying capacity and then calculate half of it? But, for lobsters, won't that be difficult, given that they live underwater? Put simply, we rarely know the value of *K*.

And the problem gets worse. For one thing, with many species, not only do we not know the carrying capacity, but we don't even know the number of individuals currently living. It is difficult to accurately count humans, so imagine how hard it is to count individuals of species that live underwater, fly, or are microscopic.

And if we were to solve the mysteries of counting individuals and knowing the carrying capacity, we would still have to figure out whether carrying capacity is stable from year to year. For instance, if it depends on levels of resources, it may be cyclic. And even if we can figure out the carrying capacity and the population size, we would not be certain which individuals to harvest. Often, not all individuals in a population are contributing equally to population growth. The post-reproductive individuals, as mentioned above, do not contribute to this growth.

As it turns out, harvesting natural resources for maximum efficiency generates insights into how to fight biological pests, such as cockroaches or termites (**FIGURE 14-12**). The problem is just turned on its head: which individuals in the pest population would you concentrate on killing so as to

The concept of maximum sustainable yield generates insights into fighting biological pests such as cockroaches.

FIGURE 14-12 **Population explosion—pests!** Population-growth models don't just guide us to maximum production of valuable resources. They can help us efficiently reduce pest populations, too.

most effectively slow population growth? Those at the age of maturity, with the highest reproductive value, contribute most to the growth of the population. Similarly, because a population can still grow very rapidly with only a few males—given that any one male can fertilize a large number of females—it is effective to focus pest control on females.

Patterns of population growth and the environmental features that influence them can lead to evolutionary changes in populations, shaping features such as life span, age of reproduction, number of offspring produced, and amount of offspring care—the topics we turn to next.

TAKE-HOME MESSAGE 14·6

Based on models of population growth, it might seem easy to utilize natural resources efficiently and sustainably. In practice, however, difficulties such as estimating population size and carrying capacity complicate the implementation of such strategies.

14·7–14·10
A life history is like
a species summary.

*Closely related to goats and sheep, the aoudad
is native to North Africa.*

14·7 Life histories are shaped by natural selection.

Some animals reproduce with a "big bang." *Antechinus* is an Australian mouse-sized marsupial, and the males are classic big-bang reproducers (**FIGURE 14-13**). At one year of age, they enter a two- to three-week period of intense mating activity, copulating for as long as 12 hours at a time. Shortly after this, the males undergo rapid physical deterioration—they lose weight, much of their fur falls out, their resistance to parasites falls—and then they die. (Although occasionally a female will live for two or three years, most die following the weaning of their first litter.)

Other animals are a bit—but only a bit—more restrained in their reproduction. The house mouse reaches maturity in about one month and produces litters of 6–10 offspring nearly every month, sometimes generating more than a hundred offspring in its first year of life.

And some animals could not be farther from big-bang reproducers. The little brown bat is also mouse-sized, yet does not reach maturity until one year of age and typically produces only a single offspring each year. It can, however, live more than 33 years in its natural habitat.

Why all the variation? And is one strategy better than the others, evolutionarily? One of the most important recurring themes in biology is that evolution nearly always finds more than one way to solve a problem. As the marsupial mouse, house mouse, and little brown bat illustrate, there are many possible responses to the challenge of when to reproduce, how often to reproduce, and how much to reproduce in any given episode. The answers to these questions make up an organism's **life history**—the species' vital statistics, including age at first reproduction, probabilities of survival and reproduction at each age, litter size and frequency, and longevity. Life histories tell us as much about a species as possible in a small amount of information.

Life histories vary from the big-bang reproductive strategy in *Antechinus,* in which the male's **reproductive investment**—all of the material and energetic contribution that an individual will make to its offspring—is made in a single episode, to strategies, such as that of humans, in which organisms have repeated episodes of reproduction. Plants also have life histories, and, like animals, they have a range of strategies. Annuals, such as corn, wheat, rice, peas, marigolds, and cauliflower, usually reproduce once and then die. Perennials, such as apples, raspberries, grapes, ginger, and garlic, on the other hand, reproduce repeatedly, often living for many years.

Antechinus

House mouse

Little brown bat

BIG-BANG REPRODUCTION
• Reaches sexual maturity at one year
• Mates intensely over a three-week period
• Males die shortly after mating period
• Females usually die after weaning their first litter

**FAST, INTENSIVE
REPRODUCTIVE INVESTMENT**
• Reaches sexual maturity at one month
• Produces litters of six to ten offspring every month

**SLOW, GRADUAL
REPRODUCTIVE INVESTMENT**
• Reaches sexual maturity at one year
• Produces about one offspring per year

FIGURE 14-13 **Reproductive strategies.**

Returning to the question posed above—which life history strategy is best?—we need to consider two questions.

1. What is the cost of the reproductive investment during any reproductive episode? Producing offspring is risky in several ways. The act of mating can be risky (by increasing the likelihood of an individual's being eaten by a predator, for example). And the wear and tear of reproduction on an individual's body takes its toll. Thus, the number of offspring an organism produces can only go so high before it becomes detrimental. An organism might produce four offspring in an episode, for example, without significantly increasing its risk of dying from the effort. But if, by chance, it produces five or six, this may take such a toll that the individual is unlikely to live to have another litter. Thus, for many organisms, a smaller litter is favored by natural selection because it enables the individual to have additional litters in the future, maximizing its *lifetime* reproductive success.

2. What is an individual's likelihood of surviving to have future reproductive episodes? If predation rates or other sources of mortality are very high, an individual might not be alive in one month. If so, it makes less sense to defer reproduction: why save for a future unlikely to come?

Q Why do humans defer reproducing so much longer than do cats or mice?

Taken together, the answers to these questions help us understand the variety of life history strategies we see in nature.

Rodents are toward the fast extreme of reproductive strategy. They experience very high mortality rates in the wild, and natural selection has consequently favored early and heavy reproductive investment. Rodents provide relatively little parental care but produce many litters, each with many offspring. Organisms with this type of life history tend to have relatively poor competitive ability but a high rate of population growth; each individual's likelihood of surviving is low, but so many are produced that the odds of two offspring surviving are high.

At the slow extreme, humans have such a low probability of dying in any given year that they can defer reproduction, or at least reproduce in small amounts—once every few years—without much risk, enabling them to have just one offspring at a time but to invest significant parental effort in maximizing its likelihood of surviving. Organisms with this type of life history generally have evolved in such a way that their competitive ability—the likelihood of an individual surviving—is high, although

their rate of population growth is low. That is, not many offspring are produced, but those that are have a high likelihood of surviving (FIGURE 14-14).

TAKE-HOME MESSAGE 14·7

An organism's investment pattern in growth, reproduction, and survival is described by its life history. Very different strategies can achieve the same outcome in which a mating pair of individuals produces at least two surviving offspring.

Humans are at the slow extreme of life histories, waiting two and sometimes several more decades before reproducing. (And they generally have just one offspring at a time.)

14·8 There are trade-offs between growth, reproduction, and longevity.

If you were designing an organism, how would you structure its life history for maximum fitness? (Recall from Chapter 8 that fitness is an organism's reproductive output relative to that of other individuals in the population.) Ideally, you would create an organism that (1) produces many offspring (beginning just after birth and continuing every year), while (2) growing tremendously large, to reduce predation, and (3) living forever.

Unfortunately, evolution operates with some constraints. Some of these traits are just not possible, because selection that changes one feature tends to adversely affect others. When evolution increases one life history characteristic, another characteristic is likely to decrease. Life history trade-offs can be better understood in the context of the three areas to which an organism allocates its resources: growth, reproduction, and survival. When resources are limited, increased allocation to one of these areas tends to reduce

Q With one simple surgical procedure, most men could add more than 10 years to their life span. Why doesn't anyone opt for it?

allocation to one or both of the others. Some examples of the best-studied trade-offs follow.

1. **Reproduction and survival.** Among big-bang reproducers such as the marsupial mouse (*Antechinus*) and salmon, investment in reproduction is exceedingly high, followed shortly by death (FIGURE 14-15). If individuals are physically prevented from reproducing, however, they can live many more years.

2. **Reproduction and growth.** Beech trees grow more slowly during years when they produce many seeds than in years when they produce no seeds. Similarly, female red deer kept from reproducing in their first year of life grow significantly larger during that year.

3. **Number and size of offspring.** The females of one lizard species, *Uta stansburiana,* can lay more eggs if the eggs are smaller, but a higher proportion of the offspring survive if the eggs are larger. Evolution has led to a

REPRODUCTION AND SURVIVAL
Big-bang reproducers such as salmon make a single, exceptionally high investment in reproduction, then die shortly afterward.

FIGURE 14-15 **Life history trade-offs.**

REPRODUCTION AND GROWTH
Beech trees grow much more slowly in the years when they produce many seeds than they do in years when they produce few seeds.

NUMBER AND SIZE OF OFFSPRING
Female lizards of the species *Uta stansburiana* produce medium-size eggs as a compromise between a large number of small eggs (with poor survival of offspring) and few large eggs (with better survival of offspring).

medium-size egg that tends to maximize the number of surviving offspring.

There are many other trade-offs, such as the number of offspring produced and amount of parental care given. Additionally, among mammals, litter size increases with latitude (distance from the equator). This trend may be due to the trade-off between litter size and frequency. Closer to the equator, animals can breed for a longer portion of the year (year-round near the equator). Farther from the equator, the breeding season is shorter. To compensate for

having fewer litters at higher latitudes, mammals produce more offspring per litter.

Can you think of other trade-offs?

TAKE-HOME MESSAGE 14·8

Because constraints limit evolution, life histories are characterized by trade-offs between investments in growth, reproduction, and survival.

? **14·9 THIS IS HOW WE DO IT**

Life history trade-offs: rapid growth comes at a cost.

Sometimes it's tricky to collect experimental evidence supporting a theoretical prediction, even when it seems extremely likely that the prediction is accurate. This is the case for the idea that there must be a trade-off between growth and longevity.

Collecting the appropriate evidence is challenging due to several confounding factors.

1. If you look at the slowest-growing organisms in a population, they may turn out not to have the greatest longevity simply because their slow growth is due to poor nutrition—which tends to reduce longevity.

2. If you look at the fastest-growing organisms, they may have the greatest longevity simply because their faster growth reflects access to better nutrition.

3. And, because growth rate is often positively correlated with adult body size—which is, in turn, positively correlated with longevity—the fastest-growing organisms may end up having the greatest longevity.

| Why is it useful to randomize subjects to experimental treatments?

In each of these cases, the data aren't appropriate for evaluating whether there is a trade-off specifically between growth and longevity. Testing that prediction requires extremely well-controlled experiments.

In a 2013 paper in the *Proceedings of the Royal Society of London,* some researchers reported a clever experimental approach that enabled them to evaluate, in a rigorous way and for the first time, the relationship between growth rate and life span. Here's how they did it.

Working with small (~5 cm long) fish called three-spined sticklebacks (*Gasterosteus aculeatus*), the researchers were able to alter the growth trajectories of the fish by manipulating the temperature of their aquarium water. The sticklebacks are ectothermic, meaning that, unlike us, they don't maintain a constant body temperature. Rather, their body temperature decreases or increases with the temperature of the water in which they are living.

The researchers randomly assigned groups of fish to one of three temperature conditions: normal (6° C), warmer (10° C), or cooler (4° C). They maintained them at this temperature for 4 weeks, designating this as Period 1. All of the groups were then maintained at 6° C (normal temperature) for the remainder of the fishes' lives, designated Period 2. Because the median life span of these sticklebacks is about two years—with many living three years or longer—Period 1 represented less than 4% of the median life span.

| What happened to fish growth when their
| water was warmer or cooler than usual?

The effect of the temperature manipulation was to reduce growth at the cooler temperature and increase growth at the warmer temperature during Period 1. As expected, at the end of this period there were significant differences in the average size of the sticklebacks in the three experimental groups.

The temperature perturbations, however, had another effect: they altered growth rates during Period 2, when all were kept at the normal temperature of their environment. Sticklebacks from the populations maintained at a cooler temperature in Period 1 (which had grown more slowly) experienced more rapid, "catch-up" growth, and those maintained at a warmer temperature in Period 1 exhibited slowed-down growth. As a result of the catch-up or slowed-down growth, after 3–4 months there were no differences among the three groups in the average size of the fish.

| Is there a cost to growing more quickly?
| What must an organism give up in exchange?

As noted above, the researchers maintained all of the sticklebacks until they died, and they monitored their longevity. When they compared the average life spans across the three groups, the results were dramatic.

• Groups that spent Period 2 in catch-up growth had a decreased median life span—14.5% lower than that of the normal-temperature group.

• Groups that spent Period 2 in slowed-down growth had an increased life span—30.6% higher than that of the normal-temperature fish.

| What can we conclude from these results?

The researchers concluded that the results demonstrated the existence of a growth–life span trade-off.

They also reported additional measurements that supported their conclusion and provided evidence that the experiment was very well-controlled. The life span differences were independent of (1) the eventual size attained by the fish, and (2) the reproductive investment made by the fish during Period 2. Also, life span was unrelated to growth rate during the temperature manipulation (Period 1).

In discussing their results, the researchers noted that the sticklebacks grew more quickly or slowly during Period 1 as a consequence of digesting and processing food more quickly or slowly. And that depended solely on the temperature. Why didn't this affect longevity? Because the differences in growth rate were not a consequence of diverting resources away from maintaining their bodies.

During Period 2, however, all the fish were living at the same temperature, so those growing more slowly were able to divert more resources to maintenance or reproduction. Those growing more quickly, on the other hand, could achieve that increased growth only by diverting valuable resources away from maintenance and reproduction.

The experimental design did not allow the researchers to determine the mechanisms for the longevity differences. They speculated, however, that more rapid growth during Period 2 may have led to more cellular damage, resulting from stresses associated with increased metabolic activity.

TAKE-HOME MESSAGE 14·9

Three-spined sticklebacks exposed to relatively cold or warm temperatures early in life have, respectively, reduced or increased growth rates. Returned to normal temperature, they exhibit catch-up or slowed-down growth. Catch-up growth reduces longevity, while slowed-down growth increases longevity, reflecting a trade-off between growth and life span.

Populations can be described quantitatively in life tables and survivorship curves.

Biologists have learned some important lessons from insurance agents. Because insurance companies need to estimate how long an individual is likely to live, they invented something called the **life table** (FIGURE 14-16). It ought to be called a death table, though, because the table tallies the number of people in a population within a certain age group, say 10–20 years old, and the number of individuals within that age range who die. From these numbers, insurers can predict an individual's likelihood of dying or surviving within a particular age interval. For the insurance companies, making money hinges on making accurate life span predictions. For biologists, a life table is a quick window into the lives of the individuals of a population, showing how long they're likely to live, when they'll reproduce, and how many offspring they'll produce.

From life tables, biologists create **survivorship curves,** graphs showing the proportion of individuals of a particular age that are now alive in a population. Survivorship curves indicate an individual's likelihood of surviving through a particular age interval. And they reveal a huge amount of information about a population, such as whether most offspring die shortly after birth—think back to the five million eggs produced by some fish—or whether most survive to adulthood and are likely to live long lives, such as humans in the United States.

Three distinct types of survivorship curves are shown in FIGURE 14-17 . The curves plot the proportion of individuals

surviving at each age, across the entire range of ages seen for that species.

Type I. At the top of Figure 14-17, in blue, is the survivorship curve seen in most human populations, and shared by the giant tortoise. We and the tortoise have a very high probability of surviving every age interval until old age, then the risk of dying increases dramatically. Species with a type I survivorship curve, which includes most large mammals, usually have a few features in common. They have few natural predators, so they are likely to live long lives. They tend to produce only a few offspring—after all, most will survive—and they invest significant time and effort in each offspring.

Type II. In the middle of the figure, in green, is a survivorship curve seen in many bird species, such as the common kingfisher, and in small mammals such as squirrels. The straight line indicates that the proportion alive in each age interval drops at a steady, regular pace. In other words, with a type II survivorship curve, the likelihood of dying in any age interval is the same, whether the bird is between 1 and 2 years old, or between 10 and 11.

Type III. At the bottom, in purple, is a survivorship curve in which most of the deaths occur in the youngest age groups. Common in most plant and insect species, as well as in many marine species such as oysters and fish, the type III curve describes populations in which the few individuals

		LIFE TABLE			
AGE (beginning of interval)	**NUMBER ALIVE** (beginning of interval)	**PROPORTION ALIVE** (beginning of interval)	**DEATHS DURING INTERVAL**	**PROBABILITY OF DYING DURING INTERVAL**	
0	210	1.00	140	0.67	
3	70	0.33	28	0.40	
6	42	0.20	28	0.67	
9	14	0.07	10	0.71	
12	4	0.02	4	1.00	
15	0	0.00	n/a	n/a	

Cactus ground finch

FIGURE 14-16 **Summarizing life and death in a table.** Shown here: a life table for the cactus ground finch on one of the Galápagos Islands. (The finch has one of the three types of "survivorship curves" discussed above.)

In species with type II survivorship curves, individuals experience approximately the same risk of death at any age.

SURVIVORSHIP CURVES

Survivorship curves show the proportion of individuals of a particular age that are alive in a population.

TYPE I

High survivorship until old age, then rapidly decreasing survivorship

Giant tortoise

TYPE II

Survivorship decreases at a steady, regular pace

Common kingfisher

TYPE III

High mortality early in life, but those that survive the early years live long lives

Mackerel

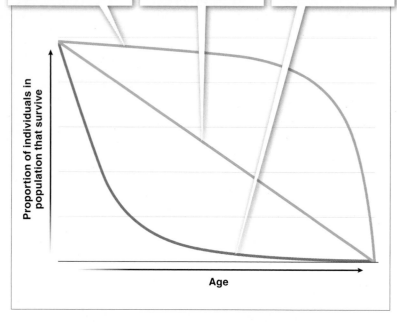

FIGURE 14-17 **Three types of survivorship.**

lucky enough to survive the first few age intervals are likely to live a much longer time. Species with this type of survivorship curve tend to produce very large numbers of offspring, because most will not survive. They also tend not to provide much parental care, if any. A classic example is the mackerel. One female might produce a million eggs! Obviously, most of these (on average, 999,998) do not survive to adulthood, or the planet would be overrun with mackerel.

In reality, most species don't have survivorship curves that are definitively type I, II, or III: they may be anywhere in between. These three types, though, represent the extremes and help us make predictions about reproductive rates and parental investment without extensive observations of individual behavior. Survivorship curves can also change over time or location. Humans in developing countries, for example, tend to have higher mortality rates in all age intervals—particularly in the earliest intervals—relative to individuals in industrialized countries.

TAKE-HOME MESSAGE 14·10

Life tables and survivorship curves summarize the survival and reproduction patterns of the individuals in a population. Species vary greatly in these patterns: the highest risk of mortality may occur among the oldest individuals or among juveniles, or mortality may strike evenly at all ages.

Protected from predators by their quills, North American porcupines are extremely long-lived.

14·11−14·13
Ecology influences the evolution of aging in a population.

14·11 Things fall apart: what is aging and why does it occur?

Picture your grandparents or people in their seventies or eighties. Clearly they have aged. But what exactly does that mean? And does it differ from the way that people in their thirties or forties have aged? Sometimes aging involves sagging skin. Sometimes it involves weakened muscles and bones. Sometimes it involves senility.

The difficulty comes when we realize that no two people age in exactly the same way, yet without fail everyone ages (**FIGURE 14-18**). So we must retreat to a somewhat vague definition. For an individual, aging is the gradual breakdown of the body's machinery. With this definition, we see how each individual can experience aging differently. But aging can be seen much more clearly by examining a population as a whole.

From a population perspective, **aging** emerges as a definitive and measurable feature: it is simply an increased risk of dying with increasing age, after reaching the age of maturity. For example, among humans, people are more likely to die between the ages of 70 and 71 than between 60 and 61. Similarly, they are more likely to die between the ages of 44 and 45 than between 34 and 35. The causes

of death at these ages often differ, but the shuffle toward death becomes more and more dire, with no respite. This is the most useful definition of aging for ecologists, and it doesn't apply only to humans. Aging can be measured and assessed for any species in the same way.

Jeanne Louise Calment lived for more than 122 years, the longest life span ever documented for a human. While this is spectacularly long relative to most organisms on earth (and is significantly longer than the average human life span in the United States, 78 years), it pales in comparison with some of the longest-lived organisms. Bristlecone pine trees are the longest-lived, surviving for thousands of years. The growth rings of one bristlecone pine showed that it was more than 7,000 years old!

Among animals, lobsters and quahogs (a type of clam) are the longest-lived invertebrates (up to 100 and 200 years, respectively). Long-lived vertebrates include striped bass, which can exceed 120 years, and tortoises, which can reach 150 years. Toward the other end of the spectrum, rodents may live just a couple of years, and fruit flies and many other insects generally live and die within the span of a few weeks.

POPULATION ECOLOGY   LIFE HISTORIES   **EVOLUTION OF AGING**   HUMAN POPULATION GROWTH

589

Aging differs from one individual to another, but we recognize individuals of different ages without any difficulty. Shown here: group portraits of the same 16 women, taken 26 years apart. (The inscription in the upper left reads "For the future.")

1973

1999

FIGURE 14-18 **What does "aging" mean?**

Interestingly, bird and bat species live 5–10 times longer than similarly sized rodents. These data prompt the questions: Why is there so much variation? And what controls how long the individuals of a species live? We explore those questions in the next section, but first we consider the more general question of why organisms age at all.

> " Things fall apart; the centre cannot hold;
> Mere anarchy is loosed upon the world,
> The blood-dimmed tide is loosed, and everywhere
> The ceremony of innocence is drowned;
> The best lack all conviction, while the worst
> Are full of passionate intensity. "
> — WILLIAM BUTLER YEATS, *The Second Coming,* 1919

Aging, in fact, is not a mystery. The question of why we age is one of the great solved problems in biology. Oddly, though, it may also be one of the best-kept secrets in science. Let's investigate.

It is easier to *state* the evolutionary explanation for why we age than to *understand* it. But that's where we start. The explanation is based on one fundamental fact: the force of natural selection lessens with advancing age. So what does this mean?

1. Imagine a mutation that causes a person carrying it to die at age 10. Will that person pass the mutation on to many offspring? Of course not. The carrier of that mutant allele will die before she gets a chance to pass it on. Alternative versions of the gene that don't cause death will be the only ones that persist.

2. Now imagine a mutation that causes a person carrying it to die at the age of 150. Will that person pass it on? Yes! The carrier of this fatal allele will already have had children, passing on the mutation to them long before she even knows she carries it. In fact, she will no doubt die long before this mutation has the opportunity to exert its disastrous effect. The same thing happens if the mutation has its negative effect at age 100 . . . or 70 . . . or even 50 (though it may be the cause of death at these "younger old" ages).

Q Many genetic diseases kill old people, but almost none kill children. Why not?

In the first example, we see that natural selection "weeds out" mutant alleles that cause sickness and death early in life. Those alleles never get passed on (**FIGURE 14-19**). No one inherits them, and they disappear from the population without fail. Individuals with the alternative versions of the gene pass on the "good" versions at a higher rate. Consequently, those "good" alleles end up as the only versions of the gene in the population. These are the genes that we inherited from our ancestors, and this is why very few people have genetic diseases that kill them in their teens. We sometimes think of natural selection as only acting positively, to increase or improve a trait—the length of a giraffe's neck, for example. But natural selection can also select *against* a trait—in this case, disease.

Natural selection isn't so effective with some other genes. The second example describes mutations that arise and only later in life make their carriers more likely to die. Examples include mutations that increase the risk of cancers, heart disease, or other ailments. Unfortunately, the carriers of these mutant alleles will already have reproduced and passed

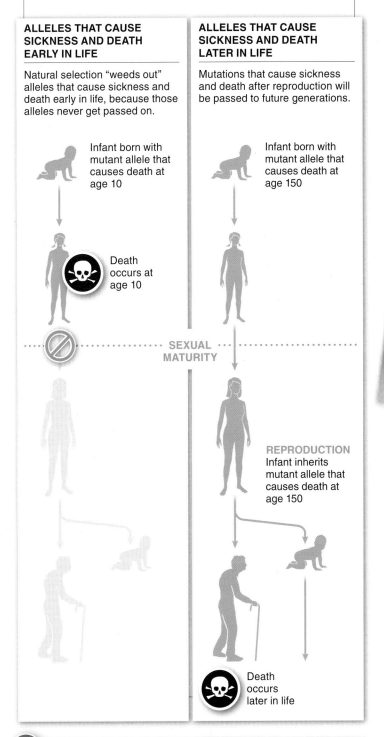

GENETIC DISEASE AND AGING

Natural selection can't weed out genetic defects that wait until we've already reproduced to harm our health. Those mutations that act early in life will be passed on to subsequent generations at a greatly reduced frequency.

ALLELES THAT CAUSE SICKNESS AND DEATH EARLY IN LIFE

Natural selection "weeds out" alleles that cause sickness and death early in life, because those alleles never get passed on.

Infant born with mutant allele that causes death at age 10

Death occurs at age 10

SEXUAL MATURITY

ALLELES THAT CAUSE SICKNESS AND DEATH LATER IN LIFE

Mutations that cause sickness and death after reproduction will be passed to future generations.

Infant born with mutant allele that causes death at age 150

REPRODUCTION
Infant inherits mutant allele that causes death at age 150

Death occurs later in life

💡 *The cumulative effects of the build-up of late-acting harmful alleles are responsible for what we see as aging.*

FIGURE 14-19 **Carrying a time bomb.**

on the alleles before they have had their negative effect. In other words, it doesn't matter whether you carry one of these mutants or an alternative allele that does not harm you, the number of offspring you produce—your **reproductive output**—is the same. Consequently, these mutants are never "cleaned" out of a population. The practical result is that we all have inherited many of these mutant alleles that have arisen over the past hundreds and thousands of generations.

These mutant alleles that have their adverse health effects later in life are responsible for aging. The later in life they have their effect—and therefore the more likely it is that they are passed on to offspring—the more common they will be in the population. This is why there are *many* causes of death, and why it seems as if all of our bodily systems fall apart as we get older.

Q A cure for cancer may be discovered, but not for aging. Why the difference?

TAKE-HOME MESSAGE 14·11

Natural selection cannot weed out harmful alleles that do not diminish an individual's reproductive output. Consequently, these mutant alleles accumulate in the genomes of individuals of nearly all species. This leads to the physiological breakdowns that we experience as we age.

What determines the average longevity in different species?

Can individuals live forever? No. Just because they are not at risk from internal, genetic sources of mortality that doesn't mean there aren't things in the world around them that might kill them—for example, predators. lightning, drought, or disease. And for different species, these risks are more or less serious. Rodents, for example, are at *very* high risk of predation nearly every minute of every day, so they are likely to be dead within a few years (**FIGURE 14-20**).

Because death from some external source is likely to come so early, organisms living in high-risk worlds must reproduce early: natural selection favors this because if they didn't, they wouldn't leave any descendants. Now, when a harmful mutation arises that causes its damage at an early age—say, one or two years of age—the rodent has probably already reproduced and passed the mutation on. And so that harmful allele increases in frequency in the population, causing individuals to age earlier.

Tortoises, by contrast, live in relatively low-risk worlds. Because they have armor-like shells that protect them from danger, death from external sources is low. Early, intensive investment in reproduction is not necessary and so has not been favored by natural selection. In such species, when a harmful mutation arises that has its effect at 5 or 10 years of age, the individual carrying it is likely to die before it has reproduced. Consequently, the mutation isn't passed on; natural selection weeds it out of the population.

The age at time of reproduction, then, is a key factor determining longevity. Factors that favor early reproduction also favor early aging; factors that don't favor early reproduction—or actively favor later reproduction—favor later aging. But then, what determines when an organism reproduces?

We can think of each population as evolving in a world with a specific **hazard factor** (see Figure 14-20). This factor

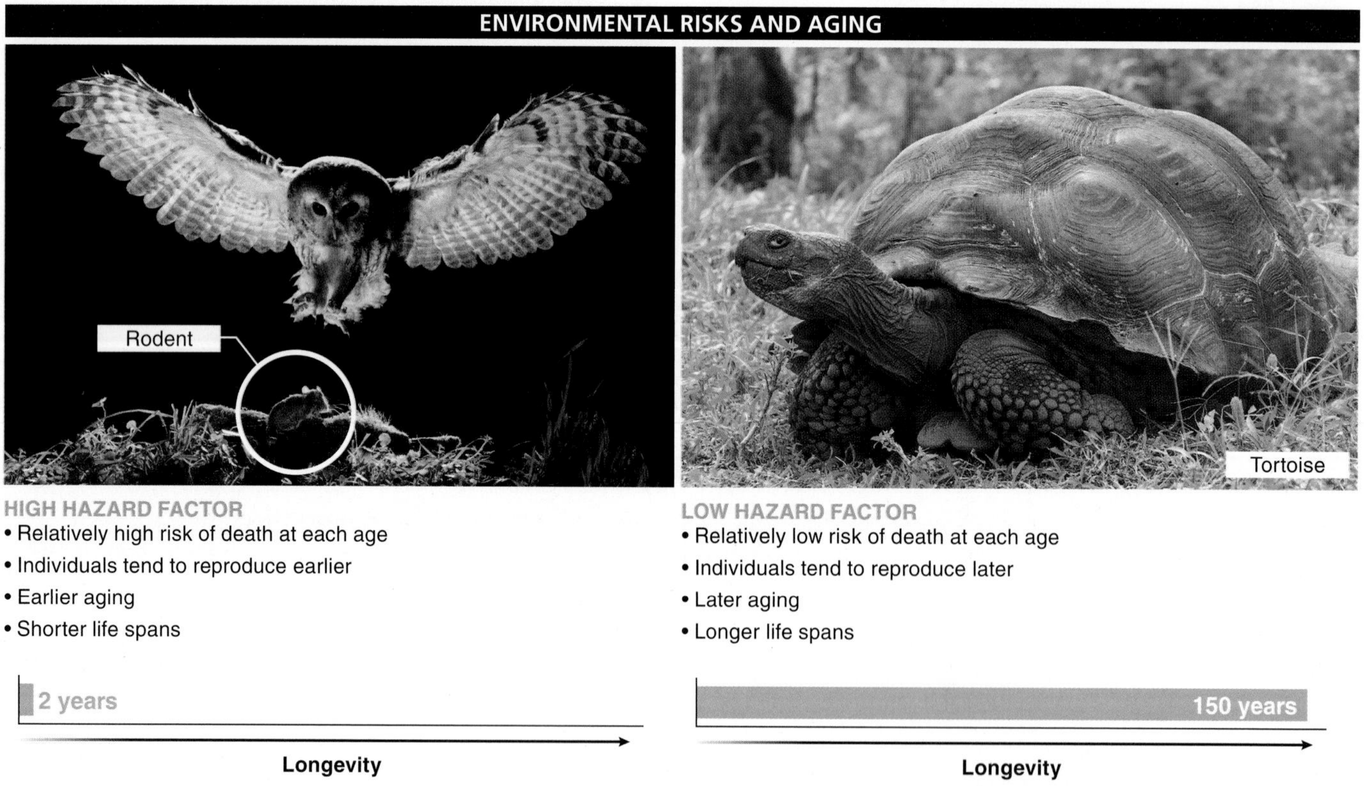

ENVIRONMENTAL RISKS AND AGING

Rodent

Tortoise

HIGH HAZARD FACTOR
• Relatively high risk of death at each age
• Individuals tend to reproduce earlier
• Earlier aging
• Shorter life spans

2 years

Longevity

LOW HAZARD FACTOR
• Relatively low risk of death at each age
• Individuals tend to reproduce later
• Later aging
• Longer life spans

150 years

Longevity

FIGURE 14-20 Hazard factor. The mouse and the tortoise face different risks of dying.

 In environments characterized by low mortality risk, populations of slowly aging individuals with long life spans evolve. In environments characterized by high mortality risk, the opposite occurs.

includes the risk of death from all types of external forces. When the hazard factor is low, as it is for tortoises and humans, individuals of a species tend to reproduce later, and so natural selection can weed out all harmful mutations except those that have their effects late in life. Individuals of these species will live a long time before they succumb to aging. A high hazard factor, on the other hand, will lead to earlier reproduction and therefore early aging and shorter life span.

Which species should age more slowly in captivity, a porcupine or a guinea pig? The difference is large, revealing unambiguously that it's good to be a porcupine. Few predators want to eat you. Or rather, few predators *can* eat you. Because of their sharp quills, porcupines have evolved in a world with little risk of predation and with a lower hazard factor.

Meanwhile, guinea pigs, who have no protective quills, live with a high risk of predation. Because they've been evolving

in a world characterized by a high hazard factor, they reproduce much sooner, age more quickly, and die younger. A guinea pig in captivity lives less than 10 years, whereas a captive porcupine can live 28 years. In the wild the difference is even greater: 15 years for the porcupine and only 3 or 4 years for the guinea pig.

What difference in longevity would you expect between poisonous and non-poisonous snakes, or between bats (which can fly) and mice?

TAKE-HOME MESSAGE 14·12

The rate of aging and pattern of mortality are determined by the hazard factor of the organism's environment. In environments characterized by low mortality risk, populations of slowly aging individuals with long life spans evolve. In environments characterized by high mortality risk, populations of early-aging, short-lived individuals evolve.

14·13   Can we slow down the process of aging?

Life extension is possible. Real, honest-to-goodness doubling of the normal life span. Not only is it possible, researchers have demonstrated it repeatedly . . . in fruit flies.

The results of these life-extension studies are astonishing, but the methods by which they were achieved are simple. Researchers kept several large populations of fruit flies in cages in the laboratory. The flies would feed and lay eggs on a small plate of food. Every few days, a fresh plate would be put in the cage and the old one discarded. At 2 weeks, eggs were collected and used to start the next generation of flies. Propagated in this way, the fly populations were maintained for hundreds of generations. When the longevity of flies in these populations was measured, the researchers found that the flies started to experience the physiological breakdowns associated with aging after a few weeks, and after about a month the flies died.

One researcher thought that changing the force of natural selection on the flies might produce an evolution in their aging pattern. So, instead of collecting eggs at 2 weeks to

Q Is it possible, with our current knowledge, to double longevity in humans?

start the new generation, he cleverly waited longer and longer (**FIGURE 14-21**). The first new generation had parents that had lived for 3 weeks. If any fly carried a mutation that caused it to die between 2 and 3 weeks, that fly did not contribute to the next generation and that mutation was not passed on.

After several generations, eggs were collected at 4 weeks instead of 3 weeks. At this point, all the flies making up the new generation had parents that had survived for 4 weeks. Any flies that died before 4 weeks of age didn't contribute to the new generation. Natural selection was now able to reduce the prevalence of mutant genes responsible for early aging.

The experiment continued, progressively increasing the generation time to 5 weeks, then 6, 7, and ultimately 10 weeks—equivalent to decreasing the hazard factor of the fly population. At this point, only those flies with a genome free of mutations that might cause death in the first 10 weeks of life could contribute to the new generation.

THE EXPERIMENT

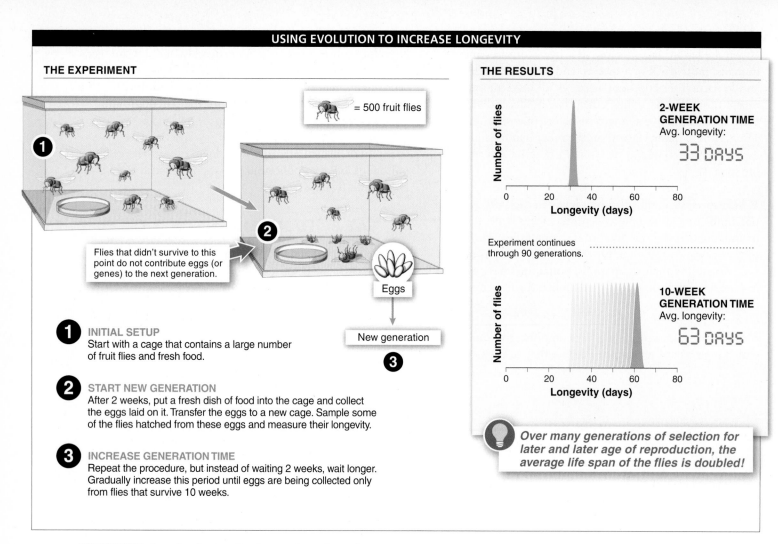

= 500 fruit flies

1

2

Flies that didn't survive to this point do not contribute eggs (or genes) to the next generation.

Eggs

New generation

3

1 INITIAL SETUP
Start with a cage that contains a large number of fruit flies and fresh food.

2 START NEW GENERATION
After 2 weeks, put a fresh dish of food into the cage and collect the eggs laid on it. Transfer the eggs to a new cage. Sample some of the flies hatched from these eggs and measure their longevity.

3 INCREASE GENERATION TIME
Repeat the procedure, but instead of waiting 2 weeks, wait longer. Gradually increase this period until eggs are being collected only from flies that survive 10 weeks.

THE RESULTS

Number of flies

Longevity (days)

2-WEEK GENERATION TIME
Avg. longevity:

33 DAYS

Experiment continues through 90 generations.

Number of flies

Longevity (days)

10-WEEK GENERATION TIME
Avg. longevity:

63 DAYS

Over many generations of selection for later and later age of reproduction, the average life span of the flies is doubled!

FIGURE 14-21 **Creating longer-lived organisms through evolution in the laboratory.**

The net result was the creation of a population of flies with much "cleaner" genomes. When put under ideal conditions outside the cages, these "super" flies did not experience aging until long after the original flies would have died. In fact, the average life span of the flies hatching from eggs harvested from parents of 10 weeks of age was more than 60 days—double that of the original flies!

TAKE-HOME MESSAGE 14·13

By increasing the strength of natural selection later in life, it is possible to increase the mean and maximum longevity of individuals in a population. This occurs in nature (as in porcupines and bats) and has also been achieved under controlled laboratory conditions.

14·14–14·16
The human population is growing rapidly.

High-density human habitations in Hong Kong.

14·14 Age pyramids reveal much about a population.

Q What is the baby boom? Why is it bad news for young people today?

The "baby boom." It happened half a century ago, yet young people in the United States today may end up paying a price for it. What was it? And why does it still matter? Increasing birth rates caused the baby boom. Beginning just after World War II, and continuing through the early 1960s, families in the United States had about 30% more babies per month than they would have had if they had followed their parents' pattern. Then the birth rates began to return to their earlier levels, where they have remained since. As the babies from the boom years grew to school age, schools had to increase in size to accommodate all the children. Then the schools had to downsize once the baby boomers had graduated. Now, as these individuals approach retirement age, the question of how their retirement and health care needs will be met is one of the biggest issues facing American society.

Populations often vary across space—dense in the cities, more sparse in rural areas. The baby boom illustrates that populations have an "age distribution" as well. Just as there may be more individuals in some areas and fewer in others, there may be more individuals in certain age groups and fewer in others.

Describing populations in terms of the proportion of individuals in each age group reveals interesting population features. A population can be divided into the percentages of individuals that are in specific age groupings, called *cohorts*, such as 0–4 years, 5–9 years, 10–14 years, and so on, in an "age pyramid." People have radically different likelihoods of dying or reproducing, for instance, depending on their age. A 10-year-old isn't going to produce any offspring, but a 30-year-old has a relatively high likelihood of reproducing. And a 10-year-old is less likely to die than an 80-year-old.

It can be useful, therefore, for a society to know the relative numbers of 10-, 30-, and 80-year-olds in its population. The age data can be used to determine whether a society would be better off investing in new schools or in fertility wards or in convalescent homes, among other social issues. Companies, too, rely on such demographic data to best

The demographic transition is a pattern of population growth that is experienced as a country industrializes. It is characterized by slow growth, then fast growth, and then slow growth again.

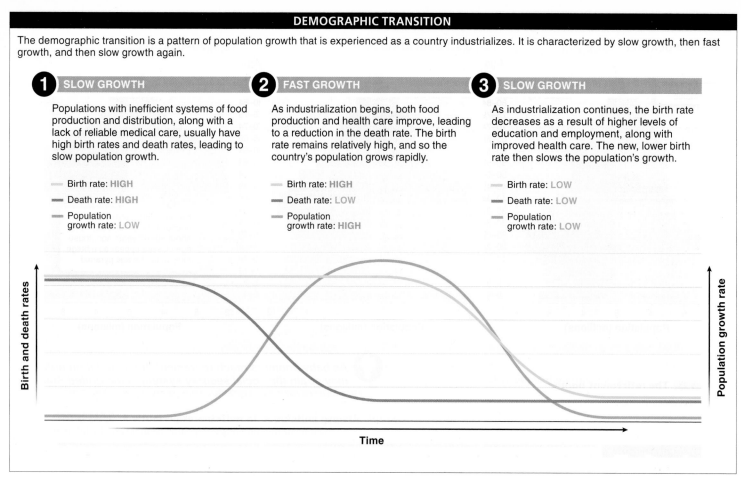

1 SLOW GROWTH

Populations with inefficient systems of food production and distribution, along with a lack of reliable medical care, usually have high birth rates and death rates, leading to slow population growth.

— Birth rate: HIGH
— Death rate: HIGH
— Population growth rate: LOW

2 FAST GROWTH

As industrialization begins, both food production and health care improve, leading to a reduction in the death rate. The birth rate remains relatively high, and so the country's population grows rapidly.

— Birth rate: HIGH
— Death rate: LOW
— Population growth rate: HIGH

3 SLOW GROWTH

As industrialization continues, the birth rate decreases as a result of higher levels of education and employment, along with improved health care. The new, lower birth rate then slows the population's growth.

— Birth rate: LOW
— Death rate: LOW
— Population growth rate: LOW

FIGURE 14-24 **With industrialization, death rates drop and, later, birth rates drop, too.**

Q Population growth is alarmingly slow in Sweden and alarmingly fast in Mexico. Why this difference?

The demographic transition can take decades to complete, so it is not always easy to identify it as it occurs. A survey of countries around the world reveals that many are at different points along the transition. Sweden, for example, has a low fertility rate (1.7 children born per woman) and a low death rate (10.4 deaths per 1,000 people). At the other extreme, Nigeria has a very high fertility rate (5.5 children born per woman) and death rate (17.2 deaths per 1,000). Mexico, in the midst of a clear demographic transition, has a moderately high fertility rate (2.45 children born per woman) yet a very low death rate (4.7 deaths per 1,000).

In the past few decades, the demographic transition has been completed in Japan, Australia, the United States, Canada, and most of Europe, leading to a slowing of population growth. In Mexico, Brazil, Southeast Asia, and most of Africa, on the other hand, the transition

is not complete and population growth is still dangerously fast.

The demographic transition illustrates how health, wealth, and education can lead to a reduction in the birth rate without direct government interventions. But because more than three-quarters of the world's population lives in developing countries and less than a quarter in developed countries, the slowed population growth that generally accompanies the demographic transition is unlikely to be sufficient to keep the world population at a manageable level. Instead, world population growth will continue to rise quickly. What are the potential consequences of such explosive growth? We explore this next.

TAKE-HOME MESSAGE 14·15

The demographic transition tends to occur with the industrialization of countries. It is characterized by an initial reduction in the death rate, followed later by a reduction in the birth rate.

14·14–14·16
The human population is growing rapidly.

High-density human habitations in Hong Kong.

14·14 Age pyramids reveal much about a population.

Q **What is the baby boom? Why is it bad news for young people today?**

The "baby boom." It happened half a century ago, yet young people in the United States today may end up paying a price for it. What was it? And why does it still matter? Increasing birth rates caused the baby boom. Beginning just after World War II, and continuing through the early 1960s, families in the United States had about 30% more babies per month than they would have had if they had followed their parents' pattern. Then the birth rates began to return to their earlier levels, where they have remained since. As the babies from the boom years grew to school age, schools had to increase in size to accommodate all the children. Then the schools had to downsize once the baby boomers had graduated. Now, as these individuals approach retirement age, the question of how their retirement and health care needs will be met is one of the biggest issues facing American society.

Populations often vary across space—dense in the cities, more sparse in rural areas. The baby boom illustrates that

populations have an "age distribution" as well. Just as there may be more individuals in some areas and fewer in others, there may be more individuals in certain age groups and fewer in others.

Describing populations in terms of the proportion of individuals in each age group reveals interesting population features. A population can be divided into the percentages of individuals that are in specific age groupings, called *cohorts,* such as 0–4 years, 5–9 years, 10–14 years, and so on, in an "age pyramid." People have radically different likelihoods of dying or reproducing, for instance, depending on their age. A 10-year-old isn't going to produce any offspring, but a 30-year-old has a relatively high likelihood of reproducing. And a 10-year-old is less likely to die than an 80-year-old.

It can be useful, therefore, for a society to know the relative numbers of 10-, 30-, and 80-year-olds in its population. The age data can be used to determine whether a society would be better off investing in new schools or in fertility wards or in convalescent homes, among other social issues. Companies, too, rely on such demographic data to best

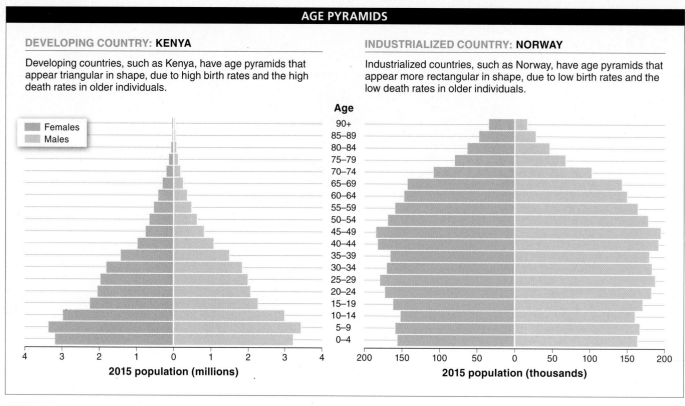

DEVELOPING COUNTRY: KENYA

Developing countries, such as Kenya, have age pyramids that appear triangular in shape, due to high birth rates and the high death rates in older individuals.

Females
Males

2015 population (millions)

INDUSTRIALIZED COUNTRY: NORWAY

Industrialized countries, such as Norway, have age pyramids that appear more rectangular in shape, due to low birth rates and the low death rates in older individuals.

Age

90+
85–89
80–84
75–79
70–74
65–69
60–64
55–59
50–54
45–49
40–44
35–39
30–34
25–29
20–24
15–19
10–14
5–9
0–4

2015 population (thousands)

FIGURE 14-22 **A visual representation of population growth.** Age pyramids go from triangular to rectangular when population growth stabilizes.

plan their strategies for producing the goods that people will want and need.

Around the world, countries vary tremendously in the age pyramids describing their populations. If two populations are the same size but have different age distributions, they will have some very different features. Most industrialized countries are growing slowly or not at all; most of the population is middle-aged or old. In these countries, such as Norway, the age pyramid is more rectangular than pyramid-shaped (**FIGURE 14-22**). Because birth rates are low, the bottom of the pyramid is not very wide. And because death rates are low, too, the higher age classes in the pyramid don't get significantly narrower. Instead, the cohorts remain about the same size all the way into the late sixties and seventies, at which point high mortality rates finally cause them to narrow. This gives the pyramid a more or less rectangular shape.

A more triangular shape is seen in the age pyramids of developing countries. Kenya, for example, has very high birth rates, reflected as a large base in its age pyramid (see Figure 14-22). But high mortality rates, usually due to poor health care, cause a large and continuous reduction in the

proportion of individuals in older age groups. In these countries, most of the population is in the younger age groups.

Notice the "bulge" in the U.S. age pyramid (**FIGURE 14-23**). The shape of the U.S. age pyramid has economists worried that the social security system (including Social Security and Medicare) will not be able to offer older citizens sufficient benefits in the next 10–30 years. Because of the unusually large number of babies born about 50–65 years ago, an unusually large number of people are now reaching retirement age. Since the baby boomers were born, there haven't been any years with such a large cohort. This means that the current numbers of working individuals who contribute to the social security system are not sufficient to cover the payouts promised to the large number of retirees, and the baby boomers will be expensive to support as they reach retirement.

TAKE-HOME MESSAGE 14·14

Age pyramids show the number of individuals in a population within any age group. They allow us to estimate birth and death rates over multi-year periods.

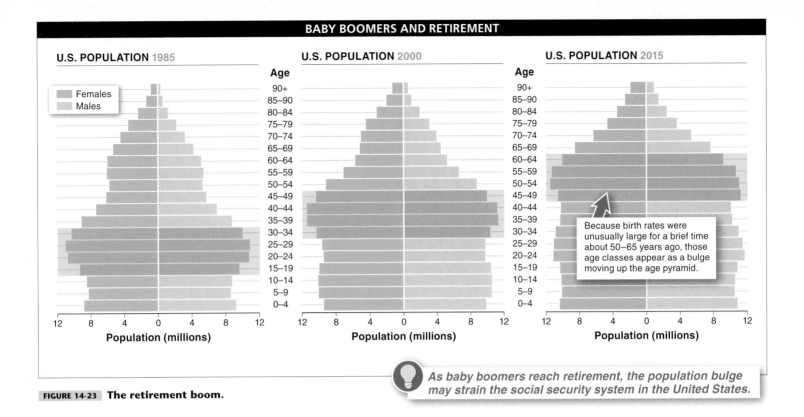

U.S. POPULATION 1985

Age
90+
85–90
80–84
75–79
70–74
65–69
60–64
55–59
50–54
45–49
40–44
35–39
30–34
25–29
20–24
15–19
10–14
5–9
0–4

Females
Males

12   8   4   0   4   8   12
Population (millions)

U.S. POPULATION 2000

Age
90+
85–90
80–84
75–79
70–74
65–69
60–64
55–59
50–54
45–49
40–44
35–39
30–34
25–29
20–24
15–19
10–14
5–9
0–4

12   8   4   0   4   8   12
Population (millions)

U.S. POPULATION 2015

Age
90+
85–90
80–84
75–79
70–74
65–69
60–64
55–59
50–54
45–49
40–44
35–39
30–34
25–29
20–24
15–19
10–14
5–9
0–4

Because birth rates were unusually large for a brief time about 50–65 years ago, those age classes appear as a bulge moving up the age pyramid.

12   8   4   0   4   8   12
Population (millions)

As baby boomers reach retirement, the population bulge may strain the social security system in the United States.

FIGURE 14-23   The retirement boom.

14·15 As less-developed countries become more developed, a demographic transition often occurs.

Populations can sometimes seem as strange and stubborn as individuals. Around the world, governments encourage (or sometimes even coerce) their citizens to reduce their reproductive rates in efforts to check population growth, but they are rarely successful. Yet it seems that when governments stop trying, their countries' population growth rates slow down all on their own.

Many countries have undergone industrialization, with increases in health, wealth, and education. And in the process, their patterns of population growth follow common paths, marked first by periods of faster growth and later by slower growth. The sequence of changes is remarkably consistent.

Start with a country prior to industrialization. Such countries usually have high birth rates and high death rates, resulting from poor and inefficient systems of food production and distribution, along with a lack of reliable medical care. Food production and health care typically improve as industrialization begins. These improvements

inevitably lead to a reduction in the death rate. The birth rate, however, remains relatively high, so the country's population grows rapidly.

As industrialization continues, further changes occur. Most importantly, the standard of living increases. This results from higher levels of education and employment, and, in conjunction with improved health care, this finally causes a reduction in the birth rate. The new, lower birth rate then slows the population's growth. The progression from

1. high birth rates and high death rates (slow population growth) *to*

2. high birth rates and low death rates (fast population growth) *to*

3. low birth rates and low death rates (slow population growth)

is called the **demographic transition** (**FIGURE 14-24**).

POPULATION ECOLOGY   LIFE HISTORIES   EVOLUTION OF AGING   HUMAN POPULATION GROWTH

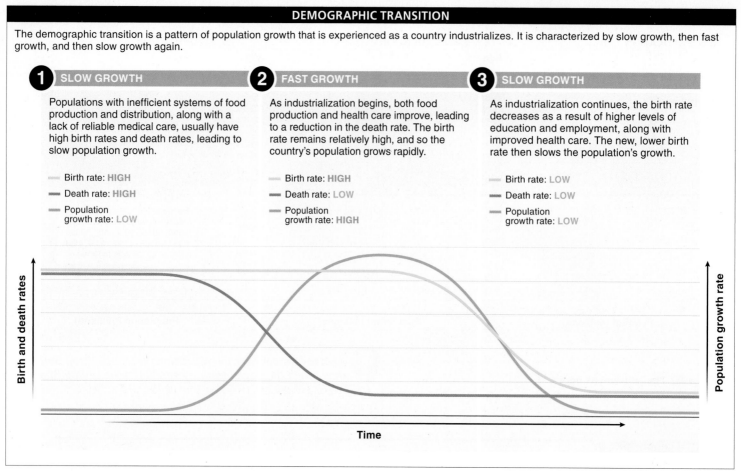

DEMOGRAPHIC TRANSITION

The demographic transition is a pattern of population growth that is experienced as a country industrializes. It is characterized by slow growth, then fast growth, and then slow growth again.

1 SLOW GROWTH

Populations with inefficient systems of food production and distribution, along with a lack of reliable medical care, usually have high birth rates and death rates, leading to slow population growth.

— Birth rate: HIGH
— Death rate: HIGH
— Population growth rate: LOW

2 FAST GROWTH

As industrialization begins, both food production and health care improve, leading to a reduction in the death rate. The birth rate remains relatively high, and so the country's population grows rapidly.

— Birth rate: HIGH
— Death rate: LOW
— Population growth rate: HIGH

3 SLOW GROWTH

As industrialization continues, the birth rate decreases as a result of higher levels of education and employment, along with improved health care. The new, lower birth rate then slows the population's growth.

— Birth rate: LOW
— Death rate: LOW
— Population growth rate: LOW

FIGURE 14-24 With industrialization, death rates drop and, later, birth rates drop, too.

Q Population growth is alarmingly slow in Sweden and alarmingly fast in Mexico. Why this difference?

The demographic transition can take decades to complete, so it is not always easy to identify it as it occurs. A survey of countries around the world reveals that many are at different points along the transition. Sweden, for example, has a low fertility rate (1.7 children born per woman) and a low death rate (10.4 deaths per 1,000 people). At the other extreme, Nigeria has a very high fertility rate (5.5 children born per woman) and death rate (17.2 deaths per 1,000). Mexico, in the midst of a clear demographic transition, has a moderately high fertility rate (2.45 children born per woman) yet a very low death rate (4.7 deaths per 1,000).

In the past few decades, the demographic transition has been completed in Japan, Australia, the United States, Canada, and most of Europe, leading to a slowing of population growth. In Mexico, Brazil, Southeast Asia, and most of Africa, on the other hand, the transition

is not complete and population growth is still dangerously fast.

The demographic transition illustrates how health, wealth, and education can lead to a reduction in the birth rate without direct government interventions. But because more than three-quarters of the world's population lives in developing countries and less than a quarter in developed countries, the slowed population growth that generally accompanies the demographic transition is unlikely to be sufficient to keep the world population at a manageable level. Instead, world population growth will continue to rise quickly. What are the potential consequences of such explosive growth? We explore this next.

TAKE-HOME MESSAGE 14·15

The demographic transition tends to occur with the industrialization of countries. It is characterized by an initial reduction in the death rate, followed later by a reduction in the birth rate.

14·16 Human population growth: how high can it go?

Humans are a phenomenally successful species. There are more than 7 billion people alive today and we add 80 million people to the total each year, because birth rates greatly exceed death rates (**FIGURE 14-25**). But for all of our success, the laws of physics and chemistry still apply. We all need food for energy and space to live. We need other resources, too, and we need the capability of processing and storing all of the waste our societies generate. Because of these limits to perpetual population growth, we may become victims of our own success.

This we know for certain: human population growth cannot continue forever at the current rate. As is the case for every other species, our environment has a carrying capacity beyond which the population cannot be maintained indefinitely. The question is, how high can it go? What is the earth's carrying capacity for humans? This, unfortunately, is a difficult question to answer.

For more than 300 years, biologists have been making estimates, starting when Antonie van Leeuwenhoek (the inventor of the microscope) made an estimate of just over 13 billion people. The median of all the estimates is just

over 10 billion, and the United Nations suggests that the carrying capacity is somewhere between 7 and 11 billion.

There is huge variation from one biologist's estimate to the next. Why is it so hard to figure this out? After all, it's important that we know so that we can work to avoid a global catastrophe. The problem, it seems, goes back to the reasons behind the tremendous human success in the first place. We are so clever that we seem to keep increasing the carrying capacity before we ever bump into it. We do this in a variety of ways, all made possible by various technologies that we invent. In particular, we make advances on three fronts (**FIGURE 14-26**).

1. **Expand into new habitats.** With fire, tools, shelter, and efficient food distribution, we can survive almost anywhere on earth.

2. **Increase the agricultural productivity of the land.** With fertilizers, mechanized agricultural methods, and selection for higher yields, fewer people can now produce much more food than was previously thought possible.

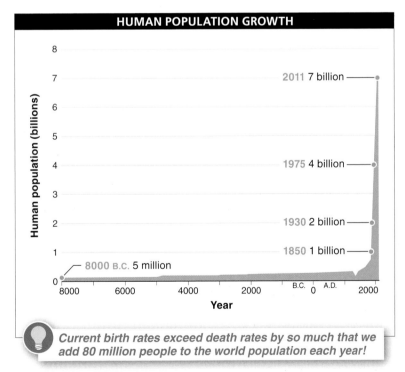

HUMAN POPULATION GROWTH

Current birth rates exceed death rates by so much that we add 80 million people to the world population each year!

FIGURE 14-25 **The world's human population: a slow start, but rapidly picking up steam.**

One reason that the carrying capacity for the human population is difficult to estimate is that we can increase it in a variety of ways.

Dubai, United Arab Emirates

EXPANDING INTO NEW HABITATS
With fire, tools, shelter, and efficient food distribution, we can survive almost anywhere on earth.

Combine harvesting wheat

INCREASING THE AGRICULTURAL PRODUCTIVITY OF THE LAND
With fertilizers, mechanized agricultural methods, and selection for higher yields, fewer people can now produce much more food than was previously thought possible.

Skyscrapers in Hong Kong

FINDING WAYS TO LIVE AT HIGHER DENSITIES
Public health and civil engineering advances make it possible for higher and higher densities of people to live together with minimal problems from waste and infectious diseases.

FIGURE 14-26 **Three ways of increasing the carrying capacity for the human population.**

3. **Circumvent the problems that usually accompany life at higher densities.** Public health and civil engineering advances make it possible for higher and higher densities of people to live together with minimal problems from waste and infectious diseases.

But the question remains: how high can the population go? The most difficult problem in determining the earth's carrying capacity for humans may be assessing just how many resources each person needs. It is possible to estimate the **ecological footprint** of an individual or an entire country by evaluating how much land, food, water, and fuel, among other things, are used. This method reveals that although some countries (e.g., New Zealand, Canada, and Sweden) have more resources available than are required to support the needs of their population, resource use in the world as a whole is significantly greater than the resources available, implying that we are already at our planet's carrying capacity. The populations of many countries—including the United States, Japan, Germany, and England—currently consume an unsustainable level of resources (**FIGURE 14-27**).

Another difficulty in estimating the earth's carrying capacity is that populations can and do occasionally alter their fundamental growth properties. Relatively small changes in the worldwide birth rate can have dramatic effects on the ultimate population size of the planet.

The difficulties in estimating the earth's carrying capacity illuminate an important issue. We must be more specific when we ask how high it can go. We must add: and at what level of comfort, security, and stability, and with what impact on the other species on earth? There are trade-offs.

> ❝There is in every American, I think, something of the old Daniel Boone—who, when he could see the smoke from another chimney, felt himself too crowded and moved further out into the wilderness.❞
> — HUBERT HUMPHREY, 38th U.S. Vice President, 1966

The ecological footprint of the 1.25 billion people currently living in India, for example, is much smaller than that of the populations of Japan, Norway, or Australia. It is possible, we know, to live with much less impact on the

environment per person. But how much do we want to sacrifice to enable a larger number to live stably on the planet? Is our goal to maximize the number? Or to reduce the number and increase the resources available to each person? Or should our goal be something else entirely? The answers to these questions will influence whether the ultimate carrying capacity is even higher than 11 billion, or perhaps lower than the current 7.3 billion people now alive.

The problem gets even more difficult, though, because the world population currently has significant momentum. That is, even if, today, we immediately and permanently reduced our fertility rate to the replacement rate of just two children per couple, the world population would continue to grow for more than 40 years, putting us up to at least

8 billion. This is because there are so many young people alive that, each year, more and more of them will enter the reproductive population. As we get closer to (or exceed) the earth's carrying capacity for humans, we should be better able to recognize it, but by then it may be too late to take preemptive measures—to avoid the resource depletion that may doom us to the much more unpleasant experience of natural population controls.

TAKE-HOME MESSAGE 14·16

The world's human population is currently growing at a very high rate, but limited resources will eventually limit this growth, most likely at a population size between 7 and 11 billion.

StreetBIO   KNOWLEDGE YOU CAN USE

Life History Trade-Offs and a Mini-Fountain of Youth

What Is the Relationship Between Reproduction and Longevity?

It's well documented that there is a trade-off between reproduction and longevity. In a wide variety of animals, decreasing reproductive effort has been shown to increase longevity.

Q: Does this have any practical applications? Yes: spay or neuter your pet to increase its life span!

Q: Does this work? Yes! Data from more than 1,000 cat autopsies showed that non-sterilized females lived a mean of 3.0 years, whereas sterilized females lived significantly longer, with a mean life span of 8.2 years. The results for males were similar. (Perhaps the most dramatic example of the reproduction–longevity trade-off is that of the marsupial mouse, *Antechinus stuartii,* in which castration led to a doubling or even tripling of the usual life span. Similarly, following castration, Pacific salmon lived up to 8 years, double their usual life span.)

Q: Not that you asked . . . In a study in the early 1900s of men institutionalized for mental retardation, those who were castrated lived 13 years longer, on average, than non-castrated men, matched for age and intelligence, at the same institution— 69.3 years vs. 55.7 years.

Q: Does a vasectomy have the same effect as castration? No! The life-extending effect of sterilization occurs only when the ovaries or testes are removed; "tying the tubes" and vasectomy leave the gonads intact. And there is no longevity increase. Why do you think that is the case?

Q: What is responsible for the trade-off between reproduction and longevity? If it were a question of sterilized individuals living longer simply because they don't have to expend energy on reproduction, perhaps similar life span increases could be achieved by simply giving animals access to more energy. In practice, however, this doesn't work (and usually decreases life span). Rather, it seems that a significant part of the "cost" of reproduction, at least in mammals, is an increased incidence of cancer, caused by the higher levels of circulating hormones in "reproductively ready" (that is, fertile and receptive) animals.

Possibilities to think about. The link between maintaining reproductive readiness and reduced longevity has led some researchers to contemplate the design of birth control pills that might have the additional effect of increasing longevity. Stay tuned.

Check Your Knowledge

Thinking critically about visual displays of data

1. What is the purpose of this graph?

2. What factors contribute to the blue line becoming flat?

3. When the population reaches the carrying capacity, it appears to persist indefinitely. How is that possible? Do organisms stop dying?

4. What data are represented in the graph?

5. What could be added to this graph to improve it?

6. Describe two assumptions about population growth that are implicit in this figure.

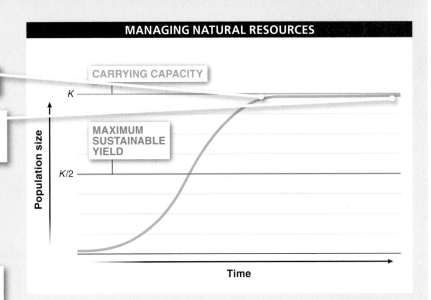

MANAGING NATURAL RESOURCES

CARRYING CAPACITY

K

MAXIMUM SUSTAINABLE YIELD

$K/2$

Population size

Time

7. Create an alternative graph showing the relationship between population size (on the *x*-axis) and growth rate (on the *y*-axis). Label carrying capacity and the point of maximum sustainable yield on your graph.

See answers at the back of the book.

Key Terms in Population Ecology

aging, p. 589
carrying capacity, *K*, p. 576
demographic transition, p. 597
density-dependent factors, p. 576

density-independent factors, p. 576
ecological footprint, p. 600
ecology, p. 572
exponential growth, p. 575
growth rate, p. 575

hazard factor, p. 592
life history, p. 582
life table, p. 587
logistic growth, p. 576
maximum sustainable yield, p. 580

population density, p. 576
population ecology, p. 573
reproductive investment, p. 582
reproductive output, p. 591
survivorship curve, p. 587

ABOUT THE CHAPTER OPENING PHOTO

Because of the limited availability of land for housing, Hong Kong is home to thousands of tall sky scrapers that house its densely packed population.

14·1–14·6 Population ecology is the study of how populations interact with their environments.

Population ecology examines features that cannot be studied on an individual organism, such as population size and growth rates.

WHAT IS ECOLOGY?

Ecology is the study of interactions between organisms and their environments. It can be studied at many levels, including:

INDIVIDUALS
Individual organisms

POPULATIONS
Groups of individual organisms that interbreed with each other

COMMUNITIES
Populations of different species that interact with each other within a locale

ECOSYSTEMS
All living organisms, as well as non-living elements, that interact in a particular area

1. Which type of interaction does population ecology, a subfield of ecology, focus on?

2. Why can't we study ecological processes within an individual?

? Check Your Knowledge

1. Ecology is best defined as the study of:

a) the relationships between all living organisms and their environments.

b) the relationships between parasites and their hosts.

c) aquatic organisms.

d) interactions between predator and prey populations.

e) the preservation of habitats.

0 — 10 — 100
EASY — HARD

2. On average, does a pair of elephants or a pair of rabbits leave more offspring that survive to become adults and reproduce?

a) The pair of elephants, because elephants live much longer and have more breeding seasons.

b) The pair of rabbits, because they have so many more offspring per breeding season than do elephants.

c) The pair of rabbits, because they reach sexual maturity more rapidly.

d) The pair of elephants, because any individual elephant that is born is more likely to survive to become an adult and reproduce.

e) If both populations are stable, the pair of elephants and the pair of rabbits will leave the same number of offspring that survive to become adults and reproduce.

0 — 63 — 100
EASY — HARD

3. In a population exhibiting logistic growth, the rate of population growth is greatest when *N* is:

a) 0.5 *K*.

b) 0.

c) above the carrying capacity.

d) *K*.

e) All of the above are correct; the rate of population growth is constant in logistic growth.

0 — 62 — 100
EASY — HARD

POPULATION GROWTH

Populations tend to grow exponentially until limited resources cause the growth to slow to logistic rates.

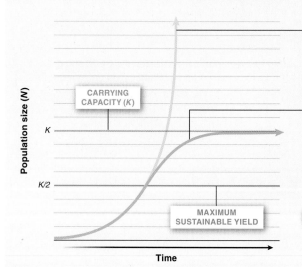

Population size (*N*)

CARRYING CAPACITY (*K*)

K

K/2

MAXIMUM SUSTAINABLE YIELD

Time

EXPONENTIAL GROWTH
Exponential growth describes a population growing at a rate proportional to its size; in other words, the bigger the population the faster it grows. Exponential growth cannot continue indefinitely (population sizes would quickly exceed available resources).

LOGISTIC GROWTH
Logistic growth describes population growth that is gradually reduced as the population nears the environment's carrying capacity.

3. If a mating pair of a particular species produced one surviving offspring, what would you expect to happen to the population size? Why?

LIMITING POPULATION GROWTH

A population's growth can be reduced both by **density-dependent factors** related to crowding, such as food supply, habitat for living and breeding, parasite and disease risk, and predation risk, and by **density-independent factors** such as natural or human-caused environmental calamities.

4. Explain the differences between density-dependent and density-independent factors that can limit a population's growth. Give an example of each.

POPULATION OSCILLATIONS

Although the logistic growth pattern is better than any other model for describing the general growth pattern of populations, some populations cycle between periods of rapid growth and rapid shrinkage.

5. Describe two examples of natural populations whose growth patterns deviate from the more typical logistic growth.

MAXIMUM SUSTAINABLE YIELD

Efficient and sustainable management of natural resources requires the determination of a population's maximum sustainable yield, the point at which the maximum number of individuals are being added to the population (and so can be harvested or utilized).

6. Explain the difference between maximum yield and maximum sustainable yield in a population.

14·7–14·10 A life history is like a species summary.

An organism's pattern of investment in growth, reproduction, and survival is described by its life history.

VARIATION IN LIFE HISTORIES

Strategies for reproducing vary widely across different species. These range from investing in just one (very intensive) episode of reproduction to making numerous smaller investments over a long period of time.

Antechinus

House mouse

Little brown bat

BIG-BANG REPRODUCTION
- Reaches sexual maturity at one year
- Mates intensely over a three-week period
- Males die shortly after mating period
- Females usually die after weaning their first litter

FAST, INTENSIVE REPRODUCTIVE INVESTMENT
- Reaches sexual maturity at one month
- Produces litters of six to ten offspring every month

SLOW, GRADUAL REPRODUCTIVE INVESTMENT
- Reaches sexual maturity at one year
- Produces about one offspring per year

? **7.** What information do you need to know to determine an organism's life history?

LIFE TABLES

Life tables summarize life and death of a species. Shown here: a life table for the cactus ground finch.

AGE (beginning of interval)	NUMBER ALIVE (beginning of interval)	PROPORTION ALIVE (beginning of interval)	DEATHS DURING INTERVAL	PROBABILITY OF DYING DURING INTERVAL
0	210	1.00	140	0.67
3	70	0.33	28	0.40
6	42	0.20	28	0.67
9	14	0.07	10	0.71
12	4	0.02	4	1.00
15	0	0.00	n/a	n/a

Cactus ground finch

The data here reveal an approximately constant probability of dying during each age interval, characteristic of a type II survivorship curve.

SURVIVORSHIP CURVES

Survivorship curves show the proportion of individuals of a particular age that are alive in a population.

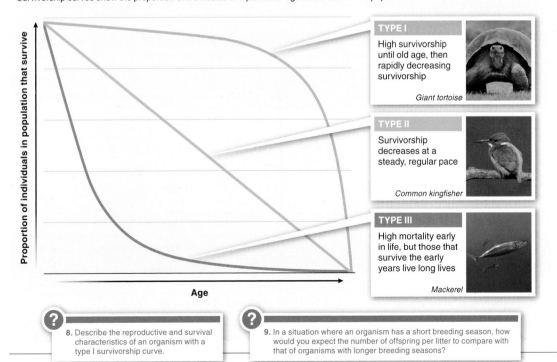

TYPE I
High survivorship until old age, then rapidly decreasing survivorship
Giant tortoise

TYPE II
Survivorship decreases at a steady, regular pace
Common kingfisher

TYPE III
High mortality early in life, but those that survive the early years live long lives
Mackerel

(y-axis) **Proportion of individuals in population that survive**
(x-axis) **Age**

? **8.** Describe the reproductive and survival characteristics of an organism with a type I survivorship curve.

? **9.** In a situation where an organism has a short breeding season, how would you expect the number of offspring per litter to compare with that of organisms with longer breeding seasons?

?

4. In a population, as *N* approaches *K*, the logistic growth equation predicts that:
a) the carrying capacity of the environment will increase.
b) the growth rate will approach zero.
c) the population will become monophyletic.
d) the population size will increase exponentially.
e) the growth rate will not change.

0 —— 46 —— 100
EASY   HARD

5. The number of individuals that can be supported in a given habitat is:
a) the innate capacity for increase.
b) one-half of the maximum sustainable yield.
c) a density-independent effect.
d) generally increasing over time.
e) the carrying capacity.

0 —— 19 —— 100
EASY   HARD

6. Which of the following statements about maximum sustainable yield is incorrect?
a) The maximum sustainable yield for a population is the population's growth rate at *K*/2.
b) The maximum sustainable yield for a population can be difficult to determine because it is not always possible to accurately measure *N*.
c) The maximum sustainable yield for a population can be difficult to determine because it is not always possible to accurately measure *K*.
d) The maximum sustainable yield for a population is a useful management guideline for harvesting plant products such as timber, but is not helpful for managing animal populations.
e) The concept of maximum sustainable yield can generate useful information for fighting the growth of pest species.

0 —— 58 —— 100
EASY   HARD

7. Dr. David Reznick has studied the evolution of life history in guppies that live in streams in Trinidad. Guppies are found in two different types of habitat: sites where predation is high and sites where predation is low. Which of the following life history characteristics would you expect to evolve in a guppy population living in a high predation site?
a) bright colors and courtship displays
b) increased egg number
c) a female-biased sex ratio
d) increased egg size
e) delayed sexual maturation

0 —— 33 —— 100
EASY   HARD

8. The death rate of organisms in a population exhibiting a type III survivorship curve is:

a) unrelated to age.

b) usually correlated with density-independent causes.

c) higher in the post-reproductive than in the pre-reproductive years.

d) lower after the organisms survive beyond the earliest age groups.

e) more or less constant throughout their lives.

0   **49**   100
EASY   HARD

9. Which of the following is a major trade-off in life histories?

a) size of offspring for amount of parental investment

b) size of offspring for number of reproductive events

c) growth for reproduction

d) size for life span

e) number of reproductive events for number of offspring per reproductive event

0   **65**   100
EASY   HARD

10. Natural selection:

a) does not influence aging, because aging is determined by an individual's environment.

b) cannot reduce the frequency of alleles that cause mortality among individuals who have not yet reached maturity.

c) cannot weed out from a population any alleles that do not reduce an individual's relative reproductive success, even if these alleles increase an individual's risk of dying.

d) can influence aging but not longevity.

e) leads to an increase in the frequency of any illness-inducing alleles that have their effect when an organism can reproduce.

0   **37**   100
EASY   HARD

11. Which of the following statements about the hazard factor of a population is incorrect?

a) It is a measure of organisms' risk of death from external sources.

b) It is lower for a population of porcupines than for a population of guinea pigs.

c) It is a measure of the ratio of mortality risk from external (environmental) causes relative to internal (genetic) causes.

d) It is responsible for the rate of aging among individuals in the population.

e) It is a measure of how quickly the individuals in a population age.

0   **55**   100
EASY   HARD

14·11–14·13 Ecology influences the evolution of aging in a population.

From a population perspective, aging is an increased risk of dying with increasing age, after reaching the age of maturity.

NATURAL SELECTION AND AGING

Natural selection cannot "weed out" harmful alleles that do not diminish an individual's reproductive output relative to other individuals in a population. Consequently, those harmful alleles accumulate in the genomes of individuals of nearly all species. This leads to the multiple physiological breakdowns that we see as aging.

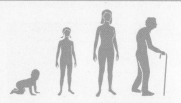

10. Explain how natural selection "weeds out" alleles that cause sickness and death early in life.

ENVIRONMENTAL RISKS AND AGING

The rate of aging and pattern of mortality are determined by the hazard factor of the organism's environment. In environments characterized by low mortality risk, populations of slowly aging individuals with long life spans evolve. In environments characterized by high mortality risk, populations of early-aging, short-lived individuals evolve.

HIGH HAZARD FACTOR
- Relatively high risk of death at each age
- Individuals tend to reproduce earlier
- Earlier aging
- Shorter life spans

Rodent

2 years

Longevity →

LOW HAZARD FACTOR
- Relatively low risk of death at each age
- Individuals tend to reproduce later
- Later aging
- Longer life spans

Tortoise

150 years

Longevity →

11. Why do some species have individuals with very long lives, while others do not? How could the life span of any species be extended?

12. Describe the process by which researchers were able to increase the life span of fruit flies.

14·14–14·16 The human population is growing rapidly.

In humans, current birth rates exceed death rates by so much that we add 80 million people to the world population each year!

AGE PYRAMIDS

Age pyramids show the number of individuals in a population within each age group at one point in time. They give us a snapshot of the age structure of a population and allow us to estimate birth and death rates over multi-year periods.

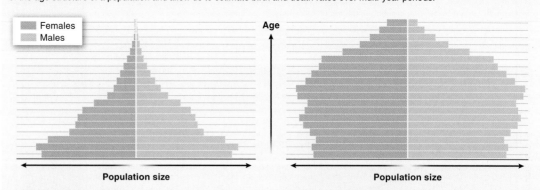

Females
Males

Age

Population size

Population size

DEVELOPING COUNTRY
Developing countries have age pyramids that appear triangular in shape, due to high birth rates and the high death rates in older individuals.

INDUSTRIALIZED COUNTRY
Industrialized countries have age pyramids that appear more rectangular in shape, due to low birth rates and the low death rates in older individuals.

13. Why is it useful for a society to know the relative numbers of 10-, 30-, and 80-year-olds in its population?

DEMOGRAPHIC TRANSITION

The demographic transition is a pattern of population growth that is experienced as a country industrializes. It is characterized by slow growth, then fast growth, and then slow growth again.

 14. What is meant by a "demographic transition"?

1 SLOW GROWTH

Populations with inefficient systems of food production and distribution, along with a lack of reliable medical care, usually have high birth rates and death rates, leading to slow population growth.

- Birth rate: HIGH
- Death rate: HIGH
- Population growth rate: LOW

2 FAST GROWTH

As industrialization begins, both food production and health care improve, leading to a reduction in the death rate. The birth rate remains relatively high, and so the country's population grows rapidly.

- Birth rate: HIGH
- Death rate: LOW
- Population growth rate: HIGH

3 SLOW GROWTH

As industrialization continues, the birth rate decreases as a result of higher levels of education and employment, along with improved health care. The new, lower birth rate then slows the population's growth.

- Birth rate: LOW
- Death rate: LOW
- Population growth rate: LOW

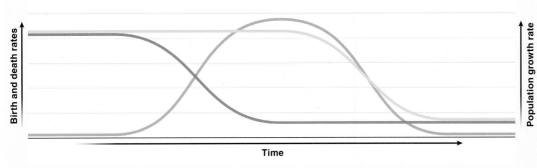

HUMAN POPULATION GROWTH

The world's human population is currently growing at a very high rate, but limited resources will eventually limit this growth, most likely at a population size between 7 and 11 billion.

2011 7 billion

1975 4 billion

1930 2 billion

1850 1 billion

8000 B.C. 5 million

Year

INCREASING THE CARRYING CAPACITY FOR THE HUMAN POPULATION

One reason that the carrying capacity for the human population is difficult to estimate is that we can increase it in a variety of ways.

Dubai, United Arab Emirates

Combine harvesting wheat

Skyscrapers in Hong Kong

EXPANDING INTO NEW HABITATS
With fire, tools, shelter, and efficient food distribution, we can survive almost anywhere on earth.

INCREASING THE AGRICULTURAL PRODUCTIVITY OF THE LAND
With fertilizers, mechanized agricultural methods, and selection for higher yields, fewer people can now produce much more food than was previously thought possible.

FINDING WAYS TO LIVE AT HIGHER DENSITIES
Public health and civil engineering advances make it possible for higher and higher densities of people to live together with minimal problems from waste and infectious diseases.

 15. Why is the ecological footprint of the 1.25 billion people currently living in India much smaller than that of the population of Japan?

---

12. Life extension:

a) is not possible, because natural selection cannot weed out disease-causing alleles that have an effect only at an age when reproduction is no longer possible.

b) can be achieved by selectively breeding those individuals that have the earliest age of maturity.

c) works in insects but could not work in humans.

d) has been achieved using laboratory selection for delayed reproduction.

e) None of the above are true.

0 ——34—— 100
EASY   HARD

13. A population pyramid:

a) represents the number of individuals in various age groups in a population.

b) can be constructed from the data in a life table.

c) directly predicts future age distributions of the population.

d) shows the current birth and death rates of a population.

e) predicts survival and mortality rates for an individual at a given age.

0 ——25—— 100
EASY   HARD

14. A primary difference between the age pyramids of industrialized and developing countries is that:

a) mean longevity is significantly greater in developing countries.

b) in developing countries, much larger proportions of the population are in the youngest age groups.

c) developing countries show a characteristic "bulge" that indicates a baby boom.

d) in developing countries, females live significantly longer than males, whereas in industrialized countries the reverse is true.

e) developing countries have significantly more individuals than industrialized countries.

0 ——20—— 100
EASY   HARD

15. Which statement best describes expectations for the world's human population by the year 2015?

a) It will exceed 10 billion if the current rate of increase continues.

b) Negative growth in the United States and Europe will counterbalance positive growth in the developing countries.

c) It will exceed 7.5 billion if the current rate of increase continues.

d) It will drop below 6 billion if the current rate of decrease continues.

e) The problem is too complex to make any predictions.

0 ——30—— 100
EASY   HARD

15 | Ecosystems and Communities

ORGANISMS AND THEIR ENVIRONMENTS

Ecosystems have living and non-living components.

Interacting physical forces create weather.

Energy and chemicals flow within ecosystems.

Species interactions influence the structure of communities.

Communities can change or remain stable over time.

Flamingoes feeding at Lake Nakuru National Park, Kenya.

15·1 What are ecosystems?

Picture a lush nature scene: some greenery, a bit of rotting wood, and abundant wildlife. Grazing animals abound, while predators feed on other animals and their eggs. Parasites are poised, looking for hosts, and, just below the surface, scavengers find meals among the organic detritus. It would seem to be the quintessential **ecosystem,** a community of biological organisms plus the non-living components with which the organisms interact. But now imagine that the entire scene gets up and walks away! The "camera" in your mind pulls back to reveal that the scene is playing out on the back of a beetle no more than 2 inches (5 cm) long (**FIGURE 15-1**).

The host of this mini-ecosystem is a beetle from New Guinea called the large weevil. The weevil is camouflaged from its predators by lichens—which consist of fungi and photosynthetic algae living together—while the lichens are given a safe surface on which to live. And the garden of lichens supports a wide range of other organisms, from tiny mites to a variety of other microscopic invertebrates, some free-living and others parasitic.

Not all ecosystems are the obvious assemblages of plants and animals that we usually picture—ponds, deserts, or tropical forests. A similar scene can just as easily be found in your large intestine, where several hundred bacterial species flourish. These ecosystems are contained within ecosystems that are contained within ecosystems. The scale can vary tremendously. The closer you look, the more you find.

What is important is that the two essential elements of an ecosystem are present: the biotic environment and the physical (abiotic) environment (**FIGURE 15-2**).

1. The **biotic** environment consists of all the living organisms within an area and is often referred to as a **community.**

💡 *Ecosystems are found not just in obvious places like ponds and tropical rain forests, but also in some unexpected places, like on the back of a beetle!*

FIGURE 15-1 **"Be it ever so humble . . ."** A small-scale ecosystem can exist on the large weevil of New Guinea.

ECOSYSTEMS

All ecosystems share two essential features:

BIOTIC ENVIRONMENT	PHYSICAL (ABIOTIC) ENVIRONMENT
• The living organisms within an area	• The chemical resources and physical conditions within an area
• Often referred to as a community	• Often referred to as the organisms' habitat

FIGURE 15-2 What makes up an ecosystem?

Biologists view communities of organisms and their habitats as "systems" in much the same way engineers might, hence the term eco*system.* Biologists monitor the inputs and outputs of the system, tracing the flow of energy and various molecules as they are captured, transformed, and utilized by organisms and later exit the system or are recycled. They also study how the activities of one species affect the other species in the community—whether the species have a conflicting relationship, such as predator and prey, or a complementary relationship, such as flowering plants and their pollinators. On a small scale, such as the back of a beetle, making some of these measurements can be easy. But there are also some well-studied giant ecosystems, such as the Hubbard Brook Experimental Forest in New Hampshire, which covers 7,600 acres. The same principles apply to the study of ecosystems regardless of size: observe and analyze organisms and their environments, while monitoring everything that goes into and out of the system.

Why should researchers bother with such a methodical—and tedious—analysis of ecosystems? Using the principles of scientific thinking and carrying out experimental tests of hypotheses have led to numerous valuable discoveries, from understanding how the clear-cutting of forests dramatically reduces soil quality to understanding the link between the use of fossil fuels and the creation of destructive acid rain. As we see in this chapter and the next, humans, perhaps more than any other species in history, are significantly affecting most of the ecosystems on earth. And, in addition to improving our understanding of environmental issues, ecosystem studies can also tell us about the microbial ecosystems living inside humans and thus lead to advances in public health.

2. The physical, or **abiotic,** environment, often referred to as the organisms' **habitat,** consists of:

    • the *chemical resources* of the soil, water, and air, such as carbon, nitrogen, and phosphorus, and

    • the *physical conditions,* such as the temperature, salinity (salt level), moisture, humidity, and energy sources.

TAKE-HOME MESSAGE 15·1

An ecosystem is a community of biological organisms plus the non-living components in the environment with which the organisms interact. Ecosystems are found not just in obvious places such as ponds, deserts, and tropical rain forests but also in some unexpected places, such as the digestive tracts of organisms or the shell of a beetle.

15·2 Biomes are large ecosystems that occur around the world, each determined by temperature and rainfall.

Dense vegetation surrounds you. Above you is a canopy of evergreen trees, 30–40 meters (100–130 feet) tall. Climbing vines hang from virtually all the trees. And dozens or even hundreds of species of insects are

flourishing around you. Even if you've never been there, the description of a tropical rain forest is easy to recognize. But where exactly would you be if you were in this scenario? It could be South America or Africa or Southeast Asia. The species are different, but the general pattern of life forms is the same. The same holds for arctic tundra: whether you were in northern Asia or North America, the view would be similar. These are examples of the largest of the earth's ecosystems, the **biomes.**

Biomes cover huge geographic areas of water or land—the deserts that stretch almost all the way across the northern part of Africa, for example. Terrestrial (land) biomes are defined and usually described by the predominant types of plant life in the area. But looking at a map of the world's terrestrial biomes, it is clear that they are mostly determined by the weather. Specifically, when defining terrestrial biomes, we ask four questions about the weather:

1. What is the average temperature?

2. What is the average rainfall (or other precipitation)?

3. Is the temperature relatively constant or does it vary seasonally?

4. Is the rainfall relatively constant or does it vary seasonally?

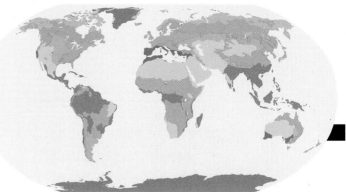

TERRESTRIAL BIOMES

Biomes are large ecosystems that cover huge geographic areas. The nine chief terrestrial biomes are determined by the temperature and amounts of precipitation in conjunction with the magnitude of seasonal variation in these factors.

TROPICAL FOREST

DESERT

SAVANNA

TEMPERATE GRASSLAND

TEMPERATE DECIDUOUS FOREST

CHAPARRAL

CONIFEROUS FOREST

TUNDRA

POLAR ICE

FIGURE 15-3 **Terrestrial ecosystem diversity.**

Aquatic biomes are determined by physical features, including salinity, water movement, and depth.

LAKES AND PONDS

RIVERS AND STREAMS

ESTUARIES AND WETLANDS

OPEN OCEANS

CORAL REEFS

FIGURE 15-4 **Aquatic ecosystem diversity.**

For example, where it is always moist and the temperature does not vary across the seasons, **tropical rain forests** develop. And where it is hot but with strong seasonality that brings a "wet" season and a "dry" season, **savannas** or **tropical seasonal forests** tend to develop. At the other end of the spectrum, in dry areas with a hot season and a cold season, **temperate grasslands** or **deserts** develop. **FIGURE 15-3** shows examples of the nine chief terrestrial biomes; all are determined, in large part, by the precipitation and temperature levels.

Aquatic biomes are defined a bit differently, usually based on physical features such as salinity, water movement, and depth. Chief among these environments are (1) lakes and ponds, with non-flowing fresh water; (2) rivers and streams, with flowing fresh water; (3) **estuaries** and wetlands, where salt water and fresh water mix in a shallow region characterized by exceptionally high productivity; (4) open oceans, with deep salt water; and (5) coral reefs, highly diverse and productive regions in shallow oceans (**FIGURE 15-4**).

If the terrestrial biomes of the world are determined by the temperature and rainfall amounts and seasonality, what determines those features? In other words, what makes the weather? We investigate next how the geography and landscape of the planet—the shape of the earth and its orientation to the sun, and patterns of ocean circulation—cause the specific patterns of weather that create the different climate zones and the biomes characteristic of each. Then we'll see how energy and chemicals are made available for life to flourish in these biomes.

TAKE-HOME MESSAGE 15·2

Biomes are the major ecological communities of earth, characterized mostly by the vegetation present. Different biomes result from differences in temperature and precipitation, and the extent to which both vary from season to season.

15·3–15·5
Interacting physical forces create weather.

Storm clouds gather over a Utah cornfield.

15·3 Global air circulation patterns create deserts and rain forests.

> *There was a hot desert wind blowing that night. It was one of those hot dry Santa Anas that come down through the mountain passes and curl your hair and make your nerves jump and your skin itch. On a night like that every booze party ends in a fight. Meek little wives feel the edge of a carving knife and study their husbands' necks. Anything can happen.*
>
> — RAYMOND CHANDLER, *Red Wind,* 1938

Temperature and rainfall, that's all. The type of terrestrial biome depends on little else besides these two aspects of the weather. But what determines the temperature and rainfall in a particular part of the world? Differences in both ultimately result from one simple fact: the earth is round. As we'll see, the earth's curvature influences temperature, and rainfall patterns are an inevitable consequence of the variation in temperature across the globe.

Wherever you are, begin walking toward the equator. As you get closer, it gets hotter. Nearly everyone is aware of this universal trend (and the reverse as well: it gets colder as you move away from the equator and approach the North or South Pole). What is responsible for the increased warmth at the equator?

The sun shines most directly on the equator (**FIGURE 15-5**). At the equator, solar energy hits the earth and spreads out

over a relatively small area. Away from the equator, the earth curves toward the North and South Poles. Because of this curvature, the same amount of solar energy hitting the earth at the Poles is spread out over a much larger area and also travels a greater distance through the atmosphere,

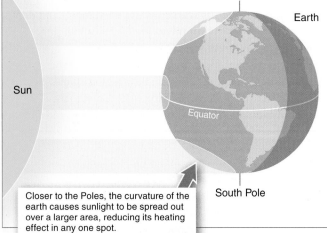

THE DISTRIBUTION OF SOLAR ENERGY

The sun shines most directly on earth's equator, where the solar energy is spread out over a smaller area than the same amount of energy hitting near the Poles. This leads to warmer temperatures at the equator.

North Pole

Earth

Sun

Equator

Closer to the Poles, the curvature of the earth causes sunlight to be spread out over a larger area, reducing its heating effect in any one spot.

South Pole

FIGURE 15-5 **Why is it warmer at the equator than the poles?**

FORMATION OF RAIN

1 AIR IS HEATED AND RISES
When solar heat hits the earth, it warms the air at that point. The heated air rises.

2 RISING AIR COOLS
As hot air rises, getting farther from the warm earth, it cools.

3 COOLING AIR LOSES MOISTURE
Because cold air holds less moisture than warm air, clouds form and rain falls.

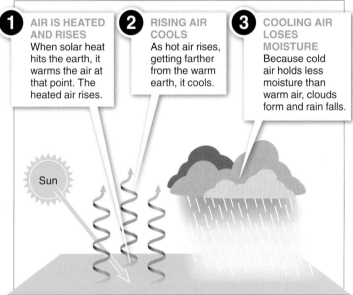

FIGURE 15-6 **Rainmaking.**

The high, cold air in the atmosphere over the equator expands outward, to about 30° north and south. Here, the cold air, which is heavier than warm air, begins to fall downward toward the earth and is warmed by heat radiated off the earth's surface. As the air gets warmer, it can hold more and more moisture, rather than releasing it as rain. In fact, the falling and rapidly heating air can hold so much moisture that it sometimes sucks up moisture from the land itself. For this reason, at about 30° north and south, around the world, there is very little rainfall, the ground is very dry, and deserts form (**FIGURE 15-7**). The Atacama Desert of South America,

Q Nearly all of the world's deserts occur a third of the way from the equator toward the Poles, at 30° latitude. Is this just a coincidence?

which absorbs or reflects much of the heat. With the energy dispersed over a large area, there is less warmth at any one point on the earth's surface. It's similar to the fact that at noon, the sun's rays hit the earth at a more direct angle than they do later in the day. That is why the sun provides less warmth late or early in the day. It is also why the risk of sunburn and skin cancer is greatest around noon and the nearer you are to the equator.

Global patterns of rainfall can be predicted just as easily, by taking into account that warm air holds more moisture than cold air. We'll start at the equator again, where the greatest warming power of the sun hits the earth, and some of that energy radiates back, warming the air. This starts a three-step process: (1) hot air rises; (2) as it rises high into the atmosphere, it cools; and (3) because cool air holds less moisture, as the air rises, clouds form, and the moisture that can no longer be held in the air falls as rain (**FIGURE 15-6**). The equator is hot, but it is also very wet.

THE FORMATION OF DESERTS

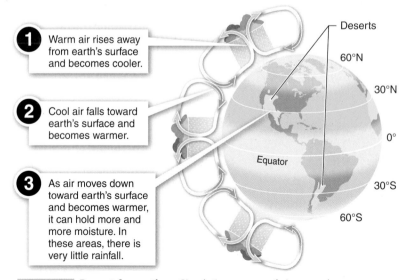

Sand dunes in Namibia

1 Warm air rises away from earth's surface and becomes cooler.

2 Cool air falls toward earth's surface and becomes warmer.

3 As air moves down toward earth's surface and becomes warmer, it can hold more and more moisture. In these areas, there is very little rainfall.

Deserts
60°N
30°N
0°
Equator
30°S
60°S

FIGURE 15-7 **Desert formation.** Circulating masses of air, caused by solar energy hitting different parts of the earth at different angles, determine rainfall patterns throughout the world.

which lies at approximately 30° south of the equator, is an extreme example. In some parts of this large desert, *no rainfall* has ever been recorded. Other great deserts of the world are also at 30° latitude, including the Sahara, Kalahari, Mojave, and Australian deserts (although some deserts do occur at other latitudes). These areas stand in stark contrast to the equator, where it is not uncommon for an area to receive more than 3 meters (almost 120 inches) of rain in a year.

As the air falls near 30° north and south and hits the earth, it spreads equally toward the equator and toward the Poles, gradually getting warmer, until, at around 60° latitude, it begins to rise because of its accumulated heat. Again, the rising air loses its moisture as rainfall. So at 60° latitude, two-thirds of the way toward the Poles from the equator,

it's not particularly warm but there is a great deal of rain. Not surprisingly, at these latitudes lie huge temperate forests with extensive plant growth. And finally, around the Poles, air masses again descend. As they do, because they can hold more moisture, very little rain falls. The Poles are cold, but with little precipitation—they resemble frozen desert.

TAKE-HOME MESSAGE 15·3

Global patterns of weather are largely determined by the earth's round shape. Solar energy hits the equator at a more direct angle than at the Poles, leading to warmer temperatures at lower latitudes. This temperature gradient generates atmospheric circulation patterns that result in heavy rain at the equator and many deserts at 30° latitude.

15·4 Local topography influences the weather.

Why is it so windy on the sidewalk around tall buildings? Why does it rain all the time on one side of some mountain ranges, while deserts form right on the other side? And is it actually warmer in the city than in the country? The answers reflect how the physical features of land, its **topography**—including features created by humans—can have dramatic but predictable effects on temperature, precipitation, and wind.

High altitudes have lower temperatures. With increasing elevation, the air pressure drops, because the weight of the atmosphere decreases as altitude increases. And when pressure is lower, the temperature drops. For each 1,000 meters above sea level, the temperature drops by about 6° C. This is why the changes in weather and vegetation that you see while climbing a mountain are similar to those you would see as you moved

FORMATION OF RAIN SHADOWS

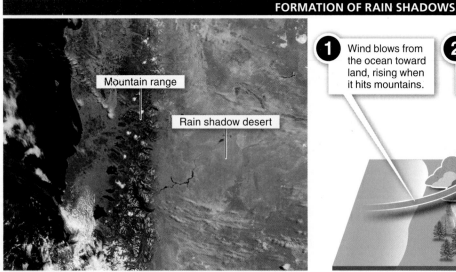

The Andes Mountains, near the west coast of South America, cause a rain shadow effect.

FIGURE 15-8 **The rain shadow effect.**

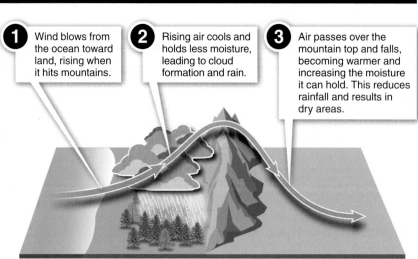

1 Wind blows from the ocean toward land, rising when it hits mountains.

2 Rising air cools and holds less moisture, leading to cloud formation and rain.

3 Air passes over the mountain top and falls, becoming warmer and increasing the moisture it can hold. This reduces rainfall and results in dry areas.

Tall buildings force wind downward.

Asphalt, cement, and building tops absorb heat, raising the temperature.

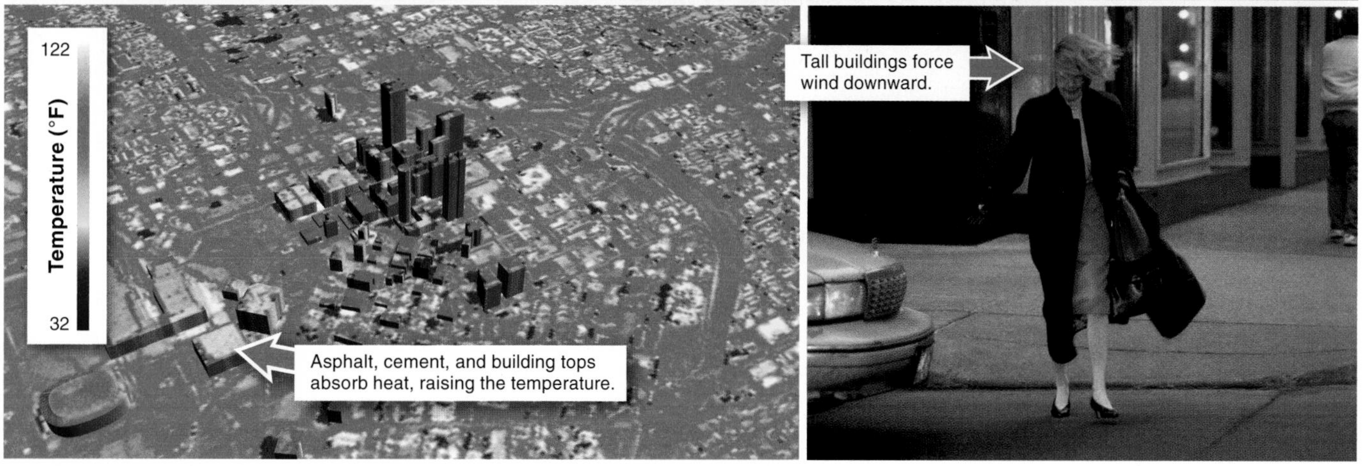

FIGURE 15-9 **Human engineering.** Unintended consequences can occur when humans alter the land, such as changes in temperature and in wind speed and direction.

farther and farther away from the equator—it gets colder in both cases.

Rain shadows create deserts. Air rises to get over the top of a mountain. But because the air cools as it rises, it can't hold much moisture. So clouds form and rain falls. As the air eventually passes over the top of the mountain, it falls and warms. And because warm air holds onto moisture, there is rarely any rain on the back side of the mountain. In fact, often the air will pull moisture from the ground, intensifying the already dry conditions, creating **rain shadow** deserts (**FIGURE 15-8**). Along the west coast of the United States, the Sierra Nevada and the Cascade mountain range are responsible for the Mojave Desert in California and the Great Sandy Desert in Oregon.

Q Is it warmer or cooler in urban areas relative to nearby rural areas?

Asphalt, cement, and tops of buildings absorb heat, raising the temperature. Modern landscapes also influence the weather, creating "urban heat islands." When energy from the sun hits concrete, pavement, or the dark roof of a building, some of it is reflected, heating the air around it, and most of the rest is absorbed by these man-made surfaces and held until night. In the darkness, the surfaces lose heat to the sky, which is colder by comparison, through radiation. In contrast, when sunlight hits trees or other plant life, the solar energy evaporates water in the leaves. The ground surface doesn't get much hotter, and neither does the air. It's not surprising, then, that cities tend to be 1° to 6° C warmer than surrounding rural areas. And not only is it hotter in

cities, but the rising warm air also alters rainfall patterns both in and around cities.

A variety of steps are being taken to create "greener" cities that absorb and release less heat from the sun, and require less energy for air conditioning. These methods include the creation of buildings with lighter-colored rooftops, the planting of trees around buildings and along roads, and the development of rooftop gardens rich with vegetation.

Tall buildings channel wind downward. Tall buildings are responsible for the perpetual winds you feel when walking on a city sidewalk. Here's why. Winds blow more strongly when they are higher above the earth, freed from the earth's frictional drag (from plants, dirt, rocks, and water) that slows wind as it gets near the surface. When these strong, elevated winds suddenly encounter tall buildings, they are deflected. Some of the wind goes up and over the building, but much of it is pushed downward, reaching double or even triple its initial speed by the time it reaches street level (**FIGURE 15-9**).

Q Why is it so windy on streets with tall buildings?

TAKE-HOME MESSAGE 15·4

Local features of topography influence the weather. With higher altitude, the temperature drops. On the windward side of mountains, rainfall is high; on the back side, descending air reduces rainfall, causing rain shadow deserts. Urban development increases the absorption of solar energy, leading to higher temperatures, and creates wind near the bottom of tall buildings.

Ocean currents affect the weather.

Weather is not just affected by circulating air masses. It is also affected by the oceans, which are vast and deep, warmed almost exclusively by the sun. And water is continuously moving and mixing, due to a combination of forces, including wind, the earth's rotation, the gravitational pull of the moon, temperature, and salt concentration. These forces create several large, circular patterns of flow in the world's oceans, as illustrated in **FIGURE 15-10** .

Q Why are beach communities cooler in summer and warmer in winter than inland communities?

Much of water's effect on weather stems from its great heat capacity. For equal volumes of water and air, water can absorb and hold 10,000 times more heat than can air. This means that temperatures fluctuate much more in air than in water. At the beach, the air temperature can go from mild to very hot and back to mild over the course of a day, while the heat of the sun will have a tiny, almost negligible effect on the water temperature. Thus, during summers, much of the heat of the sun is absorbed and held by the ocean water in coastal towns, rather than causing hot air temperatures. Conversely, during cold winter months, heat from the water can reduce the coldness of the air.

One of the strongest ocean currents is the Gulf Stream. It travels north through the Caribbean, carrying warm water up the east coast of the United States and across the Atlantic Ocean toward Europe. Because the current begins in a warm part of the globe, close to the equator, it is still warm when it reaches the east coast of the United States and then Europe. The warm water also warms the climate in these areas. In fact, if it weren't for the Gulf Stream, much of Europe—given its high latitudes—would have a climate more like Canada's. Because all ocean currents in the northern hemisphere rotate in a clockwise direction, water reaching the beaches of California, unlike water reaching east coast beaches, has just come from the north, near Alaska, where it gets very cold.

Every two to seven years, a dramatic climate change driven by ocean currents, called **El Niño,** causes a sustained surface temperature change in the central Pacific Ocean. It is blamed for flooding, droughts, famine, and a variety of other extreme climate disruptions. We can more easily understand El Niño if we contrast it with the more common climate pattern, as illustrated in **FIGURE 15-11** .

These predictable changes during an El Niño event can start a global pattern of unusual weather. Although the mechanisms aren't clear, in El Niño years, more rain falls in

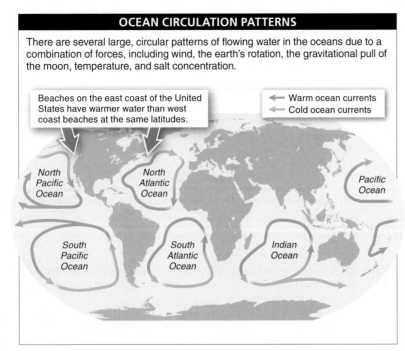

OCEAN CIRCULATION PATTERNS

There are several large, circular patterns of flowing water in the oceans due to a combination of forces, including wind, the earth's rotation, the gravitational pull of the moon, temperature, and salt concentration.

Beaches on the east coast of the United States have warmer water than west coast beaches at the same latitudes.

← Warm ocean currents
← Cold ocean currents

North Pacific Ocean

North Atlantic Ocean

Pacific Ocean

South Pacific Ocean

South Atlantic Ocean

Indian Ocean

FIGURE 15-10 Ocean currents influence water temperatures.

El Niño occurs every two to seven years and is blamed for flooding, droughts, famine, and a variety of other extreme climate disruptions.

USUAL CONDITIONS

1 Steady winds blow westward across the Pacific Ocean from South America toward Southeast Asia.

2 These winds push warm surface water away from the coast of South America, heating the air above it, which causes air to rise and produce tropical rainstorms.

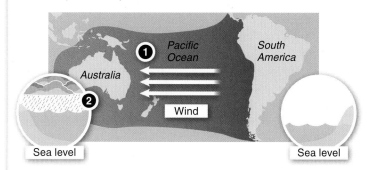

3 Off the coast of South America, colder water upwells from the ocean depths. The cold water cools the air above, causing extremely dry weather.

4 Water from the depths of the ocean brings up rich nutrients, enabling plankton to flourish and feed the huge populations of fish.

FIGURE 15-11 **Domino effect.** A reduction in the usual east-to-west ocean breeze can cause a cascade of disastrous weather.

EL NIÑO CONDITIONS

1 The usual South America–to–Southeast Asia winds ease just a bit.

2 Without the push of the wind, warm water flows back toward South America. The cooled water around Australia and the Philippines cools the air and causes dry weather that can lead to droughts.

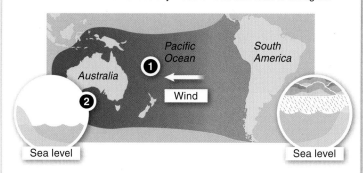

3 Without the warm surface water being blown west, colder water from the ocean depths cannot upwell. The warm surface water warms the air, resulting in rainfall.

4 With no upwelling of cold water and nutrients, plankton levels drop dramatically and fish stocks are wiped out.

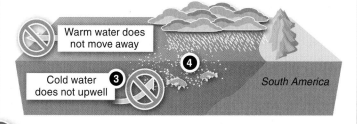

Changes during an El Niño event start a chain reaction of unusual weather around the globe: increased rainfall in the U.S. Midwest, storms in coastal California, and warmer weather in South Africa and Japan.

the Midwestern United States and storms form off the coast of California, while South Africa, Japan, and Canada enjoy warmer weather than usual. The counterpart of the El Niño phenomenon is called **La Niña.** During La Niña periods, ocean surface temperatures are lower than usual, and the climate effects are approximately opposite to El Niño effects.

TAKE-HOME MESSAGE 15·5

Oceans have global circulation patterns. Disruptions in these patterns occur every few years and can cause extreme climate disruptions around the world.

A Parson's chameleon (Calumma parsonii) *eating a grasshopper in Madagascar.*

15·6 Energy flows from producers to consumers.

All life on earth is made possible because energy flows perpetually from the sun to the earth. Looking at an ecosystem such as a desert savanna, for example—trying to understand how all the species, from grasses and trees to birds and mammals and worms, interact with one another, and what role each plays within the system—can seem overwhelming. But if we focus on just one aspect of the ecosystem—the pathways energy takes as it flows through the system—a simple and logical order becomes clear.

The sun is where our pathway of energy flow begins. Most of the energy is absorbed or reflected by the earth's atmosphere or surface, but about 1% of it is intercepted and converted to chemical energy through photosynthesis. That intercepted energy is then transformed again and again by living organisms, making about four stops as it passes through an ecosystem. Let's examine what happens at each of the stops, known as **trophic levels** (FIGURE 15-12).

First stop: producers. When it comes to energy flow, all the species in an ecosystem can be placed in one of two groups: producers and consumers. Plants (along with some algae and bacteria) are the **producers.** They convert the sun's light energy into chemical energy through photosynthesis, as discussed in Chapter 4. We use another word to describe that chemical energy: food. The amount of organic material produced in a biome is called its **primary productivity** level.

Second stop: primary consumers—the herbivores. Cattle grazing in a field, gazelles browsing on herbs,

insects devouring the leaves of a crop plant—these are the **primary consumers** in an ecosystem, the animals that eat plants. Plant material such as cellulose can be difficult to digest. Consequently, most **herbivores,** the animals that eat plants, need a little help in digesting their food. Primary consumers, from termites to cattle, often have bacteria living in their digestive system. These microorganisms benefit the organism in which they live by breaking down the cellulose, enabling the herbivore to harness the energy held in the chemical bonds of the plants' cell walls.

Third stop: secondary consumers—the carnivores. The energy that the herbivore harnesses fuels its growth, reproduction, and movement, but that energy doesn't remain in the herbivore forever. **Carnivores,** such as cats, spiders, and frogs, are animals that feed on herbivores. They are also known as **secondary consumers.** As they eat their prey, some of the energy stored in the chemical bonds of carbohydrate, protein, and lipid molecules is again captured and harnessed for their own movement, reproduction, and growth.

Fourth stop: tertiary consumers—the "top" carnivores. In some ecosystems, energy makes yet another stop: the **tertiary consumers,** or "top carnivores." These are the "animals that eat the animals that eat the animals that eat the plants." They are several steps removed from the initial capture of solar energy by a plant, but the general process is the same. A top carnivore, such as a tiger, eagle, or great white shark, consumes other carnivores, breaking down their tissues and releasing energy stored in the chemical bonds of

Energy from the sun is intercepted and converted into chemical energy, which passes through an ecosystem in about four stops.

Sun

1 PRODUCERS
Plants convert light energy from the sun into food through photosynthesis.

2 PRIMARY CONSUMERS
Herbivores are animals that eat plants.

3 SECONDARY CONSUMERS
Carnivores are animals that eat herbivores.

4 TERTIARY CONSUMERS
Top carnivores are animals that eat other carnivores.

The food chain is a simplified pathway. In actuality, **food webs** are often a better representation, because many organisms can occupy more than one position in the chain.

FIGURE 15-12 **Follow the fuel.**

The food chain pathway from photosynthetic producers through the various levels of animals is a slight oversimplification. In actuality, food chains are better thought of as **food webs,** because many organisms are **omnivores** and can occupy more than one position in the chain (see Figure 15-12). When you eat a simple meal of chicken and vegetables, after all, you're simultaneously a carnivore and a herbivore. On average, about 30% to 35% of the human diet comes from animal products and the remaining 70% to 65% from plant products. Many other animals, from bears to cockroaches, also have diets that involve harvesting energy from multiple stops in the food chain.

In every ecosystem, as energy is transformed through the steps of a food chain, organic material accumulates in the form of animal waste and dead plant and animal matter. **Decomposers,** usually bacteria or fungi, and **detritivores,** including scavengers such as vultures, worms, and a variety of arthropods, break down the organic material, harvesting energy still stored in the chemical bonds (**FIGURE 15-13**). Decomposers are distinguished from detritivores because the decomposers are able to break down a much larger range of organic molecules. But both groups release many important chemical components from the organic material,

the cells. As in each of the previous steps, the top carnivores harness this energy for their own physiological needs.

This path from producers to tertiary consumers is called a **food chain.** We see later in this chapter why a food chain almost never extends to a fifth stop.

When they die, organisms from every level in the food chain provide sustenance for decomposers and detritivores, and important chemical components are recycled through the food chain.

Mold decomposes an orange.

The dung beetle (a detritivore) feeds on decomposing matter.

 Decomposers and detritivores break down organic wastes, releasing chemical components that can then be reused by plants and other producers.

FIGURE 15-13 **Nothing is wasted.**

which can eventually be recycled and used by plants and other producers.

Energy flows from one stop to the next in a food chain, but not in the way that a baton is passed by runners in a relay race. The difference is that at every step in the food chain, much of the usable energy is lost as heat. An animal that eats five pounds of plant material doesn't convert that into five new pounds of body weight. Not by a long shot. In the next section, we'll see how this inefficiency of energy transfers ensures that most food chains are very short.

TAKE-HOME MESSAGE 15·6

Energy from the sun passes through an ecosystem in several steps known as trophic levels: (1) producers convert light energy to chemical energy in photosynthesis; (2) herbivores then consume the producers; (3) the herbivores are consumed by carnivores; and (4) the carnivores may be consumed by top carnivores. Detritivores and decomposers extract energy from organic waste and the remains of dead organisms. At each step in a food chain, some usable energy is lost as heat.

15·7 Energy pyramids reveal the inefficiency of food chains.

Look out of the nearest window. What organisms can you see? Almost without fail, you will see green plant life. Maybe some trees, possibly bushes and grasses as well. You'll have to look longer and harder to see any animals, but you'll probably see a few, most likely small animals and various insects that eat plants. On the other hand, you might stare out of the window all day and not see any animals (other than some fellow humans) that eat other animals. Why? And why are big, fierce animals so rare? Also, why are there so many more plants than animals?

The answers to these questions are closely related to our observation that an animal consuming five pounds of plant material does not gain five pounds in body weight from its meal. The actual amount of growth such a meal can support is far, far less—about 10%—and this is fairly consistent across all levels of the food chain. So the herbivore consuming five pounds of plant material is likely to gain only about half a pound in new growth, while the remaining 90% of the meal is either expended in cellular respiration or lost as feces. Similarly, a carnivore eating the herbivore converts only about 10% of the mass it consumes into its own body mass. Again, 90% is lost to metabolism and feces. Additionally, non-predatory deaths reduce the transfer of energy from one trophic level to the next. And the same inefficiency holds for a top carnivore as well. Let's explore how this 10% rule limits the length of food chains and is responsible for the rarity of big, fierce animals outside your window and across the world

Biomass is the total weight of living or non-living organic material in a given volume, such as a single organism, or, on a larger scale, the weight of all plant and animal matter

in an ecosystem. Given the 10% efficiency with which herbivores convert plant biomass into their own biomass, how much plant biomass is necessary to produce a single 1,200-pound (500 kg) cow? On average, that cow would need to eat about 12,000 pounds (5,000 kg) of grain in order to grow to weigh 1,200 pounds. But that 1,200-pound cow, when eaten by a carnivore, could only add about 120 pounds of biomass to the carnivore, and only 12 pounds to a top carnivore. That's a huge amount of plant biomass required to generate a tiny amount of our top carnivore, which explains why big, fierce animals are so rare (and why vegetarianism is more energetically efficient than meat-eating). Multiply that 5,000 kg of grain by several hundred—or more appropriately, thousands—and you can see that millions of kilograms of grain are required to support only a few top carnivores.

Q Why are big, fierce animal species so rare in the world?

How much plant biomass would be required to support an even higher link on the food chain? Ten times as much—so much that there might not be enough land in the ecosystem to produce enough plant material. And even if there were, the area required would be so large that the "top, top carnivores" might be so spread out and so busy trying to eat enough that they'd be unlikely to encounter each other in order to mate. Hence, the 10% rule limits the length of food chains.

We can illustrate the path of energy through the organisms of an ecosystem with an **energy pyramid,** in which each layer of the pyramid represents the biomass of a trophic

GRAPHIC CONTENT

Thinking critically about visual displays of data
Turn to p. 643 for a closer inspection of this figure.

ENERGY PYRAMID

"The 10% rule": only about 10% of the biomass from each trophic level is converted into biomass in the next trophic level. The rest of the available energy is lost to the environment, a consequence of several factors, including non-predatory deaths, incomplete digestion of prey/food, and respiration.

TERTIARY CONSUMERS

SECONDARY CONSUMERS ————— 10% converted to biomass

PRIMARY CONSUMERS ————— 10% converted to biomass

PRODUCERS ————— 10% converted to biomass

Inefficiencies in the transfer of energy from one trophic level to the next explain why there are so many more plants than animals.

FIGURE 15-14 **The 10% rule.**

VARIATIONS IN PRIMARY PRODUCTIVITY

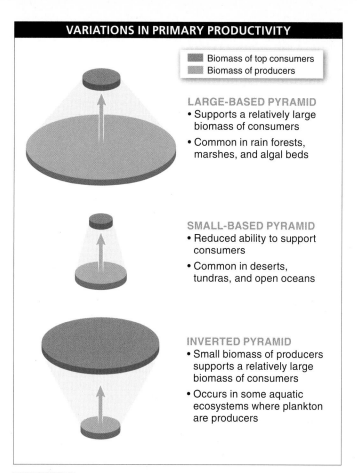

■ Biomass of top consumers
■ Biomass of producers

LARGE-BASED PYRAMID
• Supports a relatively large biomass of consumers
• Common in rain forests, marshes, and algal beds

SMALL-BASED PYRAMID
• Reduced ability to support consumers
• Common in deserts, tundras, and open oceans

INVERTED PYRAMID
• Small biomass of producers supports a relatively large biomass of consumers
• Occurs in some aquatic ecosystems where plankton are producers

FIGURE 15-15 **Relative biomass of producers and consumers.** Across ecosystems, there is huge variation in primary productivity.

level. In **FIGURE 15-14** , we can see that for terrestrial ecosystems, the biomass (in kilograms per square meter) found in photosynthetic organisms, at the base of the pyramid, is reduced significantly at each step, given the incomplete utilization by organisms higher up the food chain. **FIGURE 15-15** illustrates the huge variation in primary productivity across a variety of ecosystems. It is highest in tropical rain forests, marshes, and algal beds, and lowest in deserts, tundra, and the open ocean. In each case, the shapes of the energy and biomass pyramids are similar. With a smaller base, though, the ability of an ecosystem

to support higher levels in the food chain is reduced. One dramatic exception is seen in some aquatic ecosystems where the producers are plankton. Because plankton have such short life spans and rapid reproduction rates, a relatively small biomass can support a large biomass of consumers, giving rise to an inverted pyramid (see the bottom pyramid in Figure 15-15). However, if you quantified the amount of energy available to consumers (rather than measuring biomass), the pyramid would resemble those seen in terrestrial ecosystems.

TAKE-HOME MESSAGE 15·7

Energy pyramids reveal that the biomass of producers in an ecosystem tends to be far greater than the biomass of herbivores. Similarly, the biomass transferred at each successive step in the food chain tends to be only about 10% of the biomass of the organisms consumed. Due to this inefficiency, food chains rarely exceed four levels.

Essential chemicals cycle through ecosystems.

What is necessary for life? Energy and some essential chemicals top the list. New energy continually comes to earth from the sun, fulfilling the first need. And everything else is already here. The chemicals cycle around and around, using the same pathway taken by energy—the food chain. Plants and other producers generally take up the molecules from the atmosphere or the soil. Then as animals consume the plants or other animals, the chemicals move up the food chain. As the plants and animals die, detritivores and decomposers return the chemicals to the abiotic environment. From a chemical perspective, life is just a continuous recycling of molecules.

Chemicals cycle through the living and non-living components of an ecosystem. Each chemical is stored in a non-living part of the environment called a "reservoir." Organisms acquire the chemical from the reservoir, the chemical cycles through the food chain, and eventually it is returned to the reservoir.

We can get a deeper appreciation of the functioning of ecosystems and the ecological problems that can occur when they are disturbed—particularly by humans—by investigating a few of these cycles in more detail. Here

we explore three of the most important chemical cycles: carbon, nitrogen, and phosphorus.

Carbon Carbon is found largely in four compartments on earth: the oceans, the atmosphere, terrestrial organisms, and fossil deposits. Plants and other photosynthetic organisms obtain most of their carbon from the atmosphere, where carbon is in the form of carbon dioxide (CO_2). As we saw in Chapter 4, photosynthetic organisms use carbon dioxide in photosynthesis, separating the carbon molecules from CO_2 and using them to build sugars and other macromolecules. Carbon then moves through the food chain as organisms eat plants and are themselves eaten (**FIGURE 15-16**). The oceans contain most of the earth's carbon. Here, many organisms use dissolved carbon to build shells (which later dissolve back into the water, after the organism dies).

Most carbon returns to its reservoir as a consequence of organisms' metabolic processes. Organisms extract energy from food by breaking carbon-carbon bonds, releasing the energy stored in the bonds, and combining the released carbon atoms with oxygen. They then exhale the end product as CO_2.

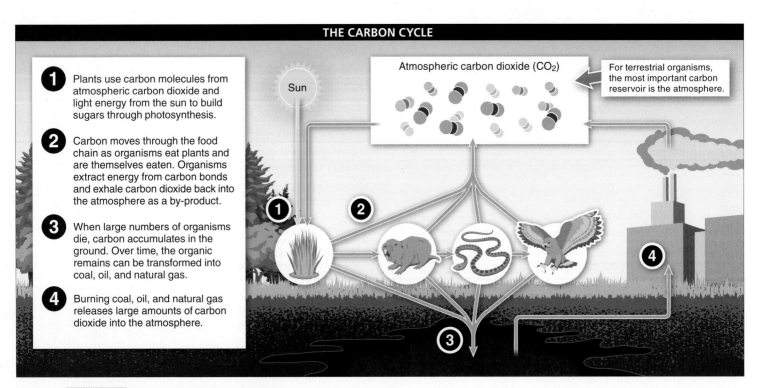

THE CARBON CYCLE

1 Plants use carbon molecules from atmospheric carbon dioxide and light energy from the sun to build sugars through photosynthesis.

2 Carbon moves through the food chain as organisms eat plants and are themselves eaten. Organisms extract energy from carbon bonds and exhale carbon dioxide back into the atmosphere as a by-product.

3 When large numbers of organisms die, carbon accumulates in the ground. Over time, the organic remains can be transformed into coal, oil, and natural gas.

4 Burning coal, oil, and natural gas releases large amounts of carbon dioxide into the atmosphere.

Atmospheric carbon dioxide (CO_2)

Sun

For terrestrial organisms, the most important carbon reservoir is the atmosphere.

FIGURE 15-16 Element cycling: carbon.

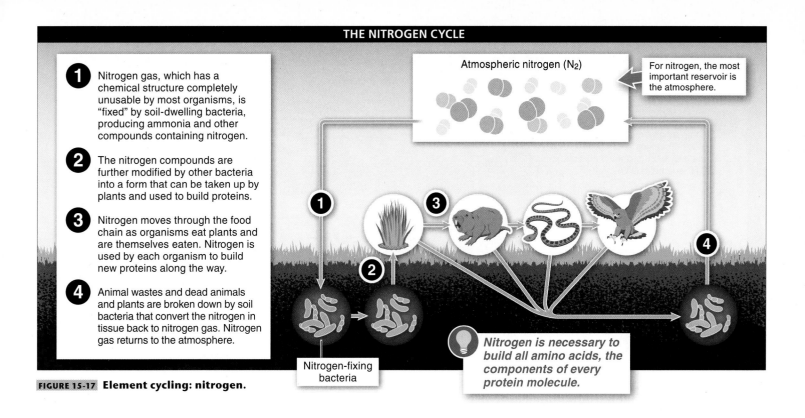

1 Nitrogen gas, which has a chemical structure completely unusable by most organisms, is "fixed" by soil-dwelling bacteria, producing ammonia and other compounds containing nitrogen.

2 The nitrogen compounds are further modified by other bacteria into a form that can be taken up by plants and used to build proteins.

3 Nitrogen moves through the food chain as organisms eat plants and are themselves eaten. Nitrogen is used by each organism to build new proteins along the way.

4 Animal wastes and dead animals and plants are broken down by soil bacteria that convert the nitrogen in tissue back to nitrogen gas. Nitrogen gas returns to the atmosphere.

Atmospheric nitrogen (N_2)

For nitrogen, the most important reservoir is the atmosphere.

Nitrogen-fixing bacteria

Nitrogen is necessary to build all amino acids, the components of every protein molecule.

FIGURE 15-17 **Element cycling: nitrogen.**

Q Why are global CO_2 levels rising?

The burning of fossil fuels is adding significantly to the atmospheric carbon reservoir. Fossil fuels are created when large numbers of organisms die and are buried in sediment lacking oxygen. In the absence of oxygen, at high pressures, and after very long periods of time, the organic remains are transformed into coal, oil, and natural gas. Trapped underground or in rock, these sources of carbon played little role in the global carbon cycle until humans in industrialized countries began using fossil fuels to power their technologies. Burning coal, oil, and natural gas releases large amounts of carbon dioxide, thus increasing the average CO_2 concentration in the atmosphere—the current level of CO_2 in the atmosphere is the highest it has been in almost half a million years. This has potentially disastrous implications, as we'll see in Chapter 16.

Q Global CO_2 levels are rising overall, but they also exhibit a sharp rise and fall over the course of each year. Why?

Although, on average, the level of CO_2 in the environment is increasing, there is a yearly cycle of ups and downs in the CO_2 levels in the northern hemisphere. This is due to fluctuations in the ability of plants to absorb CO_2. Many trees lose their leaves each fall and, during the winter months,

relatively low rates of photosynthesis lead to low rates of CO_2 consumption, causing an annual peak in atmospheric CO_2 levels. During the summer, trees have their leaves, sunlight is strong, and photosynthesis (consuming CO_2) occurs at much higher levels, causing a drop in the atmospheric CO_2 level.

Nitrogen Nitrogen is necessary to build a variety of molecules critical to life, including all amino acids, the components of every protein molecule. The chief reservoir of nitrogen is the atmosphere, but even though more than 78% of the atmosphere is nitrogen gas (N_2), for most organisms this nitrogen is completely unusable. The problem is that atmospheric nitrogen consists of two nitrogen atoms bonded tightly together, and these bonds need to be broken to make the nitrogen usable for living organisms. Only through the metabolic magic (chemistry, actually) of some soil-dwelling bacteria, the nitrogen-fixers, can most nitrogen enter the food chain (**FIGURE 15-17**).

These bacteria chemically convert, or "fix," nitrogen by attaching it to other atoms, including hydrogen, to produce ammonia and related compounds. These compounds are then further modified by other bacteria into a form that can be taken up by plants and used to build proteins. And once nitrogen is in plant tissues, animals acquire it in the same way they acquire carbon: by eating the plants. Nitrogen returns to the atmosphere when animal

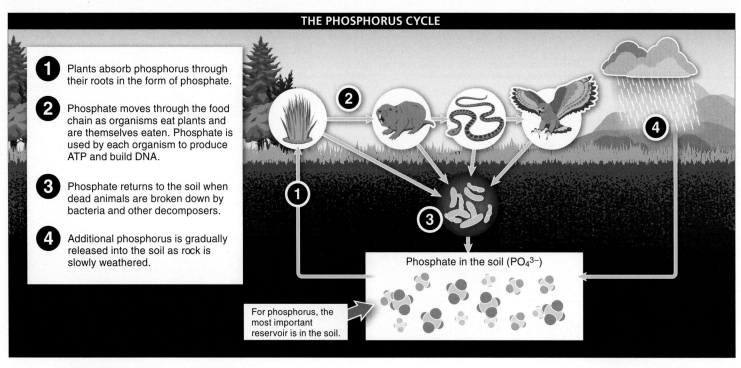

1 Plants absorb phosphorus through their roots in the form of phosphate.

2 Phosphate moves through the food chain as organisms eat plants and are themselves eaten. Phosphate is used by each organism to produce ATP and build DNA.

3 Phosphate returns to the soil when dead animals are broken down by bacteria and other decomposers.

4 Additional phosphorus is gradually released into the soil as rock is slowly weathered.

Phosphate in the soil (PO_4^{3-})

For phosphorus, the most important reservoir is in the soil.

FIGURE 15-18 **Element cycling: phosphorus.**

wastes and dead animals are broken down by soil bacteria (decomposers) that convert the nitrogen compounds in tissues back to nitrogen gas.

The availability of nitrogen is like a bottleneck limiting plant growth. Because it is necessary for the production of every plant protein, and because all nitrogen must first be made usable by bacteria, plant growth is often limited by nitrogen levels in the soil. For this reason, most fertilizers contain nitrogen in a form usable by plants.

Phosphorus Every molecule of ATP and DNA requires phosphorus. But no (or barely any) phosphorus is available in the atmosphere. Instead, soil serves as the chief reservoir. Like nitrogen, phosphorus is chemically converted— "fixed"—into a form usable by plants (phosphate) and then absorbed by their roots. It cycles through the food chain as herbivores consume plants and carnivores consume herbivores. As organisms die, their remains are broken down by bacteria and other organisms, returning the phosphorus to the soil. The pool of phosphorus in the soil is also influenced by the much slower process of formation of rock on the seafloor, its uplifting into mountains, and its eventual weathering, releasing its phosphorus (**FIGURE 15-18**).

Like nitrogen, phosphorus is often a limiting resource in soils, constraining plant growth. Consequently, fertilizers usually contain large amounts of phosphorus. This is beneficial in the short run, but it can have some disastrous unintended consequences. As more and more phosphorus (and nitrogen) is added to soil, some of it runs off with water and ends up in lakes, ponds, and rivers. It also acts as a fertilizer in these habitats, making spectacular growth possible for algae. But eventually the algae die and sink, creating a huge source of food for bacteria. The bacterial population increases and may use up too much of the dissolved oxygen, causing fish, insects, and many other organisms to suffocate and die.

«An atom of phosphorus 'X' had marked time in the limestone ledge since the Paleozoic seas covered the land. *Time, to an atom locked in a rock, does not pass.* The break came when a bur-oak root nosed down a crack and began prying and sucking. In the flash of a century the rock decayed, and X was pulled out and up into the world of living things. He helped build a flower, which became an acorn, which fattened a deer, which fed an Indian all in a single year.»

— ALDO LEOPOLD, *A Sand County Almanac,* 1949

The increase in nutrients, particularly nitrogen and phosphorus, in an ecosystem, is called **eutrophication.** The

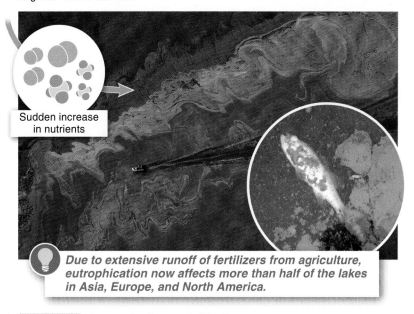

EUTROPHICATION

Eutrophication—the increase in nutrients in an ecosystem, particularly nitrogen and phosphorus—often leads to the rapid growth of algae and bacteria in aquatic ecosystems. These organisms then consume much of the oxygen, leading to large die-offs of animals.

Sudden increase in nutrients

Due to extensive runoff of fertilizers from agriculture, eutrophication now affects more than half of the lakes in Asia, Europe, and North America.

FIGURE 15-19 **Too much of a good thing?**

consequent rapid growth of algae and bacteria, followed by large-scale die-offs of other organisms, is increasingly a problem in both small and large bodies of water, including more than half of the lakes of Asia, Europe, and North America (**FIGURE 15-19**). Lake Erie, on the U.S.-Canadian border, for example, has experienced eutrophication as a result of all the phosphorus- and nitrogen-containing wastewater that drains into the lake from the extensive surrounding farmlands. A eutrophication-caused algae bloom there in August 2014 was responsible for the growth of toxic bacteria in municipal water supplies. The governor of Ohio was forced to declare a state of emergency, banning half a million people from using water for drinking, cooking, or even bathing. Given the lower use of fertilizers in South America and Africa, eutrophication is less common there.

TAKE-HOME MESSAGE 15·8

Chemicals essential to life—including carbon, nitrogen, and phosphorus—cycle through ecosystems. They are usually captured from the atmosphere, soil, or water by growing organisms, passed from one trophic level to the next as organisms eat other organisms, and returned to the environment through respiration, decomposition, and erosion. These cycles can be disrupted as human activities significantly increase the amounts of chemicals released to the environment.

15·9–15·15
Species interactions influence the structure of communities.

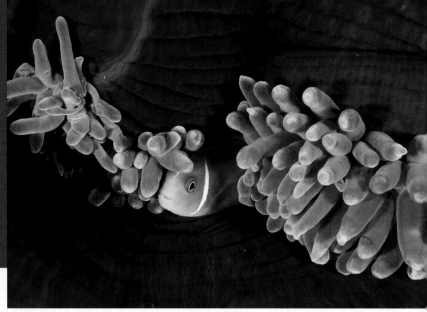

The anemone (green) and the pink anemonefish live together. The anemone's sting warns predators away, while the fish eats parasites and drives off fish that feed on anemones.

15·9 Each species' role in a community is defined as its niche.

Within a society, most humans seem to find their niche. Each person plays a particular role, defined by the nature of his or her work, activities, and interactions with others. Other species do the same thing. Within their *communities*—geographic areas defined as loose assemblages of species, sometimes interdependent, with overlapping ranges—each species has its own niche.

We can think of an organism's **niche** as its place in the community. More than just a *place* for living, however, a niche is a complete *way* of living. In other words, an organism's niche encompasses (1) the space it requires, (2) the type and amount of food it consumes, (3) the organisms for which it is a food source, (4) its influence on competitors, (5) the timing of its reproduction (its life

NICHE

NICHE FEATURES
- The space an organism requires
- The type and amount of food an organism utilizes
- The timing of an organism's reproduction
- An organism's temperature and moisture requirements and other necessary living conditions
- The organisms for which it is a food source
- Its influence on competitors

FIGURE 15-20 A way of living. "Niche" describes all of the ways in which an organism utilizes the resources of the environment.

REALIZED NICHE vs. FUNDAMENTAL NICHE

An organism's fundamental niche—the full range of conditions under which it may live—may be larger than its realized niche, due to competition or other factors.

For example, urban areas reduce suitable habitat within the geographic range of a bald eagle. Likewise, interference from other species can limit access to habitat within its geographic range.

■ Bald eagle range

history), (6) its temperature and moisture requirements, and almost every other aspect that describes the way the organism uses its environment and influences the other organisms in that environment (**FIGURE 15-20**).

Although a niche describes the role a species *can* play within a community, the species doesn't always get to have that exact role. It is common for species to find themselves competing with other species for parts of a niche, and for one of the species to be restricted from its full niche (see Figure 15-20). Consider the rats of Boston. Until the 1990s, they lived in relative peace in the sewers beneath the city's streets. But when the city embarked on the largest underground highway construction project in U.S. history, engineers displaced and forcibly drove out thousands of rats from much of their habitat. In essence, there was suddenly an overlap between the rat niche

and the human niche, and the rats were now restricted to just some portions of the sewers. As a result, the rats' **realized niche,** the narrower role that they may occupy in a community, is now just a subset of their **fundamental niche,** the full range of environmental conditions under which they can live.

TAKE-HOME MESSAGE 15·9

A population of organisms in a community fills a unique niche, defined by the manner in which the population utilizes the resources and influences the other organisms in its environment. Organisms do not always completely fill their niche; competition with other species within overlapping niches can reduce their range.

15·10 Interacting species evolve together.

There is a moth in Madagascar with a tongue that is 11 inches long! This might seem absurd—until the moth approaches a similarly odd-looking orchid. The orchid's flower has a very long tube, also about 11 inches long, with a bit of nectar at the very bottom. The moth's tongue, although usually rolled up, straightens out as fluid is pumped into it, and the moth can then insert it into the long nectar tube. As its tongue reaches the bottom, the moth slurps up the nectar. The moth also gets a bit of pollen stuck to it, which gets brushed onto the reproductive parts of the next orchid flower it visits (**FIGURE 15-21**).

Q Which came first, the long-tongued moths or the long-tubed flowers?

It's clear that having an 11-inch tongue is useful, even necessary, to extract nectar from an 11-inch nectar tube. And it's clear that putting nectar at the bottom of an 11-inch tube is a strategy to restrict access to only those pollinators that will reliably pass pollen from plant to plant of the same species. But how did such a system originate? Each trait seems to make sense only if the other already exists. The answer is that the two traits evolved—**coevolved**—together.

As a consequence of natural selection, populations of organisms commonly become better adapted to their environment. As long as there is variation for a trait and

the trait is heritable, differential reproductive success will lead to a change in the population. It is easy to imagine populations becoming more and more efficient at making use of non-living resources. But natural selection does not distinguish between biotic and abiotic resources as selective forces. Either can cause individuals with certain traits to

FIGURE 15-21   A perfect match? Long-tubed orchids and long-tongued moths in Madagascar each influence the evolution of the other.

reproduce at a higher rate than others, so either can cause evolution: small changes in the moth lead to small changes in the orchid, which select for further small changes in the moth . . . and the process continues. In the end, species become adapted not just to their physical environment but to the other species around them.

TAKE-HOME MESSAGE 15·10

In producing organisms better adapted to their environment, natural selection does not distinguish between biotic and abiotic resources as selective forces.

15·11 Competition can be hard to see, yet it influences community structure.

Interacting species' agendas are not always aligned, as they are for plants and their pollinators. Often, species interact because both are trying to exploit the same resources. In such cases—that is, when species have similar niches—conflict usually occurs. This competition doesn't last forever, though. Inevitably, one of two outcomes occurs: competitive exclusion or resource partitioning (**FIGURE 15-22**).

1. In **competitive exclusion,** two species battle for resources in the same niche until the more efficient one wins and the other species is driven to extinction in that

location ("local extinction"). In the 1930s, this was demonstrated in simple laboratory experiments using *Paramecium,* a single-celled organism. Populations of two similar *Paramecium* species were grown either separately or together in test tubes containing water and their bacterial food source. When grown separately, each species thrived. When grown together, one species always drove the other to extinction.

2. Resource partitioning is an alternative outcome of niche overlap. Individual organisms and species can adapt to changing environmental conditions, and resource partitioning can result from an organism's behavioral change or a change in its structure. When this occurs, one or both species become restricted in some aspect of their niche, dividing the resource. In other experiments with *Paramecium,* for example, researchers used one of the two species from the initial experiment and a new, third species. Again, each species thrived when grown alone. But when the two species were grown together in the same test tube, they ended up dividing the test tube "habitat." One species fed exclusively at the bottom of the test tube, and the other fed only at the top. Simple behavioral change made coexistence possible.

In many situations, resource partitioning is accompanied by **character displacement,** an evolutionary divergence in one or both of the species that leads to a partitioning of the niche. A clear example occurs in two species of seed-eating finches on the Galápagos Islands. On islands where both species live, their beak sizes differ significantly. One species has a deeper beak, better for large seeds, while the other has a shallower beak, better for smaller seeds, and they do not compete. On islands where either species occurs alone, beak size is intermediate between the two sizes (**FIGURE 15-23**).

Competition between species has one very odd feature: it is hard to actually see it happening because it causes itself to disappear. After only a short period of competition, one of the species becomes locally extinct or leaves the area where the niches overlap, or character displacement

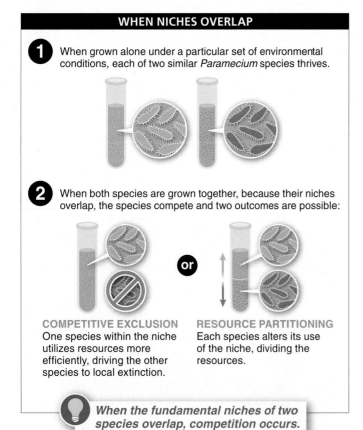

WHEN NICHES OVERLAP

1 When grown alone under a particular set of environmental conditions, each of two similar *Paramecium* species thrives.

2 When both species are grown together, because their niches overlap, the species compete and two outcomes are possible:

or

COMPETITIVE EXCLUSION
One species within the niche utilizes resources more efficiently, driving the other species to local extinction.

RESOURCE PARTITIONING
Each species alters its use of the niche, dividing the resources.

When the fundamental niches of two species overlap, competition occurs.

FIGURE 15-22 Overlapping niches: competitive exclusion and resource partitioning.

Character displacement occurs when natural selection reduces the competition between two species by producing an evolutionary divergence in one or both species.

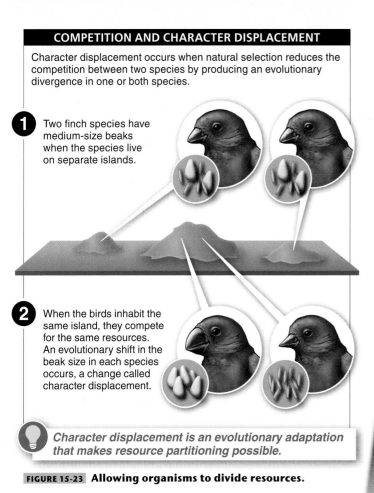

1 Two finch species have medium-size beaks when the species live on separate islands.

2 When the birds inhabit the same island, they compete for the same resources. An evolutionary shift in the beak size in each species occurs, a change called character displacement.

Character displacement is an evolutionary adaptation that makes resource partitioning possible.

FIGURE 15-23 **Allowing organisms to divide resources.**

occurs, largely reducing the competition. In either case, the net result is that the level of competition is significantly reduced or wiped out altogether. For this reason, biologists often have to look for character displacement—the "ghost of competition past"—to identify areas where competition once took place. Moreover, even while it is occurring, competition tends to be indirect, rather than head-to-head battles. Like a game of musical chairs, it's more a question of both species trying to use a particular resource, with one being a bit better at it.

Biologists do have the opportunity to see competition in occasional natural experiments in which a new species is introduced to a community. For example, the American mink, a weasel-like mammal, was introduced into Europe, where previously only the European mink had lived. Within 10 years, in the area where their ranges overlapped, the European mink increased in size while the American mink decreased in size, as biologists documented the evolution. In the next chapter, we investigate in greater depth why newly introduced species frequently win the competition with native species.

TAKE-HOME MESSAGE 15·11

Populations with completely overlapping niches cannot coexist forever. Competition for resources occurs until one or both species evolve in ways that reduce the competition, through character displacement, or until one species becomes extinct in that location.

15·12 Predation produces adaptation in both predators and their prey.

Some words of advice in case you ever think about quickly approaching a horned lizard: be afraid. Be very afraid. Here's why: as you get close to the lizard, it may zap you with streams of blood squirted from its eyes. In all likelihood, you will flinch. And as you flinch, the horned lizard will scurry away. The display is shocking, but the fact that evolution has produced extreme and effective anti-predator adaptations is not.

Predation—an interaction between two species in which one species eats the other—is one of the most important forces shaping the composition and abundance of species in a community. Predation, though, isn't restricted to the obvious interactions involving one animal chasing down and killing another. Herbivores' eating of

Q *Why do exotic species often flourish when released into novel habitats, even though natural selection has not adapted them to this new environment?*

leaves, fruits, or seeds is a form of predation, even though it doesn't necessarily kill the plant. And predators are not necessarily physically imposing. Each year more than a million humans die as a result of disease from mosquito bites, compared with fewer than a dozen from shark attacks.

Predators are a potent selective force: organisms eaten by predators tend to have reduced reproductive success. Consequently, in prey species, a variety of features have evolved (and continue to evolve)—including the blood-squirting-eyeball effect—that reduce the organisms' predation risk. But as prey evolve, so do predators. This coevolution is a sort of arms race with ever-changing and escalating predation-effectiveness adaptations causing more effective predator-avoidance adaptations, and vice versa. In this light, it may seem unexpected that exotic species often flourish when released into novel habitats, even though natural selection has not adapted them to their new environment. As it turns out, just as these species are not fully adapted to their new environment, they also have few predators there. And with low predation risk, they often can flourish. We'll examine some common adaptations of both predators and prey.

Prey adaptations for reducing predation There are two broad categories of defenses against predators: physical and behavioral.

Physical defenses include mechanical, chemical, warning coloration, and camouflage mechanisms (FIGURE 15-24).

1. Mechanical defenses. Predation plays a large role in producing adaptations such as the sharp quills of a porcupine, the prickly spines of a cactus, or the tough armor protecting an armadillo or sow bug. These, as well as claws, fangs, stingers, and other physical structures, can reduce predation risk.

2. Chemical defenses. Further prey defenses can include chemical toxins that make the prey poisonous or unpalatable. Plants can't run from their predators, so chemical defenses are especially important to them. Almost all plants produce some chemicals to deter organisms that might eat them. The toxins can be severe, such as

strychnine from plants in the genus *Strychnos,* which kills most vertebrates, including humans, by stimulating nonstop convulsions and other extreme and painful symptoms leading to death. At the other end of the spectrum are chemicals toxic to some insects but relatively mild to humans, such as those found in cinnamon, peppermint, and jalapeño peppers. Ironically, many plants that evolved to produce such chemicals to deter predators are now cultivated and eaten specifically *for* these chemicals. One organism's toxic poison is another's spicy flavor.

Some animals can also synthesize toxic compounds (or perhaps, more commonly, can safely acquire them from the organisms they eat). The poison dart frog has poison glands all over its body, making it toxic to the touch. The fire salamander is also toxic, with the capacity to squirt a strong nerve toxin from poison glands on its back. Some animals, including milkweed bugs and monarch butterflies, can safely consume toxic chemicals from plants and sequester them in their tissues, becoming toxic to predators who try to eat them.

3. Warning coloration. Species protected from predation by toxic chemicals frequently have bright color patterns to warn potential predators, essentially carrying a sign that says "Warning: I'm poisonous—keep away." To make it as easy as possible for predators to learn, different poisonous species often have the same color patterns.

In a clever twist on this, some species that are perfectly edible to their predators also have the same bright colors, in a phenomenon known as **mimicry.** As long as they are

---

PHYSICAL DEFENSES FOR REDUCING PREDATION

MECHANICAL DEFENSES
Physical structures—such as the quills of the Cape porcupine—can help protect an organism from predation.

CHEMICAL DEFENSES
Toxins—such as those found within the strawberry poison dart frog—can make an organism poisonous or unpalatable to a predator.

WARNING COLORATION
Organisms that produce toxic chemicals—such as monarch butterflies—frequently have bright color patterns, warning potential predators to stay away.

CAMOUFLAGE
Cryptic coloration—such as that seen in the praying mantis—can enable an organism to blend into its surroundings and avoid predation.

FIGURE 15-24 **Prey defenses: some physical means for avoiding predation.**

HIDING OR ESCAPING
Some prey species excel at hiding or running, or both, effectively avoiding predation. A variation of this strategy comes from safety in numbers: many species, including schooling fish, travel in large groups to reduce their predation risk.

FIGHTING BACK
Some prey species fight back against their attacker, effectively avoiding predation. For example, the fulmar, a seabird, defends its nest from attacks with projectile vomiting aimed at the intruder.

FIGURE 15-25 **Prey defenses: behavioral means for avoiding predation.**

relatively rare compared with the toxic individuals they mimic—reducing the chance that predators might catch on to their trickery—the evolutionary ruse is quite successful.

4. Camouflage. An alternative to warning coloration, and one of the most effective ways to avoid being eaten, is simply to avoid being seen. An adaptation in many organisms is patterns of coloration that enable them to blend into their surroundings. Examples include insects that look like leaves or twigs and hares that are brown for most of the year but turn white when there is snow on the ground.

Behavioral defenses include both seemingly passive and active behaviors: hiding or escaping, and alarm calling or fighting back (**FIGURE 15-25**).

1. Hiding or escaping. Anti-predator adaptations need not involve toxic chemicals, physical structures, or special coloration. Many species excel at hiding or running, or both. With vigilance, it is possible to get advance warning of impending predator attacks, and then quickly and effectively avoid the predator. A variation of this strategy comes from safety in numbers: many species, including schooling fish and emperor penguins, travel in large groups to reduce their predation risk.

Q **On islands, animals frequently have no fear of humans. Why?**

On many islands, animals show no fear of humans. Rather than encountering skittish lizards, for example, a human visitor to the Galápagos Islands must be careful not to step on them. The animals are not afraid because, given the small size of most islands, the number of predators is restricted. As a consequence, there has been no selection for skittishness and hypervigilance, traits that normally evolve in response to predation risk.

2. Alarm calling and fighting back. In many species, especially birds and mammals, individuals warn others with an alarm call. Although risky for the caller, such alarm calling can give other individuals—often close kin that are nearby—enough advance warning to escape (recall, from Chapter 9, the warning calls of Belding's ground squirrels). Some prey species also turn the tables, mobbing predators to keep them from successfully completing their task. This category might also include the blood-squirting lizard described above, or the fulmar, a seabird that defends its nest from attacks with projectile vomiting aimed at the intruder.

Predator adaptations for enhancing predation

Just as prey use physical and behavioral features to reduce their risk of predation, predators evolve in parallel ways to increase their efficiency. As milkweed plants have evolved to produce toxic chemicals, sequestered within the structures of the plant, milkweed bugs have evolved to be able to eat the toxic plants without suffering harm. And as prey have become better at hiding and escaping, predators have developed better sensory perception to help them detect hiding prey and faster running ability to catch them. Predators, too, make use of mimicry. The angler fish, tasseled frogfish, and snapping turtle all have physical

In the angler fish, a structure has evolved that tricks prey organisms into coming right to the predator, where they are eaten.

FIGURE 15-26 **A better predator.** Some adaptations enhance predation.

structures that mimic something—usually a food item—of interest to potential prey. As a prey animal comes closer to inspect, the predator snaps it up (**FIGURE 15-26**).

Although natural selection leads to predators with effective adaptations for capturing prey, the adaptations are rarely so efficient that the prey species are driven to extinction. The explanation for this is referred to as the "life-dinner hypothesis." Selection for "escape ability" in the prey is stronger than selection for "capture ability" in the predator. When, for example, a rabbit can't escape from a fox, the cost to the rabbit is its life—and it will never reproduce again. On the other hand, when a fox can't keep up with a rabbit, all it loses is a meal; it can still capture prey and reproduce in the future. In other words, the cost of losing in the interaction is much higher for the rabbit.

> **Q** Why don't predators become so efficient at capturing prey that they drive the prey to extinction?

TAKE-HOME MESSAGE 15·12

Predators and their prey are in an evolutionary arms race; as physical and behavioral features evolve in prey species to reduce their predation risk, predators develop more effective and efficient methods of predation. This coevolutionary process can result in brightly colored organisms, alarm calling, and many types of mimicry.

15·13 Parasitism is a form of predation.

For most people, the word "predation" conjures images of a large animal such as a cheetah, chasing and killing another large animal such as a gazelle. But there is another world of predation, largely unseen but equally deadly: **parasitism,** defined as a symbiotic relationship—that is, a close and commonly long-term relationship between individuals of different species—in which one organism (the **parasite**) benefits while the other (the **host**) is harmed. In fact, this hidden world of parasites is thriving with activity: parasites are probably the most numerous species on earth, with three or four times as many species as non-parasites!

Two general types of parasites make life difficult for most organisms. *Ectoparasites* (*ecto* = outside) include organisms such as lice, leeches, and ticks. One species of Mexican parrot is all too aware of ectoparasites: it has 30 different species of mites living on its feathers alone. And many of the parasites even have parasites themselves! *Endoparasites*

are parasites that live inside their hosts (*endo* = inside). They are equally pervasive. Endoparasites infecting vertebrates include many different phyla of both animals and protists, the single-celled eukaryotes (**FIGURE 15-27**). In all of these parasite-host interactions, as with all predator-prey interactions, the predator or parasite benefits and the prey or host is harmed.

Even though they are considered predators, parasites have some unique features and face some unusual challenges in comparison to other predators. The most obvious of these features is that the parasite generally is much smaller than its host and stays in contact with it for extended periods of time, normally not killing the host but weakening it as the parasite uses some of the host's resources. Being located right on your food source all the time can be advantageous. But this also leads to what is perhaps the greatest challenge that parasites face: how to get from

Parasites are predators that benefit from a symbiotic relationship with their hosts.

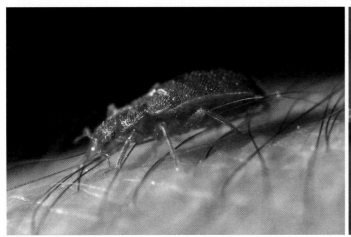

ECTOPARASITES

Ectoparasites—such as the bedbug (*Cimex lectularius*), an ectoparasite of humans—live *on* their host.

FIGURE 15-27 **Parasites: predators dwarfed by their prey.**

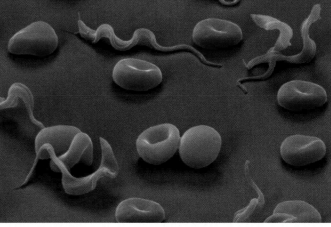

ENDOPARASITES

Endoparasites—such as *Trypanosoma brucei*, an endoparasite that invades a human host and is the cause of African sleeping sickness—live *inside* their host.

💡 *There may be three to four times as many parasitic species as non-parasitic species on earth!*

one individual host to another—after all, a parasite can't survive long once its host dies. The methods by which parasites accomplish such dispersal are surprising. Many of their complicated life cycles involve passing through two (or more) different host species (and could have come about only through coevolution with each of the host species). These life cycles are likely to give us a new appreciation for the ingenuity of these microorganisms—or rather, for the evolutionary process that produced them. Let's look at a few representative examples.

Case 1: Parasites can induce foolish, fearless behavior in their hosts. During their evolution, rats have developed a protective wariness of cats, as well as areas in which cats have been roaming. *Toxoplasma* is an organism that changes this. This single-celled parasite of rats must also spend part of its life cycle in cats. It does this by altering the brain of its rat host so that the rat no longer fears cats. In fact, *Toxoplasma*-infected rats not only lose their fear of cats, they become attracted to them. Is this an accident? No. This behavioral change, while quite dangerous for the rat, is exactly the change that increases the likelihood that the rat will be attacked by a cat, spreading the parasite in the process.

Case 2: Parasites can induce inappropriate aggression in their hosts. Rabid animals don't behave normally. They froth at the mouth and become unusually aggressive (**FIGURE 15-28**). Is this an accident? No. Rabies is caused by a virus that

infects warm-blooded animals, mostly raccoons, skunks, foxes, and coyotes. It is passed from one host to another via saliva. Inducing these "rabid" behaviors, of course, is

Rabies parasite is more effectively passed from one animal to another by causing its host to foam at the mouth and aggressively bite.

💡 *Parasites have surprising ways of solving a challenging problem: how to get from one individual host to another.*

FIGURE 15-28 **Spreading disease.** Rabies symptoms aid the disease-causing virus in spreading to new hosts.

WHAT ARE ECOSYSTEMS?   WEATHER   ENERGY AND CHEMICAL FLOW   SPECIES INTERACTIONS   COMMUNITIES

exactly the change in behavior that will most help the virus to spread.

Case 3: Parasites can induce bizarre and risky behavior in their hosts. The lancet fluke is a parasitic flatworm. It has also been described as a "zombie-maker." This fluke spends most of its life in sheep and goats, but the fluke's eggs pass into snails that graze on vegetation contaminated by sheep and goat feces. Once inside the snail, the fluke eggs grow and develop, eventually forming cysts that the snail surrounds with mucus and then excretes. Continuing on their complex life cycle, the fluke cysts find their way into ants that eat the snail mucus. The flukes' journey back to a sheep or goat is now expedited by the so-called zombie-making. In an infected ant, the flukes grow into the ant's brain, altering its behavior. Whereas ants normally remain low to the ground, when infected by the lancet fluke, an ant climbs to the top of a grass blade or plant stem and clenches its mandibles on a leaf. This behavioral change puts the ant at greater risk of

being eaten by a grazing mammal—a bad outcome for the ant, but just what the parasite needs to complete its life cycle.

These parasite-induced behavior modifications are among the most dramatic, but there are many others that are more subtle yet equally effective at allowing a parasite to thrive attached to or within its host's body. This subset of predation is an active and exciting area of ecological and physiological investigation.

TAKE-HOME MESSAGE 15·13

Parasitism is a symbiotic relationship in which one organism benefits while the other is harmed. Parasites face some unusual challenges relative to other predators, particularly in how to get from one individual host to another, and some complex parasite life cycles have evolved.

15·14 Not all species interactions are negative: mutualism and commensalism.

It's easy to get the idea that all species interactions in nature are harsh and confrontational, marked with a clear winner and loser. That is largely the case when it comes to competition and predation. However, not all species

interactions are combative. Every flower you see should be a reminder that evolution produces beneficial species interactions as well. These types of interactions fall into two categories: mutualism and commensalism.

NON-NEGATIVE SPECIES INTERACTIONS

MUTUALISM
In mutualism, both species benefit from an interaction. For example, the swamp aster flower is pollinated by a bumblebee that is nourished by the plant's nectar.

COMMENSALISM
In commensalism, one species benefits from an interaction and the other neither benefits nor is harmed. For example, the cattle egret feeds on insects stirred up by a grazing Cape buffalo. The buffalo is neither helped nor harmed in the interaction.

FIGURE 15-29 Not always "red in tooth and claw." Mutualisms and commensalisms abound in nature.

Mutualism: everybody wins Corals that gain energy from photosynthetic algae living inside their tissues. Termites capable of subsisting on wood, but only with the assistance of cellulose-digesting microbes living in their digestive system. Flowers pollinated by animals that are nourished by nectar. Each of these relationships is an example of **mutualism,** an interaction in which both species gain and neither is harmed (FIGURE 15-29). Mutualism is common in almost all communities. Plants, particularly, have numerous such interactions: with nutrient-absorbing fungi, nitrogen-fixing bacteria, and animals that pollinate them and disperse their fruits.

Mutualistic interactions are so widespread and ecologically important that they are described in significant detail throughout this book. The relationship between plants and mycorrhizae bacteria, for example, is discussed in Section 12-15. Plants and their rich variety of mutualistic relationships with animal pollinators and fruit dispersers are covered throughout Sections 12-9 and 12-11. Lichens, the symbiotic association of fungi and photosynthetic algae or bacteria, are discussed in Section 12-15.

You need not look farther than your own gut for an important mutualistic relationship. Living within your large intestine are huge populations of *E. coli* bacteria that synthesize a significant portion of the vitamin K—essential for several metabolic processes and blood clotting—you need each day.

Commensalism: an interaction with a winner but no loser Some species interactions are one-sided. The cases in which one species benefits and the other neither benefits nor is harmed are called commensal relationships, or **commensalism.** Cattle egrets have just such a relationship with grazing mammals such as buffalo and elephants (see Figure 15-29). As the large mammals graze through grasses, they stir up insects. The birds, which feed near the mammals—particularly near the forager's head—are able to catch more insects with less effort this way. The grazers are neither helped nor harmed by the presence of the birds.

TAKE-HOME MESSAGE 15·14

Not all species interactions are combative: beneficial species interactions evolve as well. Mutualism, in which both species benefit from the interaction, is widespread and critically important to all ecosystems. In commensalism, one species benefits and the other is neither harmed nor helped.

❓ 15·15 THIS IS HOW WE DO IT

Investigating ants, plants, and the unintended consequences of environmental intervention.

Mutualisms between plants and animals are hugely important to the functioning of ecosystems and to maintaining biodiversity. But the complexity of the webs of interactions within ecosystems makes it difficult to predict the impact of a change in one part of the system. This has important implications as we try to protect valuable environments and natural resources.

Consider a situation that many natural resource managers encounter. Suppose that a population of organisms with great value appears to be threatened by the predatory behavior of another species. To protect the valuable population, the natural resource manager might be tempted to intervene, blocking the predator from harming the valuable population. This seems like a reasonable plan. But how can we be sure?

In fact, our plan, with all its good intentions, may not work. Worse, it may have exactly the opposite effect of what we intend. Let's look at a situation that, surprisingly (but unambiguously), revealed a scary truth: seemingly helpful, straightforward manipulations may have significant, unintended, and negative consequences.

It started when a biologist visiting Africa observed an intervention that was in place to help a plant species—acacia trees. The intervention: large groups of these trees were fenced off to protect them from the destructive herbivory of elephants and giraffes.

> **What do you think was the intended consequence of putting enclosures around acacia trees?**

The natural resource managers assumed that the enclosures, by protecting the trees from herbivores, should benefit the plants. The acacia trees ought to grow faster and live longer. But the visiting biologist noticed that the several hundred plants within the enclosures—

distributed across six 10-acre plots of land that had been fenced off for more than 10 years—weren't doing very well. Compared with unprotected acacia plants in plots adjacent to the enclosures, most of the protected plants looked unhealthy, and many were dying. A group of researches decided to find out why.

> **With an intervention already in place, how could the researchers figure out what was responsible for the poor health of the enclosed acacia trees?**

The researchers began by observing acacia trees closely to figure out what normally occurs (that is, outside enclosures) in these plants. Here's what they found out:

1. Acacia plants and certain types of ants have a mutualistic relationship. The plant makes hollow thorns, in which the ants live, and produces sugary nectar, which the ants consume.

2. When something touches the tree—such as an elephant eating leaves—the ants swarm out and defend the tree. This reduces herbivory on the plant.

The biologists' next step was to make careful observations of the acacia trees within the enclosures to figure out what was happening as a result of the intervention to keep out elephants and giraffes. Each piece of information they collected was like one piece of a puzzle. Here's what they found for the enclosed trees:

1. **Reduced predation → reduced investment in defense against predators.** When protected from elephants and giraffes, the acacias somehow detected the reduced herbivory and reduced their investment in supporting the symbiotic ants: they decreased by one-third the number of nectar-producing sites, and they produced fewer swollen thorns in which the ants could live.

2. **Reduced support of ants → reduced defense behavior by ants.** When the nectar reward dwindled, the ants played a lesser role in plant defense: there was a 30% reduction in the number of ants remaining on the plants, and among the remaining ants, 50% fewer individuals responded to disturbances of the plant—and those that did were significantly less aggressive in their response.

> **These two changes, on their own, should not have been bad for the enclosed plants. Why?**

3. **Reduced number of ants → opportunity for competitors to get in.** When the number of mutualistic ants dwindled, this created an opportunity for competing

(non-mutualistic) species of ants to take up residence in the acacias. The proportion of trees occupied by these competing, non-mutualistic ants doubled.

4. **Increased number of competitor ants → increase in stem-boring beetles and plant damage.** The competitor, non-mutualistic ants encouraged beetles to infect the plants: the beetles bored cavities in the acacia stems, where the competing ants could lay eggs. And when beetles bore cavities, plants suffer. The researchers documented that acacia growth was much slower, and mortality was doubled.

In just a few short steps, the researchers were able to piece together what was happening:

1. Big herbivores eat acacia plants.

2. Restricting herbivores causes acacia plants to invest less in mutualistic ants.

3. Reduced support of the ants helps competing ants and stem-boring beetles to invade.

4. Formation of bore holes by the beetles slows plant growth and increases plant mortality.

In other words, the researchers learned why restricting big herbivores by placing enclosures ultimately harmed the acacia trees.

> **What can we conclude from these results?**

Beyond this specific message, the researchers noted that their results illustrate the extreme complexity of community interactions.

> **What are the implications for natural resource managers?**

The disappointing but unavoidable implication is that, within an ecosystem, seemingly helpful, straightforward manipulations may have significant, negative consequences. Or, in the words of poet Robert Burns, "The best-laid schemes o' mice an' men / Gang aft agley [often go awry]."

> **What does this study suggest about the possible consequences of the loss of large herbivores from the world's ecosystems? Why?**

TAKE-HOME MESSAGE 15·15

Study of acacia-ant mutualisms and attempts to manipulate them reveals that community interactions are very complex. Within an ecosystem, seemingly helpful, straightforward manipulations may have significant, negative consequences.

Buffalo grazing in the grasses.

15·16 Many communities change over time.

Human "progress" and development can transform an environment. A patch of barren land, for example, may be turned into a productive agricultural field. Or an urban landscape may obliterate any signs of the nature that was once there. This is why it can be surprising (and heartening) to observe what happens when humans abandon an area. Little by little, nature reclaims it. The area doesn't necessarily recover completely, and change is slow. Still, this process is almost universal and virtually unstoppable. Nature responds similarly to other disturbances, too, from a single tree falling in a forest, to a massive flood or fire, to massive volcanic eruptions.

The process of nature reclaiming an area and of communities gradually changing over time is called **succession.** It is defined as a change in the species composition over time, following a disturbance. There are two types of succession. *Primary succession* is when the process starts with no life and no soil. *Secondary succession* is when an already established habitat is disturbed, but some life and some soil remain.

Primary succession Primary succession can take thousands or even tens of thousands of years, but it generally occurs in a consistent sequence. It always begins with a disturbance that leaves an area barren of soil and life. The huge volcanic eruption on Krakatau, Indonesia, in 1883, for example, completely destroyed several islands and wiped out all life and soil on others. Primary succession has also begun, in a less dramatic fashion, in regions where glaciers have retreated, such as Glacier Bay, Alaska.

Although succession does not occur in a single, definitive order, several steps are relatively common (**FIGURE 15-30**).

- The first arrivals—called **colonizers** or pioneer species—in a lifeless, soil-less area are usually bacteria or fungi or other photosynthetic microorganisms, floating in on the air.

- Lichens—symbiotic associations between a fungus and a photosynthetic alga or bacterium—are also common first colonizers. They can grow on rocks and can generate energy through photosynthesis. As they grow, lichens change their environment, secreting acids that break down the rocks and release useful minerals, making the terrain more hospitable for other organisms.

- Mosses also tend to arrive in the early stages, using many of the nutrients freed up by lichens. They trap moisture and can provide a hospitable site for germination of the seeds of other plants.

- Following mosses, small herbs often arrive. Later, some shrubs may begin to thrive, shading out the mosses and herbs.

- Shrubs, in turn, are eventually outcompeted and shaded out by small trees.

- And the first trees generally are outcompeted by taller, faster-growing trees.

An important feature of the colonizers seen in the earliest stages of primary succession—whether plants, animals, or other species—is that while they are good **dispersers,**

Succession is the change in species composition in a community over time.

COLONIZING COMMUNITY   **INTERMEDIATE COMMUNITIES**   **CLIMAX COMMUNITY**

1 Fungi, bacteria, lichens, and seeds are often among the earliest colonizers.

2 Mosses begin to grow and trap moisture, allowing the seeds of other plants to germinate.

3 Small herbs arrive, and shrubs grow, shading out the mosses and herbs.

4 Small trees grow and outcompete the shrubs.

5 Longer-living, larger species outcompete the initial colonizers and persist as a stable and self-sustaining community.

PRIMARY SUCCESSION
Primary succession begins after a disturbance leaves an area barren of soil and with no life.

SECONDARY SUCCESSION
Secondary succession is like primary succession with a head start. It begins when a disturbance opens up part of a community to the development and growth of species previously outcompeted by other species in the area.

Disturbance is a fundamental part of most ecosystems and can repeatedly set a community back to an earlier stage of succession.

FIGURE 15-30 **The species composition of a community changes over time.**

able to move away from their original home (hence their early arrival in a new locale), they are not particularly good *competitors*. That's why they are gradually replaced.

Ultimately, it is the longer-living, larger species that tend to outcompete the initial colonizers and persist within a stable and self-sustaining community, called a **climax community.** The particular species present in the climax community depend on physical factors such as temperature and rainfall.

Secondary succession Secondary succession is much faster than primary succession, because life and soil are already present. Rather than the thousands of years that primary succession may take, secondary succession can happen in a matter of centuries, decades, or even years. It frequently begins with organisms colonizing the decaying remains of dead organisms. This may involve fungi establishing themselves in the decaying trunk of a fallen tree, and these being replaced over time by different species of fungi.

Secondary succession may also begin with weeds springing up in formerly cultivated land. If the weeds are allowed to grow, they are eventually outcompeted and replaced by perennial species, then shrubs, and eventually larger trees. The process is similar to primary succession, but with a head start. Ultimately, if undisturbed, secondary succession also leads to establishment of a stable, self-sustaining climax community.

Given that primary and secondary succession both progress toward climax communities, it might

Q Why doesn't succession occur on front lawns?

seem surprising that, at any given time, so many communities are in states far from climax. This paradox reflects the fact that succession leads to climax communities in the *absence of disturbance.* But disturbance is a fundamental part of most ecosystems. Massive disturbances, such as a wildfire that blazed across more than 500,000 acres in Oregon in 2012 or the volcanic eruptions of Mount St. Helens, can send a community back to square one. Minor disturbances, on the other hand, may undo just a small degree of the succession that has already occurred. Interestingly, it is at intermediate levels of disturbance that communities tend to support the greatest number of species. With very high or very low disturbance levels, the community is restricted to the top colonizers—those with the best dispersal or fastest reproductive rates—or the top competitors, respectively. At the intermediate level, a larger variety of species with intermediate combinations of features can persist.

TAKE-HOME MESSAGE 15·16

Succession is the change in the species composition of a community over time, following a disturbance. In primary succession, the process begins in an area with no life and no soil; in secondary succession, the process occurs in an area where life is already present. In both, the process usually takes place in a predictable sequence.

15·17 Some species are more important than others within a community.

In this chapter, we have examined the ways in which species interact with one other and with their environments, noting the dependence of species on each other. But not all species have equal dependence on or influence over others. Within a community, the presence of some species, called **keystone species,** greatly influences which other species are present and which are not (FIGURE 15-31). If a keystone species is removed, the species mix in the community changes dramatically. The removal of other species causes relatively little change.

Bison are a keystone species. Experiments were set up in which a herd of bison was allowed to graze in certain parts of a prairie in Kansas, while being excluded from other areas. The areas without the bison were soon dominated by a single species of tall grass that was like a bully to other plant species. The areas where the bison grazed, on the other hand, had significantly greater species diversity. When kept in check by the grazing bison, the bully grass was unable to outcompete the other plant species, which were able to thrive.

Sea stars, too, are a keystone species (see Figure 15-31). On the rocky seashores of the Pacific Northwest, they consume mussels, which tend to crowd out other species. In a five-year study, sea stars were removed and kept out of certain areas of the intertidal zone but allowed in other areas. As with the bison study, species diversity dropped dramatically where sea stars were excluded. In fact, ultimately, just the mussels remained. In the areas with sea stars, mussels also remained but in reduced numbers. More importantly, 28 other species of animals and algae were also able to live there.

Keytone species make it possible to get more "bang for your buck" when your aim is to conserve biodiversity. Consequently, identifying keystone species is an important part of conservation biology. Besides bison and sea stars, some other keystone species are dam-building beavers, elephants of the African savanna, and lichens in the desert. We explore these questions of conservation in greater detail in the next chapter.

Q When it comes to conservation, are some species more valuable than others? Why?

TAKE-HOME MESSAGE 15·17

Keystone species have a relatively large influence on which other species are present in a community and which are not. Unlike other species, when a keystone species is removed from the community, the species mix changes dramatically. For this reason, protecting keystone species is an important strategy in preserving biodiversity.

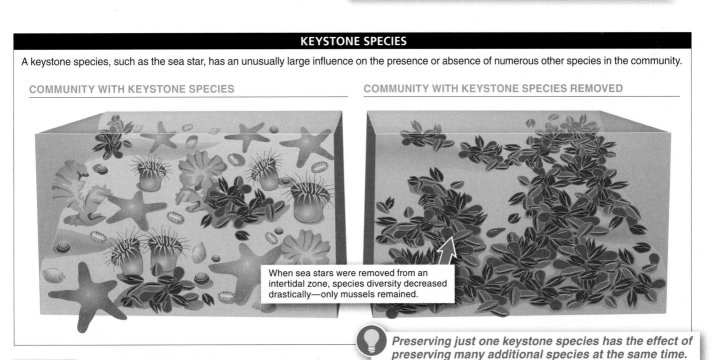

KEYSTONE SPECIES

A keystone species, such as the sea star, has an unusually large influence on the presence or absence of numerous other species in the community.

COMMUNITY WITH KEYSTONE SPECIES

COMMUNITY WITH KEYSTONE SPECIES REMOVED

When sea stars were removed from an intertidal zone, species diversity decreased drastically—only mussels remained.

Preserving just one keystone species has the effect of preserving many additional species at the same time.

FIGURE 15-31 **Preserving biodiversity.** Keystone species can keep aggressive species in check, allowing more species to coexist.

Life in the Dead Zone

In boosting plant productivity on farms, we've created a "dead zone" in the Gulf of Mexico bigger than Connecticut.

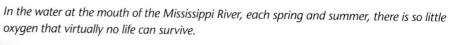

In the water at the mouth of the Mississippi River, each spring and summer, there is so little oxygen that virtually no life can survive.

Q: **Which chemical almost always limits plant growth?** Plant growth requires the production of proteins, all of which contain nitrogen. Access to nitrogen nearly always determines how much a plant grows.

Q: **If you're a farmer, how can you increase the productivity of your crops?** Add nitrogen. This, in fact, is the chief component of all plant fertilizers.

Q: **Where does the unused nitrogen end up?** Spring and summer rains wash nutrients, dissolved organic materials, and other runoff (much of which comes from human sources) from the middle of the United States into nearby rivers and thus, eventually, into the Mississippi River. From there they flow into the Gulf of Mexico.

Q: **What happens when large amounts of nitrogen are dumped into a body of water?** Organisms such as plankton and algae living near the mouth of the Mississippi grow like crazy with all the extra nitrogen and nutrients that are washed into the Gulf of Mexico. But their rapid increase in productivity disrupts the food chain as they decay, increasing the organic matter that sinks to the bottom and feeds the bacteria there. As the bacteria thrive, they consume ever-increasing amounts of oxygen.

Q: **What happens when bacteria use up too much oxygen?** Excessive bacterial growth depletes the supply of dissolved oxygen in the water, starving all other life of oxygen—including fish, crabs, oysters, and shrimp, all of which die or move out of the area. This creates a "dead zone" and ruins local fisheries, with a significant impact on the economies they support.

Q: **What can you conclude?** Disrupting food chains, even increasing the productivity of certain trophic levels, can have unintended effects throughout an ecosystem, and even thousands of miles away.

Q: **What can be done?** Reducing and reversing the Gulf of Mexico's dead zone (and other dead zones throughout the world) can be accomplished by (1) reducing fertilizer use, (2) preventing animal wastes from getting into waterways, and (3) controlling the release of other sources of nutrients (phosphorus as well as nitrogen) from industrial facilities into rivers and streams.

Check Your Knowledge

1. Why is the drawing of the hawk so much smaller than those of the other organisms?

2. What does the size of each "pie" represent? Why aren't they all the same size?

3. What does the slice of the pie represent in each drawing?

4. Why isn't there a slice taken from the top pie?

5. Would this figure be different if it included data from consumers that ate plants and other animals? How?

6. What can you conclude from this figure about the relative numbers of predators versus herbivores in any environment?

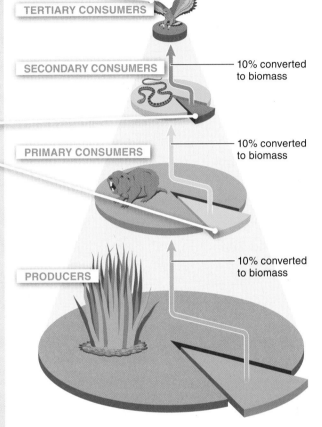

ENERGY PYRAMID

TERTIARY CONSUMERS

SECONDARY CONSUMERS — 10% converted to biomass

PRIMARY CONSUMERS — 10% converted to biomass

PRODUCERS — 10% converted to biomass

See answers at the back of the book.

Key Terms in Ecosystems and Communities

ABOUT THE CHAPTER OPENING PHOTO

An aerial view of the Landmannalaugar region in southern Iceland, known for its beautiful and unusual geological features.

Check Your Knowledge

1. An ecosystem consists of _____ in a given area.

a) all the photosynthetic organisms

b) all the living organisms

c) all the abiotic factors that influence living organisms

d) all the living organisms and non-living materials

e) the plant life and climate

0 — (17) — 100
EASY   HARD

2. Earth's largest terrestrial ecosystems, the biomes, are defined primarily by:

a) the average rainfall.

b) the average temperature.

c) the seasonal variability in temperature.

d) the seasonal variability in rainfall.

e) All of the above.

0 — (11) — 100
EASY   HARD

3. Why is it hotter at the equator than at the Poles?

a) The angle at which sunlight hits the earth leads to solar energy being spread over a smaller area at the equator than at the Poles.

b) The increased rotational speed of the earth at the Poles creates strong winds that increase radiant cooling.

c) Greater cloud cover at the equator leads to a greater "greenhouse" effect than at the Poles.

d) The water in the oceans has a high heat capacity, and there is significantly more water at the equator than at the Poles.

e) All of the above are responsible for the higher average temperature at the equator than at the Poles.

0 — (23) — 100
EASY   HARD

15·1–15·2 Ecosystems have living and non-living components.

An ecosystem is a community of biological organisms plus the non-living components in the environment with which the organisms interact.

ECOSYSTEMS

All ecosystems share two essential features:

BIOTIC ENVIRONMENT	PHYSICAL (ABIOTIC) ENVIRONMENT
• The living organisms within an area	• The chemical resources and physical conditions within an area
• Often referred to as a community	• Often referred to as the organisms' habitat

? 1. Describe the two elements of an organism's physical or abiotic environment.

TERRESTRIAL BIOMES

Terrestrial biomes—large ecosystems that cover huge geographic areas—are determined by the temperature and amounts of precipitation in conjunction with the magnitude of seasonal variation in these factors.

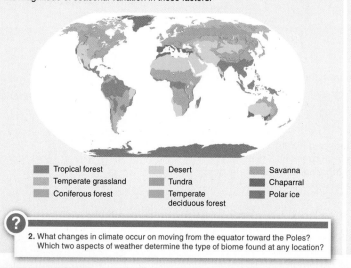

- ☐ Tropical forest
- ☐ Temperate grassland
- ☐ Coniferous forest
- ☐ Desert
- ☐ Tundra
- ☐ Temperate deciduous forest
- ☐ Savanna
- ☐ Chaparral
- ☐ Polar ice

? 2. What changes in climate occur on moving from the equator toward the Poles? Which two aspects of weather determine the type of biome found at any location?

AQUATIC BIOMES

Aquatic biomes are determined by physical features, including salinity, water movement, and depth.

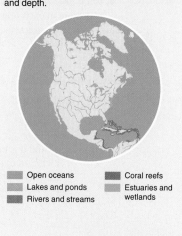

- ☐ Open oceans
- ☐ Lakes and ponds
- ☐ Rivers and streams
- ☐ Coral reefs
- ☐ Estuaries and wetlands

15·3–15·5 Interacting physical forces create weather.

Global patterns of weather are largely determined by the earth's round shape.

THE DISTRIBUTION OF SOLAR ENERGY

The sun shines most directly on earth's equator, where the solar energy is spread out over a smaller area than the same amount of energy hitting near the Poles. This leads to warmer temperatures at the equator.

North Pole

Earth

Sun

Equator

South Pole

? 3. Why are areas at lower latitudes typically warmer than high-latitude areas?

THE FORMATION OF RAIN

1 AIR IS HEATED AND RISES
When solar heat hits the earth, it warms the air at that point. The heated air rises.

2 RISING AIR COOLS
As hot air rises, getting farther from the warm earth, it cools.

3 COOLING AIR LOSES MOISTURE
Because cold air holds less moisture than warm air, clouds form and rain falls.

Sun

THE FORMATION OF DESERTS

Large circulating masses of air are responsible for many of the large-scale rainfall patterns throughout the world. Air heated by the sun rises, expands northward and southward, and moves back toward earth as it cools. As the air moves down toward earth's surface and becomes warmer, it can hold more and more moisture, causing these areas to have very little rainfall.

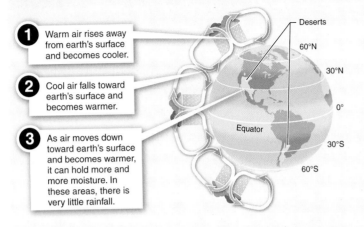

1 Warm air rises away from earth's surface and becomes cooler.

2 Cool air falls toward earth's surface and becomes warmer.

3 As air moves down toward earth's surface and becomes warmer, it can hold more and more moisture. In these areas, there is very little rainfall.

Deserts

60°N
30°N
0°
Equator
30°S
60°S

THE RAIN SHADOW EFFECT

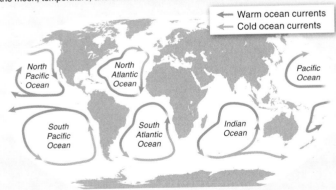

1 Wind blows from the ocean toward land, rising when it hits mountains.

2 Rising air cools and holds less moisture, leading to cloud formation and rain.

3 Air passes over the mountain top and falls, becoming warmer and increasing the moisture it can hold. This reduces rainfall and results in dry areas.

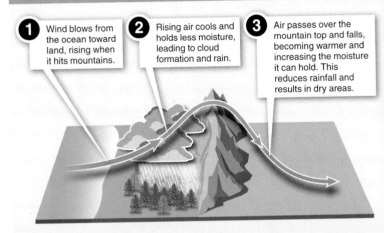

Mountain range

Rain shadow desert

The Andes Mountains, near the west coast of South America, cause a rain shadow effect.

OCEAN CIRCULATION PATTERNS

There are several large, circular patterns of flowing water in the oceans due to a combination of forces, including wind, the earth's rotation, the gravitational pull of the moon, temperature, and salt concentration.

← Warm ocean currents
← Cold ocean currents

North Pacific Ocean
North Atlantic Ocean
Pacific Ocean
South Pacific Ocean
South Atlantic Ocean
Indian Ocean

These circular patterns are responsible for the water off the west coast of the United States being much colder than the water off the east coast.

MODERN LANDSCAPES

Urban development increases the absorption of solar energy, leading to higher temperatures, and creates wind near the bottom of tall buildings.

? 4. You are charged with creating a "greener" city that will be cooler and use less air conditioning. Describe three strategies with which you could help achieve this goal.

? 5. Why do strong winds often blow on street corners with tall buildings?

EL NIÑO

Every two to seven years, a dramatic climate change driven by ocean currents, called El Niño, causes a sustained surface temperature change in the central Pacific Ocean. Changes during an El Niño event start a chain reaction of unusual weather around the globe: increased rainfall in the U.S. Midwest, storms in coastal California, and warmer weather in South Africa and Japan.

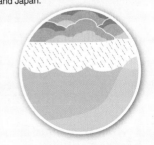

?

4. **Why do beaches on the west coast of the United States have colder water than beaches on the east coast?**

a) The ocean currents in the northern hemisphere rotate in a clockwise direction, so water off the coast of California has just come from the cold north, while water off the coast of New York has just come from the warm south.

b) West coast beaches are not colder than east coast beaches. Ocean temperatures depend only on latitude.

c) The Pacific Ocean is deeper than the Atlantic Ocean, so more of the sun's heat is absorbed by cold, deep water.

d) Trade winds blowing north from the Tropic of Cancer in the Pacific Ocean sweep the warm air away from the west coast, taking with it much of the warmth of the water.

e) Deep-water upwellings bring cool water to west coast beaches, while the surface currents on the east coast cause downwellings, which reduce heat loss due to evaporation.

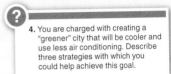
0 · · · 27 · · · 100
EASY · · · HARD

5. **When a moist ocean wind blows onshore toward a mountain range:**

a) as the air rises, it pulls moisture from the ground, causing the higher elevations to be drier.

b) as the air goes over the top of the mountains and falls back down toward lower elevations, it holds less moisture, creating a "rain shadow" zone of unusually high precipitation.

c) as the air rises, it holds less moisture, causing dissipation of all the clouds.

d) as the air goes over the top of the mountains and falls back down to lower elevations, it holds even more moisture, creating a "rain shadow" desert.

e) None of the above are correct.

0 · · · 42 · · · 100
EASY · · · HARD

6. **"Top" carnivores:**

a) are more common than secondary consumers.

b) rely directly on producers for energy.

c) consume primarily herbivores.

d) consume primarily carnivores.

e) rely on symbiotic bacteria to help them digest cellulose.

0 · · · 26 · · · 100
EASY · · · HARD

7. **The 10% rule of energy-conversion efficiency:**

a) explains why big, fierce animals are so rare.

b) explains why the biomass of herbivores must exceed that of carnivores.

c) limits the length of food chains.

d) suggests that 90% of what an organism eats is used in cellular respiration or is lost as feces.

e) All of the above are correct.

0 · · · 32 · · · 100
EASY · · · HARD

Left column questions

8. Nitrogen enters the food chain:

a) primarily through soil-dwelling bacteria that "fix" nitrogen by attaching it to other atoms.

b) from the atmosphere when "fixed" by the photosynthetic machinery of plants.

c) when rocks dissolved by rainwater become soil, which is then utilized for plant growth.

d) through soil erosion followed by runoff into streams and ponds.

e) through methane, produced by herbivores as a by-product of the breakdown of plant material.

0 ——— **34** ——————— 100
EASY   HARD

9. Which of the following statements about an organism's niche is incorrect?

a) It encompasses the space the organism requires.

b) It includes the type and amount of food the organism consumes.

c) It is not always fully exploited.

d) It may be occupied by two species, as long as they are not competitors.

e) It reflects the ways in which the organism utilizes the resources of its environment.

0 ——————— **47** ————— 100
EASY   HARD

10. The "ghost of competition past" refers to the fact that:

a) competition often leads to character displacement, which remains even after direct competition is reduced.

b) competition cannot be seen in nature.

c) competition inevitably leads to extinction of one of the competitors.

d) competition inevitably leads to extinction of both competitors.

e) the fossil record is a record of competitive interactions.

0 ——————————— **60** — 100
EASY   HARD

11. Chemical defenses are more common among plants than animals because:

a) plants cannot move to escape predators and so must develop other deterrents.

b) the cell wall can contain the chemicals more effectively than a simple plasma membrane.

c) mechanical defenses against predators can evolve only in animals.

d) parasite loads are significantly higher in plants than in animals.

e) All of the above are correct.

0 ——— **30** ——————— 100
EASY   HARD

Main content

15·6–15·8 Energy and chemicals flow within ecosystems.

Energy from the sun flows through ecosystems, while chemicals repeatedly cycle from the physical environment through living organisms and back into the environment.

ENERGY FLOW THROUGH THE ECOSYSTEM

Energy from the sun is intercepted and converted into chemical energy, which passes through an ecosystem in about four stops, known as trophic levels. At each stop, some usable energy is lost as heat.

Sun

PRODUCERS
Plants convert light energy from the sun into food through photosynthesis.

PRIMARY CONSUMERS
Herbivores are animals that eat plants.

SECONDARY CONSUMERS
Carnivores are animals that eat herbivores.

TERTIARY CONSUMERS
Top carnivores are animals that eat other carnivores.

This linear pathway is overly simplified. In actuality, food webs are often a better representation; many organisms (including you, if you eat vegetables and meat) can occupy more than one position in the chain.

6. Why is the term "food web" more accurate than "food chain" when describing the pathway of energy flow?

INEFFICIENT ENERGY FLOW

Only about 10% of the biomass from each trophic level is converted into biomass in the next trophic level. The rest of the available energy is lost to the environment, a consequence of several factors, including non-predatory deaths, incomplete digestion of prey/food, and respiration.

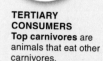

7. Describe the energy flow through an ecosystem. How does this energy flow influence the number of top carnivores?

CHEMICAL CYCLES

Chemicals essential to life—including carbon, nitrogen, and phosphorus—cycle through ecosystems. They are usually captured from the atmosphere, soil, or water by growing organisms, passed from one trophic level to the next as organisms eat other organisms, and returned to the environment through respiration, decomposition, and erosion.
These cycles can be disrupted as human activities significantly increase the amounts of chemicals added to soils for use by plants or released to the environment.

8. Why do levels of CO_2 in the northern hemisphere undergo ups and downs throughout the year?

15·9–15·15 Species interactions influence the structure of communities.

Interacting species in a community coevolve in a variety of ways, some antagonistic and others mutually beneficial.

NICHE

A population of organisms in a community fills a unique niche, defined by the manner in which the population utilizes the resources and influences the other organisms in its environment. Organisms do not always completely fill their niche; competition with other species within overlapping niches can reduce their range.

COMPETITION: WHEN NICHES OVERLAP

When the fundamental niches of two species overlap, the species compete, and two outcomes are possible:

 **or**

COMPETITIVE EXCLUSION
One species within the niche utilizes resources more efficiently, driving the other species to local extinction.

RESOURCE PARTITIONING
Each species alters its use of the niche, dividing the resources.

9. Give an example to demonstrate that "natural selection does not distinguish between biotic and abiotic resources as selective forces."

10. What is a niche? What are the potential fates of two species that occupy the same niche?

11. When the niches of different species overlap, the result is competition. Describe the two possible outcomes from competition.

PHYSICAL DEFENSES FOR REDUCING PREDATION

MECHANICAL DEFENSES
Physical structures can help protect an organism from predation.

CHEMICAL DEFENSES
Toxins can make an organism poisonous or unpalatable to a predator.

WARNING COLORATION
Organisms that produce toxic chemicals frequently have bright color patterns, warning potential predators to stay away.

CAMOUFLAGE
Cryptic coloration can enable an organism to blend into its surroundings and avoid predation.

? 12. Why would you expect plants to be more likely than animals to use chemical defense mechanisms to reduce predation?

PARASITISM

Parasitism is a symbiotic relationship in which one organism benefits while the other is harmed.

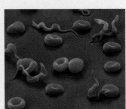

ECTOPARASITES
Parasites that live on their host

ENDOPARASITES
Parasites that live inside their host

? 13. Describe a feature of rabies that reflects the mechanism by which the disease is transmitted from one host to the next.

BEHAVIORAL DEFENSES FOR REDUCING PREDATION

HIDING OR ESCAPING
Some prey species run, hide, or find safety in numbers, effectively avoiding predation.

FIGHTING BACK
Some prey species fight back against their attacker, effectively avoiding predation.

NON-NEGATIVE SPECIES INTERACTIONS

In commensalism, one species benefits and the other is neither harmed nor helped. In mutualism, both species benefit from the interaction.

? 14. Why is commensalism described as an interaction with a winner but no loser?

15·16–15·17 Communities can change or remain stable over time.

Most communities change over time. The patterns of succession depend on the rate and magnitude of disturbances.

SUCCESSION

Succession is the change in the species composition of a community over time, following a disturbance.

COLONIZING COMMUNITY　**INTERMEDIATE COMMUNITIES**　**CLIMAX COMMUNITY**

❶ Fungi, bacteria, lichens, and seeds are often among the earliest colonizers.

❷ Mosses begin to grow and trap moisture, allowing the seeds of other plants to germinate.

❸ Small herbs arrive, and shrubs grow, shading out the mosses and herbs.

❹ Small trees grow and outcompete the shrubs.

❺ Longer-living, larger species outcompete the initial colonizers and persist as a stable and self-sustaining community.

PRIMARY SUCCESSION
Primary succession begins after a disturbance leaves an area barren of soil and with no life.

SECONDARY SUCCESSION
Secondary succession is like primary succession with a head start. It begins when a disturbance opens up part of a community to the development and growth of species previously outcompeted by other species in the area.

? 15. Which feature of colonizers is crucial in the earliest stages of primary succession?

?

12. In a commensal relationship:
a) one species pollinates the other.
b) neither species benefits or is harmed, but the community itself benefits.
c) one species provides nutrients, usually cellulose, for the other.
d) one species benefits while the other neither benefits nor is harmed.
e) the "loser" has reduced reproductive output.

0 —— **26** —————————— 100
EASY   HARD

13. Coevolution:
a) is responsible for all the beautiful flowers in the world.
b) is responsible for nectar production by plants.
c) reveals that both biotic and abiotic resources can serve as selective forces.
d) can produce an insect with a tongue as long as its body.
e) All of the above are correct.

0 ——————— **57** ———— 100
EASY   HARD

14. The chief difference between primary and secondary succession is that:
a) primary succession occurs among the plants in a habitat, whereas secondary succession occurs among the animals.
b) primary succession begins with no life and no soil, whereas secondary succession begins with both.
c) secondary succession alters the biotic environment, whereas primary succession alters the abiotic environment.
d) primary succession occurs more quickly than secondary succession.
e) primary succession can occur in terrestrial and aquatic habitats, but secondary succession can occur only in terrestrial habitats.

0 —— **29** —————————— 100
EASY   HARD

15. Keystone species:
a) occur only in intertidal zones.
b) play an unusually important role in determining the species composition in a habitat.
c) can be removed from a habitat without any impact on the remaining species in the community.
d) are producers and therefore usually are plants.
e) are more expendable than commensal species, from a conservation perspective.

0 — **11** ——————————— 100
EASY   HARD

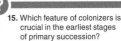

16 Conservation and Biodiversity

HUMAN INFLUENCES ON THE ENVIRONMENT

Biodiversity—of genes, species, and ecosystems—is valuable in many ways.

Extinction reduces biodiversity.

Human activities can have disruptive environmental impacts.

We can develop strategies for effective conservation.

Biodiversity—of genes, species, and ecosystems—is valuable in many ways.

Endangered Nilgiri tahr (Nilgiritragus hylocrius) visit a grassy alpine plateau in India.

16·1 Biodiversity can have many types of value.

In the Pacific Northwest of the United States, a medium-size tree grows. This conifer, the Pacific yew tree (*Taxus brevifolia*), isn't the biggest tree in the forest, nor is it the most common. But within the bark of the Pacific yew there is a chemical that is pretty remarkable. It's called taxol, and it acts as an anti-cancer agent, interfering with the division of cells that come into contact with it (**FIGURE 16-1**). Although we don't fully understand the role taxol plays for the Pacific yew—it may reduce the rate at which other organisms feed on the plant—in humans, taxol is effective in the treatment of ovarian cancer, breast cancer, and lung cancer (generating more than $1 billion a year in pharmaceutical sales).

Consider also the following:

- The chemicals vinblastine and vincristine, isolated from the Madagascar periwinkle (*Catharanthus roseus*), are so effective in treating leukemia and Hodgkin's lymphoma that both diseases, formerly incurable, are now curable in the vast majority of people.

- The chemical ancrod, found only in the Malayan pit viper snake (*Agkistrodon rhodostoma*), dissolves blood clots and is effective in treating some heart attack and stroke patients.

- Epibatidine, a poisonous compound in the saliva of a small frog (*Epipedobates tricolor*) that lives in Ecuador, is 200 times more effective than morphine in relieving pain and is non-addictive.

These are just a few examples—there are many, many more—illustrating that medically important compounds often come from other organisms living in a wide variety of locations around the world. Toxin production commonly evolves in these organisms as a method of protecting them from other organisms. Co-opting the

Q Why would another species produce a chemical that fights cancer in humans?

A chemical within the bark of the Pacific yew, called taxol, is effective in the treatment of ovarian cancer, breast cancer, and lung cancer.

Pacific yew tree (Taxus brevifolia)

FIGURE 16-1 **Taxol is cancer medicine derived from a tree.**

chemicals for human use reveals that one important value of the diverse plants and animals around the world is as a sort of universal medicine cabinet.

Plants and animals aren't the only species with great value to humans. Microbes, too, have great utility. In the next section, for example, we see that hydrocarbon-consuming bacteria can play a significant role in cleaning up oil spills (**FIGURE 16-2**).

Medicinal compounds derived from plants and animals, as well as the activities of oil-removing microbes, are just a few examples of how humans use the chemicals and metabolic processes that occur in other organisms. The plants, animals, and microbes in which these compounds and processes occur are examples of **biodiversity,** which is defined as the variety of genes, species, and ecosystems on earth. And, as we've seen, there is tremendous value that can come from earth's biodiversity, and it often comes from unexpected places. For these reasons, the loss of biodiversity—including genes, species, and ecosystems—can be hugely detrimental to humans.

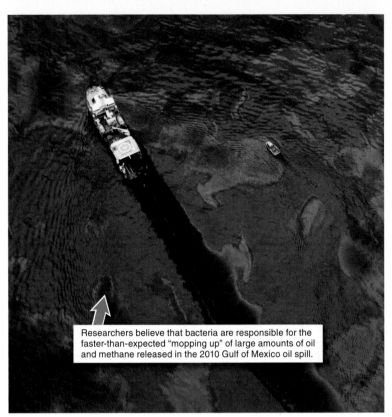

Researchers believe that bacteria are responsible for the faster-than-expected "mopping up" of large amounts of oil and methane released in the 2010 Gulf of Mexico oil spill.

FIGURE 16-2 **Can bacteria help clean up an oil spill?**

But viewing the value of biodiversity only in the context of our abilities to extract medicines or to better process the messes we make is to take much too narrow a perspective. Taking a philosophical perspective, the United Nations has asserted that biodiversity has an intrinsic value—that is, an inherent value independent of its value to humans—stating that "every form of life is unique, warranting respect regardless of its worth to man."

And, in describing the extrinsic or utilitarian value of biodiversity, scientists and economists use the concept of **ecosystem services** to highlight the benefits to humans that come from biodiversity and natural ecosystems. In a 2005 United Nations study that involved more than 1,300 scientists, the researchers identified four distinct categories of ecosystem services (**FIGURE 16-3**).

- **Provisioning services.** These include the many useful products that humans obtain from nature, including food, spices, minerals, and pharmaceutical and industrial products.

- **Cultural services.** Much of the value that humans gain from biodiversity isn't easily assigned a dollar value—but that value can be extremely important just the same. We think of a field of flowers or a deer in the

woods as beautiful to look at and pleasant to experience, for example. Or we make use of images of animals and plants to convey meaningful abstract ideas: the dove as a symbol of peace; the oak as a symbol of strength; the butterfly as a symbol of transformation; and the owl as a symbol of wisdom. Beyond the aesthetic, symbolic, and spiritual values of biodiversity, cultural services also include recreational experiences, as well as scientific and other educational value.

- **Regulating services.** Ecosystems provide numerous services relating to the regulation of our environment. These include climate regulation through carbon storage in plants; pollination; waste decomposition and detoxification; protection from natural disasters; and pest and disease control.

- **Habitat services.** Significant value from biodiversity and natural ecosystems also flows to humans in the form of soil formation, photosynthesis, and nutrient cycling.

When we recognize that people can view the same organism or gene or ecosystem and value it in different ways, we are on our way to making wiser conservation decisions. This knowledge, for example, can help us

THE VALUES OF BIODIVERSITY

INTRINSIC VALUE
The inherent value of biodiversity, independent of its value to humans

EXTRINSIC (UTILITARIAN) VALUE
The value of biodiversity to humans, often described as ecosystem services, can be grouped in four categories of services:

PROVISIONING SERVICES
The many useful products that humans obtain from nature including food, pharmaceutical, and industrial products

CULTURAL SERVICES
The aesthetic, symbolic, and spiritual values of biodiversity, as well as scientific and educational value

REGULATING SERVICES
The values that come from the regulation of our environment, including climate regulation, waste decomposition, protection from natural disasters, and pest and disease control

HABITAT SERVICES
The value to humans from soil formation, photosynthesis, and nutrient cycling

FIGURE 16-3 Ecosystem services categorize the many different ways that biodiversity has value to humans.

anticipate potential conflict when it comes time to invest limited conservation resources in conserving biodiversity and can help us seek out methods for balancing conflicting desires. We explore some strategies for making these difficult decisions later in this chapter.

TAKE-HOME MESSAGE 16·1

Biodiversity—the variety of genes, species, and ecosystems on earth—has intrinsic value as well as extrinsic value, or value to humans. The value of biodiversity to humans is often described in terms of four categories of ecosystem services: provisioning, cultural, regulating, and habitat services. These categories help distinguish among various utilitarian values, such as the production of food and medicines, aesthetic and symbolic value, and the regulation and support of our environments.

? 16·2 THIS IS HOW WE DO IT

When 200,000 tons of methane disappears, how do you find it?

Even in the midst of an ecological catastrophe, there can be opportunities for scientific thinking and problem solving. Such situations, however, can require fast thinking about how to approach a problem, particularly when a real understanding of the issues is necessary to reduce further impact of the catastrophe and to guide management strategies.

An example of just such a catastrophe occurred in April 2010. An oil rig in the Gulf of Mexico was drilling a well about a mile under the surface of the water when it exploded and caused a massive leak of crude oil. It took nearly three months to cap the well, during which time more than 200 million gallons of oil poured into the Gulf.

The oil spill caused extensive damage to the coastal and marine environments and was responsible for killing thousands of animals, including birds, dolphins, sea turtles, mollusks, and crustaceans. The longer-term consequences of the oil spill for the region—including the eight U.S. national parks threatened—are not yet clear.

Methane was the most abundant hydrocarbon dumped into the Gulf—with almost 200,000 tons released by the spill. As scientists worried about the impact of the gas and oil on the ecosystem, they needed to monitor the area in order to map where the methane was and where it was headed.

| **Can you propose a way to keep track of oil a mile below the water surface?**

Researchers set out to build a three-dimensional map of the Gulf in the regions above and around the site where the drilling had been. At 207 locations, they lowered sensors into the water and collected water samples at a large number of depths, from the surface down to the Gulf floor. This enabled them to identify where in the water column the methane was and how concentrated it was.

Writing in the journal *Science,* the team reported something quite unexpected. Just a couple of months after the leak had been sealed, they discovered that a huge proportion of the methane released in the spill was gone.

| **Most of the methane was gone from the area of the spill. What could have happened to it? What were the possibilities?**

Two explanations for the methane disappearance seemed possible. Perhaps, as the researchers tried to track the flowing gas and oil underwater, they had simply lost it. Or, they wondered, could populations of bacteria be consuming the methane?

| **How could the scientists figure out whether bacteria were eating the methane or it was just lost?**

The researchers made a testable prediction. If bacteria were consuming the methane, there would be a chemical trace of their metabolic activities. Specifically, the amount of dissolved oxygen in the water should have decreased significantly. One consequence of such metabolic activity—particularly on such a huge scale—would be that the bacteria would consume large amounts of oxygen as they burned the methane for fuel. If, on the other hand, the methane had simply flowed somewhere else in the Gulf, there would be no drop in the dissolved oxygen.

Sure enough, as the researchers sampled the water and evaluated oxygen concentrations, they noted an unprecedented, significant drop in oxygen saturation—down from 67% to 59%. This suggested that bacterial populations were indeed proliferating rapidly and "mopping up" the oil.

Making some calculations, the researchers found that the amount of missing oxygen was in fact almost exactly equal to the amount required by bacteria to consume the amount of methane that had leaked from the well. "The math worked out scary good," is how one researcher put it.

| **The diversity of bacteria in the world is huge. How could the scientists figure out which species were responsible?**

Twenty years ago, researchers would have had to culture any bacteria they found in order to identify them. But scientists of today were able to bypass this difficult step. Using DNA sequencing methods, they were able to test the microbes (and pieces of microbes) in the Gulf. In these sequences, they found genetic fragments previously identified in methane-eating microbes.

Moreover, the bacterial DNA they found was related to several previously observed oil-eating bacterial species from the Gulf of Mexico. These bacteria can live in the Gulf because natural oil—equivalent to as much as a million barrels of oil per year—seeps out of the ground. The researchers hypothesized that because the bacteria were already present in the Gulf, the populations of microbes were able to respond extremely quickly to the spill.

| **Should we be confident that we can rely on microbes to clean up any oil that we spill?**

By virtue of having methane-consuming microbes already in place, the Gulf of Mexico was particularly suited to a rapid, effective response to an oil spill. In other places, however, such as the numerous Arctic drilling locations, there is not the same sort of natural oil seepage. And with no similar populations of methane-eating bacteria already in place, the impact of an oil spill on these habitats might be quite different.

TAKE-HOME MESSAGE 16·2

Following a massive oil leak in the Gulf of Mexico, researchers noted a rapid disappearance of methane. Finding a simultaneous drop in the oxygen saturation of the water, the researchers were able to determine that populations of bacteria already living in the area grew rapidly in response to the new food source and consumed the methane.

Biodiversity occurs at multiple levels.

Which habitat has greater biodiversity, one with 3 or 4 species each of birds, reptiles, mammals, plants, and insects, or one with 38 unique species of fruit flies, or one with 100 radically different species of prokaryotes (**FIGURE 16-4**)?

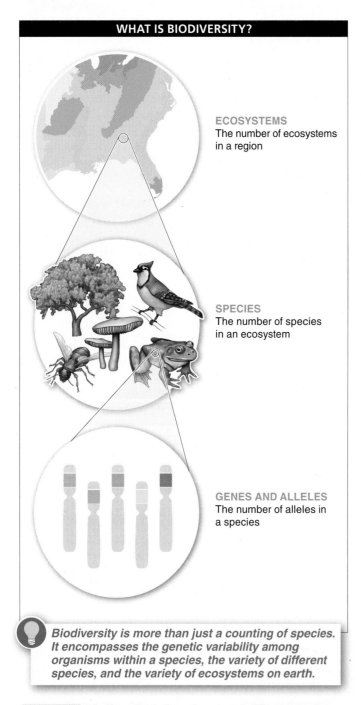

WHAT IS BIODIVERSITY?

ECOSYSTEMS
The number of ecosystems in a region

SPECIES
The number of species in an ecosystem

GENES AND ALLELES
The number of alleles in a species

Biodiversity is more than just a counting of species. It encompasses the genetic variability among organisms within a species, the variety of different species, and the variety of ecosystems on earth.

FIGURE 16-4 **Biodiversity is found on many different scales.**

Conservation biology is a multidisciplinary science that addresses how to preserve natural resources and protect biodiversity.

FIGURE 16-5 **Preserving the earth's natural resources.**

If you were given the task of overseeing the conservation of biodiversity in a region where these three habitats occur, which habitat would you protect first? In making your decision, you might prioritize the number of different classes or families of organisms in a habitat, or maybe the number of distinct species in the habitat, which is, in fact, the most commonly used measure of biodiversity. In doing so, however, you'd be acknowledging that biodiversity can be prioritized in many ways and that any assessment is a function of your values, biases, and interests. But this recognition helps us to discuss, manage, use, and conserve biodiversity in a practical and less ambiguous way.

Conservation biology is the interdisciplinary field that addresses how to understand and preserve the natural biological resources of earth, including its biodiversity at all levels (**FIGURE 16-5**). And the complexities of managing biodiversity directly affect the decisions that conservation biologists make. In this chapter, we explore how the actions of humans can have significant, even disastrous, consequences for biodiversity and natural resources, and we also explore the strategies that have been developed for effective conservation.

Underpinning many of the difficulties in conservation biology is that biodiversity, as we've noted, can have value for humans in so many different ways. Throughout the chapter, we'll see how this complexity makes it difficult to balance competing interests when they are in conflict.

TAKE-HOME MESSAGE 16·3

Assessing biodiversity can be difficult because it must be considered at multiple levels, from entire ecosystems to species to genes and alleles. In practice, biodiversity is most often defined as the number of distinct species in a habitat, which has important implications for conservation biology—the field that addresses questions of how to preserve the natural resources of earth.

16·4 Where is most biodiversity?

GRAPHIC CONTENT
Thinking critically about visual displays of data
Turn to p. 681 for a closer inspection of this figure.

If you were to stand at the equator in South America and identify the number of land mammals, you would count about 400 different species. If you then started walking southward, away from the equator, and assessed the diversity of land mammals when you got to 20° latitude, you would find about 350 different species. Continuing your long walk south, when you got to 30° latitude you would find only 200 different species. And the trend would continue for as far south as you could go, with 100 species at 40° latitude and fewer than 50 species at 50° latitude. On a similar walk northward from the equator, through North America, you would observe the same trend, with more than 300 species of land mammals at 20° latitude, reduced to only 100 species in northern Canada (**FIGURE 16-6**). Biodiversity is not evenly distributed throughout the world. And perhaps the strongest trend in terrestrial biodiversity distribution is this: as you move away from the equator in either direction, diversity is reduced with increasing latitude.

Q Why are there more species in an acre of tropical rain forest than in an acre farther from the equator, such as in a temperate forest or prairie?

The strong biodiversity gradient from the tropics to the Poles occurs not just in land mammals but in nearly any group of plants or animals observed—recent surveys have documented the trend in woody plants, herbaceous plants, and mangrove trees. Interestingly, even the diversity of organisms in the oceans follows this trend. One survey of marine copepods—tiny crustacean animals, including plankton—found the same trend moving northward from the tropical region of the Pacific Ocean (80 species present) to the

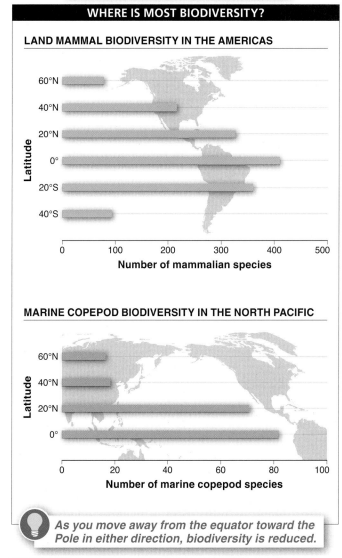

WHERE IS MOST BIODIVERSITY?

LAND MAMMAL BIODIVERSITY IN THE AMERICAS

Number of mammalian species

MARINE COPEPOD BIODIVERSITY IN THE NORTH PACIFIC

Number of marine copepod species

As you move away from the equator toward the Pole in either direction, biodiversity is reduced.

FIGURE 16-6 **Biodiversity is unevenly distributed.**

Arctic Ocean (fewer than 10 species). Similar patterns have been documented in fishes, mollusks, marine and terrestrial arthropods, corals, and marine protists.

Numerous factors that influence **species richness**— that is, the number of species in an area—are responsible for producing the tropical-to-temperate-to-polar gradient in biodiversity. Three, in particular, play strong roles (**FIGURE 16-7**).

1. Solar energy available. Perhaps the simplest predictor of species diversity is the amount of solar energy available in an area. Solar energy is, after all, the ultimate fuel for nearly all life. As we saw in Figure 15-5, the sun shines most directly on the equator, and, as we move away from the equator, the earth's curvature causes the solar energy hitting the earth to be spread out over increasingly large areas. This leads to less solar energy at any one point as latitude increases. In a variety of species of plants and animals, researchers have documented strong relationships between energy availability and species richness.

2. Evolutionary history of an area. Communities diversify over time. Consequently, the more time that passes, generally speaking, the greater is the diversity in an area. A high level of biodiversity in an area, however, can be knocked back down by climatic disasters such as glaciations. Thus, for example, biodiversity declines due to glaciations may have disproportionately occurred in temperate and polar regions (and on the tops of some mountains), which have been affected more frequently than tropical regions as glaciers have advanced and retreated over hundreds of millions of years.

3. Rate of environmental disturbance. Over time, the best competitor for a resource is expected to outcompete other species, excluding them from the community and reducing species richness. However, communities may be kept from reaching this species-depleted state. Factors such as predation, or environmental disturbances such as floods, fires, and volcanic eruptions that may wipe out most species in an area, can prevent any one species from excluding others. For this reason, a habitat with an *intermediate* amount of environmental disturbance—that is, not so high that many or all species are wiped out, and not so low that organisms with a competitive advantage drive all of the less competitive species from the community—is expected to have the greatest species richness.

Conservation biologists are increasingly interested in **biodiversity hotspots,** those regions of the world having significant reservoirs of biodiversity that are under threat of destruction. Twenty-five biodiversity hotspots have been identified around the globe (**FIGURE 16-8**). These hotspots cover less than 1% of the world's area but have 20% or more of the world's species. Habitats included among the biodiversity hotspots are tropical rain forests, coral reefs, and islands. While not considered a hotspot, the deepest regions of the oceans have recently proved to hold great biodiversity.

Tropical Rain Forests These habitats are common near the equator in Asia, Australia, Africa, South America, and Central America, as well as on many Pacific islands. An important, but still poorly studied, region of tropical rain forests is the canopy, the uppermost region of a forest, where nearly all of the sunlight is intercepted. The canopy, which can be up to 30 feet (almost 10 m) thick, is the most diverse part of the tropical rain forests; estimates suggest that as many as half of all species on earth reside in the canopy of these rain forests. Due to its inaccessibility—the canopy is 65–135 feet (20–40 m) above the ground—the canopy is not easily studied, and the extent of biodiversity there is still not fully known.

Coral Reefs Built by corals, members of the animal kingdom, coral reefs occur in ocean areas with low levels of nutrients. The physical structures created by the corals provide a home and habitat for huge numbers of other species. The Great Barrier Reef of Australia (which is not listed among the 25 biodiversity hotspots because it currently is a focus of extensive conservation efforts) is an example of just how much diversity can be supported by coral reefs. It is the largest coral reef system in the world and covers such a large area (almost the size of California)

FACTORS THAT INFLUENCE BIODIVERSITY

SOLAR ENERGY AVAILABLE
Greater access to solar energy, the fuel for all life, provides increased species richness.

EVOLUTIONARY HISTORY OF AN AREA
Communities diversify over time. The more time that passes without a climatic event, such as an ice age, the greater the diversity in an area.

RATE OF DISTURBANCE
A habitat with an intermediate amount of disturbance tends to have the greatest species richness.

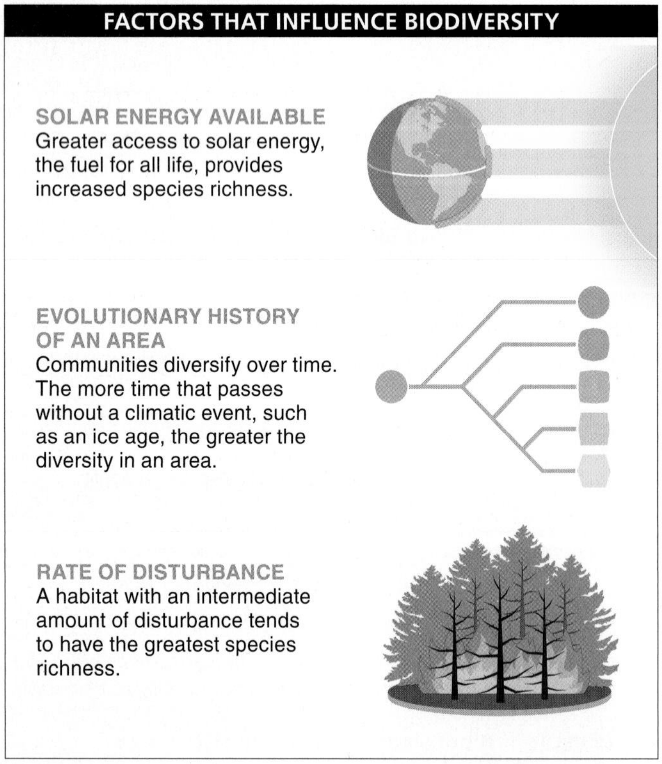

FIGURE 16-7 **What determines species richness in an area?**

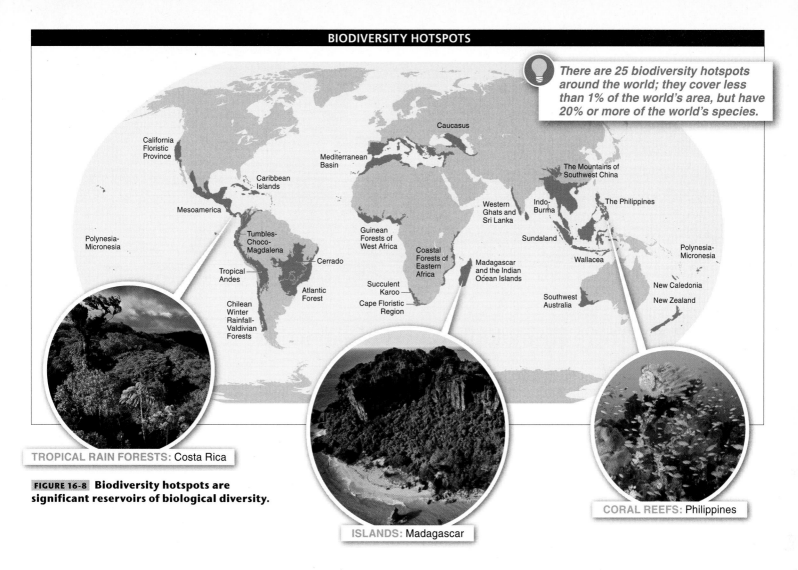

There are 25 biodiversity hotspots around the world; they cover less than 1% of the world's area, but have 20% or more of the world's species.

California Floristic Province

Mesoamerica

Polynesia-Micronesia

Tumbles-Choco-Magdalena

Tropical Andes

Chilean Winter Rainfall-Valdivian Forests

Caribbean Islands

Cerrado

Atlantic Forest

Caucasus

Mediterranean Basin

Guinean Forests of West Africa

Succulent Karoo

Cape Floristic Region

Coastal Forests of Eastern Africa

Madagascar and the Indian Ocean Islands

Western Ghats and Sri Lanka

The Mountains of Southwest China

Indo-Burma

Sundaland

Southwest Australia

The Philippines

Wallacea

Polynesia-Micronesia

New Caledonia

New Zealand

TROPICAL RAIN FORESTS: Costa Rica

FIGURE 16-8 **Biodiversity hotspots are significant reservoirs of biological diversity.**

ISLANDS: Madagascar

CORAL REEFS: Philippines

off the northeast coast of Australia that it can be seen from space. It is home to numerous species of whales, dolphins, sea turtles, sharks, and stingrays, as well as 1,500 species of fishes and 5,000 species of mollusks. Five hundred species of algae are found there, too. Above the Great Barrier Reef, more than 200 species of birds are supported by the rich diversity of life below. Other coral reefs are not as well protected by laws and conservation efforts, and the world's second largest coral reef, in New Caledonia in the southwestern Pacific, is one of the 25 biodiversity hotspots.

Islands Because of their generally smaller size, which limits the population size of organisms living on them, islands can be regions where biodiversity is at risk. Madagascar, located in the Indian Ocean off the southeastern coast of Africa, is the fourth largest island in the world (just a bit smaller than Texas). It is home to 5% of the world's species, 80% of which are **endemic** (exclusively native to a place)—a higher percentage of endemic plants and animals than in any comparably sized area on earth. One of the 25 biodiversity hotspots, Madagascar is an area unusually rich in species diversity.

Deep Oceans The oceans, covering two-thirds of the earth's surface, can be much more difficult to explore than terrestrial habitats. Deep ocean regions, in particular, are much less studied than the terrestrial parts of the world. Once thought to be completely devoid of life, the deepest regions of the oceans have recently been found to be teeming with biodiversity. One survey of the deepest parts of the Antarctic Ocean, sampling at a depth of more than 18,000 feet (6,000 m), discovered 585 new species of crustaceans alone.

TAKE-HOME MESSAGE 16·4

Biodiversity is not evenly distributed over the earth. For nearly all groups of plants and animals, both marine and terrestrial, biodiversity is greatest near the equator and falls progressively toward the North and South Poles. Factors that influence the species richness in an area include the amount of solar energy available, the area's evolutionary history, and the rate of environmental disturbance. Biodiversity hotspots are regions of significant biodiversity under threat of destruction.

IDENTIFYING BIODIVERSITY   EXTINCTION   HUMAN INTERFERENCE   CONSERVATION STRATEGIES

16·5–16·6
Extinction reduces biodiversity.

The Bengal tiger has been hunted, captured, and poisoned to near extinction.

16·5 There are multiple causes of extinction.

Imagine that you are on a deserted island. Alone. It would be hard and it would be lonely. The hope of rescue someday, however, would probably make things bearable. "Martha," a passenger pigeon living in the Cincinnati zoo in the early 1900s, found herself in a situation of this sort—minus the hope. Passenger pigeons in the wild had been completely wiped out by a combination of factors, including hunting of the birds for meat, loss of their habitat due to deforestation in North America, and possibly disease. And so, except for the small flock that Martha was part of, the species was gone. One by one, those birds died as well, and when the second-to-last passenger pigeon died in 1912, Martha was all alone. She lived for two more years, with no possibility for ever reproducing, and died on September 1, 1914 (**FIGURE 16-9**).

On average, species persist for about 10 million years, but there is huge variation. Some may last for hundreds of millions of years, but for many others the number of years is much lower than the average. An extinction occurs when all individuals in a species have died. And whether you look at biodiversity from a utilitarian, aesthetic, or symbolic perspective, extinction is a tragic loss.

As we discussed in Chapter 10, extinctions can be divided into two general categories: mass extinctions and background extinctions (**FIGURE 16-10**), which are distinguished by the numbers of species affected. A mass extinction occurs when a large proportion of species on earth are lost in a short period of time (typically less than two million years, and in some cases much less)—such as when an asteroid struck earth 65 million years ago and about 75% of all species were wiped out. These extinctions

Once numbering in the millions, passenger pigeons were hunted to extinction. The last passenger pigeon on earth, named Martha, died at the Cincinnati zoo in 1914.

FIGURE 16-9 **The only way to see one today.** Shown here is a stuffed passenger pigeon. Inset: the aftermath of a pigeon-hunting expedition; in a single day, as many as 50,000 birds could be killed.

TWO CATEGORIES OF EXTINCTIONS

Extinctions generally fall into two categories:

- **MASS EXTINCTIONS**
  A large number of species (even entire families) become extinct over a short period of time due to extraordinary and sudden environmental change.

- **BACKGROUND EXTINCTIONS**
  These extinctions occur at lower rates during times other than mass extinctions.

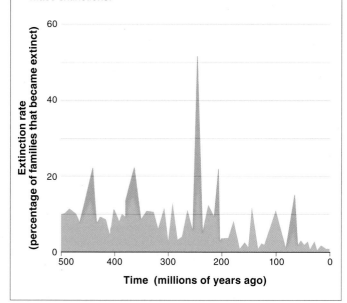

FIGURE 16-10 **Mass and background extinctions.**

FACTORS THAT INFLUENCE THE RISK OF BACKGROUND EXTINCTION

GEOGRAPHIC RANGE:
EXTENSIVE vs. RESTRICTED
Species restricted in their range are more vulnerable than those with extensive ranges.

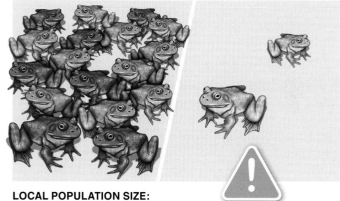

LOCAL POPULATION SIZE:
LARGE vs. SMALL
Species with small population sizes are at increased risk of extinction.

HABITAT TOLERANCE:
BROAD vs. NARROW
Species with narrow habitat tolerances are at greater risk of extinction than species with broader habitat tolerances.

FIGURE 16-11 **Range, population size, and habitat tolerance influence risk of extinction.**

---

are above and beyond the normal rate of extinctions that occur in any given period of time, referred to as the background extinction rate.

Q *Does extinction only happen to weak, unfit species?*

Both background and mass extinctions result in the same outcome for the species involved, but the causes tend to differ. In catastrophic mass extinction events, such as an asteroid impact, for example, the particular features of a species' biology—its biochemistry, physiology, behavior—don't necessarily play a role, in which case it's more like really bad luck. Background extinctions, by contrast, tend to be a consequence of one or more aspects of a species' biology.

For any species, there is always a risk of extinction. But this risk can be larger or smaller, depending on certain features. Here we look at three important aspects of a species' biology that can influence its extinction risk (**FIGURE 16-11**).

1. Geographic range: extensive versus restricted. Species restricted in their range—including those limited to small

bodies of water and those confined to islands—are more vulnerable than species with extensive ranges. The Tasmanian devil is a marsupial carnivore about the size of a dog. Although these animals once thrived in Australia, they now are confined to the island of Tasmania, smaller than the state of Maine. Unable to expand their range, Tasmanian devils are more vulnerable to extinction than if their range was not limited. The house mouse, on the other hand, is found nearly everywhere in the world and so has a much lower risk of extinction.

2. Local population size: large versus small. Tigers and California condors, along with *Welwitschia,* a slow-growing, long-lived plant in southwest Africa, are examples of species that—as a consequence of their small population sizes—are at increased risk of extinction. With a small population size, a species is more susceptible to extinction due to the variety of factors that can kill individuals, including fire, diseases, habitat destruction, and predation. Put simply, the more individuals that are alive, the more likely it is that some of them will survive such events. In addition, having more individuals to breed with typically leads to greater genetic variation. When a population is small, inbreeding is increased, which generally reduces the fertility and longevity of individuals.

3. Habitat tolerance: broad versus narrow. "Habitat tolerance" describes the breadth of habitats in which a species can survive. Some plant species, for example, can tolerate large swings in water availability or soil pH or temperature. Others are limited to very narrow habitat ranges. The now-extinct passenger pigeons could only build nests in a specific type of forest and needed large numbers of individuals, breeding communally, in order to breed successfully. As forests were cut down and as the birds were hunted, the size of their flocks diminished.

Their narrow habitat tolerance then made them extremely vulnerable to extinction. In general, because species with narrow habitat tolerance cannot adapt in the face of habitat degradation and loss, they are at greater risk than species with broader habitat tolerance.

One species that is enormously successful, thanks in part to its extensive geographic range, large local population sizes, and broad habitat tolerance, is our own. Humans are growing and expanding their range at a staggering rate, consuming resources in an unprecedentedly voracious fashion. By any measure, humans are one of the most successful species in earth's history. This success, unfortunately, is having an increasingly negative effect on the species with which we share the planet, causing extinctions at a significantly higher rate than ever before. We next explore in more detail the conflict between humans' success as a species and the survival of other species.

TAKE-HOME MESSAGE 16·5

Extinctions occur for fundamentally different reasons. Mass extinctions, which can destroy many or all of the species in an area, may reflect bad luck more than the particulars of a species' biology, including its biochemistry, physiology, and behavior. Background extinctions, on the other hand, tend to be a consequence of one or more features of the species' biology. Small population size, limited habitat range, and narrow habitat tolerance contribute to background extinctions.

16·6   We are in the midst of a mass extinction.

Sometimes you can be so close to something that it's difficult to really see it. That may be the situation that humans are in right now when it comes to the global loss of biodiversity. But we are becoming increasingly aware that we are in the midst of a mass extinction. In a recent survey by the American Museum of Natural History, 70% of biologists indicated that they believed this is true and that steps must be taken by governments and individuals to stop this massive loss of species.

Data on current rates of extinction in every well-studied group of plants and animals support the hypothesis that a

Q If mass extinctions are a natural part of earth's history, should we be concerned about our effects on extinction rates?

mass extinction is under way. Among the mammals, more than 10% of all species are currently *endangered,* under imminent threat of extinction, and approximately 21% are endangered or *threatened,* characterized as vulnerable to extinction. Among birds, 4% of all species are endangered and approximately 13% are endangered or threatened. Close to 50% are in decline. Almost a third of all amphibian species are endangered or threatened. Among fishes, mollusks, insects, fungi, and plants, too few species have been evaluated to make an accurate estimate of the proportion of all species that are endangered or threatened. But for the species that have been evaluated, the situation is similar.

Historically, evidence from the fossil record suggests that background extinction rates are about one extinction per million species per year. Currently, as the numbers above suggest, extinction rates may be 1,000 times (or more) greater than this. It is difficult to know the exact magnitude of the problem because, as we saw in Chapter 10, biologists don't have much of an idea about how many species there are on earth—the estimates range from 5 million to 100 million.

Unlike the last mass extinction event, in which the dinosaurs and most other species were wiped out in the wake of an asteroid's collision with earth, this current mass extinction seems to be the result of the activity of one species—humans. Ironically, it is the unprecedented success of humans that is responsible for this situation. The resource needs of so many people have inevitably interfered with the ability of other species to coexist with us.

The chief reason for the loss and impending loss of so many species is habitat loss and habitat degradation. Particularly harmful is habitat loss in earth's tropical rain forests, where biodiversity is greatest. In the past 25 years, half of the world's tropical rain forests have been destroyed, usually by burning to make way for agricultural use of the land or by logging (**FIGURE 16-12**). Urban development, too, is responsible for destruction of rain forests, as the growing human populations continue to expand. Intensive agriculture, livestock grazing, and the development of urban centers have led to the destruction and fragmentation of habitats worldwide.

The loss of biodiversity-supporting habitat is not restricted to the tropics. In the Pacific Northwest, as well, logging of forests has occurred at an unsustainable rate, leading to significant reductions in a diverse range of populations.

Bonobos, the closest living relative of humans, have been driven close to extinction, in part due to hunting.

FIGURE 16-13 **Exploitation of species: the bonobo.**

Beyond habitat loss, another factor that reduces biodiversity is overexploitation of species. Such overexploitation includes the killing of animals for food, pelts, tusks or other body parts, and medicinal products. Bonobos (*Pan paniscus*), for example, a type of chimpanzee and our closest living relatives, are almost never seen in the wild. As a consequence of habitat loss and their exploitation as a source of food in parts of Central Africa, they have been driven to near extinction (**FIGURE 16-13**).

Additionally, the introduction of exotic species into habitats where they would not naturally be found is also having a significantly adverse effect on biodiversity. We explore this in Section 16-8. The consequences of the loss of biodiversity, beyond the values of biodiversity discussed earlier, are largely unknown, but most likely include a serious reduction in the capacity of the environment to provide clean air and water and to recover from environmental and human-induced disasters.

Later in this chapter, we discuss the strategies by which the current high rate of biodiversity loss can be reduced.

DEFORESTATION

Deforestation in the tropics is a leading cause of biodiversity loss.

FIGURE 16-12 **Removing habitat through deforestation.**

TAKE-HOME MESSAGE 16·6

Most biologists believe that we are currently in the midst of a mass extinction, that it is the result of human activities, and that it poses a serious threat to the future survival of humans.

A dense green jungle surrounds Hong Kong.

16·7   Some ecosystem disturbances are reversible, others are not.

When you fly over southern New England today, you look down on an undulating carpet of forest that covers hills and valleys. Roads traverse the forest, and here and there you see towns and cities or cultivated land and pastures, but most of what you see is the tops of trees.

If you could have flown over the same route 200 years ago, things would have looked very different. By the early 1800s, most of the forest in southern New England had been cleared and the land was in cultivation. Homesteads consisting of farmhouses, barns, stables, and other outbuildings were scattered across the countryside. Stone walls extended across hills and valleys, separating cropland from pastures.

Going back further, the same flight 400 years ago would have taken you over mostly forested land. Here and there, Native American villages would have stood among cleared fields, but most of what you saw then would have looked similar to what you see today.

The change from forest 400 years ago to cleared land 200 years ago was a major ecosystem disturbance, but now the area has returned to forest. The species compositions are not identical—today's forests are not exactly the same as those of 400 years ago. Nonetheless, the stone fences found deep in the New England woods today are the only obvious trace of the agricultural history of the area. The fences still snake their way up and down hills, but now they run through solid forest instead of between fields and pastures.

By the middle 1800s, New England settlers were abandoning farms on marginal land and moving west. And, once abandoned, the fields previously cultivated

began a process of ecological succession that returned them to forests (see Section 15-15).

In the first year after a field is abandoned, its bare soil provides a harsh environment for plants. Sunlight blazes down, heating the ground surface and baking moisture from the soil. It takes a tough plant to sprout and grow under those conditions, but there are plants that can do it. We call those pioneering species of plants "weeds," and we see them growing in disturbed habitats—along roadsides, in vacant lots, and even in cracks in pavement.

As weeds grow and die, year after year, their organic matter enriches the soil and helps it retain moisture. And every year, the plants grow more densely so that they shade the ground surface, which no longer gets quite so hot and dry. Additional species of plants can grow in these milder conditions. These plants are taller, so they provide still more shade, and the ground becomes still cooler and more moist. At this stage, the seeds of bushes and small trees can sprout. As these woody plants grow, they make the ground cool and shady enough to allow the seeds of forest trees (oak, maple, or beech) to sprout. More saplings of forest trees sprout and grow every year, and eventually they are so closely packed that little sunshine penetrates their leafy canopies to reach ground level. Undisturbed, the forest stops changing at this point, because the mature trees have created conditions in which only their own seeds can sprout and grow.

Q Once land is cultivated or developed, can it ever return to its previous state?

An ecosystem disturbance is reversible when the disturbance alters the biotic (living organisms) and abiotic (decaying

MOUNT ST. HELENS, MAY 1980
Seen here on the day before the eruption, the volcano is surrounded by coniferous forests.

MOUNT ST. HELENS, SEPTEMBER 1980
The eruption flattened surrounding vegetation and left the area coated in a layer of ash.

MOUNT ST. HELENS, AUGUST 2011
Shrubs and small trees have returned to the plain at the foot of the active volcano.

FIGURE 16-14 **The forest returning.** Just two decades after an explosive eruption of Mount St. Helens, in Washington State, shrubs and small trees had returned to the plain at the foot of this active volcano. The recovery continues today.

organisms, soil, rocks) nature of the habitat but does not result in the complete extinction of any species, making it possible for species to re-establish their populations. The example of the New England forests is based on a disturbance that was caused by humans—clearing land for agriculture in the 1700s—but wind storms, forest fires, floods, and volcanic eruptions have been clearing forests throughout the history of life (**FIGURE 16-14**).

> "Any fool kid can step on a beetle. But all the professors in the world can't make one."
> — ARTHUR SCHOPENHAUER, German philosopher (1788–1860)

If an ecosystem disturbance involves the complete loss of a species to extinction, however, the disturbance is irreversible.

Each species is the result of a long and uninterrupted evolutionary history, involving an interplay of random and selective forces and producing a unique genome. Once lost, a species can never exist again (**FIGURE 16-15**). For this reason, ecosystem disturbances that involve the loss of species can have more devastating consequences than those that do not lead to extinctions, no matter how great the changes to the abiotic environment might be.

TAKE-HOME MESSAGE 16·7

An ecosystem disturbance is reversible as long as the disturbance does not include the complete extinction of any species, so they can re-establish their populations. An ecosystem disturbance that involves the complete loss of a species to extinction is irreversible because a species, once lost, can never exist again.

EXTINCT
Tasmanian wolves

EXTINCT
Chinese river dolphin

ENDANGERED
Blue whale

FIGURE 16-15 **The end of a species.**

 Once lost, a species will never exist again. The blue whale—the largest animal ever to have lived on earth—is at risk of being lost forever.

IDENTIFYING BIODIVERSITY   EXTINCTION   HUMAN INTERFERENCE   CONSERVATION STRATEGIES

16·8 Human activities can damage the environment: 1. Introduced non-native species may wipe out native organisms.

In the rain and sleet of a Chicago winter, the idea of colorful birds flying about is very appealing to some people, and you might see exactly that, in the form of the monk parakeet. These small green members of the parrot family are about a foot (30 cm) from head to tail tip. And if they look out of place in Chicago, it's because they are. Monk parakeets are just one example of an **exotic species** (also called **introduced species**), which refers to species introduced by human activities, intentionally or accidentally, to areas other than the species' native range.

Native to southern South America, monk parakeets were imported to the United States as pets in the 1960s and 1970s. Some of those pet birds escaped, survived, and bred, and free-living populations of monk parakeets are now found in 11 states across the country, including Oregon, Florida, and New York, in addition to those mentioned above, in Illinois (FIGURE 16-16).

Monk parakeets were imported to the United States as pets. Some of those birds escaped, survived, and bred. They now are a nuisance in cities, where they build huge nests on utility poles.

FIGURE 16-16 **A parakeet in the Midwest.**

The phrase "exotic species" suggests colorful birds and butterflies winging through a forest, but the ecological reality of such species can be much darker. Although some introduced species do not cause harm to their new habitat, in many cases they do—including economic or environmental harm, or harm to human health. When species are introduced and cause harm, they are referred to as **invasive species.**

The appearance of the free-living monk parakeets in the United States was greeted with alarm because they are regarded as agricultural pests in South America. Records from the Inca civilization that flourished before Pizarro's invasion of Peru in 1532 attest to crop damage by monk parakeets. And when Charles Darwin visited Uruguay in 1833, he was told that monk parakeets attack fruit orchards and grain fields. Fortunately, the dire predictions that these parakeets would devastate crops in the United States have not been fulfilled as yet, but monk parakeets do make a nuisance of themselves in cities such as Chicago, where they build nests on utility poles, transformers, and floodlights. There is nothing delicate about a monk parakeet nest—it is a large mass of sticks and twigs. And a nest that has been soaked with rain is heavy enough to bring down wires, interrupting electrical service to entire neighborhoods.

While monk parakeets were initially introduced to the United States intentionally, in many cases the introductions of exotic species have been unintentional. This was the case with the brown tree snake, which reached Guam sometime before 1952 by stowing away in shipments of military equipment just after the Second World War. In either case—intentional or unintentional—two species characteristics, in particular, can lead to introduced species becoming harmful, invasive species (FIGURE 16-17).

> **Q** Why should we worry about exotic species?

1. Exotic species may have no predators or pathogens in their new habitat, so their populations may grow unchecked.

2. Native plants and animals may have no mechanisms to compete with or defend themselves against invading exotic species.

Guam, which lies in the middle of the South Pacific Ocean, had no native snakes and no predators specialized

DISRUPTIVE IMPACTS | INVASIVE SPECIES

PROBLEM
Introduced species can harm habitats and their native populations.

CAUSE
When non-native species are introduced–accidentally or intentionally–and cause harm, they are called invasive species. Invasive species can multiply, unchecked by predation, overwhelming competitors and irreversibly altering ecosystems.

STRATEGIES FOR SOLUTION
Better regulation and restriction of intentional introductions; better vigilance against accidental introductions

FIGURE 16-17 **Understanding the impact of invasive species.**

to eat snakes. But Guam did have a magnificent fauna of native birds that had evolved in isolation on the island for thousands of years. Because the species of birds on Guam had never experienced predation by snakes, they had no defense mechanisms against them—and brown tree snakes have eradicated most of the native species of birds. Extraordinary efforts are being made to prevent brown tree snakes from spreading to other Pacific islands, such as Hawaii. Like Guam, these islands are home to species of birds that occur nowhere else in the world, and the islands have no native species of snakes and no native predators of snakes.

One current promising strategy, used by the U.S. Department of Agriculture, is to airdrop dead mice with tablets of acetaminophen—the active ingredient in Tylenol—packed into their bodies. In addition to killing

live animals, brown tree snakes scavenge dead animals such as mice. And it turns out that acetaminophen interferes so much with the oxygen-carrying capability of the snake's hemoglobin that a child's dose of Tylenol will kill a brown tree snake. While initial results are promising, it is too early to predict the long-term success of this program.

We don't have to look beyond the borders of the continental United States, though, to find examples of invasive species that are responsible for massive ecological shifts and economic costs. Purple loosestrife is an attractive flowering plant that is native to Eurasia. It was imported to the United States in the 1800s as a garden plant. Once here, it rapidly escaped from cultivation and invaded wetlands in every state except Florida. It produces thousands of seeds a year and also spreads by sending out underground stems. This aggressive growth overwhelms native grasses, sedges, and flowering plants, replacing diverse wetland communities with monocultures of loosestrife that provide poor-quality habitats for bog-dwelling insects, birds, reptiles, amphibians, and mammals (**FIGURE 16-18**).

The title of "Most Destructive Invaders in North America" may belong to the zebra mussels and quagga mussels. These thumbnail-size freshwater mussels are native to eastern Europe and western Asia. In the early 1800s, the mussels spread to western Europe. They unexpectedly came to North America in the 1980s—in the ballast water of ships traveling through the Saint Lawrence Seaway on their way from the Atlantic Ocean to the Great Lakes. Female mussels produce more than a million eggs per year, and the larvae settle on any solid surface. Sometimes they settle so thickly that they

INVASIVE SPECIES: ACCIDENTAL INTRODUCTIONS

BROWN TREE SNAKE
• Introduced to the island of Guam

• Eradicated most of the native bird species, which had never been preyed upon by snakes and had not developed defense mechanisms

PURPLE LOOSESTRIFE
• Introduced to the U.S. as a garden plant

• Invades wetlands and outgrows native grasses, sedges, and flowering plants in every state except Florida

ZEBRA MUSSELS
• Introduced to the Great Lakes

• Cause damage to industrial facilities by clogging water-intake pipes, and deplete resources available to small fish, eliminating the food supply for larger game fish

FIGURE 16-18 **Unwanted guests.** Invasive species sometimes get into natural habitats accidentally. The results can be devastating.

clog the intake pipes of water-treatment plants and factories and the cooling systems of power plants.

The ecological threat that zebra mussels represent is far more serious than the damage they cause to industrial facilities. The Great Lakes fisheries are based on game species (lake trout, salmon, walleye) and commercial species (whitefish, perch, herring), and they produce revenues of more than $1.5 billion annually. Most of the commercially valuable fish feed on small fish species, such as smelt, which, in turn, feed on microscopic plants and animals (phytoplankton and zooplankton). Zebra and quagga mussels are exceptionally efficient at filtering phytoplankton and zooplankton from the water—so good that they are depleting the resources available to the small fish. Without enough food, the populations of small fish are crashing and eliminating the food supply for the large game and commercial fish. The Great Lakes contain nearly one-fifth of the fresh water in the world, but this enormous ecosystem is being irreversibly degraded by just two exotic species.

One legendary pest is the cane toad. Ironically, it was initially imported into Australia to control a native sugarcane pest, the cane beetle. Unfortunately, not only were the cane toads unsuccessful at controlling the pest populations, but the cane toad populations exploded—preying on numerous species of amphibians, reptiles, mammals, and even birds. They spread throughout much of the continent and now number more than 200 million. Cane toads have become a nuisance in much of the country, as conservation workers struggle to determine how to control their numbers (**FIGURE 16-19**). The cane toad illustrates a common problem encountered when species are intentionally introduced, say, as a method of biologically, "naturally" controlling pests: the effects of a species on an ecosystem in which it does not naturally occur are unpredictable and often terrible.

Controlling or eradicating invasive species once they have become established and spread can be difficult and costly. Consequently, the best approach when dealing with invasive species is early detection and a rapid response.

INVASIVE SPECIES: INTENTIONAL INTRODUCTION

Introduced to Australia to control agricultural pests, cane toad populations exploded and have become an ecological nightmare, killing indigenous snakes, lizards, birds, and even crocodiles.

FIGURE 16-19 **Cane toad infestation.**

Toward these goals, in the United States the National Invasive Species Council has been established and spends approximately $1.3 billion per year coordinating the efforts of 38 federal departments and agencies to prevent and control invasive species.

TAKE-HOME MESSAGE 16·8

Exotic or introduced species are species intentionally or accidentally introduced into a new habitat. They are considered invasive species if they cause harm in their new habitat. Exotic species often come to be considered invasive because, in the new habitat, they often have no natural predators to reduce their population size, and they often encounter prey that have few or no defenses. Invasive species can dominate and irreversibly alter communities and entire ecosystems.

16·9 Human activities can damage the environment: 2. Acid rain harms forests and aquatic ecosystems.

"What goes up must come down." That's a saying that applies to molecules just as much as it does to larger objects. The difference is that when molecules come down, they may not be in the same form as when they went up. And in one dangerous instance of this, chemicals that rise

into the atmosphere as gases return to earth as acidic precipitations, in the form of fog, sleet, snow, or rain.

We use the term "fossil fuels" to describe oil, natural gas, and coal. Composed largely of carbon and hydrogen, these substances form carbon dioxide (CO_2) and water (H_2O) when they burn. Although they are referred to as hydrocarbons, fossil fuels are not pure hydrocarbons. In addition to carbon and hydrogen, they also contain substantial quantities of other elements, including sulfur and nitrogen (**FIGURE 16-20**).

When fossil fuels are burned, the gases sulfur dioxide (SO_2) and nitrogen dioxide (NO_2) are produced. These gases react with water vapor in the atmosphere to produce sulfuric acid (H_2SO_4) and nitric acid (HNO_3), and these acids fall to earth as acid precipitation (**FIGURE 16-21**).

In North America, precipitation has an average pH as low as 4.3 in some parts of the Northeast, more than 10 times

DISRUPTIVE IMPACTS \ ACID RAIN

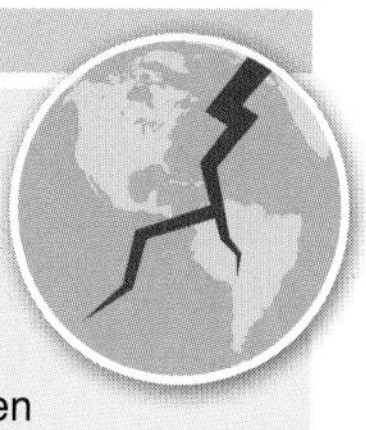

PROBLEM
Acid precipitation kills plants and aquatic animals directly, and also acts indirectly via changes in soil and water chemistry.

CAUSE
Burning fossil fuels releases sulfur dioxide and nitrogen dioxide. The compounds form sulfuric and nitric acids when combined with water vapor.

STRATEGIES FOR SOLUTION
Tighter regulation and reduction of sulfur dioxide and nitrogen dioxide emissions

FIGURE 16-21 **Acid precipitation: the problem and its cause.**

as acidic as clean rain, which has a pH of about 5.6. (Clean rain is more acidic than pure water because carbon dioxide in the air dissolves in water to form carbonic acid, making the rain slightly acidic.) High concentrations of sulfuric and nitric acids in the atmosphere are to blame, and several factors play a role. The Northeast has more people per square mile than the rest of the United States, and that means there are more houses, factories, and automobiles burning the oil, coal, and gasoline that create acid precipitation (**FIGURE 16-22**).

Q Why is acid rain an international issue?

But not all of the pollution that causes acid precipitation is local. In the United States, the Midwest and Southeast

THE CHEMISTRY OF ACID RAIN

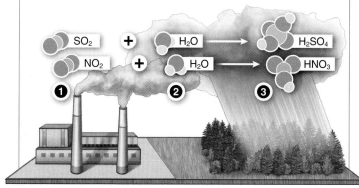

1 Burning fossil fuels releases the gases sulfur dioxide and nitrogen dioxide.

2 When combined with water vapor in the atmosphere, these compounds form sulfuric acid and nitric acid.

3 The acids fall to earth as acid precipitation.

FIGURE 16-20 **Formation of acid rain.**

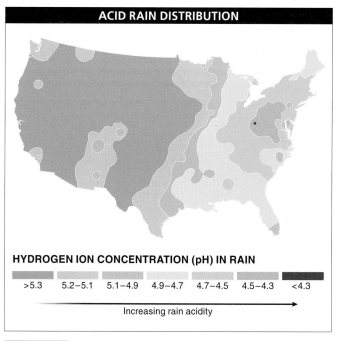

ACID RAIN DISTRIBUTION

HYDROGEN ION CONCENTRATION (pH) IN RAIN

| >5.3 | 5.2–5.1 | 5.1–4.9 | 4.9–4.7 | 4.7–4.5 | 4.5–4.3 | <4.3 |

Increasing rain acidity →

FIGURE 16-22 **Acid precipitation across the United States.** Pure water has a neutral pH of 7.

IDENTIFYING BIODIVERSITY — EXTINCTION — HUMAN INTERFERENCE — CONSERVATION STRATEGIES

have a large number of electric power plants that burn coal, and wind currents carry the sulfur dioxide and nitrogen dioxide produced from this combustion across the Northeast. The precipitation in the western states is not as acidic as that in the Northeast (but it's still significantly more acidic than clean rain). Much of this region's acid precipitation comes from sulfur dioxide and nitrogen dioxide from Asia, which are converted to acids as they blow eastward across the Pacific Ocean. Precipitation in Europe and Asia is also acidified by local and distant sources of pollution.

Both terrestrial and aquatic organisms are harmed by acid precipitation, and the effects of acid precipitation can be direct or indirect. Direct effects result from contact of living tissues with acidic water, whereas indirect effects are produced by interactions between non-living and living systems.

Dramatic evidence of the direct effects of acid precipitation on vegetation can be found in mountain forests that are often blanketed by fog. When trees are exposed to acid fog, year after year, their leaves are damaged, their rates of photosynthesis decrease, and eventually, many of the trees die (**FIGURE 16-23**).

The indirect effects of acid precipitation on forests are not as conspicuous as the direct effects, but they are more far-reaching. As acid falls on the soil, it carries away calcium, potassium, magnesium, and sodium ions. These ions are essential nutrients for plants, and when they are leached out of the soil, the rate of plant growth is reduced. Some forest animals also feel the impact of this change in soil chemistry—when calcium is depleted, for example, the

shells of snails are thinner than usual. As a result of the thinning of snail shells, birds that eat snails receive less calcium than normal, and these birds lay eggs with shells so fragile that they can crack while the parents are incubating them.

Lakes and streams are also affected directly and indirectly by acid precipitation, and some lakes in the northeastern United States and in northern Europe have pH values as low as 4.0–4.3, which is at least 100 times more acidic than most lakes (which typically have a pH between 6 and 8). Very few species of insects, mollusks, crustaceans, fishes, or amphibians can live in water that acidic. Still worse, acid precipitation dissolves aluminum in soil and carries the dissolved metal (aluminum ions)—which is toxic to animals—into lakes and streams, where it can kill most forms of aquatic animal life.

Acid rain is a solvable problem. The deposition of acids from acid rain can be both prevented and cleaned up. The best approach is to prevent or reduce the emissions of sulfur dioxide and nitrogen dioxide. Preventive measures include improving energy efficiency, reducing the use of coal, switching to the use of natural gas, and increasing the use of renewable energy resources, such as wind energy and solar energy. In the United States, many coal-burning plants have now installed special "scrubbers" on their smokestacks. The scrubbers extract hot gases from the power plant and function like sponges, removing sulfur dioxide before it is released into the atmosphere and converting it to a less hazardous form that can be physically removed from the tower and disposed of.

In the United States, regulations required reductions in sulfur dioxide and nitrogen oxides by 2010 to levels approximately one-half of the emissions levels in 1980. Implementation of these regulations has led to significant improvements in air quality. Ambient sulfate concentrations during 2007–2009, for example, were 40% lower in all regions of the country relative to 1989–1991 levels.

EFFECTS OF ACID PRECIPITATION

Trees exposed to acid rain and acid fog have damaged leaves and decreased rates of photosynthesis. Eventually, many of the trees die.

FIGURE 16-23 **Acid precipitation can kill trees.**

TAKE-HOME MESSAGE 16·9

Burning fossil fuels releases the gases sulfur dioxide and nitrogen dioxide, and these compounds form sulfuric and nitric acids when they combine with water vapor in the atmosphere. Rain, fog, sleet, and snow that contain these acids can be more than 10 times more acidic than clean rain. Acid precipitation kills plants and aquatic animals directly, by contact with living tissues, and indirectly, through changes in soil and water chemistry.

16·10 Human activities can damage the environment: 3. The release of greenhouse gases can influence the global climate.

Not long ago, debates about global climate change focused on two questions: "Is it real?" and "Is it caused by human activities?" Now, the vast majority of scientists believe that the answer to the first question is *yes* and that it is very likely that the answer to the second question is also *yes.* As a consequence, the scientific debate has increasingly moved to questions such as "What effect will global climate change have on the world we know?" and "How can we stop it?" Nonetheless, in part because of the significant political implications of global climate change—including, but not limited to, global warming—and because of the complex nature of the data establishing its occurrence, global climate change and the role of human activities remains a controversial topic.

Some of the evidence that the earth's atmosphere is warming comes from weather records that extend back into the 1700s. Those records show that the average temperature has increased rapidly during the past 50 years and that 18 of the 20 hottest years on record have occurred since 1990 (**FIGURE 16-24**). Three centuries is a substantial time from the perspective of an individual human, but it's less than the blink of an eye in the history of the earth. Data from a report on Antarctic ice cores, published in the journal *Science* in 2007, extend the temperature records back to more than 800,000 years ago.

These ice cores reveal regular cycling between periods of cold climate ("glaciations") and warmer climate on earth and show that we are currently nearing the end of a colder

GREENHOUSE EFFECT

Energy from the sun

② Escaped heat

①

Reflected energy

③

Earth's atmosphere

Trapped heat

Earth's surface

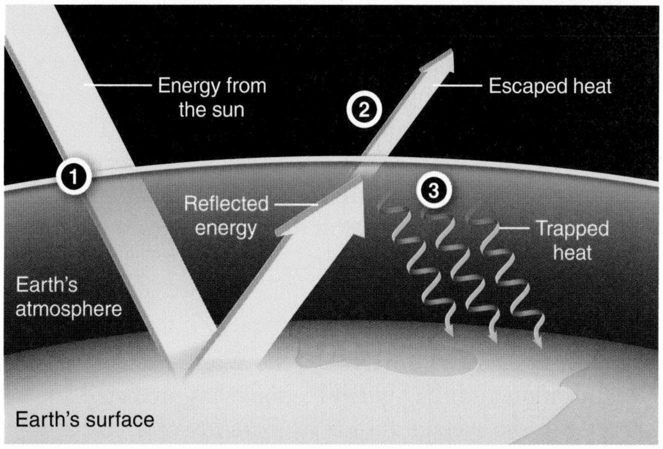

1 Energy from the sun passes easily through the atmosphere to warm earth's surface.

2 Some energy is reflected back toward space and escapes the atmosphere.

3 Some energy is absorbed by greenhouse gases and remains trapped in the atmosphere, heating the air.

Greater concentrations of greenhouse gases trap more heat and lead to higher temperatures.

FIGURE 16-25 **The earth is warmed by the greenhouse effect.**

period. But while they document that the current temperatures are not the highest on record, the Antarctic ice cores reveal some dramatic information about certain "greenhouse gases" in the atmosphere that are associated with global warming. In particular, they show that the current levels of carbon dioxide and methane in the atmosphere are far higher than *any* that have occurred during the 800,000-year span of the cores.

The term "greenhouse effect" describes the process by which energy from the sun warms the earth's atmosphere. The light we see lies in the visible portion of the light spectrum, and visible wavelengths of light pass easily through the atmosphere to warm the earth's surface. Some energy in the infrared part of the spectrum flows from earth back toward space, and several gases in the atmosphere absorb this energy, heating the air (**FIGURE 16-25**). These gases act like the glass panels that make up the roof of a greenhouse: they allow

DISRUPTIVE IMPACTS **INCREASED GREENHOUSE GAS EMISSIONS**

PROBLEM
The average temperature has increased rapidly during the past 50 years, affecting both the physical environment and the biological world.

CAUSE
Burning fossil fuels and clearing land to cultivate crops have significantly increased levels of greenhouse gases in the atmosphere.

STRATEGIES FOR SOLUTION
Reduced emissions of greenhouse gases (particularly from the burning of fossil fuels)

FIGURE 16-24 **Increasing greenhouse gases: the problem and its cause.**

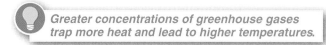

IDENTIFYING BIODIVERSITY   EXTINCTION   HUMAN INTERFERENCE   CONSERVATION STRATEGIES

669

Q Why do we blame global warming on humans?

light energy to pass through, but they keep heat from escaping—hence the name "greenhouse gases." Carbon dioxide and methane are among the most important of the greenhouse gases. Without them, heat would escape into space and the earth's average temperature would be lower.

Humans have increased the amount of greenhouse gases in the atmosphere. And we've done it for primarily one reason: we rely on the burning of fossil fuels—coal, oil, gasoline, and natural gas—for much of the energy we use. The burning of fossil fuels, which have been stored deep in the earth for millions of years, releases carbon dioxide. Clearing land and plowing soil to cultivate crops, too, releases both carbon dioxide and methane.

The Intergovernmental Panel on Climate Change, a scientific body set up in 1988 by the World Meteorological Organization and the United Nations, has predicted an increase in the average annual temperature of 2° to 12° F (1° to 6.4° C) during the 21st century. Sea levels will rise by 7–24 inches (18–59 cm) as warmer temperatures accelerate the melting of glaciers and ice sheets near the North and South Poles. Rainfall is predicted to decrease overall, and droughts will become more frequent and more severe. The impact of droughts will be most severe in areas already on the borderline of aridity, such as Australia, the American Midwest, and sub-Saharan Africa.

Both environmental and biological evidence already reveal the effects of global warming. The Arctic Ocean ice cap is melting, and the glaciers that cover Antarctica and Greenland have also been melting at unprecedented rates— and those rates are increasing (**FIGURE 16-26**). During 1996, 21.6 cubic miles (90 cubic kilometers) of the Greenland ice sheet melted, and by 2010, the annual rate of melting was more than 57 cubic miles per year—faster than in any of the 53 years over which measurements have been made. This melting will cause an increase in ocean levels, creating flooding problems for many of the world's population centers in coastal areas, including New York City, Tokyo, and Amsterdam, and in the world's small island nations.

Biological systems are also showing the effects of climate change. Trees and flowers in northern latitudes are blooming earlier in the spring than they used to; migratory birds are arriving at their summer ranges earlier than they did even a decade ago; and birds and butterflies are extending their geographic ranges northward.

The changes to the physical environment caused by global warming, such as the thinning and shrinking of the Arctic

ICE CAP MELTING

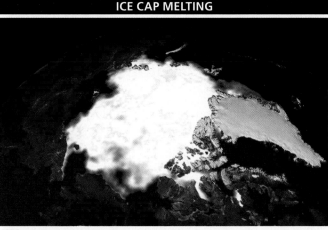

1979

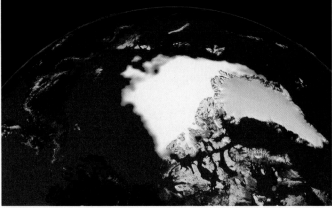

2012

FIGURE 16-26 **Disappearing ice pack.** The satellite image at the top shows the minimum concentration of Arctic sea ice in 1979. The lower image shows the concentration of sea ice recorded on September 16, 2012.

Ocean ice or the flooding of coastal estuaries, can have profound effects on biodiversity at the species and ecosystem levels. For example, in 2004, a research ship in the Arctic Ocean found nine walrus calves swimming alone in deep water. Normally, adult female walruses leave their calves on ice floes in shallow water while they dive to the seafloor to feed on clams and crabs, and then return to the ice to nurse their calves. But the Arctic ice cap over shallow water has melted, driving the female walruses to the remaining ice, which is in deep water. The mother walruses cannot dive deep enough to reach the sea bottom, and they seem to be abandoning their calves.

Other consequences of global warming reveal the tremendous interconnectedness of so many ecosystem elements. Many bird and butterfly species' migratory patterns have been altered as they search for lower temperatures. Many populations of small mammals living near high-altitude peaks are increasingly constrained as they attempt to move toward milder temperatures during the summer months—if they move to lower altitudes it gets hotter; moving up to higher altitudes brings them to milder temperatures, but if they are already at the peak, they've got nowhere else to go when it gets too hot. Massive croplands throughout the world may no longer be able to support the crops previously grown there.

Infectious disease experts have warned that global climate change is likely to alter the distribution of disease vectors—especially arthropods such as mosquitoes and ticks, which are particularly sensitive to temperature—changing and potentially increasing the incidence of diseases such as malaria and dengue fever.

Global warming is a difficult issue to tackle because it is so large in scope; it encompasses every biome, every ecosystem, every species, and every body of water, and its solutions require individual efforts in all nations as well as international cooperation among governments. Fundamentally, however, the science is straightforward: carbon emissions must be reduced significantly. This can be done by individuals who choose to cut down on their own fossil fuel use, improve the energy efficiency of their homes and cars, and support the development of carbon-free renewable energy sources. Internationally, it can be done through efforts to reduce deforestation while increasing the replanting of forests, and by developing and implementing transportation and industrial innovations that reduce dependence on the fuels that produce greenhouse gases. Innovative technologies, such as the underground sequestration of carbon, may also contribute to the reversal of global warming.

TAKE-HOME MESSAGE 16·10

Carbon dioxide and methane are called "greenhouse gases" because they trap heat in the atmosphere. As humans burn fossil fuels and clear forests, the concentrations of greenhouse gases have been increasing and global temperatures have been rising. Ecological changes in plant and animal communities have already been observed and are likely to become more serious unless there is a global reduction in emissions of greenhouse gases.

16·11 Human activities can damage the environment: 4. Deforestation of rain forests causes loss of species and the release of carbon.

Towering trees, colorful birds and butterflies, maybe a glimpse of Tarzan swinging past on a hanging vine—that is the popular image of a tropical rain forest. And it is a reasonably accurate picture (minus the Tarzan part). But it is also a picture that is rapidly fading, as agriculture, logging, gold mines, and oil wells destroy tropical forests (**FIGURE 16-27**).

Tropical rain forests grow in a region extending just a bit north and south of the equator, in South America, Africa, Asia, and Australia. As recently as a few centuries ago, that belt of rain forest covered about 6 million square miles (15.5 million square kilometers). More than half of that forest has already been destroyed, most of it in just the past 200 years (**FIGURE 16-28**).

The destruction of a tropical rain forest means an enormous loss of species, because these forests are the most diverse terrestrial habitats on earth. More than 170,000 species of plants grow in tropical rain forests—more than 60% of the total number of living plant species in the world. A survey of a Brazilian rain forest found 487 different species of trees in just 2.5 acres (1 hectare), an area less than half the size of one city block in Manhattan. To put that number in perspective, consider that only 700

DISRUPTIVE IMPACTS   DEFORESTATION

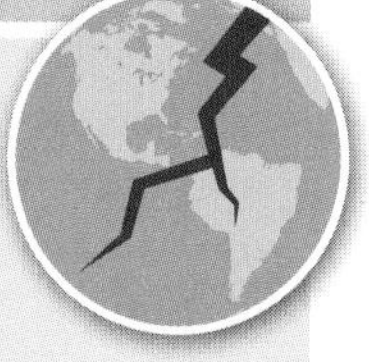

PROBLEM
Tropical rain forests are being cleared at unprecedentedly high rates, endangering countless species and increasing the concentration of greenhouse gases in the atmosphere.

CAUSE
The land is cleared for agriculture, logging, gold mines, and oil wells.

STRATEGIES FOR SOLUTION
Reduced destruction of high-biodiversity habitats, particularly tropical rain forests

FIGURE 16-27 **Deforestation: the problem and its cause.**

TROPICAL RAIN FOREST COVER BEFORE HUMAN INFLUENCE

CURRENT TROPICAL RAIN FOREST COVER

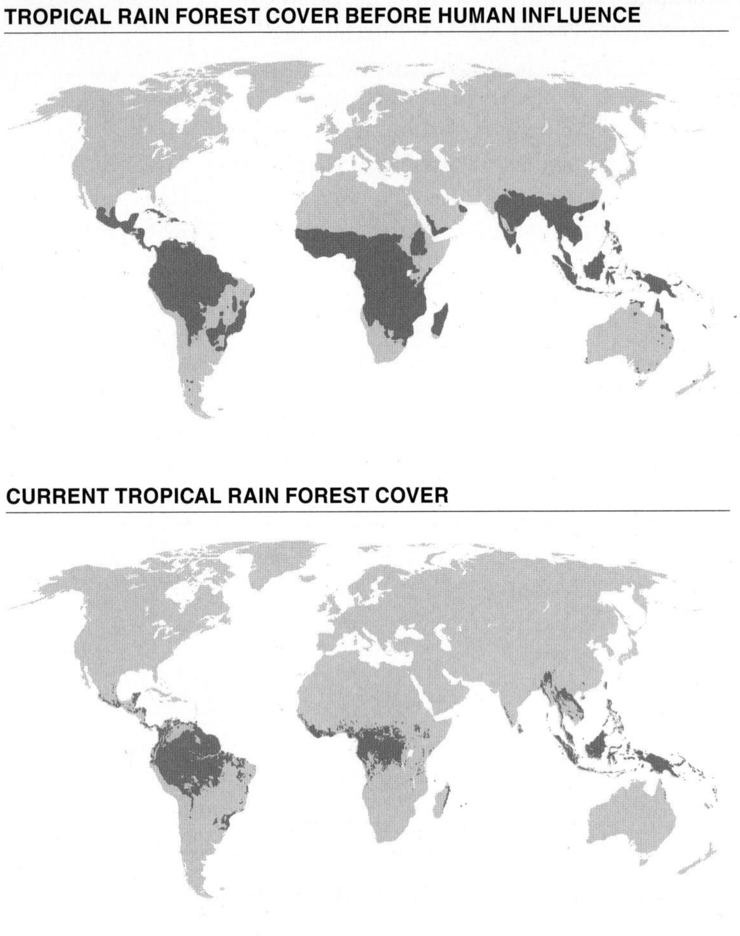

FIGURE 16-28 **Worldwide loss of tropical rain forests.**

species of trees are found in the 5 *billion* acres (2 billion hectares) of the United States and Canada combined!

Agriculture is responsible for the greatest loss of tropical forests, and it takes a variety of forms. Sometimes, a relatively small area is affected. In "slash-and-burn agriculture," trees are cut and burned, and crops are planted in the newly opened area. Usually, just a few acres of forest are cleared, but the land is fertile for only two or three years after it has been cleared. Then that plot is abandoned and more forest is cleared. So the cumulative effect of slash-and-burn agriculture is substantial. At the other end of the size scale, multinational corporations clear hundreds of acres of forest to plant bananas, coffee, or oil palms, and clear thousands of acres of forest at a time to create pastures for cattle.

Pollution from oil wells and mining in tropical rain forests has an impact that can extend far beyond the areas that are actually cleared. Leaking oil contaminates streams and

groundwater, and acidic water drains from mines. Even worse, gold miners use mercury to extract gold, and some of the mercury enters streams, where it is converted to methyl mercury—a toxic compound that accumulates in plant and animal tissues as it moves up the food chain.

Agriculture and mining both require roads to bring equipment to the sites and take crops or minerals to market. Roads have an impact on forests that greatly exceeds the relatively small area they occupy, because they make access to the forest easy. Traveling through a virgin tropical rain forest is difficult—bogs and natural tree falls often prevent travel in a straight line; slopes are often steep and the wet soil is slippery; and the ground-level vegetation can be both dense and thorny. No wheeled vehicle larger than a motorcycle can penetrate most rain forests, and traveling by motorcycle is typically slower than walking. These difficulties protect rain forests, but as soon as a road is bulldozed through a forest, people flock to it. They and their activities then spread into the forest on both sides of the road, creating new clearings that spread farther and farther into the forest.

Destruction of tropical rain forests has two serious environmental impacts: reducing the earth's biodiversity and increasing the concentration of greenhouse gases in the atmosphere (which, in turn, affects global climate change).

1. Reducing biodiversity. Tropical rain forests of Africa, Asia, the Pacific region, and Central and South America, as we've seen, contain an unusually large number of species of plants and animals and probably other groups of organisms. Half of the world's biodiversity hotspots are in tropical rain forests.

2. Increasing greenhouse gases. Photosynthesis in tropical rain forests removes an estimated 610 billion tons (550 trillion kg) of carbon dioxide from the atmosphere each year. As we learned earlier, accumulation of carbon dioxide in the atmosphere is the major cause of global warming. Thus, the continued photosynthetic activity of tropical rain forests is important in slowing the rate of warming. The huge quantity of carbon stored in rain forests can have a downside, however. When forests are cleared and burned, that carbon is released into the atmosphere. And tropical forests are being cleared at a frightening rate—more than 50,000 square miles (an area larger than the state of New York!) each year between 2000 and 2010.

Because deforestation alters the characteristics of land cover, changing the proportion of solar radiation reflected and absorbed, destruction of the rain forests can have additional significant effects on climate—including rainfall amounts and seasonality. These changes can further accelerate climate change, with potentially significant impacts on ecosystems.

The problem of tropical deforestation is one of the most difficult environmental problems to solve, and solving it will rely on international cooperation and involve multiple strategies: (1) identifying and protecting the most diverse areas; (2) addressing the poverty that drives the need to destroy rain forests for human activities; (3) developing alternative sources of food and income; (4) reducing population growth; and (5) making education about the value of biodiversity a central part of these solutions.

In spite of these varied and complex difficulties, numerous programs around the world are beginning to show some success (**FIGURE 16-29**). In the state of Pará, in the Brazilian Amazon, for example, encouraging progress has been made. As of 2012, this region, which is three times the size of California, has nearly eliminated illegal deforestation by adopting several new forest management practices.

- **Accountability and protection:** establishing a local system requiring landowners to register their holdings and an environmental police force to identify and punish illegal deforestation.

- **Education:** creating training programs for ranchers and farmers to help them make more efficient use of already-deforested land.

- **Better forest management:** developing environmentally sound infrastructure, such as reducing the building of new roads, long associated with increasing deforestation, in favor of improving existing roads and thus enhancing the ability to bring renewable forest products to market.

The initial transition is difficult, particularly because it can be accompanied by job losses. But efforts are under way to offset the loss of jobs associated with deforestation with increased economic development in non-deforestation-related areas. These areas include ecosystem services such as water management and non-wood forest products, including medicinal plants. The governor of Pará is realistic about the

Planting new trees can undo some deforestation damage.

FIGURE 16-29 **Cultivating seedlings for reforestation in the Amazon rain forest.**

challenge, but still hopeful. In a 2012 speech about his state's goal of "zero net deforestation" by 2020, he identified an essential element of the program, saying, "Additional reduction in deforestation is going to be very difficult. Locals have to really feel they are part of the solution."

TAKE-HOME MESSAGE 16·11

Tropical rain forests are being destroyed at an alarming rate. This deforestation is devastating, for two main reasons. First, tropical rain forests contain more species of plants and animals than all other terrestrial habitats combined, and half of the earth's biodiversity hotspots are in these forests. Second, tropical rain forests remove more carbon dioxide from the atmosphere than any other terrestrial habitat and, as a result, these forests are enormously important in limiting global warming. Deforestation further influences climate change by altering land cover characteristics. Programs addressing the complex difficulties of slowing tropical deforestation are beginning to see some encouraging results.

A worker carries a tree to plant at the edge of Maowusu Desert, China. Trees are planted to prevent the desert from expanding.

16·12 Reversing ozone layer depletion illustrates the power of good science, effective policymaking, and international cooperation.

You are familiar with the oxygen you breathe—it comes in molecules formed by two oxygen atoms and is represented by the chemical formula O_2. Ozone is a different molecular arrangement of oxygen that packs three oxygen atoms into a molecule; its chemical formula is O_3. Ozone irritates the respiratory pathways and lungs, causing coughing and wheezing, and exposure to ozone can induce asthma attacks.

If you live in a city, you are likely to hear radio and television warnings of high ozone levels during the summer. These warnings announce that ozone is expected to exceed safe levels that day. Children and adults with lung disease are advised to remain indoors during an ozone alert.

If ozone makes people sick, why would anyone worry about *depletion* of ozone in the atmosphere? Isn't that a *good* thing? The answer is, in the words of the old saying about what determines the value of real estate, "Location, location, location." At ground level, ozone is bad for you, but ozone in the lower part of the stratosphere, which is 30 miles (50 km) above the earth's surface, protects you from ultraviolet radiation, the rays that cause sunburn and skin cancer. Ozone is a Jekyll-and-Hyde molecule, and one set of environmental regulations tries to reduce the formation of ozone at ground level while a different set of regulations tries to protect ozone in the stratosphere.

There are two types of "ozone depletion": the general reduction in the amount of ozone in the stratosphere and the formation of areas with very low ozone concentration

(called "ozone holes") over the North and South Poles every winter. Synthetic chemicals known as chlorofluorocarbons (CFCs) are the villains in both forms of ozone depletion. CFCs, which were developed in the 1930s, have a wide range of applications, including use as coolants in refrigeration systems. When CFC molecules leak into the atmosphere, they rise to the stratosphere, where sunlight knocks a chlorine atom off the CFC molecule. These free chlorine atoms then catalyze the breakdown of ozone to oxygen (**FIGURE 16-30**).

In the 1970s, scientists noted that the amount of ozone in the stratosphere was decreasing by about 4% per decade. The cold temperatures and circular flows of air that develop over the Poles during the winter concentrate CFCs, forming

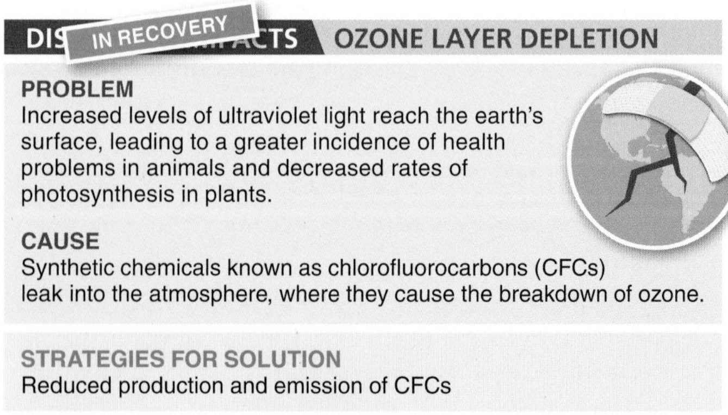

DISTRIBUTION IMPACTS **IN RECOVERY** **OZONE LAYER DEPLETION**

PROBLEM
Increased levels of ultraviolet light reach the earth's surface, leading to a greater incidence of health problems in animals and decreased rates of photosynthesis in plants.

CAUSE
Synthetic chemicals known as chlorofluorocarbons (CFCs) leak into the atmosphere, where they cause the breakdown of ozone.

STRATEGIES FOR SOLUTION
Reduced production and emission of CFCs

FIGURE 16-30 Depletion of the ozone layer: the problem and its cause.

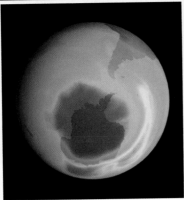

Ozone depletion has slowed, and by 2012 the atmospheric levels of ozone had stabilized.

September 1979　　September 1988　　September 2000　　September 2012

FIGURE 16-31 **Monitoring the ozone hole.** The satellite images show changes in the ozone hole over the Antarctic during the past 30 years. The darkest blue color indicates the lowest ozone concentration.

Depletion of the ozone layer and the formation of an ozone "hole" over earth's polar regions have occurred as a result of human activities.

ozone holes in those locations. The Antarctic ozone hole, which lasts for several months, covers the entire continent of Antarctica and extends northward to include the southern tips of South America and Australia (**FIGURE 16-31**). The Arctic ozone hole is smaller than the Antarctic hole and does not last as long, but it is large enough to extend southward into northern Europe, Asia, and North America.

Q **Should we be concerned about an ozone hole in our atmosphere?** Ozone depletion allows short-wavelength ultraviolet light (UVB light, with wavelengths of 270–315 nanometers) to reach the earth's surface. The 4%-per-decade reduction in global stratospheric ozone levels in the 1970s and earlier increased the intensity of UVB radiation everywhere on earth, particularly under the ozone holes. This raised concern because a 1% increase in UVB intensity increases the incidence of skin cancer by 2% to 3%. Exposure to UVB radiation also promotes the formation of cataracts and reduces the effectiveness of our immune system. Domestic animals suffer the same kinds of damage that humans do, and a few studies indicate that wild animals are also affected.

Beyond the health risks associated with the higher UVB radiation, the damage to ecosystems is an even more serious consequence of ozone depletion. UVB radiation reduces the rate of photosynthesis by plants, and reduced photosynthesis in agricultural ecosystems means that crop yields decline.

Fortunately, the worldwide response to the dangers accompanying creation of the ozone holes is an encouraging example of effective integration of science and policymaking. When CFCs were first invented, they were considered harmless, and so their use as coolants, as aerosol propellants, and in the production of materials such as Styrofoam became widespread. But the recognition that CFCs eventually reach the highest levels of the atmosphere and catalyze the destruction of the protective ozone layer prompted relatively quick reaction and efforts to find solutions (**FIGURE 16-32**) With the adoption by most countries, in the 1980s, of an agreement to discontinue the

REDUCING CFCs

Products made from ozone-depleting CFCs, such as Styrofoam hamburger boxes and hair sprays with damaging propellants, have been phased out in the hope of spurring a recovery of the ozone layer.

FIGURE 16-32 **Taking simple steps to reduce ozone depletion.**

use of CFCs, ozone depletion was slowed. And in 2010, scientists found that the atmospheric levels of ozone had stabilized. A full recovery could occur by 2050.

It's important to note that depletion of the ozone layer is not the mechanism of global warming. Ultraviolet radiation represents less than 1% of the sun's energy. And the greenhouse gases we discussed earlier do not reduce the atmosphere's ozone layer. Global warming and ozone depletion are largely unrelated environmental issues, linked only by their common cause: human activities. The response to ozone depletion, however, represents an encouraging example of how nations can work together to address and correct human-caused environmental problems.

> ### TAKE-HOME MESSAGE 16·12
>
> Ozone in the stratosphere prevents short-wavelength ultraviolet light (UVB) from reaching the earth's surface, but for many decades the amount of ozone was decreasing, largely due to synthetic chemicals called chlorofluorocarbons (CFCs). An increase in UVB light reaching the earth's surface can seriously damage ecosystems and has adverse effects on health, including increasing the incidence of skin cancer in humans and other animals. A decline in the production and use of CFCs is reducing atmospheric ozone depletion, and a full recovery of the ozone layer appears possible.

16·13   With limited conservation resources, we must prioritize which species should be preserved.

Is it preferable to save one beautiful, well-studied bird species or 1,000 species of bacteria, none of which have even been described or named? And what about the 1,200 species of beetles in Panama that live in the evergreen tree *Luehea seemannii,* almost 200 of which live *only* in this one species of tree—how should we prioritize these species for conservation relative to, say, a single primate species? From locale to locale the particulars of these questions may change, but the underlying issue remains: in a world where species are being driven to extinction faster than we can save them, which should be singled out for preservation and which should we leave to almost certain extinction (**FIGURE 16·33**)? This is a question that biologists, policymakers, and, ultimately, citizens must address.

When setting conservation priorities, it is essential to articulate a goal. In an ideal world, the goal might be to protect "all biodiversity." Barring that, options include protecting "most biodiversity"—that is, the most *diverse* subset of biodiversity (in terms of genes, species, and ecosystems)—or protecting the most *valuable* biodiversity. There are numerous arguments in favor of each goal, and each has significant flaws as well. Once a goal is decided on, a plan is needed that outlines priorities for achieving the goal. Frequently, such a plan involves an assessment of the degree to which various components of biodiversity are threatened. We can rank species, for example, as relatively intact, relatively stable, vulnerable, endangered, or critical. These rankings can then be used, in conjunction with

WHAT SHOULD WE PROTECT?

Egyptian tortoise hatchling

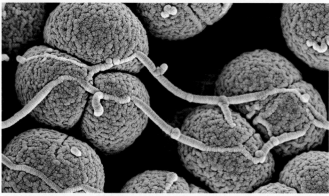

Soil bacteria

FIGURE 16·33 **Prioritizing conservation efforts.** How do we decide among the many options?

FIGURE 16-34 **Which to preserve?** Limited resources and great demand make preservation choices difficult.

Conservation efforts can focus on preserving individual species (such as orangutans) or important habitats (such as rain forests).

measures of the biological value of the biodiversity to humans, in formulating a plan.

As we saw earlier in the chapter, biodiversity can be valued in many different ways, and its worth is not easily quantified or weighed. Sooner or later, many difficult and subjective decisions must be made in the goal-setting and prioritizing process. The undiscovered bacterial species may harbor the metabolic secrets to curing a devastating medical condition in humans, but the beautiful bird species may be much loved by bird-watchers or may serve as a powerful icon of strength or freedom, inspiring generations of people. The decisions are difficult, but not facing them is, in most cases, the same as making a decision.

In the United States, much conservation policy involves response to the **Endangered Species Act (ESA),** a law that defines **endangered species** as those in danger of extinction throughout all or a significant portion of their range. The law is designed to protect these species from extinction. As we noted earlier in the chapter, species that are dwindling are listed as endangered or threatened, according to an assessment of their risk of extinction. Once a species is listed, legal tools are available to help rebuild the population and protect the habitat critical to its survival.

Seemingly straightforward, the ESA has had the effect of focusing most conservation efforts on the preservation of individual species (and populations), sometimes at the

expense of other elements of biodiversity and sometimes at the expense of efforts to reduce the loss of ecologically important habitats (**FIGURE 16-34**). Other difficulties are also associated with the ESA. Consider the task of determining the critical population size below which a population is endangered—is it 500 individuals or 5,000 or 50,000? With its emphasis on preservation of single species, the ESA does not effectively address ecosystem decline, which is an equally urgent and serious problem.

In spite of these difficulties, the species-focused approach to conservation has had some success in the United States. Since it became law in 1973, approximately 40 species that were listed as endangered—including the bald eagle, the peregrine falcon, the gray whale, and the grizzly bear—have been taken off the list as their numbers have recovered. In the final section, we examine other strategies that have been used in efforts to preserve biodiversity.

TAKE-HOME MESSAGE 16·13

Effective conservation requires the setting of goals on the elements of biodiversity (genes, species, or ecosystems) that should be conserved and priorities among those elements. The U.S. Endangered Species Act has focused much conservation effort on the preservation of species.

16·14 There are multiple effective strategies for preserving biodiversity.

In a survey asking which animal they would like to be, men and women gave very different answers. Among men, the top answer was eagle, followed by tiger, lion, and dolphin. As their first choice, women chose cat, followed by butterfly, swan, and swallow. It's just a silly survey, but the differences in male and female selections parallel the fact that people have very different preferences when it comes to preserving biodiversity. One person may view as unthinkable the loss of a particular species or habitat, while another may view other species or habitats as much more valuable.

Most approaches to conservation biology, as we've seen, have focused on the preservation of individual species. Increasingly, however, conservation biology is shifting toward the preservation of important habitats, focusing on conserving communities and ecosystems. This habitat-conserving approach also leads to the preservation of individual species, but it has the added benefit of preserving, at the same time, many different species within a habitat, including many that have not yet been identified, particularly microbes and fungi.

One example of an effort aimed at the preservation of important habitat is the World Wildlife Fund's "Global 200," an identification of all of the most biologically distinct habitats on earth. The identification of these habitats, which include terrestrial, freshwater, and marine habitats, is part of an innovative strategy, often called **landscape conservation,** that is geared toward conserving not just species but habitats and ecological processes (e.g., large-scale migrations and predator-prey interactions). In an age of scarce conservation resources and limited time, this approach prioritizes the conservation of biologically unique habitats on earth.

The habitat conservation approach is not new. The very first attempts at conservation in the United States were the creation of Yellowstone Park in 1872 and Yosemite in 1890, both large-scale efforts geared not simply toward the preservation of one or a few species but rather toward the preservation of the "wilderness." In the time since then, significant efforts have been made to establish national parks, wilderness areas, and recreation areas (FIGURE 16-35).

As conservation biologists have started to better understand population dynamics and biogeography, the design of natural preserves has evolved. Modern preserve design focuses not simply on maximizing the variety of habitats and biodiversity preserved but on using several

FIRST ATTEMPTS AT PRESERVING WILDERNESS

Yosemite was one of the first efforts in the United States at preserving wilderness.

FIGURE 16-35 **The natural beauty of Yosemite.**

design features that maximize the efficiency of the preserves (FIGURE 16-36).

- Larger, rather than smaller, preserves (including a preference for a single, undivided preserve over several smaller preserves)

- Circular, rather than linear, preserves to reduce the negative effects at the preserve edges

- Corridors—even just relatively narrow strips of land—that connect larger natural preserves, allow gene flow, and reduce inbreeding among distinct populations in the different preserves

- Buffer zones (which permit limited amounts of human activity) around core areas (which contain the habitat to be conserved)

As conservation plans aimed at preserving habitats increase, efforts focusing on individual species still continue to be effective. Several strategies targeting individual species for conservation have been particularly successful at preserving large amounts of biodiversity beyond that single species.

1. Flagship species. Some species, because they are particularly charismatic, distinctive, vulnerable, or otherwise appealing, can engender significant public

DESIGNING EFFECTIVE NATURE PRESERVES

The design of natural preserves has evolved. Modern preserves focus on the use of several design features that maximize their efficiency, including larger, rather than smaller, preserves; circular, rather than linear, preserves; and:

CORRIDORS
Strips of land that allow gene flow and reduce inbreeding among distinct populations in different larger natural preserves

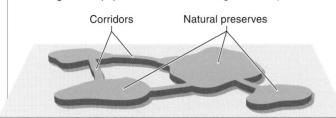

Corridors   Natural preserves

BUFFER ZONES
Areas where limited amounts of human use are permitted that surround a core natural preserve

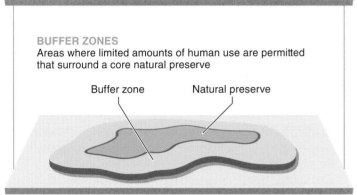

Buffer zone   Natural preserve

FIGURE 16-36 **Carefully planned nature preserves focus on features that maximize biodiversity.**

support. Examples include the giant panda of China, the golden lion tamarin of Brazil's coastal forest, the mountain gorilla of Central Africa, the orangutan of Southeast Asia, the leatherback sea turtle, the Indian tiger, and the African elephant. Preservation of these species, given their habitat needs, can serve to preserve many other species as well.

2. Keystone species. As we saw in Chapter 15, keystone species have a disproportionate effect on the biodiversity of a community. Their removal can lead to massive changes in the composition of species in an ecosystem, often causing huge loss of biodiversity. Examples include kelp, California mussels, grizzly bears, beavers, and sea stars.

3. Indicator species. These are species whose presence within an ecosystem indicates the presence or well-being of a large range of other species. For example, the presence or absence of lichens is an indicator of air quality, because they are sensitive to sulfur dioxide, a component of industrial fumes. When lichens disappear from trees, it is usually an indicator of increased pollution, which endangers the entire ecosystem. Conversely, when lichens are present and healthy, they indicate that the ecosystem is healthy.

4. Umbrella species. These include wide-ranging species with such large needs for habitat and other resources that their preservation ends up protecting the numerous other species within that same habitat, without having to identify these other species as conservation targets. Umbrella species tend to be large, wide-ranging vertebrates.

For the most critically endangered species and degraded habitats, all hope is not necessarily lost. Captive breeding programs and habitat restoration have had some success in bringing back biodiversity from the brink of extinction. Zoos and botanical gardens have taken the lead in many of these efforts. In the 1980s, for example, when the population size of the California condor had dropped to 22 individuals, due to poaching, lead poisoning, and habitat loss, all of the birds were caught and taken to zoos, where a captive breeding program began. With the success of these breeding programs, by 2011 the population had reached 400, including close to 200 birds in the wild and the remainder in zoos. Breeding programs are not a complete conservation solution on their own, however. It is essential that the species' habitats are not destroyed or altered if the species are to flourish again under natural conditions.

Restoration ecology, which uses the principles of ecology to restore degraded habitats to their natural state, has also been an important tool of conservation biologists. Wetlands that have been degraded by dredging and development, in particular, have benefited. Reintroduction of the native plant species and restoration of water and stream flow patterns can be instrumental in restoring these habitats.

Taken together, the strategies used by conservation biologists represent an important step toward reducing the adverse effects of one hugely successful (from a population growth perspective) species—humans—on other species. Although we are far from living with minimal disturbance of the environment in which we live, continued conservation efforts offer our best hope for a sustainable future.

TAKE-HOME MESSAGE 16·14

Conservation biology has focused, in the past, on preserving individual species. Increasingly, there is a shift toward the preservation of important habitats, focusing on conserving communities and ecosystems. Several methods focusing on single species remain useful, however, particularly when preserving the selected species requires the preservation of an amount and type of habitat that simultaneously preserves many other species.

The perils of (exotic) pets!

Some significant threats to biodiversity may come from your own neighborhood. And they're often aided by well-meaning small businesses: pet stores!

Q: **How might otherwise-responsible citizens wreak havoc on their enviornment?** The purchase of exotic, non-native animals as pets can have unintended and harmful consequences if those organisms are released or escape into new habitats.

Q: **What's wrong with keeping an exotic fish in your aquarium or a tropical bird in a cage?** In an aquarium or in a cage, non-native species can be kept safely without harming any natural environment. But problems can occur if they escape, get flushed down toilets, are "set free" by owners who no longer want them, or, in some other way, get into the environment. And it's not their native habitat.

Q: **What happens when non-native species are introduced to a novel environment?** In some cases, non-native species do no harm. But in many others, non-native or exotic species can alter habitat and crowd out, outcompete, or directly prey on beneficial native species. They can spread exotic diseases, for which native species have no resistance. And they can disrupt and harm fisheries, crops, and other valuable industries, costing millions of dollars. Consider just a few common examples.

Fish. Catfish can alter river vegetation and shorelines, making them uninhabitable for native species. Lionfish from home aquariums have been released into the ocean off the coast of Florida. Expert predators, they kill a wide variety of smaller fishes, mollusks, and other invertebrates. In 2008, their population densities in some new habitats were found to have increased by 700% in just five years. China's giant snakehead fish is a voracious eater and preys on numerous species of smaller fishes and amphibians. Traced to the aquarium industry, non-native populations off the east coast of the United States have wiped out numerous populations of native species. The snakehead fish can even crawl from one pond to another, surviving as long as four days out of water! Numerous other types of popular aquarium fish, including emperor angelfish, the yellow tang, and the orbicular batfish, have been found thriving in non-native habitats in the United States.

Birds. More than 150 non-native bird species have been documented in Florida, including 30 different parrot species, all of which were former pets, and many of which now cause significant crop damage, consuming fruits, berries, flowers, and seeds. Monk parakeets often build nests on electrical transformers and lampposts and have been responsible for power outages and other damage.

Amphibians and Reptiles. Terrapins (a type of turtle) and bullfrogs often outgrow their aquariums. If released into ponds, they can wipe out native fish, amphibian, and even small mammal populations.

Plants. Numerous popular aquarium plant species, such as the fast-growing seaweed *Caulerpa,* have become invasive in the Mediterranean Sea and in southern California. Freed from their natural predators, they thrive and have caused millions of dollars in damage.

Q: **What can you do?** You can let your representatives in government know of the potential dangers of non-native species and the importance of closely monitoring the import of such species into the country—an issue that is not always high on their list of priorities. The best advice here, however, is—rather than simply avoiding purchasing exotic pets (or using non-native plant species when landscaping or planting a garden)—to become educated. It turns out, for example, that programs encouraging the export of tropical fishes from the Amazon can serve to reduce deforestation by giving local people the financial incentive to develop markets for renewable natural resources rather than carrying out activities that lead to deforestation.

Check Your Knowledge

GRAPHIC CONTENT

Thinking critically about visual displays of data

1. What are the axes for these graphs? What are the variables being observed?

2. Which variable is independent? Which is dependent? Do these graphs conform to typical norms for the placement of the independent and dependent variables? Why or why not?

3. At what latitude do the largest numbers of species occur?

4. Why are data presented both for mammalian species and for marine copepod species (rather than just one or the other)? Why aren't data for all species presented?

5. At 20° north latitude, are there more mammalian or marine copepod species? What aspect of the graphs may be responsible for someone answering this question incorrectly?

6. Bar graphs often have error bars, indicating some measure of variance in the data. Why do the data bars in these graphs have no such error bars?

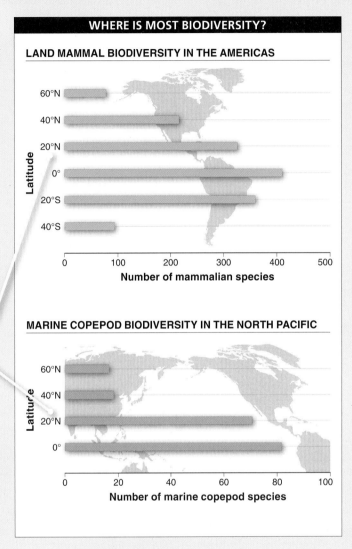

WHERE IS MOST BIODIVERSITY?

LAND MAMMAL BIODIVERSITY IN THE AMERICAS

Latitude / Number of mammalian species

MARINE COPEPOD BIODIVERSITY IN THE NORTH PACIFIC

Latitude / Number of marine copepod species

See answers at the back of the book.

Key Terms in Conservation and Biodiversity

biodiversity, p. 651
biodiversity hotspots, p. 656
conservation biology, p. 654
ecosystem services, p. 651
endangered species, p. 677
Endangered Species Act (ESA), p. 677

endemic species, p. 657
exotic species, p. 664
introduced species, p. 664
invasive species, p. 664
landscape conservation, p. 678
species richness, p. 656

ABOUT THE CHAPTER OPENING PHOTO

Baobab trees (*Adansonia grandidieri*) along a dirt road in Madagascar. One of six species of baobab trees that are endemic to Madagascar, they are endangered as a consequence of habitat loss due to agriculture.

REVIEW & REHEARSE

16

CONSERVATION and BIODIVERSITY

16·1–16·4 Biodiversity—of genes, species, and ecosystems—is valuable in many ways.

Biodiversity has intrinsic value as well as extrinsic value (that is, utilitarian value) to humans.

❓ Check Your Knowledge

1. Biodiversity is considered important because of:

a) the direct economic benefits to humans, such as medicines and food.

b) the symbolic value that elements of the natural world can provide.

c) the aesthetic value it holds.

d) its potential for helping humans understand the processes that gave rise to the diversity we see.

e) All of the above are correct.

0 — **40** — 100
EASY — HARD

2. Biodiversity hotspots are defined by which two criteria?

a) species richness and ecosystem integrity

b) size and distance from nearest alternative hotspot

c) species endemism and degree of threat

d) species richness and size

e) ecological diversity and species diversity

0 — **40** — 100
EASY — HARD

3. Madagascar is unusually important to conservation because it:

a) has more species per unit area than any other place on earth.

b) is the fourth largest island in the world.

c) is home to more endangered species than any other country.

d) has a higher percentage of endemic plants and animals than any comparably sized area on earth.

e) is the native habitat of the rosy periwinkle, a rain forest plant that helps cure childhood lymphocytic leukemia.

0 — **37** — 100
EASY — HARD

THE VALUES OF BIODIVERSITY

INTRINSIC VALUE
The inherent value of biodiversity, independent of its value to humans

EXTRINSIC (UTILITARIAN) VALUE
The value of biodiversity to humans, often described as ecosystem services, can be grouped in four categories of services:

PROVISIONING SERVICES
The many useful products that humans obtain from nature, including food, pharmaceutical, and industrial products

CULTURAL SERVICES
The aesthetic, symbolic, and spiritual values of biodiversity, as well as scientific and educational value

REGULATING SERVICES
The values that come from the regulation of our environment, including climate regulation, waste decomposition, protection from natural disasters, and pest and disease control

HABITAT SERVICES
The value to humans from soil formation, photosynthesis, and nutrient cycling

❓ 1. Following the oil leak in the Gulf of Mexico, scientists observed a drop in the oxygen saturation of the water and a disappearance of methane. What were they able to determine from this observation?

WHAT IS BIODIVERSITY?

Biodiversity is more than just a counting of species. It encompasses the genetic variability among organisms within a species, the variety of different species, and the variety of ecosystems on earth.

ECOSYSTEMS
The number of ecosystems in a region

SPECIES
The number of species in an ecosystem

GENES AND ALLELES
The number of alleles in a species

❓ 2. How is biodiversity defined?

❓ 3. What makes studying biodiversity particularly difficult?

FACTORS THAT INFLUENCE BIODIVERSITY

SOLAR ENERGY AVAILABLE
Greater access to solar energy, the fuel for all life, provides increased species richness.

EVOLUTIONARY HISTORY OF AN AREA
Communities diversify over time. The more time that passes without a climatic event, such as an ice age, the greater the diversity in an area.

RATE OF DISTURBANCE
A habitat with an intermediate amount of disturbance tends to have the greatest species richness.

❓ 4. Describe three factors that influence the species richness in an area.

BIODIVERSITY HOTSPOTS

Biodiversity hotspots are regions of significant biodiversity under threat of destruction. There are 25 biodiversity hotspots around the world; they cover less than 1% of the world's area, but have 20% or more of the world's species.

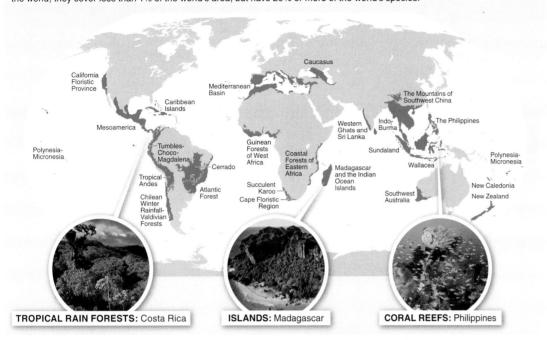

TROPICAL RAIN FORESTS: Costa Rica

ISLANDS: Madagascar

CORAL REEFS: Philippines

16·5–16·6 Extinction reduces biodiversity.

An extinction occurs when all individuals in a species have died.

TWO CATEGORIES OF EXTINCTIONS

Extinctions generally fall into two categories:

- **MASS EXTINCTIONS**
  A large number of species (even entire families) become extinct over a short period of time due to extraordinary and sudden environmental change.

- **BACKGROUND EXTINCTIONS**
  These extinctions occur at lower rates during times other than mass extinctions.

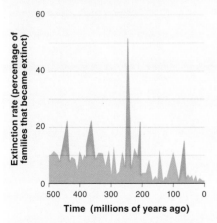

CURRENT MASS EXTINCTION

Most biologists believe that we are currently in the midst of a mass extinction, that it is the result of human activities, and that it poses a serious threat to the future survival of humans.

FACTORS THAT INFLUENCE THE RISK OF BACKGROUND EXTINCTION

GEOGRAPHIC RANGE:
EXTENSIVE vs. RESTRICTED
Species restricted in their range are more vulnerable than those with extensive ranges.

LOCAL POPULATION SIZE:
LARGE vs. SMALL
Species with small population sizes are at increased risk of extinction.

HABITAT TOLERANCE:
BROAD vs. NARROW
Species with narrow habitat tolerances are at greater risk of extinction than species with broader habitat tolerances.

5. Describe a feature of tigers that renders their species susceptible to extinction.

6. What is the chief reason for the loss and impending loss of so many species?

---

4. A habitat with _____ disturbance is expected to have the greatest species richness.
   a) a great deal of
   b) an intermediate amount of
   c) predictable
   d) very little
   e) sporadic

   0 — 79 — 100 EASY / HARD

5. What bird, once the most abundant in North America, was hunted to extinction by shooting and trapping during the 1800s?
   a) California condor
   b) blue-footed booby
   c) dodo
   d) passenger pigeon
   e) great auk

   0 — 53 — 100 EASY / HARD

6. How many species do biologists estimate to be currently existing on earth?
   a) fewer than 1 million
   b) between 3 million and 5 million
   c) more than 1 billion
   d) between 5 million and 100 million
   e) between 1 million and 2 million

   0 — 45 — 100 EASY / HARD

7. Which of the following is currently the leading cause of extinction?
   a) pollution
   b) habitat loss
   c) disease
   d) exotic species
   e) overexploitation

   0 — 29 — 100 EASY / HARD

8. Currently, the major threat to most large land mammals is:
   a) invasive species.
   b) loss of genetic variation.
   c) overexploitation.
   d) habitat loss.
   e) mutation.

   0 — 33 — 100 EASY / HARD

9. Exotic species can disrupt ecosystems because:

a) they frequently have no predators in their new habitat and grow unchecked.

b) they are favored by ecotourists.

c) they have no natural prey items and so must rely on humans for their survival.

d) they have better dispersal capability than endemic species.

e) All of the above are correct.

0 — 42 — 100
EASY   HARD

10. Even though there is a carbon cycle, carbon dioxide levels around the world seem to be rising. Which of the following best explains why this is so?

a) Animals give off carbon dioxide during their normal metabolism.

b) As the atmosphere heats up, it can contain more carbon dioxide.

c) The destruction of coral reefs leads to increased levels of carbon dioxide.

d) More carbon dioxide is being given off by ocean waters as they heat up.

e) The burning of fossil fuels releases more carbon dioxide into the atmosphere.

0 — 19 — 100
EASY   HARD

11. Increased exposure to short-wavelength ultraviolet radiation resulting from depletion of stratospheric ozone:

a) increases the acidification of lakes and streams.

b) reduces the rate of photosynthesis by plants.

c) has lowered the average daily temperature on the earth's surface by 4% per decade.

d) increases the rate of photosynthesis by plants.

e) has raised the average daily temperature on the earth's surface by 4% per decade.

0 — 75 — 100
EASY   HARD

12. A charismatic species that can engender significant public support for conservation of other species and the ecosystem they all inhabit is called:

a) an indicator species.

b) a keystone species.

c) a flagship species.

d) a phylogenetic species.

e) an endangered species.

0 — 52 — 100
EASY   HARD

16·7–16·11 Human activities can have disruptive environmental impacts.

Disruptions of ecosystems can be disastrous.

SOME DISTURBANCES ARE REVERSIBLE

An ecosystem disturbance is reversible when the disturbance does not include the complete extinction of any species, so they can re-establish their populations.

MOUNT ST. HELENS, MAY 1980
Seen here on the day before the eruption, the volcano is surrounded by coniferous forests.

MOUNT ST. HELENS, SEPTEMBER 1980
The eruption flattened surrounding vegetation and left the area coated in a layer of ash.

MOUNT ST. HELENS, AUGUST 2011
Shrubs and small trees have returned to the plain at the foot of the active volcano.

SOME DISTURBANCES ARE IRREVERSIBLE

An ecosystem disturbance that involves the complete loss of a species to extinction is irreversible because a species, once lost, can never exist again.

EXTINCT   EXTINCT   ENDANGERED

Tasmanian wolves   *Chinese river dolphin*   *Blue whale*

7. Why are ecosystem disturbances that involve the loss of species considered so devastating?

EXOTIC SPECIES

PROBLEM
Introduced species can harm habitats and their native populations.

CAUSE
When non-native species are introduced–accidentally or intentionally–and cause harm, they are called invasive species. Invasive species can multiply, unchecked by predation, overwhelming competitors and irreversibly altering ecosystems.

STRATEGIES FOR SOLUTION
Better regulation and restriction of intentional introductions; better vigilance against accidental introductions

8. Describe two reasons why we should be concerned about exotic species.

ACID RAIN

PROBLEM
Acid precipitation kills plants and aquatic animals directly, and also acts indirectly via changes in soil and water chemistry.

CAUSE
Burning fossil fuels releases sulfur dioxide and nitrogen dioxide. The compounds form sulfuric and nitric acids when combined with water vapor.

STRATEGIES FOR SOLUTION
Tighter regulation and reduction of sulfur dioxide and nitrogen dioxide emissions

9. Which gases are produced when fossil fuels are burned?

INCREASED GREENHOUSE GAS EMISSIONS

PROBLEM
The average temperature has increased rapidly during the past 50 years, affecting both the physical environment and the biological world.

CAUSE
Burning fossil fuels and clearing land to cultivate crops have significantly increased levels of greenhouse gases in the atmosphere.

STRATEGIES FOR SOLUTION
Reduced emissions of greenhouse gases (particularly from the burning of fossil fuels)

10. Why are carbon dioxide and methane called "greenhouse gases"?

GREENHOUSE EFFECT

Greater concentrations of greenhouse gases trap more heat and lead to higher temperatures.

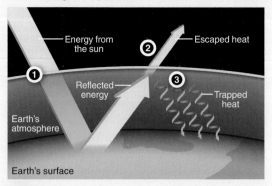

Energy from the sun — Escaped heat
Reflected energy — Trapped heat
Earth's atmosphere
Earth's surface

1 Energy from the sun passes easily through the atmosphere to warm earth's surface.

2 Some energy is reflected back toward space and escapes the atmosphere.

3 Some energy is absorbed by greenhouse gases and remains trapped in the atmosphere, heating the air.

PROBLEM

Tropical rain forests are being cleared at unprecedentedly high rates, endangering countless species and increasing the concentration of greenhouse gases in the atmosphere.

CAUSE

The land is cleared for agriculture, logging, gold mines, and oil wells.

STRATEGIES FOR SOLUTION

Reduced destruction of high-biodiversity habitats, particularly tropical rain forests

11. Describe two important ecological effects of deforestation.

DEFORESTATION: TROPICAL RAIN FORESTS

TROPICAL RAIN FOREST COVER BEFORE HUMAN INFLUENCE

CURRENT TROPICAL RAIN FOREST COVER

13. In conservation biology, what does the term "corridor" refer to?

a) a section of habitat that organisms use to travel between two or more otherwise isolated patches of habitat

b) the edge of a given patch of habitat

c) the minimum range required by a keystone species in any habitat

d) the path a conservation biologist takes through a habitat to minimize any negative effects on the habitat's biodiversity

e) the ancestral geographic range of a species

0 ————————31———————— 100
EASY   HARD

16·12–16·14 We can develop strategies for effective conservation.

Effective conservation requires setting goals on which elements of biodiversity (genes, species, or ecosystems) to conserve and setting priorities among those elements.

OZONE LAYER DEPLETION

PROBLEM

Increased levels of ultraviolet light reach the earth's surface, leading to a greater incidence of health problems in animals and decreased rates of photosynthesis in plants.

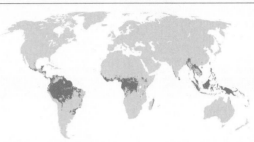
IN RECOVERY

CAUSE

Synthetic chemicals known as chlorofluorocarbons (CFCs) leak into the atmosphere, where they cause the breakdown of ozone.

STRATEGIES FOR SOLUTION

Reduced production and emission of CFCs

12. What has been one effect of the decline in the production and use of CFCs?

PRESERVING SPECIES vs. PRESERVING HABITATS

Most approaches to conservation biology have focused on the preservation of individual species. Increasingly, however, conservation biology is shifting toward the preservation of important habitats, focusing on conserving communities and ecosystems. This habitat-conserving approach also leads to the preservation of individual species, but it has the added benefit of preserving, at the same time, different species within a habitat, including many that have not yet been identified.

13. Describe one impact of the difficulty in defining biodiversity on conservation efforts.

DESIGNING EFFECTIVE NATURE PRESERVES

The design of natural preserves has evolved. Modern preserves focus on the use of several design features that maximize their efficiency, including larger, rather than smaller, preserves; circular, rather than linear, preserves; and:

CORRIDORS

Strips of land that allow gene flow and reduce inbreeding among distinct populations in different larger natural preserves

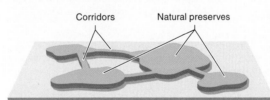
Corridors   Natural preserves

BUFFER ZONES

Areas where limited amounts of human use are permitted that surround a core natural preserve

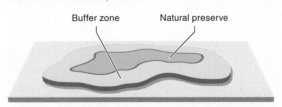
Buffer zone   Natural preserve

14. Describe the significance of keystone species such as kelp.

Periodic Table

ELEMENT CATEGORIES

- Alkali metals
- Alkaline earth metals
- Transition metals
- Lanthanides
- Actinides
- Post-transition metals
- Metalloids
- Other nonmetals
- Halogens
- Nobel gases
- Unknown chemical properties

Key:
- Atomic number
- Element symbol
- Element name
- Atomic mass

GROUP / **PERIOD**

Group	Element
1	1 H Hydrogen 1.00794
1	3 Li Lithium 6.941
2	4 Be Beryllium 9.01218
1	11 Na Sodium 22.9898
2	12 Mg Magnesium 24.3050
1	19 K Potassium 39.0983
2	20 Ca Calcium 40.078
3	21 Sc Scandium 44.9559
4	22 Ti Titanium 47.867
5	23 V Vanadium 50.9415
6	24 Cr Chromium 51.9961
7	25 Mn Manganese 54.9380
8	26 Fe Iron 55.845
9	27 Co Cobalt 58.9332
10	28 Ni Nickel 58.6934
11	29 Cu Copper 63.546
12	30 Zn Zinc 65.409
13	5 B Boron 10.811
13	13 Al Aluminum 26.9815
13	31 Ga Gallium 69.723
14	6 C Carbon 12.0107
14	14 Si Silicon 28.0855
14	32 Ge Germanium 72.64
15	7 N Nitrogen 14.0067
15	15 P Phosphorus 30.9738
15	33 As Arsenic 74.9216
16	8 O Oxygen 15.9994
16	16 S Sulfur 32.065
16	34 Se Selenium 78.96
17	9 F Fluorine 18.9984
17	17 Cl Chlorine 35.453
17	35 Br Bromine 79.904
18	2 He Helium 4.00260
18	10 Ne Neon 20.1797
18	18 Ar Argon 39.948
18	36 Kr Krypton 83.798

37 Rb Rubidium 85.4678	38 Sr Strontium 87.62	39 Y Yttrium 88.9059	40 Zr Zirconium 91.224	41 Nb Niobium 92.9064	42 Mo Molybdenum 95.94	43 Tc Technetium [98]	44 Ru Ruthenium 101.07	45 Rh Rhodium 102.906	46 Pd Palladium 106.42	47 Ag Silver 107.868	48 Cd Cadmium 112.411	49 In Indium 114.818	50 Sn Tin 118.710	51 Sb Antimony 121.760	52 Te Tellurium 127.60	53 I Iodine 126.904	54 Xe Xenon 131.293

| 55 Cs Cesium 132.905 | 56 Ba Barium 137.327 | 57–71 | 72 Hf Hafnium 178.49 | 73 Ta Tantalum 180.948 | 74 W Tungsten 183.84 | 75 Re Rhenium 186.207 | 76 Os Osmium 190.23 | 77 Ir Iridium 192.217 | 78 Pt Platinum 195.078 | 79 Au Gold 196.967 | 80 Hg Mercury 200.59 | 81 Tl Thallium 204.383 | 82 Pb Lead 207.2 | 83 Bi Bismuth 208.980 | 84 Po Polonium [209] | 85 At Astatine [210] | 86 Rn Radon [222] |

| 87 Fr Francium [223] | 88 Ra Radium [226] | 89–103 | 104 Rf Rutherfordium [261] | 105 Db Dubnium [262] | 106 Sg Seaborgium [266] | 107 Bh Bohrium [264] | 108 Hs Hassium [277] | 109 Mt Meitnerium [268] | 110 Ds Darmstadtium [281] | 111 Rg Roentgenium [272] | 112 Cn Copernicium [285] | 113 Uut Ununtrium [284] | 114 Uuq Ununquadium [289] | 115 Uup Ununpentium [288] | 116 Uuh Ununhexium [292] | 117 Uus Ununseptium ? | 118 Uuo Ununoctium [294] |

Lanthanides

57 La Lanthanum 138.906	58 Ce Cerium 140.116	59 Pr Praseodymium 140.908	60 Nd Neodymium 144.24	61 Pm Promethium [145]	62 Sm Samarium 150.36	63 Eu Europium 151.964	64 Gd Gadolinium 157.25	65 Tb Terbium 158.925	66 Dy Dysprosium 162.5	67 Ho Holmium 164.930	68 Er Erbium 167.259	69 Tm Thulium 168.934	70 Yb Ytterbium 173.04	71 Lu Lutetium 174.967

Actinides

89 Ac Actinium [227]	90 Th Thorium 232.038	91 Pa Protactinium 231.036	92 U Uranium 238.029	93 Np Neptunium [237]	94 Pu Plutonium [244]	95 Am Americium [243]	96 Cm Curium [247]	97 Bk Berkelium [247]	98 Cf Californium [251]	99 Es Einsteinium [252]	100 Fm Fermium [257]	101 Md Mendelevium [258]	102 No Nobelium [259]	103 Lr Lawrencium [262]

Answers

CHAPTER 1

Review and Rehearse

Check Your Knowledge
1. b; 2. b; 3. d; 4. b; 5. c; 6. a; 7. a; 8. a; 9. b; 10. a; 11. d; 12. a; 13. b; 14. e; 15. a; 16. d; 17. a

Short-Answer Questions

1. Two claims were made, requiring two double-blind experiments, since only one variable can be changed or tested at a time. *Experiment 1:* Measure intestinal transit time (the time it takes for consumed foods and other substances to travel through the digestive system) for two groups. *Group 1* (control group) does not receive Activia yogurt and instead receives an inert yogurt substitute. *Group 2* (experimental group) receives Activia yogurt. Intestinal transit times are measured for both groups. Neither the participants nor the administrators know which participants are consuming Activia and which are consuming the substitute. Data from group 2, the experimental group, are compared with data from group 1, the control group, to determine the effectiveness of Activia. If the data show that intestinal transit time is shorter in the experimental group, the statement by the Dannon Company is supported. *Experiment 2:* Determine whether DanActive dairy drink helps prevent colds and flu, again using two groups. *Group 1* (control group) does not receive DanActive dairy drink and instead receives an inert dairy drink substitute. *Group 2* (experimental group) receives DanActive dairy drink. The incidence of colds and flu is measured over a finite period of time, such as one year, for the two groups—the same period of time for both groups. Neither the participants nor the administrators know which participants are consuming DanActive dairy drink and which are consuming the substitute. Data collected from the experimental group are compared with data from the control group. If the incidence of colds and/or flu is lower in the experimental group, the statement made by the Dannon Company is supported.

2. Biological literacy can be defined in several ways: (a) The ability to use the process of scientific inquiry to think creatively about real-world issues that have a biological component. Examples could include using the scientific method to answer questions about which plant fertilizers are more effective or to evaluate whether companies' claims about herbal remedies are legitimate and supported by properly designed experiments. (b) The ability to communicate these thoughts to others. Examples could include discussing with a friend the importance of consuming adequate amounts of calcium or properly making tables and graphs for data collected in experiments. (c) The ability to integrate these ideas into decision making. Examples could include interpreting data from experiments and using this information to make decisions. For instance, you could read and evaluate an article about the benefits of eating a diet rich in fruits and vegetables for reducing certain cancers, then decide whether you and others would benefit from this.

3. "Empirical" refers to knowledge gathered through observations and experiments that are rational, testable, and repeatable. In contrast to this empirical approach, knowledge can also be gained through reason, reflection, and observation in the absence of experimentation.

4. Predictions are made based on a formulated hypothesis and are tested with experiments. If a hypothesis is not supported by the experimental results, it can be revised, new predictions made, and these predictions tested with new experiments. Experiments can be repeated many times by many different scientists, and the hypothesis can again be revised and tested if conflicting results arise. In this way the scientific method, done properly, is self-correcting.

5. When a phenomenon can be measured, it can be tested through experimentation. To be testable by the scientific method, all examples would need to comply with this requirement. If a phenomenon cannot be measured, it cannot be tested using the scientific method; examples include determining the existence of God or the beauty of Shakespeare's sonnets.

6. The null hypothesis would state that exercise does not affect the amount of acne a person develops.

7. The prediction would be that a group of people who eat fresh fruits will have a lower rate of illness than another group of people who do not consume fresh fruits.

8. Key features of an experiment: *Treatment:* The condition that is changed for members of the experimental group. It is the variable (e.g., temperature, pressure, light) that is changed to test a hypothesis. Only the experimental group experiences this treatment; otherwise, comparison with the control group will not be possible and the hypothesis cannot be properly evaluated. *Experimental group:* The group that receives the treatment, which is the variable being manipulated to test the hypothesis. *Control group:* The group for which all variables, with the exception of the treatment, are the same as for the experimental group. This makes it possible to determine differences between the experimental group and control group that result from the variable being tested. *Variables:* Characteristics that can be changed, such as temperature, quantity and type of food, light, or pressure, and can potentially illicit a change. *Blind experimental design:* For studies involving human beings, a study design in which the placebo effect and bias can be minimized by preventing members of the experimental group from knowing whether they are part of the experimental (treatment) or control group. However, the experimenters know who is a member of the experimental group or control group. *Double-blind experimental design:* For studies involving human beings, a study design in which the placebo effect and bias can be minimized by preventing members of the experimental and control groups *and* the experimenters from knowing who is part of the experimental (treatment) and control groups. *Randomized:* The random selection of participants to be in either the experimental or control group.

9. (a) Include a control group that does not drink green tea and instead drinks water or other inert drink. (b) Conduct this as a double-blind experiment. (c) Include a large number of people who would be representative of the population advised to drink green tea as a weight-loss method. (d) Stipulate a certain amount of green tea of the same brand to be consumed by study participants. (e) Ensure that only one variable, green tea consumption, differs between the experimental and control groups. Neither group should alter their consumption of food and beverages or change their total calories burned or consumed. (f) Perform a statistical analysis of collected data to determine whether weight loss is significant.

10. If the experimental conclusion does not support the hypothesis, the hypothesis can be revised and retested. If the hypothesis is supported by the experimental conclusion, the experiment can be repeated by other scientists. This will ensure the process is self-correcting.

11. Both theory and hypothesis are attempts to understand a natural phenomenon. A hypothesis has been proposed but not yet widely tested, whereas a theory has been widely tested by many different scientists with repeated results to support it. Both hypotheses and theories can be revised as new information arises from research and experimentation.

12. The study should be a double-blind, controlled experiment utilizing a sugar pill or other inert substance for the control group. All participants in the experiment should be informed of all the potential risks of the drug(s) being tested and the potential consequences of taking part in this study.

13. *Randomized:* Participants are randomly selected to be in either the experimental or control group. *Controlled:* There is a control group in which all variables, with the exception of the treatment, are the same as for the treatment group. This makes it possible to determine whether differences between outcomes for the experimental group and the control group result from the variable being tested. *Double-blind:* For studies involving human beings, the placebo effect and bias can be minimized by preventing members of the experimental and control groups and the experimenters from knowing who is part of the experimental (treatment) group and who is in the control group.

14. A randomized, controlled, double-blind experiment can be used to try to eliminate biases.

15. Statistics can help scientists determine whether differences between the control and experimental groups are a result of random chance or are due to the experimental variable (treatment).

16. No conclusion can be drawn. The fact that the male students who signed up for French tutoring have blue eyes could be due to chance or a result of some biological differences. Without a randomized, controlled, double-blind experiment, there is no way to determine whether these observations result from chance or not.

17. This does not disprove that Scandinavians are usually fair-skinned and blonde. There is a greater likelihood of people from this area being fair-skinned and blonde, but there are always going to be some people with genetic variation that includes darker skin and/or hair. This could result from variations within the population or migration of individuals into the area.

18. The study of life is unified by the themes of hierarchical organization and the power of evolution.

Graphic Content Questions

1. The green portion represents the proportion of all papers published in the journal *Behavioral Ecology* that had a female first author. The blue portion represents the proportion of all papers published in *Behavioral Ecology* that had a male first author. Together, the green and blue portions, making up the whole pie, represent all papers published in *Behavioral Ecology* (that is, 100%).

2. The pie chart on the left presents the proportion of all papers published in *Behavioral Ecology* that had a female first author (and, in blue, the proportion that had a male first author) during the years 1997–2001, when the journal had a "single-blind" review process in which reviewers knew the authors' identities (and, presumably, their sex). The pie chart on the right presents the same information, but for the years 2002–2005, when the journal had a "double-blind" review policy in which reviewers did not know the authors' identities or sex.

3. When the journal changed its review policy so that reviewers did not know the sex of the authors, the proportion of accepted papers that had female first authors increased from 23.7% to 31.6%. This suggests that when reviewers did know the sex of the authors, they were less likely to accept a paper with a female first author. This would seem to indicate that the reviewers for *Behavioral Ecology* at that time were biased against female authors.

4. It could be that, beginning in 2002, there was an increase in the proportion of papers submitted to *Behavioral Ecology* that had a female first author. This explanation seems unlikely, however, given the information in the text that a similar journal that did not change its review policy had no change in the proportion of published papers with a female first author. Another possible explanation is that the proportion of women in the field of behavioral ecology increased in the early 1990s, so that, during the second period measured, 2002–2005, there were more women at a more advanced stage in their careers, so the quality of the papers they submitted increased over time.

5. The increase in proportion of published papers with female first authors was $31.6\% - 23.7\%$, or 7.9 percentage points. It could also be described as a $7.9\%/23.7\% = 33\%$ increase (if you take one-third of 23.7% and add it to 23.7%, you get 31.6%).

6. The information given in the figure tells us what percentage of the papers published had a female first author. We cannot tell from this information whether 50% of the papers submitted had female first authors. If that were true, it would be reasonable to be concerned that the proportion published wasn't also 50%. But what if only 10% of the people working as behavioral ecologists are women? If that were the case, we might expect that only 10% of the papers published in the journal would have female first authors, and we might be concerned that with such a small proportion of female behavioral ecologists, more than one-quarter of the papers published in this journal had female first authors.

7. The graphs do not prove a general bias against female scientists. Because the data are drawn from just one journal in one field of science, we cannot be certain that they reflect a similar bias across all journals in biology, or in chemistry, physics, or other scientific fields. Nonetheless, the findings do suggest such a bias may exist. It would be valuable to collect additional data on this topic across all of the sciences.

CHAPTER 2

Review and Rehearse

Check Your Knowledge

1. a; 2. e; 3. d; 4. b; 5. a; 6. c; 7. a; 8. d; 9. c; 10. b; 11. a; 12. a; 13. a; 14. a; 15. a; 16. a; 17. e; 18. e; 19. d

Short-Answer Questions

1. The mass of an electron is so small, less than one-twentieth of one percent of the weight of a proton, that its mass can be ignored when calculating the mass of an atom.

2. Atoms are most stable when their outermost shell, called the valence shell, is full. Bonding with another atom will not necessarily create a more stable atom than currently exists and could potentially produce a more unstable atom with a partially full valence shell.

3. Atoms are most stable when their outermost shell, called the valence shell, is full. Since the first shell holds a maximum of two electrons, a hydrogen bonded covalently to another hydrogen will fill its valence shell as each atom shares its single valence electron, forming more stable atoms within the H_2 molecule.

4. Covalent and ionic bonding involve the sharing or transfer of electrons, respectively. Hydrogen bonding involves the attraction between hydrogen and another molecule as a result of weak opposite charges, which is a much weaker attraction.

5. Water is a polar molecule; the region of the hydrogen atoms is slightly positive and the region of the oxygen is slightly negative. Other polar molecules and ionic substances will readily dissolve in water because charged regions or ions are attracted to the oppositely charged regions of the water molecules.

6. Vomit tastes sour because the stomach produces hydrochloric acid, which causes the stomach contents to have a pH of 1–3.

7. The chemical bonds of carbohydrates are readily broken, releasing energy that the cell can use to perform cellular activities. Excess carbohydrates can be stored for future use.

8. The three possible fates of glucose are: (a) breakdown for immediate energy needs, (b) storage as glycogen, or (c) conversion to fats.

9. An enzyme in saliva breaks the bonds of starch to produce sugar, which tastes sweet. Since it takes time for the enzyme to break these bonds, the degree of sweetness will increase over time as the sugar concentration increases.

10. Humans, as well as other mammals, do not produce the enzymes capable of breaking the chemical bonds of cellulose, preventing it from being digested and absorbed.

11. Lipids contain many more carbon–hydrogen bonds than carbohydrates. These bonds are rich in energy.

12. In animals, including humans, a strong taste preference for fats has evolved because fats store so much energy.

13. Phospholipids form the lipid bilayer of the plasma membrane.

14. Any two of the following functions are acceptable answers. *Structural:* Proteins are used to build structures such as hair and nails, among others. *Protective:* Proteins are used to form antibodies in an immune response, as well as forming the components of blood clots. *Regulatory:* Proteins form some types of hormones, which are chemical messengers, and almost all enzymes, the biological catalysts that speed up chemical reactions. *Contractile:* Proteins form contractile fibers used in the contraction of muscle cells and movement of cilia and flagella. *Transport:* Proteins assist with the transport of substances around the body, such as transport of oxygen by hemoglobin in the blood.

15. Essential amino acids cannot be synthesized by the body, but must be consumed in the diet.

16. The amino acid sequence determines the three-dimensional shape of a protein, and the shape of a protein determines the function of the protein. As the shape of a protein changes, so does its function.

17. If the shape of an enzyme changes, the active site may no longer be able to bind its specific substrate, preventing a chemical reaction from occurring.

18. Nucleic acids store information by varying which base is attached at each position in the backbone of the DNA molecule. The specific sequence of nucleotides determines the particular protein produced.

19. The base sequence determines the genetic information contained in a segment of DNA; if the base sequence changes, so does the genetic information it contains. Researchers working on the Human Genome Project presented the sequence of only one DNA strand because the complementary base sequence of the other DNA strand can be inferred from this strand.

20. (a) The sugar in DNA is deoxyribose, whereas the sugar in RNA is ribose. (b) DNA forms double strands, whereas RNA generally forms single strands. (c) The base thymine is found only in DNA, and the base uracil only in RNA.

Graphic Content Questions

1. Blue dots indicate that an essential amino acid is present at optimal levels. White dots indicate that an essential amino acid is not present at optimal levels.

2. Some foods (primarily plant products) contain some of the essential amino acids, and others (primarily animal products) contain all the essential amino acids.

3. The shading indicates an example of how different foods (in this case, lentils and rice) that alone lack one or more essential amino acids can be combined to include all the essential amino acids.

4. Foods missing one or more essential amino acids at optimal levels should not be avoided. Instead, these foods can be combined with other foods containing the missing amino acids to ensure that protein needs are met.

5. Although apples are missing several essential amino acids, when combined with white rice, which contains the missing amino acids, all the essential amino acids are provided. If lentils and almonds are combined for a meal, the essential amino acids methionine and cysteine—which are missing from both—will not be supplied in adequate amounts.

6. There is no information in this figure that defines optimal levels. We can assume that consuming these foods will provide amino acids as shown in the figure, but how much of each food is needed is not indicated.

7. It would be helpful to provide the serving sizes needed to meet the daily needs for each essential amino acid.

CHAPTER 3
Review and Rehearse

Check Your Knowledge

1. e; 2. b; 3. e; 4. a; 5. d; 6. d; 7. c; 8. a; 9. d; 10. d; 11. b; 12. d; 13. c; 14. e; 15. b; 16. b; 17. e; 18. e

Short-Answer Questions

1. Cells are the smallest unit of life that can perform all the activities necessary for life, and all living organisms consist of one or more cells.

2. All prokaryotes share four structural features: (a) a plasma membrane composed of a phospholipid bilayer, which encloses the cellular contents, including the DNA, ribosomes, and cytoplasm; (b) cytoplasm, which is a jelly-like fluid contained within the plasma membrane; (c) DNA organized into one or more circular loops containing the cell's genetic material; and (d) many ribosomes, which are the sites of protein synthesis, directed by the cell's genetic material.

3. *Endosymbiosis:* According to this theory, mitochondria and chloroplasts are thought to have originally been prokaryotes capable of cellular respiration and photosynthesis, respectively, that were engulfed by a cell that developed into a eukaryotic cell. *Invagination:* In this theory, the cell membrane folded in on itself to form inner compartments that specialized to become the membranous organelles.

4. Phospholipids have a hydrophilic polar head containing a glycerol molecule attached to a phosphorus-containing molecule, resulting in regions of partial positive and partial negative charge. Phospholipids also have two nonpolar hydrophobic tails, each containing a chain of carbon and hydrogen. As a result of their chemical structure, phospholipids associate to form two layers (bilayers), with the hydrophobic tails facing the center region and the hydrophilic heads facing the outer, water-containing regions. Phospholipids can float around within the same layer, but generally do not mix with the other side of the bilayer. The bilayer is semipermeable, allowing some materials to pass through but not others.

5. The plasma membrane behaves like a *fluid* because many membrane molecules can move around within their respective layer of the phospholipid bilayer, and it is described as a *mosaic* because it is composed of a variety of proteins, lipids, and carbohydrates.

6. Individuals with cystic fibrosis have a genetic disorder in which faulty transmembrane proteins improperly pump chloride ions across the cell membrane. This occurs primarily in cells lining the respiratory and digestive systems, and results in accumulation of chloride ions in cells and the production of very thick mucus.

7. Common ways in which HIV can be transmitted from an infected person to a non-infected person include: (a) the transfer of blood; (b) the transfer of semen; (c) the transfer of vaginal fluid; and (d) the transfer of milk, from infected mother to nursing child. In each case, to infect exposed cells, the virus must have access to cells that carry the CD4 marker.

8. None of these methods of transport require the expenditure of additional energy. All are limited to movement of substances down their concentration gradients, from regions of higher concentration to regions of lower concentration.

9. An isotonic solution has the same concentration of water as another solution—in this case, the contents of the cell. There is no *net* movement of water between the solution and the cell: the same amount of water that moves into the cell will also move out of the cell.

10. Energy is required to move substances against their concentration gradient.

11. The plasma membrane surrounds an object outside the cell, forming a pocket. The pocket and its contents are moved into the cell to form a vesicle.

12. Cells lining the small intestine have tight junctions between them, preventing the passage of bacteria and nutrients from the intestine into the body cavity.

13. (a) The nucleus controls the activities of the cell by controlling the type and quantity of molecules produced. (b) The nucleus acts as the cell's genetic storehouse.

14. The three types of protein fibers making up the cytoskeleton are: *microfilaments*, solid rod-like fibers involved in cell contraction and cell division; *intermediate filaments*, a rope-like system of overlapping proteins that gives the cell great strength; and *microtubules*, thick, hollow tubes that act as tracks along which molecules and organelles move.

15. Most of the cell's energy is produced in the mitochondria, which extract energy from food. This energy conversion requires oxygen.

16. Lysosomes contain acid and about 50 different enzymes, used to break down certain materials brought into the cell, unwanted materials, and worn-out organelles.

17. Smooth endoplasmic reticulum contains enzymes involved in detoxifying chemicals. This is one of the many important functions of the liver.

18. Proteins and lipids.

Graphic Content Questions

1. The variables are the different cell types observed and the number of mitochondria they contain.

2. It would be helpful to see data for more cell types, such as sperm cells, egg cells, or brain cells. Data on variation in the number of mitochondria among cells of a particular type in the same individual would also be helpful. In other words, do all liver cells have the same number of mitochondria, or does the number vary a lot? And does the number of mitochondria in a specific cell type of an individual vary over time—such as when the person gets older, or in response to exercise?

3. Cells with higher rates of metabolic activity seem to have more mitochondria.

4. First, perhaps "number of mitochondria per unit of cell volume" would be better, which would be a measure of mitochondrial density. This would make it easier to compare energy-generating capacity for cells that differ greatly in size. Second, perhaps mitochondria can shrink or swell. If that's the case (and some researchers believe it is), the energetic needs of a cell might be met by a few big mitochondria *or* by a large number of small mitochondria—in which case, the *number* of mitochondria wouldn't be the best measure. A measure of total mitochondrial *volume* might be better.

5. Some other cells with high metabolic activity might also have relatively large numbers of mitochondria. Sperm cells, for example, might have larger numbers. Similarly, for one cell type, such as muscle cells, more active cells (such as those in leg muscles) might have more mitochondria than less used muscles (such as those controlling your baby toe).

CHAPTER 4

Review and Rehearse

Check Your Knowledge

1. a; 2. e; 3. e; 4. c; 5. a; 6. c; 7. a; 8. b; 9. d; 10. c; 11. b; 12. e; 13. b; 14. b; 15. a; 16. e; 17. d

Short-Answer Questions

1. Fossil fuels are made of plant and animal remains. They require millions of years to form, which essentially makes them a non-renewable resource: any fossil used today will not be replaced for a great deal of time—possibly long after humans have become extinct.

2. The ball has the greatest amount of *potential energy* at the top of the ramp. As it rolls down the ramp, the amount of potential energy decreases. This energy is converted to the energy of motion, or *kinetic energy*, and to *heat*, from friction between the ramp and the ball and between the air and the ball. Once it reaches the bottom of the ramp and stops moving, the ball has the least amount of potential energy and no longer has kinetic energy.

3. Some energy is always converted to heat, which is the least useful form of energy. Energy conversions in which a high percentage of energy forms heat are considered less efficient than energy conversions in which a smaller percentage of energy is converted to heat.

4. Each of the three phosphate groups of ATP carries a negative charge, causing repulsion between them. Bonds between these three phosphate groups readily break, releasing energy that can be utilized by the cell to power energy-consuming processes. Cells can easily regenerate ATP by attaching a phosphate group to an adenosine diphosphate (ADP), with the input of energy. These characteristics make ATP an ideal molecule to carry and store energy.

5. Oxygen is formed during photosynthesis as water is split in the water-splitting photosystem to supply electrons to the electron transport chain and to supply hydrogen ions. Animals inhale oxygen and use it in cellular respiration (as an electron acceptor). If oxygen were not available, there would be no final electron acceptor for electrons that have passed down the mitochondrial electron transport chain. In this case, only glycolysis would take place and relatively little ATP would be produced.

6. Most chloroplasts are located in the leaves, particularly in cells close to the leaf surface. Leaves are designed to capture light; many leaves have flat blades that face the sun, and they are often located in the highest parts of the plant.

7. Chlorophyll, found in chloroplasts, is a green pigment, meaning that it reflects light in the green part of the electromagnetic spectrum. The observed color of any object is a result of the light it reflects. Pigments are chemicals with the ability to absorb specific portions of the electromagnetic spectrum.

8. Energy moves through the cell as electrons, which carry energy, are passed from one substance to another. The substance with a stronger "pull" receives the electron as the substance with less of a pull loses the electron. In chloroplasts, electrons originate in the water-splitting photosystem from the splitting of water molecules: oxygen is a by-product, and hydrogen is used by the chloroplasts. Electrons are energized by light and pass down an electron transport chain, where ATP is produced, and enter the NADPH-producing photosystem, where they are again energized by light—but this time they are picked up by the electron carrier NADPH. The NADPH shuttles the electrons to the Calvin cycle, where they are used in the synthesis of glucose. In mitochondria, glucose enters the cellular respiration pathway, where electrons are removed and ultimately pass down the electron transport chain in the inner membrane of the mitochondria. The energy is used to concentrate hydrogen ions (protons) in the intermembrane space, and this proton gradient is used to make ATP. The electrons that pass down the electron transport chain are accepted by oxygen, which combines with hydrogen to form water.

9. An electron in a photosynthetic pigment that is excited to a higher energy state generally has one of two fates: (a) the electron returns to its resting, unexcited state, releasing energy in the process, some of which may bump electrons in a nearby molecule to a higher energy state, or (b) the excited electron itself is passed to another molecule.

10. Rubisco helps plants pluck off (remove) the carbons from carbon dioxide for use in the building of carbohydrates and other organic molecules. Carbons are used in the Calvin cycle to synthesize glucose.

11. When stomata are closed to conserve water, the exchange of gases—specifically, carbon dioxide and oxygen—is disrupted. If carbon dioxide is prevented from entering the leaf, it will not be available for fixation in the Calvin cycle. As a result, photosynthesis is halted.

12. Energy is derived from food by the breaking of chemical bonds. When the chemical bonds of food molecules, such as glucose, are broken, energy is released and captured in ATP. The chemical bond between the second and third phosphate groups of ATP stores energy for use by the cell.

13. Glycolysis takes place in the cytoplasm, outside the organelles.

14. Mitochondria are the site of aerobic cellular respiration, where a majority of the cell's ATP is produced. Glycolysis occurs in the cytoplasm, outside the mitochondria. During glycolysis, only 2 ATP molecules are produced per glucose molecule. Cellular respiration in the mitochondria, requiring oxygen, includes the Krebs cycle (citric acid cycle) and the electron transport chain and produces an additional 30–32 ATP molecules per glucose molecule. Therefore, the vast majority of the cell's ATP is produced in the mitochondria, making them the "ATP factories."

15. In cellular respiration, electrons removed from glucose ultimately pass down the electron transport chain in the inner membrane of the mitochondria. The energy derived from electrons as they pass down the electron transport chain is used to concentrate hydrogen ions (protons) in the mitochondrion's intermembrane space. This proton gradient in then used to make ATP, and the electrons that pass down the electron transport chain are accepted by oxygen, which combines with hydrogen to form water. It is possible to concentrate hydrogen ions in the intermembrane space because the mitochondrial membranes are impermeable to ions; the concentration gradient provides the potential energy that can be used to make ATP.

16. Oxygen acts as an electron acceptor at the end of the electron transport chain. Imagine that the electron transport chain is like a set of stairs. Every time an electron falls down one step, some energy is lost from the electron and is used for ATP production. Once each electron makes its way to the end of the electron transport chain (the bottom of the stairs), it is accepted by oxygen, combining with hydrogen to form water. If no oxygen were available, electrons would have no place to go and would collect at the bottom of the stairs. This would prevent other electrons from moving down the electron transport chain, halting aerobic cellular respiration and the majority of the cell's ATP production. Glycolysis would continue, but would produce a net of only two ATP molecules per glucose molecule—not enough energy to support many cellular activities.

17. Dietary lipids are broken down into fatty acids and glycerol molecules. Glycerol molecules can enter glycolysis, and fatty acids can be converted to acetyl-CoA, eventually entering the cellular respiration pathway.

Graphic Content Questions

1. The *y*-axis, or vertical axis, indicates the amount of pigment molecules present in the leaves. The *x*-axis, or horizontal axis, indicates the pigment types: chlorophyll *a,* chlorophyll *b,* and carotenoids.

2. (a) There are three variables: the *type* and *relative amount of pigments,* including chlorophyll *a,* chlorophyll *b,* and carotenoids, and the *season,* either fall or spring. (b) The amount of pigment in the leaves was measured. There are no numerical values. The bars indicate only the relative amount of each pigment. (c) Each color represents a different pigment; chlorophyll *a* is represented by dark green, chlorophyll *b* by yellowish green, and carotenoids by orange.

3. The source of the data is not indicated; if it were, this might tell us something about the reliability of the information presented. Since these data are presented in a textbook written and reviewed by experts in their fields, we can assume that the information is from a reliable source. If these same data were presented in a publication not written or reviewed by experts in the field, we would have some concerns about the reliability and accuracy of the data.

4. The two graphs indicate relative amounts of pigments during two different seasons: the graph on the left, the spring, and the graph on the right, the fall.

5. The amounts of chlorophyll *a* and chlorophyll *b* are lower in the fall than in the spring, whereas the amount of carotenoids stays the same.

6. It would be helpful if the figure indicated: (a) units of measurement for the pigments; (b) types of plants for which the pigments were measured; (c) whether the amounts shown are averages or minimum/maximum amounts of pigments; (d) the source of the data; (e) the geographic region where samples were collected; and (f) the dates the samples were obtained.

7. This is an informational graph. No hypothesis has been stated, and the information is not presented or used to test a hypothesis. If the data were presented differently—especially if numerical measurements were given and a hypothesis stated—they could be seen as data from an experiment.

CHAPTER 5

Review and Rehearse

Check Your Knowledge

1. e; 2. e; 3. b; 4. c; 5. d; 6. d; 7. e; 8. b; 9. c; 10. b; 11. b; 12. a; 13. e; 14. c; 15. d; 16. a; 17. d

Short-Answer Questions

1. When biological evidence exists, DNA analysis can be used to identify criminals with accuracy and to clear suspects and exonerate persons mistakenly accused or convicted of crimes.

2. In this DNA molecule, 30% of the bases would be thymine. Cytosine bases are complementary to guanine bases, so 20% of the bases are cytosine and 20% are guanine, for a total of 40%. Adenine and thymine make up the remainder of the bases, or 60%. Since adenine and thymine are complementary bases, they occur in equal quantities: 30% adenine and 30% thymine.

3. The sugar-phosphate backbone is the structure that holds the bases in sequence; only the base sequence carries genetic information.

4. The exact function of much of this DNA is not known. It consists of non-coding DNA sequences within genes, called introns, and other non-coding DNA outside genes; these sequences of non-coding DNA are made up of duplicate versions of genes, gene fragments, pseudogenes, and segments that act as switches. "Junk" is not an appropriate term because scientists are only just beginning to understand the functions, or lack of function, of this DNA.

5. Transcription is the process in which DNA is used as a template for the production of a messenger RNA molecule in the nucleus. The nucleotide sequence of the mRNA can then be used to encode the amino acid sequence of a protein, in the presence of ribosomes, in the cytoplasm. Since DNA is restricted to the nucleus, mRNA is needed to transfer information from the nucleus to the cytoplasm for translation.

6. For translation to occur, mRNA, synthesized in the nucleus, must move to the ribosomes, which are the cytoplasmic organelles involved in protein synthesis. Ribosomes are either attached to the endoplasmic reticulum or free-floating in the cytoplasm.

7. Transfer RNA molecules move specific amino acids, corresponding to a tRNA's anticodon sequence, to the growing amino acid chain being produced on an mRNA-ribosome complex. Each tRNA molecule binds (through its anticodon) to a complementary RNA sequence, called a codon, on the mRNA to ensure the correct sequence of amino acids making up the new protein.

8. When *E. coli* inhabits an environment with lactose but without glucose, lactose binds to the repressor protein that is produced by the regulatory gene, located outside the lac operon. This binding causes a change in the shape of the repressor and prevents it from binding to the operator. This allows RNA polymerase to bind to the promoter and begin transcription, producing the mRNA that codes for the enzymes needed for lactose metabolism.

9. Mutations allow evolution to occur. With new alleles added to the genome, new traits may provide reproductive advantages not present in the past.

10. This inherited disease is caused by a genetic mutation that affects an enzyme involved in lipid digestion in lysosomes. As a result, undigested lipid accumulates in lysosomes, and the lysosomes swell until they fill up the cell, resulting in death of the cell. Eventually, a Tay-Sachs sufferer will die from this disorder.

11. (a) Several types of crops have been genetically engineered with insecticide genes inserted into their genomes, which has reduced the need for application of pesticides. This has reduced costs for farmers and reduced crop loss. (b) Some plants have been genetically engineered by insertion of a gene that provides resistance to herbicides. Farmers can now apply herbicides to fields of these crop plants; the crop is unharmed and the weeds are killed. The need for tilling is reduced and crop yield is increased.

12. As farmers use more genetically modified crops, there is much less genetic diversity. The loss of genetic diversity makes crops more susceptible to environmental changes, such as drought, insects, and fungal infections. Plant populations no longer have the genetic potential—that is, the variety of alleles—for imparting resistance to these stressors.

13. *Improvements in agriculture:* Plants and animals can be genetically altered to improve yield, to resist environmental pressures, such as drought and insects, and to provide additional nutrients. The net result is the potential to provide more food and improve nutritional health for more people. *Treatment of diseases and production of medicines:* Many of today's medicines have resulted from biotechnology, improving or saving many lives. For example, the gene for insulin is inserted into the DNA of bacteria, which then produce insulin for use by people with diabetes. Although holding much promise, gene therapy has not yet produced the benefits hoped for. *Cloning:* Agriculture has used cloning in animals to ensure that desired traits are maintained from one generation to the next. Cloning of tissues and organs provides the potential for treating many human ailments. When used responsibly, cloning helps improve human health and nutrition.

14. Since many genetic diseases are recessive diseases, potential parents not showing symptoms of the disease might be carriers of the gene. DNA testing can show whether they are carriers. If both potential parents are carriers, they can seek genetic counseling and medical alternatives.

15. Producing animals with the traits that farmers desire takes a long time, and conventional breeding techniques may not maintain the desired traits in successive generations. Cloning of animals ensures that the desired traits are maintained, and therefore is an appealing alternative.

Graphic Content Questions

1. According to the pie chart, 86% of corn crops grown in the United States are genetically modified.

2. The pie charts indicate only the proportion of genetically modified crop, not the quantity, for each type of crop. Therefore it is impossible, from interpreting the pie charts, to determine whether more of one crop is grown than another. If the pie charts indicated the amount of each crop grown, we could determine whether more genetically modified cotton or corn is grown.

3. The majority of cotton, corn, and soybean crops grown in the United States are genetically modified. Opinions should be based on data.

4. It is much easier to interpret the data by presenting the proportion of each crop than by showing their absolute amounts. Key to these pie charts is illustrating that a majority of these crops grown in the United States are genetically modified. Presentation of their magnitude would only detract from this point.

5. With this information, the reader might consider it interesting to investigate why a smaller proportion of these crops are genetically modified in other parts of the world.

6. Several different types of charts could be produced: (a) a series of pie charts showing the proportion of genetically modified crops for each year; (b) a bar graph, with each bar representing the proportion for each year, the *x*-axis representing the year and the *y*-axis representing the proportion of genetically modified crop grown; or (c) a line graph, with the *x*-axis representing the year and the *y*-axis representing the proportion of genetically modified crop grown, with a line connecting the values for each year. The clearest conclusion from these data is the dramatic increase in the proportion of genetically modified corn grown in the United States over the years shown.

CHAPTER 6
Review and Rehearse

Check Your Knowledge
1. c; 2. b; 3. d; 4. a; 5. c; 6. a; 7. b; 8. e; 9. d; 10. b; 11. c; 12. d; 13. d; 14. e; 15. c; 16. e; 17. d

Short-Answer Questions
1. Rebuilding telomeres could result in cells becoming cancerous. Individuals would have a greater chance of dying from cancer by undertaking therapy of this kind.

2. Prokaryotes typically have circular chromosomes attached to the cell membrane, whereas eukaryotes have linear, free-floating chromosomes in the nucleus. Eukaryotic cells have more genetic material than prokaryotic cells. Eukaryotic cells typically have two or more copies of their chromosomes, contributed by maternal and paternal gametes. Since prokaryotic cells do not take part in sexual reproduction, they do not typically have multiple copies of chromosomes.

3. The main phases of the eukaryotic cell cycle are interphase, during which the cell grows and prepares for division, and mitosis (or M phase), during which the nucleus and genetic material divide, followed by division of the rest of the cellular contents. Interphase consists of several sub-phases. During Gap 1, the cell grows and performs normal functions, such as protein synthesis and waste removal. Cells spend most of their time in Gap 1. Some cells may enter a resting phase (sometimes for days or even years) called G_0, where no cell division occurs. Cells destined for mitosis enter S phase (DNA synthesis), in which chromosomes are replicated, resulting in each chromosome having an exact copy. In Gap 2, cells exhibit a high rate of growth and protein synthesis in preparation for division. In M phase, or mitosis, the parent cell's nucleus (including its chromosomes) divides. This generally is followed by cytokinesis, a process that leads to the division of cytoplasm between the two daughter cells.

4. Making a copy of the DNA ensures that daughter cells have the same genetic information as the parent cell that produced them. Cells lacking a complete copy of the genetic information would be unable to carry out normal activities, maintenance, and reproduction.

5. Mitosis makes it possible for existing cells to generate new, genetically identical cells, allowing organisms to grow and to replace damaged or dead cells.

6. (a) The chromosomes replicate, becoming two identical strands of linear DNA (sister chromatids), held together at the centromere. (b) The sister chromatids condense, coiling tightly and becoming compact.

7. During the S phase (also referred to as DNA synthesis) of interphase, the cell replicates its DNA, creating a duplicate copy of each chromosome (sister chromatids). Once this process is complete, the cell is ready to divide into two identical daughter cells.

8. Cancer cells are unlike normal cells in that: (a) they lack contact inhibition, (b) they can divide indefinitely, and (c) they have reduced "stickiness." Often, death occurs when cancer cells crowd out normal, healthy cells, causing organs and organ systems to fail.

9. During meiosis, two cell divisions occur. The first division separates the homologous chromosomes equally between two daughter cells. The second division separates the sister chromatids equally into four daughter cells. These four daughter cells, called gametes, are haploid, meaning that they contain a single copy of each chromosome. When haploid gametes fuse at fertilization (the union of two gametes), the fertilized egg returns to the diploid state—in other words, it has two copies of each chromosome. If gametes were not haploid, fertilization would result in greater numbers of chromosomes.

10. Homologous chromosomes separate in meiosis I, producing two cells with chromosomes made up of two sister chromatids. Sister chromatids separate during meiosis II, producing four haploid cells (gametes).

11. The larger female gamete contains greater amounts of cytoplasm, with the nutrients needed for development of the organism after fertilization. The sperm contributes only genetic material as it enters the cytoplasm of the egg.

12. Sexual reproduction produces genetic variation, for the following reasons: (a) alleles from two different parents are contributed during fertilization, producing offspring with a genetic makeup different from that of either parent; (b) crossing over during prophase I of meiosis results in a mixture of maternal and paternal genetic information on the sister chromatids; and (c) homologous chromosomes are randomly distributed to daughter cells during meiosis I, resulting in a mixing of maternal and paternal chromosomes.

13. During crossing over, sister chromatids swap segments of equal size between the homologous pairs of maternal and paternal chromosomes, thereby producing sister chromatids that are no longer genetically identical. This results in unique genetic combinations and offspring that are genetically different from their parents—variations on which natural selection can act in the process of evolution.

14. As we learned in Chapter 3, when fertilization occurs, the egg contributes not only chromosomes but mitochondria, which contain their own DNA. The sperm contributes only its chromosomes and no cytoplasm, and hence no mitochondrial DNA. Therefore, the female gamete always contributes more genetic material than the male gamete.

15. Non-sex chromosomes contain genetic information regarding traits other than gender-specific characteristics. The sex chromosomes contain genetic information that instructs the body to develop into one sex or the other.

16. The gender with two different sex chromosomes (e.g., in humans, the male) determines the sex of the offspring. During meiosis in human males, half of the sperm produced inherit the X chromosome and half inherit the Y chromosome. When an egg is fertilized, it is fertilized by a sperm bearing either an X chromosome or a Y chromosome, thus determining the sex of the offspring.

17. Unlike the X chromosome, the Y chromosome does not contain essential genetic information needed for normal development. Information contained on the much smaller Y chromosome is limited to directing the development of male gonads.

Graphic Content Questions

1. According to the graph, the probability of a 35-year-old woman giving birth to a baby with Down syndrome is about 1 in 250 live births (assuming 4 per 1,000 live births). The graph could also be interpreted as indicating 3 babies with Down syndrome per 1,000 live births, with a probability of giving birth to a baby with Down syndrome of 1 in 333.3 live births.

2. For women aged 45, the incidence of live births of babies with Down syndrome is 20 per 1,000 live births, which means the woman is more likely to give birth to a baby without Down syndrome than with Down syndrome. The probability of having a baby with Down syndrome is 1 in 50 live births.

3. No. The graph indicates only the incidence of live births of babies having Down syndrome per 1,000 live births. The graph does not indicate the total number of babies born with Down syndrome for women of any age. Note: Most babies with Down syndrome were born to women younger than 35, since the total of all babies born is greatest for this age group.

4. The exclusion of women over 45 may be due to the lack of data, since a much smaller number of women have children after the age of 45. Note: Some research indicates that the risks are slightly reduced after age 47, but more research is needed.

5. The graph indicates data on live births. Fetuses lost through early miscarriage, late miscarriage, or abortion that have an extra copy of chromosome 21 are not included in the data.

CHAPTER 7

Review and Rehearse

Check Your Knowledge

1. a; 2. b; 3. e; 4. e; 5. a; 6. c; 7. e; 8. a; 9. c; 10. e; 11. e; 12. d; 13. d; 14. b; 15. d; 16. a

Short-Answer Questions

1. Offspring resemble their biological parents because they receive genes from their parents. Genes code for traits that determine physical characteristics that parents and offspring may share.

2. Since offspring inherit one copy of each gene from each parent, if the gene for a single-gene trait is inherited and the trait is expressed, it is easy to determine which parent it came from. Inheritance patterns for multigene traits are more difficult to follow, because their expression can look very different, depending on which combination of genes is acquired and expressed.

3. Gregor Mendel wanted to understand how traits are passed from one generation to the next. For his research, he chose easily observable traits,

which he could categorize and count, in organisms (pea plants) that he could manipulate experimentally to understand their pattern of inheritance.

4. The cells of diploid organisms, with the exception of gametes, have two copies (alleles) of each gene, one allele on each of a pair of homologous chromosomes. For offspring to inherit the same (diploid) number of chromosomes and genes, they must receive only one copy in each parent's gamete, which combine to form a diploid zygote.

5. "Phenotype" is the term used to describe the expression of alleles as physical, physiological, or behavioral characteristics. "Genotype" is the term used to describe the alleles an organism has inherited, which determine what phenotypes are expressed.

6. The term "homozygous" indicates that an organism has only one version of an allele for a specific trait. All of its gametes will contain this allele for the trait.

7. If an individual exhibits a dominant trait but has an unknown genotype, a test-cross can reveal the genotype. The individual is crossed with an individual that is homozygous recessive for that trait. If all the offspring express the dominate trait, then the unknown genotype is homozygous dominant. If half the offspring exhibit the recessive trait and half exhibit the dominant trait, then the individual with the unknown genotype must be heterozygous for that trait.

8. A "carrier" is an individual who carries one copy of a recessive allele, which is not expressed. There is no outward evidence that the individual is carrying the recessive allele.

9. In incomplete dominance, an individual that is heterozygous for a trait expresses a phenotype that is intermediate between the phenotypes associated with the two alleles. An example is the snapdragon flower. If a true-breeding red-flowered plant is crossed with a true-breeding white-flowered plant, all their offspring have pink flowers—intermediate between red and white.

10. Assuming that an individual is diploid, it can have two different alleles for each gene, one carried on each of two homologous chromosomes. An individual cannot have more than two alleles for each gene, unless it has more than two sets of chromosomes.

11. A trait is described as polygenic if many genes control the trait. As result, there is a lot of variation in the expression of this trait. An example is height in humans.

12. An individual who is heterozygous for sickle-cell anemia has resistance to malaria because the parasite that causes malaria does not survive well in sickled cells, the red blood cells with the defective hemoglobin.

13. Both the X and Y chromosomes are sex chromosomes. There are many more genes on the much larger X chromosome, with no corresponding alleles on the Y chromosome. If a recessive allele on the X chromosome is inherited by a male (XY), that allele will always be expressed.

14. An individual with PKU inherits a defective allele for an enzyme needed to convert phenylalanine to tyrosine; phenylalanine builds up in cells, eventually increasing to toxic levels. If the person modifies his or her diet—that is, modifies the environment—to largely exclude phenylalanine, this amino acid will not increase to toxic levels in the body.

15. Mendel's second law states that the inheritance of one trait does not influence the inheritance of another trait. However, this is not true for traits determined by alleles located on the same chromosome, especially if they are close together. When a chromosome is inherited, groups of alleles, especially those close together, are inherited together.

Graphic Content Questions

1. A novelty-seeking score was measured for individuals with different genotypes. Measurement was based on the individuals' responses on a questionnaire. While this might be a fast and easy way to estimate one person's propensity for novelty-seeking relative to another person's propensity, it would have been better to observe individuals actual behaviors. If you are interested in behaviors, it is always better to see how someone behaves than to ask them how they think they would behave.

2. *CC* refers to a genotype for the promoter polymorphism –521C/T, which alters the transcription rate of the dopamine receptor gene. This gene influences the activity of the brain's pleasure centers. There are two different alleles, *C* and *T,* and a person can have one of each or two copies of *C* or two copies of *T.* Having one or two copies of the *T* allele (that is, *CT* or *TT*) seems to lead to reduced transcription of the dopamine receptor gene, while the *CC* genotype is associated with greater gene activity and thus greater novelty-seeking behavior.

3. The data reported for the initial study provide information on how much individuals with genotypes in the two groups (first group: *CC,* second group: *CT* or *TT*) varied in their score on the novelty-seeking questionnaire (since everyone would not be expected to score exactly the same). This information is useful because it helps us see that not only was the average novelty-seeking score higher among those with the *CC* genotype, but those with the lowest novelty-seeking scores in the *CC* genotype group had higher scores than those with the highest novelty-seeking scores in the *CT/TT* group. The smaller the variation observed in each group, the more likely it is that differences between the two groups reflect true differences in the measure between the groups.

4. A meta-study combines the results of several or many studies that address a related research hypothesis. This is commonly done because the results from a group of studies can allow more accurate data analysis, whereas a single study doesn't necessarily have enough power to reveal a true effect.

5. The 11 different studies did not include exactly same subjects. In fact, they may have used groups of subjects that differed in significant ways, such as age, sex, profession, or ancestry. They may also have measured different numbers of subjects, or they may have used slightly different versions of novelty-seeking questionnaires.

6. The results of any one study showing that individuals with the *CC* genotype score higher on a novelty-seeking questionnaire than individuals with the *CT* or *TT* genotype might be interesting, but unconvincing. Questions might remain about the population from which the subjects were chosen, the number of subjects used, the laboratory methods used in determining genotypes, the conditions under which the questionnaire was administered, and a whole range of unconscious, inadvertent, or otherwise hidden biases that the researchers may have had. But it is powerful when the results of all 11 studies that tested this hypothesis are reported and most seem to show similar results. This means that you can be more certain about the conclusions drawn in any one of the studies. It's important that *all* of the studies conducted on this hypothesis were collected and analyzed, rather than the meta-study simply selecting studies that supported or rejected the hypothesis. We can have more confidence in concluding that an individual's genotype for this gene influences novelty-seeking behavior. (In fact, though, without information about how well a person's score on this questionnaire predicts actual behavior, it might be safer to simply conclude that an individual's genotype influences his or her score on a novelty-seeking questionnaire. This conclusion gives rise to numerous new hypotheses that could be tested.)

CHAPTER 8

Review and Rehearse

Check Your Knowledge

1. e; 2. a; 3. a; 4. d; 5. e; 6. a; 7. e; 8. b; 9. a; 10. c; 11. d; 12. b; 13. b; 14. e; 15. d; 16. c

Short-Answer Questions

1. Fruit flies reach maturity in about two weeks, so a scientist can include many generations during an experiment such as that on starvation resistance. With an organism having a late age of maturity, it would require many years to complete a study such as the starvation resistance experiment.

2. It was thought that: (a) as in the Biblical account, the earth was only about 6,000 years old; (b) the earth had not changed very much over time, with the exception of the occasional earthquake, flood, or volcanic eruption; and (c) all species, including humans, were created at the same time, and these species would never change or die out.

3. Darwin found many fossils that were very similar to species still living in the same area. He also noticed that finches on the Galápagos Islands, which he assumed were all of the same species, had different physical characteristics on different islands. Biologists at the Zoological Society in London later discovered that they were 13 different finch species. All of these species closely resembled a single species of finch on the closest mainland, in Ecuador.

4. Darwin noticed a striking similarity between fossils found in an area and the animals living in that area. An example was the fossils of the extinct glyptodont, which was estimated to be 10 feet long and 4,000 pounds, and lived in the same area that was now occupied by armadillos. This suggested to Darwin that glyptodonts were extinct relatives of the armadillos, contradicting the belief in unchanging species.

5. In 1858, Darwin received a letter from Alfred Russel Wallace, a biologist studying in Malaysia, that presented a clear description of evolution. A joint presentation of Darwin's and Wallace's work at the Linnaean Society of London soon followed. In 1859, Darwin published *The Origin of Species,* which was an immediate success and ensured Darwin's place in history as the father of evolutionary theory.

6. Evolution is a genetic change in a population over time. A change in allele frequency in the population occurs as a result of natural selection, mutations, genetic drift, and migration.

7. Three causes of mutations are: (a) exposure to high-energy radiation; (b) exposure to certain chemicals in the environment; and (c) spontaneous mutation—that is, a random genetic change occurring by chance.

8. Genetic drift is a random change in allele frequency in a population that is not influenced by the allele's impact on fitness. Smaller populations have fewer individuals with each type of allele, so any random loss of individuals is more likely to cause a change in allele frequency in that population. An example is the founder effect, in which individuals leave one population and create a new, smaller population in a different geographic region. This new population is likely to have a different allele frequency than the originating population.

9. Fixation for an allele occurs as a result of genetic drift, when an allele frequency becomes 100% in a population. As a result, there is no genetic variation for that gene, which can be detrimental because the population may be less able to adapt to environmental changes.

10. Evolution is genetic change in a population over time—specifically, a change in allele frequency over time. Natural selection is a cause of evolution in which alleles that increase reproductive success become more prevalent in a population, while those that do not favor reproductive success become less prevalent in a population.

11. The frequency of a recessive allele does not change in a population as long as the allele does not affect reproductive success. If a recessive allele does affect reproductive success, its frequency will change. Those recessive alleles that adversely affect reproduction will decrease in frequency, and those that improve reproductive success will increase in frequency.

12. Fitness is a measure of the reproductive success of individuals. (a) An organism's fitness depends on the environment in which it lives. The organism may be fit in one environment, but less fit in another environment. (b) Fitness is relative to that of other organisms in the same environment. The fitness of one individual is measured against that of other individuals with specific genotypes and phenotypes. (c) The relative fitness of an individual is measured against that of other individuals of the same species in the same population. Individuals who reproduce more frequently than others are more fit, irrespective of the age at which they die.

13. The use of an antibiotic creates environmental pressure, and bacteria possessing one or more alleles imparting resistance to that antibiotic are more likely to reproduce than those that do not possess alleles for antibiotic resistance. Over time, bacteria with antibiotic-resistance alleles become more frequent in the population. Should another antibiotic be introduced, other alleles present in the population that impart resistance to that antibiotic will become more frequent over time.

14. (a) Natural selection requires time, sometimes more time than is allowed by the rate at which changes are occurring in the environment. The world's environments are ever changing and require continual evolutionary change if organisms are to maintain fitness. No single, perfect organism for all environments will ever evolve. (b) Since mutations occur regardless of the needs of the organism, new mutations (i.e., new alleles) may not be those required for the organism to adapt to changes in its environment. (c) There may be more than one allele suitable for selection in an environment, and thus there is no single optimal adaptation.

15. Answers can include the selective breeding of a wide range of animals and plants, such as cats, dogs, chickens, pigs, horses, cows, sheep, and goats, and apples, tomatoes, potatoes, cucumbers, corn, lettuce, strawberries, grapes, peppers, and cucumbers. Each organism chosen as an example should have been modified by humans from its natural state so that it possesses traits desired by humans.

16. Normally, babies weigh between 7 and 8 pounds at birth, as a result of stabilizing selection. Modern medicine has enabled very-low-weight, premature babies to survive, and very large babies to be delivered through Caesarean births. This has prevented selection against extremes in the size of babies born and may lead to a greater number of babies of extreme size (small or large) becoming more prevalent in the future.

17. The forelimbs of mammals have been modified to form, for example, wings in bats, flippers in porpoises, forelegs developed for running in horses, and arms with hands for grasping in humans. These are referred to as homologous structures, since the bones of the forelimbs of mammals have been modified to form a number of different structures to perform vastly different functions.

18. (a) The soft parts of animals are not well preserved in fossils. (b) There are very few environments where erosion and decay do not occur, and thus very few with the conditions required for the formation of fossils. The result is a large number of gaps in the fossil record.

19. Scientists in the field of biogeography have observed that species will migrate to nearby locations and adapt to their new habitats. Different species living less than 100 miles apart and occupying drastically different habitats resemble each other more than they resemble species many thousands of miles away that occupy very similar habitats. This suggests shared common ancestors among species that are close together geographically. An example is found on the continent of Australia, occupied by marsupials. Many niches in Australia that are occupied by marsupials would otherwise have been occupied by placental mammals. For example, the Tasmanian wolf, now thought to be extinct, occupied the same type of habitat as gray wolves. However, the Tasmanian wolf is much more closely related to other marsupials living in Australia than to gray wolves found thousands of miles away.

20. Analogous structures are structures having similar functions that developed as a result of convergent evolution but were formed from different starting materials in unrelated species. For example, the wings of insects, birds, and bats are all used for flying, but developed independently in each group of animals. Homologous structures develop in related species in which the same structures are modified to perform different functions. An example is the modification of the bones making up the forelimbs of mammals. All mammals have a common ancestor, but the bones of the forelimbs have been modified for different functions, such as flying, swimming, and walking.

21. A molecular clock is a way for biologists to determine how long two different species have been evolving separately. The longer two different species have been evolving separately, the greater the differences in their genetic sequences. For example, humans and chimpanzees share many genes, while the fruit fly and humans share much fewer genes.

Graphic Content Questions

1. The curvy line represents the average beak length in millimeters.

2. Birds with longer/larger beaks are more likely to survive during drought conditions, whereas birds with shorter beaks are more likely to survive during very wet conditions.

3. This information shows the influence of drought and wet conditions on average beak size. The average beak size is larger during drought conditions than during wet conditions.

4. If data on average precipitation were presented, it would be possible to give a qualitative prediction of average beak size. Beaks will be, on average, longer during drought conditions and shorter during wet conditions.

5. There is no optimal beak size, as illustrated by the graph. Beak size varies with level of precipitation. If there was a single, optimal beak size, there would be no variation in average beak size with changes in precipitation.

6. Average beak size varies according to precipitation on the Galápagos Islands for the finches surveyed in this study.

CHAPTER 9

Review and Rehearse

Check Your Knowledge
1. b; 2. b; 3. c; 4. b; 5. d; 6. b; 7. c; 8. a; 9. d; 10. e; 11. c; 12. a; 13. a; 14. b; 15. e; 16. b

Short-Answer Questions
1. If a trait positively affects an individual's relative reproductive success, the frequency of the allele for that trait is likely to increase in the population.

2. This is an example of an innate behavior that does not need to be learned. It is fixed, occurring in exactly the same pattern each time the behavior is started, and normally is not altered.

3. Behaviors that improved our ancestors' reproductive success and survival are more likely to be easily learned.

4. The sight of a large, egg-like object acts as a stimulus to a goose and compels it to roll the object back into its nest, regardless of whether there are any smaller egg-like objects, including its own eggs, outside the nest. Only after the goose has rolled the largest egg-like object into its nest will it roll other, smaller, egg-like objects into the nest.

5. Altruistic-appearing behaviors benefit the interests of closely related individuals, with whom an individual is more likely to share genes. It could be said that genes are acting in their own best interest.

6. The squirrel producing the alarm call is likely to be related to many of the other squirrels in the colony. When the squirrel sounds the alarm, it is helping to ensure the safety of others in the colony, many of whom share many of its genes. This is an example of inclusive fitness.

7. Any two of the following three conditions are correct: (a) Individuals must interact with others who could potentially be recipients or donors of altruistic-appearing behavior. (b) The benefit to the recipient of an altruistic-appearing behavior must be much greater than the cost to the donor. (c) Recipients of altruistic-appearing acts that do not reciprocate at a later time will be recognized and punished.

8. An example given in the text illustrates a maladaptive behavior. People often donate to a cause that benefits people who live in distant reaches of the planet—people the donor is unlikely to meet, so the behavior is unlikely to be reciprocated. The origins of this behavior go back to when humans lived in small communities in which altruistic-like behaviors were directed to other members of the community and were likely to be reciprocated. Humans of today live in much larger communities but behave as if they still lived in small ones.

9. No. Natural selection will favor behaviors that benefit the individual, not necessarily the group. In group selection, the reproductive fitness of individuals is decreased, and this type of selection cannot be sustained.

10. The female produces and retains the fertilized egg, and the offspring develops within her body. The male initially contributes only sperm. After birth of the offspring, the male may play a large role—and in some cases, an equal role—in rearing young. Because offspring develop within the female, her energetic investment will initially always be larger than the male's.

11. The sex that makes the greater reproductive investment is choosier about who it will mate with. Females make the greatest initial reproductive investment, but their overall reproductive investment may or may not be greater than that of the male. In cases where the male has a greater overall reproductive investment, it is in his best interest to ensure that his mate is carrying offspring with his genes and that these offspring grow to maturity. If the female makes the greater overall reproductive investment, she will try to find a mate that best ensures the success of her offspring. The degree of investment determines the risk: the greater the investment, the greater the potential risk.

12. Antler size indicates to the female the relative quality of the male as a potential mate.

13. In these mating systems, females are choosey about which males they mate with, choosing males with preferred physical traits or superior resources. Males of these systems must compete with other males for available females and may mate with multiple females. Males who cannot compete for mates, due to undesirable physical characteristics or poor access to resources, will mate with few if any females.

14. *Monogamy:* Two individuals mate and remain with each other. *Polygamy:* An individual has multiple mates; in *polygyny,* the male mates with multiple females; in *polyandry,* the female mates with multiple males.

15. The male moorhen probably provides the most care. The female makes a smaller reproductive investment; she must compete with other females for males and is physically larger and more aggressive than the males. Distinct differences between the sexes indicate sexual dimorphism, which is typically associated with polygamous mating systems.

16. An "honest signal" is one that cannot be faked. Both the animal making the signal and the animal receiving it have the same interests.

Graphic Content Questions

1. A female with one mate produces 60 offspring. This is the same number as for a male mating with one female. The two numbers are the same because they are measures of the same thing: one male mating with one female. In both graphs, the number of offspring produced is recorded in the leftmost bar.

2. A female with two mates produces 50 offspring. It is not clear why females with two mates would produce fewer offspring than females with one mate. At least two explanations are possible. (a) Because the numbers are measured for different females—that is, a female's reproductive output can be measured when she has one mate or two mates, but not both—this may just be a consequence of natural variation among females. Maybe those in the two-mate group just happened to have fewer offspring, unrelated to the number of mates they had. (b) Perhaps the presence of two males created stress, which somehow reduced either the female's or the males' fertility. For additional data, it would be useful to have error bars. These would give us a sense of the extent of variation in reproductive output among females given access to two males and females given access to one male. If there is much variation within those two groups, then the lower number seen here may just be an artifact of the individual females measured. In that case, if greater numbers of flies were used in the study, the average reproductive output would be the same for females given access to one mate or two mates.

3. Females with four mates would most likely still have about 60 offspring, while males with four mates would have about 150 offspring. These numbers assume continuation of the trends shown in the graphs.

4. The females do not seem to be limited by access to mates. Rather, they seem to be limited by their own production of eggs. Males' reproductive output, however, does seem to be constrained by the number of mates they have. If a male has more mates, he has greater reproductive output.

5. To make comparisons between the data in the two graphs, it is essential not only that they are measuring the same thing but that the data are presented using the same scale. Otherwise, a comparison between the two graphs could be misleading.

CHAPTER 10

Review and Rehearse

Check Your Knowledge

1. a; 2. d; 3. e; 4. e; 5. a; 6. d; 7. d; 8. a; 9. e; 10. a; 11. d; 12. d; 13. e; 14. b; 15. c; 16. a; 17. c

Short-Answer Questions

1. Bacteria-like prokaryotic cells were the first life forms. Fossils of these organisms, 3.4 billion years old, have been found in Australia and South Africa.

2. The development of a membrane enabled the separation of a group of chemicals from the environment and the evolution of cells. Later, the segregation of chemicals within cells by intracellular membranes allowed chemical reactions of different kinds to occur in different parts of the cell.

3. According to the biological species concept, a species is a population of organisms that interbreed or can possibly interbreed with each other under natural circumstances and cannot interbreed with organisms from other groups.

4. The specific epithet, the second part of the species name, indicates the species within a genus and is the narrowest classification of an organism. The specific epithet alone, however, is not used to name an organism; the same specific epithet can be used with different genus names to name different species.

5. The biological species concept is based on reproductive isolation, due to prezygotic or postzygotic barriers. The morphological species concept relies on physical characteristics or features. The biological species concept is applicable to plants and animals that use sexual reproduction. It is not applicable to organisms that use only asexual reproduction, or to fossilized organisms, since they are no longer living and reproducing. In both of these cases, the morphological species concept is applicable. There is also difficulty in applying the biological species concept to species that arise over long periods of time, to ring species such as greenish warblers, and to hybridizing species such as butterflies of the genus *Heliconius*.

6. Reproductive isolation enables organisms to have independent evolutionary fates, since there can no longer be genetic exchange between the two populations and each population can evolve differently, depending on the specific environmental conditions it encounters.

7. A phylogenetic tree not only shows relatedness among organisms but also allows biologists to hypothesize about the evolutionary history of life on earth. As new knowledge is discovered, new branches are added to the tree and a more complete understanding of the evolutionary history of life on earth is constructed.

8. Until the 1980s, evolutionary trees were constructed by comparing physical characteristics between species. Species that shared traits also shared evolutionary heritage. With advances in the late 20th century, biologists began using molecular sequencing, including DNA sequencing, in place of physical traits when constructing evolutionary trees. This method is based on the transmission of DNA from one generation to the next. Organisms that share the same genes also share evolutionary heritage.

9. Analogous traits are produced through convergent evolution. They do not share a common ancestry but have developed as similar traits evolved to serve similar purposes, through natural selection. Problems arise when constructing evolutionary trees based on similarity of structures. Although similar structures are often a sign of shared heritage, for analogous structures this is not the case.

10. Microevolution occurs as a result of changes in allele frequencies in a population; this process modifies the genetic makeup of a population simply by changing allele frequency. Macroevolution results in the development of whole new groups of species and may involve many new alleles.

11. The development of long, flat teeth and a more square jaw is an evolutionary innovation that enabled horses to inhabit and adapt to a wide range of environments where they could find suitable food. When an evolutionary innovation appears, adaptive radiation allows the evolution of many species.

12. Fossils are an invaluable source of transitional species that lived in the past. Study of the physical features of fossilized organisms allows comparisons with other fossilized organisms and with existing organisms. More recently, genetic analysis has made it possible to examine the genetic relatedness of species, often presenting a clearer picture than anatomical comparisons alone.

13. Background extinctions occur at lower rates and typically result from natural selection. Mass extinctions do not result from natural selection; they are caused by extraordinary changes in the environment over relatively short periods of time. Typically, mass extinctions are global events, while background extinctions are more local.

14. Prior to the 1970s and 1980s, classification of organisms depended on comparisons of physical characteristics. When molecular sequencing methods became available, genetic information could be used for classification.

15. The biological species concept depends on the ability of organisms to reproduce sexually and produce fertile offspring. Bacteria do not reproduce sexually and cannot be defined according to the biological species concept.

16. Archaeal cell membranes contain polysaccharides not found in bacteria or eukaryotes, and archaeal ribosomes, enzymes, and membranes are more similar to those of eukaryotes than those of bacteria.

17. The domain eukarya contains single-celled and multicellular organisms with cells containing a membrane-enclosed nucleus.

Graphic Content Questions

1. The background extinction rate ranged from a few percent to approximately 14 percent of all families of species during the time period represented.

2. Background extinction rates (in green) are lower than rates of mass extinctions (in blue). Mass extinctions occur over relatively short periods and are caused by sudden and unusual environmental changes. Background extinctions occur constantly but at much lower rates.

3. There are several possible reasons. (a) Fossils found thus far represent only a small percentage of all species present at any time in the distant past. However, fossils are representative of families of species at that time, and each may represent a number of species. (b) Scientists do not know how many species existed at any given time in the distant past. Even today, we don't know how many living species there are. (c) It may not be possible to determine whether multiple fossils are of the same species, but perhaps they can be assumed to belong to the same family.

4. Since the causes of extinction vary over time, background extinction rates will change.

5. The fossil record may or may not represent the number of species that existed at any time in the distant past. Some species leave behind fossils, others do not. Only organisms that leave fossils can be identified as extinct organisms.

6. Scientists know what the background extinction rate should be, and any indication of extinction rates higher than expected may be a sign that human activities are causing another mass extinction. Human activities could then be modified to halt the progression of extinctions.

CHAPTER 11
Review and Rehearse

Check Your Knowledge

1. d; 2. e; 3. c; 4. e; 5. d; 6. c; 7. a; 8. b; 9. e; 10. b; 11. d; 12. a; 13. c; 14. b; 15. e; 16. d

Short-Answer Questions

1. (a) Animals cannot make their own food and therefore must take in food to survive. (b) Animals are made up of more than one cell. A multicellular body enables them to have specialized cells with unique properties. (c) Animals are capable of moving at some point in their lives. Some can move for their entire life, while others can move only during a larval stage or other life stage.

2. The degree of complexity of an organism is not related to how well that organism is adapted to its environment. There is a tendency to judge animals that are more like humans as "higher organisms," but animals perfectly adapted to their environment, though they may not be at all like humans, are not "lower organisms."

3. The common ancestor of all animals is thought to have been a unicellular ancestral protist resembling a sperm in shape and size. This conclusion is based on DNA and RNA analyses.

4. Sponges can reproduce both sexually and asexually. Sponges are hermaphrodites, but individuals produce sperm or eggs at different times. Sperm swim to the eggs contained in another sponge, and the fertilized eggs develop into larvae, which are released from the sponge and float to another location to start life. Asexual reproduction occurs by the formation of buds on the body surface. These buds can break off, float away, and form new sponges, after landing on a suitable surface.

5. Cnidarians can exist as a sessile (non-mobile) polyp, which resembles a stalk with arm-like projections at its top surface, or as a bell-shaped, free-floating medusa, with arm-like extensions at its perimeter. Both forms are radially symmetrical and have stinging cells.

6. Roundworms are probably the most abundant animals on earth. Some roundworms, also called nematodes, are parasites that cause damage to plants and animals.

7. A major characteristic of segmented worms is the grooves that run around their body, marking their body segments. Terrestrial slugs do not show segmentation. Slugs have an extended foot at the base of their body, which is not present in earthworms. Finally, unlike earthworms, when slugs are threatened they produce copious amounts of slime that coats their body.

8. Mollusks have a sandpaper-like tongue called a radula, which is used for feeding. The radula's rough surface assists with pulling food into the mouth.

9. It is difficult to define what intelligence is, and even more difficult to apply this concept to non-human animals. Animals have evolved in response to the selective forces of their environments, and our concept of intelligence cannot be applied objectively to non-human animals. In other words, all animals have adapted to manipulate their environment to meet their needs in ways that may not conform to our concept of intelligence.

10. Commonly, an octopus's ability to carry out manipulations such as unscrewing a bottle to get at a fish inside is seen as a form of intelligence—a type of intelligence that most people can relate to and compare to our concept of intelligence.

11. An exoskeleton is a hard outer covering that provides an attachment for muscles, protection for the body, and reduction in water loss. The exoskeleton cannot expand as the animal grows. The animal must shed its old exoskeleton and form a new one as it increases in size.

12. You could estimate the number of beetle species by the method used by Terry Erwin. Determine the number of beetle species in one species of tree in a temperate region, count the total number of tree species in that region, then estimate the total number of beetle species in the region. For example, if 100 beetle species are found in one tree species and there are 10 species of trees, the number of beetle species can be estimated as $10 \times 100 = 1,000$. This would be a rough estimate.

13. *Benefits:* The exoskeleton provides protection, much like a suit of armor, and helps prevent water loss, keeping the organism from drying out. *Costs:* An exoskeleton prohibits growth; animals with exoskeletons must go through developmental stages or shed their exoskeleton and grow a new one as they increase in size.

14. (a) Insects' ability to fly enables them to escape from predators, find food and mates, and occupy a wide variety of habitats. (b) Their ability to pass through different life stages (larva, pupa, and adult) allows them to grow in size, even though they have an exoskeleton.

15. Echinoderms are deuterostomes, meaning that they undergo embryonic development in which the gut develops from the back of the body toward the front; this developmental characteristic is shared by vertebrates. Echinoderm larvae show bilateral symmetry, in which the body has two halves that are essentially anatomically identical; this is another trait shared by vertebrates. These similarities in anatomical characteristics, developed under the force of natural selection, indicate an evolutionary history shared, in the distant past, with vertebrates.

16. (a) The notochord is a rod of tissue that extends from the head to the tail and stiffens when muscles contract during movement. It is present in all vertebrates during at least some stage of life. (b) The dorsal hollow nerve cord is located along the back, or dorsal side, of the animal and extends from the head to the tail. It forms the brain and spinal cord of the central nervous system. (c) Pharyngeal slits, located between the back of the mouth and the top of the throat, are present during at least some stage of vertebrate development. (d) A post-anal tail is present during at least some stage of vertebrate development; a post-anal tail is located behind the anus, which is the end of the digestive system.

17. Two evolutionary innovations in mammals were: (a) long legs set vertically beneath the trunk of the body, enabling animals to run faster and for a longer time to capture prey; and (b) endothermy, a warm-blooded body, in which warm body temperature can be maintained, regardless of environmental temperature, because of an increased metabolism, which is due to the muscle contraction required for running.

18. Lungs most likely evolved from the air sacs of lobe-finned fishes; the evolution can be traced through amphibians, reptiles (including birds), and mammals. Air sacs also evolved into the air bladders of ray-finned fishes.

19. Amphibians must live close to water because most amphibians lay their eggs in water. Reptiles do not lay their eggs in water and can live in very dry locations.

20. Endotherms produce body heat as a result of cellular respiration. Ectotherms depend on outside sources of heat, such as from sunlight, to heat their body and do not depend on cellular respiration. Endotherms use much more energy to produce heat and maintain their body temperature, and thus must consume more food, more frequently, than ectotherms.

21. The placenta is rich in blood vessels and enables the transfer of nutrients, respiratory gases, and metabolic wastes between the mother's circulatory system and that of the developing fetus. Finger-like projections of the placenta enter the uterine wall, which is also rich in blood vessels, so blood vessels of both structures are in close proximity. This arrangement allows an efficient exchange between the circulatory systems of the mother and developing fetus.

22. The presence of smaller teeth suggests that tools were used for the initial preparation of food. Stone tools were found in the same deposits as the fossils with smaller teeth, supporting this idea. With the use of tools, food could be cut into smaller pieces before being eaten, doing the work previously done by larger teeth.

23. (a) The brain is larger in humans. (b) Humans walk on two legs, whereas chimpanzees mostly walk on four legs. (c) Body size is larger in humans.

24. Fossil evidence of healed broken bones suggests that: (a) there were organized social groups that worked together to support individuals with severe injuries; without social support, it is unlikely that individuals would have been able to recover from injuries; and (b) Neandertals collectively hunted larger prey in close proximity, using short spears to jab their prey; this conclusion comes from comparing the injuries seen in fossils with typical injuries received when taking part in rodeos, which also involve people being in close proximity to large animals.

Graphic Content Questions

1. We are of the species *Homo sapiens,* which has *Homo heidelbergensis* as an ancestor, which may have had the following ancestors, in order of most recent to most distant: *Homo ergaster, Homo habilis, Australopithecus* (gracile), and *Ardipithecus.* We seem less likely to be direct descendants of *Homo neanderthalensis* and *Australopithecus* (robust), which are dead-end branches that split off from our common ancestors and became extinct.

2. *Homo* appeared about 2.5 million years ago. It seems to have had a significantly larger skull (and so brain size) than its *Australopithecus* ancestors.

3. The solid lines represent transitions from one species to another that are well supported by fossils showing gradual changes. The dotted lines represent potential relationships that are not as well supported by transitional fossils. The discovery of additional fossils may allow better clarification of the relationships.

4. About 1.6 million years ago, there were probably four different human species living at the same time: *Australopithecus* (robust), *Homo erectus, Homo ergaster,* and *Homo habilis.*

5. It's not long ago that our species became the only living representative of the genus *Homo.* From this figure, it seems to be approximately 10,000 years ago.

CHAPTER 12

Review and Rehearse

Check Your Knowledge

1. a; 2. a; 3. b; 4. c; 5. a; 6. b; 7. e; 8. c; 9. c; 10. c; 11. c; 12. d; 13. b; 14. d; 15. e; 16. b; 17. d

Short-Answer Questions

1. *Sexual reproduction:* Getting sperm to eggs for fertilization can be a challenge for many plants living on land, especially if the plants are large or live in dry locations. *Resisting predators:* As a result of their inability to move, plants cannot run or otherwise escape from predators, and they have had to develop other means of defense, including thorns and poisonous chemicals.

2. The earliest plants grew close to the ground to minimize the effects of gravity on the plant body—a major adaptation necessary for life on land. Buoyancy provided by water enables aquatic organisms to resist gravity; in organisms living on land, including the plants, traits to resist gravity had to evolve.

3. Mosses produce motile sperm with flagella, which require water so they can swim to the eggs and fertilize them. Since mosses grow low to the ground, areas they inhabit become temporarily flooded as rain falls or snow melts, providing the needed environment for sperm to reach eggs.

4. Non-vascular plants do not have vascular tissue, which limits their height to only a few centimeters; their gametophyte is the dominant generation. Vascular plants have vascular tissue that transports water and nutrients over great distances within the plant body, enabling them to grow very tall; the sporophyte is the dominant generation. Vascular plants are more successful in areas where the surface of the ground dries out between rains. Non-vascular plants and mosses can grow in deserts.

5. The growing embryo and young seedling derive nutrients from the endosperm, a specialized nutritive tissue contained in the seed. The endosperm provides the nutrients and energy required for the embryo to grow before the seedling establishes roots and leaves and the ability to photosynthesize.

6. Seeds contain nutrients and an embryonic plant within the protective seed coat. Spores contain only a cell with DNA, RNA, and a small amount of nutrients. For seed plants, growing from an embryo provides a head start and can improve chances for success.

7. There are more species of conifers, which number about 600, than any other group of gymnosperms, and they inhabit more regions of the world, including every continent except Antarctica. They grow larger and live longer than other gymnosperms.

8. A flower contains: (a) petals, the colorful part of the flower; (b) sepals, which form the structure that covers the flower while in bud; (c) male reproductive structures: stamens, each with an anther, which produces pollen and is supported by the stalk; and (d) female reproductive structures: the carpel, which encloses the ovary, where eggs develop, and the style, which extends from the ovary and has a sticky tip, called the stigma.

9. Plants and animals have evolved together, depending on each other for survival. If animals became extinct, there would be no bees or other pollinators, and many plant species would no longer be able to reproduce. And for many plants, there would be no way for seeds to be dispersed away from the parent plant. Altogether, this would result in many plant species becoming less common, if not extinct. Plants that do not depend on animals for pollination and/or seed distribution would dominate.

10. A pollen grain lands on the stigma, a pollen tube grows down the style into the ovary, and two sperm travel down the pollen tube. One sperm fertilizes the egg to form a zygote, and the other fuses with the two central nuclei in the embryo sac to form endosperm. The endosperm grows into a food source for the developing embryo after germination. Thus, double fertilization forms not only a zygote but also a significant food source. Gymnosperms do not produce endosperm, so angiosperms have the advantage when it comes to nourishing the embryo.

11. The fleshy fruit provides a bribe to entice an animal into eating the fruit and transporting the fruit along with its seeds to another location, away from the parent plant. After the seeds have passed through the digestive system, they are deposited with some nutrient-rich manure to help nourish the seedling.

12. There are two categories of plant defenses: anatomical structures and chemical compounds. Plant structures that provide protection include thorns on rose bushes, spines on acacia trees and palm fronds, and the microscopic hairs of the stinging nettle, which cause pain when bare skin rubs against them.

13. Unlike other eukarya, fungi have cell walls containing chitin and are decomposers, digesting food externally prior to absorption of nutrients.

14. Answers could include: (a) fungi are a source of antibiotics, (b) they are a food source, and (c) they make possible the production of alcoholic beverages and bread.

15. Fungi are associated with the roots of many plants, forming mycorrhizae, in which fungal hyphae wrap around the plant's root hairs. The plant benefits by getting minerals such as phosphorus and nitrogen from the fungus. The fungus benefit by getting small amounts of sugar from the plant,

16. Since fungi are not capable of photosynthesis, they depend on outside food sources for nutrients. An association with plant roots enables them to obtain food easily (without having to externally digest and then absorb it), since the plant supplies them with sugar.

Graphic Content Questions

1. At some point between 20 and 47 days, plants with mycorrhizal fungi had a significantly higher survival rate than those without mycorrhizal fungi.

2. To answer this question, two considerations must be kept in mind. First, the percentage survival rates are given in intervals of 20%, so only estimations between these intervals can be made; second, error bars are given for survival data at 47 days. A best estimate is that the percentage survival rate is in the high 50s, perhaps 57% or 58%, up to about 62%, for offspring of parents with mycorrhizal fungi, and about 21% to 25% or 26% for those having parents without mycorrhizal fungi.

3. At 10 and 20 days, no plants died, so there could be no errors or variations in measurement. At 47 and 96 days, some plant deaths had occurred and thus there was a chance of measurement errors or variations.

4. "Statistically significant difference" means a difference greater than a difference due only to an error or variation in measurements.

5. A useful additional point of time would be about halfway between days 20 and 47. At some point between these dates, a difference in survival rates between the two groups arose. By adding an additional data point, it might be possible to determine exactly when this change occurred. However, many additional data points might be required to better determine the actual point in time when a benefit was derived from this mutualistic plant-fungal relationship.

CHAPTER 13

Review and Rehearse

Check Your Knowledge

1. b; 2. b; 3. c; 4. e; 5. d; 6. a; 7. a; 8. d; 9. b; 10. a; 11. a; 12. b; 13. a; 14. b; 15. d; 16. e; 17. e; 18. b

Short-Answer Questions

1. Amoebas are tiny cells. Their small size allows the exchange of gases, including oxygen and carbon dioxide, across their cell membrane at a high enough rate to meet their metabolic needs, so they do not require a respiratory system. Larger, multicellular organisms do require a respiratory system because, without it, gas exchange cannot occur at a rate fast enough to meet their metabolic needs.

2. Microbes are genetically very diverse. They live in just about every type of environment imaginable, including at extreme temperatures and pressures, and they live on just about any type of food available. Thus, it is difficult to make any generalizations about microbes. There are huge numbers of microbes living on earth, with a huge number of species, many of which are still to be identified.

3. Bacteria have a cell envelope made up of a plasma membrane and, in most cases, a cell wall, enabling them to control what moves into and out of the cell. These structures also allow bacteria to live in harsh environments and still maintain internal conditions within a range that allows them to survive.

4. *Conjugation:* A copied piece of DNA is transferred between bacteria, even if they are not of the same species. *Transduction:* A virus inadvertently transfers bacterial DNA from one bacterium to another. This occurs as a virus picks up DNA from one host and, as it infects another host, transfers the DNA to that cell. *Transformation:* As bacteria die they release their DNA into the environment, and it can be scavenged by other bacteria. This new DNA may code for traits the cell does not already have.

5. Aerobic bacteria require oxygen to carry out cellular respiration, whereas anaerobic bacteria do not. Bacteria that can live aerobically when oxygen is available and anaerobically when oxygen is not available are called facultative anaerobes.

6. The characteristic taste and texture of yogurt result from the use of bacteria in yogurt production. The process uses bacteria that metabolize milk sugar, called lactose, for energy and produce lactic acid as a by-product. Lactic acid has a low pH and causes milk proteins to denature, resulting in the characteristic taste and texture.

7. The source of this cholera outbreak was studied in 1854 by Dr. John Snow, who identified the source as the Broad Street pump, which was delivering contaminated water. The handle of the pump was removed so no one else could use it as a source of water. It was Dr. Snow who persuaded the authorities to remove the handle, and as a result the number of new cholera cases fell dramatically.

8. Some strains of *Staphylococcus aureus* have become resistant to most of the antibiotics available today. Untreatable staph infections have spread from hospitals into the larger community and are killing more people than HIV infections.

9. (a) Many people are unaware that they are infected with an STD because symptoms may be mild or absent, and they may inadvertently pass the disease to another person. (b) It is nearly impossible to completely eliminate all STDs because both partners need to be treated simultaneously, and this is hard to arrange if people are unaware of their infection.

10. Any two of the following answers are possible. Bacteria and archaea have chemical and/or structural differences in: (a) their DNA, (b) their plasma membranes, (c) their cell walls, and (d) their flagella.

11. "Extremophile" refers to the ability of these organisms to tolerate extremes in the physical and chemical environments.

12. Because many archaea live in extreme environments, it can be challenging to replicate these conditions in the laboratory so as to successfully grow and study these organisms.

13. The protists were the first organisms to develop a nucleus, a structure with a double membrane, called the nuclear envelope, that encloses the cell's chromosomes. It is thought that the double membrane could have formed from the fusion of infoldings of the cell's plasma membrane.

14. The ciliates, including *Paramecium,* are a group of protists considered to be animal-like, since they move and hunt other microorganisms, much like animals.

15. *Plasmodium* goes through a series of developmental stages, each of which has different surface proteins. These proteins are what an immune system uses to identify foreign organisms. It takes time for an immune system to recognize a foreign organism by its surface proteins, so the constant change in surface proteins with each stage of *Plasmodium* development means the immune system has to repeatedly identify and react to new proteins.

16. A virus is composed of a piece of DNA or RNA surrounded by a protein coat. Unlike living organisms, a virus is not made up of cells, cannot replicate on its own, carries out no metabolic activities within its protein coat, lacks the structures found in cells, and cannot detect and respond to stimuli in its environment.

17. The genetic material making up a DNA virus is relatively stable over time, since there are enzymes present to correct replication mistakes. There are no comparable enzymes that correct replication errors for RNA, therefore RNA viruses have a higher rate of mutation, making them more difficult for the immune system to fight.

18. The types of glycoprotein present on the surface of the virus determine which organism(s) it can infect and which tissue(s) of a multicellular organism it can infect.

Graphic Content Questions

1. A majority of the cells in a human body are microbial cells living in and on the human—about 90 trillion microbial cells, compared with 10 trillion human cells.

2. There are about 10 trillion human cells.

3. There are about 90 trillion microbial cells living in and on the human body.

4. Microbial cells are much smaller than the human cells making up the body.

5. It is not clear how the data were obtained for this graph. It would be helpful to indicate the age, gender, and number of the people in the represented sample. A breakdown of cell types might also be useful.

6. Any type of graph that compares the numbers of bacteria and human cells is acceptable. (a) A pie chart could be divided into two wedges to show the numbers of bacterial and human cells. A strength of this presentation is that it conveys nicely that most of the "pie" is bacterial rather than human. A weakness is that it conveys information on how each component represents some proportion of the whole, but no information on the absolute numbers of each cell type. (b) A bar graph could be used, with one bar representing the number of bacteria and the other bar representing the number of human cells. One strength of this approach is that it makes it easier to compare the two numbers. A weakness is that it de-emphasizes that the two components, together, make up the entire count of cells in a human body.

CHAPTER 14

Review and Rehearse

Check Your Knowledge

1. a; 2. e; 3. a; 4. b; 5. e; 6. d; 7. b; 8. d; 9. c; 10. c; 11. e; 12. d; 13. a; 14. b; 15. c

Short-Answer Questions

1. Population ecology is a subfield of ecology that studies the interactions of populations of species with their environment—specifically, study of the growth of a population and how other species and the environment influence that growth.

2. A significant part of ecology deals with the interaction of populations of organisms with other organisms and with their environment. Study of a single individual does not meet the study objectives of ecology.

3. The population size would decrease. For population size to stay the same (or increase), each mating pair must produce two (or more) offspring that reach sexual maturity.

4. Density-dependent factors are pressures on a population's growth that result from population size itself. These factors include limitations on resources such as food, water, and space. As the population size increases, fewer resources are available to support the greater number of organisms. Density-independent factors do not result from population size, but usually result from geological or meteorological forces, such as earthquakes or hurricanes, or from human activity that degrades the environment. These factors lower the environment's potential to support populations of organisms.

5. (a) Populations of desert locusts in northwest Africa sometimes undergo explosive growth. These locusts usually live solitary lives, but on occasion, for reasons not entirely understood, they form large migrating swarms that cause extensive crop damage. (b) The lynx and snowshoe hare populations of Canada undergo periodic oscillations. As the snowshoe hare population increases, the lynx, which feed on snowshoe hares, begin to increase in number. As the lynx population increases, more and more snowshoe hares are hunted and eaten by the lynx, causing the snowshoe hare population to decrease. Eventually, the decrease in number of snowshoe hares causes the lynx population to decrease, and the snowshoe hare population again increases. This pattern occurs repeatedly.

6. Maximum yield is obtained by the harvesting or removal of all or almost all available organisms of a specific kind to be used as food or for other purposes. Not enough of the organisms remain to allow normal population growth. Maximum sustainable yield is the maximum number of organisms that can be harvested or removed but still allow the normal growth of the population.

7. To determine the organism's life history, you would need to know when the organism reproduces, how often it reproduces, and how many offspring it produces.

8. Organisms with a type I survivorship curve produce a small number of offspring and have a high probability of surviving every age interval until they reach old age. Examples include most mammals and the giant tortoise.

9. Organisms with a shorter breeding season produce more offspring per reproductive event than organisms with a longer breeding season. Generally, the closer an organism lives to the equator, the longer the breeding season.

10. Sickness and death early in life prevent reproduction and the passing of alleles to the next generation, and thus the prevalence of alleles that cause disease and death early in life will be reduced in a population over time.

11. Species that have evolved in environments that present a high risk of dying at a relatively early age live shorter lives, reproduce earlier, and have more offspring per reproductive event than species that have evolved in environments that present a relatively low risk of dying. If a species that evolved in a high-risk environment were transferred to a low-risk environment, it could eventually evolve to live longer, breed later, and have fewer offspring.

12. Researchers began with a large number of fruit flies, and after 2 weeks they supplied them with a fresh dish of food and collected the eggs laid in the dish. They measured the longevity of some flies hatched from these eggs (average life span 33 days), and repeated the procedure with this new generation of flies—but this time waiting longer than 2 weeks to collect eggs. They repeated the procedure a number of times, gradually increasing the time until eggs were collected up to 10 weeks, and in this way almost doubled the average life span of the flies to 63 days.

13. These data can be used for planning and meeting the needs of people in different age groups, called cohorts. For instance, based on the current size of particular cohorts, the data would be useful for planning when and how many schools or other child services might be needed, or hospital maternity services, or assisted living facilities for the elderly.

14. "Demographic transition" refers to the change in the make-up of a region's population as it progresses from high birth rates and death rates, to high birth rates and low death rates, then finally to low birth rates and death rates.

15. The ecological footprint of a population is a description of the amount of resources it uses. Average resource use by each person is so much greater in Japan than in India, which is a much poorer nation, that, overall, the smaller population of Japan has a larger ecological footprint than the larger population of India. Generally, as nations become richer, their ecological footprint increases.

Graphic Content Questions

1. This graph illustrates the growth of a population over time, its maximum sustainable yield, and the environment's carrying capacity. With this information, it is possible to better manage a resource by understanding how much can be harvested/hunted/utilized sustainably.

2. The blue line becomes flat as the population reaches the environment's carrying capacity, the point at which all available resources are being utilized by the organisms in the population. No further increase in population size is possible.

3. A population can stay the same size (plateau) as long as the number of individuals arriving (through birth or immigration) is exactly the same as the number of individuals leaving (through death or emigration). This situation can persist indefinitely, even as organisms age and die.

4. The data contained in this graph include: (a) the size of the population, with no units given; (b) time, with no units given; (c) maximum sustainable yield, the population size at half the carrying capacity; and (d) the carrying capacity of the environment.

5. The following information could be added: (a) units of time and units of population size, to provide quantitative data; (b) the species the population data represent; (c) the environment the data represent; and (d) the season when the data were obtained.

6. The graph assumes that: (a) the population will grow exponentially until it reaches the environment's carrying capacity and will neither fall below nor exceed this rate over time, and (b) the environment's carrying capacity will not change over time, due to the degradation/reduction of resources or increased resource availability.

7. The graph will have an inverted "U" shape: growth rate starts out very fast; increases to a maximum rate, reaching maximum sustainable yield at half the carrying capacity ($K/2$); then slows down, approaching zero as the population size approaches the carrying capacity (K) of the environment.

CHAPTER 15

Review and Rehearse

Check Your Knowledge

1. d; 2. e; 3. a; 4. a; 5. e; 6. d; 7. e; 8. a; 9. d; 10. a; 11. a; 12. d; 13. e; 14. b; 15. b

Short-Answer Questions

1. The two elements of an organism's abiotic environment are: (a) chemical resources, such as soil, water, carbon, nitrogen, and phosphorus; and (b) physical conditions, including factors such as temperature, salt levels, moisture, humidity, sunlight, and air pressure.

2. The temperature generally decreases on moving from the equator toward the Poles. The type of biome located in a geographic region is determined by: (a) temperature: average temperature and whether it varies seasonally or is constant; and (b) precipitation: average amount and whether it varies seasonally.

3. At lower latitudes, closer to the equator, the sun's rays hit the earth more directly, and the solar energy spreads out over a smaller area (is more concentrated). Closer to the Poles, energy from the sun's rays is spread out over a greater area, The greater the concentration of sunlight reaching the earth, the higher the temperature.

4. By reducing the amount of solar radiation they absorb, cities can become cooler without the expenditure of energy. Steps that could be taken include: (a) making rooftops lighter in color to reflect sunlight, (b) planting trees and plants around buildings to provide shade and cool areas, and (c) creating rooftop green areas and gardens.

5. Winds blow more strongly when they are higher above the earth, freed from the earth's frictional drag. When strong elevated winds encounter tall buildings, they are deflected, and much of the wind is pushed downward, increasing speed as it reaches street level.

6. Many organisms are omnivores, which eat both plants and animals and thus can occupy more than one trophic level in a food chain. These multiple connections between organisms, not apparent in a food chain, are better represented by a food web.

7. Energy flows from producers to consumers through several levels called trophic levels. Producers, the photosynthesizers, are at the base of the food chain and outnumber all consumers. Primary consumers, the herbivores, feed on the producers. Energy then passes to secondary consumers as they feed on the primary consumers. Finally, at the very top of the food chain are the tertiary consumers: the top carnivores. Only about 10% of the energy at each trophic level is available for consumers at the next higher trophic level. Very little energy originally absorbed by producers is available to top carnivores, so a relatively small number of these animals can be supported.

8. Carbon dioxide levels reflect fluctuations in plant growth during different seasons. Photosynthetic rates decrease during the winter, resulting in higher atmospheric carbon dioxide levels. The carbon dioxide levels decrease during the summer as photosynthetic levels increase.

9. Natural selection selects for characteristics that improve reproductive success, regardless of whether the trait evolves in response to the selective pressure of physical resources or of biotic resources. For instance, an animal may develop a thick fur coat in response to cold temperatures and develop a light-colored coat to improve camouflage as a defense against predators. Evolution of a thick fur coat is in response to an abiotic resource; camouflage is in response to biotic resources.

10. An organism's niche can be described as its place in a community. The features of a niche include space and food requirements; the time of year the organism reproduces; its temperature, moisture, and other living requirements; the plants or other organisms it feeds on; and how it influences competitors. If two species occupy the same niche, two outcomes are possible. Competitive exclusion occurs when the two species compete until the one that uses resources more efficiently drives the other specie to extinction. Resource partitioning occurs when the two species alter their use of the niche through behavioral or structural changes, so that one or both species become restricted to just part of their niche.

11. *Competitive exclusion:* The two species compete until the one that uses resources more efficiently drives the other specie to extinction. *Resource partitioning:* The two species alter their use of the niche through behavioral or structural changes, so that one or both species become restricted to just part of their niche.

12. Unlike animals, plants are immobile and unable to move away from danger. They have developed chemical defenses that can injure or kill organisms that try to eat or otherwise harm them.

13. Rabies is spread through the transfer of saliva from one mammal to another. This disease causes an infected animal to become aggressive and to foam at the mouth, which increases the chance of its biting and transferring saliva to another animal.

14. Commensalism is a relationship involving one species that benefits and another species that neither benefits nor is harmed. In other words, one species is a winner, while the other derives no benefit from the relationship, but does not lose anything either.

15. A crucial feature of a colonizer in primary succession is its ability to disperse far from its site of origin. Because of their ability to travel far, these species can reach remote environments that have been disturbed. Generally, they face few if any competitors at first and can flourish, but in later stages of succession they do not compete well with other species.

Graphic Content Questions

1. The drawings are roughly proportional in size to the biomass of each level in the energy pyramid—producer, primary consumer, secondary consumer, tertiary consumer—for this particular type of ecosystem. The tertiary consumers (top predators) are fewest in number and have the lowest biomass.

2. The size of the "pie" at each trophic level represents the biomass in an ecosystem of organisms fulfilling a particular role at that level. The pies are different sizes because the biomass of producers is typically 10 times the biomass of primary consumers, and the biomass of each successive level higher up in the pyramid is typically one-tenth of that at the previous level. This decreasing biomass is due to the inefficiencies of energy flow and conversion.

3. The pie slice, typically 10% of the total pie, represents the conversion efficiency of biomass at one trophic level into biomass at the next level up the pyramid.

4. At the level of tertiary consumers, there is not sufficient biomass to support a population of quaternary consumers, given the 10% rule.

5. A level of secondary or tertiary consumers that also eat plants could have greater biomass than shown here, because these consumers could draw energy not just from the animals they eat, but also from plants. The expected biomass of such consumers would be 10% of the total biomass of the organisms they consume—both plants and animals. This would be much greater than for any of the consumers shown here.

6. There are likely to be many fewer predators than prey organisms (particularly herbivores), because the total biomass of predators is estimated to be one-tenth the total biomass of prey organisms.

CHAPTER 16

Review and Rehearse

Check Your Knowledge

1. e; 2. c; 3. d; 4. b; 5. d; 6. d; 7. b; 8. d; 9. a; 10. e; 11. b; 12. c; 13. a

Short-Answer Questions

1. The observed drop in oxygen level correlated with the proliferation of bacteria that were using up methane. This confirmed scientists' suspicion that bacteria were the cause of the disappearance of methane. Scientists also discovered, through DNA testing, that the bacteria responsible for consuming oil were also found at other oil spills. These bacteria live in the Gulf at all times, because there are naturally occurring oil leaks in this area.

2. Biodiversity can be defined as the diversity of species in an area, the diversity of genes and alleles for a particular species, and the number of different ecosystems in an area.

3. Studying biodiversity can be difficult because more than just the number of species must be studied—including the diversity of genes and alleles and the diversity of ecosystems.

4. Three primary factors contribute to species richness. (a) The solar energy available: More available solar energy correlates with greater species diversity. The sun's solar energy is more concentrated in areas closer to the equator and less concentrated in areas farther away from the equator. The more energy available, the greater the number of organisms and species that can be supported. (b) The evolutionary history of the area: The greater the time that has elapsed in an undisturbed inhabited area, the greater the biodiversity. Climatic disasters in an area diminish the number of species and lower biodiversity. (c) The rate of environmental disturbance: Environments with an intermediate amount of disturbance have the greatest biodiversity. Species that would outcompete (have a competitive advantage over) many other species may be lost from the environment due to disturbance, enabling greater species diversity.

5. The fewer the individuals making up a species, the greater the chances of the species becoming extinct. As the population of tigers has decreased, their genetic/allele diversity has decreased. Should tigers be faced with a severe environmental disturbance, disease, or other change, the population may lack the alleles that would permit some individuals to survive and reproduce.

6. The chief reason for species loss is habitat loss and habitat degradation due to human activities, particularly in tropical rain forests. It is estimated that half the world's tropical rain forests have been destroyed over the past 25 years. These areas are biologically very diverse, and their loss represents a significant loss in biodiversity. Overexploitation of species and the introduction of exotic species are additional reasons for the loss of so many species.

7. Once a species is lost, it can never exist again. Each species has formed as the result of a unique, uninterrupted, never to be repeated evolutionary history that gave rise to a genome unlike any other.

8. Exotic species may: (a) become invasive, since they often have no natural predators, leading to uncontrolled population growth, and (b) find prey that have no natural defenses against them, causing endangerment or extinction of these species. Communities or whole ecosystems can be altered as a result of invasive species. This can cause degradation of the environment and species loss.

9. Fossil fuels, which include oil, coal, and natural gas, are composed primarily of carbon and hydrogen, with smaller amounts of nitrogen and sulfur. As fossil fuels are burned, these elements react with oxygen in the air to form carbon dioxide, water, nitrogen dioxide, and sulfur dioxide.

10. Carbon dioxide and methane in the air absorb infrared energy radiating from the earth, preventing its escape into space and thus heating up the atmosphere. With carbon dioxide and methane added to the atmosphere, more heat is retained and contributes to global rises in temperature. This effect is comparable to that of the glass panes of a greenhouse, which have the same effect as carbon dioxide and methane in retaining more infrared energy and warming the internal environment of the greenhouse.

11. (a) Reducing biodiversity: A disproportionate percentage of the world's species live in the rain forests. With the disappearance of these forests, many known and unknown species are lost. (b) Reducing photosynthesis and increasing greenhouse gases: Each year, an estimated 610 billion tons (550 trillion kg) of carbon dioxide, a greenhouse gas, are removed from the atmosphere during photosynthesis by the plants found in tropical rain forests—an amount that will be reduced as rain forests are destroyed. Additionally, when rain forests are burned, carbon dioxide is released, adding more greenhouse gas to the atmosphere.

12. As a result of the reduced production and use of CFCs, atmospheric ozone levels have stabilized and a full recovery is thought possible. Images of the ozone hole over the Antarctic revealed that it was increasing in size until 2010, when it was found to have stopped enlarging as atmospheric levels of ozone stabilized.

13. The primary impact of the difficulty in defining biodiversity is identifying which organisms to include in conservation goals, given that conservation resources are limited. Identifying which species are to be protected can be very challenging, and this is made more difficult by the fact that many species and their value have yet to be identified.

14. A keystone species, such as kelp, has a disproportionate effect on the biodiversity of a community, as is revealed if that species should die out or otherwise be removed. When a keystone species is lost from a community, a relatively large number of other species, which depend on it, directly or indirectly, are also lost.

Graphic Content Questions

1. There are two graphs, each presenting data on the number of species found at different latitudes, from the equator toward the Poles. The *x*-axis shows the number of species: in the top graph, the number of mammalian species; in the bottom graph, the number of marine copepod species. The *y*-axis shows the latitude. In the top graph, data are presented for the numbers of species at 40° and 20° south, the equator, and 20°, 40°, and 60° north. In the bottom graph, data are presented for the equator and 20°, 40°, and 60° north.

2. The independent variable is latitude. The dependent variable is species number. Typically, the *x*-axis is used for the independent variable and the *y*-axis for the dependent variable. In these graphs, the variables are placed on the opposite axes. This is done because the data are overlaid on a world map, which aids in identifying exactly where in the world the selected latitudes fall. This makes it easy to see, for example, that 20° north latitude includes Central America, while 60° north latitude cuts across southern Alaska and northern Canada.

3. For mammalian species and for marine copepods, the largest numbers of species occur at 0°, which is the equator.

4. The value of having data both for mammalian species and for marine copepod species is that it enables us to see that the trend of decreasing species diversity toward the Poles holds true for a *terrestrial* species and for a *marine* species. This suggests that the phenomenon is robust and not just a quirk of one group of organisms. It would be helpful if data were presented for *all* species, but such data do not exist. (And, as we saw in Chapter 10, for prokaryotes and many other organisms it is difficult to even define what a species is.) Only a fraction of the world's species have been described. In this figure, it is more valuable to include animal groups for which most species have been described. Also important is that the figure uses large groups of organisms, for which any bias in the proportion of total species described at the equator versus the proportion described at higher latitudes is unlikely.

5. There are more mammalian species than marine copepod species at 20° north latitude. Someone might answer this question incorrectly because the bar representing the number of copepod species is longer than the bar representing mammalian species—but this is because the scales for the axes differ. There are more than 300 mammalian species and about 70 copepod species at 20° north.

6. The bars here represent exact counts rather than data from repeated measures of a variable, such as height, that would have an associated variance.

Glossary

A note about notation: The word or phrase being defined is in boldface type; it is followed by the number (in parentheses) of the chapter or chapters where it is discussed. In some cases, derivations are given. Abbreviations: Gk., Greek; Lat., Latin; *sing.,* singular; *pl.,* plural; *dim.,* diminutive (a smaller version of the object named); *pron.,* pronounced.

A

abiotic (15) Relating to the physical and chemical components in an environment. These include the chemical resources of the soil, water, and air (such as carbon, nitrogen, and phosphorus) and physical conditions (such as temperature, salinity, moisture, humidity, and energy sources).

acid (2) Any fluid with a pH below 7.0, indicating the presence of more H^+ ions than OH^- ions in solution. [Lat., *acidus,* sour]

acquired immune deficiency syndrome (AIDS) (13) Infectious human disease caused by a retrovirus, HIV (human immunodeficiency virus), which compromises the immune system by attacking T cells, leaving an individual susceptible to infections, as well as cancers.

activation energy (2) The minimum energy needed to initiate a chemical reaction (regardless of whether the reaction releases or consumes energy).

activator (2) A chemical within a cell that binds to an enzyme, altering the enzyme's shape or structure in a way that causes the enzyme to catalyze a reaction.

active site (2) The part of an enzyme to which reactants (or substrates) bind and undergo a chemical reaction.

active transport (3) Molecular movement that depends on the input of energy, which is necessary when the molecules to be moved are large or are being moved against their concentration gradient.

adaptation (8) The process by which, as a result of natural selection, a population's organisms become better matched to their environment; also, a specific feature, such as the quills of a porcupine, that makes an organism more fit.

adaptive radiation (10) The rapid diversification of a small number of species into a much larger number of species, able to live in a wide variety of habitats.

additive effects (7) Effects from alleles of multiple genes that all contribute to the ultimate phenotype for a given characteristic.

adenosine triphosphate (ATP) (4) A molecule that temporarily stores energy for cellular activity in all living organisms. ATP is composed of an adenine, a sugar molecule, and a chain of three negatively charged phosphate groups.

adult (11) (pertaining to insect developmental stages) In complete metamorphosis, the third and final stage of insect development.

aging (14) An increased risk of mortality with increasing age; generally characterized by multiple physiological breakdowns.

alleles (5, 7) Alternative versions of a gene. [Gk., *allos,* another]

allopatric speciation (10) Speciation that occurs as a result of a geographic barrier between groups of individuals that leads to reproductive isolation and then genetic divergence. [Gk., *allos,* another + *patris,* native land]

altruistic behavior (9) A behavior that comes at a cost to the individual performing it and benefits another. [Lat., *alter,* the other]

amino acid (2) One of 20 molecules built of an amino group, a carboxyl group, and a unique side chain. Proteins are constructed of combinations of amino acids linked together.

amino group (2) A nitrogen atom attached by single bonds to hydrogen atoms.

amniotes (11) Terrestrial vertebrates—reptiles, birds, and mammals—that produce eggs (called amniotic eggs) that are protected by a water-tight membrane and a shell.

amphibians (11) Members of the class Amphibia; ectotherms (that is, they are cold-blooded), with a moist skin, lacking scales, through which they can fully or partially absorb oxygen. They were the first terrestrial vertebrates. The young of most species are aquatic, and the adults are true land animals. [Gk., *bios,* life + *amphi,* on both sides]

analogous traits (10) Characteristics (such as bat wings and insect wings) that are similar because they were produced by convergent evolution, not because they descended from a common structure in a shared ancestor. [Gk., *analogos,* proportionate]

anaphase (6) A phase in mitosis and meiosis in which chromosomes separate. In mitosis, it is the third phase, in which the sister chromatids are pulled apart by the spindle fibers, with a full set of chromosomes going to opposite sides of the cell. In meiosis, the homologues separate in anaphase I and the sister chromatids separate in anaphase II. [Gk., *ana,* up + *phasis,* appearance]

anecdotal observation (1) Observation of one or only a few instances of a phenomenon.

angiosperms (12) Vascular, seed-producing flowering and fruit-bearing plants, in which the seeds are enclosed in an ovule within the ovary. [Gk., *angeion,* vessel, jar + *sperma,* seed]

animals (11) Members of the kingdom Animalia; eukaryotic, multicellular, heterotrophic (that is, they cannot produce their own food) organisms. Many of these organisms have body parts specialized for different activities and can move during some stage of their lives. [Lat., *animal,* a living being]

annelid (11) Phylum of worms having segmented bodies; protostomes with defined tissues, which grow by adding segments rather than by molting. There are about 13,000 identified species of segmented worms, including earthworms and leeches.

anther (12) The part of the stamen, the male reproductive structure of a flower, that produces pollen. [Gk., *anthos,* blossom]

apoptosis (6) Programmed cell death, which takes place particularly in parts of the body where the cells are likely to accumulate significant genetic damage over time and are therefore at high risk of becoming cancerous. [Gk., *apoptosis,* falling away]

archaea (*sing.* **archaeon**) (10) A group of prokaryotes that are evolutionarily distinct from bacteria and that thrive in some of the most extreme environments on earth; one of the three domains of life. [Gk., *archaios,* ancient]

arthropods (11) Members of the invertebrate phylum Arthropoda; characterized by a segmented body, an exoskeleton, and jointed appendages. [Gk., *arthron,* joint + *pous,* foot]

asexual reproduction (6) A type of reproduction common in prokaryotes and plants, and also occurring in many other multicellular organisms, in which the offspring inherit their DNA from a single parent.

atom (2) A particle of matter than cannot be further subdivided without losing its essential properties. [Gk., *atomos,* indivisible]

atomic mass (2) The mass of an atom; the combined mass of the protons, neutrons, and electrons in an atom (the mass of the electrons is so small as to be almost negligible).

atomic number (2) The number of protons in the nucleus of an atom of a given element.

B

background extinctions (10) Extinctions that occur at lower rates than at times of mass extinctions; occur mostly as the result of aspects of the biology and competitive success of the species, rather than catastrophe.

base (chemistry) (2) Any fluid with a pH above 7.0, that is, with more OH⁻ ions than H⁺ ions in solution.

base (of DNA) (2, 5) One of the nitrogen-containing side-chain molecules attached to a sugar molecule in the sugar-phosphate backbone of DNA and RNA. The four bases in DNA are adenine (A), thymine (T), guanine (G), and cytosine (C); the four bases in RNA are adenine (A), uracil (U), guanine (G), and cytosine (C). The information in a molecule of DNA and RNA is determined by its sequence of bases.

base pair (5) Two nucleotides on complementary strands of DNA that form a pair, linked by hydrogen bonds. The pattern of pairing is adenine (A) with thymine (T) and cytosine (C) with guanine (G); the base-paired arrangement forms the "rungs" of the double-helix structure of DNA.

behavior (9) Any and all of the actions performed by an organism, often in response to its environment or to the actions of another organism.

bilateral symmetry (11) A body structure with left and right sides that are mirror images. [Lat., *bi-*, two + *latus*, side]

binary fission (6) A type of asexual reproduction in which the parent cell divides into two genetically identical daughter cells. Bacteria and other prokaryotes reproduce by binary fission. [Lat., *binarius*, consisting of two + *fissus*, divided]

biodiversity (16) The variety and variability among all genes, species, and ecosystems. [Gk., *bios*, life + Lat., *diversus*, turned in different directions]

biodiversity hotspots (16) Regions of the world with significant reservoirs of biodiversity that are under threat of destruction.

biofuels (4) Fuels produced from plant and animal products.

biogeography (8) The study and interpretation of distribution patterns of living organisms around the world. [Gk. *bios*, life + *geo-*, earth + *graphein*, to write down]

biological literacy (1) The ability to use scientific inquiry to think creatively about problems with a biological component, to communicate these thoughts to others, and to integrate these ideas into one's decision making.

biological species concept (10) A definition of species described as populations of organisms that interbreed, or could possibly interbreed, with each other under natural conditions and that cannot interbreed with organisms outside their own group.

biology (1) The study of living things. [Gk., *bios*, life + *logos*, discourse]

biomass (15)) The total mass of all the living organisms in a given area. [Gk., *bios*, life]

biomes (15) The major ecological communities of earth. Terrestrial biomes, such as rain forest or desert, are defined and usually described by the predominant types of plant life in the area, which are mostly determined by the weather; aquatic biomes are usually defined by physical features such as salinity, water movement, and depth. [Gk., *bios*, life]

biotechnology (5) The modification of organisms, cells, and their molecules for practical benefits. [Gk., *bios*, life + *technologia*, systematic treatment]

biotic (15) Relating to living organisms; the biotic environment, or community, consists of all the living organisms in a given area. [Gk., *bios*, life]

bivalve mollusks (11) Mollusks with two hinged shells; examples are clams, scallops, and oysters. [Lat., *bi-*, two]

blind experimental design (1) An experimental design in which the subjects do not know what treatment (if any) they are receiving.

bond energy (2) The strength of a bond between two atoms, defined as the energy required to break the bond.

bottleneck effect (8) A phenomenon by which genetic drift can occur; a sudden reduction in population size (often due to famine, disease, or rapid environmental disturbance) that can lead to changes in the allele frequencies of a population.

bryophytes (12) Three groups of plants (the liverworts, hornworts, and mosses) that lack vascular tissue and move water and dissolved nutrients by diffusion. [Gk., *bruon*, tree-moss, liverwort + *phytas*, plant]

buffer (2) A chemical that can quickly absorb excess H⁺ ions in a solution (preventing it from becoming too acidic) or quickly release H⁺ ions (to counteract increases in OH⁻ concentration).

C

C4 photosynthesis (4) A method (along with C3 and CAM photosynthesis) by which plants fix carbon dioxide, using the carbon to build sugar; serves as a more effective method than C3 for binding carbon dioxide under low carbon dioxide conditions, such as when plants in warmer climates close their stomata to reduce water loss.

Calvin cycle (4) In photosynthesis, a series of chemical reactions in the stroma of chloroplasts in which sugar molecules are assembled. [From the name of one of its discoverers, Melvin Calvin, 1911–1997]

CAM (crassulacean acid metabolism) (4) Energetically expensive photosynthesis in which the stomata are open only at night to admit CO_2, which is bound to a holding molecule and released to enter the Calvin cycle to make sugar during the day. In this type of photosynthesis, found in many fleshy, juicy plants of hot, dry areas, water loss is reduced because the stomata are closed during the day.

cancer (6) Unrestrained cell growth and division. [Lat., *cancer*, crab]

capsid (13) The protein container surrounding the genetic material (DNA or RNA) of a virus. [Lat., *capsa*, box, case]

capsule (13) A layer surrounding the cell wall of many bacteria; it may restrict the movement of water out of the cell and thus allow bacteria to live in dry places, such as the surface of the skin. The capsule contributes to the virulent characteristics of some bacteria, making them resistant to phagocytosis by the host's immune system. [Lat., *capsula*, small box or case]

carbohydrate (2) One of the four types of biological macromolecules, containing mostly carbon, hydrogen, and oxygen. Carbohydrates are the primary fuel for cellular activity and form much of the cell structure in all life forms. [Lat., *carbo*, charcoal + *hydro-*, pertaining to water]

carboxyl group (2) A functional group characterized by a carbon atom double-bonded to one oxygen atom and single-bonded to another oxygen atom. Amino acids are made up of an amino group, a carboxyl group, and a side chain.

carnivores (15) Predatory animals (and some plants) that consume only animals. [Lat., *carnis*, of flesh + *vorare*, to devour]

carotenoids (4) Pigments that absorb blue-violet and blue-green wavelengths of light and reflect yellow, orange, and red wavelengths of light. [Lat., *carota,* carrot]

carpel (12) The female reproductive structure of a flower, including the stigma, style, and ovary. [Gk., *karpos,* fruit]

carrier (7) An individual who carries one allele for a recessive trait and does not exhibit the trait; if two carriers mate, they may produce offspring that do exhibit the trait.

carrying capacity (*K*) (14) The ceiling on a population's growth imposed by the limitation of resources for a particular habitat over a period of time.

cartilaginous fishes (11) Fish species characterized by a skeleton made completely of cartilage, not bone. [Lat., *cartilago,* gristle]

cell (3) The smallest unit of life that can function independently; a three-dimensional structure, surrounded by a membrane and, in prokaryotes and most plants, by a cell wall, in which many of the essential chemical reactions of the life of an organism take place. [Lat., *cella,* room]

cell cycle (6) In a cell, the alternation of activities related to cell division and those related to growth and metabolism.

cell-cycle control system (6) A system—made up of molecules, mostly proteins—by which the events of the cell cycle are coordinated and cell division is regulated. At critical points in the cell cycle, called checkpoints, progress to the next phase may be blocked until specific signals trigger continuation of the process.

cell theory (3) A unifying and universally accepted theory in biology that holds that all living organisms are made up of one or more cells and that all cells arise from other, preexisting cells.

cell wall (3) A rigid structure, outside the cell membrane, that protects and gives shape to the cell; found in many prokaryotes and plants.

cellular respiration (4) The process by which all living organisms extract energy stored in the chemical bonds of molecules and use it for fuel for their life processes.

cellulose (2) A complex carbohydrate, indigestible by humans, that serves as the structural material for a huge variety of plant structures. It is the single most prevalent organic compound on earth. [Lat., *cellula,* dim. of *cella,* room]

central vacuole (3) See **vacuole.**

centriole (6) A structure, located outside the nucleus in most animal cells, to which the spindle fibers are attached during cell division.

centromere (6) After replication, the region of contact between sister chromatids, located near the center of the two strands. [Gk., *centron,* the stationary point of a pair of compasses, thus the center of a circle + *meris,* part]

cephalopods (11) Mollusks in which the head is prominent and the foot has been modified into tentacles; examples are octopuses and squids. Cephalopods have a reduced or absent shell and possess the most advanced nervous system of the invertebrates. [Gk., *kephale,* head + *pous,* foot]

character displacement (15) An evolutionary divergence in one or more of the species that occupy the same niche that leads to a partitioning of the niche between the species. Changes in characteristics, such as behavior or body plan, of two or more very similar species that have overlapping geographic locations result in a reduction of competition between the species.

checkpoint (6) One of several critical points in the cell cycle at which progress is blocked—and cells are prevented from dividing—until specific signals trigger continuation of the process.

chemical energy (4) A type of potential energy in which energy is stored in chemical bonds between atoms or molecules.

chemolithotrophs (13) Bacteria that can use inorganic molecules such as ammonia, hydrogen sulfide, hydrogen, and iron as sources of energy. [Gk., *lithos,* stone, rock + *trophē,* food]

chemoorganotrophs (13) Bacteria that consume organic molecules, such as carbohydrates, as an energy source. [Gk., *trophē,* food]

chiasmata (*sing.* **chiasma**) (6) The sites at which chromatids exchange genetic material during recombination.

chitin (2) (*pron.* KITE-in) A complex carbohydrate, indigestible by humans, that forms the rigid outer skeleton of most insects and crustaceans. [Gk., *chiton,* undershirt]

chlorophyll (4) A light-absorbing pigment molecule in chloroplasts. [Gk., *chloros,* pale green + *phyllon,* leaf]

chlorophyll *a* (4) The primary photosynthetic pigment. Chlorophyll *a* absorbs blue-violet and red light; because it cannot absorb green light and instead reflects those wavelengths, we perceive the reflected light as the color green.

chlorophyll *b* (4) A photosynthetic pigment similar in structure to chlorophyll *a.* Chlorophyll *b* absorbs blue and red-orange wavelengths and reflects yellow-green wavelengths.

chloroplast (3, 4) The organelle in plant cells in which photosynthesis occurs. [Gk., *chloros,* pale green + *plastos,* formed]

cholesterol (2, 3) One of the sterols, a group of lipids important in regulating growth and development; an important component of most cell membranes, helping the membrane maintain its flexibility. [Gk., *chole,* bile + *stereos,* solid + *-ol,* chemical suffix for an alcohol]

chromatid (6) One of the two strands of a replicated chromosome. [Gk., *chroma,* color]

chromatin (3, 6) A mass of long, thin fibers consisting of DNA and proteins in the nucleus of the cell. [Gk., *chroma,* color]

chromosomal aberration (5) A type of mutation characterized by a change in the overall organization of genes on a chromosome, such as the deletion of a section of DNA; or the moving of a gene from one part of a chromosome to elsewhere on the same chromosome or to a different chromosome; or the duplication of a gene, with the new copy inserted elsewhere on the chromosome or on a different chromosome. [Lat., *aberrare,* to wander]

chromosome (5, 6) A linear or circular strand of DNA with specific sequences of base pairs. The human genome consists of two copies of each of 23 unique chromosomes, one from the mother and one from the father. [Gk., *chroma,* color + *soma,* body]

cilia (*sing.* **cilium**) (3) Short projections from the cell surface, often occurring in large numbers on a single cell, that beat against the extracellular fluid to move the fluid past the cell. [Lat., *cilium,* eyelid]

class (10) In the hierarchical taxonomic system developed by Carolus Linnaeus (1707–1778), a classification of organisms consisting of related orders.

climax community (15) A stable and self-sustaining community that results from ecological succession.

clone (5) A genetically identical DNA fragment, cell, or organism produced from a DNA fragment or single cell or organism. [Gk., *klon,* twig]

clone library (5) A collection of cloned DNA fragments; also known as a gene library.

cloning (5) The production of genetically identical cells, organisms, or DNA molecules.

code (5) In genetics, the base sequence of a gene; information encoded within the genetic information can be translated into proteins.

codominance (7) The case in which the heterozygote displays characteristics of both alleles.

codons (5) Three-base sequences in mRNA that link with complementary tRNA molecules, which are attached to amino acids that are specified by that codon; a codon with yet another sequence ends the process of assembling a protein from amino acids.

coevolution (15) The concurrent appearance and modification over time, through natural selection, of traits in interacting species that enable each species to become adapted to the other; an example is the 11-inch-long tongue of a moth that feeds from the 11-inch-long nectar tube of an orchid.

colonizers (15) Species introduced into an area that has been disturbed and is undergoing the process of either primary or secondary succession. The identity of a colonizing species varies, depending on the stage and type of succession.

commensalism (15) A symbiotic relationship between species in which one benefits and the other neither benefits nor is harmed. [Lat., *com,* with + *mensa,* table]

communication (9) An action or signal on the part of one organism that alters the behavior of another organism.

community (in ecology) (15) The biotic environment; a geographic area defined as a loose assemblage of species with overlapping ranges.

competitive exclusion (15) The case in which two species battle for resources in the same niche until the more efficient of the two wins and the other is driven to extinction in that location.

competitive inhibitor (2) A chemical that binds to the active site of an enzyme, blocking substrate molecules from the site and thereby reducing the enzyme's ability to catalyze a reaction.

complementarity (6) A characteristic of double-stranded DNA in which the base on one strand always has the same pairing partner, or complementary base, on the other strand.

complementary base (6) A base on a strand of double-stranded DNA that is a pairing partner to a base on the other strand: adenine (A) is the complementary base to thymine (T), and guanine (G) is the complementary base to cytosine (C).

complete metamorphosis (11) A developmental process in which, after hatching, an organism's life is divided into three completely different stages: the larva, pupa, and adult; occurs in about 83% of insect species, including beetles and butterflies, and is an important factor in the broad diversification of insects.

complex carbohydrate (2) A carbohydrate that contains multiple simple carbohydrates linked together; examples are starch, which is the primary form of energy storage in plants, and glycogen, which is the primary form of short-term energy storage in animals.

condensation (6) In the cell cycle, just before mitosis, the process in which sister chromatids coil tightly and become compact—in contrast to the uncondensed and tangled state of the chromosomes prior to replication, during most of interphase.

conjugation (13) The process by which a bacterium transfers a copy of some or all of its DNA to another bacterium, of the same or another species. [Lat., *coniugatio,* connection]

conservation biology (16) An interdisciplinary field, drawing on ecology, economics, psychology, sociology, and political science, that studies and devises ways of preserving and protecting biodiversity and other natural resources.

contact inhibition (3) The limiting factor of cell growth that occurs when normal cells come into contact; in cancer cells, contact inhibition does not take place and the cells continue to divide.

control group (1) In an experiment, the group of subjects not exposed to the treatment being studied but otherwise treated identically to the experimental group.

convergent evolution (8, 10) A process of natural selection in which features of organisms not closely related come to resemble each other as a consequence of similar selective forces. Many marsupial and placental species resemble each other as a result of convergent evolution. [Lat., *con-,* together with + *vergere,* turn + *evolvere,* to roll out]

coral reef (11) Underwater structures that are assemblies of giant calcium carbonate skeletons and corals, small cnidarians that secrete the calcium carbonate to create hard shells on which numerous individual polyps can live as a colony. Coral reefs provide an environment that is home to a greater diversity of species than any other marine habitat.

covalent bond (2) A strong bond formed when two atoms share electrons; the simplest example is the H_2 molecule, in which each of the two atoms in the molecule shares its lone electron with the other atom. [Lat., *con-,* together + *valere,* to be strong]

critical experiment (1) An experiment that makes it possible to determine decisively between alternative hypotheses.

cross (7) The breeding of organisms that differ in one or more traits.

crossing over (6) The exchange of some genetic material between a paternal homologous chromosome and a maternal homologous chromosome, leading to a chromosome carrying genetic material from each; also called recombination.

cuticle (12) A waxy layer produced by epidermal cells and found on leaves and shoots of terrestrial plants, protecting them from drying out. [Lat., *cuticula, dim.* of *cutis,* skin]

cytokinesis (6) In the cell cycle, the stage following mitosis in which cytoplasm and organelles duplicate and are divided into approximately equal parts, and the cell separates into two daughter cells. In meiosis, two diploid daughter cells are formed in cytokinesis following telophase I, and four haploid daughter cells are formed in cytokinesis following telophase II. [Gk., *kytos,* container + *kinesis,* motion]

cytoplasm (3) The contents of a cell contained within the plasma membrane, including a jelly-like fluid, called the cytosol, and, in eukaryotes, the organelles. [Gk., *kytos,* container + *plasma,* anything molded]

cytoskeleton (3) A network of protein structures in the cytoplasm of eukaryotes (and, to a lesser extent, prokaryotes) that serves as scaffolding, adding support and, in some cases, giving animal cells of different types their characteristic shapes. The cytoskeleton serves as a system of tracks guiding the intracellular traffic flow and, because it is flexible and can generate force, gives cells some ability to control their movement.

cytosol (3) The jelly-like fluid that fills the inside of the cell.

D

daughter cells (6) Cells produced by the division of a parent cell.

decomposers (12, 15) Organisms, including bacteria, fungi, and detritivores, that break down and feed on once-living organisms.

demographic transition (14) A pattern of population growth characterized by the progression from high birth and death rates (slow growth) to high birth rates and low death rates (fast growth) to low birth and death rates (slow growth).

denaturation (2) The disruption of protein folding, in which secondary and tertiary structures are lost, caused by exposure to extreme conditions in the environment such as heat or extreme pH.

density-dependent factors (14) Limitations on a population's growth that are a consequence of population density.

density-independent factors (14) Limitations on a population's growth without regard to population size, such as floods, earthquakes, fires, and lightning.

deoxyribonucleic acid (DNA) (2, 5) A nucleic acid that carries information, in the sequences of its nucleotide bases, about the production of particular proteins.

dependent variable (1) A measurable entity that is created by the process being observed and whose value cannot be controlled; generally represented on the *y*-axis in a graph and expected to change in response to a change in the independent variable.

desert (15) A type of terrestrial biome; a type of dry climate, with very little rainfall, in which water loss through evaporation exceeds water gain through precipitation; typically found at 30° north and south latitudes.

desmosomes (3) Irregularly spaced connections between adjacent animal cells that, much like Velcro, hold the cells together by multiple attachments but are not water-tight. They provide mechanical strength and are found in muscle tissue and in much of the tissue that lines the cavities of animal bodies. [Gk., *desmos*, bond + *soma*, body]

detritivores (15) Organisms that break down and feed on once-living organic matter; this group includes scavengers such as vultures, worms, and a variety of arthropods. [Lat., *detritus*, worn out + *vorare*, to devour]

deuterostomes (11) Bilaterally symmetrical animals with defined tissues in which the gut develops from back to front; the anus forms first, and the second opening formed becomes the mouth of the adult animal. [Gk., *deuteros*, second + *stoma*, mouth]

differential reproductive success (8) The situation in which some individuals have greater reproductive success than other individuals in a population; along with variation and heritability, differential reproductive success is one of the three conditions necessary for evolution by natural selection.

diffusion (3) Passive transport in which a particle (the solute) is dissolved in a gas or liquid (the solvent) and moves from an area of higher solute concentration to an area of lower solute concentration. [Lat., *diffundere*, to pour in different directions]

dihybrid (6) An individual that is heterozygous at two genetic loci.

dihybrid cross (6) A cross between two individuals that are heterozygous for the same two genetic loci.

diploid (6) Describes cells that have two copies of each chromosome (in many organisms, including humans, somatic cells are diploid). [Gk., *diplasiazein*, to double]

directional selection (8) Selection that, for a given trait, increases fitness at one extreme of the phenotype and reduces fitness at the other, leading to an increase or decrease in the mean value of the trait.

disaccharides (2) Carbohydrates formed by the union of two simple sugars, such as sucrose (table sugar) and lactose (the sugar found in milk). [Gk., *di-*, two + *sakcharon*, sugar]

discontinuous replication (6) In DNA replication, the synthesis of new DNA on the lagging strand that produces short DNA fragments.

dispersers (15) Organisms able to move away from their original home.

disruptive selection (8) Selection that, for a given trait, increases fitness at both extremes of the phenotype distribution and reduces fitness at middle values.

DNA helicase (6) In DNA replication, the enzyme that unwinds the coiled DNA and separates the two complementary strands.

DNA polymerase (6) In DNA replication, the enzyme that adds new DNA nucleotides to the 3′ end of the primer as it builds a strand complementary to the template strand.

DNA probe (5) A short sequence of radioactively tagged single-stranded DNA that contains part of the sequence of the gene of interest, used to locate that gene in a gene library. The probe binds to the complementary base pair on a gene in the library, which is identified by the radioactive tag on the probe.

DNA synthesis phase (6) In the cell cycle, the phase during which the cell prepares for cell division by creating an exact duplicate of each chromosome by replication; also called S phase.

domain (10) In modern classification, the highest level of the hierarchy; there are three domains, Bacteria, Archaea, and Eukarya.

dominant (7) Describes an allele that masks the phenotypic effect of the other, recessive allele for a trait; the phenotype shows the effect of the dominant allele in both homozygous and heterozygous genotypes. [Lat., *dominari*, to rule]

dorsal hollow nerve cord (11) The central nervous system of vertebrates, consisting of the spinal cord and brain. [Lat., *dorsum*, back]

double-blind experimental design (1) An experimental design in which neither the subjects nor the experimenters know what treatment (if any) individual subjects are receiving.

double bond (2) The sharing of two electrons between two atoms; for example, the most common form of oxygen is the O_2 molecule, in which two electrons from each of the two atoms of oxygen are shared.

double fertilization (12) In angiosperms, the fertilization process in which two sperm are released by a pollen grain, and one fuses with an egg to form a zygote, while the other fuses with two nuclei to form a triploid endosperm.

double helix (2) The spiraling ladder-like structure of DNA composed of two strands of nucleotides; the bases protruding from each strand like "half-rungs" meet in the center and bind to each other (via hydrogen bonds), holding the ladder together. [Gk., *heligmos*, wrapping]

E

ecological footprint (14) A measure of the impact of an individual or population on the environment by calculation of the amount of resources—including land, food and water, and fuel—consumed.

ecology (14) The study of the interaction between organisms and their environments, at the level of individuals, populations, communities, and ecosystems. [Gk., *oikos*, home + *logos*, discourse]

ecosystem (15) A community of biological organisms and the non-living environmental components with which they interact.

ecosystem services (16) The utilitarian value of biodiversity to humans described in terms of four distinct categories of services—provisioning, cultural, regulating, and habitat services.

ectotherms (11) Organisms that rely on heat from an external source to raise their body temperature and seek the shade when the air is too warm. [Gk., *ektos*, outside + *thermē*, heat]

El Niño (15) A sustained surface temperature change in the central Pacific Ocean that occurs every two to seven years; this event can start a chain reaction of unusual weather across the globe that can result in flooding, droughts, famine, and a variety of extreme climate disruptions. [Spanish, *the child;* a reference to the Christ child, because of the appearance of the phenomenon at Christmastime]

electromagnetic spectrum (4) The range of wavelengths that produce electromagnetic radiation, extending (in order of decreasing energy) from high-energy, short-wave, gamma rays and X rays, through ultraviolet light, visible light, and infrared light, to very long, low-energy, radio waves. [Lat., *specere*, to look at]

electron (2) A negatively charged particle that moves around the atomic nucleus.

electron transport chain (4) The path of high-energy electrons moving from one molecule within a membrane to another, coupled to the pumping of protons across the membrane, creating a concentration gradient that is used to make ATP; occurs in mitochondria and chloroplasts.

element (2) A pure substance that cannot be broken down chemically into any other substances; all atoms of an element have the same atomic number. [Lat., *elementum*, element, or first principle]

empirical (1) Describes knowledge that is based on experience and observations that are rational, testable, and repeatable. [Gk., *empeiria,* experience]

endangered species (16) As defined by the Endangered Species Act, species in danger of extinction throughout all or a significant portion of their range.

Endangered Species Act (ESA) (16) A U.S. law that defines "endangered species" and is designed to protect those species from extinction.

endemic (16) Describes species peculiar to a particular region and not naturally found elsewhere. [Gk., *en,* in + *demos,* the people of a country]

endocytosis (3) A cellular process in which large particles, solid or dissolved, outside the cell are surrounded by a fold of the plasma membrane, which pinches off, forming a vesicle, and the enclosed particle moves into the cell. The three types of endocytosis are phagocytosis, pinocytosis, and receptor-mediated endocytosis. [Gk., *endon,* within + *kytos,* container]

endomembrane system (3) A system of organelles (the rough endoplasmic reticulum, the smooth endoplasmic reticulum, and the Golgi apparatus) that surrounds the nucleus; it produces and modifies necessary molecules, breaks down toxic chemicals and cellular by-products, and is thus responsible for many of the fundamental functions of the cell. [Gk., *endon,* within + Lat., *membrana,* a thin skin]

endosperm (12) Tissue of a mature seed that stores certain carbohydrates, proteins, and lipids that fuel the germination, growth, and development of the embryo and young seedling. [Gk., *endon,* within + *sperma,* seed]

endosymbiosis theory (3) Theory of the origin of eukaryotes. For photosynthetic eukaryotes, the theory holds that, in the past, two different types of prokaryotes engaged in a close partnership and eventually one, capable of performing photosynthesis, was subsumed into the other, larger prokaryote. The smaller prokaryote made some of its photosynthetic energy available to the host and, over time, the two became symbiotic and eventually became a single, more complex organism in which the smaller prokaryote had evolved into the chloroplast of the new organism. A similar scenario can be developed for the evolution of mitochondria. [Gk., *endon,* within + *symbios,* living together]

endotherms (11) Organisms that use the heat produced by their cellular respiration to raise and maintain their body temperature above air temperature. [Gk., *endon,* within + *thermē,* heat]

energy (4) The capacity to do work, which is the moving of matter against an opposing force. [Gk., *energeia,* activity]

energy pyramid (15) A diagram that illustrates the path of energy through the organisms of an ecosystem; each layer of the pyramid represents the biomass of a trophic level.

enzymatic protein (3) See **enzyme.**

enzyme (2, 3) A protein that initiates and accelerates a chemical reaction in a living organism. Enzymes are found throughout the cell; enzymatic proteins take part in chemical reactions on the inside and outside surfaces of the plasma membrane. [Gk., *en,* in + *zyme,* leaven]

estuary (15) A tidal water passage, linked to the sea, in which salt water and fresh water mix; characterized by exceptionally high productivity. [Lat., *aestus,* tide]

ethanol (4) The end product of fermentation of yeast; the alcohol in beer, wine, and spirits. [Contraction of the full chemical name, ethyl alcohol]

eukaryote (3) An organism composed of eukaryotic cells. [Gk., *eu,* good + *karyon,* nut, kernel]

eukaryotic cell (3) A cell with a membrane-surrounded nucleus containing DNA, membrane-surrounded organelles, and internal structures organized into compartments.

eutrophication (15) The process in which excess nutrients dissolved in a body of water lead to rapid growth of algae and bacteria, which consume much of the dissolved oxygen and, in time, can lead to large-scale die-offs of other species. [Gk., *eu,* good + *trophē,* food]

evolution (8) A change in allele frequencies of a population. [Lat., *evolvere,* to roll out]

exocytosis (3) A cellular process in which particles within the cell, solid or dissolved, are enclosed in a vesicle and transported to the plasma membrane, where the membrane of the vesicle merges with the plasma membrane and the material in the vesicle is expelled to the extracellular fluid for use throughout the body. [Gk., *ex,* out of + *kytos,* container]

exoskeleton (11) A rigid external covering such as is found in some invertebrates, including insects and crustaceans. [Gk., *ex,* out of]

exotic species (16) Species introduced by human activities to areas other than the species' native range.

experimental group (1) In an experiment, the group of subjects exposed to a particular treatment; also known as the treatment group.

exponential growth (14) Growth of a population at a rate that is proportional to its current size.

extinction (10) The complete loss of all individuals in a species. [Lat., *extinguere,* to extinguish] [Lat., *extremus,* outermost + Gk., *philios,* loving]

F

facilitated diffusion (3) Diffusion of molecules through the phospholipid bilayer of the plasma membrane that takes place through a transport protein (a "carrier molecule") embedded in the membrane. Molecules that require the assistance of a carrier molecule are those that are too big to cross the membrane directly or are electrically charged and would be repelled by the middle layer of the membrane.

family (10) In the system developed by Carolus Linnaeus (1707–1778), a classification of organisms consisting of related genera.

fatty acid (2) A long hydrocarbon (a chain of carbon-hydrogen molecules); fatty acids form the tail region of triglyceride fat molecules.

female (9) In sexually reproducing organisms, a member of the sex that produces the larger gamete.

fermentation (4) The process by which glycolysis occurs in the absence of oxygen; the electron acceptor is pyruvate (in animals) or acetaldehyde (in yeast) rather than oxygen.

fertilization (6) The fusion of two reproductive cells.

filament (12) In angiosperms, the supporting stalk of the anther of a stamen in flowers. [Lat., *filum,* thread]

first law of thermodynamics (4) A physical law that states that energy cannot be created or destroyed; it can only change from one form to another.

fitness (8) A relative measure of the reproductive output of an individual with a given phenotype compared with the reproductive output of individuals with alternative phenotypes.

fixation (8) In genetics, the point at which the frequency of an allele in a population is 100%, and thus there is no more variation in the population for this gene.

fixed action pattern (9) An innate sequence of behaviors, triggered under certain conditions, that requires no learning, does not vary, and once begun runs to completion; an example is egg-retrieval in geese.

flagellum (*pl.* **flagella**) (3) Long, thin, whip-like projection from the cell body of a prokaryote that aids in cell movement through the medium in which the organism lives; in animals, the only cell with a flagellum is the sperm cell. [Lat., *flagellum,* whip]

flatworms (11) Worms with flat bodies that are members of the phylum Platyhelminthes; characterized by well-defined head and tail regions, with some having clusters of light-sensitive cells for eyespots. Most are hermaphroditic and are protostomes that do not molt. Examples are tapeworms and flukes.

fleshy fruit (12) A fruit that consists of the ovary and some additional parts of the flower; fleshy fruits are an attractive food, and when eaten by animals, the seeds may be widely dispersed.

flower (12) The part of an angiosperm that contains the reproductive structures; consists of a supporting stem with modified leaves (the petals and sepals) and usually contains both male and female reproductive structures.

fluid mosaic (3) A term that describes the structure of the plasma membrane, which is made up of several different types of molecules, many of which are not fixed in place but float, held in their proper orientation by hydrophilic and hydrophobic forces.

food chain (15) The path of energy flow from producers to tertiary consumers.

food web (15) A more precisely described path of energy flow from producers to tertiary consumers than the food chain, reflecting the fact that many organisms are omnivores and occupy more than one position in the chain.

fossil (8) The remains of an organism, usually its hard parts such as shell, bones, or teeth, that have been naturally preserved; also, traces of such an organism, such as footprints. [Lat., *fossilis,* that which is dug up]

fossil fuels (4) Fuels produced from the decayed remains of ancient plants and animals; include oil, natural gas, and coal.

founder effect (8) A phenomenon by which genetic drift can occur. The isolation of a small subgroup of a larger population can lead to changes in the allele frequencies of the isolated population, because all the descendants of the smaller group will reflect the allele frequencies of the subgroup, which may differ from those of the larger source population.

fundamental niche (15) The full range of environmental conditions under which an organism can live.

G

G_0 (6) A quiescent or "resting" phase, outside the cell cycle, in which no cell division occurs.

gamete (6) A cell (often haploid) that will combine with another (haploid) cell at fertilization to produce offspring; also called a reproductive cell or sex cell. [Gk., *gamete,* wife]

gametophyte (12) The structure in land plants and some algae that produces gametes (sperm and eggs); the haploid life stage of plants and some algae, which may be either male (producing sperm) or female (producing eggs). [Gk., *gamete,* wife + *phytas,* plant]

Gap 1 (G_1) (6) The phase of the cell cycle during which the cell may grow and develop, as well as performing its various cellular functions. Most cells spend most of their time in this phase.

Gap 2 (G_2) (6) The phase of the cell cycle that follows the DNA synthesis phase, characterized by significant growth and high rates of protein synthesis in preparation for cell division.

gap junction (3) A junction between adjacent animal cells in the form of a pore in each of the plasma membranes that is surrounded by a protein, linking the two cells and acting like a channel between them, thus allowing materials to pass between the cells.

gastropods (11) Mollusks that are members of the class Gastropoda; most have a single shell, a muscular foot for locomotion, and a radula used for scraping food from surfaces; examples are snails and slugs. [Gk., *gaster,* belly + *pous,* foot]

gene (5) The basic unit of heredity; a sequence of DNA nucleotides on a chromosome that carries the information necessary for making a functional product, usually a protein or an RNA molecule. [Gk., *genos,* race, descent]

gene expression (5) The process by which information in a gene's sequence is used to synthesize a gene product (commonly a protein, but also RNA).

gene flow (8) A change in the allele frequencies of a population due to movement of some individuals of a species from one population to another, changing the allele frequencies of the population they join; also known as migration.

gene library (5) A collection of cloned DNA fragments; also known as a clone library.

gene regulation (5) The processes by which cells "turn on" or "turn off" genes, influencing the amount of gene products formed.

gene therapy (5) The process of inserting, altering, or deleting one or more genes in an individual's cells to correct defective versions of the gene(s).

genetic drift (8) A random change in allele frequencies over successive generations; a cause of evolution.

genetic engineering (5) The manipulation of an organism's genetic material by adding or deleting genes or transplanting genes from one organism to another.

genetic recombination (6) See **recombination.**

genome (5) The full set of DNA present in an individual organism; also can refer to the full set of DNA present in a species.

genotype (5, 7) The genes that an organism carries for a particular trait; also, collectively, an organism's genetic composition. [Gk., *genos,* race, descent + *typos,* impression, engraving]

genus (*pl.* **genera**) (10) In the system developed by Carolus Linnaeus (1707–1778), a classification of organisms consisting of closely related species. [Lat., *genus,* race, family, origin]

glycerol (2, 3) A small molecule that forms the head region of a triglyceride fat molecule. [Gk., *glykys,* sweet + *-ol,* chemical suffix for an alcohol]

glycogen (2) A complex carbohydrate consisting of stored glucose molecules linked to form a large web, which breaks down to release glucose when it is needed for energy. [Gk., *glykys,* sweet + Gk., *genos,* race, descent]

glycolysis (4) In all organisms, the first step in cellular respiration, in which one molecule of glucose is broken into two molecules of pyruvate. For some organisms, glycolysis is the only means of extracting energy from food; for others, including most plants and animals, it is followed by the Krebs cycle and the electron transport chain. [Gk., *glykys,* sweet + *lysis,* releasing]

Golgi apparatus (3) (*pron.* GOHL-jee) An organelle, part of the endomembrane system, structurally like a flattened stack of unconnected membranes, each known as a Golgi body. The Golgi apparatus processes molecules synthesized in the cell and packages those molecules that are destined for use elsewhere in the body. [From the name of the discoverer, Camillo Golgi, 1843–1926]

gonads (6) The ovaries and testes in sexually reproducing animals. [Gk., *gonē,* food offspring]

Gram stain (13) A dye test used by microbiologists in identifying an unknown bacterium; the dye stains the layer of peptidoglycan outside the cell wall purple, but bacteria in which the layer of peptidoglycan is covered by a membrane are not colored by the dye. Bacteria that take a Gram stain are known as Gram-positive bacteria; those that do not are known as Gram-negative bacteria. [From the name of the inventor, Hans Christian Gram, 1853–1938]

group selection (9) The process, extremely uncommon in nature, that brings about an increase in the frequency of alleles for traits (e.g., behaviors) that are beneficial to the persistence of the species or population but are detrimental to the fitness of the individual possessing the trait (or engaging in the behavior).

growth factors (6) Chemical signals that trigger transitions to subsequent phases in the cell cycle, typically by providing feedback about the cell's environment.

growth rate *(r)* (14) The birth rate minus the death rate in a population; the change in the number of individuals in a population per unit of time.

gymnosperms (12) Vascular plants that do not produce their seeds in a protective structure; seeds are usually found on the surface of the scales of a cone-like structure. The gymnosperms include conifers, cycads, gnetophytes, and ginkgo. [Gk., *gymnos,* naked + *sperma,* seed]

H

habitat (15) The physical environment of organisms, consisting of the chemical resources of the soil, water, and air, and physical conditions such as temperature, salinity, humidity, and energy sources. [Lat., *habitare,* to dwell or inhabit]

hair (11) Dead cells filled with the protein keratin that collectively serve as insulation covering the body or a part of the body; present in all mammals.

haploid (6) Describes cells that have a single copy of each chromosome (in many species, including humans, gametes are haploid). [Gk., *haploeides,* single]

hazard factor (14) An external force on a population that increases the risk of death.

herbivores (15) Animals that eat plants; also known as primary consumers. [Lat., *herba,* grass + *vorare,* to devour]

heredity (7) The greater resemblance of offspring to parents than to other individuals in the population, a consequence of the passing of characteristics from parents to offspring through their genes. [Lat., *heres,* heir]

heritability (8) The transmission of traits from parents to offspring via genetic information; also known as inheritance.

hermaphrodite (6) An organism that produces both male and female gametes. [From the names of the Greek god Hermes and goddess Aphrodite]

heterozygous (7) Describes the genotype of a trait for which the two alleles an individual carries differ from each other. [Gk., *heteros,* other + *zeugos,* pair]

histones (6) Proteins around which the long, linear strands of DNA are wrapped; serve to keep the DNA untangled and to enable an orderly, tight, and efficient packing of the DNA within the cell.

homologous feature (10) A feature inherited from a common ancestor. [Gk., *homologia,* agreement]

homologous pair (homologues) (6) The maternal and paternal copies of a chromosome. [Gk., *homologia,* agreement]

homologous structure (8) Body structures in different organisms that, although they may have been modified over time to serve different functions in different species, are derived through inheritance from a common evolutionary ancestor.

homozygous (7) Describes the genotype of a trait for which the two alleles an individual carries are the same. [Gk., *homos,* same + *zeugos,* pair]

honest signal (9) A signal, which cannot be faked, that is given when both the individual making the signal and the individual responding to it have the same interests; it carries the most accurate information about an individual or situation.

horizontal gene transfer (10) The transfer of genetic material directly from one individual to another, not necessarily related, individual; common among bacteria.

host (13, 15) An organism in or on which a parasite lives.

human immunodeficiency virus (HIV) (13) The virus responsible for AIDS, a deadly disease that destroys the human immune system. HIV is a retrovirus, an RNA-containing virus that is thought to have been introduced to humans from chimpanzees.

hybridization (10) The interbreeding of closely related species. [Lat., *hybrida,* animal produced by two different species]

hybrids (10) Offspring of individuals of two different species. [Lat., *hybrida,* animal produced by two different species]

hydrogen bond (2) A type of weak chemical bond formed between the slightly positively charged hydrogen atoms of one molecule and the slightly negatively charged atoms (often oxygen or nitrogen) of another. Hydrogen bonds are important in building complex molecules, such as large proteins and DNA, and are responsible for many of the unique and important features of water.

hydrophilic (2, 3) Attracted to water, as, for example, polar molecules that readily form hydrogen bonds with water. [Lat., *hydro-,* pertaining to water; Gk., *philios,* loving]

hydrophobic (2, 3) Repelled by water, as, for example, nonpolar molecules that tend to minimize contact with water. [Lat., *hydro-,* pertaining to water + Gk., *phobos,* fearing]

hypertonic (3) Of two solutions, that with a higher concentration of solutes. [Gk., *hyper,* above + *tonos,* tension]

hyphae (*sing.* **hypha**) (12) (*pron.* HIGH-fee) Long strings of cells that make up the mycelium of a multicellular fungus. [Gk., *hypha,* web]

hypothesis (*pl.* **hypotheses**) (1) A proposed explanation for an observed phenomenon. [Gk., *hypothesis,* a proposal]

hypotonic (3) Of two solutions, that with a lower concentration of solutes. [Gk., *hypo,* under + *tonos,* tension]

I

inclusive fitness (9) The sum of an individual's indirect and direct fitness.

incomplete dominance (7) The case in which the heterozygote has a phenotype intermediate between those of the two homozygotes; an example is pink snapdragons, with an appearance intermediate between homozygous for white flowers and homozygous for red flowers.

incomplete metamorphosis (11) A process of growth and development in which, rather than passing through completely different stages, an organism passes through several stages in which the juvenile form resembles a smaller version of the adult—with no pupal stage; occurs in about 17% of insects, including grasshoppers and crickets.

independent variable (1) A measurable entity that is available at the start of a process being observed and the value of which can be changed as required; generally represented on the *x*-axis in a graph.

inheritance (8) The transmission of traits from parents to offspring via genetic information; also known as heritability.

inhibitor (2) A chemical that binds to an enzyme or substrate molecule and in doing so reduces the enzyme's ability to catalyze a reaction.

innate behaviors (9) Behaviors that do not require environmental input for their development. These behaviors are present in all individuals in a population and do not vary much from one individual to another or over an individual's life span; also known as instincts. [Lat., *innatus,* inborn]

instincts (9) Behaviors that do not require environmental input for their development. Instincts are present in all individuals in a population and do not vary much from one individual to another or over an individual's life span; also known as innate behaviors. [Lat., *instinctus,* impelled]

intermediate filaments (3) One of three types of protein fibers (the others are microtubules and microfilaments) that make up the eukaryotic cytoskeleton, providing it with structure and shape; a durable, rope-like system of numerous different overlapping proteins.

intermembrane space (3) In a mitochondrion, the region between the inner and outer membranes. [Lat., *inter,* between + *membrana,* a thin skin]

interphase (6) In the cell cycle, the phase during which the cell grows and functions; during this phase, replication of DNA occurs in preparation for cell division. [Lat., *inter,* between + Gk., *phasis,* appearance]

intron (5) A non-coding region of DNA.

invagination (3) The folding in of a membrane or layer of tissue so that an outer surface becomes an inner surface. [Lat., *in,* in + *vagina,* sheath]

invertebrates (11) Animals that do not have a backbone; although commonly used in organizing the animals, invertebrates are not a monophyletic group.

ion (2) An atom that carries an electrical charge, positive or negative, because it has either lost or gained an electron or electrons from its normal, stable configuration. [Gk., *ion,* going]

ionic bond (2) A bond created by the transfer of one or more electrons from one atom to another; the resulting atoms, now called ions, are charged oppositely and so attract each other to form a compound.

ionic compound (2) A chemical compound in which ions of two or more elements are held together by ionic bonds.

isotonic (3) Refers to solutions with equal concentrations of solutes. [Gk., *isos,* equal to + *tonos,* tension]

isotopes (2) Variants of atoms that differ in the number of neutrons they possess; isotopes do not differ in charge, because neutrons have no electrical charge, but the atom's mass changes with the loss or addition of a particle in the nucleus.

K

karyotype (6) A visual display of an individual's full set of chromosomes. [Gk., *karyon,* nut or kernel + *typos,* impression, engraving]

keystone species (15) A species that has an unusually large influence on the presence or absence of numerous other species in a community.

kin selection (9) "Kindness" toward close relatives, which may evolve as apparently altruistic behavior toward them, but which in fact is beneficial to the fitness of the individual performing the behavior.

kinetic energy (4) The energy of moving objects, such as legs pushing the pedals of a bicycle or wings beating against the air. [Gk., *kinesis,* motion]

kinetochore (6) During metaphase of mitosis and metaphase II of meiosis, the disk-like group of proteins that develops at the centromeres of a pair of sister chromatids to which the spindle fibers attach and, in anaphase, will pull the chromatids to opposite poles of the cell.

kingdom (10) In the system developed by Carolus Linnaeus (1707–1778), one of the three categories—animal, plant, and mineral—into which all organisms and substances on earth were placed. In modern classification, there are six kingdoms: bacteria, archaea, protists, plants, animals, and fungi.

Krebs cycle (4) The second step of cellular respiration, in which energy is extracted from sugar molecules as additional molecules of ATP and NADH are formed. [From the name of the discoverer, Hans Adolf Krebs, 1900–1981]

L

La Niña (15) The counterpart of the El Niño phenomenon, characterized by the reverse effects, including a decrease in surface temperature and air pressure at the surface of the western Pacific Ocean; the resulting climate effects are approximately opposite to El Niño effects.

lagging strand (6) In DNA replication, the template strand that is replicated discontinuously as a series of DNA fragments; nucleotides are added to the new strand in the 5′ to 3′ direction—opposite to the direction in which the replication fork is moving along the DNA.

landscape conservation (16) The conservation of habitats and ecological processes, as well as species.

language (9) A type of communication in which arbitrary symbols represent concepts and grammar; a system of rules dictates how the symbols can be manipulated to communicate and express ideas.

larva (11) In complete metamorphosis, the first stage of insect development; the larva is hatched from the egg and eats to grow large enough to enter the pupa stage. The larva (for example, a caterpillar) looks completely different from the adult (a butterfly or moth). [Lat., *larva,* ghost]

leading strand (6) In DNA replication, the template strand that is replicated by adding nucleotides to a continuously growing new strand in the 5′ to 3′ direction—the same direction in which the replication fork is moving along the DNA.

learning (9) The alteration and modification of behavior over time in response to experience.

lichen (12) Symbiotic partnership between fungi and chlorophyll-containing algae or cyanobacteria, or both.

life (10) A physical state characterized by the ability to replicate and the presence of metabolic activity.

life history (14) The vital statistics of a species, including age at first reproduction, probabilities of survival and reproduction at each age, litter size and frequency, and longevity.

life table (14) A table presenting data on the mortality rates within defined age ranges for a population; used to determine an individual's probability of dying during any particular year.

light energy (4) A type of kinetic energy made up of energy packets called photons, which are organized into waves.

linked genes (7) Genes that are close to each other on a chromosome and so are more likely than others to be inherited together.

lipid (2) One of the four types of biological macromolecules, insoluble in water and greasy to the touch. Lipids are important in energy storage and insulation (fats), membrane formation (phospholipids), and regulating growth and development (sterols). [Gk., *lipos,* fat]

lobe-finned fishes (11) Fish species characterized by two pairs of sturdy lobe-shaped fins on the underside of the body.

locus (*pl.* **loci**) (5) The location or position of a gene on a chromosome.

logistic growth (14) A pattern of population growth in which initially exponential growth levels off as the environment's carrying capacity is approached.

lysosome (3) A round, membrane-enclosed, enzyme- and acid-filled vesicle in the cell that digests and recycles cellular waste products and consumed material. [Gk., *lysis,* releasing + *soma,* body]

M

macroevolution (10) Large-scale evolutionary change involving the origins of new groups of organisms; the accumulated effect of microevolution over a long period of time. [Gk., *macros,* large + Lat., *evolvere,* to roll out]

macromolecule (2) A large molecule, made up of smaller building blocks or subunits; the four main types of biological macromolecules are carbohydrates, lipids, proteins, and nucleic acids. [Gk., *macros,* large + *dim.* of *moles,* mass]

male (9) In sexually reproducing organisms, a member of the sex that produces the smaller gamete.

mammary glands (11) Glands in all female mammals that produce milk for the nursing of young. [Lat., *mamma,* breast]

marsupials (11) Mammals in which, in most species, after a short period of embryonic life in the uterus, the young complete their development in a pouch in the female. [Gk., *marsipos,* pouch]

mass (2) The amount of matter in a given sample of a substance.

mass extinctions (10) Extinctions in which a large number of species become extinct in a short period of time, usually because of extraordinary and sudden environmental change. [Lat., *extinguere,* to extinguish]

mate guarding (9) Behavior by an individual that reduces the opportunity for its mate to interact with other potential mates.

mating system (9) The pattern of mating behavior in a species, ranging from polyandry to monogamy to polygyny; mating systems are influenced by the relative amounts of parental investment by males and by females.

maximum sustainable yield (14) The point at which the maximum number of individuals can be removed from a population without impairing its growth rate; it occurs at half the carrying capacity.

meiosis (6) In sexually reproducing organisms, a process of nuclear division in the gonads that, along with cytokinesis, produces reproductive cells that have half as much genetic material as the parent cell and that all differ from each other genetically. [Gk., *meioun,* to lessen]

membrane enzymes (3) Proteins, embedded within a phospholipid bilayer membrane, that catalyze chemical reactions.

Mendel's law of independent assortment (7) The principle that allele pairs for different genes separate independently in meiosis, so the inheritance of one trait generally does not influence the inheritance of another trait (the exception, unknown to Mendel, occurs with linked genes). [From the name of its discoverer, Gregor Mendel, 1822–1884]

Mendel's law of segregation (7) The principle that during the formation of gametes, the two alleles for a gene separate, so that half the gametes carry one allele, and half the gametes carry the other. [From the name of its discoverer, Gregor Mendel, 1822–1884]

messenger RNA (mRNA) (5) The ribonucleic acid that "reads" the sequence for a gene in DNA and then carries the information from the nucleus to the cytoplasm, where the next stage of protein synthesis will take place.

metamorphosis (11) The rebuilding of molecules from the larva to the adult stage, resulting in a change of form. *Complete metamorphosis* is the division of an organism's life history into three completely different stages; *incomplete metamorphosis* is the pattern of growth and development in which an organism does not pass through separate, dramatically different life stages. [Gk., *metamorphoun,* to transform]

metaphase (6) The second phase of mitosis, in which the sister chromatids line up at the center of the cell; in meiosis, the homologues line up at the center of the cell in metaphase I and the sister chromatids line up in metaphase II. [Gk., *meta,* in the midst of + *phasis,* appearance]

microbe (10, 13) A microscopic organism; not a monophyletic group, since it includes protists, archaea, and bacteria. [Gk., *micros,* small + *bios,* life]

microevolution (10) A slight change in allele frequencies in a population over one or a few generations. [Gk., *micros,* small + Lat., *evolvere,* to roll out]

microfilaments (3) One of three types of protein fibers (the others are intermediate filaments and microtubules) that make up the eukaryotic cytoskeleton, providing it with structure and shape, These are the thinnest elements in the cytoskeleton; long, solid, rod-like fibers that help generate forces, including those important in cell contraction and cell division.

microsphere (10) A membrane-enclosed, small, spherical unit containing a self-replicating molecule and carrying information, but no genetic material. Microspheres may have been an important stage in the development of life. [Gk., *micros,* small + *sphaira,* ball]

microtubules (3, 6) One of three types of protein fibers (the others are intermediate filaments and microfilaments) that make up the eukaryotic cytoskeleton, providing it with structure and shape. These are the thickest elements in the cytoskeleton; they resemble rigid, hollow tubes, functioning as tracks to which molecules and organelles within the cell may attach and be moved along; also help pull chromosomes apart during cell division.

migration (8) A change in the allele frequencies of a population due to the movement of some individuals from one population to another; an agent of evolutionary change caused by the movement of individuals into or out of a population. [Lat., *migrare,* to move from place to place]

mimicry (15) The evolution of an organism to resemble another organism or object in its environment to help conceal itself from predators. [Gk., *mimesis,* imitation]

mismatch (9) The phenomenon of organisms finding themselves in a situation where the environment they are in differs from the environment to which they are evolutionarily adapted. If this occurs, we expect (and see) behaviors that appear to be (and are) not evolutionarily adaptive.

mitochondrial matrix (3, 4) The space enclosed by the inner mitochondrial membrane, where the carriers NADH and $FADH_2$ begin the electron transport chain by carrying high-energy electrons to molecules embedded in the inner membrane.

mitochondrion (*pl.* mitochondria) (3) The organelle in eukaryotic cells that converts the energy stored in food, in the chemical bonds of carbohydrate, fat, and protein molecules, into a form usable by the cell for all its functions and activities. [Gk., *mitos,* thread + *chondros,* cartilage]

mitosis (6) The division of a nucleus into two genetically identical nuclei that, along with cytokinesis, leads to the formation of two identical daughter cells. [Gk., *mitos,* thread]

mitotic phase (M phase) (6) The phase of the cell cycle during which first the genetic material and nucleus and then the rest of the cellular contents divide.

molecule (2) A group of atoms held together by covalent bonds. [Lat., *dim.* of *moles,* mass]

monogamy (9) A mating system in which most individuals mate with and remain with just one other individual. [Gk., *monos,* single + *gamete,* wife]

monophyletic (10) Describing a group containing a common ancestor and all of its descendants. [Gk., *monos*, single + *phylon*, race, tribe, class]

monosaccharides (2) The simplest carbohydrates and the building blocks of more complex carbohydrates; they cannot be broken down into other monosaccharides; examples are glucose, fructose, and galactose; also known as simple sugars. [Gk., *monos*, single + *sakcharon*, sugar]

monotremes (11) Present-day mammals that retain the ancestral condition of laying eggs. Monotremes are so called because they have a single duct, the cloaca, into which the reproductive system, the urinary system, and the digestive system (for defecation) open. [Gk., *monos*, single + *trema*, hole]

morphological species concept (10) A concept that defines species on the basis of physical features such as body size and shape. [Gk., *morphē*, shape]

multiple allelism (7) The case in which a single gene has more than two possible alleles.

mutation (5, 8) An alteration in the base-pair sequence of an individual's DNA; may arise spontaneously or following exposure to a mutagen. [Lat., *mutare*, to change]

mutualism (15) A symbiotic relationship in which both species benefit and neither is harmed. [Lat., *mutuus*, reciprocal]

mycelium (12) A mass of interconnecting hyphae that make up the structure of a multicellular fungus. [Gk., *mykes*, fungus]

mycorrhizae (12) (*pron.* my-ko-RYE-zee) Root fungi; symbiotic associations between roots and fungi in which fungal structures are closely associated with fine rootlets and root hairs. [Gk., *mykes*, fungus + *rhiza*, root]

N

NADPH (4) A molecule (nicotinamide adenine dinucleotide phosphate) that is a high-energy electron carrier involved in photosynthesis, which stores energy by accepting high-energy protons. It is formed when the electrons released from the splitting of water are passed to NADP$^+$.

natural selection (8) A mechanism of evolution that occurs when, for a heritable variation of a trait, individuals with one version of the trait have greater reproductive success than individuals with a different version of that trait.

neutron (2) An electrically neutral particle in the atomic nucleus. [Lat., *neutro*, in neither direction]

niche (15) The way an organism utilizes the resources of its environment, including the space it requires, the food it consumes, and timing of reproduction.

node (phylogeny) (10) The point on an evolutionary tree at which species diverge from a common ancestor. [Lat., *nodus*, knot]

noncompetitive inhibitor (2) A chemical that binds to part of an enzyme away from the active site but alters the enzyme's shape so as to alter the active site, thereby reducing or blocking the enzyme's ability to bind with substrate.

nondisjunction (6) The unequal distribution of chromosomes during cell division; can lead to Down syndrome and other disorders caused by an individual's having too few or too many chromosomes.

nonpolar (3) Electrically uncharged.

non-vascular plants (12) Plants that do not have vessels to transport water and dissolved nutrients, but instead rely on diffusion; bryophytes are non-vascular plants. [Lat., *vasculum*, dim. of *vas*, vessel]

notochord (11) A rod of tissue extending from head to tail that stiffens the body when muscles contract during locomotion. Primitive chordates retain the notochord throughout life, but in advanced chordates it is present only in early embryos and is replaced by the vertebral column. [Gk., *notos*, back + Lat., *chorda*, cord]

nuclear membrane (3) A membrane enclosing the nucleus of a cell, separating it from the cytoplasm; consists of two bilayers and is perforated by pores, enclosed in embedded proteins, that allow the passage of large molecules between the nucleus and the cytoplasm; also called the nuclear envelope.

nucleic acid (2, 5) One of the four types of biological macromolecules, involved in information storage and transfer. The nucleic acids DNA and RNA store genetic information in unique sequences of nucleotides. [Lat., *nucleolus*, dim. of *nucleus*, kernel, small nut]

nucleotide (2, 5) A molecule containing a phosphate group, a sugar molecule, and a nitrogen-containing molecule. Nucleotides are the individual units that together, in a unique sequence, constitute a nucleic acid.

nucleus (cell biology) (3) A membrane-enclosed structure in eukaryotic cells that contains the organism's genetic information as linear strands of DNA in the form of chromosomes. [Lat., *nucleus*, dim. of *nux*, nut]

nucleus (chemistry) (2) The central and most massive part of an atom, usually made up of two types of particles, protons and neutrons, which move about the nucleus. [Lat., *nucleus, dim.* of *nux*, nut]

null hypothesis (1) A hypothesis that proposes a lack of relationship between two factors. [Lat., *nullus*, none]

nuptial gift (9) A food item or other item presented to a potential mate as part of courtship. [Lat., *nuptialis*, of marriage]

O

omnivores (15) Animals that eat both plants and other animals and thus can occupy more than one position in the food chain. [Lat., *omnis*, all + *vorare*, to devour]

operon (5) A group of several genes, along with the elements that control their expression as a unit, all within one section of DNA.

order (10) In the system developed by Carolus Linnaeus (1707–1778), a classification of organisms consisting of related families.

organelles (3) Specialized structures in the cytoplasm of eukaryotic cells, with specific functions; include, among others, the rough and smooth endoplasmic reticulum, Golgi apparatus, and mitochondria. [Gk., *organon*, tool]

origin of replication (6) The site where the coiled, double-stranded DNA molecule unwinds and separates into two strands, like a zipper unzipping, and where duplication of the DNA molecule begins. In prokaryotes, there is a single origin of replication; eukaryotes may have multiple origin sites on each chromosome.

osmosis (3) A type of passive transport in which water molecules move across a membrane, such as the plasma membrane of a cell; the direction of osmosis is determined by the relative concentrations of all solutes on either side of the membrane. [Gk., *osmos*, thrust]

ovary (12) In plants, an enclosed chamber at the base of the carpel of a flower that contains the ovules; the female gonad. [Lat., *ovum*, egg]

ovule (12) The structure within the ovary of flowering plants that gives rise to egg cells. [Lat., *ovum*, egg]

Oxygen Revolution (13) In earth's history, the accumulation in the atmosphere of oxygen released by cyanobacteria and other photosynthetic organisms.

P

pair bond (9) A bond between an individual male and female in which they spend a high proportion of their time together, often over many years, sharing a nest or other refuge and contributing equally to the care of offspring.

parasite (13, 15) An organism that lives in or on another organism, the host, and damages it. [Gk., *para,* beside + *sitos,* grain, food]

parasitism (15) A symbiotic relationship in which one organism (the parasite) benefits while the other (the host) is harmed.

parent cells (6) Cells that divide to form daughter cells, which are genetically identical to the parent cell.

passive transport (3) Molecular movement that occurs spontaneously, without the input of energy; the two types of passive transport are diffusion and osmosis.

paternity uncertainty (9) Describes the fact that among species with internal fertilization in the female, a male cannot be 100% certain that any offspring a female produces are his. [Lat., *pater,* father]

pathogenic (13) Disease-causing.

pedigree (7) In genetics, a type of family tree that maps the occurrence of a trait in a family, often over many generations.

peptide bond (2) A bond in which the amino group of one amino acid is bonded to the carboxyl group of another; two amino acids so joined form a dipeptide, several amino acids so joined form a polypeptide. [Gk., *peptikos,* able to digest]

peptidoglycan (13) A glycoprotein that forms a thick layer on the outside of the cell wall of a bacterium; in some bacteria, the layer of peptidoglycan is covered by a membrane and so is not colored by a Gram stain.

periodic table (2) A tabular display in which all the known chemical elements are arranged in the order of their atomic number and on the basis of other aspects of their atomic structures.

pH (2) A logarithmic scale that measures the concentration of hydrogen ions (H^+) in a solution, with decreasing values indicating increasing acidity. Water, in which the concentration of hydrogen ions (H^+) equals the concentration of hydroxyl ions (OH^-), has pH = 7, the midpoint of the scale. [Abbreviation for "power of hydrogen"]

phagocytosis (3, 13) One of the three types of endocytosis, in which relatively large solid particles are engulfed by the plasma membrane, a vesicle is formed, and the particle is moved into the cell.

pharyngeal slits (11) Slits in the pharyngeal region, between the back of the mouth and the top of the throat, for the passage of water for breathing and feeding. [Gk., *pharynx,* throat]

phenotype (5, 7) The manifested structure, function, and behaviors of an individual; the expression of the genotype of an organism. [Gk., *phainein,* to cause to appear + *typos,* impression, engraving]

pheromones (9) Molecules released by an individual into the environment that trigger behavioral or physiological responses in other individuals. [Gk., *pherein,* to carry]

phospholipid (2, 3) A type of lipid that is the major component of the plasma membrane. Phospholipids are structurally similar to fats, but contain a phosphorus atom and have two, not three, fatty acid chains.

phospholipid bilayer (3) The structure of the plasma membrane; two layers of phospholipids, arranged tail to tail (the tails are hydrophobic and so avoid contact with water), with the hydrophilic head regions facing the watery extracellular fluid and intracellular fluid.

photoautotroph (13) Chlorophyll-containing bacteria, or other organisms, that use the energy from sunlight to convert carbon dioxide to glucose by photosynthesis. [Gk., *phos,* light + *autos,* self + *trophē,* food]

photon (4) The elementary particle that carries the energy of electromagnetic radiation of all wavelengths. [Gk., *phos,* light]

photosynthesis (4) The process by which some organisms, including plants and some protists and bacteria, are able to capture energy from the sun and store it in the chemical bonds of sugars and other molecules. [Gk., *phos,* light + *syn,* together with + *tithenai,* to place or put]

photosystems (4) Two arrangements of light-absorbing pigments, including chlorophyll, within the chloroplast that capture energy from the sun and transform it first into the energy of excited electrons and ultimately into ATP and high-energy electron carriers such as NADPH. [Gk., *phos,* light + *systema,* a whole compounded of parts]

phylogenetic tree (5) A grouping of organisms in a hierarchical system that reflects the evolutionary history and relatedness of the organisms. [Gk., *phylon,* race, tribe, class + Lat., *genus,* race, family, origin]

phylogeny (10) The evolutionary history of organisms.

phylum (10) In the system developed by Carolus Linnaeus (1707–1778), a classification of organisms consisting of related classes. [Gk., *phylon,* race, tribe, class]

pigment (4) In photosynthesis, molecules that are able to absorb the energy of light of specific wavelengths, raising electrons to an excited state in the process. [Lat., *pigmentum,* paint]

pilus (*pl.* **pili**) (3) A thin, hair-like projection that helps a prokaryote attach to surfaces. [Lat., *pilus,* a single hair]

pinocytosis (3) One of the three types of endocytosis, in which dissolved particles and liquids are engulfed by the plasma membrane, a vesicle is formed, and the material is moved into the cell. The vesicles formed in pinocytosis are generally much smaller than those formed in phagocytosis. [Gk., *pinein,* to drink + *kytos,* container]

placebo (1) An inactive substance used in controlled experiments to test the effectiveness of another substance; the treatment group receives the substance being tested, and the control group receives the placebo. [Lat., *placebo,* I shall please]

placebo effect (1) A frequently observed and poorly understood phenomenon in which there is a positive response to treatment with an inactive substance.

placenta (6) The organ formed during pregnancy (and expelled at birth) that connects the developing embryo to the wall of the uterus and allows the transfer of gases, nutrients, and waste products between mother and fetus; the placenta is so called from its shape. [Lat., *placenta,* a flat cake]

placentals (11) Mammals in which the developing fetus takes its nourishment from the transfer of nutrients from the mother through the placenta, which also supplies respiratory gases and removes metabolic waste products. [Lat., *placenta,* a flat cake]

plants (12) Members of the kingdom Plantae; multicellular eukaryotes that have cell walls made primarily of cellulose, contain true tissues, and produce their own food by photosynthesis. Plants are sessile, and most inhabit terrestrial environments.

plasma membrane (3) A complex, thin, two-layered membrane that encloses the cytoplasm of the cell, holding the contents in place and regulating what enters and leaves the cell; also called the cell membrane. [Gk., *plasma,* anything molded]

plasmid (5, 13) A circular DNA molecule found outside the main chromosome in bacteria.

plasmodesma (*pl.* **plasmodesmata**) (3) In plants, microscopic tube-like channels connecting the cells and enabling communication and transport between them. [Gk., *plassein,* to mold + *desmos,* bond]

pleiotropy (7) A phenomenon in which an individual gene influences multiple traits. [Gk., *pleion,* more + *tropos,* turn]

point mutation (5) A mutation in which one base pair in DNA is replaced with another, or a base pair is either inserted or deleted.

polar (3) Having an electrical charge.

polar body (6) In gamete formation, one of the two cells formed when a primary oocyte divides; it gets almost no cytoplasm and eventually disintegrates.

pollen grain (12) A structure that contains the male gametophyte of a seed plant. [Lat., *pollen,* fine dust]

pollination (12) The transfer of pollen from the anther of one flower to the stigma of another flower. [Lat. pollen, *fine dust*]

polyandry (9) A polygamous mating system in which individual females mate with multiple males. [Gk., *polys,* many + *aner,* husband]

polygamy (9) A mating system in which, for one sex, some individuals attract multiple mates while other individuals of that sex attract none; among the opposite sex, all or nearly all of the individuals are able to attract a mate. [Gk., *polys,* many + *gamete,* wife]

polygenic (7) Describes a trait that is influenced by many different genes. [Gk., *polys,* many + *genos,* race, descent]

polygyny (9) A mating system in which, among the males, some individuals attract multiple mates while others attract none; among the females, all or nearly all of the individuals are able to attract a mate. [Gk., *polys,* many + *gune,* woman]

polymerase chain reaction (PCR) (5) A laboratory technique in which a fragment of DNA can be duplicated repeatedly. [Gk., *polys,* many + *meris,* part]

polyploidy (10) The doubling of the number of sets of chromosomes in an individual. [Gk., *polys,* many + *ploion,* vessel]

polysaccharides (2) Complex carbohydrates formed by the union of many simple sugars. [Gk., *polys,* many + *sakcharon,* sugar]

population (8) A group of organisms of the same species living in a particular geographic region.

population density (14) The number of individuals of a population in a given area.

population ecology (14) A subfield of ecology that studies the interactions between populations of organisms of a species and their environment.

positive correlation (1) A relationship between variables in which they increase (or decrease) together. [Lat., *com,* with + *relatio,* report]

post-anal tail (11) A tail that extends beyond the end of the trunk, a point that is marked by the anus; a characteristic of chordates. [Lat., *post,* after]

postzygotic barrier (10) A barrier to reproduction caused by the infertility of hybrid individuals or the inability of hybrid individuals to survive long after fertilization. [Lat., *post,* after]

potential energy (4) Stored energy; the capacity to do work that results from an object's location or position, as in the case of water held behind a dam. [Lat., *potentia,* power]

predation (15) An interaction between two species in which one species eats the other. [Lat., *praedari,* to plunder]

prepared learning (9) Behaviors that are learned easily by all, or nearly all, individuals of a species.

prezygotic barrier (10) A barrier to reproduction caused by the physical inability of individuals to mate with each other, or the inability of the male's reproductive cell to fertilize the female's reproductive cell. [Lat., *prae,* before]

primary active transport (3) Active transport using energy released directly from ATP.

primary consumers (15) Herbivores, which consume the output of producers.

primary electron acceptor (4) In photosynthesis, a molecule that accepts excited, high-energy electrons from chlorophyll *a,* beginning the series of electron handoffs known as an electron transport chain.

primary productivity (15) The amount of organic matter produced by living organisms, primarily through photosynthesis.

primary structure (2) The sequence of amino acids in a polypeptide chain.

primase (6) One of several proteins that make up the replication complex, binding to each of the exposed DNA strands at the replication fork and playing a role in initiating the synthesis of complementary DNA strands.

probiotic therapy (13) A method of treating infections by introducing benign bacteria in numbers large enough to overwhelm harmful bacteria in the body. [Lat., *pro,* for, on behalf of + Gk., *bios,* life]

producers (15) The organisms responsible for primary productivity, such as grasses, trees, and agricultural crops, which convert light energy from the sun into chemical energy (that is, food) through photosynthesis.

prokaryote (3) An organism consisting of a prokaryotic cell (all prokaryotes are one-celled organisms). [Gk., *pro,* before + *karyon,* nut, kernel]

prokaryotic cell (3) A cell bound by a plasma membrane enclosing the cell contents (cytoplasm, DNA, and ribosomes); there is no nucleus or other organelles.

promoter site (5) Part of a DNA molecule that indicates where the sequence of base pairs that makes up a gene begins.

prophase (6) The first phase of mitosis, in which the nuclear membrane breaks down, sister chromatids condense, and the spindle forms. In meiosis, homologous pairs of sister chromatids come together and cross over in prophase I, and the chromosomes in daughter cells condense in prophase II. [Gk., *pro,* before + *phasis,* appearance]

protein (2) One of the four types of biological macromolecules, constructed of unique combinations of 20 amino acids that result in unique structures and chemical behavior. Proteins are the chief building blocks of tissues in most organisms. [Gk., *proteion,* of the first quality]

protein synthesis (5) The construction of a protein from its constituent amino acids, by the processes of transcription and translation.

prothallus (12) The free-living haploid life stage of a fern; produces haploid gametes. [Gk., *pro,* before + *thallia,* twig]

protists (10) In modern classification, one of the four eukaryotic kingdoms; includes all the single-celled eukaryotes. Phylogenetic analyses show that they are not a monophyletic group.

proton (2) A positively charged particle in the atomic nucleus; it is identical to the nucleus of the hydrogen atom, which lacks a neutron and has atomic number 1. [Gk., *protos,* first]

protostomes (11) Bilaterally symmetrical animals with defined tissues in which the gut develops from front to back; the first opening formed becomes the mouth of the adult animal. [Gk., *protos,* first + *stoma,* mouth]

pseudoscience (1) Hypotheses and theories not supported by trustworthy and methodical scientific studies. [Gk., *pseudes,* false + Lat., *scientia,* knowledge]

punctuated equilibrium (10) A theory in biology about the tempo of evolution, suggesting that in many taxa, long periods in which there is relatively little evolutionary change are punctuated by brief periods of rapid evolutionary change.

Punnett square (7) A diagram showing the possible outcomes of a cross between two individuals; the possible crosses are shown in the manner of a multiplication table. [From the name of its designer, Reginald C. Punnett, 1875–1967]

pupa (11) In complete metamorphosis, the second stage of insect development, in which the larva is enclosed in a case and its body structures are broken down into molecules that are reassembled into the adult form. [Lat., *pupa,* a little girl]

pyruvate (4) The end product of glycolysis.

Q

quaternary structure (2) Two or more polypeptide chains bonded together in a single protein; an example is hemoglobin. [Lat., *quaterni*, four each]

R

radial symmetry (11) A body structure like that of a wheel, or pie, in which any cut through the center would divide the organism into identical halves. [Lat., *radius*, spoke of a wheel + Gk., *symmetria*, symmetry]

radioactive (2) The property of some elements or isotopes of having a nucleus that breaks down spontaneously, releasing tiny, high-speed particles that carry energy.

radiometric dating (8) A method of determining both the relative and the absolute ages of objects such as fossils by measuring both the radioactive isotopes they contain, which are known to decay at a constant rate, and their decay products.

rain shadow (15) An area in the lee of a mountain where there is no or reduced rainfall, because the air passing over the mountain falls, becoming warmer and thus increasing the amount of moisture it can hold.

random assortment (6) During anaphase of meiosis I, the process in which homologues are separated and pulled apart toward opposite poles of the cell, with either the maternal or the paternal homologue (consisting of two sister chromatids) going to one pole and the other homologue to the opposite pole. Because the process is random, the pairs of sister chromatids at each pole (which will separate in meiosis II) are a mix of maternal and paternal sister chromatids.

randomized (1) In a research study, describes a manner of choosing subjects and assigning them to groups on the basis of chance—that is, randomly.

ray-finned fishes (11) Fish species characterized by rigid bones and a mouth at the apex of the body; they are so called because their fins are lined with hardened rays.

realized niche (15) The environmental conditions in which an organism is living at a given time.

receptor-mediated endocytosis (3) One of the three types of endocytosis, in which receptors on the surface of a cell bind to specific molecules; the plasma membrane then engulfs both molecule and receptor and draws them into the cell.

receptor protein (3) A protein in the plasma membrane that binds to specific chemicals in the cell's external environment to regulate processes within the cell; for example, cells in the heart have receptor proteins that bind to adrenaline.

recessive (7) Describes an allele whose phenotypic effect is masked by a dominant allele for that trait. [Lat., *recessus*, retreating]

reciprocal altruism (9) Costly behavior directed toward another individual that benefits the recipient, with the expectation that, at some later time, the recipient will behave in a similar manner, "returning the favor." [Lat., *reciprocare*, to move backward and forward + *alter*, the other]

recognition protein (3) A protein in the plasma membrane that provides a "fingerprint" on the outside-facing surface of the cell, making it recognizable to other cells. Recognition proteins make it possible for the immune system to distinguish the body's own cells from invaders that may produce infection; they also help cells bind to other cells or molecules.

recombinant DNA technology (5) Technology that depends on the combination of two or more sources of DNA into a product; an example is the production of human insulin from fast-dividing transgenic *E. coli* bacteria that contain, through genetic manipulation, the human DNA sequence that codes for insulin.

recombination (6) The exchange of some genetic material between a paternal homologous chromosome and a maternal homologous chromosome, leading to a chromosome carrying genetic material from each; also called crossing over.

replication (6) The process in both eukaryotes and prokaryotes by which DNA duplicates itself in preparation for cell division.

replication fork (6) In DNA replication, the structure formed by the unwinding and separating of the two DNA strands, which will be used as templates.

reproductive cells (6) Haploid cells from two individuals that, as sperm and egg, will combine at fertilization to produce offspring; also called gametes or sex cells.

reproductive investment (9, 14) Energy and material expended by an individual in the growth, feeding, and care of offspring.

reproductive isolation (10) The inability of individuals from two populations to produce fertile offspring together.

reproductive output (14) The number of offspring an individual or population produces.

resource partitioning (15) A division of resources that occurs when species overlap some portion of a niche in which one or more species differ in behavior or body plan in a way that divides the resources of the niche between the species.

restriction enzymes (5) Enzymes that recognize and bind to different specific sequences of four to eight bases in DNA and cut the DNA at that point. Restriction enzymes are important in biotechnology, because they permit the cutting of short lengths of DNA, which can be inserted into other chromosomes or otherwise utilized.

retrovirus (13) A virus containing RNA and the enzyme reverse transcriptase, which uses the viral RNA as a template to synthesize a single strand of DNA. [Lat., *retro*, backward + *virus*, slime]

ribonucleic acid (RNA) (2) A nucleic acid that serves as a middleman in the process of converting genetic information in DNA into protein, as well as other functions. Messenger RNA (mRNA) takes instructions for production of a given protein from DNA to another part of the cell; transfer RNA (tRNA) interprets the mRNA code and directs the construction of the protein from its constituent amino acids.

ribosomal subunits (5) The two structural parts of a ribosome, which function together to translate mRNA to build a chain of amino acids that will make up a protein.

ribosomes (3) Granular bodies in the cytoplasm, consisting of protein and RNA, that are involved in using information in messenger RNA (which was transcribed from DNA) to synthesize proteins.

ring species (10) Populations that can interbreed with neighboring populations but not with populations separated by larger geographic distances. Because the non-interbreeding populations are connected by gene flow through geographically intermediate populations, there is no clear point at which one species stops and another begins; for this reason, ring species are problematic for the biological species concept.

root (12) The part of a vascular plant, usually below ground, that absorbs water and minerals from the soil and transports them through vascular tissue to the rest of the plant, and anchors the plant in place. The overall structure of a plant's roots is called the root system.

rough endoplasmic reticulum (rough ER) (3) An organelle, part of the endomembrane system, structurally like a series of interconnected, flattened sacs connected to the nuclear envelope; called "rough" because its surface is studded with ribosomes. [Gk., *endon,* within + *plasma,* anything molded; Lat., *reticulum, dim.* of *rete,* net]

roundworms (11) A worm phylum with members characterized by a long, narrow, unsegmented body and growth by molting. Roundworms, also called nematodes, are protostomes with defined tissues; there are some 90,000 identified species.

rubisco (4) An enzyme (ribulose 1,5-bisphosphate carboxylase/oxygenase) involved in photosynthesis; it fixes carbon atoms from CO_2 in the air, attaching them to an organic molecule in the stroma of the chloroplast. This fixation is the first step in the Calvin cycle, in which molecules of sugar are assembled. Rubisco is the most abundant protein on earth.

S

S phase (6) See DNA synthesis phase.

saturated fat (2) A fat in which each carbon in the hydrocarbon chain forming the tail region of the molecule is bound to two hydrogen atoms; saturated fats are solid at room temperature.

savanna (15) A type of terrestrial biome; a tropical or subtropical grassland with scattered woody plants, characterized by a hot climate and distinct wet and dry seasons (rainfall is less than in the tropical seasonal forest biome).

science (1) A body of knowledge based on observation, description, experimentation, and explanation of natural phenomena. [Lat., *scientia,* knowledge]

scientific literacy (1) A general, fact-based understanding of the basics of biology and other sciences, the scientific method, and the social, political, and legal implications of scientific information.

scientific method (1) A process of examination and discovery of natural phenomena that involves making observations, constructing hypotheses, testing predictions, experimenting, and drawing conclusions and revising them as necessary.

second law of thermodynamics (4) A physical law that states that every conversion of energy is not perfectly efficient and invariably includes the transformation of some energy into heat.

secondary active transport (3) Active transport in which there is no direct involvement of ATP (adenosine triphosphate); the transport protein simultaneously moves one molecule against its concentration gradient while letting another flow down its concentration gradient.

secondary consumers (15) Animals that feed on herbivores; also known as carnivores.

secondary structure (2) The corkscrew-like twists or folds of a protein that are held in place by hydrogen bonds between amino acids in the polypeptide chain.

seed (12) An embryonic plant with its own supply of water and nutrients, encased within a protective coating.

segmented worms (11) A worm phylum with members characterized by grooves around the body that mark divisions between segments. Segmented worms, also called annelids, are protostomes with defined tissues and do not molt; examples are earthworms and leeches.

sessile (11) Describes organisms that are fastened in place, such as adult mussels and barnacles. [Lat., *sedere,* to sit]

sex-linked trait (7) A trait controlled by a gene on a sex chromosome.

sexual dimorphism (9) The case in which the sexes of a species differ in size or appearance. [Gk., *di-,* two + *morphē,* shape]

sexual reproduction (6) A type of reproduction in which offspring are produced by the fusion of gametes from two distinct sexes.

sexual selection (8) The process by which natural selection favors traits, such as ornaments or fighting behavior, that give an advantage to individuals of one sex in attracting mating partners.

sexually transmitted disease (STD) (13) A disease passed from one person to another through sexual activity.

shoot (12) The above-ground part of a plant, consisting of stems and leaves, and sometimes flowers and fruits; the stem contains vascular tissue and supports the leaves, the main photosynthetic organ of the plant; also called a shoot system.

sign stimulus (9) An external signal that triggers the innate behavior called a fixed action pattern.

simple diffusion (3) Diffusion of molecules directly through the phospholipid bilayer of the plasma membrane, without the assistance of other molecules; oxygen and carbon dioxide, because they are small and carry no charge that would cause them to be repelled by the middle layer of the membrane, can pass through the membrane in this way.

simple sugars (2) Monosaccharide carbohydrates, generally containing three to seven carbon atoms, which store energy in their chemical bonds and can be broken down by cells; they cannot be broken down into other simple sugars. Examples are glucose, fructose, and galactose.

single-gene trait (7) A trait that is determined by instructions on only one gene; examples are a cleft chin, a widow's peak, and unattached earlobes.

sister chromatids (6) The two identical strands of a replicated chromosome.

smooth endoplasmic reticulum (smooth ER) (3) An organelle, part of the endomembrane system, structurally like a series of branched tubes; called "smooth" because its surface has no ribosomes. Smooth ER synthesizes lipids such as fatty acids, phospholipids, and steroids. [Gk., *endon,* within + *plasma,* anything molded; Lat., *reticulum, dim.* of *rete,* net]

solute (3) A substance that is dissolved in a gas or liquid; for example, in a solution of water and sugar, sugar is the solute. [Lat., *solvere,* to loosen]

solvent (3) The gas or liquid in which a substance is dissolved; for example, in a solution of water and sugar, water is the solvent. [Lat., *solvere,* to loosen]

somatic cells (6) The (usually diploid) cells of the body of an organism (in contrast to the usually haploid reproductive cells). [Gk., *soma,* body]

speciation (10) The process by which one species splits into two distinct species. The first phase of speciation is reproductive isolation; the second is genetic divergence, in which two populations evolve over time as separate entities with physical and behavioral differences.

speciation event (10) A point in evolutionary history at which a given population splits into independent evolutionary lineages.

species (10) Natural populations of organisms that can interbreed and are reproductively isolated from other such groups; in the Linnaean system, the species is the narrowest classification for an organism. [Lat., *species,* kind, sort]

species richness (16) The number of different species in a given area.

specific epithet (10) In the system developed by Carolus Linnaeus (1707–1778), a noun or adjective added to the genus name to distinguish a species; For example, in the name *Homo sapiens, Homo* is the genus name and *sapiens* is the specific epithet. [Gk., *epithetos,* added]

spindle (6) A part of the cytoskeleton of a cell, formed in prophase (in mitosis) or prophase I (in meiosis), from which extend the fibers that organize and separate the sister chromatids.

spindle fibers (6) Fibers that extend from one pole of a cell to the other, which pull the sister chromatids apart in anaphase of mitosis or anaphase II of meiosis.

sporangia (*sing.* **sporangium**) (12) In many ferns, the structures on the underside of the leaves in which the spores are produced. [Gk., *spora*, seed + *angeion*, vessel]

spore (12) A reproductive structure of non-vascular and some vascular plants that have an alternation of generations. Spores are typically haploid and unicellular and develop into either a male (producing sperm) or female (producing eggs) gametophyte. The eggs and sperm produced by gametophytes unite to produce the diploid generation (sporophyte). [Gk., *spora*, seed]

sporophyte (12) The multicellular diploid structure in non-vascular plants, some vascular plants, and some algae that produces asexual spores; the diploid life stage in organisms exhibiting alternation of generations.

stabilizing selection (8) Selection that, for a given trait, produces the greatest fitness at the intermediate point of the phenotypic range.

stamen (12) The male reproductive structure of a flower, consisting of a head-like anther on a stalk-like filament. Most flowers have several stamens. [Lat., *stamen*, thread]

starch (2) A complex polysaccharide carbohydrate consisting of a large number of monosaccharides linked in line; in plants, starch is the primary form of energy storage.

statistics (1) A set of analytical and mathematical tools designed to further the understanding of numerical data.

stem cells (5) Undifferentiated cells that have the ability to develop into any type of cell in the body; this property makes stem cells useful in biotechnology.

sterol (2) A lipid important in regulating growth and development. The sterols include cholesterol and the steroid hormones testosterone and estrogen; all are modifications of a basic structure of four interlinked rings of carbon atoms. [Gk., *stereos*, solid + *-ol*, chemical suffix for an alcohol]

stigma (12) The region of the carpel, the female reproductive structure of a flower, that has a flat, sticky surface that functions as a landing pad for pollen.

stomata (*sing.* **stoma**) (4) Small pores, usually on the undersides of leaves, that are the primary sites for gas exchange in plants; carbon dioxide (for photosynthesis) enters and oxygen (a by-product of photosynthesis) exits through the stomata. [Gk., *stoma*, mouth]

stroma (3, 4) In the leaf of a green plant, the fluid in the inner compartment of a chloroplast, which contains DNA and protein-making machinery. [Gk., *stroma*, bed]

style (12) The long, thin part of the carpel, the female reproductive structure of a flower, that holds up the stigma for pollen capture and through which a pollen tube will grow down into the ovary.

substrate (2) The molecule on which an enzyme acts. The active site on the enzyme binds to the substrate, initiating a chemical reaction; for example, the active site on the enzyme lactase binds to the substrate lactose, breaking it down into the two simple sugars glucose and galactose. [Lat., *sub-*, under + *stratus*, spread]

succession (15) The change in the species composition of a community over time, following a disturbance.

superstition (1) The irrational belief that actions not related by logic to a course of events can influence an outcome.

surface protein (3) A protein that resides primarily on the inner or outer surface of the phospholipid bilayer that constitutes the cell's plasma membrane.

survivorship curves (14) Graphs showing the proportion of individuals of particular ages now alive in a population; indicate an individual's likelihood of surviving through a given age interval.

sympatric speciation (10) Speciation that results not from geographic isolation but from polyploidy or hybridization and allopolyploidy; this type of speciation is relatively uncommon in animals but common in plants. [Gk., *syn*, together with + *patris*, native land]

systematics (10) The modern approach to classification, with the broader goal of reconstructing the evolutionary history, or phylogeny, of organisms. [Gk., *systema*, a whole compounded of parts]

T

telomere (6) A non-coding, highly repetitive section of DNA at the tip of every eukaryotic chromosome that shortens with every cell division; if it becomes too short, additional cell division can cause the loss of functional, essential DNA and almost certain cell death. [Gk., *telos*, end + *meris*, part]

telophase (6) The fourth and last phase of mitosis, in which the chromosomes begin to uncoil and the nuclear membrane is reassembled around them. In meiosis, in telophase I, the sister chromatids arrive at the cell poles and the nuclear membrane reassembles around them; in telophase II, the sister chromatids have been pulled apart and the nuclear membrane reassembles around haploid numbers of chromosomes. [Gk., *telos*, end + *phasis*, appearance]

tertiary consumers (15) Animals that eat animals that eat herbivores; also known as top carnivores. [Lat., *tertius*, third]

tertiary structure (2) The unique and complex three-dimensional shape of a protein formed by multiple twists of its secondary structure as amino acids come together to form hydrogen bonds or covalent sulfur-sulfur bonds. [Lat., *tertius*, third]

test-cross (7) A mating in which a homozygous recessive individual is bred with individuals of unknown genotype that have the dominant phenotype; this type of cross can reveal the unknown genotype by the observed characteristics, or phenotypes, of the offspring.

tetrapod (11) An organism with four limbs; all terrestrial vertebrates are tetrapods. [Gk., *tetra-*, four + *pous*, foot]

theory (1) An explanatory hypothesis for a natural phenomenon that is exceptionally well supported by empirical data. [Gk., *theorein*, to consider]

thermodynamics (4) The study of the transformation of energy from one type to another, such as from potential energy to kinetic energy. [Gk., *thermē*, heat + *dynamis*, power]

thylakoids (3, 4) Interconnected membranous structures in the stroma of a chloroplast, where light energy is collected and converted to chemical energy in photosynthesis. [Gk., *thylakis, dim.* of *thylakos*, bag]

tight junction (3) A continuous, water-tight connection between adjacent animal cells. Tight junctions are particularly important in the small intestine, where digestion occurs, to ensure that nutrients do not leak between cells into the body cavity and so become lost as a source of energy.

tonicity (3) For a cell in solution, a measure of the concentration of solutes outside the cell relative to that inside the cell. [Gk., *tonos*, tension]

topography (15) The physical features of a region, including those created by humans. [Gk., *topos*, place + *graphein*, to write down]

total reproductive output (9) The lifetime number of offspring produced by an individual.

trait (5, 8) Any characteristic or feature of an organism, such as red petal color in a flower.

trans fat (2) An unsaturated fat that has been partially hydrogenated (meaning that hydrogen atoms have been added to make the fat more saturated and to improve a food's taste, texture, and shelf-life). The added hydrogen atoms are in a trans orientation, which differs from the cis ("near") orientation of hydrogen atoms in the unsaturated fat. [Lat., *trans,* on the other side of]

transcription (5) The process by which a gene's base sequence is copied to mRNA.

transduction (13) A method of lateral transfer of DNA from one bacterial cell to another by means of a virus containing pieces of bacterial DNA picked up from its previous host; the virus infects the recipient bacterium and passes along new genes to it. [Lat., *trans,* on the other side of + *ducere,* to lead]

transfer RNA (tRNA) (5) The type of RNA molecules in the cytoplasm that link specific triplet base sequences on mRNA to specific amino acids.

transformation (13) A method of lateral transfer of DNA from one bacterial cell to another in which a bacterial cell scavenges DNA released from burst bacterial cells in the environment. [Lat., *trans,* on the other side of + *formare,* to shape]

transgenic organism (5) An organism that contains DNA from another species.

translation (5) The process by which mRNA, which encodes a gene's base sequence, directs the production of a protein.

transmembrane protein (3) A protein that penetrates the phospholipid bilayer of a cell's plasma membrane.

transport protein (3) A transmembrane protein that provides a channel or passageway through which large or strongly charged molecules can pass. Transport proteins are of various shapes and sizes, making possible the transport of a wide variety of molecules.

treatment (1) In a research study, any condition applied to subjects— those in the treatment group— that is not applied to subjects in a control group.

triglyceride (2) A fat having three fatty acids linked to the glycerol molecule. [Gk., *tri-* three + *glykys,* sweet]

trophic level (15) A step in the flow of energy through an ecosystem. [Gk., *trophē,* food]

tropical rain forest (15) A type of terrestrial biome, found between the Tropic of Cancer (23.5° north latitude) and Tropic of Capricorn (23.5° south latitude) and characterized by constant moisture and temperature that do not vary across the seasons; vegetation is dense.

tropical seasonal forest (15) A type of terrestrial biome characterized by hot climate and distinct wet and dry seasons; trees shed their leaves in the dry season.

true-breeding (7) Describes a population of organisms in which, for a given trait, the offspring of crosses of individuals within the population always show the same trait; for example, the offspring of pea plants that are true-breeding for round peas always have round peas.

turgor pressure (3) In plants, the pressure of the contents of the cell against the cell wall, which is maintained by osmosis as water flows into the cell when it contains high concentrations of dissolved substances. Turgor pressure allows non-woody plants to stand upright, and its loss causes wilting. [Lat., *turgere,* to swell]

U

unsaturated fat (2) A fat in which at least one carbon in the hydrocarbon chain forming the tail region of the molecule is bound to only one hydrogen atom; unsaturated fats are liquid at room temperature.

V

vacuole (central) (3) A membrane-enclosed, fluid-filled, multipurpose organelle prominent in most plant cells (but also present in some protists, fungi, and animals); its functions vary but can include storing nutrients, retaining and degrading waste products, accumulating poisonous materials, containing pigments, and providing physical support. [Lat., *vacuus,* empty]

variables (1) The characteristics of an experimental system subject to change; for example, time (the duration of treatment) or specific elements of the treatment, such as the substance or procedure administered or the temperature at which it takes place. [Lat., *variare,* to vary]

vascular plants (12) Plants that transport water and dissolved nutrients by means of vascular tissue, a system of tubes that extends from the roots through the stem and into the leaves. [Lat., *vasculum, dim.* of *vas,* vessel]

vertebrates (11) Chordate animals that have a backbone (made of cartilage or hollow bones) and, at the front end of the organism, a head containing a skull, brain, and sensory organs.

vesicle (3) A small, membrane-enclosed sac within a cell. [Lat., *vesicula, dim.* of *vesica,* bladder]

vestigial structure (8) A structure, once useful to organisms, but which has lost its function over evolutionary time; an example is molars in bats that now consume an exclusively liquid diet. [Lat., *vestigium,* footprint, trace]

viruses (10) Diverse and important biological entities that can replicate but can conduct metabolic activity only by taking over the metabolic processes of a host organism, and therefore fall outside the definition of life. [Lat., *virus,* slime]

viviparity (11) The characteristic of bearing live young, giving birth to babies (rather than laying eggs). [Lat., *vivus,* alive + *parere,* to bear]

W

waggle dance (9) Behavior of scout honeybees that indicates, by the angle of the body relative to the sun and by physical maneuvers of various duration, the direction to a distant source of food.

wax (2) A lipid similar in structure to fats but with only one long-chain fatty acid linked to the glycerol head of the molecule; because the fatty acid chain is highly nonpolar, waxes are strongly hydrophobic.

X

X and Y chromosomes (6) The human sex chromosomes.

Photo Credits

Legend: *L*, left; *C*, center; *R*, right; *T*, top; *M*, middle; *B*, bottom

CHAPTER 1 pp. 0–1: Hugo Van Lawick/National Geographic Stock. **p. 2:** © Tim Laman (www.TimLaman.com). **p. 3:** *Figure 1-1 (L)* AP Photo, *Figure 1-1 (C)* Mike Fuentes/The New York Times/Redux Pictures, *Figure 1-1 (R)* The Photo Works. **p. 4:** *Figure 1-2 (L)* 23andme.com, *Figure 1-2 (C)* Universal Pictures/Photofest, *Figure 1-2 (R)* Discovery/Photofest. **p. 5:** *Figure 1-3* age fotostock/SuperStock. **p. 6:** *Figure 1-4 (TL)* Steven Puetzer/ Getty Images, *Figure 1-4 (TC)* © Tetra Images/Corbis, *Figure 1-4 (BR)* Mark Scott/Getty Images, *Figure 1-4 (TR)* © Blue Jean Images/Corbis, *Figure 1-4 (BL)* © Ricardo Azoury/Corbis. **p. 7:** David Doubilet/National Geographic Creative. **p. 8:** *Figure 1-6 (L)* Tim Boyle/Getty Images; *Figure 1-7 (R)* Digital Vision Photography/Veer. **p. 9:** *Figure 1-8 (inset)* AP Photo/ Elaine Thompson, *Figure 1-8* Comstock Images/Getty Images. **p. 10:** *Figure 1-9* © LWA/Dann Tardif/Blend Images/Corbis. **p. 12:** *Figure 1-12* © Peter Frank/Corbis. **p. 16:** © Ocean/Corbis. **p. 18:** *Figure 1-15* © Topham/The Image Works. **p. 21:** *Figure 1-16 (R)* Courtesy Julia Phelan, *Figure 1-16 (L)* Courtesy Julia Phelan. **p. 23:** B. Boissonnet/age fotostock. **p. 26:** *Figure 1-20 (L)* Jennifer Graylock/FilmMagic/Getty Images, *Figure 1-21 (C)* Image Source Black/Alamy, *Figure 1-21 (R)* Macmillan Education Archives. **p. 27:** *Figure 1-22* Jay Phelan. **p. 28:** *Figure 1-23* Matthew Staver/Landov. **p. 30:** *Figure 1-24* Rex Features via AP Images. **p. 31:** Patrick Landmann/Science Source. **p. 32:** Jan Komsta/National Geographic Creative.

CHAPTER 2 pp. 38–39: Miroslaw Swietek (table of contents and about the chapter opening photo: Miroslaw Swietek/SN/Landov). **p. 40:** *Figure 2-1 (L)* Optimarc/Shutterstock, *Figure 2-1 (C)* Jeffrey Coolidge/Getty Images, *Figure 2-1 (R)* Zhangyang/Shutterstock; *(T)* David Nicholls/Science Source. **p. 44:** *Figure 2-7* FOX/Photofest. **p. 46:** *Figure 2-10 (inset)* Custom Medical Stock Photo/Getty Images, *Figure 2-10* Judith Collins/Alamy. **p. 48:** *Figure 2-13 (B inset)* R. B. Suter, Vassar College, *Figure 2-13 (BL)* Courtesy of Steve Krichten; *(T)* Adam Jones/Science Source. **p. 55:** *Figure 2-20 (B)* Dorling Kindersley/Getty Images; *(T)* Keren Su/Getty Images. **p. 57:** *Figure 2-22 (inset)* Frans Lemmens/Alamy, *Figure 2-22* Dmytro Mykhailov/Shutterstock. **p. 59:** *Figure 2-25 (TL)* Valentina Razumova/ Shutterstock, *Figure 2-25 (TR)* Alison Parks-Whitfield/Alamy, *Figure 2-26 (B inset)* D. Roberts/Science Source, *Figure 2-26 (B)* © Lake County Museum/Corbis. **p. 60:** *Figure 2-27* Chassenet/age fotostock. **p. 61:** *Figure 2-28 (B)* Bryan & Cherry Alexander Photography/Alamy; *(T)* Daisy Gilardini/Getty Images. **p. 63:** *Figure 2-30 (T)* © Gary Gladstone/Corbis, *Figure 2-31 (BL)* Ellen Isaacs/Alamy, *Figure 2-31 (BR)* Masterfile. **p. 64:** *Figure 2-32* © Christine Schneider/Corbis. **p. 66:** *Figure 2-34* Hulton Archive/Getty Images. **p. 67:** *Figure 2-36 (B, L-R)* Steve Bloom Images/Alamy; Steve Gschmeissner/Science Source; SPL/Science Source; Thomas Deerinck, NCMIR/Science Source; Deco/Alamy; *(T)* Jeff R. Clow/ Getty Images. **p. 69:** *Figure 2-38* Ildi Papp/Shutterstock. **p. 71:** *Figure 2-41 (L)* Richard Nowitz/National Geographic Creative, *Figure 2-41 (R)* WIN-Initiative/Getty Images. **p. 75:** Vivek Mittal. **p. 78:** © KidStock/Blend Images/Corbis.

CHAPTER 3 pp. 84–85: Clark Sumida. **p. 86:** *Figure 3-1 (BL)* Jason Edwards/National Geographic/Robert Harding Picture Library, *Figure 3-1 (BC)* © Frans Lanting/www.lanting.com, *Figure 3-1 (BR)* David Scharf/

Science Source; *(T)* Marshall Sklar/Science Source. **p. 88:** *Figure 3-3 (inset)* Courtesy Reproductive Solutions, *Figure 3-3 (T)* © John Springer Collection/Corbis, *Figure 3-3 (C)* Rosenfeld/age fotostock, *Figure 3-3 (B)* © Clouds Hill Imaging Ltd./Corbis. **p. 89:** *Figure 3-4* Alfred Pasieka/Science Source. **p. 90:** *Figure 3-5 (L)* © Frans Lanting/www.lanting.com, *Figure 3-5 (C)* © Frans Lanting/www.lanting.com, *Figure 3-5 (R)* Matt Meadows/ Peter Arnold Images/Getty Images. **p. 91:** *Figure 3-6 (L)* © Dennis Kunkel Microscopy, Inc. *Figure 3-6 (R)* James Cavallini/Science Source. **p. 94:** *Figure 3-9 (BR)* Caro/Alamy, *Figure 3-9 (BL)* Kevin MacKenzie, University of Aberdeen, Wellcome Images; *(T)* imagebroker/Alamy. **p. 98:** *Figure 3-13 (L)* Mauro Fermariello/Science Source, *Figure 3-13 (R)* Brendan Fitterer/ © Tampa Bay Times/Zuma Press. **p. 99:** *Figure 3-14b* Image Source/Veer. **p. 101:** *Figure 3-16* AP Photo/Jeffrey Boan. **p. 102:** Courtesy of Dr. G. M. Gaietta & T. J. Deerinck, CRBS/NCMIR, University of California, San Diego. **p. 106:** *Figure 3-21* Richard Keppel-Smith/Getty Images. **p. 107:** *Figure 3-22* Biophoto Associates/Science Source. **p. 108:** *Figure 3-24 (L)* SPL/Science Source, *Figure 3-24 (R)* SPL/Science Source. **p. 109:** *Figure 3-25* SPL/Science Source. **p. 110:** Lorenz Britt/Alamy. **p. 111:** *Figure 3-26 (L)* © Dennis Kunkel Microscopy, Inc., *Figure 3-26 (C)* Dennis Kunkel Microscopy, Inc., *Figure 3-26 (R)* Courtesy of Dr. G. D. Griffin, Oak Ridge National Laboratory. **p. 112:** *Figure 3-27 (B)* Mary Ellen Mark; *(T)* © Dennis Kunkel Microscopy, Inc. **p. 113:** *Figure 3-28* CNRI/Science Source. **p. 114:** *Figure 3-29 (C)* James King-Holmes/Science Source, *Figure 3-29 (T and B)* Dr. Jan Schmoranzer/Science Source. **p. 115:** *Figure 3-30 (L)* Dr. Yorgos Nikas/Phototake, *Figure 3-30 (R)* Juergen Berger/ Science Source. **p. 116:** *Figure 3-31* Keith R. Porter/Science Source. **p. 119:** *Figure 3-33* Dr. Gopal Murti/Science Source. **p. 121:** *Figure 3-35 (T)* MedImage/SPL/Science Source; *Figure 3-36 (B)* Omikron/Science Source. **p. 123:** *Figure 3-37* © Dennis Kunkel Microscopy, Inc. **p. 124:** *Figure 3-39* Omikron/Science Source. **p. 125:** *Figure 3-40* Biophoto Associates/Science Source. **p. 126:** *Figure 3-41* Biology Pics/Science Source. **p. 128:** Marco Garcia/Getty Images.

CHAPTER 4 pp. 134–135 Eye of Science/Science Source. **p. 136:** *Figure 4-1 (B)* Pat LaCroix/Getty Images; *(T)* © Frans Lanting/National Geographic Creative. **p. 138:** *Figure 4-3 (TL)* David Madison/Getty Images, *Figure 4-3 (TC)* Ron R. Bielefeld, *Figure 4-3 (BL)* © Frans Lanting/www.lanting.com, *Figure 4-3 (TR)* Bernhard Edmaier/Science Source, *Figure 4-3 (CR)* AP Photo/Alessandro Trovati; *Figure 4-4 (BR)* © Halfdark/fstop/Corbis. **p. 139:** *Figure 4-5 (L)* Ed Gifford/Masterfile, *Figure 4-5 (R)* Jay Phelan. **p. 140:** © Toshiaki Ono/amanaimages/Corbis. **p. 144:** *Figure 4-10 (far L)* © Dennis Kunkel Microscopy, Inc., *Figure 4-10 (L)* © Dennis Kunkel Microscopy, Inc., *Figure 4-10 (R)* © Dennis Kunkel Microscopy, Inc., *Figure 4-10 (far R)* © Ralph A. Clevenger/Corbis. **p. 145:** *Figure 4-12 (T)* Value Stock Images/Fotosearch, *Figure 4-12 (inset)* © Dennis Kunkel Microscopy, Inc. **p. 154:** *Figure 4-23 (T)* AP Photo/Rajesh Kumar Singh, *Figure 4-24 (BL)* Dr. Jeremy Burgess/Science Photo Library/ Science Source, *Figure 4-24 (BR)* Dr. Jeremy Burgess/Science Photo Library/Science Source. **p. 155:** *Figure 4-25 (L)* Tommy Moorman, *Figure 4-25 (C)* Beth Perkins/Getty Images, *Figure 4-25 (R)* © PBNJ Productions/Corbis.

p. 157: *Figure 4-27 (BL)* © Frans Lanting/www.lanting.com, *Figure 4-27 (BC)* © Frans Lanting/www.lanting.com, *Figure 4-27 (BR)* Volker Steger/Science Source; *(T)* Joe Petersburger/National Geographic Creative. p. 159: *Figure 4-29 (L)* age fotostock/SuperStock, *Figure 4-29 (C)* © Frans Lanting/www.lanting.com, *Figure 4-29 (R)* A. Barry Dowsett/Science Source. p. 162: *Figure 4-33* Hybrid Medical Animation/Science Source. p. 167: *Figure 4-36 (B)* Tobias Bernhard/Getty Images; *(T)* Martin Harvey/Getty Images. p. 168: *Figure 4-38 (inset)* SciMAT/Science Source, *Figure 4-38 (T)* Cephas Picture Library/Alamy, *Figure 4-38 (B)* Romilly Lockyer/Getty Images. p. 169: *Figure 4-39 (L)* © Corbis, *Figure 4-39 (C)* © IMAGEMORE/Imagemore Co., Ltd./Corbis, *Figure 4-39 (R)* © Chris Alack/the food passionates/Corbis. p. 170: © JGI/Blend Images/Corbis.

CHAPTER 5 pp. 176–177: Courtesy HudsonAlpha Institute for Biotechnology. p. 178: *Figure 5-1(B)* Vasna Wilson/Virginian-Pilot; *(T)* Cris Benton. p. 179: *Figure 5-2* Jay Phelan. p. 180: *Figure 5-3 (L)* A. Barrington Brown/Science Source, *Figure 5-3 (R)* Science Source. p. 182: *Figure 5-5 (TL)* Jay Phelan, *Figure 5-5 (BL)* Biophoto Associates/Science Source, *Figure 5-5 (TC)* Sian Irvine/Getty Images, *Figure 5-5 (TR)* Eye of Science/Science Source, *Figure 5-5 (BR)* © John White. p. 185: *Figure 5-8 (T-B)* Eye of Science/Science Source, *Figure 5-8* Jay Phelan, *Figure 5-8* Sian Irvine/Getty Images, *Figure 5-8* © John White, *Figure 5-8* Biophoto Associates. p. 186: *Figure 5-9 (T-B)* Jay Phelan, *Figure 5-9* Eye of Science/Science Source, © Dennis Kunkel Microscopy, Inc., © Fancy/Veer/Corbis, © Dennis Kunkel Microscopy, Inc. p. 188: Tek Image/Science Source. p. 194: *Figure 5-16* © Zave Smith/Corbis. p. 197: © Uli Westphal 2008- p. 198: *Figure 5-20 (L)* Andrew Syred/Science Source, *Figure 5-20 (R)* Ted Kinsman/Science Source. p. 200: *Figure 5-22 (L)* © Mika/zefa/Corbis, *Figure 5-22 (C)* © Radius Images/Corbis, *Figure 5-22 (R)* © Ocean/Corbis. p. 203: *Figure 5-23 (L)* Marni Rolfes, *Figure 5-23 (R)* Marni Rolfes. p. 204: Juan Mabromata/AFP/Getty Images. p. 208: *Figure 5-30* © Ed Young/Corbis, *Figure 5-30 (inset)* Photo by John Doebley. p. 209: *Figure 5-31 (TL)* Martin Ruegner/Digital Vision/Getty Images, *Figure 5-31 (TR)* Scottish Crop Research Institute, Dundee, *Figure 5-31 (BL)* Professor Peter Beyer/Humanitarian Board for Golden Rice/www.goldenrice.org, *Figure 5-31 (BR)* Professor Peter Beyer/Humanitarian Board for Golden Rice/www.goldenrice.org. p. 210: *Figure 5-32 (L)* BSIP/Science Source, *Figure 5-32 (C)* © Lance Nelson/Corbis, *Figure 5-32 (R)* Daniel Acker/Bloomberg via Getty Images. p. 211: *Figure 5-34* Garry D. McMichael/Science Source. p. 212: *Figure 5-35* Photo by Dr. Garth Fletcher. p. 213: *Figure 5-36 (T)* © Reuters/Corbis; *Figure 5-37 (B)* Alex Milan Tracy/Sipa USA via AP Images. p. 215: © Remi Benali/Corbis. p. 216: *Figure 5-38 (T)* © B. Boissonnet/BSIP/Corbis; *Figure 5-39 (B)* MGM/courtesy Everett Collection. p. 217: *Figure 5-40* © Rick Friedman/Corbis. p. 218: *Figure 5-41* Baylor Medical Center. p. 220: *Figure 5-43 (L)* AP Photo/John Chadwick, *Figure 5-43 (C)* AP Photo/Ben Margot, *Figure 5-43 (R)* © RIA Novosti/Alamy. p. 221: *Figure 5-44 (L)* Neville Chadwick/Science Source *Figure 5-44 (R)* David Parker/Science Photo Library/Science Source. p. 224: *(B)* Lauren Nicole/Getty Images; *(T)* Anthony Marsland/Getty Images.

CHAPTER 6 pp. 230–231: © Steve Schapiro/Corbis Outline. p. 232: Steve Gschmeissner/© Science Photo Library/Alamy. p. 233: *Figure 6-1 (T)* Science Photo Library/Science Source; *Figure 6-2 (B)* Sven Dillen/Reporters/Redux. p. 235: *Figure 6-4* CNRI/Science Source. p. 242: *Figure 6-10 (BL)* Dan Guravich/Science Source, *Figure 6-10 (BL inset)* Biology Media/Science Source, *Figure 6-10d (BR)* Alex Blackburn Clayton/Photolibrary/Getty Images, *Figure 6-10 (BR inset)* SPL/Science Source;

(T) Indeed/Getty Images. p. 243: *Figure 6-11* © Micro Discovery/Corbis. p. 244: *Figure 6-13* David Scharf/Getty Images. p. 246: *Figure 6-15 (all photos)* Courtesy of T. Wittman. p. 247: *Figure 6-15 (all photos)* Courtesy of T. Wittman. p. 248: *Figure 6-17* GJLP/CNRI/Science Source. p. 249: *Figure 6-19* Kevin Laubacher/Taxi/Getty Images. p. 250: Thierry Berrod, Mona Lisa Production/Science Source. p. 251: *Figure 6-21* AP Photo. p. 255: *Figure 6-25* Don Fawcett/Science Photo Library/Science Source. p. 258: *Figure 6-28* Margot Granitsas/Science Source. p. 259: *Figure 6-29 (B)* James H. Robinson/Science Photo Library/Science Source; *Figure 6-30 (T)* Dr. Kari Lounatmaa/Science Source. p. 260: *Figure 6-31 (B)* Biophoto Associates/Science Source; *(T)* Davi Russo/© The Weinstein Company/Courtesy Everett Collection. p. 262: *Figure 6-33 (L)* © Frans Lanting/www.lanting.com, *Figure 6-33 (M)* Audun Bakke Andersen/Getty Images, *Figure 6-33 (R)* © Daniel Heuclin/Biosphoto. p. 265: L. Willatt/East Anglian Regional Genetics Service/Science Source. p. 266: *Figure 6-35 (TL)* CNRI/Science Source; *Figure 6-36 (TR)* Saturn Stills/Science Photo Library/Science Source; *Figure 6-37 (BL)* © Mika/zefa/Corbis, *Figure 6-37 (BR)* L. Willatt/East Anglian Regional Genetics Service/Science Source. p. 270: Fresh Air Fund/WireImage/Getty Images.

CHAPTER 7 pp. 276–277: © Vincent Kessler/Reuters/Corbis. p. 278: *Figure 7-1 (B)* © Felix Wirth/Corbis, *(T)* © Frans Lanting/www.lanting.com. p. 280: *Figure 7-4* Jonathan Hordle/Rex Features, *Figure 7-4 (R)* © Taamallah/Splash News/Corbis. p. 281: *Figure 7-5 (T)* Newspix/Rex USA; *Figure 7-6 (CL)* © Michael Weber/imagebroker/Corbis, *Figure 7-6 (CR)* Aspen Photo/Shutterstock, *Figure 7-6 (BR)* Razvan Losif/Shutterstock, *Figure 7-6 (BL)* Satapat/Shutterstock. p. 283: *Figure 7-8* James King-Holmes/Science Source. p. 284: *Figure 7-9 (L)* R. L. Webber/Shutterstock, *Figure 7-9 (R)* Sally Reed/age fotostock. p. 286: *Figure 7-11* Thomas & Pat Leeson/Science Source. p. 288: Ryan McVay/Getty Images. p. 290: *Figure 7-15* Richard Ellis/Alamy. p. 292: *Figure 7-17* Courtesy Michael Comte. p. 293: © The Photo Works/Alamy. p. 294: *Figure 7-19* Rene Maltete/Rapho/Gamma-Rapho via Getty Images. p. 297: *Figure 7-23* Big Cheese Photo LLC/Alamy. p. 299: *Figure 7-24 (R)* Jackie Lewin, Royal Free Hospital/Science Source, *Figure 7-24 (L)* Lennart Nilsson/TT. p. 302: *Figure 7-26* Tommy Moorman. p. 303: *Figure 7-27 (T)* Nick Moulin Photography/Getty Images, *Figure 7-27 (B)* Juniors Bildarchiv/Alamy. p. 304: Joel Meyerowitz/Gallery Stock. p. 306: *Figure 7-29* © Armando Gallo/Retna Ltd./Corbis. p. 308: Patrick Miller.

CHAPTER 8 pp. 314–315: © Manfred Danegger/Corbis. p. 316: *Figure 8-1 (B)* Graphic Science/Alamy; *(T)* © Frans Lanting/www.lanting.com. p. 319: *(T)* Joel Sartore/National Geographic Creative; *Figure 8-3 (L-R)* © Leonard de Selva/Corbis; Sheila Terry/Science Source; The Granger Collection, New York, *Figure 8-3 (BR)* The Granger Collection, New York. Mary Evans Picture Library/Alamy. p. 321: *Figure 8-5* Marc Schlossman/Getty Images. p. 323: *Figure 8-7* The Granger Collection, New York. p. 324: *Figure 8-8* Photo by Louie Fasciolo/Bernard J. Shapero Rare Books. p. 325: Jerzy Gubernator/Science Source. p. 326: *Figure 8-9* © Renee Lynn/Corbis. p. 327: *Figure 8-11 (L)* © Fabrice Lerouge/Onoky/Corbis, *Figure 8-11 (R)* © Yann Arthus Bertrand/Corbis. p. 329: *Figure 8-12* J. D. Talasek/Phototake. p. 330: *Figure 8-13 (T)* © Lester V. Bergman/Corbis; *Figure 8-14 (B)* © Frans Lanting/www.lanting.com. p. 332: *Figure 8-16 (TL)* Stephen Simpson/Rex Features via AP, *Figure 8-16 (TC)* Catherine Ledner/Getty Images, *Figure 8-16 (TR)* © Deanne Fitzmaurice/Corbis, *Figure 8-16 (BL)* XPACIFICA/National Geographic Creative, *Figure 8-16 (BC)* © yellowdog/Corbis, *Figure 8-16 (BR)* AP Photo/The Canadian Press,

Jeff McIntosh. **p. 333:** *Figure 8-17 (T)* Max Nash/AFP/Getty Images; *Figure 8-18 (B)* Benn Mitchell/Riser/Getty Images. **p. 338:** Brian J. Skerry/National Geographic Creative. **p. 339:** *Figure 8-22* Jason Edwards/National Geographic Creative. **p. 340:** *Figure 8-23* J. Sneesby/B. Wilkins/Getty Images. **p. 342:** *Figure 8-26* © Minnesota Historical Society/Corbis. **p. 343:** *Figure 8-27 (T)* Owen Humphreys/PA Wire URN:18316334 Press Association via AP Images; *Figure 8-28 (B)* Walter B. McKenzie/Getty Images. **p. 344:** *Figure 8-29* Tom & Pat Leeson/Science Source. **p. 346:** *Figure 8-30* Will & Deni McIntyre/Science Source. **p. 347:** *Figure 8-31* Medford Taylor/National Geographic Creative. **p. 348:** © Tim Laman (www.TimLaman.com). **p. 349:** *Figure 8-32* Jason Edwards/National Geographic/Getty Images. **p. 351:** *Figure 8-36 (BL)* Eric VanderWerf, *Figure 8-36 (BC)* Eric VanderWerf, *Figure 8-36 (BR)* Jack Jeffrey, *Figure 8-36 (T)* George Grall/National Geographic Creative. **p. 352:** *Figure 8-37 (TL)* ANT Photo Library/Science Source, *Figure 8-37 (TC)* © Jean Paul Ferrero/Mary Evans Picture Library/age fotostock, *Figure 8-37 (TR)* Dave Watts/Alamy, *Figure 8-37 (BL)* Nicholas Bergkessel, Jr./Science Source, *Figure 8-37 (BC)* © Frans Lanting/www.lanting.com, *Figure 8-37 (BR)* John W. Warden/age fotostock. **p. 353:** *Figure 8-38 (TL)* From: *The Sharks of North America* by José I. Castro, © 2011, *Figure 8-38 (TR)* Laboratory of Evolutionary Morphology, *Figure 8-38 (CL)* Omikron/Science Source, *Figure 8-38 (CR)* S. K. Ackerley, Department of Integrative Biology, University of Guelph, Guelph, Ontario, Canada; *Figure 8-39 (B, L to R)* PBNJ Productions/PBNJ/age fotostock; Juniors Bildarchiv/age fotostock; Malcolm Schuyl/Alamy; Juniors Bildarchiv/age fotostock. **p. 354:** *Figure 8-40 (TL)* Bruce Dale/National Geographic Creative; *Figure 8-41 (TR)* Joe Petersburger/National Geographic Creative, *Figure 8-41 (BR)* Kevin Schafer/Getty Images. **p. 355:** *Figure 8-42 (L to R)* Getty Images; Arco Images/Alamy; © Andrew Grant/Corbis; Danita Delimont/Alamy; Ulrike Klenke and Zeev Pancer, Center of Marine Biotechnology, Baltimore, Md.

CHAPTER 9 pp. 364-365: Juniors Bildarchiv/Alamy. **p. 366:** *(T)* C. S. Ling; *Figure 9-1 (B)* AP Photo/Independence Daily Reporter/Rob Morgan. **p. 367:** *Figure 9-2 (T)* Radius Images/Alamy; *Figure 9-3 (B)* Isabel M. Smallegange. **p. 368:** *Figure 9-4 (L)* © Ralph Reinhold/age fotostock, *Figure 9-4 (TR)* © FLPA/Dave Pressland/age fotostock, *Figure 9-4 (BR)* A. Hartl/age fotostock. **p. 370:** *Figure 9-6 (T)* Stacy Gold/National Geographic Creative, *Figure 9-6 (BL)* Brooke Whatnall/National Geographic Creative, *Figure 9-6 (BR)* Courtesy Everett Collection. **p. 373:** *(T)* © tbkmedia.de/Alamy; *Figure 9-8 (B)* Dr. Theodore Evans. **p. 375:** *Figure 9-9* Richard R. Hansen/Science Source. **p. 376:** *Figure 9-11* © Paul Damien/National Geographic Society/Corbis. **p. 378:** *Figure 9-12* Courtesy of Gerald Carter. **p. 379:** *Figure 9-13* Courtesy of Jay Phelan, Julia Phelan, Alon Ziv, Joshua Malina, Terry Burnham, Jennifer Yousem, Janvi Patel, Michael Coopersman, and Jeffrey Tambor. **p. 380:** *Figure 9-14* Peter Biro/IRC. **p. 382:** *Figure 9-16 (B)* David M. Phillips/Science Source; *(T)* © Tim Laman (www.TimLaman.com). **p. 384:** *Figure 9-18 (L)* © Frank Parker/age fotostock, *Figure 9-18 (C)* © Thomas Marent/arde/age fotostock, *Figure 9-18 (R)* © Frans Lanting/Corbis. **p. 385:** *Figure 9-19 (TL)* AP Photo/A24 Films, *Figure 9-19 (TR)* © Columbia/Courtesy Everett Collection; *Figure 9-20 (B)* George Grall/National Geographic/Getty Images. **p. 387:** *Figure 9-21 (T-B)* Anthony Mercieca/Science Source; © Paul Zahl/National Geographic Creative; Graphic Science/Alamy; © Frans Lanting/www.lanting.com. **p. 388:** *Figure 9-22* age fotostock/Alamy. **p. 389:** *Figure 9-23* James H. Robinson/Science Source. **p. 391:** *Figure 9-24* © Frans Lanting/www.lanting.com. **p. 392:** *Figure 9-25* M. Botzek/age fotostock. **p. 393:** *Figure 9-27 (L)* © Frans Lanting/www.lanting.com, *Figure 9-27 (R)* © Frans Lanting/www.lanting.com. **p. 395:** © Frans Lanting/www.lanting.com. **p. 396:** *Figure 9-28 (TL)* Joel Sartore/National Geographic Creative, *Figure 9-28 (TC)* © DLILLC/Corbis, *Figure 9-28 (TR)* © Ocean/Corbis; *Figure 9-29 (BL)* Scott Camazine/Alamy, *Figure 9-29 (BC)* AP Photo/J. Pat Carter, *Figure 9-29 (BR)* Monkey Business Images/Shutterstock. **p. 397:** *Figure 9-30* Werner Bollmann/age fotostock. **p. 398:** Titus Lacoste/Getty Images.

CHAPTER 10 pp. 404-405: Kellar Autumn, Lewis & Clark College. **p. 406:** *Figure 10-1 (B)* Courtesy of Smithsonian Institution. Painting by P. Sawyer. Peter Sawyer, NMNH, Smithsonian Institution; *(T)* © Frans Lanting/www.lanting.com. **p. 407:** *Figure 10-2* © Roger Ressmeyer/Corbis. **p. 408:** *Figure 10-3* Courtesy of Andrew Knoll. **p. 411:** *Figure 10-5 (B, L-R)* Michael Regan/Getty Images; AP Photo/Evan Agostini; Plantography/Alamy; The Collection by Jan Smith Photography/Alamy; © Frans Lanting/www.lanting.com; © Frans Lanting/www.lanting.com; *(T)* © YellowPaul/Alamy. **p. 412:** *Figure 10-6 (TL)* © Frans Lanting/www.lanting.com; *Figure 10-7 (TR)* Udo Richter/EPA/Newscom; *Figure 10-8 (BL)* Juniors Bildarchiv/Alamy, *Figure 10-8 (BR)* Masterfile. **p. 417:** *Figure 10-11 (L)* John Cancalosi/age fotostock, *Figure 10-11 (R)* R. Höelzl/Wildlife/Juniors/age fotostock. **p. 418:** *Figure 10-12 (T)* Michael Woodruff, *Figure 10-12 (BR)* Ralph Lee Hopkins/National Geographic Creative, *Figure 10-12 (BL)* © D. Parer & E. Parer-Cook/ardea.com, *Figure 10-12 (MC)* Frans Lanting/National Geographic Creative, *Figure 10-12 (ML)* David Hosking/Alamy, *Figure 10-12 (MR)* Joel Sartore/National Geographic Creative. **p. 420:** © Radius Images/Corbis. **p. 423:** *Figure 10-17 (L-R)* © DLILLC/Corbis; Nigel Dennis/age fotostock; James Gritz/Getty Images; Altrendo/Getty Images. **p. 424:** *Figure 10-19* © Dietmar Nill/Nature Picture Library/Corbis. **p. 425:** *Figure 10-20 (L)* David Hosking/Alamy, *Figure 10-20 (C)* Galen Rathbun, © California Academy of Sciences, www.afrotheria.net, *Figure 10-20 (R)* Birgit Koch-Wolf **p. 426:** *Figure 10-21 (B)* Ralph Lee Hopkins/National Geographic Creative; *(T)* Courtesy Jonathan Sequeira Barboza. **p. 431:** *Figure 10-26* © Reuters/Corbis. **p. 432:** © Frans Lanting/www.lanting.com. **p. 435:** *Figure 10-29 (B)* Radius Images/Alamy; *Figure 10-30 (T)* Steve Gschmeissner/Science Source. **p. 436:** *Figure 10-31 (inset)* Derek Lovley/Kazem Kashefi/Science Source, *Figure 10-31* Massimo Lupidi. **p. 437:** *Figure 10-32 (T)* Paul Zahl/National Geographic/Getty Images, *Figure 10-32 (ML)* Paul Sutherland/National Geographic Creative, *Figure 10-32 (MR)* Adam Jones/Science Source, *Figure 10-32 (B)* Nitin Prabhudesai. **p. 438:** Reed Hutchinson.

CHAPTER 11 pp. 444–445: © Frans Lanting/www.lanting.com. **p. 446:** *Figure 11-1 (BL)* Picture Hooked/Malcolm Schuyl/Alamy, *Figure 11-1 (BC)* © Frans Lanting/www.lanting.com, *Figure 11-1 (BR)* Georgette Douwma/Science Source; *(T)* © Frans Lanting/www.lanting.com. **p. 447:** *Figure 11-2 (L)* Philippe Clement/Nature Picture Library, *Figure 11-2 (R)* © Frans Lanting/www.lanting.com. **p. 450:** *Figure 11-4 (T, L-R)* Alexis Rosenfeld/Science Photo Library/Science Source; Dirk Wiersma/Science Source; Sinclair Stammers/Science Source; Wayne G. Lawler/Science Source; *(B, L-R)* Matthew Oldfield, Scubazoo/Science Source; CNRI/Science Source; © George Grall/National Geographic Society/Corbis; Mark Conlin/Alamy. **p. 452:** *Figure 11-5 (BL)* Natural History Museum, London/Science Photo Library/Science Source, *Figure 11-5 (BR)* © Robert Yin/Corbis; *(T)* Poul Beckmann. **p. 453:** *Figure 11-6* Sam Hodge/Alamy. **p. 454:** *Figure 11-7 (T)* Melissa Fiene Photography, *Figure 11-7 (BL)* Geophoto/Natalia Chervyakova/imagebroker/Alamy, *Figure 11-7 (BR)* © Chris Howes/Wild Places Photography/Alamy. **p. 455:** *Figure 11-9 (T)* © Tim Laman (www.TimLaman.com), *Figure 11-9 (B)*

Howard Hall Productions. **p. 457:** *Figure 11-10 (TL)* Alamy, *Figure 11-10 (TC)* Sinclair Stammers/Science Source; *Figure 11-11 (TR)* London School of Hygiene & Medicine/Science Source, *Figure 11-11 (BR)* © Dennis Kunkel Microscopy, Inc. **p. 458:** *Figure 11-12 (T)* Alexander Semenov/Science Photo Library/Science Source, *Figure 11-12 (C)* Jeanne White/Science Photo Library/Science Source, *Figure 11-12 (B)* ARCO/H. Reinhard/age fotostock. **p. 460:** *Figure 11-13 (TL)* David B. Fleetham/SeaPics.com, *Figure 11-13 (TR)* © Wayne Lynch/All Canada Photos/Corbis, *Figure 11-13 (BL)* Georgette Douwma/Nature Picture Library, *Figure 11-13 (BR)* © Hal Beral/Corbis. **p. 461:** *Figure 11-14 (L)* David B. Fleetham, *Figure 11-14 (C)* © Reuters/Corbis, *Figure 11-14 (R)* © Image Source/Corbis. **p. 462:** *Figure 11-15* © Tomasz Machnik. **p. 463:** *Figure 11-16 (TL)* blickwinkel/Alamy, *Figure 11-16 (TR)* George Grall/Getty Images, *Figure 11-16 (BL)* Patrick Lynch/Alamy, *Figure 11-16 (BR)* © George Karbus Photography/cultura/Corbis. **p. 464:** *Figure 11-17 (L-R)* Alex Wild; Simon D. Pollard/Science Photo Library/Science Source; Thomas Shahan/Thomasshahan.com; Steve Allen/Science Photo Library/Science Source. **p. 467:** *Figure 11-18 (L-R)* © Frans Lanting/Corbis; Alamy; Alamy; Ron Rowan Photography/Shutterstock; Dirk Seifert. **p. 469:** *Figure 11-19 (T)* Paul Sutherland/National Geographic Stock, *Figure 11-19 (CL)* Biosphoto/Superstock, *Figure 11-19 (CR)* Tim Laman/National Geographic, *Figure 11-19 (B)* © Frans Lanting/www.lanting.com. **p. 470:** © Frans Lanting/Corbis. **p. 471:** *Figure 11-20 (TL)* © Frans Lanting/www.lanting.com, *Figure 11-20 (TR)* Ross Armstrong/Getty Images, *Figure 11-20 (BL)* © Tim Laman (www.TimLaman.com), *Figure 11-20 (BC)* Custom Life Science Images/Alamy, *Figure 11-20 (BR)* Images&Stories/Alamy. **p. 473:** *Figure 11-23 (L)* ARCO/H. Reinhard/age fotostock, *Figure 11-23 (R)* Masa Ushioda/SeaPics.com, *Figure 11-23 (inset)* Darlyne A. Murawski/Getty Images. **p. 474:** *Figure 11-24 (L)* © Paul Souders/Corbis, *Figure 11-24 (C)* © Steve Jones/Stocktrek Images/Corbis, *Figure 11-24 (R)* © Rudie Kuiter/OceanwideImages.com. **p. 476:** *Figure 11-26 (BL)* Stephen Dalton/Nature Picture Library, *Figure 11-26 (BC)* Duncan McEwan/Nature Picture Library, *Figure 11-26 (BR)* GmbH/Alamy; *(T)* © Frans Lanting/www.lanting.com. **p. 478:** *Figure 11-28 (L)* John Cancalosi/age fotostock, *Figure 11-28 (R)* © Tim Laman (www.TimLaman.com). **p. 479:** *Figure 11-29* Chuang Zhao & Lida Xing/AFP/Newscom. **p. 480:** *Figure 11-31 (L)* Dave Watts/Alamy, *Figure 11-31 (C)* Martin Rugner/Getty Images, *Figure 11-31 (R)* Denis-Huot/Nature Picture Library. **p. 482:** *Figure 11-34* © Paul Springett 07/Alamy. **p. 486:** © Ariel Skelley/Blend Images/Corbis.

CHAPTER 12 pp. 492-493: © Katinka Matson. **p. 494:** *(T)* © Frans Lanting/Corbis; *Figure 12-1 (BL)* © Frank Krahmer/Corbis, *Figure 12-1 (BC)* Wendy Foden, *Figure 12-1 (BR)* Phil Schermeister/National Geographic Creative. **p. 495:** *Figure 12-2 (TL)* © Kevin Schafer/Corbis; *Figure 12-3 (TR)* © Creasource/Corbis, *Figure 12-3c (M inset)* George Shepherd, *Figure 12-3 (M)* Rich Reid/National Geographic Creative, *Figure 12-3 (B)* Herbert Hopfensperger/age fotostock. **p. 497:** *Figure 12-5 (T)* © Frans Lanting/www.lanting.com; *(B)* Charles F. Delwiche. **p. 498:** *Figure 12-6 (L)* Jouan & Rius/Nature Picture Library, *Figure 12-6 (R)* Gallo Images/Getty Images. **p. 499:** *Figure 12-7 (TL)* Alistair Dove/Alamy, *Figure 12-7 (BL)* Jacques Rosès/Biosphoto, *Figure 12-7 (BR)* Daniel Vega/age fotostock; *Figure 12-8 (R)* Barry Turner/Alamy. **p. 500:** *Figure 12-9 (TL)* © The Irish Image Collection/Design Pics/Corbis, *Figure 12-9 (BL)* © Hemis/Alamy; *Figure 12-10 (TR)* David Wall/Alamy. **p. 501:** *Figure 12-11 (T)* © Frans Lanting/www.lanting.com, *Figure 12-11 (BR)* Pascal Goetgheluck/Science Source, *Figure 12-11 (BL)* © Luciano Gaudenzio/

Grand Tour/Corbis. **p. 502:** *Figure 12-12 (L)* Snowflake Studio/Getty Images, *Figure 12-12 (R)* © Bob Gibbons/Alamy. **p. 503:** *(T)* Rob Kesseler; *Figure 12-13 (B)* Kevin Phelan. **p. 505:** *Figure 12-14 (CL)* Fletcher & Baylis/Science Source, *Figure 12-14 (TL)* Design Pics Inc./Alamy, *Figure 12-14 (TR)* © Bohemian Nomad Picturemakers/Corbis, *Figure 12-14 (CR)* Jouan & Rius/Nature Picture Library; *Figure 12-15 (B, L-R)* Francesco de Marco/Shutterstock; Alan & Linda Detrick/Science Source; Scott S. Warren/National Geographic Stock; blickwinkel/Alamy. **p. 506:** *Figure 12-16 (L)* Dr. Keith Wheeler/Science Source, *Figure 12-16 (R)* © Tetra Images/Corbis. **p. 507:** *Figure 12-18 (T)* Tommy Moorman, *Figure 12-18 (B)* Philippe Bourseiller/JH Editoral/Getty Images. **p. 508:** © Clive Nichols/Eurasia Press/Corbis. **p. 509:** *Figure 12-19 (BL)* © Frank Krahmer/Corbis, *Figure 12-19 (TL)* blickwinkel/Alamy, *Figure 12-19 (TR)* © Frans Lanting/www.lanting.com, *Figure 12-19 (BR)* Roger Standen/Science Photo Library/Science Source. **p. 510:** *Figure 12-21 (L)* Dr. John Brackenbury/Science Source, *Figure 12-21 (R)* Topic Photo Agency/age fotostock. **p. 511:** *Figure 12-22 (T, L-R)* Craig K. Lorenz/Science Source; Rolf Nussbaumer/Nature Picture Library; Amanda Jastremski/National Geographic Stock; *Figure 12-22* © Richard Bryant/Arcaid/Corbis, *Figure 12-22 (BL)* © W. P. Armstrong, *Figure 12-22 (BR)* George Grall/National Geographic Creative. **p. 512:** *Figure 12-23* © Paul Souders/Corbis. **p. 514:** *(T)* © Ocean/Corbis; *Figure 12-25 (L)* © Tim Graham/Tim Graham LLP/Corbis, *Figure 12-25 (C)* James L. Stanfield/National Geographic Stock, *Figure 12-25 (R)* © Lothar Lenz/Corbis. **p. 516:** *Figure 12-26b (TL)* John Burcham/National Geographic Creative, *Figure 12-26 (TR)* Wildlife GmbH/Alamy, *Figure 12-26 (C)* Howard Grey/Getty Images, *Figure 12-26 (BL)* Gertrud & Helmut Denzau/Nature Picture Library, *Figure 12-26 (BR)* Dr. Keith Wheeler/Getty Images. **p. 517:** *Figure 12-27* Ed Reschke/Getty Images. **p. 518:** *(T)* © Frans Lanting/www.lanting.com; *Figure 12-28 (BL)* Elliott Neep/Getty Images, *Figure 12-28 (BC)* © Frans Lanting/www.lanting.com, *Figure 12-28 (BR)* Raul Touzon/National Geographic Creative. **p. 519:** *Figure 12-30 (L)* imagebroker.net/SuperStock, *Figure 12-30 (C inset)* Gregory G. Dimijian/Science Source, *Figure 12-30 (C)* © Dennis Kunkel Microscopy, Inc., *Figure 12-30 (R inset)* Julia Phelan, *Figure 12-30 (R)* E. Gueho/CNRI/Science Source. **p. 520:** *Figure 12-31* © Elliott Neep. **p. 521:** *Figure 12-32 (T)* Joerrn M/Photolibrary/age fotostock America; *Figure 12-33 (B)* D. Hurst/Alamy. **p. 522:** *Figure 12-34* Dr. Jeremy Burgess/Science Source. **p. 523:** *Figure 12-35 (T)* Paul Redfearn, the Ozarks Regional Herbarium, Missouri State University; *Figure 12-36 (B)* Dan Hanscom/Shutterstock. **p. 526:** Pius Otomi Ekpei/AFP/Getty Images.

CHAPTER 13 pp. 532–533 © Dennis Kunkel Microscopy, Inc. **p. 534:** *(T)* Derren Ready, Wellcome Images; *Figure 13-1 (ML)* Eye of Science/Science Source, *Figure 13-1 (MC)* Alfred Pasieka/Science Source, *Figure 13-1 (B)* Eye of Science/Science Source, *Figure 13-1 (MR)* Astrid & Hanns-Frieder Michler/Science Source. **p. 536:** *Figure 13-2 (TL)* © Dennis Kunkel Microscopy, Inc., *Figure 13-2 (TCL)* © Dennis Kunkel Microscopy, Inc., *Figure 13-2 (TCR)* © Dennis Kunkel Microscopy, Inc., *Figure 13-2 (TR)* Astrid & Hanns-Frieder Michler/Science Source; *Figure 13-3 (BL inset)* Phototake/Alamy, *Figure 13-3 (BL)* CNRI/Science Source, *Figure 13-3 (BC inset)* © Dennis Kunkel Microscopy, Inc., *Figure 13-3 (BC)* © Marc Moritsch/National Geographic Society/Corbis, *Figure 13-3 (BR inset)* © Dennis Kunkel Microscopy, Inc., *Figure 13-3 (BR)* © Scott T. Smith/Corbis. **p. 537:** *Figure 13-4* Courtesy of Jay Phelan. **p. 538:** *Figure 13-5 (B)* © Dennis Kunkel Microscopy, Inc.; *(T)* CNRI/Science Source. **p. 539:** *Figure 13-6 (L)* mediacolor/Alamy, *Figure 13-6 (R)* Biophoto Associates/

Science Source. **p. 542:** *Figure 13-9 (L)* © Dodie Ulery, *Figure 13-9 (C)* © Ralph White/Corbis, *Figure 13-9 (R)* Reino Hanninen/Alamy. **p. 543:** *Figure 13-10 (B)* © Mediablitzimages/Alamy; *(T)* © Dennis Kunkel Microscopy, Inc. **p. 546:** *Figure 13-11 (L inset)* Justin Cormack, *Figure 13-12 (R)* ImageState Royalty Free/Alamy. **p. 549:** *Figure 13-15* © Gaetan Bally/Keystone/Corbis. **p. 550:** *Figure 13-16 (B)* Eye of Science/Science Source; *(T)* Frans Lanting/National Geographic Creative. **p. 551:** *Figure 13-17(L)* © James Forte/National Geographic Society/Corbis; *Figure 13-18 (R)* Carrie Vonderhaar/Ocean Futures Society/National Geographic Creative. **p. 552:** *Figure 13-19* Tim Mantoani/Masterfile. **p. 553:** *Figure 13-20* UCL Geology Collections; *(T)* Steve Gschmeissner/Science Source. **p. 554:** *Figure 13-21 (L)* © Dennis Kunkel Microscopy, Inc., *Figure 13-21 (C)* Paul Zahl/National Geographic Creative, *Figure 13-21 (R)* Gregory Ochocki/Science Source. **p. 555:** *Figure 13-22 (B)* Frans Lanting/National Geographic Creative, *Figure 13-22 (C)* David Littschwager/National Geographic Creative, *Figure 13-22 (T)* © Inaki Relanzon/Nature Picture Library/Corbis. **p. 556:** *Figure 13-23 (B)* Phototake/Alamy. **p. 557:** *Figure 13-24 (L)* SPL/Science Source, *Figure 13-24 (R)* Hazel Appleton, Health Protection Agency Centre for Infections/Science Source; *(T)* Biozentrum, University of Basel/Science Source. **p. 559:** *Figure 13-26 (L)* Science Source, *Figure 13-26 (C)* AP Photo/Charles Tasnadi, *Figure 13-26 (R)* © Reuters/Corbis. **p. 561:** *Figure 13-28* © Bettmann/Corbis. **p. 563:** *Figure 13-30* Scott Camazine/CDC/Science Source. **p. 564:** Tommy Moorman.

CHAPTER 14 pp. 570–571: Michael Wolf/Laif/Redux. **p. 572:** *Figure 14-1 (B)* James L. Stanfield/National Geographic Creative; *(T)* © Frans Lanting/www.lanting.com. **p. 574:** *Figure 14-3* © Frans Lanting/www.lanting.com. **p. 575:** *Figure 14-4* Juan Carlos Muñoz/age fotostock. **p. 576:** *Figure 14-5* © Frans Lanting/www.lanting.com. **p. 577:** *Figure 14-7* John McColgan, Bureau of Land Management, Alaska Fire Service. **p. 578:** *Figure 14-8 (T)* George Steinmetz/National Geographic Creative, *Figure 14-9 (B)* Reuters/Pierre Holtz/Landov. **p. 579:** *Figure 14-10* Jeffrey Lepore/Science Source. **p. 580:** *Figure 14-11* Dirk Anschutz/Getty Images. **p. 581:** *Figure 14-12* © National Museum of Natural History, Smithsonian Institution, photo by Chip Clark. **p. 582:** © Frans Lanting/www.lanting.com. **p. 583:** *Figure 14-13 (L)* © Fisher/Dickman/Splash/Splash News/Corbis, *Figure 14-13 (C)* Stephen Dalton/NHPA/Photoshot, *Figure 14-13 (R)* Michael Durham. **p. 584:** *Figure 14-14* Photo by Greg Katsoulis, courtesy Terry Burnham. **p. 585:** *Figure 14-15 (L)* Gunter Marx Photography/Corbis, *Figure 14-15 (C)* Norbert Rosing/National Geographic Creative, *Figure 14-15 (R)* © Jack Goldfarb/Design Pics/Corbis; *(inset)* iStockphoto.com/jeridu. **p. 587:** *Figure 14-16* Michel Gunther/Science Source. **p. 588:** *Figure 14-17 (L)* © Frans Lanting/www.lanting.com, *Figure 14-17 (C)* © Bernd Zoller/imagebroker/Corbis. *Figure 14-17 (R)* © Mark Carwardine/Nature Picture Library/Corbis. **p. 589:** Arco Images/Robert Harding Picture Library. **p. 590:** *Figure 14-18* Them #6, photography, 78 × 200 cm, 1998–2000 (left-hand original, 2003). © Hai Bo, Courtesy Pace Beijing. **p. 592:** *Figure 14-20 (L)* Renaud Visage/Getty Images, *Figure 14-20 (R)* photoiconix/Shutterstock. **p. 595:** Michael Wolf/Laif/Redux. **p. 600:** *Figure 14-26 (T)* Yadid Levy/Getty Images, *Figure 14-26 (C)* Kletr/Shutterstock, *Figure 14-26 (B)* Michael Wolf/laif/Redux. **p. 601:** *Figure 14-27* Deng Peibo—Imaginechina via AP Images. **p. 602:** © Lane Oatey/Blue Jean Images/Corbis.

CHAPTER 15 pp. 608–609 © LUMI Images/Corbis. **p. 610:** © Frans Lanting/www.lanting.com, *Figure 15-1 (B)* © Tim Laman (www.TimLaman.com); **p. 611:** *Figure 15-2* Steve Allen/Getty Images. **p. 612:**

Figure 15-3 (TL) © Frans Lanting/www.lanting.com, *Figure 15-3 (TC)* © Andrey Nekrasov/imagebroker/Corbis, *Figure 15-3 (TR)* Fritz Poelking/Photolibrary/age fotostock, *Figure 15-3 (ML)* John Warburton-Lee Photography/Alamy, *Figure 15-3 (MC)* Michael Melford/Getty Images, *Figure 15-3 (MR)* Stephen P. Parker/Science Source, *Figure 15-3 (BL)* © Dan Lamont/Corbis, *Figure 15-3 (BC)* Peter Arnold/Alamy, *Figure 15-3 (BR)* Paul Nicklen/National Geographic Creative. **p. 613:** *Figure 15-4 (TL)* © James Randklev/Corbis, *Figure 15-4 (TR)* © Frans Lanting/Corbis, *Figure 15-4 (BL)* Rich Reid/National Geographic Creative, *Figure 15-4 (BC)* Troy Plota/Getty Images, *Figure 15-4 (BR)* Mark Conlin/V&W/Image Quest Marine. **p. 614:** © Steve Besserman. **p. 615:** *Figure 15-6 (L)* Mike Boyatt/AGStockUSA; *Figure 15-7 (R)* © Frans Lanting/www.lanting.com. **p. 616:** *Figure 15-8* Jacques Descloitres, MODIS Land Rapid Response Team, NASA/GSFC. **p. 617:** *Figure 15-9 (L)* Image courtesy of NASA/Goddard Space Flight Center Scientific Visualization Studio, *Figure 15-9 (R)* © Sandy Felsenthal/Corbis. **p. 620:** © Frans Lanting/www.lanting.com. **p. 621:** *Figure 15-13 (L)* Dr. Jeremy Burgess/Science Source, *Figure 15-13 (R)* © James Hager/Robert Harding World Imagery/Corbis. **p. 627:** *Figure 15-19* © Peter Essick/Aurora Photos/Corbis, *Figure 15-19 (inset)* Dr. Jennifer L. Graham, U.S. Geological Survey. **p. 628:** *Figure 15-20 (B)* Klaus Nigge/National Geographic Creative, *(T)* David Doubliet/National Geographic Creative. **p. 629:** *Figure 15-21* Hermann Eisenbeiss/Science Source. **p. 632:** *Figure 15-24 (L-R)* © Anthony Bannister, Gallo Images/Corbis; George Grall/National Geographic Creative; George Grall/National Geographic Creative; Medford Taylor/National Geographic Creative. **p. 633:** *Figure 15-25 (L)* David Doubilet/National Geographic Creative, *Figure 15-25 (R)* © Simon Wagen/J. Downer Product/Nature Picture Library. **p. 634:** *Figure 15-26* Dante Fenolio/Science Source. **p. 635:** *Figure 15-27 (TL)* Stephen Dalton/Science Source, *Figure 15-27 (TR)* Eye of Science/Science Source; *Figure 15-28 (B)* Evan Kafka/Getty Images. **p. 636:** *Figure 15-29 (L)* Darlyne A. Murawski/National Geographic Creative, *Figure 15-29 (R)* Arco Images GmbH/Alamy. **p. 639:** © Eric Raptosh Photography/Blend Images/Corbis. **p. 642:** NASA Image by Robert Simmon, based on Landsat data provided by the UMD Global Land Cover Facility.

CHAPTER 16 pp. 648–649: DEA/Dani-Jeske/Getty Images. **p. 650:** *Figure 16-1 (inset)* © Brad Nelson/Phototake, *Figure 16-1 (B)* Inga Spence/Getty Images; *(T)* © Frans Lanting/www.lanting.com. **p. 651:** *Figure 16-2* © Daniel Beltra for Greenpeace/Nature Picture Library. **p. 652:** *Figure 16-3 (T-B)* Dafna Ben Nun; Jupiterimages/Getty Images; Jack Phelan; © Frans Lanting/www.lanting.com, *Figure 16-3 (B)* © Ocean/Corbis. **p. 654:** *Figure 16-5b* Dennis Macdonald/age fotostock. **p. 657:** *Figure 16-8 (L)* age fotostock/Robert Harding Picture Library, *Figure 16-8 (C)* © Frans Lanting/www.lanting.com, *Figure 16-8 (R)* © Image Source/Corbis. **p. 658:** *(T)* © Gallo Images/Corbis, *Figure 16-9 (inset)* Courtesy of the Public Libraries of Saginaw, *Figure 16-9 (B)* Tom Uhlman/Alamy. **p. 661:** *Figure 16-12 (B)* © Frans Lanting/www.lanting.com; *Figure 16-13 (inset)* Courtesy Bonobo Conservation Initiative, *Figure 16-13 (T)* © Frans Lanting/www.lanting.com. **p. 662:** Simon Weller/Getty Images. **p. 663:** *Figure 16-14 (TL)* USGS Photo by Harry Glicken, *Figure 16-14 (TC)* USGS Photo by Harry Glicken, *Figure 16-14 (TR)* Tom Nevesely; *Figure 16-15 (BL)* E. J. Keller/National Zoo/Smithsonian Institution, *Figure 16-15 (BC)* Mark Carwardine/Getty Images, *Figure 16-15 (BR)* © Denis Scott/Corbis. **p. 664:** *Figure 16-16 (inset)* Gale S. Hanratty/Alamy, *Figure 16-16* Jany Sauvanet/Science Source. **p. 665:** *Figure 16-18 (L)* William Mullins/Alamy, *Figure 16-18 (C)* Michael P. Gadomski/Science

Source, *Figure 16-18 (R)* Peter Yates/Time & Life Pictures/Getty Images, *Figure 16-18 (inset)* © Stephen Barnes/Marine/Alamy. **p. 666:** *Figure 16-19* Jack Picone/Alamy. **p. 667:** *Figure 16-20* © Dennis MacDonald/Alamy. **p. 668:** *Figure 16-23* © Will & Deni McIntyre/Corbis. **p. 670:** *Figure 16-26 (T)* NASA/Goddard Space Flight Center Scientific Visualization Studio, Thanks to Rob Gerston (GSFC) for providing the data; *Figure 16-26 (B)* NASA/Goddard Space Flight Center Scientific Visualization Studio, The Blue Marble data courtesy of Reto Stockli (NASA/GSFC). **p. 673:** *Figure 16-29* Florian Kopp/© imagebroker/Alamy. **p. 674:** © Jason Lee/Reuters/Corbis. **p. 675:** *Figure 16-31 (T, far L)* NASA Goddard Space Flight Center Scientific Visualization Studio, *Figure 16-31 (TL)* NASA Goddard Space Flight Center Scientific Visualization Studio, *Figure 16-31 (TR)* NASA Goddard Space Flight Center Scientific Visualization Studio, *Figure 16-31 (T, far R)* NASA/Goddard Space Flight Center Scientific Visualization Studio; *Figure 16-32 (inset)* Jim Spellman/WireImage/Getty Images, *Figure 16-32 (B)* AP Photo/Elise Amendola. **p. 676:** *Figure 16-33 (T)* Tony Gentile/Reuters/Landov, *Figure 16-33 (B)* Martin Oeggerli/Science Source. **p. 677:** *Figure 16-34 (R)* © Frans Lanting/www.lanting.com, *Figure 16-34 (L)* © Frans Lanting/www.lanting.com. **p. 678:** *Figure 16-35* Tommy Moorman. **p. 680:** Corbis/SuperStock.

Index

boldface indicates a definition; *italics* indicates a figure.

Living organisms and systems, common characteristics of, 31
Living systems, pH and, 51–53, *52, 53*
Lobe-finned fishes, **474,** *474*
Loci, **183**
Locust swarms, 578, *578*
Logistic growth, **576,** *577*
Loneliness, health consequences of, 194
Longevity, 589–590
   determinants of, *592,* 592–593
   tradeoff with reproduction, 2
Lucy (australopithecine), *482,* 483
Luehea seemannii, 465, 676
Lungs, evolution of, 475, *475*
Lyell, Charles, 319, 320, 323
Lysosomes, **118,** 118–119, *119, 127*

M

M phase. *See* Mitotic phase (M phase)
Macroevolution, **426,** *426,* 426–431, *427*
   adaptive radiations and, 428–429, *429*
   mass extinctions and, *430,* 430–431, *431*
   uneven pace of evolution and, 427–428, *428*
Macromolecules, **55**
Mad cow disease, 558
Madagascar periwinkle, 650
Malaria, 556, *556*
Malayan pit viper, 650
Male-pattern baldness, 300–301
Males, **382**
Malthus, Thomas, 323
Mammals, *479,* 479–480, *480*
Mammary glands, **479**
Marsupials, **480,** *480*
Martha (passenger pigeon), 658, *658*
Mass, **41**
Mass extinctions, 429, *429,* **430,** *430,* 430–431, *431,* 658–659
   current, 660–661, *661*
Mate guarding, **388,** *388,* 388–390, *389*
Mating systems, **391,** *391,* 391–393, *392*
Maximum sustainable yield, **580,** *580,* 580–581, *581*
Mechanical defenses, 632, *632*
Meiosis, 233, **250,** 250–259
   gamete production by, *252,* 252–256, *254–256*
   sexual reproduction and, 250–251, *251,* 258–259
   variation created by, 257, *257*
Membrane enzymes, **97,** *97*
Mendel, Gregor, 282–285, 303, 307, 368
Mendel's law of independent assortment, 304–305, **305,** *305*
Mendel's law of segregation, 284–285, **285,** *285*
Messenger RNA (mRNA), **187**
   bacterial, 540
   transcription and, 188–190, *189*
   translation and, 190–191, *190–192,* 193
Metabolic pathways, 72
Metabolic plasmids, 540
Metabolism, 72
Metamorphosis, *467,* 467–468
Metaphase, **245**
   in meiosis, 253, 254, *254*
   in mitosis, 245, *246*
Metastasis, 248, *248*
Methane
   bonds in, *45,* 46
   consumption by bacteria, 652–653
   as greenhouse gas, 670
Methanogens, 436
Methuselah tree, 507
Methyl jasmonate (MJ), 516–517

Microbes, **432, 534.** *See also* Archaea; Bacteria; Protists; Viruses
   abundance of, 535, 536–537, *537*
   diversity of, *534,* 534–535
   habitats of, 535–536, *536*
   normal flora, 543
   size of, 535, *536*
Microevolution, **426,** 426–427, *427*
Microfilaments, **114,** *114*
Microspheres, **409**
Microsporidia, 519, *519*
Microtubules, **114,** *114,* **245**
Migration, **331**
   as mechanism of evolution, 331, *331*
Milk of Magnesia, osmosis and, 105
Miller, Stanley, 407, *407,* 409
Millipedes, 464, *464*
Mimicry, **632,** 632–633
Mismatches, **380**
Mitochondria, 91–92, **92,** *93,* **115,** 115–117, *116, 127*
   DNA of, 117
   structure of, 162, *162*
Mitochondrial matrix, **117, 162,** 162–163
Mitosis, 233, **237,** 242–249
   cancer and, 246–249, *248, 249*
   events preceding, 243–244, *244, 245*
   need for, *242,* 242–243
   rate of, 243, *243*
   steps in, 245–246, *246–247*
Mitotic phase (M phase), **236,** *236,* 236–238, *237*
MJ. *See* Methyl jasmonate (MJ)
MMR vaccine, autism and, *28,* 28–29
Mojave Desert, 616
Molecular biology, common genetic sequences revealed by, 355, *355*
Molecular structure, models representing, 45
Molecules, **45**
   bonds in, 45–47
   hydrophobic and hydrophilic, 62
   nonpolar, 51
   organic, formation of, *406,* 407, *407*
   passive transport across membranes, *102,* 102–103, *103*
   polar, 47
   self-replicating, formation of, 408, *408*
Molluscs, 459–461, *460, 461*
Monk parakeets, 664, *664*
Monkeys, fear of snakes in, 369, *369*
Monogamy, **391,** 392, *392*
Monophyletic groups, **423,** *423*
Monosaccharides, **55,** 55–56
Monotremes, **480,** *480*
Mono-unsaturated fats, 64
Monsanto Company, 213
Morphological species concept, **416**
Morphospecies, 524
Mosses, 498–500, *499, 500,* 639
Moths, mate guarding in, 388
mRNA. *See* Messenger RNA (mRNA)
MRSA infections, 548
Multigeneration experiments, *356,* 356–357, *357*
Multiple allelism, **295,** *295,* 295–297, *297*
Mussels, 460
Mutations, **197,** 197–200, *198, 199,* **326,** 326–328, *327*
   chemical-induced, 200
   point, 198, *199*
   radiation-induced, 198–200
   spontaneous, 198
Mutualism, **636,** 636–638, *637*

Mycelium, **520**
Mycorrhizae, **522,** *522,* 522–523, *523*

N

NADH, 159, 160, 161, 162, 163, 167
   jet lag and, 165–166
NADPH, **150,** 152
Naked chickens, 212, *214*
National Invasive Species Council, 666
Natterjack toads, call of, *397,* 397
Natural preserves, design of, 678–679, *679*
Natural selection
   adaptation to environment and, 339–341, *340*
   aging and, 590–591, *591*
   artificial selection as special case of, 341
   coevolution and, 629
   conditions for occurrence of, 332–334, *332–335*
   directional selection and, *342,* 342–343, *343,* 344–346
   disruptive selection and, 344, *344*
   evolution of complex traits and behaviors and, *346,* 346–347, *347*
   life histories and, 582–584, *583, 584*
   as mechanism of evolution, 15, 325–326
   stabilizing selection and, *343,* 343–344
   theory of evolution by, 15
Nautiluses, 461
Neutrons, *41*
Niches, **628,** *628,* 628–629
Nitrogen cycle, 625, 625–626
Nitrogen dioxide, acid rain and, *666,* 666–668, *667*
Nodes, 421, 421–422
Non-amniotes, **476**
Noncompetitive inhibitors, **73,** 73–74
Nondisjunction, **267,** *267*
Nonpolar head, **95,** *95*
Nonpolar molecules, 51
Non-vascular plants, **496,** *496,* 498–500, *499, 500*
Normal flora, 543
Norway spruces, 507
Notochord, **470,** *471*
Novelty-seeking individuals, 308
Nuclear membrane (nuclear envelope), **112,** *113*
Nuclear pores, 112–113
Nucleic acids, **75,** 75–77, **180.** *See also* Deoxyribonucleic acid (DNA); Ribonucleic acid (RNA)
   information stored by, *75,* 75–76
Nucleolus, **113,** *113*
Nucleotides, **75,** *75,* 75–76, **180**
Nucleus (atomic), **41,** *41*
Nucleus (of cell), **91,** *112,* 112–113, *113, 127*
Nudibranches, *460*
Null hypothesis, **9,** 9–10
Nuptial gifts, **387**

O

Observations
   anecdotal, 27, 28
   in scientific method, 8–9, *9*
Ocean currents, weather and, *618,* 618–619, *619*
Oceans
   deep, 657
   open, 613, *613*
Octopuses, 461, *461*
   intelligence of, 462, *462*
Offspring, number and size of, 584–585, *585*
Oils, hydrogenation of, 64, *64,* 78
Olestra, 65
Open oceans, 613, *613*
Operons, **195**

Oral herpes, 559
Orders, **413,** *414*
Organelles, **91**
Organic molecules, formation of, *406,* 406–407, *407*
Origin of replication, **239**
The Origin of Species (Darwin)
   on classification, 421
   on evolution of behavior, 409, 421, 428
   on natural selection, 340, 357
   on pace of evolution, 428
   publication of, 323–324, *324*
   survival of the fittest and, 338
   on variation in nature, 332
   on "warm little pond," 409
Osmosis, **104,** *104,* 104–105, *105*
Outbreeding, 513
Ovaries, 508, *509*
Ovules, **504**
Oxygen, bonding of, *45,* 45–46
Oxygen Revolution, **542**
Oysters, 460
Ozone layer depletion, *674,* 674–676, *675*

P

Pacific yew tree, 650, *650*
Pair bonds, **392**
Pan paniscus, 661, *661*
Pará (Brazil), 673
Paramecium, 554
Parasites, **556,** *556,* **634**
Parasitism, **634,** 634–636, *635*
Parent cells, **235,** *235,* 243, *244*
Parenting, by males versus females, 383
Passive transport, **102,** *102,* 102–103, *103*
Paternity uncertainty, **384,** 386, 389–390
Pathogenic bacteria, **545,** 545–547, *546*
Pauling, Linus, 21, 179–180
PCR. *See* Polymerase chain reaction (PCR)
Pea plants, Mendel's experiments using, 282–285
Peacocks, mate selection in, *387,* 391–392
Peat, 500, *500*
Peat moss, 500
Pedigrees, **291,** *291,* 291–292, *292*
Penicillium, 519, *519,* 522
Peptide bonds, **70**
Peptidoglycans, **539**
Periodic table, **42**
Persian buttercup, *411*
Pesticides, containing *Bacillus thuringiensis*, 210–211, *211*
pH, **51,** 51–53, *52, 53*
   enzymes and, *73,* 73
Phagocytosis, **107,** *107,* **554**
Pharyngeal slits, **470,** *471*
Phenotype, **187, 286**
   genotype and, *286,* 286–287, *287*
   genotype translation into, 293–303
Phenylalanine, 302, *302*
Phenylketonuria (PKU), 302, *302*
Pheromones, **395**
Phospholipid bilayer, **95,** *95,* 95–96
Phospholipids, **66,** *66,* **95**
Phosphorus cycle, *626,* 626–627
Photoautotrophs, **542,** *542*
Photons, **146**
Photosynthesis, 92, **137,** *137, 143,* 143–156, *144*
   Calvin cycle and, *152,* 152–153, *153*
   CAM and, 155–156
   C4, 154–155, *155*
   chlorophyll and. *See* Chlorophyll
   chloroplasts and, 145, *145*
   C3, 154

"photo" segment of, 144, *144,* 150, *151,* 152
photosystems and, 149–150, *149–151*
sunlight and, *146,* 146–147, *147*
"synthesis" segment of, 144, *144, 152,*
152–153, *153*
Photosystems, **149,** 149–150, *149–151,* 152
Phyla, **413,** *414*
Phylogeny, **421**
Physical defenses, **632,** *632,* 632–633
Phytoplankton, 555
Pigments, **146.** *See also* Chlorophyll
Pili, **90**
Pines, 504, *505,* 506, *506,* 507
Pinocytosis, **107**
Pitchfork, Colin, 220–221, *221*
PKU. *See* Phenylketonuria (PKU)
Placebo effect, **18**
Placebos, **13**
Placentals, **480,** *480*
Plants, **494,** *494–496,* 494–517
adapted to water scarcity, 153–156, *154–156*
carnivorous, 517, *517*
cell wall of, *124,* 124–125, *127*
chloroplasts of, *126,* 126–127, *127*
crop. *See* Agriculture
first, 497–502
flowering. *See* Angiosperms
fruits of, *514,* 514–515
fungal symbionts of, *522,* 522–523, *523*
fungi and, 518–519, *519, 522,* 522–524, *523*
land, first, 497–498, *498*
with medicinal uses, 516
non-vascular, *496, 496,* 498–500, *499, 500*
photosynthesis and. *See* Chlorophyll;
Photosynthesis
relationship with animals, 514–517
resistance to predation, 515–517, *516, 517*
seeds and, *503,* 503–504, 503–507, *514,*
514–515
vacuoles of, *125,* 125–126, *127*
vascular, **501,** 501–502, *502*
Plasma membrane, **89,** *89, 94,* 94–101
active transport across, 106, *106*
diffusion across, *102,* 102–103, *103*
endocytosis and exocytosis and, 107–109,
107–109
faulty, diseases due to, *98,* 98–99, *99*
"fingerprint" of, *100,* 100–101, *101*
functions of, 95
molecules embedded in, *96,* 96–97, *97*
origin of life and, 408–409, *409*
osmosis across, *104,* 104–105, *105*
phospholipid bilayer of, *95,* 95–96
Plasmids, **206,** *206,* 540
Plasmodesmata, *124,* **125**
Plasmodium, 556, *556*
Platypus, 480, *480*
Pleiotropy, **298,** 298–299, *299*
Point mutations, **198,** *199*
Polar bodies, **256,** *256*
Polar head, *95,* 95
Polar ice, *612. See also* Ice caps
Polar molecules, 47
Pollen grains, **504**
Pollination, **504,** 506, *510,* 510–511, *511*
Polyandry, **391**
Polychaetes, 458
Polydactyly, among Amish people, 329, *330*
Polygamy, **391**
Polygenic traits, **297,** *297,* 297–298
Polygyny, 391, **391,** *391,* 392
Polymerase chain reaction (PCR), **205,** *206*
Polypeptides, 70
Polyploidy, **418**

Polysaccharides, **58,** *58,* 58–59
Polyunsaturated fats, 64
Ponds, 613, *613*
Population bottleneck effect, 329, *330,* 331
Population density, **576**
Population ecology, 570–607, **573**
aging and, 589–594, 602
environmental limitation on population
growth and, *576,* 576–577, *577*
growth rate of populations and, 574–575, *575*
human population growth and, 595–601
life histories and, 582–588
maximum sustainable yield and, *580,*
580–581, *581*
populations cycling between large and small
and, *578,* 578–579, *579*
Population growth
bacterial, 259, *259*
environmental limit on, *576,* 576–577, *577*
rapid, cost of, 585–586
rate of, 574–575, *575*
Population size, extinction risk and, 660
Populations, *573*
human. *See* Human population
Porcupines, hazard factor and, 593
Portuguese man-o'-war, 454
Positive correlation, **27**
Post-anal tails, **470,** *471*
Postzygotic barriers, 412, *412,* **413**
Potential energy, **138,** *138*
Praying mantises, 259
Precipitation. *See* Rainfall
Predation
adaptations produced by, 631–634, *632–634*
parasitism as, 634–636, *635*
plant defenses against, 515–517, *516, 517*
Predictions, testable, devising, 10–11, *11*
Pregnancy, gestational diabetes and, *376,*
376–377
Prepared learning, **370**
Preseucoila imallshookupis, 413
Presley, Elvis, 413
Prezygotic barriers, *412,* 412–413, **413**
Prilosec, 54
Primary active transport, **106,** *106*
Primary consumers, **620,** *621*
Primary electron acceptor, **149**
Primary productivity, **620**
Primary structure of proteins, **70,** *71*
Primary succession, 639–640, *640*
Primase, **240**
Primates. *See also specific primates*
evolution of, *481,* 481–483, *482*
Principles of Geology (Lyell), 320, 321
Prions, 558
Probiotic therapy, 543–544
Producers, **620**
energy flows to consumers from, 620–622,
621
Prokaryotes, **89.** *See also* Archaea; Bacteria
cell division in, 233
gene regulation in, *194, 195,* 195–196
replication in, 234–235, *235*
structural features of, *89,* 89–90
Prokaryotic cells, **89,** *89,* 89–90, 432
chromosomes of, 234, *234*
Promoter sites, **188,** 195
Prophase, **245**
in meiosis, 253, 254, *254, 255*
in mitosis, 245, *246*
Protein synthesis, **193**
RNA and, 77, *77*
Proteins, **67,** 67–74
amino acids and, 68, *68*

in cytoskeleton, 114
denaturation of, 71, *71*
in diet, 68–69, *69*
energy from, 169, *169*
enzymatic, 68, 72–74, *72–74,* 97
functions of, *67,* 67–68
receptor, 96–97, *97*
recognition, 97, *97*
structure of, *70,* 70–71, *71*
surface, 96, *96*
transmembrane, 96, *96*
transport, 97, *97,* 103
Proteus, 538
Prothallus, **502**
Protists, **423,** 437, 448, 553–556
animal-like, 554
first eukaryotes as, *553,* 553–554
fungus-like, *554,* 554–555
pathogenic, 556, *556*
plant-like, *554, 555,* 555
Protons, *41*
number of, *41,* 41–42, *42*
Protostomes, **449**
Pseudoscience, **27,** *27,* 27–28
Punctuated equilibrium, **427,** *427–428, 428*
Punnett, R. C., 335
Punnett square, **286,** 286–287, *287*
Pupa, **467,** *467,* 467–468
Purple loosestrife, 665, *665*
Pyruvate, **158**

Q
Quagga mussels, 665–666
Quaternary structure of proteins, **70,** *71*

R
Rabies, *635,* 635–636
Rabies virus, 560
Racial differences, 438
Radial symmetry, **448**
Radiation therapy, 249
Radiation-induced mutations, 198–200
Radioactive atoms, *42*
Radiometric dating, **348,** 348–349, *349*
Rain shadows, **616**
Rainfall
acid rain, 666–668, *667, 668*
day of the week and, 32
global patterns of, *615,* 615–616
Random assortment, **253**
Randomization, 18
Rats
running in maze, *346,* 346–347
superstitious behavior in, 5, *5*
Ray-finned fishes, **473,** *473–474, 474*
Realized niche, **628,** *629*
Receptor proteins, **96,** 96–97, *97*
Receptor-mediated endocytosis, **107,** 107–108,
109
Recessive traits, 284, **284**
frequency in population, 335–337, *336*
Reciprocal altruism, **374,** 377–378, *378, 379*
Recognition proteins, **97,** *97*
Recombinant DNA technology, **208,** 208–209
Recombination, **253**
Red deer
mate selection in, 387
tradeoff between reproduction and growth
and, 584
Redwoods, 504, *505,* 506, *506*
Regeneration, in photosynthesis, 152–153
Regulating services, **651,** *652*
Repeatability, of experiments, *21,* 21–22

Replication, **234,** 238–241, *239, 240*
of chromosomes, 244, *252,* 252–253, *254*
discontinuous, 241
origin of, 239
in prokaryotes, 234–235, *235*
Replication forks, **239**
Reproducibility, of experiments, *21,* 21–22
Reproduction
asexual. *See* Asexual reproduction
difficulties classifying species and, 415, *415*
growth and, 584, *585*
sexual. *See* Sexual reproduction
survival and, 584, *585*
tradeoff with longevity, 2
Reproductive cells, **236.** *See also* Eggs; Gametes;
Sperm
Reproductive investment, **382,** 382–394, **582,**
582–583, *583*
competition and courtship and, 386–388, *387*
differing patterns in males and females, *384,*
384–386
mate guarding and, *388,* 388–390, *389*
monogamy versus polygamy and, *391,*
391–393, *392*
paternity uncertainty and, 384, 386, 389–390
sexual dimorphism and, *393,* 393–394
Reproductive isolation, **411**
Reproductive output, **591**
Reproductive success
differential, natural selection and, *333,*
333–334
fitness and, 338–339, *339*
of humans, *392,* 392–393
Resistance plasmids, 540
Resource partitioning, **630,** *630*
Respiration, cellular. *See* Cellular respiration
Restoration ecology, 679
Restriction enzymes, **205,** *205*
Retroviruses, **562**
Rh blood group, 297
Ribonucleic acid (RNA), **75**
messenger. *See* Messenger RNA (mRNA)
origin of life and, 408, *408*
protein synthesis and, 77, *77*
structure of, 77, *77*
transfer, *190,* **191**
viral, 557, 558, *558*
Ribosomal subunits, **190**
Ribosomes, **89, 90,** 113
Rice, golden, *209,* 209–210, *210*
Ring species, **416**
difficulties classifying, *415,* 415–416
Rivers, 613, *613*
RNA. *See* Ribonucleic acid (RNA)
RNA bases, 75–76
RNA viruses, 559
Rodents
hazard factors and, 592, *592*
reproductive strategy of, 583, *583*
Ronaldo, Cruistiano, 411
Roots, **495**
Rose, 411
Rough endoplasmic reticulum (rough ER), **120,**
121, 127
Roundworms, **456,** *457,* 457–458
Rubisco, **152**
Ruffin, Julius, 178, *178*

S
S phase, **237**
Sahara Desert, 616
Salamanders, DNA of, 184
Salmon, tradeoff between reproduction and
survival in, 584, *585*